Nanosensors

Series in Sensors

Series Editors
Barry Jones and Haiying Huang

Other recent books in the series:

Nanosensors
Physical, Chemical, and Biological, *2nd Edition*
Vinod Kumar Khanna

Advanced Chromatic Monitoring
Gordon R. Jones, Joseph W. Spencer

Semiconductor Radiation Detectors
Alan Owens

Radiation Sensors With 3D Electrodes
Cinzia Da Vià, Gian-Franco Dalla Betta, Sherwood Parker

A Hands-On Course in Sensors Using the Arduino and Raspberry Pi
Volker Ziemann

Electrical Impedance: Principles, Measurement, and Applications
Luca Callegaro

Compound Semiconductor Radiation Detectors
Alan Owens

Biosensors and Molecular Technologies for Cancer Diagnostics
Keith E. Herold, Avraham Rasooly

Sensors for Safety and Process Control in Hydrogen Technologies
Thomas Hübert, Lois Boon-Brett, William Buttner

Resistive, Capacitive, Inductive, and Magnetic Sensor Technologies
Winncy Y. Du

Semiconductor X-Ray Detectors
B. G. Lowe, R. A. Sareen

Portable Biosensing of Food Toxicants and Environmental Pollutants
Dimitrios P. Nikolelis, Theodoros Varzakas, Arzum Erdem, Georgia-Paraskevi Nikoleli

Optochemical Nanosensors
Andrea Cusano, Francisco J. Arregui, Michele Giordano, Antonello Cutolo

Metal Oxide Nanostructures as Gas Sensing Devices
G. Eranna

Nanosensors
Physical, Chemical, and Biological
Vinod Kumar Khanna

Handbook of Magnetic Measurements
Slawomir Tumanski

Structural Sensing, Health Monitoring, and Performance Evaluation
D. Huston

Chromatic Monitoring of Complex Conditions
Gordon Rees Jones, Anthony G. Deakin, Joseph W. Spencer

Principles of Electrical Measurement
Slawomir Tumanski

Novel Sensors and Sensing
Roger G. Jackson

Hall Effect Devices, *2nd Edition*
R.S. Popovic

Nanosensors

Physical, Chemical, and Biological

Second Edition

Vinod Kumar Khanna

CRC Press
Taylor & Francis Group
Boca Raton London New York

CRC Press is an imprint of the
Taylor & Francis Group, an **informa** business

Second edition published 2021
by CRC Press
6000 Broken Sound Parkway NW, Suite 300, Boca Raton, FL 33487-2742

and by CRC Press
2 Park Square, Milton Park, Abingdon, Oxon, OX14 4RN

First edition published by CRC Press 2012

CRC Press is an imprint of Taylor & Francis Group, LLC

Library of Congress Cataloging-in-Publication Data

Names: Khanna, Vinod Kumar, 1952- author.
Title: Nanosensors : physical, chemical, and biological / Vinod Kumar Khanna.
Description: Second edition. | Boca Raton : CRC Press, 2021. | Series: Series in sensors | Includes bibliographical references and index.
Identifiers: LCCN 2020038701 | ISBN 9780367457051 (hardback) | ISBN 9781003025559 (ebook)
Subjects: LCSH: Nanostructured materials. | Detectors.
Classification: LCC TA418.9.N35 K48 2021 | DDC 681/.25--dc23
LC record available at https://lccn.loc.gov/2020038701

ISBN-13: 978-0-367-45705-1 (hbk)
ISBN-13: 978-0-367-51480-8 (pbk)
ISBN-13: 978-1-00-302555-9 (ebk)

Visit the Taylor & Francis Web site at
http://www.taylorandfrancis.com

and the CRC Press Web site at
http://www.crcpress.com

Typeset in Times
by Deanta Global Publishing Services, Chennai, India

Dedicated to the loving memories of my late father,

Shri Amarnath Khanna,

and my late mother, Shrimati Pushpa Khanna,

for nurturing my educational career.

Also dedicated to my grandson Hansh,

my daughter Aloka, and my wife Amita

for their affection and support.

Contents

Part III Physical Nanosensors

Part IV Chemical and Biological Nanosensors

Part V Emerging Applications of Nanosensors

Part VI Powering, Networking, and Trends of Nanosensors

Preface to the Second Edition

We live in the nanotechnology era. Many phenomena take place at the nanoscale which do not have any equivalent in bulk matter. These nano-specific phenomena bestow tremendous opportunities to build advanced sensing devices with enhanced capabilities. A revolution has been fostered by nanometer-sized structures in the realm of sensors. Research on such nanosensors is being aggressively pursued worldwide.

During the period that elapsed since the first edition of the book, *Nanosensors: Physical, Chemical, and Biological*, published in 2012, enormous developments have taken place in the field of nanosensors, making it extremely important to draw level with these quick strides. Bearing in mind the goal set for the book to serve as an all-in-one resource on nanosensors, it felt necessary to publish a revised edition incorporating this recent progress, and the outcome of this revision is the book in your hands.

In its present revised form, this book looks at nanosensors from three different angles: first, from a scientific view based on nanosensor classification into physical, chemical, and biological regimes; secondly, from an applications point-of-view, where nanosensors are treated by an approach focusing on what they can do for mankind in terms of the betterment of human life; and thirdly, from significant differences of nanosensors from conventional sensors, regarding their powering and networking.

Taxonomical treatment has a foundational value from which the subject develops and is pedagogically mandatory. But the applications and powering/networking aspects are no less important. The applications are the beneficial outputs of nanosensor development, the implementation of which lead to improved human welfare. Power saving is a high-priority requirement at the moment because energy saved is equivalent to energy earned or generated, and energy utilization efficiency of all appliances must be raised to the maximum possible extent. Here, the emphasis is placed on the differences between the tiny nanosensors and their larger macro/microscopic counterparts, in the respect that it will be very convenient if these devices can work without batteries and without any connecting wires. Freedom of the nanosensor from a tethered external power source provides it with enormous flexibility in applications, because it allows easy networking. To meet this need, nanogenerators can act as robust, self-sustainable, and self-sufficient energy sources, supplying the required microwatt/nanowatt level power to drive nanosensors.

Following the three-pronged revision approach outlined above and noting that the classification perspective has been successfully presented in the 1st edition, five new chapters, which constitute Chapters 10–14 of the revised book, have been added. Among these, Chapters 10–12 cover the applications of nanosensors in environmental monitoring, namely, air and water pollution control, and in various enterprises, such as the food industry, healthcare sector, the automotive and aerospace industries, and homeland security. The remaining two chapters are on nanosensor powering and networking. These topics form the contents of Chapters 13 and 14 on self-powered nanosensors and wireless nanosensor networks, respectively. The concluding chapter, Chapter 15, reviews the present-day overall scenario in the realm of nanosensors and articulates the directions for future progress which researchers are pursuing from where we stand today.

The book is systematically subdivided into six parts as follows:

- Part I: Fundamental Concepts of Nanosensors (Chapter 1)
- Part II: Nanomaterials and Micro/Nanofabrication Facilities (Chapters 2–3)
- Part III: Physical Nanosensors (Chapters 4–7)
- Part IV: Chemical and Biological Nanosensors (Chapters 8–9)
- Part V: Emerging Applications of Nanosensors (Chapters 10–12), and
- Part VI: Nanosensor Powering, Networking, and Trends of Nanosensors (Chapters 13–15)

Each part focuses on a subdiscipline of nanosensors, and the parts are seamlessly interwoven to offer a unified, cohesive treatment.

In a nutshell, the 1st edition presented nanosensors from the classification standpoint alone. In this revised edition, the treatment is extended by paying attention to nanosensor applications, powering, and arrangement into networks. Thus, the revised edition provides an extensive coverage of nanosensor technologies, including nanoscopic detection mechanisms, nanomaterial selection, nanofabrication, and novel applications in addition to nanogenerators and nanocommunication networks. Each device is looked upon as a fusion of particular nanoscale property(s) with specific sensor structure.

It is earnestly hoped that this revised edition will continue to enjoy popularity among the readers, as happened with the first edition. The author will consider his efforts amply rewarded if the revised edition receives the same warmth and affection from its readership as its predecessor did. Wishing you all an enjoyable trip through this exciting and interesting knowledge domain.

Vinod Kumar Khanna
Chandigarh, India

Preface to the First Edition

Objectives and Goals

Interesting developments have taken place in the field of sensors in recent years. Nanotechnology holds a great potential for revolutionizing the sensor arena. Traditional sensors are being reengineered and recast from a nanotechnological perspective, and new sensor designs are being introduced. Sensitivity, detection range, and response times of sensors have shown remarkable improvement by using nanotechnological methods. Faster, better, cheaper, and smaller sensors are becoming available.

Sensor development in the nanotechnology age, the focal theme of this book, is one of the hottest topics in contemporary research. This book presents a critical appraisal of the new opportunities provided by nanotechnologies and the nanotechnology-enabled advances in the field of sensors.

Rationale

Being a new topic, information on nanosensors is scattered across research journals and scientific reports. There is a paucity of books in this area. This book intends to fill this long-felt need. Bringing together the widely spread information between the covers of an easy-to-read book will be helpful to all those working on nanosensors, for instant reference and updates.

Topical coverage

Chapter 1 provides an introductory survey of nanotechnology and introduces preliminary background material to help the reader's understanding of the rest of the book. It thus lays the groundwork for the book.

Chapter 2 describes the materials utilized for the fabrication of nanosensors, such as nanoparticles made of metals, semiconductors and insulators, quantum dots, and carbon nanotubes. These constructional nanomaterials constitute the building blocks of nanosensors.

In Chapter 3, the vision of a laboratory working on nanosensor fabrication and characterization is sketched. Such a laboratory has a broad scope and pools resources from vastly different areas. Integration of these facilities under one roof, though, is an uphill task.

The remaining chapters address various categories of nanosensors. Chapter 4 is dedicated to mechanical nanosensors, e.g., MEMS and NEMS resonators, single-electron transistor-based displacement sensors, electron tunneling accelerometers, CNT force, pressure and flow sensors, etc.

Chapter 5 deals with temperature sensors such as miniaturized thermocouples, CNT resistors, ratiometric fluorescence intensity sensors, hot-electron bolometers, and so on.

Chapter 6 provides a detailed description of optical nanosensors, like those based on surface plasmon resonance, SERS scattering, colorimetric sensors, fiber-optic nanosensors, PEBBLEs, QD devices with special emphasis on FRET-type sensors, etc. Optical nanobiosensors are also covered in this chapter.

Magnetic nanosensors, the subject of Chapter 7, have received a boost with the discovery of the giant magnetoresistance effect. Many landmark nanobiosensors have been developed and some of these are discussed in this chapter. Magnetic relaxation is another phenomenon that has been widely applied in designing nanobiosensors. This is briefly covered here.

Chapter 8 describes the coveted "nanobiosensors," which unify the living world with physical and chemical domains. The "marriage" of biomolecules and nanotechnology often yields novel sensor concepts. Such biosensors have a marked impact in different fields of application, such as medicine, food technology, environment, biochemistry, and biotechnology, as well as in information processing. These nanosensors, as well as all others, have been the subject of intensive research globally in recent years. There has been a veritable deluge of research publications in this vital area because it is directly related to human health and well-being. Electrochemical nanobiosensors are comprehensively surveyed. Many kinds of nanoparticle-modified electrodes are also reported. Apart from these nanosensors, mechanical nanosensors, such as those based on micro- or nanocantilevers, are outlined, along with additional optical nanosensors that use aptamers.

Chapter 9 deals with chemical nanosensors primarily aimed at detecting gaseous compounds, elements, and parameters such as pH of aqueous solutions. Metallic nanoparticles, metal oxide nanoparticles, and CNTs, pristine or functionalized, have been used to develop a family of sensors for chemical analysis, which will be useful for analytical chemists. *Chemiresistors*, fabricated on a MEMS micro-hotplate with a nanostructured tin oxide film coating, provide great sensitivity for gas detection. Progress in nanomaterials has created tremendous opportunities for improved gas sensor devices. In nanoparticle-based semiconductor gas sensors, the detection thresholds for gases have been lowered severalfold, compared with current commercialized sensors.

The concluding chapter, Chapter 10, reviews and introspects on the main developments in the different varieties of nanosensors that have been described in this book. The present scenario is assessed, and, on the basis of this panoramic view, the main research areas attracting attention are highlighted. Core problems facing researchers, notably the interfacing of nanosensors with the macro world and related issues, are discussed.

Style and Presentation

Nanosensors, like their conventional macro- and micro-counterparts, are an interdisciplinary field that attracts scientists, engineers, and students alike. To cater to this wide audience, a thoroughly grassroot approach has been adopted in this book. Professionals in one area are not expected to be conversant with the terminology of another discipline, and, therefore, the responsibility to explain the key terms of different areas has been clearly borne in mind. It is earnestly hoped that the subject matter will be understandable to personnel specializing in different fields. Another noteworthy feature of this book is that it is not written in a dull, monotonous vein but in a question/answer format to keep the reader's interest alive. Thought-provoking questions are raised time and again to arouse the reader's interest.

Throughout the book, end-of-chapter exercises are appended to rekindle the reader's interest. Further, worked examples are sprinkled throughout the book to supplement the text.

Intended Audience

It is anticipated that the book will provide a useful forum for a wide cross section of readers, including scientists, engineers, teachers, and students. It will serve as a useful resource, for both research and teaching purposes.

Vinod Kumar Khanna
CSIR-CEERI, Pilani, Rajasthan, India

Acknowledgments

For writing this book, a vast literature has been studied, and I am indebted to the authors of all books, journals, and internet sites whose original work has been reported in this book. Many of these excellent research papers, articles, books, and web pages are cited in the references at the end of each chapter. However, some references may have been inadvertently missed; I trust this may be forgiven.

I wish to thank the Director of CSIR-Central Electronics Engineering Research Institute (CEERI), Pilani, for relentless encouragement in my research and academic pursuits.

I am thankful to Dr John Navas, senior acquisitions editor, physics, Taylor & Francis/CRC Press, and Amber Donley, project coordinator, editorial project development, Taylor & Francis Books, for their keen interest in this work. The project started on April 29, 2009, with Dr Navas inviting me to consider writing or editing a book on *Nanosensors: Physical, Chemical, and Biological*. For me, it was a dream come true, as I was fascinated and enthusiastic about nanosensors.

I am thankful to my daughter and my wife for their love and affection and for bearing with my busy schedule during this work over the past two years. The writing time was obviously stolen from them.

Author's Profile

Brief Introduction

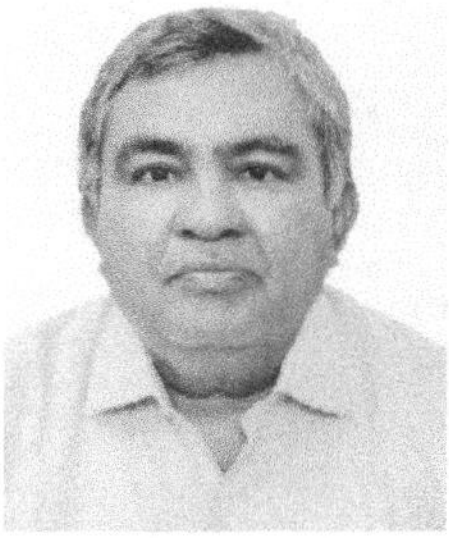

Vinod Kumar Khanna (born November 25, 1952, Lucknow, India) is a former scientist emeritus, CSIR (Council of Scientific and Industrial Research) and professor emeritus, AcSIR (Academy of Scientific and Innovative Research), India. He is a retired Chief Scientist and Head, MEMS and Microsensors Group, CSIR-CEERI (CSIR-Central Electronics Engineering Research Institute), Pilani (Rajasthan) and Professor, AcSIR, India.

Academic Qualifications

The author was awarded a national scholarship by the Ministry of Education and Youth Services, Government of India, in 1970, and he received his MSc degree in Physics in 1975 from University of Lucknow, India and his PhD (Physics) degree from Kurukshetra University, India, in 1988.

Research and Development Projects, and Significant Contributions

During his tenure of work, spanning almost 38 years at CSIR-CEERI, Pilani, the author worked on a large number of CSIR and sponsored projects on power semiconductor devices, notably high-current and high-voltage rectifiers, high-voltage transistors, fast-switching inverter grade thyristors, power DMOSFET and IGBT. Another key research area included the device and process design, and fabrication of micro- and nanoelectronics and MEMS- technology-based sensors and dosimeters. In the sensors area, his contributions focused primarily on the development of technology and characterization of moisture and microheater-embedded gas sensors, ion-sensitive field-effect transistor pH sensors, MEMS acoustic sensors and capacitive MEMS ultrasonic transducers, PIN diode neutron dosimeters, and PMOSFET gamma ray dosimeters.

He has also been instrumental in the setting-up and maintenance of diffusion/oxidation facilities, annealing and metal-sintering facilities, reactive sputtering for gate dielectric deposition of extended-gate FETs, in addition to molybdenum-silicon alloying, beveling and edge contouring for surface electric field control in power devices, and minority carrier lifetime measurements for process-induced influence studies in semiconductors.

Visits Abroad

The author is widely traveled and has worked at Technische Universität Darmstadt, Germany (1999) and Kurt-Schwabe-Institut für Mess- und Sensortechnik e.V., Meinsberg, Germany (2008). He also visited the Institute of Chemical Physics, Novosibirsk, Russia, in 2009, as a member of the Indian delegation, and Fondazione Bruno Kessler (FBK), Trento, Italy, in 2011, under the India-Trento Programme of Advanced Research (ITPAR). He participated and presented research papers at the IEEE-IAS Annual Meeting at Denver, Colorado, USA, in 1986.

Research Publications and Books

Prof. Khanna is the author/co-author of 192 research papers in prestigious peer-reviewed journals and in refereed national/international conference proceedings. In addition, he has five patents to his credit, including two US patents. Prior to this book, he has authored 16 books. He has also published six chapters in edited books.

Membership of Learned Societies

Prof. Khanna is a life member of several leading professional societies (Indian Physics Association, Semiconductor Society-India and Indo-French Technical Association), and a fellow life member of the Institution of Electronics & Telecommunication Engineers (IETE), India.

Present Interests

After superannuating as Head, MEMS and Microsensors Group, CSIR-CEERI in November 2014, and completing his tenure as Scientist emeritus, CSIR, in November 2017, Prof. Khanna currently resides at Chandigarh, India. He is a passionate author, avidly reading and writing.

About the Book (2nd Edition)

Over and above the classification-based contents of the first edition on physical, chemical, and biological nanosensors, this second edition provides information on the following:

- Nanosensors for air quality control
- Nanosensors for water quality monitoring
- Nanosensors for foodborne pathogens
- Nanosensors for foodborne toxins
- Nanosensors for cancer diagnostics
- Nanosensors for diagnosing infectious diseases
- Nanosensors for trace explosives
- Nanosensors for bioterrorist agents
- Tribo- and piezoelectric nanogenerators
- Nanosensors powered by triboelectricity
- Nanosensors powered by piezoelectricity
- Nanosensor communication networks
- Internet of nanothings

With 336 attractive diagrams and 78 tables, the new edition presents an across-the-board assessment of nanosensor technologies, presenting authoritative information in a concise package, and making it immensely useful to researchers, industry professionals and students.

Abbreviations and Acronyms

A	Ampere
Å	Angstrom unit
AA	Ascorbic acid; anti-adenosine aptamer
AAO	Anodic aluminum oxide
AAP	Ascorbic acid-2-phosphatase
Ab	Antibody
Ab-AuNPs	Antibody-conjugated gold nanoparticles
AC	Alternating current, anti-cocaine aptamer
AChE	Acetycholinesterase
ACN	Acetonitrile
AF	Alexa Fluor
AF-488	Alexa Fluor 488 (dye)
AFB	Acid-Fast Bacilli
AFB1	Aflatoxin B1
AFM	Atomic force microscope
AFM1	Aflatoxin M1
Ag	Antigen
AIS	$AgInS_2$
AK	Anti-potassium-specific G-quartet
ALCVD	Atomic layer chemical vapor deposition
ALD	Atomic layer deposition
ALE	Atomic layer epitaxy
ALP	Alkaline phosphatase
AM	Amplitude modulation
aM	Attomolar
AMI-25	Ferumoxide superparamagnetic iron oxide particles
amol	Attomole
AMR	Anisotropic magnetoresistance
AMU	Atomic mass unit
aN	Attonewton
ADV-5	Adenovirus-5
Anti-D-AA	Anti-D-amino acid
Anti IFN-γ	Anti-interferon-gamma (antibodies)
AP	Ammonium peroxydisulfate, anti-parallel
APD	Antibody phage display
AP enzyme	Alkaline phosphatase
APCVD	Atmospheric pressure CVD
APD	Antibody phage display (technique)
apo-GOx	Glucose oxidase (GOx) apoprotein, without the FAD/FADH2 cofactor (apo=away from or off)
APP	Application
Aptasensor	Biosensor made by combing an aptamer with a transducer
APTES	Aminopropyltriethoxysilane
APTMS	Aminopropyltrimethoxysilane
Apts	Aptamers
AQMS	Anthraquinone monosulfonate
A-RAM	Advanced random access memory
Arg	Arginine
ASDPV	Anodic stripping differential pulse voltammetry
ASV	Anodic stripping voltammetry
ATP	Adenosine triphosphate
4-ATP	4-Aminothiophenol
AU	Arbitrary unit
Au_{coll}	Colloidal gold
AuE	Gold electrode
AuNDs	Gold nanodendrites; Gold nanodots
AuNIs	Gold nanoislands
AuNPs	Gold nanoparticles
AuNPsA	Gold nanoparticles amalgam
BaP	Benzo[α]pyrene
BARC	Bead ARray Counter (Biosensor)
BAWs	Bulk acoustic waves
BCA	Bio-barcode assay
BCSP	Bovine conceptus secretory protein
BDT	Dithiol-1,4-benzenedithiol
BEF	Built-in electric field
BFM	Bacterial flagellar motors
Biotin-HPDP	*N*-(6-(Biotinamido)hexyl)-3′-(2′-pyridyldi thio)-propionamide
Biotin-SS-NHS	Succinimidyl 2-(biotinamido)-ethyl-1,3′-dithiopropionate
Botin-PEG	Biotin-poly(ethylene glycol)
BJT	Bipolar junction transistor
BK7	High optical quality crown glass (produced by German company SCHOTT)
$BMIMPF_6$	1-Butyl-3-methylimidazolium tetrafluorophosphate
BnOH	Benzyl alcohol
BoLcA	Botulinum Serotype A Light Chain
BoNT/A	Botulinum neurotoxin serotype A
BODIPY	Boron-dipyrromethene
BOX	Buried oxide
bp	Base pair (of double-stranded DNA), boiling point
BPT	Benzopyrene tetrol
BRCA1	BReast CAncer gene 1
BSA	Bovine serum albumin
BSG	Borosilicate glass
BTC	Benzene tricarboxylate, benzene-1,3,5-tricarboxylic acid
BTD	5-Bromothienyldeoxyuridine
B. thuringiensis	*Bacillus thuringiensis*
BTX	Benzene, toluene, xylene
BY2	Bright yellow-2
C	Coulomb
CAb	Capture antibody
Cal	Calorie
CANTOR	Cantilever-based airborne nanoparticle detector
Cat	Catalase

CB	Conduction band
CBD	Chemical bath deposition
CBF	Carbofuran
CCD	Charge-coupled device
CD	Cyclodextrin
Cd	Cadmium; Candela
CD10	Cluster of differentiation 10
CdA	Cadmium arachidate
CDI	1,1′-carbonyldiimidazole
cDNA	Complementary DNA; control DNA (probe)
CdS NPs	Cadmium sulfide nanoparticles
CEA	Carcinoembryonic antigen
CFP	Cyan fluorescent protein
CFU	Colony-forming unit
CG	Colloidal gold
chemFET	Chemical field-effect transistor
Chi-AgNPs	Chitosan-stabilized silver nanoparticles
Chit, CHIT, CS	Chitosan
CHP	Cholesterol-bearing pullulan
CIP	Current-in-plane
C. jejuni	*Campylobacter jejuni*
CL	Cross-linker; chemiluminescence
C-line	Control line
CLIO	Cross-linked iron oxide
CLIO-d-Phe	Cross-linked iron oxide-D-phenylalanine
CLIO-NH_2	Amidated cross-linked iron oxide (nanoparticles)
cm	Centimeter
CMLNPS	Carboxylate-modified latex nanoparticles
CMNPs	Chitosan-modified magnetic nanoparticles
CMOS	Complementary metal-oxide semiconductor
CMP	Chemical mechanical polishing
CNPs	Carbon nanoparticles
CNTs	Carbon nanotubes
COVID-19	Corona virus disease-2019
CPE	Carbon paste electrode
CPNPs	Conjugated polymer nanoparticles
CPP	Current-perpendicular-to-plane
CPPs	Cell-penetrating peptides
CRP	c-reactive protein
CSAFM	Current-sensing AFM
CSMAC	Carrier-sense multiple access control
CSNPs	Chitosan nanoparticles
CSPE	Chlorosulfonated polyethylene
c-strand DNA	Complementary single stranded DNA
CT	Computerized tomography; constant temperature; charge-transfer (interaction)
Ct	Catalase
CTAB	Cetyltrimethylammonium bromide
CTB	Cholera toxin B-subunit
Cu NPs	Copper nanoparticles
CV	Cyclic voltammetry
CVD	Chemical vapor deposition
CW	Continuous wave
Cy	Cyanine
Cy3B	Cy3B-*N*-hydroxysuccinimidine monoester
Cys, Cyst	Cysteamine
cyt c	Cytochrome *c*
C-Z process	Czochralski process
0D	Zero-dimensional
1D	One-dimensional
2D	Two-dimensional
2DEG	Two-dimensional electron gas
2,4-D	2,4-dichlorophenoxyacetic acid
3D	Three-dimensional
DA	Dopamine
Da	Dalton
DAb	Detection antibody
DACH	Diaminocyclohexane
dB	Decibel
DC	Direct current
DDA	Dichlorodiphenylacetic acid or 2,2-*bis*(chlorophenyl) acetic acid
dDNA	Capture DNA probe specific Dengue type 1 virus
DDT	Dichlorodiphenyltrichloroethane
DDTC	Diethyldithiocarbamate
DEP	Dielectrophoresis
Der f2	*Dermatophagoides farinae* (American house dust mite)
DIMP	Di-isopropyl methylphosphonate
DI	Deionized
DIW	Deionized water
DM	Deltamethrin
DMF	*N*, *N*′-dimethyl formamide
DMMP	Dimethyl methylphosphonate
DMS	Dimethylsuberimidate
DMSA	Meso-2,3-dimercaptosuccinic acid
DMSO	Dimethyl sulfoxide
DNA	Deoxyribonucleic Acid
DNS	Dansyl hydrazine
DNPs	Dynamic nanoplatforms
DNT	2,4-Dinitrotoluene
DoF	Depth of field
DOS	Density of states
DPA	Diphenylamine
DPASV	Differential pulse anodic stripping voltammetry
DPBS	Dulbecco's phosphate-buffered saline
DPC	1,5-Diphenylcarbazide
D-Phe	D-Phenylalanine
DPN	Dip-pen nanolithography
dppz	Dipyridophenazine
DPV	Differential pulse voltammetry
DRAM	Dynamic random access memory
DRIE	Deep reactive-ion etching
dsDNA	Double-stranded DNA
DSM	Deep submicron
DT	Decanethiolate
DTCs	Dithiocarbamates
DTD	Diode temperature detector
DTSSP	3,3′-Dithiobis(sulfosuccinimidyl propionate)
DTT	Dithiothreitol
dUTP	Deoxyuridine triphosphate
DUV	Deep ultraviolet
e^-	electron

EA, EtOAc	Ethyl acetate
EBL	E-beam or electron-beam lithography
EB-CVD	Electron-beam chemical vapor deposition
EBOV	Ebola virus
ECG	Electrocardiogram
ECL	Electrochemiluminescence, electrogenerated chemiluminescence
E. coli	*Escherichia coli*
ECR-CVD	Electron cyclotron resonance chemical vapor deposition
EDC, EDAC	1-Ethyl-3-(3-dimethylaminopropyl)carbodi imide
EDC-HCl	1-Ethyl-3-(3-dimethylaminopropyl) carbodiimide-hydrochloride
EDP	Ethylenediamine pyrocatechol
EDX	Energy dispersive X-ray (spectroscopy)
EEG	Electroencephalogram
EGDN	Ethylene glycol dinitrate
EGFET	Extended-gate field-effect transistor
EGMRA	Ethylene glycol maleic rosinate acrylate
EGP	Ebola glycoprotein
EIS	Electrochemical impedance spectroscopy
ELISA	Enzyme-linked immunosorbent assay
EMC	Electromagnetic communication
EMF	Electromotive force
EM transceiver	Electromagnetic transceiver
ENIA	Europium nanoparticle immunoassay
erfc	Complementary error function
ES	Electrospinning; electrochemical sensing
ESA	Electrostatic self-assembly
e-skin, E-skin	Electronic skin
ET	Electron-transfer
ETH	Eidgenössische Technische Hochschule
Eu-DT	Eu-tris (Dinaphthoylmethane)-*bis*-(trioctylp hosphine oxide)
eV	Electron volt
***Exo*III**	*Exonuclease* III (restriction enzyme)
F	Farad
F1	Fraction 1
F2	Coagulation factor II
1/*f* noise	Flicker or pink noise with 1/*f* power spectral density
FA	Folic acid
fA	Femtoampere
F-actin	Filamentous-actin
FAD	Flavin adenine dinucleotide
FADH	Reduced flavin adenine dinucleotide
FAM	Fluorescein amidite
FBCVD	Fluidized-bed chemical vapor deposition
Fc	Fragment crystallizable region of the antibody
FcSH	6-(Ferrocenyl)hexanethiol
FDS-NPs	Fluorescent dye doped-silica nanoparticles
FEP	Fluorinated ethylene propylene
FePt-NPs	Iron–Platinum nanoparticles
FET	Field-effect transistor
FI	Fluorescence intensity
FIB	Focused ion-beam
FIB-CVD	Focused ion-beam chemical vapor deposition
FIR	Fluorescence intensity ratio, far-infrared
FITC	Fluorescein-5-isothiocyanate
FL	Fluorescence
FM	Frequency modulation
fM	Femtomolar
FMF	Few-mode fiber
fm/√Hz	Femtometer per square root hertz
fmol	Femtomole
fN	Femtonewton
FRET	Fluorescence resonance energy transfer; Förster resonance energy transfer
ftDNA	Fragmented DNA
F. tularensis	*Francisella tularensis*
fW	Femtowatt
FWHM	Full width at half-maximum
F-Z process	Float Zone process
g	Gram
GB	Gigabyte
GC	Glassy carbon
GCE	Glassy carbon electrode
GC-SPR	Grating-coupling surface plasmon resonance
GDH	Glucose dehydrogenase
GF	Gauge factor
GFP	Green fluorescent protein
GHz	Gigahertz
GMNP	Glycan-functionalized magnetic nanoparticle
GMR	Giant magnetoresistance
GNE	Gold nanostructured electrode
GNPs	Gold nanoparticles
GNRs	Graphene nanoribbons
GOD, GO$_x$	Glucose oxidase
GOPS	Glycidoxypropyltrimethoxysilane
GP	Graphite
GPa	Gigapascal
gQD	Green-emissive QD
GR	Graphene
GSs	Graphene sheets
GSH	Reduced glutathione
G-rich DNA	Guanine-rich DNA
G/T duplex	Double-stranded DNA formed by hybridization of G-strand DNA with T-strand DNA
GQDs	Graphene quantum dots
h	Hour
h$^+$	Hole
Hb	Hemoglobin
HBsAg	Hepatitis B surface antigen
HBV	Hepatitis B virus
HCB	Hexachlorobenzene
HCG-β	Human chorionic gonadotrophin-β
HDA	Hexadecyl amine
HEB	Hot-electron bolometer
HEDD	Hot-electron direct detector
HEL	Hot embossing lithography
HEMT	High-electron-mobility transistor
HEPES	4-(2-Hydroxyethyl)-1-piperazineethanesulfonic acid)
HER2	Human epidermal growth factor receptor 2
HIgG	Human immunoglobulin

High-κ material	High dielectric constant material
HiPco	High-pressure carbon monoxide (process)
HIV-1	Human immunodeficiency virus-1
HMTA	Hexamethylenetetramine
HMX	High-melting explosive; Chemical name: Octahydro-1,3,5,7-tetranitro-1,3,5,7-tetrazocine
HOFs	Hollow-core optical fibers
HOMO	Highest occupied molecular orbital
HOPG	Highly oriented pyrolytic graphite
HPC	Hydroxypropyl cellulose
HPDP	Hexyl-3′-(2′-pyridyldithio)propionamide
HPV	Human papillomavirus
HRP	Horseradish peroxidase
hsp70	Heat-shock protein 70
HSV-1	Herpes simplex virus-1
Hz	Hertz
IBM	International Business Machines Corporation
IC	Integrated circuit
IDAMs	Interdigitated array microelectrodes
IDEs	Interdigitated electrodes
IDOS	Integrated density of states
IDTs	Interdigitated transducers
IgG	Immunoglobulin
IL	Ionic liquid
IL-6	Interleukin-6
IL6-sR	Interleukin 6-soluble receptor
IMNPs	Immunomagnetic nanoparticles
INS	Integrated nanosensor system
IONPs	Iron oxide nanoparticles
IoNT	Internet of nanothings
IP	In-plane
IR	Infrared
ISFET	Ion-sensitive field-effect transistor
ITO	Indium tin oxide
J	Joule
JEV	Japanese encaphilitis virus
K	Kelvin
kA	Kiloampere
kDa	Kilodalton
keV	Kilo electron volt
kg	Kilogram
km	Kilometer
kPa	Kilopascal
kV	Kilovolt
L	Liter
LANs	Local area networks
LB	Langmuir-Blodgett
LBL	Layer-by-layer
LC	L-carnitine
L-C circuit	Inductance-capacitance circuit
LCAO-MO	Linear combination of atomic orbitals-molecular orbitals
lcrV protein	Low calcium response locus protein V (Recombinant virulence-associated V antigen)
LCST	Lower critical soluble temperature
LDL	Lowest detection limit
LDR	Linear dynamic range
LDV	Laser Doppler vibrometer
LED	Light-emitting diode
LEDIT	Layout editor
LFB	Lateral flow biosensor
LIGA	Lithographie, Galvanoformung, und Abformung (German): Lithography, Electroplating, and Molding (English)
LiTf	Lithium triflate
L-J potential	Lennard-Jones potential
L. monocytogenes	*Listeria monocytogenes*
LOD, LoD	Limit of detection
LOQ	Limit of quantification
LPCVD	Low-pressure chemical vapor deposition
LPFGs	Long-period fiber gratings
LPS	Lipopolysaccharide
LRET	Luminescence resonance energy transfer
LSPR	Localized surface plasmon resonance
LSV	Linear sweep voltammetry
LTCC	Low-temperature co-fired ceramics
LUMO	Lowest unoccupied molecular orbital
M	Molar; a metal
m	Meter
MA	Maleic acid
MAA	Methacrylic acid
mAb or MAb	Monoclonal antibody
MAC	Multiple access control (protocol)
Maneb	Manganese ethylene-1, 2-bisdithiocarbamate (fungicide)
MarR	Multiple antibiotic resistance regulator
MBE	Molecular beam epitaxy
MBP	Maltose-binding protein
MBs	Magnetic beads
MC	Molecular communication
McAbs	Monoclonal antibodies
MCF-7	Cell named after Michigan Cancer Foundation, the institute where the cell line was isolated and established
MCH	6-Mercapto-1-hexanol
MDA-MB	MD Anderson metastatic breast (breast cancer line)
MEA	Mercaptoethylamine
MEMS	Micro-electromechanical systems
meV	Millielectron volt
MF	Magnetic fluid
MfnS	Magneto-fluorescent nanosensor
Mga	MAX gene-associated protein, MAX=MYC-associated factor X; *MYC* is a regulator gene family
6-MHO	6-Mercaptohexan-1-ol
MHPC	1-Myristoyl-2-hydroxy-sn-glycero-3-phosp hocholine
mHz	Millihertz
MHz	Megahertz
MION	Monocrystalline iron oxide nanoparticles
MIP	Molecularly imprinted polymer

miRNA-106a	Micro-ribonucleic acid-106a
mL	Milliliter
MLT	Malathion (insecticide)
mM	Millimolar
MMP	Magnetic microparticle
MMP-9	Matrix metalloproteinase-9
MNPs	Magnetic nanoparticles
MO	Molecular orbital
MOCVD	Metal-organic chemical vapor deposition
MOF	Metal-organic framework
mol	Mole
MOSFET	Metal-oxide field-effect transistor
MOVPE	Metal-organic vapor phase epitaxy
MOX NWs	Metal oxide nanowires
MOXs	Metal oxides
mp	Melting point
MP-3	Moving picture experts group (MPEG)-1 audio layer-3
MPA, 3-MPA	3-Mercaptopropionic acid
MPa	Megapascal
mPa	Millipascal
MPCVD	Microwave plasma chemical vapor deposition
MPECVD	Microwave plasma-enhanced chemical vapor deposition
mPEG	Methoxypoly(ethylene glycol)
MR	Magnetic resonance
MRI	Magnetic resonance imaging
mRNA	Messenger RNA
MR S-phone	Smartphone-based microwell reader
MRSws	Magnetic relaxation switches
ms	Millisecond
MTBE	Methyl-*tert*-butyl ether
MT devices	Multiple tube devices
MTJ	Magnetic tunnel junction
MUA	11-Mercaptoundecanoic acid
MUDA	Mercaptoundecanoic acid
mV	Millivolt
MW	Molecular weight
mW	Milliwatt
MWCNT	Multi-walled carbon nanotube
N	Newton
NA	Numerical aperture
nA	Nanoampere
NaAc	Sodium acetate
NAD	Nicotinamide adenine dinucleotide
NADH	Reduced nicotinamide adenine dinucleotide
Na_2EDTA	Disodium ethylenediaminetetraacetic acid
NALC	*N*-acetyl-L-cysteine
NBs	Nanobelts
NC	Nitrocellulose
nC	Nanocoulomb
NC100150	Superparamagnetic iron oxide nanoparticle
NCBA	(Magnetic) nanoparticle-based colorimetric biosensing assay
NDT	1,9-Nonanedithiol
NED	Naphthylethylene diamine
NEMS	Nano-electromechanical systems
NETD	Noise-equivalent temperature difference
NG	Nitroglycerine
ng	Nanogram
nG/$\sqrt{}$Hz	Nano gravitational acceleration per square root of Hertz
NHS	*N*-hydroxysuccinimide
Ni-NTA	Ni^{2+} ion coupled to nitrilotriacetic acid
NIR	Near-infrared
nM	Nanomolar
nm	Nanometer
nMgO	Nanostructured magnesium oxide
NMM	*N*-Methyl mesoporphyrin IX
NMR	Nuclear magnetic resonance
NRs	Nanorods
NSs	Nanosensors
NPs	Nanoparticles
NR	Nanorod
NSET	Nanomaterials surface energy transfer, nanosurface energy transfer
NSL	Nanosphere lithography
NSOM	Near-field scanning optical microscopy
NSPs	Nanoscale particles
NT	Nanotube
NTA	Nitrilotriacetic acid
4-NTP	4-Nitrothiophenol
NW	Nanowire
nW	Nanowatt
OASN	*N*-Octyl-4-(3-aminopropyltrimethoxysilane)-1,8 -naphthalimide
ODP	On-device percentage
ODT	1,8-octanedithiol
Oe	Oersted
Oligo	Oligonucleotide
Oligo-AuNP	Oligonucleotide-activated gold nanoparticle
OMR	Ordinary magnetoresistance
ONO	Oxide-nitride-oxide
OOP	Out-of-plane
OP	Organophosphate
OTA	Ochratoxin A
OTMS	Trimethoxy(octadecyl)silane
P	Parallel
PA	Picric acid; protective antigen
P_A	Probe for the anti-adenosine aptamer
Pa	Pascal
PAA	Poly(acrylic acid)
PAAm	Polyacrylamide
pAb	Polyclonal antibody
pag	Protective antigen
PAH	Poly(allylamine hydrochloride)
PAL	Peptidoglycan-associated protein
PAM	Polyacrylamide, pulse-amplitude modulation
PANI	Polyaniline, poly-2-aminoaniline
PAPBA	Poly(3-aminophenylboronic acid)
***Para*-MBA**	*Para*-mercaptobenzoic acid
PBN	*N*-*tert*-butyl-a-phenylnitrone
PBS	Phosphate-buffered saline
PBST	PBS-Tween= PBS + Tween (nonionic surfactant)
PC	Polycarbonate, personal computer
P_C	Probe for anti-cocaine aptamer
pcAb	Polyclonal antibody

PCR	Polymerase chain reaction
P(Cz-*co*-ANI)	Poly(carbazole-aniline)
PDMA	Poly (decyl methacrylate)
PDMS	Polydimethylsiloxane
PdNPs	Palladium nanoparticles
PDPH	3-(2-Pyridyldithio)propionyl hydrazide
PEBBLE	Photonic explorer; probe for bioanalysis with biologically localized embedding
PECVD	Plasma-enhanced chemical vapor deposition
PEDOT	Poly(3,4-ethylenedioxythiophene)
PE-FET	Piezoelectric field-effect transistor
PEG	Polyethylene glycol
PEI	Polyethyleneimine
PEM	Polyelectrolyte-modified (gold electrode)
PENG	Piezoelectric nanogenerator
PEO	Polyethylene oxide
PET	Polyethylene terephthalate
PETN	Pentaerythritol tetranitrate
PDADMAC	Poly(diallyldimethylammonium chloride)
PDMS	Polydimethylsiloxane
pDNA	Probe DNA
PfHsp70	*Plasmodium falciparum* heat-shock protein 70
PFMI	Poly(3,3′-((2-phenyl-9H-fluorene-9,9-diyl)-*bis*-(hexane-6,1-diyl))-*bis*-(1-methyl-1H-imidazol-3-ium) bromide)
PFTS	Perfluoro-octyltrichlorosilane
PG	Pyrolytic graphite
pg	Picogram
PGE	Pencil graphite electrode
pH	Potential of hydrogen
Phen	1,10-Phenanthroline
P_K	Probe for the potassium-specific G-quartet
PL	Photoluminescence
PLA	Polylactide
PLL	Phase-locked loop
PM	Particulate matter
pM	Picomolar
PMA	Pyrenemethylamine
PMMA	Polymethylmethacrylate
PMNPs	Paramagnetic nanoparticles
PMOS	P-channel metal-oxide-semiconductor (FET)
PMT	Photomultiplier tube
P3MT	Poly(3-methylthiophene)
PNAs	Peptide nucleic acids
PnC	Pneumococcal C-polysaccharide (antibody)
PNIPA, PNIPAM	Poly(*N*-isopropylacrylamide
PNLC	Polymer network liquid crystal
ppb	Parts per billion
PPE	Personal protective equipment
PPM	Pulse-position modulation
ppm	Parts per million
PPT	Plasmonic photothermal (effect)
Ppy	Polypyrrole
PRAM	Phase-change random access memory
PrG-Thiol	Protein G-thiol
Protein A/G	A protein containing four Fc-binding domains from protein A and two from protein G
Protein G	A cell wall protein of group G streptococci
PRM	Pulse rate modulation
PS	Porous silicon, polystyrene sphere
PSA	Poly(styrene-co-acrylic acid), prostate-specific antigen
pSC$_4$R	Para-sulphonato-calix[4]resorcinarene
pSC$_4$R-AgNPs	*Para*-sulfonato-calix[4]resorcinarene capped silver nanoparticles
PSG	Phosphosilicate glass
PSOP	Protease-specific iron oxide
PSS	Polystyrene sulfonate
PTB	Pulmonary tuberculosis
PTH	Pin-through-hole
PTFE	Polytetrafluoroethylene (Teflon)
PtOEPK	Platinum(II) octaethylporphyrine ketone
PVA	Polyvinyl alcohol
PVB	Polyvinyl butyral
PVC	Polyvinylchloride
PVD	Physical vapor deposition
PVP	Poly(*N*-vinyl-2-pyrrolidone)
PWM	Pulse-width modulation
PyCD	1-Pyrenebutyl-amino-β–cyclodextrin
PyoV	Pyoverdine
PZR	Piezoresistance
PZT	Lead zirconate titanate
QD	Quantum dot
QCM	Quartz crystal microbalance
QCM-D	QCM with dissipation monitoring
***Q*-factor**	Quality factor
R	An organic substituent or an organic group
rad/s	Radian/second
RAM	Random access memory
Rano	Ranolazine
RGB	Red-green-blue
RCA	Radio Corporation of America
RC delays	Resistive-capacitive delays
rDNA	Reporter DNA (probe)
RDX	Research department explosive, royal demolition explosive; chemical name: 1,3,5-trinitro-1,3,5-triazacyclohexane, cyclotrimethylenetrinitramine
REDONs	Rare-earth-doped nanocrystals
RF	Radio frequency
rGO	Reduced Graphene Oxide
RH	Relative humidity
RRE	Rev response element
RI	Refractive index
RIE	Reactive-ion etching
RIU	Refractive index unit
RNA	Ribonucleic Acid
Rnase	Ribonuclease
RNP	Ribonucleoprotein
ROS	Reactive oxygen species
Rox, ROX	Carboxy-X-rhodamine

RPM	Revolutions per minute
rQD	Red emissive QD
RSCS	Raman scattering cross section
RTA	Rapid thermal annealing
RTDs	Resistance temperature detectors
RU	Resonance unit or response unit
$(Ru[dpp]_3)Cl_2$	Ruthenium-tris(4,7-diphenyl-1,10-phenanthroline) dichloride
$Ru[dpp(SO_3Na)_2]_3$	Ruthenium(II)-tris(4,7-diphenyl-1,10-phenanthrolinedisulfonic acid) disodium salt
Ry	Rydberg
S	Entropy, Siemen
s	Second
SA	Streptavidin
SAA1	Serum amyloid A1 (Antigen)
SA-AP	Streptavidin-modified alkaline phosphatase
SAM	Self-assembled monolayer
SARS-CoV-2	Severe acute respiratory syndrome coronavirus 2
SAW	Surface acoustic wave
SBCS	Smartphone-based colorimetric system
SCE	Saturated calomel electrode
SDT	Spin-dependent tunneling
SEB	Staphylococcal enterotoxin B
SEFP	Standard extrinsic Fabry-Perot
SEM	Scanning electron microscope
SERS	Surface-enhanced Raman spectroscopy (or scattering)
SET	Single-electron transistor
sGP	Soluble glycoprotein
SH-PEG	Thiolated polyethylene glycol
SI	Systéme International d' Unités
SIMOX	Separation by IMplantation of OXygen.
SIMS	Secondary ion mass spectrometry
SiON	Silicon oxynitride
Si NPs	Silicon nanoparticles
SiNWs	Silicon nanowires
SLED	Super luminescent light emitting diode
SMPB	Succinimidyl 4-(p-malemidophenyl)-butyrate
SMT	Surface-mount-technology
SNARF1	Seminaphtho-rhodafluor-1
S-NHS	Sulfo-*N*-hydroxysuccinimide
SNIA	Silica nanoparticle-based immunoassay
SOFs	Silica optical fibers
SOI	Silicon on insulator
SPA	*Staphylococcus* protein A
SPANI	Poly(anilinesulfonic acid)
SPB	Surface plasmon band
SPE	Screen printed electrode
S. pneumoniae	*Streptococcus pneumoniae*
SPCE	Screen-printed carbon electrode
SPDP	*N*-Succinimidyl 3-(2-pyridyldithio)-propionate
SPIO	Small particle of iron oxide
SPMs	Scanning probe microscopes
SPP	Surface plasmon polariton (waves)
SPR	Surface plasmon resonance
sPS or SPS	Syndiotactic polystyrene
S. pyogenes	*Streptococcus pyogenes*
sr	Steradian
SRAM	Static random access memory
SRH-TSOOK	Symbol rate hopping time-spread on-off keying
ssDNA	Single-stranded DNA
ssG-DNA	Single-stranded genomic DNA
ssPNA	Single-stranded peptide/nucleic acid
SQUIDs	Superconducting quantum interference detectors
STF	$SrTi_{1-x}Fe_xO_3$
STM	Scanning tunneling microscope
STP	Scanning thermal profiler
Sulpho NHS	Sulpho *N*-hydroxysuccinimide
SV	Spin-valve
S/V	Surface/volume (ratio)
SV-GMR	Spin-valve giant magneto-resistance
SWASV	Square wave anodic stripping voltammetry
SWCNT	Single-walled carbon nanotube
$SWCNT_{HIPCO}$	High-pressure carbon monoxide (HiPco)-produced single-walled carbon nanotube
SWV	Square wave voltammetry
T	Tesla
2T	Two-terminal
TA	Thioctic Acid
TAE	Tris-acetate-EDTA; EDTA= Ethylenediaminetetraacetic acid
Tat protein	Transactivator of transcription protein
TB	Terabyte, tuberculosis
TBA	Thrombin-binding aptamer
TBABr	Tetra-*N*-butylammonium bromide
TCEP	Tris(2-carboxyethyl) phosphine
TCPPs	Tetra(4-carboxylatophenyl) porphyrins
TCR	Temperature coefficient of resistance
TCs	Thermocouples
TE buffer	'T' is derived from "Tris" and 'E" from "EDTA" meaning Ethylenediaminetetraacetic acid
tDNA	Target DNA
TDPC	3,3′-Thiodipropionic acid
TED	Thermoelastic dissipation
TEA	Triethanolamine
TEM	Transmission electron microscope
TENG	Triboelectric nanogenerator
TEOS	Tetraethylorthosilicate; tetraethoxysilane
Tetryl	Tetranitro-*N*-methylamine
TES	Transition edge sensor
TG	Thioglycerol
TGA	Thioglycolic acid
THF	Tetrahydrofuran
THMA	Tris(hydroxymethyl) aminomethane
THz	Terahertz
T-line	Test line
TMAH	Tetramethyl ammonium hydroxide
TMB	Tetramethylbenzidine
TMSPA	3-(Trimethoxy-silyl)propyl aldehyde

TMR	Tunnel magnetoresistance
$(TMS)_2S$	*Bis*(trimethylsilyl) sulfide
TNT	Trinitrotoluene
TOAB or TOABr	Tetraoctylammonium bromide
TOP	Tri-*n*-octylphosphine
TOPO	Tri-*n*-octylphosphine oxide
TOP-Se	Trioctylphosphine selenide
TPA	Tripropylamine
TPa	Terapascal
TREG	Triethylene glycol
Tris	Tris(hydroxymethyl)aminomethane, trisaminomethane, tromethamine or THAM, $(HOCH_2)_3CNH_2$
TRITC	Tetramethyl rhodamine isothiocyanate
Trp	Tryptophan
TS-OOK	Time-spread on-off keying
T-strand DNA	Single-stranded template DNA
UA	Uric acid
UC	Up-conversion
UCNP	Up-converting nanoparticle
UDSM	Ultra-deep submicron
UHF	Ultra-high frequency
UHVCVD	Ultra-high vacuum chemical vapor deposition
ULSI	Ultra-large-scale integration
USPIO	Ultra-small particle of iron oxide
UTI	Urinary tract infection
UV	Ultraviolet (radiation)
UV-vis	Ultraviolet-visible
V	Volt
VING	Vertically integrated nanowire generator
VP40	Viral protein 40 kDa
V-antigen	Virulence antigen (LcrV protein)
VB	Valence band
VHF	Very-high frequency
VLSI	Very-large-scale integration
VOC	Volatile organic compound
VSOP	Very-small iron oxide particles
W	Watt
WKB	Wentzel–Kramers–Brillouin (approximation)
WNNs	Wireless nanosensor networks
6x-His tag	Hexa histidine-tag or polyhistidine tag
XPS	Photoelectron spectroscopy
XRD	X-ray (powder) diffraction
XRL	X-ray lithography
ZCME	Zeolite-modified carbon paste electrode
ZIKV	Zika virus
Zineb	Zinc ethylene bis(dithiocarbamate) (fungicide)
β-CD	β-cyclodextrin
μA	Microampere
μg	Microgram
μL	Microliter
μM	Micromolar
μm	Micrometer
μs	Microsecond
μW	Microwatt
Ω	Ohm
MΩ	Megaohm

Mathematical Notation

A	Absorbance; amplitude of the incident wave; parabolic rate constant; area/area of cross section; amplitude of oscillation of the cantilever
$A\|_{1/2}$	Concentration of acceptor at which the energy transfer efficiency E is 50% (in FRET)
A420	Absorbance at 420 nm
A_{Blank}	Absorbance of blank
A519	Absorbance at 519 nm
A_{GB}	Cross section of the grain boundary
$A_{Specimen}$	Absorbance of specimen
a, a'	Distance
a_0	Radius of nanowire
A_{drive}	Amplitude at which actuator is driven
A_H	Hamaker constant
$A(\nu)$	Frequency (ν)-dependent enhancement factor of local optical field
$\hat{a}_1$ and $\hat{a}_2$	Unit vectors
B	Amplitude of reflected wave; the bandwidth requirement; constant; magnetic induction; detection bandwidth
$\mathbf{B}$	Magnetic induction vector
B_0	External magnetic field
b	Path length
B/A	Linear rate constant
C	Capacitance, constant
c	Concentration; velocity of light = $2.99792458 \times 10^8 ms^{-1}$ (in vacuum)
C_{ext}	Cross section (Scattering)
C_G	Gate capacitor of SET
C_g	Capacitance between the gate and the nanotube
$\vec{C}_h$	Roll-up vector
$C_{minimum}$	Capacitance of the smallest tunnel junction achievable with contemporary manufacturing technology
$C(x)$	Concentration function
D	Diffusion constant of the impurity at the temperature of diffusion; grain diameter; diffusion coefficient of biological or chemical target molecules in the solution
$\mathbf{D}$	Electric displacement
D_c	Critical grain diameter
D_n	Mode-dependent spring constant
d	Closest point of the tip to the sample; distance; tunnel gap; diameter
d_1 and d_2	Thicknesses of two dielectric layers
dA	An elemental area; change in surface area dA
$dC(t)/dt$	Rate of concentration of molecules
d_{SWCNTs}	Thickness of the SWCNTs layer
dW	Reversible work of deforming a surface (elastically)
E	Energy; elastic modulus/Young's modulus; electric field strength or intensity; internal energy; energy transfer efficiency in FRET
$\mathbf{E}$	Electric field vector
E^*	Biaxial modulus
E_0	Maximum electric field value
E_B	Bandgap energy of bulk semiconductor
E_c	Charging energy
E_F	Fermi level, Fermi energy
E_g	Energy gap
E_{gap}	Energy bandgap
E^0_{gap}	Zero-strain energy gap
$\mathbf{E}_i$	Internal electric field
E_j	Energy levels of the orbitals
E_{pA}	Peak anodic voltage
E_{pC}	Peak cathodic voltage
$E_{QD}(R)$	Bandgap energy of a QD of radius R
E^*_{Ryd}	Rydberg energy of electron-hole pair
e	Electron charge = $1.602176634 \times 10^{-19}$C
F	Force; free energy
F_{int}	Internal force
$F_{minimum}$	Minimum force
f_{drive}	Frequency at which actuator is driven
$f(E)$	Fermi distribution function
$F_{induced\ dipole}$	Induced dipole-induced dipole force (dispersion force)
$F_{permanent\ dipole}$	Permanent dipole-permanent dipole force
f	frequency; fundamental resonance frequency of a doubly clamped beam
f_0	Fundamental eigenfrequency; the resonance frequency of the fundamental mode of the crystal; fundamental resonance frequency of the microcantilever
f_1, f_0	Resonance frequencies of the cantilever after loading and before loading, respectively
f_1, f_2	Resonance frequency of the quartz crystal with Ab1 antibodies attached to the gold coating; resonance frequency of the quartz crystal with all the bindings including the CD-10 antigens and the AuNPs
f_2	Second bending frequency of the cantilever
f_D	Doppler frequency
f_{GaAs}	Resonance frequency of a gallium arsenide beam
FI	Fluorescence increase
f_M	Linear surface plasmon resonance frequency
f_{Si}	Resonance frequency of a silicon beam
F_{ts}	Tip-sample force
G	Gibbs free energy; conductance of the nanotube; gravitational acceleration

G_0, G	Conductances in the absence of an applied differential pressure ΔP and after ΔP is applied, respectively
G, GF	Gauge factor or strain factor
g	Gain input for the analog multiplier
H	Hamiltonian of a system; magnetic field; enthalpy; applied tickling field
h	Planck's constant = 6.62607015×10^{-34} J-s = $4.135667696\times10^{-15}$eV-s
$\hbar$	Reduced Planck's constant = Planck's constant/2π = $1.054571817\times10^{-34}$J-s = $6.582119569\times10^{-16}$eV-s
I	Electric current; area moment of inertia; the electrochemiluminescent (ECL) intensity at a given concentration of quencher [Q]; luminescence intensity of L-carnitine (LC)-capped quantum dot in a mercury ion solution; fluorescence intensity
I_0	Initial ECL intensity; original current; baseline current
I_1, I_2	Current levels
I_{11}, I_{21}, I_{31}, I_{41}	Currents obtained for a multi-particle distribution (for I_1)
I_{12}, I_{22}, I_{32}, I_{42}	Currents obtained for a multi-particle distribution (for I_2)
i	Tunnel current
I_H	Integral intensity of the $^2H_{11/2}\rightarrow{}^4I_{15/2}$ of Er^{3+} emission band
I_{max}	Luminescence intensity of LC-capped QD in a mercury ion-free solution, maximum fluorescence intensity
I_{pA}	Peak anodic current
I_{pC}	Peak cathodic current
I_S	Integral intensity of the $^4S_{3/2}\rightarrow{}^4I_{15/2}$ of Er^{3+} emission band
I_{SC}	Short circuit current
$I_{sd\ (maximum)}$	Maximum source-drain current
I_{Total}	Total tunneling current from sample to tip becomes
I-V	Current-voltage (characteristics)
I_z	Second moment of inertia
j	Electric current density
j_0	Peak electric current density
$\mathbf{J}_D$	Displacement current density
J_{SC}	Short circuit current density
$J(x)$	Diffusive flux
K	Kelvin; spring constant; restoring force constant; radius of gyration; extinction coefficient of fluorescence
K_d	Dissociation constant
K_{sv}	Quenching constant
k	Wave number; spring constant
k^*	Effective spring constant
k_0	Momentum of light in free space
k_B	Boltzmann's constant = 1.380649×10^{-23}JK^{-1}
k_i	Component of the incident light wave vector parallel to the prism interface
k_{sp}	Propagation constant of the surface plasmon

k_t	Rate of transfer of energy
k_{ts}	Tip-sample spring constant
L, l	Length; Lagrangian of a system; self-inductance; barrier width; grain size
L_0	Decay length
L_s	Thickness of surface space-charge layer of the oxide in a gas sensor
l_d	Characteristic evanescent electromagnetic field decay length
M	Bending moment; concentrated moment; modulation
$\mathbf{M}$	Magnetization vector
MR	Magnetoresistance ratio
M_{xy}	Component of the magnetization vector in the transverse magnetic plane
m	Mass; refractive index sensitivity
m_0	Effective suspended mass of the cantilever
m^*	Effective mass of a particle
m_b	Mass of the cantilever beam
m_e	Free electron mass = $9.1093837015\times10^{-31}$kg
m_e^*	Effective mass of the electron
m_{eff}	Effective mass of the cantilever
m_h^*	Effective mass of the hole
m_r	Reduced mass of the electron–hole pair
m/z	Mass-to-charge ratio
N	Number of electrons; Avogadro's number (6.02214076×10^{23} mol^{-1})
N_d	Volumetric density of the electron donors
N_s	Surface concentration of impurity; the minimum number of molecules to be captured for detection
N_t	Surface density of adsorbed oxygen ions
$N(x, t)$	Concentration of impurity in atoms cm^{-3} at distance x at time t
n	Number of electrons per unit volume; refractive index; real part of complex refractive index; quantum number, a factor
n^*	Complex refractive index
n_1, n_2	Wrapping indices
n_0	Density of the electrons trapped on the surface by the initially adsorbed oxygen in a ZnO-based gas sensor
n_b	Free-electron density in the grain body
n_d	Free-electron density in the depleted layer
n_r	Chemisorbed gas concentration
n_t	Surface electron density
P	Power; noise power; pressure
$\mathbf{P}$	Polarization field
$P(f)$	Power spectra
p	Momentum
Peak-PI(Blank)	Photoluminescence intensity without antigen
Peak-PI (PfHsp70 antigen)	Photoluminescence intensity with antigen
Q	Total impurity in atoms cm^{-2}; heat generated
$+Q$, $-Q$	Charges developed on the aluminum electrode and the gold electrode of PDMS membrane, respectively
[Q]	Concentration of quencher
$\overline{Q}$	Extra charge

$\widetilde{Q}$ Charge at which the current is maximum
q Electron charge = $1.602176634 \times 10^{-19}$C; charge; an integer
q_0 Background charge
$q'(t)$ Deflection of the tip of the cantilever
qV_s Height of the grain-boundary potential barrier (in eV); potential barrier (in eV)
R Reflection coefficient; spherical tip radius; radius of curvature; resistance; load resistor; distance between donor and acceptor (in FRET); response
R_0 Förster distance (or radius), initial resistance without strain
$R_\uparrow$ Resistance to current flow by up-spin electrons (up-spin resistance)
$R_\downarrow$ Resistance to current flow by down-spin electrons (down-spin resistance)
$R_{\downarrow\uparrow}$ Resistance measured with antiparallel magnetization
$R_{\uparrow\uparrow}$ Resistance measured with parallel magnetization
R_a Resistance of gas sensor in the reference gas (usually, the air)
$R_{Antiparallel}$ Resistance for the oppositely directed magnetic layers
R_0, R_1 Fitting parameters
R_{ct} Charge transfer resistance
$R_{dry\ air}$ Resistance in dry air
R_g Resistance of gas sensor in the reference gas containing target gases.
R_{GB} Grain-boundary resistance
R_H Rydberg energy of the hydrogen atom
R_{max} Maximum resistance
$R_{max}(\sigma)$ Maximum resistance as a function of strain σ
R_{min} Minimum resistance
R_N Neck resistance
R_{ozone} Resistance in ozone
$R_{Parallel}$ Resistance for parallel-aligned magnetic layers
R_S Resistance in series with the junction due to the metal-carbon nanotube (CNT) contacts
R_{SWCNTs} Single-walled carbon nanotube (SWCNT) film reflectance
R_{Total} Low-bias resistance of the device
r Distance between the particles, radius of curvature
r_m Radius of the cylinder constituting the neck (in the neck and grain boundary model of the gas sensor)
S Entropy; Siemen; sensitivity; sensing response
s Width of the barrier
T Kinetic energy; temperature on Kelvin scale; transmission coefficient
$\vec{T}$ Translational vector
T_1 Longitudinal relaxation time
T_2 Spin-spin relaxation time
t Time; thickness
$|t|^2$ Transmission probability
t_0 Tight-binding overlap integral
t_{ox} Oxide thickness
t_s Average response settling time
T_C Curie temperature; critical temperature; superconducting transition temperature
U Potential energy; voltage
$\langle u_n^2 \rangle$ Average fluctuation of the sensor
$u_y(x, t)$ Displacement of the beam in y direction at a point x and time t
$U_y(x, \omega)$ Displacement of the beam in y direction at a point x and angular frequency ω
V Electrical potential; Direct current (DC) bias voltage; supply voltage; volume
V_0, V_g Open-circuit potential (V_{OC}) in dry air and in target gas, respectively
V_l, V_2 Lower and higher voltage levels, respectively
V_{beam} Biasing voltage of the beam of resonator
V_c Voltage below which Coulomb blockade restricts electron flow to an island
V_G, V_g Applied gate voltage
V_{GAP} Source–drain voltage difference between former and latter V_{SD} values of the current minima
$V_{g(AC)}$ AC component of gate voltage
$V_{g(DC)}$ DC component of gate voltage
V_{OC} Open-circuit potential
V_s Height of the grain-boundary potential barrier (in V); potential barrier (in V)
V_{SD} Source–drain voltage
$V_{Tension}$ Voltage applied to electrode near the oscillating carbon nanotube (CNT) in a CNT radio transmitter
V_{tip} Tip voltage
V_{vdW} van der Waals (force)
v Velocity
$\mathbf{v}$ Velocity vector
v_0 Peak velocity
W_0 Mechanical energy accumulated in the device per vibration cycle
w Width
X Reactance
$[X]$ Analyte concentration
x Displacement
X_C Capacitive reactance
x_d Diffraction limit
X_L Inductive reactance
Y Young's modulus
y Beam deflection
z Deflection of the cantilever
z_{max} Amplitude of the cantilever oscillations
α Energy; the ratio of effective electron mass to electron mass
β transfer integral; exponential fitting parameter
β_1, β_2 Transfer integrals used to describe interactions of orbitals separated by different distances
Γ Damping constant
ΔA_{523} Change of absorbance between the blank and the specimen, as measured at 523 nm
ΔC Variation in capacitance
$\Delta d/\Delta t$ Rate of change of distance d with time t

Δf	Frequency shift; bandwidth in Hz; frequency difference (f_2-f_1)
ΔI	Change in intensity, change in current
ΔI_D	Change in drain current
Δm	Mass change per unit area
Δm_{th}	Smallest (thermal-noise limited) detectable change in the surface density of cantilever
Δn	Change in refractive index induced by an adsorbate
ΔP	Applied differential pressure
Δq	Charge induced on the SET
ΔR	Resistance change, difference between parallel and antiparallel resistances
$\Delta R/R_0$	Normalized resistance
ΔR_{ct}	Change in charge transfer resistance
ΔR_{et}	Change in electron-transfer resistance
ΔT	Temperature difference
ΔT_2	Change in spin-spin relaxation time
ΔV	Potential difference between two points
ΔW_0	Mechanical energy dissipated in the device per vibration cycle
$\Delta\lambda$	Difference in wavelengths
$\Delta\lambda_{max}$	Shift in the wavelength of maximum absorption
$\Delta\omega$	A small frequency interval
Δz	Cantilever deflection
δf_0	Absolute fluctuation of the resonance frequency
δm	Mass resolution
$\delta Q_{minimum}$	Minimum charge that can be detected by the electrometer
δq	Change in induced charge on the nanotube
δq_1	Change in the charge caused by gate voltage variation
δq_2	Change in the charge due to alteration in capacitance between the nanotube and gate
$\delta X_{minimum}$	Smallest displacement that can be sensed by the electrometer
δx_{zp}	Intrinsic fluctuation amplitude; the "zero-point motion"
$\delta\omega$	Angular frequency shift
ε	Molar extinction coefficient; depth of the potential well; strain
ε_0	Permittivity of free space = $8.8541878128\times10^{-12}$Fm^{-1}
ε_1, ε_2	Relative permittivities of two dielectric layers
ε_d	Dielectric constant of the dielectric
ε_{ext}	Dielectric constant of the external medium
ε_f	Dielectric constant of the optical fiber
ε_m	Dielectric constant of the metal
ε_r	Relative permittivity of the semiconductor material; relative permittivity of ZnO
ε_s	Relative permittivity of the semiconductor,
ε_{SWCNTs}	Complex dielectric constant of the SWCNTs layer
ε'_{SWCNTs}	Real part of the complex dielectric constant of the SWCNTs layer indicating the degree of polarization
ε''_{SWCNTs}	Imaginary part of the complex dielectric constant of the SWCNTs layer indicating the dielectric losses
$\varepsilon(\omega)$	Frequency-dependent dielectric constant
ε_∞	High-frequency permittivity limit
θ	Chiral angle; half-angle of the maximum cone of light that can enter or exit the lens
θ_c	Critical angle
θ_i	Incident light angle
κ	Dielectric constant; decay constant; curvature of the deflected beam; imaginary part of complex refractive index (extinction coefficient)
λ	Wavelength; mean free path; wavelength in a medium
λ_0	Wavelength in vacuum
λ_{air}	Wavelength in air
λ_{max}	Peak absorption wavelength i.e., wavelength in SPR spectrum at which absorption is maximum
λ_{Probe}	Wavelength of the probe beam in ultrasound sensor
$\lambda_{protein1}$	Wavelength in protein1
$\lambda_{protein2}$	Wavelength in protein 2
λ_{Pump}	Wavelength of pump beam in ultrasound sensor
μ	Refractive index of the prism material, Poisson's ratio
μ_0	Permeability of free space = $1.25663706212\times 10^{-6}$ Hm^{-1}
μ^*	Effective mass
μ_b	Electron mobility in the neutral grain body
μ_d	Electron mobility in the depleted layer
μ_q	Shear modulus of quartz
ν	Frequency, Poisson's ratio
ρ	Density
ρ_0	Minimum detectable concentration (detection limit)
ρ_q	Density of quartz
ρ_s	Sample states for tunneling from the sample
$\rho_s(E_F)$	Local density of electronic states in the test mass
ρ_t	Tip states for tunneling to the tip
σ	Distance at which the interparticle potential is zero; conductivity; conductance; surface stress; strain; standard deviation
σ_0	Pre-exponential factor in Arrhenius equation
σ_+ , σ_-	Surface stresses on the upper and lower surfaces, respectively
$-\sigma_{Contact}$, $+\sigma_{Contact}$	Charge densities produced on two dielectric surfaces, respectively, when they are brought in contact
$+\sigma_{Induction}(z, t)$	Induced charge density produced at a distance z at time t on the side of charge density $-\sigma_{Contact}$
$-\sigma_{Induction}(z, t)$	Induced charge density produced at a distance z at time t on the side of charge density $+\sigma_{Contact}$
σ_x	Normal stress
τ	A shift in the time coordinate
τ_D	Measured lifetime of the donor in the absence of the acceptor
Φ	Work function
φ	Height of the barrier; phase shift
ϕ	Phase shift; approximate work function of the metal; carbon nanotube (CNT) chiral angle; barrier height
ϕ_i	Degenerate orbitals

$\psi_{th}(f)$	White thermal noise
$\psi(x,y,z)$	Wave function or eigenfunction of a particle
ω	Angular frequency
ω_0	Angular resonance frequency
ω_1	Fundamental frequency
ω_1	Initial resonance frequency of cantilever
ω_2	Final resonance frequency of cantilever
ω_M	Angular surface plasmon resonance frequency of electron oscillations in metal nanoparticles
ω_n	Eigenfrequencies of cantilever
ω_P	Plasmon frequency

Part I

Fundamental Concepts of Nanosensors

1

Introduction to Nanosensors

1.1 Getting Started with Nanosensors

This book deals with *nanosensors* of *three key types of signals*, namely, *physical*, *chemical*, and *biological*. So, the questions that immediately come to mind are as follows: *(1) What is nanotechnology? (2) What are nanosensors?* and *(3) What is the meaning of the terms 'physical," "chemical," or "biological" alone or when applied to nanosensors?*

These questions need to be answered at the outset. Clearly, the starting point in the study of nanosensors will be the basic concepts of physics, chemistry, biology, and nanotechnology Semiconductor electronics will also be introduced. The key terms will be defined. These will enable the reader to recapitulate the preliminary knowledge required to understand the subject matter. They will also help to build up the vocabulary of the new words of nanotechnology that the reader will come across in this book. An exhaustive revision of these fields is not intended – only examples of interesting physical, chemical, and biological materials and phenomena will be provided to highlight the interdisciplinary nature of the subject.

Then, the motivation behind advancement from macro- to micro- and to nanosensors will be explained. The reader will be introduced to the rapidly advancing field of nanosensors and acquainted with different types of nanosensors, dealing with the fundamental principles and classification of nanosensors and laying the foundation upon which the succeeding chapters will be based. The scope and organizational plan of the book will be described.

Let us begin by revising our knowledge of elementary natural sciences.

1.2 Natural Sciences

What is natural science? Natural science is the systematized knowledge of nature and the physical world. It is a science, such as biology, chemistry, physics, and earth science, that deals with objects, phenomena, and laws of nature and the physical world (Brooks 2006). Parts of natural science (physics, chemistry and biology) are described in sections 1.3–1.5 below.

1.3 Physics

1.3.1 Definition of Physics

Physics is the science of forces that exist between objects, the interrelationship between matter and energy, and interactions between the two (Daintith 2010).

1.3.2 Branches of Physics

What are the different branches of physics? Physics is grouped into traditional fields, such as mechanics, properties of matter, heat, light, sound, electricity, and magnetism, as well as in modern extensions, including atomic and nuclear physics, cryogenics, solid-state physics, particle physics, and plasma physics.

1.3.3 Matter: Its States, Materials, and Particles

Matter is something that has weight and occupies space, for example, air, water, gold, iron, and wood. It exists as either a *solid*, having a definite volume and shape, such as a block of copper; a *liquid*, having a definite volume but not a definite shape, such as milk and juice; a *gas*, having neither a definite volume nor a definite shape, for example nitrogen and oxygen; or a *plasma*, containing a mixture of positive ions and free electrons in almost equal proportions.

Materials are substances out of which things are or can be made, serving as inputs to manufacturing plants. A *particle* is a very small piece or part, a tiny portion or speck of a material.

1.3.4 Molecules, Atoms, and Atomic Structure

A *molecule* is the smallest particle of a substance, element (which cannot be broken down into simpler substances), or compound (made of two or more simpler substances) that retains the chemical properties of the substance. An *atom* is the smallest part or particle of an element, having all the chemical characteristics of that element that uniquely define it, and can take part in a chemical reaction.

Does an isolated atom or molecule possess the bulk physical properties of the element or compound from which it is made? Not necessarily, because the influence of neighboring atoms or molecules, as present in bulk form, is no longer exercised.

Is an atom the ultimately smallest particle or are there further smaller particles? An atom is not the ultimately smallest particle into which matter can be broken down. Still smaller particles exist. Then, *what is the structure of the atom?* Talking about atomic structure, an atom consists of a central nucleus surrounded by a cloud of one or more electrons (Figure 1.1). The electron has a rest mass m_e of 9.1066×10^{-28} gram (g) and a unit negative electric charge of 1.602×10^{-19} coulomb (C). The *nucleus* is positively charged and contains one or more relatively heavy particles known as *protons* and *neutrons*. A proton is positively charged. A proton has a rest mass, denoted by m_p, of approximately 1.673×10^{-27} kilogram (kg). A neutron is electrically neutral and has a rest mass, denoted by m_n, of approximately 1.675×10^{-27} kg.

The number of protons in the nucleus of an atom of an element is its atomic number (Z). The total number of protons and neutrons in the nucleus of an atom gives its *mass number, atomic weight (A)*, or *relative atomic mass*. It is the ratio of the average mass per atom of the naturally occurring form of an element to 1/12th the mass of a carbon-12 isotope; an *isotope* is one or more

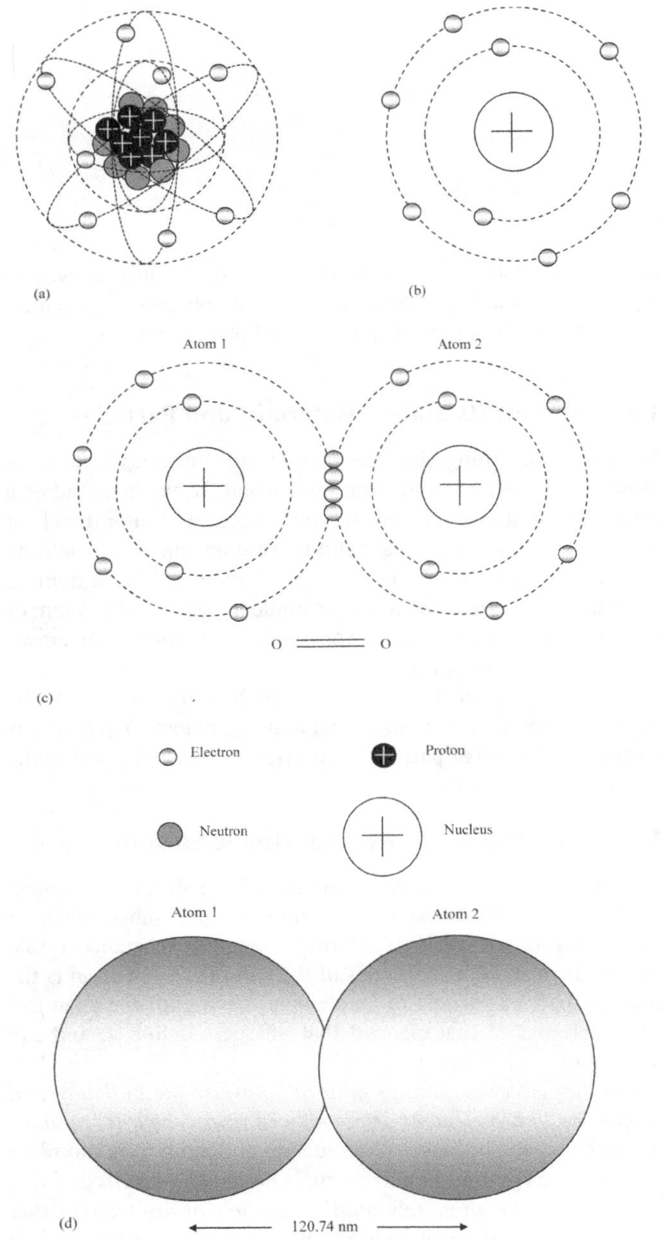

FIGURE 1.1 (a) Structure of oxygen atom (atomic number 8, atomic weight 16) containing eight protons ($1s^2\ 2p^6$) in the planetary atomic model. (b) Shell model of oxygen atom with two electrons in shell 1 and six electrons in shell 2. (c) Line diagram of oxygen molecule showing double bonds, each consisting of a pair of shared electrons. (d) Two orbitals in oxygen molecule. An orbital is the region surrounding an atomic nucleus indicating the electron's probable location.

atoms of an element having the same atomic number but a different atomic weight. One-twelfth the mass of a carbon-12 atom is called the *atomic mass unit* (amu) and is = 1.66033×10^{-27} kg.

Relative molecular mass or *molecular weight* of an element or compound is the ratio of the average mass per molecule of its naturally occurring form to 1/12th the mass of the carbon-12 atom and equals the sum of the relative atomic masses of the atoms comprising it.

1.3.5 Mechanics

Mechanics is the branch of physics that deals with forces acting on bodies and the resulting motions. The subject started with Newtonian mechanics that described the translational and rotational motion of bodies under Newton's three laws of motion, but it was extended to liquids and gases (fluids) as fluid mechanics, to molecular motion as statistical mechanics, and to subatomic particles as quantum mechanics. A *quantum* is a discrete minimum amount by which a physical quantity changes.

The *position* of a body in three dimensions is specified by its (x, y, z) coordinates. The *displacement* of a body is the distance traveled by it in a given direction. *Velocity* of a body is the distance traveled per unit of time in a given direction. The *mass* (m) of a body is the amount of matter contained in it, defined in terms of its opposition to acceleration. *Acceleration* of a body is the rate of change of its velocity. The *momentum* of a body is the product of its mass and velocity. *Force* (F) is any agent that changes the speed or direction of motion of a body.

$$\text{Force} = \text{Mass of the body} \times \text{Acceleration}. \quad (1.1)$$

The *density* of a substance is its mass per unit volume. The *viscosity* of a fluid is the force acting per unit area between its layers per unit velocity gradient.

Pressure is the force applied per unit area on a body. *Stress* is the force acting per unit cross-sectional area on a body, tending to cause its deformation. *Strain* is the ratio of change in dimension of a body (e.g., change in length) to its original dimension (original length). The ratio of stress applied to a material to the strain induced in it is called the *elastic modulus* (E) of the material.

Energy (E) is the capacity of a physical system to perform work. *Power* is the rate of doing work. *Potential energy* is the ability of a body to do work by virtue of its position, relative to others, for example, a heavy stone on the top of a mountain possesses gravitational potential energy due to its height above the ground. *Kinetic energy* is the energy possessed by a body due to its motion, such as the energy of a moving train.

The *Lagrangian* of a system (symbol L) is a function $L = T - U$, where T is the kinetic energy and U, the potential energy. The *Hamiltonian* of a system (symbol H) is a function used to express its energy in terms of momentum and positional coordinates; in simple cases, it is the sum of the kinetic and potential energies.

A *wave* is a periodic disturbance moving in space or in a medium. Waves are of two types: *longitudinal* (in which the disturbance is in the same direction as that of wave propagation) and *transverse* (in which the disturbance is perpendicular to the direction of the wave). The *wavelength* of a wave is the distance between successive compressions and rarefactions of a longitudinal wave (or crests and troughs of a transverse wave). The *frequency* of a wave is the number of cycles per second, usually expressed in Hertz (Hz). The *period* is the reciprocal of frequency. *Amplitude* is the largest difference of the disturbed quantity from its mean value.

Two fundamental differences exist between quantum and classical mechanics:

1. The parameters of a quantum system, such as the position and momentum of an electron, are affected by the act of measuring them. This is expressed in the

uncertainty principle, which states that the uncertainty in simultaneous measurement of the *x*-coordinate and the *x*-component of the momentum of the particle is greater than or equal to a minimum value defined by the equation

$$\Delta x \Delta p_x \geq \frac{h}{4\pi} \tag{1.2}$$

where h is the Planck's constant, a fundamental constant of quantum mechanics which relates the energy carried by a photon to its frequency.

2. The parameters of a quantum system are not exact or precise values but are expressed as probability distributions. This happens because electrons act as particles and waves at the same time; this concept is called *wave-particle duality*. The wavelength (λ) associated with a particle of mass m moving with a velocity v is

$$\lambda = \frac{h}{mv} \tag{1.3}$$

For an electron in an atom, the ideas of *atomic and molecular orbitals* originate from here (see Section 1.4.9).

Wave function or *eigenfunction* $\psi(x, y, z)$ of a particle is a mathematical expression involving the coordinates of the particle in space, such that the square of its absolute value $|\psi|^2$ at a point is proportional to the probability of finding the particle in a small elementary volume $dxdydz$ at that point. Each permitted wave function of a particle has a corresponding allowed energy level called the *eigenvalue*. The wave functions are the solutions of the *Schrödinger*'s *equation*, the time-independent form of which is

$$\nabla^2\psi + \frac{8\pi^2 m(E-U)\psi}{h^2} = 0 \tag{1.4}$$

where m is the mass of the particle, E is its total energy, and U is the potential energy.

1.3.6 Heat

Heat or *thermal energy* is the energy of vibration, rotation, or translation of the atoms or molecules of a body. It is transferred from one place to another by one of three modes: (i) *conduction* (through successive collisions from high-kinetic-energy molecules to neighboring low-kinetic-energy molecules); (ii) *convection* (by mass movement of high-kinetic-energy molecules from a high-temperature region to a low-temperature region); or (iii) *radiation* (by electromagnetic waves at the speed of light, not requiring any material medium).

Temperature is a measure of the degree of hotness or coldness of a body, that is, the average kinetic energy of the molecules of the body. *Absolute zero* is the theoretical lowest attainable temperature = –273.15°C or 0 K, at which all atomic motion has ceased. It cannot be attained because it requires an infinite amount of energy.

Specific heat capacity of a substance is the amount of heat required to raise the temperature of the unit mass of the substance by 1°C. *Latent heat* is the quantity of heat evolved or absorbed to change the state of a unit mass of a substance without any change in its temperature.

Heat is measured in kilocalories (kcal). The amount of heat required to raise the temperature of 1 kg of the substance by 1°C is 1 kcal.

Heat = mass × specific heat × rise in temperature.

1.3.7 Sound

Sound or *acoustic energy* is the energy of a vibrating object that is transmitted through a medium in the form of longitudinal pressure waves, consisting of compressions (high-pressure regions) and rarefactions (low-pressure regions), taking place in the same direction as that of wave propagation.

Sound requires a material medium for propagation, and its speed is different in various media. The speed of sound in a medium is the square root of the ratio of elastic bulk modulus of the medium to its density. In air, the speed of sound is 330 m s^{-1}. *Audible sound* has a frequency between 20 and 200,000 Hz. *Infrasound* has a frequency below 20 Hz. *Ultrasound* has a frequency above 20,000Hz. Ultrasonic waves are used for nondestructive flaw detection in metals. They are also used in medical diagnosis in pregnancy where X-rays could be harmful.

The decibel (dB) is a dimensionless unit of power ratio used in acoustics. For a pair of powers, P_2 and P_1, the power ratio in decibels is 10 $\log_{10}(P_2/P_1)$.

1.3.8 Light

Light energy consists of visible electromagnetic radiation with wavelengths from 400 to 750 nm. Light is propagated as waves at a speed of 3 × 10^8 m second^{-1} in vacuum. It is also conveyed as discrete energy packets or quanta, called *photons*.

Reflection is the process in which light bounces back or is deflected at the boundary between two media. *Refraction* is the process of change of direction in which light bends in crossing from one medium to another of different density. The ratio of the speed of light in a vacuum to the speed of light in a medium is called the *refractive index* of the medium. *Dispersion* is the splitting of light of mixed wavelengths into its component colors. *Diffraction* is the spreading of light when it passes through an aperture or around the edge of a barrier. *Polarization* is the process of restricting the vibrations of the electric vector of light waves to one direction; in unpolarized light, the electric field vibrates in all directions perpendicular to the direction of propagation.

Fluorescence is the emission of light by a material in response to irradiation with electromagnetic waves. It lasts for a few microseconds. *Phosphorescence* is a similar phenomenon, lasting for a few minutes.

1.3.9 Electricity

Electrical energy is a form of energy arising from the existence of charged bodies. A body is said to be *electrically charged* if, on rubbing with another body, it acquires the ability to attract light objects, like pieces of paper, fur, etc. The charge produced on a glass rod rubbed with silk is called *positive charge*, whereas that created on an ebonite rod rubbed with flannel is known as *negative charge. Unlike charges attract each other and like charges repel. Electric field* is the region of space in which force is exerted by the charge. *Electric field strength* or *intensity* (E) at a point in an electric field is defined as the force per unit charge experienced by a small charge placed at that point. *Electric current* (I) is the flow of electric charge and its magnitude is given by the rate of flow of charge, that is, the amount of charge per unit time. *Circuit* is the closed path around which electric current flows. *Electrical potential* (V) at a point in an electric field is the work done in transferring a unit positive charge from infinity to that point, whereas *potential difference between two points* (ΔV) is the work done in transferring a unit positive charge from one point to the other.

Electric displacement is the displacement of charge perpendicular to the direction of the field produced in an electric field per unit area. *Permittivity* is the ratio of electric displacement to electric field. *Relative permittivity* or *dielectric constant* of a material is its permittivity with respect to permittivity of free space. *Displacement current* is the rate of displacement of charge in an insulator (a material that does not allow current to flow through).

The steady current of constant magnitude flowing in one direction is called *direct current* (DC). *Alternating current* (AC) is a current that periodically changes its magnitude and direction.

Resistance of a body is its opposition to the flow of electric current. *Conductance* is the reciprocal of resistance. *Resistivity* of a material is the resistance of a unit cube of the material. *Conductivity* is the reciprocal of resistivity. A *Resistor* is a circuit component designed to give a specified resistance value in a circuit.

Capacitance of a conductor is the charge produced on it per unit potential applied. *Capacitor* is a circuit component for storing electric charge, designed to provide a fixed or variable value of capacitance in a circuit. A *parallel plate capacitor* consists of two metal plates separated by an insulator.

Inductance is the property by which an electromotive force (EMF) is generated in a conductor due to a changing current in itself (self-inductance) or a neighboring circuit (mutual inductance). An *inductor* is a circuit component fabricated to give a specified inductance value. It generally consists of a coil wound around a core of magnetic material.

Impedance in a circuit is the sum of contributions from resistors, capacitors (in the form of capacitive reactance, $X_C = 1/\omega C$), and inductors (as inductive reactance, $X_L = \omega L$). Here ω is the angular frequency, C is the capacitance and L is the inductance.

1.3.10 Magnetism

Magnetic energy is the energy associated with magnets. A *magnet* is a piece of iron, steel, alloy, ore, etc., showing the property of attracting iron or similar materials called *magnetic materials. Magnetic field* is the region surrounding a magnetic pole, in which the magnetic force due to it is perceived. A *magnetic pole* is each of the two regions of a magnet from which the magnetic force appears to originate. The strength and direction of the magnetic field (H) is expressed in terms of the magnetic flux density or magnetic induction, symbol (B) defined as the magnetic flux per unit area of a magnetic field perpendicular to the magnetic force. *Flux* is a measure of the quantity of magnetism, taking into account the strength and extent of the magnetic field. *Magnetic permeability* is the ratio of magnetic flux density to the magnetizing field.

Magnetism is the study of magnetic phenomena and their laws. Magnetism arises from the spinning motion of electrons so that each electron produces a small magnetic field. The magnetic effects of electrons spinning in opposite directions cancel each other out. Magnetism is of four types: diamagnetism, paramagnetism, ferromagnetism, and ferrimagnetism. In *diamagnetism*, the magnetization is opposite to the applied magnetic field, weak and temporary. Diamagnetic materials contain paired electrons. In *paramagnetism*, it is in the same direction as the applied magnetic field, but weak and temporary. In *ferromagnetism* occurring in materials like iron, cobalt, and nickel, there is an enormous increase in magnetization in the same direction as the field, due to the alignment of regions of aligned electron spin called domains. Moreover, magnetism is retained even after removal of the field. Both paramagnetic and ferromagnetic materials contain unpaired electrons. *Antiferromagnetism* is a property possessed by some metals, alloys, and salts of transition elements, such as manganese oxide (MnO), by which the atomic magnetic moments form an ordered array that alternates or spirals, so as to give no net total moment in zero applied magnetic field and hence almost no gross external magnetism. In *ferrimagnetism*, which is observed in composite materials, such as magnetite, rather than individual elements, the overall spin effect in one direction is greater than that in the other.

1.3.11 Electromagnetism

Scottish physicist J.C. Maxwell formulated four basic equations, popularly known as *Maxwell equations*, unifying electricity and magnetism, by describing the space and time dependence of electric and magnetic fields: (i) a relationship between electric field and electric charge; (ii) a relationship between the magnetic field and magnetic poles; (iii) generation of an electric current by a changing magnetic field; and (iv) production of a magnetic field by an electric current or a changing electric field.

Electromagnetic radiation is the energy emitted by an accelerated charge and its accompanying electric and magnetic fields. An *electromagnetic wave* consists of oscillating electric and magnetic fields at right angles to each other and the direction of propagation. *Electromagnetic spectrum* is the range of wavelengths or frequencies over which the electromagnetic waves extend. It covers radio waves, infrared (IR) waves, visible light, ultraviolet waves, X-rays, and gamma rays, in the order of descending wavelength.

1.3.12 SI System of Units

SI (Systéme International d′ Unités) is a coherent system of metric units for scientific usage across the world, based on seven basic units: meter (length), kilogram (mass), second (time), ampere (electric current), Kelvin (thermodynamic temperature), mole (amount of substance), and candela (luminous intensity). *Meter* (m) is the length of path traveled by light in vacuum in 1/299 792 458 of a second. *Kilogram* (kg) is the mass of the platinum-iridium (PtIr) prototype kept by the International Bureau of Weights and Measures at Sévres near Paris, France. *Second* (s) is the duration of 9,192,631,770 periods of the radiation corresponding to the transition between the two hyperfine levels of ground state of the Cesium-133 atom. *Ampere* (A) is the constant current that, when maintained in two straight, parallel, infinitely long conductors of negligible cross section, will produce a force of 2×10^7 N m^{-2} between the conductors. *Kelvin* (K) is the temperature equal to the 1/273.16 fraction of the thermodynamic temperature of the triple point of water, the temperature and pressure at which vapor, liquid, and solid phases of water are in equilibrium, = 0.01°C or 273.16 K and 611.2 Pa. *Mole* (mol) is the amount of a substance that contains as many elementary units (atoms, molecules, ions, radicals, electrons, etc.) as there are atoms in 0.012 kg of carbon-12, an isotope of carbon. *Candela* (Cd) is the luminous intensity in a given direction of a source that emits monochromatic radiation at a frequency of 540×10^{12} Hz and has a radiant intensity in this direction of 1/683 W sr^{-1} (Watt steradian^{-1}). The steradian is the unit of solid angle in the SI system.

1.4 Chemistry

1.4.1 Definition of Chemistry

Chemistry is the science of the properties, structure, and composition of substances, and how these undergo transformations (Tro 2010).

1.4.2 Elements and Compounds

Pure substances have an invariable composition and are composed of either elements or compounds. *Elements* are substances that cannot be broken down into simpler substances by chemical means, e.g., hydrogen, argon, silicon, iron, copper, silver, etc. *Compounds* can be broken down into two or more elements, e.g., water, carbon dioxide, ammonia, methane, sodium chloride, copper sulfate, etc. They are substances made by the combination of two or more different elements in fixed proportions.

1.4.3 Organic and Inorganic Compounds

Organic compounds are those that contain carbon. They contain one or more of the following elements: carbon (symbol C), hydrogen (H), oxygen (O), nitrogen (N), sulfur (S), halogens (fluorine [F], chlorine [Cl], bromine [Br], iodine [I], and astatine [At]), and occasionally phosphorus (P) and metals. Because of the ability of carbon atoms to form long chains, the number of organic compounds far exceeds that of other elements. These compounds form the basis of living matter.

Inorganic compounds are those that do not contain carbon and usually do not occur inside living organisms. These compounds are formed from elements other than carbon.

1.4.4 Subdivisions of Chemistry

What are the main subdivisions of Chemistry? Physical, organic, and inorganic chemistry. *Physical chemistry* deals with the physical principles governing the chemical phenomena, the physical structure of chemical compounds, and the physical effects of chemical reactions. *Organic* and *inorganic chemistry* are, respectively, the branches of chemistry concerned with organic and inorganic compounds. Organic chemistry is the chemistry of the element carbon, traditionally excluding its oxides and the carbonates, which are studied in inorganic chemistry.

1.4.5 Natural and Artificial Elements

Ninety-four elements, from atomic numbers 1–94, occur in nature. Elements heavier than plutonium (symbol: Pu, atomic number 94), the heaviest naturally occurring element, are called *artificial elements.* Elements up to atomic number 118 have been made in the laboratory. According to theoretical predictions, the highest atomic number lies between 170 and 210.

Hydrogen is the most abundant element in the universe (90.7% by mass), while oxygen is the most abundant element in the earth's crust (47.4% by mass). Tungsten (W) has the highest melting (mp) and boiling points (bp), namely 3407°C mp and 5657°C bp).

1.4.6 Metals, Nonmetals, and Metalloids

Elements are divided into three classes: metals, nonmetals, and metalloids. *Metals* form positively charged ions (cations) by the loss of electrons from their atoms; they have positive valency. *Nonmetals* form negatively charged ions (anions) by the gain of electrons by their atoms; they have negative valency. Metals (except mercury, Hg, which is a liquid) are solid at room temperature. In general, metals, for example, gold (Au), silver (Ag), copper (Cu), and aluminum (Al), etc., are lustrous (shiny), malleable (can be shaped by hammering), ductile (capable of being drawn into a wire, not brittle, moldable), and good conductors of heat and electricity. Examples of nonmetals are argon (Ar), bromine (Br), carbon (C), chlorine (Cl), fluorine (F), helium (He), iodine (I), krypton (Kr), neon (Ne), nitrogen (N), oxygen (O), phosphorus (P), radon (Rn), selenium (Se), sulfur (S), and xenon (Xe). Nonmetals are generally solids or gases, except for bromine, which is a liquid.

Elements such as arsenic, antimony, hydrogen, silicon, germanium, and tellurium exhibit characteristics of both metals and nonmetals, that is, properties intermediate between those of metals and nonmetals, and are called *metalloids.*

1.4.7 Periodic Table of Elements

The *Periodic Table* is a compilation of the chemical elements in ascending order of their atomic numbers to show the similarities of properties of elements. The vertical columns of the periodic table numbered I–VIII are called *groups*, while the horizontal rows 1–7 are known as *periods.* The groups contain elements that are similar to each other.

1.4.8 Chemical Change and Reaction

When water is boiled, it is converted into steam. This is a physical change because only the state has changed from liquid to gas, and can be reversed by cooling, at which point water vapor condenses. Steam and water have the same composition, consisting of H_2O molecules. But when a piece of sodium (Na) is added to water, hydrogen (H_2) gas is evolved and sodium hydroxide (NaOH) is formed. This is a chemical change in which the nature of substances is altered.

A *chemical reaction* is the interaction between one or more elements or compounds called *reactants* to form new elements or compounds termed *products*. A *chemical equation* is the symbolic representation of a chemical reaction; for example, potassium chlorate decomposes, on heating in the presence of manganese dioxide powder, into potassium chloride and oxygen. Manganese dioxide acts as a catalyst, a substance that increases the speed of a reaction without any changes in itself. Symbolically,

$$2KClO_3 = 2KCl + 3O_2\left(MnO_2\text{ catalyst}\right) \quad (1.5)$$

Enthalpy (*H*) is the heat content of a substance. *Entropy* (S) is a property of a substance related to the degree of disorder or randomness in it. *Free energy (F)* is the available energy of a chemical reaction. *Gibbs free energy G* is given by

$$G = H - TS \quad (1.6)$$

where *T* is the absolute temperature.

1.4.9 Electronic Configuration (Structure) of Elements

This is the arrangement of electrons in the different orbitals or shells of an atom of the element. The orbital or shell is the region in space surrounding the nucleus in which there is the greatest probability of finding the electron. For example, the electronic structure of oxygen is $1s^2\ 2s^2\ 2p^4$. The symbols s, p, d, f, ... represent the sub-shells. The prefixes 1, 2, ... denote the shell number while the superscript is the number of electrons in each sub-shell.

1.4.10 Chemical Bond

The chemical bond is the attractive force between the atoms in a molecule that binds together the atoms. It originates from the tendency of atoms to acquire the stable inert gas configuration. The main types of bonds are as follows: (i) *ionic* or *electrovalent bond* in which electrons are transferred from one atom to another forming positive and negative ions that stick together by electrostatic force; (ii) *covalent bond* in which the atoms are held together by sharing electron pairs; bonds formed by sharing one, two, or three pairs of electrons are called *single*, *double*, or *triple bonds* respectively; (iii) *hydrogen bond*, in which a hydrogen atom attached to one of the three elements, fluorine, oxygen, or nitrogen, is able to form a bridge with another one of these three elements; and (iv) *metallic bond*, forming in metals, which, due to their low ionization energies, lose one or more of their outer shell electrons, becoming positive ions so that the metal is pictured as a sea of free mobile electrons in which positive ions are immersed. The free electrons are said to be delocalized. *Delocalization* is the spreading of a molecule's electrons over the molecule.

1.4.11 Oxidation and Reduction

Oxidation of an atom, molecule, or ion is a reaction in which there is a loss of electrons from it, for example, in the conversion of magnesium (Mg) into magnesium oxide (MgO), Mg^{2+} and O^{2-} ions are formed. As the Mg^{2+} ion is obtained from Mg by the loss of two electrons ($2e^-$), magnesium is oxidized here. *Reduction* of an atom, molecule, or ion is the reaction in which there is a gain of electrons by it. The overall reaction is called reduction-oxidation or a *redox reaction*.

1.4.12 Acid, Base, and Salt

An *acid* is a substance that dissociates in aqueous solution to give hydrogen ions; for example, HCl and H_2SO_4. It acts as a proton donor. It is a substance that has a tendency to accept an electron pair to form a covalent bond.

A *base* is a substance, such as KOH or NaOH, that dissociates in aqueous solution to produce hydroxide ions, acting as a proton acceptor. A base is a substance that can donate an electron pair to form a covalent bond. An *alkali* is a base that is soluble in water.

A *salt* is the product of the neutralization reaction between an acid and a base formed by replacing the hydrogen of an acid with a metal or an electropositive ion, such as the ammonium (NH_4^+) ion. An example is the sodium chloride (NaCl) salt formed by the reaction between HCl (hydrochloric acid) and NaOH (sodium hydroxide). Similarly, Na_2SO_4 (sodium sulphate) is produced from H_2SO_4 (sulfuric acid).

1.4.13 Expressing Concentrations of Solutions and Gases

The concentration of a solution is expressed as the number of grams of solute per liter of solution or number of moles per liter (molar concentration). The concentration of a gas in a mixture of gases is expressed as the percentage (parts per hundred) of the gas in the mixture. It is also expressed in parts per million ($1/10^6$; ppm) or parts per billion ($1/10^9$; ppb).

1.4.14 Hydrocarbons: Saturated and Unsaturated

Hydrocarbons are compounds made of only two elements, carbon and hydrogen. Hydrocarbons are classified as follows: (i) *aromatic hydrocarbons*, such as benzene (Figure 1.2), in which carbon atoms are arranged in an unsaturated ring with delocalized electrons above and below; and (ii) *aliphatic hydrocarbons*, in which carbon atoms are arranged in an open chain with no delocalized electrons; these are further subdivided into alkanes, alkenes, alkynes, and alicyclic compounds.

Alkanes (e.g., methane (CH_4), ethane (C_2H_6), and propane (C_3H_8)) are hydrocarbons with the general formula C_nH_{2n+2}. *Alkenes* (e.g., ethane (C_2H_4), propene (C_3H_6), and butene (C_4H_8)) are hydrocarbons having the general formula C_nH_{2n}. *Alkynes* (e.g., ethyne (C_2H_2), propyne (C_3H_4), and butyne (C_4H_6)) are hydrocarbons with the general formula C_nH_{2n-2}. *Alicyclic compounds* (e.g., cyclobutane (C_4H_8), cyclopentane (C_5H_{10}), and cyclohexane (C_6H_{12})) are the class of cyclic hydrocarbons that closely resemble open chain hydrocarbons.

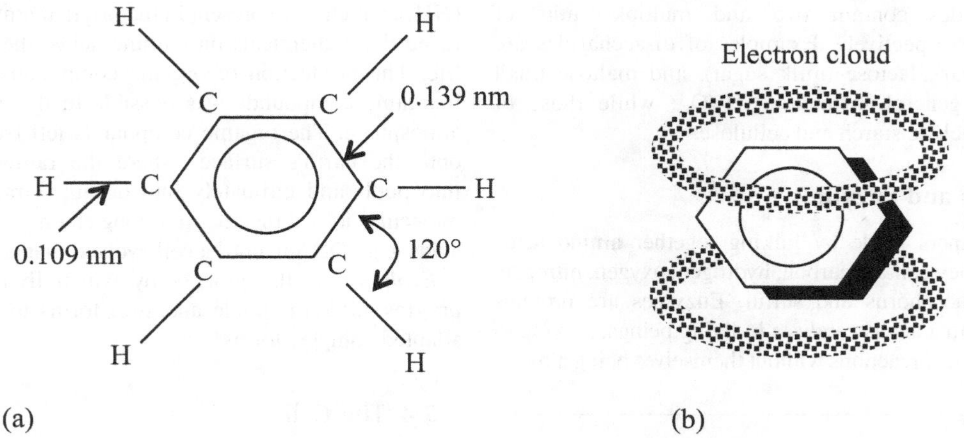

FIGURE 1.2 Benzene ring: (a) conventional symbol; (b) delocalization of electrons causing high electron density above and below the ring.

Organic compounds containing single bonds are said to be *saturated compounds;* those containing multiple (e.g., double or triple) bonds are *unsaturated.* As alkanes have single carbon-to-carbon atom bonds, they are saturated hydrocarbons. Alkenes, having double carbon–carbon bonds, and alkynes, with triple carbon–carbon bonds, are examples of unsaturated hydrocarbons.

1.4.15 Alkyl and Aryl Groups

An alkyl group (symbol R) is a group formed from an alkane by removal of a single atom of hydrogen ($-CH_3$, methyl group; $-CH_3CH_2$, ethyl group, etc.). An aryl group is a group formed from an aromatic compound by removing a single hydrogen atom, such as the C_6H_5- group from benzene.

1.4.16 Alcohols and Phenols

An alcohol is one of a class of organic molecules containing an alkyl group connected to a hydroxyl group, ROH, such as methyl alcohol or methanol (CH_3OH) and ethyl alcohol or ethanol (CH_3CH_2OH). A phenol is one of a family of aromatic organic compounds containing an aryl group attached to a hydroxyl group, for example, C_6H_5OH, hydroxybenzene or phenol.

1.4.17 Carboxylic Acids

These are weak organic acids containing the group –COOH; for example, HCOOH, methanoic or formic acid and CH_3COOH, ethanoic or acetic acid.

1.4.18 Aldehydes and Ketones

Aldehydes are organic compounds containing an alkyl group, aryl group, or a hydrogen attached to a carbonyl group, =CO, with an adjacent hydrogen, –CHO; for example, methanal (formaldehyde), HCHO; ethanal (acetaldehyde), CH_3CHO.

Ketones are also organic molecules containing a carbonyl group, =CO, attached to two alkyl groups, with the structure: R–CO–R; for example, CH_3COCH_3, propanone.

1.4.19 Amines and Amino Acids

These are alkyl or aryl derivatives of ammonia (NH_3), with the structure $R-NH_2$; for example, CH_3NH_2, methylamine; $C_2H_5NH_2$, ethylamine; and $C_6H_5NH_2$, phenylamine. The group $-NH_2$ is called the amino group.

Amino acids are compounds containing both an acidic carboxyl group, –COOH, and a basic amino group, $-NH_2$.

1.4.20 Lipids

These are water-insoluble biomolecules, such as fats, oils, and waxes. Fats are organic compounds that are made up of carbon, hydrogen, and oxygen. They are a source of energy in foods. Saturated fats are found in animal products, such as butter, cheese, whole milk, ice cream, and fatty meats, but also some plant products, such as coconut oil. Unsaturated fats are almost always plant based. Mostly, but not all, liquid vegetable oils are unsaturated fats. Fats usually imply substances that are solids at normal room temperature while oils generally refer to fats that are liquids at normal room temperature. Oils also mean substances that do not mix with water and have a greasy feel, such as essential oils. Hydrogenated oils are oils that are hardened, such as hard butter and margarine. Waxes are various natural, oily, or greasy heat-sensitive substances, such as beeswax.

1.4.21 Carbohydrates

Carbohydrates are a class of biological molecules composed of carbon with hydrogen and oxygen, in the ratio of two O atoms and one H atom per C atom. Carbohydrates are represented by the general formula $C_x(H_2O)_y$, where x is any number between 3 and 8, and y represents the number of water molecules. The formula shows that a carbohydrate molecule contains x carbon atoms attached to y water molecules. This explains the origin of the name 'carbohydrates', which translates to carbon water or watered carbon; hence carbohydrates literally mean hydrated carbon. Carbohydrates are of three types: monosaccharides {general formula $(CH_2O)_n$ where $n \geq 3$}; disaccharides; and polysaccharides. Monosaccharides are simple sugars like glucose (blood sugar) and fructose (fruit sugar). Disaccharides and polysaccharides contain two and multiple units of monosaccharides, respectively. Examples of disaccharides are sucrose (cane sugar), lactose (milk sugar), and maltose (malt sugar), with the general formula $C_{12}H_{22}O_{11}$; while those of polysaccharides include starch and cellulose.

1.4.22 Proteins and Enzymes

Proteins are polymers made by linking together amino acids (Section 1.4.19). They contain carbon, hydrogen, oxygen, nitrogen, and sometimes phosphorus and sulfur. Enzymes are proteins produced by cells that act as catalysts in living beings, serving to accelerate biochemical reactions without themselves being altered.

1.5 Biology

1.5.1 What Is Biology?

Biology is the science of life and of living organisms, including their processes and phenomena (Campbell and Reece 2004). An *organism* is an individual living being, capable of growth and reproduction.

1.5.2 Branches of Biology

What are the main branches of biology? Botany, zoology, and microbiology. *Botany* is the scientific study of all aspects of plant kingdom. *Zoology* deals with all aspects of the animal kingdom. *Microbiology* deals with all aspects of the life of microbes, organisms invisible to the naked eye, including bacteria, viruses, fungi, and algae.

1.5.3 Origin and Evolution of Life

According to modern theory, life began from simple chemical compounds such as water (H_2O), methane (CH_4), and ammonia (NH_3), which were present in the original atmosphere. Lightning fused these chemicals into amino acids, the building blocks of life. The production of organic compounds from such simple inorganic compounds was possible in the early hydrogen-rich atmosphere. The organic compounds fell from the atmosphere onto the earth's surface, where the rainwater washed them into pools and ultimately into oceans. Gradually, the organic molecules joined together into long chains, proteins, and nucleic acids (e.g., DNAs), until a cell evolved that could replicate itself.

Evolution is the process by which living organisms have progressed from simple ancestral forms to present-day better-adapted complex forms.

1.5.4 The Cell

The *cell* is the basic structural and functional unit of all organisms (Figure 1.3). The *protoplasm* is the living part of a cell consisting of a nucleus embedded in a membrane-enclosed *cytoplasm*, is the latter being the protoplasm excluding the nucleus and the membrane. The *nucleus*, which houses the DNA (deoxyribonucleic acid: the double-stranded, helical molecular chain, carrying genetic information), is contained within a membrane and separated from other cellular structures. Its functions are cell regulation and reproduction. *Chromatin* consists of masses of DNA and associated proteins. A *chromosome* is a rod-like structure in the nucleus of the cell that carries the *genes* (segments of DNA that are the basic units of heredity) of the cells and performs an important role in cell division and transmission of hereditary characters. The nucleus contains a small body, the *nucleolus*, in which ribosomes, where proteins are assembled, are made out of another nucleic acid (ribosomal ribonucleic acid, rRNA) and ribosomal proteins. *Mitochondria* provide energy to the cell by carrying out cellular respiration. *Lysosomes* are dark spherical bodies in the cytoplasm, which contain enzymes that break down complex

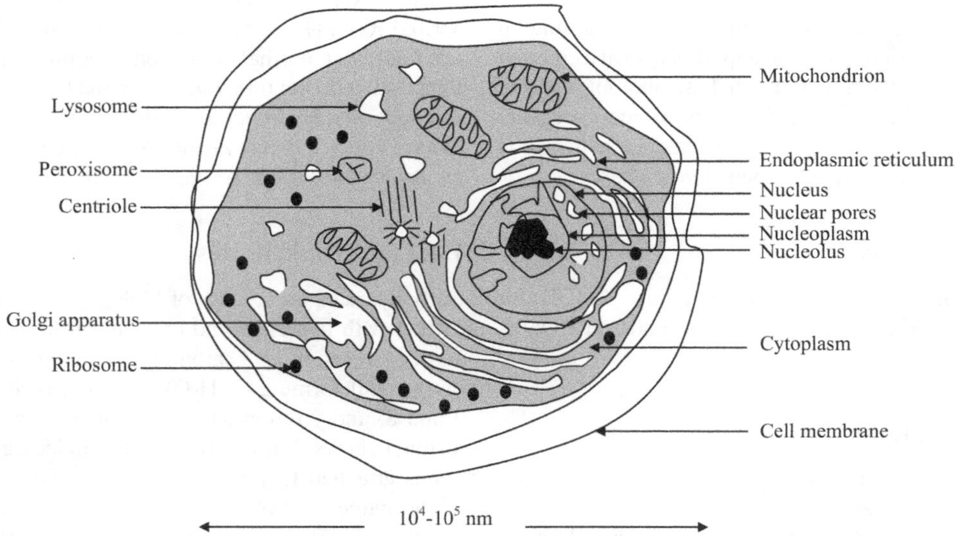

FIGURE 1.3 Generalized structure of animal cell. The cell is surrounded by a membrane made up of two fluid layers of fatty molecules. The cytoplasm comprises the total cell contents including the subcellular structures and excluding the nucleus. The nucleus contains the genetic information. Mitochondria release energy in a useful form through cellular respiration. The endoplasmic reticulum is the site of protein and lipid production. The Golgi apparatus modifies proteins and lipids and distributes them to the rest of the cell.

compounds into simpler subunits. The *endoplasmic reticulum* (ER) is the site of protein synthesis and transport and lipid metabolism. The *Golgi apparatus* modifies proteins and lipids and distributes them to the rest of the cell. *Plastids* are spherical bodies (e.g., leucoplasts store starch, while chloroplasts contain chlorophyll and carry out photosynthesis) specific to plants and photosynthetic microbes. *Vacuoles* are fluid-filled sacs for storage, which are unique to plant cells. *Microtubules* are tiny tubes associated with transport inside cells. *Centrioles* are cylindrical bodies concerned with cell division and movement. *Cilia and flagella* are motile hairs projecting from cells, and linked to movements of cells and substances.

What are the basic types of cells? There are two basic types of cells: those with membrane-bound nuclei (*eukaryotic*: eukaryotic organisms are plants, animals, algae, and fungi) and those without nuclei (*prokaryotic*: prokaryotic organisms including bacteria and archaea). All of the cells in a human being are eukaryotic.

1.5.5 Differences between Bacteria and Viruses

Two common types of microorganism are bacteria and viruses. Bacteria (singular: *bacterium*) are relatively large, unicellular microorganisms, typically a few micrometers long, having many shapes including curved rods, spheres, rods, and spirals. Viruses (from the Latin noun *virus*, meaning toxin or poison) are submicroscopic particles (ranging in size from 20 to 300 nm) that can infect the cells of an organism.

Bacteria and viruses are clearly distinguishable. (i) Viruses are the smallest and simplest life form known. They are 10–100 times smaller than bacteria. (ii) All viruses are parasitic and must have a living host, like a bacterium, plant or animal, in which to multiply, whereas, though some bacteria are parasitic, causing disease (*pathogenic bacteria*), most bacteria obtain their food from dead tissue or by consuming other micro-organisms. (iii) Bacteria are living organisms, whereas viruses are not, needing their hosts to carry out viral functions such as growth and reproduction. (iv) Some bacteria are useful but all viruses are harmful. (v) Antibiotics can kill bacteria but not viruses. (vi) An example of a disease caused by bacteria is *strep throat*, whereas a human viral disease is influenza (the *flu*).

1.5.6 Heredity, Chromosomes, Genes, and Related Terms

Heredity is the transmission of characteristics from one generation to successive generations. A *chromosome* is a thread-like body in the nucleus of eukaryotic cells, which is composed of DNA and protein, and carries the genes.

The gene, the fundamental unit of inheritance located on a specific chromosome, is a specific section of DNA that codes for a recognizable cellular product, either an intermediate form of RNA (messenger RNA, mRNA) or a polypeptide, or a protein (a long chain of 20 amino acids). *Genetic code* is the information for directing synthesis of a specific gene product, encoded as a linear sequence of four nitrogenous bases that form the structure of DNA. The *genome* is the full set of chromosomes of an individual or the total number of genes in such a set. *Genetics* is the study of heredity and the gene.

1.6 Semiconductor Electronics

1.6.1 What Is Semiconductor Electronics?

Electronics deals with the control of movement of electrons in solid-state materials, a vacuum, and gases to perform specialized tasks for computing, power conversion, regulation, communications, etc. *Semiconductors* are materials having conductivities (10^5–10^{-7} S m^{-1}) intermediate between metals (as high as 10^9 S m^{-1}) and insulators (as low as 10^{-15} S m^{-1}). Siemen is the derived unit of electrical conductance in SI system equal to inverse ohm or reciprocal of ohm, i.e., ampere/volt; it is also called mho. *Semiconductor electronics* is the aspect of electronics dealing with the use of semiconductors for controlling the flow of electrons.

1.6.2 Energy Bands in Conductors, Semiconductors, and Insulators

When isolated atoms come together to form solids, the discrete energy levels of atoms split up into multiple levels. Due to this splitting, solids are described by energy bands instead of energy levels. Electrons in the valency shells of solids are said to reside in the valence band, while the electrons that acquire energy by some means, such as heat, light, etc., are dislodged from their chemical bonds and become free to take part in electrical conduction. These are in the conduction band.

An *energy band diagram* is a plot of energy of the electrons with respect to distance through the solid (Figure 1.4). It contains two bands: valence and conduction bands. The energy difference between the top of the valence band and bottom of the conduction band is called *the forbidden energy gap* or *bandgap*.

In metals, the energy bands either overlap each other or a very small bandgap exists between the conduction and valence bands. This means that approximately all of the electrons lie in the conduction band and contribute to conduction. Semiconductors have a medium bandgap while insulators have a large bandgap. Popular semiconductors, such as germanium (Ge), silicon (Si), and gallium arsenide (GaAs) have bandgap values of 0.66, 1.1, and 1.42 eV (electron volts), respectively.

One *electron volt* is the energy gained by an electron when it is accelerated by a potential difference of 1 V (=potential difference between two points of a conductor carrying a current of 1 A when the power dissipation is 1 W). Insulators like silicon dioxide (SiO_2) and silicon nitride (Si_3N_4) have bandgaps of 9 and 5 eV, respectively.

1.6.3 Interesting Properties of Semiconductors

What are the special properties of semiconductors that make their study important to us? The electrical conductivity of semiconductors can be altered by the controlled addition of a small amount of impurities. This controlled addition of impurities to a pure semiconductor material, called an intrinsic semiconductor, is known as *doping*.

Silicon used for device fabrication is in a *single-crystal form*. A crystal consists of an ordered arrangement of atoms arranged at fixed distances repeated throughout the material in a systematic fashion.

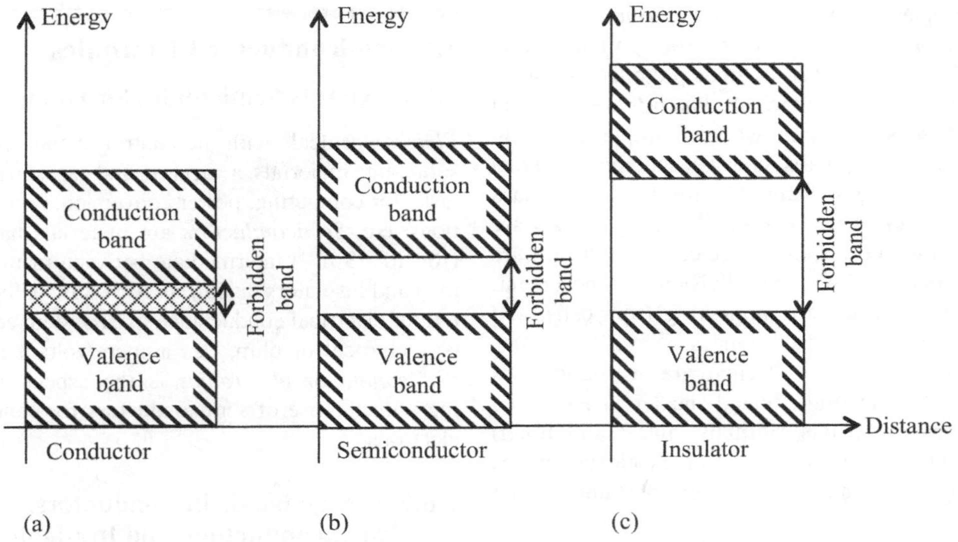

FIGURE 1.4 Energy band diagrams for: (a) conductor; (b) semiconductor; and (c) insulator.

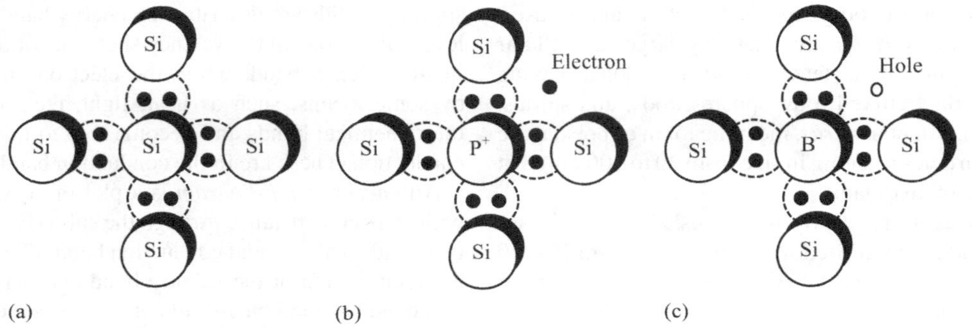

FIGURE 1.5 Bond diagrams: (a) intrinsic semiconductor; (b) N-type semiconductor; and (c) P-type semiconductor.

Silicon (Si: valency 4) has four electrons in its outermost shell. Every atom in a single crystal is surrounded by four atoms bound to it by covalent bonds, formed by sharing of electrons (Figure 1.5a). Each shared electron pair consists of one electron from the central silicon atom and one electron from the surrounding silicon atom. Thus, four pairs of electrons add up to eight electrons completing the stable inert gas configuration.

Suppose an intrinsic silicon crystal is doped with a *pentavalent impurity* such as phosphorus (P), arsenic (As), or antimony (Sb) (Figure 1.5b). When a silicon atom in the crystal is substituted by a phosphorus atom, the four electrons of the phosphorus atom take part in covalent bonding, as in silicon, but the fifth electron is available for conduction. Thus, one free electron is donated by each phosphorus atom. Phosphorus is therefore called a *donor impurity* and, because of the resulting negatively charged electrons present in a phosphorus-doped silicon crystal, the material is referred to as an *N-type semiconductor.*

If doping is done by a trivalent impurity like boron (B) (Figure 1.5c), indium (In), or gallium (Ga), three bonds around the impurity atom are complete but the fourth bond has one missing electron. An electron from an adjacent silicon atom jumps to fill and complete this bond. The electron vacancy left behind in the silicon atom is called a *hole* and acts like a positively charged particle. As one hole is produced per impurity atom, the silicon crystal has the same number of holes as the number of impurity atoms. As these impurities accept electrons from the silicon, they are termed *acceptor impurities.* Due to the presence of a large number of positively charged holes in boron-doped silicon, it is called a *P-type semiconductor.*

In an N-type semiconductor, the number of electrons is larger than the number of holes. Electrons carry the major part of the current and are called *majority carriers.* Holes are termed the *minority carriers.* In a P-type semiconductor, holes are majority carriers and electrons are minority carriers.

Thus, by deliberate incorporation of known amounts of phosphorus or boron atoms into semiconductors, conductivities of resulting N-type and P-type semiconductors can be varied within wide limits. However, both N-type and P-type semiconductors are electrically neutral as a whole, because the number of electrons = the number of positively charged donor ions in an N-type semiconductor and the number of holes = the number of negatively charged acceptor ions in a P-type semiconductor.

The distribution of electrons at different energies obeys the *Fermi-Dirac distribution function,* which expresses the probability that a state at energy E is occupied by an electron:

$$f(E) = \left[1 + \exp\left(\frac{E - E_F}{kT}\right)\right]^{-1} \quad (1.7)$$

where

E_F is the Fermi level and

T is the absolute temperature

The Fermi level is the energy at which there is a 50% probability of it being occupied by an electron. In an intrinsic semiconductor, the *Fermi level* is located approximately midway between the conduction and valence bands. When a semiconductor is doped with a donor impurity, this probability increases and therefore the Fermi level shifts upward. On doping with an acceptor impurity, the situation reverses and Fermi level shifts downward.

What happens when a semiconductor material is heated or when we shine light on the semiconductor? Conductivities of semiconductors are altered by the action of heat or light. Heat or light energies break the chemical bonds, leading to the formation of electron–hole pairs. These electron–hole pairs participate in electrical conduction, increasing the conductivity of the material.

1.6.4 P–N Junction

At a junction between P-type and N-type semiconductors, holes diffuse from the P-side to the N-side and electrons diffuse from the N-side to the P-side (Figure 1.6). *Diffusion* is the movement of particles along a concentration gradient, that is, from a high-concentration region to a low-concentration region. The above diffusion of carriers constitutes the *diffusion current* and leaves negatively charged acceptor ions on the P-side and positively charged donor ions on the N-side. Thus, an electric field is created around the junction. This electric field is directed from the N- to the P-side. It sets up an electric current in the reverse direction to the diffusion current flowing along the concentration gradient. This current set up by the electric field is called the *drift current.* Ultimately, the number of carriers crossing the junction from one side to the other by diffusion equals the number crossing in the reverse direction, due to the electric field. Then, a *dynamic equilibrium* is established. The potential difference associated with the electric field at the junction is called the *built-in potential.*

When a power supply is connected across a P–N junction, with its positive terminal on the P-side and its negative terminal on the N-side, the P–N junction is said to be *forward biased.* As long as the applied voltage is less than the built-in potential, there is no current flow across the junction. But, as soon as this voltage exceeds the built-in potential, the aforesaid dynamic equilibrium is disturbed and the current starts flowing by diffusion. Thus, current flowing across a forward-biased diode is a *diffusion current.*

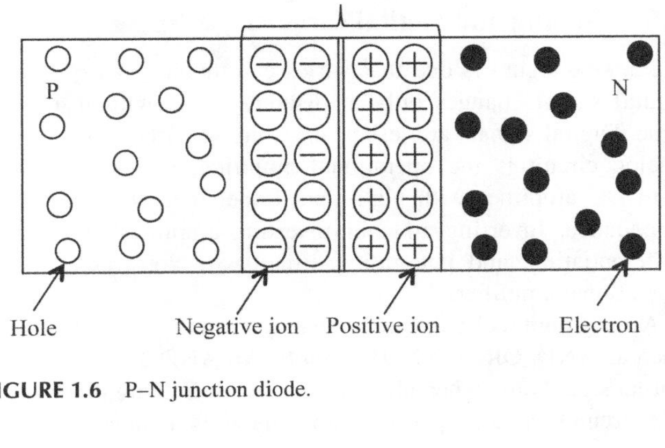

FIGURE 1.6 P–N junction diode.

On changing the connections of the power supply, with a positive voltage on the N-side and a negative voltage on the P-side, the diode is said to be *reverse biased.* Here, the depletion region is widened, and the thickness of the depletion region depends on the applied voltage. Any minority carriers in the depletion region, or close to it, such as electrons on the P-side or holes in the N-side, are carried and swept away by the applied field, leading to a *reverse leakage current.* Thus, the reverse current is essentially a *drift current.*

At high voltages, one of the following two phenomena takes place: (i) avalanche multiplication or (ii) punch-through breakdown. In *avalanche multiplication*, the carriers, electrons, and holes are accelerated to high velocities. They collide with electrons in other atoms, dislodging them from their orbits. In this way, an avalanche of carriers builds up by *impact ionization*, producing high currents and causing breakdown. In the *punch-through breakdown*, the depletion region enlarges to the full thickness of the lightly doped side touching the contact. Then, a very high current starts flowing. Another breakdown mode is the *Zener breakdown*, which occurs if both the N- and P-sides of the junction are heavily doped. The depletion region is very thin, so that carriers *tunnel* across this region, causing breakdown. It usually takes place at around 6 V.

1.6.5 Bipolar Junction Transistor

Bipolar junction transistors are of two types: NPN and PNP. The *NPN transistor* (Figure 1.7a and b) contains three regions: a heavily doped N region (emitter), a lightly doped thin P region (base), and an even more lightly doped N region (collector). On forward biasing, the emitter-base junction and, on reverse biasing, the collector-base junction, electrons are injected from the emitter into base. Very few electrons are lost by recombination in the thin base region. In practical terms, more than 99% of the injected electrons reach the collector junction. Although the collector current is slightly smaller than the emitter current, the important observation is that the transistor is able to pass current from a low-resistance forward-biased emitter-base junction to a high-resistance reverse-biased collector-base junction; hence the name "transfer resistor" or "transistor." The significant advantage here is that a high resistance can be connected in the collector circuit across which a large voltage is developed, yielding a high voltage gain. This is *voltage amplification.* The term "amplification" means strengthening a signal by drawing power from a source other than that of the signal. Clearly, the transistor does not generate extra energy nor is the principle of conservation of energy violated. The extra power comes from the supplies in the circuit.

1.6.6 Metal-Oxide-Semiconductor Field-Effect Transistor

This transistor is of two kinds: enhancement or normally-off mode device without a built-in channel or depletion or normally-on mode device, having a built-in channel.

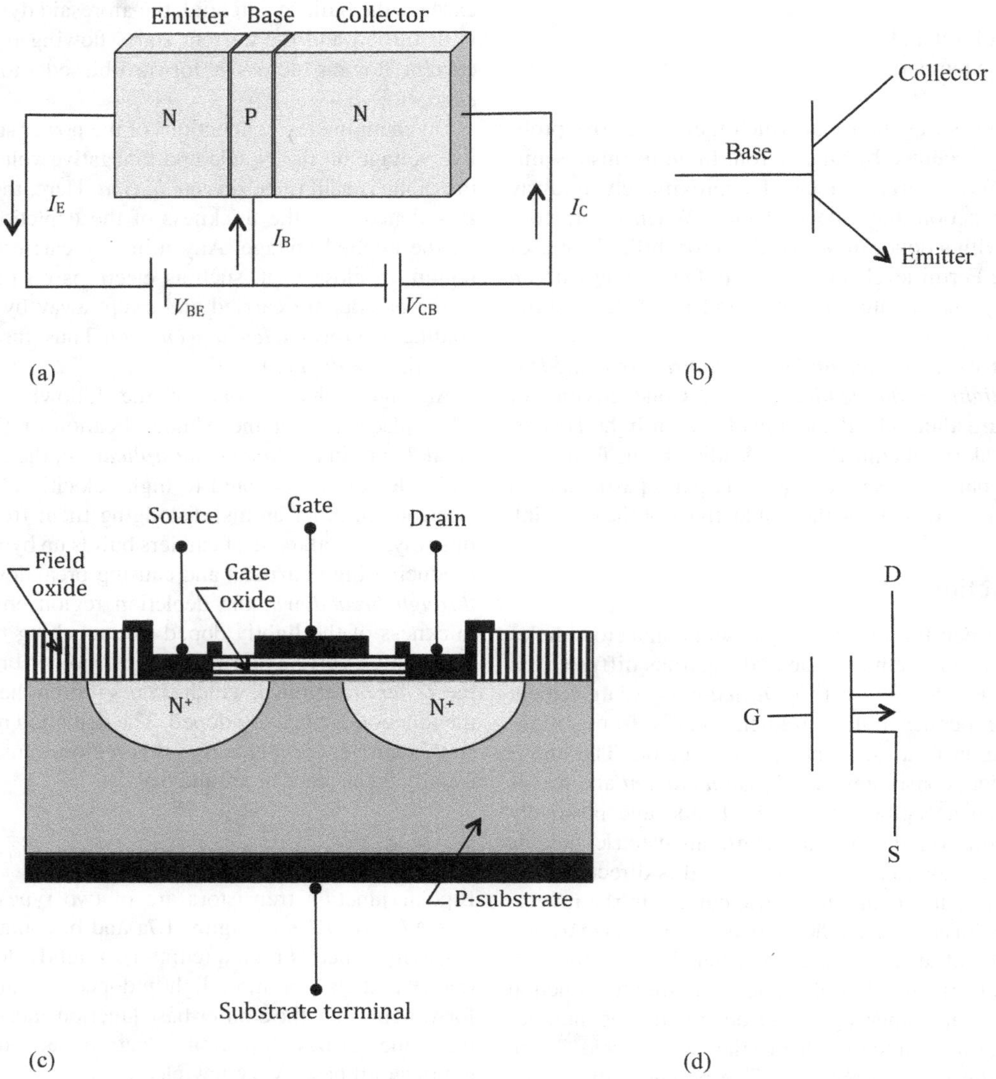

FIGURE 1.7 Transistors: (a) BJT, and (b) its circuit symbol; (c) MOSFET and (d) its circuit symbol.

The N-channel enhancement mode device (Figure 1.7c and d) contains two heavily doped N regions called source and drain in a P-substrate. The device has four terminals: source, drain, gate, and substrate. A positive voltage is applied between the source and the drain, with the source and the substrate grounded. The source-substrate junction is shorted but the drain-substrate junction is reverse biased. Therefore, there is no current flow between source and drain. But when a positive voltage is applied to the gate, by field effect through the underlying oxide layer, a depletion region is formed at the surface of the substrate below the gate. This is due to repulsion of holes by the positive gate voltage. On increasing the gate voltage, to a certain gate voltage called the *threshold voltage*, minority carrier electrons are attracted toward the surface of the P-substrate below the gate, forming an *inversion region* called the *channel*. This conducting channel connects the source and the drain, and the current starts flowing between the source and the drain.

Unlike the bipolar junction transistor (BJT), which is a current-controlled device, the metal-oxide-semiconductor field-effect transistor (MOSFET) is a voltage-controlled device. Also, MOSFET has higher input impedance than BJT. Moreover, MOSFET is a *unipolar device*, in which one type of carriers, either electrons or holes, carries current, whereas, in BJT, both the electrons and holes are responsible for current flow; hence the name "bipolar."

1.6.7 Analog and Digital Circuits

An analog signal is one that varies continuously with time. A digital signal changes only at discrete levels with respect to time. Digital signals are less prone to noise. A commonly used analog circuit is the operational amplifier, which is a high-gain DC amplifier with high-input impedance and low-output impedance. Inverting and non-inverting amplifiers, summing, differentiating, and integrating circuits are constructed using operational amplifiers.

Among digital circuits, mention may be made of logic circuits, such as AND, OR, and NOT circuits. An AND circuit gives an output signal only when all its inputs, A, B, C ... are present. An OR circuit gives an input when any one of its input signals, A or

B or C or ... is applied. A NOT circuit produces an output when there is no signal at the input. Combinations of NOT and AND circuits, and NOT and OR circuits are called NAND and NOR circuits, respectively. By suitable combinations of logic circuits, mathematical operations, like addition, multiplication, division, and subtraction, are performed.

The different analog and digital circuits are fabricated on a single chip of silicon and are available as integrated circuits (ICs).

1.7 Nanometer and Appreciation of Its Magnitude

Is "Nano" an English word? What does it mean? The word "nano" is derived from Greek, meaning dwarf, a person shorter in height than normal, and standing for *one-billionth*. It is a unit of length = one billionth of a meter = 1×10^{-9} m.

How can one imagine the sizes on such a small scale and consider magnitudes of such small dimensions? To visualize the nanometer scale, dimensions of familiar objects must be compared in nanometers (Table 1.1). The average height of a man is 1.7×10^9 nm. The thickness of a book page is ~10^5 nm. The diameter of a hair strand is 7.5×10^4 nm. The minimum length that the human eye can see is 1×10^4 nm. So, the human eye can see the hair strand. The dimensions of protein molecules are in the range 1–20 nm, the width of the DNA molecule is 2 nm while that of a carbon nanotube (CNT) is 1.3 nm. The diameter of a gold atom is 0.33 nm, that of a water molecule is 0.29 nm, and that of a hydrogen atom is 0.1 nm.

A CNT, also called a *buckytube*, is a large molecule of pure carbon that is long and thin and shaped like a tube, hundreds to thousands of nanometers long. It is 10^2 times stronger than steel but is one-sixth of its weight.

Example 1.1

Convert the average human height (5 ft 7 in.) into nanometers.

$$5\,\text{ft}\,7\,\text{in.} = 5 \times 12 + 7\,\text{in.} = 67\,\text{in.} = 67 \times 2.54\,\text{cm}$$

$$= (67 \times 2.54)/100\,\text{m}$$

$$= \{(67 \times 2.54)/100\,\text{m}\} \times 10^9\,\text{nm} = 1.7 \times 10^9\,\text{nm}$$

Example 1.2

How many hydrogen atoms can be lined up end-to-end to make 1 nm length?

Since the diameter of a hydrogen atom is 0.1 nm, 1/0.1 or 10 such atoms will be necessary to add up to 1 nm in length.

1.8 Nanoscience and Nanotechnology

In his visionary 1959 lecture at Caltech, Richard R. Feynman foresaw the potential of the ability to manipulate matter at the atomic scale.

Let us define nanoscience and nanotechnology. *Nanoscience* is the science of matter near or a little above the atomic and molecular level, imbued with special electrical and chemical properties, at dimensions less than 100 nm. *Nanotechnology* is a branch of applied science or engineering concerned with the design, synthesis, characterization, control, manipulation, and application of materials, devices, and systems at the nanoscale (1–100 nm) or having at least one physical dimension of the order of 100 nm or smaller (Booker and Boysen 2008). *How small is 100nm?* To imagine, it is about one hundred-thousandth of the diameter of a human hair (which is 50–100 μm or approximately 7.5×10^4 nm). The size of a bacterium, which is about the limit of what is visible through most optical microscopes, is about 1 μm (1000 nm). But the size of a virus is about 100 nm, one-tenth the size of a bacterium.

Table 1.2 presents the landmarks in the progress of nanotechnology while Table 1.3 traces the evolution of present-day CNT gate field-effect transistors from the triode valve.

Since such materials/devices/systems are found near the atomic or molecular scales, nanotechnology deals with objects close to these scales. Furthermore, since properties of matter at nanoscales differ from those in the bulk state, nanotechnology is engaged in the study of unusual phenomena and interesting properties of matter that make their appearance because of their small size.

In this respect, nanotechnology involves the extension of natural sciences (physics, chemistry, and biology) into the nanoscale. Nanosciences comprise nanophysics, nanochemistry,

TABLE 1.1

Small and Big Things

S. No.	Object	Diameter	Reference
1.	Electron	$<10^{-18}$ m	http://hypertextbook.com/facts/2000/DannyDonohue.shtml
2.	Proton	1×10^{-15} m	Christensen (1990)
3.	Nucleus	1.6 fm (10^{-15} m) (for a proton in light hydrogen) to about 15 fm (for the heaviest atoms, such as uranium)	http://wiki.answers.com/Q/What_is_the_diameter_of_nucleus
4.	Atom	0.1–0.5 nm	hypertextbook.com/facts/MichaelPhillip.shtml
5.	Water molecule	0.282 nm	http://www.falstad.com/scale/
6.	Polio virus	30 nm	http://users.rcn.com/jkimball.ma.ultranet/BiologyPages/V/Viruses.html
7.	Human red blood cell	7.5×10^3 nm	http://wiki.answers.com/Q/What_is_the_Diameter_of_red_blood_cell
8.	*Amoeba* cell	3×10^5 nm	http://www.answers.com/topic/amoeba
9.	Planet Earth (Equatorial)	12,756.32 km	geography.about.com/library/faq/blqzdiameter.htm
10.	Sun	1,392,000 km	http://www.daviddarling.info/encyclopedia/S/Sun.html

TABLE 1.2

Historical Developments in Nanotechnology

Year	Milestone
1959	Caltech physicist Richard R. Feynman gives his reputed lecture "There is Plenty of Room at the Bottom" on the prospects of atomic engineering
1981	Invention of scanning tunneling microscope by G. Binnig and H. Rohrer of Zurich Research Laboratory, Rüschlikon, Switzerland
1985	Discovery of buckyballs by R. R. Curl Jr., H. W. Croto, and R. E. Smalley

TABLE 1.3

Journey from Triode Valve to Carbon Nanotube Transistor

Year	Landmark
1906	Patenting of triode valve (Audion vacuum tube) by Lee De Forest
1947	Invention of transistor by William Shockley, John Bardeen, and Walter Brattain at Bell Laboratories
2003	Filing of patent on carbon nanotube gate field effect transistor by Jeng-hua Wei et al.

nanobiology, nanoelectronics, nanomechanics, nanomachining, and nanomaterials.

1.9 Nanomaterials and the Unusual Behavior at Nanoscales

What type of objects are studied in nanotechnology? The objects under study in nanotechnology are the *nanomaterials*, also called *nanostructured materials* (Bréchignac et al. 2007). As we know, all materials are composed of grains, which, in turn, are made of molecules and atoms. Nanomaterials are those having grain sizes in the range of nanometers (Ghuo and Tan 2009). These include CNTs, nanowires, nanoscaled thin films, etc. Table 1.4 summarizes the potential applications of nanotechnology, while Table 1.5 lists the drawbacks of nanomaterials.

How are nanomaterials classified? Nanomaterials are divided into three main classes (Table 1.6): (i) *nanoparticles or zero-dimensional (0D) nanomaterials*, for example, atom clusters with particle diameters below 100 nm; (ii) *one-dimensional (1D) nanomaterials* such as nanowires, nanotubes, and nanocables, having a width of less than 100 nm; and (iii) *two-dimensional (2D) nanomaterials* like nanofilms and superlattices, with layer thickness in the nano-range (Cao 2004).

Are nanomaterials found only in the form of elements? No, nanomaterials occur in elemental and also in composite form as compounds or their mixtures and alloys.

What unusual and interesting properties of matter emerge at nanoscales? These properties will be talked about very often in this book. To mention a few interesting properties here, as examples, nanomaterials offer greater reactivity, optical absorption, catalytic efficiency, superplasticity (the unusual ability of some metals and alloys to elongate uniformly by thousands of percentage points at elevated temperatures) and superparamagnetic characteristics, unique and important aspects of magnetism in nanoparticles. Superparamagnetism is a phenomenon by which magnetic materials may exhibit a behavior similar to paramagnetism even when at temperatures below the Curie or the Néel temperature. The Curie temperature (T_C), or Curie point, is the temperature at which a ferromagnetic or a ferrimagnetic material becomes paramagnetic on heating. The Néel temperature or magnetic ordering temperature, T_N, is the temperature above which an antiferromagnetic material becomes paramagnetic. Superparamagnetism occurs when the material is composed of very small crystallites (1–10 nm). In this case, even when the temperature is below the Curie or Néel temperature (and hence the thermal energy is not sufficient to overcome the coupling forces between neighboring atoms), the thermal energy is able to change the direction of magnetization of the entire crystallite. The resulting fluctuations in the direction of magnetization enable the magnetic field to average to zero so that the material behaves in a manner similar to paramagnetism.

TABLE 1.4

Promises and Expectations from Nanotechnology

Field	Application
Resources	(a) *Energy*: Lighter cars with stronger engines and frames, and more efficient fuels. House lighting using quantum dots to prevent wastage of energy as heat. Cost-effective solar cells and hydrogen fuel cells, (b) *Water*: More efficient water purification and cheaper desalination plants for supplying water from oceans
Healthcare	(a) *Diagnostics*: Point-of-care lab-on-chip systems for analyzing ailments. New contrast agents transported in the blood stream to light up tumors and for DNA mapping; (b) *Drug delivery*: Delivering the correct amount of medicine precisely at the site of disease, or nanoshells attaching to cancerous cells and destroying them following laser illumination
Security	(a) *Lightweight but stronger materials* for bridges, aircraft, and skyscrapers; (*b*) *Cheap, disposable but highly sensitive chemical nanosensors* for warning against explosives at airport or anthrax-laced letters; (c) *Quantum cryptography* of data for computers to provide unbreakable security for business and government

TABLE 1.5

Harmful Effects of Nanomaterials

Nanomaterial	Harmful Effect
Buckyballs	Single-walled carbon nanotubes produce lesions in the lungs of rats
Nanoparticles	They can interact with living cells in unanticipated ways. Effects of their inhalation are not known

Nanoparticles show greater electrical conductivity, hardness, wear resistance, strength, and toughness than bulk matter. Gold nanoparticles are used as biosensor labels, and Pt nanoparticles are superior catalysts. Polymer nanoparticles are used as carriers to target cancerous cells in medicine.

Magnetic nanoparticles serve as the new generation of nuclear magnetic resonance (NMR) contrast agents. NMR is the use of magnetic fields and radio waves (instead of the X-rays employed in the computerized tomography or CT scan) to visualize the functioning of body structures. A powerful

TABLE 1.6

Classification of Nanomaterials According to Dimensionality

S.No.	Number of Dimensions (-D)	Name of the Nanomaterial	Examples
1.	0	Nanoparticle	Buckyball, metallic nanoparticles, e.g., Au, Ag, and Fe
2.	1	Nanowire	Carbon nanotube, silicon nanowire
3.	2	Thin film	Metallic, semiconducting, or insulating films, e.g., of Au, Si, and SiO_2

magnet, large enough to enclose the body, forces the atomic nuclei in most material, such as soft tissue, to align themselves with the magnetic field. Radio waves are aimed at the selected area to excite the atoms, which, on stoppage of the waves, emit signals that are converted to computer-generated pictures. NMR is used in the detection of diseases of soft tissues, heart, brain, and spinal cord.

Zhu et al. (2009) found that the silicon nanowires deformed in a very different way from bulk silicon. Bulk silicon is very brittle and has limited deformability, meaning that it cannot be stretched or warped very much without breaking. But the silicon nanowires are more resilient and can sustain much greater deformation. Young's modulus of Si nanowires (NW) was close to the bulk value (187 GPa for <111> orientation; 1 gigapascal = 10^3 MPa = 9870 atm = 10^4 bar) when the diameters were larger than 30 nm. As the nanowire diameter decreased, the Young's modulus decreased, whereas the fracture strength increased up to 12.2 GPa. However, the softening trend was obvious when the diameters of the NW were less than 30 nm; the Young's modulus of the NWs decreased with the decreasing diameter. These experiments also showed that the fracture strain and strength of SiNWs increased as the NW diameter decreased.

More details about nanomaterials will be given in the ensuing chapters. But the curious reader must be already inquisitive about the reasons for the astonishing behavior already described of matter at the nanoscale. These explanations will be provided as we move through the pages of this book.

1.10 Moving toward Sensors and Transducers: Meaning of Terms "Sensors" and "Transducers"

In our efforts to review the basics, let us enquire into the meanings of the terms "sensors" and "transducers" (Wilson 2004). The word "sensor" is derived from the Greek word *sentire*, meaning to perceive (Sze 1994). It is a device that converts a physical stimulus such as mechanical motion, heat, light, sound, magnetic, electric, or radiant effect into an electrical signal, which is measured or recorded by an observer or an instrument. It is used for various purposes including measurement or information transfer. For example, a mercury-in-glass (Hg-in-glass) thermometer converts the measured temperature into expansion and contraction of a liquid, which is read on a calibrated glass tube. A thermocouple converts temperature to an output voltage, which is read by a voltmeter.

Similarly, the word "transducer" has originated from the Greek word *transducere*, which means "lead across." It is a device, usually electrical, electronic, electromechanical, electromagnetic, photonic, or photovoltaic, that converts power from one system to another in either the same or a different form. A transducer is a device that is actuated by energy from one system and supplies energy, usually in another form, to a second system. For example, a motor converts electrical energy to mechanical energy; hence, it is an example of a transducer. A loudspeaker is a transducer that transforms electrical energy signals into sound energy. An ultrasound transducer transforms electrical signals into ultrasonic waves or *vice versa*. Similarly, a light-emitting diode (LED) converts electric energy into light energy.

What are the differences between a sensor and a transducer? A sensor differs from a transducer in that a transducer converts one form of energy into other forms, whereas a sensor converts the received signal into an electrical form only. A sensor collects information from the real world. A transducer only converts energy from one form to another. The words transducer and sensor are used synonymously or interchangeably by many engineers. Some devices, for example, the microphone, can act as both. The thermostat in a home refrigerator takes temperature as the input variable and turns it into mechanical motion by unfolding a bimetallic strip (a strip consisting of two metals of different coefficients of expansion welded together, which buckles on heating), which turns a dial calibrated to read in units of temperature. In this way, the thermal energy associated with a specific temperature is transduced into mechanical motion.

On the contrary, many devices do not play the roles of both the sensor and the transducer. A photoresistor changes resistance as a function of impinging light without any energy conversion. A piezoresistive pressure sensor shows changes in piezoresistance value with pressure. The capacitance of a capacitive pressure sensor is altered by pressure.

A plausible way to distinguish between sensors and transducers is to use the term "sensor" for the sensing component itself and "transducer" for the sensing component together with the associated circuitry. Transducer = sensor + actuator. Thus, all transducers contain a sensor and mostly, although not always, sensors will also be transducers.

Since all transducers include sensors, nanotransducers will therefore also include nanosensors. Nanotransducers also come under the purview of this book, with the intention of knowing about their integral sensing parts, that is, nanosensors.

1.11 Definition of Sensor Parameters and Characteristics

Many terms related to sensors will be repeatedly used in the text. It is therefore useful to briefly recapitulate the definitions of these terms.

1. *Sensitivity*: the change in output value per unit change in the input variable (the measurand).
2. *Selectivity*: the ability to detect the measurand in the presence of similar species. It is a measure of the discriminating ability of the sensor with respect to interfering analytes, which are sources of noise in the output.
3. *Resolution*: the smallest measurable change in analyte (measurand) value that can be detected by the device.
4. *Response time*: the time taken by the sensor to reach $1/e$ or 63% of the final value of the sensed variable (output value).
5. *Calibration characteristic*: the curve obtained by plotting the sensor output along the abscissa and the measurand (analyte values) along the ordinate.
6. *Linearity*: the degree to which the calibration curve of the sensor matches with a specified straight line approximating the same.
7. *Repeatability*: the reproducibility of sensor output readings at given measurand values.
8. *Stability*: the degree to which the calibration curve of the sensor remains unchanged over a period of time so that the sensor need not be recalibrated.
9. *Drift*: the shift or translation in calibration curve of the sensor, with respect to time.
10. *Allowed ambient parameters*: the permissible maximum values of ambient parameters such as temperature, pressure, relative humidity, light, or illumination, etc., under which the device can operate satisfactorily.

1.12 Evolution of Semiconductor-Based Microsensors

Semiconductor-based sensors, mainly fabricated on silicon substrates, are made either in silicon or on silicon by depositing sensitive materials upon it. The earliest semiconductor sensor was a point-contact rectifier in 1904 for detecting radio waves, converting a radiation signal into an electrical one. After the invention of the transistor in 1947, the study of properties of silicon (Si) and germanium (Ge), notably the piezoresistive effect, received a boost. It was found that the resistance changed by two orders of magnitude more than in metallic strain gauges. This led to the use of a silicon bar as a resistive strain gauge by measuring its changes in resistance in response to applied strain. These bars had to be adhesively bonded to metal diaphragms. In subsequent modifications, the piezoresistors were diffused in silicon diaphragms, with the diaphragm bonded to a constraint to provide package isolation. In improved versions, the diaphragms were formed by selective anisotropic etching processes, reducing both sensor sizes and costs, and thus enabling batch fabrication. Wet and dry micromachining techniques were developed, along with silicon wafer-to-wafer and wafer-to-glass bonding methods. Thus, microelectronics was combined with micromachining to fabricate microelectromechanical systems (MEMS)-based sensors in which there were actual moving parts, like diaphragms, membranes, or moving bars, like cantilevers, etc.

The power requirements of these miniaturized sensors were considerably lower than their traditional counterparts so that signal-processing circuits had to be redesigned for proper interfacing. For a long period, sensor developments lagged behind the incredible growth in microelectronics. Therefore, compatibility of the new sensors with circuits, especially on the same chip, was not achieved. A hybrid approach followed, in which the signal-processing circuit was fabricated by thin- and thick-film technologies on a ceramic substrate, with active components die-mounted and wire-bonded onto the substrate.

But the main advantage accruing from silicon microelectronics and MEMS technology-based sensors was realized by integrating the sensor and the signal-processing electronic circuitry on the same chip. Thus, the semiconductor-based sensors were miniature in size and required lower operating power. They offered bulk-manufacturing capability and therefore lower costs when mass produced. The sizes of these microsensors ranged from hundreds of microns to several millimeters. Note that $1\ \mu m = 10^{-6}$ m.

1.13 From the Macrosensor to the Microsensor Age and the Necessity for Nanoscale Measurements

From the beginning, imperatives for sensor miniaturization were clear (Figure 1.8).

1.13.1 A Miniaturized Sensor Can Accomplish Many Tasks That a Bulky Device Cannot Perform

A little thought will explain why macro- and microsensors have many limitations. They cannot be used in many situations. Small-sized and lightweight sensors help in making portable instruments, which are essential for military and aerospace applications, in addition to mobile and handheld consumer products.

To consider an example, a predicted application of nanosensors is in the detection of cancerous cells in the human body by injecting quantum dots. These quantum dots are crystals of semiconductor materials, having sizes on the nanometer scale and emitting fluorescent radiation. These crystals are made of cadmium selenide (CdSe), cadmium sulfide (CdS), or indium gallium phosphide (InGaP), and are coated with suitable polymers that safeguard human cells from the toxic effects of cadmium and also the attachment of molecules that enable the tracking of cell processes and cancers. Obviously, such an application could not be imagined without the development of nanosensors. Macro- or microsensors cannot be inserted into the human body without harming or disturbing its normal functioning. So, it is the small size of nanosensors that is advantageously exploited

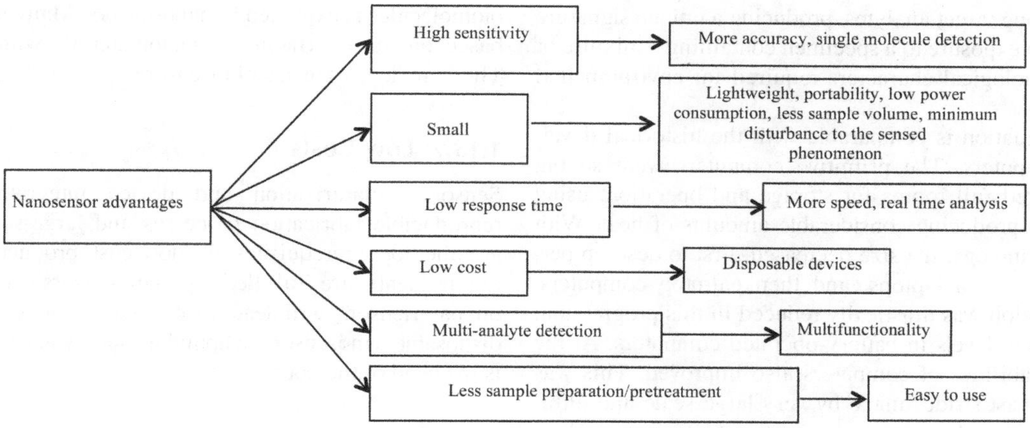

FIGURE 1.8 Advantages of nanosensors.

here. Futuristic advances can open unprecedented perspectives for the use of nanosensors as molecular-level diagnostic and treatment instruments in medicine and as networks of nanorobots for real-time monitoring of physiological parameters of a human body.

Systems biology, which is currently taking off as a research discipline to explore the basic principles of living systems by quantitative modeling of inter- and intracellular processes, depends on nanosensors and associated tools to provide data for model verification. Implantable devices, like autonomous nanorobots or multifunctional endoscopes (medical devices consisting of a long, thin, flexible or rigid tube that has a light and a video camera, whereby images of the inside of the patient's body can be seen on a screen), to achieve minimal invasive diagnostics, health monitoring, drug delivery, and many other intra-corporal (within a corpus, the body) tasks need ultraminiature sensors to fulfill their missions while minimizing invasiveness. As limitations in the downscaling of conventional systems are foreseeable, new materials with new properties at the nanoscale will emerge to fulfill sensor tasks in ultra-miniaturized sensor systems.

1.13.2 The Issue of Power Consumption

Conventional sensors were bulky and power-hungry. For example, consider a metal oxide gas sensor. An earlier version consisted of a ceramic (hard, brittle, heat-resistant, and corrosion-resistant material) tube with a heater coil fixed axially inside. A paste made from tin oxide (SnO_2) powder + palladium (Pd) was applied over the outer surface of the ceramic tube. The heater raised the temperature of the tin oxide layer to 200°C–500°C, at which point the sensitivity to inflammable gases increases. The sensing principle was the decrease in resistance of tin oxide in the presence of a combustible gas under ambient conditions. Obviously, the sensor consumed several tens of watts of power during operation. The next generation of sensors consisted of a ceramic substrate. On the bottom side of the substrate, a meander-shaped Pt heater was deposited. On the top side, the tin oxide paste was applied by screen printing (a printing technique that uses a woven mesh to support an ink-blocking stencil; a roller or squeegee is moved across the screen stencil, forcing ink past the threads of the woven mesh in the open areas). This sensor had lower power consumption than its predecessor, requiring only a few watts.

It is evident that gas sensing requires neither a large volume of material (only the surface area must be large) nor a thick layer of sensing material. Therefore, most of the power used in heating is wasted. Considerable power saving could be achieved by reducing the thermal mass of the sensor. So, the next improvement consisted of fabricating a miniaturized hotplate by silicon micromachining. This hotplate, fabricated by the MEMS technology, was as small as 100 μm × 100 μm. The thickness of the supporting plate (silicon dioxide + silicon nitride) was 2 μm, and the hotplate was suspended over a cavity to minimize conduction losses. Such a hotplate and, therefore, the resulting sensor barely consumed milliwatts of power. However, reduction in power consumption is only one of the reasons for downsizing sensors.

1.13.3 Low Response Times

Small-sized sensors require less time to reach equilibrium with the sensed environment because the signals take less time to traverse shorter path lengths. Therefore, these sensors can implement the desired operations at higher speeds.

What benefit is derived from a high speed? Measurements are performed in real time and data are immediately analyzed for appropriate action.

1.13.4 Multi-Analyte Detection and Multifunctionality

If one has to detect several gases, an array of gas sensors will be required. To construct such an array, consisting of hundreds or thousands of sensors, will lead to an enormously bulky and costly sensor, which will be virtually impractical to fabricate and use. The multiplexing capability offered by nanosensors is a vast improvement for real-time gas composition monitoring. An array of thousands of nanosensors, each coated with a different functional group and hence specific to a prefixed analyte, can be used in a single device. Such an ultraminiaturized, low-power device, supplemented with signal-processing and pattern-recognition algorithms, will exhibit effective discrimination capabilities among target analytes, producing a unique signature or fingerprint on exposure to a specimen containing a mixture of chemical and biological characters required for environmental pollution control.

The above situation is comparable with the historical development of computers. The primitive computers were so big that they required full rooms for storage and operation, using high power and producing considerable amounts of heat. With advances in technology, the size decreased, first to desktop personal computers, then laptops, and then palmtop computers. Power consumption was drastically reduced in this progression, reaching very low levels in battery-operated computers. At the same time, capabilities of computers also improved. This was enabled by the vast strides made by very-large-scale and ultra-large-scale ICs, very-large-scale integration (VLSI) and ultra-large-scale integration (ULSI).

Multifunctionality is a versatile feature in nature. The human tongue is a taste sensor and also used in speech. The human nose is used for smelling and breathing. Similarly, many other organs serve multiple functions. Multifunctionality is also expected from nanosensors.

1.13.5 Sensitivity Considerations and Need for Functionalization

Many sensors are based upon the adsorption of target analytes, and adsorption depends on surface area and surface chemistry. Making a sensor out of a nanoporous or nanocrystalline material increases the surface area tremendously; for example, the surface area of a single-walled carbon nanotube (SWCNT) is 1600 m^2 g^{-1}. The sensitivity of a gas sensor is augmented several-fold by increasing the surface area. By using tin oxide (SnO_2), indium oxide (In_2O_3), antimony oxide (Sb_2O_3), or zinc oxide (ZnO) in nanoparticle form, the sensitivity is considerably enhanced.

On their own, the CNTs have poor sensitivity toward analytes but, by surface modification, such as by suitably coating the CNTs or by doping palladium (Pd) atoms, they are made selective toward particular species.

Many types of microelectronic platforms are made. Notable examples are micro-hotplate, ion-sensitive field-effect transistor (ISFET), and microcantilever. The ISFET is a MOSFET without the metal layer on the gate and with a gate dielectric made of materials like Si_3N_4, $A1_2O_3$, or Ta_2O_s. The ISFET is a pH sensor that is converted into a specific ion sensor or a biosensor by coating suitable membranes onto the gate.

The above platforms are inherently not sensitive to any particulate analyte or sensitive to specific analytes only, and therefore need to be coated with suitable materials to make them sensitive, thereby enabling the development of families of sensors. Thus, each platform is a source of a sensor family.

1.13.6 Interfacing with Biomolecules

'Biomolecule' or biological molecule is a term used for molecules found in living organisms. Nanomaterials have sizes comparable with those of biomolecules, like proteins, viruses (small infectious agents), cells, nucleic acids, etc., and thus can readily form interfaces between biomolecules and readout instruments. Similarity or comparability of sizes of nanomaterials and biomolecules is exploited in nanosensors. Many nanosensors are based on the interfacing of nanomaterials with biomolecules. These are known as nanobiosensors.

1.13.7 Low Costs

Sensor miniaturization and device integration, based on reproducible fabrication processes and large-scale production, are the top prerequisites for low-cost production, and these requirements are fulfilled by nanosensors, so that research on nanosensors will lead to cheaper devices. Availability of disposable nanosensors in abundant supply at throw-away prices is the need of the hour.

1.13.8 Possibility of a New Genre of Devices

Nanosensors allow for building an entirely new class of devices, that provide the elemental base for "intelligent sensors", capable of data processing, storage, and analysis. These sensors will provide high accuracy, ultrahigh sensitivity, extreme specificity, and real-time *in vivo* information with greater speed, having multi-analyte options, requiring smaller quantities of sample, and minimal sample preparation, and being durable, safe, and portable. Some members of the nanosensor generation of devices have already qualified in laboratory tests, and are starting to appear in the marketplace. But there remains a long way to go, and several issues to be solved. Nonetheless, the potential is enormous.

1.14 Definition and Classification of Nanosensors

The definition of nanosensors must be clear. In fact, any sensor that uses a nanoscale phenomenon for its operation is a nanosensor. A nanosensor translates various input stimuli into readable signals, utilizing the distinctive properties of materials at ultra-small dimensions, thereby enabling the reliable, rapid, sensitive, and selective detection of the stimuli.

Any sensor fabricated by nanotechnological methods is a nanosensor, that is, a nanosensor is a nanotechnology-based sensor. To define the term, let us agree that any sensor characterized by one of the following properties will be labeled as a nanosensor: (i) the size of the sensor is at nanoscale; (ii) the sensitivity of the sensor is at the nanoscale; or (iii) the spatial interaction distance between the sensor and the object is measured in nanometers. In this book, whenever a nanosensor is talked about, the implication is that at least one of these criteria is valid.

From definition (i), nanosensors involve signal transformation from the environment, using nanostructures, that is, structures having at least one dimension in the lateral direction being less than 100 nm and the other less than 1 μm. Hence, in this definition, sensors must comply with nanotechnology, based on geometrical dimensions of the sensor (Figure 1.9).

A nanowire is a wire of a material, such as a metal, the diameter of which is less than 100 nm. Nanofibers are submicron-sized fibers whose diameter is 50–500 nm. Nanotubes are hollow cylinders only a few nanometers wide, made of one element, such as carbon. Nanobelts are nanostructures in the form of belts. Nanoprobes are optical devices for viewing extremely small objects. Quantum dots are nano-sized fluorescent semiconductor

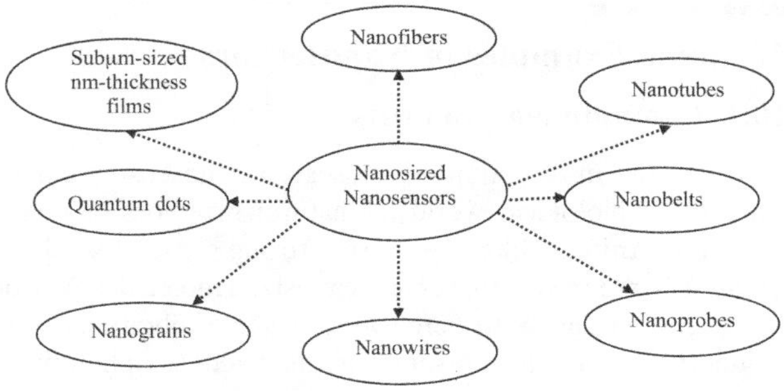

FIGURE 1.9 The nanosized nanosensor family tree.

crystals. Thin films of interest here are those having thicknesses in the nanometer range.

Many conventional techniques for nanometer-precise measurements or position control do not fulfill the criterion of definition (i). But they are in conformance with the criteria of definitions (ii) and (iii). Many authors exclude them from the domain of nanosensors. But in this book, all such sensors that are not necessarily small in size, which may be having micro- or macro-dimensions, will be categorized as nanosensors if their sensitivities fall in nano-units, for example, nm for position or displacement sensor, nN for force sensor, nK for temperature sensor, nM for solution concentration sensor, and so on. Moreover, any such sensors in which spatial interaction distance between the investigating electrode and the examined object is a few nanometers, such as in atomic force microscope (AFM), are also nanosensors (Figure 1.10). Thus, in this book, interpretation of the term "nanosensors" is broad and this will provide a wide perspective of the field, encompassing a wide diversity of nanosensors in the scope and coverage of the book.

Based on earlier discussions, nanosensors can be broadly classified into two groups, namely those that have nanoscale dimensions and those that perform nanoscale measurements but do not necessarily have nanoscale dimensions. In the former group are two subgroups: sensors having all three dimensions in nanometers, like quantum dots (fluorescent nanosized semiconductor crystals) and nanoscale thin films, like the 100–500 Å thick films of metal oxides, palladium, etc., used for gas sensing, which have only one nanoscale dimension.

Furthermore, like all sensors, nanosensors can be classified as active (those that require an energy source, for example, a thermistor [thermally sensitive resistor], a resistive strain gauge, etc.) or passive (those that do not require an energy source, such as a thermocouple, a photodiode [a P–N diode in which the reverse leakage current changes with illumination], a piezoelectric sensor [a device that uses the piezoelectric effect to measure pressure, acceleration, strain, or force], etc.) nanosensors.

Another classification scheme consists of two categories: absolute and relative nanosensors, including an absolute pressure sensor which measures pressure with reference to zero pressure (vacuum), whereas a relative pressure sensor does the same with respect to an arbitrarily chosen reference, such as atmospheric pressure. An absolute sensor detects a stimulus with respect to an absolute physical scale independent of measurement conditions, whereas the relative sensor produces a signal in reference to some special chosen case.

Proceeding further, nanosensors (like sensors in general) are classified according to the form of energy signal detected, such as physical (mechanical, thermal, optical), chemical, or biological (Table 1.7).

Advancing still further, sensor classification is based on the measurand, like a mechanical variable, such as position or displacement, or a chemical variable, like concentration of a sample (Table 1.8).

1.15 Physical, Chemical, and Biological Nanosensors

Let us elaborate on the sensor classes: physical, chemical, and biological. *Physical sensors* are used for measuring properties like temperature, pressure, flow, stress, strain, position, displacement, or force. *Chemical sensors* are meant for determining concentration or identity of a chemical substance like ethanol (C_2H_5OH), carbon monoxide (CO), gasoline or petrol, or other molecules. Biosensors are useful for dealing with biologically active substances, whether they are cellular, like the toxic plague bacterium (*Yersinia pestis*, found mainly in rodents, particularly rats) or anthrax spores (the *Bacillus anthracis* bacterium causes anthrax, forming tough dormant endospores ("spores"), which allow bacteria to

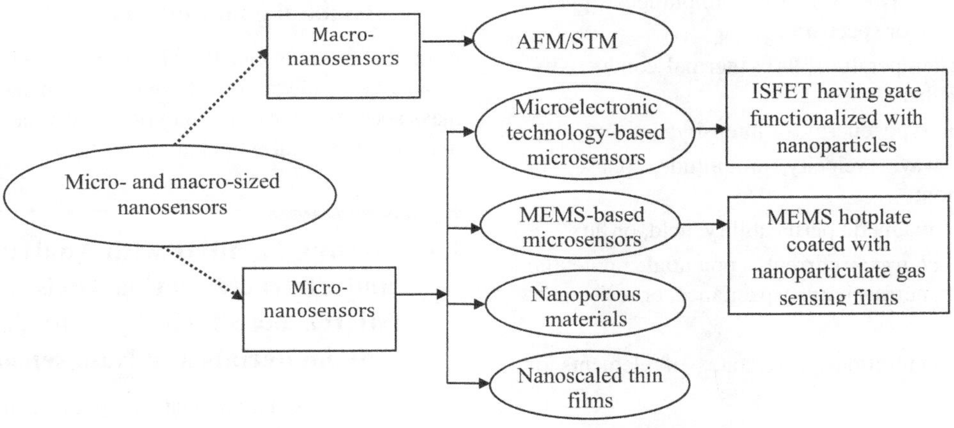

FIGURE 1.10 Micro- and macro-sized nanosensors.

TABLE 1.7
Conversion Phenomena of Sensors

Stimuli	Phenomena
Physical	Thermoelectric, photoelectric, photomagnetic, magnetoelectric, electromagnetic, thermoelastic, elastoelectric, thermomagnetic, thermo-optic, photoelastic
Chemical	Chemical transformation, physical transformation, electrochemical process, spectroscopy
Biological	Biological transformation, physical transformation, effects on test organisms, spectroscopy

TABLE 1.8
Stimuli and Measurands

Stimulus	Properties
Mechanical	Position, acceleration, stress, strain, force, pressure, mass, density, viscosity, moment, torque
Acoustic	Wave amplitude, phase, polarization, velocity
Optical	Absorbance, reflectance, fluorescence, luminescence, refractive index, light scattering
Thermal	Temperature, flux, thermal conductivity, specific heat
Electrical	Charge, current, potential, dielectric constant, conductivity
Magnetic	Magnetic field, flux, permeability
Chemical	Components (identities, concentrations, states)
Biological	Biomass (identities, concentrations, states)

go dormant under unfavorable conditions, and then multiplying rapidly upon inhalation, ingestion, or coming into contact with a skin lesion on a host), or supra-molecular (the collective behavior of organized ensembles of molecules, like influenza viruses), or molecular, like protein toxins such as staphylococcal enterotoxin B (SEB), an enterotoxin excreted by the *Staphylococcus aureus* bacterium. Sometimes, biosensors are considered to be a subset of chemical sensors. In this book, the following classification of nanosensors will be used. The measurands for physical, chemical, and biological sensors are as follows:

1. *Physical*
   a. *Mechanical*: mass, density, viscosity, position, velocity, flow rate, acceleration, force, stress, strain, moment, or torque;
   b. *Acoustic*: wave velocity, amplitude, phase, polarization, or spectrum;
   c. *Thermal*: temperature, flux, thermal conductivity, or specific heat;
   d. *Radiation*: type, energy, or intensity;
   e. *Optical*: wave velocity, amplitude, phase, or polarization;
   f. *Magnetic*: magnetic permeability, field, or flux;
   g. *Electric charge*: current, potential, potential difference, permittivity, capacitance, or resistance.
2. *Chemical*
   Identities, concentrations, and states of elements or compounds.
3. *Biological*
   Identities, concentrations, and states of biomolecules.

1.16 Some Examples of Nanosensors

1.16.1 Common Nanosensors

At present, the most commonly observed nanosensors exist in nature in the biological world, as natural receptors of outside stimulation. Animals like dogs have a strong sense of smell that operates using receptors that detect nanosized molecules. Various fishes use nanosensors to perceive minuscule vibrations in the surrounding water, while several insects detect sex pheromones using nanosensors. Similar to animals, many plants also use nanosensors to detect sunlight. In the artificial world, most film cameras have used nanosized photosensors for years. Traditional photographic film uses a layer of silver ions that become excited by solar energy and cluster into groups, as small as four atoms apiece in some cases, scattering light and appearing dark on the frame. Various other types of films can be made using a similar process to detect other specific wavelengths, such as IR, ultraviolet, and X-rays.

1.16.2 Carbon Nanotube-Based Nanosensors

Because of the ability to functionalize CNTs to detect specific molecules and the changes in their dielectric constants and permittivities with the adsorption of gases or vapors, a family of nanotube-based sensors has emerged.

1.16.3 Nanoscaled Thin-Film Sensors

Nanoscaled thin films of polymers and certain oxides have allowed preconcentration of analytes, such as heavy metals (metals with a specific gravity greater than about 5.0, especially those that are toxic, e.g., antimony [Sb], arsenic [As], bismuth [Bi], cadmium [Cd], cerium [Ce], chromium [Cr], cobalt [Co], copper [Cu], gallium [Ga], gold [Au], iron [Fe], lead [Pb], manganese [Mn], mercury [Hg], nickel [Ni], platinum [Pt], silver [Ag], tellurium [Te], thallium [Tl], tin [Sn], uranium [U], vanadium [V], and zinc [Zn]), using functionalized nanoporous thin films for X-ray fluorescence analysis, extending the limits of detection from ppm to ppb level.

1.16.4 Microcantilever- and Nanocantilever-Enabled Nanosensors

Arrays of suitably functionalized cantilevers are used to detect explosives as lower-cost alternatives to mass spectrometers; the mass spectrometer is an analytical instrument that measures the mass-to-charge ratio of charged particles.

1.17 Getting Familiar with Analytical and Characterization Tools: Microscopic Techniques to View Nanomaterials and Nanosensors

Nanomaterials and many features of nanosensors are invisible to the naked eye. But several instruments have helped in viewing and understanding these devices.

As already mentioned in Section 1.3.8, light is an electromagnetic wave, comprising electric and magnetic fields and representing energy, not matter; hence, it can travel through a vacuum. High-frequency light has more energy (in a familiar water analogy, more waves strike the seashore per second, delivering greater energy). Light is not visible to the human eye below red (IR) or beyond violet (ultraviolet). IR light has the lowest frequency and energy, whereas ultraviolet light has the highest frequency and energy. The continuum of light frequencies and energies constitutes its electromagnetic spectrum.

Does light behave as waves only? Light also behaves, in some ways, as if it consists of discrete particles. These apparent particles or energy packets have been designated as *photons*. According to wave-particle duality concept, objects show wave- or particle-like behavior, depending on the experiment being performed. When an object is heated, it becomes red hot and then white hot. Electrons are excited from normal (ground state) orbitals to excited states. Falling back to the ground states, they emit photons or light.

What is a microscope, and what is meant by microscopy? A microscope is an instrument that uses an optical lens or a combination of lenses, electronic or other processes, to produce magnified images of small objects, in particular those objects that are too small to be seen by the unaided eye. A microscope augments the power of the eye to see small objects. Microscopy is the technical field of using microscopes to view samples or objects. It deals with the examination of minute objects by means of a microscope, and is the science of interpretive use and application of microscopes for research.

Which is the most common microscope? Is it useful for nanotechnology? The most familiar and widely used microscope is the optical or light microscope but it is of little use in nanotechnology. Microscopes for nano-regime exploration will be described later with respect to their operating principles, applications, and limitations. Use of these microscopes requires special attention to sample preparation. Some microscopes can probe conducting samples only, whereas others can probe both conducting and insulating specimens. Some can work in vacuum only, whereas others can operate in air or even under liquid. Therefore, the operational and other salient features of microscopes are presented.

1.17.1 Scanning Electron Microscope

How does a scanning electron microscope (SEM) produce an image of the surface of an object? In an SEM, a stream of electrons called the *primary electrons* bombards objects under examination under vacuum to create their images (Goldstein et al. 2003). These primary electrons dislodge electrons from the sample, and the emitted secondary electrons are collected by a positively charged grid (Figure 1.11). As the primary electron beam scans the surface of the sample, secondary electrons are obtained from different portions of the sample. The secondary electrons thus gathered are used to construct the images of small invisible objects at a scale of 10 nm on a computer monitor.

What information is provided by SEM pictures? SEM images provide information on the topography of the surface, such as its texture and morphology, including the shape, size, and positions of particles on the surface of the sample.

What are the limitations of SEM? SEM is restricted to surface studies only and is not useful for exploring the interior of samples except, e.g., by freeze-fracture imaging technique in which a frozen biological specimen is broken for revealing its internal structure; or by cutting an object at right angles to its surface to take a cross-sectional or sideways view of the cut-away portion. *What will happen if the sample is an insulator?* The electron cloud produced on the sample surface will be stored there. This is called the charging effect and hinders the imaging of insulating surfaces. For discharging the surface, a thin conducting gold film, with a thickness of a few hundred angstroms, is deposited by sputter coating on the sample surface. This helps in imaging an insulating sample by removing the charge buildup by the electron cloud.

1.17.2 Transmission Electron Microscope

How does transmission electron microscope (TEM) differ from SEM in producing an image? In contrast to SEM, which scans a surface to produce its image, in a TEM, the electron beam is transmitted through a sample to observe the density of its constituents and thus produce an image on the basis of density variations. SEM works by surface bombardment with electrons and secondary electron collection. In TEM, electrons penetrate and cross to the opposite side of the sample.

In TEM, a beam of electrons passes through a sample to produce a projected image on a phosphor screen. Many electrons are able to pass through less-dense regions of the sample. These are observed as light areas. On the other hand, a smaller number of electrons can cross the denser regions of the sample, giving darker areas in the image. Thus, an *inverse density image* of the sample is obtained.

How do SEMS and TEMS differ in terms of resolution, sample preparation, costs, and applications? TEMs give greater image resolution (i.e., the details an image holds) than SEMs but TEM sample preparation procedure is extremely difficult. Also, TEMs are more expensive instruments. Therefore, SEMs are used as workhorses for routine work, whereas TEMs are engaged in more specialized tasks.

1.17.3 Scanning Tunneling Microscope

Unlike Newtonian mechanics, wave mechanics allows the flow of current, called tunneling current, across a thin barrier. If the surface to be probed is conducting, the tunneling current flow between a conducting tip and surface of the sample at small separations of a few nanometers between them helps in constructing images of the surface (Figure 1.12). This is the principle of the scanning tunneling microscope (STM).

How to examine insulating specimens? Obviously, other microscopes like an atomic force microscope (AFM) must be resorted to.

1.17.4 Atomic Force Microscope

How does an AFM produce an image? In the AFM, the tip of a cantilever moves on the surface of the sample. Interatomic van der Waals forces (weak electric forces that attract neutral molecules to one another, caused by a temporary change in dipole

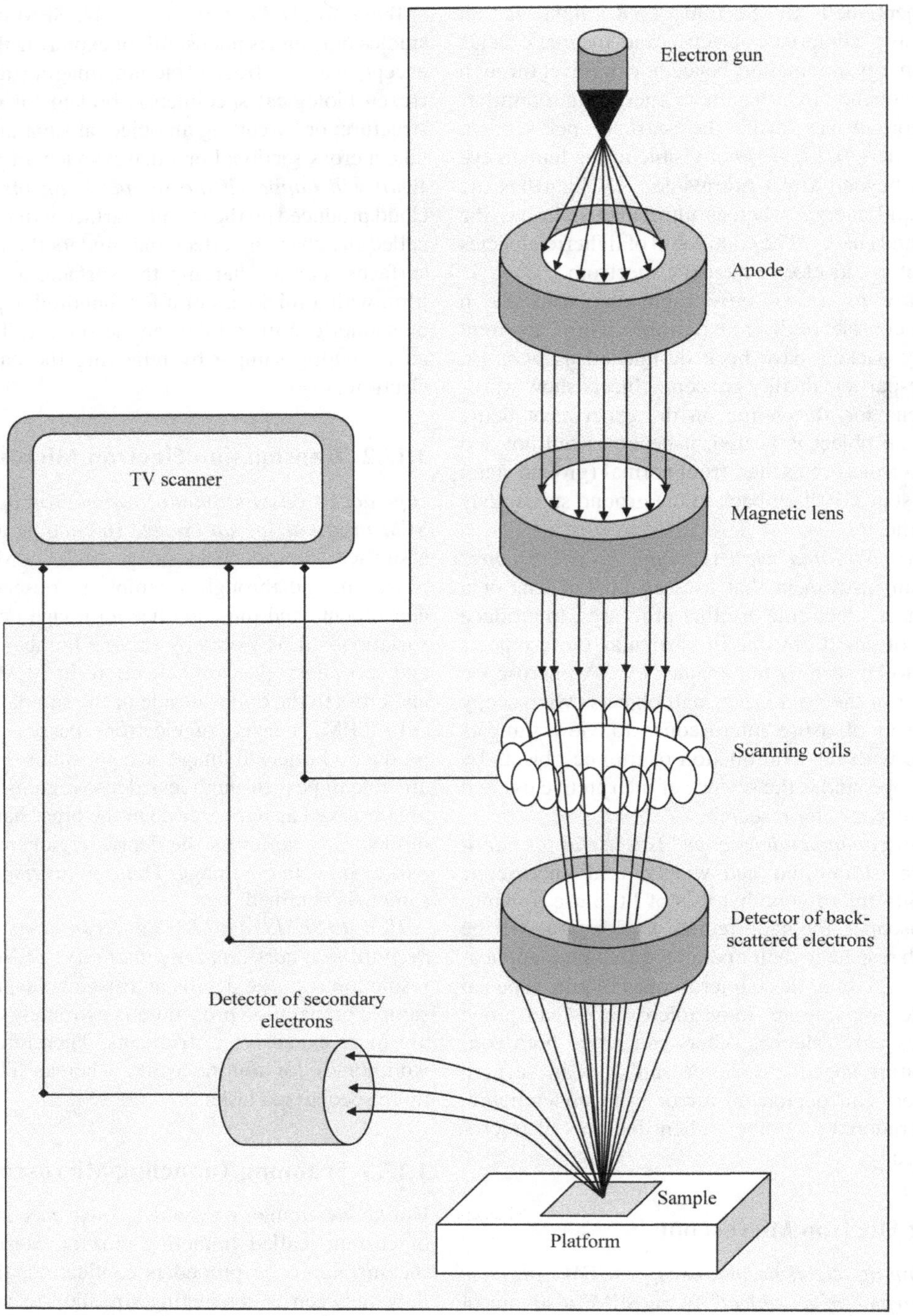

FIGURE 1.11 Image construction process by SEM. (https://www.purdue.edu/ehps/rem/laboratory/equipment%20safety/Research%20Equipment/sem.html).

moment arising from a brief shift of orbital electrons to one side of one atom or molecule, creating a similar shift in adjacent atoms or molecules) act between the tip and the sample (Figure 1.13). As the tip moves across the hills and valleys on the surface, a laser beam is reflected from the back of the cantilever into an array of photodetectors. This reflected laser beam generates a profile of the surface of the sample (Figure 1.14).

What are the distinguishing features of AFM in relation to other microscopes? Unlike SEM, AFM can operate in air and even underneath liquids. This is obviously a great advantage. A vacuum is used in AFM to prevent contamination of the sample, if necessary. Also, AFM produces a three-dimensional (3D) image of the sample, in contrast to 2D images obtained by SEM. This must be carefully noted. Furthermore, AFM is used to study conducting, semiconducting, or insulating samples. As previously remarked, in SEM, a thin conducting film must be deposited over an insulating sample to prevent charge buildup on its surface. In an STM, the sample itself must be conducting.

How does the tip of AFM affect its resolution? The sharper the AFM tip, the better is the resolution. A CNT serves as good

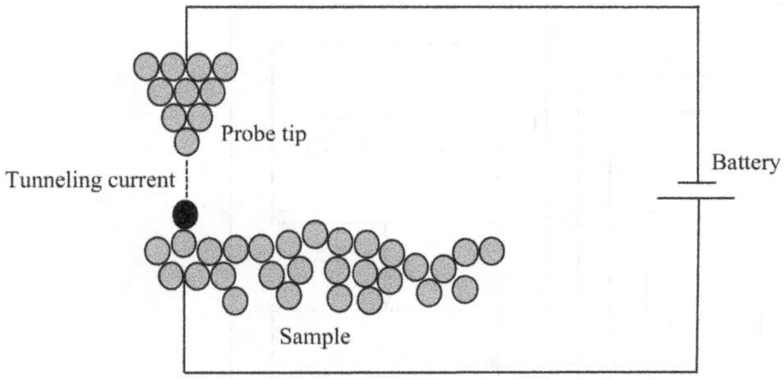

FIGURE 1.12 Principle of STM.

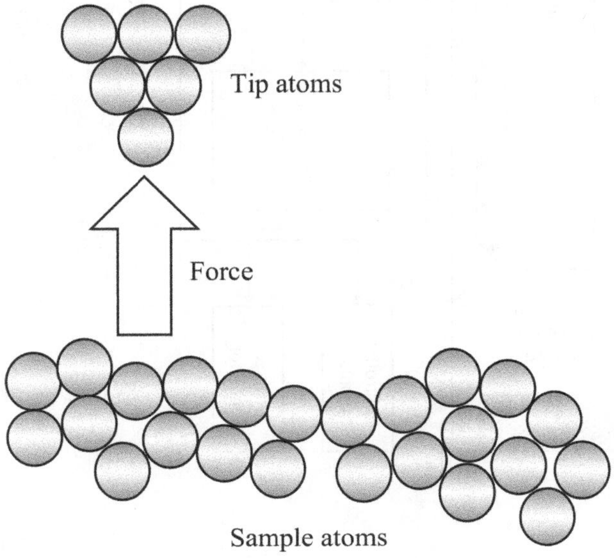

FIGURE 1.13 Operating principle of AFM.

material for an AFM tip. As it is cylindrical in shape, it provides a more accurate profile of the sample surface because it can penetrate into deep crevices on the surface where a pyramidal-shaped tip cannot reach; a pyramid is a structure where the outer surfaces are triangular and converge to a point. Moreover, it is more flexible and stronger than common silicon or silicon nitride tips.

Table 1.9 summarizes the characteristic features of various microscopes. The reader should take notice of and remember these features.

1.18 Spectroscopic Techniques for Analyzing Chemical Composition of Nanomaterials and Nanosensors

What is a spectrum? A spectrum is the distribution of a characteristic of a physical system or phenomenon, such as the distribution of energy or mass, measured as intensity, emitted by a radiant source in order of wavelength, frequency, or some other property. Examples are an electromagnetic spectrum, a visible light spectrum, and an ultraviolet light spectrum. A mass spectrum is represented as a vertical bar graph, in which each bar represents an ion having a specific mass-to-charge ratio, while the length of the bar indicates the relative abundance of the ion.

What is meant by spectroscopy? Spectroscopy is concerned with identifying elements and compounds and elucidating atomic and molecular structure by measuring the radiant energy absorbed or emitted by a substance at characteristic wavelengths of the spectrum, because such emission or absorption behavior is a unique property useful for recognizing the presence of the element or compound.

What does a spectrometer do? A spectrometer is used for producing spectral lines and measuring their intensities and wavelengths (Perkampus and Grinter 1992).

1.18.1 Infrared Spectroscopy

What are IR rays? IR radiation consists of electromagnetic waves. It occupies a part of the electromagnetic spectrum, with wave numbers from 13,000 to 10 cm^{-1}, or wavelengths between 0.78 and 1000 μm; wave number is the reciprocal of wavelength.

What is IR spectroscopy? It is a type of vibrational spectroscopy concerning the absorption measurement of different IR frequencies by a sample placed in the path of an IR beam (Hollas and Hollas 2004).

What is the principle of IR spectroscopy? The basis of IR spectroscopy is that IR radiation is in the same frequency range as a vibrating molecule. This is essentially the reason for choosing IR radiation for this purpose. Therefore, when a vibrating molecule is bombarded with IR rays, it absorbs those frequencies in the incident rays which exactly match the frequencies of the different harmonic oscillators making up that molecule. When this radiation is absorbed, the little oscillators in the molecule continue vibrating at the same frequency, but, since they have absorbed the energy of the light, the amplitude of vibration is greater. The remaining unabsorbed light by any of the oscillators in the molecule is transmitted through the sample to a detector.

IR absorption information is generally presented as a spectrum in which wavelength or wave number (the number of complete wave cycles of a wave that exists in one meter [1 m] of linear space; it is expressed in reciprocal meters [m^{-1}]), and is plotted as the abscissa, whereas the absorption intensity (the ratio of the radiant flux absorbed by a body to that incident upon it) or percentage transmittance (% incident light at a specified wavelength that passes through a sample) is plotted as the ordinate. A computer analyzes this spectrum and determines the absorbed frequencies. Using a Fourier transform (a mathematical operation that breaks a signal down into its constituent frequencies) algorithm (a set of instructions for solving a problem), one can bombard the molecule with every frequency of IR radiation at one time, and obtain a spectrum (Smith 1995).

How are the compounds identified by IR spectroscopy? IR spectrum of a given compound is unique and therefore serves as a *fingerprint* for this compound. IR spectroscopy is useful for identifying certain functional groups (collections of atoms in a molecule that participate in characteristic reactions) in molecules. But the identification of an unknown compound by IR spectroscopy alone is rarely possible.

What are the applications and limitations of IR spectroscopy? It can identify polar molecules (molecules that have mostly

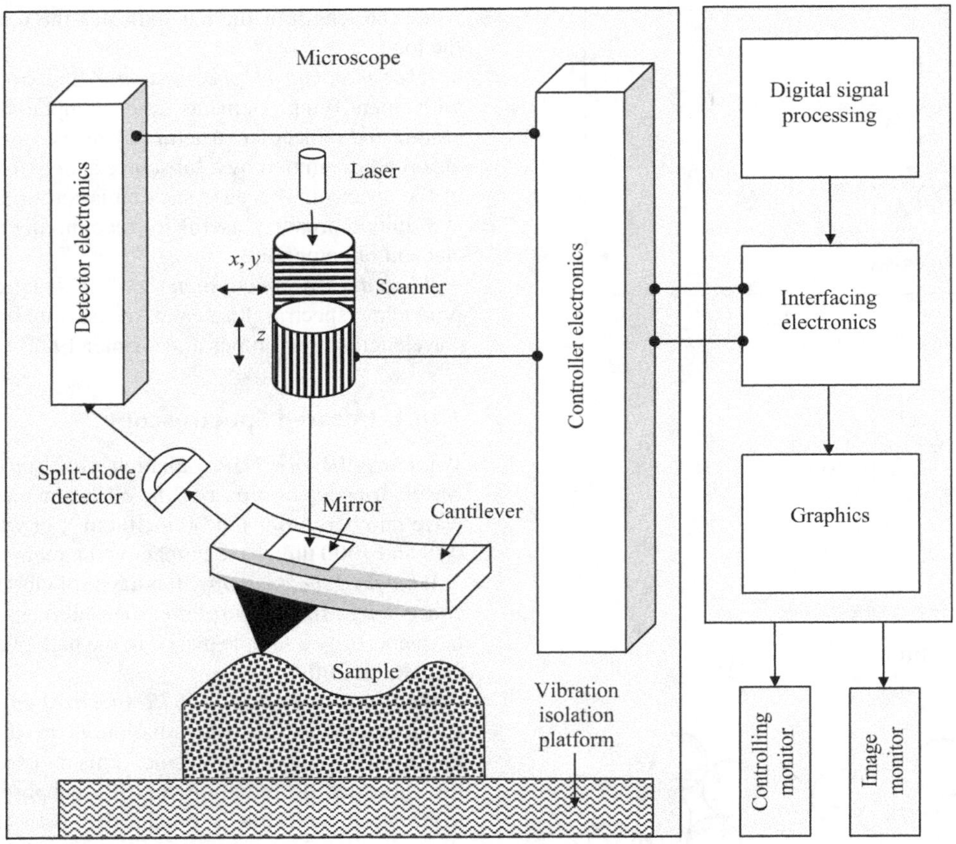

FIGURE 1.14 Parts of AFM.

TABLE 1.9
A Look at the Features and Capabilities of Different Microscopes

	Optical	SEM	STM	AFM
Imaging environment	Air, liquid, vacuum	Vacuum	Air, liquid, vacuum	Air, liquid, vacuum
Sample preparation requirement	Easy	Little to considerable	Easy	Easy
Sample restrictions	Must not be transparent to wavelength of light used	Charge buildup on the sample is to be avoided	(i) Conducting only (ii) Height variations in the sample should be <10 μm	Height variations in the sample should be <10 μm
In situ imaging possibility	Yes	No	Yes	Yes
In-fluid imaging possibility	Yes	No	Yes	Yes
Depth of field	Small	Large	Medium	Medium
Depth of focus	Medium	Large	Small	Small
Resolution (x, y)	1.0 μm	5 nm	0.1 nm	2–10 nm
Resolution z	Not applicable	Not applicable	0.05 nm	0.05 nm
Effective magnification	$1–2 \times 10^3$	$10–10^6$	$5 \times 10^2–10^8$	$5 \times 10^2–10^8$

positive charge on one side and mostly negative charge on the other; this charge difference allows the positive end of a molecule to attract to the negative end of another) but is not useful for identifying nonpolar molecules (molecules that share electrons equally and do not have oppositely charged ends). It is used for recognizing certain functional groups in molecules but it is rarely, if ever, possible to identify an unknown compound by using IR spectroscopy alone. It is a handy tool for studying SWCNTs.

1.18.2 Ultraviolet-Visible Spectroscopy

Do molecules absorb only IR radiation? No, molecules absorb visible and even UV radiation (UV = 200–400 nm, visible = 400–800 nm). Ultraviolet-visible (UV-vis) is spectroscopy that employs a UV and/or a visible light source for irradiating the sample. Electrons in different atoms absorb certain frequencies of light, reaching an "excited state." The excited atom gives off a photon to return to its original or ground state.

How does the principle of UV-vis spectroscopy differ from IR spectroscopy? Unlike IR spectroscopy (which views vibrational motions), UV-vis spectroscopy looks at electronic transitions. Applying Beer's law, $A = \varepsilon bc$ (where A = absorbance, ε = molar extinction coefficient, b = path length, and c = concentration), and from the absorbance information, one is able to determine either the concentration of a sample if the molar extinction coefficient (a property of chemical species that measures the strength with which it absorbs light at a particular wavelength per molar concentration) is known, or the molar absorptivity, from knowledge of its concentration.

What property helps in the identification of compounds? As molar extinction coefficients are specific to particular compounds, UV–vis spectroscopy is helpful in determining the identity of an unknown compound. The spectrometer measures the frequencies of radiation passing through the sample and the missing frequencies reveal the identities of atoms and molecules present in the sample.

For what purpose is UV-vis spectroscopy especially useful? UV-vis spectroscopy is commonly used to study samples that IR spectroscopy cannot identify.

1.18.3 Raman Spectroscopy

The reader will have noted that, starting with the use of vibration motions of atoms and molecules in IR spectroscopy, there is a progression toward applying electronic transitions and photon emissions in UV-vis spectroscopy. Raman spectroscopy is related to photon emission in the Raman effect.

How does Raman spectroscopy utilize the Raman effect and what is this effect? Raman spectroscopy is a spectroscopic technique based on *inelastic scattering of monochromatic light* from a laser source (Ferraro 2003). In inelastic scattering, the frequency of photons in monochromatic light (consisting of radiation of a single wavelength or of a very small range of wavelengths) changes upon interaction with a sample. Photons of the laser beam are absorbed by the sample and then reemitted. Frequency of the re-emitted photons is shifted toward higher (Stokes scattering) or lower values (anti-Stokes scattering) in comparison with original monochromatic frequency, which is called the *Raman effect*. There are two types of Raman scattering, namely Stokes scattering and anti-Stokes scattering. In Stokes scattering, the electron decays to a state higher than its original state whereas, in anti-Stokes scattering, the electron decays to a state lower than the original state.

Re-radiation of energy at the same frequency as the original light is known as *elastic Rayleigh scattering*. Stokes and anti-Stokes scattering spectra are mirror images on opposite sides of the Rayleigh line. Stokes scattering spectra are less temperature sensitive so they are generally used.

Raman spectroscopy is used to study solid, liquid, and gaseous samples. Unlike IR spectroscopy, Raman spectroscopy is capable of identifying nonpolar molecules in which the electrons are shared equally between the nuclei. Their charge distribution is spherically symmetrical, when averaged over time. Consequent to this even distribution of charge, the force of attraction between different molecules is small.

What kind of irradiation source is used in Raman spectroscopy? The Raman effect is very weak. To obtain meaningful results, an intense laser source is necessary for irradiating the sample.

1.18.4 Energy-Dispersive X-Ray Spectroscopy (EDX)

EDX is a technique based on the use of X-rays generated during the SEM examination of a sample. The X-rays produced on irradiating the sample (10–20 keV) in SEM give information about the chemical composition of the sample (Goldstein et al. 2003). In this method, called *energy dispersive X-ray (EDX) spectroscopy*, a detector collects X-rays and produces current pulses proportional to the energies of X-rays, enabling the acquisition of an image of each element in the sample.

How are these current pulses used to recognize elements? There are peaks in X-ray energies corresponding to different elements. By reading the positions of the energies of these peaks for a given sample, one is able to identify the elements present in the sample by comparing the energies of the peaks obtained in the sample under study with a chart of samples of known compositions and peak positions.

1.18.5 Auger Electron Spectroscopy

Electrons of energy 3–20 keV bombard a sample (Briggs and Grant 2006). A transference of energy occurs, exciting a core electron into an orbital of higher energy. On reaching this excited state, the atom has two possible modes of relaxation: emission of an X-ray or an Auger electron. In X-ray emission, the energy is given off as a single X-ray photon. In the Auger process, the electron transfers its energy to an inner orbital electron and this latter electron, called the "Auger electron," is ejected. It carries a characteristic kinetic energy.

How do the processes described earlier help in recognition of species? In both relaxation processes, the emitted particle has an energy peculiarity (trait) of the parent element. An energy spectrum of the detected electrons therefore shows peaks assignable to the elements present in the sample.

What are the uses and limitations of the Auger electron spectroscopy (AES) method? AES is a popular technique for determining the composition of the top few layers of a surface. It is unable to detect hydrogen or helium, but is sensitive to all other elements. It is most sensitive to the low atomic number elements.

1.18.6 X-Ray Diffraction

X-ray powder diffraction (XRD) is a rapid, analytical technique, primarily used for phase identification of a crystalline material. A diffraction pattern is obtained when X-rays interact with a crystalline substance (phase). The diffraction pattern is the distinctive pattern of light and dark fringes, rings, etc., formed by diffraction phenomena. It is an interference pattern produced when a wave, or a series of waves, undergoes diffraction, as when passing through a diffraction grating or the lattices of a crystal. This pattern provides information about the frequency of the wave and the structure of the material causing the diffraction, i.e., the X-ray diffraction pattern of a pure substance is like a fingerprint of the substance. The X-ray powder diffraction technique identifies components in a sample by a search/match procedure.

1.18.7 X-Ray Photoelectron Spectroscopy or Electron Spectroscopy for Chemical Analysis

In X-ray photoelectron spectroscopy (XPS), a high-energy X-ray source (*hν*) is used for the excitation and ejection of core-level electrons. Al K-alpha (1486.6 eV) or Mg K-alpha (1253.6 eV) are frequently chosen as the photon energies. The energy of the photoelectrons (electrons released or ejected from a substance by photoelectric effect, i.e., under irradiation from photons of sufficiently low wavelength) leaving the sample is determined, giving a spectrum with a series of photoelectron peaks. The binding energy of the peaks is characteristic of each element. XPS is insensitive to hydrogen (H) or helium (He), but detects all other elements.

1.18.8 Secondary Ion Mass Spectrometry

The surface of the sample is bombarded with high-energy ions. As a result, ions and atoms are sputtered from the surface. Sputtering is the process of ejection of ions from the surface by the impact of the bombarding ions. These sputtered ions are detected and mass analyzed to determine the elements from which the sample is made. Thus, bombardment of a sample surface with a primary ion beam, followed by mass spectrometry of the emitted secondary ions constitutes secondary ion mass spectrometry (SIMS).

In order to measure the characteristics of individual molecules, a mass spectrometer converts them to ions so that they can be moved about and manipulated by external electric and magnetic fields. Ion formation and manipulation is conducted under a high vacuum. By changing the strength of the magnetic field, ions of different masses are focused on a detector fixed at the end of a curved tube under a high vacuum (10^{-5}–10^{-8} torr). A mass spectrum is a vertical bar graph. In this bar graph, each bar represents an ion having a specific mass-to-charge ratio (*m/z*), and the length of the bar indicates the relative abundance of the ion.

Mass spectrometers can easily distinguish ions differing by only a single amu. Thus, they provide accurate values for the molecular mass of a compound. The highest-mass ion in a spectrum is considered to be the molecular ion. Lower-mass ions are fragments from the molecular ion, assuming that the sample is a single pure compound.

1.19 The Displacement Nanosensor: STM

1.19.1 Principle of Operation

Ever since the atomic structure of matter was envisioned, imaging individual atoms remained a cherished, though elusive, goal until the introduction of the STM. Gerd Binnig and Heinrich Rohrer developed the first working STM in 1981 while working at Zurich Research Laboratories in Switzerland (Chen 2008). This humble instrument provided a breakthrough in human ability to investigate matter on the atomic scale: for the first time, the individual surface atoms of flat samples could be made visible in real space. This instrument won Binnig and Rohrer the Nobel Prize for Physics in 1986.

STM works by the united application of several principles: quantum-mechanical tunneling (a quantum-mechanical phenomenon, whereby a particle tunnels through a barrier that it classically could not surmount because its total kinetic energy is lower than the potential energy of the barrier); piezoelectric effect (the ability of a material to produce electricity when subjected to mechanical stress); and feedback loop (in which information about the result of a transformation or an action is sent back to the input of the system in the form of input data, allowing self-correction) (Bowker and Davies 2010). The spectacular spatial resolution of the STM, along with its intriguing simplicity, launched a broadscale research effort, significantly impacting surface science. A large number of metals and semiconductors have been investigated on the atomic level and marvelous, beautiful images of the world of atoms were created within the first few years after the inception of the STM. Today, the STM is an invaluable asset in the toolbox of a surface analysis scientist.

Thinking in classical terms, a particle is not allowed to move over a barrier unless it has sufficient energy to cross that barrier. If two metallic objects are connected to a battery, one to each end, and the circuit is completed by touching the two objects together, a current starts to flow between the objects. If the objects are pulled apart and separated, the current ceases to flow because of inability of electrons to cross the air gap between the objects. It is said that the air gap represents a "potential energy barrier" to the electrons that cannot be compensated by the battery voltage. But when distances are measured at atomic scales, classical rules are not applicable. In quantum mechanics, electrons have wavelike properties (Figure 1.15). A tunneling current occurs when electrons move across a barrier that is classically impenetrable to them. This becomes obvious from the realization that the electron waves do not terminate abruptly at a wall or barrier, but taper off quickly on encountering a barrier. For a very thin barrier, the probability function extends into the next region through the barrier. Because of the small but nonzero probability of the presence of an electron on the opposite side of

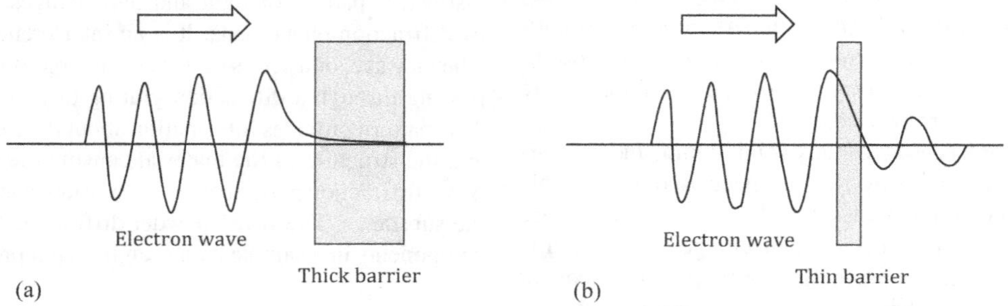

FIGURE 1.15 Interaction of electron wave with (a) thick potential barrier and (b) thin barrier. In (a), the wave does not end abruptly but tapers off exponentially However, it does not cross the barrier. A variable is said to increase or decrease exponentially if its rate of change is expressible by an exponential function. A graph of such a rate would appear as a curve, not as a straight line. In (b), the wave manages to penetrate the barrier and some electrons appear on the opposite side of the barrier.

the barrier, if a large number of electrons is available on one side of the barrier, some electrons will successfully move through the barrier and appear on the other side of the same. When electrons move through the barrier in this fashion, the phenomenon is called tunneling through the barrier, a case that is forbidden in classical physics.

The sample is scanned by a very fine metallic tip, mechanically connected to the scanner, an *XYZ* positioning device realized by means of piezoelectric materials (Figure 1.16). The scanner allows 3D positioning in the *X*, *Y*, and Z directions with subatomic precision. The tunneling tip is typically a wire that has been sharpened by chemical etching or mechanical grinding. W, PtIr, or pure Ir are often chosen as the tip material. The sample is positively or negatively biased so that, when the distance between tip and sample is in the range of several angstroms, a small current, the tunneling current, flows with the tip very close but not in contact with the sample. This feeble tunneling current is amplified and measured. This current is used as the feedback signal in a Z-feedback loop. With the help of the tunneling current I_t, the feedback electronics keep the distance between the tip and the sample constant. In the topographic mode, images are created by scanning the tip in the *XY* plane and recording the *Z* position required to keep I_t constant. If the tunneling current exceeds its preset value, the distance between tip and sample is increased. When it falls below this value, the feedback decreases the distance. The tip is scanned line by line above the sample surface, following the topography of the sample. In the *constant-height mode*, the probe scans rapidly so that the feedback cannot follow the atomic corrugations. The atoms are then apparent as modulations of I_t, which are recorded as a function of *X* and *Y*. The scanning is usually performed in a raster fashion, with a fast scanning direction (sawtooth or sinusoidal signal; named sawtooth because of its resemblance to the teeth on the blade of a saw) and a slow scanning direction (sawtooth signal). A computer controls the scanning of the surface in the *XY* plane while recording the Z position of the tip (topographic mode) or I_t (constant-height mode). Thus, a 3D image Z ($X, Y, I_t \sim$ const) or I_t ($X, Y, Z \sim$ const) is created.

1.19.2 Transmission Coefficient

Schrödinger's equation of quantum mechanics is applied to predict the increase in tunneling current between the tip and the sample with the decrease in separation between two metals. Here, the energy of all the electrons in the metal is lower than the height of the potential wall. Let the difference between the most energetic electron and the vacuum energy be denoted by the symbol φ, which is called the work function of the material. The movement and shape of the electron wave is governed by Schrödinger's equation. Although the complete description of the tunneling process requires a solution of the 3D form of Schrödinger's equation, a 1D analysis is sufficient to describe the main features of the phenomenon. The 1D Schrödinger's equation is given by:

$$\left(-\frac{\hbar^2}{2m}\right)\nabla^2\psi(x,t)+U(x)\psi(x,t)=i\hbar\left[\frac{\partial\psi(x,t)}{\partial t}\right] \quad (1.8)$$

where $U(x)$ is the potential energy of the electron as a function of its position. The plane wave (a constant-frequency wave the wavefronts of which, i.e., the surfaces of constant phase, are infinite parallel planes of constant amplitude normal to the phase velocity vector) for an electron wave function of wave number

$$k=\frac{2\pi}{\lambda} \quad (1.9)$$

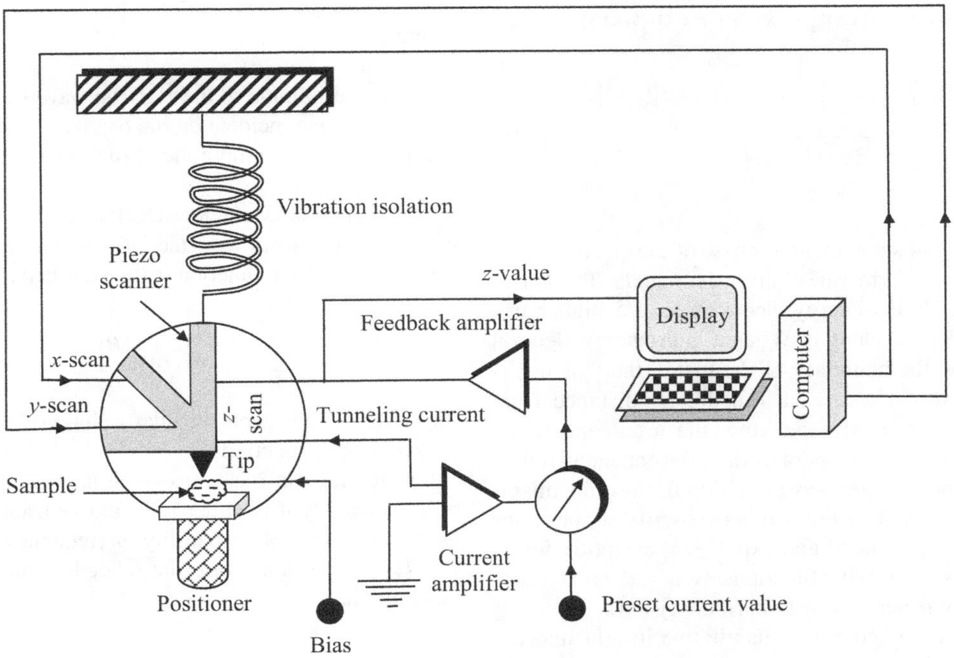

FIGURE 1.16 Scanning tunneling microscope. (Chen, C. J., *Introduction to Scanning Tunneling Microscopy*, Oxford University Press, New York, 2008.)

and angular frequency ω (a scalar measure of rotation rate in radians per second, where radian is a unit of angle similar to degrees), is represented by the equation

$$\psi(x,t) = A\exp\{-i(kx+\omega t)\} + B\exp\{i(kx-\omega t)\} \quad (1.10)$$

Electrons of energy $E(x, t) = E$ bumping into a uniform potential barrier of height $U(x, t) = U(x)$ continuously flow from one metal to the other. A steady-state time-independent situation is postulated. Under this assumption, it is necessary to solve only the 1D steady-state Schrödinger's equation, given by:

$$\left(\frac{-\hbar^2}{2m}\right)\left\{\frac{\partial^2\psi(x)}{\partial x^2}\right\} + U(x)\psi(x) = E\psi(x) \quad (1.11)$$

where

E is the kinetic energy of the electron.

$U(x)$ is the potential energy of the electron, as a function of position, and is smaller than the electron energy in the metals and larger than the electron energy in the barrier

For the sake of simplicity, it is assumed that

$$U(x) = U_0 \quad (1.12)$$

is a constant in the barrier. The general solutions to the above equation in the Metal 1, Metal 2, and barrier are given, respectively, by the equations (University of Wisconsin 2007)

$$\text{Metal 1} \quad \psi(x) = A\exp\{-(ikx)\} + B\exp\{(ikx)\}, \quad k = \frac{\sqrt{2m(E-U_0)}}{\hbar} \quad (1.13)$$

$$\text{Metal 2} \quad \psi(x) = E\exp\{-(ikx)\} + F\exp\{(ikx)\} \quad (1.14)$$

$$\text{Barrier} \quad \psi(x) = C\exp\{-(\mu x)\} + D\exp\{(\mu x)\}, \quad \mu = \frac{\sqrt{2m(U_0-E)}}{\hbar} \quad (1.15)$$

Equations 1.13 and 1.14 show that the phase of the electron wave function undergoes uniform variation in the metals. The wavelength is $\lambda = 2\pi/k$. Higher-energy electrons have a higher frequency and a smaller wavelength. When a high-energy electron wave stumbles upon the boundary of the metal, the "intensity" of the electron wave declines as a function of distance from the boundary. Mathematically speaking, the argument of the exponential function, i.e., its operand or independent variable, becomes real and the electron wave function decays (for imaginary arguments, the wave function will have oscillatory behavior because $\exp(\pm i\theta) = \cos\theta \pm i\sin\theta$ and $\exp(i\theta \pm 2\pi) = \exp(i\theta)$ due to 2π periodicity of cosine and sine functions. Hence, the imaginary exponential function is periodic with periodicity $2\pi i$).

As the objective is to acquire a quantitative insight into the electron tunneling phenomenon, it is necessary to derive an expression for the transmission coefficient (the amplitude or the intensity of a transmitted wave, relative to an incident wave), which indicates the transmitted flux from the sample to the tip through the barrier of width L. The barrier is considered to be wide but finite, such that the electron wave function decays significantly as it passes through the barrier. Furthermore, the energy and mass conservation requirements impose the condition that electron wave function and its first derivative must be continuous, that is, they join smoothly at the sample-barrier and tip-barrier boundaries. Let us set up a coordinate system in which the surface of the sample (Metal 1) is at $x = 0$ and the tip (Metal 2) is at $x = L$. Then, applying the boundary conditions (the set of conditions specified for the behavior of the solution to a set of differential equations at the boundary of its domain) for continuity, we get

$$A + B \approx C \quad (1.16)$$

$$ik(A-B) \approx -\mu C \quad (1.17)$$

(at the sample surface, $x = 0$) where D, the amplitude of the reflected wave function at the tip-barrier boundary, is ignored, since $D << A, B, C$. However, D is not small enough to be *negligible* at the tip-barrier boundary. At the tip-barrier boundary, $x = L$, continuity requires the following:

$$C\exp\{-(\mu L)\} + D\exp\{(\mu L)\} = F\exp\{(ikL)\} \quad (1.18)$$

$$-\mu C\exp\{-(\mu L)\} + \mu D\exp\{(\mu L)\} = ikF\exp(ikL) \quad (1.19)$$

Solving for B/A at $x = 0$, by obtaining the solution for C and substituting for it, we have

$$\frac{B}{A} = -\frac{1+ik\delta}{1-ik\delta} \quad (1.20)$$

where

δ is $1/\mu$

A is the amplitude of the electron wave function in the sample surface incident on the barrier

B represents the amplitude of the reflected wave function

The reflection coefficient (R: the ratio of the amplitude of a wave reflected from a surface to the amplitude of the incident wave) for the wave function is then defined as:

$$R \approx \left|\frac{B}{A}\right|^2 \quad (1.21)$$

Physically, it represents the relative intensities of the incident and reflected wave functions.

For an electron incident on the barrier, there are two possibilities, namely it is either reflected or transmitted through the barrier. In terms of probability or frequency of occurrence, $R + T = 1$, where R and T are the reflection and transmission coefficients. Thus,

$$R = \left|\frac{B}{A}\right|^2 = \left|-\frac{1+ik\delta}{1-ik\delta}\right|^2 \approx 1 \quad (1.22)$$

and therefore

$$T = 1 - R \approx 0 \tag{1.23}$$

which indicates that, for an infinitely wide barrier, the number of electrons found in the barrier region will be zero. Nevertheless, division of the first of the sample vacuum-barrier boundary conditions by A results in the equation

$$1 + \frac{B}{A} \approx \frac{C}{A} \tag{1.24}$$

The probability of finding an electron in the barrier region at $x = 0$, due to quantum-mechanical tunneling, is given by the equation

$$\left|\frac{C}{A}\right|^2 \approx \frac{4(k\delta)^2}{1+(k\delta)^2} \tag{1.25}$$

For determination of the effective tunneling transmission coefficient, $|F/A|^2$, that is, the relative probability for the frequency of occurrence of an electron tunneling out of the sample surface, across the sample-tip-barrier region, and into the tip, the tip-barrier boundary equations (at $x = L$) and Equation 1.19 are combined to get

$$\left|\frac{F}{A}\right| \approx -\frac{4ik\delta}{(1-ik\delta)^2}\exp\left[-L(\mu+ik)\right] \tag{1.26}$$

which produces the desired quantitative result

$$T(E) = \left|\frac{F}{A}\right|^2 \approx \left[\frac{4k\delta}{\left\{1+(k\delta)^2\right\}}\right]^2 \exp\left(\frac{-2L}{\delta}\right) \propto \exp\left(-2L\sqrt{\frac{2m\Phi}{\hbar^2}}\right) = \exp\left(-\frac{L}{L_0}\right) \tag{1.27}$$

where

$$k^2 = \frac{2mE}{\hbar^2},\ (k\delta)^2 = E(U_0 - E) = E\Phi \text{ and } L_0 = \frac{\hbar}{\sqrt{8m\Phi}} \tag{1.28}$$

Substituting typical values: $\Phi = 5\times10^{-19}$ J, $m = 9.11 \times 10^{-31}$ kg, and $\hbar = 1.05 \times 10^{-34}$ J s, results in a decay length $L_0 = 0.55$ Å so:

$$T(E) \approx \exp(-2L) \tag{1.29}$$

where L is expressed in angstroms. This equation shows that the probability of an electron tunneling across the barrier decreases by an order of magnitude for each angstrom change in separation. This demonstrates mathematically that tunneling current is an extremely sensitive measure of the distance between the tip and sample.

Example 1.3

Determine the tunneling transmission coefficients $T(E)$ for barrier widths (L) of 1, 2, 3, 5, 8, and 10 Å and plot T as a function of L.

From Equation 1.29, $T(E) = \exp(-2 \times 1) = \exp(-2) = 0.135$ for $L = 1$Å. Similarly, for $L = 2$, 3, 5, and 10 Å, $T(E) = 0.0183$, 0.00248, 4.54×10^{-5}, and 2.06×10^{-9}, respectively. The $T(E)$ versus L plot is shown in Figure 1.17.

In the STM, one of the metals is the sample being investigated and the other metal is the probe. The sample is usually flatter than the probe. Because the probe is formed of atoms, if it is sharpened into a tip, it will most likely have one atom at the end of the tip. The spacing between atoms is about 3Å. Therefore, any tunneling through atoms, that are one atom back from the closest atom, is a fraction exp (–2)(3) of tunneling through the atom at the tip. Virtually all of the tunneling electrons will pass through the single atom closest to the surface. Most of the tunneling current is carried by the atom that is closest to the sample (the "front atom"). If the sample is very flat, this front atom remains the atom that is closest to the sample during scanning in X and Y, and even relatively blunt tips yield atomic resolution easily. The sample surface is scanned with a single atom! This feature produces the atomic resolution capabilities of the microscope. STM, in fact, still provides the best resolution available. STM resolution can be as high as 0.1 nm. Z resolution is about 0.1 nm for a well-designed STM.

The spectacular spatial resolution and relative ease of obtaining atomic resolution by scanning tunneling microscopy are based on three properties of the tunneling current: (i) as a consequence of the strong distance dependence of the tunneling current, even with a relatively blunt tip, the probability is high that a single atom protrudes far enough out of the tip that it carries the main part of the tunneling current; (ii) typical tunneling currents are in the nanoampere range – measuring currents of this magnitude can be performed with a very good signal-to-noise ratio, even with a simple

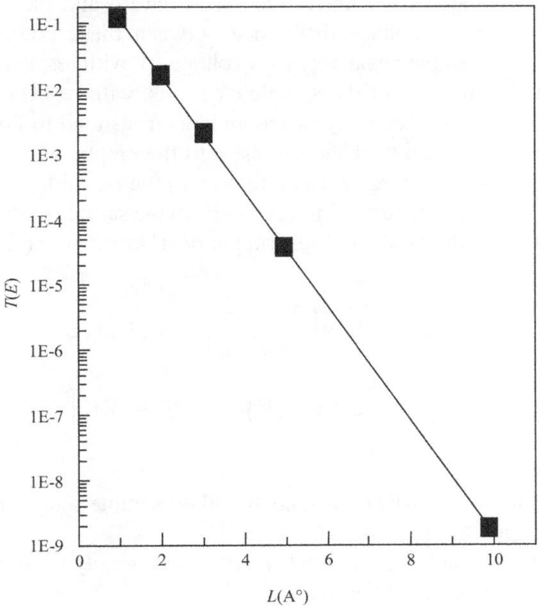

FIGURE 1.17 Effect of tip–sample spacing on transmission coefficient.

experimental setup; and (iii) because the tunneling current is a monotonic function (a function that is either entirely non-increasing or non-decreasing) of the tip–sample distance, it is easy to establish a feedback loop that controls the distance so that the current is constant.

Although the STM itself does not need vacuum to operate (it works in air as well as under liquids), ultrahigh vacuum is required to avoid contamination of the samples from the surrounding medium. Unfortunately, there are some limitations, the most important of which is that the surface of the sample must be conducting or semiconducting. This limits the materials that can be studied by STM. But even conductors – except for a few special materials, like highly oriented pyrolytic graphite (HOPG: a unique form of graphite grown on a substrate by decomposition of a hydrocarbon gas at very high temperature in a vacuum furnace, permitting the graphite to crystallize; the result is an ultrapure product that is near theoretical density and extremely anisotropic) – cannot be studied under ambient conditions by STM but have to be investigated in an ultrahigh vacuum. Under ambient conditions, the surface layer of solids constantly changes by adsorption and desorption of atoms and molecules. An ultrahigh vacuum is required for clean and well-defined surfaces. This limitation led to the invention in 1986 of the first atomic force microscope (AFM). Because electrical conductivity of the sample is not required in atomic force microscopy, the AFM can image virtually any flat solid surface without the need for surface preparation. Consequently, thousands of AFMs are in use in universities and public and industrial research laboratories all over the world. Most of these instruments are operated under ambient conditions.

1.19.3 Tunneling Current

The tunneling current flowing between the tip and the sample depends on the voltage difference between these bodies. On biasing the sample negatively by a voltage $-V$ with respect to the tip, the Fermi level of the sample electrons, with respect to the tip electrons, is effectively raised and electrons tend to flow out from the filled states of the sample into the empty states of the tip. For states of energy E (with respect to the Fermi level of the sample), the elastic tunneling current from the sample to the tip is expressed as the product of several factors (Hoffman Lab 2010a):

$$I_{\text{sample-tip}} = -2E \times \left(\frac{2\pi}{\hbar}\right)\left|\left(\frac{F}{A}\right)\right|^2 \times \rho_s(E) f(E) \times \left[\rho_t(E+qV)\{1-f(E+qV)\}\right] \tag{1.30}$$

where

- the factor $\rho_s(E)f(E)$ represents the filled sample states for tunneling from the sample,
- the factor $[\rho_t(E + qV)\{1 - f(E + qV)\}]$ is the empty tip states for tunneling to the tip,
- the factor of 2 accounts for electron spin,
- $-q$ is the electron charge,
- $|(F/A)|^2$ is a matrix element; and
- $f(E)$ is the Fermi distribution function

$$f(E) = \left\{1 + \exp\left(\frac{E}{k_B T}\right)\right\}^{-1} \tag{1.31}$$

Though the predominant tunneling current flow for negative sample voltage $-V$ is directed from sample to tip, there exists a smaller reverse tunneling current of electrons from tip to sample. The governing equation for this current is written in an identical manner. By summation of these currents, and integration over all the energies E, the total tunneling current from sample to tip becomes

$$I_{\text{total}} = -\left(\frac{4\pi E}{\hbar}\right)\int_{-E_F}^{\infty}\left|\left(\frac{F}{A}\right)\right|^2 \rho_s(E)\rho_t(E+qV) \times \left[f(E)\{1-f(E+qV)\} - \{1-f(E)\}\{f(E+qV)\}\right]dE \tag{1.32}$$

This basic equation can be simplified by applying various realistic approaches to obtaining meaningful solutions. First, suppose the measurements are performed at low temperatures. In this case, the Fermi function is sharply cut off at the Fermi surface with a cut-off width of k_BT, which is only 0.36 meV for $T = 4.2$ K. In the approximation of an ideally abrupt cut-off, the integral is subdivided into three parts, named I, II, and III. Therefore, for finding the tunneling current, the relevant range of E over which integration must be performed is reduced to region II only, where $-qV < E < 0$. (Likewise, if a positive bias voltage V had been applied to the sample, the range of integration is $0 < E < qV$.) This results in considerable saving of effort. Hence, approximately

$$I_{\text{total}} \approx -\left(\frac{4\pi E}{\hbar}\right)\int_{-qV}^{0}\left|\left(\frac{F}{A}\right)\right|^2 \rho_s(E)\rho_t(E+qV)dE \tag{1.33}$$

Second, let us choose a tip material having a flat density of states (DOS; the number of states per interval of energy at each energy level that are available to be occupied) within the energy range of the Fermi surface (an abstract constant-energy interface that defines the allowable energies of electrons in a solid; it is useful for characterizing and predicting the thermal, electrical, magnetic, and optical properties of crystalline metals and semiconductors) to be studied. As an example, if the sample DOS is to be studied within 200 meV of the Fermi surface, then the measured tunneling current is a convolution (a mathematical operation on two functions, producing a third function that is typically viewed as a modified version of one of the original functions; an integral that expresses the blending or amount of overlap of one function as it is shifted over another function) of the DOS of the tip and sample in this energy range. Then, a tip material is selected such that it has a flat DOS in this range, so that $\rho_t(E + qV)$ is treated as a constant and taken outside the integral.

$$I_{\text{total}} \approx \left(\frac{4\pi E}{\hbar}\right)\rho_t(0)\int_{-qV}^{0}\left|\left(\frac{F}{A}\right)\right|^2 \rho_s(E)dE \tag{1.34}$$

From the basic theory for vacuum tunneling, under realistic assumptions, the following is evident: (i) the tip and the sample each have their own independent DOS; (ii) each of their wave functions falls exponentially to zero in the tunneling barrier; and (iii) the overlapping is small enough (i.e., tip–sample separation is sufficiently large) that each side is feebly influenced by the tail of the wave function from the other side. Under these conditions, the matrix element for tunneling is independent of the energy difference between the two sides of the barrier. To a reasonable approximation, the matrix element is taken outside the integral and treated as a constant.

$$I_{\text{total}} \approx \left(\frac{4\pi E}{\hbar}\right)\left|\left(\frac{F}{A}\right)\right|^2 \rho_t(0)\int_{-qV}^{0} \rho_s(E)\,dE \qquad (1.35)$$

But the matrix element arises from the assumption that both tip and sample wave functions fall off exponentially into the vacuum gap. Basically, a square vacuum barrier is assumed, and a WKB (Wentzel–Kramers–Brillouin) approximation (a method for finding approximate solutions to linear partial differential equations with spatially varying coefficients) is carried out. In reality, there will be some tilting to the top of the barrier, but the tilt will be the applied voltage (around 100 meV), whereas the height of the barrier equals the energy required to remove an electron from a metal, i.e.,, the work function, has a value of several electron volts. Therefore, the tilt of the barrier will be much smaller than the height of the barrier, and can be ignored. According to WKB, the tunneling probability through a barrier is

$$\left|\left(\frac{F}{A}\right)\right|^2 = \exp(-2\gamma) \qquad (1.36)$$

where

$$\gamma = \int_0^s \sqrt{\frac{2m\varphi}{\hbar^2}}\,dx = \frac{s}{\hbar}\sqrt{2m\varphi} \qquad (1.37)$$

where m is the mass of the electron
s is the width of the barrier (the tip–sample separation)
φ is the height of the barrier, which is actually a combined effect of the work functions of the tip and the sample

The work function is measured by recording the tunneling current as a function of the tip–sample separation.

$$I \propto \exp\left(-\frac{2s}{\hbar\sqrt{2m\varphi}}\right) \qquad (1.38)$$

Therefore, φ is obtained from the slope of the natural logarithm of tunneling current versus barrier width, that is, a graph of I versus s . Typically, φ is ~ 3-4 eV. The higher the value of φ, the more the tunneling current I varies for a given change in s; therefore, a higher φ provides a tip with a superior resolution.

However, due to the exponential fall-off, measurement of the absolute value of s is not possible. This is troublesome, because there is no way to ensure that measurements are being made at constant tip–sample separations. So, if variations are noticed from one point on the sample surface to another, it is not clear whether the variation is due to intrinsic non-homogeneities in the sample at the specific energy of measurement, or due to a varying tip–sample separation, i.e., it is difficult to distinguish between the two causes.

In summary, the tunneling current is fairly well approximated by the equation

$$I_{\text{total}} \approx \left(\frac{4\pi q}{\hbar}\right)\exp\left(-s\sqrt{\frac{8m\phi}{\hbar^2}}\right)\rho_t(0)\int_{-qV}^{0} \rho_s(E)\,dE \qquad (1.39)$$

1.19.4 Measurements with STM

1.19.4.1 Topography

The common mode of STM measurement employed by research groups worldwide is "topography" (Hoffman Lab 2010b). In this mode, the tip is rastered across the surface at a fixed sample bias voltage, V_{set}. A feedback loop is employed for controlling the voltage across the z piezoelement for maintaining a constant value of the tunneling current I_{set}. By recording the voltage applied to the z piezo, the height of the surface is effectively mapped.

The meaning of the "height of the surface" is unclear. One obvious suggestion is that some contour of constant charge density is implied. However, as can be seen from Equation 1.38 for I_{total}, the tunneling current is not dependent on the total charge density, but depends only on the charge density within qV below the Fermi surface, where $-V$ is the applied bias.

An arbitrarily large applied voltage will enable more charge density to be captured, but there are two problems: (i) some of the samples are fragile compounds with weak bonds, so that if a large voltage is applied locally, pieces of the surface will literally rip off; and (ii) if V is too high (approaching the work function φ), the tunneling approximation is invalid.

So, the "height of the surface" is somewhat arbitrarily defined as the tip–sample separation for which tunneling current is fixed at a particular constant value, I_{set}, for a particular applied bias voltage, V_{set}. In practical terms, a fixed current is chosen at −100 pA, for a bias voltage of −100 mV. This is arbitrary but gains support from the fact that, in accordance with expectations, atoms and other structural features are not seen, even over a wider range of choices of I_{set} and V_{set}. The most widely varying DOS features of the superconducting samples studied so far appear to be within 75 meV of the Fermi level. Superconductivity is the occurrence of zero electrical resistance in certain materials below a characteristic temperature.

1.19.4.2 Density of States

From the tunneling equation mentioned earlier, it is easy to see that, if the tip–sample separation is held constant, at a given (x, y) location, and a negative bias voltage $-V$ is applied on the sample, we have

$$I = I_0\int_{-qV}^{0} \rho_s(E)\,dE \qquad (1.40)$$

In other words, we can measure the integral of the DOS, down to any energy $-qV$, by varying $-V$. For a negative bias voltage on the sample, electrons are tunneling from sample to tip, and the integrated density of full states below the Fermi level in the sample is being measured. For a positive bias voltage on the sample, electrons are tunneling from tip to sample, and the integrated density of empty states above the Fermi level in the sample is measured. Thus, the integrated density of states (IDOS) is obtained. But it would be useful to find the DOS. After plotting an IDOS *versus* V curve, a numerical derivative (a technique of numerical analysis to produce an estimate of the derivative of a mathematical function using values from the function and other knowledge about the same) of the data is taken to obtain the DOS. But it will be easier to measure the derivative directly. So, a lock-in amplifier (a type of amplifier that can recover signals in the presence of overwhelming background noise or can provide high-resolution measurements of relatively clean signals over several orders of magnitude) is employed to modulate the bias voltage by dV (typically a few millivolts) around a DC voltage V of interest. Due to the voltage modulation dV, a current modulation dl can be measured. This dl/dV is termed the conductance $g(V)$, so that we can write

$$g(V) \equiv \frac{dI}{dV} \propto \text{DOS}(\text{eV}) \tag{1.41}$$

Therefore, by using a lock-in amplifier and varying V, an entire DOS curve can be mapped.

The energy resolution is limited by the amplitude of the wiggle until the modulation becomes less than approximately $k_B T = 0.36$ meV at $T = 4.2$ K. So ideally, the voltage modulation can be made smaller than 0.36 mV. But in practical terms, adequate signal-to-noise ratio (the dimensionless ratio of the signal power to the noise power corrupting the signal; it provides a comparison of the amount of a particular signal with the amount of background noise) cannot be obtained at this low amplitude without using prohibitively long averaging times. Most of the data are measured with a 2-mV root-mean-square (RMS) modulation, therefore blurring the energy resolution by approximately 5.6 meV.

1.19.4.3 Linecut

In the previous section, a single DOS curve at a single location was discussed. The possibility of (x, y) control over the location of the tip, using the piezoelectric tube scanner, enables the measurement of DOS curves at any desired location. Some samples, like good metal samples (without impurities), will have a completely homogenous DOS everywhere. But other, more interesting samples are nonhomogeneous. A full DOS curve can be measured at every point along a straight line, spaced a few angstroms apart, and a "linecut" can be seen.

1.19.4.4 DOS Map

Basically, a 3D data set is being discussed, with two spatial dimensions x and y (by varying the position of the tip) and one energy dimension (by varying V). This 3D data set can be viewed as a series of DOS-*versus*-energy curves at every location (x, y), or as a series of 2D DOS-maps at each energy qV. Mapping the DOS at a specific energy provides a good visual representation to see the non-homogeneities in the DOS.

1.20 The Force Nanosensor: AFM

1.20.1 Operating Principle

AFM is the most versatile member of the family of scanning probe microscopes (SPMs) (Morris et al. 2004). The AFM is closely related to the STM, and they share key components, except for the probe tip. X and Y topographic resolution for most AFMs is typically 2–10 nm. The AFM works in the same way as our fingers touch and probe the environment when we cannot see things in darkness. By using fingers to "visualize" an object, our brain can deduce its topography while touching it. *The AFM generates images by feeling the specimens, as opposed to seeing, using optical microscopes.* Its closest predecessor is the stylus profiler, an instrument used to measure the profile of a surface, in order to quantify its roughness, step height, or thin film thickness. AFM technology uses sharper probes and lower forces than stylus profilers to provide higher-resolution information without sample damage. The essential part of an AFM is a silicon (Si) or silicon nitride (Si_3N_4) cantilever with a sharp tip at its end, having a tip radius of the order of nanometers. AFM works by bringing the cantilever tip in contact with the surface to be imaged. The AFM is similar to an STM, except that the tunneling tip is replaced by a force sensor. The potential energy between the tip and sample V_{ts} causes a Z component of the tip–sample force $F_{ts} = -\partial V_{ts}/\partial z$ and a tip-sample spring constant $k_{ts} = -\partial F_{ts}/\partial z$. Depending on the mode of operation, the AFM uses F_{ts}, or some entity derived from F_{ts}, as the imaging signal. The spring constant is the restoring force of a spring per unit length. It is given by the change in the force exerted by the spring, divided by the change in deflection of the spring, i.e., the gradient of the force *versus* deflection curve.

Unlike the tunneling current, which has a very short range, F_{ts} has long- and short-range contributions. We can classify the contributions by their range and strength. In vacuum, there are short-range chemical forces (fractions of nm) and van der Waals, electrostatic, and magnetic forces with a long range (up to 100 nm). Under ambient conditions, meniscus forces formed by adhesion layers on tip and sample (water or hydrocarbons) can also be present. The meniscus is the free surface of a liquid, which assumes a flat, convex or concave shape, depending on the solid and liquid surface. The forces acting on a liquid molecule at the free surface are: (i) the weight of the molecule, acting vertically downward; (ii) the force of adhesion; and (iii) the force of cohesion. Cohesion is the term for molecules of a substance sticking together. Liquid molecules are attracted not only to each other, but to any molecule with positive or negative charges. When a molecule is attracted to a different substance, it is termed "adhesion."

1.20.2 Lennard-Jones Potential and the Van der Waals Forces

Before describing the operation of AFM, it is essential to understand the fundamental mechanisms responsible for interactions between particles. A simple mathematical description of the interaction between two particles is given by the Lennard-Jones potential (also referred to as the L-J potential, 6-12 potential or, less commonly, 12-6 potential). The Lennard-Jones potential function is a reasonably accurate model of interactions between noble gas atoms. The Lennard-Jones potential (Figure 1.18) is given by the expression

$$\Phi_{LJ}(r) = 4\varepsilon\left\{\left(\frac{\sigma}{r}\right)^{12} - \left(\frac{\sigma}{r}\right)^{6}\right\} \tag{1.42}$$

for interaction between a pair of atoms. The choice of parameters, ε and σ, is conducted to fit the physical properties of the material. The binding energy, ε, is the depth of the potential well, σ is the (finite) distance at which the interparticle potential is zero, and r is the distance between the particles.

This potential has an attractive tail at a large r. It reaches a minimum around 1.122σ, and becomes strongly repulsive at shorter distances, passing through 0 at $r = \sigma$ and increasing steeply as r is further decreased. The term $\sim 1/r^{12}$, dominating at short distances, models the repulsion between atoms when they approach very close to one another. Its physical origin is related to the Pauli exclusion principle, namely no two electrons in an atom can be in the same quantum state or configuration at the same time, i.e., they cannot have the same set of four quantum numbers; this prevents matter from collapsing into an extremely dense state. As soon as the electronic clouds surrounding the atoms start overlapping, the energy of the system increases abruptly. In the Lennard-Jones potential expression, the exponent 12 was selected solely on a practical basis. In fact, an exponential behavior is more appropriate on physical grounds.

Thus, the interaction between neutral atoms and molecules can be broken down into two different forces: an attractive force at long distances (the van der Waals forces) and a repulsive force at short distances (due to overlap between electron wave functions). Two interacting neutral atoms are subject to two opposing forces: firstly, they are weakly attracted by van der Waals forces, and secondly, they are repelled by Pauli repulsion. It is known that

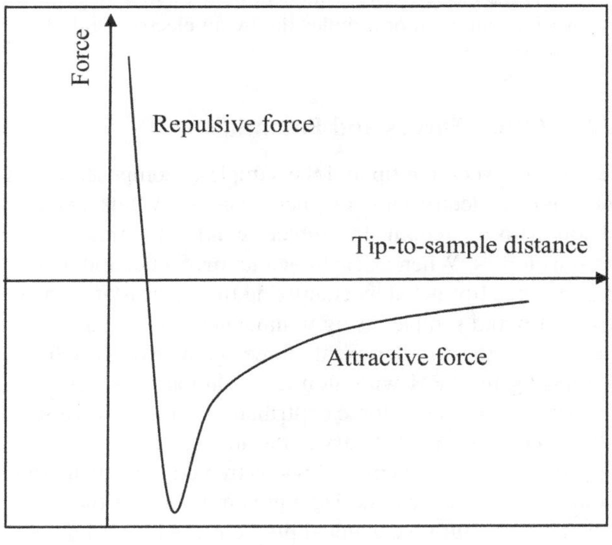

FIGURE 1.18 Graph of Lennard-Jones potential.

the van der Waals forces decay proportionally to the 6th power of the separation and that the effects of the Pauli repulsion decay exponentially. The exponent 12 was chosen exclusively because of the ease of computation. This form of the potential has little theoretical justification, but sometimes matches reality to an acceptable standard. Its use greatly simplifies numerical calculations based on the interatomic potential.

Elaborating on the attraction element, two factors contribute to this part:

1. *Dipole-dipole interaction*, i.e., electrostatic attraction between two molecules with permanent dipole moments. An electric dipole is a separation of positive and negative charges. The dipole moment is defined as the product of the total amount of positive or negative charges and the distance between their centroids. If both the molecules have permanent dipoles, they will tend to align directions so as to produce a relatively strong attraction. Thus, the main contribution to the force between the two molecules will not depend on momentary induced fluctuations. This is called a permanent dipole-permanent dipole force. It is an electrostatic interaction and decreases with the 4th power of distance

$$F_{\text{permanent dipole}} = -\text{constant} \times R^{-4} \tag{1.43}$$

where the constant of proportionality depends only on the identity of the molecules involved.
2. *Dipole-induced dipole interactions*, in which the dipole of one molecule polarizes a neighboring molecule, that is, the presence of one dipole (whether permanent or due to random fluctuation of charge density) exerts a force on the electrons of the other molecule causing a brief shift of orbital electrons to one side of one atom or molecule, creating a similar shift in adjacent atoms or molecules, thus giving it, in turn, a temporary dipole moment. Thus, weak attractive forces exist between atoms or nonpolar molecules. Essentially, the attraction between the molecules seen as electric dipoles is attraction between electron-rich regions of one molecule and electron-poor regions of another. These forces, called *London forces* or *dispersion forces*, hold together molecules with no permanent dipole moment.

The dispersion force is the weakest intermolecular force. It is a temporary attractive force that results when the electrons in two adjacent atoms occupy positions that make the atoms form temporary dipoles. It is sometimes called dipole-induced dipole attraction. This attractive force causes nonpolar substances to condense to liquids and to freeze into solids when the temperature is lowered sufficiently.

Clarifying further, the dispersion effect is the interaction between the instantaneous dipoles formed in the atoms by their orbiting electrons. The very rapidly changing dipole of one atom produces an oscillating electric field that acts upon the polarizability of a neighboring atom. The microscopic electric polarizability of an atom refers to its ability to respond to an external electric field by shifting its charges so as to create a dipole oriented favorably in the external field (negative end of the dipole closer to the positive pole of the external field). For an isolated atom this may occur by biasing the time-average distribution of electrons among ground and excited states to favor polar states. If excitation is easy, polarizability is high. For molecules in the gas or liquid state, skeletal vibrational states and rotational states may also contribute to the polarizability. Molecules that bend easily to produce a dipolar configuration have high polarizability. In the solid state, the rotational states are usually suppressed. If the atoms exist in a vacuum, the induced dipole of the neighboring atom moves in phase with the original dipole, producing an attractive atomic interaction.

Dispersion forces can also be thought of as very weak intermolecular covalent bonds, resulting from momentary dipoles forming due to asymmetry in electron cloud density. This electron cloud polarization causes a perturbation of electron clouds of nearby molecules, and the interaction of these temporary dipoles results in a net attractive interaction. Were it not for these intermolecular forces, molecules with no permanent dipole moment would exist solely in the gaseous phase.

These weak interactions dominate the bonding character of closed-shell systems, such as noble gases Ar or Kr. Molecules such as dinitrogen (the diatomic molecule of nitrogen: N_2), methane (CH_4), and fullerene (a closed graphitic form of carbon with exactly 12 pentagons) are held in liquid and solid states by *dispersion forces*.

These forces (which also, in fact, occur between permanent dipoles) are weaker than permanent dipole forces and decrease with the 7th power of distance

$$F_{\text{induced dipole}} = -\text{constant} \times R^{-7} \tag{1.44}$$

The (different) constant of proportionality again depends on the molecules involved.

Curiously, the inverse 7th power van der Waals forces turn out to be very closely related to the Casimir effect, which tells us that a force acts between two close, parallel, uncharged, conducting plates in a vacuum, due to quantum fluctuations in the electromagnetic field.

Thus, the forces acting between the tip of the AFM cantilever and the sample surface are attractive or repulsive forces between molecular entities (or between groups within the same molecular entity), other than those due to bond formation or to the electrostatic interaction of ions or of ionic groups with one another or with neutral molecules. The attraction is much weaker than a chemical bond. At short distances, the forces are repulsive in nature, which prevents the collapsing of molecules. As entities move closer to one another, these repulsions dominate.

Example 1.4

An oxygen atom has an atomic radius of 60 pm. Apply the equation for Lennard-Jones potential to show that the interatomic force changes from repulsive to attractive as the interatomic distance increases.

Considering that the distance between two oxygen atoms is minimum when the distance between the centers of two oxygen atoms = atomic diameter = 120 pm, for the closest approach of two oxygen atoms, σ = 120 pm. At $r < 120$ pm, σ/r is >1. Then, the first term in the Lennard-Jones (L-J) potential is larger than the second term, because the 12th power of a number >1 has a greater magnitude than the 6th power of the same number. Consequently, the potential has a positive value corresponding to repulsive force. But for $r >$ 120pm, σ/r is <1, i.e., a fraction. The twelfth power of a fractional number is smaller than its 6th power. The first term of the L-J potential is smaller than the second term, giving a negative value for potential; hence, the force is attractive. Potential energy is relative to some datum or reference level where the potential energy is zero at $r = \sigma$. If one extracts potential energy from the charge by letting it move with the attractive force, then the potential energy becomes smaller than the datum value, and hence negative. If one adds work to the charge to move it against the repulsive forces (from fields), its potential energy increases with respect to that datum and the potential energy is positive.

Assuming additivity and disregarding the discrete nature of matter by replacing the sum over individual atoms by an integration over a volume, with a fixed number density of atoms, the van der Waals interaction between macroscopic bodies can be calculated by the Hamaker approach (Hamaker 1937). This approach does not account for retardation effects due to the finite speed of light and is therefore only appropriate for distances of up to several hundred angstroms. For a spherical tip with radius R next to a flat surface (z is the distance between the plane connecting the centers of the surface atoms and the center of the closest tip atom), the van der Waals potential is given by (Israelachvili 1991)

$$V_{\text{vdW}} = -\frac{A_{\text{H}} R}{6z} \tag{1.45}$$

The Hamaker constant A_{H} depends on the type of materials (atomic polarizability and density) of the tip and sample. For most solids and interactions across a vacuum, A_{H} is of the order of 1 eV. Atomic polarizability is the electric dipole moment induced in a system, such as an atom or a molecule, by an electric field of unit strength.

1.20.3 Other Forces and Potentials

The force between the tip and the sample is composed of many contributions: electrostatic, magnetic, van der Waals, and chemical forces in a vacuum. In ambient conditions, there are also meniscus forces. Whereas electrostatic, magnetic, and meniscus forces can be eliminated by equalizing the electrostatic potential between tip and sample, using nonmagnetic tips, and operating in a vacuum, the van der Waals forces cannot be switched off. For imaging by AFM with atomic resolution, it is desirable to filter out the long-range force contributions and to measure only the force components that vary at the atomic scale.

A prototype of the chemical bond is treated in many textbooks on quantum mechanics: the H_2^+ ion is a model for the covalent bond. This quantum-mechanical problem can be solved analytically and gives interesting insights into the character of chemical bonds. The Morse potential (Israelachvili 1991), a convenient model for the potential energy of a diatomic molecule, describes a chemical bond with bonding energy E_{bond}, equilibrium distance a, and a decay length κ. Although the Morse potential can be used for a qualitative description of chemical forces, it lacks an important property of chemical bonds: anisotropy. Chemical bonds, especially covalent bonds, show an inherent angular dependence on the bonding strength. Using the Stillinger–Weber potential (one of the first attempts to model a semiconductor with a classical model, it is based on a two-body term and a three-body term), one can explain subatomic features in Si images. With increasing computer power, it becomes more and more feasible to perform *ab initio* calculations for tip–sample forces.

1.20.4 Force Sensor (Cantilever) and Force Measurement

The central element of a force microscope and its major instrumental difference from an STM is the spring that senses the force between tip and sample. For sensing normal tip–sample forces, the force sensor should be rigid in two axes and relatively soft in the third axis. This property is fulfilled with a cantilever beam, and, therefore, the cantilever geometry is typically used for force detectors. For a rectangular cantilever with dimensions *w* (width), *t* (thickness), and *L* (length), the spring constant *k* is given by (Chen 2008)

$$k = \frac{Ywt^3}{4L^3} \tag{1.46}$$

where *Y* is Young's modulus. The fundamental eigenfrequency (vibration frequency of a quantum system) f_0 is given by (Chen 2008)

$$f_0 = 0.162\left(\frac{t}{L^2}\right)\sqrt{\frac{Y}{\rho}} \tag{1.47}$$

where ρ is the mass density of the cantilever material. Force is calculated from the amount of bending of the cantilever, measured by a laser spot reflected onto a split photodiode detector. It is calculated by Hooke's law, by measuring the deflection of the cantilever and from the knowledge of the spring constant of the material of the cantilever. The properties of interest are the stiffness (*k*: the resistance of an elastic body to deformation by an applied force, measured by the ratio of a steady force acting on the body to the resulting displacement), the eigenfrequency f_0, the quality factor (*Q*: a parameter that describes how underdamped an oscillator is; a higher *Q* value indicates a lower rate of energy loss relative to the stored energy of the oscillator), the variation of the eigenfrequency with temperature $\partial f_0/\partial T$, and the chemical and structural composition of the tip.

The first AFMs were mostly operated in the static contact mode (see Section 1.20.5), and, for this mode, the stiffness of the cantilever should be less than the interatomic spring constants of atoms in a solid, which amounts to $k \leq 10$ N m^{-1}. This constraint on *k* was assumed to hold for dynamic atomic force microscopy as well. In the dynamic mode, the cantilever is externally oscillated at, or close to, its fundamental resonance frequency or an integer multiple of the fundamental frequency (harmonic). However, it turned out later that, in dynamic atomic force microscopy, *k* values exceeding hundreds of N m^{-1} help to reduce noise and increase stability. The *Q* factor depends on the damping mechanisms present in the cantilever. For micromachined cantilevers operated in air, *Q* is mainly limited by viscous drag (force arising from a viscous effect of a fluid, i.e., due to resistance to flow), which typically amounts to a few hundred, while, in a vacuum, internal and surface effects in the cantilever material are responsible for damping, and *Q* reaches hundreds of thousands.

The most common cantilevers in use today are built completely from silicon with integrated tips pointing in a [001] crystal direction. In crystallography, any vector direction is specified between two points of which one point is usually the origin of the coordinate system. The lengths of the projections of the vector on the three axes *X*, *Y*, *Z* are determined in terms of unit cell vectors *a*, *b* and *c*. These lengths are multiplied or divided by a common factor to reduce them to the smallest integer values denoted by u, v and w. The numbers u, v, w are enclosed in square brackets to denote the direction of the vector as [uvw].

Self-sensing cantilevers with integrated tips have a built-in deflection-measuring scheme, utilizing the piezoresistive effect in silicon. In dynamic atomic force microscopy, some requirements for the force sensor are similar to the desired properties of the time-keeping element in a watch: utmost frequency stability over time and temperature changes and little energy consumption.

Quartz tuning forks (small two-pronged devices) have many attractive properties, but their geometry gives them marked disadvantages for use as force sensors. The great benefit of the fork geometry is the high *Q* factor, which is a consequence of the presence of an oscillation mode in which both prongs oscillate opposite to one another. The dynamic forces necessary to keep the two prongs oscillating cancel out exactly in this case. However, this only works if the eigenfrequencies of the two prongs match precisely. The mass of the tip mounted on one prong and the interaction of this tip with a sample breaks the symmetry of the tuning fork geometry. This problem can be avoided by fixing one of the two beams and turning the fork symmetry into a cantilever symmetry, where the cantilever is attached to a high-mass substrate with a low-loss material.

For atomic-resolution AFM, the front atom of the tip should ideally be the only atom that interacts strongly with the sample. In order to reduce the forces caused by the shaft of the tip, the tip radius should be as small as possible. Cantilevers made of silicon with integrated tips are typically oriented so that the tip points in the [001] crystal direction. Due to the anisotropic etching rates of Si and SiO_2, these tips can be etched so that they develop a very sharp apex. Not only the sharpness of a tip, but also the coordination of the front atom is important for atomic force microscopy. The tip and sample can be viewed as two giant molecules. In chemical reactions between two atoms or molecules, the chemical identity and the spatial arrangements of both partners play a crucial role. For AFM with true atomic resolution, the chemical identity and bonding configuration of the front atom is therefore critical. In [001]-oriented silicon tips, the front atom exposes two dangling bonds (if bulk termination is assumed) and has only two connecting bonds to the rest of the tip; dangling bonds are unsatisfied valences associated with atoms in the surface layer of a solid. If we assume bulk termination, it is immediately evident that tips pointing in the [111] direction are more stable, because then the front atom has three bonds to the rest of the tip.

The force is given by

$$F = -kz \tag{1.48}$$

where

k is the spring constant and
z is the deflection of the cantilever

While scanning the tip across the sample surface, the force is kept constant. Then the vertical movement of the tip follows the surface profile and is recorded as the surface topography by the AFM. To form a map of the measured property relative to the X–Y position, these data are collected as the probe is scanned in a raster pattern (a set of pixels arranged in rows and columns) across the sample. The information is fed to a computer, which generates a map of topography and/or other properties of interest. Areas ranging from less than 100 nm^2 to as large as about 100 μm^2 are imaged.

1.20.5 Static and Dynamic Atomic Force Microscopy

In AFM, the force F_{ts} that acts between the tip and sample is used as the imaging signal. In the static mode of operation, the force translates into a deflection $q' = F_{ts}/k$ of the cantilever. Because the deflection of the cantilever should be significantly larger than the deformation of the tip and the sample, restrictions on the useful range of k apply. In the static mode, the cantilever should be much softer than the bonds between the bulk atoms in the tip and sample. Interatomic force constants in solids are in a range from 10 to about 100 N m^{-1} – in biological samples, they can be as small as 0.1 N m^{-1}. Force constant is the ratio of the force acting to restrain the relative displacements of nuclei in a molecule to its deformation from the equilibrium position. Thus, typical values for k in the static mode are 0.01–5 N m^{-1}. The eigenfrequency f_0 should be significantly higher than the desired detection bandwidth (the width of the range of frequencies between upper cut-off and lower cut-off frequencies), that is, if 10 lines are recorded per second and each line has a width of say 100 atoms, then during this imaging f_0 should be at least $10 \times 2 \times 100$ s^{-1} = 2 kHz in order to prevent resonant excitation of the cantilever.

In the dynamic operation modes, the cantilever is deliberately vibrated. The cantilever is mounted on an actuator to allow the external excitation of an oscillation. There are two basic methods of dynamic operation: amplitude modulation (AM) and frequency modulation (FM). AM is the encoding of a wave by variation of its amplitude in accordance with an input signal. FM is a method of altering a waveform by changing the instantaneous frequency. In AM-AFM, the actuator is driven by a fixed amplitude A_{drive} at a fixed frequency f_{drive}, where f_{drive} is close to but different from f_0. When the tip approaches the sample, elastic and inelastic interactions cause a change in both the amplitude and the phase (relative to the driving signal) of the cantilever; the phase is a particular point or any distinct time period in the time of a cycle, measured from some arbitrary zero and expressed as an angle. These changes are used as the feedback signal. The change in amplitude in the AM mode does not occur instantaneously with a change in the tip–sample interaction, but on a time scale of $\tau_{AM} = 2Q/f_0$. With Q factors reaching 100,000 in vacuum, the AM mode is very slow. This problem was solved by introducing the FM mode, in which the change in the eigenfrequency occurs within single oscillation cycle on a time scale of $\tau_{FM} = 1/f_0$. Both AM and FM modes were initially meant to be "noncontact" modes, i.e., the cantilever was far away from the surface and the net force between the front atom of the tip and the sample was clearly attractive. The AM mode was later used very successfully at a closer distance range under ambient conditions involving repulsive tip–sample interactions. Using the FM mode in vacuum improved the resolution dramatically. Finally, atomic resolution was obtained.

In static atomic force microscopy, the imaging signal is given by the DC deflection of the cantilever, which is subject to $1/f$ noise. Pink noise, or $1/f$ noise, is a signal or process with a frequency spectrum, such that the power spectral density is inversely proportional to the frequency. In dynamic atomic force microscopy, the low-frequency noise (noise that has a frequency between 20 and 100–150 Hz) is discriminated if the eigenfrequency f_0 is larger than the $1/f$ corner frequency (the frequency at which the $1/f$ noise spectral density equals the white noise, a random signal or process with a flat power spectral density). With a bandpass filter (an electronic device or circuit that allows signals between two specific frequencies to pass, but rejects or attenuates signals at other frequencies) with a center frequency (midpoint in the pass band) of around f_0, only the white noise density is integrated across the bandwidth B of the bandpass filter.

1.20.6 Classification of Modes of Operation of AFM on the Basis of Contact

1.20.6.1 Contact Mode

In the contact-AFM mode, the tip makes soft "physical contact" with the surface of the sample. It either scans at a constant low height above the surface or under the conditions of a constant force. In the constant-height mode, the height of the tip is fixed, whereas, in the constant-force mode, the deflection of the cantilever is fixed and the motion of the scanner in Z-direction is recorded.

For contact mode AFM imaging, it is necessary to have a cantilever that is soft enough to be deflected by very small forces yet has a high enough resonant frequency to avoid being susceptible to vibrational instabilities. Silicon nitride (Si_3N_4) tips are used for contact mode. In these tips, there are several cantilevers with different geometries attached to each substrate, resulting in different spring constants. The spring constant is the constant of proportionality k, which appears in Hooke's law: $F = -kx$, where F is the applied force and x is the displacement from equilibrium. It has units of force per unit length and is a measure of how stiff the spring is.

To avoid problems caused by capillary forces (the forces involving molecular adhesion, by which the surface of a liquid in a tube is either elevated or depressed, depending on the cohesiveness of the liquid molecules; they arise from intermolecular attractive forces between the liquid and surrounding solid surfaces), which are generated by a liquid contamination layer usually present on surfaces in air, the sample can be studied while immersed in a liquid. This procedure is especially beneficial for biological samples.

The advantages of contact mode are (a) high scan speeds, (b) atomic resolution is possible, and (c) easier scanning of rough samples with extreme changes in vertical topography.

The disadvantages of this mode are (a) lateral forces can distort the image, (b) capillary forces from a fluid layer can cause large forces normal to the tip–sample interaction, and(c) combination of these forces reduces spatial resolution and can cause damage to soft samples.

1.20.6.2 Noncontact Mode

In this mode, the probe operates in the attractive force region and the tip–sample interaction is minimized. The use of noncontact mode allowed scanning without influencing the shape of the sample by the tip–sample forces. In most cases, the cantilever of choice for this mode is one having high spring constant of 20–100 N m^{-1}, so that it does not stick to the sample surface at low amplitudes. The tips mainly used for this mode are silicon probes.

An advantage of the noncontact mode is that low force is exerted on the sample surface and no damage is caused to soft samples.

The disadvantages of noncontact mode are (i) lower lateral resolution, limited by tip-sample separation, (ii) slower scan speed to avoid contact with fluid layer, and (iii) usually only applicable to extremely hydrophobic (repelling, tending not to combine with, or being incapable of dissolving in water) samples with a minimal fluid layer.

1.20.6.3 Tapping Mode (Intermittent-Contact Mode)

In tapping-mode AFM the cantilever is oscillating close to its resonant frequency. An electronic feedback loop ensures that the oscillation amplitude remains constant, such that a constant tip–sample interaction is maintained during scanning. Forces that act between the sample and the tip will cause not only cause a change in the oscillation amplitude, but also a change in the resonant frequency (natural frequency of vibration is determined by the physical parameters of a vibrating object; it is easy to make an object vibrate at its resonant frequencies, but difficult to cause vibrations at other frequencies) and phase (the fraction of a wave cycle that has elapsed relative to an arbitrary point) of the cantilever. The amplitude is used for the feedback and the vertical adjustments of the piezoscanner are recorded as a height image. Simultaneously, the phase changes are presented in the phase image (topography). Phase imaging refers to recording the phase shift signal in intermittent-contact AFM. It is a powerful technique for producing contrast on heterogeneous samples. The phase shift can be thought of as a "delay" in the oscillation of the cantilever as it moves up and down and in and out of contact with the sample. Phase of the cantilever oscillations, φ, is measured relative to the drive signal oscillations. Hence, phase imaging implies the monitoring of the phase lag between the signal driving the cantilever oscillations and the cantilever oscillation output signal. Therefore, changes in the phase lag reflect changes in the mechanical properties of the specimen surface.

Phase images complement topography images by mapping the various regions of the sample surface, each of which interacts with the tip in a slightly or significantly different way. This difference is sometimes so subtle that it is barely noticeable in the topography image, but clearly visible in the contrast variations in the phase image.

Reverting to the mainstream discussion on the tapping mode, an advantage of the tapping mode is that there is almost no lateral force. Elimination of a large part of permanent shearing forces (internal forces in any material, that are usually caused by external forces acting perpendicular to the material, or forces that have a component acting tangentially to the material, leading to cutting and separating of the material through its cross section) create less damage to the sample surface, even with stiffer probes. Different components of the sample, that exhibit different adhesive and mechanical properties, will show a phase contrast and therefore even allow a compositional analysis. For a good phase contrast, larger tip forces are advantageous, while minimization of this force reduces the contact area and facilitates high-resolution imaging. So, in applications, it is necessary to choose the correct values matching the objectives. Silicon probes are used primarily for tapping-mode applications. Higher lateral resolution is achieved (1–5 nm).

A disadvantage of this mode is its slower scan speed than occurs in contact mode.

The tip of the AFM is used (i) for measuring forces (and mechanical properties) at the nanoscale; (ii) for imaging; and (iii) as a nanoscale tool, that is, for bending, cutting, and extracting soft materials (such as polymers, DNA, and nanotubes), at the submicron scale under high-resolution image control. AFM is used as nanorobot for manipulating and controlling nano-objects, as shown in Figure 1.19.

1.20.7 Frequency-Modulation Atomic Force Microscopy

In frequency-modulation atomic force microscopy (FM-AFM) (Giessibl 2003), a cantilever with eigenfrequency f_0 and spring constant k works under controlled positive feedback (in which a portion of the output is combined in phase with the input), which ensures its oscillation with constant amplitude A, as shown in Figure 1.20. The deflection signal enters a bandpass filter, the output from which is divided into three branches. The first branch is shifted in phase, routed through an analog multiplier (a device that produces an output voltage or current that is proportional to the product of two or more independent input voltages or currents), and fed back to the cantilever via an actuator, a mechanical device for moving or controlling a mechanism or system. The second branch is used to compute the actual oscillation amplitude – this signal is applied to calculate a gain input g for the analog multiplier. The third branch feeds a frequency detector. In frequency detectors, frequency-modulated oscillations are first converted into amplitude-modulated oscillations, which are then detected by an amplitude detector. The frequency f is decided by the eigenfrequency f_0 of the cantilever and the phase shift ϕ between the mechanical excitation generated at the actuator and the deflection of the cantilever. If $\phi = \pi/2$, the loop oscillates at $f = f_0$.

Forces between the tip and sample cause a change in $f = f_0 + \Delta f$. The eigenfrequency of a harmonic oscillator = $(k^*/m^*)^{0.5}/(2\pi)$,

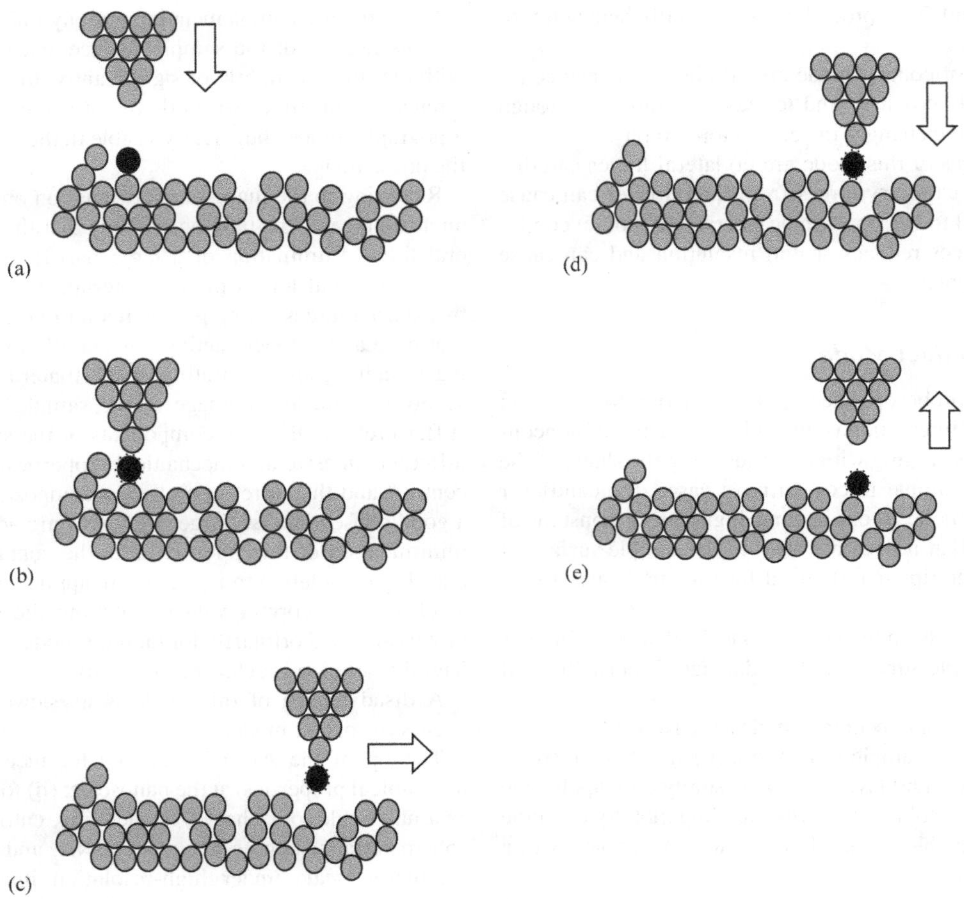

FIGURE 1.19 Manipulation of an atom with an AFM probe: (a) probe tip approaching the atom; (b) probe tip touching the atom; (c) probe tip transferring the atom to the new location; (d) probe tip placing the atom at the new position; and (e) withdrawal of the probe tip.

where k^* is the effective spring constant and m^* is the effective mass; a harmonic oscillator is a system that, when displaced from its equilibrium position, experiences a restoring force, F, proportional to the displacement, x. If the second derivative of the tip–sample potential, $k_{ts} = \partial^2 V_{ts}/\partial z^2$, is constant for the whole range covered by the cantilever, $k^* = k + k_{ts}$. If $k_{ts} << k$, the oscillating square root is expanded as a Taylor series (series expansion of a function as an infinite sum of terms calculated from the values of its derivatives at a single point) and the shift in eigenfrequency is approximately

$$\Delta f = \left(\frac{k_{ts}}{2k}\right) f_0 \tag{1.49}$$

By measuring the frequency shift Δf, the tip–sample force gradient is determined.

1.20.8 Generic Calculation

The oscillation frequency is the main observable in FM-AFM, and it is useful to formulate a relation between frequency shift and the forces acting between the tip and sample (Giessibl 2003). Whereas the frequency can be calculated numerically, an analytic calculation helps in finding the functional relationships between operational parameters and the physical tip–sample forces. The motion of the cantilever (spring constant k, effective mass m^*) is described by a weakly disturbed harmonic oscillator. The deflection of the tip of the cantilever is $q'(t)$. It oscillates with an amplitude A, at a distance of $q(t)$ from a sample. The closest point to the sample is $q = d$ and $q(t) = q'(t) + d + A$. The Hamiltonian of the cantilever is

$$H = \frac{p^2}{2m^*} + \frac{kq'^2}{2} + V_{ts}(q) \tag{1.50}$$

where $p = m^* dq'/dt$. The unperturbed motion is represented by

$$q'(t) = A\cos(2\pi f_0 t) \tag{1.51}$$

and the frequency is

$$f_0 = \left(\frac{1}{2\pi}\right)\sqrt{\frac{k}{m^*}} \tag{1.52}$$

If the force gradient, $k_{ts} = -\partial F_{ts}/\partial z$, is constant during the oscillation cycle, the frequency shift is calculated by the equation derived in Section 1.20.7. However, in classic FM-AFM, k_{ts} varies by orders of magnitude during one oscillation cycle, and a perturbation approach (mathematical methods that are used to find an approximate solution to a problem that cannot be solved

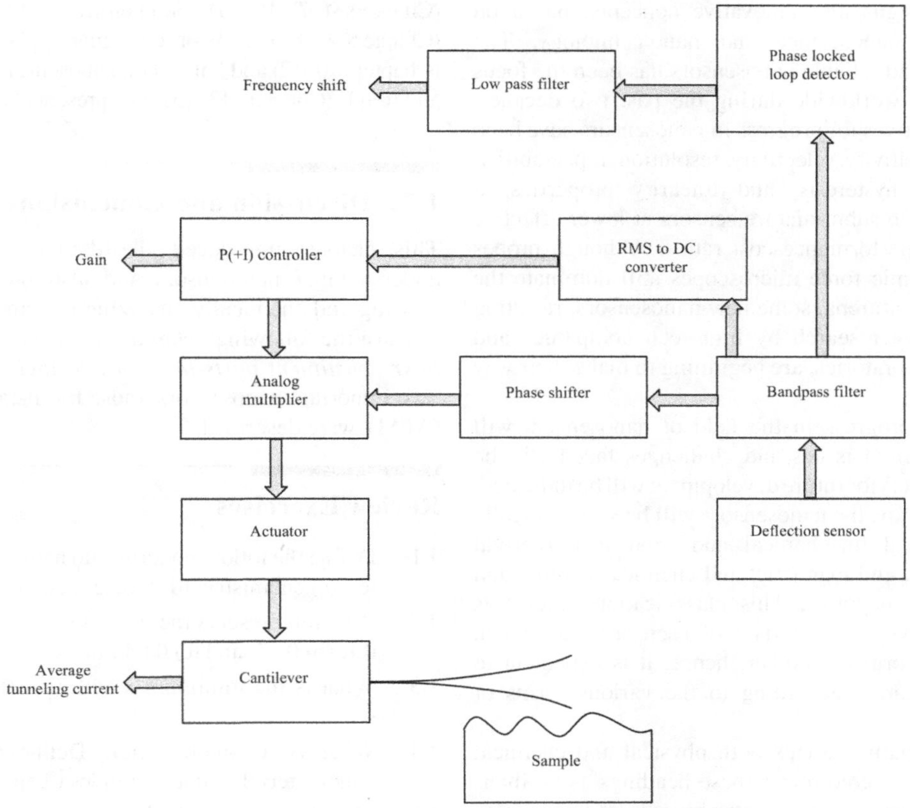

FIGURE 1.20 Frequency-modulation AFM feedback loop.

exactly, by starting from the exact solution of a related problem) is employed for the calculation of the frequency shift. This approach derives the magnitude of the higher harmonics and the constant deflection of the cantilever. The method involves the solution of Newton's equation of motion for the cantilever (effective mass μ^*, spring constant k):

$$\mu^*\left(\frac{d^2q'}{dt^2}\right) = -kq' + F_{ts}(q') \tag{1.53}$$

The cantilever motion is assumed to be periodic and therefore is expressed as a Fourier series (an expansion of a periodic function in terms of an infinite sum of sines and cosines) with fundamental frequency f:

$$q'(t) = \sum_{m=0}^{\infty} a_m \cos(m2\pi ft) \tag{1.54}$$

Inserting this into Newton's equation yields

$$\sum_{m=0}^{\infty} a_m\left\{-(m2\pi f)^2 \mu^* + k\right\}\cos(m2\pi ft) = F_{ts}(q') \tag{1.55}$$

Multiplication by $\cos(l2\pi ft)$ and integration from $t = 0$ to $t = 1/f$ gives

$$a_m\left\{-(m2\pi f)^2 \mu^* + k\right\}\pi(1+\delta_{m0}) = 2\pi f \int_0^{1/f} F_{ts}(q')\cos(m2\pi ft)\,dt \tag{1.56}$$

by applying the orthogonality of the angular functions

$$\int_0^{2\pi} \cos(mx)\cos(lx)\,dx = \pi\delta_{ml}(1+\delta_{m0}) \tag{1.57}$$

Two functions, f and g, are said to be orthogonal if their inner product is zero for $f \neq g$. The *inner product* (or dot product, or scalar product) is an operation on two vectors that produces a scalar. For a weak perturbation, $q'(t) = A\cos(2\pi ft)$ with $f = f_0 + \Delta f$, $f_0 = (1/2\pi)\sqrt{(k/\mu^*)}$ and $|\Delta f| << f_0$. In a first-order approximation, the frequency shift is given by

$$\Delta f = -\left(\frac{f_0^2}{kA}\right)\int_0^{1/f_0} F_{ts}(q')\cos(m2\pi f_0 t)\,dt = -\left(\frac{f_0}{kA^2}\right)\langle F_{ts}q'\rangle \tag{1.58}$$

which equals the result of the Hamilton-Jacobi method, a reformulation of classical mechanics and, thus, equivalent to other formulations such as Newton's laws of motion, Lagrangian mechanics, and Hamiltonian mechanics.

1.21 Outline and Organization of the Book

Sensor technology, a key technology in healthcare instrumentation, environmental monitoring, homeland security, aerospace and automotive sectors, and in industrial process control, has received a boost from new materials, and novel device structural designs and innovative concepts, based on the foundations of nanoscience and nanotechnology. The development of fast and reliable nanosensors has been the focus of intensive research worldwide during the past two decades. Prime identifiable motives for progress in nanosensors have been the quest for high sensitivity, selectivity, resolution, repeatability, reliability, desirable hysteresis, and linearity properties in devices, and to develop subminiature sensors at lower effective costs, yielding high performance-cost ratios. Although probes for scanning and atomic force microscopes still dominate the commercial nanosensor arena, some new nanosensors, resulting from multidisciplinary research by high-tech companies and academic research laboratories, are beginning to make their way to the marketplace.

In this book, the progress in the field of nanosensors will be outlined, fundamental issues and challenges faced will be addressed, and prospects for future development will be indicated. For convenience of study, the nanosensors will be systematically classified into physical (mechanical and acoustical, thermal and radiation, optical, and magnetic) and chemical (atomic and molecular energies) categories. This classification scheme is convenient because a key characteristic of a sensor is conversion of energy from one form to another; hence, it is expedient to organize the classification according to the various forms of energy.

The biological domain overlaps both physical and chemical domains and will be covered under these headings as an intermixed or interdisciplinary area, because biosensors incorporate biological recognition elements with a physical or chemical sensor. Micro-/nanocantilevers, for instance, overlap both nanomechanical and nanobiosensors. In addition, nanobiosensors will be exclusively discussed in length in a separate chapter (Chapter 8) because of their enormous potential.

The current chapter (Chapter 1) makes the reader conversant with nanosensor classification and fundamental terms. The properties of some important nanomaterials used in nanosensor fabrication will be outlined in Chapter 2. Nanotechnologies that are widely used in nanosensor fabrication will be briefly recapitulated in Chapter 3. Chapters 4 through 8 will describe the representative sensors belonging to sensor families characterized by the input signal to which they respond, whereas Chapter 9 will be devoted to nanobiosensors. Applications of nanosensors for societal benefits, in industry, and for homeland security will be addressed in Chapters 10–12. Chapter 13 will present nanogenerators for self-powering nanosensors, as well as nanosensors with built-in nanogenerators. In Chapter 14, ideas of wireless nanosensor networks will be discussed. These networks are essential to increase the capabilities of a solo nanosensor. The book will conclude in Chapter 15 by summarizing the present overall nanosensor scenario and discussing future trends.

Thus, the proposed book will serve as a complete, definitive, and authoritative guide to nanosensors, offering information on fabrication, properties, and operating mechanisms of nanosensors. Applications of nanosensors, their operation without batteries, and their interconnections to form networks will be dealt with in extensive detail.

The book consists of six parts: Part I on fundamental concepts of nanosensors (Chapter 1), Part II on nanomaterials and micro/nanofabrication (Chapters 2–3), Part III on physical nanosensors (Chapters 4–7), Part IV on chemical and biological nanosensors (Chapters 8–9), Part V on emerging applications of nanosensors (Chapters 10–12) and Part VI on nanosensor powering, networking and trends (Chapters 13–15). The present chapter falls under Part I.

1.22 Discussion and Conclusions

This chapter introduced the definitions and concepts for understanding nanosensors and also presented the tools for viewing and chemically analyzing nanomaterials. It enabled us to learn the following: *what are nanosensors? how can they or their constituent parts be seen and their composition known?* Two important nanosensors, those for distance (STM) and force (AFM), were described.

Review Exercises

1.1 Define the following terms: (a) natural science; (b) physics; (c) chemistry; (d) biology; and (e) nanotechnology.

1.2 Which represents the diameter of a hydrogen atom: (a) 0.1; (b) 0.32; and (c) 0.04 nm?

1.3 What is the minimum length that the human eye can see?

1.4 What is a nanomaterial? Define 0D, 1D, and 2D nanomaterials. Give examples of each type.

1.5 Mention some unusual properties of matter that emerge at the nanoscale.

1.6 Distinguish between sensors and transducers, giving examples. Justify the statement, “All transducers contain a sensor and mostly, although not always, sensors will also be transducers.”

1.7 List and describe the important parameters and characteristics of a sensor.

1.8 Mention two advantages of miniaturization of sensors. Explain and exemplify the statement: “A miniaturized sensor can accomplish many tasks that a bulky device cannot perform.”

1.9 Define a nanosensor. Give one example of a nanosensor that performs nanoscale measurements but does not have dimensions in the nano range.

1.10 Give two examples of natural nanosensors.

1.11 Which of the following microscopes does not need a vacuum environment for its operation: (i) SEM; (ii) TEM; and (iii) AFM.

1.12 Explain the statement “SEMs are used as workhorses for routine work whereas TEMs are engaged in specialized tasks.”

1.13 How does EDX complement the operation of SEM?

1.14 What is the advantage derived by using vacuum with AFM or STM?

1.15 Describe the contact, noncontact, and intermittent modes of operation of AFM, and point out their relative advantages and disadvantages.

1.16 Write the equation and draw the graph of Lennard-Jones potential for interaction between a pair of atoms.

1.17 IR absorption information is generally presented as a spectrum. What are the parameters plotted on the *X*- and *Y*-axes?

1.18 Can IR spectroscopy identify nonpolar molecules? What is the Raman effect? Explain the principle of Raman spectroscopy.

1.19 Prepare a comparative chart of IR and UV-vis spectroscopy in terms of principle of operation, instrument construction, and fields of application.

1.20 Explain: (i) X-ray diffraction; (ii) XPS; and (iii) SIMS.

REFERENCES

Booker, R. and E. Boysen. 2008. *Nanotechnology.* New Delhi, India: Wiley India Pvt. Ltd.

Bowker, M. and P. R. Davies (Eds.) 2010. *Scanning Tunneling Microscopy in Surface Science, Nanoscience and Catalysis.* Weinheim, Germany: Wiley-VCH.

Bréchignac, C., P. Houdy, and M. Lahmani (Eds.). 2007. *Nanomaterials and Nanochemistry.* Berlin, Germany: Springer, 753 pp.

Briggs, D. and J. T. Grant. 2006. *Surface Analysis by Auger and X-Ray Photoelectron Spectroscopy.* Manchester, UK: SurfaceSpectra Ltd.

Brooks, K. (Ed.). 2006. *Chambers Science Factfinder.* Edinburgh, UK: Chambers Harrap Publishers Ltd.

Campbell, N. A. and J. B. Reece. 2004. *Biology.* San Francisco, CA: Benjamin Cummings, 312 pp.

Cao, G. 2004. *Nanostructures and Nanomaterials: Synthesis, Properties and Applications.* London, UK: Imperial College Press, 452 pp.

Chen, C. J. 2008. *Introduction to Scanning Tunneling Microscopy.* New York: Oxford University Press.

Curtis, L. J. 2003. *Atomic Structure and Lifetimes: A Conceptual Approach.* Cambridge, UK: Cambridge University Press, 282 pp.

Daintith, J. (Ed.). 2010. *A Dictionary of Physics.* New York: Oxford University Press, 624 pp.

Ferraro, J. R. 2003. *Introductory Raman Spectroscopy.* San Diego, CA: Academic Press, 434 pp.

Ghuo, Z. and L. Tan. 2009. *Fundamentals and Applications of Nanomaterials.* Norwood, MA: Artech House, 249 pp.

Giessibl, F. J. 2003. Advances in atomic force microscopy. *Reviews of Modern Physics* 75: 949–983.

Goldstein, J., D. E. Newbury, D. C. Joy, C. E. Lyman, P Echlin, E. Lifshin, L. Sawyer, and J. R. Michael. 2003. *Scanning Electron Microscopy and X-Ray Microanalysis.* New York: Springer, 689 pp.

Hamaker, H. C. 1937. The London-van der Waals attraction between spherical particles. *Physica* 4:1058–1072.

Hoffman Lab. 2010a. STM: More technical details, http://hoffman.physics.harvard. edu/research/STMtechnical.php.

Hoffman Lab. 2010b. STM measurement types, hoffman.physics .harvard.edu/research/STMmeas.php.

Hollas, J. M. and M. J. Hollas. 2004. *Modern Spectroscopy.* New York: Wiley, 480 pp.

Israelachvili, J. 1991. *Intermolecular and Surface Forces,* 2nd edn. London, UK: Academic Press.

Morris, V. J., A. R. Kirby, and A. P. Gunning. 2004. *Atomic Force Microscopy for Biologists.* London, UK: Imperial College Press, 324 pp.

Perkampus, H.-H. and H.-Charlotte Grinter. 1992. *Uv-Vis Spectroscopy and Its Applications (Springer Laboratory).* Berlin, Germany: Springer, 234 pp.

Smith, B. C. 1995. *Fundamentals of Fourier Transform Infrared Spectroscopy.* Boca Raton, FL: CRC Press, 224 pp.

Sze, S. M. (Ed.) 1994. *Semiconductor Sensors.* New York: Wiley.

Tro, N. 2010. *Principles of Chemistry: A Molecular Approach.* Upper Saddle River, NJ: Pearson/Prentice Hall, 888 pp.

University of Wisconsin. 2007. Scanning Tunneling Microscope (revised 1/9/07), http://www.hep.wisc.edu/~prepost/407/istm/istm.pdf.

Wilson, J. 2004. *Sensor Fechnology Handbook.* London, UK: Newnes, 704 pp.

Zhu, Y, F. Xu, Q. Qin, W. Y. Fung, and W. Lu. 2009. Mechanical properties of vapor-liquid-solid synthesized silicon nanowires. *Nano Letters* 9(11): 3934–3939.

Part II

Nanomaterials and Micro/Nanofabrication Facilities

2

Materials for Nanosensors

2.1 Introduction

When constructing a house, several materials are required, such as cement, concrete, bricks, tiles, iron and steel bars, flooring materials, and so on. Similarly, a large variety of materials are used to fabricate nanosensors. In fact, numerous! But some materials, like nanoparticles (NPs), quantum dots (QDs), carbon nanotubes (CNTs), inorganic nanowires, thin films, and nanoporous materials, find widespread usage in many nanosensors. One repeatedly stumbles across their names while perusing nanosensor literature.

Low-dimensional nanometer-sized materials and systems have defined a new research area in condensed-matter physics (the field of physics dealing with the physical properties of condensed phases of matter, such as solids and liquids) within the past 20 years. These will be described in this chapter. Apart from the aforesaid categories of materials, there are innumerable materials of different types used for fabricating nanosensors. These will be mentioned in specific examples.

Information presented in this chapter will help the reader to appreciate the roles and importance of nanomaterials for nanosensor fabrication.

2.2 Nanoparticles or Nanoscale Particles, the Importance of the Intermediate Regime between Atoms and Molecules, and Bulk Matter

The first term in the discussion of nanomaterials is the "nanoparticle." A nanoparticle (NP) is a particle of a metal, polymer, or oxide, smaller than 100 nm in diameter (Rotello 2003; Yang 2003; Schmid 2004; Hosokawa et al. 2007; Eftekhari 2008; Gubin 2009). It is essentially a zero-dimensional nanostructure.

Will a fiber thinner than 100 nm be considered a nanoparticle even if it is several micrometers long? No, it is a nanowire. This is essentially the case for carbon nanotubes which are definitely not "nano" in length, but have a diameter in the order of 3 nm for a single-walled tube, and hence can be described as nanowires.

What is a one-dimensional nanostructure? A nanowire or nanorod, which is less than 100 nm in diameter but can be several micrometers long. Examples are nanotubes, nanocables, nanowhiskers, nanofibers, etc.

What is a two-dimensional nanostructure? A thin film with a thickness below 100 nm but may extend to hundreds of microns in the other two dimensions.

Are nanoparticles visible? NPs, like viruses (small infectious agents that can replicate only inside the cells of other organisms), are invisible even through the best light microscope, because they are smaller than wavelengths of light (700 nm in the red to 400 nm in the violet). They can be seen only with higher-resolution instruments, such as a scanning electron microscope.

Is a single sugar molecule, g 1 nm in size, a nanoparticle? No, because NPs are aggregates of atoms bridging the continuum between *small molecular clusters*, of a few atoms and dimensions of 0.2–1 nm, and *bulk solids*, containing millions of atoms and having the properties of macroscopic bulk material. They have properties which are neither those of atoms (the smallest unit into which an element can be divided and still retain the chemical properties of the element), molecules (smallest division of a substance, element or compound, that still exhibits all the chemical properties of the substance; it is a group of similar or dissimilar atoms held together by chemical forces), or those of bulk materials (an assembly of solid particles that is large enough for the statistical average of any property to be independent of the number of particles) but somewhat intermediate between these properties.

In nanotechnology, not only NPs but also one- and two-dimensional nanostructures are included. The interesting properties displayed by all these nanostructures differ from atomic- or molecular-scale behavior as well as bulk characteristics. The fabrication, properties, and applications of these structures, *which are neither like atoms and molecules, nor like bulk matter*, is what nanotechnology is all about.

It must be emphasized that nanotechnology is concerned with this *intermediate or in-between zone* between these two extreme limits. While bridging the gap between an atomic state and bulk phases of materials, nanoscale materials reveal novel physical and chemical properties, which are completely different from those observed in either state of the materials. Size-dependent optical responses, exotic structural configurations, reaction catalysis (acceleration of a chemical reaction by the presence of a material that is chemically unchanged at the end of the reaction), etc., are some novel properties exhibited by various classes of nanomaterials.

2.3 Classification of Nanoparticles on the Basis of Their Composition and Occurrence

How are the NPs broadly classified? There are two main classification schemes of NPs, based on their composition and occurrence. NPs are classified, according to their chemical composition, into five categories, depending on whether they are based on metals, carbon, semiconductors, polymers, or composite materials (Table 2.1). From the viewpoint of their occurrence, they are subdivided into three classes, namely natural, incidental (occurring as an unpredictable or minor accompaniment, as a subordinate or by chance), or engineered (Table 2.2).

TABLE 2.1
Classification of Nanomaterials According to Composition

Sl. No.	Name of the Class	Examples
1	Metal-based	Metallic NPs, e.g., Au, Ag, etc.; metal oxides, such as zinc or titanium oxides
2	Carbon-based	Fullerenes, buckyballs, carbon nanotubes
3	Semiconductor-based	QDs, used in exploratory medicine or in the self-assembly of nano-electronic structures
4	Polymer-based	Dendrimers (branched polymers)
5	Composite	Nanoclays, naturally occurring plate-like clay particles that strengthen or harden materials or make them flame-retardant. DNA molecules may be combined with various nanomaterials to make a nanosized biocomposite

TABLE 2.2
Classification of Nanomaterials According to Origin or Occurrence

Sl. No.	Name of the Class	Explanation	Remarks
1	Natural	Present in the environment as volcanic dust, lunar dust, mineral composites, etc.	May have irregular or regular shapes
2	Incidental (waste or anthropogenic particles)	Result from man-made industrial processes such as diesel exhaust, coal combustion, and welding fumes,	May have irregular or regular shapes
3	Engineered	Produced either by milling or lithographic etching of a large sample to obtained nanosized particles or by assembling smaller subunits through crystal growth or chemical synthesis to grow NPs of the desired size and configuration	Most often have regular shapes, such as tubes, spheres, rings, etc.

2.4 Core-/Shell-Structured Nanoparticles

What are structured nanoparticles and what is the advantage of structural configuration? Recently, core/shell NPs have been finding widespread applications (Sounderya and Zhang 2008). Structurally, core/shell NPs are nanostructures that have the core made of one material coated with another material. These NPs have a size range of 20–200 nm. The necessity to shift to core/shell NPs arises from the quest for improvement in the properties of the base NPs. Taking into consideration the size of the NPs, the shell material is chosen such that the agglomeration of particles is prevented. This implies that the monodispersity (having the same size and shape) of the particles is improved.

Besides the prevention of agglomeration, does the core/shell structure improve any other property of the base nanoparticle? The core/shell structure also enhances the thermal and chemical stability of the NPs, improves their solubility, makes them less cytotoxic (poisonous to living cells), and allows conjugation (the state of being joined together) of other molecules to these particles. In some cases, the shell also prevents the oxidation of the core material.

When a core NP is coated with a polymeric layer or an inorganic layer like silica (SiO_2), synergistical functions can be envisioned, because the polymeric or inorganic layer would endow the hybrid structure with an additional function/property on top of the function/property of the core.

2.4.1 Inorganic Core/Shell Nanoparticles

What is the composition of inorganic core/shell nanoparticles, in terms of structure? The core or the shell or both are made of inorganic materials.

What are metallic core/shell nanoparticles? In metallic core/shell NPs, the core is a metal/metal oxide or silica, while the shell is silica or a metal or metal oxide. The most widely used core/shell nanocomposites are gold or silver core with a silica shell. The gold/silica NPs are used in optical sensing, and the thickness of the silica coat alters the optical properties of gold NPs.

What are semiconductor core/shell nanoparticles? In semiconductor NPs, the core is made of semiconductor material, a semiconductor alloy, or a metal oxide, with the shell being made of semiconductor material, metal oxide, or an inorganic material like silica.

What are the specifics of these structures? These structures can be binary (consisting of two parts or components), with a core and shell, or a ternary (having three elements, parts, or divisions) structure, with a core and two shells. The most common binary structure, known by the name quantum dots (QDs) is an alloy of group III and group V metals or group IV and group VI metals, namely, CdSe/CdS, CdSe/ZnS, ZnSe/ZnS, CdTe/CdS, etc., with the shell thickness determining the emission range of these particles. They fall under the category of binary NPs (Li et al. 2009).

What are lanthanide nanoparticles? Lanthanide (rare earth: any element of the lanthanide series [atomic numbers 57–71])

NPs have a core which contains one or more lanthanide group elements, surrounded by a shell made of inorganic material like silica or a lanthanide material. Aqueous colloids of rhabdophane ((Ce, Y, La, Di)(PO_4)·H_2O), brown, pinkish, or yellowish-white mineral, consisting of a hydrated phosphate of cerium, yttrium, or rare earths, are one of these categories, that show a green luminescence. These are Ce-, Tb-doped core particles with an $LnPO_4$–xH_2O shell. These particles can further be coated with silica to enhance their luminescent properties. They have potential applications in electronics and bioimaging.

2.4.2 Organic–Inorganic Hybrid Core/Shell Nanoparticles

What are the main subdivisions of hybrid nanoparticles? These include: (i) organic core and inorganic shell NPs: polyethylene ((–CH_2–CH_2–)$_n$)/silver (Ag), poly-lactide (PLA: $C_3H_6O_3$)/gold (Au); (ii) inorganic core and organic shell NPs: SiO_2/PAPBA(poly(3-aminophenylboronic acid), Ag_2S/PVA(polyvinyl alcohol), CuS/PVA, Ag_2S/PANI(polyaniline), and TiO_2/cellulose; 3-aminobenzenebo-ronic acid monohydrate = $C_6H_8BNO_2 \cdot H_2O$; PVA = $(C_2H_4O)_x$; aniline = $C_6H_5NH_2$; cellulose = $(C_6H_{10}O_5)_n$; and (iii) polymeric core/shell NPs: polymethylmethacrylate (PMMA:$(C_5O_2H_8)_n$)-coated antimony trioxide (Sb_2O_3) compounded with polyvinylchloride (PVC:CH_2 = CHCl)/antimony trioxide composites. The interaction between PMMA and the PVC along with antimony trioxide increases the toughness and strength of PVC.

2.5 Shape Dependence of Properties at the Nanoscale

Engineered nanomaterials with identical chemical composition have a variety of shapes such as spheres, tubes, fibers, rings, and planes. *Do these particles of different shapes exhibit dissimilar properties?* Yes, every one of these shapes may have different physical properties, because the pattern of molecular bonds differs, even though they are composed of the same atoms, such as the properties of fullerenes (molecules composed entirely of carbon; buckminsterfullerene = C_{60}) differ from those of its other two allotropic forms of carbon, namely, diamond and graphite.

In diamond, every C-atom is sp^3 hybridized and covalently bound to four other C-atoms. The binding arms of these C-atoms point to the corners of a tetrahedron. There is an angle of 109° between each arm, resulting in the unique structure of diamond. In graphite, every C-atom is sp^2 hybridized and covalently bound to three other C-atoms. Thus, a plane of continuous hexagons results, with feeble van der Waals forces acting between the hexagons. In fullerenes, a spherical network exists. In contrast to diamond and graphite, which are made of expanded three-dimensional structures, fullerenes form closed molecular systems with a sp^2 hybridized C-atom as a common building block. They contain pentagons and hexagons; the pentagons are responsible for the bend and the football structure of the fullerenes. As a result, they dissolve in various solvents, which makes their chemical manipulation easier. The available fullerenes are C_{60}, C_{70}, C_{76}, C_{78}, C_{84}, C_{90}, C_{94}, and C_{96}.

2.6 Dependence of Properties of Nanoparticles on Particle Size

Does only the nanoparticle shape matters? At the NP scale, both size and shape of the particle are important. One reason for size dependence is the increase in surface area with decreasing particle size. Surface area plays a definitive role here, because most chemical reactions involving solids happen at the surfaces, where chemical bonds are incomplete. Thus, collections of nanoscale particles (NSPs), with their enormous surface areas, are exceptionally reactive (unless a coating is applied), because more than a third of their chemical bonds are at their surfaces. For example, NPs of silver have been found to be an effective bactericide (a substance that kills bacteria), inspiring several companies to design reusable water-purification filters, using nanoscale silver fibers.

Example 2.1

A simple thought experiment reveals why NPs have such high surface area per unit volume. A solid cube of a material 1cm × 1cm or 1cm^2 on each side – about the size of a sugar cube – has 6 cm^2 of surface area (Figure 2.1). But if that volume of 1 cm^3 were filled with cubes 10 nm on each side, there would be $(10^{-2} \times 10^{-2} \times 10^{-2})\ m^3/(10 \times 10^{-9} \times 10 \times 10^{-9} \times 10 \times 10^{-9})\ m^3 = 10^{18}$ cubes of 10 nm-sized cubes (volume = 10 nm × 10 nm × 10 nm), each one of which has a surface area of $6 \times 10\ nm \times 10\ nm = 6 \times 10 \times 10^{-9} \times 10 \times 10^{-9} = 6 \times 10^{-16}\ m^2$, giving a total surface area $= 6 \times 10^{-16} \times 10^{18}\ m^2 = 6 \times 10^2\ m^2 = 6 \times 10^2 \times 10^2 \times 10^2\ cm^2 = 6 \times 10^6\ cm^2$. Hence, the surface area increases by a factor of $(6 \times 10^6\ cm^2)/6\ cm^2 = 1 \times 10^6$.

2.7 Surface Energy of a Solid

How is the surface energy of a solid defined? It is the sum total of energies of all the atoms or molecules present at the surface of a solid.

A solid is formed when individual atoms come closer and distribute themselves to satisfy their bonds on all four sides. In this way, the total energy of atoms decreases to reach a more stable state. But this condition of satisfied bonds does not hold for the surface atoms because their bonds toward the surface side are

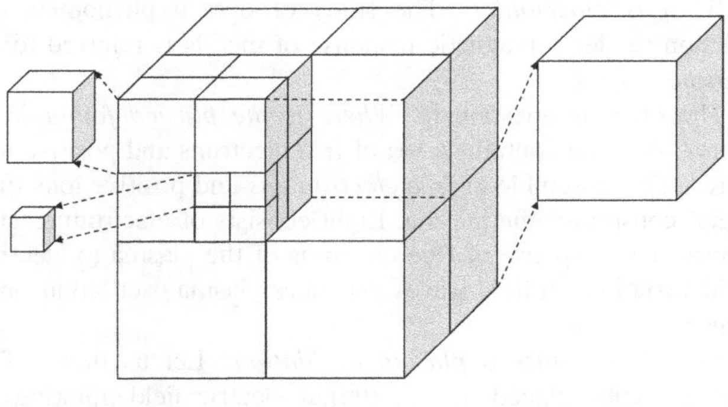

FIGURE 2.1 Subdivision of a cube into smaller cubes.

left dangling. Thus, surface atoms have more energy than do subsurface atoms. This extra energy will be released, and the surface will acquire a more stable state when the surface bonds are satisfied, for example, by the attachment of adsorbed atoms. The excess energy of surface atoms above those present in a bulk solid is collectively called the *surface energy of the solid.*

Surface energy is defined as the energy required to create a unit area of new surface.

After noting that the surface atoms of a solid possess more energy than those in the bulk solid, let us enquire about the unbalanced interatomic bond forces near the surface. *What is the effect of these forces on bond lengths*? Due to the dangling bonds at the surface, the surface atoms experience an inward-directed force toward the bulk solid, resulting in shortening of the bond lengths of surface atoms. Thus, there is a decrease in bond lengths and lattice constants at the surfaces of solids.

2.8 Metallic Nanoparticles and Plasmons

Why have metallic nanoparticles aroused interest? NPs of gold, silver, and copper have attracted attention due to their attractive electronic and optical properties. These particles have the characteristic feature of displaying beautiful bright colors due to a property called the surface plasmon band (SPB).

Are SPBs unique to nanoparticles? This phenomenon is not unique to NPs but is also observed on the surfaces of bulk metals, where it is called the surface plasmon resonance (SPR). *What is the confusion in terminology regarding SPR and SPB*? There is a little confusion in terminology because, *prima facie,* it appears that SPB is not a resonance phenomenon but SPR is. In reality, both are resonance phenomena, and SPB is sometimes referred to as *localized surface plasmon resonance* (LSPR) to distinguish it from SPR (Hutter and Fendler 2004; Dahlin et al. 2006).

What is a plasmon and how is this term coined? In general, a *plasmon* is a quantum of plasma oscillations; a plasma is a quasi-neutral gas of charged and neutral particles, which exhibits collective behavior. The coinage of the term "plasmon" is similar to calling a quantum of light a "photon" or calling a quantum of lattice vibrations a "phonon." The plasmon is the short form of a *plasmon polariton.*

Is plasmon a quasi-particle? Yes, it is a quasi-particle, a long-lived single-particle excitation in which the excitations of individual particles are modified by their interactions with the surrounding medium. Excitation is the promotion to a higher-energy quantum state.

What is plasmonics? The study of optical phenomena in relation to electromagnetic response of metals is referred to as *plasmonics.*

The obvious question is: *Where is the plasma found in a metal?* A metal contains a sea of free electrons and positive ion cores. This ensemble of *free electron gas* and positive ions in a metal constitutes the plasma. Light consists of electromagnetic waves. The response of free electrons of the plasma to electric field variations in light waves generates plasma oscillations and hence plasmons.

How to visualize a plasma oscillation? Let us think of a metallic cube placed in an external electric field pointing in the right-handed direction. Under the influence of this electric field, electrons move toward the left-hand side, thereby uncovering positive ions on the right-hand side. This continues until they cancel the field inside the metal. Now if the electric field is switched off, the electrons move to the right-hand side, repelled by each other and attracted to the positive ions left bare on the right-hand side. Thus, the electrons oscillate back and forth at the plasma frequency (the frequency at which plasma oscillations occur; clearly, there is a different plasma frequency for each species) until the energy is dissipated in some kind of resistance or damping. Plasmons are a quantization of this kind of oscillation.

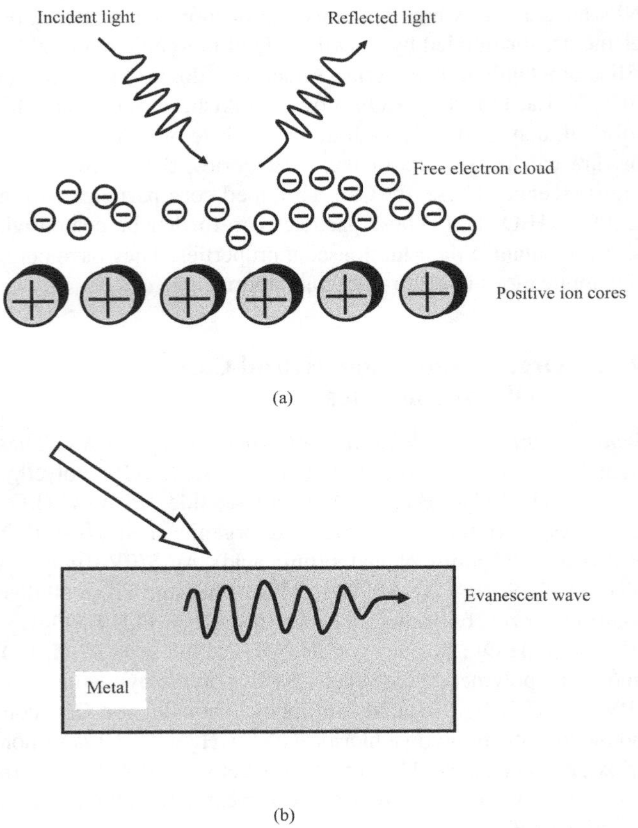

FIGURE 2.2 (a) Reflection of light by an electron cloud in metals at lower frequencies than plasma frequency. (b) Production of evanescent waves by excitation of bulk metal.

What role is played by plasmons in the optical properties of metals? Plasmons play a key role in the optical properties of metals. Metals have a shiny appearance due to reflection of light from their surfaces back to the eye. The reason for this reflection is associated with the electron cloud surrounding the metals. Because the electron cloud in the metals screens the electric field of the light, photons having a frequency of less than the plasma frequency are reflected (Figure 2.2).

Light of a frequency higher than that of the plasma frequency is transmitted, because the electrons cannot respond fast enough to screen it. Generally, for metals, the plasma frequency is in the ultraviolet region (7.5×10^{14}–3×10^{16} Hz; 10–400 nm), making them shiny or reflective in the visible portion of spectrum (wavelengths from approximately 400–720 nm). But metals like copper and gold have electronic interband transitions in the visible range. Hence, specific light energies (colors) are absorbed, resulting in the appearance of their distinct color. In semiconductors, the plasma frequency of valence band electrons is usually in the deep ultraviolet (DUV: <300 nm), so that they are also reflective.

2.8.1 Surface Plasmon Resonance on Bulk Metals

Surface plasmons are those plasmons that are confined to surfaces and which interact strongly with light, producing a polariton. They occur at the interface of a vacuum or a material with a positive relative permittivity and a negative relative permittivity (usually a metal or a doped dielectric).

Consider the excitation of a piece of bulk metal by electromagnetic waves (Figure 2.2b). Evanescent waves are produced. Evanescent waves (plasmons) are waves that are decaying in amplitude, having crossed into a medium attenuating them.

Surface plasmon waves are electromagnetic waves propagating at optical frequencies on an interface between a metal, typically gold or silver, and a dielectric. They are evanescent waves: their field intensity is concentrated in a very thin layer (a few tens of nanometers) across the interface. They are characterized by a propagation constant k_{sp}. The propagation or wave vector of a wave is a vector quantity that defines the magnitude and direction of the wave; its magnitude is $2\pi/\lambda$. For a flat interface, the propagation vector of surface plasmon waves is given by

$$k_{sp} = k_0\sqrt{\frac{\varepsilon_m\varepsilon_d}{\varepsilon_m + \varepsilon_d}} \tag{2.1}$$

where

ε_m and ε_d are the dielectric constant of the metal and the dielectric, respectively

k_0 is the momentum of light in free space at the same frequency

According to Equation 2.1, the propagation constant of surface plasmons is strongly dependent on the variations of permittivity at the interface. Therefore, SPR is a *surface-sensitive optical technique*, based on changes in refractive index at a metal–dielectric interface.

Surface plasmon resonance sensors are usually constructed by using prism coupling of incident light onto an optical substrate that is coated with a semitransparent (partially transparent: not perfectly or completely transparent) noble metal under conditions of total internal reflection (complete reflection of a ray of light at the boundary between two transparent media, which occurs if the angle of incidence is greater than a certain limiting angle, called the critical angle). Therefore, a condition of total internal reflection (Figure 2.3a) must exist at the interface. Total internal reflection will exist for incident angles greater than the critical angle θ_c; at this angle, some of the light is refracted across the interface. At the point of reflection at the interface, an evanescent field (standing wave) penetrates the exit medium to a depth of the order of one-quarter of the incident light wavelength. A standing wave, also known as a stationary wave, is a wave that remains in a constant position; standing waves are produced wherever two waves of identical frequency interfere with one another while traveling in opposite directions along the same medium.

Variables μ_1 and μ_2 are the refractive indices of the prism and the exit medium, respectively. The refracted beam at the critical angle is shown as a directed dotted line. All light is reflected at incident angles $>\theta_C$. If a semitransparent noble metal film is placed at the interface (Figure 2.3b), then, under conditions of total internal reflection, SPR occurs. This is commonly known as the *Kretschmann configuration.*

The conditions of *surface plasmon resonance* will occur when the following criteria are satisfied. The incident wave vector (a vector which helps describe a wave and the direction of which is the direction of wave propagation) is given by the equation (ICX Nomadics)

$$k_i = \left(\frac{2\pi}{\lambda}\right)\mu \sin\theta_i \tag{2.2}$$

where

k_i is a component of the incident light wave vector parallel to the prism interface

θ_i is the incident light angle

λ is the wavelength of the incident light

μ is the refractive index of the prism material

The wave vector of the plasmon mode is described by the equation

$$k_{sp} = \left(\frac{2\pi}{\lambda}\right)\sqrt{\frac{\varepsilon_m\varepsilon_d}{\varepsilon_m + \varepsilon_d}} \tag{2.3}$$

where

k_{sp} is the surface plasmon wave vector

ε_m and ε_d are the relative permittivity constants of the metal film and the dielectric exit medium, respectively

SPR takes place at the equality of the wave vectors

$$k_i = k_{sp} \tag{2.4}$$

How do evanescent waves interact with incoming waves? The evanescent waves present at the surface of the metal interact with the incoming electromagnetic waves, causing a perturbation of the reflected signal. The intensity of the reflected light will decrease at SPR, thereby leading to a well-defined minimum in the reflectance intensity.

How is resonance observed? At a given wavelength, extinction of the reflected signal is observed, which indicates the occurrence of resonance. Thus, resonance is observed through extinction of the signal, with extinction being the condition of being extinguished.

If the incident angle is fixed and polychromatic light (described by many different frequencies) is reflected from the surface, then light will be adsorbed by the resonance at particular wavelengths. This will give rise to a typical plasmon resonance minimum in the reflectance spectrum.

If monochromatic light (light of one color) is reflected from the surface over a range of incident angles, then a similar reflectance minimum is observed with respect to the angle of incidence. The reflectance minimum, that is produced from plasmon resonance, is caused by the phase difference (the time interval

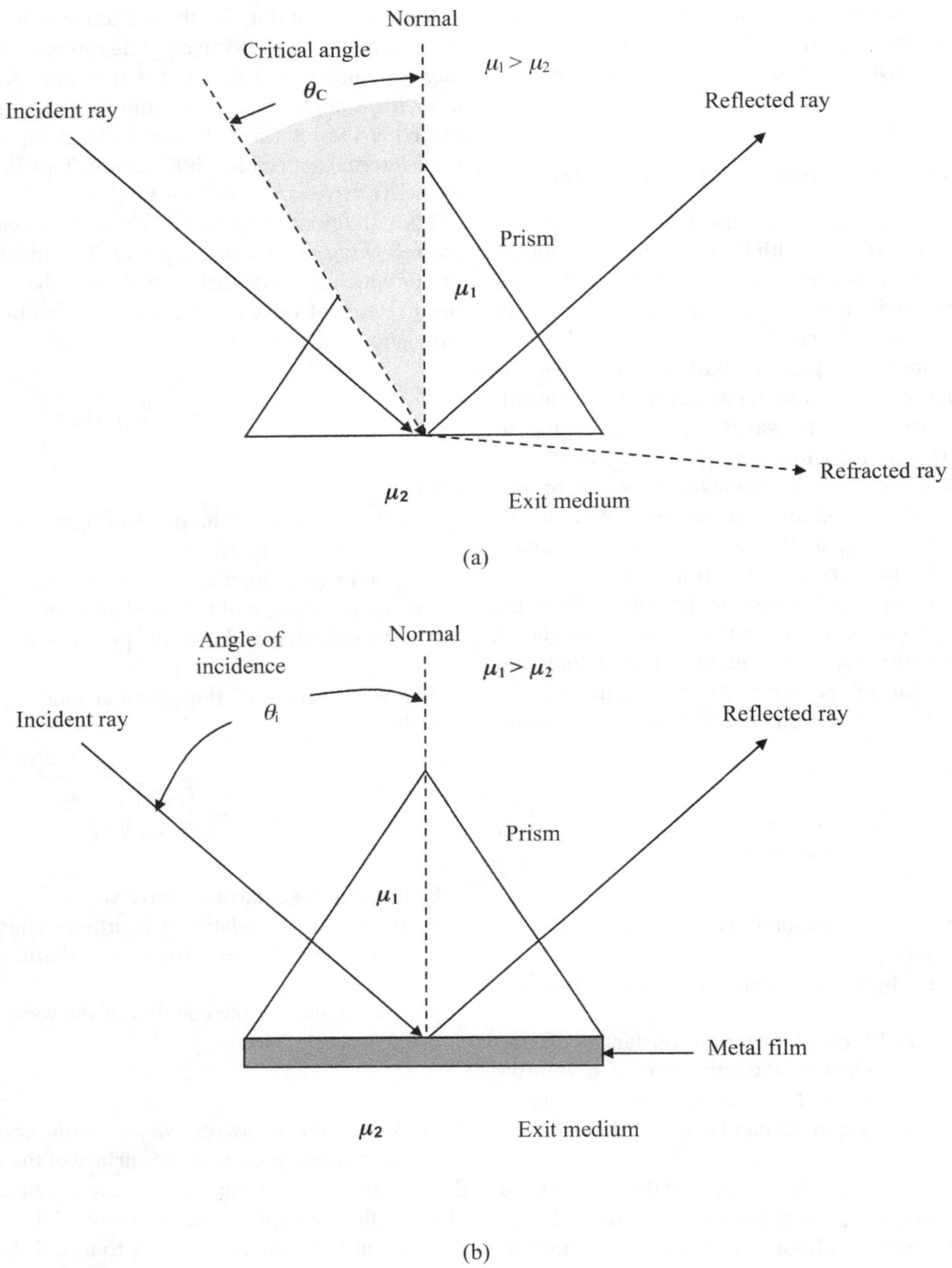

FIGURE 2.3 (a) Total internal reflection at an interface. (b) Use of a prism coupling configuration for SPR.

or phase angle by which one wave leads or lags another) of the surface mode with respect to the incident photon field. Below the resonance, the phase difference is 0° and it approaches 180° above the resonance. Therefore, the photons that are reflected from the metal-prism interface undergo destructive interference (out-of-phase superposition, yielding zero intensity) with the photons emitted by the excited plasmons that are 180° out of phase immediately above the maximum resonance. As a consequence, the characteristic reflectance minimum is produced.

How is an SPR experiment performed in the laboratory? A practical SPR set-up consists of a laser source to illuminate the metal surface (Figure 2.4). This illumination is done either at a fixed angle and a variable wavelength or at a variable angle and a fixed wavelength. The reflected signal is collected and measured by a detector.

When resonance takes place, the reflectance is minimum. Therefore, the situation of minimum reflectance is ascertained, and the corresponding wavelength or angle is recorded.

For sensor applications, it is this change in the refractive index of the dielectric exit medium at the metal surface that is of interest. Therefore, if the refractive index of the prism is constant, then a change in the resonance condition may be linked to changes in the refractive index of the exit medium. In this way, it is possible to monitor the accumulation of films on the metal

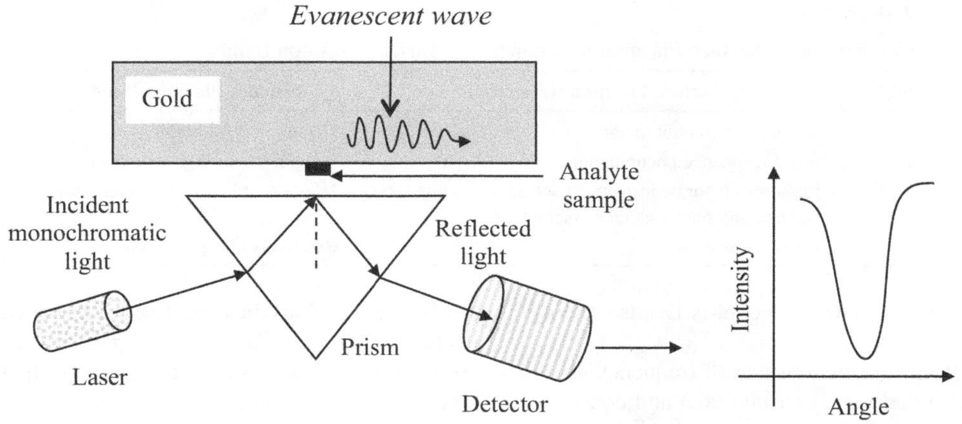

FIGURE 2.4 Schematic layout of experimental set-up for conducting SPR experiments.

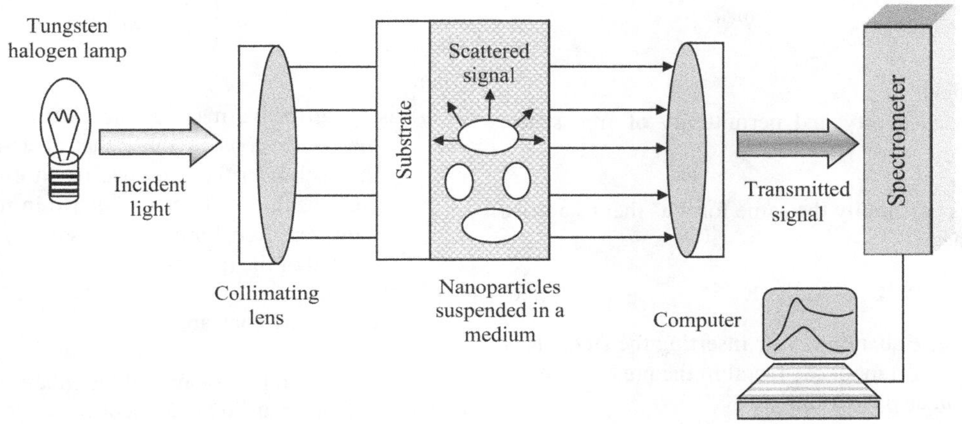

FIGURE 2.5 Schematic layout of experimental set-up for SPB experiments.

surface in order to measure the binding of molecules to a surface that has been coated with an affinity ligand (an ion, a molecule, or a molecular group that binds to another chemical entity to form a larger complex); affinity meaning "natural liking or fondness" is the force attracting atoms to each other and binding them together in a molecule). From these considerations, the SPR configuration may be regarded as a *surface-sensitive refractometer* (an instrument for measuring the refractive index of a substance), in which the sensitivity depth is defined by the penetration depth of the evanescent field (~200 nm).

2.8.2 Surface Plasmon Band Phenomenon in Metal Nanoparticles

Here the oscillation of free-electron gas in the metal NPs takes place in resonance with the incident optical signal. The resulting absorption and scattering phenomena (in which the direction, frequency, or polarization of the wave is changed when the wave encounters discontinuities in the medium, or interacts with the material at the atomic or molecular level) are known as surface plasmon band (SPB).

What is the main dissimilarity between SPB and SPR? The strikingly distinguishing feature of SPB, with respect to SPR, is the simplicity of the SPB instrument, in which a laser source is neither mandatory nor is a complex detector necessary (Figure 2.5). The phenomenon is easily observed by the naked eye. A UV-visible spectrometer is able to provide the desired information. Thus, the instrument is portable, besides being low cost and user friendly. Naturally, therefore, the SPB instrument has received the favorable attention of analytical scientists for carrying out their experiments.

Table 2.3 shows a comparison between SPR and SPB.

2.9 Optical Properties of Bulk Metals and Metallic Nanoparticles

2.9.1 Light Absorption by Bulk Metals and Metallic Nanoparticles

What models are needed to explain light absorption by bulk metals and metallic nanoparticles? The absorption of light by bulk metals and NPs is interpreted by a model for describing waves, along with a separate model for particles. The wave model is needed for electromagnetic waves and the particle model for electrons.

What are the suitable available models for this description? The appropriate wave model is Maxwell's electromagnetic wave

TABLE 2.3

Comparison of Surface Plasmon Resonance and Surface Plasmon Band

Sl. No.	Surface Plasmon Resonance	Surface Plasmon Band
1	Takes place in bulk metals	Occurs in metallic NPs
2	It is a resonance phenomenon	It is a resonance phenomenon
3	Requires elaborate instrument set-up, including laser source and photodetector assembly	Needs simpler instrumental setup
4	Costly instrument	Relatively cheap instrument

theory, and the appropriate particle model is Drude's theory of free electrons.

The equation of electromagnetic waves of frequency ω propagating in a metal of a dielectric constant ε(ω) and conductivity σ(ω) is (Moores and Floch 2009):

$$\nabla^2\mathbf{E} = -\mu_0\varepsilon_0\omega^2\left\{\varepsilon_\infty + \frac{i\sigma(\omega)}{\omega\varepsilon_0}\right\}\mathbf{E} \tag{2.5}$$

where

E is the electric field

μ_0, ε_0 are the permeability and permittivity of free space, respectively

This equation has essentially the same form as that of a wave traveling in free space:

$$\nabla^2\mathbf{E} = -\mu_0\varepsilon_0\omega^2\mathbf{E} \tag{2.6}$$

Equation 2.6 becomes Equation 2.5 by inserting the frequency-dependent dielectric constant of the metal in the presence of the wave in this equation; ε(ω) is written as

$$\varepsilon(\omega) = \varepsilon_\infty + \frac{i\sigma(\omega)}{\omega\varepsilon_0} \tag{2.7}$$

Implicit here is the fact that the dielectric constant of the metal is affected by the incident wave, that is, the incident wave alters the behavior of the metal. The influence of the wave on the metal is thus accounted for. Another important fact to be understood from Equation 2.7 is that ε(ω) is a complex quantity containing real and imaginary components, signifying that the incident wave and the dielectric response are not necessarily in phase.

Whereas the free-space problem of Equation 2.6 is easy to solve, the solution of Equation 2.5 is difficult. Although experimental approaches have been followed, theoretical models were proposed. Drude's theory is used for this purpose. In this theory, the loosely bound conduction electrons are considered to be free and independent. They are treated classically as being free to move within the crystal but colliding with each other and with the ion cores, which are assumed to be immobile. The motion of an electron cloud is represented by the sum total of the motions of the individual electrons. All the electrons move in phase so that inter-electron coupling is maximized.

How do we write the equation of motion of an electron? For a single electron, the equation of motion is written by equating the force acting on the electronic charge q in the applied electric field E, i.e., qE, with the sum of the forces responsible for the motion of the electron and the opposition to this motion by damping caused by various factors, like inelastic collisions (collisions between two particles in which part of their kinetic energy is transformed to another form of energy) of free electrons with metal ions, electron-phonon coupling, and scattering of free electrons at impurities and defects in the metallic structure. The equation of electron motion is

$$m_e^*\left(\frac{dv}{dt}\right) + m_e^*\Gamma v = qE \tag{2.8}$$

where

m_e^* is the effective mass of the electron (a parameter with the dimensions of mass that is assigned to electrons in a solid; in the presence of an external electromagnetic field, the electrons behave in many respects as if they were free, but with a mass equal to this parameter, instead of the true mass)

v is its velocity

Γ is the damping constant

(Note: see Example 6.4 regarding enhancement of the effective mass of electrons in noble metals)

It may be noted that the force applied by the magnetic field component of electromagnetic waves has been ignored, being very small as compared to the force due to electric field component of the same. *How is this justified*? Justification for this approximation is the very small velocity of electrons, compared with the velocity of light.

For a sinusoidal electric field with maximum field value E_0,

$$E = E_0\exp(-i\omega t) \tag{2.9}$$

The electron velocity will also vary sinusoidally as

$$v = v_0\exp(-i\omega t) \tag{2.10}$$

where v_0 is the peak velocity. Hence, Equation 2.8 takes the form

$$\left(-i\omega m_e^* + m_e^*\Gamma\right)v_0\exp(-i\omega t) = qE_0\exp(-i\omega t) \tag{2.11}$$

giving

$$v_0 = \frac{qE_0}{m_e^*\Gamma - i\omega m_e^*} \tag{2.12}$$

The resulting electric current density is

$$j = qnv \tag{2.13}$$

where n is the number of electrons per unit volume. In terms of the peak current density value j_0, the electric current density j is expressed as

$$j = j_0 \exp(-i\omega t) \tag{2.14}$$

where

$$j_0 = qnv_0 = \frac{q^2 n \mathbf{E}_0}{m_e^* \Gamma - i\omega m_e^*} = \sigma(\omega) E_0 \tag{2.15}$$

Here, $\sigma(\omega)$ is the conductivity of the metal and Equation 2.15 expresses Joule's law. Equations 2.7 and 2.15 are combined together to relate the dielectric constant $\varepsilon(\omega)$ in terms of known constants $\varepsilon(\infty)$, n, q, m_e, ε_0, frequency ω, and damping constant Γ as (Moores and Floch 2009):

$$\varepsilon(\omega) = \varepsilon_\infty - \frac{\omega_P^2}{\omega^2 + i\omega\Gamma} = \varepsilon_\infty - \frac{\omega_P^2}{\omega^2 + \Gamma^2} + \frac{i\omega_P^2 \Gamma}{\omega(\omega^2 + \Gamma^2)} \tag{2.16}$$

where, for simplification, we have put

$$\omega_P^2 = \frac{nq^2}{\varepsilon_0 m_e^*} \tag{2.17}$$

How is Γ determined? Γ is determined experimentally.

Equation 2.16 is valid for bulk metals. *What are the special considerations that must be taken into account for nanoparticles?* For NPs, the following conditions apply: (i) current density in the metal NPs is zero because electron density is not uniform in small particles; (ii) because of the infinitesimally small size of the NPs relative to the wavelength of incident electromagnetic waves, at any given instant of time, all the electrons confined in an NP see and hence experience the same electric field. As a result, the electric field acting on the electrons in an NP is position independent. Therefore, all these electrons behave identically and in harmony, in response to the electromagnetic waves. Under the influence of the electric field, the electron clouds in the NP are displaced from their respective positions. In this way, charges are created on the surfaces of NPs through the imbalancing of net charges in the particles. These surface charges are negative at locations where there are higher concentrations of electrons (Figure 2.6). Conversely, the surface charges are positive where lower concentrations of electrons exist. Hence, the surfaces of NPs play important roles in this phenomenon. So, it is natural to call it by the name "surface plasmon band."

Furthermore, it must be noted that the electrons act in a coordinated fashion or in unison, i.e., collectively. All the electrons present in an NP are undergoing collective motion. The collective oscillations of electrons in response to the incident waves are known as plasmon polaritons or plasmons in contrast to the free plasmons in bulk metals. This explains the use of the word "plasmon" in this phenomenon.

As explained earlier, there is a dipolar charge separation in the NPs, so that the NPs act as dipoles (a pair of electric charges, of equal magnitude but, of opposite sign, separated by a small distance). *What is the effect of these dipolar forces?* Dipolar forces try to oppose the effects of incident waves and restore the electron clouds to their original positions. *How is the influence of dipolar forces observed?* This is observed as a restitution or damping effect on the electron oscillations. Accordingly, Equation 2.8 is modified for the NP motion as

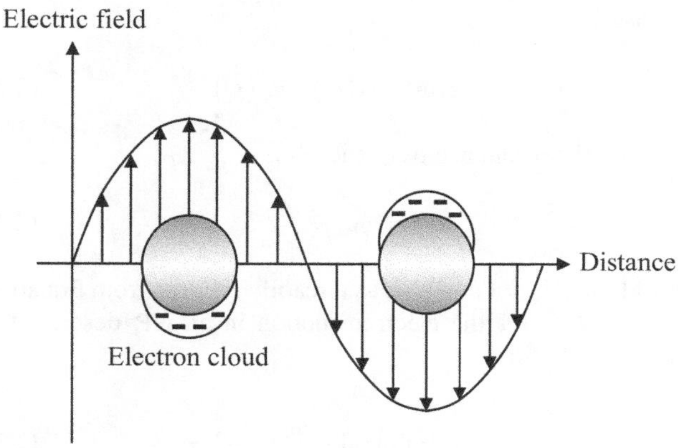

FIGURE 2.6 Electron cloud displacements in spherical metallic NPs with respect to nuclei, produced by an incident electromagnetic wave. (Moores, A. and Floch, P. L., The metal nanoparticle plasmon band as a powerful tool for chemo- and biosensing, In: *Biosensing Using Nanomaterials,* Merkoçi, A. (Ed.), John Wiley & Sons, Inc., New Jersey, 137, 2009.)

$$m_e^* \left(\frac{dv}{dt}\right) + m_e^* \Gamma v + Kx = qE \tag{2.18}$$

where
K is the restoring force constant
x is the position of an electron in the electron cloud

In terms of x, Equation 2.18 is recast as

$$m_e^* \left(\frac{d^2x}{dt}\right) + m_e^* \Gamma \left(\frac{dx}{dt}\right) + Kx = qE \tag{2.19}$$

Thus, the motion of an electron in the electron cloud of an NP is analogous to that of a damped harmonic oscillator, that is, a classical mechanical oscillator. A harmonic oscillator is a system which, when displaced from its equilibrium position, experiences a restoring force proportional to the displacement from the mean position; damping is any effect that tends to reduce its amplitude of oscillation.

Looking at the electric field seen by the NP, it is the one applied externally but altered by the polarizability (the electric dipole moment induced in a system, such as an atom or molecule, by an electric field of unit strength) of the medium of the NP. Applying the boundary condition in a spherical particle, the internal field in an NP, surrounded by vacuum, is expressed as

$$\mathbf{E}_i = \mathbf{E}_0 \left\{\frac{3}{\varepsilon(\omega) + 2}\right\} \tag{2.20}$$

Equation 2.20 gives the condition of resonance which occurs when $\mathbf{E}_i$ is maximum, that is, when $|\varepsilon(\omega) + 2|$ is minimum. In other words, this takes place when

$$|\varepsilon_1(\omega) + 2|^2 + |\varepsilon_2(\omega)|^2 = \text{minimum} \tag{2.21}$$

because

$$\varepsilon(\omega) = \varepsilon_1(\omega) + i\varepsilon_2(\omega) \tag{2.22}$$

Thus, for the resonance frequency ω_M,

$$\varepsilon_1(\omega_M) = -2 \tag{2.23}$$

Considering $\varepsilon_\infty = 1$ and $\Gamma \ll \omega$, it readily follows, from Equations 2.16 and 2.23 for the electron motion in an NP, described by Equation 2.19, that

$$\varepsilon_1(\omega_M) = 1 - \frac{\omega_P^2}{\omega_M^2} = -2 \tag{2.24}$$

from which we have

$$\omega_M = \frac{\omega_P}{\sqrt{3}} \tag{2.25}$$

Substituting for ω_P from Equation 2.17, we get

$$\omega_M = (q)\sqrt{\left(\frac{n}{3\varepsilon_0 m_e^*}\right)} \tag{2.26}$$

To summarize, the electron cloud in the NP oscillates in the incident light. Surface charges are produced at the edges of the NP, leading to the formation of dipoles and thereby providing the restoring force. Resonance takes place at the frequency ω_M given by Equation 2.25.

Was the aforementioned treatment oversimplified for nanoparticles? Yes, the treatment of the phenomenon presented was overly simplified because only one isolated NP in a vacuum was considered and several assumptions were made, including $\varepsilon_\infty = 1$ and $\Gamma \ll \omega$. Nonetheless, the physical picture of the phenomenon occurring in the NP could be clarified. Three forces, namely those due to the electric field, and the restoring and damping effects acted on the NP.

Does an example of the exploitation of metallic nanoparticle effects occur in history? Yes. The ancient Romans exploited the size dependence of the plasmon resonance frequency of gold at the nanoscale when they added gold salts during glass making to obtain red-colored glass due to the presence of gold nanocrystals. The remarkable Lycurgus cup in the British Museum is an example of this technique.

2.9.2 Light Scattering by Nanoparticles

What is the main idea of this approach? Here, we note that, under photoexcitation (the process of exciting the atoms or molecules of a substance by the absorption of radiant energy), the electrons in the electron cloud of the NP undergo acceleration. Since *an accelerated charge radiates energy in space in the form of electromagnetic waves*, the accelerated electrons themselves emit electromagnetic waves in all directions. Interaction takes place between the emitted waves and the incoming waves. The emitted waves scatter the energy of the incoming unidirectional wave. Scattering is a general physical process, where light is forced to deviate from a straight trajectory by one or more localized nonuniformities in the medium through which it passes. This also includes deviation of reflected radiation from the angle predicted by the law of reflection. Reflections that undergo scattering are often called diffuse reflections.

At the resonance condition, the scattering of energy is maximum. Hence, most of the incident waves are scattered and not transmitted. Therefore, the transmitted signal reaches a minimum. Thus, the absorbance spectrum (a spectrum of radiant energy, the intensity at each wavelength of which is a measure of the amount of energy at that wavelength that has passed through a selectively absorbing substance) of plasmonic particles is explained. The wavelength of the scattered light is the same as that of the incident light.

Scattering theory was the first approach applied to explain the phenomenon of SPB. Mie applied Maxwell's equations to spherical NPs embedded in a medium and chose adequate boundary conditions to derive the equation for cross-section C_{ext} given by (Mie 1908)

$$C_{ext} = \frac{24\pi^2 R^3 \varepsilon_m^{2/3}}{\lambda} \frac{\varepsilon_2}{(\varepsilon_1 + 2\varepsilon_m)^2 + \varepsilon_2^2} \tag{2.27}$$

where

λ is the wavelength of the incident wave

ε_m is the dielectric constant of the surrounding medium

Scattering cross section is a hypothetical area, which describes the probability or likelihood of light being scattered by a particle. At resonance,

$$\varepsilon_1 = -2\varepsilon_m \tag{2.28}$$

so that C_{ext} reaches a maximum. For $\varepsilon_m = 1$, Equation 2.28 reduces to

$$\varepsilon_1 = -2 \tag{2.29}$$

which is the same as Equation 2.23.

2.10 Parameters Controlling the Position of Surface Plasmon Band of Nanoparticles

What are the vital parameters affecting the position of surface plasmon band? Equations 2.16, 2.17, 2.23, and 2.24 suggest that the electron density n inside the NP, as well as the permittivity of the medium surrounding the particle (here, ε_0), are critical deciding parameters controlling the dielectric constant of the NP and the resonance frequency of the surface plasmon phenomenon, that is, the position of SPB. These properties have been extensively exploited in nanosensor fabrication and therefore require a detailed explanation.

2.10.1 Effect of the Surrounding Dielectric Medium

What are the principal approaches followed to explore this effect? Two approaches are followed here. The first approach is to change the matter surrounding the NP, and the second one is to vary the distance separating two NPs. In experiments by Underwood et al. (1994) and Liz-Marzán et al. (1996), it was shown, by using gold (Au) NPs stabilized with a polymeric comb (for preventing aggregation of the particles upon solvent transfer), that the SPB shifted toward the lower-frequency and therefore lower-energy red side when the refractive index n of the solvent and, hence its dielectric constant $\varepsilon_r = n^2$, was increased.

In further experiments (Ung et al. 2001), silica-coated gold NPs were prepared in a core–shell structure. The thickness of the silica (SiO_2) coating was changed to study its effect. In solution, the position of the SPB was found to depend on the thickness of the silica coating and, therefore, on the separation between particles.

Then, thin films of silica-coated NPs were deposited on glass substrates. The colors of these films were observed to vary with silica thickness, which determined the interparticle distance. As this distance decreased, the color of the film shifted from red to blue. The close distance between the particles causes dipole–dipole interactions (electrostatic interactions of permanent dipoles in molecules by virtue of the existence of partial charges on its atoms, e.g., attractive forces between the positive end of one polar molecule and the negative end of another polar molecule). The interaction energy depends on the strength and relative orientation of the two dipoles, as well as on the distance between the centers and the orientation of the radius vector connecting the centers with respect to the dipole vectors, leading to the SPB shift. This effect is identical to the color change observed on aggregation of NPs in solution. The number of layers of particles in the film was also found to affect the SPB position.

In solutions, chemically induced reversible aggregation and de-aggregation of NPs is possible. Grafted (to attach) gold NPs stabilized with dithiol (a compound having two thiol groups, the thiol being a sulfur-containing organic compound having the general formula RSH, where R is another element or radical, such as D-proline reductase, an enzyme) could be cleaved or rejoined (Thomas et al. 2004). Upon aggregation of the NPs, a red shift (a proportional decrease in the frequency of light toward lower energy, or longer wavelength light) of the solution was noted.

2.10.2 Influence of Agglomeration-Preventing Ligands and Stabilizers

What is the function of ligands (molecules, ions, or atoms bonded to the central metal atom of a coordination compound; substances, e.g., hormone, drug, functional group, etc., that bind specifically and reversibly to another chemical entity to form a larger complex) *and stabilizers on nanoparticle surfaces?* In order to lower their surface tension, metal NPs in solution have a strong urge to agglomerate, thereby forming bigger particles. To ensure NP stability, their surfaces are protected by covering with ligands or stabilizing agents, such as amines (e.g., RNH_2 are organic derivatives of ammonia formed by replacing the hydrogen atoms in the ammonia, one at a time, by hydrocarbon groups), thiols (organosulfur compounds structurally similar to alcohols but containing a sulfur atom in place of the oxygen atom normally found in alcohols), or phosphines (PH_3 and compounds derived from it by substituting one, two, or three hydrogen atoms, which are called primary, secondary, and tertiary phosphines, respectively, by hydrocarbyl groups R_3P, RPH_2, R_2PH, and R_3P [$R \neq H$],.

How are the ligands/stabilizers bonded to metal nanoparticles? For bonding purposes, either there is an exchange of electrons between the metal atom of the NP and the ligand through oxidation/reduction, or bonding takes place through an electrostatic mechanism. In all cases, the stabilizers form shells surrounding the metal NPs.

How is the SPB position affected? The alteration of electron density n inside the NP, resulting from the bonding of the stabilizer with the NP, changes the SPB position. It moves toward the lower-frequency red side when n falls.

2.10.3 Effect of Nanoparticle Size and Shape

How does the size of nanoparticles affect their SPB position? From Equation 2.27, it is evident that the cross section of a spherical NP is proportional to its radius. Furthermore, the mean free path of an electron in gold (Au) or silver (Ag) is around 50 nm. Hence, an electron is more likely to encounter the wall of the NP if the NP has dimensions in this range. This disturbs the system dynamics. As a result, the SPB of a noble metal NP is blue shifted with decreasing particle diameter.

How does nanoparticle shape exercise its influence on the SPB? Metal NPs are made in a variety of shapes such as nanospheres, nanorods, nanotriangles, nano-octahedra, etc.; octahedra are polyhedrons with eight faces. These shapes differ in their scattering properties. Generally, a reduction in symmetry of the NP shape is accompanied by the appearance of additional modes of resonance, e.g., the nanosphere shows one resonance peak whereas a nanorod has two. Two resonances take place: one for oscillations of electrons along the small axis of the NP (transverse mode) and the second for oscillations along the longer axis (longitudinal mode); the second one is blue shifted with respect to the first.

2.10.4 Compositional Effect

The composition of the NP strongly affects the SPB. Cu, Ag, and Au NPs show different behaviors, for example, Ag NPs provide a sharper SPB signal, a different window and four times greater extinction coefficient (a parameter defining how strongly a substance absorbs light at a given wavelength, per mass unit or per molar concentration) (Mulvaney 1996; Cao et al. 2001). A red shift is noticed in the position of SPB of an Ag-Au alloy on moving from pure Ag to pure Au (Zou and Schatz 2004; Zou et al. 2004).

2.11 Quantum Confinement

With what material properties is the quantum confinement effect related, and in what range of sizes is it observed? Quantum confinement is the change of electronic and optical properties of a material when the material sampled is of sufficiently small size, typically 10 nm or less.

How does quantum confinement originate? This phenomenon arises from the association of several physical processes or interactions with solids with a characteristic length (a convenient reference length [usually constant] of a given configuration) scale. When one dimension of the solid is comparable to or smaller than this characteristic length, the property concerned becomes sensitive to the size of the solid.

2.11.1 Quantum Confinement in Metals

The reduction of the size and dimensionality of metals results in a marked change in the electronic properties, as the spatial length scale of the electronic motion is reduced with decreasing size. When one dimension of a metallic material becomes comparable to the de Broglie wavelength (the wavelength of a particle, given by $\lambda = h/p$ where h is Planck's constant and p is the momentum), typically a fraction of a nanometer for Cu, the quantum confinement effects appear.

Are quantum confinement effects easily observed in metals? Scattering phenomena (in which the waves do not maintain a fixed and predictable phase relationship with each other over a period of time) prevent the observation of these effects in metals in the nano state, unless low temperatures or high magnetic fields are used.

2.11.2 Quantum Confinement in Semiconductors

Electrons occupy one of two bands (valence or conduction) in a crystal of a semiconductor material. The *valence band* contains electrons that occupy positions in the crystal lattice, i.e., in chemical bonding, and the *conduction band* contains those electrons participating in electrical conduction. By providing the proper stimuli, one or more electrons can be encouraged to move from the valence band to the conduction band, i.e., it is released from the chemical bond and is free to move (Figure 2.7). As an electron moves from the valence band to the conduction band, it creates a *hole*, which is positively charged. Thus, excitation of a semiconductor creates an electron–hole pair:

$$h\nu \Leftrightarrow e^-(\text{CB}) + h^+(\text{VB}) \tag{2.30}$$

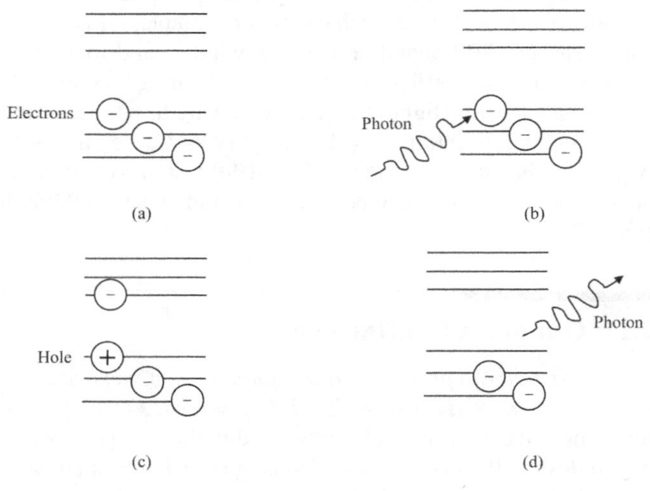

FIGURE 2.7 (a) Energy band diagram of a semiconductor at 0 K, showing electrons in the valence band. (b) A photon strikes an electron in the conduction band to create an exciton. (c) Due to the photon strike, an electron is promoted to the conduction band leaving behind a hole in the valence band. An exciton has been produced. (d) The electron drops to the valence band, recombining with a hole and emitting light.

where $h\nu$ is the energy of the photon.

Together, the hole and the electron are referred to as a *Wannier–Mott exciton* or, simply, an *exciton*. The electron and the hole in the exciton normally maintain their distance from each other, usually delocalized over a length much longer than the lattice constant. This distance is called the *exciton Bohr radius*. Delocalization is the spreading or sharing of an electron–hole pair between more than two atoms. The exciton is considered an elementary excitation of condensed matter that can transport energy without conveying net electric charge. The electron and hole may have either parallel or antiparallel spins; these spins are coupled by the exchange interaction, producing exciton fine structure.

The hole is thought of as the absence of an electron and acts as a particle with its own effective mass and charge in the solid. The spatial separation of the electron and its hole (in an "exciton") is calculated using a modified Bohr model (a planetary model of an atom, consisting of a small, positively charged nucleus orbited by negatively charged electrons, similar to the planets orbiting the Sun [except that the orbits are not planar]; the gravitational force of the solar system is mathematically akin to the Coulomb [electrical] force between the positively charged nucleus and the negatively charged electrons) in which the radius of the sphere defined by the three-dimensional separation of the electron–hole pair is given by (Murphy and Coffer 2002)

$$r = \frac{\varepsilon_0 \varepsilon h^2}{\pi m_r q^2} \tag{2.31}$$

where
ε_0 is the free-space permittivity
ε is the dielectric constant of the semiconductor
h is Planck's constant
m_r is the reduced mass of the electron–hole pair
q is the electronic charge

The reduced mass of the electron–hole pair is simply written as

$$m_r = \frac{m_e^* m_h^*}{m_e^* + m_h^*} \tag{2.32}$$

where m_e^*, m_h^* are the effective masses of the electron and the hole, respectively. The effective mass of a particle is the mass it seems to carry in the *semiclassical model* of transport in a crystal. Electrons and holes in a crystal respond to electric and magnetic fields almost as if they were particles with a mass dependence in their direction of travel, an effective mass tensor. Tensors, defined mathematically, are simply arrays of numbers, or functions, which transform according to certain rules under a change of coordinates, extending the notion of scalars, geometric vectors, and matrices to higher orders. Tensors provide a natural and concise mathematical framework for formulating and solving problems in areas of physics.

In a simplified picture, that ignores crystal anisotropies, electrons and holes behave as free particles in a vacuum, but with a different mass. This mass is usually stated in units of the ordinary mass of an electron m_e (9.11×10^{-31} kg). In these units, it is usually in the range 0.01–10 but can also be lower or higher. The effective masses of the electron and the hole have been determined for many semiconductors by ion cyclotron resonance (a phenomenon related to the movement of ions in a magnetic field; it is used for accelerating ions in a cyclotron, and for measuring the masses of an ionized analyte), and have been found to be in the range of $0.1m_e$–$3m_e$. The radius r has values such as 1 nm for CuCl, 2 nm for ZnS, 3 nm for ZnSe, 4 nm for CdS, 6 nm for CdSe, 8 nm for CdTe or InP, 20 nm for PbS, 34 nm for InAs, 46 nm for PbSe, and 57 nm for InSb (Zimmer 2006).

Example 2.2

Calculate the Bohr radii for CdS ($\varepsilon = 8.4$) and CdSe ($\varepsilon = 9.7$), given that, for CdS, the electron and hole effective masses are $m_e^* = 0.14m_e$, $m_h^* = 0.51m_e$, respectively; and for CdSe, $m_e^* = 0.11m_e$ and $m_h^* = 0.44m_e$ where m_e = electron mass = 9.1095×10^{-31} kg. Also, $\varepsilon_0 = 8.854 \times 10^{-12}$ Fm^{-1}, $h = 6.626 \times 10^{-34}$ J s, and $q = 1.60219 \times 10^{-19}$ C.

The Bohr radius is

$$r = \frac{\varepsilon_0 \varepsilon h^2}{\pi\left(m_e^* m_h^* / \left(m_e^* + m_h^*\right)\right) q^2} \tag{2.33}$$

For CdS,

$$r_{CdS} = \frac{8.854\times10^{-12}\times8.4\times\left(6.626\times10^{-34}\right)^2}{3.14\times\left(\left(0.14m_e\times0.51m_e\right)/\left(0.14m_e+0.51m_e\right)\right)\left(1.60219\times10^{-19}\right)^2}$$
$$= \frac{3.688\times10^{-39}}{9.1095\times10^{-31}} = 4.0485\times10^{-9}\,\text{m} \tag{2.34}$$

For CdSe,

$$r_{CdS} = \frac{8.854\times10^{-12}\times9.7\times\left(6.626\times10^{-34}\right)^2}{3.14\times\left(\left(0.11m_e\times0.44m_e\right)/\left(0.11m_e+0.44m_e\right)\right)\left(1.60219\times10^{-19}\right)^2}$$
$$= \frac{5.31586\times10^{-39}}{9.1095\times10^{-31}} = 5.84\times10^{-9}\,\text{m} \tag{2.35}$$

2.11.3 Bandgap Energies

How does the bandgap in a bulk semiconductor differ from that in a small nanosized crystal? In a bulk semiconductor, the bandgap is centered about the atomic energy levels, with a width proportional to the nearest-neighbor interactions. The Fermi level is located between the conduction and valence bands, and the optical behavior is controlled by the levels near the band edges. The bandgap energies are lower than in smaller-sized crystals, where quantum confinement effects become perceptible. The bandgap of bulk CdSe is 1.7eV. On shrinking the crystal size from 6 to 2 nm, the bandgap ranges between 1.9 and 2.8 eV. Likewise, the bandgap of CdS is 2.4 eV. By dimensional shrinkage, the bandgap varies between 2.5 and 4 eV.

What is the explanation for bandgap enlargement as crystal size is decreased? The increase in bandgap with decreasing crystal size is considered to be the energy cost of confining the exciton within dimensions smaller than the Bohr radius. Clearer interpretation is provided by the particle-in-a-box model of electrons discussed in Section 2.11.4.

2.11.4 Bandgap Behavior Explanation by Particle-in-a-One-Dimensional Box Model of Electron Behavior

To understand the effect of confining the particles within defined boundaries on bandgap energy, let us recall one of the simplest models of electron behavior in elementary quantum mechanics, that of a particle in a one-dimensional box (Figure 2.8). ("The particle in a one-dimensional box.") It is a mathematical machine for predicting the behaviors of microscopic particles on the basis of their dual particle-like and wave-like behaviors, departing from classical mechanics primarily at the atomic and subatomic scales.

What are the salient features of this model and the prescribed particle energies? An electron is allowed to move along an arbitrary spatial dimension, say, the x-axis. The x-axis is divided into three regions, I, II, and III, where $\{I \mid x < 0\}$, $\{II \mid 0 < x < L\}$, and $\{III \mid x > L\}$. In regions I and III, an infinite potential energy barrier is present. The potential energy of any particle within these regions becomes infinite, implying that, in order to exist there,

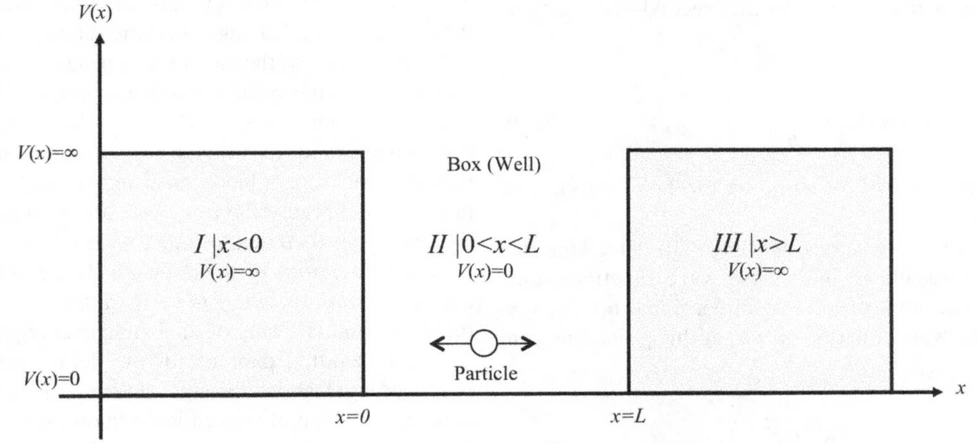

FIGURE 2.8 Particle-in-a-one-dimensional box model.

the total mechanical energy of the particle must be infinite. Since this will never be true, the particle cannot exist within regions I and III. Hence, it is restricted to region II.

What are the possible wave function values in different regions? Clearly, the wave function (a variable quantity that mathematically describes the wave characteristics of a particle; the square of the wave function gives the probability or likelihood for finding the particle at a given point in space and time) of the electron must be zero everywhere, except for $\{0 < x < L\}$. Remembering that all wave functions are continuous (the functions, the graphs of which have no breaks, gaps, or holes), the wave function at $x = 0$ and at $x = L$ must be zero; otherwise a discontinuity would occur.

What is the goal of this model and how do we achieve it? The objective is to determine the set of possible wave functions associated with the electron, and then to use such wave functions to determine the energy eigenvalues (discrete or quantized energies; the German word *eigen* means "characteristic" or "unique"). One should pay attention to the reason why such discrete (step-like) eigenvalues appear. This is because a particle motion accompanies the wave function. In other words, the eigenstates, with discrete energy eigenvalues, occur because of the *particle-wave duality*. This is extremely surprising and never happens in classical theory.

To begin, one looks at the time-independent Schrödinger equation.

$$-\frac{\hbar^2}{2m}\frac{d^2\psi}{dx^2} = E\psi \tag{2.36}$$

ignoring the potential energy term.

Why is the time-dependent wave equation not used? The time-dependent form is not necessary since the states of the electron and its surroundings are constant.

For solving Equation 2.36, it is rearranged into the form

$$\frac{d^2\psi}{dx^2} = -k^2\psi \tag{2.37}$$

by multiplying Equation 2.36 by $(-2m/\hbar^2)$ to obtain

$$\frac{d^2\psi}{dx^2} = -\left(\frac{2mE}{\hbar^2}\right)\psi \tag{2.38}$$

The general trigonometric solution to this second-order differential equation is

$$\psi = N\sin\left(\frac{\sqrt{2mE}}{\hbar}x\right) \tag{2.39}$$

which is the generalized wave function set for the electron; N is a constant.

How are the energy eigenvalues obtained? To determine the energy eigenvalues associated with these wave functions, one looks at the right-end point of the one-dimensional box, at $x = L$. Knowing that the wave function is zero at this point, one can write

$$0 = N\sin\left(\frac{\sqrt{2mE}}{\hbar}L\right) \tag{2.40}$$

Since the sine function produces an output of zero when its input values are multiples of π, we can write

$$\frac{\sqrt{2mE}}{\hbar}L = n\pi \tag{2.41}$$

where n is a member of the natural number set {1, 2, 3,...}.

What is the significance of number n? The number n is called a quantum number, a number which determines the particular state that the electron occupies.

Rearranging Equation 2.41 to isolate energy, we write

$$E = \frac{n^2\pi^2\hbar^2}{2mL^2} = \frac{n^2\pi^2(h/2\pi)^2}{2mL^2} = \frac{n^2h^2}{8mL^2} \tag{2.42}$$

Thus, the energy of a particle of mass m (= m_e for electron) confined by a one-dimensional box of dimension L of infinite potential is given by the equation

$$E_n = \frac{n^2h^2}{8m_eL^2} \tag{2.43}$$

where

n is the quantum number

h is Planck's constant

What are the implications of Equation 2.43? Equation 2.43 implies that, on decreasing the dimension L of the box, the spacing between the energy levels of the particle in the box increases (Figure 2.9). Thus, it can be said that, on constraining the charge-carrier particles within reduced dimensions, the quantum confinement effect causes a change in the density of the electronic states. For a particle that is confined in a three-dimensional box constrained by walls of infinitely high potential energy, the allowed energy states for the particle are discrete with a nonzero ground state energy. As the length of the box, which corresponds to the radius of the QD, is changed, the energy gap between the ground and the first excited state varies in proportion to $1/r^2$.

For a real nanocrystal, this means that the smaller the particle radius, the larger the energy gap to the first electronically excited state becomes. For a semiconductor material, there exists a size regime bounded by the onset of a molecular cluster structure on the smaller side and by the Bohr radius on the larger side, in which the bandgap energy varies strongly with the crystal size. As relatively small changes in dimensions produce large changes in bandgap energy, the nanocrystal behaves as a tunable bandgap material, showing variable bandgap controlled by the crystal size.

What happens for optically allowed transitions? In case of an optically allowed transition, this relates to a blue shift in absorption and emission, as has been observed in all kinds of nanocrystals. This is because the frequency and hence the energy of blue light is higher than that for other parts of the spectrum.

Is the tunability of bandgap produced here by altering the chemical composition of the semiconductor? Interestingly, this tunability and tailoring of the bandgap energy, and therefore the energy of emitted photons for the fluorescent semiconductors, is not brought about through changes in the chemistry of the material, as done in conventional fluorophores; the fluorophores are molecules or parts of molecules that emit fluorescence when excited with light.

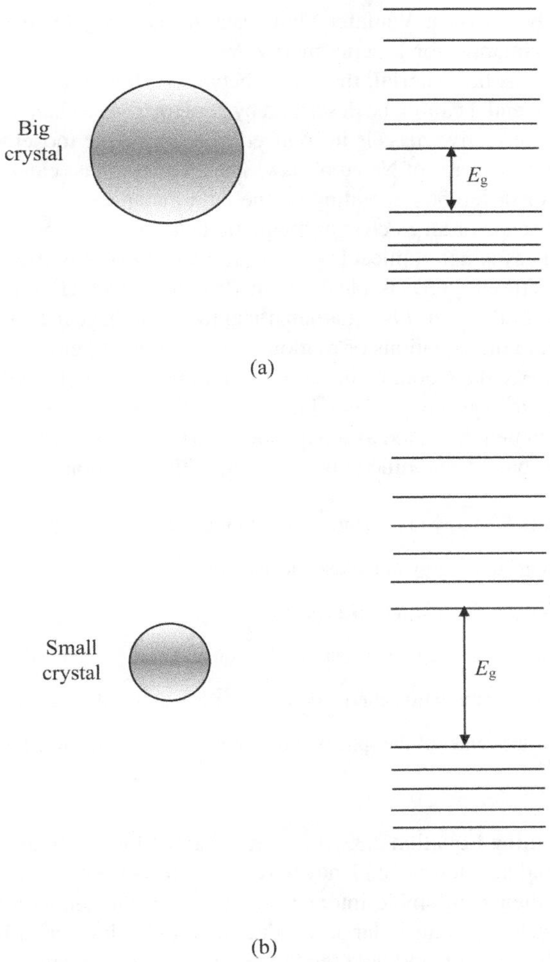

FIGURE 2.9 Effect of dimensions of the box on energy-level diagrams of the particle-in-a-box: (a) large crystal and (b) small crystal. Note the change in inter-level energy differences as well as in bandgap with decrease in size of the semiconductor crystal.

What specific advantage is derived from these nanocrystals? A series of fluorophores, with nearly identical reactivities and chemical properties, are obtained from the same material through the influence of size changes only. Compositional changes are not required.

Example 2.3

Given that m_e = free electron mass = 9.1095 × 10^{-31} kg, h = Planck's constant = 6.626 × 10^{-34} J s, calculate the energy-level differences $(E_2 - E_1)$ for semiconductor nanocrystals of sizes 2 and 20 nm. Do you notice the broadening of the energy-level difference with a decrease in size of the nanocrystal?

For L = 2 nm,

$$E_1 = \frac{n^2h^2}{8m_eL^2q}\text{eV}$$

$$= \frac{(1)^2 \times (6.626\times10^{-34})^2}{8\times9.1095\times10^{-31}\times(2\times10^{-9})^2\times1.60219\times10^{-19}} \quad (2.44)$$

$$= 0.094\,\text{eV}$$

and E_2 = 0.376 eV.

Therefore $(E_2 - E_1)$ = 0.376 − 0.094 = 0.282 eV. Similarly, for L = 20 nm,

$$E_1 = \frac{n^2h^2}{8m_eL^2q}\text{eV}$$

$$= \frac{(1)^2 \times (6.626\times10^{-34})^2}{8\times9.1095\times10^{-31}\times(20\times10^{-9})^2\times1.60219\times10^{-19}} \quad (2.45)$$

$$= 0.00094\,\text{eV}$$

and E_2 = 0.00376 eV. Therefore $(E_2 - E_1)$ = 0.00376 − 0.00094 = 0.00282 eV. It is found that $(E_2 - E_1)$ increases 100-fold for the 2-nm crystal, as compared with the 20-nm crystal.

Example 2.4

Calculate the energies of the levels E_1, E_2, E_3,... E_n for L = 1, 2, 5, and 10 nm. Then, show the changes in the bandgap energies for the electron-in-a-box model for the above dimensions L = 1, 2, 5, and 10 nm. Also, draw the energy band diagrams for different L values.

For L = 1 nm,

$$E_1 = \frac{n^2h^2}{8m_eL^2q}\text{eV}$$

$$= \frac{(1)^2 \times (6.626\times10^{-34})^2}{8\times9.1095\times10^{-31}\times(1\times10^{-9})^2\times1.60219\times10^{-19}} \quad (2.46)$$

$$= 0.376\,\text{eV}$$

Similarly, E_2, E_3, E_4, E_5, ... = 1.504, 3.384, 6.016, 9.4, ... eV. For L = 2 nm,

$$E_1 = \frac{n^2h^2}{8m_eL^2q}\text{eV}$$

$$= \frac{(1)^2 \times (6.626\times10^{-34})^2}{8\times9.1095\times10^{-31}\times(2\times10^{-9})^2\times1.60219\times10^{-19}} \quad (2.47)$$

$$= 0.094\,\text{eV}$$

and E_2, E_3, E_4, E_5, ... = 0.376, 0.846, 1.504, 2.35, ... eV.

For L = 5 nm,

$$E_1 = \frac{n^2h^2}{8m_eL^2q}\text{eV}$$

$$= \frac{(1)^2 \times (6.626\times10^{-34})^2}{8\times9.1095\times10^{-31}\times(5\times10^{-9})^2\times1.60219\times10^{-19}} \quad (2.48)$$

$$= 0.01504\,\text{eV}$$

and E_2, E_3, E_4, E_5, ... = 0.06016, 0.13536, 0.24064, 0.376, ... eV.

For $L = 10$ nm,

$$E_1 = \frac{n^2h^2}{8m_eL^2q}\text{eV}$$

$$= \frac{(1)^2 \times (6.626\times10^{-34})^2}{8\times9.1095\times10^{-31}\times(10\times10^{-9})^2\times1.60219\times10^{-19}} \quad (2.49)$$

$$= 0.00376\,\text{eV}$$

and E_2, E_3, E_4, E_5, … = 0.01504, 0.03384, 0.06016, 0.094, … eV.

Tabulating the calculated values E_1, E_2, E_3, E_4, E_5, … for L = 1, 2, 5, and 10 nm, Table 2.4 is obtained. The energy bandgap is modeled by omitting the intermediate energy level (Algar and Krull 2009). Thus, Table 2.4 becomes Table 2.5. The energy band diagrams corresponding to L = 1, 2, 5, and 10 nm are sketched in Figure 2.10.

2.12 Quantum Dots

2.12.1 Fundamentals

What are quantum dots and why are they so called? These are colloidal semiconductor nanocrystals, 1–10 in diameter, containing typically 10^3–10^4 atoms, so called because quantum confinement takes place here in all three spatial dimensions (Wang and Herron 1991, Bruchez et al. 1998; Murphy and Coffer 2002; Zimmer 2006).

Are quantum dots "artificial atoms"? In some respects, but not completely! In a QD, the electrons occupy discrete energy states as they would in an atom. Using this argument, the QDs have been called artificial atoms.

Is there a particle-in-a-box picture for quantum dots? Brus (1984) provided the analysis of the particle-in-a-box picture for QDs based on a Wannier-Mott exciton and an effective mass approximation for kinetic energy. When an electron is moving inside a solid material, the forces between other atoms affect its motion and it cannot be described by Newton's law. Then the concept of effective mass is introduced to describe the movement of electron in terms of Newton's law. The effective mass can be negative or different, depending on the circumstances. Generally, in the absence of an electric or magnetic field, the concept of effective mass is not applicable. The third element of a particle-in-a-box picture for QDs is a hydrogenic (hydrogen-like) Hamiltonian. The Hamiltonian H is a mathematical function that can be used to generate the equations of motion of a dynamic system; for many such systems, it equals the sum of the kinetic and potential energies of the system, expressed in terms of the system's coordinates and momenta, treated as independent variables. (iv) Particle-in-a-sphere basis wave functions, resulting in the equation:

Bandgap energy in a quantum dot of radius R
– Bandgap energy in bulk semiconductor
= Energy for a particle in a sphere
+ Coulomb energy between the electron and hole
+Spatial correlation energy between the electron and hole to reside near the center of the sphere to maximize dielectric stabilization. (2.50)

Examining Equation 2.33, it is noted that the loosely bound electron and hole have a tendency to reside near one another to maximize their Coulombic interaction. However, the semiconductor dielectric constant is large, producing appreciable screening so that the electron and hole tend to reside close to the center of the sphere for dielectric stability. Mathematically, Equation 2.50 is expressed as

$$E_{\text{QD}}(R) - E_{\text{B}} = \frac{\hbar^2\pi^2\left(m_e^{*-1} + m_h^{*-1}\right)}{2R^2} - 1.786\frac{q^2}{4\pi\varepsilon_0\varepsilon_r R} - 0.248E_{\text{Ryd}}^* \quad (2.51)$$

where

$E_{\text{QD}}(R)$ and E_{B} are the bandgap energies in a QD and bulk semiconductor, respectively
R is the radius of the QD
m_e^*, m_h^* are the effective masses of electron and hole
ε_r is the relative permittivity of the semiconductor material

TABLE 2.4
Calculated Energy Level Values for Different Box Sizes

Sl. No.	Energy Level Designations	Energy Values in eV For $L = 1$ nm	$L = 2$ nm	$L = 5$ nm	$L = 10$ nm
1	E_1	0.376	0.094	0.01504	0.00376
2	E_2	1.504	0.376	0.06016	0.01504
3	E_3	3.384	0.846	0.13536	0.03384
4	E_4	6.016	1.504	0.24064	0.06016
5	E_5	9.4	2.35	0.376	0.094

TABLE 2.5
Modified Table of Energy Level Values for Different Box Sizes

Sl. No.	Energy Level Designations	Energy Values in eV For $L = 1$ nm	$L = 2$ nm	$L = 5$ nm	$L = 10$ nm
1	E_1	0.376	0.094	0.01504	0.00376
2	E_2	1.504	0.376	0.06016	0.01504
3	E_g	6.106–1.504 = 4.602	1.504–0.376 = 1.128	0.24064–0.06016 = 0.18048	0.06016–0.01504 = 0.04512
4	E_4	6.016	1.504	0.24064	0.06016
5	E_5	9.4	2.35	0.376	0.094

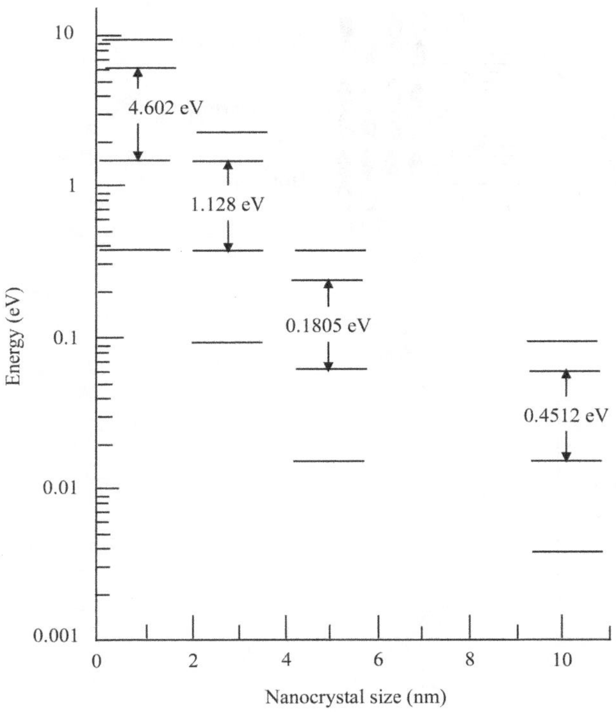

FIGURE 2.10 Energy band diagrams for different L values.

E^{*}_{Ryd} is the Rydberg energy of the electron–hole pair (= $m_r R_H/\varepsilon_r^2$ where m_r is the reduced mass of electron-hole pair given by Equation 2.32 and R_H is the Rydberg energy of the hydrogen atom = 13.6 eV = the work required to remove an electron from a hydrogen atom; 1 Rydberg [Ry], a subsidiary unit of energy used in atomic physics and optics, is equal to the ionization energy of an atom of hydrogen).

Are there any approximations used in the Brus model? Implicit in this equation is the assumption that the QDs are spherical and that the effective masses of charge carriers and the dielectric constant of the solid are constant as a function of QD size.

What is the applicability zone of the Brus model? The Brus model maps energy gap $E_g = E_{QD}(R)$ and size (radius R) well for larger QDs, but its predictions do not match well with experimental findings for very small particle sizes. Although containing the basic physics of the quantum size effect, the model cannot be expected to be quantitatively correct, especially for very small dot sizes, as has been shown by a growing number of experimental studies. This is because, for small QDs, the eigenvalues of the lowest excited states are located in a region of the energy band that is no longer parabolic (the breakdown of the effective mass approximation). Many other more complex approximations have been derived theoretically that better match values which have been experimentally determined.

For a quantitative description of the quantum size effect, one should move beyond the effective mass approximation. A better description of the energy band is obtained from the linear combination of atomic orbitals–molecular orbitals (LCAO–MO) (a quantum superposition of atomic orbitals for calculating MO in quantum chemistry) or a tight-binding approach (an approach for the calculation of electronic band structure, using an approximate set of wave functions, by superposition of wave functions for isolated atoms located at each atomic site) (Wang and Herron 1991). It provides a natural framework by which to understand the evolution of clusters from molecules to bulk and the size dependence of the lowest excited-state energy (bandgap).

2.12.2 Tight-Binding Approach to Optical Bandgap (Exciton Energy) *Versus* Quantum Dot Size

First, let us briefly review molecular orbital concepts and theory. *Molecular orbitals* (MOs) are the solutions of the Schrödinger's equation for molecules in the same way as atomic orbitals are solutions of this equation for atoms. Two atomic orbitals overlap to give two MOs: the molecular orbital at a lower energy than the overlapping atomic orbitals is called a *bonding molecular orbital,* whereas the molecular orbital at a higher energy than the overlapping orbitals is known as an *antibonding molecular orbital.*

Huckel theory is an approximate molecular orbital theory in which the neighboring atoms are postulated to overlap and bond strongly and equally, whereas atoms that are not nearest neighbors are assumed to be noninteracting.

Degenerate level is an energy level of a quantum-mechanical system that corresponds to more than one quantum state. Degenerate states are quantum states of a system having the same energy.

A simple one-dimensional analog of the bulk solid is provided by an infinite chain of carbon atoms, in which individual atoms are separated by distance a (Wang and Herron 1991) (Figure 2.11a). Moreover, each atom carries one $p\pi$ orbital (polyene, an idealized polyacetylene chain). This infinite chain is equivalent to an N-annulene, where N is infinite. Annulene is a class of monocyclic hydrocarbons (having a molecular structure with only one ring) with conjugated double bonds (two or more double bonds separated by single bonds; conjugation is possible by means of alternating single and double bonds). The main ideas behind the quantum size effects of semiconductor QDs are understood by investigating the length dependence of N-annulene or N-polyene; polyene is an unsaturated compound containing more than two double bonds.

What happens in the absence of any interaction between the carbon atoms? In this situation, an N-atom polyene has N degenerate orbitals, ϕ_i, each with on-site energy (Coulomb integral) α,$<\phi_i |H| \phi_i>$ where H is the electronic Hamiltonian.

The degeneracy is removed by considering the interaction between ϕ_i and ϕ_j, represented by the transfer integral β, α, $<\phi_i |H| \phi_j>$. The lowest energy orbital is situated at $\alpha + 2\beta$ and is bonding between all neighboring atoms (Figure 2.11b); a *bonding orbital* is a molecular orbital formed by a bonding electron, the energy of which decreases as the nuclei are brought closer together, resulting in a net attraction and chemical bonding. The highest energy orbital is located at $\alpha - 2\beta$ and is antibonding between all neighboring atoms; an *antibonding orbital* is an atomic or molecular orbital, the energy of which increases as atoms are brought closer together, indicating a net repulsion rather than a net attraction and chemical bonding. In the middle, at the energy α, there is a nonbonding orbital.

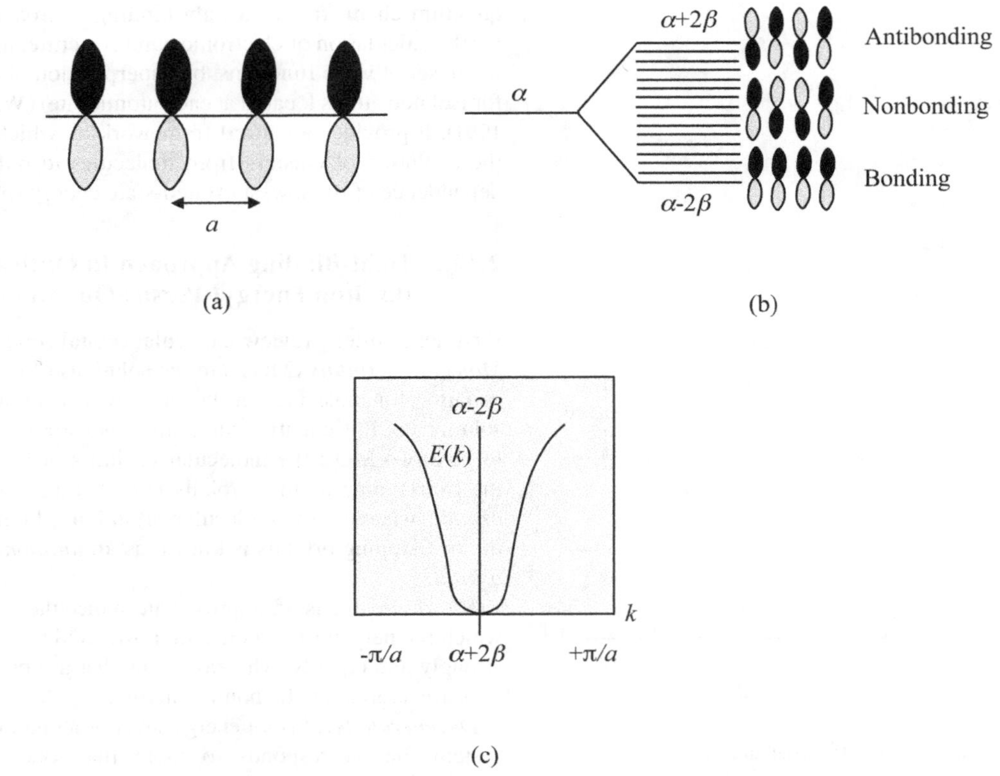

FIGURE 2.11 (a) Orbitals of a chain of carbon atoms. (b) Splitting of the energy level of the orbital into levels for bonding, nonbonding, and antibonding orbitals. (c) Plot of *E(k)* against *k*. (Wang, Y. and Herron, N., *J. Phys. Chem.*, 95, 525, 1991.)

The energy levels of these orbitals, E_j, are obtained easily on the Huckel level as

$$\text{Annulene}: \quad E(j) = \alpha + 2\beta\cos\left(\frac{2j\pi}{N}\right) j = 0, \pm 1, \ldots,$$

$$(\pm N/2 \text{ for } N \text{ even}) \text{ or } \{\pm(N-1)/2 \text{ for } N \text{ odd}\} \tag{2.52}$$

$$\text{Polyene}: E(j) = \alpha + 2\beta\cos\left(\frac{j\pi}{N+1}\right) j = 1, 2, \ldots, N \tag{2.53}$$

By defining a new index k, these equations are rewritten in a different form

$$E(k) = \alpha + 2\beta\cos(ka) \tag{2.54}$$

$k = 0$ to ($\pm\pi/a$ for annulene) or (or π/a for polyene). Here $k = 2j\pi/Na$ for annulene and $j\pi/(N+1)a$ for polyene; it is called the *wave vector*, a vector which has magnitude equal to the wave number (the number of waves per unit distance in a series of waves of a given wavelength) and points in the direction of propagation of the wave.

By plotting the $E(k)$ versus k curve (Figure 2.11c), dispersion of the energy band is obtained. For an infinite chain, the energy band is continuous from $\alpha + 2\beta$ to $\alpha - 2\beta$, with a width of the energy band = 4β but, for a chain of finite atoms, the eigenvalues are discrete.

For linear polyene, there is no degeneracy (k varies from 0 to π/a), so it is easily proven, by filling electrons into these orbitals, that the HOMO–LUMO (highest occupied molecular orbital-lowest unoccupied molecular orbital) gap increases with diminishing chain length, N.

There is no bandgap in an idealized polyene chain. By introducing two-atom repeat units and Peierls distortion (Jahn-Teller distortion) into the chain, a bandgap (the HOMO–LUMO gap) is created in the middle of the dispersion curve at $k = \pi/a'$ (Figure 2.12). The *Jahn–Teller* effect is the distortion of nonlinear molecules or complexes for avoiding degenerate electronic states. For N-annulene, the energy levels are given by

$$E(k) = \alpha \pm \sqrt{\beta_1^2 + \beta_2^2 + 2\beta_1\beta_2\cos ka'} \tag{2.55}$$

where β_1, β_2 are transfer integrals used to describe interactions of orbitals separated by different distances. At $k = \pi/a'$, the bandgap $= 2(\beta_1 - \beta_2)$.

When an electric field is applied, the electron or hole moves as if it has an effective mass of $h^2/(d^2E(k)/dk^2)$. Near the bottom of the band, where the $E(k)$ versus k curve is close to parabolic, the effective mass is $m^* = \hbar^2(\beta_1 + \beta_2)/\beta_1\beta_2 a^2$.

For a Peierls-distorted N-annulene, the HOMO–LUMO gap also increases with decreasing N when N is odd (Equation 2.55). But for even N values, the energies of the HOMO and LUMO are independent of N. However, on examination of the wave functions, it is noticed that the HOMO and LUMO are two nonbonding orbitals with zero overlap. Hence the actual observable

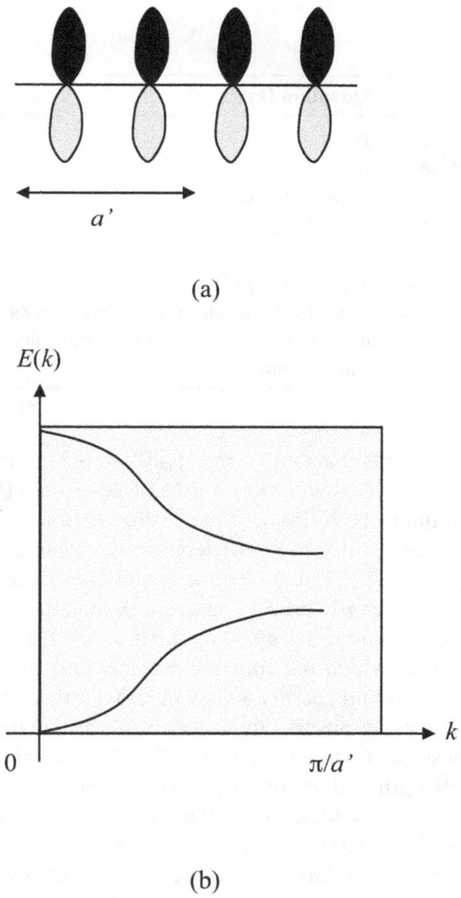

FIGURE 2.12 (a) Definition of *a*′. (b) Bandgap creation. (Wang, Y. and Herron, N., *J. Phys. Chem.*, 95, 525, 1991.)

transition takes place at the next LUMO, which is again dependent on *N*.

The LCAO–MO approach presented earlier offers a simple formulation for appreciating important solid-state concepts. The main observation is that, with a decrease in the size of the *N*-annulene or *N*-polyene, its energy levels become discrete and the distance between the eigenvalues, and HOMO–LUMO gaps, increases, as is clear from Equations 2.53 and 2.55. This is essentially the first part of the quantum size effect asserting the increase of the excited-state energy with decreasing QD size.

It may be noted that the curvature of the band and the effective mass of the electron or hole are determined by β. For large values of β, with the curvature of the band being steep, the electron or hole effective mass is small, and the width of the band is large. Consequently, a more pronounced quantum size effect is anticipated. In the limit of β approaching the zero limit, as in the cases of most molecular solids (the constituent units are molecules, which do not carry any charge; these molecules are held together by the van der Waals forces), the band is flat and the quantum size effect is absent.

Example 2.5

The forbidden energy gap of CdS (E_B) is 2.4 eV. Find the bandgap energy of a CdS QD of radius R = 2 nm, if the electron effective mass (m_e^*) = 0.14 electron mass units, the hole effective mass (m_h^*) = 0.51 electron mass units, the free electron mass (m_e) = 9.11 × 10^{-31} kg, the dielectric constant of CdS (ε_r) is 8.4, the electronic charge (q) = 1.6 × 10^{-19} C, the reduced Planck's constant ($\hbar$) = 1.054572 × 10^{-34} J s, and the Rydberg energy of hydrogen atom (R_H) = 13.6056923 eV.

The exciton Rydberg energy is

$$E_{\text{Ryd}} = \left(\frac{m_r}{m_e\varepsilon_r^2}\right)R_H = \left(\frac{m_e/9.104}{m_e\varepsilon_r^2}\right)R_H = \frac{R_H}{9.104\times\varepsilon_r^2}$$
$$= \frac{13.6056923}{9.104\times(8.4)^2} = 2.118\times10^{-2}\text{ eV} \tag{2.56}$$

where

R_H is the Rydberg energy of the hydrogen atom = 13.6056923 eV

μ is the reduced mass of electron–hole pair given by

$$\mu = \left(\frac{1}{m_e^*}+\frac{1}{m_h^*}\right)^{-1} = \left(\frac{1}{0.14m_e}+\frac{1}{0.51m_h}\right)^{-1} = \left(\frac{0.51+0.14}{0.14\times0.51m_e}\right)^{-1}$$
$$= \left(\frac{9.104}{m_e}\right)^{-1} = \frac{m_e}{9.104} \tag{2.57}$$

From Equation 2.51,

$$E_{\text{QD}}(R) = E_B + \frac{\hbar^2\pi^2\left(m_e^{*-1}+m_h^{*-1}\right)}{2R^2} - 1.786\frac{q^2}{4\pi\varepsilon_0\varepsilon_r R} - 0.248E_{\text{Ryd}}^*$$
$$= 2.4 + \frac{\left(1.054572\times10^{-34}\right)^2\times(3.14)^2\times\left(9.104/\left(9.11\times10^{-31}\right)\right)}{2\left(2\times10^{-9}\right)^2\times1.6\times10^{-19}}$$
$$-1.786\frac{\left(1.6\times10^{-19}\right)^2}{4\times3.14\times8.854\times10^{-12}\times8.4\times2\times10^{-9}\times1.6\times10^{-19}}$$
$$-0.248\times2.118\times10^{-2}$$
$$= 2.4 + \frac{1.0957857\times10^{-36}}{1.28\times10^{-36}} - \frac{4.572\times10^{-38}}{2.9892\times10^{-37}} - 5.2526\times10^{-3}$$
$$= 2.4 + 0.8561 - 0.15295 - 5.2526\times10^{-3} = 3.098\text{ eV} \tag{2.58}$$

2.12.3 Comparison of Quantum Dots With Organic Fluorophores

How do quantum dots compete with organic fluorophores? Organic fluorophore-based diagnostics is a versatile and widely practiced technology. QDs provide good substitutes for these fluorophores, serving essentially as reporters of information, but in a more reliable manner. Advantages, as well as disadvantages, of organic fluorophores and QDs are listed in Table 2.6.

The drive to measure more biological indicators simultaneously imposes new demands on the fluorescent probes used in these experiments. To better appreciate the advantages of QDs, we need to understand: *What difficulties are faced with conventional fluorophores in a multiplexed experiment?* A multiplexed experiment is performed by exciting and observing photoluminescence (PL) from many fluorophores at the same time. Due to the broad red-tailed PL and Stokes shift (the displacement of

TABLE 2.6

Organic Fluorophores and Quantum Dots

Sl. No.	Organic Fluorophores	Quantum Dots
1	Narrow and weak absorption spectra	Broad and strong absorption spectra
2	Broad red-tailed photoluminescence (PL) spectra	Narrow and symmetrical PL spectra
3	Short fluorescent lifetimes, typically nanoseconds	Longer-lived PL spectra (tens of nanoseconds)
4	Highly susceptible to photobleaching	Resistant to photobleaching
5	Susceptible to chemical degradation	Chemically stable
6	Not optimally suited for multiplexing	Superior spectral properties for multiplexing
7	Smaller in size	Larger in size (hydrodynamic radii~10–15 nm along with surface chemical modification) causing structural or functional perturbation of biomolecules
8	No blinking effect	Luminescence turns on and off upon continuous excitation

spectral lines or bands of luminescent radiation toward longer wavelengths than those of the absorption lines or bands; it is the difference in wavelength or frequency units between positions of the band maxima of the absorption and emission spectra) in the range of 10–40 nm of most organic fluorophores, spectral overlapping takes place between fluorophores causing crosstalk (a disturbance caused by one communication signal affecting a signal in an adjacent channel) between detection channels. This crosstalk is avoided by using spectrally widely spaced dyes. But an organic fluorophore absorbs strongly over only a narrow wavelength range. Hence, when using fluorophores, a separate excitation source is needed for the efficient working of each fluorophore. Moreover, rapid photobleaching (photochemical destruction) of the dyes makes their use difficult. Due to the photobleaching effect, the signal drifts over a single measurement or is lost over several measurement cycles. Moreover, conventional dye molecules impose stringent requirements on the optical systems used to make these measurements; their narrow excitation spectrum makes simultaneous excitation difficult in most cases, hindering multicolor real-time imaging (where the rapid acquisition and manipulation of information enables images to be produced almost instantaneously). Their broad emission spectrum, with a long tail at red wavelengths, retards efficient multiplexing (the process whereby multiple channels are combined for transmission over a common transmission path) by introducing spectral crosstalk between different detection channels, unless accurate emission filters are used to prevent bleed through (often termed crossover or crosstalk) to adjacent channels. Finally, synthesis of new molecules is necessary for spectrally shifting the absorption and PL spectra of a fluorophore.

In contrast to the fluorophores, the size of the QD determines its peak PL wavelength. With decreasing size of the QD, the peak PL wavelength changes from red to blue. The absorbance onset and emission maximum shift to higher energy. A QD absorbs light at any wavelength smaller than its peak PL wavelength. The excitation tracks the absorbance, resulting in a tunable fluorophore that can be excited efficiently at any wavelength shorter than the emission peak, yet will emit with the same characteristic narrow, symmetric spectrum regardless of the excitation wavelength. This enables the simultaneous excitation of many QDs having different emission wavelengths by a single wavelength source in the blue or ultraviolet region, creating a Stokes shift, ~100–300 nm, much larger than for fluorophores. Variation of the material used for the nanocrystal and also of the size of the nanocrystal affords a spectral range of 400 nm–2 μm in the peak emission, with typical emission widths of 20–30 nm (full width at half-maximum [FWHM]) in the visible region of the spectrum and large extinction coefficients in the visible and ultraviolet range (~10^5 M^{-1} cm^{-1}). The *full width at half maximum)* is a parameter commonly used to describe the width of a "bump" on a curve or function; it is given by the distance between points on the curve at which the function reaches half its maximum value. The *extinction coefficient* is a parameter that defines how strongly a substance absorbs light at a given wavelength per mass unit. Many sizes of nanocrystals may therefore be excited with a single wavelength of light, resulting in many emission colors that may be detected simultaneously. QDs combine a broad excitation spectrum with a narrow emission spectrum.

Do quantum dots have any more advantages over fluorophores? Apart from the aforementioned advantages, QDs offer better or comparable brightness, high photostability (resistance to change under the influence of radiant energy and especially of light) and low reactivity (the rate at which a substance tends to undergo a chemical reaction) than fluorophores. In addition, the fluorescence lifetimes (the lifetime refers to the average time the molecule stays in its excited state before emitting a photon) are longer or of the order of tens of nanoseconds, long enough to allow time-gated detection (monitoring the fluorescence of a sample as a function of time after excitation by a flash or pulse of light) but short enough so that the signal is not limited by low turnover rates (as is the case for lanthanides [14 rare-earth chemical elements which lie between lanthanum and lutetium on the periodic table] or metal ligand complexes, the structures consisting of a central atom or ion [usually metallic], bonded to a surrounding array of molecules or anions [ligands, complexing agents]) with millisecond lifetimes. A short laser pulse acts as a gate for the detection of emitted fluorescence and only fluorescence that arrives at the detector at the same time as the gate pulse is collected, with the detector turned on after excitation and integration of the fluorescence intensity generated. Fluorescence from single QDs has been observed as longer than from most other individual fluorophores, enabling the collection of a larger number of photons. Furthermore, the synthetic protocol of QDs with different emission wavelengths is simpler and the reaction time is the dot size-controlling variable.

What are the disadvantages of quantum dots? Disadvantages of QDs include the toxicity of the component heavy metals, Cd or Pb, used in II-VI semiconductor dots, the most widely used QDs.

Furthermore, the inorganic nature of QDs renders them inherently insoluble in aqueous media. Also, the larger (of QDs ~10–15 nm are a serious impediment in their application because they cause disturbances in biomolecular reactions. The hydrodynamic radius is the effective radius of an ion in a solution measured by assuming that it is a body moving through the solution and resisted by the viscosity of the solution; if the solvent is water, the hydrodynamic radius includes all the water molecules attracted to the ion.

Blinking or *photoluminescence intermittency effect* in QDs is their other major shortcoming. It is a near-universal aspect of nanocrystal luminescence, yet it is poorly understood. When it was first observed, "blinking" was attributed to rare events of photoionization (the physical process in which an incident photon ejects one or more electrons from an atom, ion, or molecule), that is, the events in which optically excited electrons tunnel to a nearby trap state.

2.12.4 Types of Quantum Dots Depending on Composition

How are quantum dots classified on the basis of composition? QDs are classified as primary (the first stage), binary (composed of two parts or two pieces), and ternary (composed of three items). *Primary QDs* are those made of a single element, such as silicon QDs. *Binary QDs* are those composed of two elements, for example, CdS, CdSe, etc. *Ternary QDs* contain three elements like CdSeTe, ZnCdSe, etc.

2.12.5 Classification of Quantum Dots Based on Structure

How are quantum dots classified from the viewpoint of their structures? The use of semiconductor nanocrystals in a biological context is potentially problematic because the high surface area of the nanocrystal might lead to reduced luminescence efficiency and photochemical degradation; luminescence efficiency is the ratio of the energy of the emitted photon to that of the excited photon, assuming that every excited photon yields one emitted photon. As already pointed out in Section 2.4, efforts to improve properties of base NPs have resulted in core/shell NPs. Concepts of bandgap engineering (controlling or altering the bandgap of a material by controlling the composition) borrowed from materials science and electronics have led to the development of core-shell nanocrystal samples with high room-temperature quantum yields (the ratio of the number of photons emitted to the number of photons absorbed, ~50%) and much improved photochemical stability. By enclosing a core nanocrystal of one material with a shell of another having a larger bandgap, one can efficiently confine the excitation to the core, eliminating nonradiative relaxation pathways and preventing photochemical degradation. Thus, the shell helps in isolating the carriers from the core, moving the particle surface away from the electron and hole spatially. This separation helps in avoiding the interaction of carriers with surface defects and trapping of the electron-hole in an intra-bandgap energy level, thus preventing radiative recombination of carriers and consequent lowering of the quantum yield or disallowing the production of broad red-shifted emission from a deep trap state, far away from the Gaussian emission of the band edge. Provision

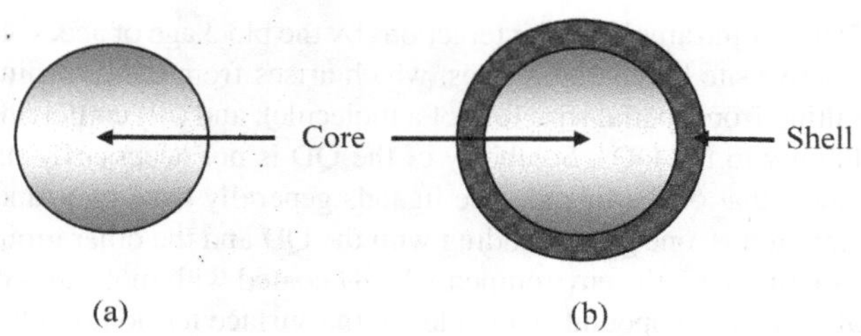

FIGURE 2.13 Structures of QDs: (a) core structure, and (b) core-shell structure.

of chemical protection to the core, particularly against oxidation, is another advantage of the shell.

Thus, there are two classes of QDs (Figure 2.13): *core nanocrystals*, consisting of a single semiconductor material, and *core-shell nanocrystals*, in which the core semiconductor material is surrounded by another semiconductor with a higher energy gap. Generally, the shell material improves the PL quantum yield and stability of the QD, but the emission wavelength of the QD is determined by the size of the core semiconductor. Core/shell structures confine the exciton, typically increasing steady-state luminescence (a form of cold body radiation: the emission of light by certain materials when they are relatively cool) and decreasing carrier trapping on the nanocrystal surface, and also reducing the blinking rate.

What are type I and type II quantum dots? In type I *quantum dots*, the shell generally has a higher bandgap than the core, with the conduction and valence bands of the shell material lying at higher and lower energies, respectively, with respect to the vacuum. However, growth of shells having overlapping or interpenetrating conduction or valence band energies is also possible. Such QDs are labeled as type II QDs. *Type II quantum dots* fluoresce at energies lower than the bandgap energies of either core or shell material. Some QDs contain more than one shell, with the first shell being an alloy of the core and a small/large band of offset material and a third outermost shell of a higher bandgap material.

What kind of matching of properties is desired between the core and the shell, and how is it achieved? In order that the shell completely passivates the dangling bonds (unsatisfied chemical bonds associated with an atom in the surface layer of a solid that do not join the atom with a second atom but extend in the direction of the solid's exterior) on the core surface and that there are no gaps separating the islands of shell material, causing exposure of the core to environmental species, there should be crystal lattice matching between the core and the shell. This is achieved by growing the shell epitaxially on the core, acquiring the crystalline structure and lattice spacing of the core. Therefore, the lattice constants of the core and the shell should be nearly the same, differing by 5% at most.

2.12.6 Capping Molecules or Ligands on the Surfaces of Quantum Dots

What are the roles of caps or ligands on QD surfaces? The caps or ligands serve three-fold purposes: (i) partial passivation of dangling bonds on the surfaces of QDs; (ii) prevention of agglomeration of QDs through steric hinderance (the prevention or retardation of inter- or intramolecular interactions by the blockage of access to a reactive site by nearby groups, which arises from the crowding resulting from spatial structure of a molecule); and (iii) conferring solubility to the QDs. Solubility of the QD is not necessarily the same as that of the ligand. The ligands generally have two functional groups, one group binding with the QD and the other group interacting with the environment. A QD coated with molecules of a surfactant (compounds which lower the surface tension of a liquid) presents hydrocarbon chains to the environment, whereas the polar head groups of the surfactant associate with the QD surface. Consequently, a surfactant-coated QD is water insoluble but is easily dispersed in nonpolar organic solvents, such as toluene (C_7H_8 or $C_6H_5CH_3$) or hexane (C_6H_{14}). Polarity refers to a separation of electric charge, leading to a molecule or its chemical groups having an electric dipole or multipole moment.

Are there any specific points to be noted for ligands? It must be emphasized that: (i) ligands allow the tuning of hydrodynamic diameters of QDs, i.e., their sizes in water, within wide limits; (ii) the thicker the ligand, the more stable is the QD to degradation; and (iii) ligands and the properties imparted by them are interchangeable after the synthesis of QDs through "ligand-exchange procedures." Simultaneous optimization of all the characteristics is difficult, so that compromising optimal solutions must be aimed at, for example, maximization of stability requires a thick ligand shell but a thicker shell may make the QD unusable in applications where size restrictions are imposed.

2.13 Carbon Nanotubes

2.13.1 What Are Carbon Nanotubes?

Conceptually, carbon nanotubes are seamless (continuous, flawless) hollow cylinders made by rolling sheets of graphene (an individual graphite layer), although they are not made in this manner. Figure 2.14 shows a graphene sheet (Ebbesen 1997; Dresselhaus et al. 2001; O'Connell 2006; Dervishi et al. 2009).

How are carbon nanotubes classified? Carbon nanotubes (CNTs) are divided into two main classes: single-wall carbon nanotubes (SWCNTs) (Figure 2.15) and multiwalled carbon

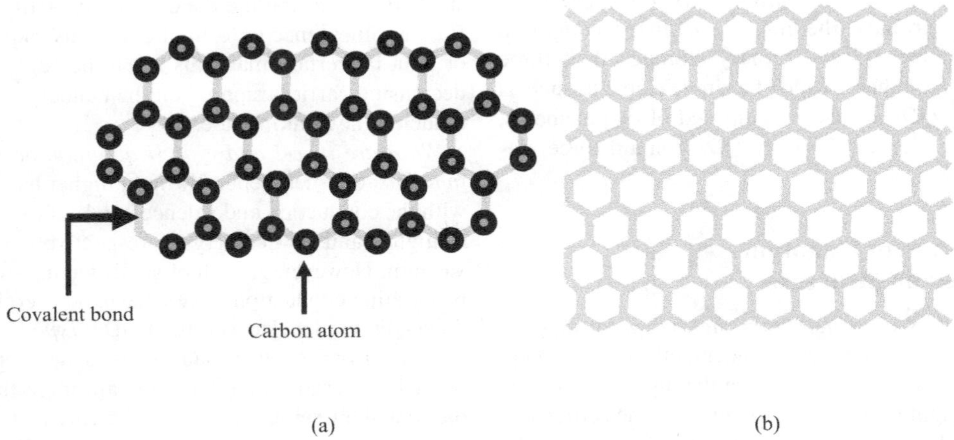

FIGURE 2.14 Atomic arrangement in a graphene sheet: (a) carbon atoms connected by covalent bonds in graphene, and (b) simpler atomic representation of graphene.

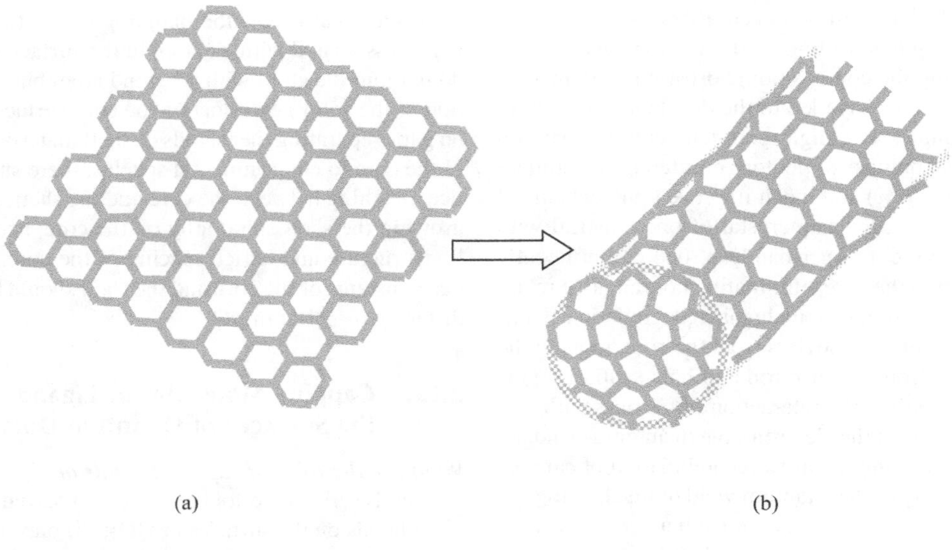

FIGURE 2.15 (a) Graphene and (b) SWCNT.

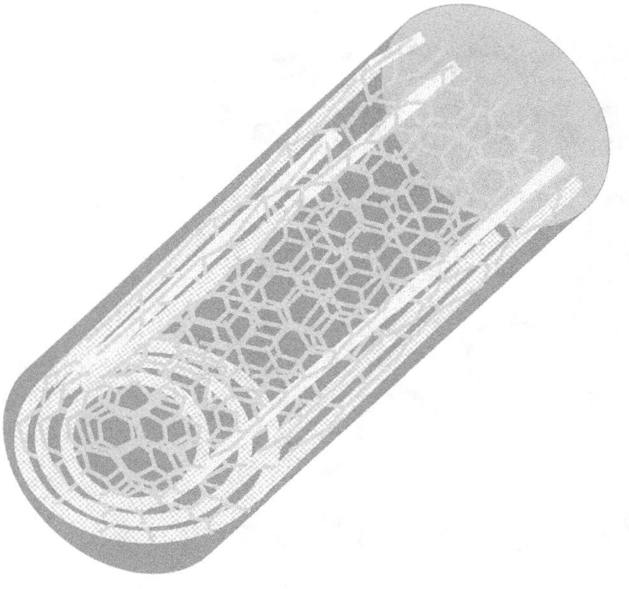

FIGURE 2.16 A MWCNT.

nanotubes (MWCNTs) (Figure 2.16). SWCNTs consist of one cylindrical graphene sheet, whereas MWCNTs comprise several nested concentric cylinders, with an interlayer spacing of 0.34–0.36 nm. They can be considered as coaxial assemblies of cylinders of SWCNTs.

What are the ranges of lengths and diameters of carbon nanotubes? The lengths of CNTs, either SWCNTs or MWCNTs, are well over 1 μm, with diameters ranging from ~1 nm for SWCNTs to ~50 nm for MWCNTs. Because of their high aspect ratio (the ratio of the longer dimension to the shorter dimension), they are often referred to as *one-dimensional nanomaterials*.

2.13.2 Structure of Graphene

Graphene is a single atomic layer of graphite. Graphite, a crystalline allotropic form of carbon, is the "mother material of graphene". What is the importance of the structure of graphene? Because a CNT is essentially rolled-up graphene, it is interesting to examine the structure of graphene, as shown in Figure 2.15a.

What are the special features of the structure of graphene and how are the conducting properties of graphene/graphite explained? The honeycomb-shaped graphene is formed by hexagons of carbon atoms bonded together through sp^2 hybrid bonds, resulting in interlocking hexagonal carbon rings. Each carbon atom has an extra electron, that is, one electron more than the number of atoms to which it is bonded. The outer-shell electrons of each carbon atom form three in-plane σ bonds and an out-of-plane π-bond (orbital). A σ-bond is the strongest type of covalent chemical bond. In a σ-bond, the electron pair occupies an orbital – a region of space associated with a particular value of the energy of the system – located mainly between the two atoms and symmetrically distributed about the line determined by their nuclei. A π-bond is a covalent bond where two lobes of one involved electron orbital overlap two lobes of the other involved electron orbital. It is a cohesive interaction between two atoms and a pair of electrons that occupy an orbital located in two regions, roughly

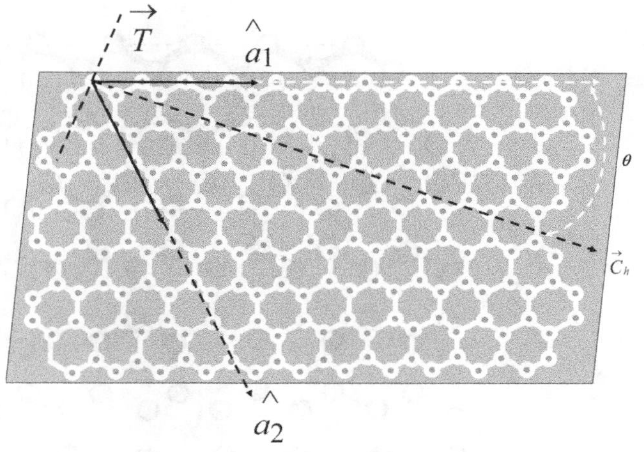

FIGURE 2.17 Chiral vector and its unit vector components. (Dervishi, E. et al., *Particul. Set. Technol.*, 27, 107, 2009.)

parallel to the line determined by the two atoms. The out-of-plane π-orbital or electron is delocalized and distributed over the entire graphene plane. This freedom of movement of the electron makes graphene/graphite thermally and electrically conducting.

Why is graphite a good lubricant and drawing tool? While the carbon atoms in the graphene plane are tightly bonded to each other through covalent linkages, the graphene sheets are feebly held together by the van der Waals forces. As a result, the graphene sheets slide easily across each other making graphite a good lubricant and drawing tool. For the same reason, the graphite found in a pencil is not so strong, although graphene, which is simply one layer of graphite, is very strong.

2.13.3 Structure of SWCNTs

The infinite number of possibilities of rolling a sheet into a cylinder leads to a variety of diameters and microscopic structures of the nanotubes. A diversity of possible configurations is found in practice, and no particular type is preferentially formed.

What is the chiral angle? Figure 2.17 shows the unrolled two-dimensional graphene sheet. The translational vector $\vec{T}$, chiral or roll-up vector $\vec{C}_h = m\hat{a}_1 + n\hat{a}_2$, the chiral angle θ, and the unit vectors $\hat{a}_1$ and $\hat{a}_2$ are indicated; *m*, *n* are integers. The *chiral angle*, θ, is the angle between the axis of its hexagonal pattern and the axis of the tube, i.e., the angle subtended by $\vec{C}_h$ relative to the direction defined by $\hat{a}_1$.

What are the different structures of carbon nanotubes? Armchair, zigzag, and chiral or helical. *How are these structures defined in terms of the tube axis and the C-C bond directions?* When the vector $\vec{T}$ (the tube axis) is perpendicular to the C-C bonds, located on opposite sides of each hexagon in the graphene sheet, the structure is referred to as "armchair" because of the 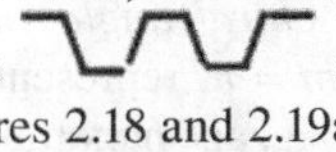 shape perpendicular to the nanotube axis (Figures 2.18 and 2.19a).

If the tube axis $\vec{T}$ is parallel to the C-C bonds, which are located on opposite sides of the hexagons, the "zigzag" structure is obtained (Figures 2.18 and 2.19b). It is named after the  shape perpendicular to the axis of the carbon nanotube.

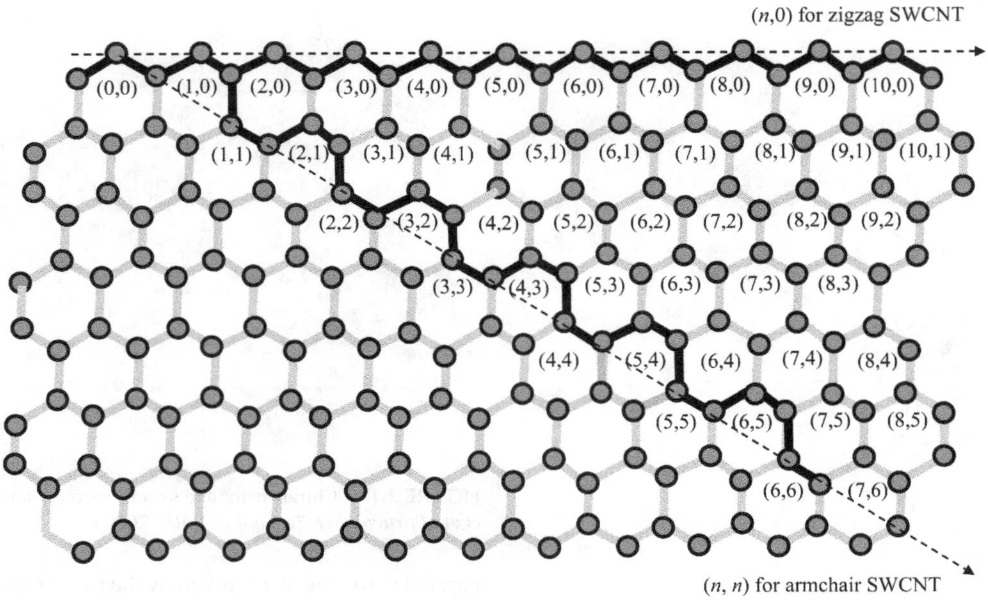

FIGURE 2.18 Chiral vector directions for zigzag and armchair SWCNTs. (Dervishi, E. et al., *Particul. Sci. Technol*, 27, 107, 2009.)

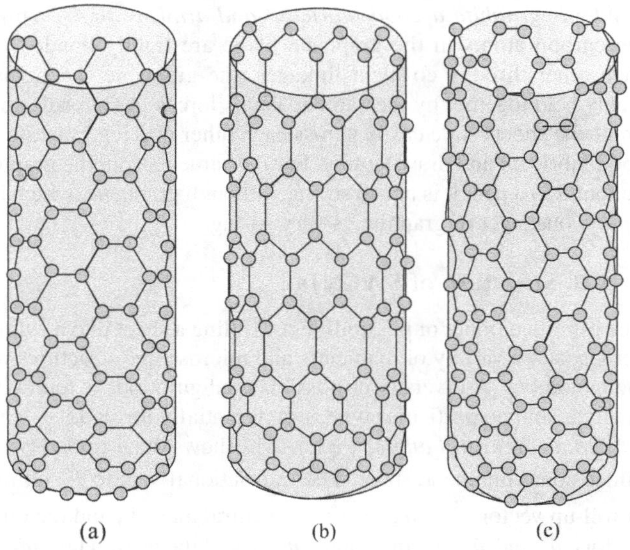

FIGURE 2.19 Carbon nanotubes: (a) armchair, (b) zigzag, and (c) chiral. (Dervishi, E. et al., *Particul. Sci. Technol*, 27, 107, 2009.)

It is also possible to roll up the sheet in a direction that differs from a symmetrical axis. In the arrangements where the vector T lies at an angle with respect to the C-C bonds, structures known as "chiral" or "helical" are formed, in which the equivalent atoms of each unit cell are aligned on a spiral (Figure 2.19c).

How are the values of integers m and n related for the three structures? What are the corresponding chiral angle values? The "armchair" CNTs are formed when $m = n$, represented as (n, n) and the chiral angle is 30°, whereas "zigzag" nanotubes are formed when either n or m is zero [which can be written as $(n, 0)$ or $(0, m)$] and the chiral angle is 0°. The rest of the nanotubes ($0 < \theta < 30°$) are called "chiral" nanotubes, i.e., all nanotubes with general (n, m) values and a chiral angle intermediate between 0° and 30° are known as chiral nanotubes.

Since the chiral angle is 0° for zigzag tubes, it is also the angle between the chiral vector and the zigzag direction. It is expressed as (Dervishi et al. 2009):

$$\theta = \tan^{-1}\left\{\frac{\sqrt{3}n}{2m+n}\right\} \tag{2.59}$$

and the diameter of the CNT is given by

$$d = \left(\frac{a}{\pi}\right)\sqrt{m^2 + mn + n^2} \tag{2.60}$$

where a, the C-C bond length in graphene, = 0.246 nm.

Are the ends of SWCNTs open or closed? Pristine (in original, pure state) SWCNTs are generally closed at both ends. An SWCNT has a well-defined spherical tip but the shape of an MWCNT cap is more polyhedral (a geometric solid in three dimensions, with flat faces and straight edges) than spherical.

What kind of defects are observed in CNT structures? Defects are observed in the hexagonal (a polygon with six edges and six vertices) lattice, usually occurring in the form of pentagons (any five-sided polygon) and heptagons (a heptagon or septagon is a polygon with seven sides and seven angles). Pentagons produce a positive curvature of the graphene layer and are frequently found at the cap. Heptagons create a negative curvature of the tube wall.

2.13.4 Mechanical Properties of CNTs

CNTs have properties that distinguish them from other allotropes of carbon; allotropes are different molecular structure modifications of the same element and can exhibit quite different physical properties and chemical behaviors. The *tensile strength or breaking strain* of individual MWCNTs, the amount of force which a specimen can withstand before tearing, is about 50 GPa (1 GPa = 10^9 Pa; Pascal = newton m^{-2}), which is 20 times stronger than steel (Seetharamappa et al. 2006).

What is the reason for this enormous tensile strength? This strength results from the covalent sp^2 bonds formed between the carbon atoms, which is stronger than the sp^3 bond in diamonds. CNTs are held together by van der Waals forces in a rope-like structure. Another reason why they are strong is that they are just one large molecule. Unlike other materials, CNTs do not have weak spots, such as grain boundaries (the internal interfaces that separate neighboring misoriented grains or single crystals in a polycrystalline solid) in steel.

Young's modulus (Y) of a material is the ratio of longitudinal stress-to-strain in that direction, a measure of how stiff or flexible an isotropic elastic material is, that is, the material's tendency to deform elastically when a force is applied to it. The value of Y for a material is therefore one of the most important properties in engineering design. Young's modulus of CNTs is independent of tube chirality but depends on its diameter.

How does the Young's modulus of CNTs vary with tube diameter? SWCNTs with diameters between 1 and 2 nm have a very high Young's modulus of about 1 TPa (10^{12} Pa), whereas MWCNTs can have a Young's modulus as high as 1.2 TPa. For comparison, the Young's modulus of steel is only about 0.21 TPa. However, the Young's modulus of SWCNTs decreases from 1 TPa to 100 GPa when the diameter of an SWCNT bundle increases from 3 to 20 nm.

2.13.5 Electrical, Electronic, and Magnetic Properties of CNTs

Depending on their diameter and the chirality, SWCNTs are either metallic or semiconducting. Some nanotubes have conductivities higher than that of copper, while others behave more like silicon.

When are SWCNTs metallic or semiconducting? SWCNTs are metallic when $2n + m = 3i$, where i is an integer; otherwise, they are semiconducting.

What is the electrical behavior of armchair and zigzag tubes? All "armchair" carbon nanotubes are metallic because they satisfy the previous equation, whereas the "chiral" and "zigzag" nanotubes can be either metallic or semiconducting.

What typically is the ratio of metallic to semiconducting SWCNTs? What are representative bandgap values? Approximately one-third of the SWCNTs are metallic and two-thirds are semiconducting. Metallic SWCNTs show a bandgap of 0 eV (a unit of energy, equal to the energy acquired by an electron falling through a potential difference of 1 V, = 1.602×10^{-19} J) and are better conductors than metals. The semiconducting SWCNTs have a bandgap of 0.4–0.7 eV.

What is the effect of a magnetic field on bandgap? Under the influence of a large magnetic field, the bandgap of semiconducting CNTs can decrease slightly.

How are CNTs doped? The main methods of doping CNTs are interstitial doping and substitutional doping. In *interstitial doping*, the compound is formed by atoms occupying empty spaces between atoms in a CNT lattice, though not at regular lattice sites, so that the lattice remains the same. In *substitutional doping*, the dopant atoms replace the carbon atoms and form sp^2 bonding in the CNT structure. Boron and nitrogen are used for P- and N-type doping, respectively. By doping, an order of magnitude increase in the electrical conductivity is observed.

In what direction does the current flow in a CNT? The dominant conduction path in a CNT is along its axis. This is because, in the radial direction, electrons are confined by the monolayer thickness of the graphene sheet. If the wavelength of the electron is not a multiple of the circumference of the nanotube, it will destructively interfere with itself. Only the wavelengths that are integral multiples of the nanotube circumference will exist.

Why are CNTs called quantum wires? CNTs are often referred to as one-dimensional "quantum wires" (electrically conducting wires in which quantum effects are affecting transport properties) due to the quantum confinement effect on the nanotube circumference.

A quantum wire is a nanostructure, having a diameter of the order of a nanometer (10^{-9} m). Alternatively, quantum wires can be defined as structures that have a thickness or diameter constrained to tens of nanometers or less and an unconstrained length. At these scales, quantum mechanical effects prevail, which is the reason for coining the term "quantum wires." Many different types of quantum wires exist, including metallic (e.g., Ni, Pt, Au), semiconducting (e.g., Si, InP, GaN, etc.), and insulating quantum wires (e.g., SiO_2, TiO_2). An example of an organic quantum wire is DNA.

2.14 Inorganic Nanowires

A variety of inorganic materials have been prepared in the form of nanowires with a diameter of a few nm and lengths going up to several microns. In order to produce the nanowires, both vapor-growth and solution-growth processes have been exploited (Meyyappan and Sunkara 2010). Examples of materials used for nanowires in sensor fabrication are silicon (Si), palladium (Pd), gallium nitride (GaN), and metal oxides such as SnO_2, ZnO, CuO, TiO_2, CeO_2, WO_3, In_2O_3, and V_2O_5. The nanowires dispersed in a suspension are deposited on a substrate and contacts are made by thermal or electron beam evaporation. For single-crystal semiconductor nanowires, having a diameter less than the Bohr radius, quantum confinement takes place affecting electron transport, band structure, and optical properties.

2.15 Nanoporous Materials

Porous materials are of scientific and technological importance because of the presence of voids of controllable dimensions at the atomic, molecular, and nanometer scales (Lu and Zhao 2004; Sayari and Jaroniec 2008). The voids endow these materials with properties to discriminate and interact with molecules and clusters. Interestingly, the notable feature about this class of materials is about the "nothingness" within them, that is, the pore space.

What are nanoporous materials? Nanoporous materials are those having pore sizes less than 100 nm. Such materials are found in abundance in the natural world, both in biological systems and natural minerals. The walls of animal cells are made of nanoporous membranes. The petroleum industry has long used nanoporous materials called *zeolites* (microporous, aluminosilicate minerals) as catalysts (Sun et al. 2008). *Aerogels* are highly porous materials manufactured with the lowest bulk density of any known porous solid. They can have densities as low as four times that of air. They are derived from a gel whose liquid component has been replaced with a gas. For silica aerogels, the distribution of pore sizes has a peak of around 5 nm radius. *Activated carbon*, a form of nanoporous carbon, is another interesting material. Advances in the capacity to view and manipulate material at the nano level have enabled the directed design of nanoporous materials, rather than to exploit their opportunistic availability. The main research challenges in nanoporous materials include the fundamental understanding of structure-property relationships and the tailor-made design of nanostructures for specific properties and applications.

2.15.1 Nanoporous Silicon

Nanoporous silicon, with a sponge-like structure and very large surface-to-volume ratio (about 10^3 m^2 cm^{-3}), is attractive for sensor applications.

What is the history of nanoporous silicon? Nanoporous Si is not a new material. Porous silicon (PS) layers, formed on monocrystalline silicon substrates by electrochemical etching, have been known for many years. However, it has been extensively studied in the past 20 years. Its PL behavior was observed by Canham (1990), who reported the discovery of significant visible PL from PS under ultraviolet illumination. Although first PS layer had been observed by Uhlir (1956) and later by Turner (1958), only Canham's report about room-temperature luminescence aroused significant interest in this material. Later, more detailed studies of this film were carried out by researchers around the world. Considerable progress has been made following the demonstration in 1990 that highly porous material could emit very efficient visible PL at room temperatures. Since that time, all features of the structural, optical, and electronic properties of the material have been subjected to in-depth scrutiny. The majority of research into PS has focused on observations of and explanations for both PL and electroluminescence from this material, and its potential optoelectronic applications. The refractive index of nanoporous silicon can be changed by light. It can emit acoustic waves through thermal stimulation. Furthermore, it produces a stream of electrons (field emission) in the absence of a vacuum. *Field emission* is the emission of electrons from a solid due to a large electric field at its surface.

What is the electrochemical etching technique? Electrochemical etching is the opposite of electroplating. Instead of metal addition, gradual and uniform erosion removes it. A basic electrochemical-etching system consists of an anode (negatively charged electrode) and a cathode (positively charged electrode) submerged a certain distance from each other in an electrolytic bath. The wafer to be etched forms the anode and a gold cathode is used. The wafer is back-coated with a metallic layer (usually aluminum) to provide a low-resistance electrical contact. The layer of PS is formed by electrochemically etching the crystalline silicon wafer, employing a mixture of hydrofluoric acid and ethanol as an electrolyte.

How does porous silicon compare with microcrystalline silicon? Typical PL intensity of PS in the visible region (1.5–2.0 eV) is larger by several orders of magnitude than the monocrystalline silicon, which gives off luminescence in the near-infrared range, (0.7–1.0) μm to 5 μm, corresponding to the 1.1 eV energy of monocrystalline Si. Silicon, being an indirect semiconductor, is the dominant material of microelectronics. However, it is a poor light emitter and therefore cannot be used in optoelectronics. PS, prepared on a Si substrate, shows the high external efficiencies of PL (a process in which a substance absorbs photons, i.e., electromagnetic radiation, and then reradiates photons) and electroluminescence (emission of light by a material in response to the passage of an electric current) and is suitable for photonic applications.

How is photoluminescence in porous silicon explained? The origin of PL in PS is still controversial. A few models have been proposed to explain the mechanism of PL. According to the model proposed by Canham (1990), radiative recombination of electron–hole pairs occurs within nanometer silicon wires and their energy gaps become larger than that of bulk Si by a quantum confinement effect. This model, modified by Koch et al. (1993), suggests that electron–hole pairs are photoexcited in nanometer silicon particles and radiatively recombined *via* Si intrinsic surface states. Another model (Brandt et al. 1992; Cullis et al. 1997) suggests that luminescence from PS was caused by some special luminescent materials, such as SiH_x complexes, polysilanes (a range of polymers having a backbone of continuous silicon atoms), or SiO_2, instead of being an intrinsic property of nanometer Si. A third model suggests that excitation of charge carriers occurs in nanometer silicon particles and the photoexcited carriers transfer into the luminescent centers (Qin and Qin 2000).

How is the structure of nanoporous silicon described? PS is composed of a silicon skeleton permeated by a network of pores. It is possible to define the characteristics of a particular PS layer in a number of ways. The methods of identification include the average pore and silicon branch widths, porosity, pore and branch orientation, and layer thickness.

How is porous silicon used in nanosensors? For sensing applications, the structures are fabricated by forming a PS layer on a silicon substrate and contacting both the PS surface and the rear face of the substrate. Using these structures, a gas sensor, based upon the current variation due to the dipole moment of the gas, and a humidity sensor, based upon the changing current with humidity, have each been demonstrated. Additionally, applications for PS in biosensing have also been demonstrated, using penicillin (a group of antibiotics derived from the *Penicillium* fungus) as an example. Coating the large surface area of the PS with a penicillin-sensitive enzyme causes the capacitance voltage curve of the junction to shift with changing concentrations of penicillin.

2.15.2 Nanoporous Alumina

What is the historical background of nanoporous alumina? Nanoporous alumina was first discovered in the 1950s. For the next 20–30 years, limitations in microscopy and other equipment limited the ability of researchers to evaluate the porosity of nanoporous alumina. Anodic oxidation was known to be a poor coating process; so, it was abandoned as a process for anodizing aluminum. The porosity of nanoporous alumina has been rediscovered and has sparked significant interest in its properties. Recent work has been performed to produce thin foils of nanopores, but these are generally aimed to be filters for both the detection of biological specimens and for use in fuel cells. These pores can be made uniform over large areas. These films are produced from relatively thick aluminum films and can be obtained as commercially available stand-alone filters.

What are the common porous alumina sensors? Porous alumina sensors are generally available for humidity sensing. These sensors work by adsorbing water molecules onto their surface. This adsorption modifies the dielectric constant, thereby changing the sensor capacitance. Anodic nanoporous films are produced, using common chemicals and electroplating power supplies. Much attention has been devoted toward nanopore ordering and nanopore shape. Highly ordered pore structures require starting with a foil of approximately 25 μm in thickness and the use of chromic acid. Chromic acid (a corrosive, oxidizing acid, H_2CrO_4) preparations are used to increase the number of pore initiation steps. Highly ordered pores also require sequencing between an etch solution and the pore-forming solution. Anodic nanoporous alumina is made from deposited aluminum, which holds promise for its direct integration into either bipolar or complementary metal-oxide-semiconductors (CMOS) integrated circuits.

2.15.3 Nano-Grained Thin Films

Thin-film gas sensors (the film thickness is typically less than 100 nm) are of interest because of their relatively small size and low power consumption. A sensor with a thin film of less than a few hundred nanometers has relatively high sensitivity, but usually shows poor stability, due to its weak mechanical strength. Consequently, optimization of the micro- or nanostructure of a tin oxide thin film should enhance sensor sensitivity and its stability. Nanocrystalline porous tin oxide thin films, obtained by the sol-gel technique, have a high CO sensitivity combined with a fast and reproducible response.

2.16 Discussion and Conclusions

This chapter provided glimpses of nanomaterials useful for designing and fabricating nanosensors. The knowledge acquired in this chapter about nanosensor constructional materials paves the way toward greater appreciation of nanosensor-enabling techniques and processes.

Review Exercises

2.1 Define nanomaterials. How are nanomaterials classified on the basis of their composition and origin or occurrence? Name two nanomaterials used in nanosensor fabrication.

2.2 What is an NP? How many dimensions does an NP have? Give one example each of: (i) a one-dimensional nanostructure; and (ii) a two-dimensional nanostructure.

2.3 What is a core/shell NP? Explain how the shell improves the properties of the NP. Classify core/shell NPs and give one example of each class.

2.4 Why do the properties of NPs vary with their sizes? Explain.

2.5 Explain the dependence of properties of NPs on their shapes.

2.6 Why do atoms on the surface of a solid possess more energy than those in the bulk? Define surface energy of a solid.

2.7 What is a plasmon? Mention two popular names of surface plasmon phenomena observed with metallic NPs. Is surface plasmon phenomenon restricted to metallic NPs?

2.8 What are evanescent waves? Differentiate SPR from LSPR.

2.9 Write and derive the equation for the plasmon resonance frequency of an NP.

2.10 How does the dielectric medium surrounding an NP affect the position of its surface plasmon band? Do ligands and stabilizers have any influence on SPB position of an NP?

2.11 What factors are responsible for non-observation of quantum confinement effects in metals at room temperature?

2.12 In a thought experiment, a semiconductor crystal is continuously subdivided into smaller pieces. Does this subdivision impact the properties of the semiconductor crystal? At what stage is any deviation from bulk behavior observed?

2.13 What is an exciton? Define Bohr exciton radius. Write the formula for Bohr exciton radius and explain the symbols used.

2.14 What is the effect of shrinking the size of a semiconductor crystal on bandgap energy? Describe the particle-in-a-box model of quantum mechanics and explain the effect of decreasing crystal size on energy level diagrams and bandgap energy of the crystal.

2.15 What is a QD? Why is it called an "artificial atom?" Write the Brus equation for the difference in bandgap in a QD and a bulk semiconductor. Explain the symbols used and provide the interpretation of the three terms in the equation. Which is the dominant term in energy change?

2.16 Provide an explanation for the advantages offered by QDs in multiplexed experiments in comparison to the organic fluorophores. Mention one disadvantage of QDs.

2.17 Give one example each of single element, binary, and ternary structure QDs.

2.18 For what purposes are capping molecules or ligands applied on the surfaces of QDs?

2.19 What is graphene? What are the two main classes of carbon nanotubes? What are typically the diameters of single and MWCNTs?

2.20 Why is graphene electrically conducting? What is the reason for graphite to act as a good lubricant?

2.21 Define chiral vector and chiral angle. When the chiral angle is 0°, name the CNT formed. What is the CNT for a chiral angle of 30°?

2.22 Why is the tensile strength of CNTs much higher than that of steel? Compare the Young's modulus of CNTs with that of steel.

2.23 State the mathematical equation which determines which SWCNTs behave metallic.

2.24 Name a P-type dopant for CNTs.

2.25 What are nanoporous materials? Give some examples of such materials.

REFERENCES

Algar, W. R. and U. J. Krull. 2009. Quantum dots for the development of optical biosensors based on fluorescence. In: *Biosensing Using Nanomaterials*. Merkoci, A. (Ed.), Hoboken, NJ: Wiley, pp. 199–245.

Brandt, M. S., H. D. Fuchs, M. Stutzmann, J. Weber, and M. Cardona. 1992. The origin of visible luminescence from porous silicon—A new interpretation. *Solid State Communications* 81(4): 307–312.

Bruchez, M., Jr., M. Moronne, P. Gin, S. Weiss, and A. P. Alivisatos. 1998. Semiconductor nanocrystals as fluorescent biological labels. *Science* 281(25): 2013–2016.

Brus, L. E. 1984. Electron-electron and electron-hole interactions in small semiconductor crystallites: The size dependence of the lowest excited electronic state. *Journal of Chemical Physics* 80: 4403–4409.

Canham, L. T. 1990. Silicon quantum wire array fabrication by electrochemical and chemical dissolution of wafers. *Applied Physics Letters* 57(10): 1046–1048.

Cao, Y., R. Jin, and C. A. Mirkin. 2001. DNA-modified core-shell Ag/Au nanoparticles. *Journal of the American Chemical Society* 123(32): 7961–7962.

Cullis, A. G., L. T. Canham, and P. D. J. Calcott. 1997. The structural and luminescence properties of porous silicon. *Journal of Applied Physics* 82: 909 (57 pp.).

Dahlin, A. B., J. O. Tegenfeldt, and F. Hook. 2006. Improving the instrumental resolution of sensors based on localized surface plasmon resonance. *Analytical Chemistry* 78(13): 4416–4423.

Dervishi, E., Z. Li, Y. Xu, V. Saini, A. R. Biris, D. Lupu, and A. S. Biris. 2009. Carbon nanotubes: Synthesis, properties, and applications. *Particulate Science and Technology* 27(2): 107–125.

Dresselhaus, M. S., G. Dresselhaus, and P. Avouris. 2001. *Carbon Nanotubes: Synthesis, Structure, Properties, and Applications*. Berlin, Germany: Springer, 447 pp.

Ebbesen, T. W. 1997. *Carbon Nanotubes: Preparation and Properties*. Boca Raton, FL: CRC Press, 296 pp.

Eftekhari, A. 2008. *Nanostructured Materials in Electrochemistry*. Weinheim, Germany: Wiley-VCH, 463 pp.

Gubin, S. P 2009 *Magnetic Nanoparticles*. Weinheim, Germany: Wiley-VCH, 466 pp.

Hosokawa, M., K. Nogi, M. Naito, and T. Yokoyama (Eds.). 2007. *Nanoparticle Technology Handbook*. Amsterdam, the Netherlands: Elsevier, 644 pp.

Hutter, E. and J. H. Fendler. 2004. Exploitation of localized surface plasmon resonance. *Advanced Materials* 16(19): 1685–1706.

ICX Nomadics: Overview of surface plasmon resonance, http://www.discoverse nsiq.com/uploads/file/support/spr/Overview:of_SPR.pdf.

Koch, F., V. Petrova-Koch, T. Muschik, A. Nikolov, and V. Gavrilenko. 1993. Some perspectives on the luminescence mechanism via surface-confined states of porous Si, In: *Micro Crystalline Semiconductors: Materials Science and Devices*, Materials Research Society Symposium Proceedings, Edited by Fauchet, P. M., Tsai, C. C., Canham, L. T., Shimizu, I., and Aoyagi, Y., November 30-December 4, 1992, Boston, Massachusetts, U.S.A, Pittsburgh, PA: Materials Research Society, Vol. 283, pp. 197–202.

Li, S., M. L. Steigerwald, and L. E. Brus. 2009. Surface states in the photoionization of high-quality CdSe core/shell nanocrystals. *ACS Nano* 3(5): 1267–1273.

Liz-Marzán, L. M., M. Giersig, and P Mulvaney. 1996. Synthesis of nanosized gold-silica core-shell particles. *Langmuir* 12(18): 4329–4335.

Lu, G. Q. and X. S. Zhao (Eds.). 2004. *Nanoporous Materials: Science and Engineering*. London, UK: Imperial College Press, 912 pp.

Meyyappan, M. and M. Sunkara. 2010. *Inorganic Nanowires: Applications, Properties and Characterization*. Boca Raton, FL: CRC Press, Taylor & Francis Group, 433 pp.

Mie, G. 1908. Contributions to the optics of turbid media, particularly of colloidal metal solutions. *Amalen der Physik* 25(3): 377–445.

Moores, A. and P. L. Floch. 2009. The metal nanoparticle plasmon band as a powerful tool for chemo- and biosensing. In: *Biosensing Using Nanomaterials*. Merkoçi, A. (Ed.), Hoboken, NJ: Wiley, pp. 137–176.

Mulvaney, P. 1996. Surface plasmon spectroscopy of nanosized metal particles. *Langmuir* 12(3): 788–800.

Murphy, C. J. and J. L. Coffer. 2002. Quantum dots: A primer. *Applied Spectroscopy* 56(1): 16A–27A.

O'Connell, M. J. 2006. *Carbon Nanotubes: Properties and Applications*. Boca Raton, FL: Taylor & Francis, 319 pp.

Qin, G. G. and G. Qin. 2000. Multiple mechanism model for photoluminescence from oxidized porous Si. *Physica Status Solidi (a)* 182(1): 335–339.

Rotello, V. 2003. *Nanoparticles: Building Blocks for Nanotechnology*. New York: Springer, 300 pp.

Sayari, A. and M. Jaroniec. 2008. *Nanoporous Materials: Proceedings of the 5th International Symposium*, Vancouver, British Columbia, Canada, May 25–28, 2008. Hackensack, NJ: World Scientific, 738 pp.

Schmid, G. (Ed.). 2004. *Nanoparticles: From Theory to Application*. Weinheim, Germany: Wiley-VCH, 444 pp.

Seetharamappa, J., S. Yellappa, and F. D'Souza. 2006. Carbon nanotubes: Next generation of electronic materials. *The Electrochemical Society Interface* 15(2): 23–25 and 61.

Sounderya, N. and Y. Zhang. 2008. Use of core/shell structured nanoparticles for biomedical applications. *Recent Patents on Biomedical Engineering* 1: 34–42.

Sun, J., D. Zhang, Z. He, S. Hovmöller, X. Zou, F. Gramm, C. Baerlocher, L. B. McCusker, A. Corma, M. Moliner, and M. J. Díaz-Cabañas. 2008. Zeolite structure determination using electron crystallography. In Zeolites and related materials, 2 volume set: Trends, targets and challenges–Proceedings of the 4th International FEZA conference, Volume 174 of *Studies in Surface Science and Catalysis*. Gedeon, A., P. Massiani, and F. Babonneau (Eds.), Paris, France, 2–6 September 2008. Amsterdam: Elsevier, pp. 799–804.

The Particle in a one-dimensional box, http://user.mc.net/~buckeroo/PODB.html.

Thomas, K. G., S. Barazzouk, B. I. Ipe, S. T. S. Joseph, and P. V. Kamat. 2004. Uniaxial plasmon coupling through longitudinal self-assembly of gold nanorods. *Journal of Physical Chemistry B* 108(35): 13066–13068.

Turner, R. D. 1958. Electropolishing silicon in hydrofluoric acid solutions. *Journal of the Electrochemical Society* 105(7): 402–408.

Uhlir, A., Jr. 1956. Electrolytic shaping of germanium and silicon. *Bell System Technical Journal* 35: 333–347.

Underwood, S. M., J. R. Taylor, and W. van Megen. 1994. Sterically stabilized colloidal particles as model hard spheres. *Eangmuir* 10(10): 3550–3554.

Ung, T., L. M. Liz-Marzán, and P. Mulvaney. 2001. Optical properties of thin films of Au@SiO_2 particles. *Journal of Physical Chemistry B* 105(17): 3441–3452.

Wang, Y. and N. Herron. 1991. Nanometer-sized semiconductor clusters: Materials synthesis, quantum size effects, and photophysical properties. *Journal of Physical Chemistry* 95: 525–532.

Yang, P. (Ed.). 2003. *The Chemistry of Nanostructured Materials.* River Edge, NJ: World Scientific, 386 pp.

Zimmer, J. P. 2006. Quantum-dot based biological materials for biomedical imaging. PhD Thesis, Department of Chemistry, Copyright Massachusetts Institute of Technology, MIT, MA.

Zou, S. and G. C. Schatz. 2004. Narrow plasmonic/photonic extinction and scattering line shapes for one and two dimensional silver nanoparticle arrays. *Journal of Chemical Physics* 121: 2606–12612.

Zou, S., N. Janel, and G. C. Schatz. 2004. Silver nanoparticle array structures that produce remarkably narrow plasmon line shapes. *Journal of Chemical Physics* 120: 10871 (5 pp.).

3

Nanosensor Laboratory

3.1 Introduction

A nanosensor laboratory is a place where nanosensors are designed, fabricated, packaged, and characterized. It also includes the facilities for building interfacing signal processing instrumentation. To cater to physical, chemical, and biological nanosensors, it is a multidisciplinary laboratory, encompassing physical and life sciences facilities. Figure 3.1 depicts the concept of a nanosensor laboratory, incorporating the various sections indicated earlier. The heart of the nanosensor laboratory is the nanotechnology laboratory, wherein the nanomaterials are grown/deposited, characterized, and improved.

3.2 Nanotechnology Division

Nanoparticle synthesis and the study of their size and properties are of fundamental importance.

3.2.1 Synthesis of Metal Nanoparticles

What are the synthetic routes for metal nanoparticles? Nanoparticles are prepared by various synthetic routes, for example: (i) the reduction of metal salts in the presence of suitable low- molecular-weight or polymeric stabilizers; (ii) electrochemical preparation; (iii) breakdown of organometallic (compounds that contain bonds between carbon and a metal) precursors (compounds that participate in a chemical reaction producing another compound); and (iv) vapor deposition methods. *What is the simplest approach of synthesis?* The simplest and the most-commonly used bulk-solution synthetic method for metal nanoparticles involves the chemical reduction of the corresponding metal salts.

3.2.1.1 Gold Nanoparticles

Gold nanoparticles are synthesized by the reduction of tetrachloroauric acid ($HAuCl_4 \cdot nH_2O$, $n=0$, 3, or 4) with trisodium citrate ($Na_3C_6H_5O_7 \cdot nH_2O$, $n=0$ or 2) (McFarland et al. 2004; Ambrosi et al. 2010). The reduction half-reaction of gold is given as follows:

$$Au^{3+}(aq) + 3e^{-} \rightarrow Au(s) \tag{3.1}$$

Materials: Hydrogen tetrachloroaurate (III) trihydrate ($HAuCl_4 \cdot 3H_2O$) and trisodium citrate ($Na_3C_6H_5O_7 \cdot 2H_2O$) (s)

Method: (i) All glassware and magnetic stir bars used in the synthesis are scrupulously cleaned in aqua regia (HCl/HNO_3 3:1, v/v) and rinsed in distilled water; (ii) they are oven-dried before use. This precaution is imperative to avoid unnecessary nucleation during the synthesis as well as aggregation of gold colloid solutions; (iii) a 0.25-mM $HAuCl_4 \cdot 3H_2O$ solution (125 mL) is boiled with forceful stirring in a 250-mL round-bottomed flask equipped with a condenser; the condenser is an apparatus used to condense vapors to maintain a constant volume of the reaction mixture; (iv) a 40-mM trisodium citrate solution (12.5 mL) is then added quickly to the boiling solution. As the citrate reduces the gold (III), the gold colloid forms slowly, showing a color change from pale yellow to dark red that signifies the formation of Au NPs; (v) the solution is maintained at boiling temperature for 10 minutes, and then withdrawn from the heating mantle; the heating mantle is a laboratory equipment to apply heat to glass containers without shattering. Stirring is continued for a further 15 minutes; (vi) for detecting the presence of a colloidal suspension, reflection of a laser beam from the particles is observed; (vii) a small quantity of the gold nanoparticle solution is poured in two test tubes. One tube is used as a color reference and 5–10 drops of NaCl solution are added to the other tube. *Does the color of the solution change as the addition of sodium chloride brings the nanoparticles near one another?* Remember that the formation of nanoparticles earlier was also indicated by a color change. As the particles de-aggregate and move farther apart, the solution turns red. Gold nanoparticle aggregation is accompanied by a change in color from red to blue. Red is the color for smaller nanoparticles and blue for larger particles.

3.2.1.2 Silver Nanoparticles

Silver nanoparticles are prepared by the chemical reduction method (Mulfinger et al. 2007), by reduction of silver nitrate using sodium borohydride. The method produces 1.2 ± 2 nm particles with plasmon absorbance near 400 nm.

Materials: Sodium borohydride ($NaBH_4$) and silver nitrate ($AgNO_3$).

Method: (i) Thirty milliliters of 0.002 M sodium borohydride ($NaBH_4$) is added to a flask. A magnetic stir bar is added. The flask is seated in an ice bath on a stirring plate. The liquid is stirred and cooled for about 20 minutes; (ii) two milliliters of 0.001 M silver nitrate ($AgNO_3$) is poured into the stirring $NaBH_4$ solution at the rate of one droplet per second. As soon as all the $AgNO_3$ has been added, the stirring is stopped; (iii) as previously mentioned for gold nanoparticles, reflection of a laser beam from the particles aids in perceiving the presence of a colloidal suspension; (iv) a small amount of the solution is transferred to a test tube. When a few drops of 1.5 M sodium chloride (NaCl) solution are added, the suspension turns darker yellow as nanoparticles aggregate. Then it turns gray with increasing aggregation of nanoparticles. *What does this color change suggest?* Color changes are correlated with the aggregation and separation of nanoparticles.

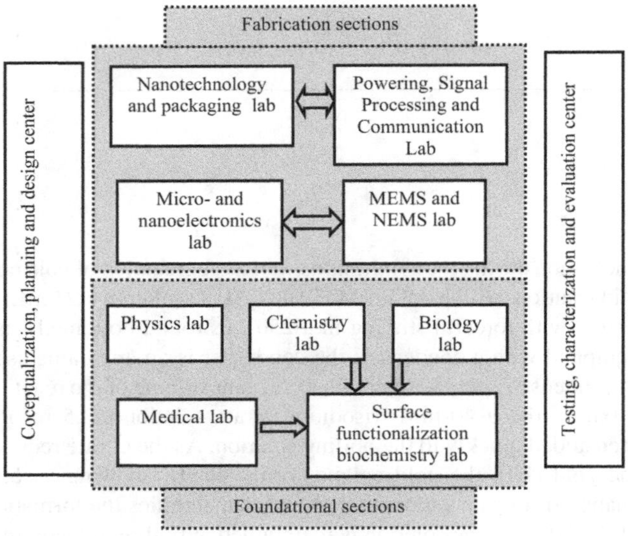

FIGURE 3.1 Divisions of a nanosensor laboratory.

3.2.1.3 Platinum Nanoparticles

The method adopted for the preparation of platinum nanoparticles is the reduction of hexachloroplatinate by refluxing a mixed solution of potassium bitartrate ($KC_4H_5O_6$), stabilizer, and hexachloroplatinate. Here, poly(*N*-vinyl-2-pyrrolidone) [$(C_6H_9NO)_n$, (PVP)], polyethylene glycol [$C_{2n+2}H_{4n+6}O_{n+2}$, (PEG)], or 3,3'-thiodipropionic acid [$C_6H_{10}O_4S$, (TDPC)] are used as the protective agents (Tan et al. 2003).

Materials: Hydrogen hexachloroplatinate (IV) hydrate ($H_2PtCl_6{\cdot}xH_2O$), PVP, TDPC, potassium bitartrate, and PEG.

Method: (i) An aliquot (20 mL) of an aqueous solution of potassium bitartrate (0.5 wt%) is brought to reflux (~100°C) while stirring, and 20 mL of an aqueous solution of H_2PtCl_6 (1.0 mM), containing the designated amount of PVP, PEG, or TDPC, is added quickly. Reflux means to boil a liquid in a vessel attached to a condenser so that the vapors continuously condense for reboiling; (ii) the concentrations of PVP, PEG, and TDPC vary (0.5, 1.0, 2.0, or 4.0 wt% (for PVP or PEG) and 0.5, 1.0, 2.0, or 4.0 mM (for TDPC). This variation is aimed at controlling the size as well as the size distribution of Pt nanoparticles; (iii) brown-colored solutions are obtained after reflux of 20 minutes and 3–24 hours in the presence of PEG and PVP, respectively. The color of the ensuing Pt colloid solutions changes from brown to light brown with increasing concentration ratio of TDPC to H_2PtCl_6.

3.2.1.4 Palladium Nanoparticles

Materials: Potassium tetrachloropalladate (II) (K_2PdCl_4), PVP, TDPC, potassium bitartrate, and potassium hydroxide (KOH) (Tan et al. 2003).

Method: (i) KOH (0.028 g) is dissolved in 20 mL of an aqueous solution of K_2PdCl_4 (1 mM) containing PVP (0.5 wt%) or TDPC (1 mM); (ii) the solution is added to 20 mL aqueous potassium bitartrate (0.5 wt%) solution, the temperature of which is raised to ~100°C under constant stirring; (iii) continued refluxing under stirring for 2 hours results in a dark brown-colored solution after the formation of palladium colloid.

3.2.2 Synthesis of Semiconductor Nanoparticles

Silica nanoparticles (SiNPs) have attracted attention because of their unrivaled optical properties. SiNPs are biocompatible. They exhibit high brightness and size-dependent tunable light emission. They also show high photoluminescence quantum efficiency (the average number of electron–hole pairs photoelectrically emitted from a photoreactive surface per incident photon of a given wavelength) and stability against photobleaching (or fading, the photochemical destruction of a fluorophore occurring when a fluorophore permanently loses the ability to fluoresce due to photon-induced chemical damage and covalent modification; it is then considered a photodegradation product) compared to organic dye molecules. The highly photoluminescent stable and non-aggregated Si nanoparticles provide a competitive, viable replacement for Cd-based quantum dots for biomedical applications. These nanoparticles are prospective non-heavy-element-containing quantum dots, well suited for applications in biology. They are useful for electronic and bioimaging purposes.

Rosso-Vasic et al. (2009) synthesized stable and brightly emitting amine (organic compounds of nitrogen, such as ethylamine, $C_2H_5NH_2$, that may be considered ammonia derivatives, in which one or more hydrogen atoms have been replaced by a hydrocarbon radical)-terminated SiNPs with different alkyl chain lengths between the Si core and the amine end group. Their procedure was as follows: (i) 1.5 g of tetraoctylammonium bromide, TOAB or TOABr, $[CH_3(CH_2)_7]_4NBr$, molecular formula: $C_{32}H_{68}BrN$, were mixed with 100 mL of dry toluene (C_7H_8 or $C_6H_5CH_3$). The mixture was sonicated (using ultrasound energy to agitate particles in a sample) for 30 minutes; this sonication was done under a flow of dry argon; (ii) 100 mL of $Si(OCH_3)_4$ (tetramethyl orthosilicate) was added through a gas-tight syringe and sonication was continued for 30 minutes allowing entry into the "micelles" (a submicroscopic aggregation of molecules, as in a droplet in a colloidal system); (iii) 2.3 mL of $LiAlH_4$ [1 M in tetrahydrofuran (THF)] was added to the above for the formation of hydrogen-terminated SiNPs; (iv) After 30 minutes sonication, dry and de-aerated methanol (30 mL) was added to react with the surplus $LiAlH_4$; (v) Alkylamine (a compound consisting of an alkyl group attached to the nitrogen of an amine; an example is ethylamine, $C_2H_5NH_2$)-terminated NPs with three different alkyl chain lengths resulted in the reactions of degassed allyl-amine (2.7 g), hex-5-en-1-amine (2.4 g) and undec-10-en-1-amine ($Cl_{11}H_{23}N$) (4.4 g) to each flask with hydrogen-terminated SiNPs obtained under argon (Ar), in the presence of 40 μL of 0.05 M H_2PtCl_6 catalyst; (vi) after 30 minutes sonication, 3-aminopropyl, 6-aminohexyl, and 11-aminododecyl SiNPs were extracted with water; 3-aminopropyl = $NH_2CH_2CH_2CH_2$; 6-aminohexyl = $C_7H_{16}N_2O_2$, carbamic acid; dodecyl = $CH_3(CH_2)_{10}CH_2$. They were washed with ethyl acetate (EtOAc), $CH_3COOC_2H_5$ or $C_4H_8O_2$) and filtered thoroughly by passing through syringe membrane filters; and (vii) the resulting SiNPs were further purified by dialysis (the separation of smaller molecules from larger molecules or of dissolved substances from colloidal particles in a solution by diffusion through a selectively permeable membrane) against water to remove any remaining unreacted aminoalkene (amino = the radical $-NH_2$; alkene = open chain hydrocarbons with one or more carbon–carbon double bonds, having the general formula C_nH_{2n}) and surfactant (a material that can greatly reduce the surface tension of a liquid when used in very low concentrations), that had remained with the NPs.

3.2.3 Synthesis of Semiconductor Nanocrystals: Quantum Dots

There are two general approaches for the preparation of quantum dots (QDs) (Drbohlavova et al. 2009): (i) *formation of nanosized semiconductor particles through colloidal chemistry*: the colloidal QD synthesis relies on rapid injection of semiconductor precursors into hot and energetically stirred specific organic solvents, containing molecules that can coordinate with the surface of the precipitated QD particles. This synthetic route is unsophisticated and can be performed in "one-pot." It constitutes the traditional route for synthesizing quantum dots; (ii) *epitaxial growth and/or nanoscale patterning*: employing lithography-based technology, this method is widely used in optoelectronics devices, such as lasers, infrared photodetectors, etc. However, it will be promising to use epitaxially grown QDs for *in situ* biosensing, mainly due to the ease of detection.

QD synthesis can be tailored to meet specific requirements, with core, shell, and coating characteristics all affecting photochemical properties. QDs may be manufactured with diameters ranging between broad limits, starting from a few nanometers to as large as a few micrometers. Size distributions are controllable within 2%, using accurate growth techniques, involving high annealing temperatures (Jamieson et al. 2007).

3.2.3.1 CdSe/ZnS Core/Shell QDs

CdSe-ZnS core/shell QDs are prepared using synthetic techniques, involving growth and annealing of organometallic precursors at high temperature. The method, according to Wang et al. (2009), is as follows: (i) a mixture of tri-*n*-octylphosphine oxide [TOPO: $OP(C_8H_{17})_3$], a tertiary alkylphosphine, and hexadecyl amine [HDA: $C_{16}H_{35}N$] is heated. Cadmium acetate $Cd(Ac)_2$ is added to the solution; (ii) the stock solution of trioctylphosphine selenide (TOP-Se) is prepared by dissolving 0.2 g selenium in 5 g of tri-*n*-octylphosphine (TOP); (iii) this stock solution is injected quickly into the reaction solution along with vigorous stirring of the same. Then, nucleation of CdSe nanocrystals takes place; (iv) $Zn(Ac)_2$ and bis(trimethylsilyl) sulfide $[(TMS)_2S]$ are added for the inorganic epitaxial growth of the shell on the surface of the core. This epitaxial layer growth is carried out for about 2 hours; and (v) the CdSe/ZnS QDs with diameters of ~3 nm, synthesized in this way, are dissolved in chloroform. They are preserved in a sealed container.

Riegler et al. (2008) described the following method: cadmium stearate Cd $(C_{17}H_{35}COO)_2$ and TOP-Se are reacted at temperatures above 200°C by fast injection of TOP-Se into a mixture of trioctylphosphine oxide (TOPO) and cadmium stearate. The CdSe-cores are passivated and annealed by growing a shell of two additional monolayers of ZnS on their surface. Diethylzinc $[(C_2H_5)_2Zn]$ and hexamethyldisilathiane $(C_6H_{18}SSi_2)$ are reacted for 12 hours with the CdSe-cores at 160°C in the presence again of TOPO and TOP. The core/shell particles obtained are repeatedly washed with methanol (CH_3OH). They are dispersed in chloroform $(CHCl_3)$ and finally stored in 50 mL chloroform as the stock solution.

3.2.3.2 CdSe/CdS Core/Shell QDs

In the method reported by Zhai et al. (2011), $Cd(Ac)_2$ and Se powder are selected as source materials. Triethylene glycol (TREG: triglycol is a colorless, odorless, viscous liquid, molecular formula = $HOCH_2CH_2OCH_2CH_2OCH_2CH_2OH$ or $C_6H_{14}O_4$) is used as the solvent due to its good hydrophilic features and high boiling point (288°C). The preparation of CdSe/CdS QDs is carried out in accordance with the following process steps: (i) 1 g poly(acrylic acid) (PAA, MW = 1800) $[(C_3H_4O_2)_n]$ and 0.5 mmol cadmium acetate $[Cd(Ac)_2]$ are dissolved in 20 mL TREG. The solution is heated to 200°C under argon flow; (ii) after 30 minutes, the solution is cooled down to room temperature, and 19 mg of Se powder is added to it; (iii) the mixture is heated to 240°C. It is kept at that temperature for a certain period of time such as 1, 5, 60, or 120 minutes; (iv) as a sulfur source, 19 mg thiourea (CH_4N_2S) is added into the aforementioned CdSe precursor solution. The excess $Cd(Ac)_2$ in the CdSe precursor solution is used as the source of cadmium; (v) the mixture is heated to 160°C for 1 hour; (vi) after reaction for 2 hours, the solution is quickly cooled to room temperature. It is precipitated by ethyl acetate; and (vii) the resultant solid products are further purified by dialysis and ultrafiltration. Ultrafiltration is a filtration process through a selectively permeable medium, in which particles of colloidal size are retained by a filter medium, while solvent and accompanying low-molecular-weight solutes are allowed to pass through.

3.2.3.3 PbS and PbS/CdS Core/Shell QDs

Silva et al. (2006) used the fusion method to produce PbS quantum dots embedded in an S-doped glass matrix (SiO_2–Na_2CO_3–Al_2O_3–PbO_2–B_2O_3:S). The sulfur-doped oxide glass matrix was prepared from high-purity powders, using SiO_2 as glass former (an oxide that can readily form a glass) and Na_2CO_3 to decrease the melting point. The mixture was melted in an alumina crucible at 1200°C for 30 min, then allowed to cool down to room temperature. Supplementary thermal treatment of the glass matrix was performed at 500°C to enhance the diffusion of Pb^{2+} and S^{2-} ions. As a consequence of this thermal treatment, PbS quantum dots were formed in the glass matrix.

Pietryga et al. (2008) described a partial cation-exchange method to synthesize PbS/CdSe core/shell QDs in a controlled manner, consisting of quenching of the reaction, separation of QDs, and their redispersal for further reaction. For PbS cores with a 5-nm diameter, the process is as follows: (i) 0.27 g (1.2 mmol) amount of PbO, 0.8 mL (2.5 mmol) of oleic acid $(C_{17}H_{33}COOH)$, 5 mL of TOP (trioctylphosohine) $(C_{24}H_{51}P)$, and 10 mL of 1-octadecene $[CH_2{=}CH(CH_2)_{15}CH_3]$ were heated to 150°C with continuous stirring under argon flow for 1 hour; (ii) the mixture was removed from the heat and a solution of 0.1 g (0.6 mmol) of hexamethyldisilathiane $[(CH_3)_3SiSSi(CH_3)_3]$ in 6 mL of 1-octadecene was immediately injected into it; (iii) the reaction was quenched after 10 seconds by adding 15 mL of cold hexane; (iv) PbS QDs were collected by precipitating with 10 mL of methanol (CH_3OH) and 15 mL of acetone $[(CH_3)_2CO]$, followed by centrifugation and removal of the decantate; (v) after redispersing in ~10 mL of toluene, the QDs were washed again by the same process. They were redispersed once again in 12 mL of toluene; and (vi) the suspended cores were subjected to further reaction without initial isolation as a solid, because this synthetic method gives essentially quantitative yield.

For CdS shell formation, the procedure is: (i) 1.0 g (7.8 mmol) of CdO, 6 mL (19 mmol) of oleic acid [$CH_3(CH_2)_7CH{=}CH(CH_2)_7COOH$ or $C_{18}H_{34}O_2$], and 16 mL of phenyl ether ($C_{12}H_{10}O$) were heated to 255°C under argon until all of the CdO had dissolved; (ii) the clear solution was cooled to ~155°C under Ar flow to remove water; (iii) the PbS core solution from above was degassed by Ar flow for ~30 minutes; (iv) immediately after the temperature of the PbS solution was set to 100°C, the cadmium oleate solution was added *via* a cannula (a flexible tube), accompanied by stirring; (v) aliquots were removed and cooled by mixing with hexane, and QDs were isolated. They were collected in a fashion analogous to that used for PbS cores.

3.2.4 Synthesis of Metal Oxide Nanoparticles

Many routes have been pursued for SnO_2 nanoparticle growth. These techniques include: (i) *thermolysis* (decomposition of a material by heating) of organometallic precursors; (ii) *sol-gel process* (chemical solution deposition: a process in which the aqueous sol (colloidal dispersion) is converted into gel spheres by partial dehydration); (iii) *oxidation* of $SnCl_2$ for tin oxide nanoparticles; (iv) *sonochemistry* (application of ultrasound to chemical reactions); and (v) *hydrothermal synthesis* (the chemical reaction of materials in aqueous solution heated in a sealed vessel).

Let us look closely at some of the methodologies adopted for synthesizing tin oxide nanoparticles of desired morphologies and sizes. Nayral et al. (1999) reported the synthesis of tin-tin oxide nanoparticles of low size-dispersion (the degree of scatter of data, usually about an average value, such as the median) through a mechanism combining the decomposition of an organometallic precursor (homoleptic tin (II) amides; metal compounds with all ligands identical) and controlled surface hydrolysis (decomposition of a chemical compound by reaction with water), as well as their oxidation into tin oxide nanoparticles without coalescence (uniting into a whole) or change in size.

Baik et al. (2000) studied the stabilization of tin oxide (SnO_2) particles by subjecting a precursor sol solution of SnO_2 to a hydrothermal treatment (hot water). They examined the effects of treating tin oxide gel hydrothermally in an ammonia solution at 200°C for 3 hours. Hydrated tin oxide (gel) was precipitated by mixing aqueous solutions of NH_4HCO_3 and $SnCl_4$. After washing with deionized water, a definite amount of precipitated gel (stannic acid, $SnO_2 \cdot nH_2O$), was suspended in an aqueous ammonia solution (NH_4OH) (stannic acid content: 10–40 mass%; stannic acid is an acidic form of tin dioxide: any of the series of acids, usually occurring as amorphous powders and varying in composition from H_2SnO_3 [alpha-stannic acid] to H_4SnO_4) to undergo a hydrothermal treatment in an autoclave at 200°C for 3 hours. Tin oxide contents of transparent sol solution were calculated as 1.8, 3.2, 6.1, and 8.6 wt%, respectively, from the weight of tin oxide after drying the specific amount of sol solution (stannic acid contents: 10, 20, 30, or 40 mass%) at 120°C. *What were the sizes of tin oxide particles*? The transparent solutions acquired were used for determining tin oxide particle size with a laser particle distribution analyzer and transmission electron microscope (TEM). The important observation was that the mean crystallite size (5–7 nm) was almost unchanged. On the other hand, the particle size was 4–6 nm in 1.8 wt% sol solution and furthermore, the size of sol particles increased with rising concentration of the consequential sol solution, that is, 5, 8, 10, or 32 nm for 1.8, 3.2, 6.1, or 8.6 wt% tin oxide sol.

Pinna et al. (2004) performed nonaqueous synthesis of nanocrystalline semiconducting metal oxides (tin and indium oxides): (i) tin (IV) *tert*-butoxide [$C_{16}H_{36}O_4Sn$] (500 mg, 1.216 mmol) or indium (III) isopropoxide [$In_5O(OC_3H_7)_{13}$] (200 mg, 0.685 mmol) was added to benzyl alcohol (abbreviation: BnOH, formula: $C_6H_5CH_2OH$) (20 mL); (ii) the reaction mixture was transferred into a teflon (PTFE: polytetrafluoroethylene $(C_2F_4)_n$) cup of 45 mL inner volume, placed in a steel autoclave, and cautiously sealed; (iii) the autoclave was withdrawn from the glovebox (an enclosed workspace in the form of a sealed compartment in which a desired atmosphere is maintained and which has openings, with long gloves attached on the sides, to enable an operator to insert hands through the gloves for manipulating the inside contents, sometimes toxic, without any harm or injury, without causing contamination of the contents and without breaking the containment). It was heated in a furnace at 220°C for 2 days; (iv) the resulting milky suspensions were centrifuged and the pellets were thoroughly washed with ethanol and dichloromethane (CH_2Cl_2); they were subsequently dried in air at 60°C. The centrifuge is a machine based on sedimentation principle which is used to separate particles of varying density dispersed in a fluid by setting the fluid in rotation about an axis.

Juttukonda et al. (2006) reported the synthesis of tin oxide nanoparticles stabilized by a variety of dendritic polymers (polymers having a branching tree-like appearance). Their technique consists of the simple oxidation of an encapsulated stannate (containing SnO_3^{2-} units) salt *via* reaction with carbon dioxide (CO_2) under ambient environment. Synthesis of tin (IV) oxide nanoparticles was accomplished *via* the reaction of carbon dioxide with stannate ions immobilized by dendritic polymers. Stirring of an aqueous or ethanolic solution containing sodium stannate and a polymeric host was first carried out for a minimum period of 2 hours to ensure saturated host-guest chelation (to firmly bind a metal ion with an organic molecule to form a ring-shaped molecular complex in which the metal is firmly bound and isolated; the resulting ring structure protects the mineral from entering into unwanted chemical reactions). Bubbling gaseous carbon dioxide through the room-temperature mixture with vigorous stirring for 30 minutes resulted in a light-yellow solution.

The best molar ratio of dendrimer: sodium stannate [$Na_2SnO_3{\cdot}3H_2O$ or $Na_2Sn(OH)_6$] was 1:4. On introducing more stannate, a white cloudy precipitate formed through reaction with CO_2 owing to the presence of free SnO_3^{2-} ions in solution. In addition to providing control over nanoparticle growth, their use of an under-supplied number of stannate ions resulted in the majority of primary amines being left behind unchelated, accessible for subsequent reaction/functionalization, if necessary.

Amongst other metal oxide NPs, let us consider silica NPs. Tabatabaei et al. (2006) synthesized silica nanoparticles from tetraethylorthosilicate (TEOS): $SiC_8H_{20}O_4$, ethanol, and deionized water. Colloidal silica nanoparticles with a constricted particle size distribution were obtained by the hydrolysis reaction of TEOS in ethanol containing water and ammonia.

Materials: TEOS ($SiC_8H_{20}O_4$), NH_4OH solution, and ethanol.

Method: (i) reagents were mixed into two starting time solutions of ethanol (EtOH): (a) TEOS/EtOH and (b) $NH_4OH/H_2O/$EtOH. By regulating the contents of solutions (a) and (b), the concentrations of TEOS, H_2O, and NH_4OH were fixed at prearranged values. The solutions were prepared in a glove box at room temperature under a controlled humidity of a few percentages; And (ii) solutions (a) and (b) were mixed with each other at 298 K. The mixture was subjected to vigorous stirring by hand for six seconds. Glycerol $[C_3H_5(OH)_3]$ was added to the $NH_4OH/H_2O/$EtOH mixture. This was followed by addition of TEOS.

Condensation reaction began 2–10 minutes after the addition of TEOS; it is a reaction in which two molecules react with the resulting loss of a molecule of water (or other small molecule): the formal reverse of hydrolysis. The condensation reaction was easily observed. The hydrolysis reaction, forming silicic acid (H_4O_4Si), is not seen. After this unseen reaction, the condensation of supersaturated silicic acid was indicated by an increasing dullness and opacity of the mixture. Then, a turbid white suspension was formed.

3.2.5 Synthesis of Carbon Nanotubes

Presently, there are three principal techniques to produce high-quality single-walled carbon nanotubes (SWCNTs): (i) laser ablation (the process of evaporating material from a target solid surface by irradiating it with pulses of laser beam); (ii) electric arc discharge (a self-maintaining direct-current electrical current flow through a gas, producing an ongoing, brightly glowing plasma discharge and characterized by a voltage drop approximately equal to the ionization potential of the gas); and (iii) chemical vapor deposition (CVD). *What are the salient features of these techniques?* Among the aforementioned techniques, laser ablation and arc discharge are physical methods. In fact, they are modified physical vapor deposition (PVD) techniques and involve the condensation of hot gaseous carbon atoms generated from the evaporation of solid carbon. However, the equipment requirements and the large amount of energy consumed by these methods make them mostly appropriate for laboratory work. The problem lies in scaling-up these methods to fabricate large quantities of CNTs to cater to larger-scale industrial applications. Both the laser ablation and arc discharge techniques suffer from limitations resulting from the volume of sample they are able to produce in comparison to the size of the carbon source. In addition, larger percentages of impurities are found in the nanotubes synthesized by these methods, in the form of amorphous carbon and catalyst particles, because of the high-temperature nature of the heat source. Since the growth is difficult to control, the final product consists mostly of multi-walled carbon nanotubes (MWCNTs) with poor alignment. Nonetheless, the overall yield of CNTs produced is sufficient for laboratory-scale experiments and they have been mostly used for academic research.

What is the favored technique for CNT manufacturing? It is CVD. *What is CVD?* CVD involves irreversible deposition of a solid from a gas or a mixture of gases through a heterogeneous chemical reaction. *Where does this reaction occur?* It takes place at the interface of a gas-solid substrate. Depending on the deposition conditions, the growth process is restrained either by diffusion or by surface kinetics.

Generally speaking, CVD of hydrocarbons over a metal catalyst is a classical method that has been used by many people in the production of various carbon materials, like carbon fibers and filaments. *What are the merits of CVD in general?* Its advantages are as follows: (i) CVD is the preferred technique for fabrication of thin layers of metals, insulators, and semiconductors on different substrates; and (ii) it offers better growth controllability because of the equilibrium nature of the chemical reactions involved.

CVD of CNT uses carbon precursor gases, such as methane (CH_4), ethylene (C_2H_4), acetylene (C_2H_2), benzene (C_6H_6), carbon monoxide (CO), and ethanol (C_2H_5OH). The process usually involves high-temperature decomposition of hydrocarbons in hydrogen over the catalyst, which is pre-deposited on the solid substrate. *What are the merits of CVD for CNTs?* Advantages of CVD for CNTs are as follows: (i) CVD has the capability to maneuver the size, shape, and alignment of the nanotubes through a cautiously designed patterning of the catalysts on the surface of the substrate; (ii) it is a continuous process and currently the best-known technique for high-yield and low-impurity production of CNTs at moderate temperatures; and (iii) Amongst CNT production techniques, this method can readily be up-scaled to industrial production. It is widely recognized as having the capability for large-scale production of CNTs, at least 10,000 tons/annum per plant (MacKenzie et al. 2008). In particular, the fluidized-bed CVD (FBCVD) technique (where the CVD reaction occurs within a fluidized bed of catalyst particles) offers the greatest expectation to produce high-quality CNTs economically and in large quantities (See and Harris 2007). A *fluidized bed* is formed when a quantity of a solid particulate substance, generally present in a holding vessel, is placed under appropriate conditions to cause the solid/fluid mixture to behave as a fluid. This is usually achieved by the introduction of pressurized fluid through the particulate medium, resulting in the medium acquiring many properties and characteristics of normal fluids, for example, the ability to free-flow under gravity, or to be pumped using fluid-type technologies.

Consequently, CVD has emerged as the most viable commercial approach for manufacturing carbon nanotubes.

3.2.5.1 Arc Discharge Method of CNT Production

Arc discharge, initially used to produce C_{60} fullerenes (molecules composed entirely of carbon, in the form of a hollow sphere, ellipsoid, or tube; the third allotropic form of carbon material after graphite and diamond), creates CNTs through arc vaporization of two carbon rods placed end to end, separated by <1 mm, in an enclosure that is usually filled with inert gas (helium, He, or argon, Ar) at low pressure (Figure 3.2). A direct current of 50–100 A, driven by a potential difference ~30 V, creates a high-temperature (>3000°C) plasma discharge between the two electrodes. The discharge vaporizes the surface of one of the carbon electrodes and forms a small rod-shaped deposit on the other electrode.

What are the shortcomings of the arc discharge method? Drawbacks of the arc discharge method are as follows: (i) the inherent design of these systems poses limitations to the large-scale production of CNTs; for example, vacuum

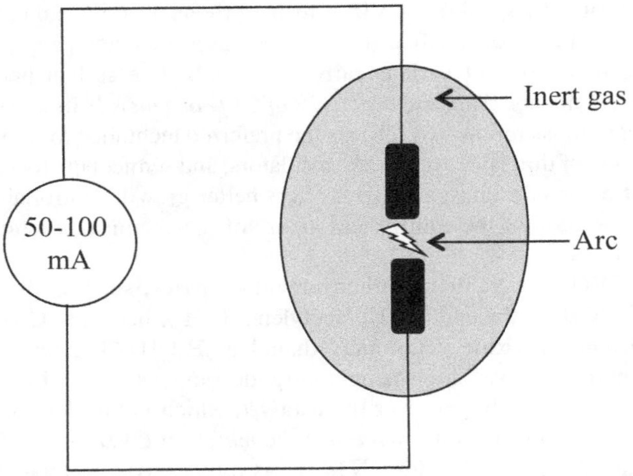

FIGURE 3.2 Arc discharge arrangement for CNT synthesis. (Singh, P. et al., *J. Optoelectron. Biomed.* M., 2, 91, 2010.)

conditions are required to prevent interference from the formation of unwanted ions at the high temperatures used in the arc discharge. But vacuum conditions are troublesome and expensive to scale up; and (ii) the graphite targets and electrodes require continuous replacement as the synthesis proceeds, and, hence, these techniques cannot be operated continuously.

3.2.5.2 Laser Ablation Method of CNT Production

A laser source is used to generate high temperatures on a carbon target (Figure 3.3). The vaporized carbon rapidly cools in a carrier gas stream, for example, helium (He), and forms CNTs and other carbonaceous secondary products. The use of two laser pulses in succession reduces the amount of carbon deposited as soot (a fine black or brown powder, consisting of impure carbon particles, that forms through incomplete combustion of a hydrocarbon) to the minimum extent. *How does this happen?* The second laser pulse breaks up the larger particles ablated (eroded by melting) by the first one and feeds them into the maturing nanotube structure. On the whole, the CNT material produced by this method appears as an entangled mat of "stout cords," 10–20 nm in diameter and up to 100 μm or more in length.

3.2.5.3 Chemical Vapor Deposition Method of CNT Production

Large amounts of CNTs can be formed by catalytic CVD of acetylene (C_2H_2) over cobalt (Co) and iron (Fe) catalysts residing on silica (SiO_2) or zeolite (microporous, aluminosilicate minerals used as commercial adsorbents) support (Singh et al. 2010). High yields of SWCNTs have been obtained by catalytic decomposition of a H_2/CH_4 mixture over well-dispersed metal particles such as cobalt (Co), nickel (Ni), and iron (Fe) on magnesium oxide (MgO) at 1000°C (Figure 3.4). The reduction produces very small particles of transition metals (an element whose atom has an incomplete *d* sub-shell, or which can give rise to cations with an incomplete *d* sub-shell) at a temperature of usually >800°C. The disintegration of CH_4 over the newly formed nanoparticles hinders their further growth and thus results in a very high proportion of SWCNTs but few MWCNTs.

Figure 3.5 depicts how the synthesis of CNT takes place. In principle, CVD of CNT is a thermal dehydrogenation (a chemical reaction that involves the elimination of hydrogen; it is the reverse process of hydrogenation) reaction by which a transition metal catalyst, for example, iron, nickel, or cobalt, is used to bring down the temperature required in order to "crack" a gaseous hydrocarbon feed into carbon and hydrogen. In several ways, it is comparable to the large-scale synthesis of hydrocarbon fuels, for example, octane (chemical formula C_8H_{18}, with a structural formula of $CH_3(CH_2)_6CH_3$), by catalytic cracking, meaning the decomposition of heavy oils to produce lighter hydrocarbons. In this field, it has already been successfully scaled up to large throughputs.

There exist a large number of CVD modalities. CVD variants include (i) fixed beds; (ii) fluidized beds; (iii) aerosols (colloidal particles dispersed in a gas); (iv) floating catalysts; and (v) combination methods, for example, plasma-enhanced (PECVD) and laser-assisted techniques, all of which are capable of producing free-standing (not supported by or adjoining another structure) CNTs. These free-standing CNTs resemble those grown using laser ablation or by arc discharge, that is, they are not grown on templates (patterns or gauges, usually as thin boards/metal pieces, for guiding cutting or drilling). A PECVD reactor is a commonly used equipment in the microelectronics industry. Figure 3.6 shows the PECVD method for CNT formation.

The fixed-bed reactor has traditionally been used for CNT synthesis via CVD. A cross-flow set-up is usually employed

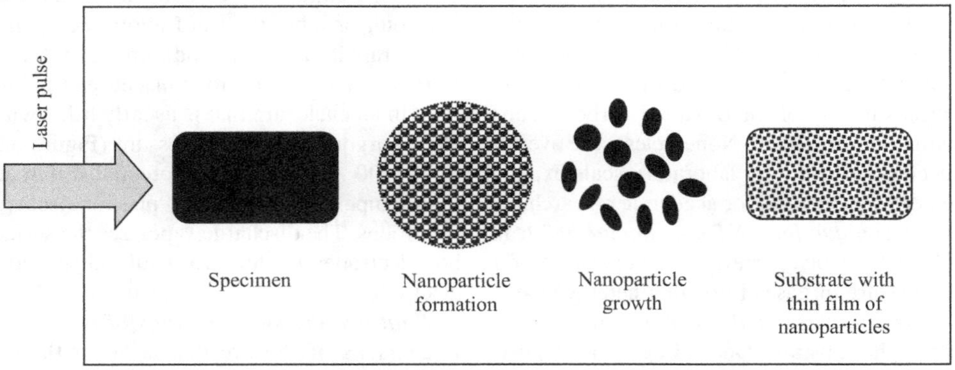

FIGURE 3.3 Laser ablation method of CNT production. (Singh, P. et al., *J. Optoelectron. Biomed.* M., 2, 91, 2010.)

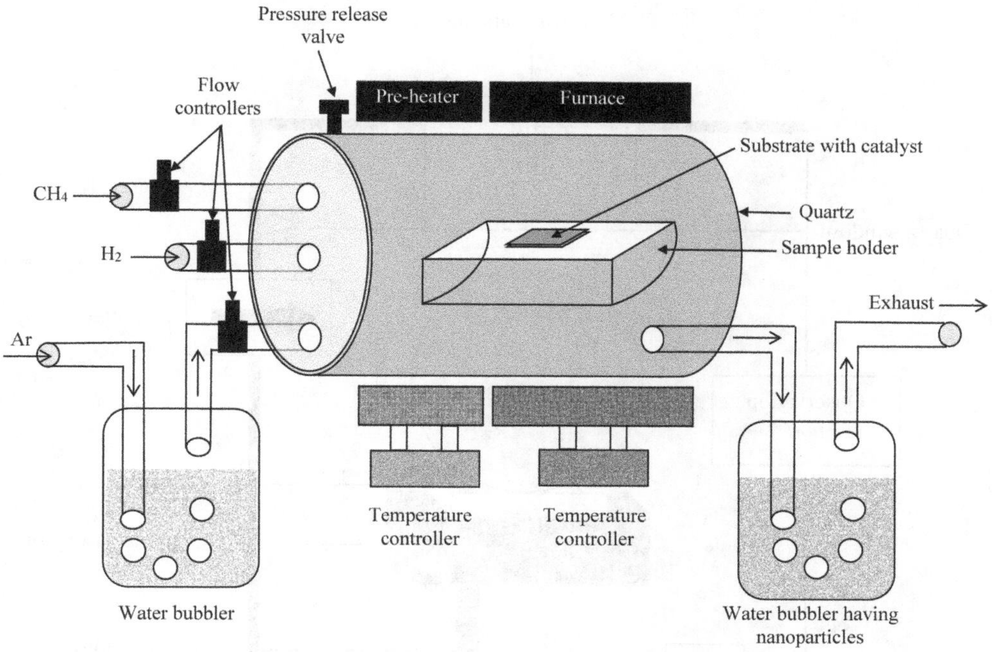

FIGURE 3.4 Chemical vapor deposition of carbon nanotubes: source of carbon atoms is usually an organic compound, the precursor (methane, ethylene, acetylene, benzene, carbon monoxide, or ethanol). The bubbler provides water vapor carried by argon gas into the reactor. Water serves as an oxidizing agent, helping to remove amorphous carbon from the surface of the catalyst particle. In addition, hydrogen and argon are added to the gas mixture for better control of the carbon concentration in the gas phase. A sample holder accommodates the substrate with patterned catalysts on its surface. Transition metals such as Fe, Co, and Ni are frequently used as catalysts. Furnace temperature ranges between 600°C and 1200°C. Carbon nanotubes are deposited on the reactor wall, which is usually constructed of quartz.

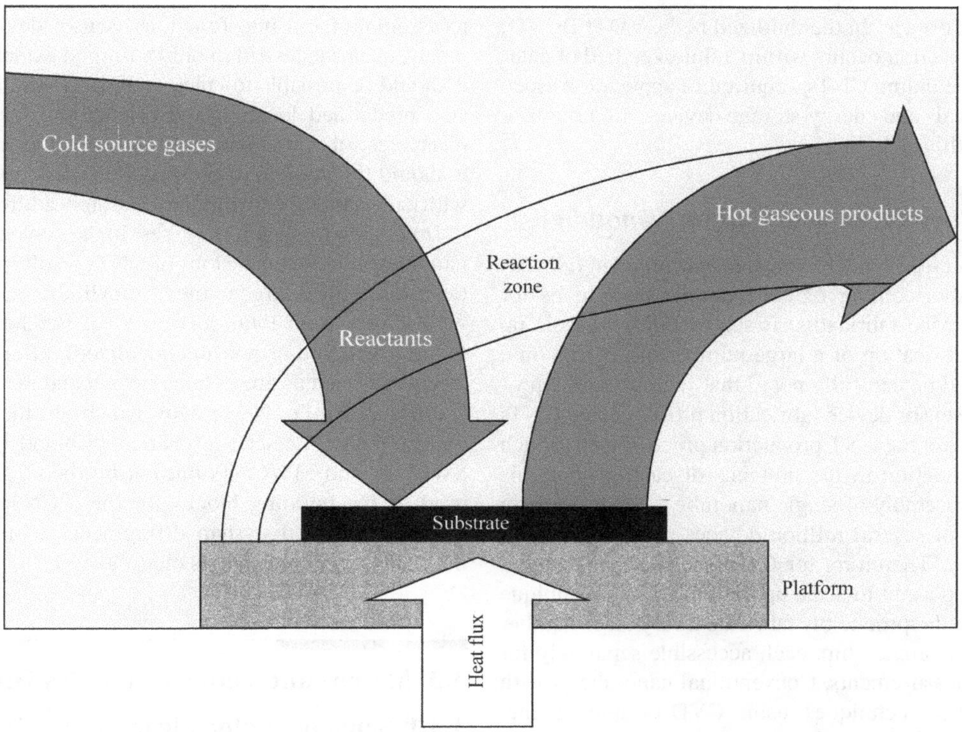

FIGURE 3.5 Mechanism of CVD of CNTs: gases enter chamber at room temperature (cooler than the reaction temperature). They are heated as they approach the substrate. They react with the substrate or undergo chemical reaction in the "reaction zone" before reacting with the substrate, forming CNT deposits. The gaseous products are then removed from the reaction chamber.

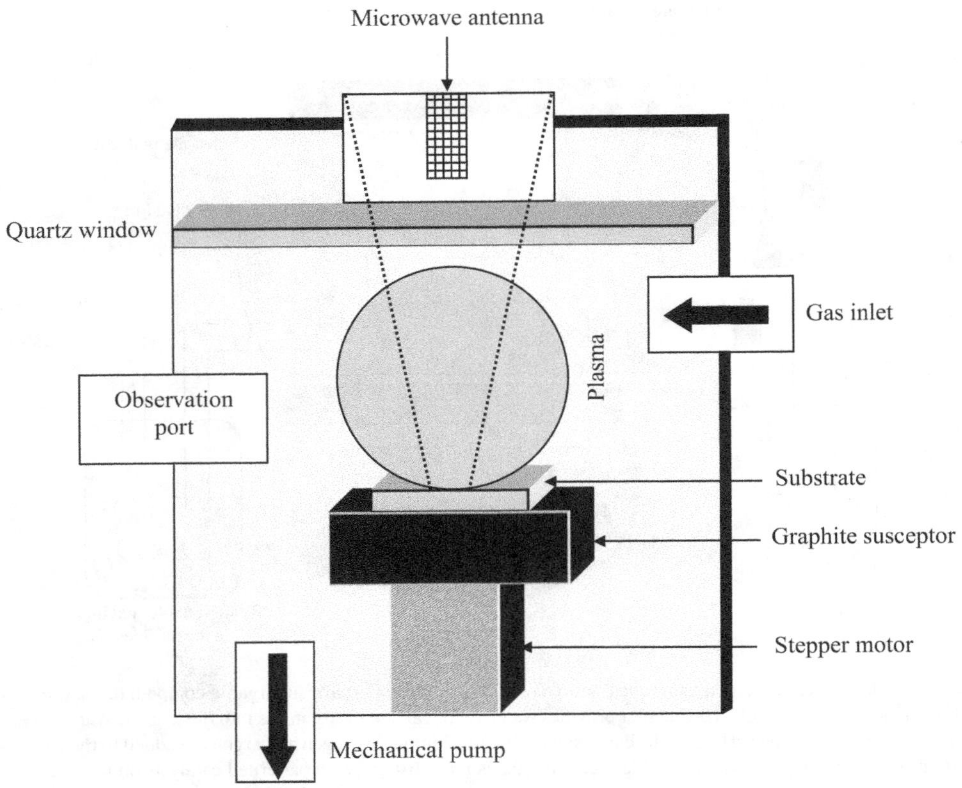

FIGURE 3.6 Plasma-enhanced CVD-based CNT synthesis: alumina substrate (Al_2O_3) is coated with ferric nitrate [$Fe(NO_3)_3{\cdot}6H_2O$] (the catalyst). Methane (CH_4) and hydrogen (H_2) are introduced into the vacuum chamber at a total pressure of 15 Torr, and the chamber is then heated to 850–900°C with an input microwave power of 600 W. CNTs are grown from the substrate material.

inside a horizontal furnace. In the fluidized bed CVD (FBCVD) process, the CVD reaction occurs within a fluidized bed of catalyst particles. Free-standing CNTs, required in applications such as composite materials and energy-storage devices, are produced cost-effectively by this technique.

3.2.5.4 Difficulties Faced with Carbon Nanotubes

What are the obstacles to CNT usage in nanosensor fabrication? One major shortcoming of conventional techniques for carbon nanotube device fabrication is the inability to scale up the processes for fabrication of a large number of devices on a single chip. It should be carefully noted that increasing the proportion of the nanosensor device fabrication process using CNTs is being referred to, not the CNT production process itself, which was the focus of attention in the previous discussion. For this purpose, accurate assembly of single nanotube devices, with an integration density of several million devices per square centimeter, is mandatory. Therefore, for CNT sensors to become a reality, it is very important to scale up the fabrication technique to simultaneously and reproducibly fabricate a very large number of such devices on a single chip, each accessible separately for precise electrical measurements. Conventional nanotube growth and device fabrication techniques, using CVD or spin casting, have failed to achieve this goal because of the lack of precise control over nanotube positioning and orientation. Due to the difficulties encountered in correctly handling and manipulating these nanoscale objects at the individual level, various attempts to assemble them into functional sensor devices have, unfortunately, been successful to only a limited extent. In the ideal case, it should be possible to "pick and place" an individual nanotube at a predefined location and orientation, forming robust, low-resistance, ohmic contacts to two metallic leads. Furthermore, it should be possible to do this at a scalable integration density, with each nanotube forming a separately addressable nanosensor.

Do CNTs exhibit toxicity? Yes. In the context of toxicology, the nanometer-scale dimensions of CNTs result in milligram quantities possessing a large number of cylindrical, fiber-like particles with a very high total surface area. Batches of pristine CNT (nonpurified and/or nonfunctionalized), after synthesis, contain impurities such as amorphous carbon and metallic nanoparticles (catalysts: Co, Fe, Ni, or Mo), which are the sources of severe toxic effects. Platelet aggregation in blood is induced by both SWCNTs and MWCNTs but not by the C_{60} fullerenes that are used as the building blocks for these CNTs. Platelets are the small, colorless, disc-shaped fragments, with no nucleus, found in blood and involved in its clotting.

3.3 Micro- and Nanoelectronics Division

3.3.1 Semiconductor Clean Room

A clean room is an enclosed clean space in which semiconductor manufacturing takes place. Airborne particles are reduced to a technically feasible minimum. A clean room is defined as a room with air containing N particles of size <0.5 μm per cubic foot (0.02832 cubic meters), where $N \leq 1$, 10, 100, 1,000, or 10,000; accordingly, it is designated as Class 1, Class 10, Class 100, Class 1000, or Class 10000. *What are the temperature and humidity guidelines for semiconductor fabrication clean rooms?* Temperature (20°C–25°C) and humidity (40%–50%) of ambient air are strictly controlled, both for the efficient running of equipment and to maintain the critical ambient requirements for carrying out the fabrication processes effectively. Noncompliance with these conditions will invariably lead to equipment breakdown and device processing faults.

3.3.2 Silicon Single-Crystal Growth and Wafer Production

What is the starting material for integrated circuit fabrication? Integrated circuits are built on single-crystal silicon substrates that possess a high level of purity and crystalline perfection. *How is such high-grade Si material obtained*? The acquisition of such high-grade starting silicon material involves two major steps: (i) refinement of raw material (such as quartzite, a hard metamorphic rock, that was originally sandstone) into electronic-grade polycrystalline silicon (EGS), using a complex multistage process; and (ii) growing of single-crystal silicon from this EGS, by either the Czochralski (C-Z) or Float Zone (F-Z) process. Silicon is readily available through the treatment of silica, SiO_2, with pure graphite (as coke) in an electric furnace.

$$SiO_2 + 2C \rightarrow Si + 2CO \tag{3.2}$$

Very pure silicon is made by the reaction of $SiCl_4$ with hydrogen, followed by zone refining of the resultant silicon. Zone refining is also known as zone purification, a technique to purify materials, in which a narrow molten zone is moved slowly along the complete length of the specimen to cause impurity segregation, and which depends on the differences in composition of the liquid and solid in equilibrium) Material used in single-crystal growth is EGS (99.999999999% pure).

C-Z crystal growth involves the crystalline solidification of atoms from a liquid phase at an interface. The silicon charge, taken in a fused silica crucible, is melted (Si melting point = 1421°C), and a seed of crystal silicon with precise orientation is introduced into the molten silicon. The seed crystal is withdrawn at a controlled rate, and the seed crystal and the crucible are rotated in opposite directions. By controlling the temperature gradients, the rate of pulling and the speed of rotation, a large, single-crystal, cylindrical ingot (a material, usually metal that is cast into a shape suitable for further processing) is extracted from the melt.

The F-Z process entails the passing of a molten zone through a polysilicon rod that has approximately the same dimensions as the final ingot. A polycrystalline rod is passed through an radio-frequency (RF) heating coil, creating a localized molten zone from which the crystal ingot is grown. A seed crystal (a small piece of single-crystal material, from which a large crystal of the same material is typically grown) is used at one end to initiate the growth process. The purity of an ingot produced by the F-Z process is higher than that produced by the C-Z process. These impurities are mainly introduced by the material of the crucible (a refractory container used for metal, glass, etc., production), which holds the silicon melt. As such, C-Z silicon has much higher oxygen and carbon impurities than does F-Z silicon.

What are the typical operating parameters of the C-Z *and* F-Z *processes?* In a representative C-Z process, induction heating (heating an electrically conducting object, usually a metal, by electromagnetic induction) at 0.4 MHz melts the Si held in a stationary clear fused quartz crucible nested in a graphite susceptor (material used for its ability to absorb electromagnetic energy and convert it to heat). The ambient atmosphere is argon at 0.1 bar above atmospheric pressure. Typically, the seed crystal is ~0.5 cm in diameter and 10 cm long. The growth rate is 1.5 mm min^{-1} with a crystal rotation rate of 14 rpm. F-Z growth is carried out on Si using RF heating with a stationary one-turn coil operating at 2 MHz in an argon ambient atmosphere at 0.3 bar above atmospheric pressure, to minimize heater or crucible sources of O, C, and other impurities. The typical growth rate is 3 mm min^{-1} with a 13–16 rpm crystal rotation rate and a 2–3 rpm feed rod rotation rate (crystal on the bottom side, moving downward).

3.3.3 Molecular Beam Epitaxy

Epitaxy or *epitaxial growth* is the process of depositing a thin layer (0.5–20 μm) of single-crystal material over a monocrystalline substrate. Molecular beam epitaxy (MBE) is a technique for achieving epitaxial growth *via* the interaction of one or several molecular or atomic beams that occur on the surface of a heated crystalline substrate. The solid source materials are housed in evaporation cells to provide an angular distribution of atoms or molecules in a beam. The substrate is heated to the necessary temperature. When needed, it is continuously rotated to improve the homogeneity of growth. The molecular beam condition that the mean free path λ (the average distance covered by a moving particle between successive impacts) of the particles should be larger than the geometrical size of the chamber is easily fulfilled if the total pressure does not exceed 10^{-5} Torr. Also, the condition for growing a sufficiently clean epilayer must be satisfied. It is necessary that the monolayer deposition times of the beams t_b and the background residual vapor t_{res}, should obey the relation $t_{res} < 10^{-5}\, t_b$. Thus, MBE is a technique of growing single crystals, in which beams of atoms or molecules bombard a single-crystal substrate in a vacuum, giving birth to crystals whose crystallographic orientation is related to that of the substrate. In MBE, a source material is heated to produce an evaporated beam of particles. These particles propagate and disseminate through a very high vacuum (10^{-8} Pa) to the substrate, where they condense. MBE has lower throughput than other forms of epitaxy.

3.3.4 Mask Making

Based on the electrical specifications of the given circuit, a geometrical layout for the same is made, defining the areas for diffusion windows, contact window openings, metal patterns, etc., for resistors, capacitors, diodes, transistors, and other devices. Then, in accordance with the fabrication process sequence, a series of masks, in the form of aligned patterns, are designed. Mask making consists of two steps: (i) layout generation (defining the pattern that will appear on the mask); and (ii) pattern transferring to the mask.

Layout is the process of defining the patterns that will be transferred to the masks, and as such will define the geometry of the devices. Layout is typically performed in a graphical editing tool such as LEDIT, Cadence, etc. Distinct layer names are used for each layer in the layout; the layer designation for each layer should be remembered. The polarity of the mask must be indicated by either bright-field (space left open in the layout is transparent) or dark-field examination (space left open in the layout is opaque).

For optical lithography, commonly used substrates are soda-lime glass and quartz. Quartz masks have a similar coefficient of thermal expansion to that of Si but are more expensive than soda-lime masks. The two most widespread mask-coating materials used for optical lithography are emulsion and chrome. Photographic emulsion is a light-sensitive colloid, such as silver halide crystals (silver bromide [AgBr], silver chloride [AgCl] and silver iodide [AgI]) suspended in gelatin (a protein substance derived from collagen, a natural protein present in the tendons, ligaments, and tissues of mammals). The light-exposed crystals are reduced by the developer to black metallic Ag particles, forming the image. Emulsion is much cheaper to make; however, the coating is not of as high a quality as chrome and does not last the rigors of processing as well.

For the production of chromium film masks, a thin layer of chromium (Cr), ~0.1 μm thick, is deposited onto a transparent substrate, most preferably glass, by vacuum deposition. The chromium layer formed is then selectively etched into the desired pattern by a photoetching process, which comprises the following steps: (i) applying a photoresist material, typically a natural or synthetic polymer, to the layer to be etched; (ii) selectively exposing the photoresist material, whereby certain portions thereof become soluble or insoluble upon development; (iii) removing the soluble portions that yields a negative or positive image of the desired pattern; and (iv) etching the portions of the underlying layer thus exposed.

The exposure of the resist is commonly done in one of two ways: (i) laser pattern generation; or (ii) electron-beam (E-beam). Laser writing can create features down to 500 nm, whereas E-beam reaches to 10 nm. Laser pattern generation involves the use of a rectangular shutter, which is capable of variable sizes and orientations. The shutter is used to sequentially expose the resist on the mask substrate to a laser beam until the requisite pattern is fully defined. As the mask is exposed sequentially, small intricate features and curved features may require a large number of flashes to define the pattern. A pattern with a high flash count may make this method of mask pattern transfer uneconomic.

E-beam is a technique in which an electron beam is used to selectively expose the resist on the mask. The beam is raster scanned across the wafer, defining the pattern. The beam may be made very narrow, so that very intricate features may be defined using this approach. The narrower the beam used, the longer the write time for a mask (as more scans are required), which increases the price.

3.3.5 Thermal Oxidation

It is a method to produce a thin layer of oxide (usually silicon dioxide) on the surface of a silicon wafer by exposing the wafer to an oxidizing environment of O_2 or H_2O at elevated temperature (700°C–1200°C) (Figure 3.7); accordingly, the process is called dry and wet oxidation. Silicon is consumed in the process. Oxidation occurs at the silicon–silicon dioxide interface.

$$\begin{aligned} &\mathrm{Si(solid)} + \mathrm{O_2(gas)} = \mathrm{SiO_2(solid)}; \\ &\mathrm{Si(solid)} + 2\mathrm{H_2O(gas)} = \mathrm{SiO_2(solid)} + 2\mathrm{H_2(gas)} \end{aligned} \tag{3.3}$$

This oxide layer is an excellent insulator and serves as a mask against diffusion of impurities, like phosphorus or boron, into the silicon.

The *Deal-Grove model* is a model describing the kinetics of thermal oxidation of silicon, based on a chemical reaction between silicon and the oxidizing species. It is most accurate for oxides thicker than about 30 nm but is of limited use for oxides thinner than about 10 nm. It assumes surface reaction-controlled oxide growth in the early stage of oxidation (linear regime) and controlled by diffusion of oxidizing species through the oxide during extended oxidation (parabolic regime). It is one of the best-established and celebrated models in silicon processing.

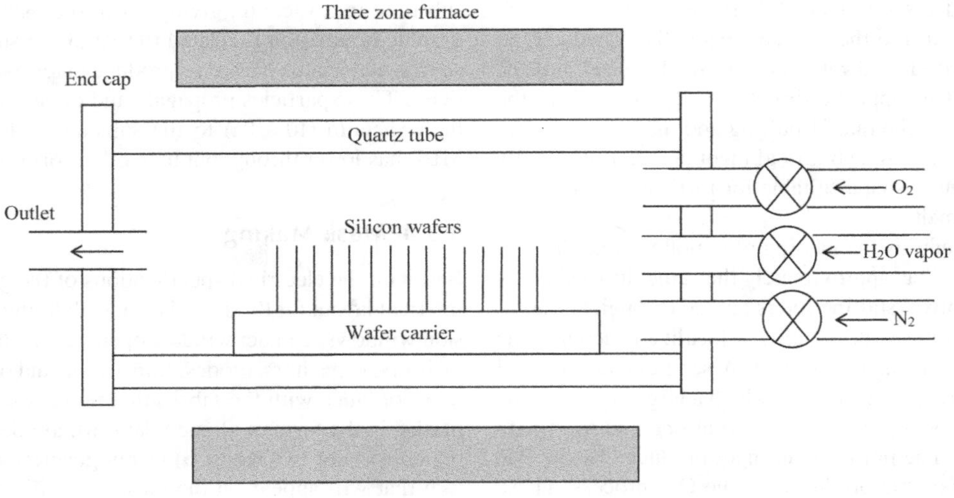

FIGURE 3.7 Typical thermal oxidation furnace for silicon wafers.

According to the Deal-Grove model, the thickness of thermal oxide is expressed as

$$t_{ox} = \left(\frac{A}{2}\right)\left\{\sqrt{1+\frac{t+\tau}{\left(A^2/4B\right)}}-1\right\} \tag{3.4}$$

where

B/A is a linear rate constant

B is the parabolic rate constant

τ denotes a shift in the time coordinate to account for the presence of the initial oxide layer

Example 3.1

Wet oxidation of silicon is carried out at 1000°C. Given the following values of parameters in the Deal-Grove model, namely $A=0.226$ µm, $B=0.287$ µm² hour⁻¹, and $\tau=0$, find the oxide thickness for growth times $t=0.01$ hour and $t=100$ hour.

In Equation 3.4, $A=0.226$ µm, $B=0.287$ µm² hour⁻¹, and $\tau=0$.

For $t=0.01$ hour, we have

$$t_{ox} = \left(\frac{A}{2}\right)\left\{\sqrt{1+\frac{t+\tau}{\left(A^2/4B\right)}}-1\right\}$$

$$=\left(\frac{0.226\times10^{-4}\text{ cm}}{2}\right)$$

$$\times\left[\sqrt{1+\frac{0.01\text{h}+0}{\left\{\left(0.226\times10^{-4}\text{ cm}\right)^2/\left(4\times0.287\times10^{-4}\times10^{-4}\text{ cm}\times\text{cm/h}\right)\right\}}}-1\right] \tag{3.5}$$

$$=1.13\times10^{-5}\left[\sqrt{1+\frac{0.01}{\left\{5.1076\times10^{-10}/1.148\times10^{-8}\right\}}}-1\right]$$

$$=1.13\times10^{-5}\left[\sqrt{1+\frac{0.01}{4.449\times10^{-2}}}-1\right] \tag{3.6}$$

$$=1.13\times10^{-5}\left[\sqrt{1.2248}-1\right]$$

$$=1.206\times10^{-6}\text{ cm}=0.01206\,\mu\text{m}$$

For $t=100$ hour, we get

$$t_{ox} = \left(\frac{A}{2}\right)\left\{\sqrt{1+\frac{t+\tau}{\left(A^2/4B\right)}}-1\right\}$$

$$=\left(\frac{0.226\times10^{-4}\text{ cm}}{2}\right)$$

$$\times\left[\sqrt{1+\frac{100\text{h}+0}{\left\{\left(0.226\times10^{-4}\text{ cm}\right)^2/\left(4\times0.287\times10^{-4}\times10^{-4}\text{ cm}\times\text{cm/h}\right)\right\}}}-1\right] \tag{3.7}$$

$$=1.13\times10^{-5}\left[\sqrt{1+\frac{100}{\left\{5.1076\times10^{-10}/1.148\times10^{-8}\right\}}}-1\right]$$

$$=1.13\times10^{-5}\left[\sqrt{1+\frac{100}{4.449\times10^{-2}}}-1\right] \tag{3.8}$$

$$=1.13\times10^{-5}\left[\sqrt{2.2487\times10^{3}}-1\right]$$

$$=5.2455\times10^{-4}\text{ cm}=5.2455\,\mu\text{m}$$

As the thickness of the gate dielectric scales below 2 nm, the leakage current due to tunneling rises markedly, leading to heavy power consumption. Materials with high dielectric constants, such as silicon oxynitride (SiON), hafnium dioxide (HfO_2), zirconium dioxide (ZrO_2), or titanium dioxide (TiO_2), permit increasing the thickness of the gate dielectric, thereby alleviating the tunneling current problem. Thus, substituting the silicon dioxide gate dielectric with a high-κ material yields the required high gate capacitance without the attendant undesirable current leakage.

3.3.6 Diffusion of Impurities in a Semiconductor

The process by which molecules spread from areas of high concentration to areas of low concentration is used for junction formation, that is, transition from P- to N-type or *vice versa.* It is typically accomplished by the process of diffusing the appropriate dopant impurities in a high-temperature furnace. At high temperature, many atoms in the semiconductor move out of their lattice site, leaving vacancies into which impurity atoms move, to take up their places. The impurities thus diffuse by this type of vacancy motion and occupy lattice positions in the crystal after it is cooled.

The behavior of diffusion particles is governed by Fick's laws, which, when solved for appropriate boundary conditions, give various dopant distributions called *profiles.* These profiles are approximated during actual diffusion processes. *Fick's first law* states that the flux, *J*, of a component of concentration C across a membrane of unit area, in a predefined plane, is proportional to the concentration differential across that plane; *Fick's second law* states that the rate of change of concentration in a volume element of a membrane, within the diffusional field, is proportional to the rate of change of concentration gradient at that point in the field. Although Fick's first law allows the diffusive flux $J(x)$ to be calculated as a function of the concentration gradient, the second law enables the concentration function $C(x)$ to be calculated as a function of time.

Fick's laws are applied to solve diffusion problems in semiconductor fabrication. Depending on boundary conditions, two types of solutions are obtained. These solutions provide two types of impurity distribution: (i) *constant or infinite source distribution,* following *complementary error function (erfc),* and (ii) *limited or finite source distribution,* following *Gaussian distribution function.* Mathematically, the constant source diffusion is described by the equation

$$N(x,t)=N_s\text{erfc}\left(\frac{x}{2\sqrt{Dt}}\right) \tag{3.9}$$

where

$N(x, t)$ is the concentration of impurity in atoms cm^{-3} at distance x at time t

N_s is the surface concentration

D is the diffusion constant of the impurity at the temperature of diffusion

For limited source diffusion, the equation is

$$N(x,t) = \frac{Q}{\sqrt{\pi Dt}} \exp\left(-\frac{x^2}{4Dt}\right) \tag{3.10}$$

Q is the total impurity in atoms cm^{-2} given by

$$Q = N_s\sqrt{\pi Dt} \tag{3.11}$$

How is diffusion performed in practice? In practical terms, diffusion of impurities in a semiconductor is performed as a sequence of two steps. The two-step diffusion consists of a deposition step and a drive-in step. The unlimited source diffusion is called the impurity *pre-deposition step.* The limited source diffusion is known as the impurity *drive-in step.*

Example 3.2

Calculate the junction depth for a constant-source boron diffusion into an N-type silicon substrate (1×10^{15} cm^{-3}), performed at 1200°C for 100 minutes. Given that the solid solubility of boron at this temperature $= 2.3\times10^{20}$ atoms cm^{-3}, diffusion coefficient $= 2.5\times10^{-12}$ cm^2 $second^{-1}$, and $erfc^{-1}$ (0.999 9957) = 3.25.

The required junction depth is obtained by applying the equation

$$N(x,t) = N_s\text{erfc}\left(\frac{x}{2\sqrt{Dt}}\right) \tag{3.12}$$

from which

$$1\times10^{15} = \left(2.3\times10^{20}\right)\text{erfc}\left\{\frac{x}{2\sqrt{2.5\times10^{-12}\times100\times60}}\right\} \tag{3.13}$$

so that

$$\frac{1\times10^{15}}{2.3\times10^{20}} = \text{erfc}\left\{\frac{x}{2\sqrt{2.5\times10^{-12}\times100\times60}}\right\} \tag{3.14}$$

or

$$4.3478261\times10^{-6} = \text{erfc}\left\{\frac{x}{2.45\times10^{-4}}\right\} \tag{3.15}$$

or

$$\text{erfc}^{-1}\left(1-4.3478261\times10^{-6}\right) = \frac{x}{2.45\times10^{-4}} \tag{3.16}$$

or

$$\begin{aligned} x &= 2.45\times10^{-4}\times\text{erfc}^{-1}\left(1-4.3478261\times10^{-6}\right) \\ &= 2.45\times10^{-4}\times\text{erfc}^{-1}\left(0.99999565\right) \\ &= 2.45\times10^{-4}\times3.25 = 7.96\times10^{-4}\text{ cm} = 7.96\ \mu\text{m} \end{aligned} \tag{3.17}$$

Example 3.3

For fabricating a PN junction containing a positive P-side (with excess holes) and negative N- side (with excess electrons), P-type diffusion was done in an N-type silicon wafer of background concentration 1×10^{15} atoms cm^{-3}. The final impurity profile was of the form exp $\{-x^2/(4Dt)\}$, where the junction depth (x) was 2 pm and the time (t) was 1 hour. A surface concentration of 1×10^{18} atoms cm^{-3} was achieved. Find the diffusivity of the dopant and the pre-deposition dose.

From the equation

$$Q = N_s\sqrt{\pi Dt} \tag{3.18}$$

we get

$$\begin{aligned} Q = N_s\sqrt{\pi Dt} &= 1\times10^{18}\times\sqrt{3.14\times D\times1\times60\times60} \\ &= 1\times10^{18}\times\sqrt{3.14\times D\times1\times60\times60} \\ &= 1\times10^{18}\times\sqrt{11{,}304D} \end{aligned} \tag{3.19}$$

Applying the equation

$$N(x,t) = \frac{Q}{\sqrt{\pi Dt}}\exp\left(-\frac{x^2}{4Dt}\right) \tag{3.20}$$

we have

$$\begin{aligned} 1\times10^{15} &= \frac{1\times10^{18}\times\sqrt{11{,}304D}}{\sqrt{3.14\times D\times1\times60\times60}}\exp\left\{-\frac{\left(2\times10^{-4}\right)^2}{4D\times1\times60\times60}\right\} \\ &= \frac{1\times10^{18}\times\sqrt{11{,}304D}}{\sqrt{11{,}304D}}\exp\left\{-\frac{4\times10^{-8}}{1.44\times10^4 D}\right\} \end{aligned} \tag{3.21}$$

yielding

$$\frac{1\times10^{15}}{1\times10^{18}} = \frac{\sqrt{11{,}304D}}{\sqrt{11{,}304D}}\exp\left\{-\frac{4\times10^{-8}}{1.44\times10^4 D}\right\} \tag{3.22}$$

or

$$10^{-3} = \exp\left\{-\frac{4\times10^{-8}}{1.44\times10^4 D}\right\} = \exp\left(-\frac{2.78\times10^{-12}}{D}\right) \tag{3.23}$$

Taking natural logarithm of both sides

$$\ln\left(10^{-3}\right) = -\frac{2.78\times10^{-12}}{D} \tag{3.24}$$

giving

$$D = -\frac{2.78 \times 10^{-12}}{\ln\left(10^{-3}\right)} = -\frac{2.78 \times 10^{-12}}{-6.907755} \tag{3.25}$$

$$= 4.0245 \times 10^{-13}\ \text{cm}^2\ \text{s}^{-1}$$

Substituting for D from Equation 3.25 in Equation 3.19

$$Q = 1 \times 10^{18} \times \sqrt{11{,}304D}$$

$$= 1 \times 10^{18} \times \sqrt{11{,}304 \times 4.0245 \times 10^{-13}} \tag{3.26}$$

$$= 6.745 \times 10^{13}\ \text{atoms cm}^{-2}$$

For phosphorus diffusion, compounds such as PH_3 (phosphine) and $POCl_3$ (phosphorus oxychloride) are used. In the case of P-diffusion, using $POCl_3$ (Figure 3.8), the reactions occurring on the silicon wafer surface are

$$Si + O_2 = SiO_2(\text{silica glass});\ 4POCl_3 + 3O_2 = 2P_2O_5 + 6Cl_2 \tag{3.27}$$

This results in the production of a glassy layer on the silicon wafers, a mixture of phosphorus glass and silica glass called phosphosilicate glass (PSG), which is a viscous liquid at the diffusion temperature. The mobility of the phosphorus atoms in this glassy layer and the phosphorus concentration is such that the phosphorus concentration at the silicon surface is maintained at the solid solubility limit throughout the time of the diffusion process. Solid solubility is the maximum concentration of impurity that can dissolve in silicon under a given temperature and pressure. This diffusion step is referred to as a pre-deposition step in which the dopant atoms deposit into the surface regions (~0.1-μm depth) of the silicon wafers. *What is the use of the PSG layer?* PSG is preferable because it protects the silicon atoms from pitting or evaporating and acts as a "getter" (a material added in small amounts during a chemical reaction to absorb impurities and contaminants). It is etched off before the next diffusion.

The pre-deposition step is followed by a second diffusion process in which the external dopant source (PSG) is removed, such that no additional dopants enter the silicon. During this diffusion process, the dopants that are already present in the silicon move further inside and are thus redistributed. The junction depth increases, and, at the same time, the surface concentration decreases. This type of diffusion is called *drive-in*.

For boron diffusion, a controlled flow of carrier gas (N_2) is bubbled through boron tribromide, which, with oxygen again, produces boron trioxide (B_2O_3) at the surface of the wafers, as per the following reaction:

$$4BBr_3 + 3O_2 = 2B_2O_3 + 6Br_2 \tag{3.28}$$

The result is deposition of a glassy layer on the silicon surface, which is a mixture of silica glass (SiO_2) and boron glass (B_2O_3) called borosilicate glass (BSG). The BSG glassy layer is a viscous liquid at the diffusion temperatures and the boron atoms move around relatively easily. The rest of the process details for boron diffusion are similar to those for phosphorus diffusion.

3.3.7 Ion Implantation

Semiconductor manufacturers today use ion implantation for almost all doping in silicon integrated circuits (ICs). *What are the common implanted species?* The most commonly implanted species are arsenic (As), phosphorus (P), boron (B), indium (In), antimony (Sb), germanium (Ge), nitrogen (N), hydrogen (H), and

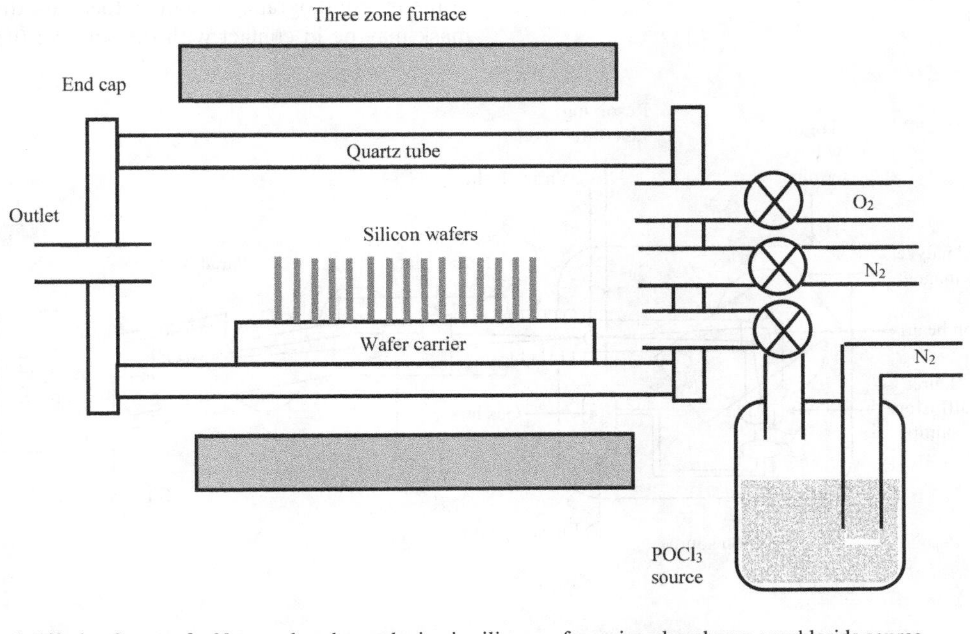

FIGURE 3.8 Thermal diffusion furnace for N-type phosphorus doping in silicon wafers using phosphorus oxychloride source.

helium (He). Ion implantation works by ionizing the required atoms, accelerating them in the electric field, selecting the correct species using an analyzing magnet, and bombarding the substrate with the ion beam in a pre-calculated manner.

The concentration and depth of the desired dopant are specified directly in the equipment settings for implant dose and energy, respectively. Therefore, the two key parameters defining the implant profile are the dose Φ (usually given in atoms cm^{-2}) and energy E (in keV). The dose is related to the beam current I. Typical beam currents and the implantation doses are in the ranges 1 μA–30 mA and 10^{11}–10^{16} atoms cm^{-2}, respectively. The lowest energies used start at the sub-keV area for ultra-shallow junctions and increase to the MeV range for deep wells.

What is the construction of an ion implanter? An ion implantation machine is constructed to deliver a beam of ions of a particular impurity at a specified energy and in a given dose to the surface of a silicon wafer. Figure 3.9 shows the schematic diagram of an ion-implantation machine. A source of gas, such as boron trifluoride (BF_3), feeds a small quantity of source gas into the ion source, in which a heated filament splits up the molecules into charged fragments, forming an ion-plasma containing the desired ions along with other ions and any accruing contamination. A high voltage ~20 kV extracts the ions and pushes them into the analyzer. Pressure in the remaining equipment is maintained at 10^{-6} Torr to avoid collisions and scattering of ions from the gas molecules. The magnetic field of the analyzer is carefully selected to allow the passage of ions of the required charge-to-mass ratio without blockage by the analyzer walls. Thus, the chosen ions of required charge-to-mass ratio continue their onward journey and enter the acceleration tube, where they are accelerated to achieve the desired implantation energy as they move from high voltage to ground. The aperture collimates the ion beam. The X- and Y-electrostatic deflection plates are employed to scan the surface of the wafer with the ion beam. A commercial ion implanter has typical dimensions of 6 m length × 3 m width × 2 m height.

3.3.8 Photolithography

Photolithography is the process of transferring geometric shapes or patterns on a mask to the surface of a silicon wafer coated with a photoresist. *What is a photoresist?* It is a light-sensitive material performing two basic functions, namely precise pattern formation and protection of the substrate from chemical attack during the etching process. Photolithography is the means by which the small-scale features of integrated circuits are created. The steps involved in the photolithographic process are wafer cleaning, barrier layer formation, photoresist application, soft baking (the step during which the solvents are removed from the photoresist), mask alignment, exposure and development, and hard baking (the step to harden the photoresist and improve adhesion of the photoresist to the wafer surface). A diffusion barrier layer is a layer of thermally grown silicon dioxide that blocks the entry of dopant impurities, like phosphorus and boron, into the silicon wafer so that these impurities enter only through the windows in the oxide layer etched after the photolithographic operation.

What is the photolithographic process? A resist is applied to the surface using a spin-coating machine. This device holds the wafer of the semiconductor, using a vacuum, and spins it at a high speed (3000–6000 rpm) for a period of 15–30 seconds. A small quantity of resist is dispensed onto the center of the spinning wafer. The rotation causes the resist to be spread uniformly across the surface of the wafer, with excess resist being thrown off by centrifugal action. Preparation of the resist is completed by a prebake step, where the wafer is gently heated in a convection oven and then a hotplate to evaporate the resist solvent, as already mentioned, and to partially solidify the resist.

The photomask (an opaque plate with transparent regions that allow light to shine through in a defined pattern) used in photolithography is created by a photographic process and developed onto a glass substrate. Chrome (vernacular for chromium) on quartz glass is used for the high-resolution deep UV (electromagnetic radiation between violet light and X-rays, having wavelengths of 200–400 nm) lithography. Depending on the design of the photolithography machine, there are three options: (i) the mask may be in contact with the surface; (ii) very close to the

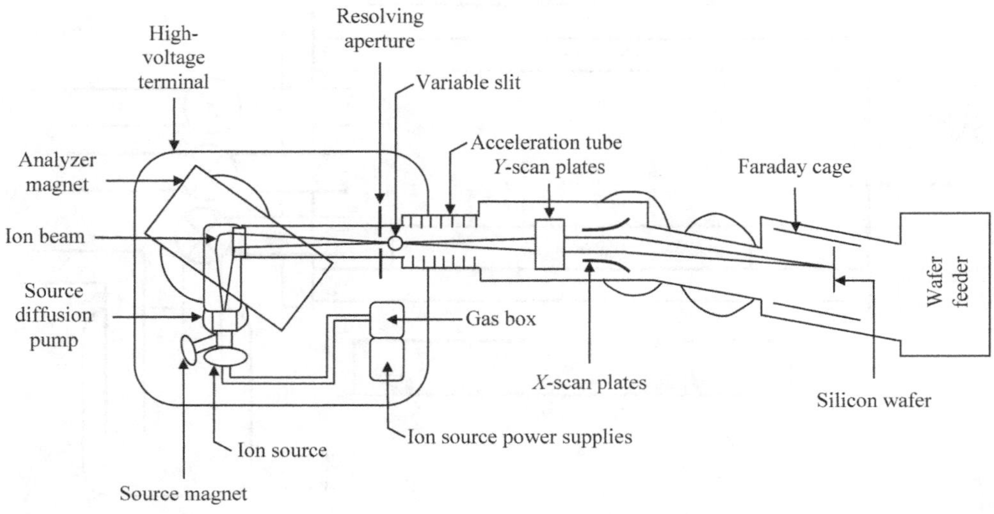

FIGURE 3.9 Parts of an ion-implantation machine.

surface; or (iii) it is used to project the image of the pattern on to the surface of the substrate. These methods are called, not surprisingly, contact, proximity, and projection, respectively. Figure 3.10 shows a schematic diagram of these methods. Option (i) provides good resolution but the mask is spoilt, sometimes excessively damaged, on repeated abrasion with the substrate. Option (ii) increases the life of the mask but the resolution is slightly sacrificed. Option (iii) is good for both the mask and the resolution, and hence is the best solution.

What happens during resist exposure? During the exposure process, the resist undergoes a chemical reaction. Depending on the chemical composition of resist, it can react in two ways when the light strikes the surface. The action of light on a positive resist causes it to become soft and easily removable, where it has been exposed to the light. Alternatively, a negative resist has the reverse property. Exposure to UV light causes the resist to become hard and difficult to remove. After the developing process (in which the exposed photoresist is dissolved), a negative image of the mask remains as a pattern of resist.

What is the purpose of post baking? The wafer undergoes a post-bake process to further harden it and to remove any residue of the developer.

What mechanisms are involved in resist usage? A positive resist consists of two components: a resin (clear to translucent, solid or semisolid, viscous compound that can be hardened with treatment) and a photoactive compound (a dissolution inhibitor), dissolved in a solvent. On UV exposure, it becomes more soluble in an aqueous developer solution. The negative photoresist also consists of two parts: a chemically inactive polyisoprene (monomer isoprene = C_5H_8) rubber, which is the film-forming component, and a photoactive agent. On exposure to UV, the photoactive agent reacts with the rubber to form cross-links (covalent or ionic) that link one polymer chain to another between the rubber molecules, making the rubber less soluble in an organic developer solvent. Therefore, the developer solvent dissolves the unexposed resist whereas the exposed resist swells as the non-cross-linked molecules are dissolved. This swelling causes distortion of the features in the pattern, limiting the resolution to 2–3 times the initial film thickness.

Which resist provides higher resolution, positive or negative, and why? A positive resist gives higher resolution. Because, in a positive photoresist, the unexposed resist regions do not swell much in the developer solution, it is evident that a higher resolution is achievable with a positive photoresist.

3.3.8.1 Physical Limits

The high spatial resolution needed in the image on the wafer necessitates large numerical aperture (NA) lenses. *Resolution* is defined as the ability of a microscope to allow the viewer to distinguish between small objects. It is the shortest distance between two points on a specimen that can be distinguished by the observer or a camera system as separate entities. The *NA* of a lens (a number that expresses the ability of a lens to resolve fine detail in an object being observed) is

$$NA = n \sin\theta \tag{3.29}$$

where

n is the refractive index of the working medium
θ is the half-angle of the maximum cone of light that can enter or leave the lens

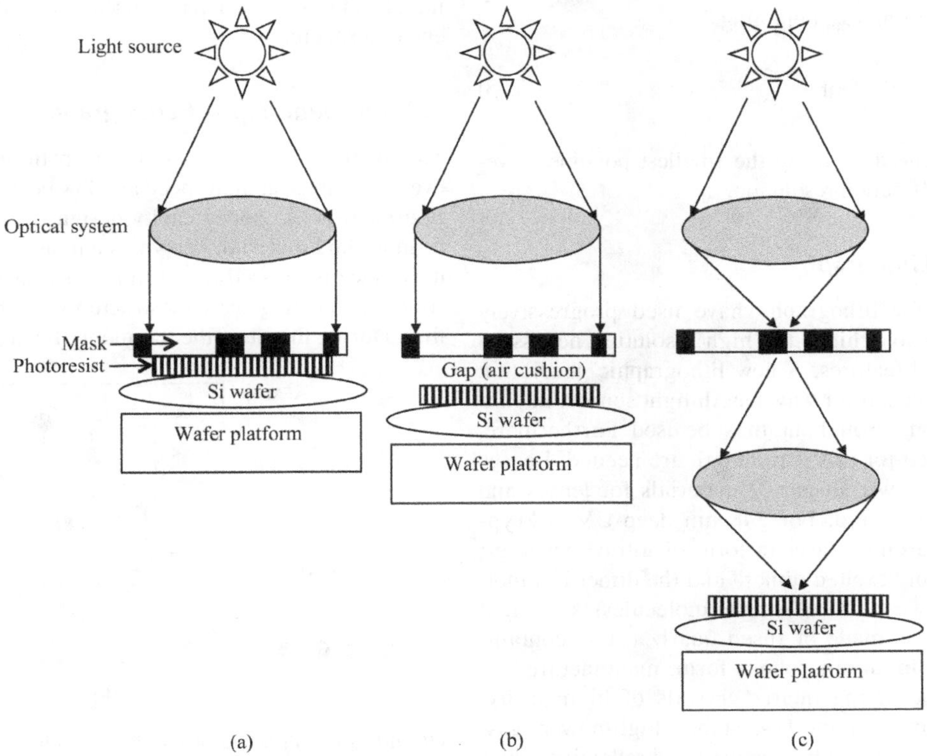

FIGURE 3.10 Photolithographic techniques: (a) contact, (b) proximity, and (c) projection.

Diffraction is the change in the directions and intensities of a group of waves after passing by an obstacle or through an aperture, the size of which is approximately the same as the wavelength of the waves. The *diffraction limit of resolution* on the wafer is

$$x_d = \frac{\eta\lambda}{NA} \quad (3.30)$$

where
λ is the wavelength of the illuminating light for exposure
η is a constant

The diffraction limit should be smaller than the minimum pattern dimension. Computerized lens designs have increased NA to values close to unity and appreciably reduced η, rendering the possibility of fabricating features with sizes considerably smaller than the exposing wavelength. However, using a high NA for increasing resolution results in a small depth of field (DoF: the distance between the nearest and farthest objects that appear in acceptably sharp focus). At a distance d from the focal plane, a point becomes an area of diameter x

$$x = NA \times d \quad (3.31)$$

In lithography, *depth of field* implies the vertical distance from the focal plane at which x is equal to the diffraction limit, $x = x_d$. Thus,

$$\text{DoF} = \frac{\eta\lambda}{NA^2} \quad (3.32)$$

and the lithographic depth of focus decreases as the square of NA. DoF may be looked upon from an alternative viewpoint by eliminating NA from x and x_d, obtaining

$$\text{DoF} = \frac{x_d^2}{\eta\lambda} \quad (3.33)$$

Hence, DoF is increased by using the smallest possible wavelength for a given diffraction resolution, x_d.

3.3.8.2 Optical Lithography

The light sources for lithography have used progressively shorter wavelengths to achieve the high resolution necessary for delineating small features. A new lithographic generation means that a different shorter wavelength light source, intense enough to permit high throughput, must be used. Furthermore, new resists optimized for this wavelength are needed. Lenses are redesigned and new transparent materials for lenses and mask supports are required. For 248 nm deep-UV, a krypton fluoride (KrF) excimer laser (a form of ultraviolet laser; "excimer" is short for "excited dimer" and the dimer is a molecule consisting of two identical simpler molecules) is the light source and the lens is made of fused quartz (glass containing primarily silica in an amorphous form, manufactured by melting naturally occurring quartz crystals of high purity) whereas, for 13.4 nm extreme-UV, stable, high-power, gas-discharge plasma is used as a light source and reflective optics are applied.

3.3.8.3 Electron-Beam Lithography

Exposure to a direct-writing scanning electron beam for producing line widths ~10–20 nm can help in overcoming the problem of light source. Additionally, it relieves the limits associated with resolution and depth of focus, the range of image distances corresponding to the range of object distances included in the DoF. However, E-beam lithography (EBL) suffers from the following disadvantages (i) limited speed, so it is therefore suited for low-volume production; (ii) expensive equipment; and (iii) as pattern generation is serial, for high throughput, parallelism is necessary to simultaneously expose a considerable area, instead of depending solely on the scanning beam.

3.3.8.4 X-Ray Lithography

X-ray lithography (XRL) (0.8–0.1 nm) allows proximity masking, to avoid the use of lenses. But it is restricted by two factors: (i) the lack of X-ray sources of sufficiently high intensity; and (ii) the exposure of resists by secondary electrons, produced by high-energy photons.

3.3.8.5 Dip-Pen Nanolithography

A scanning probe nanopatterning technique, DPN is becoming a workhorse tool for researchers interested in fabricating and studying soft and hard matter on the sub-100 nm length scale. It is a direct-write technique that provides high-resolution patterning for a number of molecular and biomolecular inks on various substrates, such as metals and semiconductors. Here, an AFM tip is used to deliver molecules to a surface *via* a solvent meniscus (the curve in the upper surface of a standing body of liquid) (Figure 3.11). This meniscus naturally forms in the ambient atmosphere.

3.3.8.6 Nanoimprint Lithography

This method, having the ability to pattern sub-25-nm patterns over large areas at high speed and low cost, comprises two steps. *Imprint step*: A shaped cavity designed used to give a definite form to fluid material, known as a mold with nanostructures on its surface is pressed into a thin resist layer coated onto a substrate, duplicating the nanostructures on the mold in the resist film. During this step, the thermoplastic (a polymer that turns to

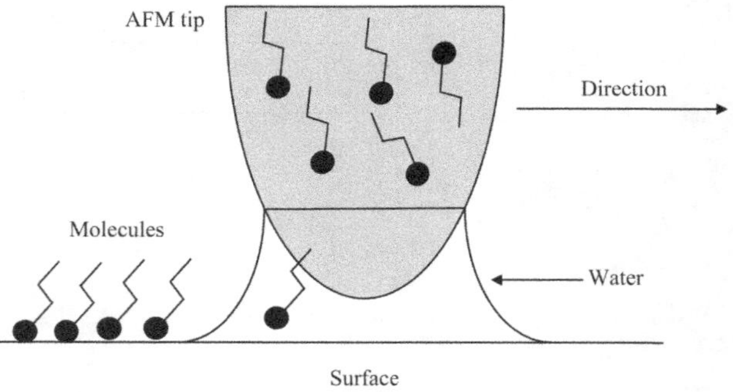

FIGURE 3.11 Principle of DIP-PEN nanolithography. (Khanna, V. K., *Proceedings of IMS-2007, Trends in VLSI and Embedded Systems*, Punjab Engineering College, Chandigarh, 19, 2007.)

a liquid when heated and freezes to a glassy state when cooled) resist is heated to a temperature above its glass transition temperature (the critical temperature at which the material changes its behavior from being "glassy" to being "rubbery") so that it can flow and can be readily deformed into the shape of the mold. *Pattern transfer step*: An anisotropic (directionally dependent) etching process, such as reactive-ion etching (RIE, see Section 3.3.11), is used to remove the residual resist in the compressed area, transferring the thickness contrast pattern into the entire resist.

3.3.8.7 Nanosphere Lithography

Localized surface plasmon resonance (LSPR) and surface-enhanced Raman scattering (SERS) are optical nano-sensing techniques. Nanoparticle size, shape, interparticle spacing, nanoparticle–substrate interactions, solvent, dielectric overlayers, and molecular adsorbates overwhelmingly affect the LSPR spectrum of nanoparticles. In addition, the SERS technique is sensitive to the working surface. This surface is often prepared with a distribution of metal nanoparticles. The shape and the size of such particles strongly affect the factor of increases in Raman scattering, because they change absorption and scattering rates on the surface.

Photolithography has not been widely applied to nanostructure fabrication in view of its diffraction-limited resolution. Nanostructure fabrication processes that have hitherto been used are EBL and XRL. EBL is able to achieve a feature size as small as 1–2 nm but is not adaptable for mass-volume fabrication, because it is not a parallel process. On the other hand, XRL has achieved a comparatively higher production volume but has resolution limits inferior to those of EBL. Nanosphere lithography (NSL) (Haynes and Van Duyne 2001) is an inexpensive, simple-to-implement, inherently parallel, high-throughput, nanostructure fabrication process, which systematically produces a two-dimensional (2D) array of periodic structures. It is a powerful fabrication technique, generating nanoparticle arrays with controlled shape, size, and interparticle spacing, using self-assembled polystyrene $[(C_8H_8)_n]$ nanospheres as templates to replace photoresist masks for fabrication of nanostructures on various substrates. The NSL masks are created by spin-coating polystyrene nanospheres on the substrate of interest. After deposition of relevant material, that is, Ag or Au, the polystyrene nanospheres are removed from the substrate by dissolving them in CH_2Cl_2, with the aid of sonication.

3.3.9 Chemical Vapor Deposition

CVD has already been described earlier (Section 3.2.5), in the context of carbon nanotubes. It is a very important process in silicon technology for depositing layers like polysilicon, silicon dioxide (SiO_2), silicon nitride (Si_3N_4), etc., and will be discussed here in that context. CVD is a generic name for a group of processes that use a chamber of reactive gas for depositing a solid material from a gaseous phase through a chemical reaction of vapor-phase precursors. Precursor gases (often diluted in carrier gases) are delivered into the reaction chamber at approximately ambient temperature. As they pass over or come into contact with a heated substrate, they undergo reaction or decomposition, forming a solid phase, which is deposited onto the substrate. The substrate temperature is critical and influences the reactions taking place.

What processes of semiconductor technology fall under the umbrella of CVD? CVD covers processes such as (i) atmospheric pressure CVD (APCVD); (ii) low-pressure or ultrahigh-vacuum CVD (LPCVD [at sub-atmospheric pressures, 10–100 Pa] or UHVCVD [10^{-6} Pa]; (iii) metal-organic CVD (MOCVD); (iv) plasma-enhanced CVD (PECVD); and (v) atomic layer deposition CVD (ALDCVD).

APCVD is performed in a reactor at temperatures up to ~400°C. As it is carried out at atmospheric pressure, it generally results in inferior film quality and conformality (the retention of angular relationships at each point) of coating. The LPCVD process (Figure 3.12) produces layers with excellent uniformity of thickness and material characteristics. The main problems with the LPCVD process are the high deposition temperature (greater than 600°C) and the relatively slow deposition rate. The PECVD process operates at lower temperatures (down to 300°C or less), due to the extra energy supplied to the gas molecules by the plasma in the reactor. However, the quality of the deposited films tends to be inferior to those from processes running at higher temperatures. In MOCVD, atoms required in the crystal are combined with complex organic gas molecules and passed over a hot semiconductor wafer. The heat breaks up the molecules and deposits the desired atoms on the surface, layer by layer. By varying the composition of the gas, the properties of the crystal are manipulated at an almost atomic scale. The reaction for GaN growth is

$$Ga\left(CH_3\right)_3 + NH_3 \rightarrow GaN + 3CH_4 \tag{3.34}$$

In what way is ALD a modification of CVD? ALD, sometimes called atomic layer epitaxy (ALE), pulsed CVD, or atomic layer chemical vapor deposition (ALCVD), is a modified version of the CVD process, having the following salient features: (i) gaseous precursors are introduced sequentially to the substrate surface (Jones and Hitchman 2009); and (ii) the reactor is purged with an inert gas, or evacuated, which removes any excess precursor molecules and volatile by-products from the reaction chamber, evaporating readily at normal temperatures and pressures, thus eliminating the possibility of occurrence of unwanted gas phase reactions.

How does ALD differ from conventional CVD? In marked contrast to traditional thermal CVD, which involves pyrolysis (thermolysis: thermochemical decomposition of materials at elevated temperatures in the absence of oxygen) of precursor molecules, ALD proceeds through surface exchange reactions, such as hydrolysis (subject to the chemical action of water), between chemisorbed metal-containing precursor fragments and adsorbed nucleophilic reactant molecules; a *nucleophile* is a chemical compound or group that is attracted to nuclei and tends to donate or share electrons. It must be noted that the chemical reactions leading to film deposition in ALD take place exclusively on the substrate at temperatures below the thermal decomposition temperature of the metal-containing precursor, with the gas-phase reactions being unimportant.

By CVD, various films required in CMOS process are deposited, notably polysilicon and different dielectric films.

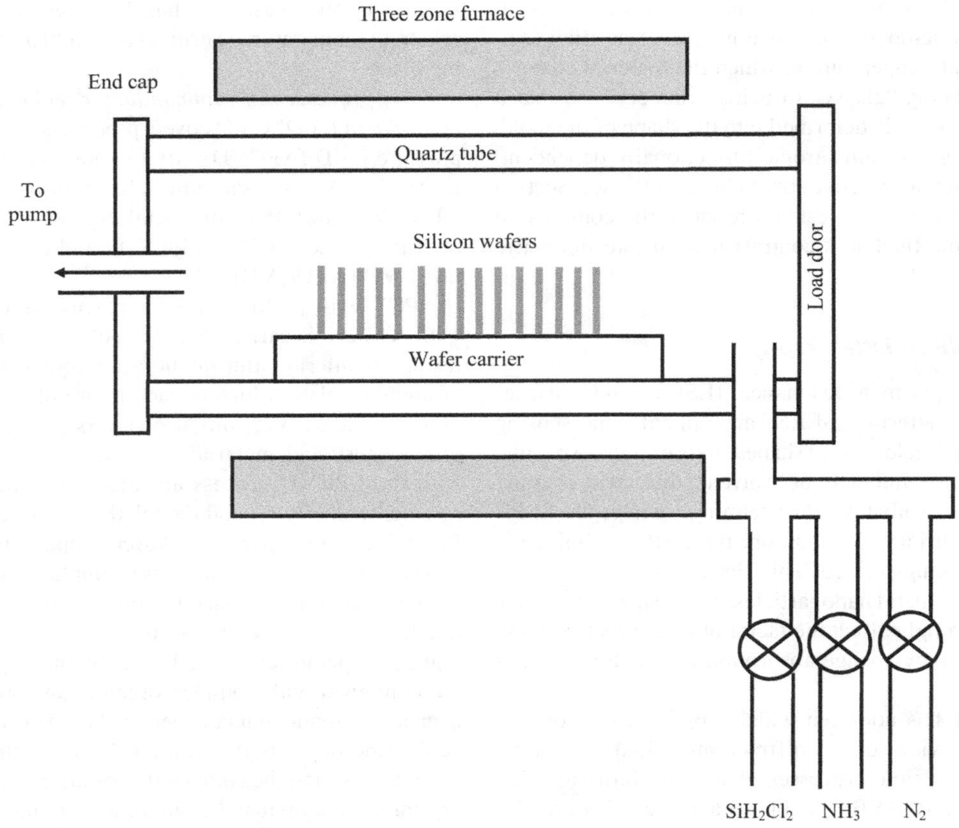

FIGURE 3.12 Low-pressure chemical vapor deposition (LPCVD) furnace for silicon nitride deposition on silicon wafers, using dichlorosilane and ammonia.

Polysilicon deposition reaction

$$SiH_4 \rightarrow Si + 2H_2 \left(575°C - 650°C\right) \quad (3.35)$$

Silicon dioxide deposition reactions

$$\begin{aligned} &SiH_4 + O_2 \rightarrow SiO_2 + 2H_2 (< 500°C);\ Si\left(OC_2H_5\right)_4 \rightarrow SiO_2 \\ &\quad + \text{by-products}\left(650°C - 750°C\right); \\ &SiCl_2H_2 + 2N_2O \rightarrow SiO_2 + 2N_2 + 2HCl \end{aligned} \quad (3.36)$$

SiH_4 is silane, $Si(OC_2H_5)_4$ is tetraethoxysilane (TEOS), and $SiCl_2H_2$ is dichlorosilane.

Silicon nitride deposition reactions

$$\begin{aligned} &3SiH_4 + 4NH_3 \rightarrow Si_3N_4 + 12H_2 \left(700°C - 900°C\right); \\ &3SiCl_2H_2 + 4NH_3 \rightarrow Si_3N_4 + 6HCl + 6H_2 \left(700°C - 800°C\right) \end{aligned} \quad (3.37)$$

What about the higher dielectric constant (high-κ) film deposition? The silicon oxynitride is the starting point toward the use of a material with high-κ and possessing satisfactory insulating properties as the gate insulator. It enables the reduction of the gate tunneling current, using materials compatible with silicon technology, and combining SiO_2 and silicon nitride (Si_3N_4) in the insulating layer. The results achieved depend on the processing of the mixture. Various mixtures of the form SiN_xO_y, where $3x+2y=4$, known as *oxynitrides*, are used as the gate insulator. A gate–substrate capacitance, equivalent to an oxide insulator about 2 nm thinner, is obtained for a given tunneling current density. Another method superimposes a layer of silicon nitride above a thin SiO_2 layer as the gate insulator, thereby preserving the desirable qualities of the contact of SiO_2 on silicon. The nitride has a dielectric constant of 7.0, as compared with 3.9 for the oxide, and the use of silicon nitride is well established in silicon technology.

What are the other dielectrics available? As already stated, other dielectrics include hafnium and zirconium silicates and oxides that are typically deposited using ALCVD. The dielectric constant of HfO_2 and SrO_2 is ~20–25, which is 5–6 times that of silicon dioxide. Rare-earth (RE) oxides such as crystalline praseodymium oxide (Pr_2O_3), deposited by MBE, have been successfully introduced into a CMOS process. RE elements are a collection of 17 chemical elements in the periodic table, specifically the 15 lanthanoids plus scandium and yttrium. From capacitance-voltage measurements, a dielectric constant of $\kappa=36$ has been calculated.

3.3.10 Wet Chemical Etching and Common Etchants

This is the simplest etching technology, using a container with a liquid solution that will dissolve the material in question. Buffered HF, a mixture of a buffering agent, such as ammonium fluoride (NH_4F), and hydrofluoric acid (HF), is commonly used for etching silicon dioxide selectively, whereas orthophosphoric acid (H_3PO_4)-based compositions are employed for aluminum etching. An etchant for gold is potassium iodide (KI) in iodine (I_2). Silicon etchants are mixtures of HF, HNO_3, and acetic acid (CH_3COOH).

3.3.11 Reactive Ion Etching

Can etching be done without liquid etchants? Yes, reactive-ion etching (RIE) is a process that uses dry chemical and physical processes to etch away desired materials. *What is the essence of RIE?* In RIE, the substrate is placed inside a reactor, in which a mixture of gases is introduced (Figure 3.13). A plasma (a gas of positive ions and free electrons containing approximately equal positive and negative charges) is struck in the gas mixture, using an RF (radio frequency) power source, breaking the gas molecules into ions. The ions are accelerated toward and react at the surface of the material being etched, forming another gaseous material. This is known as the *chemical part of RIE*. There is also a *physical part*, that is similar in nature to the sputtering deposition process. If the ions have sufficiently high energy, they can knock atoms out of the material to be etched without a chemical reaction. This dislodging of atoms by impact or collision constitutes the physical part of the etching. *Synergy between the chemical reaction and ion bombardment leads to high etch rates.*

How does RIE compare with wet etching? Compared to wet etching, RIE has higher anisotropy (having dissimilar properties in different directions), greater uniformity and control, and improved etch selectivity.

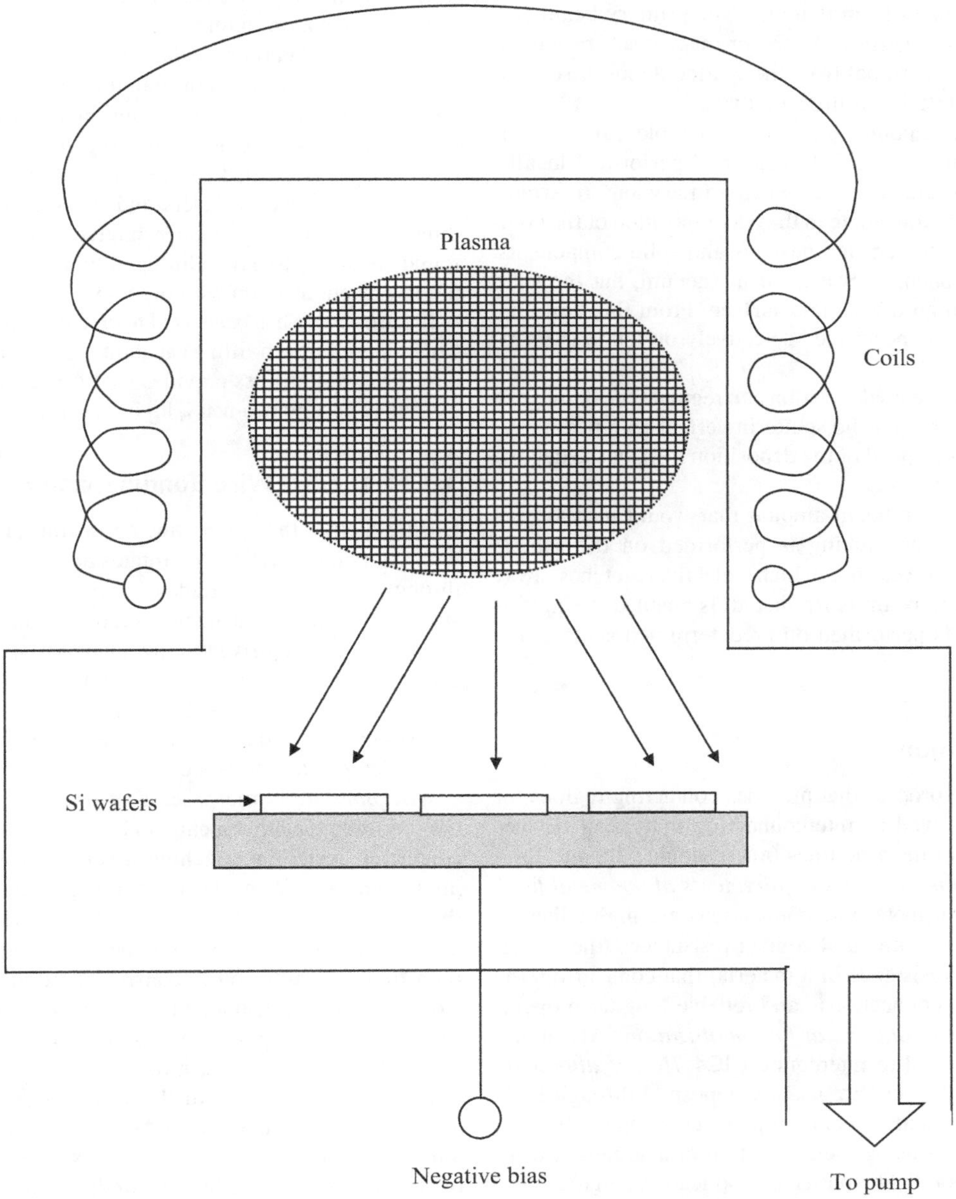

FIGURE 3.13 Transformer-coupled plasma RIE system.

3.3.12 Focused Ion Beam Etching and Deposition

Focused ion beam (FIB) technology is a widely used tool for microelectronics. An FIB system, using liquid metal ion sources, is capable of forming very small probes with high current densities. The ions strike the specimen to remove materials through a physical sputtering process. A combination of a FIB and a reactive gas also causes various physical effects on the specimen. Gas molecules above a specimen may be dissociated, resulting in a local etching or deposition of the specimen.

FIB systems operate in a similar fashion to a scanning electron microscope (SEM), with the exception that, instead of a beam of electrons, FIB systems use a sharply focused beam of gallium (Ga^+) ions, that is operated at low beam currents for imaging or high beam currents for site-specific sputtering or milling. The gallium (Ga^+) primary ion beam strikes the sample surface and sputters a small amount of material, which leaves the surface as either secondary ions or neutral atoms. The primary beam also produces secondary electrons. As the primary beam rasters on the sample surface, the signal from the sputtered ions or secondary electrons is collected, forming an image.

By spraying a compound gas on the sample surface near the ion beam irradiation area, deposition is performed locally. Secondary electrons are generated when primary ions are irradiated. Secondary ions contribute to the decomposition of the compound gas, which splits up into gaseous and solid components. The gaseous component is evacuated in vacuum, but the solid component piles up on the sample surface. From this build-up, maskless deposition is performed selectively on ion beam irradiation areas.

Although electron beam deposition can regulate the damage to the sample, because there is no sputtering effect with beam irradiation, the formation speed of the deposition film, or the deposition rate, appears to be low.

By increasing the ion beam amount that would increase the sputtered atom amount, etching is performed on the sample surface. Maskless processing, which selectively etches away those parts where ion beam is irradiated, is enabled. Using this technique, etching is performed on predetermined spots of the sample.

3.3.13 Metallization

Metallization is the process that provides contacting regions for separate devices as well as interconnecting individual devices together by means of metallic lines (microscopic wires) to form circuits. *What are the principal requirements of the metal film?* Desirable features of metal interconnections are high adhesion to the substrate, low ohmic and contact resistances (the contributions to the total resistance of a material that comes from the electrical leads and connections), and reliable long-term operation. *Which metal is most popular for metallization?* Aluminum is a popular metal used to interconnect ICs. *How is aluminum deposited?* Al metal layers are usually deposited through PVD by sputtering. In a magnetron (a high-powered vacuum tube that generates microwaves using the interaction of a stream of electrons with a magnetic field) sputtering application (Figure 3.14), the high voltage is delivered across a low-pressure gas (usually argon) to create high-energy plasma. This plasma emits a colorful halo of light often referred to as a "glow discharge" and consists of electrons and gas ions. These energized plasma ions strike a target composed of the desired coating material. The force causes atoms to eject from the target material, and strike and bond with those of the substrate. Because sputtering takes place in a high-energy environment, it creates a virtually unbreakable bond between the film and its substrate at the atomic level, creating one of the thinnest, most adherent, most uniform, and cost-effective films possible. Another method used for metallization is electron beam (E-beam) evaporation, in which a beam of electrons emitted by a tungsten filament under high vacuum is focused on the source material.

The replacement of aluminum (resistivity of $2.65 \times 10^{-6}\ \Omega$ cm), long used for wiring of integrated circuits, with lower-resistivity copper ($1.67 \times 10^{-6}\ \Omega$ cm) to reduce RC (resistive-capacitive) has been successful. Copper is also much less susceptible to electromigration, the movement of atoms carrying large electrical currents, than aluminum.

Are copper interconnections made in the same way as aluminum? No, the process of manufacturing of copper (Cu) interconnections is not the same as aluminum-based interconnections because of possible oxide poisoning and diffusion problems. A barrier layer is needed to prevent the diffusion of copper into neighboring dielectric layers and the silicon substrate. A cladding around on-chip copper interconnects is formed. It was found necessary to have a lining of tungsten (W) around the copper to prevent reaction with the SiO_2.

Electroplating has been used to create copper interconnects due to its superior trench-filling capability, greater reliability, and lower cost. The IBM process provides void-free copper electro-filling in extremely narrow geometry, high aspect ratio IC structures.

3.3.14 Dicing, Wire Bonding, and Encapsulation

During dicing, the wafer saw, consisting of a blade embedded with diamond particles that rotates at a very high speed, passes through the wafer at boundaries between dies known as *saw streets*, established during wafer fabrication. The dicing machine is programmed to drive the saw blade through the saw streets at a defined spindle speed, saw rate, and depth, separating the wafer into individual dies. The blade must be carefully aligned with saw streets to avoid any chip loss at this stage, where the wafer has undergone all the processes.

Wire bonding is the process of providing electrical connection between the silicon chip and the external leads of the semiconductor device by attaching a very thin wire, usually 25–75 μm in diameter, from one connection pad to another, completing the electrical connection in the electronic device. The most frequently used method of joining the wires is *ultrasonic welding*; a high-frequency ultrasonic acoustic vibration is locally applied to workpieces being held together. *Thermocompression wire bonding* requires the application of a high force on the surface along with a high temperature, around 300°C. It provides excellent, reliable Al-Au bonds with flexibility in the bonding direction allowed. The wire material is Au but the pad can be either Au or aluminum. Depending on their shapes, there are two basic types of wire bonds: ball/wedge and wedge/wedge. About 90% of all assemblies are produced using ball/wedge bonds and about 10% are done with wedge/wedge bonds.

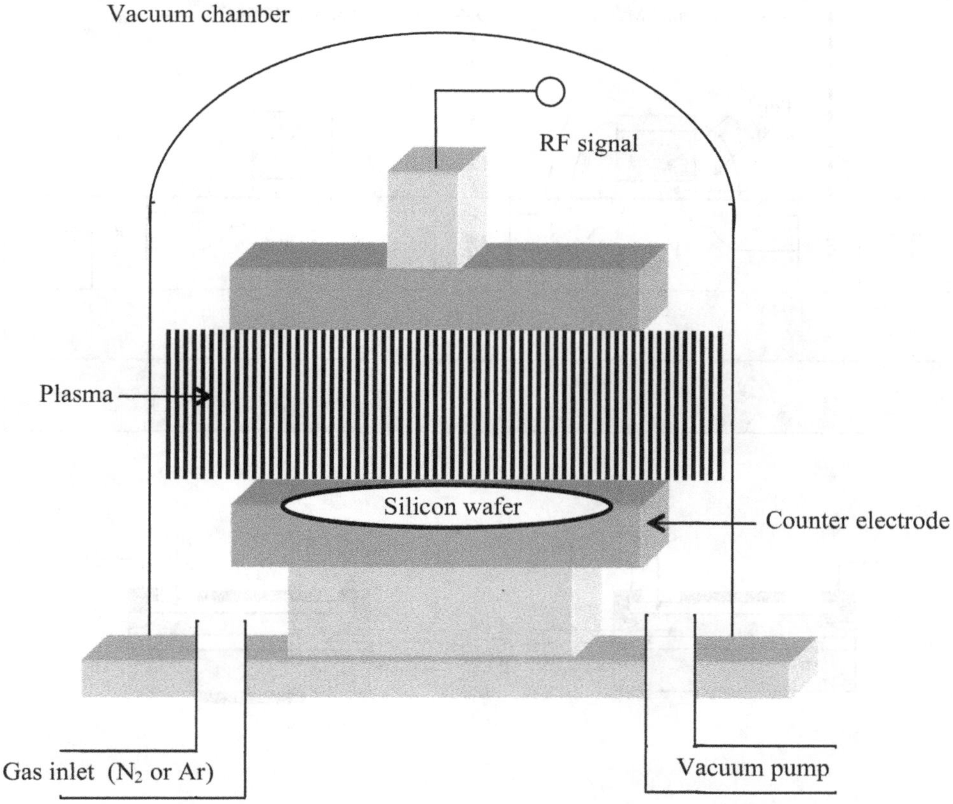

FIGURE 3.14 RF sputtering system.

Package is a case or molding forming the body of a commercial semiconductor device. *Integrated circuit encapsulation* refers to the design and manufacturing of protective packages for these circuits. IC packaging materials include plastics, ceramics, laminates (materials that are constructed by uniting two or more layers of material together), metal, etc. IC package categories are pin-through-hole (PTH), where pins are inserted through holes in the circuit board, and surface-mount-technology (SMT), in which packages have leads that are directly soldered to the exposed metal lands on the surface of the circuit board.

Reliable operation of nanosensors in harsh environments, such as elevated temperatures, high pressure, and aggressive chemical media, poses stringent requirements on packaging, which cannot be met using polymer-based technologies. Ceramic technologies, especially low-temperature co-fired ceramics (LTCC), an evolution of thick-film technology, where the multilayer substrate is co-fired with the other layers (resistors, conductors, etc.), offer a trustworthy platform on which to build stable and reliable packages. LTCC allows fabrication of complicated 3D structures, with fluidic channels for liquids and/or gases, while simultaneously permitting 3D electrical circuits in the same device.

3.3.15 IC Downscaling: Special Technologies and Processes

3.3.15.1 Downscaling Trends

During the past several decades, CMOS (complementary metal-oxide semiconductor) integrated circuits have been at the frontiers of research and development globally. Research activities in this field have expanded exponentially and a revolution is apparent, leading to reliable, low-cost ICs for consumer electronics, computer, and telecommunication sectors (Khanna 2004; 2007b). Deep submicron (DSM) technology involves the use of <0.35 μm (<350 nm) and ultra-deep submicron (UDSM) <0.25 μm (<250 nm) feature sizes in ICs. State-of-the-art nano-CMOS. technologies work in the 22 nm and 10 nm nodes which will be extended to 5 nm node (Radamson et al 2019). With reference to the *International Roadmap for Devices and Systems*: 2017 Edition, logic industry node range labelling is done as 5 nm (2021), 3 nm (2024), 2.1 nm (2027), 1.5 nm (2030) and 1.0 nm (2033); year of production is enclosed in brackets. Although the resolution of optical lithography is gradually increasing, known or proven methods for the anticipated ultra-small structure sizes are lacking. EBL or XRL can meet this goal but low-cost solutions for mass production are still elusive.

3.3.15.2 SOI-MOSFETs

The monumental growth of silicon technology is built on the foundation of new fabrication technologies and materials. Silicon-on-insulator metal-oxide-semiconductor field-effect transistors (SOI-MOSFETs) (Figure 3.15) offer advantages such as easy device isolation, absence of latch-up (a failure mechanism in the form of a particular type of short circuit that occurs in an improperly designed device or circuit), extremely low junction capacitance, improved transconductance, better subthreshold slope, suppression of the short-channel effect (an effect whereby

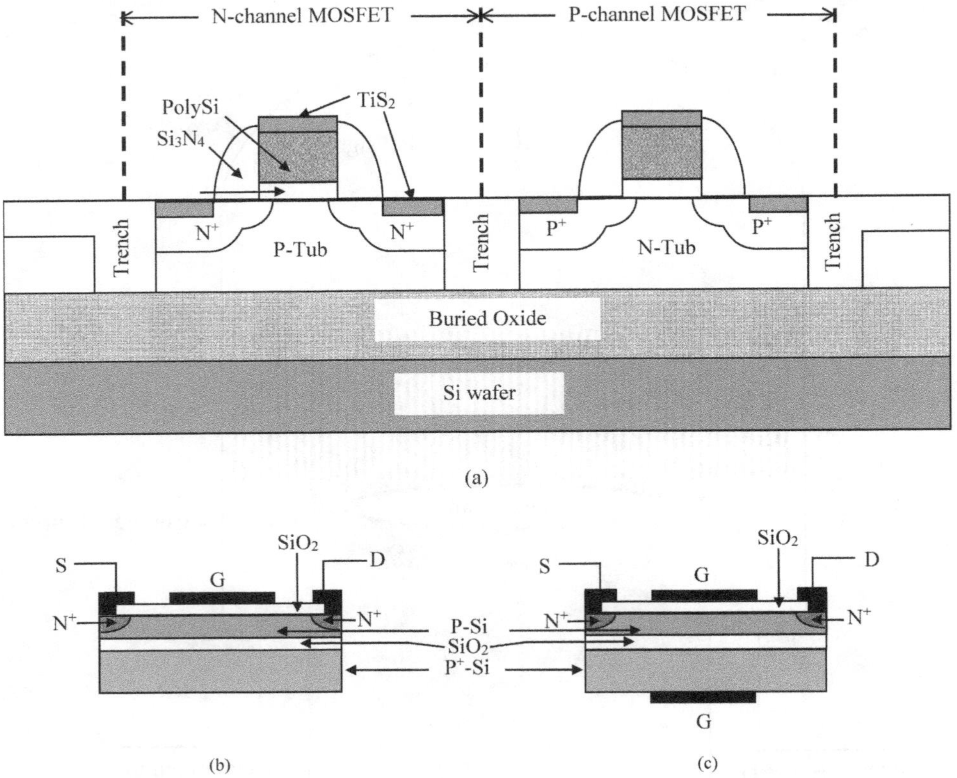

FIGURE 3.15 (a) Trench-isolated SOI CMOS structure, (b) single-gate N-channel FET, and (c) double-gate N-channel FET. (Khanna, V. K., *Proceedings of IMS-2007, Trends in VLSI and Embedded Systems*, Punjab Engineering College, Chandigarh, 19, 2007.)

a MOSFET, in which the channel length is the same order of magnitude as the depletion-layer widths of the source and drain junctions, behaves differently from other MOSFETs), and high-temperature operation capability, such as 350°C. The isolation advantage makes the SOI process sequence free from trench- or well-formation steps. Two methods of making wafers for the SOI devices commonly used in practice are *SIMOX* (separation by the implantation of oxygen) and *smart cut* processes.

3.3.15.3 SIMOX Process

The SIMOX method employs oxygen ion implantation to produce a high concentration of oxygen in a layer beneath the silicon surface. High-temperature annealing of the implanted wafer causes crystallization of a silicon layer at the surface that is separated from the bulk by a layer of SiO_2. The buried insulator is called *buried oxide* (BOX).

3.3.15.4 Smart Cut Process

In the Smart Cut Process, a layer of SiO_2 is formed on a silicon wafer, A. Proton implantation through the oxide layer formed into silicon produces a damaged layer at the end of the penetration range of protons. The oxidized surface is bonded to another silicon wafer, B. The bonded wafer pair is heated so that the two wafers split along the hydrogen-implanted plane; this procedure is termed *hydrogen slicing*. Wafer B becomes the substrate of the SOI wafer and the remaining wafer A is reused as a substrate for another SOI wafer. Wafer B is annealed and polished to obtain the required SOI wafer for device fabrication.

3.3.15.5 Strained Silicon Process

Stresses influence the bandgap and carrier mobility (drift velocity per unit applied electric field) in silicon. The pursuit of high mobility has led to the introduction of germanium into silicon. Silicon and germanium are miscible elements. They form alloys with a range of lattice parameters that increases from silicon to germanium. Silicon is grown epitaxially (e.g., by using metal-organic vapor phase epitaxy, MOVPE) on a Ge-Si alloy with a larger lattice constant (the constant distance between unit cells in a crystal lattice) and strained in such a way that current in the plane of the channel is carried by high-mobility, low-mass electrons. In particular, charge carriers in germanium have a smaller effective mass and greater mobility than those in silicon. Thus, silicon technology has returned to a thin layer of germanium as a means to obtain higher mobility.

The *IBM strained silicon process*, called the *dual stress liner*, enhances the performance of both types of semiconductor transistors, called N-channel and P-channel transistors, by stretching silicon atoms in one transistor and compressing them in the other. *Strained silicon* is a layer of Si, in which the Si atoms have been stretched beyond their normal interatomic distance. Because of the natural tendency for atoms inside compounds to align with one another, the atoms in the silicon layer align with those in the SiGe layer, where the atoms are farther apart. Hence, the links between the silicon atoms become extended, leading to strained silicon.

In strained silicon, the interatomic spacing in the plane of the wafer is greater than in unstrained Si. Biaxial (having two axes) distortion of the Si lattice alters its electronic band structure. Hence, electron and hole mobilities increase. Electrons move at 70% faster speeds, so that strained silicon devices can switch 35% faster.

It may be noted that the introduction of germanium into the critical areas of the integrated circuits provides an alternative means of improving chip performance from the traditional method of simply shrinking circuitry. This is becoming increasingly important as further miniaturization becomes more difficult and yields diminishing returns. Future silicon technology is likely to depend on locally strained silicon channels.

3.3.15.6 Top-Down and Bottom-Up Approaches

The *top-down route* of silicon technology had been running easily until its basic step of optical lithography met its physical limits (minimum feature size around that of the light wavelength). Quest for new technological options has begun. Nanotechnology has provided the answer by developing techniques to handle individual molecules. In comparison with the present downscaling of CMOS ICs, it is necessary to reverse the process flow, to a bottom-up approach.

The *bottom-up approach* is nonlithographic, in contrast to the top-down approach using expensive photolithographic, thin-film deposition, etching, and metallization processes. Its main tools are chemical synthesis and self-assembly. *Inexpensive chemistry* is used to promote self-assembly of complex mesoscopic architectures; the *mesoscopic* world lies in between the microscopic and the macroscopic world. Further, in molecular components made by organic synthesis, smaller variations are observed in structural parameters and chemical composition. Molecular species are tailored for different device applications through *surface molecular engineering.*

Molecular electronics is the stage of electronics in which the data-handling device will be coincident with the constituting material (the molecule). Instead of imagining about the ideal situation of total molecular electronics, the hybrid molecular electronics approach, in which microelectronic circuits host molecular devices grafted to the silicon surface, seems the most realistic one for future electronics. Thus, top–down and bottom–up approaches will not be antagonistic but complementary.

3.3.15.7 DNA Electronics

What striking qualities of DNA (deoxyribonucleic acid) molecules favor their application in electronics? Organic molecules, like DNA, can be used either as active components through which electron current flows, such as an active channel of an field-effect transistor (FET) or quantum device, or as passive devices, such as the charge storage medium of a memory. DNA also possesses complementarity, which is a *self-assembling ability.* DNA forms a self-assembled structure that always has hydrogen-bonded base pairs of guanine ($C_5H_5N_5O$) and cytosine ($C_4H_5N_3O$), and of adenine ($C_5H_5N_5$) and thymine ($C_5H_6N_2O_2$), the four bases found in DNA and represented by the letters G-C-A-T. This structure can be controlled by programming the base sequence. Due to the self-assembling property of DNA, it can be highly integrated, error-free, without using microfabrication technologies. In addition, DNA has an address with 0.34–0.36 nm intervals. It serves as an information material that can be used as a template for aligning molecules and clusters at a nanoscale without any ultrafine processing. DNA is also an insulator, and periodic DNA can be a wide-bandgap semiconductor. Since the electric conduction in DNA is controlled by doping, doped DNA can be regarded as a nanoscale semiconducting molecular wire. In the atmosphere, the electric conduction by DNA is dominated by ion conduction in the water layer adsorbed onto it.

Thus, DNA serves as a nanoscale electronics material that works as an electronic transport material in 1D molecular wires, if used in a vacuum, and as an ion transport material, if used in the atmosphere.

3.3.15.8 Spintronics

Spin-based electronics, often called *spin-electronics* or *spintronics*, employs the spin degrees of freedom in solid-state systems (Tanaka 2005). Spintronics harnesses an electron's spin (a quantum property with either an "up" or a "down" orientation) for data encoding and processing. *What types of controlling fields are used in electronics and spintronics?* Whereas electronics uses electrical fields to push electrons for reading information in a current, spintronics uses *magnetic fields* to transport electrons. Spintronics could lead to the development of computers using less power and with greater data storage capability.

The spin MOSFET consists of a MOS gate with an Si channel (Figure 3.16). The source and drain materials are half-metallic ferromagnets with 100% spin polarization at the Fermi energy, implying that carriers are completely spin polarized. The source/channel and drain/channel contacts are Schottky barriers (potential barriers formed at a metal–semiconductor junction that have rectifying characteristics). They allow efficient spin injection and detection, and also act as blocking contacts for stopping the drain–source current in the off-state of FET. Spin injection utilizes the

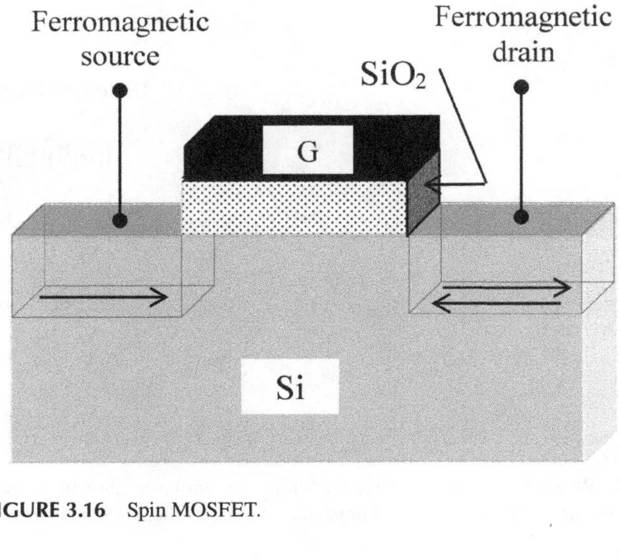

FIGURE 3.16 Spin MOSFET.

strong, short-range quantum mechanical exchange interaction of the injected spin-polarized electrons with the atomic spins on the atoms. From the half-metallic source, highly spin-polarized carriers are injected by tunneling across the Schottky barrier into the Si channel. The half-metallic drain selectively extracts the spin-polarized carriers from the channel. This happens only when the spin configuration between the ferromagnetic source and drain is parallel. Hence, the output current depends on the relative magnetization configuration of the source and the drain. Since the Si channel can be intrinsic, P-type, N-type, or both P-type and N-type, spin MOSFETs, and, therefore, CMOS logic gates can be designed. These are compatible with the CMOS technology and show promise for computing applications.

3.4 MEMS and NEMS Division

3.4.1 Surface and Bulk Micromachining

Surface and bulk micromachining processes are used to create microstructures in MEMS devices (Figure 3.17) (Khanna 2007a). *How do these processes differ?* The difference between surface and bulk micromachining is that, instead of etching the silicon substrate (as done in bulk micromachining), surface micromachining etches away layers deposited on top of the silicon substrate. Surface micromachining combines structural materials and sacrificial layers for making MEMS structures on the silicon surface. The process starts with a silicon wafer, upon which structural and sacrificial layers are deposited. *What are structural and sacrificial layers? Structural layers* are the layers that form the desired structures. *Sacrificial layers* are the layers that are etched away and are used to support the structural layers, until they are etched away. Typically, a sacrificial layer of silicon dioxide is formed by a combination of thermal and CVD processes. Phosphosilicate glass (PSG) is also often used as a sacrificial layer because of its high etch rate in hydrofluoric acid (HF). After the polysilicon structural layers are selectively deposited on top of the sacrificial layer, the silicon dioxide is etched away using HF. This process, in which polysilicon is the commonly used structural material, while SiO_2 and related glasses like PSG act as the sacrificial material, that is etched away at a later stage of the process, is useful for creating cantilever beams, bridges, and sealed cavities.

Bulk micromachining selectively etches the silicon substrate to create microstructures in MEMS devices. Typically, a layer of silicon dioxide is patterned onto a silicon substrate, using a mask. The silicon dioxide is patterned to protect certain areas of the silicon substrate from etching. Thus, in bulk micromachining, 3D structures are etched into the bulk of silicon, in contrast to building the features, layer by layer, on the surface of a silicon wafer.

3.4.2 Machining by Wet and Dry Etching Techniques

Etching is a process in which material is removed from selected regions of the substrate (Khanna 2007c). In *wet etching*, liquid etchants are used to remove material. Wafers are immersed in

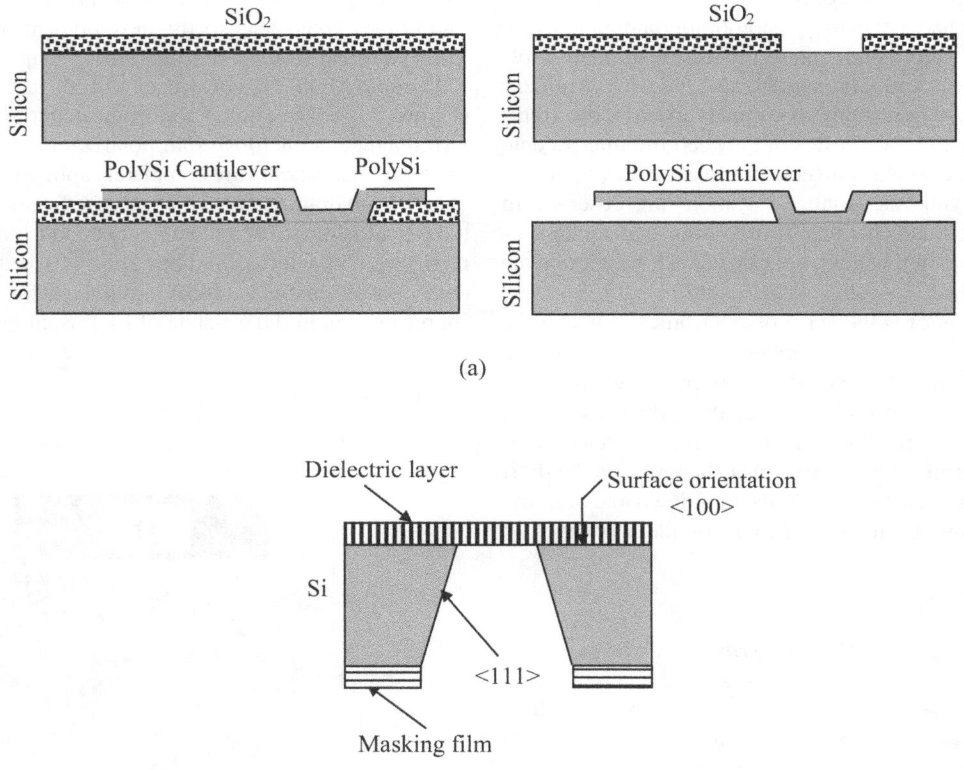

FIGURE 3.17 (a) Surface micromachining for cantilever fabrication, and (b) bulk micromachining. (Khanna, V. K., *Proceedings of IMS-2007, Trends in VLSI and Embedded Systems*, Punjab Engineering College, Chandigarh, 317, 2007.)

a tank containing the reactants. Wet chemical etching is of two types: isotropic and anisotropic. *Isotropic etchants* attack Si at the same rate in all directions. *Anisotropic etchants* attack the silicon wafer at different rates in different directions. *Which etching has more control on the resulting shapes?* Anisotropic etching is direction-sensitive and so there is more control of the shapes produced. An etching process that progresses in one direction, for example, vertical only, is termed "completely anisotropic." *What etchants are commonly used for silicon?* The common Si etchants are inorganic aqueous solutions of KOH, NaOH, RbOH, CsOH, NH_4OH, hydrazine ($(N_2H_4)_n$), and organic etchants, like EDP (ethylenediamine pyrocatechol), which consists of ethylenediamine ($C_2H_4(NH_2)_2$), pyrocatechol ($C_6H_4(OH)_2$), pyrazine ($C_4H_4N_2$), and water, and TMAH (tetramethyl ammonium hydroxide), a quaternary ammonium salt with the molecular formula $(CH_3)_4NOH$. *Which amongst these are the most popular etchants, and what are their typical etching rates?* Amongst these, the most commonly used etchants include KOH, TMAH, and EDP. KOH (50 g) in 100 mL H_2O/isopropanol (C_3H_8O) at 50°C gives an etch rate of 1 μm minute^{-1}. Biswas and Kal (2006) reported that 22 wt% KOH solution achieves maximum silicon etch rates of 89.2 and 88.1 pm hour^{-1} for N-type and P-type silicon, respectively, at 80°C. Biswas et al. (2006) also reported that undoped 5% TMAH achieves etch rates of 10–60 μm hour^{-1} and surface roughness of between 1.5 and 3.0 μm at etching temperatures varying from 50°C to 80°C. A 5% TMAH solution, doped with 38 g L^{-1} silicic acid (H_4O_4Si) and 7 g L^{-1} AP [ammonium peroxodisulfate: $(NH_4)_2S_2O_8$)] gives a high <100> Si etch rate of 70 μm hour^{-1}.

In wet etching, the choice of a particular silicon etchant for fabricating structures depends not only on the etch rate selectivity and anisotropy but also on its compatibility with integrated circuit technology. TMAH is fast gaining popularity in MEMS as an alternative to KOH and EDP. A comparison of etchants is made in Table 3.1.

How is micromachining carried out by dry etching and what are the common types of dry etching? As briefly described before, dry etching is done by a plasma without using any liquids. Three types of dry etching, namely reactive-ion etching, sputter etching, and vapor phase etching, are distinguished in Table 3.2.

3.4.3 Deep Reactive-Ion Etching

What is DRIE? A special subcategory of RIE, that is rapidly gaining reputation, is *deep RIE (DRIE)*, which has virtually changed the MEMS scenario. In this process, etch depths of hundreds of microns are achieved; at the same time, almost vertical sidewalls are ensured.

What is the remarkable feature of DRIE? It prevents lateral etching of the silicon resulting in highly anisotropic etch profiles at fast etch rates and with large aspect ratios, compared with wet chemical etching. The *aspect ratio* of a given feature, like a trench, is defined as the ratio between the depth of the trench and the width.

What is the sequence of steps in a DRIE process? In the time-multiplexed (the process of dividing up one-time slot into smaller time slots) dry etching scheme, the etching and passivating gases are made to flow independently. An etching cycle and a

TABLE 3.1

Different Silicon Etchants for Bulk Micromachining

Etchant	Features and Advantages	Disadvantages
Potassium hydroxide (KOH), an alkali metal hydroxide	Nontoxic, economical, requires simple etching setup, provides high silicon etch rate, high degree of anisotropy, moderate Si/SiO_2 etch rate ratio and low etched surface roughness.	Damages exposed aluminum metal lines very quickly. Not CMOS compatible due to the presence of K^+ alkali metal ions in it.
Ethylenediamine pyrocatechol (EDP), a diamine-based silicon etchant.	Moderate silicon etch rate, high Si/SiO_2 etch rate ratio, low degree of anisotropy and partly CMOS compatible.	Toxic, requires careful handling, special safety measures and therefore a complex etching apparatus. Ages quickly.
Tetramethyl ammonium hydroxide (TMAH), a quaternary ammonium hydroxide-based silicon etchant.	A nontoxic, CMOS- compatible organic solution, moderately high silicon etch rate and high selectivity to masking layers.	High cost and requires a complex etching setup. Gives rough etched silicon surfaces. When doped with suitable amounts of silicic acid and AP [$(NH_4)_2S_2O_8$)], provides full aluminum passivation along with smoothly etched surfaces.

TABLE 3.2

Types of Dry Etching

Sputter Etching (Physical Etching)	Vapor Phase Etching	Reactive-Ion Etching
Material is removed by bombarding the Si substrate with gas ions. Etchant is an inert gas, like argon, giving a slow etch rate, typically tens of nanometer per minute.	Si wafer to be etched is placed inside a chamber in which one or more gases are introduced, and the material is dissolved at the surface in a chemical reaction with the gas molecules. Silicon etching, using xenon difluoride (XeF_2), BrF_3, and ClF_3, is isotropic in nature. Silicon reacts with the gas, forming SiF_4. Etch rate in XeF_2 is ~0.27 micron minute^{-1}	By striking a plasma in the gas mixture (SF_6), using an RF power source, the gas molecules are ionized, which are accelerated toward, and which react at the surface of the material being etched, forming another gaseous material. This is the chemical part of RIE. It also has a physical part. If the ions have high energy, they knock atoms out of the material to be etched without a chemical reaction. Since the chemical part is isotropic and the physical part highly anisotropic, the combination forms sidewalls that have shapes from rounded to vertical. Etch rate is ~2–3 micron minute^{-1}.

passivating cycle are alternately applied to the machine. During the etching cycle of duration ≤12 seconds, a shallow trench is formed in the silicon substrate with an isotropic profile of fluorine (F)-rich glow discharge. In the ensuing passivation cycle of duration ≤10 seconds, a protective fluorocarbon (an organo-fluorine compound that contains only carbon and fluorine bonded together in strong carbon-fluorine bonds) film is formed on all the surfaces. In the subsequent etching cycle, ion bombardment removes the passivant from all the horizontal surfaces, continuing the etching process, but fails to do so on the vertical sidewalls so that the walls are not etched, allowing the etching profile to evolve in a highly anisotropic fashion. *What is this alternation of cycles known as?* This alternation of etching and passivating cycles is known as the *Bosch process* after the German company that invented and patented it (Figure 3.18).

Is there any other version of DRIE process? Yes, another variant of DRIE uses cryogenic (very low temperatures below –150°C) process. The differences between the two DRIE processes is clarified in Table 3.3.

In brief, dry processes, such as RIE, have been widely used for making narrow vertical-wall structures. The addition of C_xF_y polymerizing species to a fluorine-containing plasma is a flexible high-etching rate approach wherein the ratio of vertical to lateral etch rate is adjusted, providing fine-tuning of the profile from vertical to tapered. But it must be emphasized that these processes require relatively high initial equipment and running costs.

What factors determine the suitability of an etching technique for a particular application? Appropriateness of a particular technique for a specific application is governed by several

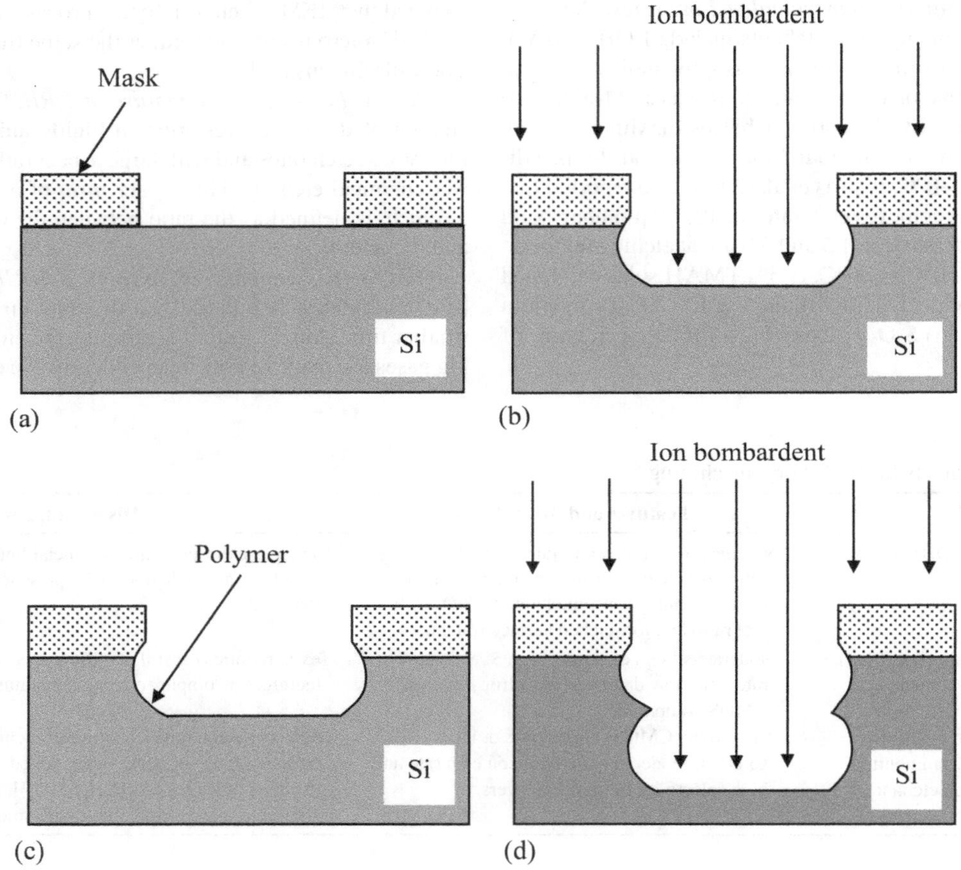

FIGURE 3.18 Bosch process sequence: (a) photolithography for etching window definition, (b) etching, (c) polymer deposition, and (d) etching to greater depth. (Khanna, V. K., *Proceedings of the All India Conference on Recent Developments in Manufacturing & Quality Management (RDMQM - 2007)*, Punjab Engineering College, Chandigarh, 46, 2007.)

TABLE 3.3
DRIE Processes

Bosch Process	Cryogenic Process
"Bosch process," a patented process developed by Robert Bosch GmbH, is based on alternating multiple steps of etching by high-density plasma (SF_6) with sidewall passivation by an etch-resistant polymer, usually a fluorocarbon with composition C_xF_y, like C_4F_8 (Figure 3.18). Etching rates ~6–12 mm minute^{-1}, which are several-fold higher than wet etching.	The mechanism is condensation of reactant gas (SF_6+O_2) on the sidewalls of the reactor by cooling, while condensation at the bottom is removed by ion bombardment. Advantage is absence of polymer contamination. Also provides lower sidewall roughness.

factors. Techno-economic aspects, such as the extent of control over etching profile *vis-à-vis* equipment cost, installation, maintenance, and infrastructural expenses are prime considerations. Anisotropic wet etching of <100> -oriented silicon is an established technique that has been widely used for realization of microstructures. For defining the crystal orientation, the Miller indices of a plane of atoms in the crystal lattice are determined. The Miller indices comprise a set of three numbers obtained by finding the intercepts of the plane with the crystallographic axis and expressing them in terms of unit cell dimensions. Then reciprocals of intercepts are taken and fractions are cleared to write the Miller indices <*hkl*>, e.g. the Miller indices of a plane parallel to two axes but intersecting the third axis at a distance equal to one edge of unit cell are <100>, <010> and <001> depending on the axis cut.

The use of dry etching is restricted by prohibitively high cost for specialized applications where other methods are inadequate. Dry etching is a high-precision and controlled process but also a high-cost enabling technology.

Under what circumstances is dry etching essential? If feature resolution in thin film structures is of primary concern, or vertical sidewalls are needed for deep etchings in the substrate, dry etching must definitely be carried out. *Under what conditions can dry etching be dispensed with?* If the above are of secondary interest and the cost of the process and device must be reduced, the use of dry etching is minimized.

3.4.4 Front- and Back-Side Mask Alignment

The fabrication of MEMS/NEMS devices often involves double-side mask alignment. Approaches to implement this kind of alignment are as follows: (i) conventional alignment of a pattern on the bottom surface of a wafer to one on the top surface, that is, front-to-back alignment, is done with an infrared mask aligner for aligning through the substrate; (ii) another method for double-sided alignment requires the use of a fixture that holds two masks in alignment with each other. The substrate is placed between the two masks and imaged; and (iii) bottom objectives are used to align the mask features to the back of the wafer by aligning a captured image of the mask with the real image of the wafer.

3.4.5 Multiple Wafer Bonding and Glass-Silicon Bonding

In MEMS/NEMS fabrication, it is frequently necessary to bond two silicon wafers together or to bond a silicon wafer with a glass wafer. For this purpose, various bonding approaches are used, as follows: *Anodic bonding*: such bonding between silicon and glass is done at 450°C at 500–1000 V to attract Na^+ ions in the glass to the negative electrode, where they are neutralized (Figure 3.19). This forms a space charge at the glass–silicon interface, producing a strong electrostatic attraction between Si and glass whereby oxygen ions move from glass to glass-silicon interface, forming silicon dioxide (SiO_2) and hence a permanent bond. The typical bond strength achievable by this method is about 5.6 kg cm^{-2}. Thus, anodic bonding is a method of hermetically and permanently joining a glass substrate with a silicon wafer without using adhesives. *Silicon direct bonding*

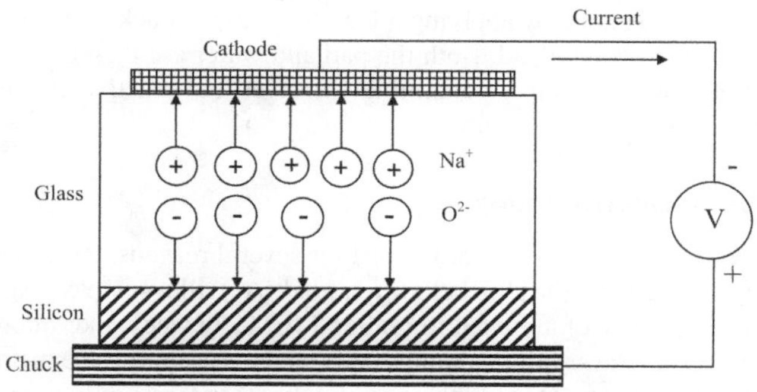

FIGURE 3.19 Glass-Si anodic bonding. (Khanna, V. K., *Proceedings of IMS-2007, Trends in VLSI and Embedded Systems*, Punjab Engineering College, Chandigarh, 317, 2007.)

or silicon fusion bonding: it is based on initial bonding by hydroxyl groups present on silicon wafer surfaces prepared by RCA (Radio Corporation of America) cleaning before bonding. Mechanical spacers are placed at the edges. Removing the spacers allows a single bonding wave to propagate from the center of the wafers. The wafers are annealed at temperatures above 1000°C, so that the hydroxyl groups from water molecules create Si-O-Si bonds as H_2 diffuses away. *Intermediate layer bonding*: it includes eutectic and glass-frit materials made from 100% crushed post-consumer recycled glass. Gold has a eutectic temperature of 363°C with silicon. Gold is evaporated or plated on one of the wafers. For eutectic bonding, the wafers are held slightly above eutectic temperature (the temperature in a two-component mixture where a liquid solution and both pure solids exist at a fixed pressure). In glass-frit bonding, a thin glass layer such as lead borate [$Pb(BO_2)\cdot 2H_2O$] is used and the temperature is melting point of glass <600°C. The typical bond strength achieved is ~9.8 kg cm^{-2}. *Thermocompression bonding*: it involves placing the two mirror-finished surfaces of the silicon in intimate contact at a high temperature. It leads to the development of *van der Waals forces* (intermolecular forces arising from polarization of molecules into dipoles or multipoles) between the two surfaces. In this case, the bond strength is weak.

3.4.6 Wafer Lapping

Back lapping is the thinning of semiconductor wafers by removing material from the backside, which is the unpolished or unprocessed face. The backside of the wafer is brought into contact with an abrasive slurry to remove material from this side. The slurry is a combination of lapping oil and silicon carbide (SiC) or aluminum oxide (Al_2O_3) particles.

3.4.7 Chemical Mechanical Polishing

Chemical mechanical polishing (CMP) is a process that is used for the planarization of semiconductor wafers. CMP takes advantages of the synergistic effect of both physical and chemical forces for polishing of wafers. Chemistry alone cannot planarize wafers because most chemical reactions are isotropic. On the other hand, mechanical grinding may achieve surface planarization, but the surface damage is very high. Therefore, wafer polishing is done by applying a load force to the back of a wafer while it rests on a pad. Both the pad and wafer are then counter rotated while a slurry containing abrasives and reactive chemicals is passed underneath.

3.4.8 Electroplating

Electroplated layers are of interest for several reasons. Patterned layers of nickel (Ni), nickel alloys, gold (Au) alloys, silver (Ag), and copper (Cu) are used to realize MEMS devices, like micro switches, relays, valves, pumps, coils, and gyroscopes. The metal layers have to meet the highest requirements in terms of homogeneity, as well as mechanical, electric, or magnetic characteristics. Electroplating has to cope with specific effects influencing the conditions in the micrometer range.

Electroplating is defined as the deposition of a metal onto a metallic surface from a solution by an electrolysis process. The electroplating process is well suited to make films of metals such as copper, gold, and nickel. The films can be made in any thickness from ~1 μm to >100 μm.

As an example, in copper plating, electroplating takes place by means of the reaction

$$Cu^{2+} + 2e \Leftrightarrow Cu \tag{3.38}$$

Cu^{2+} in this equation represents an ion that is carried to the metal surface to be plated, known as the *cathode*, from the source of the metal being plated, known as the *anode*, which is made of copper. The ions are forced to the cathode by an external source such as a battery. The electrolytic solution is a salt of the metal being plated; in the case of copper plating, it is copper sulfate, $CuSO_4{\cdot}5H_2O$.

What are the primary requirements to be fulfilled by the substrate to be electroplated? It may be noted that electrodeposition requires the following: (i) electrical contact to the substrate when immersed in the liquid bath; and (ii) furthermore, the surface of the substrate must have an electrically conducting coating before the deposition can be carried out.

3.4.9 LIGA Process

What does LIGA stand for? LIGA is an acronym representing the main steps of the process steps involved, i.e., deep XRL, electroforming (a metal-forming process that forms thin parts through the electroplating process, differing from electroplating in that the plating is much thicker and can exist as a self-supporting structure), and plastic molding (a process used in manufacturing to shape materials), in German: Lithographie, Galvanoformung, Abformung (Figure 3.20). By deep XRL, structures of lateral design with high aspect ratios are produced, that is, with heights of up to 1 mm and lateral resolution down to 0.2 μm. The transparent carrier of the mask is a thin metal foil (e.g., titanium, Ti, and beryllium, Be). The absorbers consist of a comparatively thick layer of Au. Synchrotron (cyclic particle accelerator) radiation is employed to transfer the lateral structural information into a plastic layer, generally polymethylmethacrylate (PMMA), $(C_5O_2H_8)_n$. Exposure to radiation modifies the plastic material in such a way that it becomes removable with a suitable solvent, leaving behind the structure of the unirradiated plastic as the primary structure.

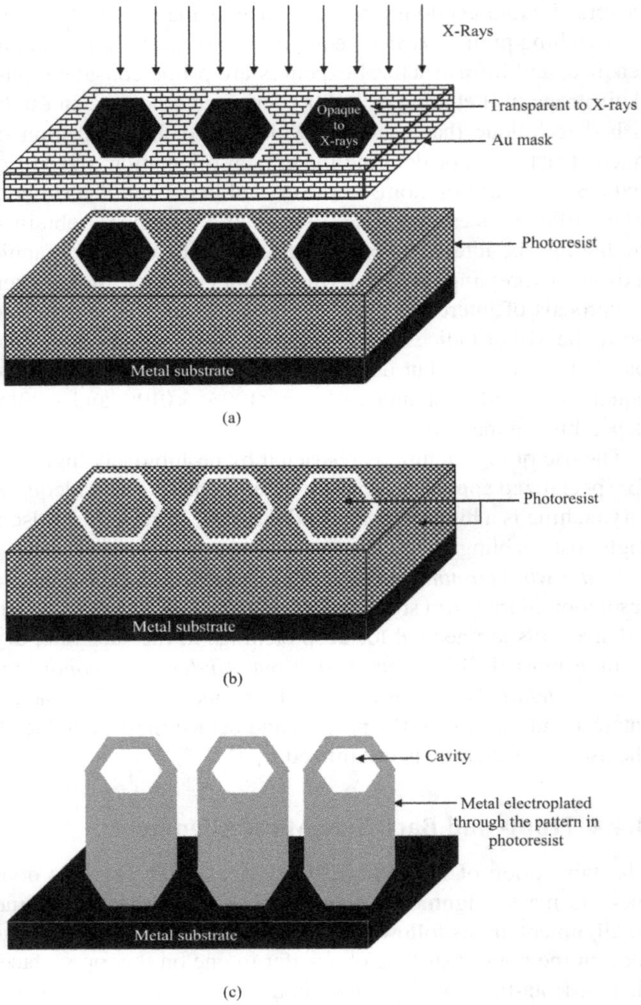

FIGURE 3.20 LIGA process: (a) exposure, (b) developing, and (c) electroplating and removing photoresist. (Khanna, V. K., *Proceedings of IMS-2007, Trends in VLSI and Embedded Systems*, Punjab Engineering College, Chandigarh, 317, 2007.)

The spaces generated by the removal of the irradiated plastic material are filled with metal using an electroforming process. *What are the uses of micro/nanostructures produced by the LIGA process?* The metal micro/nanostructures produced by deep XRL and electroforming are used as molding tools for the economical production of faithful replicas of the primary structure in large numbers.

3.4.10 Micro-Injection Molding

This is a widely adopted procedure for fabrication of microparts or micro-structured components from plastics, ceramics, or metals. The molding mass is melted in the injection molding machine and then injected into a heated and frequently evacuated tool, where it cools and solidifies into the final part. This tool is equipped with micro-structured mold inserts, manufactured by mechanical micromachining (micro-cutting), laser micromachining, X-ray or UV lithography, based on the LIGA process, or by a combination of these and other processes. Depending on the component to be molded, the injection molding process is conducted either isothermally or with heating, prior to injection, and cooling, prior to demolding. After cooling and opening of the tool, the injection-molded components are removed. This is done mostly by a handling device or robot. Nearly all thermoplastics (also known as thermo-softening plastics, are those which become soft when heated and hard when cooled), as well as thermoplastic elastomers (polymers having the elastic properties of natural rubber: ELAST(IC)+[(POLY]MER) are suitable for micro-injection molding.

Noteworthy advantages of micro-injection molding are as follows: (i) cycle times are relatively short; (ii) components are fabricated in an integrated manner from several multicomponent materials; (iii) a high degree of process automation is achieved; (iv) the process is applicable to a wide range of materials (plastics, ceramics, metals); and (v) commercially available equipment form the basis of the system and tool technologies.

3.4.11 Hot Embossing and Electroforming

Hot embossing is the stamping (making a distinctive mark or impression upon an object) of a pattern into a polymer that has been softened by raising its temperature above the *glass transition temperature* (the temperature at which the transition occurs in the amorphous regions between the glassy and rubbery state). The stamp used to define the pattern in the polymer is made by micromachining from silicon LIGA or other similar processes. *Hot embossing lithography* (*HEL*) has proved its potential for structuring resists with high-aspect ratios by thermoplastic molding. It is an important fabrication process for topographic nanostructures to a lateral resolution as low as a few tens of nanometers; topography involves studies of the surface shapes and features. The process is parallel and has the implementational advantage of high throughput at a low cost. Figure 3.21 shows the process steps for fabrication of metal stamps with nanoscale fidelity. By pressing a structured master onto a thermoplastic thin film, the latter is shaped down to sub-10 nm resolution in its viscous state.

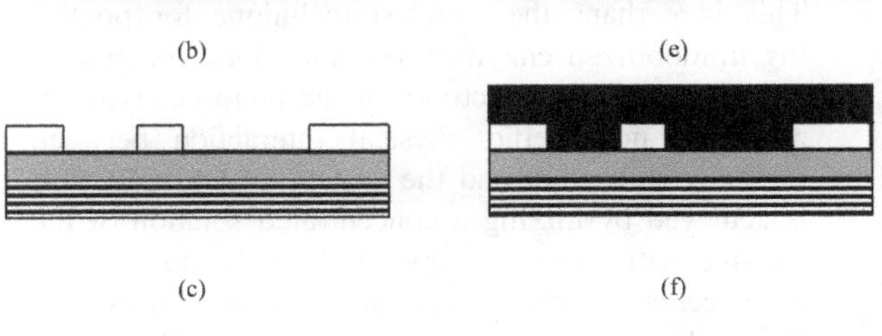

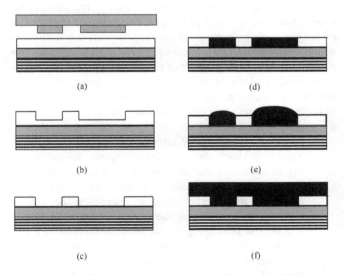

FIGURE 3.21 (a) Embossing, (b) de-molding, (c) dry etching, (d) electro-filling, (e) overplating, and (f) back-plating. (Khanna, V. K., *Proceedings of IMS-2007, Trends in VLSI and Embedded Systems*, Punjab Engineering College, Chandigarh, 317, 2007.)

Process parameters like temperature, pressure, vacuum, etc., are previously optimized to ensure complete filling of the cavities in the stamp with different aspect ratios and shapes. The polymer is hardened by cooling and then de-molded. For electroforming of hot embossed structures, a seed layer is deposited below the resist. Pattern transfer is achieved by removing the residual resist layer, using oxygen RIE, which opens the seed layer windows for plating. Depending on the extent of electroplating, the structure height is either preserved or increased.

Daughter stamps are manufactured by extending the overplating to form a sizeable supporting base. Thus, a robust metal stamp, capable of withstanding repeated use and high mechanical loads, is constructed.

3.4.12 Combination of MEMS/NEMS and CMOS Processes

The *hybrid approach* in which the MEMS/NEMS devices and CMOS circuits are fabricated independently and connected together afterward, is the most primitive method. There are three distinct approaches to fabrication of MEMS components, along with interface electronics on the same chip: *preprocessing*, in which wells are etched up to a depth equal to the height required for fabrication of MEMS devices. Then, MEMS devices are fabricated and protected with an encapsulating layer. The next step is CMOS circuit fabrication followed by the removal of the encapsulating layer to release the MEMS structures. *Postprocessing*: the CMOS circuit is first fabricated and protected by a chemically resistant layer. Then, micromachining steps are carried out, avoiding high-temperature processing. *Combined processing* utilizes the CMOS layers for MEMS device fabrication.

3.5 Biochemistry Division

Biochemistry is the study of the chemical and physicochemical processes of living organisms.

3.5.1 Surface Functionalization and Biofunctionalization of Nanomaterials

What is a functional group? A functional group is an atom or group of atoms within a molecule that is responsible for the characteristic properties of the molecule and reactions in which it takes part. *What does the functional group of a molecule tell us?* It defines the structure of a family of compounds and determines the properties of the family; for example, methanol (CH_3OH), ethanol (C_2H_5OH), and isopropanol [$(CH_3)_2CHOH$] are all classified as alcohols, because each contains a functional hydroxyl group. *What is meant by functionalization of a material?* Functionalization is the addition of functional groups onto the surface of a material by chemical synthesis methods.

What is functionalization of a material used for? Functionalization is employed for modification of surfaces of industrial materials to achieve desired surface properties. Common examples are water-repellent coatings for automobile windshields and non-biofouling, hydrophilic coatings for contact lenses. In addition, functional groups are used to covalently link functional molecules to the surfaces of chemical and biochemical devices.

What does biofunctionalization mean? Biofunctionalization is the modification of a material (especially a nanomaterial) to imbue a biological function to it. *Why is biofunctionalization done*? Biofunctionalization of gold, fluorescent, and magnetic nanoparticles, and carbon nanotubes is carried out as a routine procedure in biochemistry laboratories for fabricating various chemical and biosensors. These processes will be presented in the relevant Chapters 9–12 in reference to the particular nanosensor being discussed. *What features should a biofunctionalization scheme take care of?* The appropriate biofunctionalization of nanoparticles clearly must address the features of colloidal stability, bio-inertness, and specificity with respect to target biomolecules, and low cytotoxicity.

Colloidal particles collide with each other due to Brownian motion, convection, gravity, and other forces. Collisions may cause coagulation of the particles and destabilization of the colloid. Coagulation is the destabilization of colloids by neutralizing the electric charge of the dispersed phase particles, which leads to aggregation of the colloidal particles. Colloidal stability is achieved due to the repulsion forces balancing the attraction forces. Two mechanisms are responsible for the colloidal stability: (i) electrostatic stabilization of colloids is the mechanism by which the attractive van der Waals forces are counterbalanced by the repulsive Coulomb forces acting between the negatively charged colloidal particles; and (ii) polymeric stabilization of colloids involves the addition of polymeric molecules to the dispersion medium in order to prevent the aggregation of the colloidal particles. The polymeric molecules create a repulsive force, counterbalancing the attractive van der Waals force between particles approaching each other.

Does biofunctionalization affect (a) the detection limit of bioanalytical systems, for example, *biosensors or (b) its degree of selectivity and specificity?* It affects both parameters (a) and (b). Although the detection limit of the transducer surface is mainly determined by the detector element (transducer), the role of biofunctionalization cannot be ignored. For the same sensor, biofunctionalization, incorporating nanoparticles, gives lower limits for detection than is achieved without these nano entities.

3.5.2 Immobilization of Biological Elements

What are the prominent biorecognition elements? Recognizable biorecognition elements include: (i) enzymes, such as glucose oxidase (GO_x: an oxido-reductase that catalyzes the oxidation of glucose to hydrogen peroxide and D-glucono-δ-lactone: $C_6H_{10}O_6$), penicillinase (an enzyme secreted by bacteria to inactivate the fungus-derived antibiotic penicillin), and urease (an enzyme that catalyzes the hydrolysis of urea into carbon dioxide and ammonia); (ii) antibodies, for example, anti-human Ig (immunoglobulin) and anti-PSA (prostate-specific antigen); and (iii) nucleic acids, for example, DNA and RNA. Other biological elements are cells, tissues, organelles, etc. *How is biofunctionalization accomplished?* Biofunctionalization is carried out by immobilizing so-called capture molecules (e.g., antibodies and enzymes) or ligands on the transducer surface, which are able to bind only to the corresponding analyte molecules ("lock-and-key" principle). As an alternative, sensors can be coated with biospecific polymers, that are typically more cost-efficient and chemically more stable than capture molecules; however, they may not be available for all analytes. *What main guidelines must be followed during biofunctionalization?* (i) When biofunctionalizing the transducer surface, the capture molecules or bio-specific polymers must not be damaged, that is, the subsequent binding of analyte molecules must be guaranteed. (ii) Furthermore, the surface must be protected against binding of non-analyte molecules contained in the sample medium (e.g., serum), for instance, by suitable intermediate layers, otherwise false-positive signals may be obtained, vitiating the measurements. (iii) In addition, the immobilization procedure has to be chosen in accordance with the chemical environment provided by the transducer and housing material, as well as the detection principle, as this may limit the thickness of the sensitive layer. *What are the representative applications of preparing biosensor surfaces by biofunctionalization?* Mention may be made of detection of analyte concentrations, such as the concentration of low-molecular-weight compounds (such as penicillin in milk) and the concentration of disease markers in serum (e.g., the breast cancer marker HER2/neu and the inflammation marker c-reactive protein CRP).

What is the meaning of "enzyme immobilization"? Literally, "immobilization" means making the enzyme incapable of being moved, restricting the free movement of something. The enzymes are soluble in aqueous media. Immobilization means associating the biocatalysts with an insoluble matrix, so that it is retained in a suitable reactor geometry for its economic reuse under stabilized conditions. Immobilization thus allows decoupling of the enzyme location from the flow of the liquid carrying the reagents and products.

What are the main classes of methods for enzyme immobilization? A large number of techniques are available for the immobilization of enzymes on a variety of natural and synthetic supports. The choice of the support, as well as the technique, depends on the nature of the enzyme and the substrate, and its ultimate application. Therefore, it is not possible to recommend any single universal means of immobilization. The search must continue for matrices that provide facile, secure immobilization of enzymes, along with good interaction with substrates, and which conform in shape, size, and density to their intended use. A combination of one or more of these techniques is also sometimes applied. Methods used for the immobilization of enzymes fall into four main categories, with a combination of two or more of these techniques also being used sometimes.

1. *Physical adsorption of enzyme on an inert carrier*: This is perhaps the simplest technique for preparing immobilized enzymes, one that does not grossly or coarsely alter the activity of the bound enzyme. It relies on nonspecific physical interaction between the enzyme protein and the surface of the matrix. It is achieved by mixing a concentrated solution of the enzyme with the solid support. Usually, no reagents are necessary. Only a minimum of activation steps is required so that the adsorption is cheap, easily carried out, and is minimally disruptive to the enzyme protein. Binding is secured mainly by hydrogen bonds, multiple salt linkages, and van der Waals forces. *But these are*

weak forces? What is the effect of using such weak forces for immobilization on the enzyme function? As these binding forces would be inadequate, it is not astonishing that very often the protein desorbs, due to changes in temperature, pH, ionic strength, etc. Also, due to the deficiency in binding strength, the properties of the enzyme are sometimes altered by further adsorption of other proteins or substances, during the use of the immobilized enzyme. This variation of properties severely impairs the device function.

2. *Inclusion of enzyme in the lattice of a polymerized gel*: The hydrophilic matrix (readily mixed with or wetted by water) is polymerized in an aqueous solution of the enzyme, with the polymeric mass being broken up to the desired particle size. As no bond is formed between the enzyme and the polymer matrix, disruption of the protein molecules does not take place. However, free radicals generated during polymerization affect the activity of the entrapped enzyme molecule. Further, the method is unsuitable for those enzymes that act on large macromolecular substrates, such as ribonuclease (e.g., RNase, a type of nuclease that catalyzes the degradation of RNA into smaller components), trypsin (an enzyme that acts to degrade proteins, i.e., an example of a proteolytic enzyme, or proteinase), and dextranase (an enzyme that catalyzes the endohydrolysis of 1,6-α-glucosidic linkages in dextrans). This is because only low-molecular-weight substrates are able to diffuse rapidly to the immobilized enzyme.
3. *Cross-linking of the protein occurs with a bifunctional reagent: What is a bifunctional reagent?* It is a reagent with two reactive groups, usually at opposite ends of the molecule, that are capable of reacting with and thereby forming bridges between side chains of amino acids in proteins. This helps in identifying the locations of naturally reactive areas within proteins.

In this enzyme immobilization technique, intermolecular cross-linking of the protein is carried out. This linking is done either to other protein molecules or to functional groups on an insoluble support matrix. Enzymes are immobilized through chemical cross-linking, using homo- as well as heterobifunctional cross-linking agents. Two bifunctional reagents used are glutaraldehyde, as an amine-reactive homobifunctional cross-linker (CL) and dimethylsuberimidate (DMS), a bifunctional imidoester. *What is meant by homo- and heterobifunctional cross-linking agents?* CLs are either homo- or heterobifunctional reagents, with identical or nonidentical reactive groups, respectively, permitting the establishment of inter- as well as intramolecular cross-linkages. *Among the cross-linking agents, which substance has been most widely used and why?* Glutaraldehyde [$CH_2(CH_2CHO)_2$], which interacts with the amino groups through a base reaction, has been extensively used in view of its low cost, high efficiency, and stability.

The enzymes are normally cross-linked in the presence of an inert protein like gelatin (a colorless or slightly yellow, tasteless, transparent, brittle, water-soluble protein derived from collagen, and formed by boiling the specially prepared skin, bones, and connective tissues of animals), albumin (water-soluble protein found in egg white, milk, blood, etc.), and collagen (a fibrous protein; the main structural protein in the animal extracellular matrix and connective tissue, forming molecular cables that strengthen the tendons and vast, resilient sheets that support the skin and internal organs). As some of the protein material is inevitably always acting as a support, there is relatively low enzyme activity. Cross-linking is therefore best used in conjunction with one of the other methods.

1. *Covalent binding of enzyme to a reactive insoluble support*: This acts through formation of covalent bonds between the enzyme and the support matrix. The binding reaction must be performed under conditions that do not cause loss of enzyme activity. In addition, the active site of the enzyme must be unaffected by the reagents used. *Which functional groups of proteins are suitable for covalent binding?* The functional groups of proteins suitable for covalent binding under gentle conditions are the alpha amino groups (attached to the carbon atom immediately adjacent to the carboxylate group [the C-2, or α-carbon]; generic formula $H_2NCHRCOOH$, where R is an organic substituent) of the chain and the epsilon amino groups (the amino acid lysine has a positively charged amino group in its side chain located at the epsilon-carbon and hence is called the epsilon-amino group) of lysine (an α-amino acid well known for its antiviral properties, present in protein and essential in the diet of vertebrates, with the chemical formula $HO_2CCH(NH_2)(CH_2)4NH_2$); arginine (a non-essential amino acid, 2-amino-5-guanidinopentanoic acid, $C_6H_{14}N_4O_2$); the alpha carboxyl group of the chain end and the beta and gamma carboxyl groups of the amino acids: aspartic acid (an α-amino acid which acts as a neurotransmitter, $C_4H_7NO_4$) and glutamic acid [(2S)-2-Aminopentanedioic acid, $C_5H_9NO_4$]; the phenol ring of tyrosine (a hydrophilic amino acid which participates in the synthesis of hormones); the thiol group (a sulfhydryl group, -SH) of cysteine (a sulfur-containing proteinogenic amino acid: HO_2CCHCH_2SH or $C_3H_7NO_2S$, which is essential in the human diet); the hydroxyl groups of serine (a hydrophilic amino acid) and threonine (an essential amino acid essential for growth: $HO_2CCH(NH_2)$ $CH(OH)CH_3$); the imidazole group ($C_3H_4N_2$) of histidine [an essential amino acid: $C_6H_9N_3O_2$; synonyms: 4-(2-Amino-2-carboxyethyl) imidazole, (S)-alpha-amino-1H-imidazole-4-propanoic acid] and the indolyl group (C_8H_6N, derived from indole C_8H_7N by removal of a hydrogen atom from a ring structure) of tryptophan (an amino acid, symbol Trp or W, belongs to the family of indolyl carboxylic acids and derivatives, needed for normal growth in infants and for nitrogen balance in adults: $C_{11}H_{12}N_2O_2$).

Does covalent immobilization affect the functional groups of enzymes responsible for catalytic action? Enzymes are covalently linked to the support through those functional groups in the enzymes, which are not essential for the catalytic activity. It is often advisable to carry out the immobilization in the presence of its substrate or a competitive inhibitor (molecules that bind to the active site of the enzyme, the latter altering the catalytic action of the enzyme and consequently slowing down, or, in some cases, stopping catalysis; the inhibitor has a similar shape to the usual substrate for the enzyme, and competes with it for the active site but once it is attached to the active site, nothing happens to it because it does not react – essentially, it just gets in the way) so as to protect the active site. The covalent binding should also be optimized so as not to alter its conformational flexibility; "conformation" refers to the structural arrangement or the way a substance is formed. Some of these problems, however, can be obviated by covalent bonding through the carbohydrate moiety when a glycoprotein (any of a class of compounds consisting of a protein in combination with a carbohydrate) is concerned.

What are the favorable features for this immobilization strategy? Due to the wide variety of binding reactions, and insoluble carriers with functional groups capable of covalent coupling, or being activated to give such groups, this is a generally applicable method of insolubilization (to make an enzyme insoluble). Like cross-linking, covalent bonding provides stable, nonsolubilized enzyme derivatives that do not leach enzyme into the surrounding solution.

What are the criteria for selection of a particular method of enzyme immobilization, amongst those described in the preceding paragraphs? Knowledge of the active sites of particular enzymes enables methods to be chosen that avoid reaction with the essential groups therein. One must carefully choose a method of attachment aimed at reactive groups outside the active catalytic and binding site of that enzyme. As an alternative, these active sites can be protected during attachment as long as the protective groups are removable without loss of enzyme activity. In some cases, this protective function is fulfilled by a substrate of the enzyme or a competitive inhibitor. Several vital roles are played by the surface on which the enzyme is immobilized. These include retention of the tertiary (3D) structure of the enzyme by hydrogen bonding or the formation of electron transition complexes. Retention of the tertiary structure may also be a crucial factor in maximizing thermal stability in the immobilized state.

How does the microenvironment imposed upon enzymes by their supporting matrices and by the products of their own action affect their properties?

1. The first important property is stability. Depending on whether the carrier provides a microenvironment capable of denaturing the enzyme protein or of stabilizing it (changing the natural qualities of a protein by heating, acidity, etc.), the stability of the enzyme either increases or decreases. To reduce inactivation by autodigestion of proteolytic enzymes (proteases), the enzyme molecules are isolated from attack by immobilizing them on a matrix.
2. The second property of an enzyme is its specific activity. Generally, the specific activity of an enzyme decreases upon insolubilization. This is due to denaturation of the enzyme protein caused by the coupling process. It may arise from physical and chemical characteristics of the support matrix alone or may result from interactions of the matrix with substrates or products involved in the enzymatic reaction. It must not be forgotten that the immobilized enzyme is residing in a drastically different microenvironment from that existing in a free solution.

3.5.3 Protocols for Attachment of Antibodies on Sensors

Sensors, such as microcantilevers, require biofunctionalization for their operation. Several protocols have been developed for this purpose. As an example, Kale et al. (2007) described a detailed process to biofunctionalize silicon nitride (Si_3N_4) cantilevers. This process consists of the following steps. (i) The cantilever was subjected to sulfochromic acid (H_2SO_4/CrO_3) treatment for 10 minutes, producing silanol (Si-OH group) sites, and then dried by heating in a vacuum. (ii) The cantilever was immersed in 1% silane for 5 minutes. Silanes (compounds of silicon and hydrogen, of the formula Si_nH_{2n+2}) and other monomeric silicon compounds are able to bond inorganic materials, such as glass, metals, and metallic oxides to organic resins. The adhesion mechanism is due to two groups in the silane structure. The $Si(OR_3)$ portion reacts with the inorganic reinforcement, whereas the organofunctional (vinyl: $-CH{=}CH_2$; amino: ($-NH_2$); epoxy: an oxygen atom, joined by single bonds to two adjacent carbon atoms, thus forming the three-membered epoxide ring, etc.) group reacts with the resin. The coupling agent is applied to the inorganic materials as a pretreatment and/or added to the resin.

After silane application to the cantilever, the excess silane was removed by rinsing in ethanol, and the cantilever was heated at 110°C for 10 minutes in an argon atmosphere. (iii) The cantilever was dipped in 1% aqueous solution of glutaraldehyde [$CH_2(CH_2CHO)_2$] for 30 minutes. The linker molecules are bound to surface amine sites (molecules that contain the functional group $-NH_2$). Thus, free aldehydes were left on the surface, to which biomolecules can attach. (iv) On the modified cantilever surface, human immunoglobulin (HIgG) antigens were immobilized by incubating them on the cantilever surface for 1 hour. Loosely adsorbed biomolecules on the surfaces were subsequently removed by washing with a detergent solution and rinsing with PBS (phosphate-buffered saline: $KCl+KH_2PO_4+NaCl+Na_2HPO_4$). (v) FITC (fluorescein-5-isothiocyanate: $C_{21}H_{11}NO_5S$)-tagged goat anti-human IgG (1 mL mL^{-1} in PBS) was allowed to react with the sample surfaces for 1 hour, and the cantilever die was rinsed; FITC is a commonly used fluorescent dye. It is an amine-reactive fluorescein derivative, suitable for protein labeling. (vi) The cantilever die was inspected under a fluorescence microscope. If the FITC-tagged antibodies were attached to the corresponding antigens, they became visible under the fluorescence microscope. This microscope is a variation of the compound laboratory light microscope, which is arranged to admit ultraviolet, violet, and sometimes blue radiation to a specimen, which then fluoresces. Basically, the microscope irradiates the specimen with a specific wavelength and then separates the much weaker emitted fluorescence from the excitation light. Thus, the sample to be studied is itself the light source. First, the microscope has a filter that only allows through radiation with the desired wavelength that matches the fluorescing material. To become visible, the emitted light is separated from the much brighter excitation light by a second filter. Here, the fact that the emitted light is of a lower energy and has a longer wavelength is exploited. The fluorescing areas are observed under the microscope and shine out against a dark background with a high contrast. In a properly configured microscope, only the emission light reaches the eye or detector so that the resulting fluorescent structures are superimposed with high contrast against a very dark (or black) background. The limits of detection are generally governed by the darkness of the background, and the excitation light is typically several 10^5–10^6 times brighter than the emitted fluorescence.

It is necessary to mention here that FITC is a small organic molecule, that is typically conjugated to proteins *via* primary amines, that is,i.e., lysine residues. Generally, 3–6 FITC molecules are conjugated to each antibody molecule; higher conjugations cause solubility problems as well as internal quenching and reduced brightness. Thus, an antibody is usually conjugated in several parallel reactions to different amounts of FITC, and the resulting reagents are compared for brightness and background adherence to decide the optimal conjugation ratio. Fluorescein is typically excited by the 488 nm line of an Ar laser. The emission is collected at 530 nm.

3.5.4 Functionalization of CNTs for Biological Applications

How are CNTs functionalized for biological applications? For biological uses, CNTs are functionalized by attaching biological molecules, such as lipids, proteins, biotin, etc., to them. In this way, it is possible to solubilize and disperse carbon nanotubes in water, thus opening the path for their facile manipulation and processing in physiological environments. Then, they can usefully mimic certain biological functions, such as gene therapy and drug delivery. In biochemical and chemical applications, the carboxylic acid poly(aminobenzoic sulfonic acid), polyvinyl alcohol, amino acid derivatives, and halogens have been used to functionalize CNTs.

3.5.5 Water Solubility of Quantum Dots

For use in biological applications, quantum dots need to be water soluble. TOPO [tri-*n*-octylphosphine oxide: $OP(C_8H_{17})_3$]-stabilized quantum dots show hydrophobic surface properties (Murcia and Naumann 2005). *What surface modification schemes have been applied for dispersing TOPO-stabilized QDs in water?* To disperse TOPO-stabilized quantum dots in aqueous solution, several surface modification strategies have been pursued. (i) A common approach is to synthesize quantum dots in TOPO and replace the hydrophobic TOPO layer with bifunctional molecules containing thiol and hydrophilic moieties, separated by a molecular spacer. The thiol groups bind to the CdSe or ZnS surface, whereas the hydrophilic moieties diverge from the surface of the corresponding semiconductor. Unfortunately, thiols bind less strongly to ZnS than to Au, which leads to a dynamic equilibrium between bound and unbound thiols. This behavior degrades the long-term water solubility of ZnS-capped quantum dots. To shift the equilibrium toward bound moieties, monothiols have been replaced with molecules containing more than one thiol group. (ii) Another stabilization concept is to enhance binding *via* surface cross-linking of bound molecules. On the basis of this concept, ZnS-shelled quantum dots have been made water soluble by adding a silica shell to the nanoparticles by using alkoxysilanes ($R_nSiX_{(4-n)}$), where R is a nonhydrolyzable organic moiety that can be an alkyl, aromatic, or organofunctional group, or a combination of any of these groups, during polycondensation (a chemical condensation of a monomer having two functional groups leading to the formation of a polymer, that is, bifunctional or multifunctional monomers react to form first a dimer, then a trimer, longer oligomer, and eventually long-chain polymer, a polycondensate, along with the release of water or a similar simple substance). Two types of silanes have been used to stabilize quantum dots in aqueous solution. The first category includes silanes whose surface functional groups are positively or negatively charged at neutral pH. The second type includes silanes with polyethylene glycol $[C_{2n+2}H_{4n+6}O_{n+2}]$ chains. (iii) TOPO coatings can be made water soluble without their replacement by adding amphiphilic molecules (chemical compounds possessing both hydrophilic [*water-loving*] and lipophilic [*fat-loving*] properties), such as lipopolymers (lipids with a polymer chain covalently attached to the lipid) or amphiphilic diblock copolymers (block copolymers with two distinct blocks; copolymers are those derived from two or more monomeric species and block copolymers comprise two or more homopolymer subunits linked by covalent bonds), whose hydrophobic moiety stabilizes the TOPO coating *via* hydrophobic forces, the hydrophilic moiety of which is exposed to the solvent environment, guaranteeing water solubility. The last approach has the advantage of not exposing the sensitive surface of the quantum dot during a surface-exchange step.

3.5.6 Low Cytotoxicity Coatings

Successful application of fluorescent nanoparticles in a biological environment requires not only high dispersion stability and the suppression of nonspecific biomolecule adsorption, but also low cytotoxicity (Murcia and Naumann 2005). This issue, in particular, is of great consequence for applications involving quantum dots, where toxic heavy metal ions may be released. If properly passivated, quantum dots appear to have insignificant cytotoxicity. *What are the approaches followed for toxicity reduction of QDs?* (i) One approach is simply to passivate the quantum dot surface with binding ligands (atoms or molecules attached to a central atom, usually a metallic element, in a coordination or complex compound). Interestingly and importantly, such a passivation has a dual role because it not only lowers the cytotoxicity of QDs, but also provides the bonus of enhancing their quantum yields. (ii) Quantum dots can be further passivated by adding a protecting semiconductor shell. Such coatings also considerably lower the cytotoxicity of CdSe quantum dots, but do not completely eradicate the problem. (iii) Surface silanization is another promising approach to suppressing surface oxidation. The stability of the coating is provided by cross links within the siloxane (a compound that has alternate silicon and oxygen atoms composed of units of the form R_2SiO, where R is a hydrogen atom or a hydrocarbon group) shell. Such shells are quite stable in a biological environment. Furthermore, CdSe/ZnS quantum dots with a protective shell of cross-linked silica significantly reduce the release of Cd^{2+}, thus diminishing the cytotoxicity of fluorescent nanoparticles. (iv) Fascinatingly, the addition of a bovine serum albumin (BSA) layer further lessens the cytotoxicity of

CdSe/ZnS quantum dots. The BSA protein, added to the mercaptoacetic acid-functionalized quantum dot surface, using EDC coupling, acts as a diffusion barrier for O_2 molecules.

3.6 Chemistry Division

Surface modification is used in the fabrication of many chemical sensors for gas detection. A few examples will be cited in the following Section.

3.6.1 Nanoparticle Thin-Film Deposition

Thin films composed of nanoparticles have emerged as useful chemical sensor platforms. Numerous different techniques exist for the deposition of nanoparticle films: (i) controlled pulling; (ii) the use of charged polymers and charged nanoparticles; or (iii) slow incubation of thiol-capped nanoparticles in a dithiol (compound having two thiol groups) solution. In a report (Childs et al. 2005), a technique is described to deposit nanoparticle films of different thickness on prepared electrodes in a reproducible manner. The nanoparticles used for these studies employed 5-nm gold and 2-nm platinum particles. The process steps were as follows: (i) substrates were first silanized with a solution of tetrakis(dimethylamino) silane, $Si[N(CH_3)_2]_4$, in toluene (C_7H_8); (ii) the remaining amino functionalities were displaced with 1,8-octanedithiol (ODT) $[HS(CH_2)_8SH]$ from a toluene solution, to achieve a thiolated (reacted with a thiol) surface capable of binding nanoparticles; (iii) the substrate was then incubated in a solution of dodecylamine ($C_{12}H_{27}N$)-capped metal nanoparticles, dissolved in toluene or hexane; (iv) the film thickness was increased *via* alternating exposure to solutions of bifunctional cross-linking molecules (containing two reactive groups, to provide a means of covalently linking two target groups) and nanoparticles.

3.6.2 Polymer Coatings in Nano Gas Sensors

Here, a nanomaterial, for example, a CNT, is coated with specific polymers to make the CNT sensitive to a particular gas. For certain gases and vapors, pristine SWCNTs do not respond at all, and, in such cases, coating or doping of the nanotubes is able to elicit a signal. Li et al. (2006) coated SWCNTs, deposited on an interdigitated electrode pattern, with polymers such as chlorosulfonated polyethylene (CSPE: a synthetic rubber) or hydroxypropyl cellulose (HPC: a partially substituted, water-soluble cellulose ether) to make them sensitive to chlorine (Cl_2) or hydrochloric acid (HCl) vapor, respectively. *What was the coating procedure followed?* Chlorosulfonated polyethylene (0.804 mg) was dissolved in 25 mL of tetrahydrofuran (THF: C_4H_8O) solvent as a coating solution for sensing Cl_2. Hydroxypropyl cellulose (0.791 g) was dissolved in 25 mL of chloroform ($CHCl_3$) to coat the nanotubes to detect HCl. In each case, a 5 nL aliquot of polymer solution was drop deposited onto the SWCNT network to coat the SWCNTs .

3.6.3 Metallic Nanoparticle Functionalization of Si Nanowires for Gas Sensing Applications

Chen et al. (2007) fabricated Si nanowires (NWs). Pd nanoparticles were deposited on SiNWs for H_2 sensing. *How was this achieved?* First, the oxide covering on the As-grown SiNWs was etched away by dipping in a 5% HF aqueous solution for 10 minutes. After rinsing with distilled water, Pd particles were deposited on the surface of SiNWs by immersion in a saturated palladium chloride ($PdCl_2$) solution for 10 minutes. Consequently, the Pd^{2+} ions were reduced, and Pd metal particles were deposited onto the surface of SiNWs in accordance with the following chemical equation:

$$Pd^{2+} + 2Si - H \rightarrow Pd + H_2 + 2Si \tag{3.39}$$

The Pd-modified SiNWs thus obtained were dispersed on a silicon wafer with a 300-nm oxide layer. For contacts, gold electrodes were deposited onto the sample by thermal evaporation and current-voltage characteristics were measured with a two-probe system.

Upon exposure to 5% hydrogen, the current signal of the sensor increased by about 20 fold. The response time was 3 seconds only, which was appreciably faster than that of the macroscopic Pd wire sensor. A Pd NW arrays-based hydrogen sensor will be elaborated in the Chapter 8, Section 8.3 on chemical nanosensors.

3.7 Nanosensor Characterization Division

This division of the nanosensor research and development facility covers the methods for testing and evaluating the already-developed nanosensors in an unbiased manner. The importance of this Section cannot be overestimated, because proper evaluation of new nanosensors, a critical assessment of their properties, and the provision of feedback to the device design and processing personnel is essential to achieve improvements in their nanosensor characteristics. As a first step, the nanosensor in question should be classified as belonging to a particular category: physical, chemical, or biological. Accordingly, it is diverted to the appropriate characterization facility. As the testing methods for these nanosensors are different, it is abundantly clear that the characterization facility will have rooms in the physics, chemistry, and biology divisions to undertake the testing jobs. Obviously, a blood glucose sensor cannot be tested in a physics lab. However, its physical transducer can be tested there. Apart from general characterization, the above facility also looks into the unique characterization tests for individual nanosensors and upgrades the instrumentation periodically, as deemed necessary by the requirements.

3.8 Nanosensor Powering, Signal Processing, and Communication Division

Most of the issues in this field are currently research challenges that will be addressed by future nanosensor devices and systems (Akyildiz and Jornet 2010). Therefore, the scenario is more aspirational and imaginary than realistic. Here, we provide a brief glimpse of these topics, shelving them for a later, more detailed reconsideration. The deferred discussion will be thoroughly undertaken in Chapters 13–14, because these chapters bring about issues, which mark a revolutionary change in thinking about nanosenors, completely breaking away from old-school thought.

3.8.1 Power Unit

3.8.1.1 Lithium Nanobatteries

Nanomaterials can be used to manufacture nanobatteries with high power density. *What are nanobatteries?* These are batteries fabricated using nanoscale technology. In battery manufacture, nanotechnology offers a number of benefits. It increases the available power from a battery and decreases the time required to recharge a battery. *How is this possible?* This is achieved by coating the surface of an electrode with nanoparticles, hence increasing the surface area of the electrode and thereby allowing more current to flow between the electrode and the chemicals inside the battery. It increases the shelf life of a battery by using nanomaterials to separate liquids in the battery from the solid electrodes under the conditions where no current is being drawn from the battery. Such separation prevents the low-level discharge occurring in the battery, which increases the shelf life of this battery markedly.

Lithium ion (or Li-Ion) batteries are one of the most common types of batteries used in consumer electronic products, ranging from mobile phones to laptop computers. The origins of lithium batteries date back to the beginning of the twentieth century. Lithium nanobatteries were constructed using alumina (Al_2O_3) membranes having pores 200 nm in diameter (Vullum and Teeters 2005; Vullum et al. 2006). Each one of these pores was filled with poly(ethylene oxide) (PEO: $(OCH_2CH_2)_n$, where n represents the average number of oxyethylene groups) and lithium triflate (CF_3LiO_3S), i.e., (PEO + lithium triflate) electrolyte and capped with cathode material. V_2O_5 ambigel confined to the pores served as cathodes for individual nanobatteries. The anode was made of lithium. Thus, each pore became an effective nanobattery. Individual nanobatteries in the arrays were characterized by charge/discharge tests, using the cantilever tip of an AFM to make electrical contact with the 200-nm cathodes of the nanobatteries. The volumetric capacity for each individual nanobattery was ~23–30 μA hour^{-1} cm^{-2} μm^{-1}, which is confirmatory of its promise for delivering power to nanodevices. These nanobatteries were charged and discharged and had high capacities, proving that these small batteries could function as minute power systems.

Lithium batteries could function as viable, miniaturized power sources for future nanodevices.

One of the most important aspects of this nanobattery technology, compared with existing battery types, is its safety. Since each nanobattery comprises thousands of small batteries, even if one of these small batteries has a short circuit and fails, the entire battery can keep functioning, losing only a very small amount of power. Similar damage to a conventional Li-ion battery could result in substantial loss of power or a complete malfunction and, in extreme cases, even fire or explosion.

3.8.1.2 Self-Powered Nanogenerators

The operation of these devices is based on the transformation of the following types of energy into electrical energy (Wang 2008): (i) *mechanical energy*: sources of this energy are the human body movements, or muscle stretching, like walking, vibrations, jerks, etc.; (ii) *vibrational energy*: this is generated by acoustic waves or structural vibrations of buildings, such as the sound of human speech, vehicles, or other noise; (iii) *Hydraulic energy*: it is produced by body fluids or the blood flow.

The conversions of these energies are affected by piezoelectric effect in zinc oxide (ZnO) NWs. When these NWs are subjected to mechanical deformations, such as upon bending, a potential difference appears in the NWs.

3.8.1.3 Energy Harvesting from the Environment

Nanotubes and nanocantilevers can be designed to absorb vibrational energy at specific frequencies. A nonlinear oscillator can harvest energy from a wide spectrum of vibrations or even mechanical and thermal noise. Besides mechanical, vibrational, or hydraulic energy (the latter derived from the force or energy of moving water), it is also possible to gather energy from electromagnetic waves in the nanoscale, for example, an NEMS-based resonator can be applied to convert electromagnetic radiation into vibrational energy. This can be transformed into electricity, using ZnONWs. Alternatively, nanoscale "rectennas" can be developed, using CNTs. These are rectifying antennas that convert electromagnetic waves into DC electricity, which can be used to supply power to nanosensors.

A simple rectenna element consists of a dipole antenna with a Schottky diode placed across the dipole elements. This diode rectifies the AC current induced in the antenna by the microwaves, to produce DC electricity. The reason for choosing Schottky diodes is that they have the lowest voltage drop and the highest speed, therefore waste the least amount of power due to conduction and switching. Large rectennas comprise an array of many such dipole elements.

3.8.1.4 Synthetic Chemical Batteries Based on Adenosine Triphosphate

Adenosine triphosphate (ATP) is the energy source of cells and living organisms, which can be obtained by chemical reactions in the nanoscale, emulating cellular respiration in mitochondria. It is considered by biologists to be the energy currency of life. ATP batteries, imitating the behavior of mitochondria, are an alternative energy source for bionanodevices.

3.8.2 Signal Processing Unit

Enabling of nanoscale processors is facilitated by the development of different forms of smaller FET transistors. Transistors based on graphene, an allotrope of carbon, are not only smaller but are predicted to be faster than existing devices. As graphene shows almost ballistic transport of electrons, the electrons traverse larger distances without back-scattering, permitting the fabrication of swifter switching devices. Additionally, the reduction of the channel length contributes to a faster response of the transistor. Theoretically, the switching frequencies of graphene-based transistors are predicted to be around a few hundreds of terahertz, which is higher than any existing silicon FET. However, irrespective of the specific approach adopted to design these nanotransistors, the main challenge will be their integration into future processor architectures.

3.8.3 Integrated Nanosensor Systems

The capability of a nanosensor is noticeably increased by incorporating the signal- processing circuit and the nanosensor on the same chip. Such a system is called an integrated nanosensor system (INS). Once the performance of a nanosensor has been optimized individually, this kind of integration will be widely practiced.

Integrated nanosensors can go far beyond simple nanosensors, providing features such as standard interfaces, self-testing, fault tolerance, and digital compensation to extend overall system accuracy, dynamic range, and reliability. Another aspect that has been neglected in considering integrated nanosensors is that a variety of new nanosensors, hitherto not practical, may become feasible. There exists a wide unexplored range of possibilities for combining digital signal processing and instrumentation techniques.

3.8.4 Wireless Nanosensor Networks

Why are wireless nanosensor networks (WNSNs) such an exciting research area? This is because they are the enabling technology of many applications that will impact our society and change our daily lives, ranging from healthcare to homeland security to environmental protection. The technology of wireless sensor networks helps to run factories, optimize widely spread processes, monitor the weather, detect the spreading of toxic gases in chemical industries, and even provide precious extra time in advance of tornados and earthquakes. The widespread use of wireless sensor networks, driven by the concept of distributed sensing and computing of the indoor and outdoor environment, has revolutionized the present state of environmental protection and control.

Two main methods for communication are envisioned at the nanoscale: (i) *molecular communication*, defined as the transmission and reception of information encoded in molecules; and (ii) *nanoelectromagnetic communication*, dealing with the transmission and reception of electromagnetic radiation from components utilizing novel nanomaterials. Electromagnetic communication among nanosensors will be aided by the development of nanoantennas and the corresponding electromagnetic transceiver. To seek the nanoantenna goal (Hagerty et al. 2004), the following approaches are likely: (i) more accurate models for nanoantennas based on nanotubes and nanoribbons need to be formulated, by providing their specific bands of operation, radiation bandwidth, and radiation efficiency. All these parameters will determine the communication capabilities of nanosensor devices; (ii) novel nanoantenna designs and radiating nanostructures must be put forward by exploiting the properties of nanomaterials and new manufacturing techniques; and (iii) a new antenna theory must be framed by considering the quantum effects observed at the nanoscale.

The EM transceiver of a nanosensor device will embed the essential circuitry to perform baseband processing, frequency conversion, filtering, and power amplification of the signals to be broadcast or those received from the free space through the nanoantenna. Toward this aim, the suggestions are as follows: (i) more accurate models for the electromechanical nanotransceiver are required by accounting for the radiation bandwidth and energy efficiency of the complete process; (ii) the noise in reception needs to be characterized by recognizing which types of noise affect the electromechanical unit and how they influence the demodulated signal; and (iii) new nanoreceiver architectures, able to support more advanced, robust, or bandwidth efficient modulations, need to be developed.

3.9 Discussion and Conclusions

A nanosensor fabrication facility was conceptualized in this chapter. The pluridisciplinary character of the laboratory was envisaged. Not only the fabrication facility, but also the characterization equipment needs inputs, bordering and cutting across various fields of science and technology. A strong infrastructural support is required to keep these facilities in running condition. A nanosensor designer chooses the nanomaterial and selects the appropriate enabling fabrication technology to realize the device from amongst the wide-ranging options described.

Review Exercises

3.1 Nanoparticles of noble metals are characterized by the presence of bright colors. What is the underlying cause of these colors?

3.2 Is the position of a surface plasmon band dependent on the nanoparticle size? What is the role of nanoparticle shape in deciding the position of this band? What is the effect of interparticle distance on the band position?

3.3 Are the following statements true or false?
   (i) Larger particles show absorption that is more red shifted as compared with the smaller ones.
   (ii) Aggregates of nanoparticles have plasmon oscillations that become red shifted as compared with the individual nanoparticles, a phenomenon widely used in chemical and biological sensors.

3.4 (i) Do SiNPs exhibit size-dependent tunable light emission? (ii) Do Si nanoparticles exhibit a low inherent toxicity in comparison with the CdSe/ZnS, CdSe/CdS, PbS/CdS core/shell quantum dots?

3.5 Unlike bulk silicon, the emission from nanoscaled silicon can be attributed to radiative recombination of carriers confined in SiNPs. Do you agree?

3.6 Define the large-scale production of CNTs in terms of number of tons/annum per plant. Mention three reasons justifying why CVD has become the most important commercial approach for manufacturing CNTs. Name two other commonly used methods of CNT synthesis.

3.7 What are the control parameters for CVD nanotube synthesis? By what technique are large arrays of well-aligned CNTs synthesized?

3.8 Why should agglomerated metal oxide nanoparticles be minimized for gas sensor applications? A tin oxide nanoparticle gas sensor takes a time of 110 seconds to attain 63.2% of the initial difference in electrical resistance before and after changing the gas environment. What is the time constant of the sensor?

3.9 Do silver nanoparticles exhibit antimicrobial activity? Can these nanoparticles be used in medicine to reduce infections as well as to prevent bacteria colonization on prostheses? Describe one method of synthesizing silver nanoparticles.

3.10 A primary process for semiconductor nanofabrication is RIE, which uses chemical and physical processes to etch away desired materials. Compare RIE with wet etching in terms of anisotropy, uniformity and control, and etch selectivity. Differentiate between RIE, sputter etching, and vapor phase etching.

3.11 In the broadest sense, CVD involves the formation of a thin solid film on a substrate material by a chemical reaction of vapor-phase precursors. How do you distinguish it from PVD processes, such as evaporation and reactive sputtering?

3.12 Compare LPCVD and PECVD processes in terms of film deposition conditions and resultant film quality.

3.13 Name two processes of junction formation, that is, transition from P- to N-type or *vice versa*. What is meant by substitutional and interstitial diffusion of impurities in a semiconductor?

3.14 How are phosphorus pre-deposition and drive-in cycles carried out in a semiconductor fabrication facility? How are boron pre-deposition and drive-in performed? Write the relevant chemical equations.

3.15 Boron and phosphorus are the basic dopants of most ICs. Arsenic and antimony, which are highly soluble in silicon and diffuse slowly, are used before epitaxial processing or as a second diffusion. What are the difficulties faced in using gallium, indium, and aluminum as P-type dopants in the IC industry?

3.16 Phosphorus is useful not only as an emitter and base dopant, but also for gettering fast-diffusing metallic contaminants, such as Cu and Au, which cause junction leakage current problems. Thus, phosphorus is indispensable in very-large scale integration (VLSI) technology. Elaborate these remarks.

3.17 Gold diffuses very rapidly in silicon. What is the use of gold in silicon technology?

3.18 What are positive and negative resists used in photolithography? Differentiate between soft and hard-baking steps.

3.19 Most wet etchants are isotropic, meaning that they etch silicon equally in all directions. This causes a well-known phenomenon in the microelectronics industry called *undercut*. How is undercutting avoided in deep RIE?

3.20 Distinguish between pin-through-hole and surface mount technologies in IC packaging. Compare plastic and ceramic packaging materials.

3.21 What are the two main types of wire bond called? Of what materials are the wires, commonly used in wire bonding, made?

REFERENCES

Akyildiz, I. F. and J. M. Jornet. 2010. Electromagnetic wireless nanosensor networks. *Nano Communication Networks* 1: 3–19.

Ambrosi, A., F. Airo, and A. Merkoc. 2010. Enhanced gold nanoparticle based ELISA for a breast cancer biomarker. *Analytical Chemistry* 82: 1151–1156.

Baik, N. S., G. Sakai, K. Shimanoe, N. Miura, and N. Yamazoe. 2000. Hydrothermal treatment of tin oxide sol solution for preparation of thin-film sensor with enhanced thermal stability and gas sensitivity. *Sensors and Actuators B: Chemical* 65(1–3): 97–100.

Biswas, K., S. Das, D. K. Maurya, S. Kal, and S. K. Lahiri. 2006. Bulk micromachining of silicon in TMAH-based etchants for aluminum passivation and smooth surface. *Microelectronics Journal* 37(4): 321–327.

Biswas, K. and S. Kal. 2006. Etch characteristics of KOH, TMAH and dual doped TMAH for bulk micromachining of silicon. *Microelectronics Journal* 37(6): 519–525.

Chen, Z. H., J. S. Jie, L. B. Luo, H. Wang, C. S. Leel, and S. T. Lee. 2007. Applications of silicon nanowires functionalized with palladium nanoparticles in hydrogen sensors. *Nanotechnology* 18: 345502 (5pp), doi: 10.1088/0957-4484/18/34/345502.

Childs, K., S. Dirk, S. Howell, R. J. Simonson, and D. Wheeler. 2005. SANDIA REPORT SAND2005-6004. Functionalized nanoparticles for sensor applications, 35pp.

Drbohlavova, J., V. Adam, R. Kizek, and J. Hubalek. 2009. Quantum dots—Characterization, preparation and usage in biological systems. *International Journal of Molecular Sciences* 10: 656–673, doi: 10.3390/ijms10020656.

Hagerty, J., F. Helmbrecht, W. McCalpin, R. Zane, and Z. Popovic. 2004. Recycling ambient microwave energy with broad-band rectenna arrays. *IEEE Transactions on Microwave Theory and Techniques* 52(3): 1014–1024.

Haynes, C. L. and R. P. Van Duyne. 2001. Nanosphere lithography: A versatile nanofabrication tool for studies of size-dependent nanoparticle optics. *Journal of Physical Chemistry B* 105(24): 5599–5611, doi: 10.1021 /jp010657m.

International Roadmap for Devices and Systems. 2017. Copyright © 2018 IEEE, p. https://irds.ieee.org/images/files/pdf/2017/2017IRDS_MM.pdf

Jamieson, T., R. Bakhshi, D. Petrova, R. Pocock, M. Imani, and A. M. Seifalian. 2007. Biological applications of quantum dots. *Biomaterials* 28: 4717–4732.

Jones, A. C. and M. L. Hitchman. 2009. Overview of chemical vapour deposition. In *Chemical Vapour Deposition: Precursors, Processes and Applications*. A. C. Jones and M. L. Hitchman (Eds.), Cambridge, UK: Royal Society of Chemistry, pp. 1–36.

Juttukonda, V., R. L. Paddock, J. E. Raymond, D. Denomme, A. E. Richardson, L. E. Slusher, and B. D. Fahlman. 2006. Facile synthesis of tin oxide nanoparticles stabilized by dendritic polymers. *Journal of American Chemical Society* 128(2): 420–421, doi: 10.1021/ja056902n.

Kale, N. S., M. Joshi, P. N. Rao, S. Mukherji, and V. R. Rao. 2007. Bio-functionalization of silicon nitride-based piezoresistive microcantilevers. In *Proceedings of the 10th International Conference on Advanced Materials (ICAM)*, October 8–13, 2007. Bangalore, India: International Union of Material Research Societies (IUMRS).

Khanna, V. K. 2004. Emerging trends in ultra-miniaturized CMOS (complementary metal-oxide-semiconductor) transistors, single-electron and molecular-scale devices: A comparative analysis for high-performance computational nanoelectronics. *Journal of Scientific & Industrial Research* 63(10): 795–806.

Khanna, V. K. 2007a. Advancing frontiers of MEMS and NEMS technologies: Micro- and nanomachining. In *Proceedings of IMS-2007, Trends in VLSI and Embedded Systems*, J. N. Roy and D. Syal (Eds.), August 17–18, 2007. Chandigarh: Punjab Engineering College, pp. 317–325.

Khanna, V. K. 2007b. Retrospection of manufacturing technologies and materials for deep submicron and nanometer CMOS integrated circuits. In *Proceedings of IMS-2007, Trends in VLSI and Embedded Systems*, J. N. Roy and D. Syal (Eds.), August 17–18, 2007. Chandigarh: Punjab Engineering College, pp. 19–27.

Khanna, V. K. 2007c. Wet- and dry-etching based micromachining techniques in silicon MEMS fabrication. In *Proceedings of the* All India *Conference on Recent Developments in Manufacturing & Quality Management* (*RDMQM –2007*), Chief Editor P. B. Mahapatra, October 5–6, 2007. Chandigarh: Punjab Engineering College, pp. 46–52.

Li, J., Y. Lu, and M. Meyyappan. 2006. Nano chemical sensors with polymer-coated carbon nanotubes. *IEEE Sensors Journal* 6(5): 1047–1051.

MacKenzie, K. J., O. M. Dunens, C. H. See, and A. T. Harris. 2008. Large-scale carbon nanotube synthesis. *Recent Patents on Nanotechnology* 2: 25–40.

McFarland, A. D., C. L. Haynes, C. A. Mirkin, R. P. Van Duyne, and H. A. Godwin. 2004. Color my nanoworld. *Journal of Chemical Education* 81(4): 544A.

Mulfinger, L., S. D. Solomon, M. Bahadory, A. V. Jeyarajasingam, S. A. Rutkowsky, and C. Boritz. 2007. Synthesis and study of silver nanoparticles. *Journal of Chemical Education* 84(2): 322–325, doi: 10.1021/ed084p322.

Murcia, M. J. and C. A. Naumann. 2005. Biofunctionalization of fluorescent nanoparticles. *Biofunctionalization of Nanomaterials*. Nanotechnologies for the Life Sciences. Vol. 1. Challa S. S. R. Kumar (Ed.), Weinheim, Germany: Wiley-VCH, pp. 1–40.

Nayral, C., T. Ould-Ely, A. Maisonnat, B. Chaudret, P. Fau, L. Lescouzeres, and A. Peyre-Lavig. 1999. A novel mechanism for the synthesis of Tin/Tin Oxide nanoparticles of low size dispersion and of nanostructured SnO^2 for the sensitive layers of gas sensors. *Advanced Materials* 11(1): 61–63, doi: 10.1002/(SICI)1521-4095(199901).

Pietryga, J. M., D. J. Werder, D. J. Williams, J. L. Casson, R. D. Schaller, V. I. Klimov, and J. A. Hollingsworth. 2008. Utilizing the lability of lead selenide to produce heterostructured nanocrystals with bright, stable infrared emission. *Journal of American Chemical Society* 130(14): 4879–4885, doi: 10.1021 /ja710437r.

Pinna, N., G. Neri, M. Antonietti, and M. Niederberger. 2004. Nonaqueous synthesis of nanocrystalline semiconducting metal oxides for gas sensing. *Angezvandte Chemie International Edition* 43(33): 4345–4349.

Radamson, H. H., X. He, Q. Zhang, J. Liu, H. Cui, J. Xiang, Z. Kong, W. Xiong, J. Li, J. Gao, H. Yang, S. Gu, X. Zhao, Y. Du, J. Yu, and G. Wang. 2019. Miniaturization of CMOS. *Micromachines* 10: 293 (52 pages).

Riegler, J., F. Ditengou, K. Palme, and T. Nann. 2008. Blue shift of CdSe/ZnS nanocrystal-labels upon DNA-hybridization. *Journal of Nanobiotechnology* 6: 7, doi: 10.1186/1477-3155-6-7.

Rosso-Vasic, M., E. Spruijt, Z. Popovi, K. Overgaag, B. van Lagen, B. Grandidier, D. Vanmaekelbergh, D. Dominguez-Gutierrez, L. De Cola, and H. Zuilhof. 2009. Amine-terminated silicon nanoparticles: Synthesis, optical properties and their use in bioimaging. *Journal of Materials Chemisry* 19: 5926–5933.

See, C. H. and A. T. Harris. 2007. A review of carbon nanotube synthesis via fluidized-bed chemical vapor deposition. *Industrial & Engineering Chemistry Research* 46(4): 997–1012, doi: 10.1021 /ie060955b.

Silva, R. S., A. F. G. Monte, P. C. Morais, A. M. Alcalde, F. Qu, and N. O. Dantas. 2006. Synthesis and characterization of PbS quantum dots embedded in oxide glass. *Brazilian Journal of Physics* 36(2A): 394–396.

Singh, P., R. M. Tripathi, and A. Saxena. 2010. Synthesis of carbon nanotubes and their biomedical application. *Journal of Optoelectronics and Biomedical Materials* 2(2): 91–98.

Tabatabaei, S., A. Shukohfar, R. Aghababazadeh, and A. Mirhabibi. 2006. Experimental study of the synthesis and characterization of silica nanoparticles via the sol-gel method. *Journal of Physics: Conference Series* 26: 371–374.

Tan, Y., X. Dai, Y. Li, and D. Zhu. 2003. Preparation of gold, platinum, palladium and silver nanoparticles by the reduction of their salts with a weak reductant-potassium bitartrate. *Journal of Materials Chemistry* 13:1069–1075.

Tanaka, M. 2005. Spintronics: Recent progress and tomorrow's challenges. *Journal of Crystal Growth* 278(1–4): 25–37.

Vullum, E and D. Teeters. 2005. Investigation of lithium battery nanoelectrode arrays and their component nanobatteries. *Journal of Power Sources* 146(1–2): 804–808.

Vullum, F., D. Teeters, A. Nyten, and J. Thomas. 2006. Characterization of lithium nanobatteries and lithium battery nanoelectrode arrays that benefit from nanostructure and molecular self-assembly. *Solid State Ionics* 177(26–32): 2833–2838.

Wang, Z. L. 2008. Towards self-powered nanosystems: From nanogenerators to nanopiezotronics. *Advanced Functional Materials* 18(22): 3553–3567.

Wang, Z., Q. Xu, H.-Q. Wang, Q. Yang, J.-H. Yu, and Y.-D. Zhao. 2009. Hydrogen peroxide biosensor based on direct electron transfer of horseradish peroxidase with vapor deposited quantum dots. *Sensors and Actuators B: Chemical* 138: 278–282.

Zhai, C., H. Zhang, N. Du, B. Chen, H. Huang, Y. Wu, and D. Yang. 2011. One-pot synthesis of biocompatible CdSe/CdS quantum dots and their applications as fluorescent biological labels. *Nanoscale Research Letters* 6: 31(5pp).

Part III

Physical Nanosensors

4

Mechanical Nanosensors

4.1 Introduction

What is a physical sensor? Recalling Section 1.15, it is a device enabling precise measurements of physical variables, such as mechanical, thermal, optical, and other properties of matter. Such measurements are required every day in many applications. Next, we move to mechanical sensors. Under 'Physics' (Section 1.3), we talked about mechanics (Section 1.3.5). Mechanical sensors are those dealing with mechanical properties of material bodies related to forces and motion, like mass, displacement, acceleration, force, torque, pressure, strain, fluid flow, viscoelasticity, etc.

The center of attention here is primarily nanoscale and nanosensors. The main methods of sensing mechanical measurands have been established for many years and are therefore directly applicable to nanosensors. Two mechanical nanosensors, recognized as landmarks in the annals of nanotechnology, namely those for displacement (STM: Scanning tunneling microscope) and force (AFM: Atomic force microscope), were described in Sections 1.19 and 1.20, respectively.

Let us go over the broad canvas of mechanical nanosensors. There is a conspicuous effect that must be accounted for when considering micro-and nanoscale devices, which is, of course, *scaling*. Some physical effects favor the typical dimensions of smaller devices, whereas others do not, e.g., as the linear dimensions of an object are reduced, other parameters do not cooperatively decrease in the same manner. Consider a simple cube of material of a given density. If the length l is reduced by a factor of 10, the volume (and hence mass) will be reduced by a factor of 1000 (l^3). Besides physical phenomena that can be extrapolated to nanoscale, many new phenomena that occur only at the nano level, such as quantum-mechanical tunneling, single-electron transistor (SET) action, and so on, are exploited in the operation of mechanical nanosensors.

Scaling laws illustrate that, on shrinking a body, there is not only a size reduction, but the physical effects are also influenced (Wautelet 2001). Gravitational force between bodies dominates in the macroworld, but, at the microscopic level, adhesion forces are dominant, and these are primarily the van der Waals forces between atoms and molecules. At the nanoscale, the gravitational force may be ignored in the analysis. As another example, the lowest frequency v of a vibrating beam or string corresponds to a state where the length of the beam/string equals one-quarter or one-half of the wavelength λ. The well-known equation for phase velocity of the wave, $\upsilon = v\lambda$, means that resonance frequencies are larger in smaller systems. Citing the case of a cantilever beam, the deflection of a cantilever loaded by its own weight varies inversely to the square of its length so that a 10^3 times smaller beam bends 10^6 times more, due to its own weight.

4.2 Nanogram Mass Sensing by Quartz Crystal Microbalance

The quartz crystal microbalance (QCM) is a mass-sensing device capable of measuring mass changes in the nanogram range (Kanazawa and Cho 2009). It is basically a quartz crystal resonator with the ability to measure very small mass changes in real time.

A *resonator* is a mechanical structure designed to resonate, that is, to exhibit resonance or resonant behavior, naturally oscillating at some particular frequencies, called its *resonance frequencies*, with greater amplitude than at others. Resonators can be fabricated from a range of single-crystal materials, with micron-sized dimensions using various micromachining processes. The high dimensional stability and low-temperature coefficient (the relative change of a physical property when the temperature is changed by 1 K) of quartz makes it a good resonator, keeping the resonant frequency constant.

The sensitivity of the QCM is approximately 100-fold higher that of an electronic balance (sensitivity~0.1 mg). Elaborating further this comparison of QCM with other balances, depending on their maximal load, the balances used today in laboratories, e.g., beam, cantilever, torsion, or spring balances, can detect down to 10^{-10} kg. On the other hand, the QCM can detect down to 10^{-16} kg. This has the implication that QCM is capable of measuring mass changes as small as a fraction of a monolayer or a single layer of atoms.

What are the applications of QCM? Since the QCM was first introduced by Sauerbrey in 1959, it has become a widely used instrument for small mass measurements in vacuum, gas, and liquid phases. The high sensitivity and the real-time monitoring of mass changes on the sensor crystal render QCM a very attractive technique for a wide range of applications.

What is the construction of QCM? The heart of the QCM is the piezoelectric AT-cut quartz crystal, sandwiched between a pair of electrodes (Figure 4.1). Crystal blanks are cut at different orientations from the bar of quartz to realize specific desirable characteristics. The cut is determined by the specifications of the crystal. The keyhole-shaped electrodes on both major faces of the quartz resonator are vacuum-deposited gold or silver films, having a thickness of about 150 nm.

How does QCM work? When the electrodes are connected to an oscillator and an AC voltage is applied across these electrodes, the quartz crystal vibrates and thickness shear acoustic waves called bulk acoustic waves (BAWs), arising from the piezoelectric effect, undergo constructive interference, such that resonances occur at particular frequencies. Essentially, the quartz crystal starts oscillating at its resonance frequency, acting as a piezoelectric resonator.

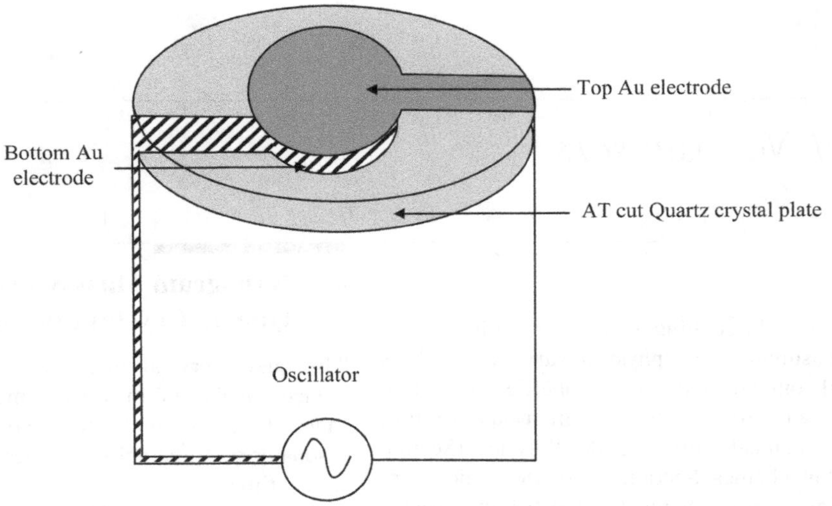

(a)

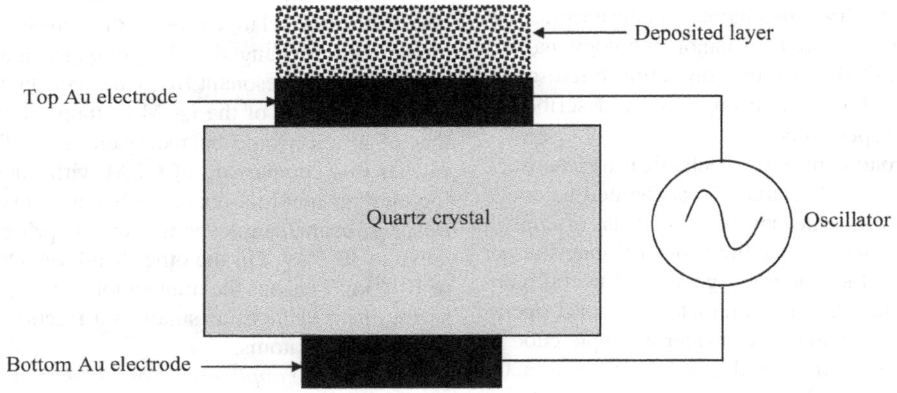

(b)

FIGURE 4.1 Quartz crystal microbalance: (a) schematic diagram and (b) cross section.

How is mass measured with a QCM? If a rigid layer is evenly deposited on one or both the electrodes, the resonance frequency decreases proportionally to the mass of the adsorbed layer, according to the Sauerbrey equation:

$$\Delta f = -\frac{2 f_o^{\,2}\,\Delta m}{A\sqrt{\rho_q \mu_q}} \tag{4.1}$$

where

Δf is the measured frequency shift

f_o is the resonance frequency of the fundamental mode of the crystal

Δm is the mass change per unit area (g cm^{-2})

A is the piezoelectrically active area of the device

ρ_q is the density of quartz, 2.648 g cm^{-3}

μ_q is the shear modulus of quartz, 2.947×10^{11} g cm^{-1} s^{-2}

The resonance frequency changes as a linear function of the mass of material deposited on the crystal surface. The mass-sensitive area is situated in the central part of the resonator, covering approximately the area where the two electrodes overlap.

Under what conditions does the Sauerbrey equation apply? The Sauerbrey equation is derived by treating the deposited mass as though it were an extension of the thickness of the underlying quartz. Hence, it is only strictly applicable to uniform, rigid, thin-film deposits. In certain situations, the Sauerbrey equation does not hold; citable instances are when the added mass is (i) not rigidly deposited on the electrode surface(s), (ii) slips on the surface, or (iii) is not deposited evenly on the electrode(s).

On the other hand, because of the method of derivation of Sauerbrey's equation, the mass-to-frequency correlation, as determined by this equation, is largely independent of electrode geometry. This has the benefit of allowing mass determination without calibration, making the set-up desirable from a cost- and time-investment standpoint.

Sauerbrey also developed a method for measuring the characteristic frequency and its changes by using the crystal as the frequency-determining component of an oscillator circuit. His method still remains the primary tool in QCM experiments for conversion of frequency to mass and is valid in nearly all applications.

The QCM is a BAW device. For comparison, a surface acoustic wave (SAW) gas sensor is shown in Figure 4.2 (Ho et al. 2003). It consists of the gas-sensitive material deposited between the input and output interdigitated transducers (IDTs). The input IDT creates electric fields alternating in time and space, which penetrate into the piezoelectric substrate and generate mechanical stresses through the piezoeffect. In this way, the transducer produces SAWs, which propagate on the surface. These waves are detected by the output transducer.

The gas gets adsorbed onto the sensitive film when the SAW sensor is exposed to it at room temperature and/or at elevated temperature, which, in turn, changes the conductivity of the film. This conductivity change affects the velocity of the SAW traveling across the film and gives rise to a frequency change, which corresponds to the percentage of the gas molecules adsorbed by the film.

A *delay line* is a circuit designed to introduce a calculated delay into the transmission of a signal. An *acoustic delay line* delays the propagation of a sound wave by circulating it through a liquid or solid medium. *SAW delay lines* are used as sensitive mass detectors by virtue of changes in SAW velocity.

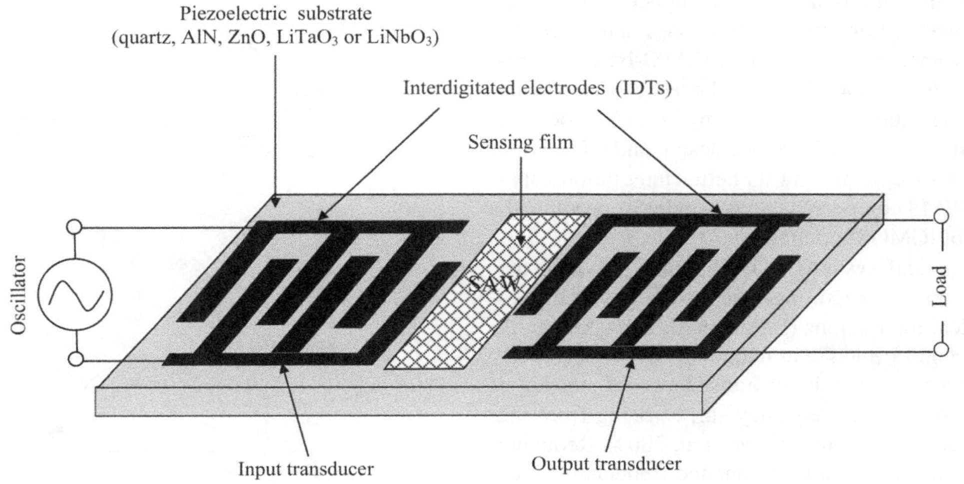

(a)

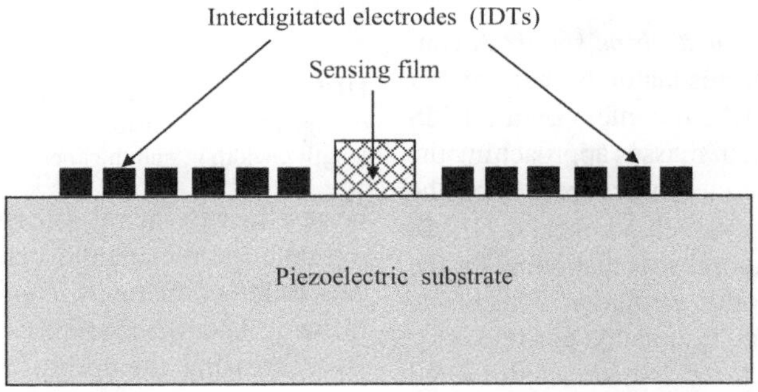

(b)

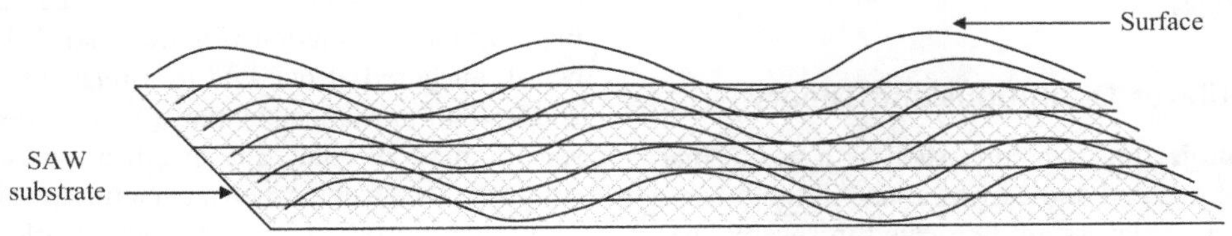

(c)

FIGURE 4.2 Surface acoustic wave (SAW) device: (a) schematic layout, (b) cross-sectional diagram, and (c) SAWs.

Decreasing the SAW device area increases the operating frequency, f, and dramatically improves the mass sensitivity (the mass detection limit is proportional to $1/f^3$, assuming that noise is linear in f). The operating frequency of SAW devices is inversely proportional to the spatial periodicity of the transducers, so that central frequencies in the GHz range for the sensor require submicron or nanometric lithography.

4.3 Attogram (10^{-18}g) and Zeptogram (10^{-21}g) Mass Sensing by MEMS/NEMS Resonators

Sensor designs based on a QCM have been refined over decades and are approaching their theoretical limitations. Some of the drawbacks of quartz crystals are their relatively large sizes, their sensitivity to shock and vibration, and their higher failure rates than silicon components. For ultrasensitive mass detection, the silicon micro-electromechanical system (MEMS)-based microresonator is better than a QCM sensor because of its much smaller minimum detectable mass limit, by several orders of magnitude, its greater flexibility in device design and fabrication for various kinds of specimens, and its better integration capability with other MEMS components and complementary metal-oxide semiconductor (CMOS) circuits.

Nanoelectromechanical systems (NEMS) have provided the unsurpassed opportunity for gaining insight into physical phenomena and chemical interactions (Ekinci et al. 2004; Dai et al. 2009). The fundamental principle in nanomechanical detection, using a NEMS resonator, is the direct transduction of molecular adsorption into the resonance frequency shift, arising from the atomic mass of adsorbed molecules (Sone et al. 2004). Reducing the size of a resonator increases its resonance frequency. Since frequencies in the gigahertz (GHz) range are possible, NEMS resonators are envisioned for radio-frequency (RF) or microwave communication.

Why should the mass of the resonator itself be small? A consequence of the smaller mass of the resonator is that smaller changes in the mass are detectable, which confers on a NEMS resonator the capability for perception of masses approaching the level of single atoms. The smaller the mechanical resonator, the greater sensitivity it exhibits. Based on such a concept, NEMS resonators have been suggested as nanosensors that allow for the sensitive detection of molecules at atomic resolution. This is one field where nanotechnology has an unquestionable position.

Various types of resonator structures include cantilevered resonators (Li et al. 2003; Jenkins et al. 2004; Lin et al. 2005; Hansen and Thundat 2005; Ikehara et al. 2007; Rinaldi 2009), clamped-clamped beam, free-free beam (Figures 4.3 and 4.4), etc. (Beeby et al. 2004).

4.3.1 Microcantilever Definitions and Theory

A *cantilever* is a projecting structure, such as a beam, that is supported rigidly at one end and carries a load at the other end or along its length. Cantilevers are fabricated in several shapes (Yang et al. 2000; Ansari et al. 2009): rectangular, two-legged, or V-shaped cantilevers (Figure 4.5). The V-shaped cantilever is the predominant shape manufactured and is used currently in AFM. The popular V-shaped cantilever is intuitively thought to

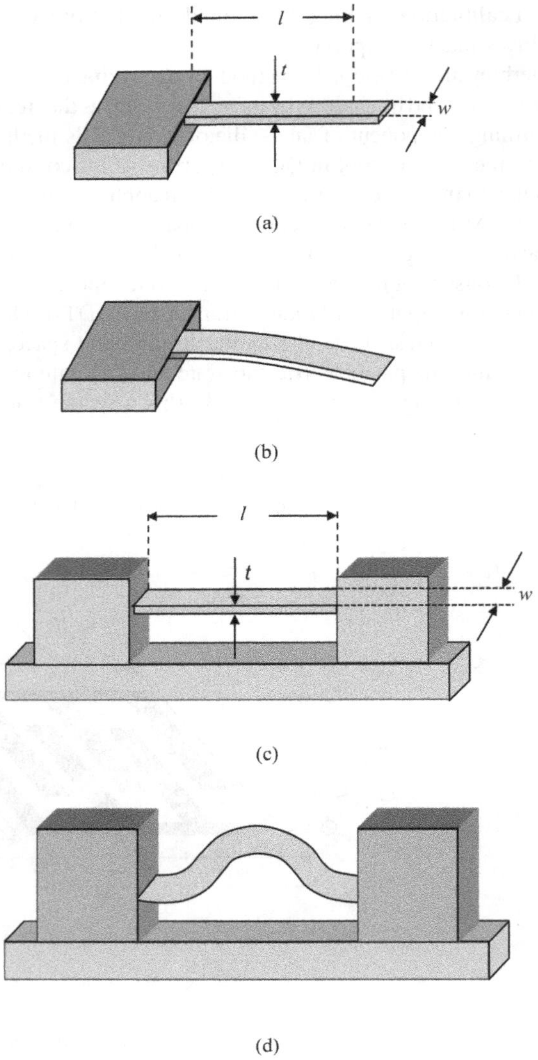

FIGURE 4.3 (a) Straight cantilever, (b) bent cantilever, (c) straight doubly clamped beam resonator, and (d) vibrating doubly clamped beam. Symbols: length l, width w, and thickness t.

be resistant to lateral forces and therefore less prone to twisting than the rectangular beam. But V-shaped cantilevers make calibration of the microscope and interpretation of the data more difficult. The two-leg type cantilever reduces the spring constant. By decreasing the spring constant, a significant performance improvement is achieved, through the corresponding increase in sensitivity.

A microcantilever is the miniaturized counterpart of a diving board that moves up and down at a regular interval, undergoing a vibrational motion (Figure 4.6a). It is a miniature diving board, anchored at one end to a relatively large mass (Vashist 2007). The movement of the microcantilever changes when a specific mass of analyte is specifically adsorbed on its surface, similar to the change when a person steps onto the diving board. Molecules adsorbed onto a microcantilever cause deflection (Figure 4.6b) or vibrational frequency changes of the microcantilever (Figure 4.6c). Many sensor applications require arrays comprising several cantilevers, each of which is assigned a specified task (Figure 4.7).

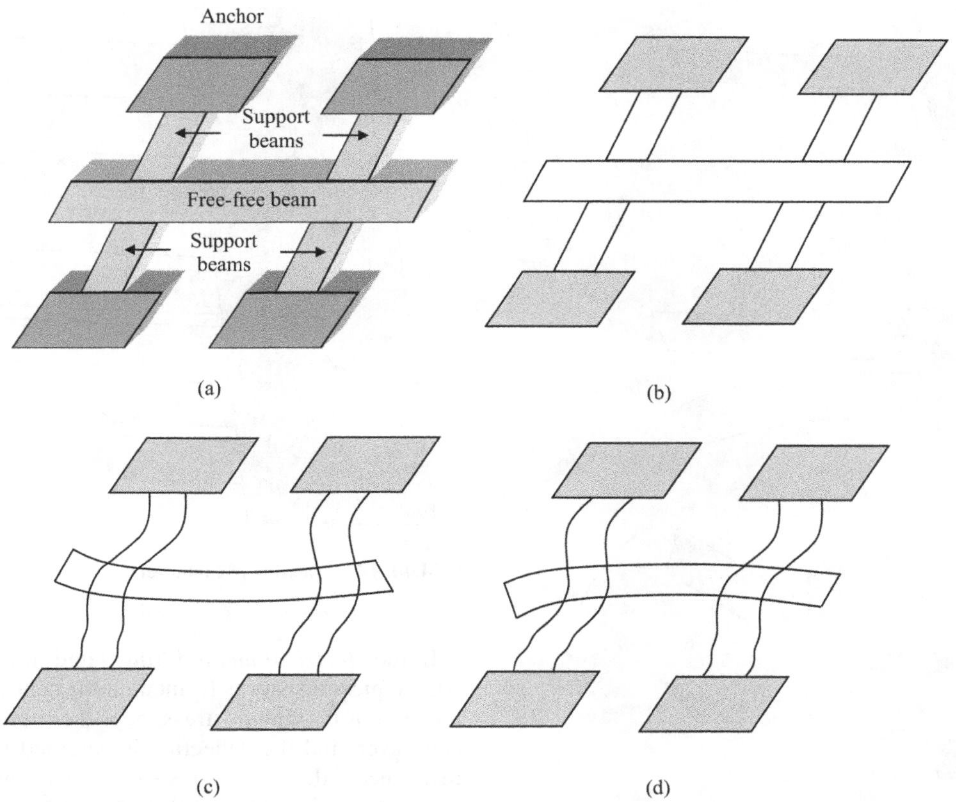

FIGURE 4.4 Free-free beam: (a) 3D and (b) 2D views in equilibrium; (c) 2D and (d) 3D vibrating positions.

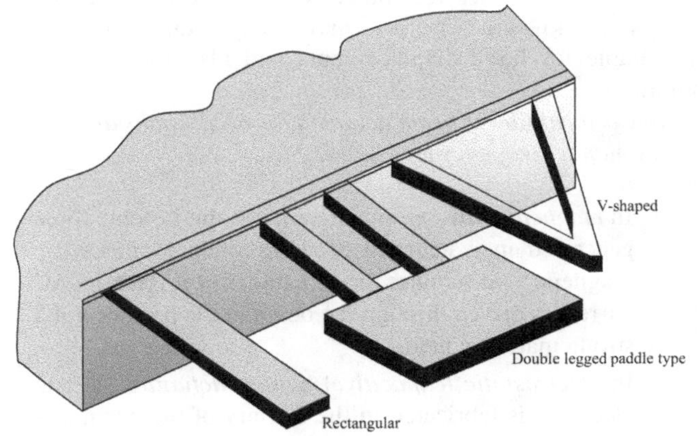

FIGURE 4.5 Cantilevers of different shapes (Vashist, *Journal of Nanotechnology Online* 3: 1–15, 2007.)

What are the operational modes of a cantilever? Modes of cantilever operation can be divided into static and dynamic modes from the viewpoint of the measured parameter (Lavrik et al. 2004). In the static mode, the static bending or deflection of the cantilever is measured while, in the dynamic mode, its dynamic properties, like resonance frequency, are determined. Each of these modes, in turn, is associated with different transduction scenarios. Static cantilever deflections are caused by one of the two primary causes: either external forces exerted on the cantilever, as in AFM, or intrinsic stresses generated on the cantilever surface or within the cantilever. Whereas cantilever microfabrication technology is capable of producing nearly stress-free suspended beams, additional intrinsic stresses may subsequently be produced from several factors, like thermal expansion, interfacial processes, and physicochemical changes.

How do viscous properties of the medium influence cantilevers in their two operating modes? Cantilever sensors operating in the dynamic mode are essentially mechanical oscillators, resonance characteristics of which depend upon both the attached mass and the viscoelastic properties of the medium, e.g., adsorption of analyte molecules onto a resonating cantilever results in lowering of its resonance frequency, due to the increased suspended mass of the resonator (Chen et al. 1995). The damping effects (effects that tend to reduce the amplitude of oscillations) of a liquid medium reduce resonance responses of cantilever devices. In most common liquids, such as aqueous solutions, the amplitude of the cantilever oscillations at the resonance is orders of magnitude lower than that of the same resonating cantilever operating in air. On the other hand, operation in the static mode is unaffected by viscous properties of the medium. Therefore, microcantilever sensors operating in the static mode are especially attractive as platforms for nanomechanical biochemical assays and other biomedical applications.

How is the small deflection of a cantilever, often in nm, measured? There are two methods, namely optical and electrical (Figure 4.8). In the *optical method*, shown in Figure 4.8a, a laser beam, reflected from a gold film-coated spot on the surface of the cantilever, is received by a photodetector, both in the unbent and bent states of the cantilever. The deflection is determined from the difference in position of the reflected beam for the two

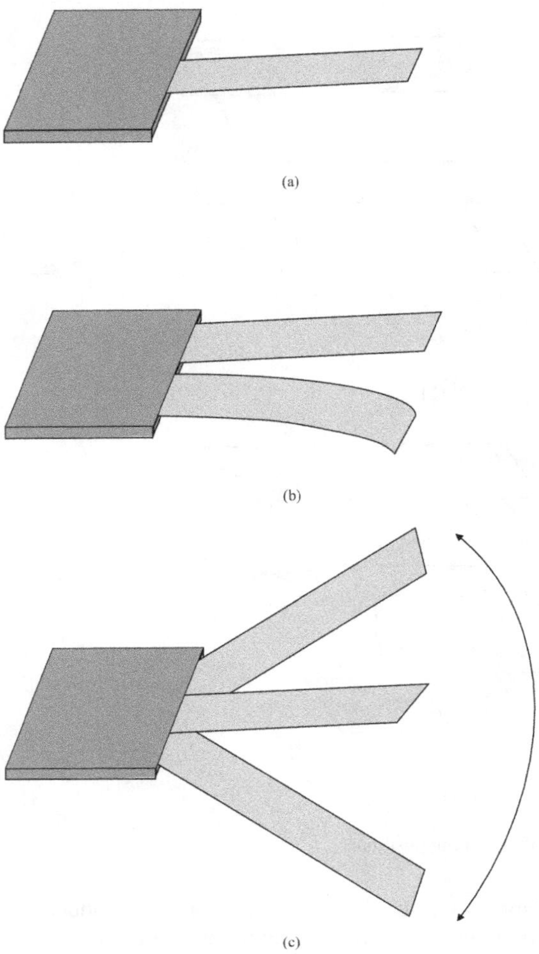

FIGURE 4.6 Cantilevers: (a) straight cantilever, (b) bent cantilever, and (c) vibrating cantilever.

cases, i.e., for the bent beam with respect to the unbent beam. This method provides sub-nanometer accuracy. Another optical method is based on interferometry (an investigative technique, involving the design and use of optical interferometers), utilizing the interference between a reference laser beam and the one reflected from the cantilever, but is less suitable for liquids and hence finds limited usage in biosensing. In addition to the aforementioned methods, optical diffraction gratings and charged-coupled devices are also used for deflection measurements. The *diffraction grating* is a component with a periodic structure, which splits and diffracts light into several beams traveling in different directions. The *charge-coupled device (CCD)* is a light-sensitive integrated circuit consisting of an array of closely spaced metal-oxide semiconductor (MOS) capacitors that can store and transfer information, using packets of electric charge. Each capacitor in the array represents a pixel so that, by applying an external voltage to the top plates of the MOS structure, charges (electrons or holes) can be stored in the resulting potential well. These charges can be shifted from one pixel to another by digital pulses applied to the top plates (gates), and, in this way, the charges can be transferred, row by row, to a serial output register. The picture represents the display of the electron distribution.

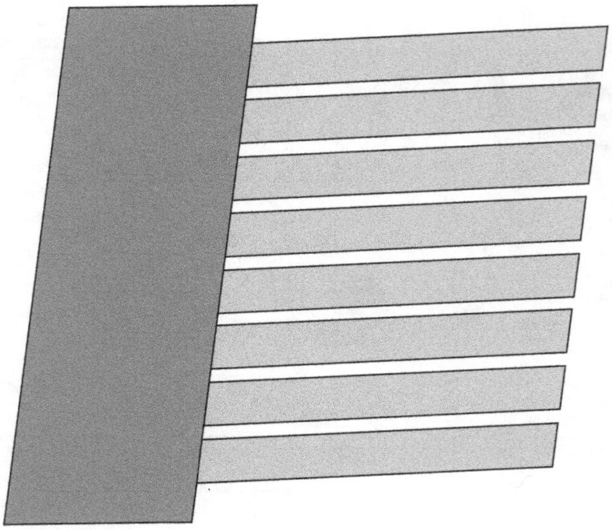

FIGURE 4.7 An array of cantilevers.

In the *electrical method*, illustrated in Figure 4.8b, a polysilicon piezoresistor is formed on the cantilever surface at the position of maximum stress, near the supporting point of the cantilever, and the deflection is obtained from the change in resistance value of this resistor as determined by the upsetting of the balance of the Wheatstone bridge; the *Wheatstone bridge* is a circuit containing four resistances, a constant voltage input, and a voltage gauge, that is used to determine the value of an unknown resistance, when the other three resistances are known. The alternative capacitance method is unsuitable for large displacements and also for electrolytic solutions.

How is a cantilever beam actuated for oscillations and how is its resonance frequency measured?

1. In *magnetomotive motion actuation*, the Lorentz force generated upon a current-carrying conductor in a static magnetic field supplies the actuation force. Here, an AC current is driven through the beam in the presence of a strong magnetic field.
2. In *electrostatic (capacitive) motion actuation*, a gate electrode is fabricated in the vicinity of the beam and a voltage is applied between the gate electrode and the device. The actuation force depends non-linearly upon the excitation voltage and has components present at higher harmonics. To obtain a large contribution to the drive frequency, a voltage having a large DC and a small AC component is usually applied.
3. An important actuation technique is based on piezoelectricity. A piezoelectric material responds to an externally applied electric field by deforming. Among various types of micro-resonators, piezoelectric thin-film transduced micro-resonator offers competitive applications because of its unique characteristics of low power consumption, small driving voltage, self-actuation self-sensing capability, and impedance matching with CMOS circuits.

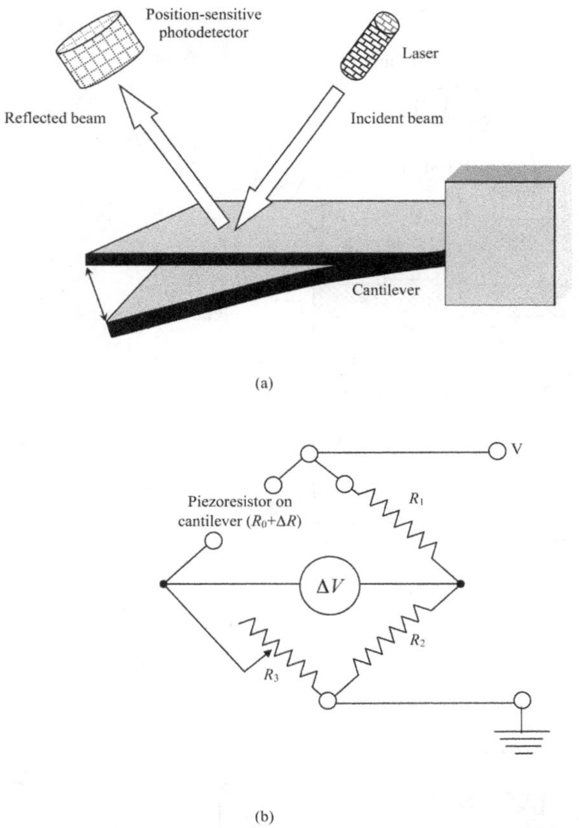

FIGURE 4.8 Measurement of cantilever deflection by: (a) optical deflection system and (b) Wheatstone bridge circuit, in which the piezoresistor on a cantilever located near the end connected to its support is put on one arm of the bridge.

How do these cantilevers interface with CMOS circuitry? Piezoelectric micro-resonators provide better power handing capability than do capacitive resonators, as a low driving voltage of several 100 mV is adequate for resonator actuation, and this facilitates the integration of micro-resonators with CMOS circuits. Lu et al. (2007) fabricated a partially lead zirconate titanate [PZT, $Pb(Zr_{0.52},Ti_{0.48})O_3$] film-covered single-crystal silicon cantilever for ultrasensitive mass detection applications. The PZT film was integrated only at the fixed end of the cantilever to suppress energy dissipation and other negative effects, as well as to improve the mechanical properties of the cantilever. The 30-μm wide 100-μm long cantilever (Figure 4.9) was fabricated on silicon-on-insulator wafers. The PZT was used as both the micro-actuator to drive the microcantilever and as a microsensor to determine its resonance frequency.

How is vibration analysis of a cantilever (or in general, any mechanical microstructure) performed? The *Laser Doppler Vibrometer (LDV)* is a state-of-the-art instrument for vibration analysis of a wide range of mechanical systems (Figure 4.10) (Yang et al. 2000). The LDV is a noncontact technique for velocity and displacement measurement of vibrating structures, so that the tested device is uninfluenced by the measurement process, giving a reliable assessment of its properties. The laser beam from the LDV is directed at the object of interest, and its vibration amplitude and frequency are extracted from the Doppler shift of the laser beam frequency due to the motion of the object. The *Doppler effect* describes the relative change in wavelength and frequency of a wave when the source and observer are in relative motion with respect to each other.

A vibrometer is essentially a two-beam laser interferometer that measures the frequency (or phase) difference between an internal reference beam and a test beam. A coherent laser beam (in which waves have a constant difference in phase) is divided into object and reference beams by beam splitter 1. The object beam strikes a point on the moving or vibrating object and light reflected from that point travels back to beam splitter 2 and mixes or interferes with the reference beam at beam splitter 3. If the object is moving or vibrating, this mixing process produces an intensity fluctuation in the light. Whenever the object has moved by one-half of the wavelength, $\lambda/2$, which is 0.3169 μm in the case of a He-Ne laser, the intensity goes through a complete dark-bright-dark cycle. A detector converts this signal to a voltage fluctuation. The Doppler frequency f_D of this sinusoidal cycle is proportional to the velocity υ of the object. Instead of detecting the Doppler frequency, the velocity is directly obtained by a *digital quadrature demodulation method*, which involves translating a given bandpass signal to a baseband signal and producing from it the in-phase and quadrature components. In the quadrature demodulator, the modulated carrier is passed through an inductance-capacitance (LC) tank circuit, that shifts the signal by 90° at the center frequency. This phase shift is either greater or less than 90°, depending on the direction of deviation. A phase detector compares the phase-shifted signal to the original to give the demodulated baseband signal.

The Bragg cell, which is an acousto-optic modulator to shift the light frequency by 40 MHz, is used for identifying the sign of the velocity.

What are the advantages of an LDV over similar measurement devices, such as an accelerometer? The LDV can be directed at targets that are difficult to access, or that may be too small or too hot to attach a physical transducer. Also, the LDV makes the vibration measurement without mass-loading the target, which is especially important for MEMS devices.

Returning to the discussion on cantilevers, what benefits accrue from using cantilevers? One great advantage of the cantilever technique is that its five response parameters (resonance frequency, phase, amplitude, Q-factor, and deflection) can be measured simultaneously. A compelling feature of the cantilever-based sensors operating in the resonance mode is that four response parameters (resonance frequency, phase, amplitude, and Q-factor, measured simultaneously) may provide complementary information about the interactions between the sensor and the environment. The bending and resonance frequency shifts of a cantilever are measurable to sub-angstrom resolution, using current techniques perfected for the AFM, such as optical beam deflection, piezoelectric, piezoresistive, and capacitance measurement methods.

Describing the theory of microcantilever, simple beam theory is restricted to a prismatic (equal cross section), homogeneous, straight, and untwisted structure (Figure 4.11). The beam is a structural member, usually horizontal, whose main function is to carry loads transverse to its longitudinal axis. These loads

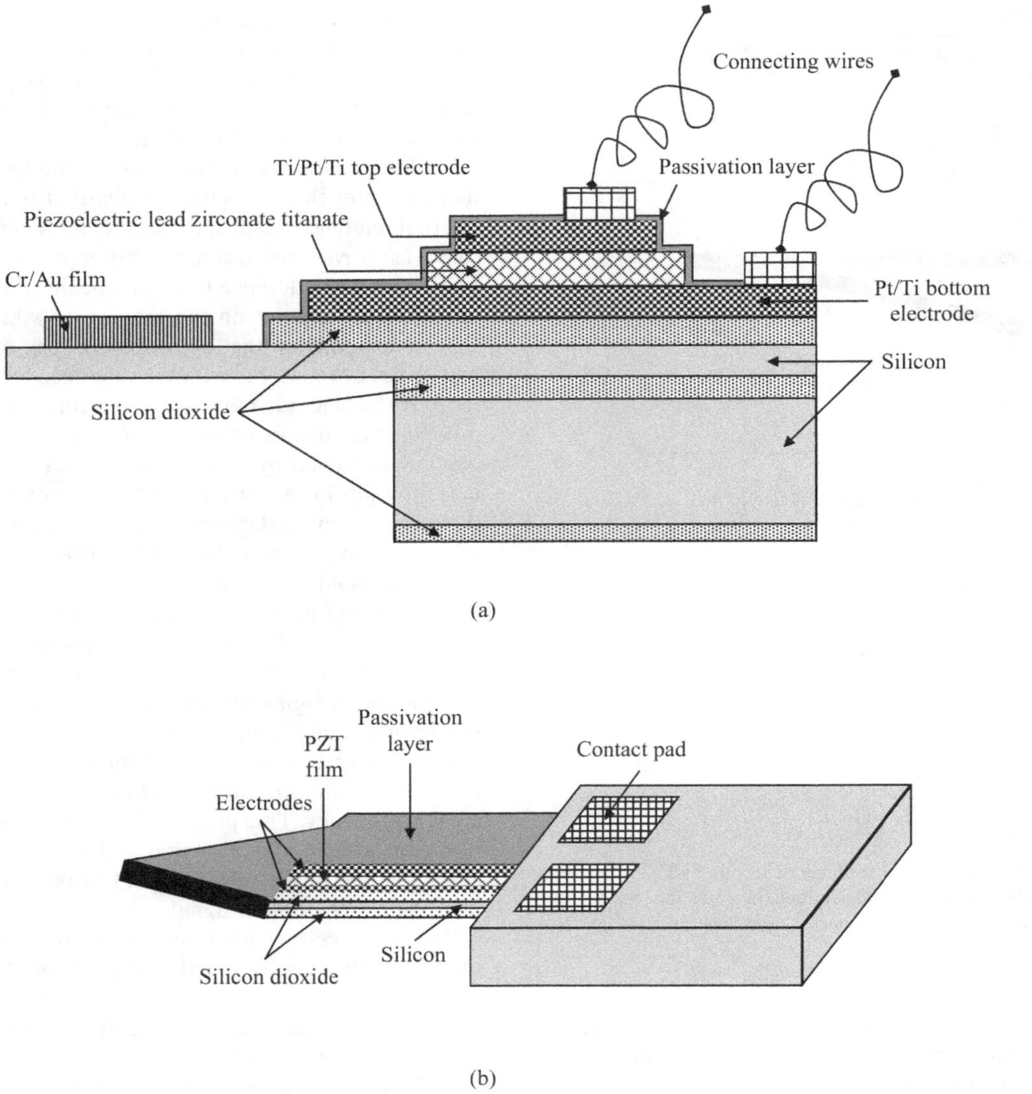

FIGURE 4.9 PZT-transduced cantilever for mass sensing: (a) cross section and (b) top and side view. (Lu, J. et al., *Jpn. J. Appl. Phys.*, 46, 7643, 2007.)

usually cause bending of the beam member. The beam is defined as a structure having two of its dimensions much smaller than the third. The thickness (t) and width (w) of the cantilever are small compared to the length (L), which reduces the analysis to a one-dimensional problem along the length of the beam. Additionally, it is presumed that the normal stresses (σ) in the y and z direction are negligible. The following derivation only holds if the maximum deflection (in y) is smaller than the radius of gyration (K). The *radius of gyration* of an object describes its dimensions. The radius of gyration of an object or body about a given axis is computed in terms of the moment of inertia around its center of gravity or a specified axis, and the total mass. It equals the square root of the ratio of the moment of inertia of the body about the given axis to its mass. Moment of inertia is the name given to rotational inertia, the rotational analog of mass for linear motion. It plays the same role in rotational dynamics as mass in linear dynamics. The measure of the inertia in the linear motion is the mass of the system, and its angular counterpart is the so-called moment of inertia. The moment of inertia of a body is not only related to its mass but also to the distribution of the mass throughout the body. It must be specified with respect to a chosen axis of rotation. For a point mass, the moment of inertia is simply the mass times the square of the perpendicular distance to the rotation axis.

Because K represents the root-mean-square distance of the infinitesimal parts constituting the object from the axis, if all the mass of the body is considered to be placed a distance K from the axis, the moment if inertia would be the same. Hence, it is a measure of how far the mass of the body is concentrated from its center of mass.

In beam theory, K is the distance from the neutral axis of a section to an imaginary point at which the whole area of the section could be concentrated and still have the same moment of inertia. It equals the square root of the moment of inertia divided by the area of the section.

If the maximum deflection approaches K, nonlinear terms have to be considered. The coordinate system used for the following derivation is shown in Figure 4.6b.

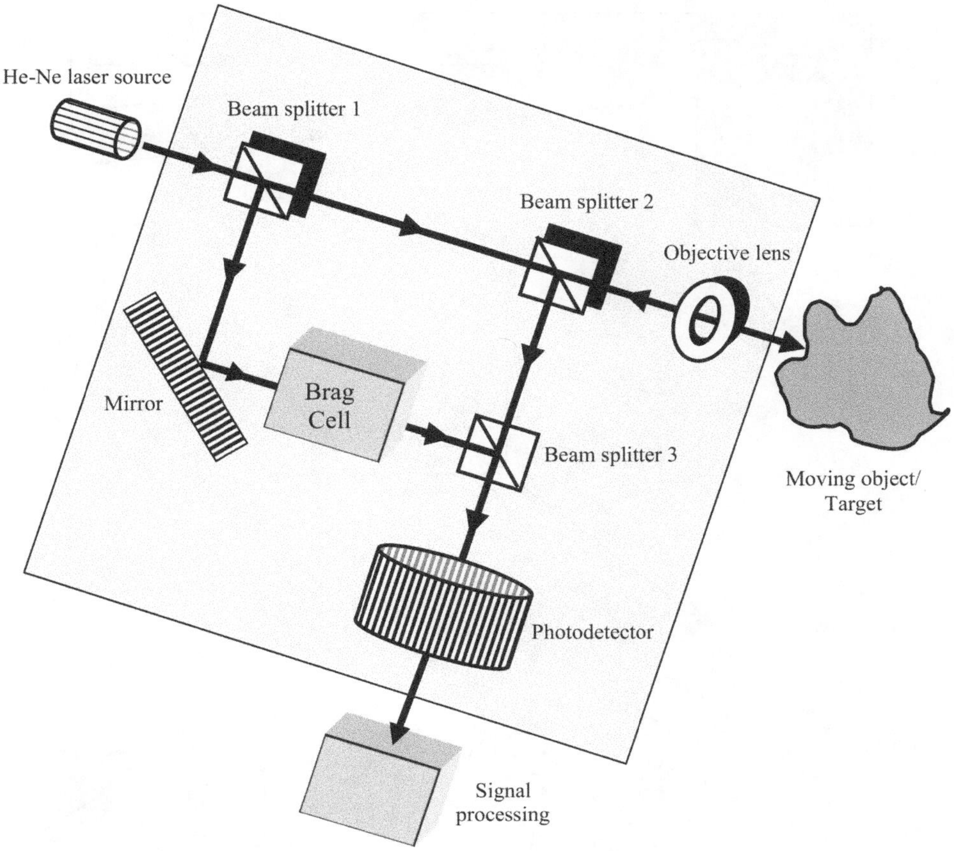

FIGURE 4.10 Modules of a laser Doppler vibrometer.

4.3.1.1 Resonance Frequency Formula

In Figure 4.11, the beam tip is under a uniformly distributed loading, σ. Its unit is Newton meter^{-1}, not Pascal, and $\sigma>0$ is tensile, and known as the surface stress. This surface stress effect is modeled as a concentrated moment applied at the free end of the cantilever beam (Duemling 2002).

The problem is simplified if all shear (a force tending to cause deformation of a material by slippage along a plane parallel or tangential to the imposed stress) and rotational forces are insignificant. The formulation is called the *Euler-Bernoulli beam theory*, also known as the engineer's beam theory or classical beam theory. It covers the case for small deflections of a beam which is subjected to lateral loads only.

The only remaining normal stress σ_x can be written as

$$\sigma_x = ky \tag{4.2}$$

where
k is a constant
$y=0$ lies in the center of the beam

The total internal force has to be zero, and it is given by

$$F_{\text{int}} = \int_A \sigma_x dA = 0 \tag{4.3}$$

A bending moment is a term used to describe the force or torque exerted on a material and leads to the event of bending or flexure within that material. Moments, internal or external, are rotation equivalents of forces, equal to force times distance. The bending moment in a beam is the moment, internal to the beam, necessary to counteract externally applied moments. The bending moment at a section through a structural element is defined as the sum of the moments about that section of all external forces acting to one side of that section. Without application of any external momentum, the total bending moment = the moment due to internal forces, which are only nonvanishing in the z direction:

$$M = M_z = \int_A y\sigma_x dA = k\int_A y^2 dA \tag{4.4}$$

The *area moment of inertia*, also known as the *second moment of inertia*, is a property of shape that is used to predict deflections and stresses in beams. The area moment of inertia of a cross-sectional area of a beam measures the ability of the beam to resist bending. The larger the moment of inertia, the less the beam will bend. This moment of inertia is a geometrical property of a beam and depends on a reference axis. The smallest moment of inertia about any axis passes through the *centroid*. The following is the mathematical equation to calculate the second moment of inertia:

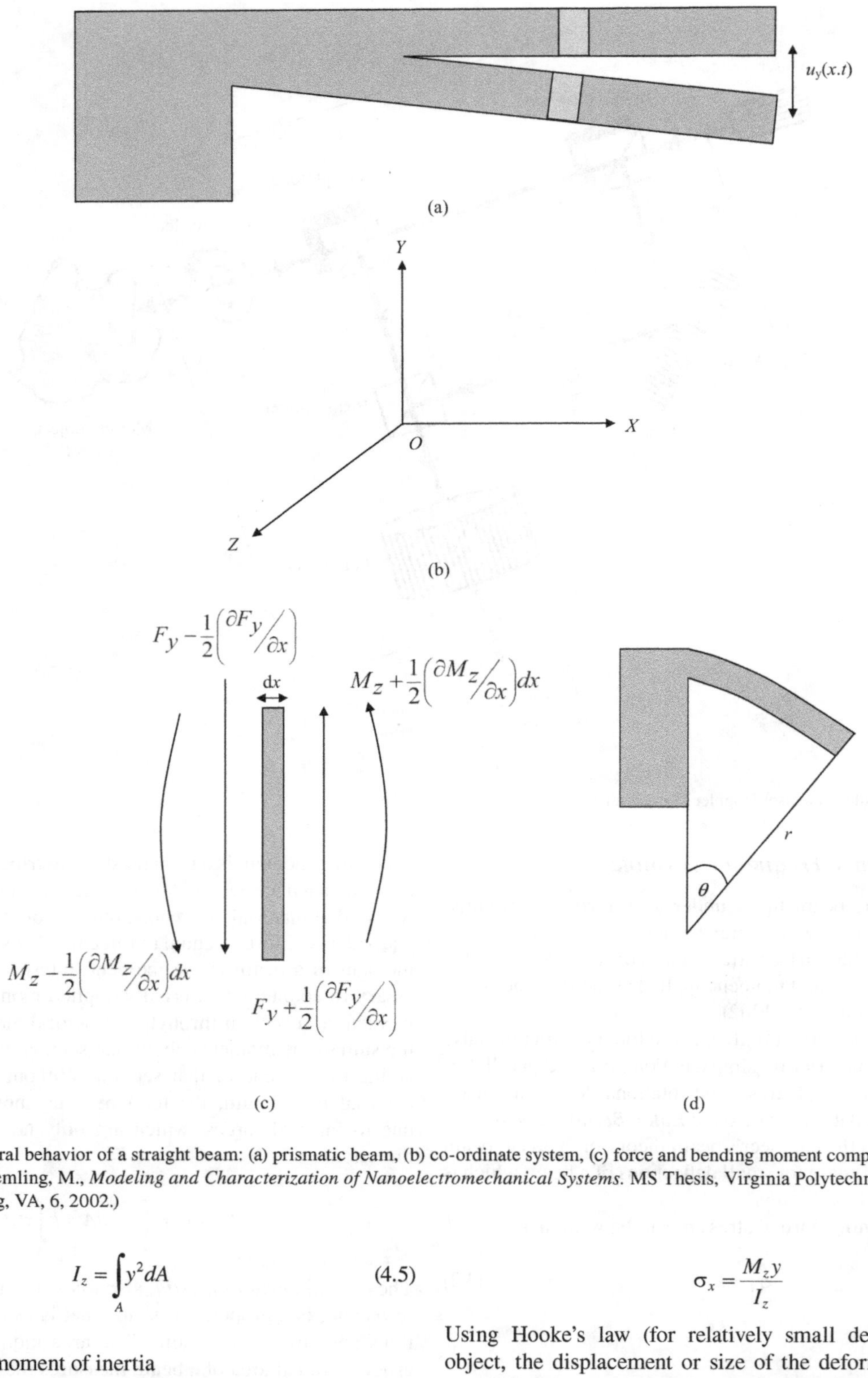

FIGURE 4.11 Flexural behavior of a straight beam: (a) prismatic beam, (b) co-ordinate system, (c) force and bending moment components, and (d) strain in the cantilever. (Duemling, M., *Modeling and Characterization of Nanoelectromechanical Systems.* MS Thesis, Virginia Polytechnic Institute and State University, Blacksburg, VA, 6, 2002.)

$$I_z = \int_A y^2 dA \tag{4.5}$$

where
- I_z is the second moment of inertia
- y is the perpendicular distance from axis z to the element dA
- dA is an elemental area
- k is defined by substituting Equation 4.5 into Equation 4.4 as

$$k = \frac{M_z}{I_z} \tag{4.6}$$

The stress of the cross section is therefore given by

$$\sigma_x = \frac{M_z y}{I_z} \tag{4.7}$$

Using Hooke's law (for relatively small deformations of an object, the displacement or size of the deformation is directly proportional to the deforming force or load), the strain (ε) is calculated as

$$\varepsilon_x = \frac{\sigma_x}{E} \tag{4.8}$$

where E is Young's modulus (longitudinal stress/longitudinal strain).

If $u_y(x, t)$ is the displacement of the beam in the y direction and the deflection is small ($du_y/dy \ll 1$), then the second derivative of the deflection is ≈ the inverse of the radius of curvature r

$$\frac{\partial^2 u_y(x,t)}{\partial x^2} = \frac{1}{r} \tag{4.9}$$

The strain can be found as

$$\varepsilon = \frac{dL - dL_o}{dL_o} = \frac{(r-y)d\theta - rd\theta}{r\sin d\theta} = -\frac{y}{r} \tag{4.10}$$

Combining Equations 4.8 through 4.10 gives the *Euler–Bernoulli law* of elementary beam theory

$$M_z = -EI_z\left\{\frac{\partial^2 u_y(x,t)}{\partial x^2}\right\} \tag{4.11}$$

In the absence of any external forces or bending moments acting on the beam, the equation of motion is

$$m\left\{\frac{\partial^2 u_y(x,t)}{\partial t^2}\right\} = \sum F_{\text{int}} \tag{4.12}$$

and the total moment has to be zero

$$\sum M_{\text{int}} = 0 \tag{4.13}$$

The total force is calculated

$$\sum F_{\text{int}} = \left\{F_y + \frac{1}{2}\left(\frac{\partial F_y}{\partial x}\right)dx\right\} - \left\{F_y - \frac{1}{2}\left(\frac{\partial F_y}{\partial x}\right)dx\right\} = \left(\frac{\partial F_y}{\partial x}\right)dx \tag{4.14}$$

and the sum of the bending moments is

$$\begin{aligned}\sum M_{\text{int}} &= \left\{M_z + \frac{1}{2}\left(\frac{\partial M_z}{\partial x}\right)dx\right\} - \left\{M_z - \frac{1}{2}\left(\frac{\partial M_z}{\partial x}\right)dx\right\} \\ &\quad + \left\{F_y + \frac{1}{2}\left(\frac{\partial F_y}{\partial x}\right)dx\right\}\frac{dx}{2} - \left\{F_y - \frac{1}{2}\left(\frac{\partial F_y}{\partial x}\right)dx\right\}\left(-\frac{dx}{2}\right)\end{aligned} \tag{4.15}$$

Combination of Equations 4.13 and 4.15 give the relationship interlinking the bending moment and the force

$$F_y = -\frac{\partial M_z}{\partial x} \tag{4.16}$$

Using Equations 4.16, 4.14, and the mass

$$m = \rho A dx \tag{4.17}$$

where
- ρ is the density
- A is the cross section

the equation of motion is expressed as

$$\rho A \frac{d^2 u_y(x,t)}{dt^2} + EI_z \frac{\partial^4 u_y(x,t)}{\partial x^4} = 0 \tag{4.18}$$

This harmonic linear fourth-order differential equation is solved using the *separation of variables* method. For a partial differential equation in a function $\Phi(x, y, \ldots)$ and variables $x, y, \ldots$, separation of variables can be applied by making a substitution of the form $\Phi(x, y, \ldots) \equiv X(x)Y(y), \ldots$, breaking the resulting equation into a set of independent ordinary differential equations, solving these for $X(x)$, $Y(y)$, …, and then plugging them back into the original equation. This technique works because, if the product of functions of independent variables is a constant, each function must separately be a constant.

Here, the complete solution is not of interest to us, but only the natural resonance frequency of the beam, which is easily obtained by using a Fourier transformation. The Fourier transform, in essence, breaks down or separates a function into sinusoids of different frequencies, which sum to generate the original waveform represented by that function. The Fourier transform is based on the fact that it is possible to take any periodic function and resolve it into an equivalent infinite summation of sine waves and cosine waves, with frequencies that start at 0 and increase in integer multiples of a base frequency.

Applying a Fourier transformation to Equation 4.18 with

$$\phi\{u_y(x,t)\} = U_y(x,\omega) \tag{4.19}$$

yields

$$\rho A (i\omega)^2 U_y(x,\omega) + EI_z \frac{\partial^4 U_y(x,\omega)}{\partial x^4} = 0 \tag{4.20}$$

The calculation becomes simpler if the equation is rewritten as

$$-\alpha^4 \omega_y^2 U_y(x,\omega) + \frac{\partial^4 U_y(x,\omega)}{\partial x^4} = 0 \tag{4.21}$$

with

$$\alpha = \sqrt[4]{\frac{\rho A}{EI_z}} \tag{4.22}$$

The solution of this differential equation is

$$\begin{aligned}U_y(x,\omega) &= A_1 \exp\left(\alpha x\sqrt{\omega}\right) + A_2 \exp\left(-\alpha x\sqrt{\omega}\right) \\ &\quad + A_3 \exp\left(i\alpha x\sqrt{\omega}\right) + A_4 \exp\left(-i\alpha x\sqrt{\omega}\right)\end{aligned} \tag{4.23}$$

where A_1, A_2, A_3, and A_4 are complex constants which are determined from the boundary condition. The relationship between the trigonometric functions and the complex exponential function, Euler's formula for the complex exponential states that, for any real number θ, $e^{i\theta} = \cos\theta + i\sin\theta$, where e is the base of the natural logarithm, i is the imaginary unit, and cos and sin are the trigonometric functions cosine and sine). Using the Euler equations and the comparable equations for sinh and cosh [sinh

$x=(e^x-e^{-x})/2$ and $\cosh x=(e^x+e^{-x})/2$], Equation 4.23 can be transformed to an equation with real constants B_1, B_2, B_3, B_4, as

$$U_y(x,\omega)=B_1\sin\left(\alpha x\sqrt{\omega}\right)+B_2\cos\left(\alpha x\sqrt{\omega}\right)+B_3\sinh\left(\alpha x\sqrt{\omega}\right)+B_4\cosh\left(\alpha x\sqrt{\omega}\right) \quad (4.24)$$

By applying the boundary conditions of the problem to Equation 4.24, the resonance frequency is found. In the following discussion, a cantilever will be considered. Let us focus our attention on the clamped side and free end of the beam. Because, at the clamped side of the cantilever ($x=0$), no displacement takes place and the beam is straight, the boundary conditions are given by

$$U_y(0,\omega),\quad \frac{dU_y(0,\omega)}{dx}=0 \quad (4.25)$$

At the free end of the beam ($y=L$), there is no bending moment and no shear forces act on the beam:

$$\frac{d^2U_y(l,\omega)}{dx^2}=0,\quad \frac{d^3U_y(l,\omega)}{dx^3}=0 \quad (4.26)$$

Applying the first two boundary conditions, we have $B_2=B_4$ and $B_1=-B_3$. The last two boundary conditions reduce Equation 4.24 to

$$\frac{2+2\cos\left(\alpha l\sqrt{\omega}\right)\cosh\left(\alpha l\sqrt{\omega}\right)}{\sin\left(\alpha l\sqrt{\omega}\right)-\sinh\left(\alpha l\sqrt{\omega}\right)}=0 \quad (4.27)$$

A nontrivial solution (a solution of a set of equations in which at least one of the variables has a value different from zero) is found if

$$\cos\left(\alpha l\sqrt{\omega}\right)\cosh\left(\alpha l\sqrt{\omega}\right)=-1 \quad (4.28)$$

There is no analytical solution (one that can explicitly be written down, also called a closed-form solution) for this equation but it is solved numerically by using the substitution

$$\beta=\alpha l\sqrt{\omega} \quad (4.29)$$

The values for β_i are given in Table 4.1. From Equations 4.22 and 4.21, the natural resonance frequency and its harmonics (frequencies that are integral multiples of the fundamental frequency) are calculated:

$$\omega_i=\left(\frac{\beta_i^2}{l^2}\right)\sqrt{\frac{EI_z}{\rho A}} \quad (4.30)$$

TABLE 4.1

β_i^2 values for Different Beams

	Value of β_i^2 for i =				
Type of Beam	**1**	**2**	**3**	**4**	**5**
Cantilever	3.516	22.034	61.701	120.912	199.855
Clamped-clamped beam	22.373	61.678	120.903	199.860	298.526
Free-free beam	22.373	61.678	120.903	199.860	298.526

For $i=1$, $\beta_i^2=3.516$; and for a rectangular cantilever

$$I_z=\frac{wt^3}{12} \quad (4.31)$$

Hence, from Equation 4.30, we get

$$\omega_1=\left(\frac{3.516}{l^2}\right)\sqrt{\frac{Ewt^3}{12\rho tw}}=1.015\sqrt{\frac{E}{\rho}}\left(\frac{t}{l^2}\right) \quad (4.32)$$

The fundamental frequency ω_1 is a function of the material parameters E and ρ, as well as the beam dimensions t and l. High frequencies are achieved by reducing the overall cantilever scale, by choosing stiffer (not easily bent or changed in shape) and lighter materials, and by reducing the aspect ratio l/t.

Example 4.1

Nine cantilevers are fabricated from three materials: silicon, silicon nitride, and silicon carbide. The geometrical dimensions of the cantilevers are: (200 μm [length] × 100 μm [width] × 10 μm [thickness]; 6 μm [length] × 1 μm [width] × 0.17 pm [thickness]; 10 μm [length] × 0.6 μm [width] × 0.75 μm [thickness]). The fabricated set of nine cantilevers cantilevers consists of three silicon cantilevers, one of each specified length × width × thickness; similarly three silicon nitride cantilevers, one of each given size; and three silicon carbide cantilevers, one of each size. The properties of these materials are presented in Table 4.2. Calculate the angular resonance frequencies of these cantilevers.

For cantilever length = 200 μm, width = 100 μm, thickness = 10 μm, from Equation 4.32,

$$\omega_1=1.015\sqrt{\frac{E}{\rho}}\left(\frac{t}{l^2}\right)=1.015\left\{\frac{10\times10^{-6}}{\left(200\times10^{-6}\right)^2}\right\}\sqrt{\frac{E}{\rho}}=250\sqrt{\frac{E}{\rho}} \quad (4.33)$$

For cantilever length = 6 μm, width = 1 μm, thickness = 0.17 μm,

$$\omega_1=1.015\sqrt{\frac{E}{\rho}}\left(\frac{t}{l^2}\right)=1.015\left\{\frac{0.17\times10^{-6}}{\left(6\times10^{-6}\right)^2}\right\}\sqrt{\frac{E}{\rho}}=4722.22\sqrt{\frac{E}{\rho}} \quad (4.34)$$

TABLE 4.2

Young's Modulus and Density of Materials (See Example 4.1)

Material	Young's Modulus (E) in N m^{-2}	Density (ρ) in kg m^{-3}
Silicon	1.7×10^{11}	2329
Silicon nitride	3.1×10^{11}	3290
Silicon carbide	4.1×10^{11}	3100

For cantilever length = 10 μm, width = 0.6 μm, and thickness = 0.75 μm,

$$\omega_1 = 1.015\sqrt{\frac{E}{\rho}}\left(\frac{t}{l^2}\right) = 1.015\left\{\frac{0.75\times10^{-6}}{\left(10\times10^{-6}\right)^2}\right\}\sqrt{\frac{E}{\rho}} = 7500\sqrt{\frac{E}{\rho}} \quad (4.35)$$

The angular resonance frequencies are shown in Table 4.3. To obtain linear frequencies *f*, they are divided by 2π.

4.3.1.2 Deflection Formula

In addition to resonance frequency change, deflection (bending) of the cantilever beam may change due to adsorption-induced surface stress, and it is interesting to identify the parameters that influence this bending. The equation governing the beam deflection is as follows:

$$E^*I\left(\frac{d^2y}{dx^2}\right) = M \quad (4.36)$$

and the boundary conditions are given as

$$y(0) = 0, \quad \frac{dy(0)}{dx} = 0 \quad (4.37)$$

Here the symbol E^* denotes the biaxial modulus, which is the modulus corresponding to biaxial stress, i.e., the stress in a structural member about two perpendicular axes at the same time. E^* takes into consideration the biaxial plane strain conditions associated with thin films and the film-substrate interface. The parameter E^* is defined as

$$E^* = \frac{E}{1-\mu} \quad (4.38)$$

where

E is Young's modulus of the beam

μ is Poisson's ratio (the ratio of transverse contraction strain to longitudinal elongation strain in the direction of the stretching force)

y is the beam deflection

I is the area moment of inertia

For a rectangular beam,

$$I = \frac{wt^3}{12} \quad (4.39)$$

w and *t* are the beam width and thickness, respectively, *M* is the concentrated moment defined as

$$M = \frac{\sigma wt}{2} \quad (4.40)$$

Equation 4.36 is solved by using integration and boundary conditions as follows:

$$y = \frac{Mx^2(1-\mu)}{2EI} \quad (4.41)$$

For linear analysis of small deflections, the curvature κ of the deflected beam approximates to d^2y/dx^2. So, the curvature κ and the radius of curvature *R* are given by

$$\kappa = \frac{1}{R} = \frac{M(1-\mu)}{EI} = \frac{6\sigma(1-\mu)}{Et^2} \quad (4.42)$$

Equation 4.42 is the *Stoney formula*, serving as a cornerstone for the analysis of curvature-based measurements. The Stoney (1909) formula has been modified by later workers many times to evaluate bilayer and multilayer structures with arbitrary layer thickness ratios. Here, it was assumed, for the sake of clarity and simplicity, that surface stress existed on only the upper surface of the beam. For more generalized cases, σ is substituted for by differential surface stress

$$\sigma = \sigma_+ - \sigma_- \quad (4.43)$$

where σ_+ and σ_- are surface stresses on the upper and lower surfaces, respectively. Also, this substitution does not affect any derivation for linear analysis. Evidently, the modeling of the surface stress effect as a concentrated moment at the free end of the cantilever beam assures that the beam curvature is uniform (constant).

The radius of curvature *R* is related to the deflection *y* and length *l* of the cantilever as

$$\frac{1}{R} = \frac{3y}{2l^2} \quad (4.44)$$

TABLE 4.3
Resonance Frequencies for Different Cantilever Structures, Fabricated Using Various Materials (See Example 4.1)

			ω_1 (Radians s^{-1})		
Material	**E/ρ (kg × ms^{-2} m^{-2} kg^{-1} × m^{-3}) = (s^{-2} m^{-2})**	**$\sqrt{(E/\rho)}$ $\sqrt{(s^{-2}\,m^{-2})} = s^{-1}\,m^{-1}$**	**l = 200 μm, t = 10 μm**	**l = 6 μm, t = 0.75 μm**	**l = 10 μm, t = 0.75 μm**
Silicon	$1.7\times10^{11}/2329 = 7.2993\times10^7$	8.54×10^3	2.135×10^6	4.0328×10^7	1.601×10^8
Silicon nitride	$3.1\times10^{11}/3290 = 9.422\times10^7$	9.71×10^3	2.4275×10^6	4.585×10^7	1.821×10^8
Silicon carbide	$4.1\times10^{11}/3100 = 1.323\times10^8$	1.15×10^4	2.875×10^6	5.43×10^7	2.156×10^8

An equation connecting the cantilever displacement with the differential surface stress is obtained as

$$y = \frac{4l^2\sigma(1-\mu)}{Et^2} \tag{4.45}$$

Therefore, for the remaining parameters which stay constant, the deflection of the cantilever is directly proportional to the adsorption-induced differential surface stress. Surface stress has units of N m^{-1} or J m^{-2}. An alternative, popular form of Equation 4.45 contains the multiplier "3" in place of "4."

Example 4.2

For a silicon cantilever of length = 5 μm, width = 1 μm, thickness = 0.1 pm, E = 1.5×10^{11} N m^{-2}, Poisson's ratio of silicon = 0.17. If the deflection of the cantilever is 5 nm, what is the stress developed in the cantilever beam?

Equation 4.45 gives the stress σ as

$$\sigma = \frac{yEt^2}{\{4l^2(1-\mu)\}} = \frac{5\times10^{-9}\times1.5\times10^{11}\times(100\times10^{-9})^2}{4\times(5000\times10^{-9})^2\times(1-0.17)} \tag{4.46}$$

$$= 0.0904 \text{ N m}^{-1}$$

4.3.2 Energy Dissipation and *Q*-Factor of Cantilever

In analogy to other types of resonators, quantification of energy dissipation in micro-electromechanical systems (MEMS)/nanoelectromechanical systems (NEMS) is commonly carried out in terms of the *quality factor* or *Q-factor.* One of the essential features of microcantilevers is that they are mechanical devices (oscillators) that accumulate and store mechanical energy. The *Q*-factor is inversely proportional to the damping coefficient, the ratio of the logarithmic decrement (defined as the natural logarithm of the ratio of two successive maxima of the decaying sinusoid) of any under-damped harmonic motion to its period, or total energy lost per cycle of vibration in a microcantilever transducer, and is defined as

$$Q = \frac{2\pi W_0}{\Delta W_0} \tag{4.47}$$

where W_0 and ΔW_0 are the mechanical energy accumulated and dissipated, respectively, in the device per vibration cycle. As the *Q*-factor critically controls both the resonance behavior of any microcantilever and its off-resonance thermal noise, it is an important parameter characterizing MEMS/NEMS sensors operating in resonance as well as in static regimes. *Thermal noise* is the noise generated by thermal agitation of electrons in a conductor. The noise power, *P*, in watts, is given by $P = k_B T \Delta f$, where k_B is Boltzmann's constant in joules per Kelvin, *T* is the conductor temperature in Kelvin, and Δf is the bandwidth in hertz. Thermal noise power, per hertz, is equal throughout the frequency spectrum, depending only on k_B and *T.*

Based on the spectral analysis, an advanced mathematical technique for studying phenomena that occur in cycles, the *Q*-factor is calculated as a ratio of the resonance frequency f_0 to the width of the resonance peak at its half amplitude. Hence, the *Q*-factor is frequently used to represent the degree of sharpness of the resonance peak.

What parameters influence the Q-factor of a microcantilever? It depends on a number of intrinsic and extrinsic parameters, such as cantilever constructional material, its geometrical shape, and the viscosity (the quantity that describes a fluid's resistance to flow) of the medium in which the cantilever is immersed. Obviously, increased damping of a microcantilever oscillator by the medium translates into lower *Q*-factor values, as compared with the same oscillator in a vacuum. Models of drag forces, surface friction, and pressure, exerted on solid bodies in fluids are used to appraise viscous damping effects.

The other damping mechanisms involve clamping loss and internal friction (the motion-resisting force between the surfaces of particles making up a substance) within the microcantilever. *Clamping loss,* occurring at the joint, grip, or support, has a negligible contribution to the total dissipation in the longer microcantilever, possessing high length-to-width and width-to-thickness ratios. However, the ultimate minimization-of-clamping loss is achieved in oscillators with double-paddle or butterfly geometries, instead of single-clamped cantilevers or double-clamped bridges. Hence, scientists delving into fundamental studies of intrinsic friction effects in MEMS frequently perform resonance measurements in double-paddle resonators. *Q*-factors as high as 10^5 have been reported for torsional butterfly-shaped resonators fabricated from single-crystal silicon. Torsion refers to deformation caused when one end of an object is twisted in one direction and the other end is held motionless or twisted in the opposite direction.

Internal friction is linked to assorted physical phenomena, especially thermoelastic dissipation (TED) motion of lattice defects, phonon-phonon scattering, and surface effects. TED is an energy loss mechanism whereby mechanical energy stored in the cantilever is irrecoverably transferred to the thermal domain; strain gradients lead to temperature gradients in the cantilever, and, if these temperature gradients are allowed to relax *via* heat flow, the mechanical energy is irrecoverable, i.e., TED has occurred, and phonon is the quantum of acoustic or vibrational energy. In any vibratory structure, the strain field causes a change in the internal energy (the total kinetic and potential energy associated with the motions and relative positions of the molecules of an object, excluding the kinetic or potential energy of the object as a whole), such that its compressed region is heated whereas its extended region becomes cold. Due to the resulting lack of thermal equilibrium between various parts of the vibrating structure, energy is dissipated when an irreversible heat flow driven by the temperature gradient occurs; this is TED, as pointed out earlier in this paragraph.

The TED limit and phonon-phonon scattering mechanisms correspond to very high *Q*-factors (10^6–10^8), which are scarcely observed experimentally due to the dominant contributions from other dissipation mechanisms affecting real MEMS devices. The assertion that surface effects constrain *Q*-factors of MEMS oscillators in vacuum is confirmed by annealing the device (treatment of a device by heating to a predetermined temperature, holding for a certain time, and then cooling to room temperature) and changing its surface in a controlled way. As the thickness of the oscillator decreases, TED becomes a trivial mechanism, causing dissipation.

What are the typical magnitudes of Q-factors of cantilevers? Q-factors of MEMS resonators under vacuum are extremely high. However, Q-factors of rectangular microcantilevers in air are typically in the range of 10–1000, whereas cantilever transducers in aqueous solutions hardly ever have Q-factors above 10.

Very intense damping in viscous liquids makes resonant operation of microcantilevers, and, in turn, measurements of adsorbed mass using microcantilever sensors, somewhat challenging. In order to surmount the difficulties of resonant cantilever operation in liquids, cantilever transducers are used as a part of a self-oscillating system with positive feedback (effect of an action returned to amplify what caused it), e.g., the signal from the microcantilever readout is amplified and fed back to a piezoelectric actuator connected to the microcantilever.

4.3.3 Noise of Cantilever and Its Mass Detection Limit

When the cantilever resides in a thermal bath with a temperature T, the free cantilever is subjected to a random time-dependent Brownian motion around its equilibrium position. This is generally known as *noise*. *Brownian motion* is the irregular motion in pseudo-random (a process that appears to be random but is not) or *stochastic* paths (involving chance or probability, as opposed to deterministic paths) of small particles in a fluid that takes place even if the fluid in question is calm. It arises from the thermal motion of the molecules of the fluid.

Noise processes in microcantilever sensors (Albrecht et al. 1991) can be distinguished into processes intrinsic to the device and those due to interactions of the cantilever with its environment, e.g., adsorption-desorption noise or noise originating from the read-out circuit. Here, the focus is on the intrinsic noise mechanisms. *Why?* Because it is these mechanisms that determine the ultimate fundamental limits of the microcantilever sensor performance and thereby establish the lower mass detection boundary of the device.

Efforts have been made to recognize the intrinsic sources of noise in a mechanical system and to identify the relationships between parameters of the mechanical system and its noise level. In contrast to a simple harmonic oscillator, a cantilever has different eigenfrequencies (resonance frequencies; *eigen*=own, particular) ω_n and a mode-dependent spring constant (the restoring force of a spring per unit of length; a measure of how stiff the spring is) D_n. For the nth mode of the cantilever, and for a small frequency interval $\Delta\omega$, on resonance, the average fluctuation $\langle u_n^2 \rangle$ of the sensor is given by (Rast et al. 2000)

$$\langle u_n^2 \rangle = \left(\frac{2k_BT}{\pi}\right)\left(\frac{Q\Delta\omega}{\omega_n D_n}\right) \tag{4.48}$$

where

k_B is Boltzmann's constant
T is the absolute temperature
Q is the quality factor
D_n is the mode-dependent spring constant

For the first four modes, D_n has values as follows: $D_1 = 3EI/L^3$, $D_2 \approx 121.3D_1$, $D_3 \approx 951.6D_1$, and $D_4 \approx 3654.3D_1$. It is easily surmised that the noise amplitude increases at higher temperatures T and decreases dramatically at higher modes, since the mode-dependent spring constant increases. *Prima facie*, it appears that the force sensitivity of an interacting cantilever can be increased by using higher eigenmodes in ambient atmosphere, due to the reduction in viscous air damping. But this seems to be only one aspect of the circumstances. The quality factors Q of commercial cantilevers were measured by Rast et al. (2000). The base pressure of the vacuum system was maintained at 10^{-5} mbar throughout the experiment. It was found that, for the second harmonic, the Q-value was more than 50% smaller than that for the first harmonic. Unequivocally, this indicated that internal friction of the sensor has a frequency-dependent behavior. Since the Q-factor decreased on the second eigenmode (a normal mode of vibration of an oscillating system), the sensitivity cannot be increased using only higher eigenmodes. To improve the sensitivity, crystalline materials with a low internal friction, such as aluminum or silicon, should be used. For small structures, the surface and bulk represent two independent dissipation channels. A native SiO_2 layer grows on silicon cantilevers when exposed to air. The internal friction of bulk SiO_2 is larger than that of crystalline Si. The sensitivity of the cantilever is therefore appreciably reduced. Promising results are obtained on removing this native oxide layer by HF etching, sputtering, and/or annealing under ultrahigh vacuum conditions.

What are the deciding factors for the ultimate mass detection limit of a cantilever? Let us try to answer this question. Upon equilibrating a microcantilever detector with the ambient thermal environment (a thermal bath), a continuous exchange of the mechanical energy which accumulated in the device occurs with the thermal energy of the environment. This exchange, dictated by the fluctuation dissipation theorem (a nonequilibrium state may have been reached either as a result of a random fluctuation or an external force, e.g., an electric or magnetic field, and that the evolution toward equilibrium is the same in both cases for a sufficiently small fluctuation), produces spontaneous oscillation of the microcantilever. Hence, the average mechanical energy per mode of cantilever oscillation is defined by thermal energy k_BT. This noise is referred to as thermally induced cantilever noise. Also called *Johnson noise, thermal noise* is the random *white* noise generated by the thermal disturbance of electrons in a conductor or electronic device. *White noise* is a random signal (or process) with a flat power spectral density, meaning that the power spectral density is nearly equal throughout the frequency spectrum. Thus, it is understandable that any cantilever in equilibrium with its thermal environment has a built-in source of white thermal noise $\psi_{th}(f)$, given by (Albrecht et al. 1991)

$$\psi_{th} = \frac{4m_0k_BT}{Q} \tag{4.49}$$

where m_0 is the effective suspended mass of the cantilever. At frequencies far below the resonance frequency, the amplitude of the thermally induced oscillation of a cantilever beam is proportional to the square root of the thermal energy and is expressed as (Lavrik et al. 2004)

$$\sqrt{\langle \delta z^2 \rangle} = \sqrt{\frac{2k_BTB}{\pi k f_0 Q}} \tag{4.50}$$

where B is the bandwidth requirement. Clearly, lower cantilever stiffness corresponds to greater amplitudes of thermal noise. Consequent upon the dynamic exchange between cantilever mechanical energy and the ambient thermal energy, the actual frequency f of thermally induced cantilever oscillations at any given moment deviates markedly from the resonance frequency f_0. The amplitude of such frequency fluctuations δf_0, due to the exchange between mechanical and thermal energy, is (Albrecht et al. 1991)

$$\delta f_0 = \left(\frac{1}{z_{max}}\right)\sqrt{\frac{2\pi f_0 k_B TB}{kQ}} \tag{4.51}$$

where z_{max} is the amplitude of the cantilever oscillations. Equation 4.51 predicts increased absolute fluctuation of the resonance frequency δf_0 as the resonance frequency f_0 increases. Nevertheless, the relative frequency instability $\delta f_0/f_0$ decreases for higher-frequency oscillators (Lavrik et al. 2004)

$$\frac{\delta f_0}{f_0} = \left(\frac{1}{z_{max}}\right)\sqrt{\frac{2\pi k_B TB}{kQf_0}} \tag{4.52}$$

By changing the physical dimension of a cantilever, its mass detection limit is affected by many orders of magnitude. For a given cantilever design, the smallest (thermal-noise-limited) detectable change in the surface density is found by combining Equation 4.52 with the equation for fundamental resonance frequency f_0 of the microcantilever:

$$f_0 = \left(\frac{1}{2\pi}\right)\sqrt{\frac{k}{m_0}} \tag{4.53}$$

for a spring constant k and an effective suspended mass m_0 which consists of both a concentrated and a distributed mass. It is expressed as (Lavrik et al. 2004)

$$\Delta m_{th} = 8\sqrt{\frac{2\pi^5 kk_B TB}{f_0^5 Q}} \tag{4.54}$$

This equation shows that the mass sensitivity of a cantilever transducer, operating in the resonance mode, increases as its dimensions are reduced. Therefore, cantilever sensors with progressively increased mass sensitivity are fabricated by simply decreasing the transducer dimensions. As the technology of nanosized mechanical structures moves forward, nanomechanical devices approach the gigahertz frequency domain, that is already widely discovered with electronic and optical devices.

How are the aforesaid small-dimension cantilevers realized practically? Two advances have been crucial to breaking the 1 GHz barrier in NEMS (Huang et al. 2003). First and foremost is the use of silicon carbide epilayers, which are of comparable density to the usual silicon but are many-fold stiffer than silicon, allowing higher frequencies to be attained for structures of similar geometry. This will be discussed in more detail in the ensuing paragraphs and further in Section 4.3.4. The second factor is the development of balanced, high-frequency displacement transducers, which enable the ubiquitous passive embedding impedances, brought into play by electrical connections to the macroworld, to be nullified. If left uncontrolled, these parasitic impedances mask the electromechanical impedance of interest, i.e., the signal, in ultra-small NEMS. These advances overcome the dual challenges of detecting tiny displacements (on the scale of femtometers) at microwave frequencies. The characteristic frequency of NEMS increases with decreasing size, but their displacement (when operating linearly) and their electromechanical impedance both simultaneously decrease.

Fundamental mechanical resonances for two nominally identical, cubic silicon carbide (3C–SiC) double-clamped beams, roughly 1.1 μm long, 120 nm wide, and 75 nm thick, were detected at 1.014 and 1.029 GHz by Huang et al. (2003).

The advantages favoring silicon carbide (SiC) require detailed explanation. SiC is an excellent material for high-frequency NEMS for the following reasons. The goal of attaining extremely high fundamental resonance frequencies in NEMS, while simultaneously preserving the small force constants necessary for high sensitivity, requires pushing against the ultimate resolution limits of lithography and nanofabrication processes. Flexural (a curve, turn, or fold) mechanical resonance frequencies for beams depend directly upon the ratio (E/ρ). The ratio of Young's modulus, E, to mass density, ρ, for SiC is markedly higher than for other commonly used semiconducting materials for electromechanical devices, such as Si and GaAs. SiC, given its larger (E/ρ), yields devices that operate at much higher frequencies for a given geometry, than would otherwise be possible using conventional materials. Furthermore, for NEMS, the Q-factor is governed by surface defects and depends on the device surface-to-volume ratio. SiC possesses excellent chemical stability. This makes surface treatments an option for higher-quality factors (Q-factors) of resonance.

4.3.4 Doubly Clamped and Free-Free Beam Resonators

The mathematical analysis carried out for the cantilever is extendable to other MEMS/NEMS structures. For a beam clamped on both sides, the boundary conditions are written as

$$U_y(0,\omega) = 0,\ \frac{dU_y(0,\omega)}{dx} = 0,\ U_y(l,\omega) = 0,\ \frac{dU_y(l,\omega)}{dx} = 0 \tag{4.55}$$

while, for a beam free on both sides, the boundary conditions are stated as

$$\begin{aligned} &\frac{d^2U_y(0,\omega)}{dx^2} = 0, \quad \frac{d^3U_y(0,\omega)}{dx^3} = 0, \\ &\frac{d^2U_y(l,\omega)}{dx^2} = 0, \quad \frac{d^3U_y(l,\omega)}{dx^3} = 0 \end{aligned} \tag{4.56}$$

The solution method is similar to the one formulated for the cantilever. The final solution is the same for the clamped-clamped and the free-free beam and only differs from the cantilever in the factor β_i, the values of which are also listed in Table 4.1.

Doubly clamped beam resonators based upon NEMS, with operating frequencies within the microwave L-band (1–2 GHz), have been achieved. Quality factors of these microwave NEMS resonators decreased as the device frequency was increased. *What could be the causes?* A major possible reason for the decrease of quality factor in these devices is the *clamping loss* intrinsic to the doubly clamped boundary condition.

Wang et al. (2000) explored the use of a free-free boundary condition to reduce this source of acoustic loss for micromachined resonators. Huang et al. (2003; 2005) implemented the in-plane (lateral) free-free beam design strategy to nanoscale mechanical resonators, to explore the possibility of reducing clamping loss. They fabricated pairs of devices consisting of a free-free beam and a doubly clamped beam, very similar in designed fundamental-mode resonance frequencies, for comparison in control experiments.

How do the quality factors for the two structures differ? The SiC free-free beam nanomechanical resonators offered extraordinary improvement in quality factor, compared with the doubly clamped beam design, operating at similar frequencies. The free-free beam had a much higher quality factor than the best value from its doubly clamped counterpart. The quality factor of the doubly clamped beam was ~4,500, whereas that of the free-free beam resonator was ~11,000. The dissipation according to clamping loss in doubly clamped beam devices increased as the aspect ratio was reduced.

What is the effect of film surface roughness on device performance? Devices were fabricated from films differing widely in surface roughness to investigate the effect of film quality on resonator performance. Devices made from films that had a low surface roughness (~2 nm or below) operated well in the ultra-high frequency (UHF) (300 MHz–3 GHz)/microwave (3–30 GHz) regime. In contrast, devices made from rougher films (up to ~7.1 nm) were operational only in the very high frequency (VHF) (30–300 MHz) range, but not higher. These results firmly indicate that it is critically important to optimize the growth processes to produce ultrasmooth thin films, if these films are to be used for nanomechanical resonators having reasonably high Q values.

By examining devices made from SiC wafers with different roughness, a strong correlation between surface roughness and quality factor was established from experiments. The quality factor decreases with increasing surface roughness.

Example 4.3

Doubly clamped beam was fabricated using silicon carbide. The dimensions were 1000 nm (length) × 125 nm (width) × 75 nm (thickness). Find the resonant frequency.

For a doubly clamped beam, from Equations 4.30 through 4.32 and Table 4.1, the resonance frequency formula is

$$\omega_1 = \left(\frac{22.373}{l^2}\right)\sqrt{\frac{Ewt^3}{12\rho tw}} = 6.46\sqrt{\frac{E}{\rho}}\left(\frac{t}{l^2}\right) \quad (4.57)$$

For silicon carbide, $\sqrt{(E/\rho)} = 1.15 \times 10^4$ s^{-1} m^{-1} (from Table 4.3). Since $t/l^2 = 75 \times 10^{-9}/(1000 \times 10^{-9})^2 = 75{,}000$ m^{-1}, therefore, $\omega_1 = 6.46 \times 1.15 \times 10^4 \times 75{,}000 = 5.572 \times 10^9$ rad second^{-1}.

4.4 Electron Tunneling Displacement Nanosensor

Let us now move to displacement nanosensors. The first of these functions on the quantum-mechanical phenomenon of electron tunneling. The tunneling transducer (Figure 4.12) is generally realized in the form of a sharp tip, placed within a fraction of a nanometer of the moving mechanical element (Ekinci 2005). By drawing a tunneling current from the surface of the mechanical element, the tip converts the displacement $x(t)$ into an electrical signal. Generally, a tunneling transducer is followed by a transimpedance amplifier (current-to-voltage converter).

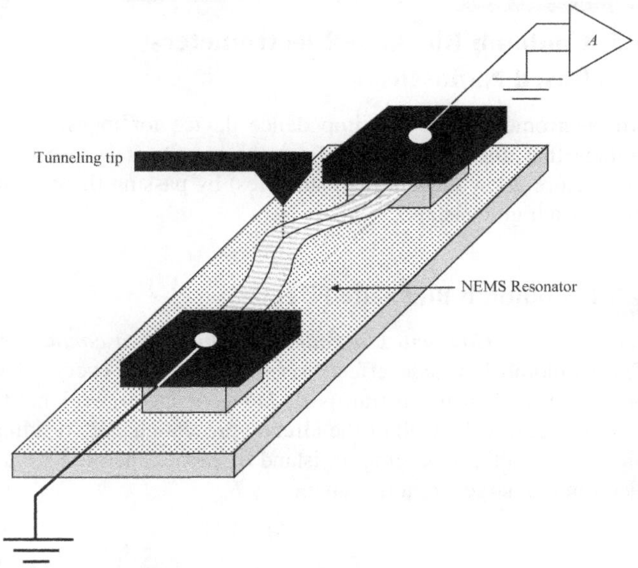

FIGURE 4.12 Electron tunneling displacement nanosensor. (Ekinci, K. L., *Small*, 1, 786, 2005.)

The tunnel current i through the junction is related to the DC bias voltage V and the tunnel gap d as

$$i = \rho_s(E_F)V\exp(-2\kappa d) \quad (4.58)$$

There are two important quantities that determine the transduction characteristics: $\rho_s(E_F)$ and κ. The former quantity, $\rho_s(E_F)$, is the local density of electronic states in the test mass. For $V \ll E_F$, where E_F is the Fermi energy (the average energy of electrons in a metal) of the metal in question, $\rho_s(E_F)$ can be assumed to be voltage-independent. The latter, κ, is the decay constant (the decrease of some physical quantity according to the exponential law) for the electron wave function within the gap and

$$\kappa = \frac{\sqrt{2m_e\phi}}{\hbar} \quad (4.59)$$

where

m_e is the mass of the electron
ϕ is the approximate work function of the metal
$\hbar$ is Planck's constant divided by 2π

$\phi \sim 3$–5 eV and $\phi \gg$ eV in most cases. Using typical values, one can determine that $\kappa \sim 0.1$ nm^{-1}. Keeping in mind the exponential dependence of i upon the tunnel gap and the fact that $\kappa \sim 0.1$ nm^{-1}, it is agreed that a tunneling transducer is extremely sensitive to the displacement of the mechanical element.

4.5 Coulomb Blockade Electrometer-Based Nanosensor

An electrometer is a high-impedance device for measuring a voltage that draws no current from the source. It is also used to measure a low current (nanoamperes) by passing the current through a high resistance.

4.5.1 Coulomb Blockade Effect

Why does the Coulomb blockade take place on a conductor? The Coulomb blockade effect arises because, for every additional charge dq which is transported to a conductor, work has to be done against the field of the already-present charges residing on the conductor. Charging an island of capacitance C with an electron of charge e requires an energy E_c

$$E_c = \frac{e^2}{2C} \quad (4.60)$$

Under these conditions, no additional electron can move into the island unless its energy is raised through the amount E_c by an external voltage

$$V_c = \frac{E_c}{e} = \frac{e}{2C} \quad (4.61)$$

What is the Coulomb blockade regime? For voltages $V < V_c$, at zero temperature, no current can flow into the island and the system is said to be in the *Coulomb blockade regime.*

How is the Coulomb blockade effect modulated? The effect is modulated by coupling a small capacitor into the island, forming a three-terminal device called the *Coulomb blockade electrometer* (White 1993). Thus, the Coulomb blockade electrometer consists of two tunnel junctions (extremely thin potential barriers to electron flow, so that the transport characteristic, the current-voltage curve, is primarily governed by the quantum-mechanical tunneling process, which permits electrons to penetrate the barrier) of capacitances C_1 and C_2, as shown in Figure 4.13, where the Coulomb island is marked by a dashed line. The electrometer is completed by a capacitor C_G weakly coupling into the central region of the device, the island, biased by a voltage U, with the island externally biased by a supply V. The capacitor C_G constitutes the gate of the device.

A background charge q_0 on the island generally produces a non-integral charge offset. *How is this background charge induced?* The background charge is induced by stray capacitances, that are not shown in the circuit diagram and impurities located near the island, which are practically always present. The effect of the gate electrode is that the background charge q_0 can be changed at will, because the gate additionally polarizes the island.

Consider the situation when both the gate and island voltages are zero. Then, electrons do not have enough energy to enter the island and current does not flow. For an electron to move onto the island, its energy must equal the Coulomb energy, $e^2/2C$. As the bias voltage between the source and the drain is increased, the energy in the system reaches the Coulomb energy, and an electron can pass through the island. The critical voltage needed to transfer an electron onto the island, equal to e/C, is called the *Coulomb gap voltage.*

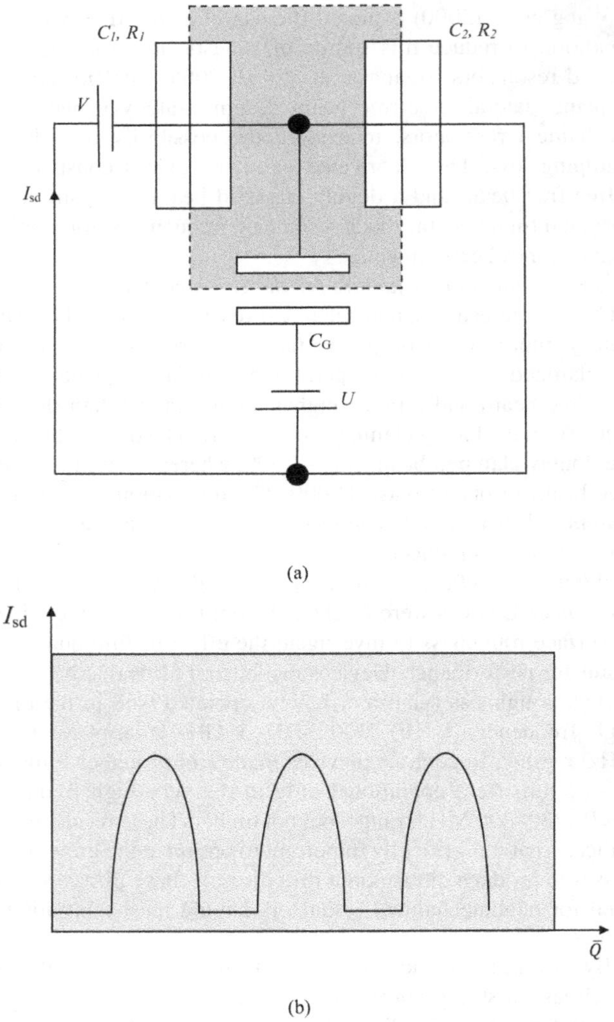

FIGURE 4.13 (a) Coulomb blockade electrometer and (b) its response to varying charge. (White, J. D., *Jpn. J. Appl. Phys.*, 32, L1571, 1993.)

Now imagine that the bias voltage is kept below the Coulomb gap voltage. When $U=0$ and $V<E_c/e$, electron transport through the island is impeded by the Coulomb blockade because there is an energy gap between the states where the island has a charge $-Ne$ and $-(N+1)e$, where N is an integer. By changing U, an extra charge $\bar{Q} = C_G U$ is induced on the island, which is added to $-Ne$. It is possible to make these states degenerate, thereby lifting the Coulomb blockade. Thus, the charging energy may be overcome by changing the source–drain voltage, as well as by changing the gate voltage.

On applying a suitable gate bias, the device starts conducting, with a maximum source–drain current $I_{sd\,(maximum)}$ given by

$$I_{sd(maximum)} \sim \frac{V}{2R} \quad (4.62)$$

where the resistance R arises from the tunneling effects across the island when an electron jumps from the left contact across the left tunnel junction (resistance R_1) to the island (enclosed inside the dashed rectangle in Figure 4.13), and from the island across the right tunnel junction (resistance R_2) to the right contact under an average potential difference $V/2$.

A closer inspection reveals that the current is maximum at a charge

$$\tilde{Q} = \frac{N|e|}{2} \tag{4.63}$$

And, when $\tilde{Q} = |e|$, the device returns to the state equivalent to the zero gate bias. Again, there is an energy gap between the states, $-(N+1)e$ and $-(N+2)e$. Thus, a cyclic process takes place with the current I_{sd} oscillating between low and high values. The period of oscillation in $\tilde{Q}$ is $|e|$. This implies that the conductivity of the island can be significantly varied by changing $\tilde{Q}$, thereby enabling measurements of $\tilde{Q}$ changes at a level lower by several orders of magnitude than the electronic charge e.

So far, $\tilde{Q}$ was changed by altering the voltage U applied to the gate. There is another technique by which $\tilde{Q}$ can be changed. *How*? By changing the value of C_G at the constant gate bias U, by varying the separation between the two plates of C_G, e.g., by keeping the plate near the island fixed and transferring the other plate relative to it. This produces an identical effect on $\tilde{Q}$.

How does the electrometer sense displacement? The electrometer acts as a motion transducer because it is able to sense the movement of the capacitor plate. The smallest displacement that can be sensed by the electrometer ($\delta X_{minimum}$) is expressed as

$$\delta X_{min} = \frac{X\delta Q_{minimum}}{C_G U} \tag{4.64}$$

where the gate is formed by suspending a conducting plate above the island at a height X, and $\delta Q_{minimum}$ is the minimum charge that can be detected by the electrometer.

When $C_G \ll$ minimum (C_1, C_2), the charge $\delta Q_{minimum}$ is estimated as (White 1993)

$$\delta Q_{minimum} = 5.4e\left\{\frac{k_B T}{\left(e^2 / C_{minimum}\right)}\right\}^{0.5} \sqrt{RC_{minimum}\Delta f} \tag{4.65}$$

where

$C_{minimum}$ is the capacitance of the smallest tunnel junction which contemporary manufacturing technology can make

Δf is the relevant bandwidth

Combination of Equations 4.64 and 4.65 yields the displacement sensitivity

$$\frac{\delta X_{minimum}}{\sqrt{\Delta f}} \approx \left(\frac{21.6}{e}\right)\sqrt{k_B TRXC_{minimum}} \tag{4.66}$$

assuming that the biasing point for the device is $\tilde{Q} = e/4$. Using electron beam technology, $C_{minimum} \sim 10^{-16}$ F. Taking X to be 10 μm, the displacement sensitivity is 5×10^{-12} mHz$^{-1/2}$ at $T = 10$ mK.

4.5.2 Comparison with Tunneling Sensors

The following are the points of dissimilarity between the electron tunneling sensor and Coulomb blockade-based sensor:

1. Although electron tunneling sensors have a higher sensitivity by at least two orders of magnitude, they have a dynamic range of the extent of tip-surface distance of around 5 Å. A larger dynamic range requires integration of the tip with a translational component, so that the final resolution is determined by the tip-translator combination. In contrast, the dynamic range of an electrometer-based sensor is the gate–island distance, which is much larger than the bare tunneling tip.
2. Moreover, establishing a reliable tunnel junction contact is not easy outside an ultrahigh vacuum environment. A Coulomb blockade sensor does not suffer from this limitation.

The surface from which the current tunnels off is never flat on the atomic scale. This means that a large change in tunneling current may occur when the tunneling tip sweeps across a small vertical step downward or upward, showing that this kind of sensor is not a one-dimensional motion sensor. The same is not true for a Coulomb blockade device.

4.6 Nanometer-Scale Displacement Sensing by Single-Electron Transistor

The SET is the most sensitive electrometer, with a demonstrated sensitivity below 10^{-5} eHz$^{-1/2}$. Presently, a promising approach for displacement sensing is to use a SET capacitively coupled to a flexural beam (a beam subjected to bending action) resonator (Figure 4.14).

Position measurements of an oscillator are ultimately limited by quantum mechanics, where "zero-point motion" fluctuations in the quantum ground state combine with the uncertainty relation, to give way to a lower limit on the measured average displacement. Average total energy of a classical simple harmonic oscillator (any physical system that is bound to a position of stable equilibrium by a restoring force or torque proportional to the linear or angular displacement from this position), in equilibrium with its environment at temperature T, is $k_B T$. The position of the oscillator fluctuates continuously with a root-mean-square displacement amplitude given by (Knobel and Cleland 2003)

$$\delta x = \sqrt{\frac{k_B T}{m\omega_0^2}} \tag{4.67}$$

for an oscillator of mass m and resonance frequency $f_0 = (\omega_0/2\pi)$. The classical displacement amplitude can be made arbitrarily small by reducing the temperature. Quantum mechanics has two implications here: (i) the quantized nature of the oscillator energy yields an intrinsic fluctuation amplitude, the "zero-point motion"

$$\delta x_{zp} = \sqrt{\frac{\hbar}{2m\omega_0}} \tag{4.68}$$

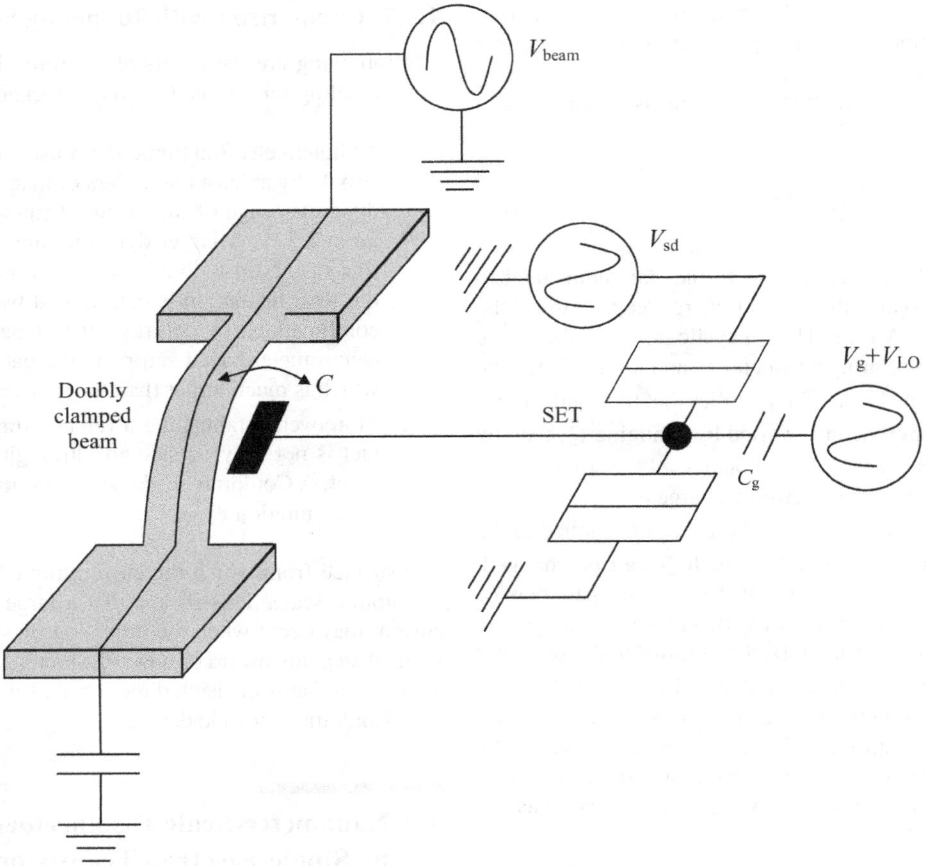

FIGURE 4.14 Capacitive coupling of the electrode on a resonator flexural beam to a SET gate electrode. (Knob el, R. G. and Cleland, A. N., *Nature*, 424, 291, 2003.)

which is achieved for temperatures T well below the energy quantum $T \ll T_Q \equiv \hbar\omega_0/k_B$; and (ii) the instrument used to measure the position of the oscillator will unavoidably perturb it, further limiting the possible measurement resolution, as quantified by the *Heisenberg Uncertainty Principle* (increasing the accuracy of measurement of one observable quantity increases the uncertainty with which another conjugate quantity may be known).

An implementation that allows near-quantum-limited sensitivity uses a SET as a displacement sensor. The exquisite charge sensitivity of the SET at cryogenic temperatures is exploited to measure motion. Knobel and Cleland (2003) demonstrated an ultrasensitive, potentially quantum-limited displacement sensor based on a SET, allowing the motion of a nanomechanical resonant beam at its resonance frequency to be read out. For a model cantilever and SET, applying the optimal bias voltage yields a displacement sensitivity of 4×10^{-16} mHz$^{-1/2}$, approaching the sensitivity needed to measure quantum effects.

What is the construction of SET and how does it work? The SET consists of a conducting island separated from leads by low-capacitance, high-resistance tunnel junctions. The current through the SET is modulated by the charge induced on its gate electrode, with a period *e*, the charge of one electron.

How is SET applied for detecting the motion of a nanomechanical resonator? This is achieved by capacitively coupling the gate of the SET to a metal electrode on a flexural beam of the resonator and biasing the electrode at a constant voltage, V_{beam}. In this scheme, the capacitance C between the SET and the beam has a coupled charge $q = V_{beam}\,C$. As the beam vibrates in the x direction, in the plane of the device, the resulting variation in capacitance modulates the charge induced on the SET, $\Delta q = V_{beam}\,\Delta C$, changing the SET source–drain current. When the voltage V_{beam} is increased, the charge modulation Δq and the sensitivity to the resonator motion increases. However, the source–drain current is due to the stochastic flow of electrons through the SET, so the voltage of the center island fluctuates randomly. This causes a fluctuating "back-action" force on the beam. Increasing the voltage applied to the beam leads to larger coupled-charge signals, but also increases the back-action coupling between the SET and the beam. The force increases as V_{beam} increases, resulting in a voltage for which the total noise is minimized. The device had a displacement sensitivity of 2.0×10^{-15} mHz$^{-1/2}$ at the 116.7 MHz resonance frequency of the mechanical beam at a temperature of 30 mK, which was limited by the noise in the conventional electronics. The sensitivity was roughly two orders of magnitude greater than the quantum limit for this oscillator.

4.7 Magnetomotive Displacement Nanosensor

Cleland and Roukes (1996) introduced the magnetomotive (a substance or phenomenon that gives rise to a magnetic flux) displacement detection technique into the NEMS domain. This

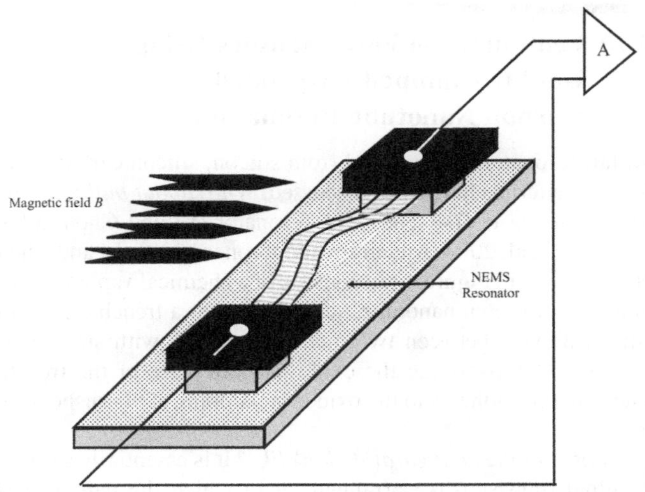

FIGURE 4.15 Magnetomotive displacement nanosensor. (Ekinci, K. L., *Small*, 1, 786, 2005.)

scheme is based upon movement of a conducting nanomechanical element through a uniform static magnetic field (Figure 4.15). The resultant time-varying flux generates an induced electromotive force (EMF) in the loop, which, in turn, is picked up by the detection circuit (Ekinci 2005).

4.8 Piezoresistive and Piezoelectric Displacement Nanosensors

These sensors do not detect the actual displacement. *How do piezoresistive/piezoelectric sensors sense displacement?* Piezoresistive and piezoelectric detection are both sensitive to the strains generated inside a material during its motion. A *piezoresistive material* is one that exhibits a change in resistivity due to strain. Likewise, a *piezoelectric material* becomes electrically polarized in the event of a deformation.

Piezoresistive sensing is realized by detecting the resistance changes through a piezoresistive NEMS device upon actuation (Ekinci 2005). There are, however, some obstacles in this method, e.g., in a P-type Si doubly clamped beam, an optimistic estimate for the resistance change ΔR, as a function of the displacement x, the beam length l, and the resistance R, is $\Delta R/R \sim x^2/l^2$, indicating that $\Delta R_{max}/R \sim 1/100$. In view of the already high resistance of a doped semiconductor beam approaching $R \sim 10\ k\Omega$, such a small resistance change is likely to be obscured at high resonance frequencies by the associated parasitics.

Polarization fields created by the strain fields within the piezoelectric nanomechanical element as it undergoes motion form the basis of the piezoelectric scheme. The detection is realized by precisely measuring the potential drop across the strained device during its motion. In most materials, the piezoelectric coefficients linking the strain to the polarization field are $\sim 10^{-12}$–10^{-9} mV^{-1}, suggesting that the piezoelectric voltage generated across a nanoscale device is very small. The high source impedance of the generated (piezoelectric) signal makes detection difficult at high frequencies.

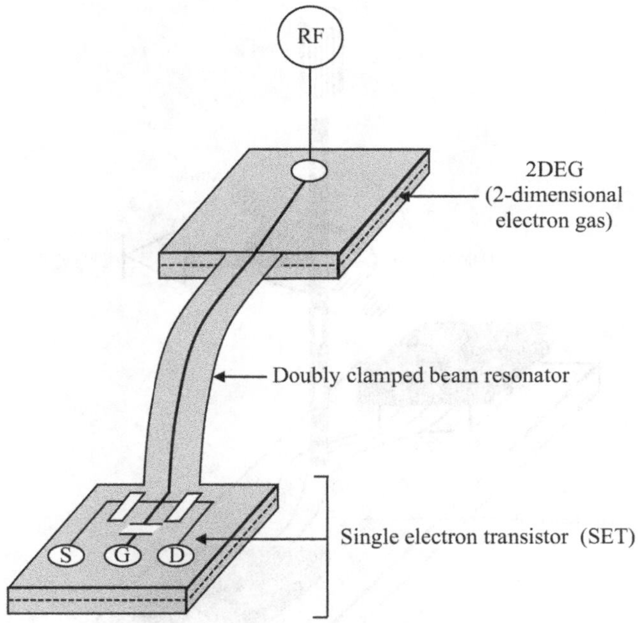

FIGURE 4.16 Piezoelectric displacement sensing using SET. (Knobel. R. and Cleland. A. N., *Appl. Phys. Lett.*, 81, 2258, 2002.). The two-dimensional electron gas (2DEG) formed at GaAs/AlGaAs interface is used as a ground plane for actuation/detection.

Knobel and Cleland (2002) proposed piezoelectric detection in NEMS using a SET (Figure 4.16). Here, the gate of the SET was placed at a position of maximum strain, and detection was realized by monitoring the consequential conductance changes in the current-biased SET. In other words, the resonator was fabricated from a piezoelectric material, such as gallium arsenide (GaAs), aluminum gallium arsenide (AlGaAs), or aluminum nitride (AIN), and the SET was configured to sense the piezoelectric voltage developed when the beam flexes. A noticeably higher displacement sensitivity was achieved here than by using capacitive displacement sensing, primarily due to the strong piezoelectric coupling strength (Figure 4.8). Knobel and Cleland (2002) calculated a noise figure of 5×10^{-17} $mHz^{-1/2}$, for a 1 GHz GaAs resonator, dominated equally at peak sensitivity by the current and back-action noise of the SET. Electromechanical coupling describes the conversion of energy by the element from electrical to mechanical form or *vice versa*. Coupling coefficient is a dimensionless number related to the effectiveness of electrical-to-mechanical energy conversion in piezoelectric devices.

Piezoelectric detection has another attractive feature because piezoelectric signals scale favorably to the small, stiff resonators needed to approach the regime, where $\hbar\omega_1 \geq k_B T$. In both the capacitive and piezoelectric schemes, the detection of deviations from the classical motion is difficult for nanomechanical resonators with $\omega_1/2\pi < \text{GHz}$.

4.9 Optical Displacement Nanosensor

Optical interferometry techniques (Krishnan et al. 2007), notably path-stabilized Michelson interferometry and Fabry-Perot

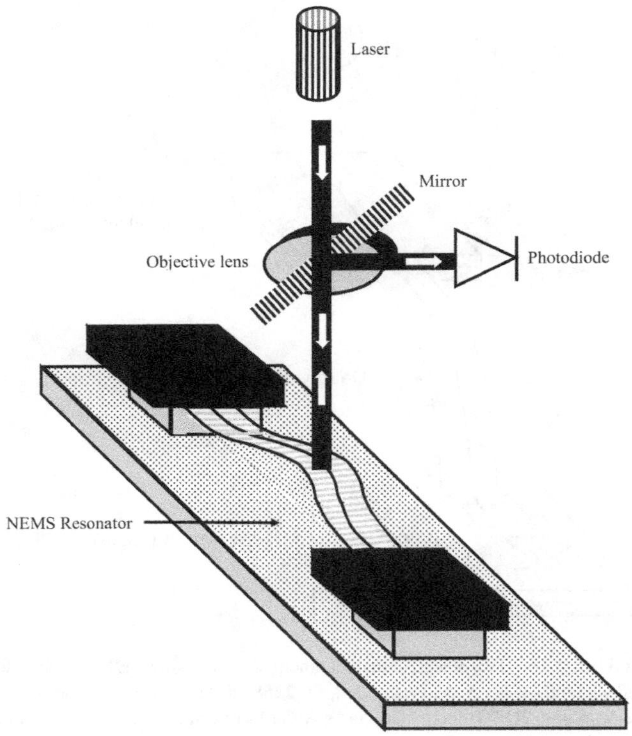

FIGURE 4.17 Optical displacement nanosensor. (Ekinci, K. L., *Small*, 1, 786, 2005.)

interferometry, have been extended into the NEMS domain. The Michelson interferometer produces interference fringes by splitting a beam of monochromatic light into two paths so that one beam strikes a fixed mirror and the other a movable mirror; when the reflected beams are recombined, an interference pattern results. On the other hand, the Fabry-Perot interferometer uses multiple reflections between two closely spaced, partially silvered surfaces; part of the light is transmitted each time, reaching the second surface, and resulting in multiple offset beams which interfere with each other, producing an interference pattern with extremely high resolution, somewhat like the multiple slits of a diffraction grating increasing its resolution. In Fabry-Perot interferometry (Figure 4.17), the optical cavity formed within the sacrificial gap of the NEMS, between the NEMS surface and the substrate, varies the optical signal and directs it onto a photodiode (a semiconductor diode in which the reverse current varies with illumination) with the movement of the NEMS device in the out-of-plane direction.

In path-stabilized Michelson interferometry, a tightly focused probing laser beam, of wavelength λ, is focused at the center of a doubly clamped beam through an objective lens (the lens of a microscope at the bottom near the sample). It is reflected from the surface of the moving NEMS device and undergoes interference with a stable reference beam.

In both the aforementioned techniques, strong diffraction effects dominate as the relevant NEMS dimensions decrease below the optical wavelength used. The light gathered by the probing lens is weakened to a small fraction of the incoming light by strong scattering.

4.10 Femtonewton Force Sensors Using Doubly Clamped Suspended Carbon Nanotube Resonators

So far, resonators constructed from silicon, silicon carbide, and related materials have been discussed. *What about building resonators using CNTs?* The CNT resonator (Li and Chou 2003, Sazonova et al. 2004; Anantram and L'eonard 2006; Li and Chou 2006) shown in Figure 4.18a consists of a chemical vapor deposition (CVD)-grown nanotube, suspended over a trench etched in silicon dioxide between two metal electrodes, with small portions of CNT overlying the oxide on both sides of the trench. The nanotube adheres to the oxide by clamping at the suspension points.

What does the configuration look like? It is essentially a three-terminal transistor-like arrangement in which the nanotube is the channel between its two metal contacts. The current in the channel is modulated by the voltage applied to the gate through capacitive coupling.

The measurement is carried out in a vacuum chamber at pressures below 10^{-4} Torr, by actuating the nanotube motion through electrostatic interaction with the gate electrode. The applied gate voltage V_g induces an additional charge, $q=C_gV_g$, on the nanotube, where C_g is the capacitance between the gate and the nanotube. The mutual attraction between the charge q and its opposite charge $-q$ on the gate produces a downward electrostatic force on the nanotube, bending the tube towards the gate.

What are the components of the gate voltage and what roles do they play? The gate voltage has two components: a static (DC) component and a small time-varying (AC) component. The DC voltage $V_{g(DC)}$ produces a static force on the nanotube that is used to vary its tension. The AC voltage $V_{g(AC)}$ produces a periodic electric force but the force is *always attractive in nature.*

How does the nanotube vibrate? When $V_{g(AC)}=-V_{g(DC)}$, i.e., the two voltages are equal in magnitude but opposite in sign, the nanotube returns to its previous upward position by virtue of its elastic resilient property, because the downward electrical force ceases to act during this period. But when $V_{g(AC)}$ changes its direction or magnitude, a net downward force again starts acting on the nanotube. Thus, the nanotube is either pulled downward or there is no force acting on it. During the latter period, the nanotube recovers its original upward position. In this way, the nanotube moves up and down, swinging into continuous oscillations.

When the driving frequency matches the natural frequency of the tube, the displacement of the nanotube becomes large, due to resonance. *How is the resonance frequency determined?* To find the resonance frequency, the resonance condition must be detected. It must be noted that the change in induced charge on the nanotube (δq) = the change in the charge caused by gate voltage variation ($\delta q_1=C_g\delta V_g$) + change in the charge due to alteration in capacitance between the nanotube and gate, arising from the continuous distance variation between them during vibration of the nanotube ($\delta q_2=\delta C_gV_g$). The resulting AC current flowing in the nanotube is expressed as

$$i_{\text{nanotube}}=\frac{\delta q_1}{\delta t}+\frac{\delta q_2}{\delta t}=C_g\left(\frac{\delta V_g}{\delta t}\right)+V_g\left(\frac{\delta C_g}{\delta t}\right) \quad (4.69)$$

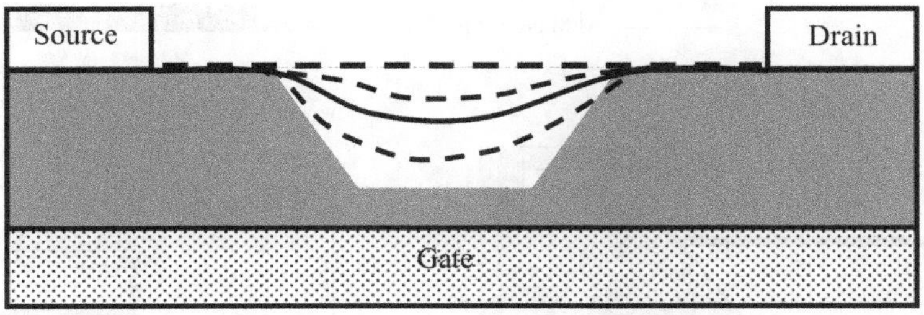

(a)

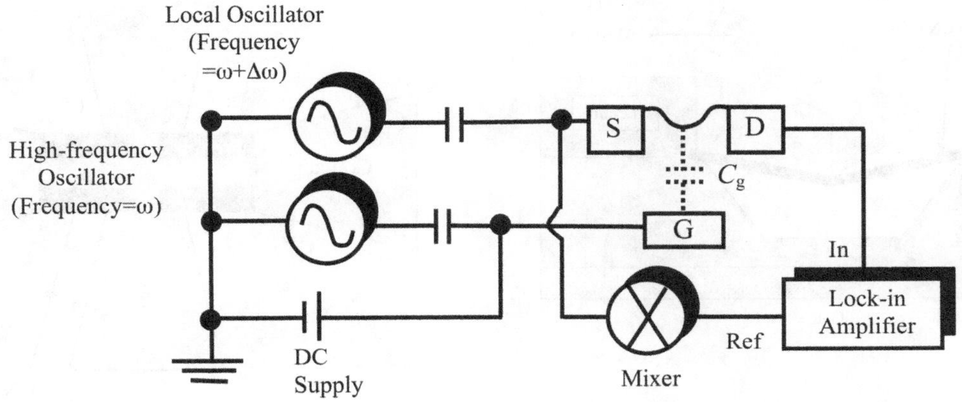

(b)

FIGURE 4.18 (a) A CNT resonator, (b) layout of experimental set-up for measurement of resonance frequency of CNT resonator. (Sazonova, V. et al., *Nature*, 431, 284, 2004.)

Only the second term is related to the distance between the nanotube and the gate, i.e., the amplitude of oscillations. The first term is independent of this amplitude. When ($\delta C_g/\delta t$) is maximum, corresponding to the maximum displacement of the nanotube from its equilibrium position, the current registers peaks in both directions. Thus, the resonant frequency is determined by observing the change in nanotube current with frequency of AC field. The frequency at which the current swings of largest magnitude are perceived pertains to the maximum amplitude of vibration of the nanotube from its mean position. This is the resonant frequency of oscillation. Taken as a whole, the resonator exhibits source–drain AC conductance enhancement when the driving frequency of the gate matches the resonance frequency of mechanical oscillations of the nanotube.

How are the resonance conditions determined practically? To detect the current or conductance change of the nanotube, it is used as a mixer (Figure 4.18b). This method helps in avoiding complications due to capacitive currents between the gate and the drain electrodes. The current equals the product of the AC voltage on the source electrode and the modulated nanotube conductance. The measured current response is fitted to a Lorentzian function (a singly peaked function used for the fitting of raw data), with a normalized linewidth, a resonance frequency of 55 MHz, and an appropriate phase difference between the actuation voltage and the force on the nanotube.

The resonance frequency increases as the DC gate voltage rises. The DC gate voltage is adjusted to tune the resonance frequency. Nanotubes of dissimilar diameters and lengths also yield different resonance frequencies, because the oscillator masses differ.

An important parameter characterizing the oscillator is the quality factor, Q, the ratio of the energy stored in the oscillator to the energy lost per cycle owing to damping. It is in the range of 40–200, with no observed frequency dependence. Maximizing Q is important for most applications.

To use a nanotube as a force/strain sensor, it is noted that the resonance frequency of a resonator is closely related to its dimensions. When a carbon nanotube is deformed, the tube length and tube diameter are changed, and its resonance frequency changes accordingly. There exists a relationship between the shift of resonance frequency of a nanotube resonator and the force the nanotube is subjected to. This is the underlying principle for carbon nanotube-based force sensors. Thus, the key issue of this study

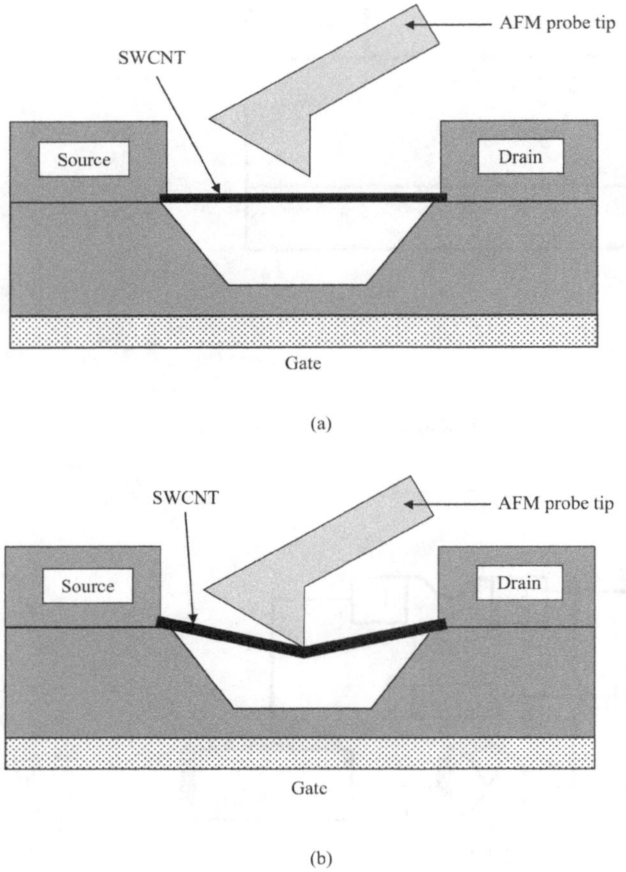

FIGURE 4.19 Using an AFM tip to apply a force on SWCNT: (a) no force applied on the tube and (b) a force applied on the tube. (Minot, E. D. et al., *Phys. Rev. Lett.*, 90, 156401, 2003.)

is to establish the relationship between the resonance frequency shift and the applied force.

How is force applied on the nanotube? An AFM tip can be used to simultaneously vary the force on the CNT and to electrostatically gate the tube (Figure 4.19) (Minot et al. 2003). The following section will illustrate two configurations of the CNT force sensor, in which a cantilever and a bridge structure are used to apply the force.

What is the ultimate force sensitivity of the sensor? The ultimate limit on force sensitivity is prescribed by the thermal vibrations of the nanotube. This force sensitivity is 20 $aNHz^{-1/2}$ for typical parameters; atto (a) = 10^{-18}. The observed sensitivity of 1 $fNHz^{-1/2}$ was 50 times lower than this limit; femto (f) = 10^{-15}. *Why was this sensitivity lower?* This is probably due to the low values of transconductance (the ratio of change in current to change in voltage that caused it) for the measured nanotubes at room temperature. At low temperatures (~1 K), the sensitivity increases by orders of magnitude due to the higher transconductance of the nanotube under these conditions.

4.11 Suspended CNT Electromechanical Sensors for Displacement and Force

In a series of papers, researchers at ETH Zurich reported electromechanical sensors based on CNTs (Stampfer et al. 2005; 2006a; 2006c). A contacted, suspended single-walled carbon

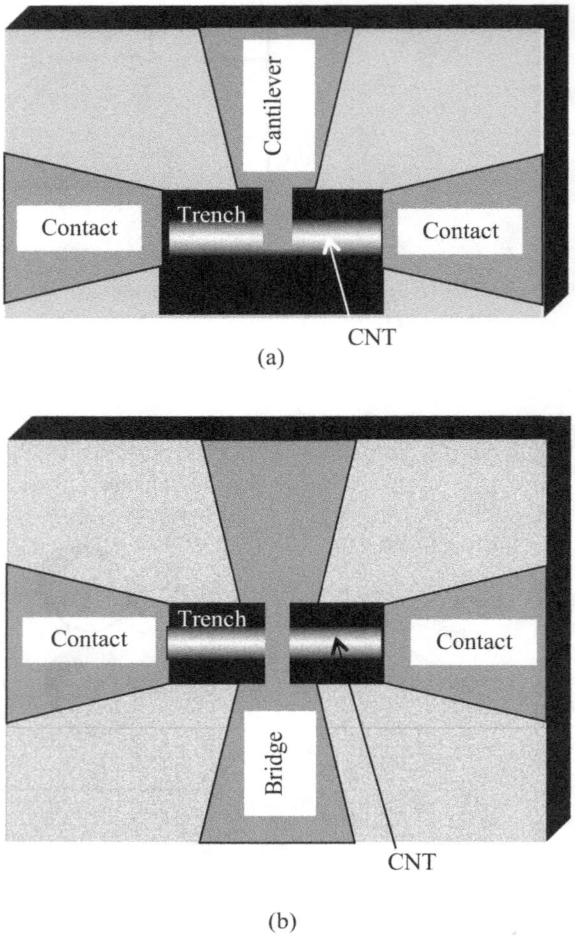

FIGURE 4.20 Configurations for building force sensors by applying force over: (a) cantilever and (b) bridge. (Stampfer, C. et al., *The 13th International Conference on Solid-State Sensors, Actuators and Microsystems*, Seoul, Korea, June 5–9, 2103, 2005; Stampfer, C. et al., *IEEE Sensors J.*, 6, 613, 2006.)

nanotube (SWCNT) is fixed underneath a freestanding cantilever (Figures 4.20a and 4.21a) or bridge (Figures 4.20b and 4.21b). On applying an external out-of-plane force on the cantilever or bridge, it undergoes deflection, producing a mechanical deformation of the underlying SWCNT. The structural change in the SWCNT yields a change of conductance, which is electrically measured. The conductance changed by a factor of 3 for approximately 120 nN of force for a metallic SWCNT with 5 MΩ base resistance.

An AFM tip introduces the external out-of-plane force for device actuation. In practical applications, the cantilever can act as a support spring, connected to a seismic (test) mass (the mass that converts the acceleration to spring displacement). Deflection of the seismic mass, e.g., by an acceleration, causes a mechanical deformation of the SWCNT, leading to an electrical signal.

Minot et al. (2003) employed an AFM tip to simultaneously vary the CNT strain and to electrostatically gate the tube. They found that, under strain, the conductance of the CNT can increase or decrease, depending on the tube. By using the tip as a gate, they showed that this is related to the increase or decrease in the bandgap of the CNT under strain.

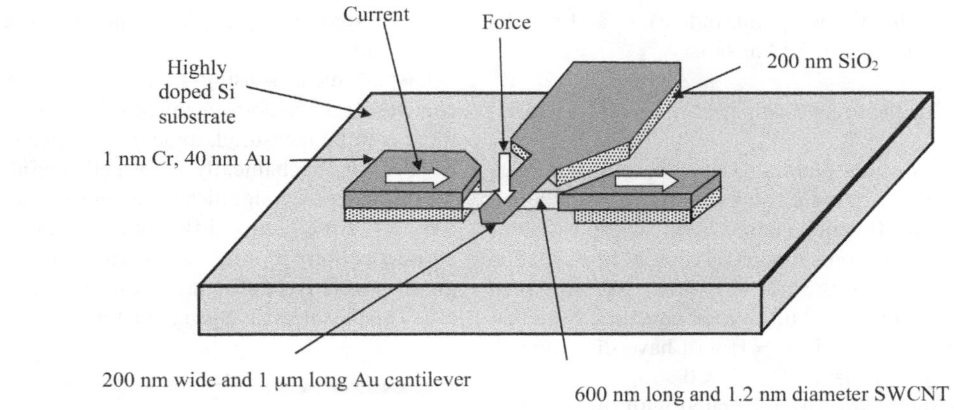

(a)

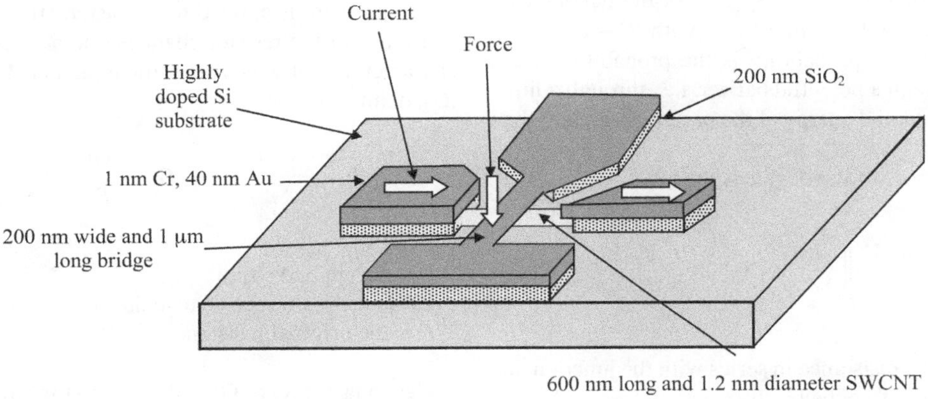

(b)

FIGURE 4.21 Suspended CNT-based electromechanical force sensors with: (a) cantilever and (b) bridge. The gap between the CNT and the underlying substrate is provided by oxide layer thickness. (Stampfer, C. et al., *The 13th International Conference on Solid-State Sensors, Actuators and Microsystems*, Seoul, Korea, June 5–9, 2103, 2005; Stampfer, C. et al., *IEEE Sensors J.*, 6, 613, 2006.)

V_{tip} was swept ~3 times per second over a range of a few volts, as strain σ was slowly increased. A CNT is either metallic or semiconducting, depending on the orientation between the atomic lattice and the tube axis. As the metallic tube was strained, an asymmetric dip, centered at V_{tip} ~ 1 V, developed in $G- V_{tip}$ (G is the conductance of the nanotube). On the other hand, the semiconducting tube showed an increase in G with strain and a reduction in the asymmetry of the dip. The maximum resistance R_{max} (σ) for each sweep of V_{tip} as a function of strain is fitted to the functional form:

$$R_{max}(\sigma) = R_0 + R_1 \exp(\beta\sigma) \tag{4.70}$$

where

β, R_0, and R_1 are fitting parameters.

β, the exponential fitting parameter, is related to the strain dependence of the gap:

$$\frac{dE_{gap}}{d\sigma} = \beta k_B T \tag{4.71}$$

Minot et al. (2003) showed that strain can open a bandgap in a metallic CNT and modify the bandgap in a semiconducting CNT. Theoretical work predicts that bandgap changes can range between ~100 meV per 1% stretch, depending on CNT chirality (Yang et al. 1999). They obtained $dE_{gap}/d\sigma = -53$ meV for the semiconducting tube and $dE_{gap}/d\sigma = +35$ meV for the metallic tube. They also estimated chiral angles ϕ ~ 19° and 23° for the two tubes, respectively. These results can be understood from the relationship between the rate of change of bandgap with strain, $dE_{gap}/d\sigma$:

$$\frac{dE_{gap}}{d\sigma} = \text{sign}(2p+1)3t_0(1+\nu)\cos 3\phi \tag{4.72}$$

where

the signum function of a real number x is defined as $\text{sign}(x>0)=1$, $\text{sign}(x=0)=0$ and $\text{sign}(x<0)=-1$

t_0 ~ 2.7 eV is the tight-binding overlap integral

ν ~ 0.2 is the Poisson ratio

ϕ is the CNT chiral angle

$p = -1$, 0, or 1, such that the wrapping indices, n_1 and n_2, (the integers *m*, *n* of Section 2.13.3) satisfy

$$n_1 - n_2 = 3q + p \quad (4.73)$$

where *q* is an integer. The tight-binding (TB) model is an approach to the calculation of the electronic band structure employing an approximate set of wave functions based upon superposition of wave functions for isolated atoms situated at each atomic site. The maximum value of $|dE_{gap}/d\sigma|$ is $3t_0\,(1+\nu) = 100$ meV, similar in magnitude to $dE_{gap}/d\sigma$ of typical bulk semiconductors. Note that half of all semiconducting CNTs ($p = 1$) will have $dE_{gap}/d\sigma > 0$, while the other half ($p = -1$) have $dE_{gap}/d\sigma < 0$.

To model the resistance $R_{max}(\sigma)$ associated with this conductance minimum, they took into account transport by thermal activation, ignoring tunneling across the depleted region. Electrons with energy *E*, such that $|E - E_F| > E_{gap}$, cross the barrier with a transmission probability $|t|^2$, while those with $|E - E_F| < E_{gap}$ are reflected; transmission probability is the probability that a particle will pass through a potential barrier, i.e., through a finite region in which the potential energy of the particle is greater than its total energy.

The low-bias resistance of the device is then

$$R_{total} = R_S + \left(\frac{1}{|t^2|}\right)\left(\frac{h}{8e^2}\right)\left\{1 + \exp\left(\frac{E_{gap}}{k_B T}\right)\right\} \quad (4.74)$$

where

the first term R_S is the resistance in series with the junction due to the metal-CNT contacts, etc.

the second term is the resistance of the junction region

From Equation 4.72, we have

$$E_{gap} = E^0_{gap} + \left(\frac{dE_{gap}}{d\sigma}\right)d\sigma \quad (4.75)$$

Additional knowledge about the device can be gained from the fitting parameter $R_1 \sim 49h/e^2$, and, hence, transmission probability $|t|^2 = 0.25$. Using an estimate of $|t|^2 = 0.25$ from the preceding text, they inferred the zero-strain energy gap, $E^0_{gap} = 160$ meV. This inferred energy gap corresponded to a tube with diameter $d = 4.7$ nm.

Displacement sensing is also obtained from conductance changes. If the nanomechanical structure, of which the displacement is to be measured, undergoes a mechanical deflection, the SWCNT is mechanically deformed (mainly axially stretched), which leads to a significant change in the conductance of the SWCNT. An effective differential sensor sensitivity of 1016 kΩ at a deflection = 35 nm was obtained for the metallic SWCNT, giving a relative differential resistance sensitivity of 27.5% per nm. The piezoresistive gauge factor was 2900.

4.12 Membrane-Based CNT Electromechanical Pressure Sensor

When pressure is applied to a resistor, strain is induced and the resistance of the resistor changes (the piezoresistive effect). The characteristic of a piezoresistor is modeled by its gauge factor (*G*), defined as

$$G\varepsilon = \frac{\Delta R}{R} \quad (4.76)$$

where

ε is the applied strain

Δ*R* is the resistance change due to the applied strain

R is the original resistance

Gauge factor (G or GF), also called the strain factor of a strain gauge, is the ratio of relative change in electrical resistance to the mechanical strain ε. When the CNT sensor is used as a piezoresistor, the bias current should be limited to a small value, so that self-heating is minimized. The threshold value of this bias current varies from sensor to sensor.

The pressure sensor (Figure 4.22) consists of an ultrathin atomic layer deposited (ALD) circular alumina (Al_2O_3) membrane, to the surface of which an SWCNT is adhered by van der Waals forces and clamped in place by two metal electrodes (Li and Chou 2004; Stampfer et al. 2006b). The pressure sensor design is based on circular membranes in contrast to rectangular

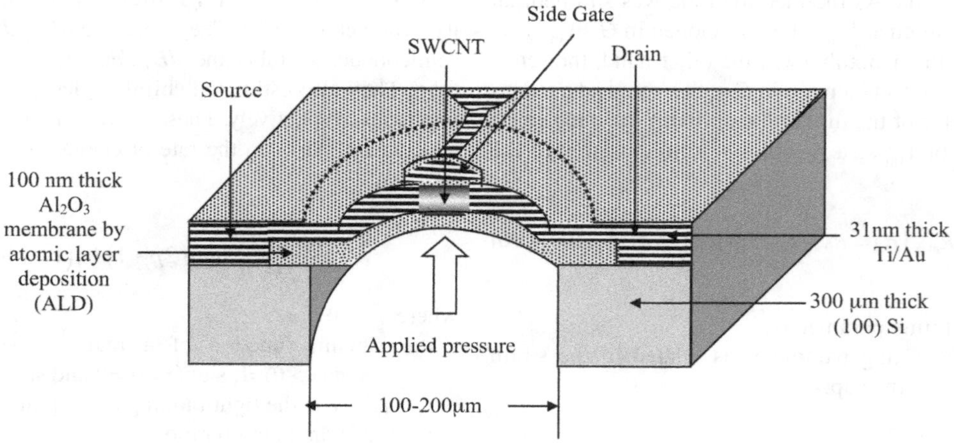

FIGURE 4.22 CNT-on-Al_2O_3-membrane type pressure sensor. (Stampfer, C. et al., *Nano Lett.* 6, 233, 2006b.)

membranes, mainly because of the isotropic nature (identical in all directions) of the strain in the central area of the membrane. Assuming that the SWCNT experiences the same stretching as the membrane, the device works as a strain gauge similar to the doped-silicon strain gauges used in MEMS pressure sensors.

The pressure sensor has been tested for a pressure range of 0–130 kPa (standard atmospheric pressure is 101,325 Pa = 101.325 kPa = 1,013.25 mbar = 760 Torr). For small pressures (up to $\Delta p \sim 70$ kPa), a monotonic increase in resistance (consistently increasing and never decreasing in value) with increasing pressure was observed. For larger pressures, there was a non-monotonic anomaly. However, the piezoresistive gauge factor of the metallic SWCNT, $= 200 \pm 8$, slightly exceeded the value of the doped-Si strain gauge (200).

4.13 Tunnel Effect Accelerometer

4.13.1 Principle of Motion Detection

As already explained in Section 1.19, the STM is a displacement nanosensor, utilizing the extreme sensitivity of tunneling current to variations in electrode spacing. Several methods have been developed for implementation of position sensors, based on electron tunneling. These methods help in position detection with sub-angstrom resolution, using simple mechanical structure and control circuits. Position or displacement detection, in turn, constitutes the basis of acceleration, force, and pressure measurements, e.g., the acceleration of a body is determined from the displacement of a "proof mass" when the sensor structure undergoes acceleration.

What are the available methods for displacement measurement, in addition to electron tunneling? What are their shortcomings? The methods available for displacement measurement rely on the changes in resistance, capacitance, or inductance, resulting from the movement of a pendulum or cantilever. To achieve high sensitivity, a large proof mass is required. The sensor is therefore bulky and requires high power consumption. The electron tunneling accelerometer provides high sensitivity along with a reduced sensor mass and lower power consumption.

4.13.2 Construction and Working

The accelerometer (Figure 4.23) consists of a thin glass plate covered by a 100-nm thick thermally evaporated Au film, which forms the Au electrode (Waltman and Kaiser 1989). The glass plate is fixed below a piezoelectric bimorph cantilever actuator; "bimorph" means an assembly of two piezoelectric crystals cemented together so that an applied voltage causes one to expand and the other to contract, converting electrical signals into mechanical energy. The Au electrode is positioned over an Au wire tunnel tip. The accelerometer is housed in an aluminum structure.

The Au electrode and the tunnel tip are suitably biased to start the tunneling current. The accelerometer functions with tunneling voltage and current over the ranges 0.01–1 V and $(0.05\text{–}10) \times 10^{-9}$ A. Coarse adjustment of the tunnel tip is done by a machine screw. The accelerometer operates through feedback control by comparing the measured tunneling current to a set point and maintaining a constant separation between the Au electrode and the tunnel tip, thereby keeping the tunneling current constant. The accelerometer has a sensitivity of 10 μg $\text{Hz}^{-1/2}$ (where $g = 9.8$ m s^{-2}) and a bandwidth of 3 kHz.

Example 4.4

The tunneling current I, flowing between a pair of planar electrodes separated by a rectangular energy barrier of height ϕ and width s, is given by

$$I = V\exp\left(-\alpha s\sqrt{\phi}\right) \tag{4.77}$$

where V is the bias voltage and $\alpha = 1.025$ $\text{eV}^{-1/2}$ Å^{-1}. The equation is valid for small bias voltages $V < \phi$.

Apply Equation 4.77 to find: (a) the factor by which the tunneling current increases for a decrease in separation between the electrodes by 1 Å for a typical barrier height $\phi = 3$ eV; (b) the decrease in separation required to change the tunneling current by a factor of 10; (c) the value of separation for a 10-fold change in tunneling current, if the barrier height is 1 eV.

(a) When the separation decreases from s Å to $(s - 1)$ Å, the tunneling current I_1 is

$$I_1 = V\exp\left\{-\alpha(s-1)\sqrt{\phi}\right\} \tag{4.78}$$

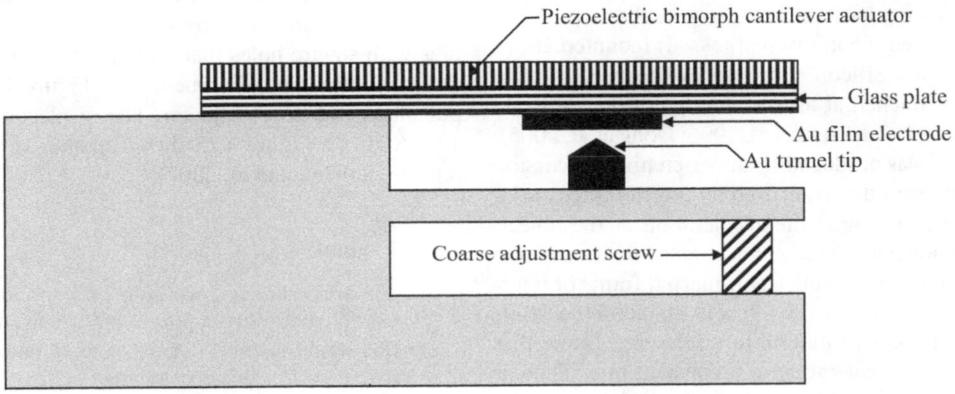

FIGURE 4.23 The electron tunneling accelerometer. (Waltman, S. B. and Kaiser, W. J., *Sensor. Actuator.*, 19, 201, 1989.)

so that the ratio I_1/I is

$$\frac{I_1}{I}=\frac{V\exp\{-\alpha(s-1)\sqrt{\phi}\}}{V\exp\{-\alpha(s)\sqrt{\phi}\}}$$

$$=\exp\{-\alpha(s-1-s)\sqrt{\phi}\} \quad (4.79)$$

$$=\exp\{\alpha\sqrt{\phi}\}=\exp\{(1.025)\sqrt{3}\}$$

$$=\exp\{(1.025)\times 1.732\}=5.90$$

(b) If the decrease in separation required to change the tunneling current by one order of magnitude is denoted by x Å:

$$10=\frac{V\exp\{-\alpha(s-x)\sqrt{\phi}\}}{V\exp\{-\alpha(s)\sqrt{\phi}\}}=\exp\{-\alpha(s-x-s)\sqrt{\phi}\} \quad (4.80)$$

$$=\exp(\alpha x\sqrt{\phi})$$

Taking the natural logarithm of both sides,

$$\ln(10)=\alpha x\sqrt{\phi} \quad (4.81)$$

from which

$$x=\frac{\ln(10)}{\alpha\sqrt{\phi}}=\frac{\ln(10)}{1.025\sqrt{3}}=\frac{2.303}{1.7753}=1.297\approx 1.3\,\text{Å} \quad (4.82)$$

(c) $$x=\frac{\ln(10)}{\alpha\sqrt{\phi}}=\frac{\ln(10)}{1.025\sqrt{1}}=\frac{2.303}{1.025}\approx 2.25\,\text{Å} \quad (4.83)$$

4.13.3 Micromachined Accelerometer

Piezoelectric actuators are sensitive to temperature variations, hysteresis (the lagging of an effect behind its cause), and creep (the plastic deformation of a material under a constant load or force at elevated temperature) in the response of the materials. Therefore, micromachined sensors employing electrostatic actuators were fabricated (Figure 4.24). These are immune to thermal drifts (drifts caused by internal heating of material during normal operation or by changes in external ambient temperature) and creep effects, and are easily miniaturized. A folded cantilever spring with integrated tip was resiliently (marked by the ability of readily bouncing back or recovering its shape, position, etc., after being stretched, bent, or compressed) mounted above a counter-electrode on a silicon substrate by means of a strip (beam) or rectangle attached at an end or edge of the substrate (Kenny et al. 1991; 1992; Rockstad et al. 1996; Dong et al. 2005). The strip or rectangle was produced by under-etching. Excursion of the resilient, cantilevered part, caused by inertial forces arising from acceleration, indicated the acceleration of the object. *Inertial force*, also known as effective force, is the fictitious force acting on a body as a result of using a non-inertial frame of reference, i.e., any force invoked by an observer to maintain the validity of Newton's second law of motion in a reference frame that is rotating or otherwise accelerating at a constant rate. Thus, it is an imaginary force supposed to act upon an accelerated body, equal in magnitude and opposite in direction to the resultant of the real forces. Examples are the centrifugal and Coriolis forces that appear in rotating coordinate systems.

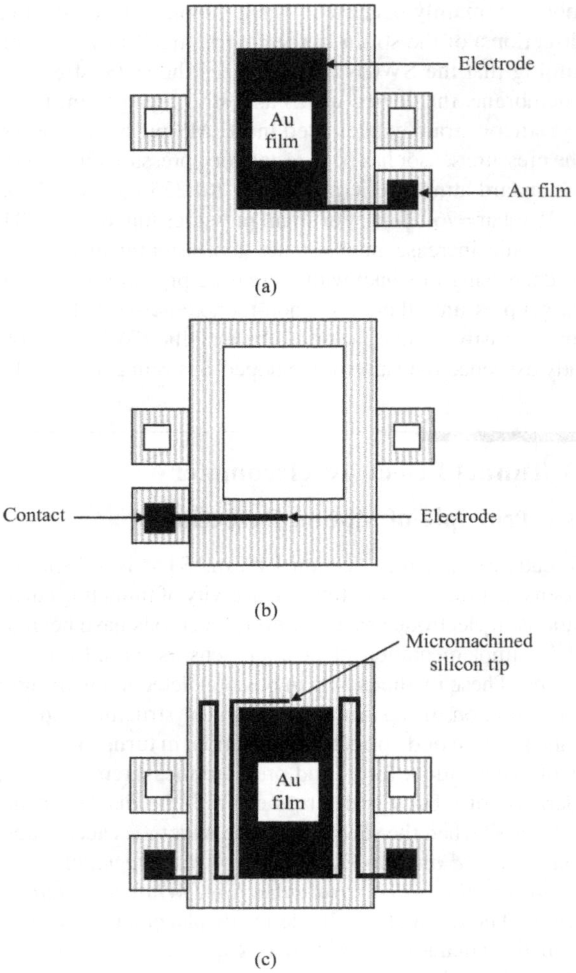

FIGURE 4.24 Three components of micromachined silicon tunnel sensor: (a) deflection counter electrode, (b) tunneling counter electrode, and (c) folded cantilever spring and tip. The tabs, having square holes, allow the components to be aligned properly. The inner rectangular area of part (c) can be deflected downward with respect to its outer segments by applying a voltage between the large electrode and corresponding deflection counter-electrode in part (a). (Kenny, T. W. et al., *Appl. Phys. Lett.*, 58, 100, 1991.)

The accelerometer was constructed by assembling separate components, fabricated by lithographic techniques and having tabs with square holes that allowed the components to be laterally constrained by alignment pins. Figure 4.24 shows the components of the accelerometer. The displacement sensitivity was 10^{-4} Å $Hz^{-1/2}$. Figure 4.25 shows another version of the accelerometer (Krishnan et al. 2007).

Example 4.5

An accelerometer, working on capacitive detection, uses a cantilever-supported proof mass with a deflection sensitivity of 500 Å G^{-1} of acceleration. The capacitive signal of this accelerometer changes by 0.5% for a 1-G change in acceleration. Taking the barrier height as 3 eV, as in Example 4.4, find: (a) the deflection of

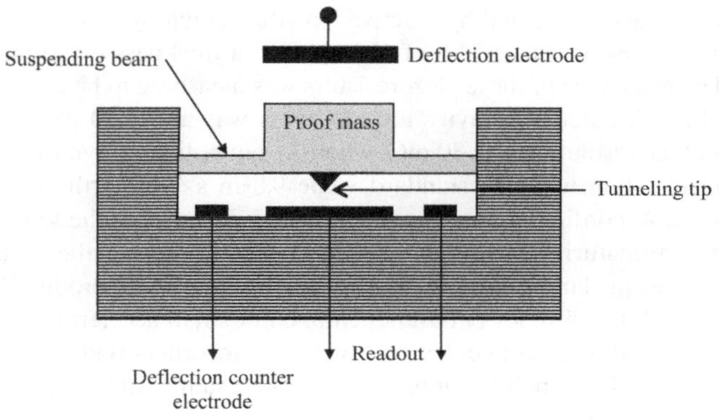

FIGURE 4.25 Tunneling accelerometer. (Krishnan, G. et al., *J. Indian. I. Sci.*, 87, 333, 2007.)

the cantilever for a 0.5% change in tunneling current; (b) the improvement in signal response to acceleration in the tunneling accelerometer. Note: G is acceleration due to gravity = 9.8 m s^{-2}; it is usually denoted by g, but here we have used G to avoid confusion with gram (also having the symbol g).

(a) 0.5% = 0.5/100 = 0.005 = 5×10^{-3}. From Example 4.4, for an increase or decrease in tunneling current of a factor of 10, the required separation is 1.297 Å. Therefore, for a change in tunneling current by a factor of 5×10^{-3}, it is $(1.297/10) \times 5 \times 10^{-3} = 6.485 \times 10^{-4}$ Å.

6.485×10^{-4} Å deflection of the tunneling accelerometer corresponds to 500 Å deflection of capacitive accelerometer. Hence, a 1-Å deflection of the tunneling accelerometer = $500/(6.485 \times 10^{-4}) = 7.7 \times 10^{5}$ Å deflection of the capacitive accelerometer. This means that the tunneling accelerometer is 7.7×10^{5} times superior to the capacitive one. In other words, the tunneling sensor technology results in an improvement of signal response by 7.7×10^{5} times. The 6.485×10^{-4} Å deflection of the tunneling accelerometer refers to an acceleration of $1/(77 \times 10^{5}) = 1.299 \times 10^{-6}$ G = 1.299 μG.

4.14 NEMS Accelerometer

NEMS devices can be generally defined as devices fabricated by microtechnologies and including at least one movable mechanical part with a dimension <100 nm. Ollier et al. (2006) designed two configurations: in-plane (IP) accelerometers with a sensitive axis included in the plane of the wafer, and out-of-plane (OOP) accelerometers with a sensitive axis perpendicular to the wafer (Figure 4.26). An acceleration causes the motion of a proof mass and the corresponding displacement is measured by means of a variation in capacitance between two sets of electrodes, which are either interdigitated combs (IP) or two parallel plates (OOP). In order to avoid parasitic deformations and undesirable vibration modes, a rigid mass was designed with typical dimensions

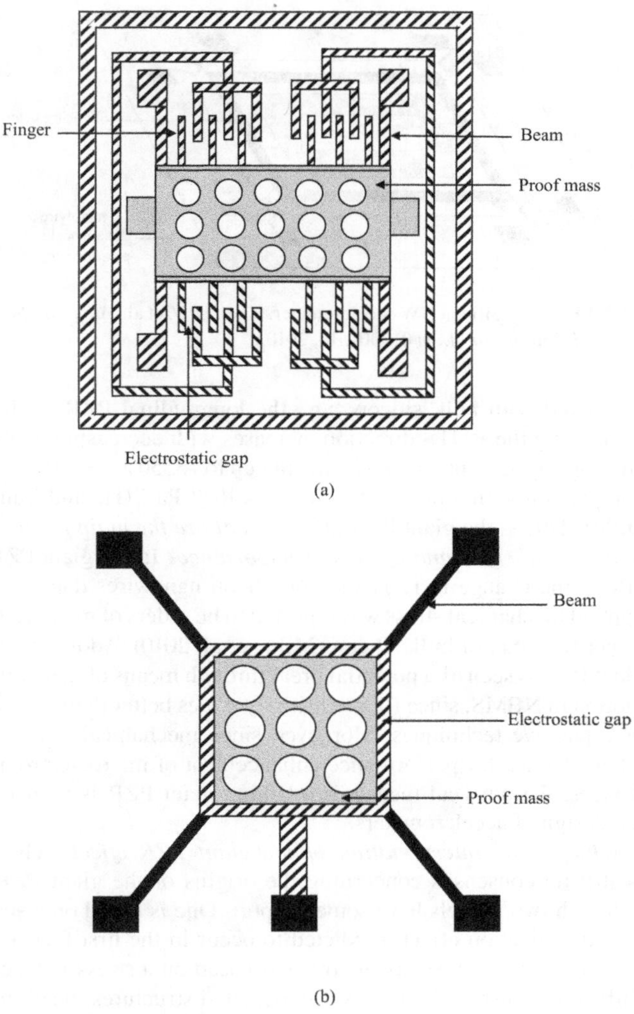

FIGURE 4.26 NEMS accelerometer: (a) in plane (IP) and (b) out-of-plane (OOP) configurations. (Ollier, E. et al., *IEEE Sensors*, EXCO, Daegu, Korea, October 22–25, 54, 2006.)

of 75 μm × 150 μm. In addition, for IP accelerometers, low cross-sensitivity implies an increase in the ratio thickness/width of the mechanical springs, leading to 50-nm wide mechanical springs. At the same time, an increase in electrical sensitivity requires a decrease in electrostatic gap to 100–400 nm. Typically, the length of springs and electrostatic teeth range from 1 to 10 μm. Some dimensions are now in the nanoscale, particularly the mechanical spring width (50 nm) and the electrostatic gaps (100–400 nm). The sensitivities are in the ranges of 100 and 500–1000 aF g^{-1} for IP and OOP accelerometers respectively; aF = 10^{-18} F.

4.15 Silicon Nanowire Accelerometer

Application of strain to a crystal causes a change in electrical conductivity due to the piezoresistance (PZR) effect. When a uniaxial stress is applied and the electric field and current are in the same direction, the longitudinal PZR coefficient can be measured. It is defined as the relative change in conductivity per unit stress. Si nanowires display an exceptionally large PZR effect,

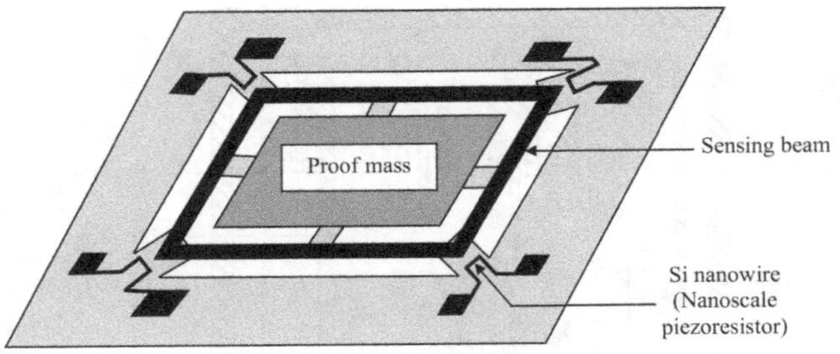

FIGURE 4.27 Silicon NW accelerometer. (Dao, D. V. et al., *Adv. Nat. Sci.: Nanosci. Nanotechnol.*, 1: 013001, 10, 2010.)

compared with bulk silicon, e.g., the longitudinal PZR coefficient along the <111> direction increases with decreasing diameter for P-type Si nanowires, reaching up to -3550×10^{-11} Pa^{-1}, in comparison with a bulk value of -94×10^{-11} Pa^{-1} (He and Yang 2006). This is the giant PZR effect. *What are the main features of this giant effect and what is its importance?* In the giant PZR effect, the change in resistance of silicon nanowires due to an applied mechanical stress was reported to be orders of magnitude larger than that of bulk silicon (Milne et al. 2010). Additionally, giant PZR is seen as a potential breakthrough means of detecting motion in NEMS, since PZR sensitivity scales better than optical or capacitive techniques. Moreover, since mechanical stress is a key element for performance enhancement of microelectronic devices, the physical mechanism behind giant PZR is vital for the design of accelerometers.

What is the interpretation of the giant PZR effect? There is still no consensus concerning the origins of the giant PZR, although two models have some support. One is based on a surface quantization effect, predicted to occur in the first few silicon monolayers, whereas the other is based on a stress-induced shift of the surface Fermi level in depleted structures, resulting from a change in surface charge. The atomic length scale of the former seems to be in disagreement with the typical wire diameters reported in the literature, which are at least several tens of nanometers, whereas the characteristic length scale of the latter is the surface depletion layer width (1–10 nm, depending on the doping density).

How is mechanical sensing performed, using the giant PZR effect? Using both atomic-level simulation and experimental evaluation methods, Dao et al. (2010) elucidated the giant piezoresistive effect in single crystalline silicon nanowires (SiNWs) along different crystallographic orientations. They used SiNWs as nanoscale piezoresistors for mechanical sensing by constructing a SiNW accelerometer. In this accelerometer (Figure 4.27), the seismic mass was suspended on four surrounding sensing beams, which were themselves connected to the frame at the ends. Silicon nanowires were placed near the fixed ends on the surface of the sensing beams. Upon acceleration of the sensor, the seismic mass moved, due to the inertial force. This motion deformed the beams so that the resistance of the nanowires was altered. Using a Wheatstone bridge, the aforementioned change in resistance was converted to a voltage change. The overall size of the accelerometer chip was 500 μm × 500 μm × 400 μm. This accelerometer has been made from multilayer separation by ion implantation of oxygen (SIMOX) "silicon on insulator" (SOI wafer) and fabricated by reactive ion silicon etching. The SiNW piezoresistor has a width of 128 nm and a thickness of 50 nm. The resistance of these piezoresistors was measured to be 20 kΩ. The calculated sensitivity for each axis was about 50 μV G^{-1}, and the resolution was 30 mG, where G symbolizes gravitational acceleration with the standard value 9.8 m s^{-2} on earth at sea level. Accordingly, by using SiNWs as piezoresistors, the sensor was miniaturized while increasing the sensitivity. Smaller chip size means larger number of chips per wafer, greater productivity, and therefore lower cost per chip. Ultra-small accelerometers are useful for portable devices, such as camcorders (video camera recorders), mobile phones, navigation systems, and entertainment devices.

4.16 CNT Flow Sensor for Ionic Solutions

Theoretically, Král and Shapiro (2001) showed the generation of an electric current in a metallic carbon nanotube immersed in a flowing liquid. They proposed a mechanism for the generation of the electric current and voltage through the transfer of momentum from the flowing liquid molecules to the acoustic phonons (quasi-particles representing the quantization of the modes of lattice vibrations of periodic, elastic crystal structures of solids) in the nanotube as the phonon quasi-momentum (a momentum-like vector). This, in turn, dragged free charge carriers into the nanotube, producing the current flow.

Ghosh et al. (2003) reported that the flow of a liquid on SWCNT bundles induced a voltage in the sample along the direction of the flow. The voltage generated fitted logarithmic velocity dependence over nearly six decades of velocity. It was observed that the induced voltage tended to saturate at low flow velocities, ~10^{-5} m second^{-1}. Furthermore, the magnitude of the voltage depended sensitively on the ionic conductivity and the polar nature of the liquid.

They put forward the notion of a flow sensor, based on SWCNTs, in which an electrical signal was created in response to a fluid flow. They believed that this sensor could be scaled down to micrometer length dimensions comparable to the length of the individual nanotubes, making it usable for very small liquid volumes. The sensor had a high sensitivity at low velocities and a fast response time (better than 1 ms).

Bourlon et al. (2007) found that individual electrolytically gated SWCNT transistors of ~2 nm diameter, when incorporated into microfluidic channels, locally sensed the change in electrostatic potential induced by the flow of ionic solutions on charged surfaces, known as the *streaming potential.* It is the potential which is produced when a liquid is forced to flow through a capillary tube or porous solid. It is the difference in electric potential at zero electric current between the capillary tube, porous solid, membrane, plug or diaphragm, and a liquid that is forced to flow through it. The streaming potential depends upon the presence of an electrical double layer at a solid–liquid interface. This electrical double layer is made up of ions of one charge type that are fixed to the surface of the solid and an equal number of mobile ions of the opposite charge, which are distributed through the neighboring region of the liquid phase.

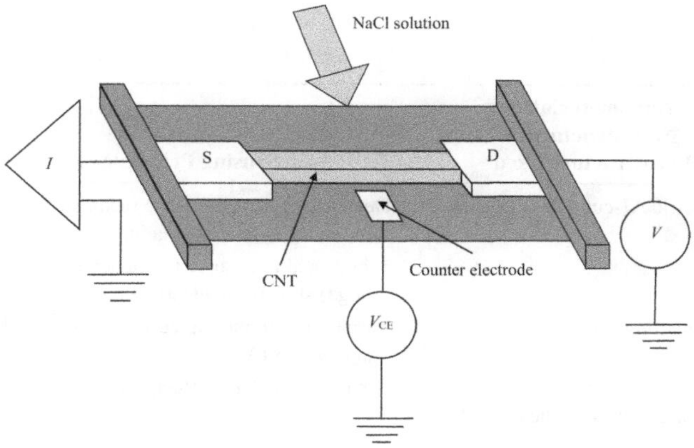

FIGURE 4.28 CNT-based flow sensor. (Bourlon, B. et al., *Nat. Nanotechnol.*, 104, 104, 2007.)

Taking advantage of the coupling between electrostatic potential and fluid dynamics, they demonstrated that individual nanotube transistors acted as fluidic flow sensors (Figure 4.28) that sensed local potentials with nanometer-scale resolution, which could be used to probe electrokinetic phenomena. These are the phenomena associated with movement of charged particles through a continuous medium or with the movement of a continuous medium over a charged surface. Examples are electrophoresis (the movement of colloidal particles or polyelectrolytes under an electric field), electro-osmosis (the motion of a liquid through an immobilized set of particles, such as a capillary or porous plug, in response to an electric field), etc., involving tangential fluid motion adjacent to a charged surface and considered as a manifestation of electrical properties of interfaces in steady-state and isothermal conditions.

How were the CNT transistors beneficial as flow sensors? Because of their ~2 nm height, nanotube devices were integrable into micro- or nanoscale fluidic circuits, making them appropriate for use in lab-on-a-chip applications, such as providing active flow sensing and high-resolution flow mapping on surfaces. Another benefit derived from this approach was that it enabled the study of nanotube conductance using the nanotube transconductance as an amplifier. In particular, this ensured that the changes induced by the flow were generated locally from only the nanotube, eliminating potential artifacts (spurious experimental results), that could originate at the electrode level.

Bourlon et al. (2007) projected that the thermodynamic limit to sensitivity in a 1-Hz bandwidth was ~ 100 nL min^{-1} for a typical 100 μm×200 μm microfluidic channel. To estimate the fundamental limits to flow sensing, using this technique, Bourlon et al. (2007) assumed that Johnson noise (the random fluctuations in the voltages due to the thermal agitation of charge carriers in a resistor; it is the mean-square EMF in conductors due to thermal agitation of the electromagnetic modes which are coupled to the thermal environment by the charge carriers) from the fluid resistance was the ultimate limit to the sensitivity. Johnson noise introduced thermal fluctuations δV_{CE} to the voltage applied to the electrolytic gate. They anticipated δV_{CE} ~100 μV Hz$^{-1/2}$ for $n = 10^{-6}$ M. Based on the measured values of α, the lowest detectable flow rate in the channel in a 1-Hz bandwidth was $Q = 100$ nL min^{-1}.

4.17 Discussion and Conclusions

Salient features of mechanical nanosensors are briefly presented in Table 4.4. Mass, displacement, acceleration, force, pressure, and fluid flow are the parameters that have been addressed for the fabrication of physical nanosensors. The quartz crystal microbalance (QCM) has long been famous as an ultrasensitive acoustic wave-based mass measurement device. But MEMS and NEMS resonators have taken the lead and improved precision for the measurement of infinitesimally small masses, such as adsorbed atoms or molecules, on surfaces, and stretched the limits far below expectations.

Displacement sensing has progressed from electron tunneling devices to Coulomb blockade electrometer and SET. In addition to magnetomotive, piezo-based and optical options are also available for displacement determination.

For measuring extremely small displacements and forces, using CNTs, another approach has been pursued, in which a bridge or cantilever presses downward on a CNT placed below it. The resulting deformation of the CNT causes a change in its resistance in accordance with the displacement or force, through which the latter can be evaluated. Such kind of CNT-based displacement and force nanosensors have been developed.

Review Exercises

4.1 Describe the electrical actuation and detection of the guitar-string-like oscillation modes of doubly clamped nanotube oscillators. How can the devices be used to transduce very small forces?

4.2 Why is SiC an extremely promising material for high-frequency NEMS? Explain why, for a given geometry, nanometer-scale SiC resonators are capable of yielding substantially higher frequencies than GaAs and Si resonators.

4.3 NEMS resonators have been suggested as nanosensors that allow the sensitive detection of molecules at atomic resolution. Explain.

4.4 SAW chemical sensors are based on the effects of adsorbed molecules on geometrical, elastic, and electric properties of gas-sensing layers and the corresponding mass-loading of the working surface of the substrate-carrying SAW. What is the SAW parameter which changes and is used for determining sensor output?

4.5 Explain how the motion of a nanomechanical resonator is detected by capacitively coupling the gate of the SET to a metal electrode placed on the resonator, and biasing the electrode at a constant voltage?

4.6 In a quartz crystal resonator, the crystal resonance frequency directly relates to the mass adsorbed; hence it is the parameter typically measured. Explain how the mass measurement is performed by inserting the crystal as the frequency controlling element of a feedback oscillator circuit.

TABLE 4.4
Mechanical Nanosensors at a Glance

Sl. No.	Name of the Nanosensor	Sensed Quantity/ Applications	Nanomaterial/ Nanostructure/ Microstructure Used	Sensing Principle
1	Quartz crystal microbalance (QCM)	Mass (up to 10^{-13} g)	Piezoelectric AT-cut quartz crystal	Decrease of acoustic wave resonance frequency with mass loading
2	MEMS/NEMS resonators	Mass (up to 10^{-18} and 10^{-21} g)	NEMS	Vibrational resonance frequency (GHz range) shift from adsorbed molecules
3	Electron tunneling displacement nanosensor	Displacement	Sharp tunneling tip	Variation of tunneling current with tunnel gap, as in STM
4	Coulomb blockade electrometer	Displacement	Two tunnel junctions with a capacitor coupling to the central region	Coulomb blockade effect
5	Single-electron transistor (SET)	Displacement	Capacitively coupled SET with a flexural beam resonator	Modulation of source–drain current by charge on gate electrode
6	Magnetomotive displacement nanosensor	Displacement	Conducting nanomechanical element	Induced EMF by a time-varying flux
7	Piezoresistive and piezoelectric displacement nanosensors	Displacement	NEMS	Resistance or current changes
8	Optical displacement nanosensor	Displacement	NEMS	Michelson and Fabry-Perot interferometry
9	Doubly clamped suspended CNT resonators	10^{-15} N force sensing	Carbon nanotube	Change in resonance frequency with applied force
10	Suspended CNT electromechanical sensors	Displacement and force	Suspended SWCNT fixed underneath a freestanding cantilever or bridge	Force acting on the cantilever or bridge deflects them and deforms the SWCNT below changing its conductance
11	Membrane-based CNT pressure sensor	Pressure	CNT adhered to an alumina membrane	Piezoresistive effect
12	Tunnel effect accelerometer	Acceleration	MEMS	Relation of deflection of proof mass (measured through tunneling current changes) with acceleration
13	NEMS accelerometer	Acceleration	MEMS and NEMS	Capacitance variation between two sets of electrodes
14	Silicon nanowire accelerometer	Acceleration	Si NWs placed near the fixed ends on the surface of the sensing beams	Giant piezoresistive effect in single, crystalline SiNWs
15	CNT flow sensor for ionic solutions	Liquid flow	CNT wall	Liquid flow produces a voltage along CNT wall depending on ionic conductivity of liquid; similar phenomenon occurs in electrolytically gated CNT transistors

4.7 When the resonance frequency shift Δf occurs due to mass change, mass change Δm is calculated using

$$\Delta m = -2\left(\frac{m}{f}\right)\Delta f \tag{4.84}$$

Applying this equation, prescribe the conditions regarding the mass of the cantilever and its frequency to detect a small Δm.

4.8 Write the Sauerbrey equation between the frequency drop and mass loading of a QCM and explain the symbols used. State the conditions of validity of the equation.

4.9 Explain the operation of self-sensing cantilevers using piezoresistive displacement transduction.

4.10 Does the flow of a liquid on SWCNT bundles induce a voltage in the sample along the direction of the flow? How does the magnitude of the voltage depend sensitively on the ionic conductivity and on the polar nature of the liquid?

4.11 What happens when two conductors are brought into extreme proximity (~1 nm) with an applied bias between them? Name the phenomenon and show how it is utilized in the accelerometer.

4.12 What is the Coulomb blockade effect? What is a Coulomb blockade electrometer? Compare the Coulomb blockade electrometer and electron tunneling transistor as displacement-sensing devices.

4.13 How are individual CNT transistors useful for probing electrokinetic phenomena and fluid flow in microfluidic channels? Can these fluidic flow sensors locally sense potentials with nanometer-scale resolution?

4.14 For cantilever devices with sizes comparable to the mean free path in air, the added benefit of preservation of a high-quality factor under ambient conditions enables unprecedented, sub-attogram-scale mass resolution at room temperature and atmospheric pressure. Elaborate this statement to explain that nanoscale mechanical sensors offer a greatly enhanced performance than is unattainable with microscale devices.

4.15 The Langmuir-Blodgett (LB) technique is a molecular engineering technique capable of fabricating materials with high structural order and sensing multilayers, monolayer by monolayer, with fine thickness control. How are SWCNT films prepared by the LB technique on a sensor surface for volatile organic compound detection?

4.16 Mention a few advantages of electron tunneling displacement transducers. Draw and explain the operation of a micromachined electron tunneling displacement transducer.

4.17 Describe two techniques for actuation of NEMS devices. Explain also how these techniques are utilized for motion detection.

4.18 An important concept in the study of electromechanical transducers is backaction. Backaction describes the perturbation of the mechanical element by the sensor and the output circuit during the detection. Define backaction noise.

REFERENCES

Albrecht, T. R., P. Grtitter, D. Horne, and D. Rugar. 1991. Frequency modulation detection using high-Q cantilevers for enhanced force microscope sensitivity. *Journal of Applied Physics* 69(2): 668–673.

Anantram, M. P. and F. L'eonard. 2006. Physics of carbon nanotube electronic devices. *Reports on Progress in Physics* 69: 507–561.

Ansari, M. Z., C. Cho, J. Kim, and B. Bang. 2009. Comparison between deflection and vibration characteristics of rectangular and trapezoidal profile microcantilevers. *Sensors* 9: 2706–2718.

Beeby, S., G. Ensell, M. Kraft, and N. White. 2004. *MEMS Mechanical Sensors*. Boston, MA: Artech House, Inc., 269 pp.

Bourlon, B., J. Wong, C. Miko, L. S. Forro, and M. Bockrath. 2007. A nanoscale probe for fluidic and ionic transport. *Nature Nanotechnology* 104(2): 104–107.

Chen, G. Y., T. Thundat, E. A. Wachter, and R. J. Warmack. 1995. Adsorption-induced surface stress and its effects on resonance frequency of microcantilevers. *Journal of Applied Physics* 77(8): 3618–3622.

Cleland, A. N. and M. L. Roukes. 1996. Fabrication of high frequency nanometer scale mechanical resonators from bulk Si crystals. *Applied Physics Letters* 69: 2653–2655.

Dai, M. D., K. Eom, and C.-W. Kim. 2009. Nanomechanical mass detection using nonlinear oscillations. *Applied Physics Letters* 95: 203104.

Dao, D. V., K. Nakamura, T. T. Bui, and S. Sugiyama. 2010. Micro/nanomechanical sensors and actuators based on SOI-MEMS technology. *Advances in Natural Sciences: Nanoscience and Nanotechnology* 1: 013001 (10pp.).

Dong, H., Y. Jia, Y. Hao, and S. Shen. 2005. A novel out-of-plane MEMS tunneling accelerometer. *Sensors and Actuators A* 120: 360–364.

Duemling, M. 2002. *Modeling and Characterization of Nanoelectromechanical Systems*. MS Thesis, Virginia Polytechnic Institute and State University, Blacksburg, VA, pp. 6–13.

Ekinci, K. L. 2005. Electromechanical transducers at the nanoscale: Actuation and sensing of motion in nanoelectromechanical systems (NEMS). *Small* 1(8–9): 786–797.

Ekinci, K. L., Y. T. Yang, and M. L. Roukes. 2004. Ultimate limits to inertial mass sensing based upon nanoelectromechanical systems. *Journal of Applied Physics* 95(5): 2682–2689.

Ghosh, S., A. K. Sood, and N. Kumar. 2003. Carbon nanotube flow sensors. *Science* 299: 1042–1044.

Hansen, K. M. and T. Thundat. 2005. Microcantilever biosensors. *Methods* 37: 57–64.

He, R. and P. Yang. 2006. Giant piezoresistance effect in silicon nanowires. *Nature Nanotechnology* 1: 42–46, doi: 10.1038/nnano.2006.53.

Ho, C. K., E. R. Lindgren, K. S. Rawlinson, L. K. McGrath, and J. L. Wright. 2003. Development of a surface acoustic wave sensor for *in-situ* monitoring of volatile organic compounds. *Sensors* 3(7): 236–247, doi: 10.3390/s30700236.

Huang, X. M. H., C. A. Zorman, M. Mehregany, and M. F. Roukes. 2003. Nanoelectromechanical systems: Nanodevice motion at microwave frequencies. *Nature* 421: 496.

Huang, X. M. H., X. L. Feng, C. A. Zorman, M. Mehregany, and M. L. Roukes. 2005. VHF, UHF and microwave frequency nanomechanical resonators. *New Journal of Physics* 7(247): 1–15, doi: 10.1088/1367-2630/7/1/247.

Ikehara, T., J. Lu, M. Konno, R. Maeda, and T. Mihara. 2007. A high quality-factor silicon cantilever for a low detection-limit resonant mass sensor operated in air. *Journal of Micromechanics and Microengineering* 17: 2491–2494.

Jenkins, N. E., L. P. DeFlores, J. Allen, T. N. Ng, S. R. Garner, S. Kuehn, J. M. Dawlaty, and J. A. Marohn. 2004. Batch fabrication and characterization of ultrasensitive cantilevers with submicron magnetic tips. *Journal of Vacuum Science and Technology B* 22(3): 909–915.

Kanazawa, K. and N.-J. Cho. 2009. Quartz crystal microbalance as a sensor to characterize macromolecular assembly dynamics. *Journal of Sensors* 2009: 824947, 17 pp., doi:10.1155/2009/824947.

Kenny, T. W., S. B. Waltman, J. K. Reynolds, and W. J. Kaiser. 1991. Micromachined silicon tunnel sensor for motion detection. *Applied Physics Letters* 58(1): 100–102.

Kenny, T. W., W. J. Kaiser, J. K. Reynolds, J. A. Podosek, H. K. Rockstad, E. C. Vote, and S. B. Waltman. 1992. Electron tunnel sensors. *Journal of Vacuum Science Technology A* 10(4): 2114–2118.

Knobel, R. and A. N. Cleland. 2002. Piezoelectric displacement sensing with a single electron transistor. *Applied Physics Letters* 81(12): 2258–2260.

Knobel, R. G. and A. N. Cleland. 2003. Nanometre-scale displacement sensing using a single electron transistor. *Nature* 424: 291–293.

Král, P and M. Shapiro. 2001. Nanotube electron drag in flowing liquids. *Physical Review Letters* 86:131–134.

Krishnan, G., C. U. Kshirsagar, G. K. Ananthasuresh, and N. Bhat. 2007. Micromachined high-resolution accelerometers. *Journal of the Indian Institute of Science* 87(3): 333–361.

Lavrik, N. V., M. J. Sepaniak, and P. G. Datskos. 2004. Cantilever transducers as a platform for chemical and biological sensors. *Review of Scientific Instruments* 75 (7): 2229–2253.

Li, C. and T.-W. Chou. 2003. Single-walled carbon nanotubes as ultrahigh frequency nanomechanical resonators. *Physical Review* B68F: 073405-1–073405–3.

Li, C.-Y. and T.-W. Chou. 2004. Strain and pressure sensing using single-walled carbon nanotubes. *Nanotechnology* 15: 1493–1496.

Li, C. and T.-W. Chou. 2006. Atomistic modeling of carbon nanotube-based mechanical sensors. *Journal of Intelligent Material Systems and Structures* 17: 247–254.

Li, X., T. Ono, Y. Wang, and M. Esashi. 2003. Ultrathin single-crystalline-silicon cantilever resonators: Fabrication technology and significant specimen size effect on Young's modulus. *Applied Physics Letters* 83(15): 3081–3083.

Lin, Y.-C., T. Ono, and M. Esashi. 2005. Fabrication and characterization of microma-chined quartz-crystal cantilever for force sensing. *Journal of Micromechanics and Microengineering* 15: 2426–2432.

Lu, J., T. Ikehara, Y. Zhan, T. Mihara, T. Itoh, and R. Maeda. 2007. High quality factor silicon cantilever transduced by piezoelectric lead zirconate titanate film for mass sensing applications. *Japanese Journal of Applied Physics* 46(12): 7643–7647.

Milne, J. S., A. C. H. Rowe, S. Arscott, and Ch. Renner. 2010. Giant piezoresistance effects in silicon nanowires and microwires. *Physical Review Letters* 105(22): 226802.

Minot, E. D., Y. Yaish, V. Sazonova, J.-Y. Park, M. Brink, and P. L. McEuen. 2003. Tuning carbon nanotube band gaps with strain. *Physical Review Letters* 90(15): 156401–156404.

Ollier, E., L. Duraffourg, M. Delaye, S. Deneuville, V. Nguyen, P. Andreucci, H. Grange, et al. 2006. Thin SOI NEMS accelerometers compatible with in-IC integration. In: *IEEE Sensors* 2006, EXCO, Daegu, Korea, October 22–25, 2006. New York: IEEE, pp. 54–57.

Rast, S., C. Wattinger, U. Gysin, and E. Meyer. 2000. The noise of cantilevers. *Nanotechnology* 11: 169–172.

Rinaldi, G. 2009. Simple and versatile micro-cantilever sensors. *Sensor Review* 29(1): 44–53.

Rockstad, H. K., T. K. Tang, K. Reynolds, T. W. Kenny, W. J. Kaiser, and T. B. Gabrielson. 1996. A miniature, high-sensitivity, electron tunneling accelerometer. *Sensors and Actuators A* 53: 227–231.

Sazonova, V., Y. Yaish, H. Ustunel, D. Roundy, T. A. Arias, and P. L. McEuen. 2004. A tunable carbon nanotube electromechanical oscillator. *Nature* 431: 284–287.

Sone, H., Y. Fujinuma, and S. Hosaka. 2004. Picogram mass sensor using resonance frequency shift of cantilever. *Japanese Journal of Applied Physics* 43(6A): 3648–3651.

Stampfer, C., A. Jungen, and C. Hierold. 2005. Nano electromechanical transducer based on single-walled carbon nanotubes. The 13th International Conference on Solid-State Sensors, Actuators and Microsystems, 2005. *Digest of Technical Papers. TRANSDUCERS '05.* (IEEE Cat. No. 05TH8791), Seoul, South Korea, Vol. 2, Publisher: IEEE, NJ, pp. 2103–2106, doi: 10.1109/SENSOR.2005.1497518.

Stampfer, C., A. Jungen, and C. Hierold. 2006a. Fabrication of discrete nanoscaled force sensors based on single-walled carbon nanotubes. *IEEE Sensors Journal* 6: 613–617.

Stampfer, C., T. Helbling, D. Obergfell, B. Schöberle, M. K. Tripp, A. Jungen, S. Roth, V. M. Bright, and C. Hierold. 2006b. Fabrication of single-walled carbon-nano-tube-based pressure sensors. *Nano Letters* 6(2): 233–237.

Stampfer, C., A. Jungen, R. Linderman, D. Obergfell, S. Roth, and C. Hierold. 2006c. Nano-electromechanical displacement sensing based on single-walled carbon nanotubes. *Nano Letters* 6(7): 1449–1453.

Stoney, G. G. 1909. The tension of metallic films deposited by electrolysis. *Proceedings of the Royal Society of London. Series A* 82(553): 172–175.

Vashist, S. K. 2007. A review of microcantilevers for sensing applications. *Journal of Nanotechnology Online* 3: 1–15.

Waltman, S. B. and W. J. Kaiser. 1989. An electron tunneling sensor. *Sensors and Actuators* 19(3): 201–210.

Wang, K., A.-C. Wong, and C. T.-C. Nguyen. 2000. VHF free-free beam high-micromechanical resonators. *Journal of Microelectromechanical Systems* 9(3): 347–360.

Wautelet, M. 2001. Scaling laws in the macro-, micro- and nanoworlds. *European Journal of Physics* 22: 601–611.

White, J. D. 1993. An ultra high resolution displacement transducer using the Coulomb blockade electrometer. *Japanese Journal of Applied Physics* 32: L1571–L1573.

Yang, L., M. P. Anantram, J. Han, and J. P. Lu. 1999. Band-gap change of carbon nanotubes: Effect of small uniaxial and torsional strain. *Physical Review B* 60(19): 13874–13878.

Yang, J., T. Ono, and M. Esashi. 2000. Mechanical behavior of ultrathin microcantilever. *Sensors and Actuators* 82:102–107.

5

Thermal Nanosensors

5.1 Introduction

In physics (Section 1.3), we talked about heat or thermal energy (Section 1.3.6). Thermal nanosensors utilize effects, such as changes in thermo electromotive force (thermo EMF) and resistance in response to temperature, and so forth. Temperature is a fundamental physical property in both the scientific and industrial fields. It is one of the primary physical quantities that is routinely measured to derive other thermodynamic quantities, such as heat, energy, or specific heat capacity (the ratio of the amount of heat, measured in calories, required to raise the temperature of 1 g of a substance by 1°C to the amount of heat required to raise the temperature of a similar mass of a reference material, usually water, by the same amount). *Thermodynamics* is the science that deals with the relationships and conversions between heat and other forms of energy, most notably mechanical work. Temperature sensing is therefore a vital requirement in many parts of industry, including aerospace (comprising the atmosphere of Earth and the surrounding space, typically referring to the industry concerned with vehicles moving through air and space), nuclear, mechanical, chemical, and medical technologies. If these sensors are small in size, they can be readily embedded in the systems without intrusion.

As semiconductor devices continue to decrease in size, and more and more applications are found for micro- and nanodevices, there is an urgent need for measuring physical quantities at submicrometer- or even nanometer-sized dimensions. The burgeoning field of nanotechnology promises to revolutionize many scientific fields.

What are the existing methods for temperature measurement? Presently, temperature is measured using thermocouples (TCs) (temperature sensors consisting of two wires made of different metals, e.g., one chromel and one constantan wire); semiconductor diodes working on the dependence of their forward voltages on temperature; resistance temperature detectors (RTDs, commonly made of platinum); thermistors (temperature-sensitive resistors made from barium titanate ($BaTiO_3$) or the oxides of the transition metals, manganese (Mn), cobalt (Co), copper (Cu), and nickel (Ni)); and infrared (IR; wavelength = 0.7–300 μm) thermometers (noncontact thermometers, determining temperature from the black-body radiation of a body; a black body is an idealized physical body that absorbs all incident electromagnetic radiation). Although infrared radiation is invisible, humans can sense it as heat. Astronomers have found that infrared radiation is particularly useful when trying to probe areas of our universe that are surrounded by clouds of gas and dust. Because of the longer wavelength of infrared radiation, it can pass right through these clouds and reveal details invisible by observing other types of radiation. Especially interesting are areas where stars and planets are forming and the cores of galaxies, where it is believed that huge black holes reside.

More advanced yet are near-field optical thermometry and other exotic methods. The spatial resolution of infrared thermometry is limited to the wavelength of collected IR radiation due to diffraction effects (the apparent bending of waves around small obstacles and the spreading out of waves past small openings). The temperature of objects inaccessible to far-field infrared thermometry is found by near-field thermometry. For ultra-high optical resolution, near-field scanning optical microscopy (NSOM) is currently the photonic instrument of choice.

The near-field region is defined as the region above a surface with dimensions less than a single wavelength of light, incident on the surface. Near-field imaging takes place when a submicron optical probe is positioned at a very short distance from the sample, and light is transmitted through a small aperture at the tip of this probe. Within the near-field region, evanescent light is not diffraction limited, so that nanoscale spatial resolution is achievable. Evanescent is defined as "temporal" or "fleeting"; unlike normal light, evanescent light is extremely limited in terms of the distance it can be propagated.

Lateral resolution of 20 nm and vertical resolution of 2–5 nm have been demonstrated. The collected radiation depends on the temperature-dependent thermal reflectance (the fraction of incident radiation reflected by a surface) of the sample.

Each of the techniques mentioned earlier has its own advantages and disadvantages, e.g., the TC is extremely simple in design and is inexpensive, whereas the semiconductor diode is unsuitable for high temperatures but is used extensively to measure cryogenic temperatures. The spatial resolution of most of these techniques is about 10 μm, except for the near-field thermometry, in which the spatial resolution can be of the order of 50 nm. But near-field thermometry involves complicated optical instrumentation and is quite a cumbersome technique.

5.2 Nanoscale Thermocouple Formed by Tungsten and Platinum Nanosized Strips

How is a nanosized thermocouple junction made? Fabrication of a nanosized sensor (Figure 5.1) (Chow et al. 2002) consists of the following two steps: (i) deposition of a nanosized strip of the first metal (tungsten) on an electrical insulator substrate by a focused ion beam (FIB) deposition process; and (ii) deposition of a nanosized strip of a second metal (platinum) on the same substrate by a FIB process in a partially overlapping portion on the first-metal nanosized strip.

The partially overlapping segment is selected from one of a ball-shaped portion, and a point-shaped configuration portion, wherein the first-metal nanosized strip and the second-metal nanosized strip both have a thickness of around 50 nm, and constitute a bimetal sensing junction from the partially overlapping

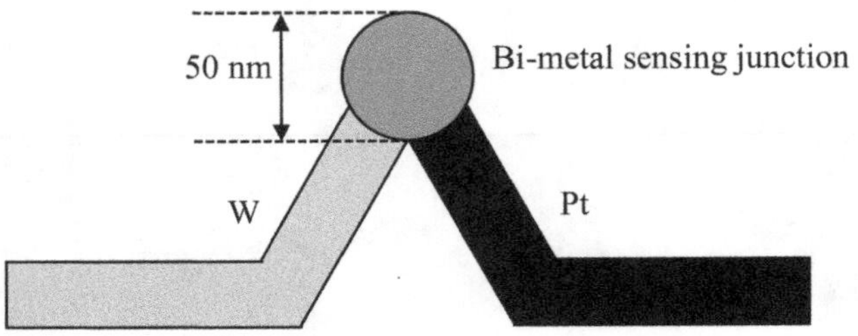

FIGURE 5.1 Overlapped nanosensor structure. (Chow, L. et al. Fabrication of nano-scale temperature sensors and heaters. US Patent 6905736, Filing Date: 02/27/2002, Publication Date: 06/14/2005.)

portion between the first and the second strips. The cross-sectional area of the bimetal sensing junction is approximated to 50 nm × 50 nm = 2500 nm^2.

The process used to deposit metal nanostrips to fabricate the sensor is ion-beam-assisted chemical- vapor-deposition (IBA–CVD). The FIB Ga^+ ion beam is used to decompose $W(CO)_6$ molecules to deposit a tungsten nanostrip on a suitable substrate. For the Pt strip, the precursors are trimethyl platinum $(CH_3)_3Pt$. Because of the use of a Ga^+ beam in the deposition, both Pt and W nanostrips contain a definite proportion of Ga impurities, which are denoted as Pt(Ga) and W(Ga), respectively. By characterization of the response of this Pt(Ga)/W(Ga) nanoscale junction, a temperature coefficient of ~5.4 mV °C^{-1} was specified. This was a factor of ~130 times larger than that of the conventional K-type TCs. The K-type TC (which is made of a nickel-chromium/nickel-aluminum junction, called chromel/alumel) is the "general purpose" TC. K-type TCs are available in the 95°C–1260°C range. Sensitivity is approximately 41 μV °C^{-1}.

5.3 Resistive Thermal Nanosensor Fabricated by Focused-Ion-Beam Chemical-Vapor-Deposition (FIB-CVD)

Can a resistive nanosensor be made by the FIB method? FIB–CVD is a very useful technique by which to make three-dimensional nanostructures and can be used to fabricate various nanodevices, such as nanosensors. Ozasa et al. (2004) produced tungsten thermal nanosensors with 3D nanostructure by FIB–CVD, using $W(CO)_6$ source gas material. Four-terminal Au electrodes were formed on a 100-nm thick SiO_2 on Si substrate by electron-beam lithography and dry etching. Then, a thermal nanosensor with 100 nm diameter was made by FIB-CVD. The exposure time was 9 min.

The temperature measurement using a nanosensor was based on the change of electrical resistivity with temperature. The resistivity of tungsten deposited by FBI-CVD was about 100 times higher than that of bulk tungsten at 25°C. The resistivity of a thermal nanosensor can be correctly measured by a four-terminal electrode arrangement as the contact resistances are ignored in this method. The four-point probe method is an electrical impedance measuring technique that uses separate pairs of current-carrying and voltage-sensing electrodes to carry out more accurate measurements than the traditional two-terminal (2T) sensing.

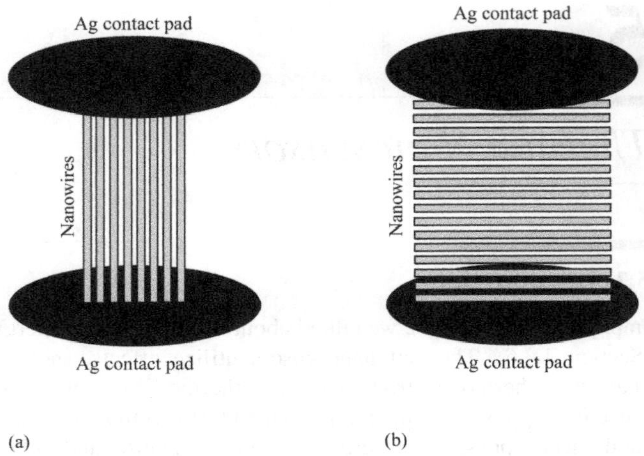

FIGURE 5.2 Carbon nanowire temperature sensor: (a) nanowires fixed parallel to contact pads and (b) nanowires fixed perpendicular to contact pads. (Zaitsev, A. M. et al., *Phys. Status Solidi* A, 204, 3574, 2007.)

The results indicated that the resistivity of a thermal nanosensor was proportional to temperature, from room temperature to 120°C.

5.4 Carbon "Nanowire-on-Diamond" Resistive Temperature Nanosensor

A temperature sensor based on an array of carbon nanowires written by a 30 keV Ga^+ FIB on a diamond substrate was developed by Zaitsev et al. (2007). *What is the reason for using carbon nanowires in place of carbon nanotubes (CNTs)?* The use of carbon nanowires in nanoelectromechanical systems (NEMS) and nanoelectronics is greatly facilitated by their far more flexible technology, compared with CNTs, the controllability of their positioning and the reproducibility of their electrical and structural parameters.

The sensor (Figure 5.2) showed an exponential increase in current in response to temperature at a rate of 0.1 dB °C^{-1} in the temperature range from 40°C to 140°C. The conductance along nanowires was high and ohmic, whereas the conductance in the orthogonal direction was much lower and super-linear. The advantages of the novel carbon nanowire sensor are light blindness, compatibility with carbon nanotechnology, reproducibility, and simplicity of fabrication.

What is the explanation for the temperature-dependent characteristic of a nanowire? The temperature sensitivity of the electrical conductance between the nanowires is interpreted by the thermal activation of charge carriers in a graphite-diamond heterostructure over an energy barrier of 0.24 eV, where a heterostructure is a structure of two different materials.

5.5 Carbon Nanotube Grown on Nickel Film as a Resistive Low-temperature (10–300 K) Nanosensor

Saraiya et al. (2006) grew CNTs on nickel film deposited on float glass substrate, using an ion- beam-deposition technique

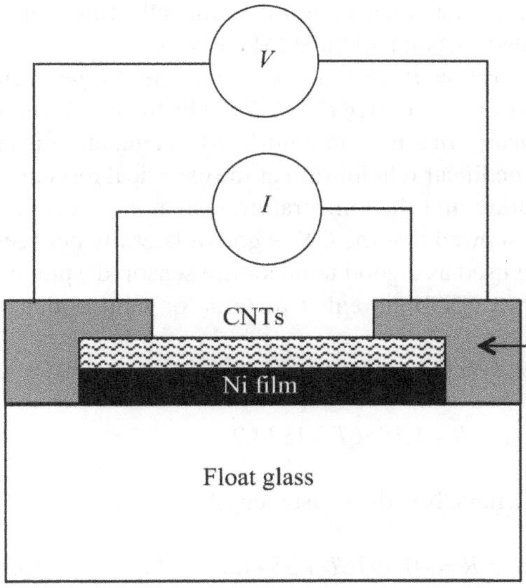

FIGURE 5.3 A CNT low-temperature sensor. (Saraiya, A. et al., *Syn. React. Inorg. Met.*, 36, 163, 2006.)

(Figure 5.3), a process of applying thin films of materials on a surface through the application of an ion beam, either directly from an ion source or by sputtering from a target bombarded by an ion beam. The percentage change of resistance of CNTs between 10 and 300 K was more than 600%. The drastic change in resistance at low temperature indicates that CNTs grown on nickel film can be used as low-temperature sensors.

5.6 Laterally Grown CNTs between Two Microelectrodes as a Resistive Temperature Nanosensor

The obvious question is: *What is the reason for the insistence on using CNTs? How is the use of CNTs as a temperature-sensing element beneficial?* In a complicated thermal flow system, the perturbation to the system from the temperature sensor becomes critically important and cannot be ignored. The advantage derived by using CNTs as a sensor is that the extremely small size of the CNT provides accurate measurements at the nanoscale size without disturbing the neighboring environment. In addition, the power consumption is very low for a small-sized sensor, while CNT can also provide an extremely small time constant, giving extremely rapid response to the temperature of the object on which measurement is done.

Kuo et al. (2007) presented direct growth or fabrication of CNTs between microelectromechanical system (MEMS) electrodes. A high-quality CNT sensor (Figure 5.4) was obtained by adopting properly controlled growth conditions. After a thin oxide film was deposited on a silicon wafer, a thin film of nickel was formed and used to make nanoparticles by sputter etching, a process whereby sputtering is used to remove atoms from a

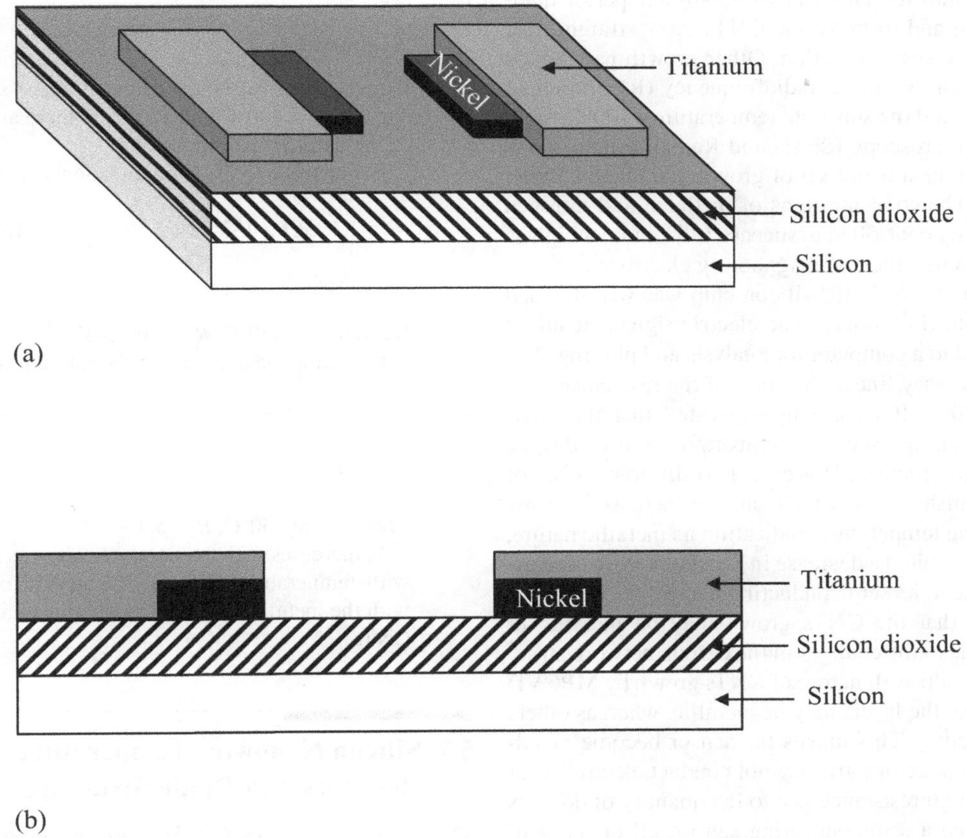

FIGURE 5.4 Electrode arrangement of a temperature nanosensor chip: (a) top view and (b) cross section. CNT (not shown here) is grown laterally between electrodes. (Kuo, C. Y. et al., *IEEE Trans. Nanotechnol.*, 6, 63, 2007.)

material. These nanoparticles were used as catalysts to grow CNTs at a later stage. Then, titanium film was deposited and patterned as electrodes and metal leads. When the titanium film was patterned, the nanoparticles were exposed on the sidewall of the electrodes and these were used to grow CNTs in the horizontal direction.

Kuo et al. (2007) grew CNTs laterally and selectively between electrodes made by MEMS techniques. The CNT growth method employed was microwave-plasma chemical vapor deposition (MPCVD). After electrodes were made, the silicon chip was transferred into an electron cyclotron resonance chemical vapor deposition (ECR-CVD) chamber, which was powered by microwaves, to grow CNTs. ECR is the resonance absorption of energy from a radiofrequency or a microwave frequency electromagnetic field by electrons in a uniform magnetic field, when the frequency of the electromagnetic field equals the cyclotron frequency of the electrons. In an ECR source, the vapor of the desired element is held in a specially designed magnetic field long enough for the elemental atoms to be ionized in collisions with electrons, which are kept in motion by microwaves.

The nanotubes grown were actually multi-wall nanotubes. The source gas used was CH_4, the concentration of which was varied from 10% to 60% during the CNT growth process. The addition of N_2 gas increased the bonding force among carbon atoms and enhanced the carbon activity when they precipitated from the catalyst. Thus, better-quality CNTs could be made with a high degree of graphitization (the formation of graphite-like material from organic compounds). Therefore, a different concentration of N_2 gas was fed into the chamber to obtain a superior quality of graphitization and to make the CNTs grow straight, like their growth in the vertical direction. Other growth parameters were chamber pressure = 30 torr, radiofrequency (RF) power of the plasma = 800 W, and the substrate temperature = 400°C. Both scanning electron microscopy (SEM) and Raman spectroscopy were used to determine a better set of growth parameters in the MPCVD process. The concentrations of methane and nitrogen gas had to be properly controlled to successfully obtain a higher-quality CNT, connecting the two neighboring electrodes.

After the growth of CNTs, the silicon chip was wire-bonded onto a polycarbonate (PC) board. The electric signals acquired at the CNT were fed to a computer for analysis and plotting. The results indicated the very linear variation of the resistance with the temperature (20°C–160°C). This suggested that the laterally grown CNTs can be used as a temperature sensor despite the seriously curled structure. However, two different types of CNTs were distinguished: one which showed increase in resistance with increasing temperature, indicating its metallic nature, whereas the other exhibited a decrease in resistance with increasing temperature, due to its semiconducting nature.

The observation that the CNTs grown may be metallic or semiconductor is due to the changes in the carbon network structure. However, the multi-wall nature of CNTs grown by MPCVD suggests that some of the layers may be metallic, whereas others may be semiconducting. This makes the sensor become metallic, since the semiconductor part may not conduct electricity, or it may have a very high resistance due to the quantity of dopants absorbed. To become a semiconducting sensor, all of the wall layers in the CNT must be semiconductors. The probability that all of the layers in the CNT become semiconductors is very low. Therefore, one does not often obtain a semiconducting sensor from a multi-walled carbon nanotube (MWCNT).

What is the current-voltage (I-V) behavior and temperature coefficient of resistance (TCR) of the CNTs? The linear *I-V* curve of the CNTs indicates that the titanium/MWCNT junction is an ohmic contact. The linear relationship of the electrical resistance with the temperature and the temperature response or recovery test of the CNTs showed that the CNTs grown laterally between electrodes can be used as a good temperature sensor, despite the fact that the grown CNTs are either metallic or semiconductor in nature

For a metallic CNT, the resistance (R) is (Kuo et al. 2007):

$$R = 1.1956T + 1834.2 \tag{5.1}$$

For a semiconducting CNT, the resistance is:

$$R = -0.8976T + 5344.3 \tag{5.2}$$

where T is the temperature in °C. The TCR for the current MWCNT sensor, which was grown at CH_4 = 20% and N_2 = 20 standard cubic centimeters per minute (sccm), obtained within the experimental growth conditions and gap widths, varied from 0.0008152 to 0.0001759. The TCR is quite low as compared with the TCR using platinum as a temperature sensor. However, the nanosize of CNT can have a very high response and sensitivity to the environmental temperature change. This is very useful for measurements in systems with very rapid temperature variations.

Example 5.1

Compare the resistances of metallic and semiconducting CNTs at 100°C and 150°C. Is there any difference in the nature of the variation?

For metallic CNT, from Equation 5.1, at 100°C:

$$\begin{aligned} R_1 &= 1.1956T + 1834.2 = 1.1956 \times 100 + 1834.2 \\ &= 1953.76\,\Omega \end{aligned} \tag{5.3}$$

Similarly, at 150°C, $R_1 = 2013.54\,\Omega$.

For semiconducting CNT, from Equation 5.2:

$$\begin{aligned} R_2 &= -0.8976T + 5344.3 = -0.8976 \times 100 + 5344.3 \\ &= 5254.54\,\Omega \end{aligned} \tag{5.4}$$

Likewise, at 150°C, $R_2 = 5209.66\,\Omega$.

R_1 increases with temperature, whereas R_2 decreases with temperature. This is expected in conformation with the metallic and semiconducting behaviors of the CNTs.

5.7 Silicon Nanowire Temperature Nanosensors: Resistors and Diode Structures

The silicon nanowires (SiNWs) are accredited as temperature sensors, because of the fact that the conductance of these semiconductor wires is highly dependent on temperature. Agarwal

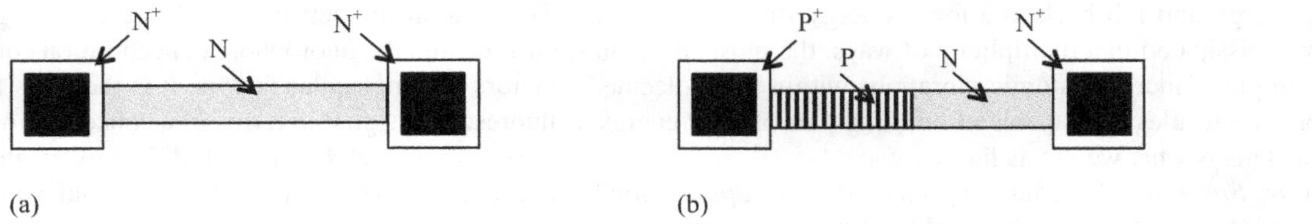

FIGURE 5.5 (a) A nanowire resistor and (b) a nanowire diode. (Agarwal, A. et al., *Sensor Actuat.* A, 145–146, 207, 2008.)

et al. (2008) presented SiNWs as nano temperature sensors. They employed two configurations, namely (i) resistance temperature detector (RTD) and (ii) diode temperature detector (DTD) types (Figure 5.5). The first configuration, of RTD-type sensors, showed TCR values ~7500 ppm K^{-1}, which were increased beyond 10,000 ppm K^{-1} by the application of back-bias. The second configuration, of DTD-type sensors, using nanowires, recorded more than one order of variation in the reverse-bias current, in the temperature range 293–373 K. Both types of nano temperature sensors were highly sensitive and could be integrated with other biochemical sensors in lab-on-a-chip devices that integrate one or several laboratory functions on a single chip only millimeters to a few square centimeters in size.

What advantages favor SiNWs for temperature sensing? The SiNW approach paved the way for realizing high-density, highly sensitive biochemical sensor chips integrated with nano temperature sensors, using mass fabrication technologies, and thus had the potential to achieve multiplexed diagnosis, using nanowire arrays (Figure 5.6), which was essential for the understanding and treatment of complex diseases. The approach also had the inherent advantage of integrating with Si-based signal processing and communication circuits. Conversely, most present techniques reported for nano temperature sensors, like assembly of CNTs as a thermal probe, face mass fabrication challenges. The mass manufacturing capability, reproducibility, and integration with other functionalities in a biochip is extremely challenging for these sensors.

How are Si nanowire sensors fabricated? SiNWs were realized on 8″ =20.32-cm diameter wafers. For single-crystal SiNWs, silicon-on-insulator (SOI) wafers were used. For polysilicon nanowires, 80-nm thick polysilicon was deposited on 50-nm thick thermal oxide grown on a P-type test wafer. The wafers were doped with P-type (boron) or N-type (phosphorus) impurities by ion implantation. The dopants were activated in a rapid thermal annealing (RTA) short-time heat treatment furnace, and

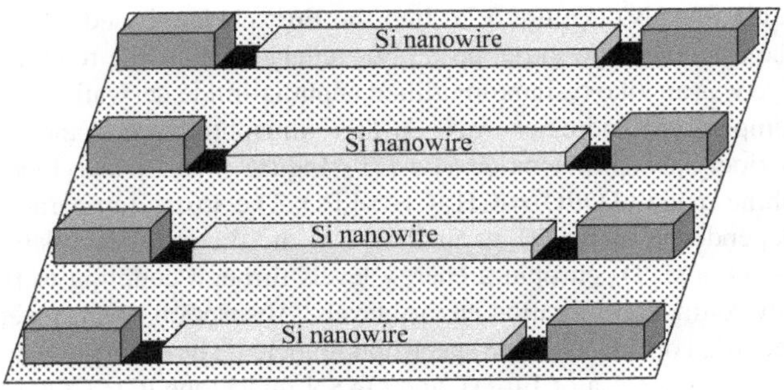

FIGURE 5.6 Nanowire array with metal lines and contacts to individual nanowires. (Agarwal, A. et al., *Sensor Actuat. A*, 145–146, 207, 2008.)

nanowire fins were patterned, using deep ultraviolet lithography in the array format. Generally, techniques for RTA fall into two main types, depending on the heating method: furnaces using steady heat sources or those employing electrical lamps, either incandescent or arc, with programmed optical output cycles.

The two ends of the nanowires were further doped to obtain N^+ or P^+ regions, followed by connection to the contact metal and alloying to realize ohmic contacts. The device was passivated by silicon nitride film, except for the active nanowire sensor area and metal pads.

How does the device measure temperature? The temperature-conductance κ (*T*) curve for several polysilicon nanowires showed a linear increase in conductance in response to temperature, having a standard deviation of ~85 pS over the set of 10 wires at 20°C. By suitable back-gate bias (V_G), applied from the back of the Si substrate, the absolute value of TCR could be increased to more than 10,000 (at $V_G = 4$ V). The relative change in conductance with temperature was appreciably more for polysilicon nanowires than for single-crystal nanowires.

How are DTDs fabricated? The process of fabricating a DTD is identical to that for an RTD, except for the different implant patterns. Diode-type temperature detectors are realized by selectively doping the N- and P-regions of the nanowires, using lithography. Similarly, the two contacts are doped to form N^+ and P^+ contact areas.

Measurement of the reverse-bias diode current was an excellent way to sense temperature as it varied by more than one order of magnitude (at −0.5 V) for a temperature change from 293 to 373 K.

5.8 Ratiometric Fluorescent Nanoparticles for Temperature Sensing

What are the advantages of fluorescence-based temperature sensors? In the realm of temperature sensing and imaging, fluorescence-based measurements have carved their niche because of the merits of non-invasiveness, accuracy, and robustness in even strong electromagnetic fields. Fluorescence is a process distinct from incandescence (i.e., emission of light due to high temperatures). Heating is generally detrimental to the process of fluorescence, and most substances, when fluorescing, produce very little heat. Due to this reason, fluorescence has commonly been referred to as "cold light." In the standard conceptual model of molecules, the electrons occupy distinct orbits and thus energy levels. When light falls on a molecule that absorbs rather than transmits, one or more of the electrons of the molecule are "kicked" into a higher energy state. All these excited electronic states are unstable, and, sooner or later, the electrons lose their excess energy and fall back to a lower energy state. This extra energy is dissipated in a multiplicity of ways, the most common being simply to increase atomic vibrations within the molecule. But some molecules are capable of emitting part of their energy as light. This is what we see as fluorescence.

Why are Europium (III) chelates preferred in this application? Europium (III) chelates display highly temperature-dependent emissions, large Stokes' shift (the difference between excitation wavelength and emission wavelength), and long lifetime (the average time the molecule stays in its excited state before emitting a photon), which make them strong candidates for fluorescent temperature sensing. A chelate is a chemical compound composed of a metal ion and a chelating agent. A chelating agent is a substance whose molecules can form several bonds to a single metal ion. In other words, a chelating agent is a multidentate ligand, i.e., a ligand capable of donating two or more pairs of electrons in a complexation reaction to form coordinate bonds. The ligand is a molecule, ion, or atom bonded to the central metal atom of a coordination compound.

Peng et al. (2010) reported fluorescent nanoparticles (NPs) prepared from a visible-light-sensitized Eu^{3+} chelate for temperature sensing and imaging over the physiological range (25°C–45°C). The NPs displayed strong temperature dependence in terms of both fluorescence intensity (FI) and lifetime.

What is the conceptual basis for measuring FI? The probability that a fluorophore will emit a photon is a product of the probability that a photon is absorbed and the probability that the excited state decays by emitting a photon (related to quantum yield). The probability of absorption is expressed by the molecular extinction coefficient. The molecular extinction coefficients of a fluorophore are somewhat sensitive to the microenvironment, which tends to influence the energy of the electronic states. To measure FI, given a source of photons with known energies (spectral characteristics), the excitation optics will select a set of wavelengths that are incident on the sample (Gaigalas et al. 2001). The excitation optics will illuminate a specific volume inside the sample. The collection optics will collect light emitted from a volume element that we call the detection volume. The total FI after the collection optics will depend on the overlap of the illumination and detection volumes, and we call these overlapping volumes, the sensing volume. Only fluorophore molecules in the sensing volume can contribute to the fluorescence signal. The FI depends directly on the concentration of fluorophore in the sensing volume, the molecular extinction coefficient, and the quantum yield. The quantum yield is defined as the ratio of the total number of emitted fluorescent photons to the total number of absorbed photons. The collection optics will direct light of a selected spectral range to the detector and provide a measurement of emission intensity over the selected range of wavelengths. Assuming that the detector is a photomultiplier tube (PMT), the PMT current is the sum of the fluorescent photons of all transmitted wavelengths. The current from the PMT will be determined by PMT gain, the quantum efficiency of the PMT photocathode, and the flux of fluorescence photons per unit wavelength. The resultant of the fluorescence flux per unit wavelength from the sensing volume with the current from the PMT gives the explicit relationship between the concentration of fluorophores and the measured current at the PMT detector. The relationship between the fluorescence signal, in the form of current, and the fluorophore concentration contains instrumental factors and molecular factors. It is the basis for interpreting the fluorescence signal in terms of a concentration of fluorophores.

How does a PMT work? PMTs convert photons into an electrical signal. They have a high internal gain and are sensitive detectors for low-intensity applications. A PMT consists of a photocathode and a series of dynodes in an evacuated glass enclosure. When a photon of sufficient energy strikes the photocathode, it ejects a photoelectron due to the photoelectric effect. The photocathode material is usually a mixture of alkali metals, which makes the PMT sensitive to photons throughout the visible region of the electromagnetic spectrum. The current produced by the incident light is multiplied by as much as 10^8 times (i.e., up to 160 dB), in multiple dynode stages, enabling the detection of individual photons when the incident flux of light is very low.

How is fluorescence lifetime measured? There are two complementary techniques of lifetime measurement: the time domain and the frequency domain. In the time domain technique, a short pulse of light excites the sample, and the subsequent fluorescence emission is recorded as a function of time. This usually takes place on the nanosecond timescale. In the frequency domain technique, the sample is excited by a modulated source of light. The fluorescence emitted by the sample has an analogous waveform. However, it is modulated and phase-shifted from the excitation curve. Both modulation (M) and phase-shift (φ) are determined by the lifetime of the sample emission. Lifetime is calculated from the observed modulation and phase-shift.

How do sensitivity and lifetime methods compare? The sensitivity of single intensity-based measurements is undermined by the distribution of probes and drifts of the optoelectronic system (lamps and detectors). Although lifetime-based sensing schemes are free of these drawbacks, the complexity and the demands on the components increase with the decreasing decay times. Comparatively, fluorescence ratiometric (two-wavelength) methods are more robust and convenient in practical applications because of the built-in calibrations provided by the simultaneous detection of two signals under one single-wavelength excitation.

How are temperature-insensitive reference dyes and temperature-sensitive probing NPs used together to make a ratiometric temperature sensor? Peng et al. (2010) introduced a temperature-insensitive dye, an alkoxysilane (a reactant used for the preparation of chemically bonded phases) in which silicon is attached by an oxygen bridge to an alkyl group, i.e., a SiOR-modified naphthalimide (1,8-naphthalimide = $C_{12}H_7NO_2$) derivative, *N*-octyl-4-(3-aminopropyltrimethoxysilane)-1,8-naphthali mide (OASN) (Figure 5.7a) into the Eu^{3+} chelate-based NPs as the reference dye to the probe Eu^{3+} chelate. Thus, the reference dye was an alkoxysilane-modified naphthalimide derivative. The temperature indicator was Eu-Tris (dinaphthoyl-methane)-*bis* -(trioctylphosphine oxide (Eu-DT) Figure 5.7b): trioctylphosphine oxide (TOPO) = $[CH_3(CH_2)_7]_3PO$. The strong temperature dependence of Eu-DT is attributed to the thermal deactivation of $^5D_1$ and $^5D_0$ europium energy levels through coupling to the environmental vibration energy levels. Absorption spectra were recorded on a UV-visible spectrophotometer. The absorption and emission spectra of Eu-DT and OASN in acetone (CH_3COCH_3) solutions showed that the absorption bands of the two dyes partly

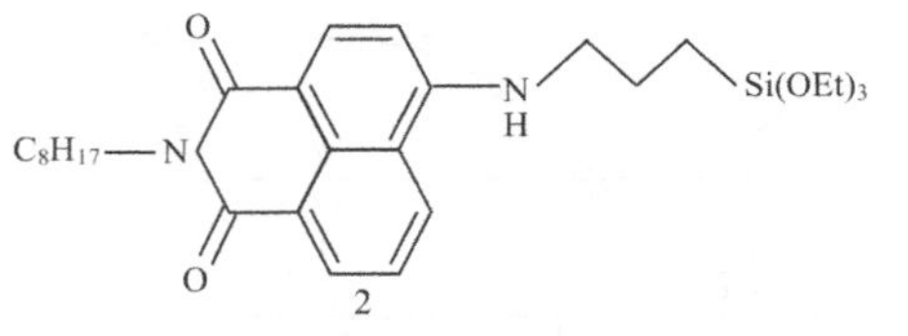

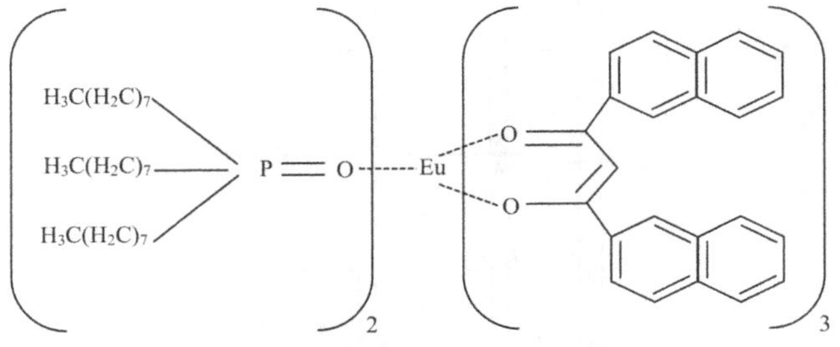

FIGURE 5.7 (a) OASN (N-octyl-4-(3-aminopropyltrimethoxysilane)-1, 8-naphthalimide 2) and (b) Eu-DT (Eu-Tris(dinaphthoylmethane)-bis-(trioctylphosphine oxide). (Peng, H -S. et al., *J. Nanopart. Res.*, doi: 10.1007/s11051-010-0046-8, 2010.)

overlapped, which ensured ratiometric fluorescence under one single-wavelength excitation, i.e., at 400 nm. Following photoexcitation, OASN gave a green fluorescence (with a peak at 497 nm), while Eu-DT gave the characteristic red emission.

How are the ratiometric NPs made? The ratiometric NPs were prepared by the encapsulation-reprecipitation method. Eu-DT, OASN, PMMA (polymethyl methacrylate): $[C_6H_{10}O_n]_n$, and BTD (5-bromothienyldeoxyuridine: $C_{13}H_{13}$ BrN_2O_5S), respectively, were used as the probe, reference dye, matrix, and silica-based encapsulation agent (Figure 5.8). The resulting NPs gave a two-wavelength emission under one single-wavelength excitation, the ratio of which was highly temperature dependent. The ratiometric NPs exhibited an intensity temperature sensitivity of –4.0% °C^{-1} from 25°C to 45°C.

What are the possible applications of these nanosensors? Given their small size (20–30 nm in diameter) and biocompatible nature (silica, SiO_2, outer layer), such types of ratiometric NPs are ideal nanosensors for sensing and imaging of cellular temperatures.

5.9 Er^{3+}/Yb^{3+} Co-Doped Gd_2O_3 Nanophosphor as a Temperature Nanosensor, Using Fluorescence Intensity Ratio Technique

Nanostructured materials for photonic applications have attracted considerable attention. One class of such materials is the rare-earth-doped nanocrystals (REDONs).

What are rare-earth elements? Rare-earth elements or rare-earth metals are a group of 17 elements in the Periodic Table, including scandium, yttrium, and 15 lanthanoids with atomic number (Z) ranging continuously from 57 to 71 (lanthanum – La, cerium – Ce, praseodymium – Pr, neodymium – Nd, promethium – Pm, samarium – Sm, europium – Eu, gadolinium – Gd, terbium–– Tb, dysprosium – Dy, homium – Ho, erbium – Er, thulium – Tm, ytterbium – Yb, and lutetium – Lu). Scandium and yttrium are considered rare earths because they tend to occur in the same ore deposits as the lanthanoids and show similar chemical properties. The term "rare earth" originates from the minerals from which they were first isolated, which were uncommon oxide-type minerals ("earths") found in gadolinite extracted from one mine in the village of Ytterby, Sweden. Nevertheless, except for the highly unstable prometheum, rare-earth elements are found in relatively high concentrations in the Earth's crust, with cerium being the 25th most-abundant element in the Earth's crust at 68 ppm.

REDON materials have been examined for use as phosphors in amplifiers and lasers. The promising optical properties of REDONs for photonic applications also led to the study of frequency up-conversion (UC). *What are the parameters on which UC efficiency depends?* The UC efficiency depends on the nanoparticle shape, the site symmetry, and the statistical distribution of active ions. In addition, the process of miniaturization of materials to the nanometer scale revealed that the radiative electronic relaxation probabilities of rare-earth ions, doped in dielectric nanoparticles, are strikingly different from their bulk counterparts.

As nanotechnology progresses, it is interesting to enquire into the potential of REDONs as nanosensors. Temperature dependency of absorption and emission properties of rare-earth-doped fluorescent materials enables their use as temperature sensors. Many researchers have considered the application

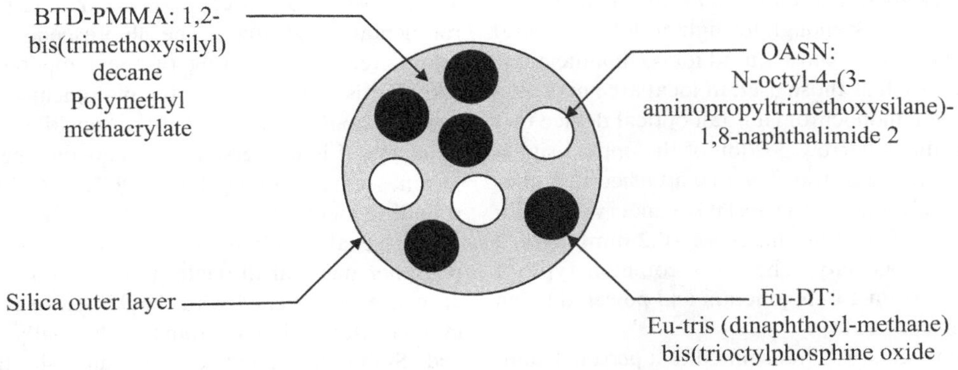

FIGURE 5.8 Ratiometric fluorescent nanoparticles. (Peng, H.-S. et al., *J. Nanopart. Res.*, doi: 10.1007/S11051-010-0046-8, 2010.)

of rare-earth-doped bulk materials for temperature measurements using different techniques. *Are there any areas where they provide the sole solutions to temperature measurement?* They are particularly useful in many areas where electrical methods become unpractical, such as in biology and optical telecommunications.

In particular, temperature sensors, based on the fluorescence intensity ratio (FIR) among different emission lines, is an important method. *What is the FIR-based method?* The FIR method involves measurements of the fluorescence intensities from two closely spaced electronic energy levels that are thermally coupled and assumed to be in a thermodynamic quasi-equilibrium state (the quasi-balanced state of a thermodynamic system near to thermodynamic equilibrium). A process is termed to be at "quasi-equilibrium" or at a "quasi-static process" if its deviation from thermodynamic equilibrium is infinitesimal. No system can be in complete equilibrium as it undergoes a real process, but some processes are in almost perfect equilibrium throughout the process.

What is the available operating range of the FIR method? The FIR method has the ability to cover a wide range (-196°C to +1350°C), with reasonable measurement resolution.

Alencar et al. (2004) reported the application of REDONs for temperature measurements, using the FIR method. They studied the use of Er^{3+}-doped $BaTiO_3$ nanocrystals for thermometry and presented ensemble (a unit or group of complementary parts that contribute to a single effect) measurements on frequency up-conversion of erbium ions (Er^{3+}) doped in $BaTiO_3$ nanocrystals to evaluate this material as a candidate for use as FIR-based temperature nanosensors. The fluorescence properties of the (Er^{3+})-doped $BaTiO_3$ nanocrystals were analyzed by changing the temperature of samples made of nanocrystallites of different sizes and Er^{3+} concentrations. *How are the measurements done?* For the fluorescence measurements, the samples were placed on a hotplate and the temperature was monitored with TCs located close to the samples. Frequency-up-converted emissions centered at 526 and 547 nm from two thermodynamically coupled excited states of Er^{3+} doped in $BaTiO_3$ nanocrystals were recorded (Figure 5.9). The samples were excited with a low-power, continuous-wave diode laser (wavelength: 980 nm; intensity: 2.0×10^3 W cm^{-2}) and the up-converted fluorescence, easily visible to the naked eye, was collected in the temperature range from 322 to 466 K, with a multimode fiber. The multimode fiber is a type of fiber optic cable that is thick enough for light to follow several paths through the code, and is mostly used for communications over short distances, such as those used in local area networks (LANs) connected to a monochromator (an optical device that separates and transmits a narrow portion of the optical signal chosen from a wider range of wavelengths) attached to a photomultiplier. The small size of the $BaTiO_3$ nanocrystal samples (typical diameter of ~3 mm; thickness ~0.2 mm) favored their temperature homogeneity. The most common type of laser diode is formed from a P–N junction and powered by an injected electric current.

The up-conversion signal was transmitted to a personal computer for processing. The behavior of the FIR for the emission lines, centered at 526 and 547 nm, was studied as a function of the temperature. The monolog plot of the experimental data

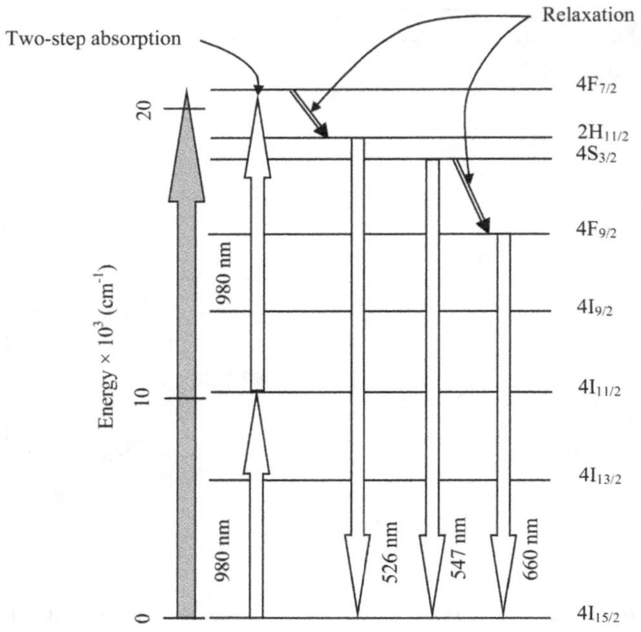

FIGURE 5.9 Simplified energy-level diagram of Er^{3+}, showing the frequency up-conversion scheme. (Alencar, M. A. R. C. et al., *Appl. Phys.* Lett., 84, 4753, 2004.)

yielded a linear dependence of the FIR with the inverse of the temperature. The data were fitted with a linear curve with parameters α and β

$$\ln(\text{FIR}) = -\left(\frac{\alpha}{T}\right) + \beta \quad (5.5)$$

The slope of the linear curve (parameter α) is related to the sensitivity of the sensor, defined as the rate at which the FIR changes with the temperature. The sensitivity found for the samples was $\leq 0.0052\ K^{-1}$. Note that α does not change, within experimental error, for samples with different Er^{3+} concentrations and of the same size but it varies when the nanoparticle size is reduced. This was explained by taking into consideration the modifications of non-radiative relaxation mechanisms (in which light is not emitted) with the size of the nanocrystals.

A temperature nanosensor, based on the FIR from a single REDON, will be a permanent device, illuminated with a low-cost diode laser, and will not require specific environmental conditions. The fluorescence detection scheme allows real-time reading of the temperature. One possible use of this nanosensor is for biological applications. In this case, it should be noted that the sensitivity changes when the nanocrystal is immersed in an aqueous medium, because the luminescence lifetime of a REDON also changes with the surrounding medium.

Singh et al. (2009) prepared Gd_2O_3: Er^{3+}/Yb^{3+} nanocrystalline phosphor material through optimized combustion and studied up-conversion excited with the 976 nm wavelength. Frequency up-converted emissions from two thermally excited states, $^2H_{11/2}$ and $^4S_{3/2}$ of Er^{3+}, centered at 523 and 548 nm in the phosphor, were recorded in the temperature range 300–900 K. They evaluated this material as a prospective candidate for use as temperature nanosensors, based on FIR.

5.10 Optical Heating of Yb^{3+}-Er^{3+} Co-Doped Fluoride Nanoparticles and Distant Temperature Sensing through Luminescence

Tikhomirov et al. (2009) found that pumping at 975 nm by a laser diode had two effects: (i) a strong up-conversion luminescence signal, and (ii) substantial heating of the Yb^{3+}-Er^{3+} co-doped nanoparticles up to several hundred degrees. The heating and up-conversion luminescence are essentially coexisting and simultaneous effects in these nanoparticles, pointing out that both effects can be used for heating of their surrounding nanovolumes and for detection of temperature rise and location of these nanovolumes.

How is the rise in temperature of nanoparticles measured? The temperature rise in the nanoparticles is estimated on the basis of measurements of the intensity ratio of up-conversion luminescence bands ${}^2H_{11/2}\rightarrow{}^4I_{15/2}$ and ${}^4S_{3/2}\rightarrow{}^4I_{15/2}$ of Er^{3+}. The up-conversion visible luminescence allowed distant optical detection of heated nanoparticles and their surrounding nanovolumes.

What is the measurement procedure? The Yb^{3+}–Er^{3+} co-doped nano powder was kept either in ampoules sealed under vacuum or was free-standing in the air, while fixed between silica glass plates. The luminescence spectra were excited by a laser diode operating at 975 nm and up to 260 mW power. When pumped by a 975 nm laser diode into the absorption band of Yb^{3+}, a laser-induced temperature rise up to 800°C was detected in the nanoparticles by measuring the ratio of the intensities of the thermalized up-conversion luminescence bands.

If I_H and I_S stand for integral intensity of the ${}^2H_{11/2}\rightarrow{}^4I_{15/2}$ and ${}^4S_{3/2}\rightarrow{}^4I_{15/2}$ of Er^{3+} emission bands, respectively:

$$\ln\left(\frac{I_H}{I_S}\right) = C - \frac{B}{T} \tag{5.6}$$

where B and C are the constants, $B \approx 1100$ K, while C varies between 1.5 and 2.5. Hence, the ratio I_H/I_S is defined by the temperature T of the emitting sample. However, the exact values of constants B and C should be experimentally found for a specific sample before Equation 5.6 is applied to evaluate the sample temperature precisely. Therefore, Tikhomirov et al. (2009) first measured a temperature dependent on the ratio I_H/I_S versus sample temperature T when these emission bands were pumped *via* the higher-lying ${}^4F_{3/2}$ absorption level of Er^{3+}.

Where are such heaters and sensors useful? The aforementioned nanoheater and temperature nanosensor may be used in medicine for local hypothermal treatment of cells and for perforation of nanoholes in organics and metals.

Example 5.2

Find the integral intensity ratio I_H/I_S at $T=700$ and 900 K for Yb^{3+}–Er^{3+} co-doped nanoparticles, assuming $B \approx 1100$K and $C=2$.

At 700 K, using Equation 5.6:

$$\frac{I_H}{I_S} = \exp\left(C - \frac{B}{T}\right) = \exp\left(2 - \frac{1100}{700}\right) = \exp(0.4286) = 1.535 \tag{5.7}$$

Similarly, at 900 K, the I_H/I_S ratio is obtained as:

$$\frac{I_H}{I_S} = \exp(2 - 1.22) = 2.18 \tag{5.8}$$

5.11 Porphyrin-Containing Copolymer as a Thermochromic Nanosensor

What is the present universal approach for obtaining desirable multicolors? The strategy is the use of nanocrystals, such as gold (Au) and silver (Ag), and semiconductor quantum dots. However, tunability of the color of these nanocrystals offers limited flexibility. Moreover, they are generally synthesized under harsh conditions and suffer from cytotoxicity and leakage into the biological system. *How is this problem solved?* The problem can be solved by using stimulus-responsive chromatic (relating to colors or color) materials as an alternative route to fulfill the color modulation. Although great efforts have been devoted to the simple- or multiple-responsive chromatic switches, the study of soluble nanosensors that have distinct color transitions and broad color-tuning range (>three kinds) is still a hot research area.

A porphyrin is any one of a class of water-soluble, nitrogenous biological pigments (biochromes), derivatives of which include the hemoproteins (porphyrins combined with metals and protein). A porphyrin is an organic compound that contains four pyrrole rings. A pyrrole is a pentagon-shaped ring of four carbon atoms with a nitrogen atom at one corner (C_4H_5N). The porphyrins are an enormous group of organic compounds, found all over the living world. They are universal, found in most living cells of animals and plants, where they perform a wide variety of functions. The special property of porphyrins is that they bind metals. The four nitrogen atoms in the middle of the porphyrin molecule act as teeth: they can grab and hold metal ions such as magnesium (Mg), iron (Fe), zinc (Zn), nickel (Ni), cobalt (Co), copper (Cu), and silver (Ag).

Based on the principle of specific chromophores (a chemical group capable of selective light absorption, resulting in the coloration of certain organic compounds), location and polymer phase induction, Yan et al. (2010) reported a full-color polymer optical nanosensor, possessing four prominent advantages: (i) metal-ion triggering of a full-spectrum-tunable sensing pattern that can enable its use as an ion detector and as a light colorimeter (a device for measurement of color) at ambient temperature; (ii) each ion-containing polymeric sensor possesses a distinct thermochromic temperature, covering a broad range as an ultrasensitive thermometer; "thermochromism" is the ability of a substance to change color due to a change in temperature; (iii) dispersive polymeric nanoparticles, with high water solubility, acting as sensing units, maintaining a discrete signal transport; (iv) full-color modulation so that the sensor is capable of sensing multichannel signals; and (v) low cytotoxicity for acceptability in biosystems.

Yan et al. (2010) designed and synthesized a specific ABC triblock copolymer. Block copolymers are material systems in which a chain of N_A units of monomer A is covalently linked to a chain of N_B units of monomer B. Difunctional initiators enable polymerization by addition at both ends of the B block, leading to triblocks, such as ABA, ABC, etc. *What interesting properties does this polymer display?* Interest in triblock copolymers has increased due to their complexity and physical properties. Below the lower critical solution temperature (LCST) of poly(*N*-isopropylacrylamide) (PNIPA) hydrogel (~32°C), a transparent red-brown solution, with a diagnostic tetra(4-carboxylatophenyl) porphyrin (TCPP) Soret band (λ_{max}=434 nm), was observed; the meso-tetra(4-carboxyphenyl) porphyrin ($C_{48}H_{30}N_4O_8$) Soret band is a very strong absorption band in the blue region of the optical absorption spectrum of a heme protein. Upon heating above 35°C, interestingly, a clear blueshift, from 434 to 406 nm, was seen and was accompanied by a visible turbid transition (transmission: 100%–20%) by UV-visible monitoring, implying an aggregated state transition around TCPP species.

To ensure the thermochromism of the copolymer, they investigated as to how to further enlarge the stimulus-responsive full-color range. Below 32°C, addition of different metal ions to the solution leads to an unprecedented full-spectrum color range, with the Soret $\pi \rightarrow \pi^*$ absorption of the TCPP species showing a large variation from 418 to 512 nm upon metal ion coordination with TCPPs. Even though the metalloporphyrin (a combination of a metal with porphyrin, as in heme) had a color different from that of the porphyrin, a full-spectrum color-tuning sensor, using a cation-stimulus TCPP-containing block copolymer, was shown for the first time. Most surprisingly, on heating, these multicolor nanosensors displayed discrete thermochromic characteristics in the temperature range of 35°C–61°C, with the phase transition point (the value of the temperature, pressure, or other physical quantity at which a phase transition or transformation [a change of a substance from one phase to another [solid-liquid-gas]] occurs) being dependent on the metal ion.

The synthesis and evaluation of responsive copolymers containing TCPP species demonstrated that they possess possible thermochromism in the absence of metal ion stimuli. Upon modification by metal ions, the available color range was hugely extended and, upon heating the solutions, color transitions at different thermochromic points were observed. It is anticipated that this polymeric optical nanosensor can give new insights into full-spectrum colorimetric and ultrasensitive thermometric arrays.

5.12 Silicon-Micromachined Scanning Thermal Profiler (STP)

The scanning tunneling microscope (STM) and the atomic force microscope (AFM) have found widespread use and are considered to be important milestones in the development of scanning probe microscopes (SPMs). The spatial resolution achievable by either one of these techniques is 1 Å, whereas the operating gap between the tip and the sample is typically 10 Å.

What is the principle of STP? Similar in principle to the aforementioned instruments, the scanning thermal profiler (STP) interacts with the sample by sensing heat conducted through the scanning probe. It provides both temperature and topographic information (as graphical representation of the surface features). Although the best reported spatial resolution for it is about an order of magnitude larger than for AFMs and STMs, it permits wide latitude in the gap size. *Why?* This is because the thermal interaction between the tip and the sample occurs over a much wider range than that associated with tunneling current and near-field forces.

What is the construction of STP and how does it function? The STP is basically a miniature TC, which is thermally biased with respect to the sample and separated from it by a small air gap (Figure 5.10) (Gianchandani and Najafi 1997). In a closed-loop operation, the voltage of the TC is stabilized by a feedback system that adjusts the gap such that the tip traces a contour of constant temperature (CT). If the variation in temperature across the sample surface is small, compared with the thermal bias, the contour provides a topographic map of the sample.

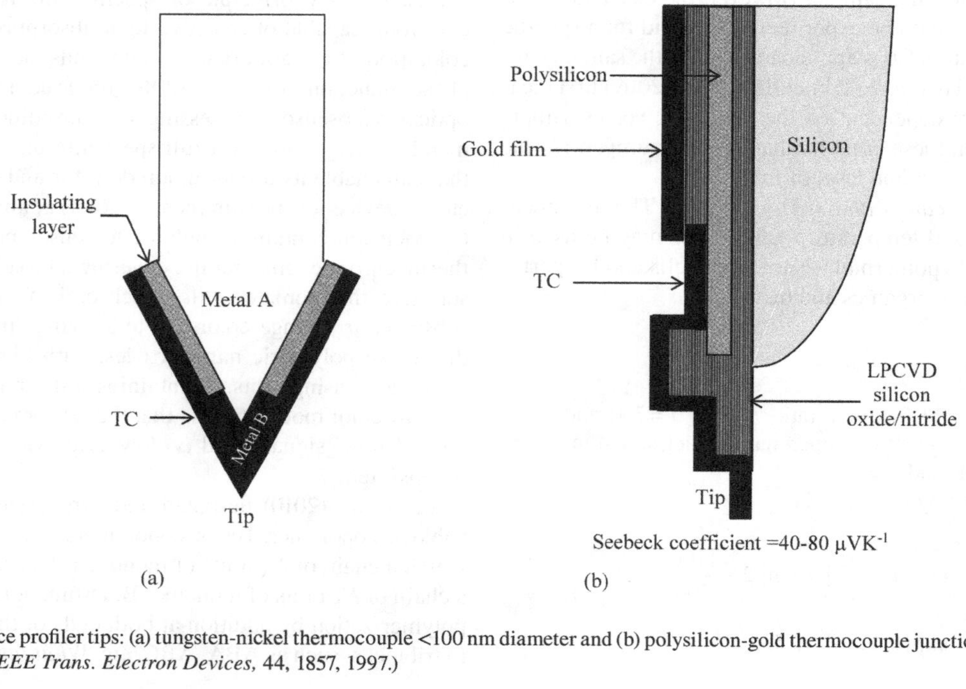

FIGURE 5.10 Surface profiler tips: (a) tungsten-nickel thermocouple <100 nm diameter and (b) polysilicon-gold thermocouple junction tip. (Gianchandani, Y. B. and Najafi, K., *IEEE Trans. Electron Devices,* 44, 1857, 1997.)

What are the operational modes of STP? There are two scan modes in which the STP is operated: the DC mode and the dithered (a state of indecisive agitation) mode. In the former, the tip is rastered across the sample while the air gap is controlled to keep the TC signal constant. In the latter, the tip is oscillated perpendicular to the sample, modulating the air gap by about 10%. *What is the advantage of an oscillating tip?* Dithering the tip reduces the sensitivity of the TC signal to fluctuations in the ambient temperature and also improves the signal-to-noise ratio, a ratio of desired signal to undesired signal (noise), serving to quantify how much a signal has been corrupted by noise.

The scanning probe overhangs the glass substrate, and is integrated with a suspension, which includes flexible support beams and electrostatic comb drives for longitudinal actuation. A thin film polysilicon–Au TC runs along the probe, with one junction near the protruding metal tip and another near the suspension. Polysilicon-gold TCs have a Seebeck coefficient (the ratio of the open-circuit voltage to the temperature difference between the hot and cold junctions) of 40–80 $\mu V\ K^{-1}$. A polysilicon resistive heater located on the lower surface of the suspension is used for generating the thermal bias between the tip and the sample. The approach selected for fabricating the SMTP is the bulk silicon micromachining technique.

What parameters determine the spatial resolution of the tip? Spatial resolution of a silicon-micromachined thermal profiler (SMTP) depends on the Seebeck coefficient of the TC and on the thermal impedance of the probe shank, as well as on operating conditions, such as temperature bias, air gap, and the sensitivity of the interface electronics. For a 1 μm × 0.5 μm tip and a 0.1 μm long air gap, the spatial resolution and the device noise equivalent temperature difference (NETD) have been theoretically estimated as ~3.33 nm and ~0.1 $mK\ Hz^{-1/2}$, respectively. NETD is a measure of the sensitivity of a detector of thermal radiation. It specifies the amount of radiation required to produce an output signal equal to the detector's own noise (due to inner component heat). Thus, it specifies the minimum detectable temperature difference. In general, detector cooling is required to limit the detector's own noise and to improve the NETD.

5.13 Superconducting Hot Electron Nanobolometers

What is a bolometer? A bolometer is a device for measuring the energy of incident electromagnetic radiation. It measures electromagnetic radiation in its various forms, from radio waves (30 kHz–300 GHz) to ultraviolet radiation (7.5×10^{14}–3×10^{16} Hz) and gamma rays (typically >10^{20} Hz). Invented by American astronomer Samuel Pierpont Langley in 1878, the first bolometer was used in conjunction with a telescope to measure infrared radiation (0.003–4×10^{14} Hz) on astronomical objects, namely, the Moon. Since the original invention, bolometers have come a long way in terms of improving the sensitivity and expanding the frequency range, from X-rays (3×10^{16} Hz upward) and optical (4–7.5×10^{14} Hz) and UV(7.5×10^{14}–3×10^{16} Hz) radiation to the submillimeter waves (3×10^{11}–3×10^{12} Hz or 3THz), the latter range containing 50% of the total luminosity of the universe and 98% of all the photons emitted since the Big Bang occurred.

What is the construction of the bolometer? A bolometer consists of an absorptive element, such as a thin layer of metal or metallic strip, connected to a heat sink (a body of CT) through a thermal link. Hence, any radiation impinging on the absorptive element, i.e., the metallic strip, raises its temperature above that of the heat sink. The more energy is absorbed, the higher is the temperature. The temperature change is measured directly or *via* an attached thermometer.

Metal bolometers, produced from thin foils or metal films, usually work without cooling. Today, most bolometers use semiconductor or superconductor materials as absorptive elements instead of metals (Romestain et al. 2004). Superconductor materials have zero electrical resistivity. Moreover, the magnetic field inside a bulk superconducting sample is zero (the Meissner effect). Superconducting photodetectors offer an interesting alternative to traditional photon-counting systems. In these superconducting devices, different mechanisms are exploited to detect the photon absorption, depending on the type of detector (transition edge sensor [TES]), superconducting tunnel junction, or hot-electron bolometer (HEB).

What is a hot-electron bolometer? An HEB is a detector for submillimeter radiation and far-nfrared radiation (Cherednichenko and Drakinskiy 2008; Gerecht and You 2008; Jiang et al. 2009). The detector is mainly of interest for detecting radiation with a frequency above 1000 GHz, an area where heterodyne SIS (supe rconducting-insulating-superconducting) receivers no longer work well. The heterodyne is a signal frequency produced by mixing two alternating current signals of different frequencies. Of the two the resultant frequencies, one is equal to the sum of the two original frequencies while the other equals their difference. Either of the two newly generated frequencies may be used in receivers by proper tuning or filtering. Heterodyning is applied to frequency shifting information of interest into a useful frequency range.

A miniscule but supersensitive sensor can help solve the mysteries of outer space. Cosmic radiation (energetically charged subatomic particles, originating from outer space, which impinge on the Earth's atmosphere), which contains the terahertz frequencies (3×10^{11}–3×10^{12} Hz) that the sensors detect, offers astronomers important information about the birth of star systems and planets.

Like the heterodyne SIS receiver, the HEB makes use of heterodyne (Greek roots *hetero-* "different," and *dyn-* "power") mixing to transform the received signal to a lower frequency. The mixed signal has a frequency of a few gigahertz. The bolometer can follow this mixed signal exactly. The frequency, phase, and power of the radiation absorbed can be deduced from the electronic signal.

What are nanobolometers? These are based on thin film metal nanostructures that work at ultralow temperatures (Vystavkin et al. 2010). HEB detectors, made of thin superconducting films (≈5 nm thick) of niobium nitride (NbN), have been found to present the best figure of merit for detecting single visible optical photons (Figure 5.11). The core of the detector comprises a small piece of superconducting NbN; detectors based on it can detect a single photon in the 3–10 μm section of the infrared spectrum. Clean superconducting contacts that are kept at a CT of −268°C (5° above absolute zero) are attached to both ends of the superconducting NbN. A miniscule gold antenna catches the terahertz radiation and sends it, *via* the contacts, to the small piece of NbN,

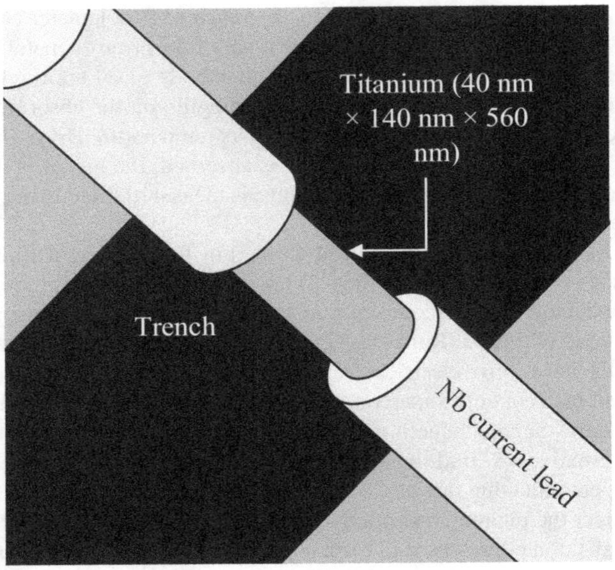

FIGURE 5.11 A Ti nanosensor. (Vystavkin, A. N. et al., *J. Commun. Technol. Electron.*, 55, 710, 2010.)

which functions as an extremely sensitive thermometer. Thus, the bolometer consists of the superconducting NbN nanobridge at the center, which connects to the on-chip gold spiral antenna *via* additional contact pads.

What does the term "hot electron" mean, in reference to the bolometer? Due to the low temperature of the strip, the incoming radiation initially heats up the electrons in the NbN nanobridge. The electron gas in the volume of the nanobolometer serves as the absorber. As the electrons of the sensor absorb radiation, the temperature of the electrons rises above that of the sensor itself. In superconducting thin films, the electrons are only loosely coupled to the crystal lattice and the electron–electron interaction is enhanced. When bombarded by photons, these electrons equilibrate at a temperature greater than that of the lattice. The electrons then cool either through phonon coupling to the underlying lattice or through diffusion across normal metal contacts. Principally, it is the length of the bridge that decides which cooling mechanism dominates. Diffusion cooling (the situation in which the average energy of an electron swarm is reduced by the diffusion of the faster electrons to the container walls) becomes prominent as the bridge length becomes smaller.

Due to the heating of free electrons by the absorption of photons, their temperature locally becomes higher than the superconducting transition temperature (the critical temperature at which the electrical resistivity of a metal drops to zero) of the NbN material ($T_C = 15.7$ K). These devices are therefore referred to as hot-electron bolometers (HEBs); the bolometer derives its name from this characteristic.

What is a transition edge sensor? Superconducting transition edge sensors (TES) are biased close to the superconducting-normal metal transition edge to detect the large increase in resistance as the photon energy is absorbed. When biased on the superconducting transition, the device has a finite electrical resistance that is less than the resistance in the fully non-superconducting state. Energy coupled to the detector increases the temperature of the superconducting material, pushing it further into the non-superconducting state and thereby increasing its electrical resistance. The nanobridge suffers from a localized loss of its superconducting characteristics and it detects the radiation used to detect very small changes in temperature, and hence in energy.

The temperature of the electrons in the strip must be able to chase the wave movement of the mixed signal exactly, more than 10^9 times per second; the electrons are therefore slightly warmed. For continued measurements, the electrons in the strip must be cooled down to the superconducting temperature extremely quickly. *Is it possible to cool the electrons?* Yes, because the electrons rapidly release their heat to the atoms of the NbN film. As the speed of this process is directly dependent on the film thickness, the NbN is extremely thin; 5 nm is just a few hundred atomic layers.

Ultrasensitive arrays based on superconducting nanobolometer sensors are used for passive radars (radars that have one or more receivers but no active transmitter, i.e., radars that only receive, instead of alternating transmission and reception) at the terahertz frequencies (0.3–10 THz), which include the imaging radiometers for the ground-based and space astronomical complexes, security systems (detection of hidden weapons, explosives, etc.), medical diagnostics, environmental control, and diagnostics of various products (including food products), among others. *What applications require the highest sensitivity?* The greatest sensitivity is needed for the space telescopes (instruments launched into orbit in outer space, which are used for observation of distant planets, galaxies, etc.), in which the fundamental limit is related to the quantum noise (the uncertainty of some physical quantity due to its quantum origin, attributable to the discrete and probabilistic nature of physical phenomena and their interactions, and representing the fundamental limit of the achievable signal-to-noise ratio of an optical communication system) of the background space signal and the received radiation. The noise per bolometer is ~10^{-21} W/Hz$^{1/2}$ in the terahertz frequency range. The nanobolometer with a working temperature of about 40 mK and dimensions (length, width, and thickness) of 1.0, 0.13, and 0.04 μm, respectively, is predicted to have a sensitivity close to the limiting sensitivity.

Wei et al. (2008) fabricated nanobolometers using electron-beam lithography and electron-gun deposition of Ti and Nb films on Si substrates covered with a film of SiO_2 or Si_3O_4. The device consisted of a titanium "island" with a volume of ~10^{-2} μm^3, flanked by niobium current leads. Low resistance of the Ti/Nb interface was ensured through *in situ* deposition of Ti and Nb at different angles through a "shadow mask." The critical temperature T_C (Ti) = 0.1–0.2 K for thin ($d \approx 40$ nm) and narrow ($w \approx 0.1$ μm) Ti nanobridges was lower than the bulk value (0.4 K). These devices operated at sub-Kelvin temperatures in the hot-electron regime at the temperature $T < T_C$ (Ti). Absorption of radiation by electrons in the titanium island increased their effective temperature T_e and caused a corresponding change in resistance, whereas the crystal lattice remained in equilibrium with the bath. Superconducting leads with high T_C severely increased the thermal isolation of electrons in the nanosensor.

5.14 Thermal Convective Accelerometer Using CNT Sensing Element

This device works as an anemometer, a device for measuring the speed of airflow in the atmosphere, in wind tunnels, and in other gas-flow applications, through temperature sensing. Structurally, the CNT-based thermal convective motion sensor (Zhang et al. 2009) consists of a sensor chip and a chamber to seal the convection medium. The sensor chip was fabricated using conventional micro fabrication methods along with manipulation of CNTs by dielectrophoresis (DEP) phenomenon. DEP is an electrostatic phenomenon in which a translational motion is induced in uncharged particles suspended in a medium, relative to the suspending medium, when the suspension is subjected to a non-uniform electric field. A 1-mm thick glass was selected as the substrate. The forced convection (the transference of heat by the flow of fluids, such as air or water, driven by fans, blowers, or pumps) in the chamber generated by vibration was sensed by the CNT bundle.

The heat-generating CNT bundle was first heated up in a micro chamber using constant current. External acceleration caused thermal convection (heat transfer by the combined mechanisms of fluid mixing and thermal conduction) within the chamber, which was detected by the fluctuation of the output voltage of the CNT temperature-sensing element.

Two temperature sensors were placed symmetrically around the heater along the sensitive axis. Under acceleration, the temperature distribution around the heater became asymmetrical with the heater, due to induced convection. The acceleration was derived since it was proportional to the measured temperature difference between the two sensors.

What are the advantages that favor this method? By using this detection method, the sensing block consumes only tens of picowatts (10^{-12} W) to function as motion sensors. Further, by using this sensing method, only micro heater and temperature detectors are needed and no moving parts (proof-mass) are used. So, this type of accelerometer is more robust to shocks compared with conventional micro motion sensors.

5.15 Single-Walled Carbon Nanotube Sensor for Airflow Measurement

The heat transfer between the CNT and the environment is influenced by the rate of fluid flow around the CNT. This heat transfer in turn affects the power required to heat the CNT to a specified temperature; hence, a CNT, carefully biased for self-heating, is usable as a fluid flow rate sensor.

What are the special advantages offered by CNT-based sensors for fluid flow studies? CNT-based fluid flow sensors are easily self-heated in that they have microwatt power consumption compared with the milliwatt power consumption of polysilicon-based resistors. Moreover, CNT fluid flow sensors can be used at a scale of hundreds of micrometers within a micro wind tunnel, a chamber through which air is forced at controlled velocities and in which airplanes, motor vehicles, etc., or their scale models, are tested to determine the effects of wind pressure on aerodynamic flow around airfoils, or other objects. Fluid flow rate inside the tunnel is increased, due to compression, and the responsivity of the sensor is thus increased.

Chow et al. (2010) developed single-walled carbon nanotube (SWCNT) sensors for airflow shear-stress measurement inside a polymethyl methacrylate (PMMA), $(C_5O_2H_8)_n$, micro wind-tunnel chip; PMMA is a transparent, thermoplastic polymer derived from methylacrylate, and used as a substitute for glass). Airflow shear stress is the form of stress on a body which is applied parallel or tangential to a face of a material, as opposed to a normal stress, which is applied perpendicularly. The shear stress causes parallel layers of the material to move relative to each another in their own planes. It has a tendency to initiate cutting rather than bending or stretching.

The developed shear-stress sensor consisted of a thermal element. The element was located on the surface of the substrate and possessed a pronounced temperature dependence of resistivity. The thermal element resided within a velocity boundary layer. In this layer, the velocity changed from zero, at the wall, to the value of the mean stress flow. When current was applied to the thermal element, Joule heating (Q=I2Rt, where Q is the heat generated by a constant current I flowing through a conductor of electrical resistance R, for a time t) took place, raising the temperature of the thermal element. But when the air flow was introduced on to the heated element, the temperature of the heated element decreased. This was caused by the interaction between the flow and the heated element. The resistance decreased if the element had positive TCR or increased if it had a negative TCR. The rate of heat loss from a heated resistive element to the ambient was dependent on the velocity profile (the shape of the velocity curve) in the boundary layer. Since the sensor was operated in constant temperature (CT) mode, when the sensor lost heat to the surrounding flow, its resistance changed, accompanied by a current flow to the sensor to keep the sensor resistance constant with an input power of ~230 μW. The voltage output of the sensor increased with the increasing flow rate in the micro wind tunnel, and the measurable volumetric flow (the volume of the fluid that passes through a given surface per unit time) was of the order of 1 × 10–5 m3 s–1.

An array of sensors was fabricated by employing dielectrophoretic technique to manipulate bundled SWCNTs across the gold microelectrodes on a PMMA substrate. After the nanotubes were manipulated across the Au microelectrodes by DEP, the sensor was annealed by a high current for several cycles. The intent was to burn off those SWCNTs that had a weak adhesion to the Au microelectrodes. The sensors were then integrated into a PMMA micro wind tunnel, which was fabricated by the SU-8 molding/hot-embossing technique. The SWCNT sensors showed a negative TCR of approximately –0.25% $°C^{-1}$.

5.16 Vacuum Pressure and Flow Velocity Sensors, Using Batch-Processed CNT Wall

Pressure and flow sensing (Choi and Kim 2010) rely on the electrothermal-thermistor effect comprising three main mechanisms: (i) negative TCR of the CNTs; (ii) the thermistor effect of an electrothermally heated CNT wall (thermistors are a type of resistor composed of solid semiconducting materials, with an electrical resistance that possesses either a negative or a positive temperature coefficient of resistivity); and (iii) the temperature-dependent tunneling rate at the CNT-silicon interface.

How does the sensor work? When the supply of constant electrical energy on the CNT wall is maintained, the temperature of the CNTs depends on the loss of thermal energy *via* heat transference through the surrounding gas molecules. Hence, the pressure and flow velocity changes cause a temperature change in the CNTs and at the interface between the CNTs and silicon, that is monitored by measuring the resistance change. The sensing capability of the CNT wall extends across a wide pressure range.

How was the sensor fabricated? For sensor fabrication, a CNT wall was synthesized using a thermal CVD on the sidewall of a microelectrode array, fabricated by a simple micromachining process on an SOI wafer. The synthesized CNT wall was a bamboo-shaped structure consisting of an MWCNT, as observed through transmission electron microscopy (TEM) analysis. The temperature dependence of the resistance of the CNT wall was measured, and the extracted TCR was –0.84% K^{-1}. The sensor responded immediately to changes in pressure and flow velocity. The maximum detectable range of pressure was between 10^5 and 3×10^{-3} Pa, and the range of flow velocity tested was from 1 to 52.4 mm s^{-1} in a nitrogen environment.

What are the benefits of the batch process and how does it differ from existing methods? The batch fabrication process, described by Choi and Kim 2010, including synthesis, integration, and assembly of CNTs, was quite simple and resulted in a high yield in comparison with existing methods for CNT-based sensors. Contrary to the proposed method, the traditional manufacturing methods for CNT sensors involve fairly low-throughput techniques, such as dielectrophoretic placement of CNTs, dispersion of pre-synthesized CNTs onto the substrate, and electron-beam lithography patterning. Although these conventional approaches are efficient in integrating an individual CNT or a few CNTs between electrodes, the labor-intensive process is serial, time-consuming, and uneconomical, thereby inhibiting the mass production of CNT-based sensors.

In the batch fabrication process, the microelectrodes were fabricated on a SOI wafer by photolithography and deep reactive-ion etching (DRIE) of the silicon device layer. The CNTs were synthesized using a thermal CVD to form CNT walls around an array of microelectrodes. During the subsequent electron-beam evaporation, the catalyst metal (Fe) was deposited on the top of the electrodes and substrate and on the sidewall of the electrodes due to imperfect directionality, even though the thickness deposited at the sidewall was thinner than on the top. By a reactive-ion etching (RIE) step, the Fe catalyst was removed from the top of the electrodes and the substrate.

Why and how is the process amenable to mass production? CNT-based devices were realized through the CNT wall formed and self-assembled between the electrodes and substrate along the sidewall of the electrodes, without additional post-assembly processing. After synthesizing the CNTs, no CNTs were observed on the top of the electrodes and substrate. This revealed the complete etching away of the catalyst metal on the topside of the electrodes during the RIE process. This synthesis of CNTs at selected locations was responsible for the bulk-production capability of the process.

Since batch processing was involved in all the fabrication steps, including the formation of the microelectrodes and the synthesis and assembly of the CNTs, high throughput and yield were achieved, in contrast with fabrication of most of the existing CNT-based sensors.

5.17 Nanogap Pirani Gauge

The miniaturization limits of classical diaphragm-based pressure sensors are determined by the technological limits on diaphragm fabrication, junction depth, etc. Although these limits have continuously improved, the physical problem of sensing very small diaphragm stresses or displacements still persists. Pirani-based pressure sensors have no moving parts, they are relatively easy to fabricate, and can be miniaturized to an extreme level. The traditional Pirani gauge (range: 1–10^{-4} torr, 10^2–10^{-2} Pa) is a thermal conductivity gauge (a gauge measuring 100–10^{-4} torr pressure and working on the principle that thermal energy dissipation is affected by the pressure and thermal conductivity characteristics of the residual gas), containing two filaments placed along the two arms of a Wheatstone bridge. One heated filament (120°C–200°C) is exposed to the vacuum in the chamber while the other, cold filament acts as a *reference or compensator* to take care of ambient temperature fluctuations. The temperatures of the two filaments are kept constant by adjusting the power delivered to the filaments. The power fed to the measuring filament is related to the pressure in the chamber, because it is inversely proportional to the number of gas molecules striking the filament and absorbing thermal energy from it.

A Pirani gauge with a range above atmospheric pressure is a potential substitute for a diaphragm-based pressure sensor at moderate atmospheric pressures. The miniaturization of a Pirani gauge by micromachining provides important advantages, including a small size, low fabrication cost, low power consumption, greater measurement sensitivity, fast response time, and improved dynamic range (the ratio between the largest and smallest possible values of a changeable quantity) of operating pressures.

A micromachined Pirani-based pressure sensor consists of a heater suspended over a heat sink (substrate) with a small gap between them; the heat sink is a component designed to lower the temperature of an electronic device by dissipating heat into the surrounding air. To extend the upper pressure limit of a Pirani gauge into the atmospheric range, the gap size of the device has to be as small as possible. The smaller gap size pushes the transition from molecular to continuum heat conduction into higher pressures. The transition pressure of the gauge is an empirical value that distinguishes the continuum and molecular heat conduction regimes of a gas. The continuum gas conduction regime occurs where the gas molecules carrying the heat collide with each other in the process of traveling. When the vacuum pressure is relatively low, heat conduction is by free molecule gas conduction. Above the transition pressure, a Pirani gauge loses its sensitivity and the gauge output gradually saturates.

Puers et al. (2002) used an FIB-based prototyping method to fabricate a nano Pirani gauge, involving a combination of FIB deposition and milling (the process of cutting away solid material). The sensor consisted of a suspended tungsten microbridge with a submicron gap from the substrate (Figure 5.12). The tungsten bridge resistor was connected to preprocessed aluminum bonding pads. The active area of the sensor was 10×1 μm. In the nano gauge fabrication process, the deposition of metallic structural layers and supporting features resulted from a local CVD reaction induced by the ion beam, while milling took place by scanning a focused beam of energetic Ga^+ ions, causing sputtering from the surface.

Khosraviani and Leung (2009) fabricated a surface micromachined Pirani pressure sensor with an extremely narrow gap (~50 nm) between its heater and the heat sink (substrate), with

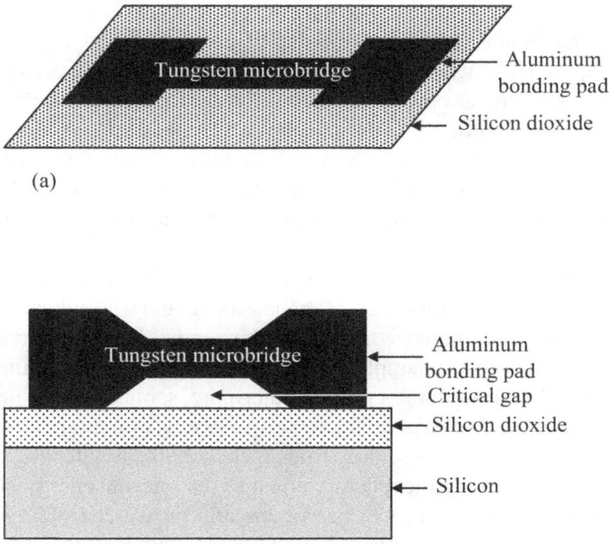

FIGURE 5.12 (a) Top view and (b) cross section of a nano Pirani gauge. (Puers, R. et al., *Sens. Actuators, A*, 97, 208, 2002.)

superior output linearity in the atmospheric pressure range. The gap size of the device had been reduced to 50 nm by using a layer of plasma-enhanced chemical vapor deposition (PECVD) amorphous silicon as a sacrificial layer and a xenon difluoride (XeF_2) gas phase-etching technique. Such a slender gap pushes the transition from molecular to continuum heat conduction to pressures beyond 200 kPa. Theoretically, when the gap size approaches the size of gas molecules, the device sensitivity decreases drastically as the gap will be devoid of the gas molecules responsible for heat conduction. This lower limit of the gap dimension is less than 1 nm, which is about two orders of magnitude smaller than the gap size achieved.

5.18 Carbon Nanotube-Polymer Nanocomposite as a Conductivity Response Infrared Nanosensor

How is the IR photoresponse of SWCNTs affected by embedding them in an insulating matrix? Pradhan et al. (2008) reported that the IR photoresponse of SWCNTs was dramatically improved by embedding SWCNTs in an electrically and thermally insulating polymer matrix, such as polycarbonate (PC, a thermoplastic polymer containing carbonate groups [–O–(C=O)–O–]) in the air at room temperature (Figure 5.13). *How does the response of embedded SWCNTs differ from those achieved without embedding?* High-pressure carbon monoxide (HiPco)-produced SWCNT ($SWCNT_{HiPCO}$) film (power intensity: 7 mW mm^{-2}; on/off time period: 200 s) exhibited two prominent features: (i) weak conductivity change (1.10%) upon IR illumination; and (ii) gradual increase and decrease in conductivity with the on/off IR illumination. In contrast to the gradual photoresponse and feeble conductivity change (1.10%) observed in a $SWCNT_{HiPco}$ film in the air at room temperature, the 5 wt% $SWCNT_{HiPco}$–PC

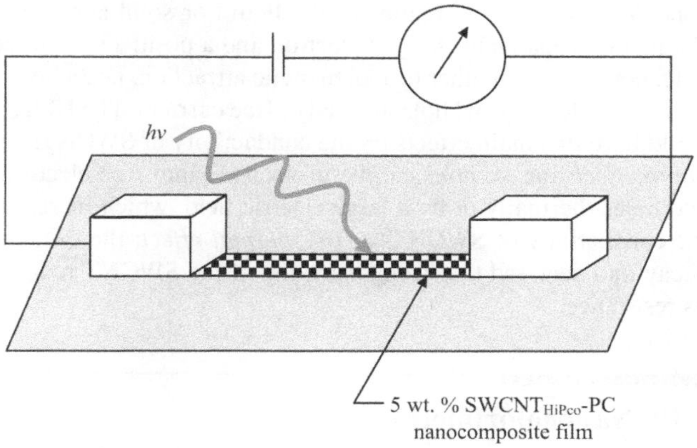

FIGURE 5.13 Set-up of a CNT-polymer nanosensor for conductivity measurements. (Pradhan, B. et al., *Nano Lett.*, 8, 1142, 2008.)

nanocomposite under the same IR illumination (power intensity: 7 mW mm^{-2}; on/off time period: 30 s) showed a sharp photoresponse and pronounced conductivity change (4.26%), which was approximately four times that (1.10%) observed in the pure $SWCNT_{HiPco}$ film under the same conditions, 21 times that (0.2%) in the pure $SWCNT_{arc}$ film in the air at room temperature (12 mW mm^{-2}), and between six and nine times that (0.7%) in the pure $SWCNT_{arc}$ film in the vacuum at 50 K. Even the 1 wt% $SWCNT_{HiPco}$–PC nanocomposite film gave a remarkable conductivity increase of 2.56% upon the same IR radiation, which surpassed that of pure $SWCNT_{HiPco}$ film (1.10% conductivity change) under the same conditions. Whereas the IR photoresponse of the pure $SWCNT_{HiPco}$ film was dominated by the thermal effect, in the IR photoresponse of the 5 wt% $SWCNT_{HiPco}$–PC nanocomposite, the photoexcitation effect dominated. Photoexcitation is the mechanism of electron excitation by photon absorption, leading to the formation of an excited state of atoms or molecules, when the energy of the photon is too low to cause photoionization.

How are the SWCNTs embedded in PC matrix? Purified $SWCNTs_{HiPco}$ were solubilized in chloroform ($CHCl_3$), with standard personal protective equipment (PPE), by vigorous shaking and/or short bath sonication. The $SWCNT_{HiPco}$ solution thus formed was mixed with a PC solution (for undoped samples) or a PC-iodine(I) solution (for doped samples) in chloroform to produce a homogeneous $SWCNT_{HiPco}$–PC composite solution. This composite solution was cast on a glass dish and dried very slowly to achieve a free-standing film after peeling from the substrate. The aforementioned process was followed to prepare doped and undoped $SWCNT_{HiPco}$–PC composites with various $SWCNT_{HiPco}$ loadings. The pure $SWCNT_{HiPco}$ film was obtained from the PTFE [polytetrafluoroethylene: $(C_2F_4)_n$] membrane after filtration of a suspension of $SWCNT_{HiPco}$ in chloroform. The typical film thickness was in the range of 25–60 μm. The mass ratios of PTFE: $SWCNT_{HiPco}$ and iodine: $SWCNT_{HiPco}$ were kept at 0.4 and 5, respectively. SEM showed the excellent dispersion of PTFE-$SWCNTs_{HiPco}$ in the PC matrix, which was vital for obtaining composites with isotropic electrical conductivity.

What are the likely effects of photoexcitation on the conductivity of SWCNTs? The photoexcitation of semiconducting SWCNTs with IR light leads to the generation of excitons. Excitons are a fundamental quantum of electronic excitation in condensed matter, i.e., matter in the liquid or solid state, consisting of a negatively charged electron and a positively charged hole, bound to each other by electrostatic attraction, i.e., a bound state of an electron and hole, instead of free carriers. The IR light could have two main effects on the conductivity of SWCNTs: (i) *photo effect*: the excitons can be dissociated into free electrons and holes thermally or by a large electric field, which increases the conductivity of SWCNTs;or (ii) *thermal effect*: the excitons decay into heat and the strong warming of the SWCNT reduces its resistance.

5.19 Nanocalorimetry

So far, attention has been focused on temperature sensing, but temperature is the effect and heat is the underlying cause. Accurate measurements of quantities of heat at the nanoscale are therefore essential, which fall under the purview of nanocalorimetry. For ease of understanding, let us approach the subject starting from calorimetry and then move toward nanocalorimetric concepts.

What is calorimetry? Derived from the Latin *calor* meaning "heat", and the Greek *metry* meaning "to measure", calorimetry is the science of measuring the amount of heat generated by chemical reactions or physical changes. *What is the scope of calorimetry?* Any process that results in heat being generated and exchanged with the environment is a subject for a calorimetric study. *What is a calorimeter?* It is a device used to measure the heat of a reaction. Various types of calorimeters have been designed for operation under different conditions, such as adiabatic (thermodynamic process taking place without loss or gain of heat) or isoperibolic (approximately isothermal). Isoperibolic calorimetry is a technique based on calorimetry for measuring thermal effects of chemical reactions or processes, where the temperature is kept constant in the calorimeter jacket.

What is a calorie? This is the amount of energy that raises the temperature of 1 mL of water by 1°C.

How does a calorimeter work? A calorimeter is a container with two chambers, one holding the reaction to be measured and the second chamber containing a measured volume of water. These two chambers are separated by a metal wall to conduct the heat from the reaction to the water. The temperature of the water is measured by a thermometer. The whole assembly is insulated from the environment so that the heat is retained inside the calorimeter. To use the calorimeter, a precisely known volume of pure water is put into the water chamber. The temperature of the water is recorded. Then a precise amount of chemicals is added to the reaction chamber. As the chemical reaction progresses, the temperature rises or falls. The maximum or minimum temperature achieved is noted. The heat of reaction (the amount of heat that must be added to or removed from during a chemical reaction in order to keep all of the substances present at the same temperature) is calculated in calories by multiplying the difference between final and initial temperatures of water by the volume of water, keeping in mind that the density of water is 1 g mL^{-1} and its specific heat is 1 cal $°C^{-1}$.

What is nanocalorimetry? Nanocalorimetry is concerned with performing thermal analysis on extremely small amounts of sample (nano- and picograms, 10^{-9} and 10^{-12} g, respectively). It is a method for measuring heats of reaction of nanoscale samples with nanojoule (10^{-9} J) sensitivity in order to detect reactions and quantify reaction thermodynamics and kinetics (the study of rates of chemical processes) in multilayer structures. Interfaces of interest include those in multilayer advanced gate stacks for microelectronic devices. Nanocalorimetry is the calorimetry for systems of nanometer dimensions, independently formed by deposition, which involves the difficulty in measuring the minuscule amount of heat.

What makes nanocalorimetry interesting? Understanding the thermodynamic properties of nanoscale materials, which are obviously different from their bulk behavior, has become increasingly important as samples of nanometer scales are routinely produced in the laboratory and furthermore, semiconductor processing technologies continue to shrink to tens of nanometers in size, and beyond. These properties include enthalpy (denoted as H, the thermodynamic quantity equal to the internal energy of a system plus the product of its volume and pressure: $H=E+PV$, where E is internal energy, P is pressure, and V is volume; a measure of the heat content of a chemical or physical system), Gibbs free energy, G (a thermodynamic property defined to predict whether a process will occur spontaneously at constant temperature and pressure: $G=H-TS$ where H, T and S are the enthalpy, temperature, and entropy, respectively; the change in G ($\Delta G=\Delta H\text{-}T\Delta S$) is negative for spontaneous processes, positive for non-spontaneous processes, and zero for processes at equilibrium), specific heat, and entropy S, (the quantitative measure of disorder in a system, i.e., the amount of thermal energy not available to do work). Accurate thermodynamic measurements are essential to understand fundamental properties of materials, providing direct insight into the thermodynamics of thin film reactions and phase transitions, and the transformation of a thermodynamic system from one phase or state of matter to another. Going forward, new classes of materials may only be synthesized as thin films, a scale at which traditional calorimetric techniques are not useful. The specific heat and other thermodynamic properties of small (nanogram, 10^{-9} g) samples can be quantified, including ultrathin and multilayered films, polymer coatings, biomaterials, and nanocrystalline and amorphous materials. Nanocalorimetry can determine the stability of multilayer thin film structures by quantifying the thermodynamics of interfacial reactions. Other applications for nanocalorimeters are the assessment of relaxation in glassy sugars ($C_{12}H_{22}O_{11}$), such as glucose, sucrose, maltose, etc.; evaluation of bulk metallic glass stability; and the quantification of reactions such as nickel-silicide (NiSi) formation in silicon-integrated circuit devices.

One particular phenomenon – particle size-dependent melting point depression – occurs when the particle size is of the order of nanometers. At these reduced dimensions, the surface-to-volume ratio is high and the surface energy substantially affects the interior bulk properties of the material, e.g., the melting point T_m of nanometer-sized Au particles can be 300 K lower than the bulk value.

Nanocalorimetry research has a great impact on several industries. Performing thermal analysis on smaller and smaller samples can lead to the development of pharmaceutics (the subject that deals with the technology of preparation of medicines) where new material production and purification is a time-limiting step and the sizes of samples are often very small. It can also foster the development of new surface-related technologies that require modification of nanometer-thick surface layers in a controlled manner.

What are the desirable features of a nanocalorimeter? The calorimeter suitable for measuring extremely small amounts of heat should have a very small heat capacity, of comparable magnitude to the nanoscale materials and films, in addition to operating over a wide range of temperatures. *How can one achieve these features?* These features can be obtained by taking advantage of the advanced thin film and membrane fabrication technologies because the thermal mass (the ability of a material to absorb and store heat, releasing it slowly) of the calorimeter is dramatically reduced. The low thermal conductance (measure of the amount of heat in calories that will pass through 1 m^2 of a material of any thickness each second for each degree centigrade difference in temperature on the opposite side of the material) and heat capacity of amorphous Si-N membranes make them ideal for thin film calorimeters. This occurs because of their intrinsic properties, such as high Debye temperature (the temperature above which a certain crystal behaves classically, i.e., the temperature above which thermal vibrations are more important than quantum effects) and the corresponding low specific heat, and their mechanical strength, whereby they can be made extremely thin with a large area. These membranes are made from a low-stress, silicon-rich film that is grown by low-pressure chemical vapor deposition (LPCVD). *What is the structure of a nanocalorimeter?* Figure 5.14 shows a schematic and cross section of the nanocalorimeter developed by Queen and Heilman (2009); the sample heaters and thermometers are also shown. *How is the sample heated?* It is heated by the Pt sample heater.

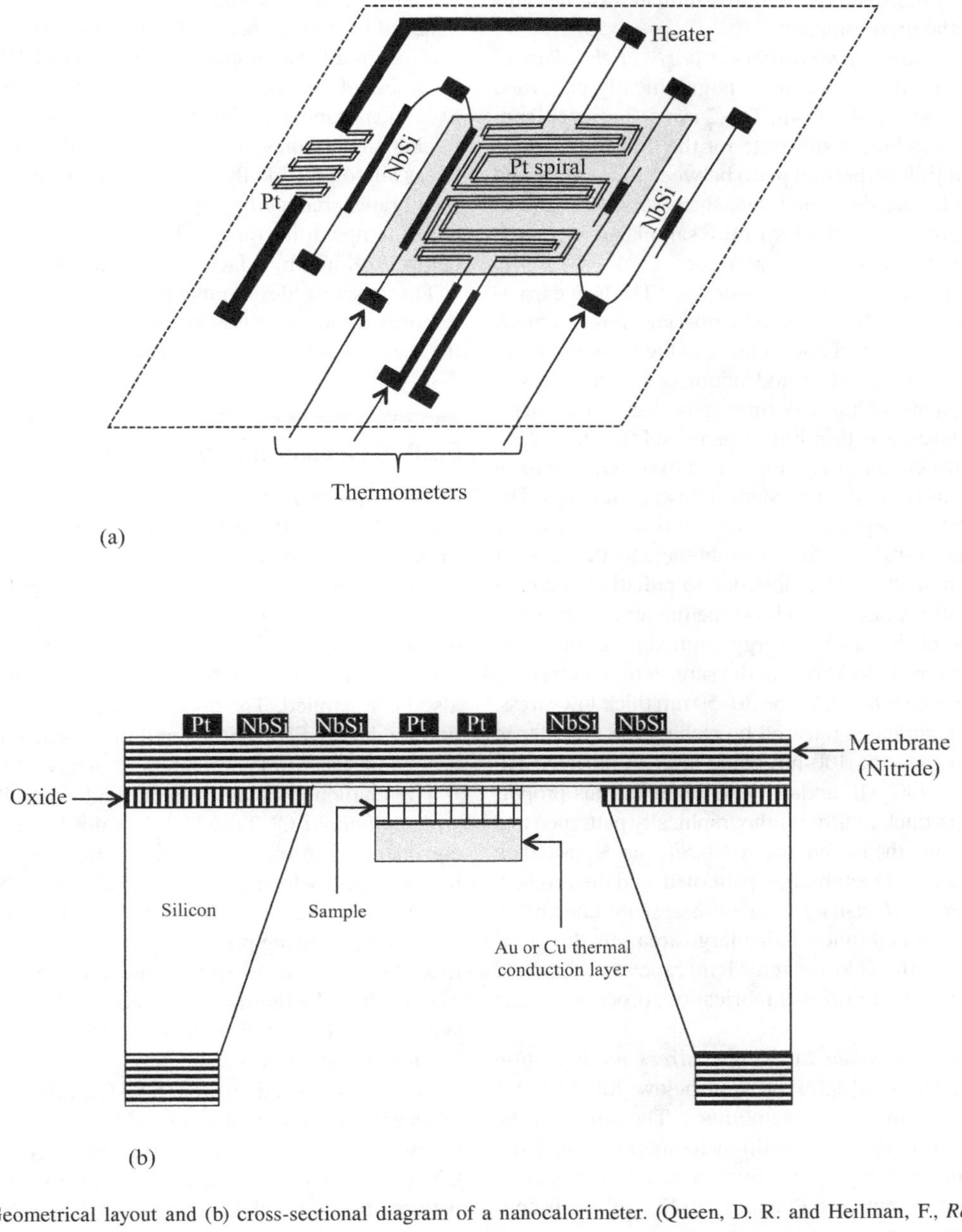

FIGURE 5.14 (a) Geometrical layout and (b) cross-sectional diagram of a nanocalorimeter. (Queen, D. R. and Heilman, F., *Rev. Sci. Instrum.*, 80, 063901-1, 2009.)

The Pt sample heater is used to raise the temperature of the sample area from fractions of a degree to several hundred degrees (300 mK to 800 K) above the frame temperature, depending on the measurement technique.

How is the temperature of the sample determined? It is measured with the help of three sample thermometers (two α-Nb_ySi_{1-y} and one Pt) located on the sample area, with impedance-matched thermometers on the frame. The resistance of each thermometer is optimized for a different temperature range. Lead resistances are <2% of the measured resistance for the heater and over the range of each thermometer. The Pt heater and thermometer are 12 µm wide. The α-Nb_ySi_{1-y} thermometers have the same doping y and resistances that differ by a factor of 12.5 due to geometry.

How is the heat conducted from the sample to the thermometers? A Cu or Au thermal conduction layer is deposited on the back of the membrane in the same central sample area to ensure that the sample is isothermal (i.e., with the temperature remaining constant) with the thermometers.

What are the functions of the different parts of the device? All of the elements on the device are lithographically patterned and their geometries are well known. The 2 mm × 2 mm × 30 nm Si-N membrane acts as both a substrate for the thin film sample and a weak thermal link (a thermal path) between the sample and the Si frame. Samples are deposited onto the reverse side and in the center of the membrane in the 1 × 1 mm^2 sample area through a separately micromachined deposition mask.

What are the contributing factors to heat flow? The heat capacities of the membrane, the thermal conduction layer, and sample, when present, are the dominant contributors to the total measured heat capacity. The background or addendum (a list of things to be added) heat capacity of the calorimeter is due to the membrane, conduction layer, and thin film heaters and thermometers. The thermal link that couples the sample area to the environment includes contributions from thermal conduction and radiation. The thermal conduction term depends on the elements that connect the sample area and the frame: the Si-N membrane and Pt leads for the heater and thermometers. Heat loss due to radiation becomes appreciable above 100 K and depends on the membrane area and emissivity (the ratio of the radiant energy emitted by a surface to the radiation emitted by a blackbody at the same temperature).

How is the membrane released? The 30–50 nm thick low-stress, amorphous Si-N membrane is released by etching the Si wafer in KOH and then removing the 100-nm oxide layer in buffered HF (prepared by mixing 49% HF and 40% NH_4F in various proportions). The 10–20-nm thick Pt film is lithographically patterned and etched prior to releasing the membrane. α-Nb_ySi_{1-y} is deposited on the wafer with the released membranes, patterned, and then etched.

What parameters are decisive for membrane size? The ability to reliably make thin membranes with a large area sets the minimum possible size for the calorimeter. Membranes as thin as 30 nm are sufficiently strong to survive fabrication, processing, and routine handling.

What precautions are taken to achieve stress minimization in deposited films? Residual stress is kept below 200 MPa for all films to reduce strain on the membranes. The stress in the LPCVD Si-N depends on the deposition temperature and the ratio of dichlorosilane (SiH_2Cl_2) to ammonia (NH_3) precursor gases. The stress in the sputtered Pt and α-Nb_ySi_{1-y} films is controlled by varying the pressure of the argon (Ar) gas.

How is the device used in heat measurements? The addendum heat capacity of the nanocalorimeter is less than 2×10^{-7} J K^{-1} at room temperature and 2×10^{-10} J K^{-1} at 2.3 K. Heat capacities of several Cu and Au films were measured. The heat capacities of thin Cu and Au films have been reported and agree with bulk values. These measurements showed that the nanocalorimeter can be used to measure the heat capacities of films as thin as 30 nm with absolute accuracy (<5%), limited by a combination of electrical noise (a random fluctuation in an electrical signal, a characteristic of all electronic circuits), film thickness uncertainty, and ≤2% systematic error from the measurement technique. A systematic error is a type of error that deviates by a fixed amount from the true value of measurement; this commonly occurs with the measuring instrument having an offset or zero setting error, multiplier, or scale factor error or by the wrong use of the instrument or by changes in the environment during the experiment.

The specific heat and thermal conductivity (the ability of a material to transfer heat, measured in Watts per square meter of surface area for a temperature gradient of 1 Kelvin (K) per unit thickness of one meter (W m^{-2}K^{-1}) of the thin amorphous silicon Si-N membrane, have been extracted and compared with data on thicker membranes. The thermal conductivity of the thin low-stress silicon nitride (Si_3N_4) is substantially smaller than thicker membranes, reduced by approximately a factor of three over the entire temperature range whereas the specific heat is enhanced below 20 K and by a factor of five at 2 K.

The ongoing development of new nanocalorimetric devices requires advances in thermometry and detailed understanding of the thermal behavior of the materials from which they are made.

5.20 Discussion and Conclusions

Table 5.1 presents the scenario of thermal nanosensors. A look at Table 5.1 reveals the efforts made in different research directions to achieve temperature sensing, by employing nanotechnology. In one approach, the FIB technique was applied to make a small TC W/Pt junction or W resistor. These types of nanosensors are useful for measuring temperatures of nanospots. Utilization of carbon nanowires or CNTs for temperature measurements has also been applied. The controllability of carbon nanowire positioning during FIB writing, and the reproducibility of their electrical and structural parameters is an advantage in their favor.

The challenge of the reliable and controllable manipulation with individual CNTs, which has not been met so far, poses a considerable problem for their practical applications. Therefore, in one study, selective growth of CNTs on Ni was achieved by CVD, using Ag contacts. Thin nickel film breaks into small nanoclusters during preheating at 750°C in a CVD furnace, due to surface tension, as well as the compressive stress due to the mismatch of the thermal expansion coefficients of the float glass substrate and nickel film. This nickel nanocluster acts as a catalyst for growing CNTs in a CVD system. In another study, a thin oxide was deposited on silicon wafer. After this step, a thin film of nickel was deposited and used to make nanoparticles by sputter etching. These nanoparticles were used as catalysts to grow CNTs at a later time. Then, a titanium film was deposited and patterned as electrodes and metal leads. A CNT was grown in the horizontal direction by MPCVD.

TABLE 5.1

Different Types of Thermal Nanosensors

Sl. No.	Name of the Nanosensor	Sensed Quantity/Applications	Nanomaterial Used	Sensing Principle
1	Nanoscale thermocouple	Thermo- electromotive force (EMF)	W/Pt	Seebeck effect
2	Tungsten resistive thermal nanosensor	Local thermal measurement	Tungsten	Change of electrical resistivity with temperature
3	Carbon-nanowire-on-diamond resistive temperature nanosensor	Medical and biological applications	Carbon nanowire	The temperature sensitivity of the electrical conductance between the nanowires
4	Carbon nanotube grown on nickel film	Low-temperature sensor (10–300 K)	Carbon nanotubes (CNTs) grown on nickel film deposited on float glass substrate	A very linear metal-like relationship between the electric resistance of CNTs and temperature (Temperature coefficient of resistance [TCR] ranging from 0.00022 to 0.00064°C^{-1})
5	Laterally grown CNT between two microelectrodes	High-quality temperature sensor	CNT	TCR varies from 0.0008152 to 0.0001759°C^{-1}
6	Silicon nanowire temperature nanosensors: resistance temperature detector (RTD) and Diode temperature detector (DTD)	For microreactors	Silicon nanowire	(i) Dependence of electrical conductance of nanowire on temperature; (ii) Variation of reverse-bias diode current with temperature.
7	Ratiometric fluorescent nanoparticles	Physiological temperature nanosensors for cellular sensing and imaging	*Temperature indicator*: Eu-DT [(Eu-Tris(dinaphthoylmethane)-*bis*-(trioctylphosphineoxide)] and *reference dye*: OASN [N-octyl-4-(3-aminopropyltrimethoxysilan e)-1, 8-naphthalimide]	Ratiometric fluorescence between a temperature indicator and reference dye introduced in a nanoparticle
8	Er^{3+}/Yb^{3+} Co-doped Gd_2O_3 nano phosphor	Optical thermometry using the green up- conversion emission centered at 523 and 548 nm	Combustion synthesized Gd_2O_3: Er^{3+}/Yb_3^+ nanocrystalline phosphor	Fluorescence intensity ratio (FIR) technique. Frequency-up-converted emissions from two thermally coupled excited states $^2H_{11/2}$ and $^4S_{3/2}$ of Er^{3+} centered at 523 and 548 nm in the phosphor pumped by near-infrared (NIR) source in temperature range 300–900 K
9	Yb^{3+}–Er^{3+} Co-doped fluoride nanoparticles	In medicine for local hypothermal treatment of cells, for perforation of nanoholes in organics and metals	Rare-earth-doped fluoride nanoparticles	Intensity ratio of up-conversion luminescence bands
10	Porphyrin-containing copolymer	Full-spectrum colorimeter and ultrasensitive thermometer	ABC triblock copolymer	Color transitions at different thermochromic points
11	Scanning thermal profiler	To provide both temperature and topographic information	Polysilicon-gold thermocouple	A scanning probe microscope with a miniature thermocouple (TC) at its tip that provides topographic and thermographic information by sensing heat conducted across a small airgap
12	Superconducting hot-electron nanobolometers	Imaging array radiometers of the terahertz frequency range	Niobium nitride (NbN) thin film	Hot-electron effect
13	Thermal convective accelerometer	Acceleration	CNT bundle	Relation of temperature change of heated CNT convectively affected by acceleration
14	Single-walled carbon nanotube (SWCNT) sensor for airflow measurement	Airflow	SWCNT	Influence of air flow on heat transference between CNT and its environment
15	Vacuum pressure and flow velocity sensor	Vacuum pressure and air flow	CNT wall	Loss of thermal energy from CNT wall *via* heat transference through the surrounding gas molecules
16	Nano Pirani gauge	Vacuum pressure	Nano gap between heater and heat sink	Pressure-dependent heat loss from the microbridge to the substrate
17	Carbon nanotube-polymer nanocomposite	Infrared sensor	SWCNT-PC	Enhancement in the IR photoresponse in the electrical conductivity of SWCNTs by embedding the SWCNTs in an insulating polymer
18	Nanocalorimeter	Heat sensor	Silicon nitride (Si-N) membrane with sample heaters	Thermal analysis of minute quantities of sample deposited on an ultrathin membrane

The SiNW fabrication process is compatible with top-down complementary metal-oxide semiconductor technology. The wires are patterned through a deep-UV photolithographic method, etched, and are formed by stress-limited oxidation; deep UV = 224–248 nm. The process uses dry oxidation at 900°C for 2–6 h and subsequent etching-away of the silicon dioxide to form the nanowires. The compatibility of this approach with complementary metal-oxide-semiconductor (CMOS) technology will be an important consideration for mass production of the devices.

The ratiometric NPs are highly temperature dependent in the physiological range (25°C–45°C). Rare-earth-doped nanocrystalline fluorescent materials have often been used as temperature sensors because of the temperature dependency of their absorption and emission properties. They are particularly useful in many areas where electrical methods are unpractical and are applicable over a wide temperature range (300–900 K). FIR has been used to measure the temperature. Yb^{3+}–Er^{3+} co-doped nanoparticles have been heated up to 800°C and the temperature has been detected by means of visible up-conversion luminescence and visual observation.

A CNT accelerometer, working on the thermal convection principle, has also been fabricated. Flow of air adjoining a heated CNT affects its temperature. This principle has been applied to the construction of air-flow sensors, using CNTs. The loss of thermal energy from the CNT wall has been utilized for estimation of both vacuum pressure and air flow velocity.

Review Exercises

5.1 Give some examples of carbon-based temperature sensors and point out their relative merits and demerits in terms of flexibility of technology, controllability of positioning of conductors, and reproducibility of electrical and structural parameters.

5.2 How does a pumping at 975 nm of the Yb^{3+}–Er^{3+} co-doped nanoparticles by laser diode result in a strong up-conversion luminescence signal, used as a distant temperature sensing property? Does this cause any heating effect? What are the possible applications?

5.3 In view of the penetration capability of terahertz radiation through clothes, dust, smoke, and biological materials being better than that of infrared or visible light, terahertz imagers are useful for detecting concealed weapons, illicit drugs, and biological materials. Are HEB devices useful for applications requiring low noise temperatures at frequencies from 0.5 to 10THz? How do they utilize the effect of electron heating by the incoming terahertz radiation? Discuss the salient features of HEB technology.

5.4 Which sensor provides topographic and thermographic information by sensing heat conducted across a small air gap? Draw and explain the working of a polysilicon-gold TC. Compare its Seebeck coefficient with a tungsten-nickel TC.

5.5 For temperature sensing and imaging, fluorescence-based measurements are useful because of their noninvasive nature, precision, and sturdiness in high-intensity electromagnetic fields. How are fluorescent NPs, prepared from a visible-light-sensitized Eu^{3+} chelate, used for temperature sensing?

5.6 Are Er^{3+}-doped $BaTiO_3$ nanocrystals suitable for use as FIR-based temperature nanosensor? Is the sensitivity of the nano-thermometer influenced by nonradiative relaxation channels? Do these relaxation mechanisms have any relation to the size of the nanocrystal?

5.7 Describe a temperature sensor fabricated by selectively growing CNTs using the CVD process and discuss its features.

5.8 Discuss the use of the Gd_2O_3:Er^{3+}/Yb^{3+} nanocrystalline phosphor as a temperature sensor in the range 300–900 K. Cite some application areas where electrical methods are unpractical.

5.9 MWCNT yarns are available in lengths of 100 m. When embedded in polymers, they would form the basis for a CNT yarn/carbon fiber composite, providing local and/or global measurements of strain and temperature. Cite some application examples of such multifunctional strain and temperature sensing.

5.10 Which of the following is a TES? (i) a CNT resistor; (ii) a superconducting bolometer; (iii) polysilicon-gold TC; or(iv) FIR thermometer.

5.11 From the viewpoint of temperature dependence, which of the following exhibits a metallic as well as a semiconducting nature? (i) carbon nanowire; (ii) CNT; (iii) gold; or (iv) alumina?

5.12 Although the best spatial resolution for STP reported is about an order of magnitude larger than for AFMs and STMs, it permits a wide latitude in the gap size. Explain this, with reference to the existence of thermal interaction between the tip and the sample over a much longer range than tunneling current and near-field forces.

5.13 The spatial resolution of infrared thermometry is limited by diffraction to dimensions close to the wavelength of the collected infrared radiation, typically 5 pm at room temperatures. Thermal property variations, temperature gradients, and defects with dimensions smaller than the diffraction limit are inaccessible to far-field infrared thermometry. How does the near-field method help in improving the spatial resolution of infrared thermometry?

5.14 Measurement of phosphor temperature in a cathode ray tube or color TV picture tube under operating conditions is of interest for establishing phosphor compositional parameters. Mention one temperature-sensitive mechanism involved in phosphor luminescence, that will be useful for temperature measurements.

5.15 What are the thermal and photo effects in the IR photoresponse of the pure $SWNT_{HiPco}$ film? Which effect dominates the IR photoresponse of $SWNT_{HiPco}$–PC nanocomposites?

REFERENCES

Agarwal, A., K. Buddharaju, I. K. Lao, N. Singh, N. Balasubramanian, and D. L. Kwong. 2008. Silicon nanowire sensor array using top-down CMOS technology. *Sensors and Actuators A* 145–146: 207–213.

Alencar, M. A. R. C., G. S. Maciel, and C. B. de Araújo. 2004. Er^{3+}-doped $BaTiO_3$ nanocrystals for thermometry: Influence of nanoenvironment on the sensitivity of a fluorescence based temperature sensor. *Applied Physics Letters* 84(23): 4753–4755.

Cherednichenko, S. and V. Drakinskiy. 2008. Low noise hot-electron bolometer mixers for terahertz frequencies. *Journal of Low Temperature and Physics* 151: 575–579, doi: 10.1007/s10909-007-9695-0.

Choi, J. and J. Kim. 2010. Batch-processed carbon nanotube wall as pressure and flow sensor. *Nanotechnology* 21: 105502 (8pp.).

Chow, L., D. Zhou, and F. Stevie. 2002. Fabrication of nano-scale temperature sensors and heaters. United States Patent 6905736, Filling Date: 02/27/2002, Publication Date: 06/14/2005.

Chow, W. W. Y., Y. Qu, W. J. Li, and S. C. H. Tung. 2010. Integrated SWCNT sensors in micro-wind tunnel for air-flow shear-stress measurement. *Microfluid Nanofluid* 8: 631–640.

Gaigalas, A. K., L. Li ,O. Henderson, R. Vogt, J. Barr, G. Marti, J. Weaver, and A. Schwartz. 2001. The development of fluorescence intensity standards. *Journal of Research of the National Institute of Standards and Technology* 106: 381–389.

Gerecht, E. and L. You. 2008. Terahertz imaging and spectroscopy based on hot electron bolometer (HEB) heterodyne detection. *Proceedings of SPIE* 6893, 689308-1–689308-8.

Gianchandani, Y. B. and K. Najafi. 1997. A silicon micromachined scanning thermal profiler with integrated elements for sensing and actuation *IEEE Transactions on Electron Devices* 44(11): 1857–1868.

Jiang, L., S. Shiba, K. Shimbo, N. Sakai, T. Yamakura, M. Sugimura, P. G. Ananthasubramanian, H. Maezawa, Y. Irimajiri, and S. Yamamoto. 2009. Development of THz waveguide NbTiN HEB mixers. *IEEE Transactions on Applied Superconductivity* 19(3): 301–304.

Khosraviani, K. and A. M. Leung. 2009. The nanogap Pirani—A pressure sensor with superior linearity in an atmospheric pressure range. *Journal of Micromechanics and Microengineering* 19: 045007 (8pp.), doi: 10.1088/0960-1317/19/4/045007.

Kuo, C. Y., C. L. Chan, C. Gau, C.-W. Liu, S. H. Shiau, and J.-H. Ting. 2007. Nano temperature sensor using selective lateral growth of carbon nanotube between electrodes. *IEEE Transactions on Nanotechnology* 6(1): 63–69.

Ozasa, A., R. Kometani, T. Morita, K. Kondo, K. Kanda, Y. Haruyama, J. Fujita, T. Kaito, and S. Matsui. 2004. Fabrication and evaluation of thermal nanosensor by focused-ion-beam chemical-vapor-deposition, *Digest of Papers. 2004 International Microprocesses and Nanotechnology Conference,* Osaka, Japan, October 26–29, 2004, 28P-7-15. New York: IEEE, pp.266-267, DOI: 10.1109/IMNC.2004.245644

Peng, H.-S., S.-H. Huang, and O. S. Wolfbeis. 2010. Ratiometric fluorescent nanoparticles for sensing temperature. *Journal of Nanoparticle Research* 12(8): 2729–2733, doi: 10.1007/s11051-010-0046-8.

Pradhan, B., K. Setyowati, H. Liu, D. H. Waldeck, and J. Chen. 2008. Carbon nanotube-polymer nanocomposite infrared sensor. *Nano Letters* 8(4): 1142–1146.

Puers, R., S. Reyntjens, and D. De Bruyker. 2002. The NanoPirani: An extremely miniaturized pressure senior fabricated by focused ion beam rapid prototyping. *Sensors and Actuators A: Physical* 97–98: 208–214.

Queen, D. R. and F. Hellman. 2009. Thin film nanocalorimeter for heat capacity measurements of 30 nm films. *Review of Scientific Instruments* 80: 063901-1–063901-7.

Romestain, R., B. Delaet, P. Renaud-Goud, I. Wang, C. Jorel, J.-C. Villegier, and J.-Ph. Poizat. 2004. Fabrication of a superconducting niobium nitride hot electron bolometer for single-photon counting. *New Journal of Physics* 6(129): 1–15.

Saraiya, A., D. Porwal, A. N. Bajpai, N. K. Tripathi, and K. Ram. 2006. Investigation of carbon nanotubes as low temperature sensors. *Synthesis and Reactivity in Inorganic, Metal-Organic and Nano-Metal Chemistry* 36: 163–164.

Singh, S. K., K. Kumar, and S. B. Rai. 2009. Er^{3+}/Yb^{3+} codoped Gd_2O_3 nano-phosphor for optical thermometry. *Sensors and Actuators A: Physical* 149(1): 16–20.

Tikhomirov, V. K., K. Driesen, V. D. Rodriguez, P. Gredin, M. Mortier, and V. V. Moshchalkov. 2009. Optical nanoheater based on the Yb^{3+}–Er^{3+} codoped nanoparticles. *Optics Express* 17(14): 11794–11798.

Vystavkin, A. N., A. G. Kovalenko, S. V. Shitov, O. V. Koryukin, I. A. Kon, A. A. Kuzmin, A. V. Uvarov, and A. S. Il'in. 2010. Hot electron superconducting nanobolometer-sensors for ultra-sensitive array radiometers of the terahertz frequency range. *Journal of Communications Technology and Electronics* 55(6): 710–715.

Wei, J., D. Olaya, B. S. Karasik, S. V. Pereverzev, A. V. Sergeev, and M. E. Gershenson. 2008. Ultrasensitive hot-electron nanobolometers for terahertz astrophysics. *Nature Nanotechnology* 3: 496–500.

Yan, Q., J. Yuan, Y. Kang, Z. Cai, L. Zhou, and Y. Yin. 2010. Dual-sensing porphyrin-containing copolymer nanosensor as full-spectrum colorimeter and ultra-sensitive thermometer. *Chemical Communications* 46: 2781–2783, doi: 10.1039/b926882k.

Zaitsev, A. M., A. M. Levine, and S. H. Zaidi. 2007. Carbon nanowire-based temperature sensor Physica Status Solidi (a) 204(10): 3574–3579.

Zhang, Y., W. J. Li, and O. Tabata. 2009. Extreme-low-power thermal convective accelerometer based on CNT sensing element. In: *The* 4th IEEE International Conference on Nano/Micro Engineered and Molecular Systems, *IEEE-NEMS* 2009, Shenzhen, China, January 5–8, 2009. New York: IEEE, pp. 1040–1042.

6

Optical Nanosensors

6.1 Introduction

Nanosensors pertaining to the sub-branch optics (Section 1.3.8) of physics (Section 1.3) are grouped under optical nanosensors (NSs). These are devices in which principally optical phenomena are used for determining (detecting and quantifying) the excitation signal. Examples are changes in refractive index, absorbance, reflectance, color effects, photoluminescence, chemiluminescence, fluorescence, etc. Optical NSs are ultra-advanced analytical tools that bring together the advantages of customary sensor technology with the flexibility of using dissolved indicators. They are defined as devices with dimensions smaller than 1000 nm that have the capability to continuously monitor chemical or biological parameters by optically converting the available information into signals useful for analytical purposes. The dimensions of optical NSs vary from a few nanometers, for example, for macromolecules with sensing properties, up to 1 μm. Many micrometer-sized core/shell systems have a sensing layer with a thickness ~ tens to hundreds of nanometers; these are also included under this grouping. These systems often perform analogously to beads of nanometer dimensions, and many of them are produced by identical procedures.

Miniaturization of many fiber-optic sensors to submicron size has been achieved. Moreover, optical fibers impregnated with nanomaterials, like carbon nanotubes, have been successfully used. Interest in surface-plasmon resonance (SPR) and localized surface plasmon resonance (LSPR) spectroscopy has been reawakened and expanded by the availability of new fabrication methods for plasmonic materials. All these topics will be included within the scope of this chapter.

Figure 6.1 shows the schematic representation of optical NSs proposed by Borisov and Klimant (2008). They classified NSs into six categories: (i) macromolecular (very large molecules commonly created by some form of polymerization) NSs; (ii) NSs based on polymer materials and sol-gels (gelatinous fluids); (iii) multifunctional core/shell systems; (iv) multifunctional magnetic beads; (v) NSs based on quantum dots (QDs); and (vi) NSs based on metal beads.

Dendrimers are spheroidal (shape of an ellipsoid) or globular (having the shape of a sphere or ball) nanostructures engineered to carry molecules encapsulated in their internal void spaces or attached to their surfaces. The polymeric shell of dendrimers not only confers water solubility to the attached indicators, but also protects them against interfering materials and markedly affects the diffusion of analyte molecules, rendering them more similar to other types of NSs than to the individual indicator molecules. As an example of attached indicators, the abundance of functional groups at the terminal extremities of the branches of polyamidoamine (PAMAM) dendrimer has been utilized for immobilizing anti-DENV II E (anti-Dengue virus II envelope) antibodies (Kamil et al. 2019). These antibodies bind with any DENV II E proteins present in the analyte, thereby serving as indicators of dengue infection. This type of sensor is called a biosensor, which is defined at the end of this section.

Polymer nanoparticles (NPs) are generally formed from polymerization of a microemulsion (an emulsion whose particles are less than 1 μm in size, usually 0.75–2 μm), which produces nanosized (~15 nm) particles smaller than those obtainable by emulsion polymerization. Particularly, polar (having a pair of equal and opposite charges) polymers confer water solubility and stability against aggregation. Alternatively, the surface of the hydrophobic (water repellant) beads can be modified with polar groups. In contrast to bulk sensor films (typically several microns thick), in NSs many indicator molecules are located near the surface, so that leaching (a natural process by which water-soluble substances are washed out from a solid) becomes a serious problem, which is avoided by using indicators with functional groups which can be cross-linked with the polymer.

Sol-gels (inorganic silica beads and organically modified silica) are popular materials for designing optical NSs, because the beads are easily manufactured by reversed microemulsion polymerization. They are porous, allowing analytes to freely diffuse inside, robust, and biocompatible, making them suitable for intracellular use (use within a cell).

NSs based on metal beads are designed by modification of the surface of a metal nanobead, such as Au, to achieve functionality. In contrast to the NSs mentioned earlier, those based on metal beads do not possess an environment that can significantly tune the sensitivity and selectivity. The main function of the metal core is the modulation of fluorescent properties of an indicator located on the surface. Mostly, the metal core acts as a quencher (suppressor of fluorescence), and, notably, many such sensors are virtually irreversible.

QDs represent promising alternatives to organic fluorophores (chemical groups responsible for fluorescence). QDs possess good brightness, high photostability (unchanged by the influence of light), broad absorption (radiation enters in a wide range of wavelengths), that allows for their simultaneous excitation, and relatively narrow emission (frequency spectrum of radiation leaving the QD is not wide), which is easily discriminated. Application of QDs in optical sensors can, however, be compromised by a high degree of nonspecific binding and by the fact that many substances act as quenchers. Susceptibility to quenching is often a serious drawback. However, it can be used to design sensors, providing that other interferences are minimized.

NSs based on other materials include NSs based on single-walled carbon nanotubes (SWCNT), silicon nanowires, and various nanomaterials. The important feature of SWCNTs is their absorption and emission in the near-infrared (NIR), the optical window that allows for subcutaneous measurements.

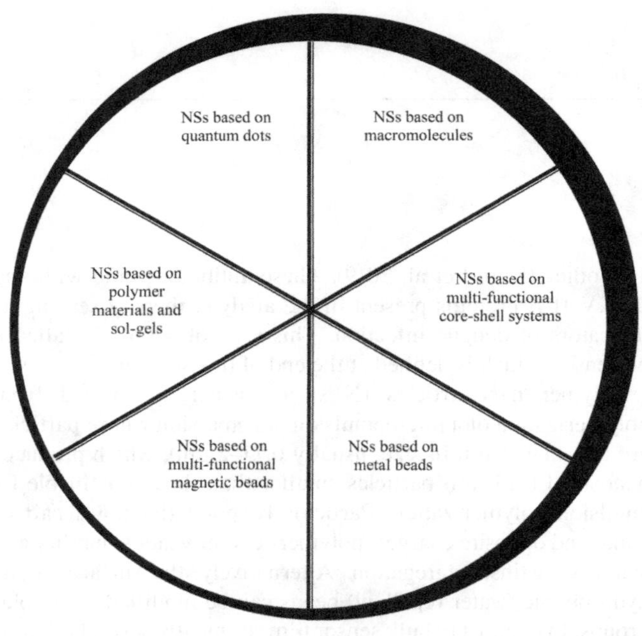

FIGURE 6.1 Different types of optical nanosensors. (Borisov, S. M. and Klimant, I., *Analyst,* 133, 1302, doi: 10.1039/b805432k, 2008.)

Multifunctional NSs are not only capable of sensing an analyte but also possess additional functionalities that make the material more versatile. Many core/shell materials are multifunctional, such as the NS for pH, in which the core of the bead is stained with an inert fluorescent dye to enable ratiometric referencing. Another type of multifunctional material includes magnetic sensor beads. In contrast to other NSs, magnetic beads can be collected, moved, rotated, or separated from the solution with the help of an external magnetic field. Since they can be collected in the proximity of a detector, the signal-to-noise ratio (a measure of signal strength relative to background noise) is often very high.

As we see, several nano-devices known in the literature as NSs act practically irreversibly and are better classified as "nanoprobes." *What is the difference between NSs and nanoprobes?* Only those systems that respond fully reversibly to the stimulus are designated as "true nanosensors." Nanoprobes do not recover from their altered states. But sometimes the term "nanosensor" is used less exactingly, and nanoprobes are also considered under NSs.

The optical NSs are extremely diverse in structure, functionality, and with respect to the materials used in their preparation (Pandana et al. 2008; Zwiller et al. 2008; Coupland et al. 2009). They rely on the same sensing principles as classical optical sensors, namely dynamic luminescence quenching, Förster resonance energy transfer (FRET), and acid-base equilibria.

What is dynamic luminescence quenching? When an atom or molecule is electronically excited by the absorption of a photon, there are a number of pathways or schemes by which that species can then return to the ground state. One mode involves the emission of a photon of light, which causes the substance to fluoresce or phosphoresce. Alternatively, species can lose their excitation by nonradiative means: (i) through collisions with other atoms or molecules; or (ii) by simply returning to the ground state along a downhill energy path, that involves several coupled vibrational and electronic energy states. The first of these nonradiative schemes is called dynamic or collisional luminescence quenching.

What is FRET? It will be elaborately explained in Section 6.11.

What are acid-base equilibria? An acid-base equilibrium system contains an acid and its conjugate base.

The main physical phenomena exploited for optical chemical sensing are absorption, fluorescence, chemical luminescence, Raman scattering, and plasmon resonance. *Optical or light absorption* refers to the process in which energy of light radiation is transferred to a medium. An *absorption* spectrum is the absorption of light as a function of wavelength or frequency. *Fluorescence* is the emission of electromagnetic radiation, especially of visible light, stimulated in a substance by the absorption of incident radiation and persisting only as long as the stimulating radiation is continued. *Chemical luminescence* is the emission of light by a substance, caused by *chemical* means. *Raman scattering* is the inelastic scattering of a photon, which creates or annihilates an optical phonon. *Plasmon resonance* is concerned with the excitation of surface plasmons by light.

Optical NSs are favored for nanoscale biological and chemical analysis, and for environmental applications (Clark et al. 1999; Aylott 2003). Features of optical NSs include the possibility of measuring in very small volumes (including single cells), suitability for three-dimensional imaging, low toxicity, high selectivity, little effect of nonspecific binding and other interferences (such as water) on sensor properties, and versatility in tuning most properties within the polymer matrix. They are divided into two main classes: optical nanobiosensors and optochemical NSs, both of which have become very widely used and are applied to many fields of science and technology, such as biology, biotechnology, environmental analysis, clinical medicine, and marine science, to mention but a few. The instruments, now at advanced stages of development in laboratories, demonstrate that optical NSs are some of the best options for scientists and physicians when continuous monitoring *in vivo* is necessary.

As we shall come across the term "biosensor" very frequently in this chapter, let us define the same. Biosensors are analytical devices for the detection of an analyte that combine a biological component with a physicochemical detector component. They are powerful tools aimed at providing selective identification of chemical compounds at ultra-trace levels in industrial products, chemical substances, environmental samples (e.g., air, soil, and water) or biological systems (e.g., bacteria, virus, or tissue components) for biomedical diagnosis.

6.2 Noble-Metal Nanoparticles With LSPR and UV-Visible Spectroscopy

When the incident photon frequency matches with that of the collective oscillation of conduction electrons, the local electromagnetic fields near the surface of the plasmonic metal film/nanostructure are strengthened (Yonzon et al. 2007). The increase in the electromagnetic fields leads to the intense signals observed in SPR and LSPR. In this section, the NS using the LSPR technique is described with reference to its conventional SPR partner, bringing out the salient features of LSPR.

For a long time, the SPR reflectivity measurements have been used to characterize the thickness or refractive index of ultrathin organic and biopolymer films at *bulk noble metal* surfaces. Refractive index (n) of an optical medium is a dimensionless number equal to the ratio of the speed of light in vacuum to that in the medium. For a given pair of materials, relative refractive index of material 2 with respect to medium 1 denoted by n_{21} is a constant defined as ratio of the speed of light in material 1 to the speed of light in material 2.

Since the introduction of the Biacore (a life science product company based in Sweden) SPR instrument, uses of SPR spectroscopy have proliferated in the fields of chemistry and biochemistry for characterization of biological surfaces and for monitoring binding events. It is a sensor chip-based analytical system where interactions between two or more biomolecules are measured by immobilizing one type of molecule on the surface of a sensor chip and passing a solution containing the other molecule over the surface under controlled conditions. The sensor chip consists of a glass surface coated with a thin layer of gold that provides the physical conditions required for the SPR phenomenon.

By monitoring the interaction between the immobilized compound, and one or more compounds in the solution, the technology provides a level of functional data that is difficult or even impossible to obtain by any other single technique. Light does not enter the sample, thereby eliminating the quenching or absorbance (a measure of the light-absorbing ability of an object, expressed as the logarithm to base 10 of the reciprocal of the internal transmittance) problems that beset all spectrophotometric (quantitative measurement of the reflection or transmission properties of a material as a function of wavelength) and fluorescent techniques.

LSPR spectroscopy is a *noble metal NP*-based optical sensing technique, effective for quantitative detection of chemical and biological targets, in which sensing is accomplished by measuring the wavelength shift in the LSPR extinction or scattering maximum (λ_{max}), that is induced by the binding of target analytes to the NP surface. The shift in λ_{max} (peak absorption wavelength, i.e., specific wavelength in the spectrum at which absorption is maximum) is quantitatively related to the concentration of target analytes. LSPR NSs have been demonstrated as sensitive platforms for the detection of streptavidin, antibiotin, concanavalin (a hemagglutinin, isolated from the meal of the jack bean seed, which reacts with polyglucosans in the blood of mammals, causing agglutination), biomarkers of diseases, and many other biorecognition events. LSPR has been used for detecting a biomarker for Alzheimer's disease (Haes et al. 2005). The Alzeimer's disease is a progressive, irreversible disease characterized by degeneration of the brain cells and commonly leading to severe dementia, a decline in intellectual functioning, including problems with memory, reasoning, and thinking.

What are the similarities in SPR and LSPR spectroscopy? The NSs based on LSPR spectroscopy operate in a manner identical to the propagation of SPR sensors, which work by transducing small changes in the refractive index near the metallic surface into a measurable wavelength shift response. LSPR NSs induce small local refractive index changes at the surfaces of metallic NPs. Variations of the reflectivity (the ability of a surface to reflect radiation, measured as the fraction of radiant energy that is reflected from it) as a function of the angle of incidence in SPR and extinction peaks (peaks in the plot of *extinction* or transmittance versus wavelength or wavenumber) in LSPR are associated with the same physical phenomenon, that is, collective oscillation of electrons in the metal. The minimum of reflection relates to a maximum of absorption. The behavioral response of SPR and LSPR sensors is described by the following equation (Barbillon et al. 2007):

$$\Delta\lambda_{\max} = m\Delta n\left\{1-\exp\left(-\frac{2d}{l_{\mathrm{d}}}\right)\right\} \tag{6.1}$$

where

$\Delta\lambda_{max}$ is the shift in the wavelength of maximum absorption
m is the refractive index sensitivity
Δn is the change in refractive index induced by an adsorbate (an adsorbed substance)
d is the effective adsorbate layer thickness
l_d is the characteristic evanescent (very short-lived) electromagnetic field decay length

This equation is valid for both SPR and LSPR, because they are intrinsically related to the same physical phenomenon, albeit applicable to different situations.

What are differences between SPR and LSPR? There are two main differentiating points, namely the propagating and decay lengths for the two techniques. The propagating length of the surface plasmon is ~10 μm for SPR but only ~10 nm for LSPR, and the decay length is much smaller for LSPR than for SPR.

This model assumes a single exponential decay (decreasing at a rate proportional to its value, following the exponential function) of the electromagnetic field normal to the planar surface, which is a simplification for the electromagnetic fields associated with metallic NPs. The simplified model enables us to optimize the response of LSPR NSs. The m factor for SPR sensors is about 2×10^6 nm/RIU (refractive index units) and is about 2×10^2 nm/RIU for LSPR sensors. RIU is the relative change in the index of refraction of the fluid medium. It is often used to describe SPR sensitivity.

The enormous difference in m factor values between SPR and LSPR sensors is largely compensated by the very low decay length offered by LSPR gold NPs. Indeed, this decay length is around 200–300 nm for SPR sensors, whereas it is only a few nanometers for LSPR NSs (for Au NPs $l_d = 15$ nm). This decay length depends on the size, shape, and composition of the NPs and is responsible for the large sensitivity of the LSPR NSs. *What is the importance of this small decay length?* This low decay length allows the detection of a very thin layer of adsorbate molecules.

Even if the sensitivity of the SPR sensors is slightly superior to that of LSPR NSs, a direct comparison is not possible because of the different mechanisms that give rise to their respective sensitivity gains. However, some advantages of LSPR NSs can be cited. The LSPR NSs do not require temperature control, compared with SPR, since the large refractive index sensitivity of SPR induces a strong dependence on the environmental temperature. No specific angular conditions of excitation are required for LSPR. There is no requirement of prism coupler-based, grating coupler-based, or optical waveguide optical accessories. In practice, SPR sensors require at least 10×10 μm^2 area for a sensing experiment, whereas, for LSPR sensing, the probed zone can be minimized to a large number of individual sensing elements up to a single NP, using confocal or near-field measurement techniques. "Confocal" means having the same focus or foci. Confocal microscopy is an optical imaging technique, offering several advantages over conventional optical microscopy, including a shallower depth of field, and elimination of out-of-focus glare. Resolution and contrast of a micrograph are increased by using point illumination and a spatial pinhole to eliminate out-of-focus light in specimens that are thicker than the focal plane. Near-field measurement techniques include near-field scanning optical microscope, in which the intensity of light focused through a pipette with an aperture at its tip is recorded, as the tip is moved across the specimen in a raster pattern at a distance of much less than a wavelength; this allows for the surface inspection with high spatial, spectral, and temporal resolving power. Finally, the extinction spectroscopy (highly sensitive optical spectroscopic technique that enables measurement of absolute optical extinction by samples that scatter and absorb light) of LSPR does not need a complex device; an ultraviolet (UV)-visible microspectrometer is sufficiently efficient to obtain the spectra.

Barbillon et al. (2007) used the electron beam lithography (EBL) for the fabrication of these NSs. The EBL system can control, with great precision, the shape, size, and also the distance between NPs, and consequently can tune LSP resonance of metallic NP arrays in the whole visible range. The metallic nanocylinders obtained have in-plane diameters of 100 nm, and a 200-nm interparticle distance, as viewed by SEM.

Barbillon et al. (2007) demonstrated the sensitivity of gold NSs by studying the influence of the concentration of 11-mercaptoundecanoic acid (MUA) [$HS(CH_2)_{10}CO_2H$] on the shift of LSPR wavelength. Additionally, to study the selectivity of NSs, the biotin/streptavidin binding system was used to detect very weak concentrations of biomolecules. The maximum LSPR wavelength shifts observed for MUA and streptavidin were $\Delta\lambda_{max} = 32.8$ and 28 nm, respectively. They noted that the surface-binding affinity was $1.1 \pm 0.2 \times 10^3\ M^{-1}$ for the binding of MUA to Au NPs and $6.1 \pm 0.3 \times 10^9\ M^{-1}$ for the binding between biotin and streptavidin. Furthermore, the limit of detection for the NS was determined to be 1.88×10^{-16} mol for MUA and 1.25×10^{-18} mol for SA, corresponding to a density of 11300 MUA molecules per NP and 75 SA molecules per NP, respectively. The total number of probed molecules was 1.13×10^8 for MUA and 7.5×10^5 for SA, with a probed zone of $30 \times 30\ \mu m^2$.

Most LSPR sensor designs described so far include only thin film on a chip or glass slide by conjugating the metallic NPs with ligands/receptors, that is, containing large aggregates of NP sensors, instead of a single NP sensor. The extinction and scattering spectra of plasmonic NPs have spectral shifts that are sensitive to small changes in the local refractive index. Most organic molecules have a higher refractive index than the buffer solution. Hence, upon binding to NPs, the local refractive index increases, causing the extinction and scattering spectrum to be red shifted.

Example 6.1

In an LSPR experiment, the observed wavelength shift ($\Delta\lambda_{max}$) was 30 nm. The refractive index sensitivity (m) is 2×10^2 nm/refractive index units, the decay length (l_d) is 15 nm, and the absorbate thickness (d) is 10 nm. Determine the change in refractive index.

Applying Equation 6.1, we have

$$\Delta n = \frac{\Delta\lambda_{max}}{m\{1-\exp(-2d/l_d)\}} = \frac{30}{2\times10^2\{1-\exp(-(2\times10)/15)\}} = \frac{0.15}{\{1-\exp(-1.33)\}} = 0.204 \text{ refractive index units (RIU)} \quad (6.2)$$

Example 6.2

If one is interested in using LSPR to detect molecular binding taking place on a NS surface, then sensitivity is appropriately expressed in surface coverage unit defined through mass, for example, pg mm^{-2}. The unit RU (resonance unit or response unit) is defined as 1 RU = 1 pg mm^{-2} and is also often used to determine surface coverage.

If a monolayer of cytochrome *c* leads to an angular shift of ~0.5°, and the corresponding mass coverage is –3000 pg mm^{-2}, find the mass sensitivity for an angular sensitivity of 0.1×10^{-3} degree.

An 0.5-degree angular shift corresponds to a mass coverage of 3000 pg mm^{-2}.

∴1 degree angular shift corresponds to mass coverage of (3000 pg mm^{-2})/0.5 = 6000 pg mm^{-2}.

∴1×10^{-3} degree angular shift corresponds to mass coverage of $6000 \times 0.1 \times 10^{-3} = 0.6$ pg mm^{-2}.

Hence, the required mass sensitivity is 0.6 pg mm^{-2} or 0.6 RU.

Example 6.3

Two SPR instruments use the same glass prisms and have similar angular sensitivity, but one uses λ_{low} light and the other uses λ_{high} light, where $\lambda_{high} > \lambda_{low}$. Which instrument is more sensitive for measuring molecular binding?

The one using λ_{low} light is more sensitive in terms of molecular binding. The penetration length of the evanescence field (typically ~200 nm), created by SPR into the fluid medium, increases with the wavelength. Longer wavelengths (e.g., NIR) have the apparent advantage of being able to probe further beyond the sensor surface. However, this results in a significant loss of surface sensitivity. Although longer wavelengths allow for slightly deeper detection into the solution bulk, this results in a significantly lower sensitivity for measuring molecular binding on the sensor surface.

6.3 Nanosensors Based on Surface-Enhanced Raman Scattering

Unlike the refractive index-based detection schemes, surface-enhanced Raman scattering (SERS) is a vibrational spectroscopic method, that yields unique vibrational signatures for small molecule analytes, as well as quantitative information (Jackson and Halas 2004; Kneipp et al. 2010a; 2010b). For applications in biosensors, easily exploitable characteristics of the SERS signal transduction mechanism include their sensitivity, selectivity, low laser power, and lack of interference from water molecules. Trace analysis of DNA, bacteria, glucose, living cells, posttranslational modification of proteins (the chemical modification of a protein after its translation; translation is the third stage of protein biosynthesis in which messenger RNA [mRNA] produced by transcription is decoded by the ribosome to produce a specific amino acid chain, or polypeptide, that will later fold into an active protein), chemical warfare agents (asphyxiating or nerve gases, poisons, defoliants, etc.), and carbon nanotubes provide some interesting examples of applications. A peptide is a compound comprising a short chain of two or more amino acids joined together by the peptide bond, i.e., an amide type of covalent chemical bond, in which carboxyl group of one amino acid is connected to the amino group of the other amino acid. A polypeptide is a longer unbranched chain of amino acids linked together by peptide bonds. It may contain up to fifty amino acids.

What is the information content of a Raman signal? The Raman spectrum is like a "fingerprint" of a molecule. The Raman scattering signal is composed of many sharp lines. The frequency shift between the excitation light and the Raman lines is determined by the energy of the molecular vibrations, $h\nu_M$, which themselves depend on the kinds of atoms and their bond strengths and arrangements in specific molecules.

What is the main advantage of Raman spectroscopy? The main advantage is its capability to provide detailed information about the molecular structure of the sample. Sophisticated data analysis techniques, based on multivariate analysis (any statistical technique used to analyze data from more than one variable) have enabled exploitation of the full information content of Raman spectra, allowing inferences to be made about the chemical structure and composition of very complex systems, such as biological materials. In this respect, Raman spectroscopy has the potential to become a key tool for health monitoring based on molecular information. The high molecular specificity and large database of chemical information from the technique makes it also a very useful tool for environmental control.

But there is a great disadvantage in any application of Raman spectroscopy. *What is this disadvantage?* It is the extremely small cross section of the effect, resulting in very weak spectroscopic signals.

Fortunately, the situation is dramatically improved when a new methodological approach is applied. *What is this new approach?* The new approach combines the interesting optical properties of metal nanostructures with modern laser spectroscopy (in which a laser is used as an intense, monochromatic light source). Due to resonances between the optical fields and the collective motion of the conduction electrons (surface plasmons) in metallic nanostructures, strongly enhanced local optical fields exist in the close vicinity of these structures. Exciting opportunities for enhancing spectroscopic signals are available when spectroscopy takes place in these enhanced fields. Raman scattering signals from molecules attached to silver or gold NPs are increased by up to 14 orders of magnitude. The effect is called surface-enhanced Raman scattering. It amplifies the Raman scattering cross section (RSCS), an important parameter in the studies with and applications of the Raman spectroscopy. It shows the molecular microscopic property as well as the frequency shift and the linewidth. It also indicates the light-scattering capacity of a particular molecule. It is usually obtained by comparing the Raman intensity for an unknown cross section to that for a standard with a known cross section.

SERS is an impressive effect for demonstrating the capabilities of this new spectroscopic direction, based on local optical fields. At present, SERS is the only way to detect a single molecule and simultaneously identify its chemical structure.

As a spectroscopic technique, SERS blends the advantages of fluorescence spectroscopy and Raman spectroscopy, as illustrated in Figure 6.2. Additionally, since SERS spectroscopy takes place in the local optical fields, the lateral confinement is determined by the confinement of the local fields, allowing the collection of spectroscopic data from volumes below 5-nm dimensions. SERS intensity has been shown to be dependent upon the excitation of the LSPR. To maximize signal strength and ensure reproducibility, it is vital to control all of the factors affecting the LSPR. These factors, including material, size, shape, interparticle spacing, and dielectric environment, must be chosen carefully to ensure that the incident laser light excites the LSPR. The term "enhancement factor" is used to describe the magnitude to which the SERS effect increases the intensity of the Raman scattering for a given experimental system. The enhancement factor is calculated by dividing the SERS spectral intensity by the normal Raman scattering intensity after normalizing both with respect to collection time, laser power, and number of molecules present in the sampling volume.

Figure 6.3 shows the principle of the technology approach for the NS based on SERS. Molecules (unfilled circles) are attached to a gold nanosphere (the big ball). The molecules respond to local optical fields, enhanced by factor $A(\nu)$, depending on frequencies ν. In addition to this *electromagnetic field enhancement*, the electronic interaction between the Raman molecule and the metal causes an increase of the Raman cross section itself, known as *chemical or electronic enhancement.* (The filled circles represent atomic-scale active sites, where this electronic or chemical interaction occurs; Figure 6.3.) *What are the relative contributions of chemical and electro-magnetically induced enhancements toward the Raman signal?* In many experiments on nanometer-scaled silver or gold structures, the chemical effect provides a contribution of 10–100 times to the total SERS enhancement, whereas the electromagnetic field enhancement gives SERS enhancement factors of up to 10–12 orders of magnitude, resulting in up to 14 orders of magnitude of net SERS enhancement.

The SERS pH sensor was developed using Ag NPs with Au nanoshell/silica core NPs (Talley et al. 2004). The sensor, made of 50–80 nm diameter silver NPs, functionalized with para-mercaptobenzoic acid (para-MBA), showed a characteristic SERS spectrum dependent on the pH of the surrounding solution; para-mercaptobenzoic acid = ($C_7H_6O_2S$). It was sensitive to pH changes in the range 6–8. The broad application of SERS technology, however, is greatly hampered by the lack of reliable and reproducible substrates (Netti and Stanford 2006). Cheap and reproducible substrates are required for using SERS as a standard analytical tool. One solution to engineering reproducible

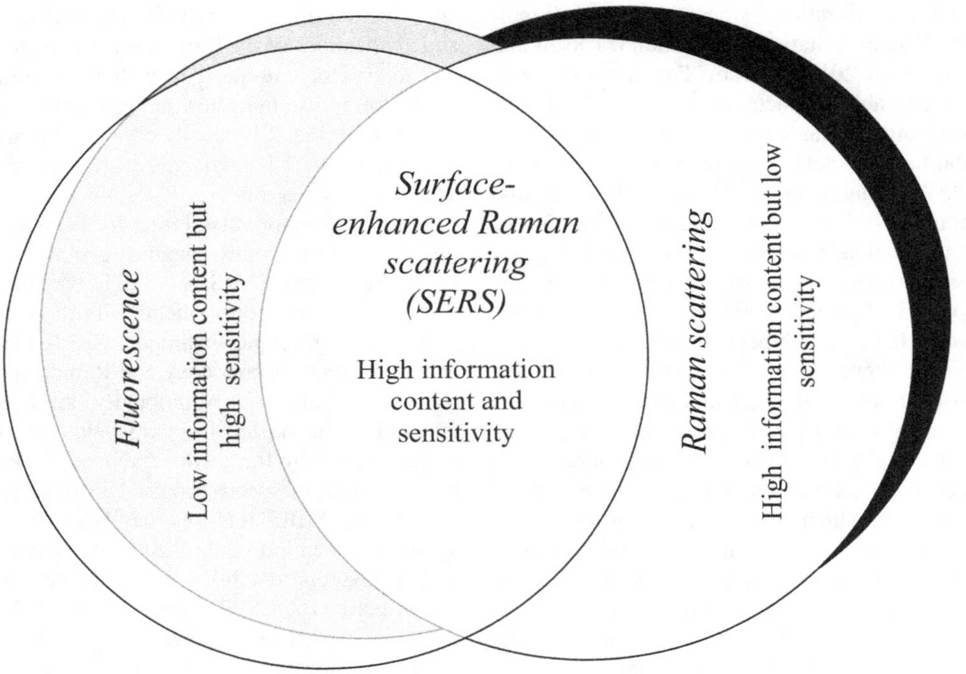

FIGURE 6.2 Coalescence of fluorescence and Raman scattering to form SERS.

SERS substrates is to blend the unique physical properties of photonic crystal devices with reliable semiconductor manufacturing techniques. Photonic crystal research has established how periodic texturing of surfaces gives rise to new optical properties, when the features have dimensions comparable with the wavelength of light ~0.1–1 μm. Metal-coated photonic crystal surfaces result in devices with remarkable dual functionality. They act as *antennae* that concentrate and localize the optical field at the individual features. They also serve as *transducers* that couple the laser beam in and out of the molecule adsorbed onto the surface. One advantageous feature of photonic crystal technology is the ability to tune the optical properties of such devices by modifying the dimensions and the geometry of the texture features and the metal. Consequently, the sensitivity of the substrates can be tuned to different wavelengths or tailored to the type of experiment.

Additionally, chemical functionality is introduced in the textured gold surface, exploiting well-proven surface chemistry protocols, for example, the coating of SERS substrates with ligand molecules having specific terminal groups is enormously effective in improving the selectivity of the SERS substrate. This functionalization of SERS substrates is particularly useful for complex molecules, such as proteins, or for molecules having low affinity to metal.

6.4 Colloidal SPR Colorimetric Gold Nanoparticle Spectrophotometric Sensor

A *colloid* is a substance consisting of particles between 1 and 1000 nm, forming a suspension in a fluid, such as smokes, fogs, foams (a substance that is formed by trapping many gaseous bubbles in a liquid or solid), aerosols (submicron to several micron particles in suspension in the atmosphere), etc., comprising a dispersed phase surrounded by a dispersion medium. A *colorimeter* is a device for measuring color, particularly hue, saturation, and luminous intensity. Hue is one of the main properties of a color, defined technically as the degree to which a stimulus can be described as similar to or different from stimuli that are described as red, green, blue, and yellow (the unique hues). Saturation refers to the dominance of hue in the color.

SPR methods have made far-reaching contributions to the quantification of biomolecular interactions. Disappointingly, conventional SPR reflectometry is difficult to realize in a large-scale array format because of the optics associated with the detection system. This latter limitation is problematical because high-throughput biochemical assays based on protein arrays are necessary to measure the protein–protein and protein–ligand interactions.

Nath and Chilkoti (2002) reported a new label-free optical sensor, which retained most of the desirable features of conventional SPR reflectometry, namely the ability to monitor the kinetics of biomolecular interactions in real time without a label, but which had several improvements: the sensor was easy to fabricate, and simpler to implement, requiring only a UV–visible spectrophotometer or a flatbed scanner. Importantly, the sensor could easily be multiplexed to enable high-throughput screening of biomolecular interactions in an array-based format.

This optical biosensor exploited colloidal SPR of *immobilized, self-assembled gold NPs on an optically transparent substrate.* The method relies on the change in the absorbance spectrum (the electromagnetic spectrum, broken by a specific pattern of dark lines or bands, presented as a graph of absorbance *(A)* versus frequency or wavelength, observed when radiation traverses a particular absorbing medium, showing the fraction of incident

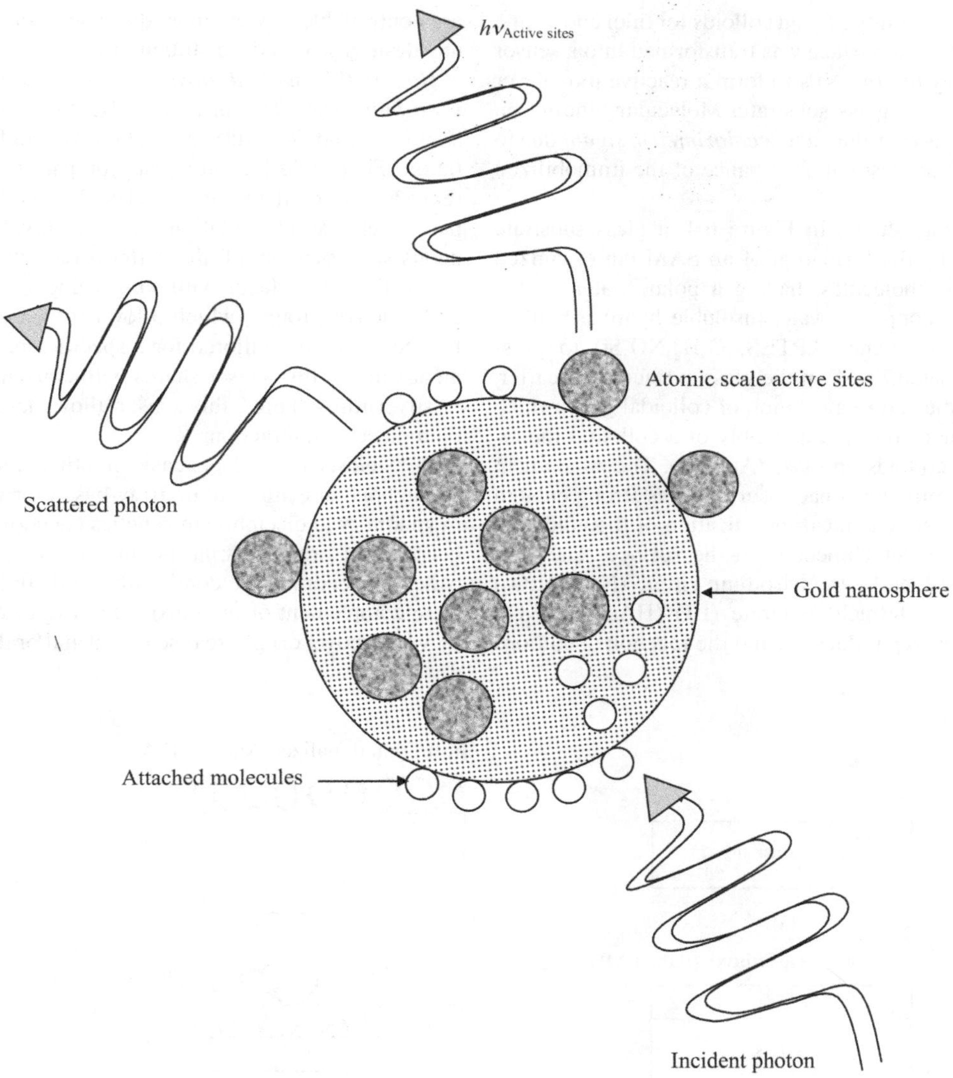

FIGURE 6.3 SERS-based nanosensor.

electromagnetic radiation absorbed by the material over a range of frequencies, a unique property that can be used to identify the material) of a self-assembled monolayer (SAM) of colloidal gold on glass, as a function of biomolecular binding to the surface of the immobilized colloids Figure 6.4). The implementation of colloidal SPR provides an experimentally simple and convenient biosensor, that is easily implemented in most laboratories. The change in the absorbance spectrum associated with biomolecular binding to the immobilized colloids is easily measured in a UV-visible spectrophotometer with an analytical sensitivity and temporal resolution that is adequate to quantify these interactions.

What are the specialties of colloidal SPR? Colloidal SPR is responsible for the intense colors of colloidal solutions of noble metals and is ascribed to the collective oscillations of surface electrons in visible light. It is an interfacial phenomenon and is used in two complementary modes for transduction of binding events at the colloid surface. First, changes in the proximity of colloids, due to their aggregation in suspension, cause a large change in the absorbance spectrum of the colloidal suspension by long-range coupling of surface plasmons. The interparticle distance-dependent color change of colloidal gold, due to aggregation of gold colloids, has been used in solution-based immunoassays. Second, the optical signal arises from the dependence of the peak intensity and the position of the surface plasmon absorbance of gold NPs upon the local refractive index of the surrounding medium, which is changed due to binding at the colloid–solution interface. This mode, resembling conventional SPR, has been utilized to determine biomolecular binding on the surface of a colloid in suspension.

Nath and Chilkoti (2002) sought to develop an SPR biosensor in a planar, chip-based format, using immobilized gold colloids on an optically transparent substrate, because gold colloids permit the transmission of light in the visible region of the electromagnetic spectrum. In principle, a chip-based colloidal SPR sensor enables SPR to be performed in transmission mode in a UV-visible spectrophotometer.

Exploiting the high affinity of gold colloids for thiol and amine functional groups, a glass surface was transformed into a sensor chip by self-assembly of gold NPs to form a reactive monolayer on an amine-terminated glass substrate. Molecular binding on the sensor surface was transduced to a *colorimetric signal* due to the changes in surface plasmon absorbance of the immobilized gold NPs.

For the experiments shown in Figure 6.4, a glass substrate was functionalized by the formation of an SAM (an organized layer of amphiphilic molecules, having a polar, water-soluble group attached to a nonpolar, water-insoluble hydrocarbon) of 3-aminopropyltriethoxysilane (APTES, $C_9H_{23}NO_3Si$) to present an amine-terminated SAM on the glass surface. Thereafter, the surface was immersed in a solution of colloidal gold, which resulted in the spontaneous self-assembly of a colloidal monolayer (CM) of gold colloids on glass (Au CM). The attachment of gold NPs to an amine-terminated surface was strong enough to withstand subsequent chemical modification of the gold NPs without causing their detachment from the surface. The gold colloids were synthesized by the trisodium citrate ($Na_3C_6H_5O_7$) reduction of hydrogen tetrachloroaurate (III) ($HAuC1_4.xH_2O$). This is a simple, one-step reduction, and the reaction conditions are controllable to yield monodisperse (monosized) gold NPs of any desired size in the 5–100 nm range.

How is this method advantageous in comparison with conventional SPR? The primary advantage of this sensor is its simplicity and flexibility at several different levels of implementation. First, gold NPs are easily prepared, and can easily and reproducibly be deposited on glass (or another optically transparent substrate) by solution self-assembly. Second, the spontaneous self-assembly of alkanethiols on gold allows convenient fabrication of surfaces with well-defined interfacial properties and reactive groups, which allows the chemistry at the interface to be readily tailored for a specific application of interest, an advantage this sensor shares with conventional SPR on gold or silver films. Third, this sensor allows label-free detection of biomolecular interactions.

The biosensor offers easy multiplexing, enabling high-throughput screening in an array-based format for applications in genomics (a discipline in genetics concerning the study of the genomes of organisms; an organism's complete set of DNA is called its genome), proteomics (the study of an organism's complete complement of proteins), and drug discovery (the process by which novel drugs are discovered and/or designed).

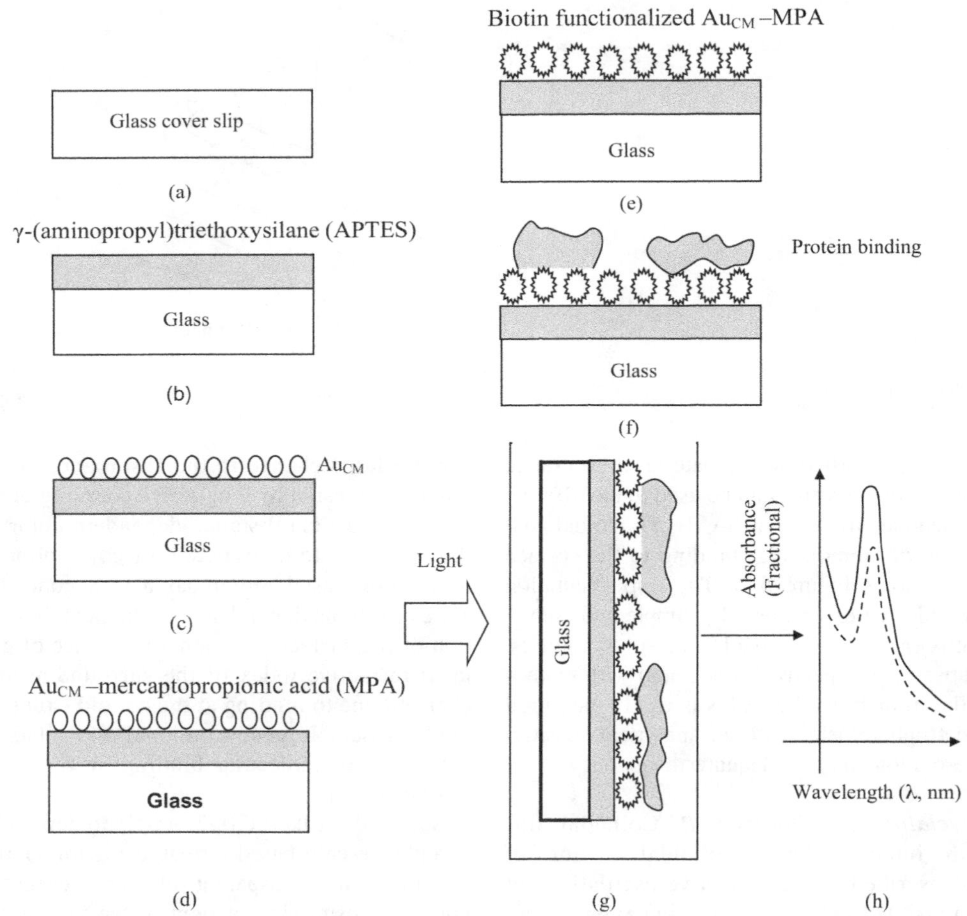

FIGURE 6.4 Steps in the fabrication of colloidal gold sensor strip: (a) glass substrate; (b) silanization; (c) self-assembled monolayer of gold colloids (Au_{CM}); (d) self-assembled monolayer (SAM) of mercaptopropionic acid (MPA); (e) biotin functionalization; (f) biomolecular binding; (g) absorbance study; (h) shift in absorbance-wavelength curve (dotted curve: before binding; full line: after binding). (Nath, N. and Chilkoti, A., *Anal. Chem.*, 74, 504, 2002.)

Example 6.4

The number of free electrons per unit volume (n) in gold is 5.9×10^{28} m^{-3}. Find the SPR frequency of gold NPs, and the corresponding wavelength in vacuum, taking the electron effective mass (m_e^*) as 10%–15% higher than the free electron mass value. Discuss the effect of the dielectric medium on the obtained values of frequencies and wavelengths.

From Equation 2.26, the angular surface plasmon frequency of gold NPs is

$$\omega_M = (q)\sqrt{\frac{n}{3\varepsilon_0 m_e^*}} \tag{6.3}$$

The concentration of free electrons in gold metal is 5.90×10^{28} m^{-3}. Applying the correction for increase in the effective mass of electrons in noble metals by 10%–15%, i.e., by 12.5%, the mass becomes $m_e^* = 9.11\times10^{-31}$ $\{1+(12.5/100)\} = 1.025\times10^{-30}$ kg.

Therefore, the angular resonance frequency is

$$\omega_M = \left(1.6\times10^{-19}\right)\sqrt{\frac{5.90\times10^{28}}{3\times8.854\times10^{-12}\times1.025\times10^{-30}}} \tag{6.4}$$
$$= 7.45\times10^{15}\ \text{rads}^{-1}$$

and the linear resonance frequency is

$$f_M = \frac{\omega_M}{2\pi} = \frac{7.45\times10^{15}}{2\times3.14} = 1.186\times10^{15}\ \text{s}^{-1} \tag{6.5}$$

The wavelength is

$$\lambda_M = \frac{c}{f_M} = \left(\frac{3\times10^8}{1.186\times10^{15}}\right)\times10^9\ \text{nm} = 253\,\text{nm} \tag{6.6}$$

where c is the velocity of light in a vacuum. The calculated frequency or wavelength in a vacuum is in the UV range. Gold NPs display a variety of colors, ranging from an intense red color (for particles less than 100 nm) to a dirty yellowish color (for larger particles). The color of a gold NP solution depends on the size of the particles. Small NPs absorb light in the blue-green portion of the spectrum (~400–500 nm), whereas red light (~700 nm) is reflected, yielding a deep red color. As particle size increases, the wavelength of the SPR-related absorption shifts to longer, redder wavelengths. This means that red light is now adsorbed, and bluer light is reflected, yielding particles with a pale blue or purple color. As particle size continues to increase toward the bulk limit, SPR wavelengths move into the IR portion of the spectrum and most visible wavelengths are reflected. This gives the NPs a clear or translucent color.

Note that the calculation presented earlier is approximate. For gold NPs, the surface plasmon wavelength is 520 nm and the bulk plasmon is at 330 nm.

6.5 Fiber-Optic Nanosensors

Optical fiber-based transduction techniques are very attractive in chemical sensing applications. *What are the characteristics by which optical fibers are favored for sensor applications?* The foremost desirable property is their unique characteristics, such as small size, low weight, immunity to electromagnetic interference (the disruption of operation of an electronic device when it is in the vicinity of an electromagnetic field), ease of multiplexing, possibility of use in harsh environments, and double function of probe and data communication channel. Another feather in their cap is the versatility of optical sensing, because it allows simultaneous collection of intensity and wavelength information and includes a wide range of techniques, for example, absorbance, reflectance, fluorescence, SPR, refractive index, and colorimetry.

Sensor technologies and nanotechnology have joined together to offer new opportunities, using nanometer-scale structures for chemical and biological applications. Ultrasmall fiber-optic devices show great value in sensing applications. Phenomena hitherto below detection levels can now be sensed by nanostructures. Fiber-optic chemical NSs or bio-NSs may achieve sensitivity down to the single molecule level. Conversely, additional problems may be encountered such as the fundamental challenge of the integration of physical structures and devices on the nanometer scale, as well as their integration into the microscale world. Sensitive coatings at nano- and sub-wavelength-scale, based on syndiotactic polystyrene (sPS; thermoplastic semicrystalline material), CNTs, and metal oxides (MOXs), are used in sophisticated fiber-optic chemical NSs.

6.5.1 Fabry-Perot Reflectometric Optochemical Nanosensor, Using Optical Fibers and SWCNTs

Diverse optical fiber devices, using chiefly standard silica optical fibers (SOFs) and hollow-core optical fibers (HOFs) in the reflectometric configuration and coated long-period fiber gratings (LPFGs: the period of the grating considerably exceeds the wavelength of radiation propagation in the fiber), have been deployed for proper integration with appropriate sensing overlays. In all cases, the basic procedure adopted derives advantage from the changes in the optical properties of the sensitive overlays, caused by chemical interaction with target analytes to produce modulated light signals. Also, a common feature adopted in all the schemes relies on the use of sensitive overlays at the nanoscale. The CNTs possess the outstanding capability for reversible adsorption of molecules of environmental pollutants, undergoing a modulation of their electrical, geometrical, and optical properties, such as resistivity, dielectric constant, and thickness.

Penza et al. (2005), Consales et al. (2006; 2007; 2008), and Cusano et al. (2008) worked on the room-temperature sensitivities and response-time analysis of SOF sensors, coated with SWCNTs, in response to aromatic volatile organic compounds (VOCs). SWCNT Langmuir-Blodgett (LB) multilayers were employed as highly sensitive nanomaterials. The LB process is a recognized technique for depositing defect-free, molecularly ordered ultrathin films with prescribed thickness and orientation, allowing fine surface modifications in a multilayered film of carbon nanotubes with a highly controlled manipulation for implementing molecularly self-organizing nanomaterials.

To construct the interferometer, thin films of SWCNTs were transferred onto the distal end of a standard silica single-mode optical fiber (Figure 6.5). The term "reflectometric configuration"

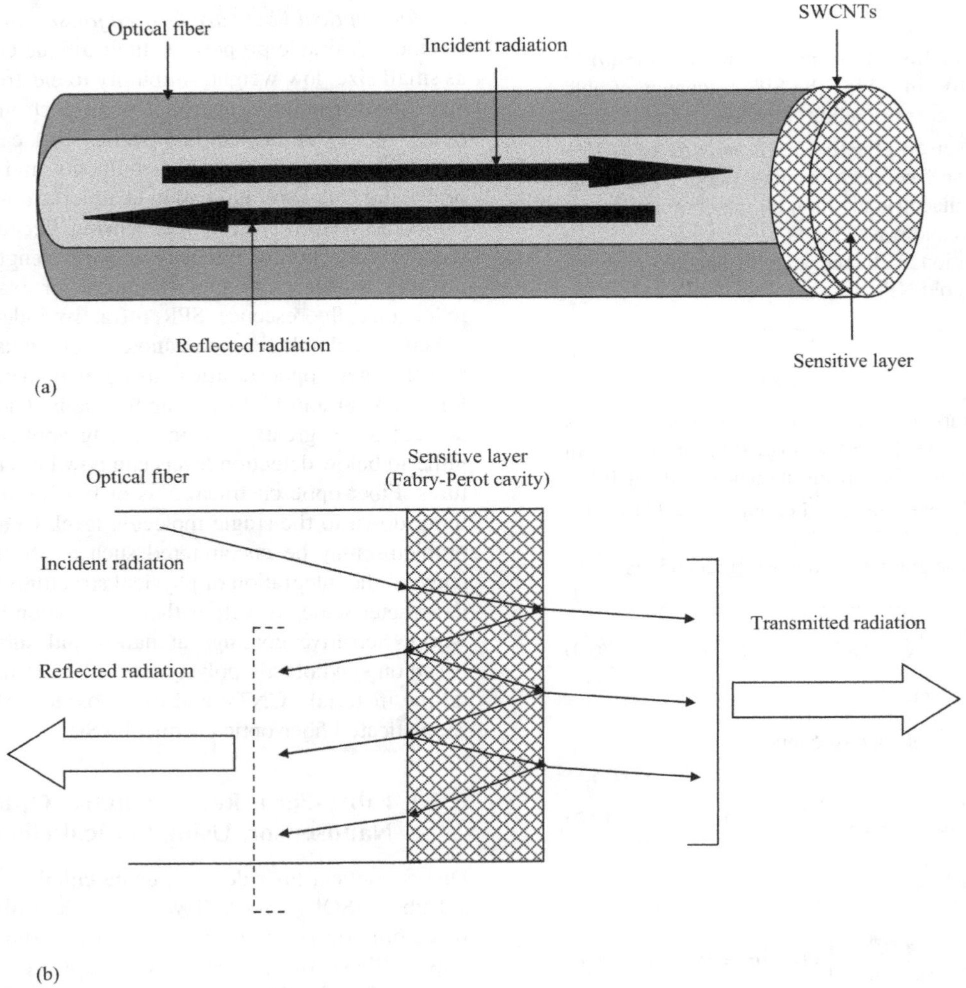

FIGURE 6.5 (a) Optical fiber configuration for vapor testing; (b) multiple reflected beams in the Fabry-Perot cavity. (Consales, M. et al., *J. Sens.*, Article ID936074, 29, doi:10.1155/2008/936074, 2008; Cusano, A. et al., *Anal. Anal. Chem.*, 4, 296, 2008.)

refers to the standard extrinsic Fabry-Perot (SEFP) configuration, working on reflectance measurements from a low-finesse (performance) and extrinsic Fabry-Perot interferometer, achieved by depositing a flat thin micro-structured sensitive film deposited at the distal end of a properly cut and prepared SOF (Cusano et al. 2008). At the fiber–film interface of this arrangement, the light propagating within the optical fiber is partially transmitted and, to some extent, reflected, depending on the optical and geometrical properties of the SWCNT overlay. Consequently, any change occurring within the sensitive layer, due to chemical adsorption of target analyte molecules, produces a corresponding variation in the film reflectance (R_{SWCNTs}):

$$R_{SWCNTs} = \left| \frac{r_{12} + r_{23}\exp\left(-i\beta_{SWCNTs}\right)}{1 + r_{12}r_{23}\exp\left(-i\beta_{SWCNTs}\right)} \right|^2, \quad r_{12} = \frac{\sqrt{\varepsilon_f} - \sqrt{\varepsilon_{SWCNTs}}}{\sqrt{\varepsilon_f} + \sqrt{\varepsilon_{SWCNTs}}},$$

$$r_{23} = \frac{\sqrt{\varepsilon_{SWCNTs}} - \sqrt{\varepsilon_{ext}}}{\sqrt{\varepsilon_{SWCNTs}} + \sqrt{\varepsilon_{ext}}}, \tag{6.7}$$

$$\beta_{SWCNTs} = \frac{2\pi\left(2\sqrt{\varepsilon_{SWCNTs}} \times d_{SWCNTs}\right)}{\lambda}$$

where

ε_f and ε_{ext} are the dielectric functions of the optical fiber and external medium, respectively

λ is the optical wavelength; further

$$\varepsilon_{SWCNTs} = \varepsilon'_{SWCNTs} - i\varepsilon''_{SWCNTs} \tag{6.8}$$

where ε_{SWCNTs} and d_{SWCNTs} represent the complex dielectric function and the thickness of the SWCNT layer, respectively.

Upon exposure of the optical probe to external pollutants, the adsorption of pollutant molecules takes place inside the sensitive layer, thereby changing its complex dielectric function and, accordingly, the optical signal reflected at the fiber–film interface. The sensing property is the dependence of the reflectance at the fiber-sensitive layer interface on the optical properties of the sensitive materials, namely its complex refractive index ($n^* = n + i{\cdot}\kappa$; the real part of the refractive index n indicates the phase speed, while the imaginary part κ indicates the amount of absorption loss when the electromagnetic wave propagates through the material; κ is often called the extinction coefficient). The interaction of the field with the chemical molecules present in the environment under investigation occurs not in the volume of the layer but on its surface, by means of the evanescent part of the field, providing a considerable improvement in performance of the sensor. The chemo-optic variations induced by the surface–chemical interactions change the film reflectance, and, in turn, the intensity of the optical signal reflected at the fiber–film interface. The abovementioned phenomenon has paved the way to a new concept of high-performance chemical sensors based on the manipulation of light through micro- and nanosized structures.

SWCNTs thin films were transferred, monolayer by monolayer, onto the substrates of the SOF sensors by means of the Langmuir–Blodgett film deposition technique. This deposition technique has been used owing to its ability to deposit ultrathin organic films with precision control over the structure of the film at the molecular level. A solution (0.2 mg mL^{-1}) of SWCNT pristine material in chloroform ($CHCl_3$) was spread onto a sub-phase consisting of deionized water of resistivity 18 megaohms (MΩs) with 10.4 M of $CdCl_2$. The sub-phase pH was 6.0 and the temperature 23°C. The monolayer was compressed with a barrier rate of 15 mm minute^{-1} up to a surface pressure of 45 mN m^{-1}. The single layer was deposited with a dipping rate of 3 mm minute^{-1}. The transfer ratio of the monolayer, from the sub-phase to the substrate surface, was in the range 0.5–0.7. The optical fibers used for the fabrication of the SWCNT-based optical probes were standard SOFs, having core and cladding diameters of 9 and 125 μm, respectively. Their effective refractive index was 1.476 at 1310 nm.

In the optical sensor interrogation system (Figure 6.6), reflectance measurements were performed by illuminating the sensing fiber with a "pigtailed" super-luminescent light-emitting diode (SLED), having a bandwidth of ~40 nm and a central wavelength of 1310 nm. For testing and making sensitivity comparisons of response and recovery times of the optical sensor toward xylene [C_8H_{10}, $C_6H_4(CH_3)_2$ or $C_6H_4C_2H_6$] and toluene (C_7H_8 or $C_6H_5CH_3$) vapors at room temperature, SOF transducers were exposed to the preselected VOC vapors, following a time-division multiplexing approach for the parallel interrogation of up to eight optical transducers. Upon target analyte exposures, the SOF sensor showed an increase of the interface reflectance, due to the change in the dielectric constant of the sensitive overlay. Optical sensors were more sensitive to xylene than to toluene vapors and provided very high reproducibility and sensitivity with resolution of a few hundreds of parts per billion. The effect of the SWCNT monolayer number on sensor sensitivity and response time was

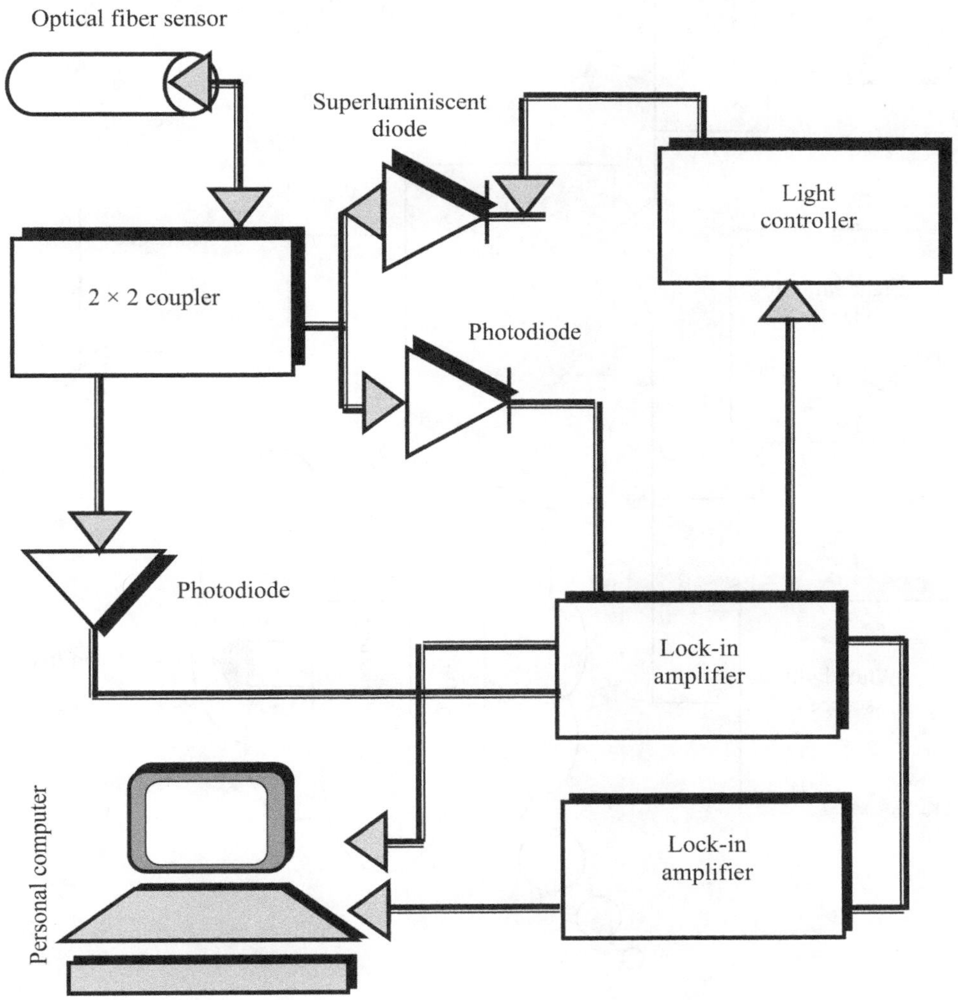

FIGURE 6.6 Optical fiber sensor system. (Penza, M. et al., *Nanotechnol.*, 16, 2536, 2005.)

also investigated, showing that the performance of the transducers could be enhanced by tailoring the geometric properties of the sensitive nanomaterial.

6.5.2 In-Fiber Nanocavity Sensor

Elosua et al. (2006) developed a VOCs NS based on Fabry–Perot interferometry, using in-fiber nanocavity doped with vapochromic material (showing pronounced and reversible changes in color and/or emission in the presence of volatile vapors). They prepared a fiber-optic NS (Figure 6.7) by constructing a nanocavity onto a cleaved-ended optical fiber pigtail by the electrostatic self-assembly (ESA) method, doping this structure with the vapochromic material belonging to a family of complexes of general formula $[Au_2Ag_2(C_6F_5)_4L_2]_n$, (where L can be pyridine (C_5H_5N), 2,2′-bipyridine, 1,10-phenanthroline ($C_{12}H_8N_2$), 1,2-diphenylacetylene ($C_{14}H_{10}$), and other ligands). It is able to change its optical properties reversibly in the presence of organic vapors, such as ethanol (C_2H_5OH), methanol (CH_3OH), isopropanol (C_3H_8O), acetic acid ($C_2H_4O_2$), and dichloromethane (CH_2Cl_2). The sensor was fabricated using a multimode optical fiber with core and cladding diameters of 62.5 and 125 μm, respectively. Employing a reflection scheme, a change in the intensity-modulated reflected signal at 850 nm was found. The interferometric response was produced by the two mirrors of the nanocavity (fiber–film and film–air interfaces). As the bilayers are deposited, the film grows, and, hence, the reflected optical power changes, following an interferometric (two-beams) pattern. The nano Fabry-Perot

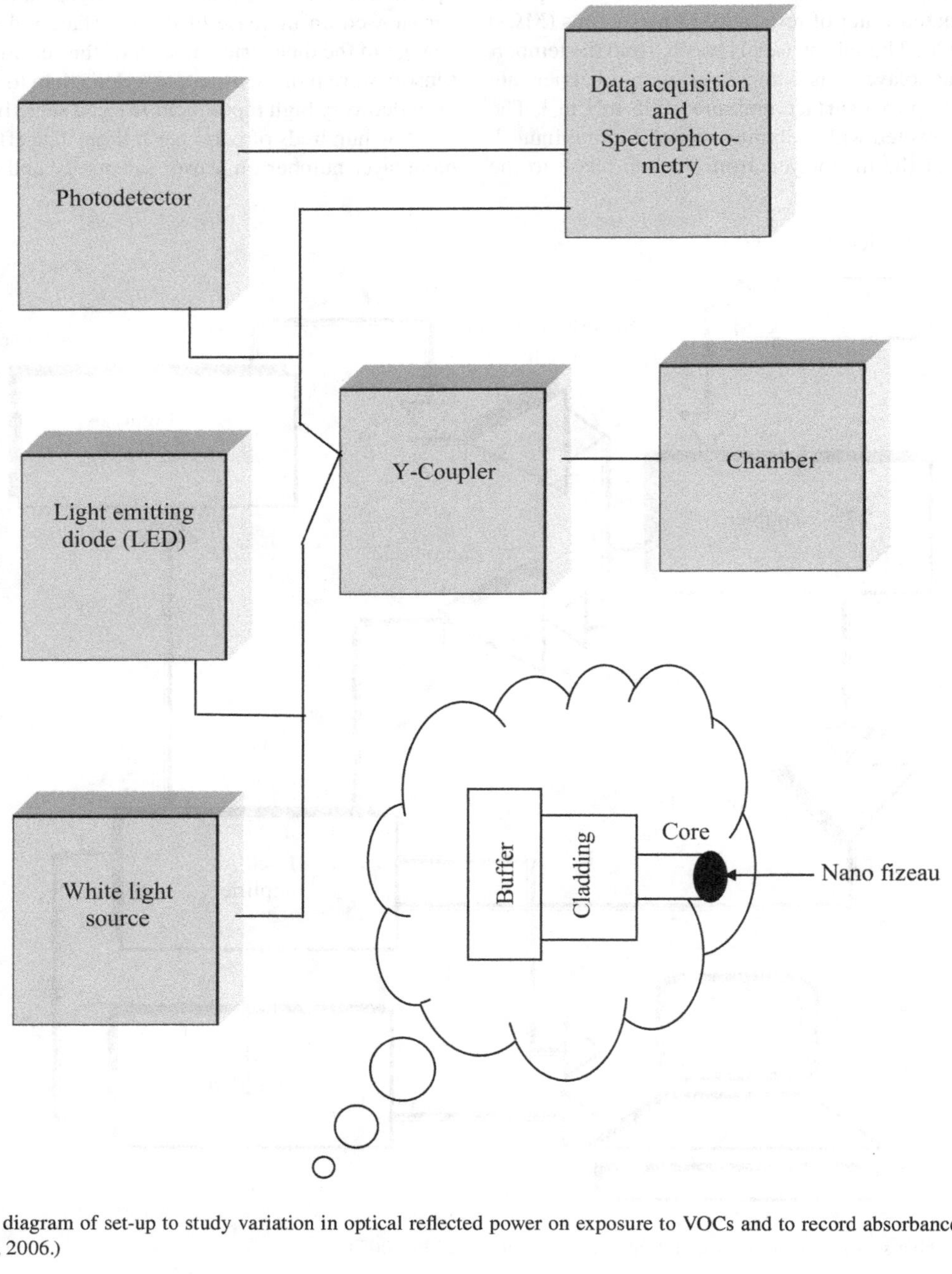

FIGURE 6.7 Block diagram of set-up to study variation in optical reflected power on exposure to VOCs and to record absorbance spectra. (Elosua, C. et al., *Sensors*, 6, 578, 2006.)

cavity consists of 25 bilayers. The length of the nanocavity was 250 nm (one-quarter of the wavelength used to excite it). The sensor head was able to distinguish among some VOCs and also could determine different concentrations of individual components. Changes of up to 1.44 dB in the reflected optical power were registered in less than 2 minutes.

6.5.3 Fiber-Optic Nanosensors for Probing Living Cells

Generally speaking, fiber-optic NSs can be defined as nanometer-scale measurement devices, consisting of a layer containing either biological or chemical recognition elements, covalently attached to the transducer. Interaction between the target analyte and the sensing layer produces a physicochemical perturbation that is convertible into a measurable electrical signal. Miniaturization by using optical fibers can make it possible to directly measure biomedical parameters by allowing the probe to be placed inside the body. Fiber-optic nanobiosensors are the tools used to investigate biological processes at the cellular level *in vivo*. These NSs have the capability to perform single-cell analysis (Figure 6.8), which is used for dynamic examination of interactions within individual living cells, which are critical for mapping and deciphering cell-signaling pathways and networks. Not only can antibodies be developed against specific epitopes (sites on the surface of an antigen molecule to which a single antibody molecule binds), but also an array of antibodies can be established to investigate the overall structural architecture of a given protein. Finally, the most noteworthy advantage of NSs for cell monitoring is the *minimal invasiveness* of the technique. The integration of advances in biotechnology and nanotechnology may usher a new generation of nanosystems with unprecedented sensitivity and selectivity for probing sub-compartments of living cells at the molecular level (Vo-Dinh and Kasili 2005). An ideal bioanalytical and biomedical sensor should achieve, in live cells *in-vivo,* real-time tracking of biological/chemical/physical processes as well as the detection of disease-related abnormal features, without interferences.

The fabrication of fiber-optic NSs is a crucial precondition for their successful deployment. It involves three steps. The first step is the fabrication of the nanotip. A fairly well-characterized method for fabricating the nanofiber tips is the so-called "heat and pull technique," involving pulling of nanotips from a larger diameter (600 μm) SOF, using a special fiber-pulling device. It is based on local heating of a glass fiber, by a CO_2 laser or a heat filament, and afterward pulling the fiber apart. The shapes of the resulting tip depend on the temperature and the timing of the procedure. The fiber is then safely put into the fiber-pulling device and the laser-heating source is focused on the median point of the fiber. The optical fiber is pulled to obtain two fibers with submicron tip diameters.

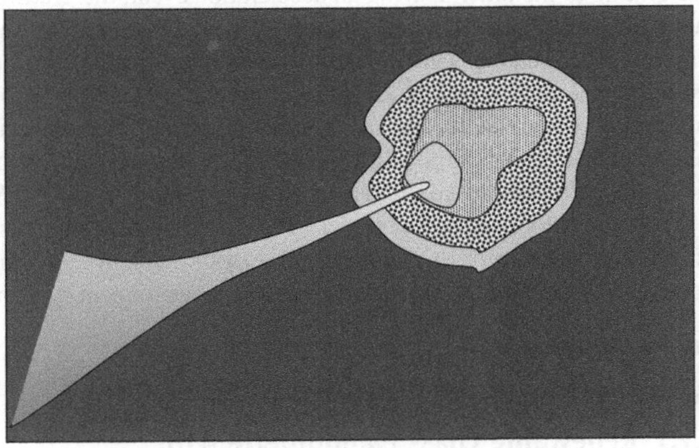

FIGURE 6.8 Fiber optic nanobiosensor inserted into a living cell. Tip diameter of the drawn fiber is 50 nm whereas that of an Ag-coated fiber is 250 nm. (Vo-Dinh, T. and Kasili, P., *Anal. Bioanal. Chem.*, 382, 918, 2005.)

The second step of the nanofabrication process involves coating the tapered side walls of the optical fiber with a thin film (100–200 nm) of silver, aluminum or gold by thermal vapor deposition, serving to restore the refractive index and enabling propagation of the excitation light down the tapered sides of the nanofiber. Because the fiber tip is pointed away from the metal source, there is no metal coating on the tip. The coating procedure is designed to leave the distal end of the fiber free from silver for subsequent derivatization (a technique used in chemistry that transforms a chemical compound into a product of similar chemical structure, called a derivative), to facilitate covalent immobilization of biorecognition molecules, such as antibodies or synthetic peptides (peptides that do not occur in nature, such as those containing unnatural amino acids; peptides are synthesized by coupling the carboxyl group or C-terminus of one amino acid to the amino group or N-terminus of another) coupled to a fluorescent molecular probe, on the exposed silica nanotip. For this purpose, derivatization of the nanotips is done by silanization with glycidoxypropyltrimethoxysilane (GOPS) (3-glycidoxypropyltrimethoxysilane, $C_9H_{20}O_5Si$, is a silane coupling agent (KH-560), a versatile epoxy functional organosilane, possessing a reactive organic epoxide and hydrolyzable inorganic methoxysilyl groups), with activation by 1,1′-carbonyldiimidazole (CDI) ((formula $(C_3H_3N_2)_2CO$), an organic, white, crystalline solid; it is often used for the coupling of amino acids for peptide synthesis). This is the third step. Because the diameter of the optical fiber tip is much less than the wavelength of light used to excite the target analyte, the photons cannot escape from the tip of the fiber by conventional optics. Instead, in a nanobiosensor, after the photons have traversed as far down the fiber as possible, evanescent fields continue to travel through the remainder of the tip, providing excitation for the fluorescent species of interest present in the vicinity of the biosensing layer. Hence, only species that are in extremely close proximity to the fiber tip (e.g., antigens bound to the antibody probes) can be excited, thereby preventing excitation of interfering fluorescent species within other locations of the sample.

Fiber-optic nanobiosensors, consisting of antibodies coupled to an optical transducer element, have been used to detect biochemical targets, benzopyrene tetrol (BPT), and benzo[α]pyrene (BaP), inside single cells; BPT is a metabolite of the carcinogen benzo[α]pyrene. *Apoptosis* or programmed cell death is a process by which cells in our tissues and organs degenerate during normal development, aging, or in disease. Fiber-optic nanobiosensors offer a strategy to monitor and measure apoptosis proteins early in the cell-death cascade. Furthermore, several different optical fiber-based chemical NSs have been studied for measurement of pH, concentrations of various ions, and other chemical species.

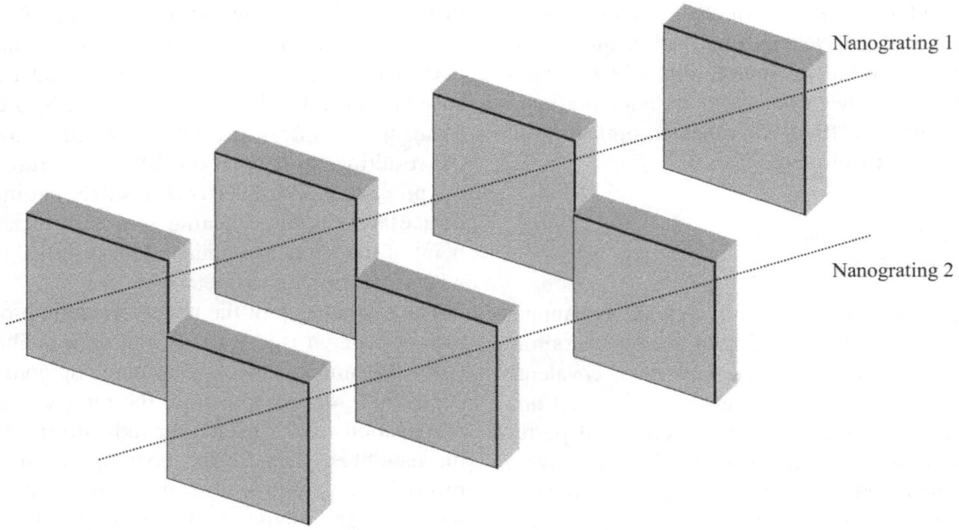

FIGURE 6.9 Nanograting. (Carr, D. W. et al., Femto-photonics: Optical transducers utilizing novel sub-wavelength dual layer grating structure. In: *Proceedings of the Hilton Head Solid-State Sensors, Actuators, and Microsystems Workshop,* Hilton Head Island, SC, 91, June 2004; Krishnamoorthy, U. et al., *Sens. Actuators, A,* 145–146, 283, 2008.)

However, as we shall see in Section 6.8, there is some degree of invasiveness still present when the fiber pierces the cell wall. Combined with the unavoidable damage associated with the readout fiber, there is a little impairment of normal cell functioning, so that further damage-minimizing techniques are necessary.

6.6 Nanograting-Based Optical Accelerometer

High-precision inertial navigation and seismic sensing for geophysical and oil-field applications require measurement of extremely small acceleration signals ($nG/\sqrt{Hz}$) at very low frequencies (<100 Hz). Challenges in building such a device were addressed by Krishnamoorthy et al. (2008). These include designing very-high-sensitivity displacement sensors with low thermomechanical noise and minimal $1/f$ noise. A practical problem is integrating a large proof mass to the displacement sensor to meet the sensitivity requirements for the device (since a larger mass with weak spring constants relates to greater sensitivity).

Why is an optical method chosen for acceleration measurements? Advantages of optical detection techniques, compared with capacitive or piezoresistive methods, include high sensitivity and performance close to the limits of Brownian noise (the kind of noise resulting from Brownian motion; hence, also called random walk noise) of the mechanical structure.

What is a diffraction grating? Diffraction grating is an optical device used to resolve the different wavelengths or colors contained in a beam of light. The device usually consists of thousands of narrow, equidistant, closely spaced, parallel slits (or grooves).

An in-plane nanophotonic resonant sensor, based on multilayer sub-wavelength optical gratings, was used (Carr et al. 2004). Through-wafer holes were incorporated under the nano-grating-based displacement sensors to accommodate transmission detection schemes. This nano-optic sensor consisted of two vertically offset layers of sub-wavelength polysilicon nano-gratings separated by an air gap that modulates the near-field intensity and polarization of an incident light source, based on the relative lateral motion of the two gratings. The nano-grating-based optical micro-electromechanical systems (MEMS) resonant sensors had extremely high lateral displacement sensitivities ($12\ fm/\sqrt{Hz}$ at 1 kHz) and greater than 120 dB of open-loop dynamic range.

Figure 6.9 shows two gratings fabricated to lie on top of one another, with a small gap separating the two. The two-layer nanogratings were fabricated in a surface micromachining process. The wafers underwent a high aspect-ratio deep reactive-ion etch (DRIE) of the substrate silicon to structurally define the proof-mass and the mechanical springs for both the proof-mass and its gimbal (a device consisting of two rings mounted on axes at right angles to each other so that an object, such as a ship's compass, will remain suspended in a horizontal plane between them, regardless of any motion of its support). The bottom grating was attached to the compliant mass and moved laterally with it. The upper grating layer was attached to the relatively stiff gimbal.

As the gratings have sub-wavelength dimensions and spacing, they create a local effective index of refraction for that region of space. The index varies as a function of the relative placement of one grating to the other. During an acceleration event, the gratings move relative to each other, modulating the effective index of the structures. This changes the intensity of the reflected and transmitted light. The light typically arrives at a normal incident angle on the grating. The reflected intensity is monitored to determine the relative displacement of the gratings, which resulted from accelerations applied to the device. Achieved device sensitivities are 590 V G^{-1} and noise floors correspond to $17\ nG/\sqrt{Hz}$ (at 1 Hz); the noise floor is the measure of the signal created from the sum of all the noise sources and undesired signals within a measurement system.

6.7 Fluorescent pH-Sensitive Nanosensors

6.7.1 Renewable Glass Nanopipette with Fluorescent Dye Molecules

Piper et al. (2006) demonstrated the detection of both ratiometric and intensity-based fluorescent changes for physiological levels of pH and sodium by using a simply fabricated fluorescent NS in the form of a 100 nm-sized probe. They displayed the concept of the nanopipette as a pH sensor by trapping a negatively charged ratiometric pH-sensitive fluorophore, SNARF-1-dextran, at the pipette tip (Figure 6.10). The sensor was able to measure bath pH from 4.0 to 9.2, responding to the addition of acid in less than 30 ms.

The stability of the fluorescence signal from the NS helped in developing an intensity-based sodium sensor using the negatively charged CoroNa Green dye (an Na^+ indicator dye that has its absorbance maximum at 492 nm). This dye (50 nM) was trapped at +5V pipette voltage, and the fluorescence intensity increased as a function of bath NaCl concentration.

The time resolution was measured to be ~2 ms, and the spatial resolution of 600 nm established here can be further improved by bringing the probe and sample nearer to each other. The method is applicable to any fluorescence-based reporter dye and therefore a broad range of analytes and concentrations.

When using free dye molecules, one must concede the problems associated with an uncovered sensing element directly making contact with the intracellular environment and the possible cytotoxic effects this has upon the biological system. In addition, the effect that the intracellular environment exerts upon the dye is detrimental to acquiring reliable measurements from within the cell. Nonetheless, the small size of the free-sensing dye molecules is a key advantage, providing high spatial resolution and allowing information throughout the cell to be collected *en masse.*

6.7.2 Ratiometric pH Nanosensor

Ratiometric measurement uses the ratio of two fluorescent peaks instead of the absolute intensity of one peak. The advantage of the ratiometric sensors is that factors such as excitation source fluctuations and sensor concentration will not affect the ratio between the fluorescence intensities of the indicator and reference dyes. Fluorescent NS particles possess several advantages over direct loading of cells with fluorescent probes, which is a classical method used for monitoring metabolic processes of living cells: (i) the biocompatible particle matrix protects the intracellular environment from any toxic dyes; (ii) the cover prevents the dyes from any potential interferences in the cellular environment, such as non-specific binding of proteins; (iii) the nanometer size minimizes the physical perturbation of the cell and the small size provides a fast response time for the sensor; and (iv) several dyes can be included in the same particle.

Sun et al. (2006) reported the synthesis of a prototype ratiometric pH NS. It contains two dyes: a pH-sensitive fluorescent dye and a pH-insensitive reference dye. The dyes are embedded in a polymer matrix, consisting of porous, highly cross-linked polyacrylamide (–$CH_2CHCONH_2$–). Microemulsion polymerization was used for synthesis of particles with diameters ~20–30 nm. To facilitate the insertion of NSs, protoplasts (plant cells without cell wall) of BY2 (Bright yellow-2) tobacco (the most popular and widely used cell line in plant research) were used as the first model system, and the NS particles were inserted into living protoplasts by gene gun bombardment. Confocal laser scanning

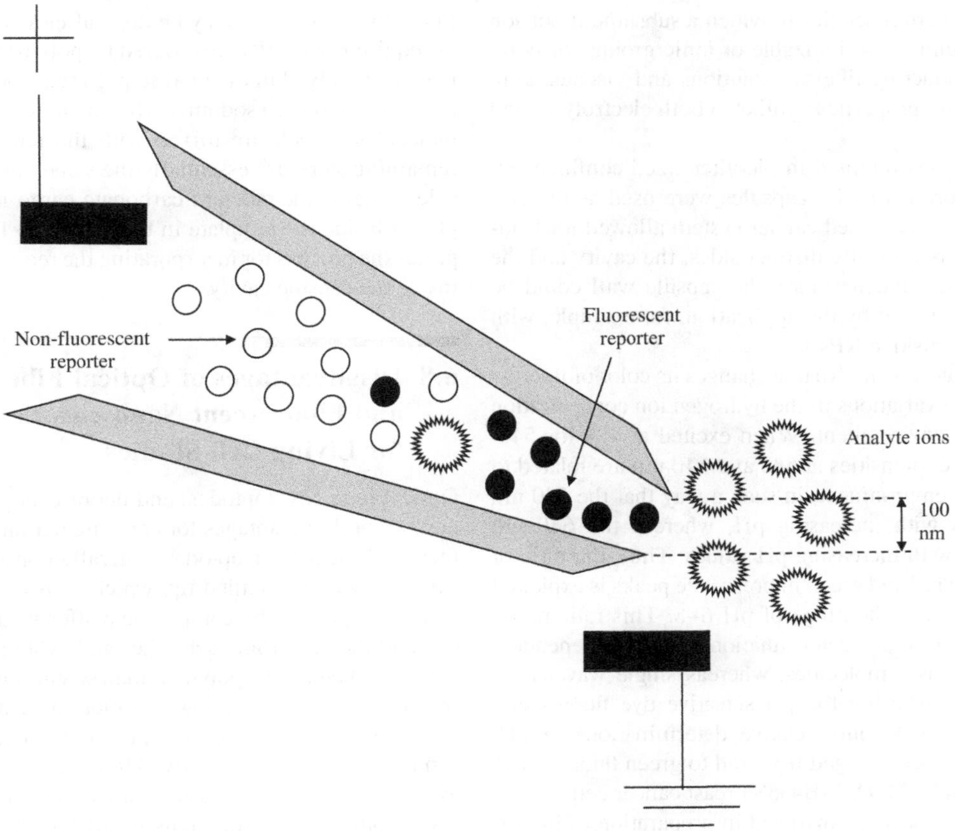

FIGURE 6.10 Glass nanopipette. (Piper, J. D. et al., *J. Am. Chem. Soc.*, 128, 16462, 2006.)

microscopy (a tool for obtaining high-resolution images with depth selectivity and 3D reconstructions by acquiring in-focus images from selected depths, a process known as optical sectioning) was used to visualize the fluorescence responses. Responses received from the pH-sensitive dye and the reference dye permitted intracellular pH measurements by fluorescence ratio imaging microscopy. In this method, two different excitation wavelengths are used and the emitted light levels compared.

Leaching of covalently attached dyes from the NPs was found to be minimal when compared to physically embedded dyes. The sensors were stable and robust because the covalent linkage between the dyes and the matrix minimized the leaching of dyes and the dyes were protected from the environment and *vice versa* as they were incorporated throughout the matrix. The sensors were capable of reflecting pH in the physiologically relevant range from pH 5.8 to 7.2.

6.7.3 pH-Sensitive Microcapsules With Nanoparticle Incorporation in the Walls

Kreft et al. (2007) introduced an alternative system for pH measurement, based on polyelectrolyte microcapsules filled with the pH-sensitive, fluorescent seminaphtho-rhodafluor-1-dye (SNARF1–1) as the local pH probe. Microcapsules are small capsules designed to release their contents when broken by pressure, or when dissolved. They are also useful for drug delivery. *What are polyelectrolytes?* These are polymers of high molecular weight, the repeating units of which bear an electrolyte group, which dissociate in aqueous solutions, releasing positive and negative ions and making the polymers charged. Naturally, the solutions of such macromolecules in which a substantial portion of the constituent units have ionizable or ionic groups, or both, are electrically conducting like salt solutions and viscous, akin to polymers, showing properties similar to both electrolytes and polymers.

The fluorophore was retained in picoliter-sized confinement, but, instead of a porous matrix, capsules were used as the carrier system. This capsule-based carrier system allowed for loading fluorophores at two locally distinct sides, the cavity and the walls of the capsules. Furthermore, the capsule wall could be functionalized as required by the application, for example, with magnetic- or light-sensitive NPs.

SNARF1-1 is a fluorescent dye that changes its color of fluorescence in response to variations in the hydrogen ion concentration of the surrounding environment. When excited at 488 (or 540) nm, the fluorescence intensities at 580 and 640 nm are related to the pH of the local environment, in such a way that the 580 nm intensity decreases with increasing pH, whereas the 640 nm intensity increases with increasing pH values. Thus, the ratio of the intensities of the red and green fluorescence peaks is exploited for pH measurements in the range of pH 6–8. This ratiometric method allows an absolute determination of pH, independent of the amount of sensor molecules, whereas single wavelength measurements (i.e., by using the pH sensitive dye fluorescein, $C_{20}H_{12}O_5$) would provide only relative determinations of pH. SNARF-loaded capsules changed from red to green fluorescence upon internalization by MDA-MB435S breast cancer cells.

Figure 6.11 shows the steps involved in preparation of microcapsules. One starts with the template (a stencil, pattern, or overlay), either positively or negatively charged. The template here is positively charged. The template is made of calcium carbonate ($CaCO_3$) polyester, or melamine formaldehyde ($C_4H_6N_6O$) latex particles (*cis*-1,4-polyisoprene or natural rubber, which is produced in the milky latex of certain plants; polyisoprene is polymer of the isoprene (C_5H_8) unit). Polyester is a category of polymers that contain the ester functional group in their main chain. They are defined as "long-chain polymers", chemically composed of at least 85% by weight of an ester, a dihydric alcohol and a terephthalic acid [$C_6H_4(COOH)_2$]. Although there are many polyesters, the term "polyester" refers most commonly to a specific material: polyethylene terephthalate (PET), $(C_{10}H_8O_4)_n$. A polyelectrolyte layer of opposite polarity to the positively charged template, the negatively charged layer, is coated on the template by electrostatic attraction. As a result, the particle has an overall negative charge. In sequence (a), succeeding layers of opposite polarities are coated at each step, that is, of opposite polarity to the preceding step. In the last step of this layer-by-layer (LBL) coating procedure, the template is removed by dissolution. The final charge of the capsule is the charge that was developed in the last step so that the finished microcapsule is either positively or negatively charged. Different layers of the microcapsule are held together by Coulombic attraction. These layers are reminiscent of the layers in an onion. The number of layers of polyelectrolyte is decided by the desired final thickness of the capsule wall. The capsule wall is a few nanometers thick, whereas the overall diameter of the capsule ranges from tens of nanometers to micrometers (the drawing is not to scale). In sequence (b), the intent is to incorporate nanometric particles inside the capsule walls. This is done here after the deposition of first polymer layer. It may be done after any other layer as well, as required. The NPs are covered by polymer layers and the template is dissolved in the final step. In sequence (c), calcium chloride is reacted with sodium carbonate to form calcium carbonate molecules, which are mixed with the template particles. The remaining steps are essentially the same but the final microcapsule contains the calcium carbonate cargo molecules that were placed inside the template in the first step. The sequence exemplifies the process for incorporating the required cargo molecules inside the capsule cavity.

6.8 Disadvantages of Optical Fiber and Fluorescent Nanosensors for Living Cell Studies

Optical fiber NSs (optodes) and fluorescent probes suffer from a number of disadvantages for carrying out intracellular measurements. Optical fiber optodes generally consist of a pulled optical fiber with a modified tip, which is inserted directly into the cell. They provide biocompatible platforms that protect both the dye and the cell from each other and avoid adverse interactions between them. The polymer matrix entrapping the chemical-sensing elements also allows for more intricate sensing schemes to be devised, due to the close proximity of several elements that can interact complementarily. All this is achieved at a cost, however, because their large size, compared with a single cell, causes severe physical perturbations to all but the largest cells. This perturbation arises from the need to collect the signal generated

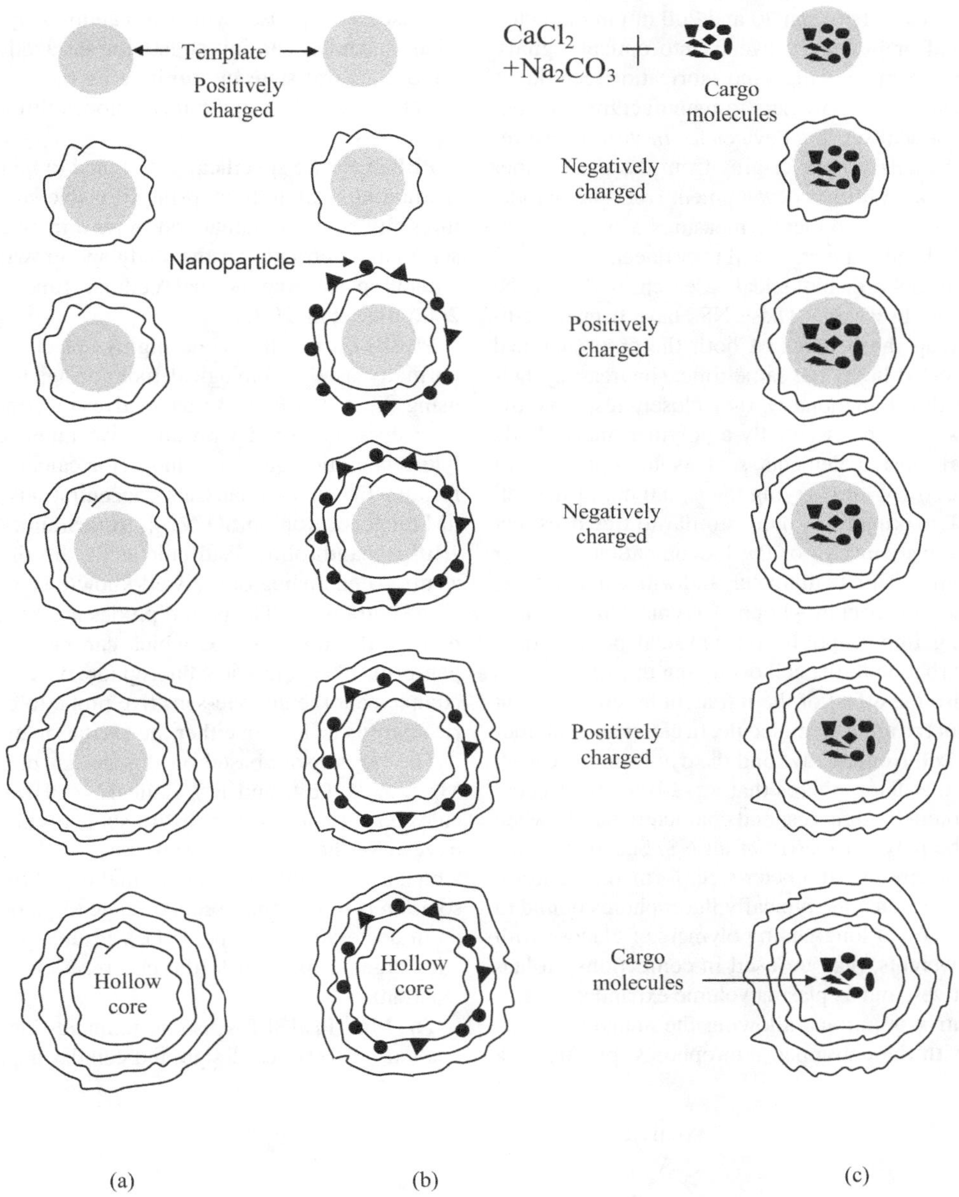

FIGURE 6.11 Steps in the preparation of microcapsules: (a) without nanoparticle incorporation; (b) with nanoparticle incorporation in the side walls; and (c) by embedding cargo molecules inside the core. (Javier, A. M. et al., In *Biosensing Using Nanomaterials,* Merkoci, A. (Ed.), John Wiley & Sons, New Jersey, 256, 2009.)

at the tip of the sensor when inserted into the cell, requiring a physical pathway to the detector. Even though optodes have submicron diameters, typical penetration to the point of the nucleus exceeds 1% of the cell volume, inducing severe perturbations and limiting cell viability. Also, the technique requires highly skilled personnel to properly position the optode in the cell; furthermore, the resultant signals are not spatially resolved. Thus, optodes are bulky and impractical for routine measurements inside living cells due to the connecting fibers, which occupy excessive space inside the cell and cause biological perturbations. Moreover, only low spatial resolution is achievable as very few optodes can be successfully inserted into a cell before it is irreversibly damaged.

Conversely, fluorescent probes, while being easier to use and offering spatially resolved images, still perturb the cell. There are a number of cytotoxic issues concerning fluorescent probes. Variations in signal intensity caused by nonspecific protein binding, and organelle sequestration or efflux of the dye by membrane proteins adversely affect the experiments. They have limited experimental time and once the fluorophore is loaded inside the cell, fluorescent probes often require extensive calibration for individual types of cells.

6.9 PEBBLE Nanosensors to Measure the Intracellular Environment

PEBBLE is the acronym for "photonic explorer or probe for bioanalysis with biologically localized embedding." It is a spherical, optical NS, ranging in size between 20 and 200 nm in diameter, transducing chemical or biological events into optical signals. PEBBLE is a generic term covering nano-fabrication techniques that utilize biologically inert polymers to manufacture nanometer-sized spherical, optical sensing devices for *in-vitro* measurements. PEBBLE NSs derive their origins from the pulled fiber technology, which was, in turn, a development from the optode, an optical sensor device that optically measures a specific substance, usually with the aid of a chemical transducer.

PEBBLEs are the tools for biological research, realizing the dream of "true silent observers." These NSs have been specifically designed to reap the benefits of both the opto-chemical sensors and free dyes while, at the same time, suppressing their negative characteristics. Functionally, they closely resemble the tip of an optode, as they are essentially a polymer matrix body entrapping chemical-sensing elements, such as fluorophores. All the components necessary for carrying the signal out of the cell have been detached, making them more similar to free dyes, but retaining the protective capacity of the biocompatible polymer matrix. Typically, only 50 nm in diameter, and with a narrow size distribution, these sensors occupy ~1 ppb of a typical mammalian cell volume, causing imperceptibly low physical perturbation. Any chemical perturbation to the cell occurring due to free dyes is annulled, as is also the effect of the intracellular environment upon the efficiency of the reporter molecule. In effect, the polymer matrix protects the cell from the dye and the dye from the cell. It has been observed that fluorophores that are adversely affected through random protein binding respond characteristically when entrapped within the polymer matrix of an NS. Succinctly, this NS is a polymeric sphere of nanometer size, formed by entrapping chemical-sensing elements, typically fluorophores bound to dextrans (any of a group of long-chain polymers of glucose with various molecular weights, that are used in confections, in lacquers, as food additives, and as plasma volume expanders).

The polymer matrix is porous, allowing the analytes to diffuse and interact with the entrapped fluorophores, producing a fluorescent response, which is captured by an optical system. Thus, scientists can investigate the intracellular environment in a more natural state by minimizing the chemical and physical effects of free dyes and bulky optodes through the use of polymeric NSs.

PEBBLEs are specifically designed to be minimally invasive, facilitating analyte monitoring in viable single cells or cell cultures (living cells maintained *in vitro* in an artificial medium of serum and nutrients for the study and growth of certain strains) without perturbing normal cellular functions (Sumner et al. 2006; Lee et al. 2009).

PEBBLEs use the same highly selective fluorescent probes normally used in biological monitoring, for example, a probe, using the Alexa Fluor series of dyes or Oregon Green as reference dyes, is paired with an active fluorescent probe for analyte recognition such as Fluo-4 for calcium (a fluorescent dye for quantifying cellular Ca^{2+} concentrations in the 100 nM to 1 μM range) or $Ru[dpp(SO_3Na)_2]_3$ [ruthenium(II) tris(4,7-diphenyl-1,10-phenanthroline disulfonic acid), disodium salt] for oxygen, a sulfonated analog of the well-known oxygen probe $(Ru[dpp]_3)Cl_2$ (Figure 6.12). The paired probes are encased in a permeable biologically inert matrix, which can be imagined as a sensing platform. The reference fluorophore is chosen specifically to complement the analyte-sensitive probe to ensure that ratiometric analysis is done in either the excitation or emission modes.

The NP sensors based on this design include ions (H^+, Ca^{2+}, Mg^{2+}, Zn^{2+}, Fe^{3+}, and K^+), radicals (•OH radical), small molecules (O_2, singlet oxygen, hydrogen peroxide, H_2O_2), etc. Singlet oxygen is the lowest excited state of the dioxygen molecule, which is less stable than the normal triplet oxygen. Its lifetime in solution is in the microsecond range (3 μs in water to about 700 μs in deuterated benzene, C_6D_6). It undergoes several reactions with organic molecules (the ene reaction and the Diels–Alder reaction).

The Mg^{2+} PEBBLE sensor is an interesting case illustrating the relationship between design and cellular application. It achieved

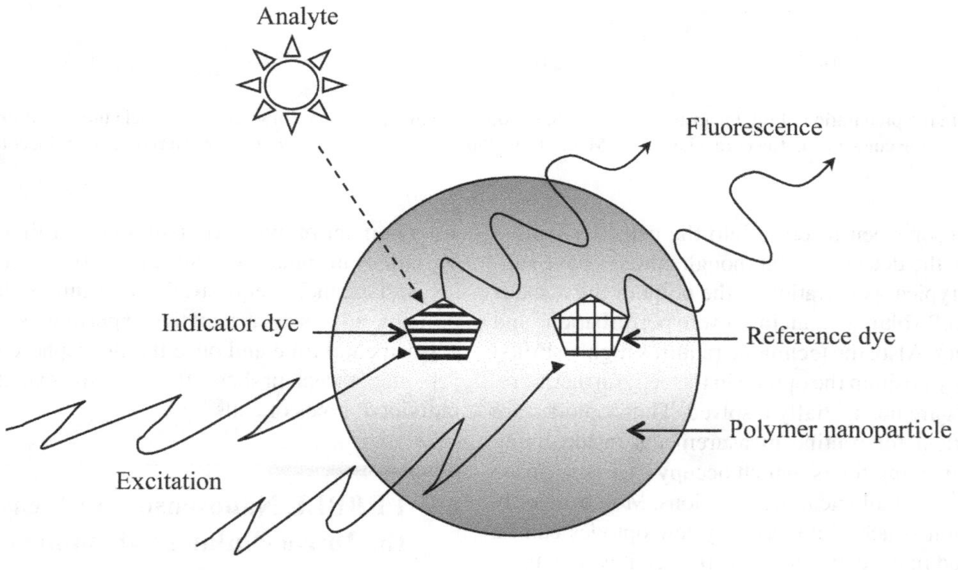

FIGURE 6.12 Polymer nanoparticle sensor using fluorescent indicator and reference dyes. (Lee, Y.-E. K. et al., *Annu. Rev. Anal. Chem.*, 2, 57, doi: 10.1146/annurev.anchem.1.031207.112823, 2009.)

both high selectivity and sensitivity by using the correct combination of NP matrix and dyes. The traditional measurement of magnesium ion concentrations in biological environments has experienced severe interference from calcium ions. Coumarin 343 is a small hydrophilic dye that is unable to penetrate the cell membrane by itself but is a very sensitive Mg^{2+} ion probe, having a greater selectivity for magnesium over calcium than any other commercially available probe. A hydrophilic material has a strong affinity for water. The contact angle of water with its surface is acute. It has a high wettability and water spreads evenly on it.

The Mg^{2+} PEBBLE sensor was constructed by encapsulating this hydrophilic dye and a commercial reference dye (Texas Red: $C_{31}H_{29}C1N_2O_7S_2$; sulforhodamine 101 acid chloride, a red fluorescent dye used in histology for staining cell specimens) in a hydrophilic polyacrylamide (PAM: $-CH_2CHCONH_2-$, white polyamide, related to acrylic acid) NP The linear response of these PEBBLEs to Mg^{2+} ions, in the range of 0.1–10 mM, is not affected even in the presence, simultaneously, of 1 mM calcium, 20 mM sodium, and 120 mM potassium, convincingly proving that this Mg^{2+} PEBBLE should serve as a reliable indicator of intracellular Mg^{2+} concentrations. These Mg^{2+} PEBBLEs were utilized to determine the role of Mg^{2+} inside human macrophage cells in the presence of invading *Salmonella* bacteria (a Gram-negative facultative rod-shaped bacterium, that is a major cause of food-borne illness throughout the world; it is transmitted to humans through consumption of contaminated food of animal origin, mainly meat, poultry, eggs, and milk) (Martin-Orozco et al. 2006).

Oxygen PEBBLE sensors with enhanced sensitivity and dynamic range were reported using the more sensitive platinum-based oxygen dyes, as well as reference dyes, embedded in a hydrophobic matrix, for example, organically modified silica (ormosil) (Koo et al. 2004) or poly(decyl methacrylate), PDMA (Cao et al. 2004). The hydrophobic matrix is more suitable for oxygen sensing than the hydrophilic one because of the higher oxygen solubility in the former matrix.

The embedded platinum(II) octaethylporphyrin ketone (PtOEPK), another oxygen-sensitive dye, exhibits infrared fluorescence, enabling these sensors to work even in human plasma samples, without being influenced by the notoriously high light scattering and autofluorescence (background fluorescence) of plasma. The oxygen PEBBLE sensors made of ormosil NPs were used for real-time imaging of oxygen inside live cells, that is, monitoring metabolic changes inside live cells of C6 glioma (a type of tumor that starts in the brain or spine; it is called a glioma because it arises from glial cells) (Koo et al. 2004).

Oxygen sensors have also been prepared by embedding ruthenium dyes into the polyelectrolyte (an electrolyte of high molecular weight) layers on commercial fluorescent NP surfaces (Guice et al. 2005) or into polymerized phospholipid vesicles (liposomes) (Cheng and Aspinwall 2006). Vesicles are phospholipid-bound globules that usually contain some kind of protein, for example, an enzyme; vesicles store the protein temporarily before it is needed. However, the sensitivities of these sensors were low, being only 60% for polyelectrolyte-coated NP sensors and 76% for sensors as polymerized liposomes (synthetic, microscopic, spherical sacs, composed of one or more concentric phospholipid bilayers and used in particular to deliver microscopic substances, such as drugs or DNA, to targeted body tissues).

A singlet oxygen (the diamagnetic form of molecular oxygen O_2) PEBBLE sensor was obtained by incorporating the singlet oxygen-sensitive 9,10-dimethyl anthracene ($C_{16}H_{14}$) or the singlet oxygen-insensitive octaethylporphyrin ($C_{36}H_{46}N_4$) within ormosil NPs (Cao et al. 2005). The NP matrix enhanced the selectivity of the indicator dye toward singlet oxygen as the matrix blocked the entry of short-lived polar reactive oxidative species (ROS: free radical molecules that contain oxygen in their molecular formula and that are formed during aerobic metabolism), like the hydroxyl radical and the superoxide anion (O_2^-) radicals. These nanoprobes were used for monitoring the singlet oxygen produced by "dynamic nanoplatforms (DNPs)" that were developed for photodynamic therapy (a treatment to kill cancer cells involving three key components: photosensitive chemicals, light in the form of laser beams controlled and directed at harmful cells, and tissue oxygen) (Reddy et al. 2006). *Nanoplatforms* are nanoscale structures designed as general platforms for multifunctional nanotechnology applications such as nanoporous oxide coatings and superpara-magnetic NPs.

A PEBBLE sensor for the OH radical was designed by covalently attaching the hydroxyl indicator dye, coumarin-3-carboxylic acid ($C_{10}H_6O_4$) onto the poly(acrylic acid) (PAA), $(C_3H_4O_2)_n$ NP surface, while encapsulating the reference dye deep inside it (King and Kopelman 2003). This design circumvents two potential problems with respect to the hydroxyl radical, which is the most reactive ROS: (i) the inability to penetrate significantly into any matrix without being destroyed; and (ii) the ability to oxidize and photobleach (the photochemical destruction of a fluorophore) most potential reference dyes.

How are PEBBLEs fabricated? These sensors are fabricated in a microemulsion (clear, stable, isotropic liquid mixture of oil, water, and surfactant, frequently in combination with a cosurfactant: a thermodynamically or kinetically stable liquid dispersion of an oil phase and a water phase), and consist of fluorescent indicators entrapped in a polyacrylamide ($-CH_2CHCONH_2-$) matrix. A generalized polymerization method is employed, permitting the production of sensors containing any hydrophilic dye or combination of dyes in the matrix. The PEBBLE matrix protects the fluorescent dye from interference by proteins, allowing reliable *in- vivo* calibrations of dyes.

The PEBBLE sensor platform provides a convenient cage that facilitates the entrapment of a number of sensing components, thus offering the ability to fabricate ratiometric sensors from complementary probes or a means to achieve complex sensing schemes with the entrapment of enzymes or ionophores (organic compounds that facilitate the transport of ions across the cell, for example, by increasing the permeability of biological membranes). The sensing components are enclosed in a permeable nano-environment, separate from the cellular environment, which offers access only to small analytes, while physically preventing interactions of larger proteins with the fluorescent probes. Ratiometric sensors reduce signal variations caused by changes in temperature, focal change shifts, or photobleaching. Moreover, the fabrication process confers homogeneous distribution of the dyes entrapped within the sensors, allowing spatial resolution and continuity to ratiometric measurements *in vitro*.

How are PEBBLEs introduced into the cells? This is an important issue. A range of methods to introduce PEBBLEs to a range of cell types have been developed, including functionalization with cell-penetrating peptides (CPPs), endocytosis (cellular process where cells absorb molecules or substances from outside the cell by surrounding and overwhelming them within the cell membrane), commercial agents for lipid transfection (the process of deliberately introducing nucleic acids into cells), cytochalasin D (a mycotoxin (fungal metabolite) that blocks cytoplasmic cleavage by blocking formation of contractile microfilament structures resulting in multinucleated cell formation, reversible inhibition of cell movement, and the induction of cellular extrusion; it specifically interferes with actin polymerization), picoinjection (the picoinjector is a robust device to add controlled volumes of reagent using electro-microfluidics at kilohertz rates), and bombardment with a gene gun (a device for injecting cells with genetic information at high pressure; the payload is an elemental particle of a heavy metal coated with plasmid DNA) bombardment (Webster et al. 2007). The *gene gun* offers high loading of PEBBLE sensors to adherent cells, while *CPP (cell-penetrating peptide) functionalization* is effective for cell cultures generally, with varying levels of sensor loading, depending on the cell type. *Lipid transfection* gives good results once the concentration of transfection reagent and sensor loading are established. *Picoinjection* allows location-specific PEBBLE delivery, although the loading is not as high as with other methods and this technique is therefore more suitable for larger cell types, where single-cell investigations are of interest. Pinocytosis (an endocytosis-like process by which liquid droplets are ingested by living cells by forming narrow channels through its membrane that pinch off into vesicles, and fuse with lysosomes that hydrolyze or break down contents) and cytochalasin D cell-permeable mycotoxin are unsuitable for preimplantation embryos, once the tight junctions have formed, because the space between the cells becomes very small, stopping the entry of nanosensors (an embryo is an unborn offspring of a multicellular organism in its early developmental stage). Both the method of delivery and cell type of interest are critical to achieving optimal intracellular delivery of NSs.

6.10 Quantum Dots as Fluorescent Labels

Labels or tags constitute the foundations of luminescence and fluorescence imaging and sensing. Labeling or tagging means to incorporate into a compound a substance ("label") that is readily detected, such as a radionuclide (an unstable form of a chemical element that radioactively decays, resulting in the emission of nuclear radiation), whereby its metabolism can be followed or its physical distribution detected. Metabolism is a broad term encompassing the sum-total of all chemical reactions taking place in the cells of a living organism that keep it alive and functioning, including the conversion of nutrients into energy for sustaining vital processes and activities, repair of cells, and elimination of wates. These biochemical reactions are divided into two classes: catabolism, the breaking down of complex molecules into simpler ones to release energy, and anabolism, the synthesis of compounds required by the cells, which uses energy. Aerobic means "with oxygen". Aerobic metabolism is an energy-generating system in the presence of oxygen using carbohydrates, lipids and proteins as energy sources. Anaerobic metabolism is oxygen independent metabolism.

What types of labels are popular at present? Presently, organic dyes (colored substances that have an affinity to the substrate to which they are being applied) are the most versatile molecular labels. But QDs have the potential to substitute for organic fluorophores (synthetic molecules such as Oregon Green and Texas Red) in many applications, which have so far been the sole province of these fluorophores. It will be meaningful to study the status of QD applications in this area *vis-à-vis* their organic counterparts. Indeed, one of the most exhilarating, yet also questionable advances, in label technology has been the appearance of QDs that have unique optical and chemical properties but complicated surface chemistry, as *in-vitro* and *in-vivo* fluorophores. The scenario has been comprehensively reviewed by Resch-Genger et al. (2008).

What are the required features of a good label? Vital properties of a good label are as follows. A suitable label should be conveniently excitable. Also, the biological matrix should not be concurrently excited and the label should be easily detectable with conventional instrumentation, it should be bright, having a high molar absorption coefficient (an index of how strongly a chemical species absorbs light at a given wavelength) at the excitation wavelength, along with a large fluorescence quantum yield (the ratio of photons absorbed to photons emitted), and it should be soluble in relevant buffer solutions, cell culture media (liquids or gels designed to support the growth of microorganisms or cells), or body fluids. It must show stability under the conditions of usage, the label should have necessary the functional groups necessary for site-specific labeling, its photophysics, such as semiconductor band picture of the solid-state photophysical properties, should be well known, and its availability at a reproducible quality must be ensured, and, where application demands, its deliverability into cells should be possible without any toxic effects, with suitability for multiplexing and compatibility for signal amplification are also sought-after features.

Typical QDs are core/shell (e.g., CdSe core with a ZnS shell) or core-only (e.g., CdTe) structures, functionalized with different coatings. Their properties depend to a considerable extent on particle synthesis and surface modification. Addition of the passivation shell often has the effect of producing a small red shift in absorption and emission relative to the core QD, due to tunneling of charge carriers into the shell. In comparison with organic dyes, QDs have the striking property of an absorption that gradually increases toward shorter wavelengths and a narrow emission band of a mostly symmetrical shape. The spectral position of absorption and emission are tunable by particle size (the so-called *quantum size effect*).

Labeling of biomolecules, such as peptides, proteins or oligonucleotides (short nucleic acid polymers), with a fluorophore requires suitable functional groups for covalent binding or for non-covalent attachment of the fluorophore. The advantage of organic dyes in this aspect is the market availability of a tool-box of functionalized dyes, in conjunction with established labeling protocols. *What is the situation with QDs?* There are no consensus methods for labeling biomolecules with QDs. The general principle for QD bio-functionalization is that, first, the QDs are made water dispersible and then they are bound to biomolecules. Binding is done by selecting from one of several methods, such as electrostatically, *via* biotin-avidin interactions, or by covalent cross-linking (e.g., carbodiimide-activated coupling between amine and carboxylic groups or maleimide-catalyzed coupling between amine and sulfhydryl groups or between aldehyde and hydrazide functions) or by binding to polyhistidine tags. *Avidin* is a biotin-binding protein produced in the oviducts of birds, reptiles, and amphibians, and deposited in the whites of their eggs. *Biotin* is a water-soluble B-complex vitamin (vitamin B_7). A *carbodiimide* or a methane-diimine is a functional group, with formula RN=C=NR, often used to activate carboxylic acids (oxoacids having the structure RC(=O)OH) toward amide formation; amides are amine (a group of organic compounds of nitrogen, such as ethylamine, $C_2H_5NH_2$, that may be considered to be ammonia derivatives) derivatives of carboxylic acids. *Maleimide* is an unsaturated imide (a functional group consisting of two carbonyl groups (C=O) bound to nitrogen) with the formula $H_2C_2(CO)_2NH$. A *polyhistidine-tag* is an amino acid motif in proteins that consists of at least five histidine (His) residues, often at the N- or C-terminus of the protein.

Alternatively, ligands present during synthesis are exchanged for biomolecules containing active groups on their surface. The latter strategy works well for labeling oligonucleotides. A *ligand*, or *complexing agent,* is a polar molecule or ion bonded to a central metal ion.

Currently, only a few standard protocols for labeling biomolecules with QDs are available, and the choice of suitable coupling chemistries depends on surface functionalization. It is difficult to prescribe general principles because QD surfaces are unique, depending on their preparation. Accordingly, for users of commercial QDs, knowledge of surface functionalization is essential.

QDs are promising in microarrays (multiplex lab-on-a-chip devices), immunoassays (laboratory techniques that make use of the binding between an antigen and its homologous antibody in order to identify and quantify), and fluorescence *in situ* hybridization (the use of labeled nucleic acid sequence probes for the visualization of specific DNA or RNA sequences). Resolution of single nucleotide polymorphisms (SNPs) on cDNA arrays has been achieved for mutation in the human oncogene p53. Single nucleotide polymorphism (SNP) is a genetic sequence variation which affects only one of the nucleotides (adenine A, thymine T, cytosine C or guanine G) in a segment of a DNA molecule, e.g. it may involve replacement of the nucleotide cytosine with thymine in a stretch of DNA. Polymorphism is the occurrence of two or more different forms called phenotypes in a population of a species. A polymorph is one of the several forms in which an organism is found. An oncogene is a sequence of DNA that has been altered or mutated from its original form, the proto-oncogene. *Onco-,* from the Greek *onkos*, meaning "bulk," or "mass," refers to the tumor-causing ability of the oncogene. The p53 gene is a tumor-suppressor gene, its activity halting the formation of tumors. Mutations in p53 are found in most tumor types. A multiplexed analysis for the hepatitis B and hepatitis C viruses has been carried out by concurrently using QDs with peak luminescence at 566 and 668 nm (Ulgar and Krull 2009). Hepatitis B and C are viral infections that cause inflammation of the liver. Different viruses are known to cause these infections. Hepatitis B virus (HBV) is a partially double-stranded circular DNA virus while Hepatitis C virus (HCV) is an enveloped positive sense single-stranded RNA (+ssRNA) virus of Flaviviridae family. Here, the apparent sensitivity of QD arrays may be less than observed with organic fluorophores, because of their poor excitation efficiency. This shortcoming is overcome by using blue-shifted (a decrease in the wavelength of radiation) source and/or red-shifted (an increase in the wavelength of radiation) LEDs, wherever possible.

The ability to track biomolecules within their native environment, e.g., on the cell surface or inside the cells, is a key characteristic for any fluorescent label and a prerequisite for assessing molecular function *in vivo. How successful are the QDs in extracellular (outside the cell) and intracellular (within a cell) labeling?* External labeling of cells with QDs is relatively straightforward, but intracellular delivery is riddled with complexity. Extracellular targeting with QDs is frequently practiced. This is usually accomplished through QD functionalization with specific antibodies to image cell-surface receptors or *via* biotin ligase-catalyzed biotinylation (the process of covalently attaching biotin to a protein, nucleic acid, or other molecule) in combination with streptavidin-functionalized QDs. Whole-cell labeling with QDs has been carried out through microinjection (a simple mechanical process in which a needle, roughly 0.5–5 μm in diameter, penetrates the cell membrane to insert substances), electroporation (also called electro-permeabilization, is a mechanical method used to introduce polar molecules into a host cell through the cell membrane, using an externally applied electrical field to increase the permeability of the cell plasma membrane; microbiologists have long been using this method to temporarily punch holes in cell membranes and hence ferry drugs into the cell) or nonspecific or receptor-mediated endocytosis (a process by which cells internalize molecules or viruses by engulfing them without passing through the cell membrane; it depends on the interaction of that molecule with a specific binding protein in the cell membrane called a receptor). The labeling specificity and efficiency is improved by using functionalized QDs.

In a milestone study (Dubertret et al. 2002), QDs encapsulated in phospholipid micelles (molecular aggregates of colloidal dimensions) were used to label individual blastomeres (the two cells produced by cleavage of a zygote or fertilized ovum during earlydevelopment) in *Xenopus* (an African clawed toad) embryos. These encapsulated QDs were stable *in vivo*, did not aggregate, and were able to label all cell types in the embryo. At the levels required for fluorescence visualization (2×10^9/cell), the QD-micelles were nontoxic to the cells, but, at concentrations of 5×10^9/cell, produced aberrations (deviations from the normal development). The QDs were restricted to the injected cell and its offspring, though inadvertent translocation to the nucleus was observed at a particular stage in the development of the embryo. Another group labeling *Dictyostelium discoideum* (a soil-living amoeba; "slime mold") mentioned that cell labeling was possible for over a week, and that QD labeling had no discernible effects on cell morphology or physiology. Differently colored QDs could also be used to label various populations in order to scrutinize the effect of starvation on *D. discoideum* development. These cells could be tracked for long duration with no conspicuous loss of fluorescence.

Zebrafish (a small freshwater tropical fish with barred, zebra-like markings) embryo blastomeres labeled with QDs and co-injected with cyan fluorescent protein (CFP), a traditionally used lineage marker (a standard cocktail of antibodies designed to remove mature hematopoietic cells from a sample), showed inheritance of QDs to daughter cells in most cases, although some cells displaying CFP fluorescence displayed no QD fluorescence; this was interpreted to be due to aggregation of QDs, causing unequal inheritance by daughter cells (the cells that result from the reproductive division of one cell during mitosis or meiosis). This is a documented problem, along with fluorescence loss and instability, in the QD structure in biological solutions.

What problems arise during intracellular labeling? Generally, owing to their size, the intracellular delivery of QDs is not effortless; compared with organic dyes, the state-of-the-art of delivery of QDs into cells and internal labeling strategies are lagging behind. There is no general procedure to achieve delivery, and individual solutions need to be empirically established. Several methods have been used to deliver QDs to the cytoplasm for staining of intracellular structures, but are successful only to some extent. Microinjection techniques have been applied to label *Xenopus* and zebrafish embryos, causing pan-cytoplasmic labeling, but this is a very arduous task, prohibiting high-volume analysis. QD uptake into cells, *via* endocytic and non-endocytic pathways, has also been corroborated, but results in only endosomal localization (endosomes are membrane-enclosed vesicles). Two novel approaches have shown pan-cytoplasmic labeling, by conjugation with Tat protein (trans-activator of transcription protein produced by a lentivirus, such as HIV, within infected cells that greatly increases the rate of viral transcription and replication and that is also secreted extracellularly, where it plays a role in increasing viral replication in newly infected cells and in enhancing the susceptibility of T cells to infection), and by encapsulation in cholesterol (a waxy steroid metabolite)-bearing pullulan (CHP), modified with amine groups; pullulan is a natural water-soluble polysaccharide polymer. Coating with a silica shell may be helpful.

Labeling of specific intracellular structures outside endocytosed vesicles (the vesicle [membrane-bound bubble within the cell] which is formed during the process of endocytosis and which contains substances being imported from outside the cell. *Endocytosis* is the process by which cells absorb molecules [such as proteins] by engulfing them) or imaging of cellular reactions in the cytoplasm or the nucleus with QDs necessitates sophisticated tools. Positively charged peptide (a molecule consisting of two or more amino acids)-transduction domains such as Tat (peptide from the cationic domain of the HIV-1 Tat protein), polyarginine (cationic peptide that promotes transfection efficiency at an appropriate concentration), polylysine (a homopolymer of L-lysine, an essential amino acid), and other specifically designed CPPs can be coated on to QDs for their delivery into cells. The adaptation of other recently developed cell penetrating agents, such as streptaphage, a synthetic ligand based on an *N*-alkyl derivative of 3β-cholesterylamine ($C_{27}H_{47}N$) that efficiently delivers streptavidin to mammalian cells by promoting non-covalent interactions, or polyproline (a sequence of several proline residues in a polypeptide or protein; it tends to generate an anomalous left-handed helix conformation) systems, equipped with cationic and hydrophobic (incapable of dissolving in water) moieties (functional groups) for QD delivery, is not yet unequivocally confirmed.

How are QDs used for labeling proteins? Labeling of F-actin fibers (filamentous actin; actin is a protein found in muscle that, together with myosin, is responsible for muscle contraction) showed that QDs could be used for labeling proteins where preservation of enzyme activity was desirable. Streptavidin-coated QDs were used to label individual isolated biotinylated (being combined with a biotin tag) filamentous-actin (F-actin) fibers. But, compared with Alexa Fluor 488 (an organic fluorophore), a smaller proportion of labeled filaments were motile (able to move spontaneously and independently using metabolic energy) mitochondria (autonomous membrane-bound cell organelles that generate energy to power the biochemical reactions of the cell). Intracellular labeling of these filaments is also practicable. In addition, QDs have been used to label mortalin (a member of the heat-shock protein 70 or Hsp70 family of proteins), and *p*-glycoprotein (a protein with covalently attached sugar units) molecules, which are important in tumor cells. Labeling with QDs gave much more photostability than with organic dyes, with a 420-fold increase in photostability relative to Alexa Fluor 488. This was utilized to image three-dimensionally the localization of *p*-glycoprotein, with the long fluorescence lifetime allowing successive *z*-sections to be imaged.

What is the application of QDs in multicolor labeling? Owing to the optical properties of organic dyes, their suitability for multicolor signaling at single-wavelength excitation is limited. QDs are the perfect contenders for spectral multiplexing at a single excitation wavelength because of their great flexibility in excitation and their very narrow and symmetrical emission bands, which simplify color discrimination. Multiple color labeling of different intracellular structures has been reported (Jamieson et al. 2007). Simultaneous labeling of nuclear structures and actin filaments with QDs of two different colors was carried out but variable labeling of nuclear structures was observed. Labeling of mitochondria and nuclear structures was carried out, producing distinct red labeling of the nucleus and green labeling of the mitochondria. Single-color labeling of *Her2* (gene encoding human epidermal growth factor receptor 2; a gene that sends control signals to cells, directing them to grow, divide, and make repairs. A healthy breast cell has two copies of the *Her2* gene. Some kinds of breast cancer start when a breast cell has more than two copies of that gene, and those copies begin overproducing the HER2 protein, so that the affected cells grow and divide much too quickly) has also been shown to be possible and worthy of attention because expression of this gene is an extrapolative and prognostic (an advance indication or portent of a future event) marker for breast cancer.

What are microenvironmental effects on dyes and QDs? Organic dyes, like fluorescein ($C_2OH_{12}O_5$) and tetramethyl rhodamine isothiocyanate (TRITC), $C_{25}H_{21}N_3O_3S$, and the majority of NIR (near-infrared) fluorophores have poor photostability. For QDs, the microenvironment effect on spectroscopic features is mainly governed by the accessibility of the core surface. In turn, this depends on the ligand (and the strength of its binding to the QD surface) and the shell quality. Provided that no ligand desorption occurs, properly shelled QDs are minimally sensitive to microenvironment polarity. Also, QD emission is scarcely responsive to viscosity, contrary to that of many organic dyes, and QDs are less susceptible to aggregation-induced fluorescence quenching (a decrease in the fluorescence intensity of a given substance). Limited dye photostability can still hamper microscopic applications requiring high-excitation light intensities in the UV-visible light region or requiring long-term imaging.

In contrast, adequately surface passivated QDs display excellent thermal and photochemical (the *effect of light in causing or modifying chemical reactions*) stability. Photo-oxidation (*oxidation* induced by light or some other form of radiant energy) is almost completely suppressed for relevant time intervals as a consequence of their additional inorganic surface layers and shielding of the core material. This is a prominent advantage over organic fluorophores for imaging applications that use intense laser excitation sources or for long-term imaging. Toxicity of organic dyes is not a major problem. In the case of QDs, the cytotoxicity of elements such as cadmium, which is present in many of these nanocrystals, is well known (cytotoxicity is the ability of certain chemicals to destroy living cells). Thus, it is critical to recognize whether these cytotoxic substances can leak out of the QD particles over time, upon illumination or oxidation, in addition to whether ligands or coatings are cytotoxic. Cytotoxicity of QDs was observed in some reports in the literature, whereas, in others, it was not found. In cases where cytotoxicity was observed, it was usually attributed to leaking of Cd^{2+}, cytotoxic surface ligands, and/or NP aggregation.

Apart from their unique potential for all bioanalytical applications requiring or benefiting from multiplexing, QDs have a bright future in NIR fluorescence *in-vivo* imaging, which requires labels that exhibit high fluorescence quantum yields in the 650–900 nm window, have adequate stability, good aqueous solubility, and low cytotoxicity in conjunction with large two-photon action cross sections, as needed for deep-tissue imaging. Moreover, QDs are attractive candidates for the development of multifunctional composite materials for the combination of two or more biomedical imaging modalities, like NIR fluorescence-magnetic resonance imaging. Yet, the routine use of QDs is, at present, strongly limited by the very small number of commercial systems available and the scanty data on their reproducibility and comparability, as well as on their potential for quantification.

6.11 Quantum Dot FRET-Based Probes

What is the principle of FRET? FRET involves the nonradiative transfer of fluorescence energy from an excited donor particle to an acceptor particle *via* dipole-dipole interaction through space. The following conditions must be satisfied by molecular FRET donor-acceptor pairs: (i) overlapping of the fluorescence emission spectrum of the donor and absorption spectrum of the acceptor; and (ii) the distance between the donor and the acceptor must be smaller than a critical radius, known as the Förster radius or distance, typically 10–100 Å. This leads to a reduction in the emission and excited state lifetime of the donor, and an increase in the emission intensity of the acceptor.

The rate of transfer of energy, dependent on the spectral overlap between the donor and acceptor, the relative orientation of donor-acceptor transition dipoles, and the donor-acceptor distance, is expressed as

$$k_t = \tau_D^{-1}\left(\frac{R_0}{R}\right)^6 \tag{6.9}$$

where

τ_D is the measured lifetime of the donor in the absence of the acceptor

R is the distance between the donor and the acceptor

R_0 is the Föster distance, the distance at which efficiency of energy transfer is 50%

It is given by the equation:

$$R_0 = \left\{\frac{3000}{4\pi N \mid A \mid_{1/2}}\right\}^{1/3} \tag{6.10}$$

where

N is Avogadro's number (=6.022×10^{23} mol^{-1})

$|A|_{1/2}$ is the concentration of acceptor at which the energy transfer efficiency E is 50%

For covalently bound donor-acceptor pair, energy transfer efficiency is written as

$$E = \frac{R_0^6}{R_0^6 + R^6} \tag{6.11}$$

It is measured experimentally by monitoring the changes in donor and/or acceptor fluorescence intensities or changes in fluorescence lifetimes of the fluorophores; the fluorophore or fluorochrome is a type of fluorescent dye.

FRET is suited to measuring changes in relative distances, instead of absolute distances, making it appropriate for measuring protein conformational changes (change in shape of a macromolecule in response to its environment or other factors; each possible shape is called a conformation, and a transition between them is called a conformational change), monitoring protein interactions and assaying enzyme activity (determining enzyme kinetics). Several groups have attempted to use QDs in FRET technologies, particularly when conjugated to biological molecules, including antibodies, for use in immunoassays. These are highly sensitive and specific chemical tests conducted for detecting or quantifying a specific substance, the analyte, in a blood or body fluid sample, using an immunological reaction. Their high specificity arises from the use of antibodies and purified antigens as reagents. An antibody is a protein (immunoglobulin) produced by B-lymphocytes (immune cells) in response to stimulation by an antigen. Immunoassays measure the formation of antibody–antigen complexes and detect them *via* an indicator reaction.

How are QDs used in FRET? QD-FRET sensors are based on the interactions between QDs serving as donors and molecular fluorophores attached to the QD surface, acting as fluorescent acceptors (Figure 6.13). QDs have been exploited as FRET donors, with organic dyes as acceptors, with the QD emission size-tuned to match the absorption band of the acceptor dye. Through the FRET mechanism, QDs respond indirectly to environmental changes without any chemical interaction that could affect their photophysical properties and degrade their brightness.

Several QD-based FRET strategies have been developed for detecting nucleic acid, proteins, and other macromolecules. Owing to the free choice of the QD excitation wavelength, cross talk is circumvented in such FRET pairs. However, the distance dependence of FRET means that both the size of the QD itself and that of the surface coating change the FRET efficiency (the percentage of the excitation photons that contribute to FRET);

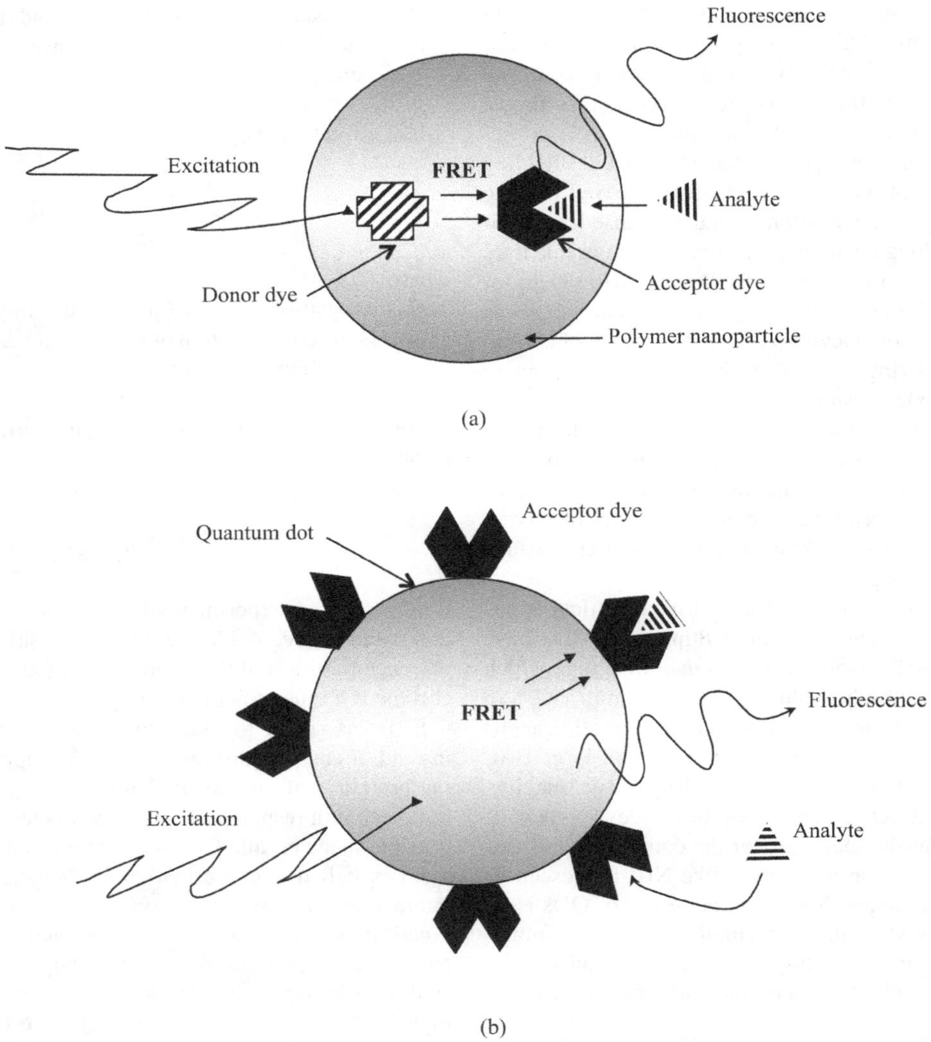

FIGURE 6.13 FRET process using (a) dyes and (b) quantum dot and dye. (Lee, Y.-E. K. et al., *Annu. Rev. Anal. Chem.*. 2. 57. doi: 10.1146/annurev.anchem.1.031207.112823. 20093).

this typically renders QD-FRET less efficient, as compared with FRET with organic dyes. Owing to the substantial size of QDs, this limitation can be partly overcome by increasing the number of adjoining small organic acceptor dyes.

Example 6.5

For tryptophan (donor)/dansyl chloride (acceptor) FRET pair, the Förster radius is 2.1 nm. Calculate the FRET efficiency at a distance of (i) 3 nm and (ii) 4 nm.

$$E = \frac{R_0^6}{R_0^6 + R^6} \tag{6.12}$$

gives

$$\text{(i)}\quad E = \frac{(2.1)^6}{2.1^6 + 3^6} \times 100\% = \frac{85.766}{85.766 + 729} \times 100\% \tag{6.13}$$
$$= 10.53\% \quad \text{at } R = 3\,\text{nm}$$

and

$$\text{(ii)}\quad E = \frac{(2.1)^6}{2.1^6 + 3^6} \times 100\% = \frac{85.766}{85.766 + 4096} \times 100\% \tag{6.14}$$
$$= 2.051\% \quad \text{at } R = 4\,\text{nm}$$

Example 6.6

In a FRET experiment, the Förster radius R_0 was 2 nm and the fluorescence lifetime was 5×10^{-9} s. What is the rate of energy transfer at (a) $R = 10$ nm and (b) 5 nm? Comment on the variation.

(a) From Equation 6.9 at $R = 10$ nm,

$$k_t = \tau_D^{-1}\left(\frac{R_0}{R}\right)^6 = \frac{1}{5 \times 10^{-9}}\left(\frac{2}{10}\right)^6 = \frac{6.4 \times 10^{-5}}{5 \times 10^{-9}} \tag{6.15}$$
$$= 1.28 \times 10^4\,\text{s}^{-1}$$

(b) At $R=5$ nm,

$$k_t = \tau_D^{-1}\left(\frac{R_0}{R}\right)^6 = \frac{1}{5\times10^{-9}}\left(\frac{2}{5}\right)^6 = \frac{4.096\times10^{-3}}{5\times10^{-9}} = 8.192\times10^5 \quad (6.16)$$

As R decreases from 10 to 5 nm, by a factor of 2, the rate of energy transfer increases by a factor of $8.192\times10^5/1.28\times10^4=64$.

6.11.1 QD-FRET Protein Sensor

QD-FRET has been used for monitoring protein interactions in the Holliday Junction (McKinney et al. 2003; Hohng and Ha 2005; Jamieson et al. 2007). The Holliday Junction is an intermediate in the recombination of DNA, that undergoes conformational change on addition of Mg^{2+} ions. QD585 was used as a donor on one arm of the DNA, and Cy5 as an acceptor on a perpendicular arm. Movement of the arms on addition of Mg^{2+} was detected as a change in the emission of both donor and acceptor. However, the changes were detected with noticeably less efficiency than with the equivalent Cy3/Cy5 FRET. Cy3 and Cy5 are reactive, water-soluble fluorescent dyes of the cyanine dye family. Cy3 dyes are red (~550 nm excitation, ~570 nm emission, and hence appear red), whereas Cy5 is fluorescent in the far red region (~650/670 nm) but absorbs in the orange region (~649 nm). These dyes are small organic molecules, and are typically conjugated to proteins *via* primary amines, that is, lysine residues.

Generally, the disadvantages of organic dyes and fluorescent proteins for FRET applications have their foundations in cross talk. This cross talk results from direct acceptor excitation due to the relatively broad absorption bands of these fluorophores. Additional difficulties are met in spectral discrimination of the fluorescence emission, owing to their relatively broad emission bands, their small Stokes shifts (the difference (in wavelength or frequency units) between positions of the band maxima of the absorption and emission spectra, such as fluorescence and Raman spectra of the same electronic transition), and the "red tails" of the emission spectra in the case of dyes like fluoresceins ($C_{20}H_{12}O_5$, yellow crystals), rhodamines (red fluorescent dyes: supplements to fluoresceins, as they offer longer wavelength emission maxima and provide opportunities for multicolor labeling or staining), *boron-dipyrromethene* (BODIPY, a class of fluorescent dyes), and cyanines. This renders tedious correction of measured signals mandatory.

6.11.2 QD-FRET Protease Sensor

Shi et al. (2006) synthesized a QD-FRET-based protease (enzyme that catalyzes the hydrolytic breakdown of proteins) sensor to measure the protease activities *in vivo.* QDs coated with unlabeled cell-binding arginine-glycine-aspartic acid-cysteine (RGDC) peptide (a molecule composed of two or more amino acids joined through amide formation, which are structurally like proteins, but smaller) molecules emit green light. When capped with rhodamine-labeled RGDC molecules, the emission color changes to orange through FRET interactions between the QDs and rhodamine molecules (serving as acceptors). Upon enzymatic cleavage of the RGDC peptide molecules, the rhodamine (acceptor) molecules no longer provided an efficient energy transfer channel to the QDs. Consequently, the emission color of the QDs changed back to the original green (Figure 6.14).

Through a unique cysteine ($C_3H_7N_1O_2S_1$), approximately 48 copies of the rhodamine-labeled collagenase (an enzyme that breaks the peptide bonds in collagen, a fibrous protein used to connect and support other body tissues) substrate RGDC were coated onto each QD (Zhou and Ghosh 2006). The emission wavelength of QD at 545 nm matched the absorption wavelength of rhodamine. Following peptide cleavage by collagenase, QD fluorescence at 545 nm increased and rhodamine fluorescence at 590 nm decreased. Ratiometric changes in fluorescence were employed in a quantitative study of protease activity in HTB126 cancer cell lines (human mammary gland; disease: carcinoma) in which collagenase is aberrantly expressed. These experiments showed that real-time measurements of protease activity were possible by using QD-FRET-based ratiometric sensors. Since activities of different proteases are measurable by changing the peptide sequences, more proteolytic enzyme (enzyme that breaks the long chainlike molecules of proteins into shorter fragments, peptides, and eventually into their components, amino acids) assays are likely to be performed, using QD-FRET-based sensors Furthermore, there are opportunities for the introduction of multiplexing (sending multiple signals or streams of information on a carrier at the same time in the form of a single, complex signal and then recovering them) and high-throughput formats.

6.11.3 QD-FRET Maltose Sensor

As QDs have broad absorption spectra and narrow emission spectra, they are usually employed as fluorescence donors. The fluorescence acceptor is often an appropriately labeled peptide (any of various natural or synthetic compounds containing two or more amino acids linked by the carboxyl group of one amino acid to the amino group of another).

A FRET biosensor based on a QD-protein/peptide conjugate was developed by Medintz et al. (2003) and Clapp et al. (2004; 2005) for detecting maltose (malt sugar, a white crystalline sugar, $C_{12}H_{22}O_{11}$, a disaccharide formed from two units of glucose joined with an $\alpha(1\rightarrow4)$ bond). Multiple copies of *Escherichia coli* maltose-binding protein (MBP) (a part of the maltose/maltodextrin system of the human gut bacterium *E. coli,* responsible for the uptake and efficient catabolism of maltodextrins) coordinated to each QD by a C-terminal oligohistidine segment (a construct in which an *oligohistidine* tag is placed at the C *terminus*) and functioned as sugar receptors. Histidine (abbreviated as His or H), $C_6H_9N_3O_2$, is an essential amino acid in humans.

Sensors were self-assembled in solution in a controlled fashion. In one configuration, a β-cyclodextrin-QSY9 dark quencher conjugate, bound in the MBP saccharide-binding site, results in FRET quenching of QD photoluminescence. Cyclodextrins are a family of compounds made up of sugar molecules bound together in a ring (cyclic oligosaccharides), produced from starch by enzymatic conversion. QSY9 is carboxylic acid, succinimidyl ester, a fluorescent dye used to stain biological specimens.

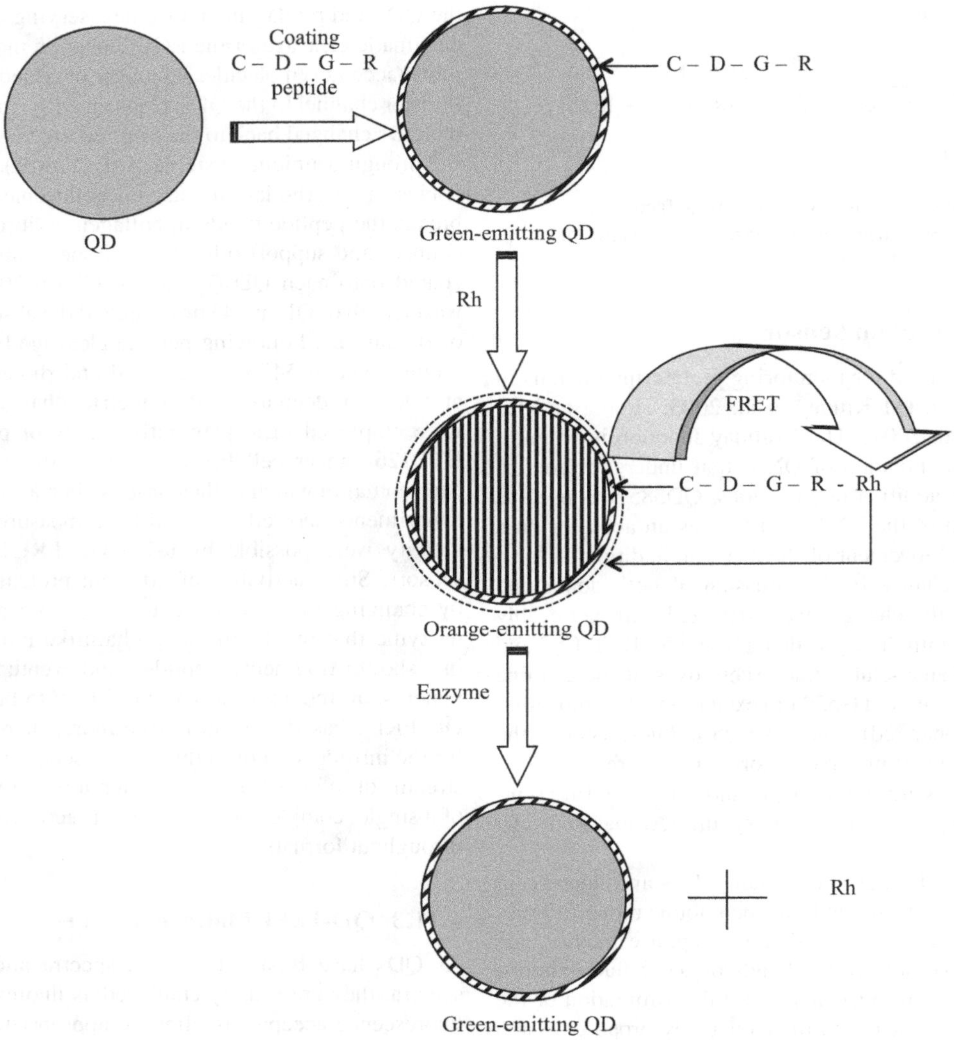

FIGURE 6.14 Scheme of QD protease biosensor. (Shi, L., et al., *J. Am. Chem. Soc.*, 128, 10378, 2006.)

Added maltose displaces the β-cyclodextrin-QSY9 and increases the fluorescence of QD in proportion to the maltose concentration.

To overcome inherent QD donor-acceptor distance limitations, the sensors were modified to operate on a two-step FRET mechanism by labeling the MBP with another fluorescence dye (Zhou and Ghosh 2006). The organic dye QSY-9 (absorption ~560 nm)-labeled β-CD (β-cyclodextrin) was used to bind to MBP attached to the QD560 (emission ~ 560 nm) surface, such that the fluorescence was quenched (Figure 6.15a). With the addition of free maltose, displacement of the β-CD-QSY9 occurred. As a consequence, there was fluorescence recovery from the QD. The optimized sensor contained ten copies of β-CD-QSY9 bound to the QD complex. Here, 75% of the QD fluorescence was quenched. Upon complete displacement of β-CD-QSY9 with maltose, a threefold increase in fluorescence was observed.

These applications are problematical due to the uncertain distance between QD and acceptor. The FRET efficiency may increase through electron transfer between QDs and organic dyes. To surmount these limitations, a maltose sensor was designed in which ten copies of Cy3 (absorption 556 nm and emission 570 nm) labeled MBP were first incorporated on the QD530 (emission 530 nm) surface Figure 6.15b). The second step involved binding of the Cy3.5 (absorption ~575 nm and emission ~595 nm) labeled β-CD. Fluorescence energy was first transferred from the QD530 to MBP-Cy3. Subsequently, the emission energy of Cy3 was transferred to β-CD-Cy3.5. The displacement of β-CD-Cy3.5 from MBP by maltose produced an associated fluorescence fall from Cy3.5 and a rise from Cy3. This geometry permitted sufficient energy to transfer to Cy3. However, it minimized direct energy transfer to Cy3.5. The abovementioned approaches showed how appropriately designed QD complexes, with peptide immobilization tags, are useful in determining small molecule concentrations in the 100 nM–10 μM range.

6.11.4 Sensor for Determining the Dissociation Constant (K_d) between Rev and RRE

Rev is a vital HIV-1 regulatory protein within the *env* gene (a *gene* that encodes a protein precursor for the envelope proteins, found in the retroviral genome) of HIV-1 RNA genome (the entirety of the organism's hereditary information). The HIV-1 regulatory protein Rev is expressed early in the virus life cycle and thus may be an important target for the immune control of

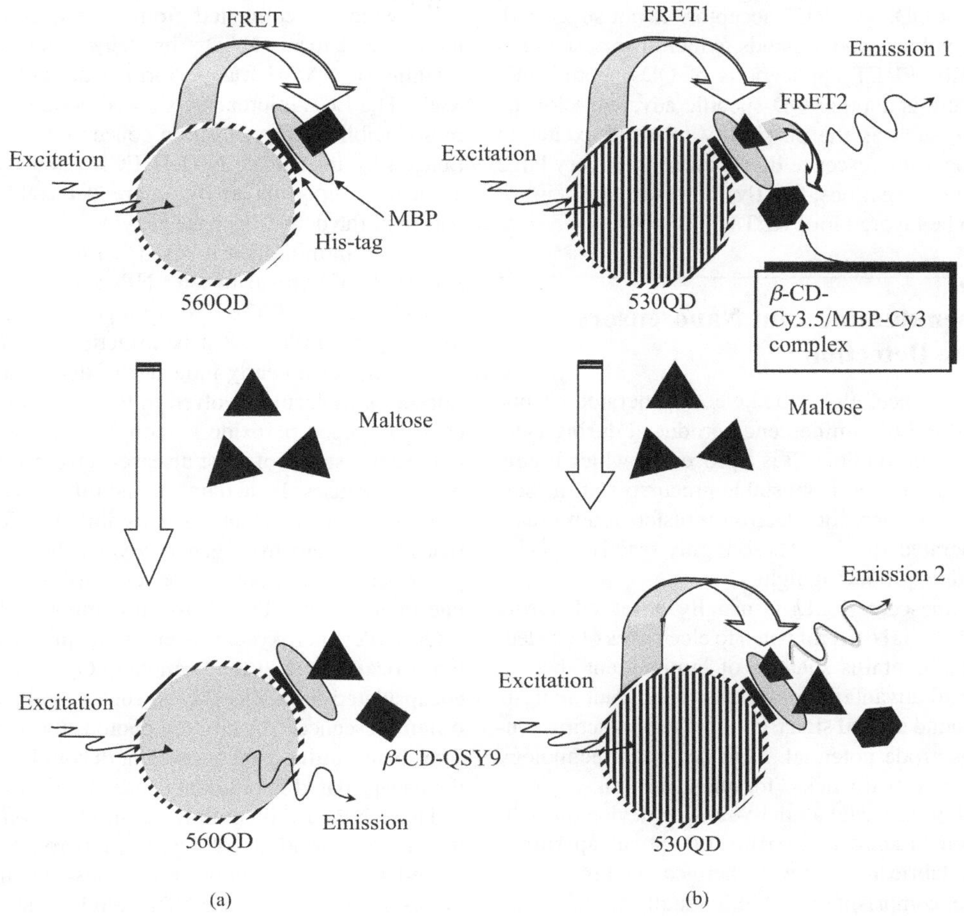

FIGURE 6.15 QD-based maltose nanosensor: (a) *above*: β-CD-QSY-9/MBP complex is bound to QD through a peptide His-tag. QD fluorescence is quenched by QSY-9. *below*: The displacement of β-CD-QSY-9 with maltose results in the recovery of QD fluorescence; (b) *above*: β-CD-Cy3.5/MBP-Cy3 complex is bound to QD through a peptide His-tag. Cy3.5 fluorescence is emitted through a two-step FRET process; *below*: the displacement of p-CD-Cy3.5 with maltose results in Cy3 fluorescence emission. (Zhou, M. and Ghosh, I., *Pept. Sci.*, 88, 325, doi: 10.1002/bip, 2006.)

HIV-1 infection and for effective vaccines. Zhang and Johnson (2006) used QDs instead of organic dyes to develop a sensor to determine the dissociation constant (K_d) between Rev and RRE (the *Rev* response element is a region in the RNA molecule of the HIV *env* gene), based on FRET. The QDs (emission ~605 nm) were functionalized with streptavidin, a tetrameric protein that binds very tightly to the small molecule, biotin ($C_{10}H_{16}N_2O_3S$). The streptavidin-coated QDs functioned as both a nano-scaffold and a FRET donor in this QD-based NS. Streptavidin was bound to biotin-labeled RRE. Nanoscaffolding is a medical process used to regrow tissue and bone, including limbs and organs.

The 605QD is an excellent energy donor with Cy5 for several reasons: (i) absence of cross talk between the emission spectra of 605QD and that of Cy5; (ii) absence of direct excitation of Cy5 at the wavelength of 488 nm; (iii) 605QD has a high quantum yield (-0.6), whereas Cy5 has a high extinction coefficient (-250,000 M^{-1} cm^{-1}); and (iv) a single 605QD can efficiently couple to multiple Cy5-labeled Rev-RRE complexes around its surface, enabling efficient FRET even at distances approaching $2 \times R_0$ for this 605QD/Cy5 FRET pair.

The RRE-functionalized QDs effectively complexed multiple copies of a Cy5-labeled 17-mer peptide (a peptide with 17 amino acid residues) containing the arginine ($C_6H_{14}N_4O_2$)-rich region of Rev; arginine also called L-arginine, symbolized as Arg or R, is an amino acid (2-amino-5-guanidinovaleric acid) frequently found at the active sites in proteins and enzymes.

This reduced the QD fluorescence and increased Cy5 fluorescence. QD fluorescence was recovered by titrating unlabeled Rev peptide into the FRET complex. The K_d of the Rev-RRE complex was estimated by monitoring the fluorescence change of QD and Cy5 in this titration process. Identical results were obtained from the titration curves from the two fluorophores (loss in FRET emission and increase in non-FRET emission), giving K_d ~ 11–14 nM. This afforded a simple method for monitoring Rev–RRE interaction and possibly any peptide/RNA, peptide/DNA, or peptide/protein interaction.

This QD-based NS offers a convenient approach for sensitive Rev–RRE interaction assay due to its exceptional characteristics of noninterference in the native interaction of Rev with RRE, improved FRET efficiency, high sensitivity, and FRET-related two-parameter detection, using the instrumentation of standard fluorescence spectrometer and available streptavidin-coated QDs. It is used to study the effect of an inhibitor on Rev–RRE interaction, and likely to have more applications in the development of new drugs against HIV-1 infection.

The application of QDs as FRET acceptors is not suggested, because of their broad absorption bands, which favor excitation cross talk. Generally, FRET applications of QDs should only be considered if there is another QD-specific advantage for the system in question, such as the possibility to avoid excitation cross talk, their longer fluorescence lifetimes, or their very large two-photon action cross sections. Mostly, fluorescent proteins or organic dyes are to be favored for FRET.

6.12 Electrochemiluminescent Nanosensors for Remote Detection

Electrochemiluminescence, also called electrogenerated chemiluminescence, is a kind of luminescence produced during electrochemical reactions in solutions. It is a process in which highly reactive species are generated from stable precursors at the surface of an electrode by energetic electron transfer (redox) reactions of electrogenerated species. These highly reactive species react with one another, producing light.

Electrochemiluminescent (ECL) is usually observed during the application of potential (several volts) to electrodes of an electrochemical cell that contains solution of luminescent species. Combining analytical advantages of chemiluminescent analysis (absence of background optical signal) with ease of reaction control by applying electrode potential, ECL detection technology promises scientists new "yardsticks" for quantification.

Chovin et al. (2006) reported a micro-sensing device for ECL imaging through an ordered microarray of optical apertures. This microdevice, fabricated by wet chemical etching of an optical fiber bundle, comprising 6000 individually clad 3–4 μm diameter optical fibers, consisted of a systematically planned array of etched optical fiber cores, each tip of which was surrounded by a ring-shaped gold nanoelectrode (Figure 6.16). By this method of fabrication, the initial architecture of the optical fiber bundle was retained and thus the microarray maintained its imaging properties. An electrophoretic paint was deposited cathodically onto the surface of the nanotip array. The electrophoretic paint is a coating formed on a cathode as a result of electrophoresis (the migration of charged colloidal particles or molecules through a solution under the influence of an applied electric field) and coagulation of colloidal particles. By carefully controlling the deposition time of the paint, the array was covered with insulating paint. Heating of the resin-coated array resulted in a shrinking of the resin (clear to translucent, solid, or semisolid viscous substance) film and exposure of the metal tips of the fiber bundle. To permit light transmission through the etched fiber cores, the exposed gold layer was removed by reacting in a gold-etching solution. The gold film accomplished two functions: it served to confine light to the tip apex and it also acted as the electrode material to perform electrochemical reactions. The high-density microarray comprised 6000 NSs. Each sensor was a sub-wavelength aperture formed at the apex of an etched fiber core and surrounded by a gold nanoring electrode. Such nanostructured microdevices integrated ECL-light generation, collection, and imaging in a microarray format.

The microdevice was applied to the ECL detection of NADH. "NADH" is an abbreviation for the reduced form of NAD (nicotinamide adenine dinucleotide), a coenzyme found in all living cells, which is converted from its reduced form (NADH) to its oxidized form (NAD^+) by dehydrogenase enzymes. Direct imaging of NADH was performed through the NS microarray itself. The ECL microarray showed good temporal stability and reproducibility. The substrate concentration was monitored and imaged by measuring NADH ECL intensity. Therefore, many clinically important analytes are detectable by changing the nature of the dehydrogenase enzyme.

A chemiluminescence-based sensor for hydrogen peroxide was made of peroxalate ester NPs and encapsulated fluorescent dyes (Lee et al. 2007). Hydrogen peroxide has a reputation as a topical germ killer, but it is attracting attention in the medical community as an early indicator of disease in the body. It is an important molecule involved in the physiological signaling process. Hydrogen peroxide is thought to be accumulated by cells at the early stages of most diseases. Overproduction of hydrogen peroxide causes tissue damage and inflammation.

The sensing mechanism is as follows. When the nano particles bump into hydrogen peroxide, the latter reacts with the peroxalate ester groups, producing a high-energy dioxetanedione intermediate. The chemical compound 1,2-dioxetanedione, often called peroxyacid ester, is an unstable oxide of carbon (an oxocarbon) with the formula C_2O_4. It chemically stimulates encapsulated fluorescent dyes, such as pentacene, exciting chemiluminescence. The emitted photons (or light) can be detected. The chemiluminescence wavelength could be tuned by changing the encapsulated fluorescent dyes.

These NPs are incredibly sensitive, so one can detect nanomolar concentrations of hydrogen peroxide. Such NPs could someday be used as simple, all-purpose diagnostic tools for most diseases; for example, the NPs would be injected by needle into a certain area of the body such as the heart. If the nanoparticles encountered hydrogen peroxide, they would emit light. When a doctor notices a significant amount of light activity in the region, the doctor would know that the patient may be presenting early signs of a disease in that area of the body. Furthermore, these NPs penetrate deep into the tissue and operate at a high wavelength, making them sensitive indicators of the presence of hydrogen peroxide produced by any kind of inflammation.

6.13 Crossed Zinc Oxide Nanorods As Resistive UV-Nanosensors

UV photosensors find wide-ranging applications in medicine, ecology, space communications, high-temperature plasma research, chemical and biological sensing, and in the military for flame and missile-launching detection. UV detectors are used to monitor and determine the earth's ozone layer thickness. Although several kinds of photosensors are available now for the UV range, the development of NSs for these applications will enable more portable detectors. Usually, direct wide bandgap materials, like GaN, ZnSe, and AlGaN, are used as sensing materials. Analogous to III-V materials, ZnO also possesses high UV photosensitivity, which is most important for UV photodetection. Thus, zinc oxide is also a potentially strong UV-sensing material.

Chai et al. (2009) reported a self-assembled ZnO nanocross-based UV photosensor (Figure 6.14). They showed a simple

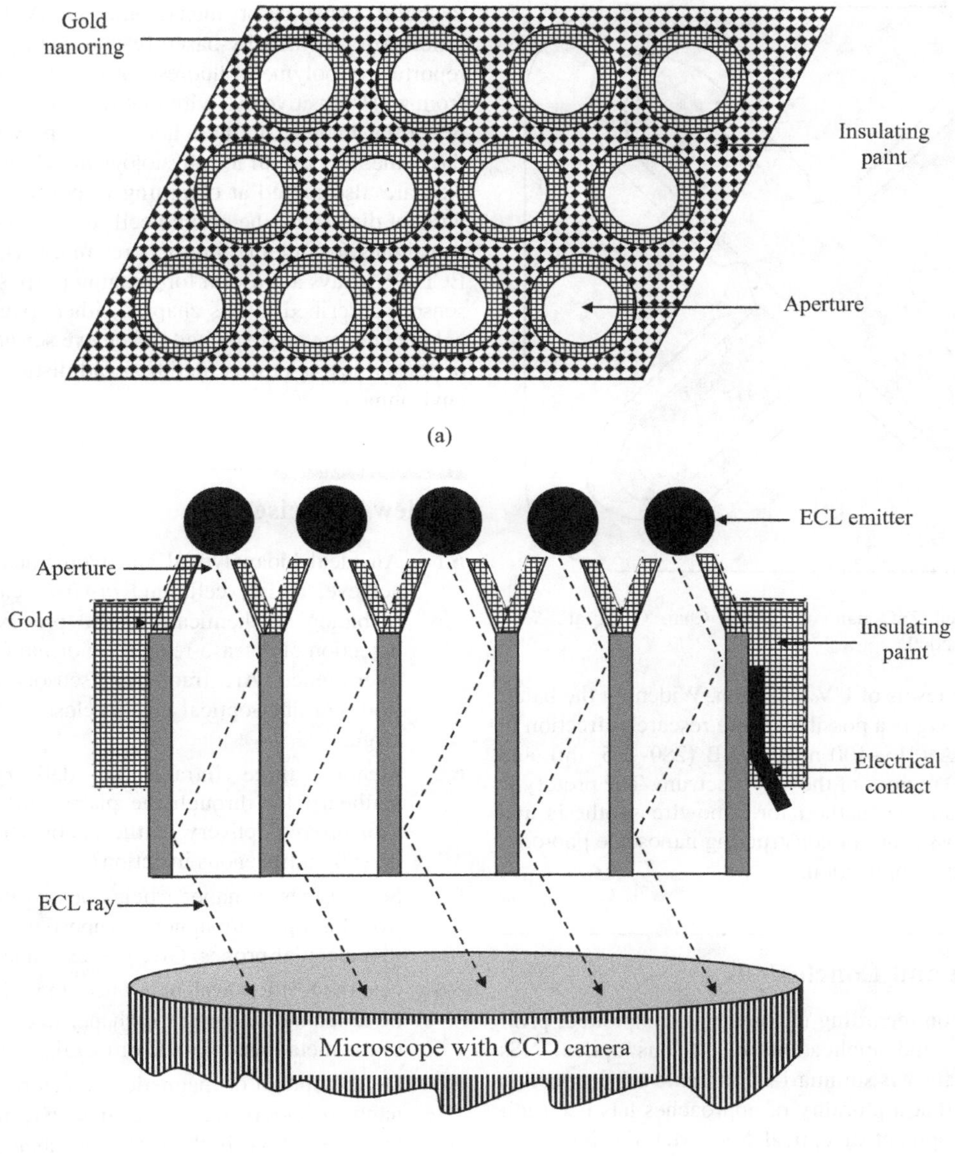

FIGURE 6.16 Microarray of electroluminescent nanosensors: (a) top view; and (b) side view. (Chovin, A. et al., *Meas. Sci. Technol.*, 17, 1211, 2006.)

and cost-effective method to fabricate a crossed ZnO nanorods device, possibly leading to the next generation of photosensors for a wide range of applications. The crossed ZnO nanorods (Figure 6.17) were synthesized in a hydrothermal reactor with an aqueous solution deposition technique. With "*hydros*" meaning water and "*thermos*" meaning heat, hydrothermal processing is defined, in general, as any heterogeneous reaction in the presence of aqueous/nonaqueous solvents or mineralizers under high pressure and temperature conditions to dissolve and recrystallize (recover) materials that are relatively insoluble under ordinary conditions. The hydrothermal technique is an important tool for advanced materials processing, particularly owing to its advantages in the processing of nanostructural materials for a wide variety of technological applications, such as electronics, optoelectronics, catalysis, ceramics, magnetic data storage, biomedical, and biophotonics. Zinc sulfate $Zn(SO_4)\cdot 7H_2O$ and ammonia solution NH_4OH (29.6%) were used as starting materials. Transmission electron microscopy (TEM) confirmed that the nanorods were crystalline. An *in-situ* lift-out technique was used to fabricate this photosensor, based on self-assembled crossed nanorods. A focused ion beam system was used to contact single crossed ZnO nanorods grown by an aqueous solution process. In order to realize the easy attachment, the nanorod was deposited and micro/nano manipulated on to a SiO_2/Si substrate already processed to have respective distribution of pre-patterned external electrodes. The UV sensitivity was measured using a two-terminal ZnO nanorod device. The current-voltage (*I-V*) characteristics showed linear behavior. The photosensor exhibited a response of ~15 mA W^{-1} for UV light (361 nm) under 1 V bias. Response measurements showed that such a photosensor

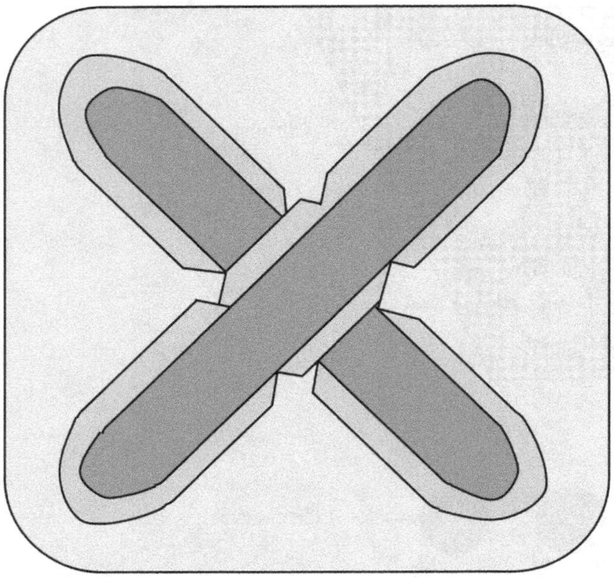

FIGURE 6.17 Crossed ZnO nanorod. (After Chai, G. et al., *Sens. Actuators, A*, 150, 184, 2009.)

was suitable for low levels of UV detection. Widening the bandgap by doping with Mg is a possible future research direction in order to cover UV-A (315–400 nm), UV-B (280–315 nm), and UV-C (100–280 nm) regions of the UV spectrum. The prototype device provided a simple method for nanowire synthesis and demonstrated the possibility of constructing nanoscale photodetectors for nano-optics applications.

6.14 Discussion and Conclusions

Important information regarding nanomaterials exploited, principles of operation, and applications of various optical NSs described in this chapter is summarized in Table 6.1. Inspection of Table 6.1 shows that a plurality of approaches has been followed for the development of optical NSs. Metallic NPs, like gold, silver, and copper, constitute interesting sensing tools by providing a physical output, such as a change in color from a chemical input, like the chemically induced modification of properties of the particle. The position of the surface plasmon band of metallic NPs in solution, that is, the color of the solution provides a vital clue but enhancement of the signal by Raman spectroscopy has given rise to a new field, known as SERS. The introduction of SERS has enabled single-molecule detection. Highly portable, user-friendly instruments are available. A simplified replacement of the SPR instrument was reported in which an optically transparent substrate was used to enable transmission mode operation. Colloidal gold NPs were immobilized on the substrate. Color changes were produced by the sample. The device was coupled to a UV-visible spectrometer.

Building noninvasive or minimally invasive NSs is the goal that all nanotechnologists strive to reach. In all these sensors, nano dimension is the main theme. Fiber optic NSs for living cell examination are an attempt to visualize the cellular phenomena in living conditions. Sensors for VOCs employ carbon nanotubes as high-sensitivity nanomaterials with optical fiber technology to build Fabry-Perot interferometers. A nanopipette giving ratiometric or intensity-based responses to pH and sodium was reported. A polymeric fluorescent pH-NS compares the signals from a pH-sensitive dye with that from a reference dye to obtain pH of the analyte. Both are housed in a polymeric casing, hence the name. It is useful for physiological pH monitoring. PEBBLE NSs are also aimed at obtaining responses from living beings without disturbing the normal cell functioning. QDs are used as substitutes for organic fluorophores in a variety of applications. ECL NS arrays are useful for imaging purposes. Thus, in all the sensors described in this chapter, either special nanoscale optical phenomena are exploited to achieve sensitivity improvement or sensor size is downscaled to avoid disturbance to the sensed environment.

Review Exercises

6.1 An ideal bioanalytical and biomedical sensor should achieve, in live cells and *in-vivo,* real-time tracking of biological/chemical/physical processes, as well as detection of disease-related abnormal features, without interference. Are traditional sensors like microelectrodes or fiber-optical sensors close to the ideal sensor? Argue.

6.2 Mention three intracellular delivery methods of nanoparticles through the plasma membrane barrier. Can *in-vivo* delivery of the nanoparticles to cells be done by intravenous injection?

6.3 Some types of nanoparticles possess unique but controllable optical/magnetic properties that are superior to molecular probes. Give two examples.

6.4 Can the LSPR wavelength of the metallic nanoparticles be tuned by changing the shape, size, and composition of the metal nanoparticle or metal shell thickness?

6.5 Do the optical or magnetic characteristics of QDs and nanoparticles themselves change in response to specific analytes? What is the effect of labeling with suitable dyes, ligands, or receptors?

6.6 FRET-based maltose sensors were developed using QD, MBP and a dye complex that forms a FRET pair with QD. The dye complex, β-cyclodextrin-acceptor dye conjugates, is initially bound to MBP, resulting in FRET quenching of the QD fluorescence. Added maltose displaces the dye complex. What is the influence of maltose concentration on the fluorescence of QD? Explain.

6.7 An important phenomenon is defined as follows: a distance-dependent energy transfer mechanism between two dye molecules by a nonradiative, long-range dipole-dipole coupling mechanism. Name the phenomenon.

6.8 In a measurement technique, a signal is measured with respect to a reference signal. What is it called?

6.9 What is the full form of: (i) LSPR; (ii) SPR; (iii) SERS; (iv) PEBBLE; and (v) SPIO?

6.10 Give reasons why ZnO has attracted great interest as a UV photosensor material?

TABLE 6.1
Optical Nanosensor Family

Sl. No.	Name of the Nanosensor	Nanomaterial Used	Operating Principle	Sensed Quantity/Applications
1.	Localized surface plasmon resonance (LSPR) spectroscopy	Noble metal nanoparticles	Refractive index-based detection: wavelength shift in the LSPR extinction or scattering maximum	Chemical and biological targets
2.	Surface-enhanced Raman scattering (SERS)	Metallic nanostructures	LSPR, fluorescence, and Raman spectroscopy	Biosensing applications
3.	Colorimetric gold nanoparticle spectrophotometric sensor	Colloidal gold on an optically transparent substrate	A colorimetric signal due to the changes in surface plasmon absorbance.	Label-free detection of biomolecular interactions
4.	Fiber-optic nanosensor	Optical fibers and SWCNTs; in-fiber nanocavity doped with vapochromic material	Absorbance, reflectance, fluorescence, surface plasmon resonance (SPR), refractive index, and colorimetry	Chemical and biological applications
5.	Nanograting-based optical accelerometer	Acceleration	Nanogratings	Modulation of the effective refractive index of the structures
6.	Fluorescent pH-sensitive nanosensor	A pH- sensitive fluorescent dye along with a pH-insensitive reference dye embedded in a polymer matrix	Responses from the pH- sensitive dye and the reference dye provide pH measurements by fluorescence ratiometric imaging microscopy	Physiological levels of pH and sodium
7.	PEBBLE nanosensors	Fluorescent indicators encased in a permeable biologically inert matrix	Ratiometric analysis	Bioanalysis
8.	Quantum dots as fluorescent labels	QD	Labeling	Biological
9.	Quantum dot FRET-based probes	QD	FRET	Biological
10.	ECL nanosensors	Microarray of optical apertures	Electrochemiluminescence	Chemical and biological
11.	ZnO nanorod ultraviolet nanosensor	ZnO nanorods	Current per watt change under voltage bias	UV photosensing

6.11 Point out the advantages of PEBBLE scheme over other means of cellular measurement with reference to the size, spatial resolution, interference from background fluorescence, protection of the recognition elements and cell, and co-location of multiple fluorophores.

6.12 Mention some methods for delivery of PEBBLEs into cells. Does the polymer matrix of PEBBLEs allow macromolecules, such as proteins, to diffuse across it?

6.13 Is SERS a vibrational spectroscopic method or a refractive index-based detection scheme? Does the SERS intensity depend upon the excitation of the LSPR?

6.14 The Langmuir-Blodgett technique is very useful for depositing ultrathin organic films, with precise control over the architecture of the films at the molecular level. In light of this, discuss its application to optical fiber sensors for SWCNT deposition on the distal face. Draw and explain the optical fiber interrogation system.

6.15 In an optical fiber nanobiosensor, the optical fiber's tip is significantly smaller than the wavelength of light used to excite the target analyte, so that the photons cannot escape from the tip of the fiber by conventional optics. After the photons have traveled as far down the fiber as possible, how is the excitation for the fluorescent species of interest provided in the vicinity of the biosensing layer? Discuss the role of evanescent fields here.

REFERENCES

Aylott, J. W. 2003. Optical nanosensors—An enabling technology for intracellular measurements. *Analyst* 128: 309–312, doi: 10.1039/b302174m.

Barbillon, G., J.-L. Bijeon, J. Plain, M. L. de la Chapelle, P.-M. Adam, and P. Royer. 2007. Biological and chemical gold nanosensors based on localized surface plasmon resonance. *Gold Bulletin* 40(3): 240–244.

Borisov, S. M. and I. Klimant. 2008. Optical nanosensors—Smart tools in bioanalytics. *Analyst* 133:1302–1307, doi: 10.1039/b805432k.

Cao, Y., Y. E. L. Koo, and R. Kopelman. 2004. Poly(decyl methacrylate)-based fluorescent PEBBLE swarm nanosensors for measuring dissolved oxygen in biosamples. *Analyst* 129: 745–750 [PubMed: 15284919].

Cao, Y., Y. E. L. Koo, S. M. Koo, and R. Kopelman. 2005. Ratiometric singlet oxygen nano-optodes and their use for monitoring photodynamic therapy nanoplatforms. *Photochemistry and Photobiology* 81(6): 1489–1498 [PubMed: 16107183].

Carr, D. W., G. R. Bogart, and B. E. N. Keeler. 2004. Femtophotonics: Optical transducers utilizing novel sub-wavelength dual layer grating structure. In: *Proceedings of the Hilton* Head Solid-State Sensors, Actuators, and Microsystems *Workshop*, Hilton Head Island, SC, June 2004, Cleveland, OH: Transducers Research Foundation, Inc., pp. 91–92.

Chai, G., O. Lupan, L. Chow, and H. Heinrich. 2009. Crossed zinc oxide nanorods for ultraviolet radiation detection. *Sensors and Actuators A* 150: 184–187.

Cheng, Z. L. and C. A. Aspinwall. 2006. Nanometre-sized molecular oxygen sensors prepared from polymer stabilized phospholipid vesicles. *Analyst* 131(2): 236–243 [PubMed: 16440088].

Chovin, A., P. Garrigue, G. Pecastaings, H. Saadaoui, and N. Sojic. 2006. Development of an ordered microarray of electrochemiluminescent nanosensors. *Measurement Science and Technology* 17: 1211–1219.

Clapp, A. R., I. L. Medintz, J. M. Mauro, B. R. Fisher, M. G. Bawendi, and H. Mattoussi. 2004. Fluorescence resonance energy transfer between quantum dot donors and dye-labeled protein acceptors. *Journal of the American Chemical Society* 126(1): 301–310.

Clapp, A. R., I. L. Medintz, H. T. Uyeda, B. R. Fisher, E. R. Goldman, M. G. Bawendi, and H. Mattoussi. 2005. Quantum dot-based multiplexed fluorescence resonance energy transfer. *Journal of the American Chemical Society* 127:18212–18221.

Clark, H. A., M. Hoyer, M. A. Philbert, and R. Kopelman. 1999. Optical nanosensors for chemical analysis inside single living cells. 1. Fabrication, characterization, and methods for intracellular delivery of PEBBLE sensors. *Analytical Chemistry* 71: 4831–4836.

Consales, M., S. Campopianoa, A. Cutoloa, M. Penza, P. Aversab, G. Cassanob, M. Giordanoc, and A. Cusano. 2006. Carbon nanotubes thin films fiber optic and acoustic VOCs sensors: Performances analysis. *Sensors and Actuators B* 118: 232–242.

Consales, M., A. Cutolo, M. Penza, P. Aversa, G. Cassano, M. Giordano, and A. Cusano. 2007. Carbon nanotubes coated acoustic and optical VOCs sensors: Towards the tailoring of the sensing performances. *IEEE Transactions on Nanotechnology* 6(6): 601–612.

Consales, M., A. Cutolo, M. Penza, P. Aversa, M. Giordano, and A. Cusano. 2008. Fiber optic chemical nanosensors based on engineered single-walled carbon nanotubes. *Journal of Sensors* 2008: 936074, 29, doi: 10.1155/2008/936074.

Coupland, P. G., S. J. Briddon, and J. W. Aylott. 2009. Using fluorescent pH-sensitive nanosensors to report their intracellular location after Tat-mediated delivery. *Integrative Biology* 1: 318–323.

Cusano, A., M. Giordano, A. Cutolo, M. Piscol, and M. Consales. 2008. Integrated development of chemoptical fiber nanosensors. *Current Analytical Chemistry* 4: 296–315.

Dubertret, B., P. Skourides, D. J. Norris, V. Noireaux, A. H. Brivanlou, and A. Libchaber. 2002. *In vivo* imaging of quantum dots encapsulated in phospholipid micelles. *Science* 298: 1759–1762.

Elosua, C., I. R. Matias, C. Bariain, and F. J. Arregui. 2006. Development of an in-fiber nanocavity towards detection of volatile organic gases. *Sensors* 6: 578–592.

Guice, K. B., M. E. Caldorera, and M. J. McShane. 2005. Nanoscale internally referenced oxygen sensors produced from self-assembled nanofilms on fluorescent nanoparticles. *Journal of Biomedical Optics* 10: 64031-1–64031-10.

Haes, A. J., L. Chang, W. L. Klein, and R. P. Van Duyne. 2005. Detection of a biomarker for Alzheimer's disease from synthetic and clinical samples using a nanoscale optical biosensor. *Journal of the American Chemical Society* 127(7): 2264–2271.

Hohng, S. and T. Ha. 2005. Single-molecule quantum-dot fluorescence resonance energy transfer. *Chemphyschem* 6: 956–960.

Jackson, J. B. and N. J. Halas. 2004. Surface-enhanced Raman scattering on tunable plasmonic nanoparticle substrates. *Proceedings of National Academy of Sciences* 101(52): 17930–17935.

Jamieson, T., R. Bakhshi, D. Petrova, R. Pocock, M. Imani, and A. M. Seifalian. 2007. Biological applications of quantum dots. *Biomaterials* 28: 4717–4732.

Javier, A. M., P. D. Pino, S. Kudera, and W. J. Parak. 2009. Chapter 8: Nanoparticle-Based Delivery and Biosensing Systems: An Example, In *Biosensing Using Nanomaterials*. Merkoci, A. (Ed.), Hoboken, NJ: Wiley, pp. 247–274.

Kamil, Y. M., S. H. Al-Rekabi, M. H. Yaacob, A. Syahir, H. Yeechee, M. A. Mahdi and M. H. Abu Bakar. 2019. Detection of dengue using PAMAM dendrimer integrated tapered optical fiber sensor. *Scientific Reports* 9: 13483 (10 pages).

King, M. and R. Kopelman. 2003. Development of a hydroxyl radical ratiometric nanoprobe. *Sensors and Actuators B* 90: 76–81.

Kneipp, J., H. Kneipp, B. Wittig, and K. Kneipp. 2010a. Novel optical nanosensors for probing and imaging live cells. Nanomedicine: Nanotechnology, Biology, and Medicine 6: 214–226.

Kneipp, J., B. Wittig, H. Bohr, and K. Kneipp. 2010b. Surface-enhanced Raman scattering: Anew optical probe in molecular biophysics and biomedicine. *Theoretical Chemistry Accounts* 125: 319–327.

Koo, Y., Y. Cao, R. Kopelman, S. M. Koo, M. Brasuel, and M. A. Philbert. 2004. Realtime measurements of dissolved oxygen inside live cells by ormosil (organically modified silicate) fluorescent pebble nanosensors. *Analytical Chemistry* 76: 2498–2505 [PubMed: 15117189].

Kreft, O., A. M. Javier, G. B. Sukhorukovac, and W. J. Parak. 2007. Polymer microcapsules as mobile local pH-sensors. *Journal of Materials Chemistry* 17: 4471–4476.

Krishnamoorthy, U., R. H. Olsson III, G. R. Bogart, M. S. Baker, D. W. Carr, T. P. Swiler, and P. J. Clews. 2008. In-plane MEMS-based nano-g accelerometer with subwavelength optical resonant sensor. *Sensors and Actuators A: Physical* 145–146: 283–290.

Lee, D., S. Khaja, J. C. Velasquez-Castano, M. Dasari, C. Sun, J. Petros, W. R. Taylor, and N. Murthy. 2007. *In vivo* imaging of hydrogen peroxide with chemiluminescent nanoparticles. *Nature Materials* 6(10): 765–769 [PubMed: 17704780].

Lee, Y.-E. K., R. Kopelman, and R. Smith. 2009. Nanoparticle PEBBLE sensors in live cells and in vivo. *Annual Review of Analytical Chemistry* 2: 57–76, doi: 10.1146/annurev.anchem.1.031207.112823.

Martin-Orozco, N., N. Touret, M. L. Zaharik, E. Park, R. Kopelman, S. Miller, B. B. Finlay, P. Gros, and S. Grinstein. 2006. Visualization of vacuolar acidification-induced transcription of genes of pathogens inside macrophages. *Molecular Biology of the Cell* 17(1): 498–510 [PubMed: 16251362].

McKinney, S. A., A. C. Declais, D. M. J. Lilley, and T. Ha. 2003. Structural dynamics of individual holliday junctions. *Nature Structural Biology* 10: 93–97.

Medintz, I. L., A. R. Clapp, H. Mattoussi, E. R. Goldman, B. Fisher, and J. M. Mauro. 2003. Self-assembled nanoscale biosensors based on quantum dot FRET donors. *Nature Materials* 2: 630–638.

Nath, N. and A. Chilkoti. 2002. A colorimetric gold nanoparticle sensor to interrogate biomolecular interactions in real time on a surface. *Analytical Chemistry* 74: 504–509.

Netti, C. and H. Stanford. 2006, June 1. Applications of reproducible SERS substrates for trace level detection. Retrieved from Spectroscopy Online: https://www.spectroscopyonline.com/view/applications-reproducible-sers-substrates-trace-level-detection

Pandana, H., K. H. Aschenbach, and R. D. Gomez. 2008. Systematic aptamer-gold nanoparticle colorimetry for protein detection: Thrombin. *IEEE Sensors Journal* 8(6): 661–666.

Penza, M., G. Cassanol, P. Aversa, A. Cusano, A. Cutolo, M. Giordano, and L. Nicolais. 2005. Carbon nanotube acoustic and optical sensors for volatile organic compound detection. *Nanotechnology* 16: 2536–2547.

Piper, J. D., R. W. Clarke, Y. E. Korchev, L. Ying, and D. Klenerman. 2006. A renewable nanosensor based on a glass nanopipette. *Journal of the American Chemical Society* 128:16462–16463.

Reddy, G. R., M. S. Bhojani, P. McConville, J. Moody, B. A. Moffat, D. E. Hall, G. Kim, Y. E. Koo, M. J. Woolliscroft, J. V. Sugai, T. D. Johnson, M. A. Philbert, R. Kopelman, A. Rchemtulla, and B. D. Ross. 2006. Vascular targeted nanoparticles for imaging and treatment of brain tumors. *Clinical Cancer Research* 12(22): 6677–6686 [PubMed: 17121886].

Resch-Genger, U., M. Grabolle, S. Cavaliere-Jaricot, R. Nitschke, and T. Nann. 2008. Quantum dots versus organic dyes as fluorescent labels. *Nature Methods* 5(9): 763–775.

Shi, L., V. De Paoli, N. Rosenzweig, and Z. Rosenzweig. 2006. Synthesis and application of quantum dot FRET-based protease sensors. *Journal of the American Chemical Society* 128: 10378–10379.

Sumner, J. P., N. M. Westerberg, A. K. Stoddard, C. A. Fierke, and R. Kopelman. 2006. Cu^{+}- and Cu^{2+}-sensitive PEBBLE fluorescent nanosensors using DsRed as the recognition element. *Sensors and Actuators B* 113: 760–767.

Sun, H., A. M. Scharff-Poulsen, H. Gu, and K. Almdal. 2006. Synthesis and characterization of ratiometric, pH sensing nanoparticles with covalently attached fluorescent dyes. *Chemistry of Materials* 18(15): 3381–3384.

Talley, C. E., L. Jusinski, C. W. Hollars, S. M. Lane, and T. Huser. 2004. Intracellular pH sensors based on surface enhanced Raman scattering. *Analytical Chemistry* 76: 7064–7068 [PubMed: 15571360].

Ulgar W. R. and U. J. Krull. 2009. Quantum dots for the development of optical biosensors based on fluorescence, In *Biosensing Using Nanomaterials*. Merkoçi A. (Ed.), Hoboken, NJ: Wiley, p. 226.

Vo-Dinh, T. and P. Kasili. 2005. Fiber-optic nanosensors for single-cell monitoring. *Analytical and Bioanalytical Chemistry* 382: 918–925.

Webster, A., P. Coupland, F. D. Houghton, H. J. Leese, and J. W. Aylott. 2007. The delivery of PEBBLE nanosensors to measure the intracellular environment. *Biochemical Society Transactions* 35(3): 538–543.

Yonzon, C. R., X. Zhang, J. Zhao, and R. P. Van Duyne. 2007. Surface-enhanced nanosensors. *Spectroscopy* 22(1): 42–56.

Zhang, C.-Y. and L. W. Johnson. 2006. Quantum-dot-based nanosensor for RRE IIB RNA-Rev peptide interaction assay. *Journal of the American Chemical Society* 128(16): 5324–5325, doi: 10.1021/ja060537y

Zhou, M. and I. Ghosh. 2006. Quantum dots and peptides: A bright future together. *Peptide Science* 88(3): 325–339, doi: 10.1002/bip.

Zwiller, V., N. Akopian, M. van Weert, M. van Kouwen, U. Perinetti, L. Kouwenhoven, R. Algra, J. Gómez Rivas, E. Bakkers, G. Patriarche, L. Liu, J.-C. Harmand, Y. Kobayashi, and J. Motohisa. 2008. Optics with single nanowires. C. R. Physique 9: 804–815.

7

Magnetic Nanosensors

7.1 Introduction

What is a magnetic sensor? It is a transducer that converts a magnetic field into an electrical signal. It relates to the sub-discipline magnetism (Section 1.3.10) of Physics (Section1.3), dealing with magnetic phenomena, notably changes in magnetoresistance (ordinary, anisotropic, giant, tunneling), magnetic relaxation times, superconductivity-based effects, etc.

A magnetic sensor is required not only for magnetic signal detection but is also necessary whenever a nonmagnetic signal is to be measured by means of an intermediary signal detection into the magnetic signal domain in a tandem transducer, for example, the detection of a current through its magnetic field or the mechanical displacement of a magnet. Therefore, two groups of magnetic sensors are distinguishable, direct and indirect.

How are magnetic sensors used? Magnetic sensors have been employed for a long time for applications, such as current sensing, encoders, gear tooth sensing, linear and rotary position sensing, rotational speed sensing, motion sensing, and so on. In particular, solid-state magnetic field sensors, which are generally used for this purpose, have an inherent advantage in terms of size and power consumption, compared with search coil, fluxgate, and more intricate low-field sensing techniques, such as superconducting quantum interference detectors (SQUIDs) and spin resonance magnetometers. This holds true even for high-frequency applications.

The SQUID is an extremely sensitive magnetic flux-to-voltage transducer. The main component of the SQUID is the Josephson junction, comprising essentially two superconductors weakly coupled through a small insulating gap or constriction. This junction has unique electrical/magnetic properties. The SQUID device consists of two superconductors separated by thin insulating layers to form two parallel Josephson junctions. Two such superconductors, separated by a thin insulating layer, experience tunneling of Cooper pairs of electrons (two electrons that are bound together at low temperatures with correlated motion and constituting a single entity) through the junction. The Cooper pairs on each side of the junction are represented by a wave function similar to a free particle wave function. In the DC Josephson effect, a current proportional to the phase difference of the wave functions flows in the junction in the absence of a voltage.

The SQUID device may be configured as a magnetometer (a scientific instrument used to measure the strength and/or direction of the magnetic field) to detect incredibly small magnetic fields. A superconducting pick-up coil couples the SQUID to the ambient magnetic field. If a constant biasing current is maintained in the SQUID device, the measured voltage oscillates with the changes in phase at the two junctions, which depends upon the change in the magnetic flux.

An electron spin resonance magnetometer utilizes the electron paramagnetic resonance for measuring the instantaneous value of the time-varying magnetic field.

Returning to our discussion on solid-state magnetic sensors, *what is the working principle of solid-state magnetic sensors?* These sensors work on the principle of converting the magnetic field into a voltage or resistance, using a DC supply. The sensing can be done in an extremely small, lithographically patterned area, with additional reduction of size and power requirements. The diminutive size of a solid-state element increases the resolution for fields that vary over small distances. It also allows the enclosure of arrays of sensors in a small package, further decreasing size and power requirements.

What has been the industrial impact of solid-state magnetic sensors? Solid-state magnetic field sensors have revolutionized measurement and control systems. Industry continues to reap the benefits of solid-state magnetic field sensing. Every day, new applications are found for solid-state magnetic field sensors.

One of the key types of solid-state magnetic field sensors is the magnetoresistive sensor.

7.2 Magnetoresistance Sensors

What is magnetoresistance (MR)? Magnetoresistance is the property of a material or system of materials that results in a change of resistance when they are exposed to a magnetic field (Freitas et al. 2007; Koh and Josephson 2009). *Is magnetoimpedance, a phenomenon involving the change of total impedance of a ferromagnetic conductor in a magnetic field, when a high-frequency alternating current flows through it, considered as magnetoresistance?* No.

The discovery of large magnetoresistive effects has led to the development of solid-state magnetic sensors that can replace high-priced wire-wound sensors in various applications. The usual parameter of value for magnetoresistance is the *magnetoresistance (MR) ratio*, traditionally defined in terms of maximum resistance R_{max} and minimum resistance R_{min} (Jander et al. 2005) as

$$\mathrm{MR}\% = \left\{\frac{R_{max} - R_{min}}{R_{min}}\right\} \times 100 \tag{7.1}$$

By this definition, can MR ratio exceed 100%? Yes.

What does the MR ratio indicate? The MR ratio indicates the maximum signal that can be obtained from the sensor.

Example 7.1

In a magnetoresistor, the resistance changed from 700 to 710 Ω. Find the MR ratio.

From Equation 7.1,

$$\mathrm{MR}\% = \left\{\frac{R_{\max} - R_{\min}}{R_{\min}}\right\} \times 100 = \left\{\frac{710 - 700}{700}\right\} \times 100 = 1.43\% \quad (7.2)$$

7.2.1 Ordinary Magnetoresistance: The Hall Effect

All conductors exhibit a weak MR effect, known as *ordinary magnetoresistance* (OMR). One such effect is the *Hall effect*: the generation of a voltage in a current-carrying conductor placed in a magnetic field; the voltage produced is perpendicular to both the electric current and magnetic field. Its origin lies in the Lorentz force. *What is the Lorentz force?* The Lorentz force is the force acting on an electrically charged particle, moving through a magnetic plus an electric field. It has two vector components, one proportional to the magnetic field and the other proportional to the electric field, to be added vectorially to obtain the total force. The strength of the *magnetic component* is proportional to the charge q of the particle, the speed v of the particle, the intensity B of the magnetic induction, and the sine of the angle between the vectors **v** and **B**. The *direction* of the magnetic component is decided by the right-hand rule: put the right hand along **v** with fingers pointing in the direction of **v** and the open palm toward the vector **B**. Then stretching the thumb of right hand, the Lorentz force acts along it, pointing from your wrist to the tip of your thumb. The *electric component* of the Lorentz force $= q \cdot \mathbf{E}$ (charge of the particle multiplied by the electric field).

The explanation of the Hall effect is as follows: disturbance of the current distribution in the conductor by the Lorentz force, due to the applied magnetic field acting on the electrons flowing in the conductor, is responsible for the creation of potential difference (voltage) across it. The deviation of the current path due to the magnetic field produces an increase in the current path length, leading to an increase in the effective resistance.

7.2.2 Anisotropic Magnetoresistance

Ferromagnetic materials exhibit a larger magnetoresistive effect known as *anisotropic magnetoresistance (AMR).* When a current is passed through a magnetic conductor, resistance changes occur, based on the relative angle between the current and the conductor's magnetization. The magnetization direction is controlled by the applied magnetic field. *Is AMR displayed by all metallic magnetic materials?* Yes.

Why is it called "anisotropic?" It is termed *anisotropic* because, in contrast to the previously known OMR, it depends on the angle between the electric current and the magnetization direction. Resistance increases when the magnetic field is applied in a direction parallel to that of current flow but decreases when the magnetic field is applied in the perpendicular direction.

On a microscopic level, the AMR effect is interpreted on the basis of a change in the scattering cross-section of atomic orbitals caused by their distortion by the magnetic field. This dependence implies that the probability of scattering of electrons is larger during their motion parallel to the direction of magnetization as compared to motion perpendicular to magnetization. As a consequence, the material exhibits maximum resistivity for current flowing along the direction of magnetization and minimum resistivity for current flowing orthogonally. A material deposited as a thin film has only one easy axis of magnetization in the plane of the film. So, in a stripe made from this film, the current flows in a direction parallel to the magnetization resulting in maximum resistance. But when a magnetic field is applied at right angles to the easy axis, the vector of magnetization rotates around its original position at some angle. Since the probability of electron scattering is maximum along the magnetization and the magnetization vector has shifted from its parallel alignment with the current flow direction, therefore a decrease in resistance of the film is observed. Hence the film can be used as a sensor for a perpendicularly applied magnetic field (Andreev and Dimitrova 2005).

Devices use permalloy (a nickel-iron magnetic alloy, generically, with about 20% iron and 80% nickel content) as sensing material deposited on Si substrates in a Wheatstone bridge configuration. *How do AMR devices compare with OMR devices?* AMR devices typically have MR ratios of 1%-2% while those of OMR devices are very feeble.

7.2.3 Giant Magnetoresistance

Like other magnetoresistive effects, giant magnetoresistance (GMR) represents the change in electrical resistance of some materials in response to an applied magnetic field. *What is the usual construction of a GMR device?* A GMR device is a sandwich structure consisting of two or more layers of ferromagnetic metal (typically Fe, Co, NiFe, CoFe, or a related transition metal alloy) separated by ultrathin, nonmagnetic metal spacer layers (Cr, Cu, Au, or Ru); these spacer layers have to be only a few nanometers in thickness and their interfaces must be of the highest quality.

It was discovered that, upon application of a magnetic field to magnetic metallic multilayers, such as Fe/Cr and Co/Cu, in which ferromagnetic layers were separated by nonmagnetic spacer layers, a few nanometers thick, the electric current experienced a strong influence on the relative orientation of the magnetizations of the magnetic layers, resulting in a significant reduction in the electrical resistance of the multilayer (about 50% at 4.2 K). This effect was found to be much larger than other magnetoresistive effects that had ever been observed in metals, and was, therefore, called "giant magnetoresistance." The term giant magnetoresistance was coined because the tremendous change in resistance found in GMR far exceeded that of any AMR devices. In Fe/Cr and Co/Cu multilayers, the magnitude of GMR can be higher than 100% at low temperatures. In modern GMR structures at room temperature, MR levels above 200% are achieved (Reig et al. 2009).

In what type of structures is GMR effect noticed? GMR response is observed in magnetic multilayer structures comprising at least a trilayer arrangement, with two magnetic layers separated by a spacer layer. *Is it found in structures containing two spacer layers and one magnetic layer?* No.

What are the two basic configurations of GMR effect? Two variants of GMR effect have been applied in devices: current-in-plane (CIP) and current-perpendicular-to-plane (CPP). GMR devices are typically operated with the sense current in the plane of the films using electrical contacts at the ends of long, often serpentine, lines. Although the magnetoresistance is reduced because of current shunting through the layers, the alternative CPP configuration has a resistance that is too low for practical circuit applications (Jander et al. 2005).

Figures 7.1a and b show the structures of the sensor and magnetization directions in the absence of the magnetic field and after applying magnetic field, respectively. In the former case, the magnetizations of the ferromagnetic layers are antiparallel (Figure 7.1a), whereas, in the latter case, they become parallel (Figure 7.1b). Here, the current flows parallel to the planes of the layers. Figures 7.1c and d represent the situation for current flowing perpendicular to the planes of layers.

The change in the resistance of the multilayer arises when the applied field aligns the magnetic moments of the successive ferromagnetic layers, as is illustrated schematically in Figure 7.2. When the field is absent, the magnetic moments (the torque exerted on a magnet within a magnetic field) are antiparallel

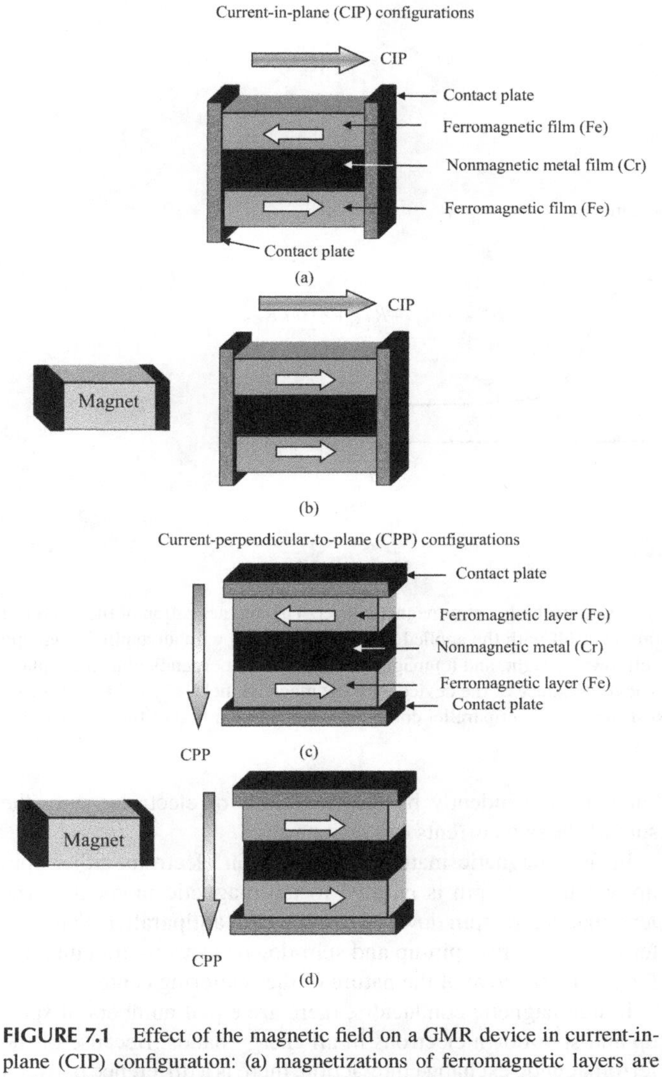

FIGURE 7.1 Effect of the magnetic field on a GMR device in current-in-plane (CIP) configuration: (a) magnetizations of ferromagnetic layers are antiparallel in the absence of applied magnetic field, and (b) the magnetizations are parallel under an applied field. Effect of the magnetic field on a GMR device in current-perpendicular-to-plane (CPP) configuration: (c) antiparallel magnetization, and (d) parallel magnetization.

(Figure 7.2a). Application of the magnetic field turns the magnetic moments through different angles and makes them perpendicular (Figure 7.2b). Finally, it aligns the magnetic moments, rendering them parallel (Figure 7.2c). The field saturates the magnetization (the quantity of magnetic moment per unit volume) of the multilayer and leads to a decrease in the electrical resistance of the multilayer; this resistance variation is depicted in Figure 7.2d. The enormous variation of the resistance is attributed to the scattering of the electrons at the interfaces between the layers.

7.2.3.1 Scientific Explanation of GMR

Electrons have the basic properties of mass and charge. The negative charge of the electron is a familiar property, as is its small mass. These properties enable it to move easily through a metal. *In addition to charge and mass, do electrons have a third property that is often overlooked?* Yes, the *spin* of electrons. The intrinsic spin angular momentum associated with an electron is a fundamental, unvarying property like electron charge and rest mass. It has become possible to take advantage of the spin of electrons for controlling their motion through thin metallic conductors. Electrons spin about their axes, like a child's top. The spin of an electron creates a tiny magnetic field that makes the electron act like a minute magnet.

How many types of spin motion are possible? Spin motion is of two types: spin up (Figure 7.3a) and spin down (Figure 7.3b). In the presence of a magnetic field, the laws of quantum mechanics affirm that an electron can spin in one of two directions: Either its own magnetic field, resulting from its spin motion, aligns with the external magnetic field, which physicists call the "up direction" or its own field aligns oppositely to the external magnetic field, which is referred to as the "down direction." In a magnetic material, electron spins are unequal, with the majority of electrons spinning in one direction (Figure 7.3c). Most electrons are aligned in the direction of the cumulative magnetic field, that is, there are more up electrons than down electrons. In a nonmagnetic material, the electron spins are opposite; the up and down electrons counterbalance each other (Figure 7.3d).

In a metal conductor, electricity is transported in the form of electrons, which move freely through the material (Figure 7.4a). The current is conducted because of the movement of electrons in a specific direction. The straighter the path of the electrons, the greater the conductance of the material. Electrons carrying a current are like balls rolling down a hill. If the slope is smooth, then the balls descend very fast. But if the slope of the hill has a lot of bumps, the balls are slowed down. Defects in the atomic lattice are like these bumps. When electrons collide with these defects, a process that physicists call "scattering," the electrons are slowed down. Therefore, scattering generates electrical resistance. Electric resistance is due to electrons diverging from their straight paths when they scatter against irregularities and impurities in the material. The more the electrons scatter, the higher the resistance is. The resistance of metals depends on the *mean free path* (the average distance covered between successive collisions) of their conduction electrons, which, in GMR devices, depends on the *spin orientation.*

The above must be emphasized. The quantum nature of electrons, that is, either spin up or spin down, is exploited in GMR

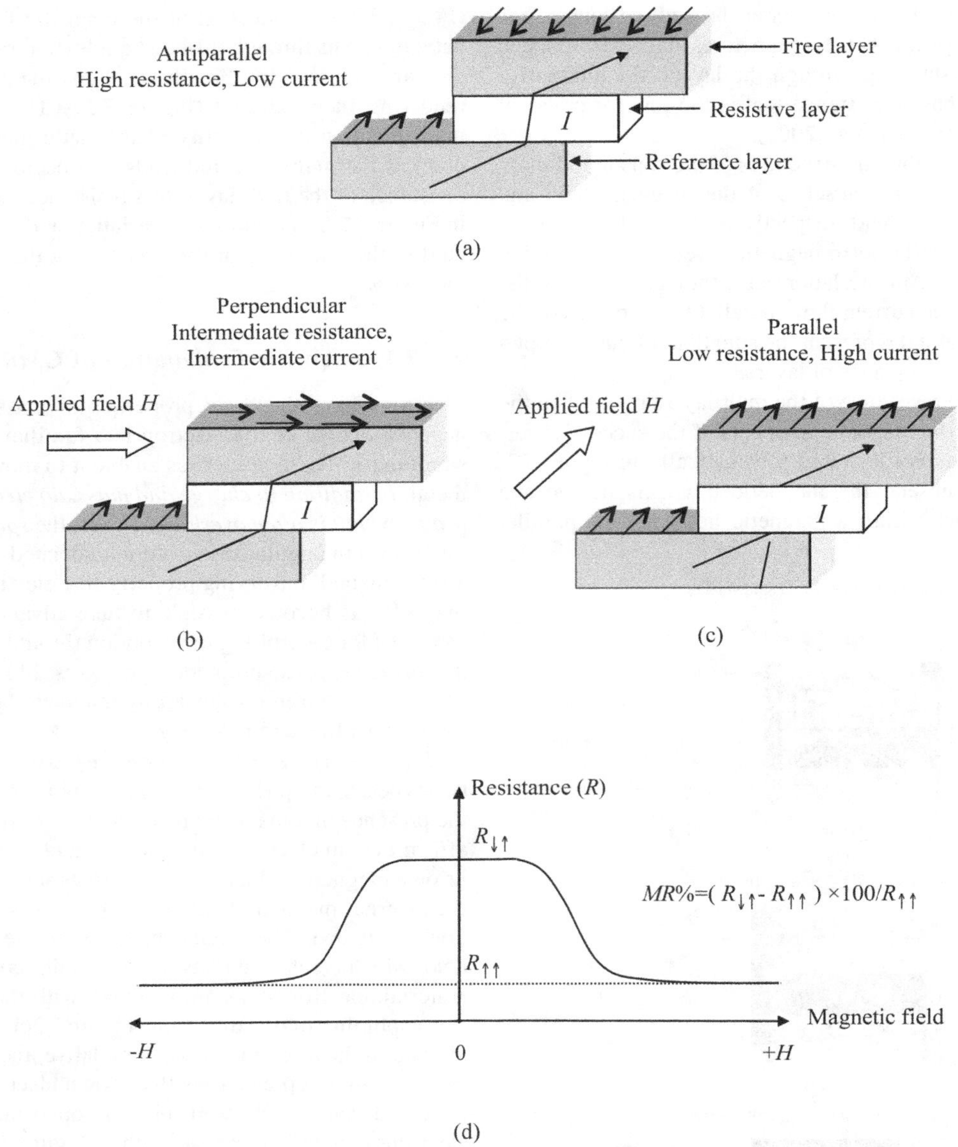

FIGURE 7.2 Resistance variation with the degree of alignment of magnetic layers sandwiching a non-magnetic layer. The magnetization of the reference layer is pinned and independent of the applied field. The free layer rotates and aligns itself with the applied tickling field H: (a) without applied magnetic field, (b) magnetic field applied along the plane of the paper and directed from left towards right, and (c) magnetic field applied perpendicular to the plane of the paper and directed downward into the paper. (d) Typical variation of electrical resistance of the device with the magnetic field. $R_{\downarrow\uparrow}$ and $R_{\uparrow\uparrow}$ are resistances measured with antiparallel and parallel magnetizations, respectively. Resistances for antiparallel configuration are much higher than for parallel configuration.

sensors. In a magnetic material, most of the spins point in the same direction (in parallel) Figure 7.4b). A smaller number of spins, however, always point in the opposite direction, antiparallel to the general magnetization. This spin imbalance gives rise not only to the magnetization as such, but also to the fact that electrons with different spins are scattered to a smaller or greater degree against irregularities and impurities, and especially at the interfaces between materials.

GMR is qualitatively understood using the *Mott model*, according to which the electrical conductivity in metals is described in terms of two largely independent conducting channels, corresponding to the spin-up and spin-down electrons. Unfortunately, there is no way to measure the electrical current carried independently by the two kinds of electrons; only the sum of the two currents can be measured.

In ferromagnetic materials, conduction electrons either spin up, when their spin is parallel to the magnetic moment of the ferromagnet, or spin down, when they are antiparallel. The scattering rates of the spin-up and spin-down electrons are quite different, irrespective of the nature of the scattering centers.

In nonmagnetic conductors, there are equal numbers of spin-up and spin-down electrons in all energy bands. Because of the ferromagnetic exchange interaction, there is a difference between the number of spin-up and spin-down electrons in the conduction bands. Quantum mechanics dictates that the probability of an electron being scattered when it passes into a ferromagnetic

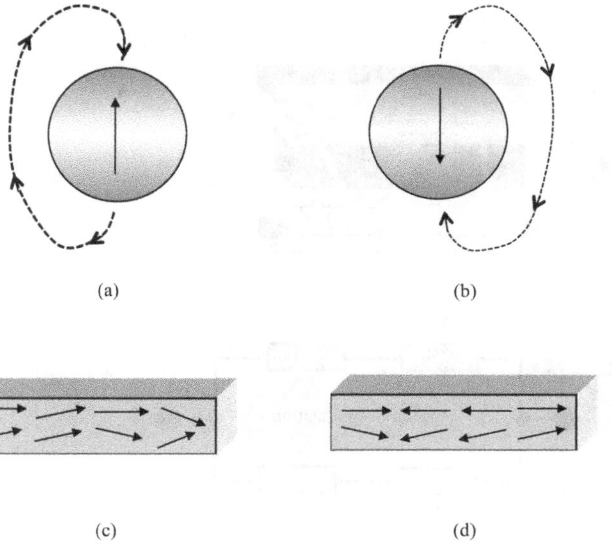

FIGURE 7.3 Spins and materials: (a) spin up, (b) spin down, (c) magnetic material, and (d) nonmagnetic material.

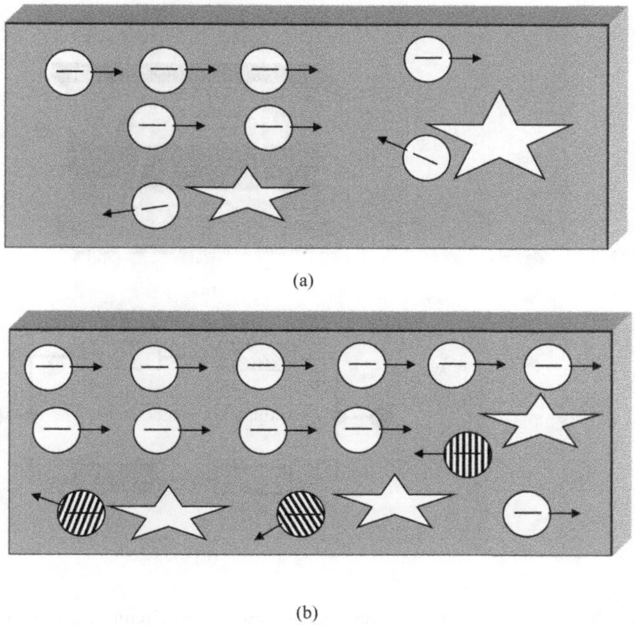

FIGURE 7.4 (a) The electrical resistance in a conductor arises when electrons scatter against irregularities in the material, so that their forward movement is obstructed. (b) In a magnetic conductor, the direction of spin of most electrons is parallel with the magnetization (empty circles with minus sign). A minority of electrons have spin in the opposite direction (vertically hatched circles with minus sign). Electrons with antiparallel spin are scattered more.

conductor depends on the direction of its spin. Conduction electrons with spin parallel to the magnetization of the material (spin up) undergo free movement, whereas the motion of those electrons with antiparallel orientation (spin down) is impeded by collisions with atoms in the material. In general, electrons with spin aligned with the majority of spins in the ferromagnets traverse a longer distance without being scattered. In other words, the scattering is strong for electrons with spin antiparallel to the magnetization direction, and is weak for electrons with spin parallel to the magnetization direction.

1. If the magnetizations are oppositely directed, then electrons originating in one layer are stopped by the adjacent layer (Figure 7.5a). For the antiparallel-aligned multilayer, both the spin-up and spin-down electrons are scattered strongly within one of the ferromagnetic layers, because, within one of the layers, their spin is antiparallel to the magnetization direction. The disruption of the free motion of the electrons results in an increase in the electrical resistance. Therefore, in this case, the total resistivity of the multilayer is high. Thus, both electron spins experience small resistance in one layer and large resistance in the other, and the total resistance is

$$R_{\text{Antiparallel}} = \left(\frac{1}{2}\right)\left(R_{\uparrow} + R_{\downarrow}\right) \tag{7.3}$$

The interfaces in a magnetic multilayer play an important role in spin-dependent transport in bulk elemental ferromagnets. If the interface separates ferromagnetic and nonmagnetic metals, the transmission is spin-dependent, due to the spin dependence of the band structure of the ferromagnetic layer; for example, there is a relatively large band mismatch between Cu and the minority spins in Co and thus the transmission of the minority spin electrons across the Co/Cu interface is anticipated to be poor. When the excess spins in the two ferromagnetic layers are in opposite directions, both the spin-up and spin-down electrons are scattered at one of the interfaces. The electrical resistivity for the aligned case is lower, because spin-up electrons in this case experience very little resistance and act like a short circuit. Let us visualize the limiting case in which spin-up electrons for the aligned case have no resistance, and spin-down electrons have resistance $2R$, with R being the contribution from each interface. Because the spin up provides a short-circuit channel, the resistance of the whole system is zero. For the case in which excess spins on two ferromagnetic layers are aligned in opposite directions, both spins up and down have resistance R, because both scatter off one interface. In this case, the total resistance is $R/2$, which is definitely greater than the value zero for the aligned case.

2. For the parallel-aligned magnetic layers, that is, when the magnetic layers are aligned in the same direction, the spin-up electrons originating in one layer cross relatively freely through the structure, almost without scattering, because their spin is parallel to the magnetization of the layers, as illustrated in Figure 7.5b. On the contrary, the spin-down electrons are scattered strongly within both ferromagnetic layers, because their spin is antiparallel to the magnetization of the two layers. Since conduction occurs in parallel for the two spin channels, the total resistivity of the multilayer is determined mainly by the highly conducting spin-up electrons and appears to be low. Thus, spin-up electrons

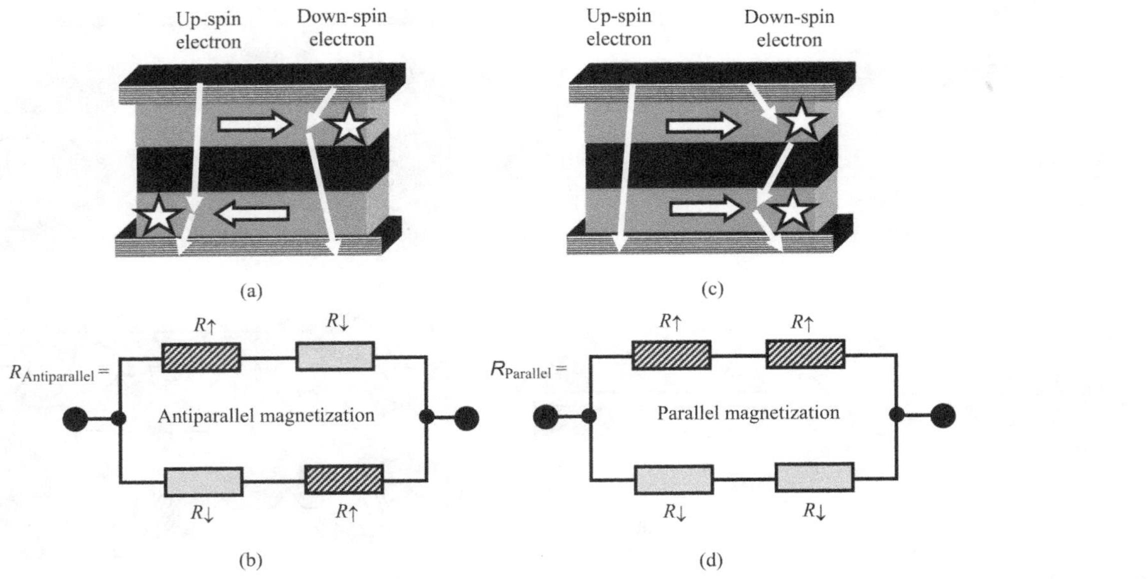

FIGURE 7.5 (a) Electron transport and (b) equivalent circuit for antiparallel magnetization. (c) and (d) The same for parallel magnetization. CPP case only is considered.

experience low resistance, the spin-down electrons experience high resistance, and the sum-total resistance is given by

$$R_{\text{Parallel}} = \frac{2R_{\uparrow}R_{\downarrow}}{R_{\uparrow}+R_{\downarrow}} \tag{7.4}$$

There are similarities between the band structure of Cu and the band structure of the majority spins in Co. This high band matching implies a high transmission probability for the majority-spin electrons across the Co/Cu interface. The spin-up electrons barely notice any difference in the number of electrons per atom in their journey from the ferromagnetic layer to the non-magnetic layer. To them, the lattice potential is smooth and more-or-less defect free. On the other hand, the spin-down electrons see a large difference in electron numbers between atoms of copper and cobalt. They perceive several bumps at the interface (because copper and cobalt atoms often intermix there), so they very likely scatter there. When the excess spins in the two ferromagnetic layers are in the same direction, spin-up electrons travel freely from one ferromagnetic (cobalt) layer across the interface to the other ferromagnetic layer, and spin-down electrons scatter strongly at both interfaces.

From Equations 7.2 and 7.3, the difference between parallel and antiparallel resistances is given by

$$\Delta R = R_{\text{Parallel}} - R_{\text{Antiparallel}} = -\frac{1}{2}\frac{\left(R_{\uparrow}-R_{\downarrow}\right)^2}{\left(R_{\uparrow}+R_{\downarrow}\right)} \tag{7.5}$$

Essentially, the interfaces of the Co/Cu multilayer act as *polarizing filters* for the spin of the electrons. When the filters are aligned, the majority-spin electrons pass through relatively easily. When the filters are not aligned, the electrons in the two spin channels are reflected at one of the interfaces. To obtain the GMR effect, the spacer layers must be thin compared to the mean free path of electrons to ensure that electrons spin polarized in one layer pass into the other layers before disturbance of their polarization by scattering. The spacer layers allow the magnetic directions of the layers to differ, while still permitting the passage of electrons.

From the aforesaid oversimplified arguments, it is evident that resistance is lower when excess spins in the magnetic layers are in the same direction. Thus, if excess spins are pointing in different directions in different magnetic layers when no magnetic field is applied, applying an external magnetic field turns them around so that they point in the same direction, resulting in a decrease in resistivity.

Example 7.2

A GMR effect-based magnetic field sensor showed a nominal resistance of 5000 Ω. Then, a magnetic field was applied and, with this field, a minimum resistance of 4400 Ω was achieved. Find (a) MR ratio, (b) the spin-up resistance $R_{\uparrow}$, and (c) the spin-down resistance $R_{\downarrow}$. (d) Verify that the obtained values satisfy Equation 7.5.

(a) Applying Equation 7.1,

$$\text{MR}\% = \left\{\frac{R_{\max}-R_{\min}}{R_{\min}}\right\}\times 100 = \left\{\frac{5000-4400}{4400}\right\}\times 100 \tag{7.6}$$

$$= 13.64\%$$

(b) and (c) Here, $R_{\text{Antiparallel}} = 5000\ \Omega$ so that Equation 7.3 gives

$$5000 = \left(\frac{1}{2}\right)(R_\uparrow + R_\downarrow) \quad (7.7)$$

$R_{\text{parallel}} = 4400\ \Omega$, so that Equation 7.4 yields

$$4400 = \frac{2R_\uparrow R_\downarrow}{R_\uparrow + R_\downarrow} \quad (7.8)$$

Combining Equations 7.7 and 7.8, we have

$$4400 = \frac{2(10{,}000 - R_\downarrow)R_\downarrow}{10{,}000 - R_\downarrow + R_\downarrow} = \frac{20{,}000R_\downarrow - 2R_\downarrow^2}{10{,}000} \quad (7.9)$$

or

$$2R_\downarrow^2 - 20{,}000R_\downarrow + 44{,}000{,}000 = 0 \quad (7.10)$$

giving

$$R_\downarrow = \frac{20{,}000 \pm \sqrt{(20{,}000)^2 - 4\times 2\times 44{,}000{,}000}}{2\times 2}$$

$$= \frac{20{,}000 \pm \sqrt{48{,}000{,}000}}{2\times 2} = \frac{20{,}000 \pm 6{,}928.203}{2\times 2} \quad (7.11)$$

$$= 6{,}732.05 \text{ or} 3{,}267.95\Omega$$

Therefore, the spin-down resistance $R_\downarrow = 6732.05$, or 3267.95 Ω, and the spin-up resistance $R_\uparrow = 3267.95$, or 6732.05 Ω.

(d) Putting one set of values of $R_\downarrow = 6732.05\ \Omega$ and $R_\uparrow = 3267.95\ \Omega$ in Equation 7.5,

$$\Delta R = -\frac{1}{2}\frac{(R_\uparrow - R_\downarrow)^2}{(R_\uparrow + R_\downarrow)} = -\frac{1}{2}\frac{(3{,}267.95 - 6{,}732.05)^2}{(6.732.05 + 3{,}267.95)} \quad (7.12)$$

$$= -\frac{11{,}999{,}988.81}{20{,}000} = -599.99\Omega \approx -600\Omega$$

The negative value is obtained because ΔR is defined as the difference between R_{Parallel} and $R_{\text{Antiparallel}}$ and $R_{\text{Antiparallel}}$ is $> R_{\text{Parallel}}$ Substituting the other set of values for $R_\downarrow = 3267.95\ \Omega$ and $R_\uparrow = 6732.05\ \Omega$, it is seen that $\Delta R = -600\ \Omega$.

7.2.3.2 Simple Analogies of GMR

To understand how GMR works on the atomic level, consider the following analogies. Let us consider a ball to be analogous to a conduction electron. If a person throws a ball between two sets of rollers turning in the same direction (like parallel spin-aligned magnetic layers), the ball tends to go through smoothly. But if the rollers turn in opposite directions, the ball tends to bounce and scatter. Alternatively, the GMR effect may be compared to light passing through optical polarizers. When the polarizers are aligned, light passes through them easily but when their optical axes are rotated with respect to each other, light is prevented from passing across.

7.2.3.3 Optimizing Parameters

What are the critical structural parameters of a GMR device that must have optimum values? Optimal layer thicknesses enhance magnetic-layer antiparallel coupling, which is necessary to keep the sensor in the high-resistance state when no field is applied. When an external field overcomes the antiparallel coupling, the moments in the magnetic layers align and reduce the resistance. If the layers do not have the proper thickness, however, the coupling mechanism can eliminate the GMR effect by causing ferromagnetic coupling between the magnetic layers.

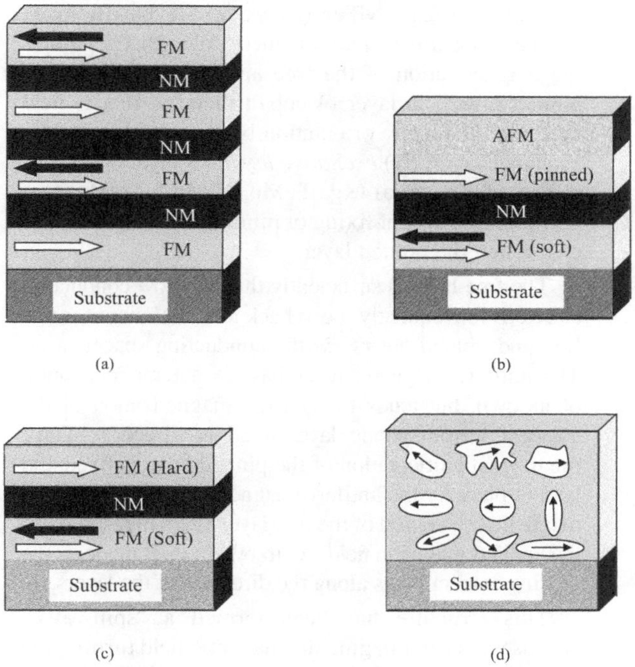

FIGURE 7.6 Different GMR structures: (a) multilayer stack, (b) spin valve, (c) pseudo spin valve, and (d) granular thin film. FM: ferromagnetic, NM: non-magnetic, and AFM: antiferromagnetic.

For spin-dependent scattering to be a significant part of the total resistance, the layers must be thinner (to a magnitude of several nanometers) than the mean free path of electrons in most spintronic materials. A typical GMR medical sensor has a conducting layer approximately 3 nm thick. For reference, this is less than 10 atomic layers of copper.

7.2.3.4 GMR Sensor Structures

Besides magnetic multilayers, GMR is exhibited by three main structures: spin valves, pseudo-spin valves, and granular solids. Features of all the four structures are presented below:

1. *Magnetic multilayers*: These work by an antiferromagnetic (materials in which the magnetism from magnetic atoms or ions oriented in one direction is canceled out by the set of magnetic atoms or ions that are aligned in the reverse direction) interlayer coupling (Figure 7.6a). Although the measured values of GMR are higher in magnetic multilayers, spin valves are more attractive from the point of view of applications, because only small magnetic fields need to be applied to change the resistance value.
2. *Spin-valve (SV) GMR sensor*: It consists of four layers (Figure 7.6b) (Pelegrí et al. 2003; Qian et al. 2003; Gupta et al. 2010): (i) *free ferromagnetic layer* – the

sensing layer (e.g., NiFe); (ii) *spacer* – typically made from copper, this is a nonmagnetic layer that separates the magnetization of the free and pinned layers; (iii) *pinned layer* – a layer of cobalt material that is held in a fixed magnetic orientation by its proximity to the exchange layer; (iv) *exchange layer* – a layer of antiferromagnetic material (e.g., FeMn, NiMn, or IrMn) that performs the task of fixing or pinning the magnetic orientation of the pinned layer.

The free layer is sufficiently thin to allow conduction electrons to frequently move back and forth between the free and pinned layers *via* the conducting spacer layer. The antiferromagnetic layer has no net magnetization of its own, but tends to hold the magnetization of the adjacent ferromagnetic layer fixed in direction. Thus, the magnetic orientation of the pinned layer is fixed and held in place by the antiferromagnetic layer, whereas the magnetic orientation of the free layer alters in response to the applied magnetic field. As in other GMR devices, the sensing current flows along the direction of the layers.

This structure has been termed a "spin-valve" because one can imagine the magnetic field turning the upper layer like a faucet valve to control the flow of spin-polarized electrons through the device.

3. *Pseudo-spin-valve*: In this structure, the antiparallel alignment is obtained due to different coercivities of the two ferromagnetic layers (Figure 7.6c). The magnetic moments of the soft and hard magnetic layers switch at different values of the applied magnetic field, providing a range of magnetic fields in which they are antiparallel and the resistance is higher.
4. *Magnetic granular solids*: Here, magnetic precipitates are embedded in a nonmagnetic metallic material (Figure 7.6d) (Arana et al. 2004; 2005). In the absence of the field, the magnetic moments of the granules are in a state of random orientation. The magnetic field aligns the moments in a certain direction and thus there is a decline in resistance.

7.3 Tunneling Magnetoresistance

Tunneling magnetoresistance (TMR) structures are similar to spin valves. One dissimilarity is that they utilize an ultrathin insulating layer to separate two magnetic layers, in place of a conductor. The thickness of this layer is fairly low, in order that electron tunneling takes place. Complete prototype sensors are fabricated using an Al_2O_3 (alumina) barrier with a thickness of 1.3 nm. The two electrodes are made of ferromagnetic metal.

TMR devices use the spin-valve arrangement of a pinned magnetic layer and a free magnetic layer. Electrons pass from one layer to the other, through the insulator, by quantum-mechanical tunneling. It is common to electrically connect multiple TMR devices in series to increase the overall resistance and limit the voltage at each tunnel barrier. Voltages above a few hundred millivolts may damage the thin insulator. The electric current tunnels through or flows in a direction perpendicular to the layers (CPP), with contacts on the top and bottom of the film stack.

Depending on the orientation of their magnetization, the resistance of the junction varies. TMR is the relative variation of the resistance, depending on the magnetization of the two electrodes.

The ease of tunneling between the two magnetic layers is modulated by the angle between the magnetization vectors in the two layers. The two possible configurations are called parallel (P) and antiparallel (AP) configuration. When the magnetization of the layers is aligned, many states are available in the bottom layer into which the spin-polarized electrons from the top layer can tunnel. When the magnetization directions are opposite, the spin-polarized electrons are prevented from tunneling because they do not have the correct orientation for entry in the bottom layer. The process is also known as *spin-dependent tunneling (SDT)* (Kasatkin et al. 2000; Freitas et al. 2007).

Figure 7.7 illustrates the configuration that exhibits switchable TMR (Cockburn 2004). A fixed ferromagnet plate is separated from a free ferromagnet plate by a very thin (e.g., ≤1.5 nm thick film) insulating barrier. The magnitude of the tunneling current through the barrier is proportional to the product of the densities of spin-aligned electron quantum states in the conduction sub-band

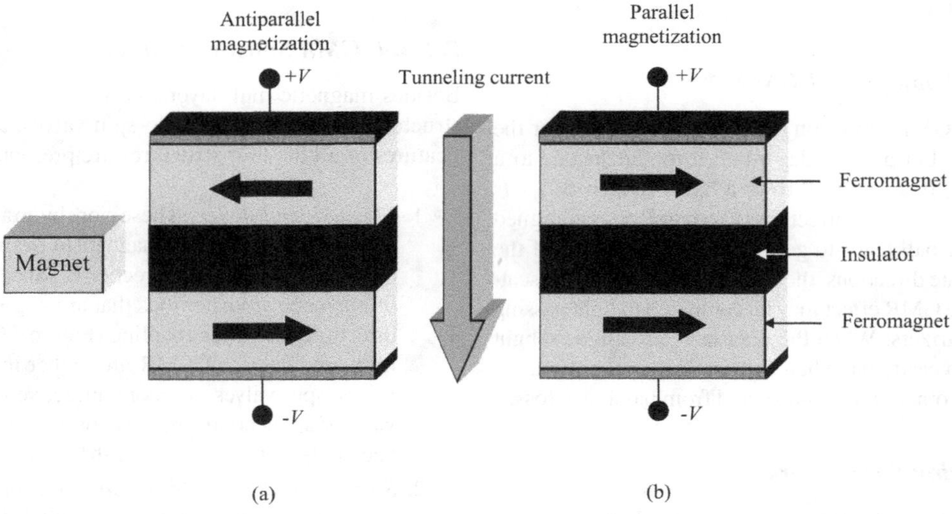

FIGURE 7.7 Tunneling magnetoresistance: (a) antiparallel and (b) parallel spins.

of the plates on either side of the barrier. If the magnetizations (and hence the majority electron spins) in the two plates are parallel, then the majority-spin-aligned electrons in the two plates tunnel more readily across the tunneling barrier. If the magnetizations (and hence majority electron spins) in the two plates are antiparallel, then the majority-spin-aligned electrons in one plate are inhibited from tunneling to the other plate (and *vice versa*) because the spin-aligned electron states are in minority there.

Magnetic tunnel junctions are made with transition-metal ferromagnetic materials and exhibit a TMR ratio of up to 50% at room temperature. Changes in resistance with a magnetic field of up to 70% have been observed in TMR structures. The field required for maximum change in resistance depends upon the composition of the magnetic layers and the method of achieving antiparallel alignment. Values of saturation field range from 0.1 to 10 kA m^{-1} (1.25-125Oe), offering, at the low end, the possibility of extremely sensitive magnetic sensors. The oersted is defined as $1000/4\pi$ ($\approx$79.5774715) A m^{-1}.

7.4 Limitations, Advantages, and Applications of GMR and TMR Sensors

7.4.1 Shortcomings

GMR and TMR sensors are nonlinear and this drawback limits their use as a replacement of Hall sensors for a large number of applications where sensitivity is not crucial. Design of the sensor and the supporting feedback electronics must be carried out to overcome this problem. *What is the major constraint for using GMR and TMR sensors for accurate measurement applications*? It is the building of hysteresis-free spin valves, which is necessary to give an efficient immunity against large random magnetic fields and to ensure reproducibility of the measurements. Hysteresis is the lagging of an effect behind its cause.

What are other constraints? A second constraint is the linearity of the sensor in the working region. Finally, the reduction of noise, and, in particular, the magnetic noise is essential for the use of GMR and TMR as sensors. At low frequencies, the inverse of frequency ($1/f$) noise dominates the thermal noise; it is related to structural defects and magnetic configuration. $1/f$ noise (occasionally called flicker noise or pink noise) is a type of noise, the power spectra $P(f)$ of which, as a function of the frequency f, behaves like: $P(f) = 1/f^a$, where the exponent a is very close to 1. Thermal noise is the electronic noise generated by the thermal agitation of the charge carriers, usually the electrons, in a conductor.

7.4.2 Advantages

Notwithstanding their drawbacks, compared to the Hall effect and AMR sensors, the many advantages of GMR sensors must be mentioned. They allow for large magnetic field-dependent changes in resistance, a larger output signal and the ability to have a larger air gap (distance from the sensor to the target), enabling tighter mechanical tolerances in system components to be achieved. GMR devices are, apparently, not susceptible to damage from large magnetic fields and also do not tend to exhibit any latch-up effect. Dense packing of magnetic sensors into very small areas is carried out using photolithography for TMR devices, with high resistance measuring only several μm on a side. Extremely small SDT devices, several μm on a side, with high resistance values, are fabricated using photolithography, allowing very dense packing of magnetic sensors into small areas, with SDT magnetic sensors having a number of attractive properties such as a pronounced spin-tunneling effect, sensitivity at the picotesla level (10–12 tesla; the tesla (T) is the SI unit of magnetic field: a particle carrying a charge of 1 C and passing through a magnetic field of 1 T at a speed of 1 m s^{-1} experiences a force of 1 N), high resistance, a small-sized junction, and low density of the sensor current (Kasatkin et al. 2000). From comparison with AMR sensors using similar materials, one should expect wide operational temperature and frequency ranges for SDT sensors.

7.4.3 Applications

The greatest technological application of GMR is in the *data storage industry*. Disk drives, based on GMR head technology, use their properties to control a sensor that responds to very small magnetic rotations on the disk. The magnetic rotation yields a very large change in sensor resistance, which, in turn, provides a signal that is collected by the electric circuits in the drive. Applications of GMR are diverse, in automotive sensors, solid-state compasses and nonvolatile magnetic memories; "nonvolatile" means that the memory is maintained even when power to the device is switched off.

Focusing on the hard disk, the GMR read-head sensor in a hard disk is manufactured using a spin- valve structure (Figure 7.8). Spin-valve resistance shows a steep change in the small field range around $H = 0$. As the magnetic bits on the hard drive pass under the read head, the magnetic alignment of the sensing layer in the spin valve changes, resulting in variation of its resistance. When the head passes over a magnetic field of one polarity, the electrons in the free layer respond by aligning with those on the pinned layer, creating a lower resistance in the head structure. When the head passes over a field of opposite polarity, the free layer electrons rotate so that they are no longer aligned with the electrons in the pinned layer. This increases the resistance of the structure. Because the resistance changes are ushered in by changes to the spin characteristics of electrons in the free layer, GMR heads are also known as *spin valves*, a term that was coined by IBM.

What are other important applications of GMR sensors? Besides data storage, GMR sensors are also widely used in biological analysis for biosensors that measure the presence of DNA or antibodies, in immunosorbent assays (Millen et al. 2005), magnetic arrays, among other applications. An immunosorbent assay is a biochemical technique used in molecular biology to detect the presence of an antibody or an antigen in a sample.

7.5 Magnetic Nanoparticle Probes for Studying Molecular Interactions

Research on nanoparticles of metals and semiconductors has been extensively pursued worldwide because of the unique electronic and optical properties of nanoparticles. When coupled to

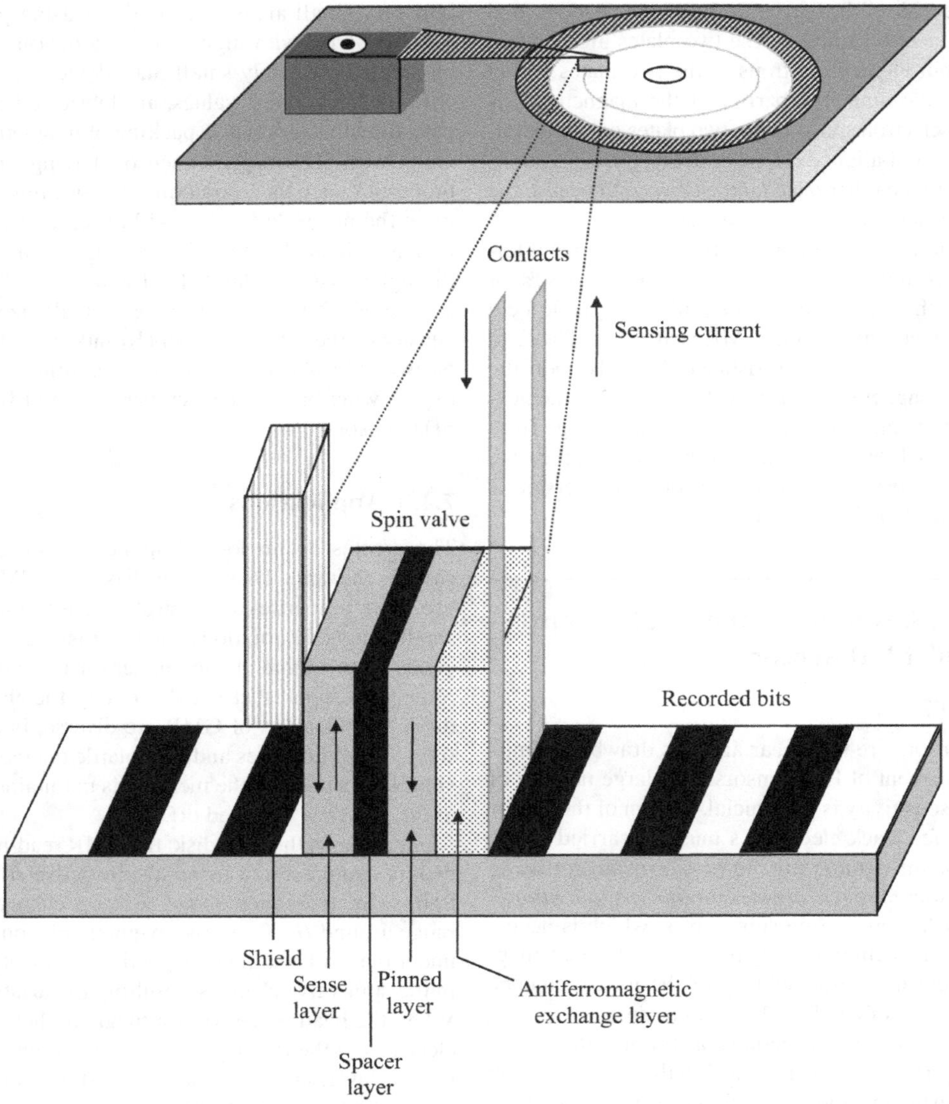

FIGURE 7.8 Application of GMR spin-valve sensor in a hard disk as a reading device.

affinity ligands, such nanoparticles function as sensitive biological nanosensors. A ligand is a substance that forms a complex with a biomolecule to serve a biological purpose. The interaction of ligands with their binding sites is characterized in terms of a binding affinity. In general, high-affinity ligand binding results from greater intermolecular force between the ligand and its receptor, whereas low-affinity ligand binding involves smaller intermolecular force between the ligand and its receptor.

Alternative detection technologies have been developed using magnetic nanosensors composed of magnetic nanoparticles, that are used to detect molecular interactions by magnetic resonance (MR), such as nuclear magnetic resonance (NMR)/magnetic resonance imaging (MRI) techniques.

What do NMR/MRI mean? NMR is a medical imaging method for visualization of body structures and their functioning, in which the nuclei of material placed in a strong magnetic field absorb radio waves supplied by a transmitter at particular frequencies. MRI is a diagnostic imaging technique that applies a magnetic field and radiowaves to produce proton density images from NMR of protons.

What is the composition, structure, and dimensions of magnetic nanosensors? Magnetic nanosensors are composed of 3–5 nm diameter monocrystalline iron oxide nanoparticles (MION), with an inverse spinel structure (cubic close-packed) of $(Fe_2O_3)_n(Fe_3O_4)_m$. Owing to their magnetic properties, iron oxide nanocores have a natural tendency to aggregate in suspensions.

What methods are adopted to prevent the particles from self-aggregation? To keep particles separated, two principal ways of particle stabilization have been developed. The first option is *steric stabilization*, where particles are coated with a thick layer of macromolecules that keep the iron oxide cores sterically apart. Dextran (complex, branched glucan [polysaccharide made of many glucose molecules]) and polyethylene glycol (PEG) polymers are extensively used for nonspecific and molecular imaging probes. These polymer coatings allow covalent coupling of chemical compounds, peptides (Schellenberger et al. 2004;

Schellenberger 2010), and proteins without impairing their steric stabilization. Therefore, sterically stabilized superparamagnetic particles have dominated molecular probe development for MRI. "Steric" relates to the spatial arrangement of atoms in a molecule. Steric effects originate from the fact that each atom within a molecule occupies a fixed amount of space. If atoms are brought too close together, there is an associated expenditure of energy due to overlapping electron clouds, and this affects the preferred shape (conformation) and reactivity of the molecule. The second option is electrostatic stabilization, with positively or negatively charged surface coatings. The advantage of electrostatic stabilization over the steric method is the possibility of developing smaller specific superparamagnetic nanoparticles with similar contrast effects. However, direct modification of the stabilized charged surface frequently results in an alteration of the surface charge, inducing destabilization and particle aggregation.

What is the average size of these nanoparticles? Depending on their preparation, the average size of the nanoparticles, surrounded by a 10 kDa dextran coating of approximately 10-nm thickness is about 25–30 nm (Perez et al. 2004). This is equivalent to a 750–1200 kDa globular protein (a protein that consists of long chains of amino acids folded up into complex shapes). 1 kDa = the weight of 1000 hydrogen atoms = 1.66×10^{-21} g.

Are these nanoparticles superparamagnetic? Yes.

What is superparamagnetism? Superparamagnetism occurs when a material is composed of very small crystallites (1–10 nm) (Gossuin et al. 2009). In this case, even though the temperature is below the Curie or Neel temperature (the temperature above which a ferromagnetic substance loses its ferromagnetism and becomes paramagnetic) and the thermal energy is not sufficient to overcome the coupling forces between neighboring atoms, the thermal energy is adequate to change the direction of magnetization of the entire crystallite. The resulting fluctuations in the direction of magnetization cause the magnetic field to average to zero. The material behaves in a manner similar to a paramagnetic material, except that, instead of each individual atom being independently influenced by an external magnetic field, the magnetic moment of the entire crystallite tends to align with the magnetic field. Essentially, superparamagnetism is a form of magnetism that occurs only in the presence of an externally applied magnetic field.

The nanoparticles become magnetized when placed in an external magnetic field and lose their magnetic moment when the field is withdrawn (Haun et al. 2010). This phenomenon arises because the nanoparticles consist of a single crystal domain and thus exhibit a net magnetic moment directed along an anisotropy axis. Aided by thermal energy, however, the magnetic moment can overcome the anisotropy barrier and spontaneously flip from one direction of anisotropy to another. Consequently, an ensemble of magnetic nanoparticles displays negligible magnetic remanence. The remanence is the strength of the magnetic field that remains in a magnetic nanoparticle after it is exposed to a strong, external magnetic field and the external field is then removed.

When placed inside an external magnetic field, the magnetic moments of the above-mentioned nanoparticles align in the direction of the field lines and enhance the magnetic flux. This behavior is similar to paramagnetism, but, since it is associated with a large particle with fixed magnetic moment, it is referred to as superparamagnetism.

How are neighboring water protons affected when these particles are placed in a magnetic field? When subjected to an external magnetic field, each superparamagnetic nanoparticle produces a large magnetic dipole. The resulting local magnetic field gradient creates an inhomogeneity in the external magnetic field, which destroys the coherent precession of nuclear spins of neighboring water protons.

What is the half-life of magnetic nanoparticles? Different MION preparations have a blood half-life of >10 h in mice. Related clinical nanoparticle preparations have circulation times of 24 h in humans.

What are amino-CLIO nanoparticles? CLIO (cross-linked iron oxide) nanoparticles (Haun et al. 2010) contain a superparamagnetic iron oxide core (3–5 nm-sized MION) composed of ferromagnetic magnetite (Fe_3O_4) and/or maghemite (γ-Fe_2O_3). Maghemite is structurally and functionally similar to magnetite but differs in one aspect, in that it contains cation vacancies in the sublattice.

How are functional groups attached to the magnetic nanoparticles? To develop more stably coated and amino-functionalized sensors, the iron oxide core is coated with dextran. The dextran coating is cross-linked with epichlorohydrin (C_3H_5ClO: an organochlorine compound and an epoxide). The dextran is successively treated with epichlorohydrin, to form stabilizing cross-links, and ammonia (NH_3), to provide primary amine group functionality. Preparations of the resulting aminated cross-linked iron oxide nanoparticle (amino-CLIO) have about 40 amino groups per particle, with an average particle size of 40–50 nm.

Amino groups in amino-CLIO react by bifunctional (having two functions) *N*-hydroxysuccinimide-based (NHS: $C_4H_5NO_3$) cross-linking (a covalent bond is formed between polymer chains, either within or across chains). This allows the attachment of a range of sulfhydryls, molecular formula, -C-SH or R-SH, where R represents an alkane, alkene, or other carbon-containing segment. The sulfhydryl group is a highly reactive and ubiquitous ligand in biological systems, found in most proteins and in some low molecular weight compounds such as cysteine (Cys) and glutathione (GSH). Sulfhydryl exists in the side chain of cysteine amino acids. The L-glutathione has a sulfhydryl group on its cysteinyl portion. These sulfhydryl groups facilitate the formation of biomolecule/amino-CLIO nanoparticle conjugates with unique biological properties. Such provision is not available with simple polymer-coated iron oxide nanoparticles.

What kind of treatments or handling can these coated magnetic nanoparticles tolerate? These nanoparticles can withstand harsh treatments, such as incubation at 120°C for 30 minutes, without any change in their size or loss of their dextran coating.

How are molecular interaction-probing experiments performed with magnetic nanoparticles? Binding of these magnetic nanoparticles with their intended molecular targets leads to the formation of stable nanoassemblies (Figure 7.8). As a result, there is a corresponding decrease in the spin-spin relaxation time (T_2) of surrounding water molecules. The T_2 of water molecules is the primary parameter of attention in these studies. Thus, the main idea behind the use of magnetic nanoparticles is to employ them as *proximity sensors* that modulate the spin-spin relaxation time of neighboring water molecules. This relaxation time is quantified using common clinical MRI scanners or benchtop NMR relaxometers.

What are the salient features of MRI scanners and benchtop relaxometers? MRI scanners have the disadvantages of (i) high operating costs, (ii) bulky size (mainly due to superconducting magnets: electromagnets made from coils of superconducting wires), and (iii) the need for relatively large sample volumes (hundreds of microliters). On the other hand, benchtop relaxometers offer a lower-cost alternative to MRI scanners, making them more accessible for use in diagnostic MR sensing. Benchtop systems operate at lower NMR frequencies (100 kHz–50 MHz), and are equipped with a permanent, low-field (<1 T) magnet for field generation. However, benchtop systems have two drawbacks, lacking the capability to perform parallel measurements and still requiring relatively large sample volumes.

It is necessary to recall here: *What is magnetic relaxation?* Magnetism is associated with angular momentum, called *spin*, because it usually arises from the spin of nuclei or electrons. The spins may interact with applied magnetic fields, the so-called *Zeeman energy* (the energy of the interaction between the magnetic moment of an atom or molecule and an applied magnetic field); with electric fields, usually atomic in origin; and with one another, through *magnetic dipole or exchange coupling* (the coupling between two nuclear spins due to the influence of bonding electrons on the magnetic field running between the two nuclei), the purported *spin-spin energy*. The *magnetic relaxation* phenomenon is defined as the approach of a magnetic system to an equilibrium or steady-state condition, over a period of time. Relaxation involves several processes by which nuclear magnetization, created in a nonequilibrium state, reverts to the equilibrium distribution. Furthermore, the relaxation is not an instantaneous phenomenon but requires time. The characteristic times involved in magnetic relaxation are known as *relaxation times*.

What are the different relaxation processes and the corresponding relaxation times? Different physical processes are responsible for the relaxation of the components of the nuclear spin magnetization vector **M**, parallel and perpendicular to the external magnetic field, $\mathbf{B}_0$ (which is conventionally oriented along the z axis). These two principal relaxation processes are designated as T_1 and T_2 relaxation times, respectively.

What is spin-lattice relaxation? It is the magnetic relaxation in which the excess potential energy associated with electron spins in a magnetic field is transferred to the lattice. (As used here, the term "lattice" does not refer to an ordered crystal but instead signifies degrees of freedom other than spin orientation, such as the translational motion of molecules in a liquid.) *What is spin-lattice relaxation time T_1?* T_1, also called the *longitudinal relaxation time*, characterizes the rate at which the longitudinal M_z component of the magnetization vector recovers. It is thus the time taken by the signal to recover around 63% [1 – (1/*e*)] of its starting value after being flipped into the transverse magnetic plane.

What is spin-spin relaxation? It is the magnetic relaxation, observed after application of a weak magnetic field, in which the excess potential energy associated with electron spins in a magnetic field is redistributed among the spins, resulting in heating of the spin system. It is a complex phenomenon corresponding to a decoherence or dephasing of the transverse nuclear spin magnetization. *What is spin-spin relaxation time? Spin-spin relaxation time*, also known as *transverse relaxation time* (T_2), is a time constant in NMR and MRI. It is named in contrast to T_1, the spin-lattice relaxation time. T_2 characterizes the rate at which the M_{xy} component of the magnetization vector in the transverse magnetic plane undergoes decay. It is the time taken by the transverse signal to fall to 37% (1/*e*) of its initial value after flipping into the transverse magnetic plane. Hence, the following formula describes the process:

$$M_{xy}(t) = M_{xy}(0)\exp\left(\frac{-t}{T_2}\right) \tag{7.13}$$

T_2 decay occurs 5–10 times more rapidly than does T_1 recovery, and different tissues have different values of T_2. Fluids have the longest T_2, ~700–1200 ms. Water-based tissues have T_2 in the 40–200 ms range, whereas fat-based tissues have T_2 in the 10–100 ms range.

It may be noted here that spin-spin relaxation does not change the total energy of the spin interactions, whereas spin-lattice relaxation modifies the same. Further, spin-spin relaxation is associated with an internal equilibrium of the spins amongst themselves. Spin-lattice relaxation is associated with the approach of the spin system to thermal equilibrium with the host material.

Returning and referring to the discussions on molecular interaction probing experiments, *what kind of assays are carried out with magnetic nanoparticles?* Sensitive homogeneous assays are performed using magnetic nanoparticles to detect a variety of different molecular interactions in biological samples, with minimal or no sample preparation. These magnetic nanosensors are able to detect specific mRNAs (messenger RNA), proteins, enzymatic activity, or pathogens (e.g., viruses) with sensitivity in the low femtomolar range (0.5–30 fmol).

What is the mechanism of magnetic relaxation-switching in the above assay experiment? Being superparamagnetic, the iron oxide crystal core in CLIO becomes magnetized when placed in an external magnetic field. The combined electron spins in the crystal produce a single large magnetic dipole, creating a local magnetic-field gradient and therefore an inhomogeneity in the external magnetic field. Water protons diffuse within the local inhomogeneity process at an off-resonance frequency, dephasing their spins and thus increasing the relaxation rate ($1/T_2$), as described by outer-sphere theory, in which $1/T_2$ is directly proportional to nanoparticle cross-sectional area.

Dephasing or decoherence is the process by which quantum-mechanical interference is destroyed so that coherence in a substance caused by perturbation decays over time, and the system returns to the state before perturbation, being the mechanism that recovers classical behavior from a quantum system.

Outer-sphere electron transfer is, by definition, intermolecular. Electron transfer is the process by which an electron moves from one atom or molecule to another atom or molecule. Outer-sphere electron transfer occurs between identical or dissimilar chemical species, differing only in their oxidation state. Here, the participating redox centers are not linked *via* any bridge during the electron event, so that the electron hops through space, from the reducing center to the acceptor.

It is hypothesized that, when individual superparamagnetic nanoparticles assemble into clusters (Figure 7.9) and the effective cross-sectional area becomes larger, the nanoassembly becomes

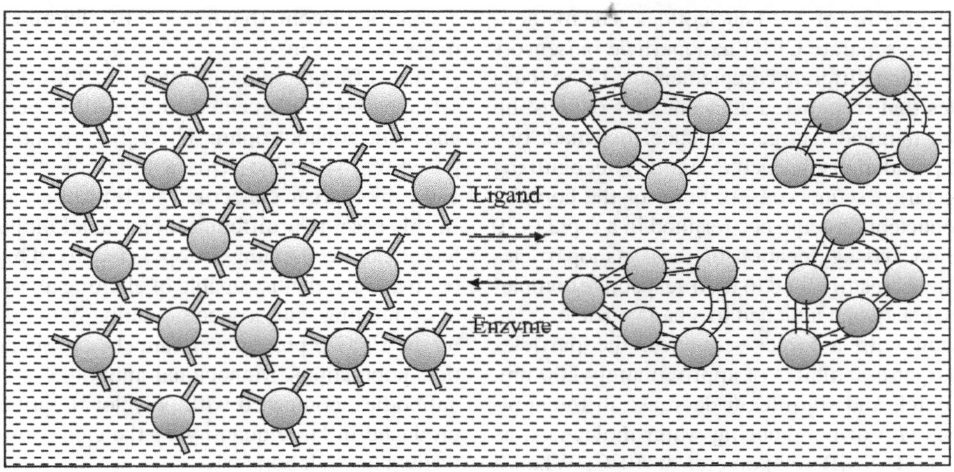

FIGURE 7.9 Self-assembly of superparamagnetic particles in the presence of a target with an associated decrease in T_2 relaxation time.

more efficient at dephasing the spins of surrounding water protons, leading to an enhancement of the relaxation rates ($1/T_2$). Hence, these magnetic nanosenors are referred to as *magnetic relaxation switches.* They act as switches through target-mediated clustering. Due to the presence of molecular targets, they switch from a dispersed state to a clustered state or *vice versa,* with changes in spin-spin relaxation time of water (T_2).

Interestingly, the spin-lattice relaxation time (T_1) is independent of nanoparticle assembly formation. Hence, this parameter is used to measure concentration in both nano-assembled and -dispersed states within the same solution. Nanoassembly formation is designed to be reversible (e.g., by temperature, chemical cleavage, pH, etc.) so that forward or reverse assays can be designed and implemented. Several examples of forward (clustering) or reverse (declustering) types of assays for detecting a large variety of biologically relevant materials can be cited.

Biosensing strategies based on magnetic nanoparticles have received considerable attention because they offer unique advantages over other techniques. *What are the special advantages of this method?* These include the following: (i) rapid detection of a target without elaborate purification of the sample or requirement of signal amplification; (ii) biological samples exhibit virtually no magnetic background, and, thus, highly sensitive measurements can be performed in turbid or otherwise visually obscured samples without further processing. The use of light, as in analysis entailing fluorescence, absorbance, chemiluminescence, etc., is avoided. The significant benefit is that light does not affect the outcome of the assay, and experiments are conducted in turbid, light-impermeable media such as blood, cell suspensions, culture media, lipid emulsions, or even whole tissue; (iii) immobilization of the sample onto a flat surface, such as microarray glass slides, is not necessary. Hence, faster hybridization and binding kinetics are observed; (iv) the assay provides the flexibility to detect various biomolecular interactions, like mRNA, protein, and enzymatic activity simultaneously. It is accomplished in a high-throughput format by using MRI and NMR; (v) At the low iron concentrations (<20 µg Fe per mL) used in the assay, the nanometer-sized clusters do not aggregate or precipitate; and (vi) magnetic nanoparticles are inexpensive to produce, are physically and chemically stable, biocompatible, and environmentally safe. Since identical iron oxide nanoparticles are commonly used in clinical studies and have shown little to no toxicity, the assay is applicable to *in-vivo sensing* of molecular targets by MRI.

Example 7.3

For tissues having $T_2 = 85$ ms, find the time by which $M_{xy}(t)$ component of magnetization decreases by 50%.

Here, $M_{xy}(0) = 100$ and $M_{xy}(t) = 50$. By Equation 7.13,

$$50 = 100 \times \exp\left(\frac{-t}{T_2}\right) \quad (7.14)$$

Or

$$\ln(0.5) = \frac{-t}{85} \quad (7.15)$$

from which $-0.69315 = -t/85$; therefore, $t = 58.92$ ms.

7.5.1 DNA Analysis

Josephson et al. (2001) and Perez et al. (2002a, b) conducted several experiments to detect oligonucleotide sequences by coupling an average of three oligonucleotides (12 base pairs) to the nanoparticles, using *N*-succinimidyl-3-(2-pyridyldithio)propionate (SPDP) as a linker. SPDP, $C_{12}H_{12}N_2O_4S_2$, is a heterobifunctional, thiol -cleavable and membrane-permeable crosslinker. A thiol (thio- + alcohol) is an organosulfur compound represented by R-SH where R is an alkyl group or other organic substituent and the –SH functional group is called thiol or sulfhydryl group, the sulfur analogue of the hydroxyl or alcohol group. SPDP contains an amine-reactive NHS ester, $C_4H_5NO_3$, that reacts with lysine ($C_6H_{14}N_2O_2$) residues to form a stable covalent amide bond; an amine is an organic compound/functional group that contains a basic nitrogen atom with a lone pair. The other end of the spacer arm is terminated in the pyridyl disulfide ($C_{10}H_8N_2S_2$) group, that will react with sulfhydryls (–C–SH or R–SH, where R = an alkane, alkene, or C-containing part of a molecule) to form a reversible disulfide bond (a covalent bond, usually derived by the coupling of two thiol groups: R–S–S–R). An oligonucleotide is a short nucleic acid polymer of 2–20 nucleotides. The methodology is as follows: two separate CLIO nanoparticle biosensors were prepared by attaching different complementary 12-bp (base pair) oligonucleotides that recognized adjacent sites on a 24-bp target sequence for each intended target sequence; they made two unique nanoparticle populations (termed P_1 and P_2), recognizing adjacent 24-bp-long target sequences. Addition of the complementary target oligonucleotide sequence resulted in a rapid (<30 min) cluster formation and aggregation of these nanoparticles because oligonucleotides readily bind to their respective complementary nucleotides, e.g., when attached to a nanoparticle.

Upon hybridization with a target sequence, the particles oligomerized into larger assemblies of approximately 200 nm. This resulted in a quick and significant decrease in the spin-spin relaxation times (T_2) of neighboring water molecules, as measured with a 0.47 T NMR relaxometer at 40°C. When a non-complementary oligonucleotide was used, however, no change in T_2 was noted. The average decrease in T_2 was linearly related to the amount of DNA added, in the concentration range displayed. Similar changes in T_2 relaxation times were observed when the experiments were conducted in turbid solutions.

The nanoparticle assemblies could be dissociated by heating. Moreover, no change in T_2 was observed when scrambled oligonucleotide sequences were used, but the inclusion of even a single mismatch in the target sequence resulted in a detectable signal change relative to the perfectly matched sequence. This technique may therefore prove valuable for mutational analysis, i.e., biochemical identification of mutational changes in a nucleotide sequence. Using this technique, researchers were successful in detecting as little as 10 fmol of DNA, with lower amounts (0.5 fmol) being detected, using a 1.5 T MR imager.

7.5.2 Protein Detection

Biosensors were formerly developed to detect soluble proteins by leveraging the specificity of antibodies (proteins produced by the body's immune system when it detects foreign substances, called antigens). The first illustration involved detection of *green fluorescent protein* (GFP) to establish a proof of the principle. GFP from the jellyfish *Aequorea victoria* is a 27 kDa monomeric protein consisting of 238 amino acids. It fluoresces green in the presence of UV light. It is a molecular tag that can be inserted into genes to make transgenic animals and plants (i.e., which contain the alien gene) glow green.

GFP-sensitive CLIO was prepared by conjugating CLIO with avidin, followed by attachment of biotinylated anti-GFP polyclonal antibody; a polyclonal antibody or antisera is the antibody derived from multiple B cells or cell lines in response to the antigen in question. Perez et al. (2002a; 2002b) prepared avidin-P1 conjugates as generic reagents for attachment of any biotinylated antibody (or peptide) to the nanosensor. Normally, each nanoparticle contained two molecules of avidin (i.e., eight binding sites). Biotinylated anti-GFP polyclonal antibody was attached to yield a GFP-sensitive nanosensor. When the nanosensors were used to investigate the GFP protein, noteworthy changes in T_2 relaxation time were found. GFP was rapidly (<30 minutes) detected in a dose-dependent manner to as low as the femtomolar range. These changes were time- and dose-dependent. Incubation with a control protein (bovine serum albumin [BSA]) showed no major changes in T_2.

In a different study (Kim et al. 2007), a slightly dissimilar scheme was implemented to detect the beta subunit of human chorionic gonadotrophin (HCG-β), a glycoprotein hormone composed of 244 amino acids with a molecular mass of 36.7 kDa, and a biomarker concerned with prostate and ovarian cancers. In this case, two different monoclonal antibodies that bind different, nonoverlapping epitopes (localized regions on the surface of an antigen that is capable of eliciting an immune response and of combining with a specific antibody to counter that response) on the HCG-β protein were attached to separate CLIO nanoparticle populations. Monoclonal antibodies are the highly specific antibodies, compared with polyclonal antibodies, which can be produced in large quantity by the clones of a single hybrid cell formed in the laboratory.

Using this system, HCG-β was quantitatively detected in a dose-dependent manner, and the sensitivity improved when an HCG dimer was used as the target, along with an increase in the number of antibodies per CLIO nanoparticle.

7.5.3 Virus Detection

Perez et al. (2003) reported the construction of magnetic nanosensors having the capability of detecting complex targets, such as intact viral particles in serum (a component of blood which is collected after coagulation). Since proteins are promptly detected by using this technique, they argued that magnetic nanoparticles, coated with antibodies raised against virus surface proteins, could be applied to identify the viral particle in solution. They conjectured that the multiple interactions occurring between a multivalent target (virus) and a multivalent magnetic nanoparticle-antibody conjugate could be utilized in an extremely sensitive assay for the detection of viruses.

The magnetic viral nanosensors developed were composed of a superparamagnetic iron oxide core with a shell of a dextran coating, on which virus-surface-specific antibodies were attached (Figure 7.10). Binding of anti-adenovirus 5 (ADV-5) or anti-herpes simplex virus 1 (anti-HSV-1) antibodies was accomplished *via* protein G coupling, attached to the dextran shell through *N*-succinimidyl-3-(2-pyridyldithio)propionate (SPDP); protein G is an immunoglobulin (IgG)-binding cell surface protein of group G streptococcal bacteria, and the SPDP protein ($C_{12}H_{12}O_4N_2S_2$) is a cleavable, water-insoluble, amino- and thiol-(sulfhydryl-) reactive heterobifunctional protein crosslinker. By and large, the particle size of the magnetic nanoparticles was 46 ± 0.6 nm.

Perez et al. (2003) first explored whether incubation of the ADV-5 viral particles with the anti-ADV-5 magnetic nanosensor would be able to form nanoassemblies in solution. Immediately after addition of 10^4 viral particles to the magnetic nanosensors, two distinct populations of particles were detected by light scattering. These pertained to magnetic nanoparticles alone (46 ± 0.6 nm) and viral particles (100 ± 18 nm). Within 30 minutes of incubation, the viral particle population became undetectable by light scattering, being replaced by a larger-sized population (494 ± 23 nm), corresponding to the virus-induced nanoassembly. The size of the nanoassembly continued to increase to some extent, reaching a plateau within 2 hours of 550 ± 30 nm. Atomic

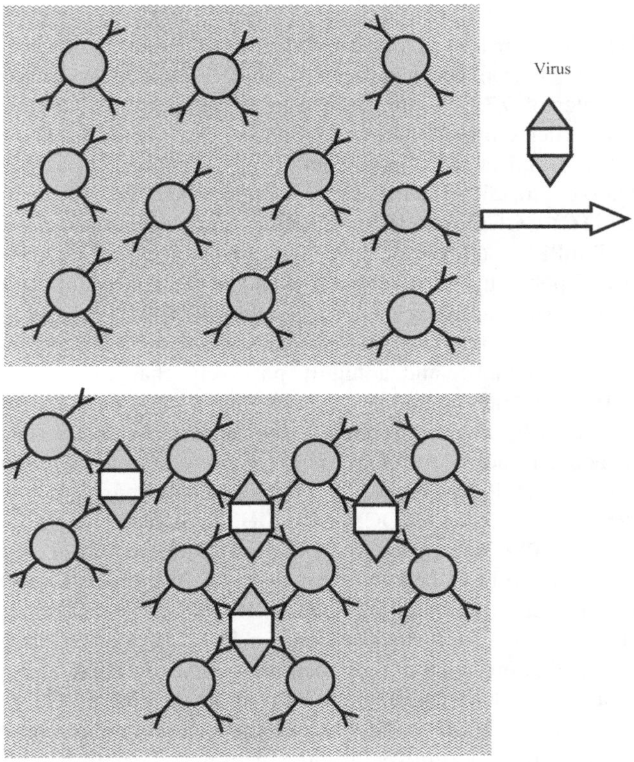

FIGURE 7.10 Magnetic viral nanosensors: clustering of nanoparticles in the presence of virus. (Perez, J. M., et al., *J. Am. Chem. Soc.*, 125, 10192, 2003.)

force microscopy (AFM) studies confirmed the presence of virus-induced nanoassemblies in solution. They investigated the effects of nanoassembly formation on the T_2 relaxation times of water at 0.47 T (20 MHz) with a tabletop relaxometer. Within 30 minutes, ΔT_2 was half-maximum and reached a plateau in less than 2 hours.

Using this approach, they were able to detect low levels (five viral particles in 10 μL) of herpes simplex virus-1 (HSV-1) and adenovirus-5 (ADV-5) in serum solutions. The method was also independent of the optical properties of the solution, permitting the detection of the virus in complex turbid media.

7.5.4 Telomerase Activity Analysis

Telomerase is a eukaryotic ribonucleoprotein (RNP) complex, containing both an essential RNA and a protein reverse transcriptase subunit. Using its intrinsic RNA as a template for reverse transcription, it helps to stabilize telomere length in human stem cells, reproductive cells, and cancer cells by adding TTAGGG (thymine-thymine-adenine-guanine-guanine-guanine) repeats onto the telomeres at the 3′ end of DNA strands in the telomere regions, found at the ends of chromosomes; this addition of repeat sequences to the telomere is catalyzed by the telomerase enzyme Telomerase activity has been found in almost all human tumors but not in adjacent normal cells. High levels of telomerase activity are found in most malignancies and are believed to play a critical role in tumorigenesis, offering an attractive target for therapeutic intervention and diagnostic or prognostic purposes. This makes telomerase a target not only for cancer diagnosis but also for the development of novel anticancer therapeutic agents. The ability to detect telomerase activity rapidly, quantitatively, and repeatedly would have significant value in cancer research, allowing the development of more efficient telomerase inhibitors, titrating treatment efficacy, and in the detection of enzyme levels for prognostic (predictive) purposes. Grimm et al. (2004) described the use of magnetic nanoparticles for investigating telomerase activity, to achieve rapid screening of telomerase activity in biological samples. The basis was hybridization of magnetic nanoparticles (consisting of iron oxide crystals core with a shell) to telomerase-associated TTAGGG repeats. Due to nanoparticle assembly formation, the relaxation time (T_2) of surrounding water changes significantly, which is readily measured by MR relaxometers or imaging systems.

Amidated cross-linked iron oxide nanoparticles (CLIO-NH_2) were conjugated to oligonucleotides *via* a stable thioether (a functional group in organosulfur chemistry with the connectivity C-S-C) linkage with *N*-succinimidyl iodoacetate ($C_6H_6INO_4$), a sulfhydryl- and amino-reactive, heterobifunctional protein cross-linking reagent. Each CLIO particle bound an average of four oligonucleotides. Aminated CLIO (amino-CLIO) nanoparticles have an average hydrodynamic diameter of 25–40 nm, with approximately 40–80 amines available per particle for conjugation of biomolecules. The hydrodynamic diameter of a particle is the effective diameter in a solution, measured by assuming that it is a body moving through the solution and resisted by the viscosity of the solution. If the solvent is water, the hydrodynamic diameter includes all the water molecules attracted to the particle. As a result, it is possible for a small particle to have a larger hydrodynamic diameter than a large particle, if it is surrounded by more solvent molecules. Simply stated, the hydrodynamic diameter is not measuring the particle, but is effectively measuring the nanoparticle and anything affiliated with its surface, for example, a surrounding double-layer, polymers or other capping agents, or nanoparticle aggregates. Thus, hydrodynamic diameter is an equivalent sphere diameter, derived from a measurement technique involving hydrodynamic interaction between the particle and the fluid.

The amine groups can then be reacted with various reagents to attach biomolecules *via* anhydride (formed from another compound by removal of water), amine (a functional group (–NH_2) formed by the loss of a hydrogen atom from ammonia), hydroxyl (–OH), carboxyl (–COOH), thiol (–SH), or epoxide (an oxygen atom joined by single bonds to two adjacent carbon atoms, thus forming the three-membered epoxide ring) groups. The resultant sensors were stable in solution over months, were monodisperse (in that the suspended particles have identical size, shape, and interaction) and had an overall size of 45 ± 4 nm. When sensors were incubated with telomeric repeats, they assembled in a linear fashion along the repeats, a feature that was not observed with control nonsense sequences.

Grimm et al. (2004) tested the efficacy of different telomerase inhibitors in crude human and murine (a term relating to a rodent of the family Muridae or subfamily Murinae, including rats and mice) samples, and showed that phosphorylation (the addition of a phosphate (PO_4) group to a protein or other organic molecule) of telomerase regulates its activity. High-throughput adaptation of the technique by MRI allowed processing of hundreds of samples within tens of minutes at ultra-high sensitivities.

They reported that, for all the cells tested, all tumor cell lines (breast, prostate, ovarian, pancreatic, lung, and liver carcinoma and glioma, a type of tumor that starts in the brain or spine; melanoma, the most dangerous type of skin cancer; lymphoma, a type of cancer involving cells of the immune system, called lymphocytes; and insulinoma, tumor of the pancreas that produces excessive amounts of insulin) and primary tumor tissue samples (liver metastasis) as well as human skin fibroblasts (the most common type of cells found in connective tissues) from newborns tested positive for telomerase, whereas primary skin melanocytes (a pigment-producing cell in the skin that determines its color) tested negative.

7.6 Protease-Specific Nanosensors for MRI

Imaging of enzyme activity is a central goal of molecular imaging. With the introduction of fluorescent "smart" probes for the *in-vivo* detection of enzyme activity, optical imaging has become the modality of choice for experimental *in-vivo* detection of enzyme activity. Protease, also called proteinase or peptidase, is a digestive enzyme that catalyzes the hydrolytic breakdown of long chain-like protein molecules comprising amino acids linked together by peptide bonds, into peptides and ultimately amino acids. Perez et al. (2002a; 2002b) introduced magnetic relaxation switches for sensing enzymatic activity of protease based on superparamagnetic high-relaxivity probes for *in-vitro* MRI. Following protease activation, the sterically stabilized cross-linked iron oxide particles (CLIO) switch from large clustered high-relaxivity aggregates to smaller low-relaxivity single particles.

The above technique is not suitable for *in-vivo* application, since the nonactivated clustered particles are too large to have a favorable bioavailability and, more importantly, activation by elevated enzyme activity in the target tissue *reduces* the relaxivity and, consequently, the contrast effect. Schellenberger et al. (2008) described a new design of protease-sensitive nanosensors (Figure 7.11), which are based on electrostatically stabilized, citrate-coated very small iron oxide particles (VSOP) with a hydrodynamic diameter of 7.7 ± 2.1 nm, instead of larger sterically stabilized nanoparticles (over 20 nm).

The synthesized fluorescein ($C_{20}H_{12}O_5$)-labeled MMP-9 (matrix metalloproteinase-9) peptide NH_2-GG*PRQITA*G-K(FITC)-GGGG-RRRRR-G-RRRRR amide was reacted with NHS-mPEG (*O*-[(*N*-succinimidyl)succinyl-amino ethyl]-*O*′-methyl-poly(ethylene glycol) at the end of the cleavage domain; the italicized amino acids correspond to the MMP-9 substrate. The peptide consists of a cleavage domain with the enzyme recognition motif, and a highly positively charged, arginine ($C_6H_{14}N_4O_2$)-rich coupling domain, the two domains being connected by a linker sequence. For analysis by fluorescence methods, a fluorescein molecule is coupled to the peptide. The resulting MMP-9-peptide-mPEG copolymer was purified by gel filtration: mPEG$=CH_3(OCH_2CH_2)_nOH$. To prepare 6×-MMP-9-PSOP (PSOP: protease-specific iron oxide particles), with six peptide-mPEG copolymers per VSOP (very small superparamagnetic iron oxide particle), in HEPES [2-[4-(2-hydroxyethyl)piperazin-1-yl] ethanesulfonic acid: $C_8H_{18}N_2O_4S$] buffer (pH$=$7.5) were mixed with peptide-mPEG in HEPES buffer and stirred immediately. A ratio of six peptide-mPEG per VSOP yielded MMP-9-PSOP with a hydrodynamic diameter of 24.9 ± 7.0 nm. Mixtures with ratios between 6 and 16 peptide-mPEG consistently yielded particles with sizes around 24 nm, whereas ratios below 6 resulted in particles over 30 nm in size.

When the sterically stabilized MMP-9-PSOPs were exposed to MMP-9, the protease cleaved the peptide at the recognition site, resulting in loss of the sterically stabilizing mPEG shell. Due to the superparamagnetic properties of the iron oxide cores and the ambivalent surface of the remaining particles with positively (coupling domains) and negatively (citrate ($C_6H_8O_7$) coat of VSOP) charged areas, the particles aggregated, driven by magnetic and electrostatic attraction.

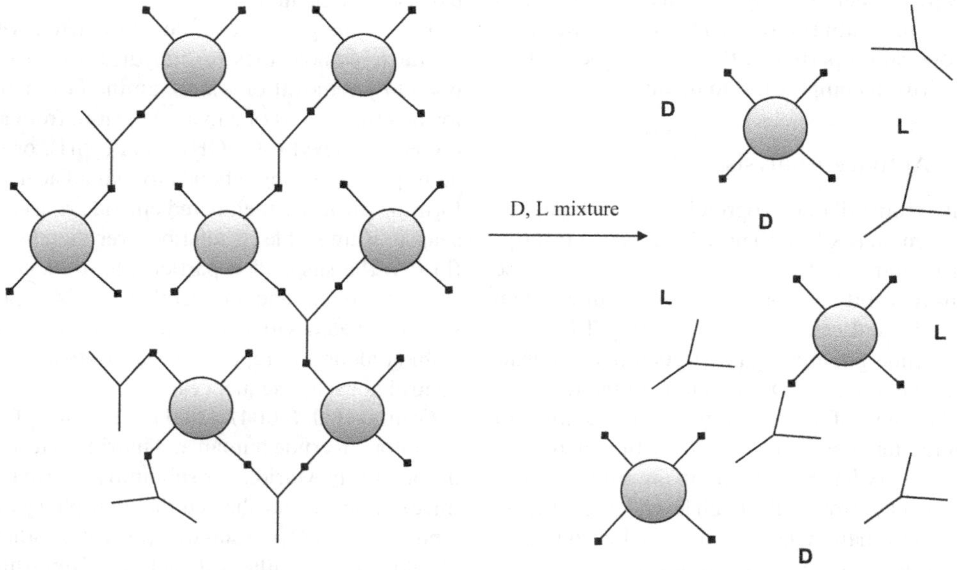

FIGURE 7.11 The addition of D-Phe impurities contained in L-Phe samples causes dispersion of CLIO-D-Phe/anti-D-AA self-assemblies and a corresponding increase in the T_2 relaxation time. (Schellenberger, E., et al., *Bioconjugate Chem.*, 19, 2440, 2008.)

This process has two important consequences. First, clustering of the superparamagnetic nanoparticles causes a substantial increase in R_2-relaxivity, called a magnetic relaxation switch. Second, the particles are converted from mPEG-covered stealth particles, avoiding detection, into highly aggregative particles with strongly charged surfaces. Conveniently, mPEG-5000 has been shown to be optimal for achieving stealth properties for nanoparticles. Consequently, once injected, the intact PSOP should remain in the blood circulation for a long time until they reach an MMP-9 expressing target tissue, where they are converted into aggregative particles and accumulate.

7.7 Magnetic Relaxation Switch Immunosensors

Tsourkas et al. (2004) developed a homogeneous enantioselective immuno-sensor that utilizes magnetic relaxation switching (MRS). An enantiomer is one of two stereoisomers that are mirror images of each other. "Enantioselective" relates to a chemical reaction in which one enantiomer of a chiral product (a molecule that is not superimposable on its mirror image) is preferentially produced. The enantioselective MRS immunosensor was based on magnetic nanoparticles labeled with a derivative of D-phenylalanine (D-Phe), $C_9H_{11}NO_2$, a form of the essential amino acid, L-phenylalanine, that, when taken as a supplement, protects the body's production of naturally occurring painkillers.

The magnetic nanoparticles consisted of a superparamagnetic iron oxide core with an aminated cross-linked dextran coating (CLIO). The CLIO-D-Phe nanoparticles were generated by first coupling tyramine ($C_8H_{11}NO$) to the primary amino groups of the CLIO nanoparticles by using a homobifunctional NHS ester ($C_4H_5NO_3$). Diazotization (a reaction between a primary aromatic amine and nitrous acid to give a diazonium compound) was then used to couple *p*-amino-D-phenylalanine ($C_9H_{12}N_2O_2$) through its side chain to the tyraminyl residues, thus preserving both the α-amino and carboxy groups attached to the stereogenic center (also known as a chiral center, it is characterized by an atom which has different groups bound to it in such a manner that its mirror image is nonsuperimposable).

When antibodies specific to D-amino acids (anti-D-AA) were added to the CLIO-D-Phe nanoparticles (Figure 7.12), the divalent nature of the antibodies resulted in the self-assembly of the nanoparticles, which led to a decrease of more than 100 ms in the T_2 relaxation time. The presence of D-Phe impurities in samples of L-Phe was then determined by performing a one-step competitive immunoassay. Upon addition of mixtures of the enantiomers (one of two stereoisomers that are mirror images of each other that are non-superposable) to the CLIO-D-Phe/anti-D-AA self-assembled structures, the presence of D-Phe impurities resulted in the dispersion of the nanoparticles by competing with the CLIO-D-Phe conjugates for antibody- binding sites. This subsequently led to an increase in the T_2 relaxation time. The presence of free D-Phe impurities could be detected within minutes, and the rate and magnitude of change in the T_2 relaxation time was dependent on the concentration of impurities.

An important attribute of magnetic relaxation switch immunosensors is their potential for the rapid determination of enantiomeric excess in a high-throughput format.

7.8 Magneto Nanosensor Microarray Biochip

7.8.1 Rationale and Motivation

Medical decision making is increasingly based on molecular testing. Some of the crucial issues involved are as follows. Detection across varied samples is difficult. A urologist (a physician who has specialized knowledge and skill regarding problems of the male and female urinary tract and the male reproductive organs) supplies a urine sample, a neurologist dealing with disorders of the nervous system gives cerebrospinal fluid (a protective nourishing fluid that circulates around the brain and spinal cord of the central nervous system), a cardiologist (a heart specialist) provides blood, or an oncologist (a physician who studies, diagnoses and treats cancerous tumors) supplies cell lysates (the cellular debris and fluid produced by lysis). The diversity of such supporting matrices has stalled the generalization and sensitivity of protein-detection platforms, thus greatly degrading their clinical utility. Another serious impediment is signal distortion that occurs in various matrices due to heterogeneity in ionic strength, pH, temperature, and autofluorescence (the fluorescence of substances other than the fluorophore of interest, increasing the background signal). In addition, sample preparation is in itself an elaborate and intricate preparatory step for any analysis. Samples need to be made in pure water or in accurately controlled salt solutions.

What is the solution to the problem? A matrix-independent method is necessary. A general sensing platform that can be ubiquitously applied to detect the range of biomolecules in varied clinical samples (e.g., serum, urine, cell lysates, or saliva) with high sensitivity and with a large linear dynamic range is the answer to the problem. The additional advantage is that the matrices of even the most complex biological samples lack a detectable magnetic background signal and produce no interference with the magnetic transduction mechanism. Therefore, a magnetic field-based detection platform is most appropriate for protein detection in clinical samples.

Reasoning on the above lines, Gaster et al. (2009) presented a magnetic nanosensing protein detection technology that overcomes the problems associated with other methodologies. They described a magnetic array biochip that is used for detecting surface binding reactions of biological molecules previously labeled with 10-100 nm superparamagnetic nanoparticles.

Is magnetic array biochip similar to DNA fluorescence-based biochip? Yes, in several respects, but magnetic labeling offers distinct advantages as compared to optical fluorescent labeling. *What are these advantages?* Some of the advantages are as follows: (i) inexpensive chip reader instrumentation, amenable to miniaturization; (ii) absence of label bleaching; (iii) improved background rejection; (iv) potentially greater sensitivity; (v) ability to simultaneously measure multiple binding reactions in homogeneous assays requiring no separation steps, with pipetting, incubating, and measuring being the only steps required; (vi) capable of seamless integration with magnetic separation methodologies; (vii) although biological molecules can exhibit autofluorescence, they have no intrinsic magnetic signal; and (viii) magnetically labeled biomolecules can be manipulated by applied magnetic fields, to pre-concentrate the analyte using microfluidic devices.

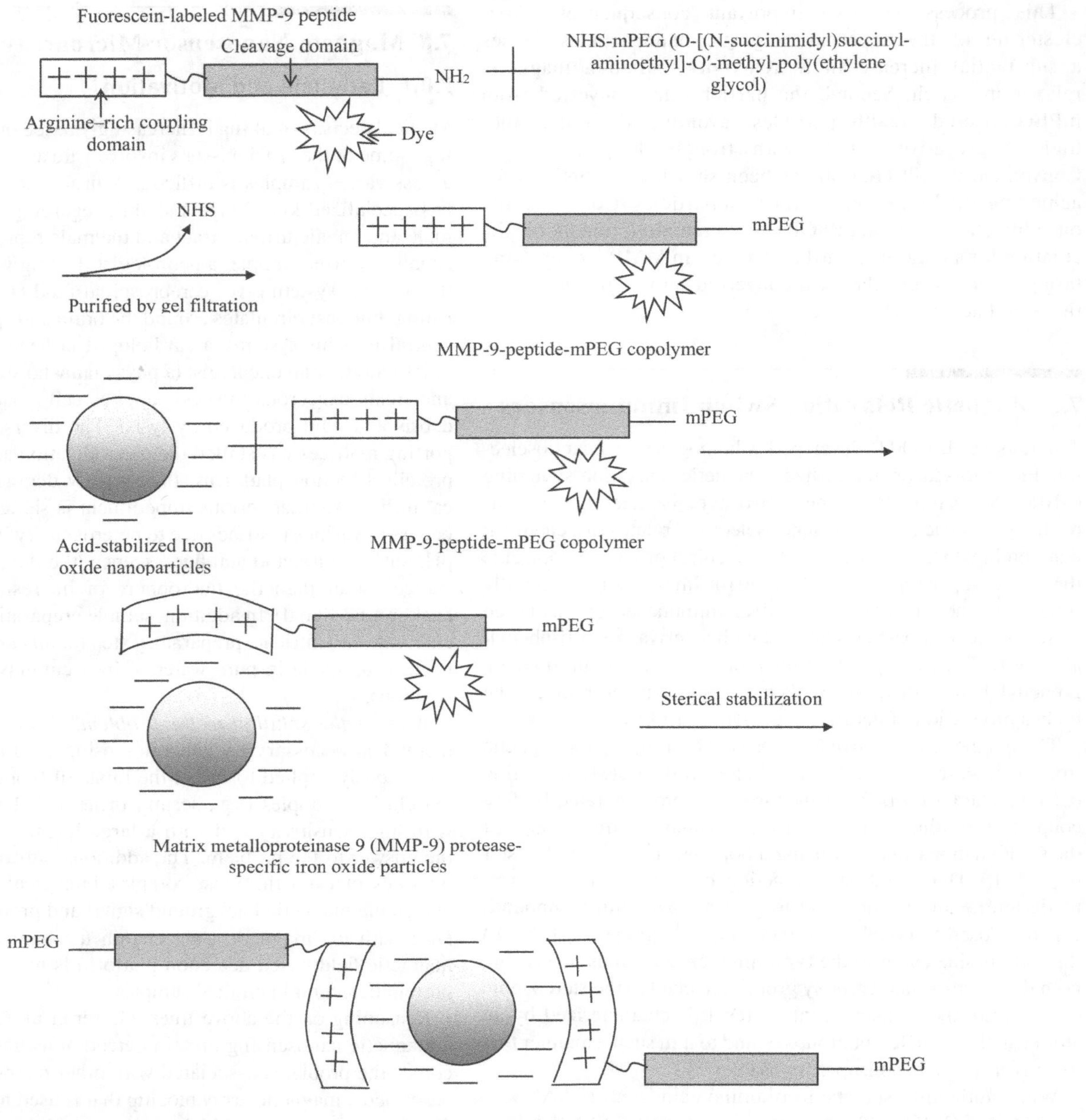

FIGURE 7.12 Synthesis and function of MMP-9-activatable protease-specific iron oxide particles. (Tsourkas, A. et al., *Angew. Chem.*, 116, 2449, doi: 10.1002/ange.200352998, 2004.)

The basic elements of magnetic array biochip are the *giant magnetoresistive (GMR) sensors* whose operation is explained in Section 7.8.2.

7.8.2 Sensor Choice, Design Considerations, Passivation, and Magnetic Nanotag Issues

As already described, and for obvious reasons, GMR sensors are favored (Kasatkin et al. 2010). These sensors, originally developed for use as read heads in hard-disk drives, are multilayer thin-film structures. They work on the basis of a quantum-mechanical effect, wherein a change in the strength and orientation of the local magnetic field induces a modification in resistance of the sensor.

The magnetoresistive thin film is sputtered on an insulating substrate, a silicon wafer, and patterned by lithographic techniques to delineate the array of several sensors. The aforementioned magnetic array biochip contained 64 such sensors in an area of 1.2 cm × 1.2 cm. Using a reasonably simple readout circuit, each sensor constituting the array can be independently and simultaneously accessed and the signal obtained.

The magnetoresistive sensor responds very quickly, on the timescale of nanoseconds. *How does the response of magnetoresistive sensors compare with conventional inductive coils?* Response of magnetoresistive sensors is a direct function of the proximate magnetic field whereas the inductive pickup coils respond only to changing magnetic fields.

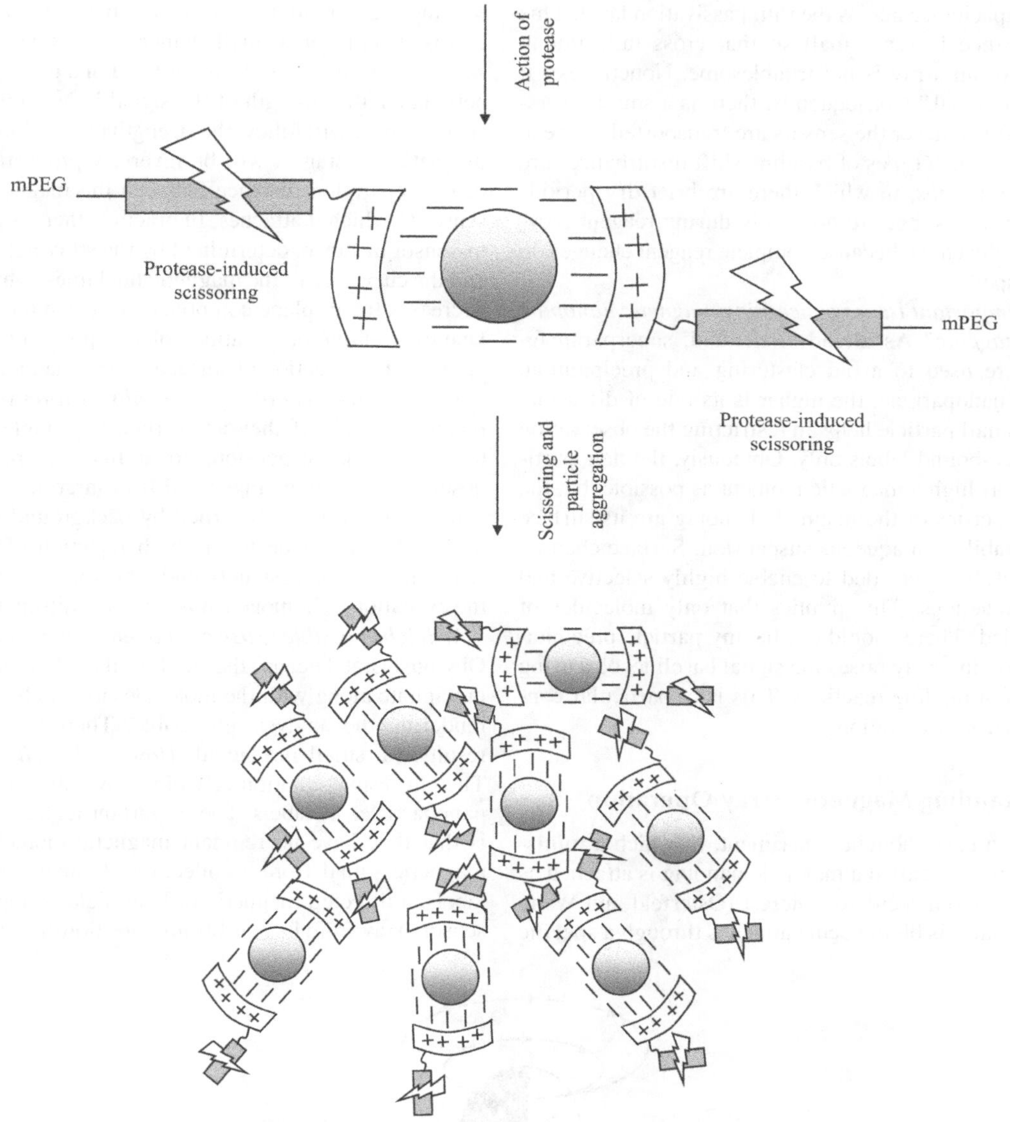

FIGURE 7.12 (Continued)

What are the design considerations for the sensor or biochips? The width of the sensor is an important performance-determining parameter. When very small magnetic nanoparticles are used, the narrower the sensor, the greater the sensitivity. But the benefit thus derived is limited because, according to theory, the edge of the spin valve produces an inverse (negative) signal, decreasing the positive signal from the reducing area of the spin valve. Attention must be paid to the magnetic domain formation of the sensor. The manner in which the free layer responds to the changing tickling field is governed by the layout of the sensor. Edges produce a local demagnetizing field, favoring the alignment of the magnetization parallel to any edges of the sensor. This can be utilized to instill a bias into the magnetization of the free layer, which will tend to align with the long axis of a linear segment. In the same way, curved or bent segments may inhibit coherent domain rotation, degrading the linearity and reproducibility of domain rotations and thereby increasing the noise in the measurements. The final resistance value obtained from the design is also a vital parameter. It should lie in a range that is easily measurable at low voltages. Higher measurable voltages mean applying more stress to the very thin passivation films isolating the sensor from the aqueous reagents.

What are the challenges facing the passivation of the magnetic arrays? It should be sufficiently durable to withstand corrosion in an aqueous environment. Leakage currents should be minimized. But it must be thin enough so that sensitivity of the chip is not decreased. An oxide-nitride-oxide (ONO) trilayer of 50-nm thickness serves as a good passivation. The silicon dioxide layers symmetrically placed on the two sides of this composite layer do not allow any residual stress to be developed, which is likely in a two-layer structure. Also, silicon nitride is a good diffusion barrier. Many spin-valve sensors show no corrosion damage as evidenced by a reproducible signal baseline in repeated measurements and stability of resistance values, when immersed in aqueous solutions or buffers. Although corrosion damage may be insignificant, alternating current leakage occurs from parasitic capacitance across the thin passivation layer. This parasitic capacitance is very small so that cross talk among various sensors of an array is not troublesome. Nonetheless, it leads to a "water signal." Consequently, there is a small reversible baseline shift whenever the sensors are transported between dry and wet states. Such types of baseline shift disturbances are found in open-well chips, in which there are brief dry periods during reagent changes, but are not likely during reagent recycling in microfluidic chips, because complete reagent changes do not involve dry spells.

What are the important issues concerning magnetic nanoparticles used for tagging? As already indicated, superparamagnetic particles are used to avoid clustering and precipitation. The smaller the nanoparticle, the higher is its rate of diffusion. Furthermore, a small particle helps in restricting the observation volume to surface-bound labels only. Obviously, the nanoparticles should have as high a magnetic moment as possible, but the most critical properties of the magnetic nanotag are its surface chemistry and stability in aqueous suspension. Surface chemistry must be carefully controlled to enable highly selective and strong binding reactions. This implies that only molecules of interest are labeled. There should not be any particle precipitation because it continuously raises the signal baseline, obscuring the equilibration of binding reactions. This is of particular concern at low analyte concentrations.

7.8.3 Understanding Magnetic Array Operation

Let us perform an easy thought experiment, in which a miniscule permanent magnet called a magnetic nanotag is attached as a tag or label to a biomolecule of interest (Osterfeld and Wang 2009). Suppose that this biomolecule attaches through a specific binding reaction to a sensor of the magnetic array. Consequent to this attachment, a small change in resistance of the particular sensor of the array will be recorded. For a properly oriented magnetic label, the strength of the signal induced in the magnetoresistive sensor will follow the strength of the dipolar field from the magnetic nanotag; it will be inversely proportional to the cube of the separation distance between the magnetic label and the sensor to which it attaches. In practice, there is an optimal label-to-sensor distance, determined by the stray field from the sensor and the curvature of the magnetic field lines, which, on approach, decreases the in-plane component to which the sensor responds. Due to the finite observation volume, properly designed sensors are ideal for detection of surface-bound magnetic nanotags.

What is the effect of this finite observation volume? Unbound magnetic labels, if they are sufficiently stable at high concentrations in the suspension, are unlikely to penetrate into this observation volume (Figure 7.13) in large numbers, so that the experiments are not disturbed by background noise from these unbound labels. Due to the high rejection of unbound labels, removal of the excess unbound labels is not required, enabling the execution of homogeneous assays, omitting the washing step.

Will it be possible to use permanent magnetic labels in reality? Obviously not, because the labels will tend to cluster together and precipitate along with the molecules to which they are attached, rendering the analysis unfeasible. Their free orientation will result in a small net signal. *How is this situation overcome?* The clustering situation is avoided by using superparamagnetic nanoparticles as labels. The important feature of these particles is that they have no remnant magnetic moment. To produce a magnetic signal from a collection of superparamagnetic nanolabels, a directed magnetic tickling field is applied, driving the sensor array from its equilibrium position. The tickling field also

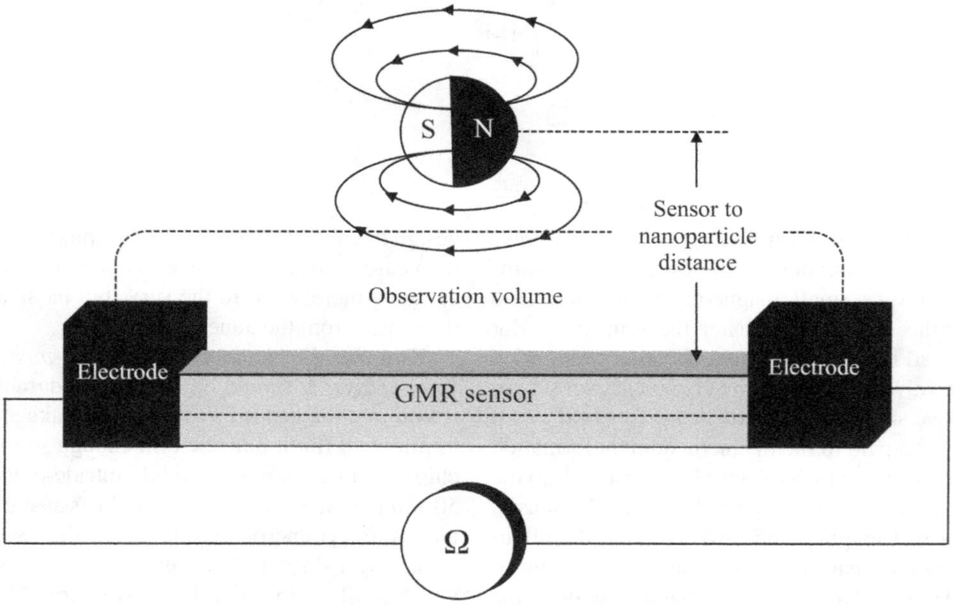

FIGURE 7.13 Observation volume: the signal from a magnetic nanoparticle decreases perceptibly as the sensor-to-nanoparticle distance increases to a few hundred nanometers, resulting in a finite observation volume covering mainly surface-bound nanoparticles. The background signal from the unbound particles is almost negligible. Rejection of this background signal avoids interferences from it. (Osterfeld, S. J. and Wang, S. X, MagArray biochips for protein and DNA detection with magnetic nanotags: Design, experiment and signal-to-noise ratio, in *Microarrays: Preparation, Micro fluidics, Detection Methods and Biological Applications*, Dill, K., R. H. Liu, and R Grodzinski (Eds.), Springer, New York, 299, 2009.)

affects the supermagnetic labels, because they become magnetized by this field and produce a small dipolar magnetic field opposite in direction to the tickling field itself. This dipolar field of magnetic labels slightly decreases the response of the sensor to the tickling field.

The magnetic array strategy employs a "sandwich" assay, in which the target antigen is sandwiched between two antibodies, one bound to the sensor and the other tagged with a superparamagnetic nanoparticle. Under an external magnetic field, the nanoparticles magnetize, and their presence or absence can be detected by the underlying GMR sensor.

What is the general methodology for conducting an assay? The main steps of the assay are as follows: (i) capture antibodies, that are complementary to a chosen antigen of interest, are immobilized onto the surface of each sensor; (ii) the complementary antigens bind to the antibodies while the noncomplementary antigens are washed off; (iii) after adding a concoction of detection antibodies, the biotinylated detection antibody complementary to the antigen of interest fastens in the sandwich structure, and the noncomplementary antibodies are washed away; (iv) a streptavidin-labeled magnetic nanoparticle tag is added to the solution, and it binds to the biotinylated detection antibody; biotin=$C_{10}H_{16}N_2O_3S$; and (v) as the magnetic tags diffuse to the GMR sensor surface and bind to the detection antibody, the magnetic fields from the magnetic nanoparticles are detected by the underlying MR sensor in real time in the presence of a small external modulation magnetic field.

Osterfeld and Wang (2009) described a systematic procedure to carry out a high-sensitivity proteomics assay for quantifying the concentration of biotinylated anti IFN-γ (anti-interferon-gamma) antibodies in phosphate-buffered saline (PBS) solution. The magnetic array is shown in Figure 7.14a. Sensor A has a thick passivation layer; therefore, it remains unaffected by magnetic nanoparticles. Because nanoparticles are prevented from entering the physical barrier of observation volume for this sensor A by the passivation layer, clearly, sensor A will not respond to any magnetic nanoparticle label. It will maintain an absolute reference level and serve as a *reference sensor* for the measurements. Sensor B is coated by IFN-γ in PBS manually by pouring a droplet over it. Such a sensor is called a *positive sensor* and is expected to interact with the analyte to be examined. Sensors like sensor C are not covered with any biomolecule and are therefore susceptible to nonspecific adsorption of IFN-γ or any molecules passing by. Hence, such sensors are termed *neutral sensors*. Sensors, like sensor D, are saturated with IL6-sR (interleukin 6-soluble receptor) and will therefore not be vulnerable to any nonspecific adsorption like sensor C. They are known as *negative sensors*. Using the procedure outlined above, portions of the magnetic array are functionalized in one of the four ways to produce reference sensors (A), positive sensors (B), neutral sensors (C), and negative sensors (D), as shown in Figure 7.14b.

After functionalizing the chip, it is rinsed with 1% BSA in PBS solution and the chip well is filled with analyte containing anti-IFN-γ in PBS. The chip is rinsed again with BSA and transferred to a measuring station for analyte quantification. Two pairs of differential signals are recorded. One pair is between the positive sensor and the reference sensor indicating the amount of specific adsorption, while the; other pair is between the negative and reference sensors, revealing the amount of nonspecific adsorption on a nonmatching functionalization, hence, achieving cross-reactivity. An additional differential pair between signals on the neutral and reference sensors will indicate the amount of nonspecific adsorption on a bare surface.

After the differential pairs have been established as desired, the chip is subjected to priming rinses in PBS several times. During the alternating wet and dry transitions between these priming rinses, the baseline shifts back and forth, but these shifts are reversible and reproducible. After stabilization, the streptavidin-coated nanoparticle solution is delivered to the reaction well of the chip and incubated. Capturing of these nanoparticles by the biotin ligand of the analyte causes their immobilization within the observation volume of the sensors, accompanied by a rise in the signal level. After the incubation period, the excess nanoparticles are flushed out of the reaction well by air, followed by rinsing with deionized water.

An important figure from the array is the *biochemical signal-to-noise ratio*, defined as the ratio of positive and negative signals, taking into account nonspecific adsorption and cross-reactivity.

7.8.4 Influence of Reaction Conditions on the Sensor

Gaster et al. (2009) investigated how the sensor itself (before the addition of the detection antibody) responded to various reaction conditions, including pH, temperature, and turbidity (the cloudiness or haziness of a fluid). In contrast to nanowires, in which a change of 0.5 pH causes considerable signal fluctuations, their sensing technology was unaffected by changes in ionic strength and pH change between pH 4 and 10. In addition, unlike microcantilevers, for which even a 0.5°C change caused substantial cantilever deflection, their sensors were unaffected by changes in the temperature of the sample, provided that a simple temperature correction algorithm was implemented. This is performed in real time without having to rely on reference sensors. Finally, optical activity or turbidity of the sample solutions had no effect on this detection platform, as it does not use optical-based detection methods, as do enzyme-linked immunosorbent assays (ELISAs), protein microarrays, and quantum dots. Protein arrays comprise a library of proteins or antibodies immobilized in an ordered manner in a 2D addressable grid on a chip. They are used to identify protein-protein interactions, to identify the substrates of protein kinases, or to identify the targets of biologically active small molecules.

7.8.5 DNA and Tumor Marker Detection

A DNA microarray (also commonly known as gene chip, DNA chip, or biochip) is a collection of microscopic DNA spots attached to a solid surface. It is used to measure the expression levels of large numbers of genes simultaneously or to genotype multiple regions of a genome.

Xu et al. (2008) developed a GMR biochip based on a spin-valve sensor array and magnetic nanoparticle labels for inexpensive, sensitive, and reliable DNA detection. The resistance of a GMR sensor changes with the magnetic field applied to the sensor, so a magnetically labeled biomolecule can induce a signal. Human papillomavirus (HPV) is a subset of papillomaviruses that infects the epithelial cells of the skin and mucus membranes in humans. Infection with HPV is associated with various forms of cancers, including cervical cancer, cancer that starts in the cervix, the lower part of the uterus (womb) that opens at the top of the vagina.

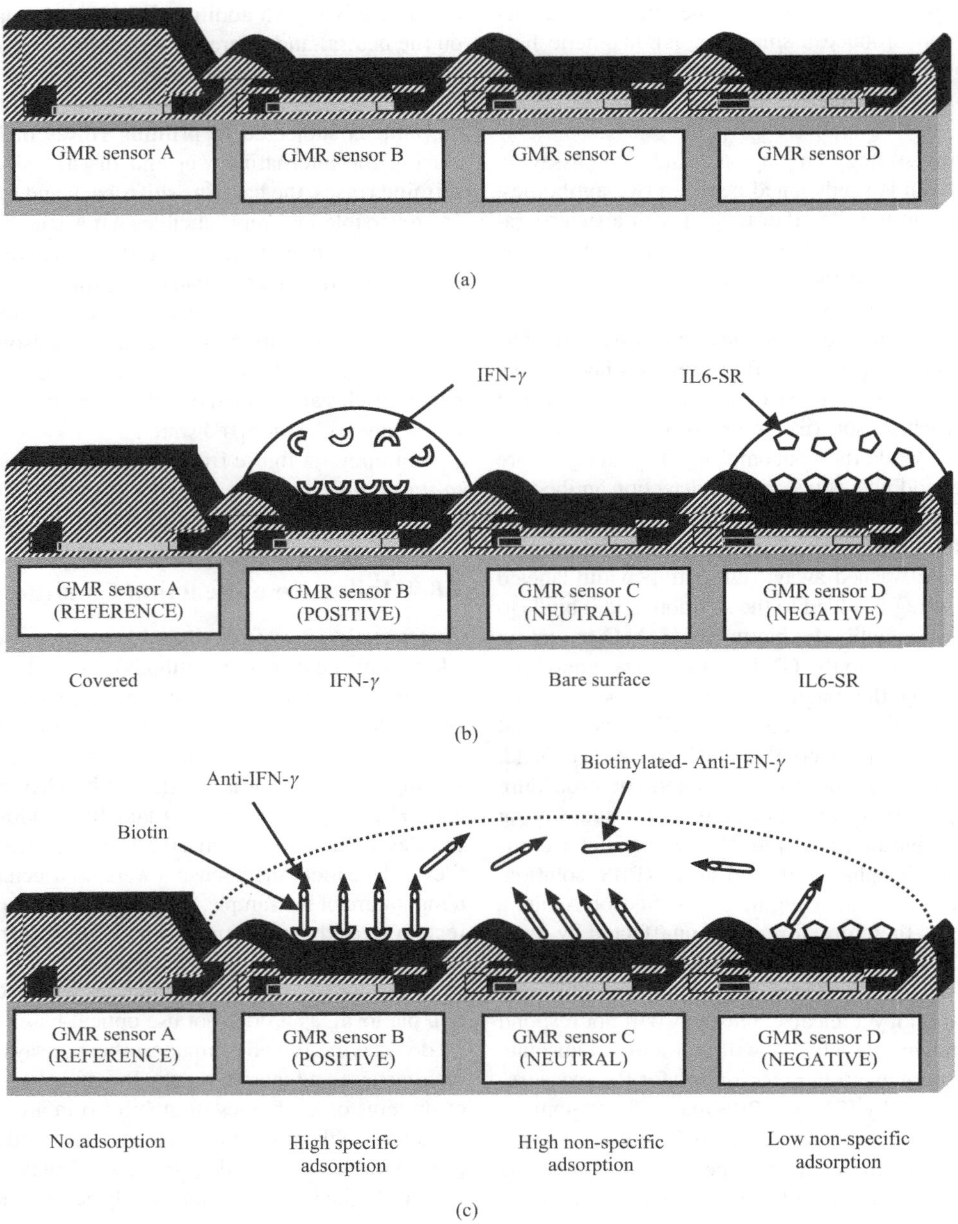

FIGURE 7.14 Procedure for conducting an assay for biotinylated anti IFN-γ: (a) magnetic array, (b) functionalization of chosen sensors with IFN-γ and IL6-sR by droplet coating, (c) incubation of the analyte, and (d) real-time magnetic measurement when magnetic nanoparticle labels get attached to analyte molecules and binding of the particles takes place. (Osterfeld, S. J. and Wang, S. X, MagArray biochips for protein and DNA detection with magnetic nanotags: Design, experiment and signal-to-noise ratio, in *Microarrays: Preparation, Microfluidics, Detection Methods and Biological Applications*, Dill, K., R. H. Liu, and P. Grodzinski (Eds.), Springer, New York, 299, 2009.)

The GMR sensor had a bottom spin-valve structure: Si/Ta(5)/seed layer/IrMn(8)/CoFe(2)/Ru/(0.8)/CoFe(2)/Cu(2.3)/CoFe(1.5)/Ta(3) – all numbers in parentheses are in nanometers. Each chip has 32 pairs of GMR sensors, which are connected to the bonding pads on the periphery by a 300-nm thick Ta/Au/Ta lead. Each sensor consists of 32 spin-valve strips in serial connection. Each strip has an electrical active area of 93 μm × 1.5 μm. To protect the sensors and leads from corrosion, two passivation layers were deposited by ion-beam sputtering: first, a thin passivation layer of SiO_2(10 nm)/Si_3N_4(20 nm)/SiO_2(10 nm) was deposited above all sensors and leads, exposing only the bonding pad area; second, a thick passivation layer of SiO_2(100 nm)/Si_3N_4(150 nm)/SiO_2(100 nm) was deposited on top of the reference sensors and leads, exposing the active sensors and bonding pad area. In the absence of an applied field, the total resistance of one sensor is about 35 kΩ. The sensors showed an accuracy of ~90%, with good signal consistency across chips.

Gaster et al. (2009) applied magnetonanosensors to simultaneously monitor real-time binding events of multiple tumor markers in several biological fluids. Their technology employs a "sandwich" assay in which the target antigen is sandwiched between two antibodies, one bound to the sensor and the other

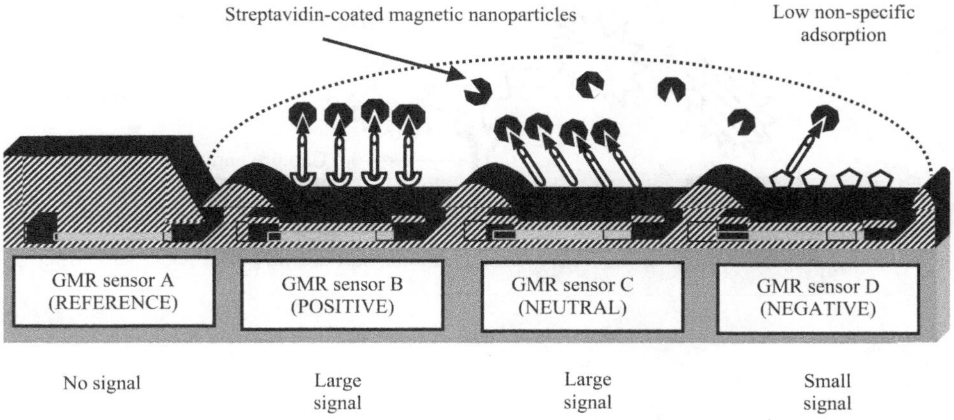

(d)

FIGURE 7.14 (Continued)

tagged with a superparamagnetic nanoparticle. The basic mechanism is to capture antigens – deleterious compounds produced and shed by the cancer cells – using antibodies that naturally tend to bond with the antigens. The antibodies, termed "capture antibodies," are applied to a sensor, so that, when the matrix of interest is placed onto the sensor chip, the appropriate antigens bind. While the antigens are held fast, another aliquot of the antibodies is applied. These antibodies are attracted to the antigens held on the sensors, and, in bonding with them, effectively seal the antigens inside an antibody sandwich. The researchers then apply a wash, containing magnetic nanoparticle tags that have been tailored to fit specific antibodies. The magnetic nanotags attach themselves to the outer antibody on the sandwich. Under an external magnetic field, the nanoparticles magnetize, where they alter the ambient magnetic field in a small but distinct and detectable way, and their presence or absence can be detected by the underlying GMR sensor (Figure 7.15).

They functionalized magnetosensors with antibodies to a representative panel of tumor markers. They were also able to perform protein detection in human serum, human urine, and human saliva (also known as spit, which is a clear liquid that is made in the mouth 24 hours a day, every day). Using chips measuring 1.2 cm × 1 cm, each containing an array of 64 GMR individually addressable magnetonanosensors, they exhibited real-time measurements of protein concentrations down to the attomolar level in a variety of clinically relevant media with a linear dynamic range (that is the concentration over which the sensor output is linearly related to the analyte concentration) of more than six orders of magnitude. The matrix insensitivity of the platform to various media demonstrates that magnetic nanosensor technology can be directly applied to a variety of settings, such as molecular biology, clinical diagnostics, and biodefense.

7.8.6 GMR-Based Detection System With Zeptomole (10^{-21} Mol) Sensitivity

Srinivasan et al. (2009) reported the development of a highly sensitive detection system, based on a GMR sensor and 12.8 nm high-moment cubic FeCo nanoparticles, which linearly detects 600–4500 copies of streptavidin based on the biotin-streptavidin interaction (Figure 7.16). They also demonstrated the feasibility of this detecting system for real biological applications, with the example of the linear detection of human interleukin 6 (IL-6, a potential lung cancer biomarker) through a sandwich-based principle. IL-6 is one of the most important mediators of fever.

The GMR sensor consists of a multilayer structure: Ta(5 nm)/Ir0.8 Mn0.2(10 nm)/Co0.9Fe0.1(2.5 nm)/Cu(3.3 nm)/Co0.9Fe 0.1(1 nm)/Ni0.82Fe0.12(2 nm)/Ta(5 nm). The GMR sensor surface was sequentially modified with 3-aminopropyltriethoxysilane (APTES), $C_9H_{23}NO_3Si$, followed by Chromalink biotin ($C_{38}H_{49}N_8NaO_{13}S_2$) *N*-hydroxysuccinimidyl ester ($C_4H_5NO_3$). High-magnetic-moment FeCo nanoparticles, with a composition ratio of 70:30, were synthesized using a sputtering gas condensation technique. In gas condensation, a metallic or inorganic material is vaporized using thermal evaporation sources such as Joule heated refractory crucibles. An alternative method employs sputtering or laser evaporation. These methods may be used instead of thermal evaporation. Sputtering is a non-thermal process in which surface atoms are physically ejected from the surface by momentum transfer from an energetic bombarding species of atomic/molecular size.

A high residual gas pressure causes the formation of ultra-fine particles (100 nm) by gas phase collision. The ultra-fine particles are produced by collision of evaporated atoms with residual gas molecules. Gas pressures >3 mPa (10 torr) are required.

The FeCo nanoparticles have an oxidation layer 1.5-nm thick. FeCo nanoparticles were first modified with amino groups on the surface by using APTES; this resulted in approximately 660 copies of APTES molecules on each nanoparticle. APTES-modified nanoparticles were subjected to streptavidin-AF (Alexa Fluor) 488 modification, using 1-ethyl-3-(3-dimeth-ylaminopro pyl)carbodiimide (EDC), $C_8H_{17}N_3$ coupling chemistry. EDC is a water-soluble carbodiimide, usually obtained as the hydrochloride form, and generally employed as a carboxyl activating agent for the coupling of primary amines to yield amide bonds. It is estimated that each nanoparticle is modified by 1.3 streptavidin AF 488 molecules. To explore the sensitivity of this detecting system based on GMR sensors and magnetic nanoparticles, varied quantities of streptavidin-AF488-modified magnetic

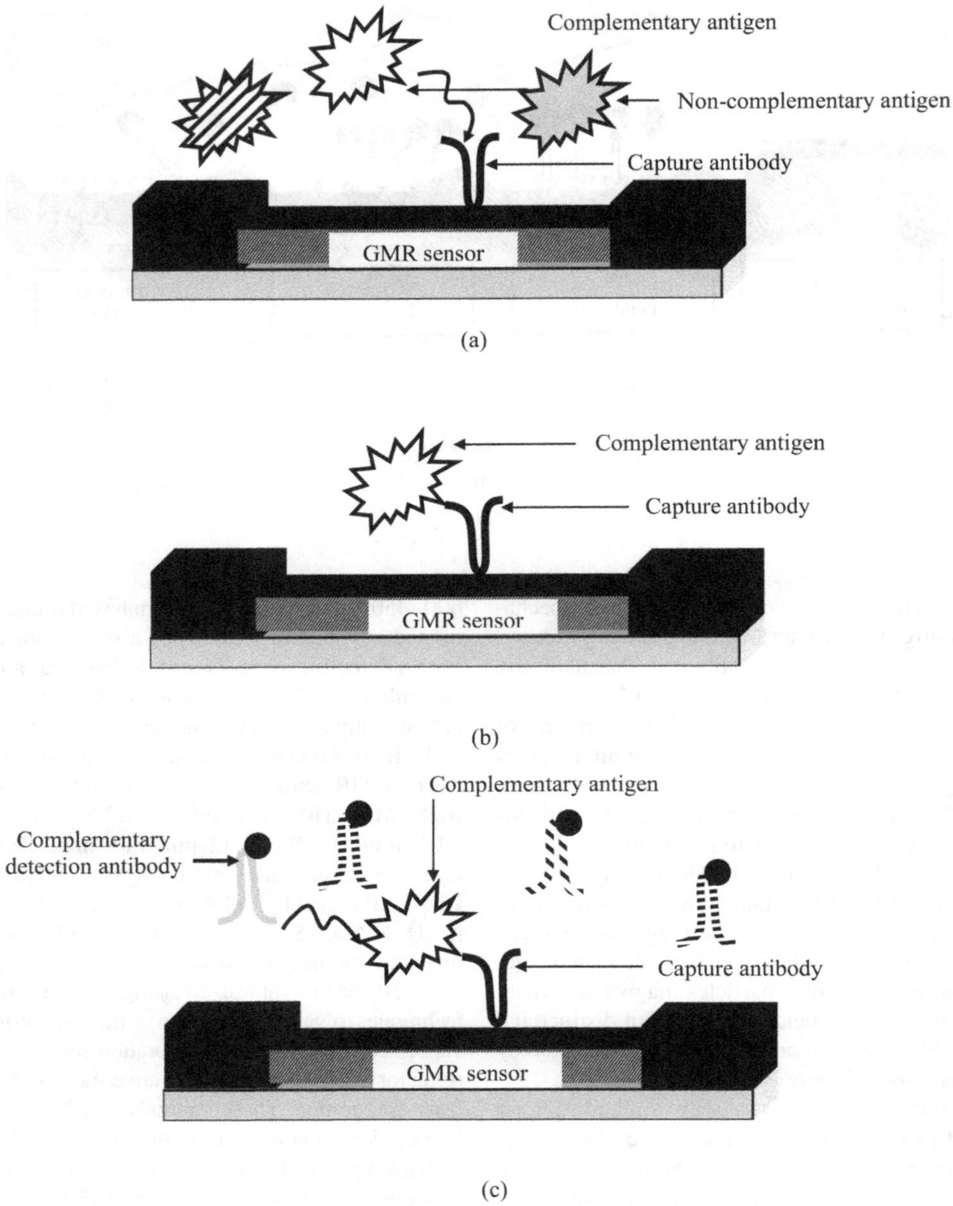

FIGURE 7.15 Sandwich assay: (a) Captured antibodies that are complementary to a chosen antigen are immobilized onto the surface of each sensor. (b) The noncomplementary antigens are subsequently washed away. (c) After adding a cocktail of detection antibodies, the biotinylated detection antibody, complementary to the antigen of interest, binds in a sandwich structure, and the noncomplementary antibodies are washed away. (d) A streptavidin-labeled magnetic nanoparticle tag is added to the solution, and it binds to the biotinylated detection antibody. (e) As the magnetic tags diffuse to the GMR sensor surface and bind with the detection antibody, the magnetic fields from the magnetic nanoparticles can be detected by the underlying GMR sensor in real time in the presence of a small external modulation magnetic field. (Gaster, R. S. et al., *Nat. Med.*, 15, 1327, 2009.)

nanoparticles were applied onto the surface of a GMR sensor, modified with Chromalink biotin. After thorough washing to remove the potentially unbound streptavidin, magnetic signals from the magnetic nanoparticles specifically retained on the GMR sensor through biotin-streptavidin interactions were measured by the GMR sensor.

The GMR sensor detected signals from as few as 600 copies ($<10^{-21}$ mol, zeptomol) of streptavidin; this sensitivity is expected to suffice for detection of all known potential biomarkers from body fluid samples of 10 nL volume or less. More importantly, there was a linear dose-response relationship between the amount of streptavidin applied and the magnetic signals detected by the GMR sensors (saturation of signal was observed with more than 20,000 copies of streptavidin). Such a dynamic range of linearity outperforms most other GMR-based detecting systems reported to date.

7.8.7 Bead ARray Counter (BARC) Biosensor

Microbeads are uniformly sized particles, typically 0.5–500 µm in diameter. Bio-reactive molecules can be adsorbed or coupled to their surface, and used to separate biological materials such as cells, proteins, or nucleic acids. Baselt et al. (1998) conducted a proof-of-concept biosensor experiment, showing the worth of using GMR sensors with magnetic microbeads in a Bead ARray

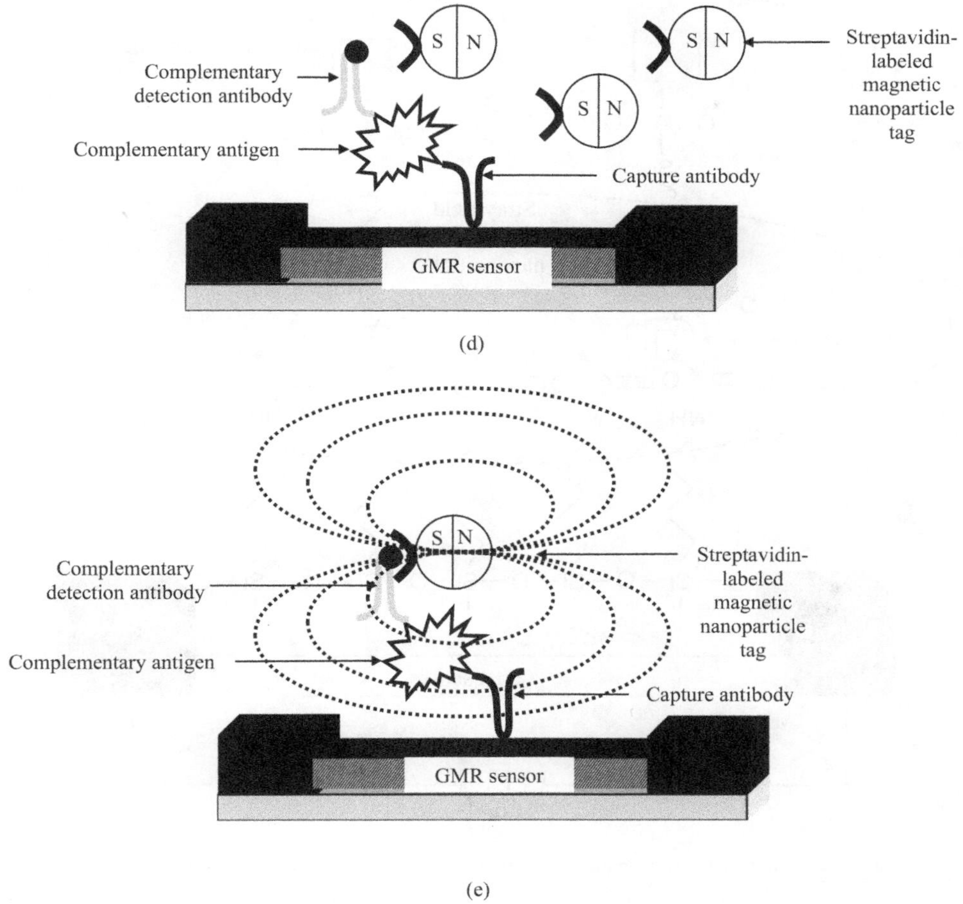

FIGURE 7.15 (Continued)

Counter (BARC). An array of 80 μm×5 μm GMR sensor elements was fabricated from sandwich GMR material. Each sensor was coated with different biological molecules that bound to different materials to be assayed. The magnetic microbeads were also coated with the materials to be analyzed. The microbeads in suspension were allowed to settle onto the GMR sensor array where specific beads will bond to specific sensors only if the materials are designed to attract each other. Non-binding beads were removed by a small magnetic field. The beads were then magnetized at 200 Hz by an AC electromagnet. The 1-μm microbeads were made up of nm-sized iron oxide particles, which had little or no magnetization in the absence of an applied field. A lock-in amplifier extracted the signal at twice the exciting frequency from a Wheatstone bridge constructed of two GMR sensor elements, one of which was used as a reference, and two normal resistors. High-pass filters were used to eliminate offset and the necessity of balancing the two GMR sensor elements. With this detection system, the presence of as few as one microbead was detectable.

7.9 Needle-Type SV-GMR Sensor for Biomedical Applications

Cancer is the most deadly disease in the world today. Hyperthermia is medical treatment method for this disease by the induction of fever, as occurs by the injection of a foreign protein or the application of heat. It is a cancer treatment method that utilizes the property that cancer cells are more sensitive to high temperature than are normal cells. This method is used for treating cancers of the prostate, liver, bladder, breast, and other organs. Magnetic hyperthermia is the name given to the treatment based on the fact that magnetic nanoparticles, when subjected to an alternating magnetic field, produce heat. Hence, if magnetic nanoparticles are put inside a tumor and the whole patient is placed in an alternating magnetic field of well-chosen amplitude and frequency, the tumor temperature will rise. This could kill the tumor cells by necrosis (the death of living cells or tissues when there is not enough blood flowing to the tissue, whether as a result of injury, radiation, or chemicals) if the temperature is above 45°C. It can improve the efficiency of chemotherapy if the temperature is raised above approximately 42°C. A hyperthermia cancer treatment based on induction heating is achieved by injecting magnetic fluid with magnetic nanoparticles into the tumor. The control of temperature is an important task in achieving success using this treatment method.

The SVGMR-based needle-type sensor (Mukhopadhyay et al. 2007; Gooneratne et al. 2008) is used to measure the magnetic flux density of the magnetic fluid inside the human body from which the temperature is estimated. The magnetic fluid (magnetite) is usually injected into human body to kill cancerous cell using hyperthermia-based treatment. To control the heat treatment, an accurate knowledge of temperature is essential. The

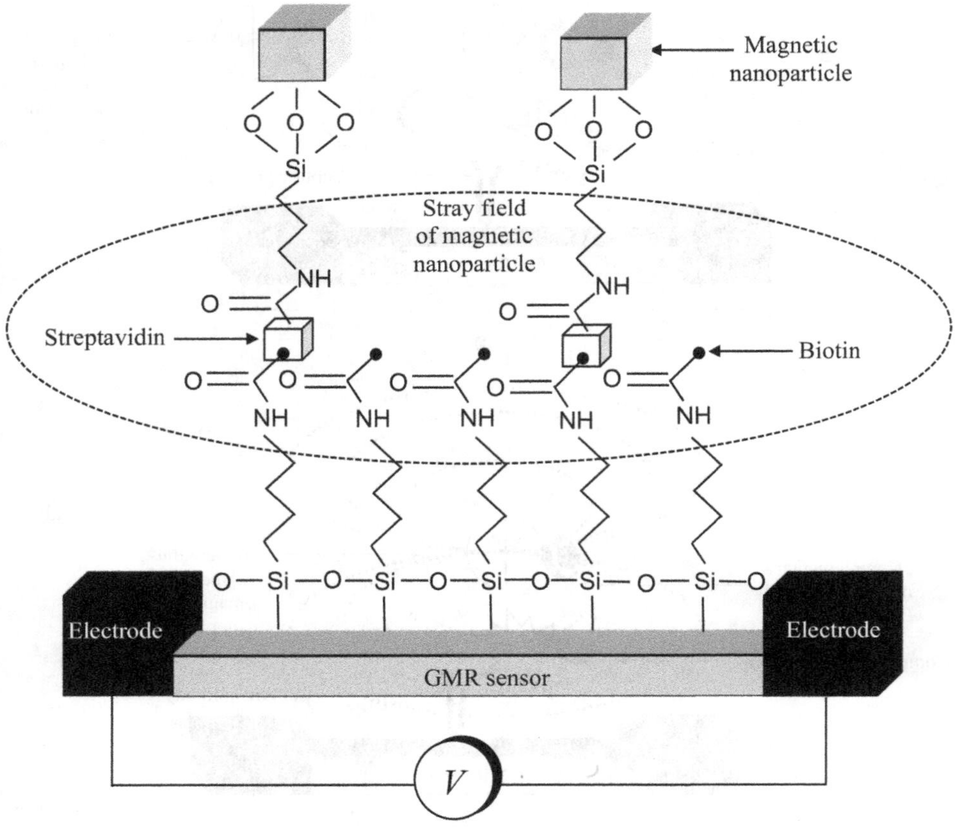

FIGURE 7.16 GMR nanosensor for detecting biotin-streptavidin interaction. (Srinivasan, B. et al., *Angew. Chem. Int. Ed.*, 48, 2764, 2009.)

SV-GMR element (Figure 7.17), with a sensing area of 75 μm × 40 μm, is on the tip of the needle. Sensing direction is parallel to the needle. The volume/weight density of the magnetic fluid is estimated by measuring and comparing the applied magnetic flux density (outside body) and the flux density in the magnetic fluid. A constant current of 0.5 mA is applied to the SV-GMR sensor. The sensitivity is approximately 10 μV μT^{-1}.

7.10 Superconductive Magnetic Nanosensor

Granata et al. (2008) fabricated a nano-SQUID (superconducting quantum interference device) for measurement of petite local magnetic signals arising from small atomic or molecular populations. They presented a nano-SQUID with an effective area of 4×10^{-2} μm^2, based on nanometric niobium (Nb) Dayem bridge junctions (Figure 7.18). The Dayem bridge is a thin-film variant of the Josephson junction, with dimensions of the order of a few micrometers or less. The sensor, having a washer shape with a hole of 200 nm and two Josephson-Dayem nanobridges of 80 nm × 100 nm, consists of a Nb(30 nm)/Al(30 nm) bilayer, patterned by electron beam lithography (EBL) and shaped by lift-off and reactive-ion etch (RIE) processes. The presence of the niobium coils, integrated on-chip and tightly coupled to the SQUID, allows the ease of excitement of the sensor in order to get the voltage-flux characteristics and to flux bias the SQUID at its optimal point. The measurements were performed at liquid helium temperature. A voltage swing of 75 μV and a maximum voltage-flux transfer coefficient (responsivity) as high as 1 mV/ϕ_0 were directly measured from the voltage-flux characteristic. The detection of the spin-flips of magnetic atom clusters is a very interesting application of nano-SQUIDs. Possible applications of this nanosensor are envisaged in magnetic detection of nanoparticles and small clusters of atoms and molecules, in the measurement of nanoobject magnetization, and in quantum computing.

7.11 Electron Tunneling-Based Magnetic Field Sensor

Here, the polymer material (PMMA) is used to fabricate inexpensive, batch-fabricated, high-yield, and highly sensitive magnetic field sensors based on a tunneling sensor platform (Wang et al. 2006). PMMA or poly(methyl methacrylate), $(C_5O_2H_8)_n$, is a clear plastic, used as a shatterproof replacement for glass. The PMMA magnetic field sensor (Figure 7.19) includes mechanical components and three electrodes. The mechanical components are composed of a membrane with a ferromagnetic film at the top surface and a tunneling tip opposite to the film at the bottom. The electrodes include a tip electrode, a counter electrode under ferromagnetic film, and a deflection electrode. The metallic tip electrode and the counter electrode form a tunneling junction. The tunneling current varies exponentially with the gap change.

The titanium/gold (Ti/Au) bilayer is chosen as the electrode metal for its inert chemical characteristics as well as the relatively high work function (the smallest energy in electron volts needed to remove an electron at 0 K to a point immediately outside the solid surface) of gold. When operating at the closed-loop

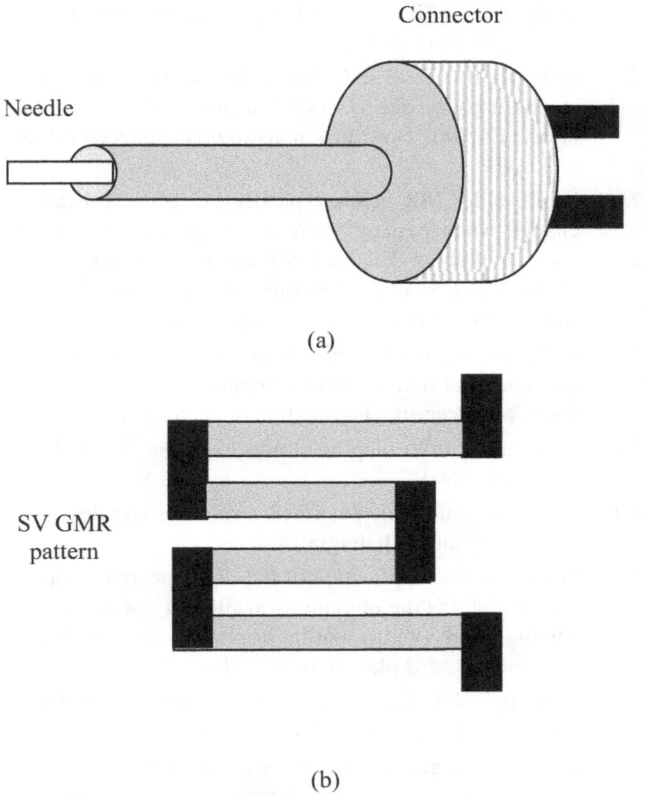

FIGURE 7.17 SV GMR needle sensor: (a) Structure of the probe and (b) spin=valve giant magnetoresistive pattern. (Mukhopadhyay, S. C. et al., *IEEE Sens.* J., 7, 401, 2007; Gooneratne, C. P. et al., *J. Sens.*, Article ID 890293, 7, doi: 10.1155/2008/890293, 2008.)

mode, the sensor maintains a constant tip-to-membrane distance by applying an electrostatic feedback force on the deflection electrode.

The displacement changes are caused by the ferromagnetic force when alternating field is applied to the membrane. The transfer function between the output voltage and the ferromagnetic force is derived. The output voltage of the sensor is proportional to the magnetic force applied to the magnetic film. The magnitude of the transfer function is determined by the radius

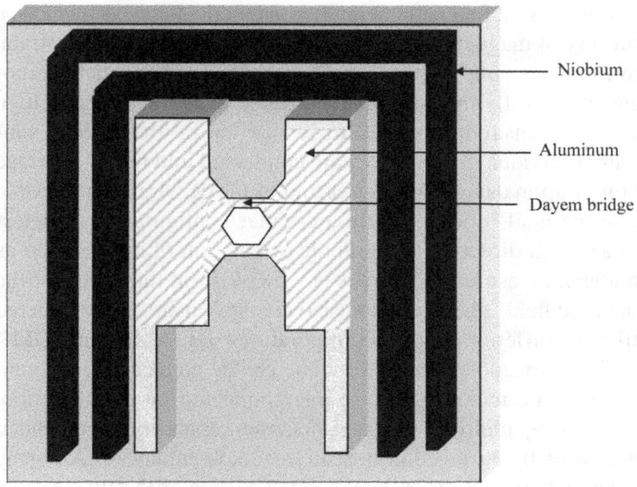

FIGURE 7.18 Nano-SQUID based on niobium Dayem nanobridges. (Granata, C., et al., *Nanotechnol.*, 19, 275501, doi: 10.1088/0957-4484/19/27/275501, 2008.)

and the thickness of the membrane. The transfer function of a PMMA sensor is substantially large, compared with a silicon sensor, if the structures remain the same. This is the main reason why the polymer is selected as the alternative material instead of silicon.

A natural frequency response of 1.3 kHz was observed, and a tunneling barrier height of 0.713 eV was tested. Due to the quadratic relationship (one variable is related to another by a function involving constant terms and first-order terms or higher) between magnetic force and the field, the sensor field response ($7\ 0 \times 10^6$ V T^{-2}) was also quadratic. The noise voltage at 1 kHz was 0.2 mV, corresponding to a magnetic field of 0.46×10^6 T. The bandwidth of this sensor was 18 kHz.

7.12 Nanowire Magnetic Compass and Position Sensor

Peczalski (2010) reported a nanowire magnetic sensor and position sensor for determining the position of a magnetic object and

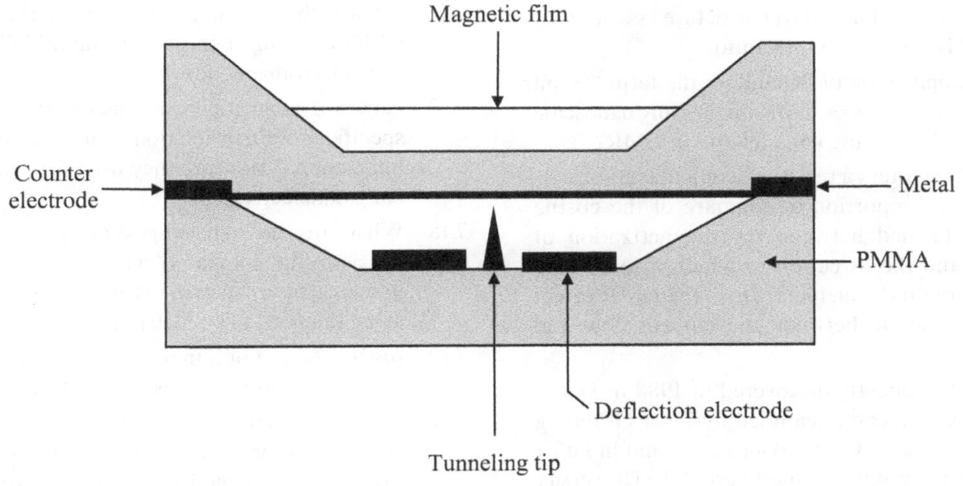

FIGURE 7.19 Micromachined magnetic field tunneling sensor. (Wang, J. et al., *IEEE Sens. J*, 6, 97, 2006.)

the direction of magnetic field. The magnetic compass includes a plurality of magnetic nanosensors printed on a flexible substrate, which covers 360° angle at equal intervals. Each magnetic nanosensor generally includes magnetoresistive nanowires with high magnetic sensitivity, printed in sets of ten on the flexible substrate. Individual nanosensors are connected into resistive bridge configuration using AMR wires and/or GMR wires. The flexible substrate is also bent to form a circular configuration to detect the azimuth direction (geographic orientation of a line given as an angle, measured in degrees clockwise from the north) of the magnetic field. The magnetoresistive nanosensors are connected utilizing different bridge configurations for AMR and GMR wires, for example, the Wheatstone bridge connection, wherein the magnetic nanosensors are interconnected by metallization. By utilizing multiple nanosensors, the capability of extended angular or linear position measurements is enhanced. An array of magnetic nanosensors is created. The size of the linear array of magnetic nanosensors and the small dimension of the magnetoresistive nanowires enables very high resolution of about 1–0.1 mdeg.

7.13 Discussion and Conclusions

As outlined in the previous sections, two main branching points of the magnetic sensor family tree are abundantly clear. On one branch are perched the nanosensors based on the GMR principle, which obviously has roots in nanotechnology. It can be considered to be a major breakthrough in physics, which has impacted our everyday lives profoundly. On the other branch reside the sensors working as proximity switches and depending on the MRI principle of T_2 variation of adjoining water molecules. Most of the research papers have tended to pursue either of the two directions. In addition, there are developments in nano-SQUID devices and other devices, based on the quantum-mechanical tunneling principle for magnetic field measurement. Table 7.1 summarizes a few of the magnetic nanosensor examples.

Review Exercises

7.1 What is a magnetoresistive effect? Is GMR a magnetoresistive effect? Is a multilayer structure essential for observing GMR? What is GMR ratio?

7.2 What is the implication of "Giant" in the term "Giant Magnetoresistance"? Does GMR involve any nanoscale phenomenon? What is the nano feature in GMR?

7.3 The AMR effect alters a resistance of a magnetoresistive element in proportion to a square of the cosine of an angle formed between the magnetization of the element and the direction in which a sense current flows through the element. Does the GMR effect depend on the angle between the current flow and magnetic field?

7.4 GMR was independently discovered in 1988 in Fe/Cr/Fe trilayers by a research team led by Peter Grünberg of the Jülich Research Centre (Germany), and in Fe/Cr multilayers by the group of Albert Fert of the University of Paris-Sud (France). What prestigious award was given to them in 2007?

7.5 Is the word "Giant" in GMR not incongruous for a nanotechnology device? Is GMR a quantum-mechanical effect? Explain. How do you distinguish between AMR and GMR?

7.6 While older MR heads typically exhibit a resistance change when passing from one magnetic polarity to another of about 2%, for GMR heads this is anywhere from 5% to 8%. What does this mean in terms of sensitivity, noise, and realizable storage capacity?

7.7 GMR heads are made up of four layers of thin material that combine into a single structure. Name these four layers and describe the function of each layer.

7.8 What is the role of pinned and exchange layers in a spin-valve GMR?

7.9 Name the four principal GMR structures and describe their operation with diagrams.

7.10 Discovery and application of the GMR phenomenon is responsible for the ubiquitous availability of economical, high-density information storage devices like compact 160 GB Mp3 players and 1TB hard drives. Explain.

7.11 What is TMR? Explain the SDT phenomenon and the resistance changes in a TMR structure.

7.12 How are magnetic nanoparticles used for detecting specific DNA sequences by structured self-assembly of the disperse particles into a stable assembly (clusters)? Why are they called magnetic nanoswitches?

7.13 Magnetic nanoparticle conjugates, that self-assemble in the presence of peroxidases, can serve as an *in-vivo* diagnostic tool to identify peroxidase-induced diseases by magnetic resonance imaging. Explain the application of peroxidase substrate nanosensors for MRI.

7.14 Give two biosensor examples in which hybridization-induced assembly of magnetic nanoparticles causes a decrease in the spin-spin relaxation time of neighboring water protons.

7.15 What are aminated cross-linked iron oxide nanoparticles (amino-CLIOs)? What is the function of amino groups in these particles?

7.16 What is the advantage of the iron oxide crystal core in CLIO in being superparamagnetic? What will happen if it is ferromagnetic?

7.17 How are magnetic nanosensors designed to detect a specific protein in solution by using antibody-mediated interactions? How are they used for measuring various enzymatic activities?

7.18 What are the advantages of a magnetic nanosensor assay as compared with an optical assay? What is meant by a "matrix-insensitive assay"? Why is the assay required to be matrix independent?

7.19 Justify the use of a magnetic nanoparticle-based assay as a matrix-insensitive protein detection assay.

7.20 How is a highly sensitive and specific multiplexed detection of protein tumor markers in a matrix-insensitive assay realized by a magnetosensor array?

TABLE 7.1

The Magnetic Nanosensor Family

Sl. No.	Name of the Nanosensor	Operating Principle
1.	Giant magnetoresistance (GMR) sensor	Magnetic-field-dependent resistivity changes due to the quantum-mechanical GMR effect, which arises from the spin-dependent scattering of conduction electrons in ferromagnetic materials. As fabricated, the magnetic moments between adjacent magnetic layers are antiparallel due to antiferromagnetic exchange coupling, and the spin-dependent electron scattering in the multilayer structure is at a maximum. This situation corresponds to the high resistance state of the device. If, however, an external magnetic field is applied in the plane of the multilayer slab, exchange coupling can be overcome and the spins in the magnetic layers brought into alignment with the external field. This alignment lowers the spin-dependent electron scattering and, therefore, the resistance of the GMR. Importantly, the degree of parallel/antiparallel alignment of the ferromagnetic layers is extremely sensitive to the strength of applied magnetic field. As a consequence, GMRs can show up to an 80% change in resistivity at room temperature and at modest magnetic fields (e.g., 10–2000 Oe)
2.	Tunneling magnetoresistance (TMR) sensor	Spin-dependent tunneling (SDT), which results in a change in effective resistance, is caused by a change in the applied field. The magnitude of the tunneling current between the two magnetic layers is modulated by the direction between the magnetization vectors in the two layers. The resistance versus field effect is similar to the usual GMR spin valve effect but of larger magnitude
3.	Magnetic nanoparticle probes	MR relaxation properties of superparamagnetic nanoparticles, such as relaxivity dependence on their degree of clustering. The relaxation induced by these particles is caused by the diffusion of water protons in the inhomogeneous magnetic field surrounding the particles. Magnetic nanoparticles are employed as *proximity sensors* to modulate the spin-spin relaxation time of neighboring water molecules
4.	Protease-specific nanosensor	Upon specific protease cleavage, the nanoparticles rapidly switch from a stable low-relaxivity stealth state to become adhesive, aggregating particles, exhibiting increased T_2-relaxivity
5.	Relaxation switch immunosensor	When antibodies specific to D-amino acids (anti-D-AA) are added to the CLIO-D-Phe nanoparticles, the divalent nature of the antibodies results in the self-assembly of the nanoparticles, leading to a decrease of more than 100 mseconds in the T_2 relaxation time
6.	Superconductive magnetic nanosensor	Nano-SQUID is based on nanometric niobium Dayem bridge junctions
7.	Tunneling-based magnetic field sensor	Electron tunneling displacement sensing and relation between magnetic field and magnetic force
8.	Nanowire magnetic compass and position sensor	Magnetoresistive nanowires are utilized as a position sensor of a magnetic object for position determination and finding the direction of the magnetic field

REFERENCES

Andreev, S. and P. Dimitrova. 2005. Anisotropic magnetoresistance integrated sensors. *Journal of Optoelectronics and Advanced Materials* 7(1): 199–206.

Arana, S., E. Castaño, and F. J. Gracia. 2004. High temperature circular position sensor based on a giant magnetoresistance nanogranular AgxCol-x alloy. *IEEE Sensors Journal* 4: 221–225.

Arana, S., N. Arana, R. Gracia, and E. Castaño. 2005. High sensitivity linear position sensor developed using granular Ag-Co giant magnetoresistances. *Sensors and Actuators A: Physical* 123: 116–121.

Baselt, D. R., G. U. Lee, M. Natesan, S. W. Metzger, P. E. Sheehan, and R. J. Colton. 1998. A biosensor based on magnetoresistance technology. *Biosensors & Bioelectronics* 13: 731–739.

Cockburn, B. F. 2004. Tutorial on magnetic tunnel junction magnetoresistive random-access memory. In: *Records of the 2004 International Workshop on Memory Technology, Design and Testing* (MTDT'04), 1087-4852/04, San Jose, CA.

Freitas, P. P., R. Ferreira, S. Cardoso, and F. Cardoso. 2007. Magnetoresistive sensors. *Journal of Physics: Condensed Matter* 19: 165221, doi: 10.1088/0953-8984/19/16/165221.

Gaster, R. S., D. A. Hall, C. H. Nielsen, S. J. Osterfeld, H. Yu, K. E. Mach, R. J. Wilson, B. Murmann, J. C. Liao, S. S. Gambhir, and S. X. Wang. 2009. Matix-insensitive protein assays push the limits of biosensors in medicine. *Nature Medicine* 15(11): 1327–1332.

Gooneratne, C. P., S. C. Mukhopadhyay, and S. Yamada. 2008. An SV-GMR needle sensor-based estimation of volume density of magnetic fluid inside human body. *Journal of Sensors* 2008: 890293 (7 pp.), doi: 10.1155/2008/890293.

Gossuin, Y., P. Gillis, A. Hocq, Q. L. Vuong, and A. Roch. 2009. Magnetic resonance relaxation properties of superparamagnetic particles. *WIREs Nanomedicine and Nanobiotechnology* 1: 299–310.

Granata, C., E. Esposito, A. Vettoliere, L. Petti, and M. Russo. 2008. An integrated superconductive magnetic nanosensor for high-sensitivity nanoscale applications. *Nanotechnology* 19: 275501 (6 pp.), doi: 10.1088/0957-4484/19/27/275501.

Grimm, J., J. M. Perez, L. Josephson, and R. Weissleder. 2004. Novel nanosensors for rapid analysis of telomerase activity. *Cancer Research* 64: 639–643.

Gupta, A., S. Mohanan, M. Kinyanjui, A. Chuvilin, U. Kaiser, and U. Herr. 2010. Influence of nano-oxide layer on the giant magnetoresistance and exchange bias of NiMn/Co/Cu/Co spin valve sensors. *Journal of Applied Physics* 107: 093910-1–093910-5.

Haun, J., T.-J. Yoon, H. Lee, and R. Weissleder. 2010. Magnetic nanoparticle biosensors. *WIREs Nanomedicine and Nanobiotechnology* 2: 291–304.

Jander, A., C. Smith, and R. Schneider. 2005. Magnetoresistive sensors for nondestructive evaluation. In: *The 10th SPIE International Symposium, Nondestructive Evaluation for*

Health Monitoring and Diagnostics, Conference 5770, Copyright © 2005, NVE Corporation, March 6–10, 2005, San Diego, CA. Washington, DC: SPIE, 13pp.

Josephson, L., J. M. Perez, and R. W. Weissleder. 2001. Magnetic nanosensors for the detection of oligonucleotide sequences. *Angewandte Chemie* 113: 3304–3306; *Angewandte Chemie—International Edition* 40: 3204–3206.

Kasatkin, S. I., P. I. Nikitin, A. M. Muravjev, V. V. Lopatin, F. F. Popadinetz, A. V. Svatkov, A. Yu. Toporov, F. A. Pudonin, and A. I. Nikitin. 2000. Spin-tunneling magnetoresistive sensors. *Sensors and Actuators* 85: 221–226.

Kasatkin, S. I., N. P. Vasil'eva, and A. M. Murav'ev. 2010. Biosensors based on the thin-film magnetoresistive sensors. *Automation and Remote Control* 71(1): 156–166.

Kim, G. Y., L. Josephson, R. Langer, and M. J. Cima. 2007. Magnetic relaxation switch detection of human chorionic gonadotrophin. *Bioconjugate Chemistry* 18: 2024–2028.

Koh, I. and L. Josephson. 2009. Magnetic nanoparticle sensors. *Sensors* 9: 8130–8145, doi: 10.3390/s91008130.

Millen, R. L., T. Kawaguchi, M. C. Granger, and M. D. Porter. 2005. Giant magnetoresistive sensors and superparamagnetic nanoparticles: A chip-scale detection strategy for immunosorbent assays. *Analytical Chemistry* 77: 6581–6587.

Mukhopadhyay, S. C., K. Chomsuwan, C. P. Gooneratne, and S. Yamada. 2007. A novel needle-type SV-GMR sensor for biomedical applications. *IEEE Sensors Journal* 7(3): 401–408.

Osterfeld, S. J. and S. X. Wang. 2009. MagArray biochips for protein and DNA detection with magnetic nanotags: Design, experiment and signal-to-noise ratio, in *Microarrays: Preparation, Microfluidics, Detection Methods and Biological Applications*. Dill, K., R. H. Liu, and P. Grodzinski (Eds.), New York: Springer, pp. 299–314.

Peczalski, A. 2010. Nanowire magnetic compass and position sensor. United States Patent Application 20100024231.

Pelegrí, J., D. Ramírez, and P. P. Freitas. 2003. Spin-valve current sensor for industrial applications. *Sensors and Actuators A: Physical* 105: 132–136.

Perez, J. M., L. Josephson, and R. Weissleder. 2002a. Magnetic nanosensors for DNA analysis. *European Cells and Materials* 3(Suppl. 2): 181–182.

Perez, J. M., L. Josephson, T. O'Loughlin, D. Hogemann, and R. Weissleder. 2002b. Magnetic relaxation switches capable of sensing molecular interactions. *Nature Biotechnology* 20: 816–820.

Perez, J. M., F. J. Simeone, Y. Saeki, L. Josephson, and R. Weissleder. 2003. Viral-induced self-assembly of magnetic nanoparticles allows the detection of viral particles in biological media. *Journal of the American Chemical Society* 125: 10192.

Perez, J. M., L. Josephson, and R. Weissleder. 2004. Use of magnetic nanoparticles as nanosensors to probe for molecular interactions. *J Chem BiolChem* 5: 261–264, doi: 10.1002/cbic.200300730.

Qian, Z., J. M. Daughton, D. Wang, and M. Tondra. 2003. Magnetic design and fabrication of linear spin-valve sensors. *IEEE Transactions on Magnetics* 39(5): 3322–3324.

Reig, C., M.-D. Cubells-Beltrán, and D. R. Muñoz. 2009. Magnetic field sensors based on giant magnetoresistance (GMR) technology: Applications in electrical current sensing. *Sensors* 9: 7919–7942, doi: 10.3390/s91007919.

Schellenberger, E. 2010. Bioresponsive nanosensors in medical imaging. *Journal of the Royal Society. Interface / the royal society* 7 Supplement 1: S83–S91, doi: 10.1098/rsif.2009.0336.focus.

Schellenberger, E. A., F., Reynolds, R. Weissleder, and L. Josephson. 2004. Surface-functionalized nanoparticle library yields probes for apoptotic cells. *ChemBioChem: a European Journal of Chemical Biology* 5: 275–279.

Schellenberger, E., F. Rudloff, C. Warmuth, M. Taupitz, B. Hamm, and J. Schnorr. 2008. Protease-specific nanosensors for magnetic resonance imaging. *Bioconjugate Chemistry* 19: 2440–2445.

Srinivasan, B., Y. Li, Y. Jing, Y. H. Xu, X. Yao, C. Xing, and J.-P. Wang. 2009. A detection system based on giant magnetoresistive sensors and high-moment magnetic nanoparticles demonstrates zeptomole sensitivity: Potential for personalized medicine. *Angewandte Chemie International Edition* 48: 2764–2767.

Tsourkas, A., O. Hofstetter, H. Hofstetter, R. Weissleder, and L. Josephson. 2004. Magnetic relaxation switch immunosensors detect enantiomeric impurities. *Angewandte Chemie* 116: 2449–2453, doi: 10.1002/ange.200352998.

Wang, J., W. Xue, N. V. Seetala, X. Nie, E. I. Meletis, and T. Cui. 2006. A micromachined wide-bandwidth magnetic field sensor based on all-PMMA electron tunneling transducer. *IEEE Sensors Journal* 6(1): 97–105.

Xu, L., H. Yu, M. S. Akhras, S.-J. Hana, S. Osterfelda, R. L. White, N. Pourmand, and S. X. Wang. 2008. Giant magnetoresistive biochip for DNA detection and HPV genotyping. *Biosensors and Bioelectronics* 24(1): 99–103, doi: 10.1016/j.bios.2008.03.030.

Part IV

Chemical and Biological Nanosensors

8

Chemical Nanosensors

8.1 Introduction

The question may be raised that, during description of physical sensors, chemical phenomena are frequently intermixed; e.g., in Sections 6.7.2 and 6.7.3, pH nanosensors were dealt with. It is well-nigh impossible to tightly compartmentalize the discussion into physical, chemical and biological domains, such that chemical and biological phenomena do not enter the domain of physical sensors. Similar remarks are applicable to chemical and biological sensors. Nor does it happen in reality or in a real-world situation, that one particular type of phenomenon occurs in total isolation. Nonetheless, it provides a description of events, in which chemical phenomena are at the forefront, with physical/biological effects in the background when talking about chemical sensors. Other phenomena may sneak in but may at times be considered less consequential. The same holds for other classes of sensors. So, a serious discussion on chemical sensors is warranted in this setting, and this exactly is the aim of this chapter.

What is a chemical sensor? A device in which chemical species, chemical stimuli, and chemical reactions play a leading role (Section 1.4) while the physical (Section 1.3) and biological effects (Section 1.5) have secondary functions.

How is a chemical nanosensor defined? A chemical nanosensor is an electronic device, consisting of a transducer and a sensitive element, whose operation relies on at least one of the physical and chemical properties characteristic of the nanostate (Francia et al. 2009).

What distinguishes a chemical nanosensor from other chemical sensors? Basically, the operation of a chemical nanosensor resembles any other chemical sensor, involving charge transference between molecules and a sensitive material, thereby producing an electrical and/or optical signal correlated to the kind and number of molecules. However, compared with macroscopic sensors, chemical nanosensors derive advantages from the unification of four different features of the nanostate: (i) the quantum confinement effect; (ii) the surface/volume ratio, S/V, typical of a specific surface termination and nanoparticle (NP) doping; (iii) the NP morphology and aggregation; and (iv) the nanomaterial agglomeration state. The term "nanoparticle" here refers, in general, to any kind of structure with at least one of its dimensions in the nanorange. Examples are nanowire (NW), nanodot, and nanotube. These properties improve the sensitivities of the constructional materials of devices, and high sensitivities have been demonstrated. The field of chemical nanosensors has received tremendous impetus from the synthesis and engineering of materials to create devices that exhibit functionalities originating specifically from their nanostates.

What kinds of materials are commonly used in chemical nanosensors? Due to their interesting structures, nanomaterials afford many opportunities for investigating their sensing behavior by making new types of nanosensing structures. The structure of nanowires (NWs) and nanorods is comparable to each other, being dominated by a wire-like structure whose diameter varies over a broad range from several tens of nanometers to a micrometer.

What is the difference between NWs and nanorods? The typical length of the NWs ranges from several tens to several hundred micrometers, whereas the nanorods are only several micrometers long.

How do nanotubes differ from NWs? Nanotubes possess wire-like nanostructures, but they have hollow cores.

What are the features of nanobelts (NBs)? The belt-like nanostructure has a rectangular cross section. Each NB has a uniform width along its entire length, and the typical widths of the NBs lie in the range of several tens to several hundred nanometers.

This chapter will provide a review on the current research status of chemical sensors based on various new types of nanostructured materials, such as nanotubes, nanorods, NBs, and NWs. These nanostructure-based sensors represent authoritative detection platforms for a broad assortment of biological, electrochemical, gas, and pH sensors. *What are the various types of nanosensing devices?* The sensing devices include individual nanostructured sensors, multi-nanostructured sensors, MOSFET-based sensors, nanostructured film sensors, and so forth. These nanosensor devices have a number of key characteristics, including high sensitivity, exquisite selectivity, fast response and recovery, and potential for integration of addressable arrays on a massive scale. These properties set them apart from currently available sensor technologies.

8.2 Gas Sensors Based on Nanomaterials

Why is the development of gas sensors currently being carried out intensively? This is because environmental pollution and toxicants in industrial and domestic environments represent acute problems facing mankind. Gas detection is obligatory in many different fields, e.g., industrial, fuel emission control, automobile exhaust emission control, household security, and environmental pollution monitoring. Gas sensors are utilized in factories, laboratories, hospitals, and almost all technical installations. *What gases have evoked special interest?* Gases of interest include CO_2, CO, NO_2, SO_2, O_2, O_3, H_2, Ar, N_2, NH_3, and H_2O, and organic vapors such as those of methanol (CH_3OH), ethanol (C_2H_5OH), isopropanol (C_3H_8O), benzene (C_6H_6), and some amines (organic compounds and functional groups that contain a basic nitrogen atom with a lone pair).

What are the gas recognition mechanisms? The mechanisms for recognizing the gases to be determined are the absorption processes, e.g., in metal oxides (MOXs) and carbon nanotubes (CNTs), and specific recognition for the formation of supramolecules (large molecules formed by grouping together or bonding several molecules together), or covalent bonds between the sensor and the analyte, as in some metal complexes. Studies have revealed that sensitivity increases or response time decreases as the film thickness or the particle size of MOXs or organic polymers decreases.

A characteristic feature of gas nanosensors is that they have a transducer system consisting of nanometric materials. On this basis, there are two kinds of device. In some devices, the sensor and the transducer are different components, whereas, in other devices, the sensor also functions as the transducer (as in semiconducting oxides and other systems).

8.3 Metallic Nanoparticle-Based Gas Sensors

While fabricating a gas sensor, why are metal NPs generally spread out on the surface of a substrate? Metal NPs are dispersed on the surface of a substrate to increase the area/volume ratio and favor the adsorption of gases. When they come into contact with the analyte, their electronic properties or the properties of the substrate itself change because the gas molecules adsorb onto the metal. The NPs allow greater interaction between the molecules and the substrate, which provides more contact area with the analyte. Generally, the deposition is made by vaporizing the metal precursor, which provides the NPs or films for a subsequent annealing (a heat treatment wherein the microstructure of a material is altered, causing changes in its properties).

Why is hydrogen sensing very important? Hydrogen, being an important source of alternative energy for the transportation, residential, and industrial sectors, needs a great deal of attention in research related to its storage, transportation, and handling. One of the major problems is that hydrogen is extremely inflammable and volatile. The presence of only 2%–4% hydrogen in air can cause an explosion. Hence, the research community is searching for a better hydrogen sensor for safety monitoring.

What are the desirable qualities of a hydrogen sensor? For a hydrogen sensor to be efficient and reliable, it should be able to detect hydrogen below its explosive limits. Only then will it be worth using. Generally, group VIII transition metals, Ni, Pd, and Pt nanostructured films, are used for hydrogen detection. *What are the drawbacks of conventional palladium-based sensors for hydrogen?* Macroscopic Pd-based hydrogen sensors suffer from two main shortcomings: (i) long response times ~0.5 seconds to several minutes, which is inordinately slow for real-time measurements; and (ii) adsorption of other gas molecules, such as CH_4, O_2, and CO, onto the sensor surfaces, causing blockage of the adsorption sites for hydrogen molecules.

How do Pd NW sensors alleviate these problems? Favier et al. (2001) and Walter et al. (2002a; 2002b) investigated Pd NW arrays as hydrogen sensors. Palladium NW hydrogen sensors overcame the above problems. NWs consist of agglomerated Pd grains separated by inter-grain nanogaps. On exposure to hydrogen, the gas diffuses into the lattice and reacts with the metal forming a metal hydride (PdH_x). As a result, the wire expands in volume, leading to a partial or total closure of the gaps. Due to reduction or shutting down of gaps between grains, a strong increase in the electrical conductivity is observed. This grain swelling is not completely reversible. On withdrawal of H_2, the original volume of grains is restored but their starting positions are not recovered.

How are Pd NWs fabricated for making a gas sensor? Palladium mesowires and NWs are electrically deposited from aqueous solutions of Pd^{2+} onto the naturally present step-edges on highly oriented pyrolytic graphite (polycrystalline carbonaceous material that is deposited from the gas phase during pyrolysis of hydrocarbons at temperatures that range from 750°C to 2400°C) surfaces; this process is known as *electrodeposition*. Freshly deposited Pd NWs are separated from the graphite surfaces and transferred onto a glass substrate by means of cyanoacrylate glue, an adhesive substance with an acrylate base that is used in industry and medicine.

8.4 Metal Oxide Gas Sensors

What is the basis of the commercially popular solid-state chemical sensors? Most commercial solid-state chemical sensors are based on appropriately structured and doped MOXs, mainly SnO_2 and ZnO. Additionally, the following oxides show a gas response in their conductivity: Cr_2O_3, Mn_2O_3, Co_3O_4, NiO, CuO, SrO, In_2O_3, WO_3, TiO_2, V_2O_3, Fe_2O_3, GeO_2, Nb_2O_5, MoO_3, Ta_2O_5, La_2O_3, CeO_2, and Nd_2O_3.

What different nanomaterial shapes have been used for building these sensors? Chemical sensors are based on nanotubes, nanorods, NBs, and NWs (Huang and Choi 2007) as follows: (i) in general, nanotube-based sensors include MOX tubes, such as Co_3O_4, Fe_2O_3, SnO_2, and TiO_2; (ii) MOX nanorods are ZnO, MoO_3, and tungsten oxide; (iii) as for NB-based sensors, most attention has been focused on MOXs, such as ZnO, SnO_2, and V_2O_5 nanosensors, especially on ZnO-NB sensors; and (iv) NWs include In_2O_3, SnO_2, ZnO, β-Ga_2O_3, etc.

How is sensitivity of a gas sensor defined? The definition of sensitivity is different for reducing and oxidizing gases: (i) for reducing gases, sensitivity (S) can be defined as the ratio R_a/R_g; and (ii) for oxidizing gases, it is the inverse ratio, $S = R_g/R_a$. The symbol R_a stands for the resistance of gas sensors in the reference gas (usually, the air) and R_g stands for the resistance in the reference gas containing the target gases.

What physicochemical phenomena take place in these sensors? For several decades, the phenomena participating in gas sensor response have been among the most interesting subjects in semiconductor gas sensors. The semiconducting nature of some MOXs renders it possible for the electrical conductivity of the material to alter when the gaseous composition of the surrounding atmosphere changes. This happens by the mechanism of charge transfer between surface complexes, such as O^-, O_2^-, H^+, and OH^-, on MOXs with interacting gas molecules. Hence, most of these devices rely on conductivity changes. In principle, they are conductimetric (relating to the measurement of conductivity) nanosensors and are aptly called *chemoresistors*, suggesting variation in resistance by chemical means.

The exact fundamental mechanisms that cause a gas-sensing response are still controversial and debatable, but, in effect, they include two phenomena: (i) trapping of electrons at adsorbed molecules; and (ii) resultant band bending, induced by these charged molecules, are responsible for a change in conductivity

Why do these gas sensors require an elevated temperature for their functioning? As the operating mechanism for gas sensors, namely the change in electrical conductivity, requires an activation energy (the least amount of energy needed by a system to initiate a particular process), the classical sensors operate only at high temperatures, in excess of 200°C, thereby increasing the power consumption, an intrinsic and inconvenient requirement tied to these sensors.

What arrangements are commonly made for heating the gas-sensitive material? The sensor (Figure 8.1) consists of a platform made of alumina ($A1_2O_3$), on the lower surface of which is a spiral platinum heater; it also serves as a temperature sensor. On the top surface lie interdigitated Pt electrodes over which the gas-sensing nanoparticle film is deposited. The ceramic (any of various hard, brittle, heat-resistant and corrosion-resistant materials made by shaping and then firing a nonmetallic mineral, such as clay) thick film hotplate consumes a few watts of power during its operation.

What other versions of hotplates are available? A power-saving hotplate, based on low temperature co-fired ceramic (LTCC) technology, is shown in Figure 8.2. The CMOS (complementary metal-oxide semiconductor)-compatible MEMS hotplate, shown in Figure 8.3, consumes only a few milliwatts of power. It has an embedded polysilicon microheater sandwiched between the bottom silicon nitride (Si_3N_4)-thermal silicon dioxide (SiO_2)

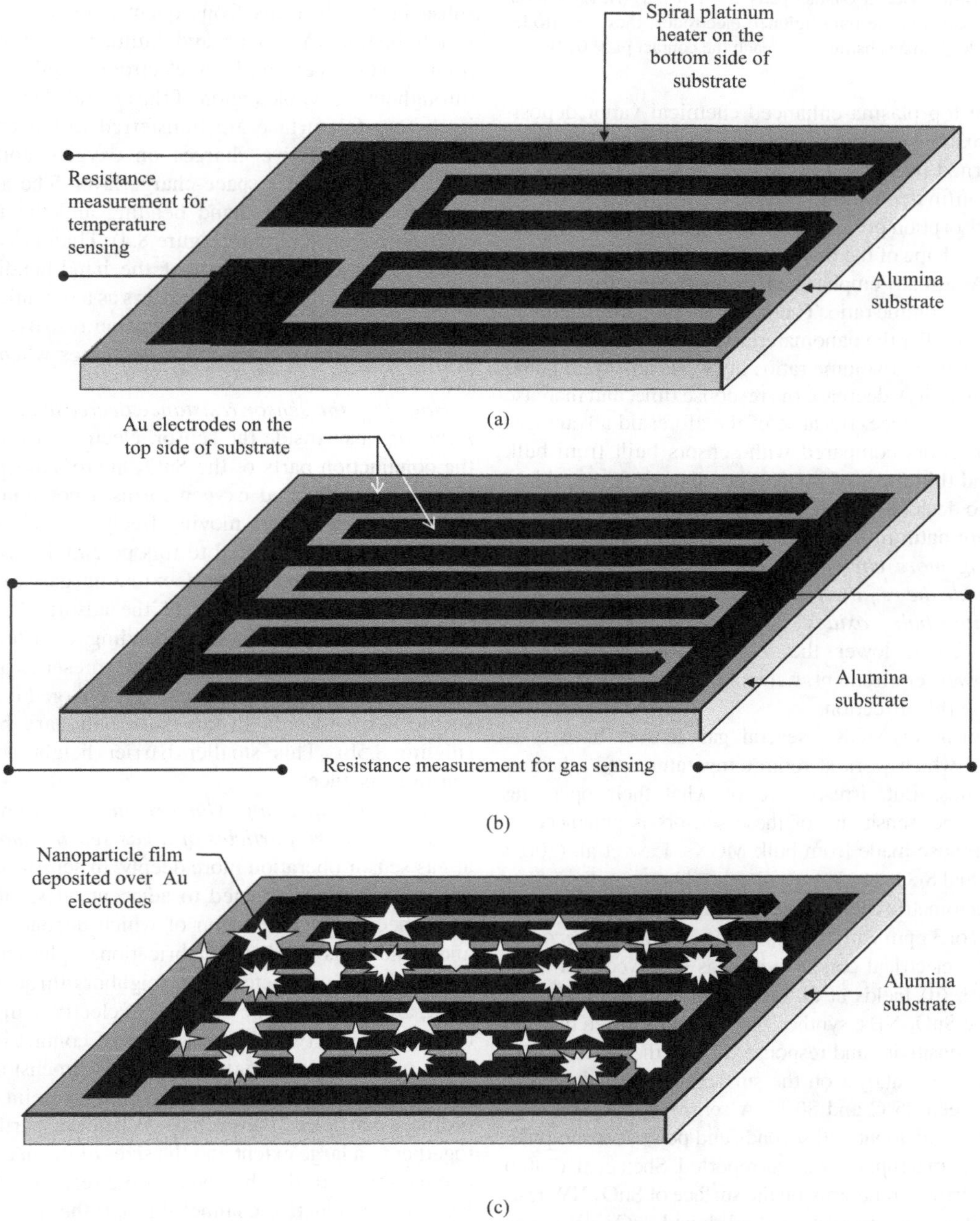

FIGURE 8.1 Gas sensor structure: (a) heater, (b) interdigitated electrodes, and (c) sensing film, containing nanoparticles deposited over the interdigitated electrodes.

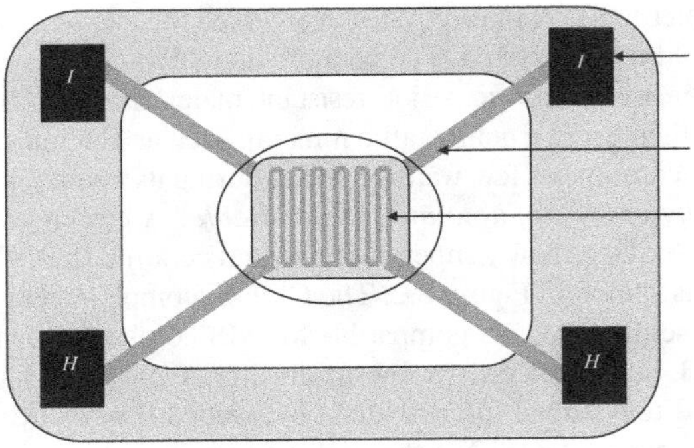

FIGURE 8.2 LTCC hotplate for gas-sensing application. The heater is suspended on a substrate. Heater contact pads are shown as (*H*, *H*). On the bottom side of the substrate lie interdigitated electrodes; these electrodes pass through *via* holes in the substrate and reach the contact pads (*I*, *I*).

platform and the top plasma-enhanced chemical vapor deposition (PECVD) oxide isolation layer. The interdigitated Pt electrodes are patterned over this PECVD oxide layer, followed by gas-sensing nanofilm deposition.

Because the adsorption process plays the main role, the physical properties and the shape of the material determine the response of the nanosensor. As already emphasized, nanosized materials have a very favorable area/volume ratio. Therefore, for the same chemical composition, the smaller the nanomaterials are, the more sensitive the sensor is. Higher area/volume ratios favor gas adsorption (and a change in conductivity), decrease the response time, and increase the sensitivity of the device. Because of the aforesaid advantages offered by nanosensors compared with sensors built from bulk material, NPs and thin films of MOXs (less than 1000-nm thick) have been used to detect a wide range of gases.

It was therefore natural to investigate *whether MOXs in their nanostates, being more chemically reactive than in the bulk form, could provide the required sensitivity at room temperature comparable with bulk oxides at elevated temperatures.* Obviously, this could lower the working temperature and minimize the power requirement. Let us look for some examples of efforts made in this direction.

Using nano forms of MOXs, several gas sensors have been reported, some of which work at room temperature and others at higher temperatures. But, irrespective of what their operating temperatures may be, sensitivity of these sensors is enhanced in comparison with those made from bulk MOXs. Law et al. (2002) fabricated individual SnO_2 single-crystal nanoribbons configured as four-probe conductometric chemical sensors. They reported a detection limit for NO_2 of 3 ppm with response/recovery times ~ seconds. The change in the electrical conductivity was observed even near room temperature. (ii) Fields et al. (2006) developed a hydrogen sensor from single SnO_2 NBs, synthesized *via* catalyst-free thermal evaporation. The sensitivity and response time of the sensors were measured without any catalyst on the surface to 2% hydrogen at temperatures between 25°C and 80°C. A sensitivity >0.3/(% H_2), with a response time of about 220 seconds and power consumption of only 10 nW at room temperature, was reported. Shen et al. (2009) distributed PdO particles randomly on the surface of SnO_2 NW testing sensors, based on both undoped and Pd-doped SnO_2 NWs. The devices showed a reversible response to H_2 at room temperature, with the signal increasing with a rise in Pd concentration.

8.4.1 Sensing Mechanism of Metal Oxide Sensors

These sensors work on the generation of a barrier potential at the MOX surface by gas adsorption. *How is the barrier potential created in a MOX sensor?* As is readily understood, the response to oxygen is the starting base of semiconductor gas sensors. The sensors are active in air, being responsive to the target gases. In a first-order elementary analysis, when a MOX crystal such as SnO_2 is heated to a certain high temperature in air, oxygen is adsorbed on the crystal surface, acquiring a negative charge, as explained in the following text. Considering oxygen adsorption on a sufficiently large N-type semiconductor crystal, it is known that, if there are traps of electrons on the surface of a semiconductor, conduction electrons are transferred from the subsurface to the traps. Consequently, an electron-depleted layer is left behind (Yamazoe and Shimanoe 2008). This electron transfer continues until an electronic equilibrium is reached throughout the whole region of the crystal. Thus, donor electrons in the crystal surface are transferred to the adsorbed oxygen, resulting in negative charges on oxygen atoms and leftover positive charges in a space-charge layer. The above-mentioned phenomenon leads to band bending and the formation of an electron-depleted region (Figure 8.4). Thickness of the electron-depleted region = the length of the band-bending region. The surface potential thus formed serves as a potential barrier against electron flow (Figure 8.5a), increasing resistivity parallel to the surface. The surface resistance decreases when a reducing gas is present.

How does the sensor resistance decrease in the presence of a reducing gas? Inside the sensor, electric current flows through the conjunction parts of the SnO_2 micro/nanocrystals. At grain boundaries, adsorbed oxygen forms a potential barrier, which prevents carriers from moving freely. The electrical resistance of the sensor is attributed to this potential barrier. Reaction of these oxygen species with deoxidizing gases or a competitive adsorption and replacement of the adsorbed oxygen by other molecules decreases the band bending, as it tends to curve the bands in the opposite direction. In the presence of a reducing gas, the surface density of the negatively charged oxygen decreases, so the barrier height in the grain boundary becomes smaller (Figure 8.5b). This smaller barrier height decreases overall sensor resistance.

Electrically speaking, what are the two different types of connections between particles of a gas-sensing material? Looking at gas sensor operation more deeply, in sensor devices, constituent crystals are connected to adjacent ones, either by contacts or by necks, the proportion of which depends on the methods and conditions of device fabrication. Spherical crystals (uniform in size), connected with neighbors through a contact or a neck, are believed to be depleted of electrons in the outer region only. There are two different types of contact between the particles of sensing materials: (i) the two-dimensional (2D) contact between necked particles; and (ii) point-to-point contact between ordinary particles (Figure 8.6). When the particles are necked together to a large extent and the sizes of the necked part become comparable with the thickness of the resistive electron-depleted layer, the conductive channel through the neck is deterministic of the total resistivity (*neck model*). When the particle sizes are much larger than the thickness of the electron-depleted layer, the

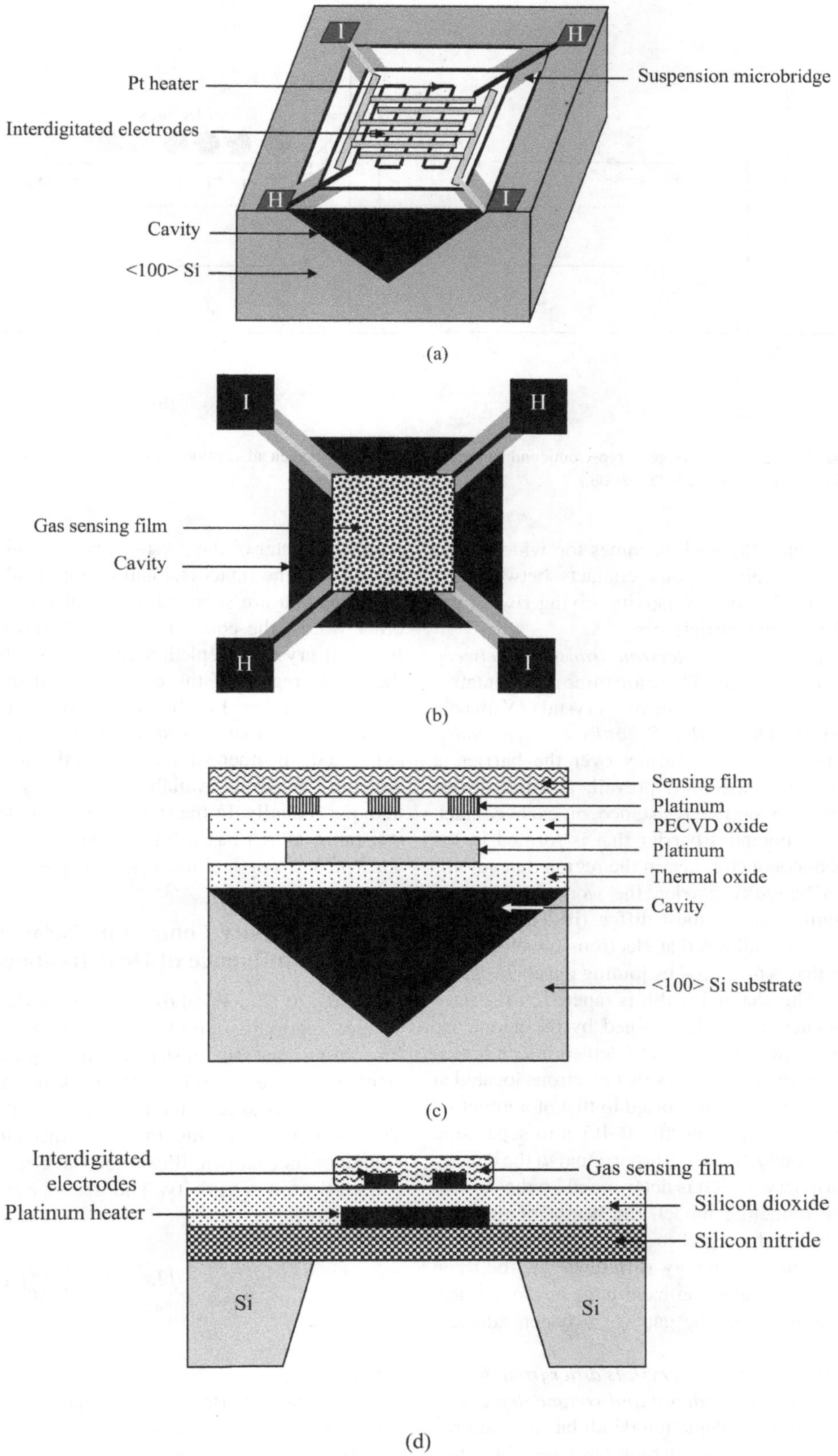

FIGURE 8.3 Schematic representation of (a) a MEMS hotplate without gas-sensing film with cavity etched from the top side of the silicon wafer, (b) its top view with gas-sensing film, (c) its cross-sectional diagram, and (d) the cross section of an alternative version of a MEMS hotplate with cavity etched from the bottom side.

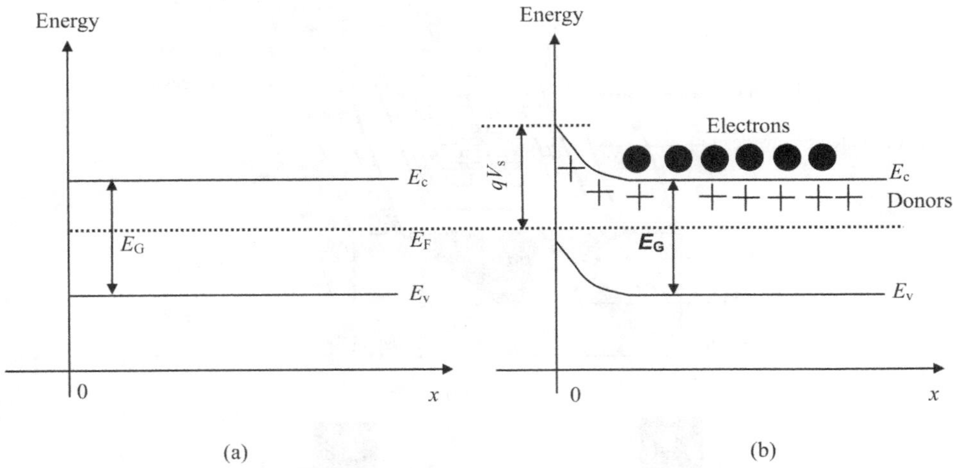

FIGURE 8.4 Energy band diagram, for a large N-type semiconductor crystal: (a) before oxygen adsorption, and (b) after oxygen adsorption. (Yamazoe, N. and Shimanoe, K., *J. Electrochem. Soc.*, 155, J85, 2008.)

conductive channel through the neck becomes too wide. Under these circumstances, the point-to-point contacts between the grain boundaries govern the total resistivity, giving rise to gas sensitivity, independent of the particle size.

What are the main models of electron transport between crystals of a gas-sensing material? There are three representative models of electron transport between nearby crystals (Yamazoe and Shimanoe 2009). (i) The *double Schottky barrier model*, which presupposes that electrons journey over the barrier at the contact. The contact connections prevail, and the device resistance equals the sum of the resistance of each contact. A Schottky barrier is a potential barrier that is formed in the contact region of a semiconductor, i.e., in the region that adjoins a metal. To obtain a Schottky barrier, the work functions of the metal and the semiconductor must differ. (ii) The *neck* (or *conduction channel model*) allows that electrons travel through a conduction channel that is produced by joining the core regions of bordering spheres. The channel width is tapered at the neck parts so that the conductance is determined by the geometric relationship between the neck size and the depletion depth. (iii) The *tunneling transport model* assumes that electrons located at the periphery of one sphere are transported to that of another by tunneling through a small gap (typically 0–0.1 nm) separating them (Figure 8.7). The conductance is proportional to the density of electrons at the periphery, which is determined by the surface barrier height. In the tunneling model, the height of the wall to tunnel through is set to be equal to the electron affinity of the crystals, i.e., the potential energy difference *(V)* between the vacuum level and the conduction band edge at the surface, whereas the thickness is set to the gap (*L*) between adjacent crystals.

How does gas adsorption on small crystals differ from that on larger ones? What is meant by regional and volume depletion? Consider the situation after the depletion depth has reached the radius of spheres or necks. Such a situation ensues easily when the spheres are more compact in size or the gas adsorption is strengthened. The gas adsorption equilibrium on small crystals is much different from that on large crystals. There are two different cases. In the case of large crystals, depletion does not reach the center of the crystals under usual conditions (*regional depletion*). The intact residual region inside acts as a source of electrons that are supplied to the adsorbates on the surface. In other words, the equilibrium is buffered by the intact region. In small crystals, depletion can advance beyond just covering the whole region of the crystals to enter into the next stage, *volume depletion*. Equilibration at this stage takes place under the conditions that almost all electrons inside the crystals are exhausted. In other words, the equilibrium is close to one, in which all electrons available in each crystal are understood to take part equally. In the stage of volume depletion, the reduced resistance (R/R_0) has a linear relationship with *y*, the reduced adsorptive strength of the adsorbing gas.

8.4.2 Sensitivity Controlling Parameters and the Influence of Heat Treatment

Ma et al. (2002) applied the equation of the semi-infinite planar surface approximation to the grain-boundary potential barrier. Presuming that the resistance in a porous ZnO-based gas element was due mainly to the neck resistance and the grain-boundary resistance, a model of gas sensitivity was developed. This model, combining the neck mechanism and the grain boundary mechanism, illustrated the effects of neck and grain boundaries on sensitivity. The grain-boundary resistance was derived as

$$R_{GB} = \left\{ \frac{\sqrt{2\pi\alpha m k_B T}}{A_{GB} q^2 n_b} \right\} \exp\left(\frac{qV_s}{k_B T} \right) \quad (8.1)$$

where

α is the ratio of effective electron mass to electron mass
m is the mass of an electron
k_B is the Boltzmann constant
T is the absolute temperature
A_{GB} is the cross section of the grain boundary
q is the electronic charge
n_b denotes the free-electron density in the grain body

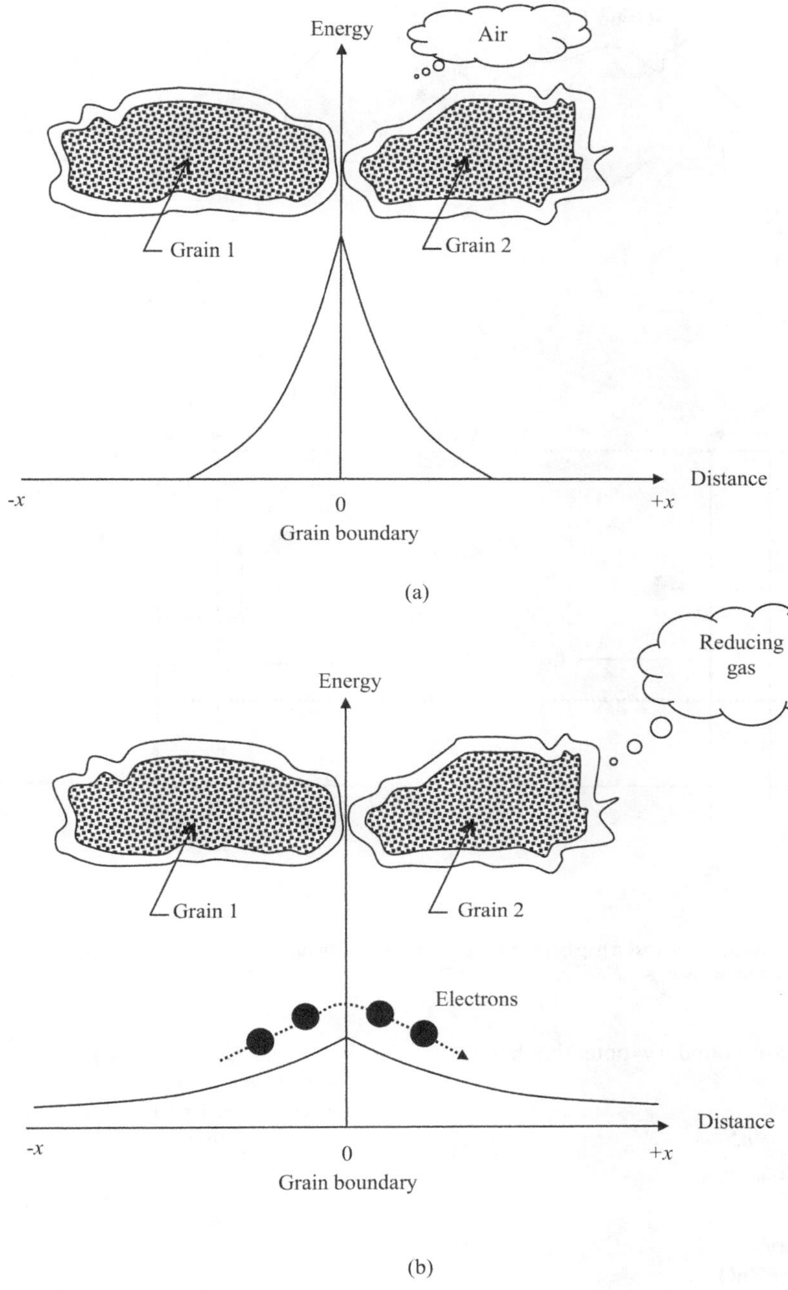

FIGURE 8.5 Intergrain potential barrier: (a) in air and (b) in the presence of reducing gas.

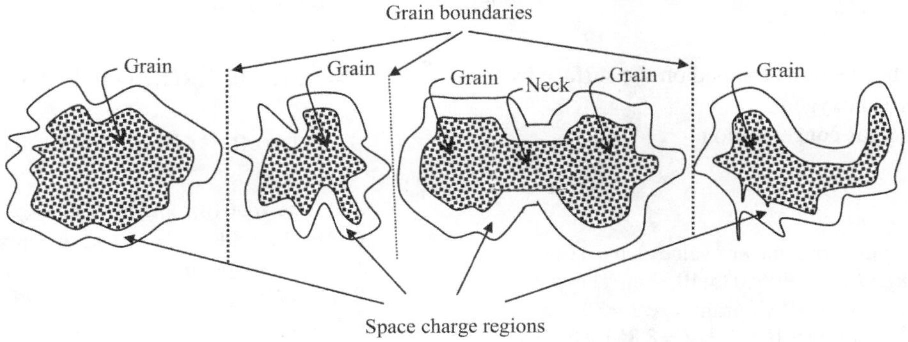

FIGURE 8.6 Neck and grain boundary model of a gas sensor. (Ma, Y. et al., *J. Wide Bandgap Mater.*, 10, 113, 2002.)

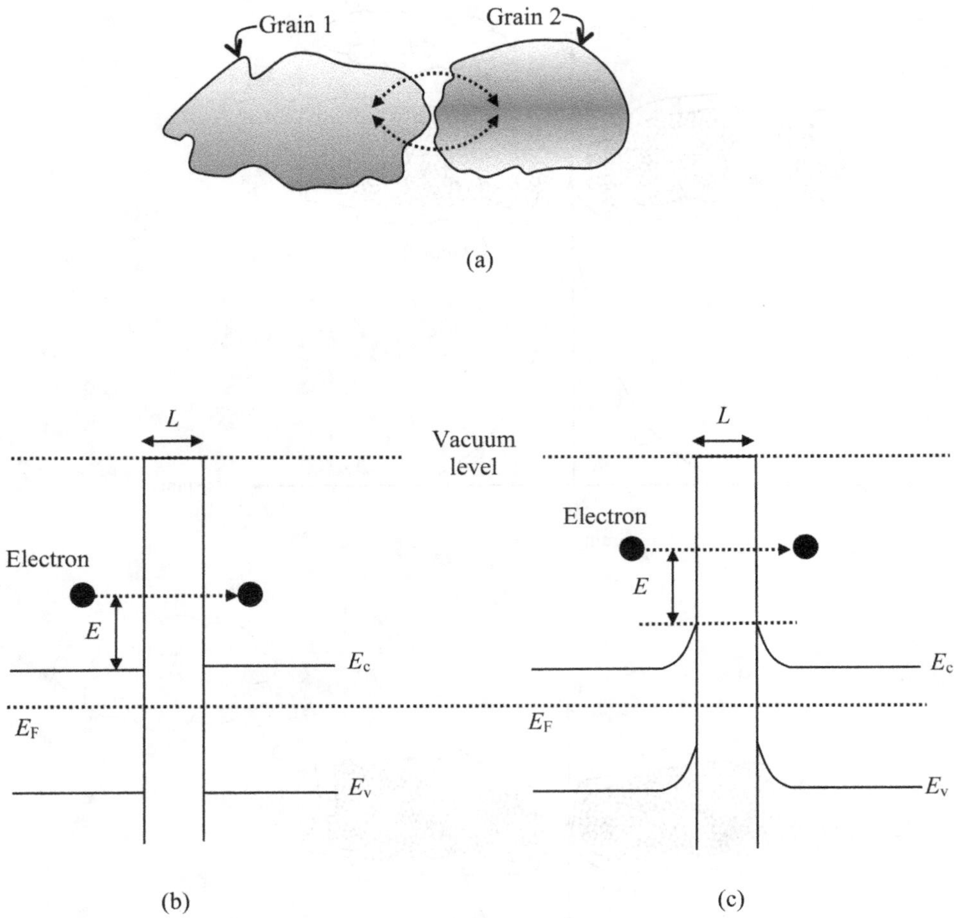

FIGURE 8.7 (a) Electron tunneling conduction across a gap between two grains. Electron tunneling models of gas-sensing material in (b) flat band and (c) bent band conditions. (Yamazoe, N. and Shimanoe, K., *J. Sens.*, Article ID 875704, 21 pages, doi: 10.1155/2009/875704, 2009.)

qV_s is the height of the grain-boundary potential barrier given by

$$qV_s = \frac{q^2 n_t^2}{2\varepsilon_0 \varepsilon_r n_b} \tag{8.2}$$

where

ε_0 is the free-space permittivity
ε_r is the relative permittivity of ZnO
n_t is the surface electron density

$$n_t = n_0 - n_r \tag{8.3}$$

where

n_0 is the density of the electrons trapped on the surface by the initially adsorbed oxygen
n_r is the chemisorbed gas concentration

Example 8.1

Given the following parameter values: $\alpha = 0.018$, $m = 9.11 \times 10^{-31}$ kg, $k_B = 1.3806503 \times 10^{-23}$ m^2 kg s^{-2} K^{-1}, $T = 570$ K, $q = 1.6 \times 10^{-19}$ C, grain size $L = 30$ nm, $n_b = 2.1 \times 10^{14}$ m^{-3}, $n_0 = 1.26 \times 10^{16}$ m^{-2}, $\varepsilon_0 = 8.854 \times 10^{-12}$ F m^{-1}, and $\varepsilon_r = 8$, find the potential barrier and grain-boundary resistances at two values of chemisorbed gas concentration: 1×10^{15} m^{-2} and 1×10^{16} m^{-2}.

From Equation 8.2,

$$V_s = \frac{qn_t^2}{2\varepsilon_0\varepsilon_r n_b} = \frac{1.6 \times 10^{-19} \times (n_0 - n_r)^2}{2 \times 8.854 \times 10^{-12} \times 8 \times 2.1 \times 10^{24}}$$
$$= 5.378 \times 10^{-34} \times \left(1.26 \times 10^{16} - n_r\right)^2 \tag{8.4}$$

At $n_r = 1 \times 10^{15}$ m^{-2},

$$V_s = 5.378 \times 10^{-34} \times \left(1.26 \times 10^{16} - 1 \times 10^{15}\right)^2$$
$$= 5.378 \times 10^{-34} \times \left(1.16 \times 10^{16}\right)^2 = 0.07237\text{ J} \tag{8.5}$$

Now, when $\alpha = 0.018$, $m = 9.11 \times 10^{-31}$ kg, $k_B = 1.3806503 \times 10^{-23}$ m^2 kg s^{-2} K^{-1}, $T = 570$ K, the cross-sectional area of the grain boundary is $A_{GB} = \pi(L/2)^2 = 3.14 \times (30 \times 10^{-9}/2)^2 = 7.065 \times 10^{-16}$ m^2, $q = 1.6 \times 10^{-19}$ C, and $n_b = 2.1 \times 10^{24}$ m^{-3},

$$R_{GB} = \left\{ \frac{\sqrt{2\pi\alpha m k_B T}}{A_{GB} q^2 n_b} \right\} \exp\left(\frac{qV_s}{k_B T} \right)$$

$$= \left\{ \frac{\sqrt{2 \times 3.14 \times 0.018 \times 9.11 \times 10^{-31} \times 1.3806503 \times 10^{-23} \times 570}}{7.065 \times 10^{-16} \times \left(1.6 \times 10^{-19}\right)^2 \times 2.1 \times 10^{24}} \right\}$$

$$\times \exp\left(\frac{1.6 \times 10^{-19} \times 0.07237}{1.3806503 \times 10^{-23} \times 570} \right)$$

$$= \frac{2.8468 \times 10^{-26}}{3.7981 \times 10^{-29}} \times \exp(1.47136) = 7.4953 \times 10^2 \times 4.355$$

$$= 3.264\,\text{k}\Omega \tag{8.6}$$

At $n_r = 1 \times 10^{16}\,\text{m}^{-2}$,

$$V_s = 5.378 \times 10^{-34} \times \left(1.26 \times 10^{16} - 1 \times 10^{16}\right)^2$$
$$= 5.378 \times 10^{-34} \times \left(2.6 \times 10^{15}\right)^2 = 6.636 \times 10^{-3}\,\text{J} \tag{8.7}$$

$$R_{GB} = \frac{2.8468 \times 10^{-26}}{3.7981 \times 10^{-29}} \times \exp\left(\frac{1.6 \times 10^{-19} \times 6.636 \times 10^{-3}}{1.3806503 \times 10^{-23} \times 570} \right)$$

$$= 7.4953 \times 10^2 \times \exp(0.1349) = 7.4953 \times 10^2\,\Omega \times 1.1444$$

$$= 0.8578\,\text{k}\Omega \tag{8.8}$$

If the neck is a cylinder of radius r_m, the neck resistance is

$$R_N = \frac{L}{\pi q \left\{ \mu_b n_b r_0^2 + \mu_d n_d \left(r_m^2 - r_0^2 \right) \right\}} \tag{8.9}$$

where
- L is the grain size
- μ_b is the electron mobility in the neutral grain body
- μ_d is the electron mobility in the depleted layer
- n_d is the free-electron density in the depleted layer

$$r_0 = r_m \left\{ 1 - \frac{2n_t}{n_b r_m} \right\}^{0.5} \tag{8.10}$$

r_m is the radius of the cylinder constituting the neck.

The gas sensitivity will increase with decreasing grain size. Gas sensors fabricated with nanometer ZnO have high sensitivity when the width of the space-charge layer of the neck in air is comparable with the neck radius. Although the grain-boundary resistance may be ≪ than the neck resistance, it cannot be ignored. It was suggested that decreasing the ratio between the numbers of grain boundaries and necks was a possible approach to the development of nano-ZnO gas sensors with a high sensitivity.

What factors control the sensitivity of a gas sensor? According to Yamazoe (2005), sensing performances, especially sensitivity, are controlled by three independent factors. These factors are receptor function, transducer function, and utility (Figure 8.8). *Receptor function* concerns the ability of the oxide surfaces to interact with the target gas. Chemical properties of the surface oxygen of the oxide itself are responsible for this function in a clean oxide device, but this function can be, for the most part, modified to induce a large change in sensitivity when an additive (noble metals, acidic oxides, or basic oxides) is loaded on the oxide surface. *Transducer function* concerns the ability to convert the signal caused by chemical interaction of the oxide surface (work function change: work function is the minimum amount of energy required to remove an electron from the surface of a metal) into electrical signal. This function is executed by each boundary between grains, to which a double Schottky barrier model is applied. The resistance depends on the barrier height and then on the concentration of the target gas. This situation is essentially unchanged with a change in the grain size (diameter, D) of the oxide, unless D is kept above a critical value (D_c), which is just equal to twice the thickness (L_s) of the surface space-charge layer of the oxide. For values of D smaller than D_c (6 nm for SnO_2), sensitivity increases sharply with decreasing D.

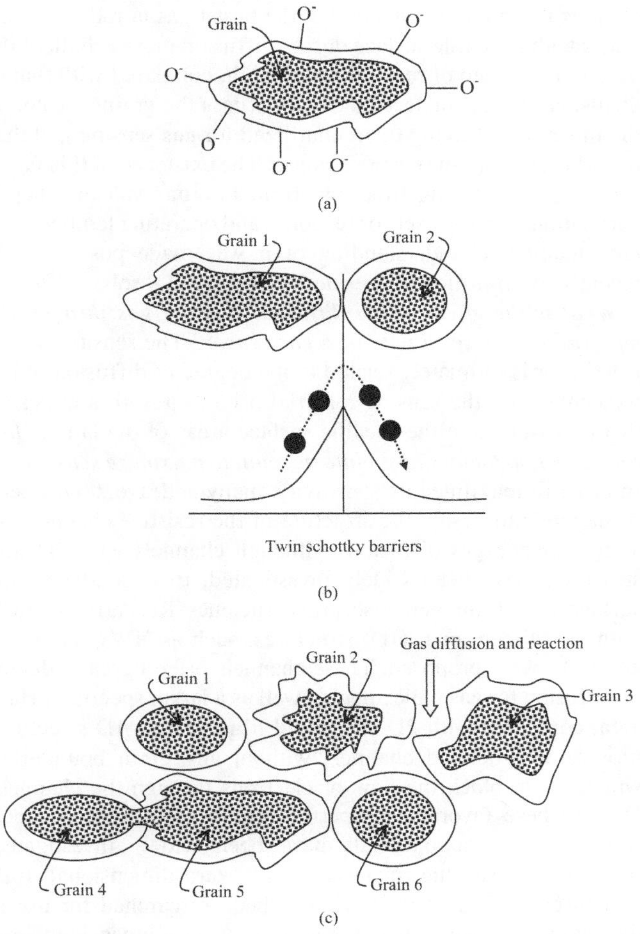

FIGURE 8.8 Three sensitivity-controlling parameters of a gas sensor [Yamazoe (2005)]: (a) receptor, (b) transducer functions, and (c) utility factor.

The last factor, *utility*, concerns the accessibility of inner oxide grains to the target gas. The importance of this factor is made obvious when one considers that the target gas (a reducing gas) reacts with the oxide surface during diffusion into the bulk of the device. If the rate of reaction is too high, compared with that of diffusion, the gas molecules cannot access the grains located at the inner sites, leaving them unutilized for gas sensing and thus resulting in a loss in sensor response. The existence of this factor was proposed a long time ago from familiar volcano-shaped correlations between sensor response and operating temperature, but quantitative understanding of it was made possible only recently for thin-film devices derived from SnO_2 sols.

What is the effect of the diffusion rates of gas through the material on the sensitivity of a gas sensor? The sensitivity of a gas sensor is intimately related to the degree of diffusion of gas molecules into the sensing material used to adsorb and desorb them, as well as to the specific surface areas of the latter. *How are these two factors taken into account to maximize sensitivity?* In order to maximize its sensitivity, many endeavors have been made lately to design the structure of the resistive channels, so as to enhance gas diffusion. Although channels with 2D film structures have been widely investigated, their small specific surface areas limit their adsorption efficiency. Recently, channels with one-dimensional (1D) structures, such as NWs, nanorods, and NBs, were proposed. These channels offer a greater degree of diffusion for gas molecules, as well as a larger specific surface area, compared with 2D structured films. These 1D structures provide well-defined channels without any grain boundaries, which might block the flow of electrons through the channels. Due to these favorable properties, gas sensors with channels made of 1D nanomaterials have been widely investigated. In addition, sensing materials with zero-dimensional (0D) structures, such as NPs, have also been researched for use in the channels, since their larger specific surface areas, compared with those of the other dimensional structures, will hopefully dramatically enhance the sensitivity characteristics of the gas sensors. However, the relatively high temperature (600°C) of the heat treatment used to stabilize the NPs brings about their agglomeration. This causes two critical problems for the gas sensors: (i) the sensitivity decreases, since the gas-sensing reaction occurs mainly at the surface of the agglomerates; and (ii) the response is sluggish, since the diffusion of gas molecules into the agglomerates becomes more difficult and slower.

Jun et al. (2009) proposed gas sensors with channels composed of nonagglomerated, necked NPs to maximize their sensitivity and response characteristics. *What are the main ideas of this approach?* Non-agglomerated, necked NPs have the characteristics of a mixture of 0D and 1D structures. If they are slightly necked with the adjacent NPs without any agglomeration, all of them participate in the gas-sensing reaction and, thus, the gas sensitivity is enhanced. In addition, the rapid diffusion of the target gas and the speedy counter-diffusion of the product gas are enabled and, thus, the response characteristics are improved. The slight necking of the NPs with their neighbors also enhances the conductivity of the channels, due to the lowering of the potential barriers present in the interfaces of the adjacent NPs. Hence, channels composed of non-agglomerated, necked NPs have the merits of both 0D structures, with the largest possible specific surface areas, and 1D structures, with well-defined channels.

How were the ideas implemented and their benefits verified? The above workers pasted the NPs across the electrodes and subsequently heated the NPs at 400°C. The sensing characteristics of the gas sensors with channels composed of the heat-treated NPs as sensing materials were then investigated and compared with those of gas sensors made with channels composed of the as-pasted NPs. When exposed to 5 ppm NO_2 gas, the response of the gas sensor with the channels composed of the heat-treated ZnO NPs was about 400, while that of the gas sensors with the channels composed of the as-pasted ZnO NPs was about 150. The higher response of the gas sensors composed of the heat-treated ZnO NPs results from the necking of the neighboring NPs and the size distribution of the NPs. The response and recovery characteristics of the gas sensors developed were remarkably fast at an NO_2 concentration of 0.4 ppm. Their response and recovery times were 13 and 10 seconds, respectively. Therefore, the heat treatment of NPs at 400°C is one of the useful methods of fabricating gas sensors having a high and fast response. These observations indicate that the non-agglomerated necking of the NPs induced by the heat treatment remarkably boosts the gas-sensing performance of the NP-based gas sensors.

8.4.3 Effect of Additives on Sensor Response

What is the effect of additives on sensor response? The ability of noble metals to act as highly effective oxidation catalysts can be used to enhance the reactions on gas sensor surfaces. A wide diversity of methods, including impregnation, sol-gel, sputtering and thermal evaporation, has been used for introducing noble metal additives into oxide semiconductors. The sol-gel process generally involves the use of metal alkoxides, which undergo hydrolysis and condensation polymerization reactions to give gels. The sol-gel process comprises solution, gelation, drying, and densification. The preparation of a silica glass begins with an appropriate alkoxide, which is mixed with water and a mutual solvent to form a solution. Hydrolysis leads to the formation of silanol groups (Si–OH). These species are only intermediates. Subsequent condensation reactions produce siloxane bonds (Si–O–Si). The silica gel formed by this process leads to a rigid, interconnected 3D network, consisting of sub-micrometer pores and polymeric chains.

Thus, sol-gel is a colloidal suspension of silica particles that is gelled to form a solid. The resulting porous gel is chemically purified and consolidated at high temperatures into high-purity silica. Removing the liquid located within the pores leads to a dried gel named xerogel.

Referring back to the discussion on the effect of additives, usually L_s is a function of the concentration of electron donors in the bulk oxide, and D_c is changed by doping the base oxide with a foreign oxide. When the oxide is loaded with a foreign additive, the additive can also modify L_s if it interacts electronically with the oxide. Indeed, such a change in L_s or barrier height elucidates the marked sensitizing effects of certain noble metals, like Pd, for the sensors of this type, e.g., in the case of Pd-loaded SnO_2, following exposure to air, Pd is oxidized into PdO, which acts as a strong acceptor of electrons from SnO_2. In this state, each grain of SnO_2 is covered with a strongly electron-deficient space-charge layer, leading to a high resistance. Whenever there is contact with a combustible gas in air, PdO is reduced to Pd, which is no longer an electron acceptor, resulting in a steep plummeting of the electrical resistance.

What are the mechanisms causing this behavior? It is noted that the sensitizing effects emerge through coupling a redox modification of the additive with a change in its electronic interaction with the oxide grains. Two concepts are hypothesized to explain the improvement of the sensing performance of NW upon Pd deposition: (i) the "electronic mechanism" proposal; and (ii) a "chemical process." The *electronic mechanism* considers that depletion zones are formed around the modified particles. Modulation of the nano-Schottky barriers (and hence the width of the conduction channel) occurs. This is due to changes in the oxidation state of the Pd, accompanying oxygen adsorption and desorption. The modulation is responsible for the sensing enhancement. Explanation of the kinetics and temperature dependence brought about by Pd functionalization is difficult through the electronic mechanism, while the chemical process does not experience such problems. The mechanism of the chemical process is based on the highly effective dissociation catalytic ability of Pd. The ionosorption of oxygen takes place at defect sites of the pristine surface. Pd is a better oxygen dissociation catalyst than tin oxide, catalytically activating the dissociation of molecular oxygen. Then, atomic products diffuse to the MOX support. Furthermore, it is not necessary for molecular oxygen to dissociate on the Pd surface only. It is believed that oxygen molecules reside briefly on an oxide support and diffuse to a catalyst particle before they have any chance of desorption. This is called "the back-spillover effect." It increases the sensitivity along with diffusion of atomic products to the MOX support.

8.5 Carbon Nanotube Gas Sensors

8.5.1 Gas-Sensing Properties of CNTs

What properties of CNTs are useful for gas sensing? CNTs are very interesting in the sensor field because they are characterized by a high, theoretically infinite surface/volume ratio (Zhang and Zhang 2009). The development of CNT-based gas sensors has attracted all-embracing interest because of their high response, sub-ppb concentration detection levels, prompt response, low operating temperature, small power consumption, miniature size for nanodevice shrinking, functionalization for specific detection of molecules, and ability to build massive sensor arrays. The development of CNT nanosized sensors opens application possibilities of multitransducer and multisensor arrays by using pattern recognition techniques for efficient chemical gas analysis.

What are the chief action lines for realizing CNT gas sensors? There are two approaches to CNT gas sensor fabrication: (i) resistive gas sensor and (ii) field-effect gas sensor. Figure 8.9a shows the adsorption of gas molecules on a CNT surface, producing a conductance change. A resistive CNT gas sensor is shown in Figure 8.9b. It is fabricated on a silicon substrate on which an insulating thermal oxide layer is grown. Bundles of CNTs are cast over the two electrodes. The resistance change of CNT bundles with ambient gas concentration is the sensing parameter. A field-effect CNT gas sensor (Figure 8.10) consists of a single semiconducting CNT, spanning over the source and drain electrodes. It serves as the channel of the field-effect transistor (FET) device. CNT conductivity varies with gas

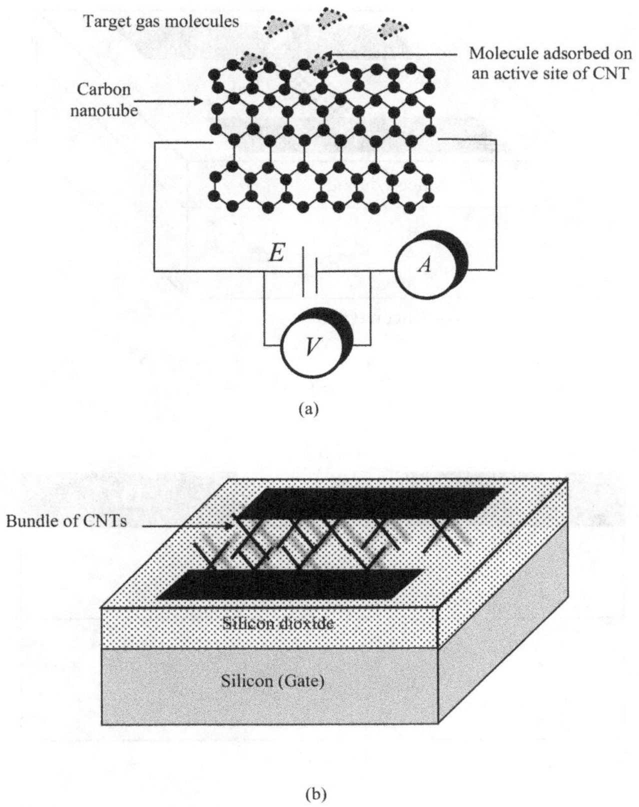

FIGURE 8.9 (a) Basic mechanism of a resistive CNT gas-sensing device and (b) a resistive gas sensor using CNT bundles.

concentration. A silicon back gate is used to produce the field-effect action. The field-effect gas sensor will be discussed in more detail in Section 8.5.4.

Several CNT-based nanosensors for gases have been investigated because interaction of molecules adsorbed on their surfaces affects their electrical properties (basically, their conductivity). Studies have been made of how they behave when exposed to gases such as NO_2, NH_3, CO, H_2O, N_2, or organic vapors. Nanotubes respond differently to different gases, that add or remove electron density from the surface of the nanotubes. Single-walled carbon nanotubes (SWCNTs) are sensitive to NO_2, NH_3 and volatile organic compounds. The adsorption of gaseous molecules either donates or withdraws electrons from the SWCNTs, leading to changes in electrical properties of the SWCNTs. The drawback of these sensors is their slow and incomplete recovery.

Li et al. (2003) proposed a simple resistive device, fabricated by casting a solution of purified SWCNT in dimethyl formamide, $(CH_3)_2NC(O)H$, on a silicon substrate. The sensor worked at room temperature and offered a very high sensitivity to NO_2 and nitrotoluene ($C_7H_7NO_2$) with detection limits of 44 and 262 ppb, respectively, in N_2.

To detect diisopropyl methylphosphonate (DIMP), $C_7H_{17}O_3P$, and dimethyl methylphosphonate (DMMP), $C_3H_9O_3P$, simulants for the G series organophosphorus compounds nerve agents, Soman ($C_7H_{16}FO_2P$) and Sarin ($C_4H_{10}FO_2P$), respectively, Cattanach et al. (2006) described the use of network films of SWCNT bundles on flexible substrates such as polyethylene

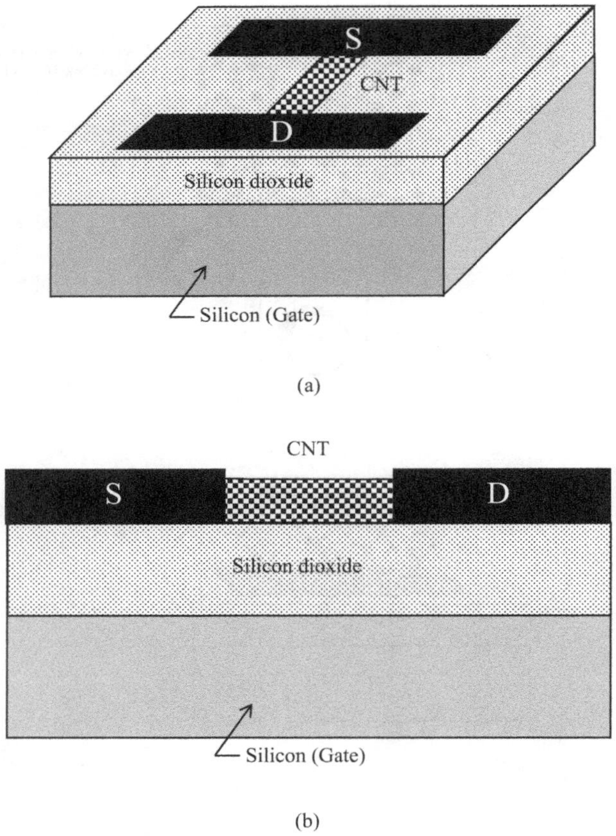

FIGURE 8.10 (a) 3D view of a gas sensor based on FET, with a single carbon nanotube as THE sensing element and (b) its cross-sectional diagram.

terephthalate (PET), $(C_{10}H_8O_4)_n$. Upon exposure to DIMP or DMMP vapors, large reproducible resistance changes (75%–150%), were observed. Concentrations as low as 25 ppm could be measured. These SWCNT/PET films were readily synthesized using line patterning, which is a simple and rapid room temperature method (involving no photolithography or printing), to obtain any desired pattern of SWCNT films from aqueous surfactant-supported dispersions of SWCNTs on flexible substrates, like plastic, paper, cloth, etc.; a surfactant is a surface-active agent that lowers the surface tension of a liquid, increasing the contact between the liquid and another substance. The sensor functioned robustly even when the SWCNT/PET sensor was bent all the way to a crease.

Network films of SWCNT/PET are also able to sense DIMP and DMMP when other interferent vapors are present. In addition to air, humidity, etc., network films of SWCNT/PET are able to detect simulant vapors in the presence of large quantities of interferent vapors commonly found in field applications, such as hexanes (a colorless flammable liquid hydrocarbon with the chemical formula C_6H_{14}) and xylenes, C_8H_{10}, $C_6H_4(CH_3)_2$, or $C_6H_4C_2H_6$, constituents of gasoline fuel (a volatile mixture of flammable liquid hydrocarbons derived chiefly from crude petroleum) and diesel (a mix of hydrocarbons typically in the C_9H_2O to $C_{12}H_{26}$ range, derived from petroleum but heavier than gasoline/petrol), respectively In order to improve the selectivity of SWCNT/PET films to simulants in a mixture containing large amounts of interferent vapors, the researchers screened a wide range of polymer coatings for chemoselectivity (the preferential reaction of a chemical reagent with one functional group in the presence of other similar functional groups). It was found that a 2-μm thick film of polyisobutylene (PIB), $-(CH_2-C_3H_6)_n-$, spin-coated) on the SWCNT sensor was an effective barrier coating. The PIB-coated sensor entirely screened out 30,000 ppm of interferent vapors. At the same time, it was able to detect 299 ppm DIMP. This showed that the use of PIB as a barrier coating polymer could screen out high concentrations of interferent vapor signatures.

8.5.2 Responses of SWCNTs and MWCNTs

Do SWCNTs, multi-walled carbon nanotubes (MWCNTs), and aligned CNTs differ in their gas-sensing behaviors? Sensors have been fabricated using single-walled, multi-walled, and well-aligned CNT arrays (Penza et al. 2004). Suehiro et al. (2005) demonstrated that the normalized response of the SWCNT sensor was higher than that of the MWCNT sensor. This is probably because SWCNTs contained more semiconducting tubes. In general, MWCNTs show a conducting (metallic) behavior at room temperature. However, MWCNTs could contain some semiconducting tubes among predominantly metallic ones. The higher normalized response of the SWCNT sensor may be attributed to greater abundance of the semiconducting tube, which is responsible for the sensor response. *Does the alignment of CNTs affect the sensor response?* Yes, nanotube alignment also affects the sensor response, e.g., the response to ammonia (NH_3) of a sensor with aligned SWCNTs was double that achieved by a sensor with disordered SWCNT (Lucci et al. 2005). This effect was probably induced by the fact that the ordered SWCNTs were more uniformly exposed to the NH_3 molecules, than in the case of placement of SWCNTs in the form of a random network.

8.5.3 Modification of CNTs

Why do CNTs need to be modified for making sensors for particular gases? Sensitivity of CNTs to the surrounding chemical environment is advantageous but, on the other hand, it implies a distinct drawback, because it indicates a lack of selectivity. Sensors based on raw CNTs are devices that operate at ambient temperature and are highly sensitive to different types of gases, although they are not very selective. To improve the selectivity of CNT-based sensors, nanotube functionalization has been carried out. For this purpose, several functionalization methods have been developed. These include doping or inserting atoms or functional groups into nanotube structures, which constitutes a type of covalent functionalization. Functionalization can also be noncovalent. This consists of depositing particles or coating the surface of the CNT with a material that interacts with the analyte of interest. In the latter case, the nanotubes still act as a transducer system, while the material deposited on the surface becomes the sensing part of the device. The sensitivity of some sensors has been dramatically enhanced from ppm to ppb level after surface modification or functionalization and doping.

What attempts have been made toward CNT functionalization? CNTs have been modified by many workers using functional groups, metals, oxides, or polymers, or by doping CNTs with other elements to enhance the response and selectivity of the sensors:

1. Lu et al. (2004) reported a SWCNT-based gas sensor, designed for methane (CH_4), with an operating range in dry air of between 6 and 100 ppm, fabricated by casting, using a suspension of purified SWCNT in deionized water (DIW), where sputtered Pd NPs were dispersed.
2. Li et al. (2006) fabricated a chemiresistor with an interdigitated electrode (IDE) configuration. Purified SWCNTs were laid across the electrodes using a solution-casting process in the form of a network. Polymer coating of SWCNTs allowed selective sensing of certain gases, as demonstrated here for chlorine (Cl_2) and hydrochloric acid (HCl) vapor. These are industrially important chemicals, and sensors for leak and spill detection are desirable. The sensor consisted of two electrodes connected by interdigitated fingers (or interdigitated electrodes, IDEs), with nanotubes bridging the finger gaps. The IDE was fabricated using conventional photolithography with a nominal finger width of 10 μm and a gap size of 8 μm. The fingers were made of thermally evaporated Ti (20 nm) and Au (40 nm) on a layer of SiO_2 thermally grown on a silicon substrate. The sensing material was bulk-produced SWCNTs that were purified to remove amorphous carbon and metal impurities. Dispersal of the purified nanotubes was done in dimethyl formamide, $(CH_3)_2NC(O)H$, to create a suspension. The polymer coatings were chlorosulfonated polyethylene (CSPE), a synthetic rubber, and hydroxypropyl cellulose (HPC), a derivative of cellulose, exhibiting both water solubility and organic solvent solubility.
3. Zhang et al. (2006b) functionalized SWCNTs by an electrochemical process with polyaniline (PANI). PANI is a versatile conducting polymer of the semi-flexible rod polymer family, easily processed by melt or solution process, and is environmentally and thermally stable. The PANI-SWCNT composite behavior was tested in NH_3, exhibiting a detection limit of 50 ppb. The response time at room temperature was ~ several minutes and the recovery time was a few hours.
4. Padigi et al. (2007) functionalized MWCNT with thiol (an organo-sulfur compound) for the detection of the first four fundamental aliphatic hydrocarbons, namely CH_3OH, C_2H_5OH, C_3H_7OH, and C_4H_9OH. High degrees of selectivity and sensitivity up to a detection concentration of 1 ppm were shown. In the presence of a chemical species, there was a shift of the resonance frequency peak.

In an attempt to decrease the working temperature of some materials that can be used as the sensing part, and to increase the sensitivity of the devices, several studies have combined CNTs with oxides or metal particles. In these studies, the functionalization is not covalent, since the CNT structure is not modified: The process incorporates NPs or films on the surface of the nanotubes, thus changing the adsorption mechanism of the analytes with respect to the unmodified nanotubes. CNTs, then, play the role of the transducer in this type of sensor.

8.5.4 CNT-Based FET-Type Sensor

Several gas-sensing devices based on FET architecture have been proposed. *What is the principle of CNT FET sensors?* As may be recalled from the viewpoint of electrical conductivity, nanotubes can be metals or semiconductors, depending on the chirality of their structure. Semiconducting nanotubes are used to construct devices such as FETs. These devices are designed to detect analytes by means of a change in the electrical conductivity of the CNTs in response to the effect of the analyte on their surface. This effect can have two consequences: (i) they may promote a charge transfer from the analyte to the nanotube; or (ii) the analyte may act as a scattering potential (i.e., potential to cause the scattering of a particle). The charge transfer increases the conductivity when the analyte adsorbed is an electron attractor (hole donor). If the analyte is an electron donor, the number of holes on the nanotubes will decrease, which leads to a decrease in the electrical conductivity. This effect can be used to modulate FET-type devices with a third gate electrode. On the other hand, if the analyte acts as a scattering center, the current will decrease without shifting the characteristics.

What CNT FET gas sensors have been fabricated? Wong et al. (2003) used MWCNTs obtained by microwave plasma-enhanced chemical vapor deposition (MPECVD) on silicon substrates to construct nano FETs. The top part of the MWCNTs was coated with a sputter-deposited thin layer of palladium (100 nm), which also served as the gate electrode. The Pd layer acts as a catalytic agent, enhancing the dissociation of the H_2 molecules into its atoms. The atomic hydrogen, adsorbed on the Pd NPs, transferred electrons to the CNTs. This decreased the number of hole carriers and, therefore, the conductivity of the device. The researchers monitored the change in the electric current by substituting air for a flow of H_2 and observed the effect of temperature. They noted that there was a remarkable increase in sensitivity at higher operating temperatures (180°C).

An NO_x and NH_3 SWCNTs-based chemical sensor was fabricated by a dielectrophoretic process (a phenomenon in which a force is exerted on a dielectric particle, when it is subjected to a nonuniform electric field; this force does not require the particle to be charged) on an FET structure (Lucci et al. 2006). It was found that NH_3 reduced the conductivity because of the charge transfer to the SWCNT, whereas NO_x induced an opposite effect.

8.5.5 MWCNTs/SnO_2 Ammonia Sensor

Hieu et al. (2008) developed an SnO_2/MWCNTs composite-based NH_3 sensor working at room temperature. The fabrication process involved the dispersion of MWCNTs in the SnO_2 dispersion. *What is the reason for combining tin oxide with CNTs?* It was intended to derive the merits of both materials in one device, i.e., blending two types of properties in one device. The SnO_2-based gas sensors can detect NH_3 gas with high sensitivity and response recovery time, but they operate only at elevated temperatures. In contrast, the CNT-based sensors can detect NH_3

gas at room temperature, but their sensitivity is low and response recovery time is excessively long. Thus, the former is unable to detect NH_3 gas at room temperature and the latter has very long recovery and response times with respect to detection of NH_3 gas at room temperature. But the composite MWCNTs/SnO_2 thin-film gas sensor amalgamates the desirable characteristics of both types of sensors, providing a high response and a faster recovery time in the detection of NH_3 gas at room temperature. It thus solves the problems of individual SnO_2-based and CNT-based sensors.

An MWCNTs/SnO_2 composite sensor behaved as a P-type semiconductor. *What is the reason for this P-type character?* A plausible explanation is that the composite has a much higher CNT content; therefore, the major charge carriers are the holes, which are mainly supplied by CNTs.

At room temperature, the optimal composite sensor exhibited much higher response and faster response-recovery (less than 5 min) to NH_3 gas at concentrations ranging from 60 to 800 ppm, in comparison with the CNT-based NH_3 sensor.

8.5.6 CNT-Based Acoustic Gas Sensor

How do mechanical gas nanosensors work? Penza et al. (2004) demonstrated the integration of SWCNTs onto a quartz-crystal microbalance (QCM) for alcohol detection at room temperature (Figure 8.11). The operating principle of QCM sensors is based on the fact that the change in the resonance frequency of a vibrating quartz-crystal resonator is proportional to the mass of the adsorbed gas molecules onto a sensitive overcoating. A conventional 10 MHz QCM was used as an acoustic sensor. It consists of a circularly shaped AT-cut quartz crystal with a diameter of 10 mm and a thickness of 0.1 mm. The aluminum electrodes deposited on both sides of the quartz were 100 nm thick and 4 mm in diameter. The active area for deposition of sensing CNTs was 12.5 mm^2. The noise of the uncoated 10 MHz QCM sensor was 0.5 Hz in 10 minutes.

The cadmium arachidate (CdA), prepared by the Langmuir-Blodgett (LB) technique, was used as multilayered buffer material on the acoustic sensor to promote the adhesion of the SWCNTs as sensing overlayers on the sensors. The Langmuir-Blodgett technique is a method of depositing crystalline films, one molecular layer at a time, by dipping the substrate into water containing a polymer, that forms a single layer of molecular chains on the surface. This layer is then transferred from the water to the substrate. The dipping is repeated to create an ordered multilayer film that does not require electrical poling (application of a strong electric field across the film) to orient the molecules. An LB film is a set of monolayers, or layers of organic material, one molecule thick, deposited on a solid substrate. It can consist of a single layer or many, up to a depth of several visible-light wavelengths.

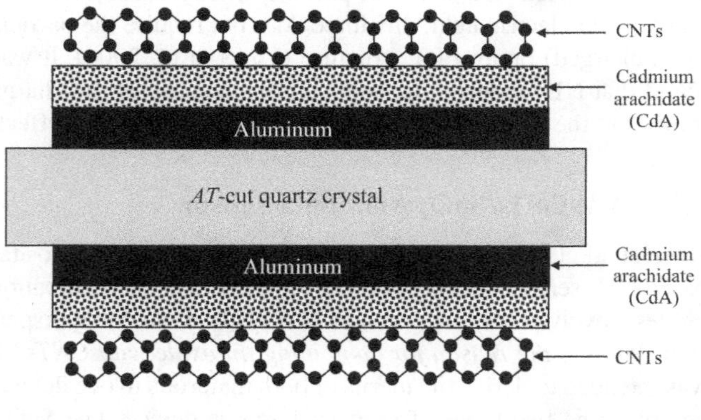

FIGURE 8.11 A quartz-crystal microbalance with linker-buffer layer of cadmium arachidate sensors to promote the adhesion of the single-walled carbon nanotubes as a sensing overlayer on the sensor. (Penza, M. et al., *Appl. Phys. Lett.*, 85, 2379, 2004.)

The CdA was chosen as a linker-buffer material with hydrophobic surface characteristics; a hydrophobic surface is a water repelling surface that resists wetting and on which the contact angle of water droplet is obtuse. LB films, consisting of tangled bundles of SWCNTs without surfactant molecules, were successfully transferred onto the QCM.

The resonance frequency was the sensor output. The frequency decreases with respect to nitrogen reference ambient in response to ethanol (ethyl alcohol, C_2H_5OH) exposure with a shift of 230 Hz with an injection of 59 mm Hg ethanol for the SWCNT-coated QCM. The resonant frequency decrease was ascribed to the mass of the alcohol molecules adsorbed onto nanosensing SWCNTs. The frequency shift of the SWCNT-coated QCM was much higher than that of the CdA-coated QCM, with good reversibility and excellent repeatability: The response increased from 60 to 230 Hz for exposure to 59 mm Hg ethanol. Hence, the influence of the CNTs on alcohol sensitivity was clearly discernible. This enhanced sensitivity could be accredited to the typical high surface area of SWCNTs.

Example 8.2

The frequency of a quartz crystal is 10 MHz. Its active area is 10 mm^2. The frequency changes by 50 Hz on adsorption of a gas in the gas-sensitive film deposited on the active area of the sensor. Find the change in mass of the crystal upon gas adsorption.

Here, $A = 10\ \text{mm}^{-2} = 10\times10^{-2}\ \text{cm}^2 = 0.1\ \text{cm}^2$, $f_0 = 10\ \text{MHz} = 10\times10^6\ \text{Hz} = 10^7\ \text{Hz}$, and $\Delta f = 50$ Hz. Recalling the Sauerbrey equation from Equation 4.1:

$$\Delta f = \frac{-2f_0^2\Delta m}{A\sqrt{\rho_q\mu_q}} \tag{8.11}$$

where

$\rho_q = 2.648\ \text{g cm}^{-3}$
$\mu_q = 2.947\times10^{11}\ \text{g cm}^{-1}\times\text{s}^{-2}$

This equation is rearranged as

$$\Delta m = \frac{-\Delta f\left(A\sqrt{\rho_q\mu_q}\right)}{2f_0^2} = -\frac{50\times0.1\times\sqrt{2.648\times2.947\times10^{11}}}{2\times\left(10^7\right)^2} \tag{8.12}$$

$$= 2.2\times10^{-8}\ \text{g} = \frac{2.2\times10^{-8}}{10^{-9}}\ \text{ng} = -22\ \text{ng}$$

This is the mass of the gas responsible for increasing the mass of the crystal. The minus sign in this equation explains that Δf decreases as Δm increases.

8.6 Porous Silicon–Based Gas Sensor

What is the difference between crystalline silicon and porous silicon (PS) in terms of gas response? The physical properties of crystalline silicon are insensitive to the environment, but, when crystalline silicon is reduced to the nanoscale, the resulting porous nanocrystalline structure reacts strongly, even explosively, with various analytes (Plessis 2008). The silicon Bohr exciton radius is about 5 nm and, upon electrochemical engraving, a PS structure with wall thickness in the range 2–5 nm is formed. The consequential nanocrystallites show a high chemical reactivity toward the environment.

The instability of the native surface termination is the major obstruction prohibiting extensive application of PS in the sensor field. Massera et al. (2004) found that the electrical performance of a PS NO_2 gas sensor was greatly improved by stabilizing its surface with long-standing exposure to high NO_2 concentrations prior to usage.

PS is sometimes used as a large surface area substrate on which NPs are deposited. *How is a PS-based hydrogen gas sensor made?* Luongo et al. (2005) developed a hydrogen sensor in which the sensing part consisted of NPs of Pd deposited on a P-type silicon substrate. Electrochemical etching of the substrate was performed in HF/ethanol/H_2O, forming nanopores with an average diameter of ~10 nm on the surface. Subsequently, approximately 4-nm thick Pd film was deposited and stabilized by annealing at 900°C for 1 hour, during which time SiO_2 and PdO were formed. This process augmented the contact surface area of the sensor. Also, the electrical resistance increased, because vacancies were produced in Si. The PS-Pd sensor was tested with a four-point probe configuration. The conductivity of the device was found to increase under nitrogen and with hydrogen concentrations of 0%–1.5%. The Pd acted as the catalyst for the dissociation of hydrogen, which transferred electrons to the silicon substrate. This relocation of electrons decreased the resistance. Since the PdO formed during the annealing was reduced at temperatures of 100°C or above, thereby degrading device performance, the sensor was operated at ambient temperature.

8.7 Thin Organic Polymer Film-Based Gas Sensors

How have polymer films contributed to gas sensing? Numerous sensors have been developed using few tens/hundreds of nanometer thick polymeric materials. Nohria et al. (2006) constructed a humidity sensor by depositing thin films of poly(anilinesulfonic acid) (SPANI) on a substrate by spin coating, from a solution of the polymer, leading to a film 90-nm thick. As a second approach, films were also deposited by the layer-by-layer (LbL) nano-assembly technique, which negatively charges the substrate to couple the layers of a polycation [poly(allylamine hydrochloride)] (PAH), $(C_3H_7N)_n{\cdot}xHCl$ and a polyanion [polystyrene sulfonate)] (PSS). Once the layers had been deposited, the PSS was replaced by the SPANI to give a 26-nm thick film. On exposure to air, resistance of the sensor decreased as the relative humidity (RH) increased, with response times of 15 s for an LbL film sensor and 27 s for a spin-coated sensor.

Zhang et al. (2006) worked on the detection of vapors of tetrahydrofuran, C_4H_8O, a clear, highly flammable liquid used as a solvent for natural and synthetic resins and in the production of nylon; benzene (C_6H_6), a colorless, flammable, aromatic hydrocarbon; toluene, C_7H_8 or $C_6H_5CH_3$, a clear, water-insoluble liquid with the typical smell of paint thinners; cyclohexane, a cycloalkane with the molecular formula C_6H_{12}; *n*-hexane (petroleum naphtha), C_6H_{14}, a colorless liquid with a slightly disagreeable odor; carbon tetrachloride, also known as tetrachloromethane, CCl_4, a colorless liquid with a sweet smell; chloroform, $CHCl_3$, a clear volatile liquid with a strong smell like ether; ethyl acetate (ethyl ethanoate), EtOAc or EA, $C_4H_8O_2$, a colorless liquid with a fruity odor; diethyl ether, $(C_2H_5)_2O$, a colorless and highly flammable liquid; acetone, C_3H_6O, a colorless, volatile, extremely flammable liquid ketone; methanol, CH_3OH, a colorless, toxic, flammable liquid, used as an antifreeze, a general solvent and a fuel; ethanol, C_2H_6O, a clear, colorless liquid with a characteristic, agreeable odor; propanol, $CH_3CH_2CH_2OH$, a clear colorless liquid used as a solvent and as an antiseptic; and isopropanol, $(CH_3)_2CHOH$, a flammable liquid. They used a polystyrene (a vinyl polymer: $(C_8H_8)_n$) film with carbon nanofibers (cylinders consisting of tightly rolled graphite sheets that have diameters of 70–200 nm and lengths of 1–100 mm) to obtain stable sensors. The vapors were noted through the change in electrical resistance shown by the device. The fact that the material is in the form of fibers gave greater stability because, when the gas was adsorbed, the NPs increased in size and had a propensity to join together.

What are the benefits of using polymers in gas sensing? (i) Organic polymers are cheap and highly sensitive to several analytes; (ii) The deposition techniques used to obtain the thin films are also relatively simple and economical. For these reasons, polymers offer a faster method of building gas sensors. *What are the problems associated with polymers?* The main problems associated with these materials are their lack of selectivity and their lower stability, with respect to time and temperature. Additive materials, such as carbon NPs or fibers, have been explored to increase the selectivity and stability of polymer-based sensors.

8.8 Electrospun Polymer Nanofibers as Humidity Sensors

Can polymer nanofibers be used for moisture measurement? Yes. Li et al. (2009) prepared nanofibers of a polymer composite of a silicon-containing polyelectrolyte and soluble polyaniline (PANI: a unique type of polymer because it can be configured to conduct across a wide range, from being utterly nonconductive, for insulation use, to highly conductive for other electrical purposes) by electrospinning (ES). They investigated the humidity sensing properties of these polymers. ES is a simple and versatile technique for the easy preparation of continuous nanofibers.

The above nanofibers, having a diameter of 250–500 nm, formed a nonwoven carpet with a highly porous structure. The polymer composite nanofibers showed an impedance change from 6.3×10^6 to 2.5×10^4 Ω with the rise in relative humidity (RH) from 22% to 97% at room temperature, exhibiting high sensitivity and good linearity of the relationship on a semi-logarithmic scale. In addition, they showed a fast and highly reversible response, characterized by very small hysteresis of ~2% RH and a short response time ($t_{90\%}$: 7 and 19 seconds for adsorption and desorption between 33% and 97% RH, respectively). The high resistance of electrospun nanofibers endorses their high porosity.

How is the humidity sensitivity of the sensor increased and what is the reason for this improvement? The modification of the underlying gold electrode with poly(diallyldimethylammonium chloride) (PDDA), prior to the deposition of nanofibers, improved the humidity response of the sensor, due to the enhancement of contact between the nanofibers and the underlying substrate. The introduction of PANI into the composite effectively decreased the impedance of the nanofibers.

8.9 Toward Large Nanosensor Arrays and Nanoelectronic Nose

What is odor made of? Odor is composed of molecules, each having a specific size and shape. All these molecules have a correspondingly sized and shaped receptor in the human nose. When a specific receptor receives a molecule, it transmits a signal to the brain and the brain identifies the smell associated with that particular molecule.

How does an electronic nose (e-nose) mimic human nose? E-noses, based on the biological model, work in a similar manner to the human nose, albeit substituting sensors for the receptors, and transmitting the signal to a program for processing, in place of the brain. An e-nose is a device, which comprises an array of electronic chemical sensors with partial specificity (whether in the form of MOX semiconductors or *via* the use of different types of polymers) and an appropriate pattern recognition system, capable of recognizing simple or complex odors. Each sensor responds slightly differently to a given odor, and altogether, the sensors give an "odor fingerprint," a characteristic response pattern, such as a series of colors, to each odor.

What are the common applications of the e-nose? The most common use for the e-nose at the present time is in the food and drink industries. In addition to this field, e-nose can be used in other areas, such as classification and degradation studies of olive oils, qualitative and quantitative analysis of petroleum, detection of explosives, development of a field-odor detector for environmental applications, and quality control applications in the automotive industry.

How is the e-nose realized? Qi et al. (2003) showed that large arrays of devices, containing multiple SWCNT bridging electrodes, could be easily produced with 100% yield by chemical vapor deposition (CVD) growth of nanotubes from a micropatterned catalyst. The ensemble of SWCNTs in each device collectively exhibits substantial electrical conductance changes to electrostatic gating, because of the high percentage of semiconducting nanotubes grown by CVD. The multiple tube devices (MT devices) are highly sensitive to chemical gating effects and, decisively, exhibit lower electrical noise than individual semiconducting SWCNT devices. Large arrays of nanotube chemical sensors with excellent characteristics are thus available. Furthermore, Qi et al. (2003) carried out polymer functionalization of the MT devices to achieve ultrahigh sensitivity for NO_2 detection and to impart selectivity to nanotube sensors. They demonstrated that microspotting allows for functionalization of nanotube sensor arrays in a multiplex fashion, enabling detection of molecules in gas mixtures.

They investigated functionalization of nanotubes in the MT devices by various polymers, to enhance sensitivity and impartselectivity to nanotube sensors, e.g., they adsorbed polyethyleneimine (PEI), $(C_2H_5N)_n$: linear form, onto SWCNTs in MT devices by simple immersion in a 20 wt% PEI/methanol solution for 2 hours, rinsing with methanol, and then baking at 50°C for 1 hour. The PEI-coated MT devices evolved into N-type, in response to electron transfer doping by high-density amine groups (a nitrogen atom with three bonds, one of which is to a carbon atom) on PEI, similar to that observed with individual semiconducting SWCNTs. The PEI functionalized N-type MT devices were ultra sensitive to NO_2, responding to as little as 100 ppt (parts per trillion) of NO_2 (versus ~10–50 ppb for as-grown MT devices) with an appreciable conductance decrease. The lowest concentrations of NO_2 reliably detected by the N-type MT devices were several ppb with a response time of ~1–2 min (defined as the time for 80% conductance change to take place). The conductance change *versus* NO_2 concentration relationship showed a linear dependence for $[NO_2] = 100$ ppt to 3 ppb and a tendency for saturation at higher $[NO_2]$.

They also explored selective detection of NH_3 with the MT devices and deciphered that coating an as-grown MT device with Nation ($C_7HF_{13}O_5S{\cdot}C_2F_4$) blocks certain types of molecules, including NO_2, from reaching the nanotubes. This allows for more selective NH_3 detection. Nation is a sulfonated tetra-fluoroethylene-based fluoropolymer-copolymer with a teflon [PTFE: $(C_2F_4)_n$] backbone and sulfonic acid (the class of organosulfur compounds with the general formula $RS(=O)_2–OH$, where R is an alkyl or aryl) side groups, well known to be permselective or preferentially permeable to –OH-containing molecules, including NH_3, that tend to react with H_2O in the environment, to form NH_4OH. A permselective membrane is one which has the capability to separate components from a mixture by allowing certain molecules or ions to pass through while others are retained.

Furthermore, they used microspotting to coat MT devices in an array with different polymers, aimed at detecting NO_2 and NH_3 in gas mixtures with multiplex-functionalized nanotube sensors. Microspotting of polymer solutions was carried out with a commercial micropin mounted on a micromanipulator, which was equipped with an optical microscope that allowed for positioning of the pin over the MT devices. It was noted that the conductance of a Nafion-coated device (spotted with 1% Nation in a water droplet) decreases (due to NH_3 electron donation to the P-type device, reducing the majority hole carriers), while a PEI-coated (0.1% in a water droplet) device on the same chip shows no response to NH_3. When introducing 1 ppm of NO_2 into the environment in the presence of 500 ppm of NH_3, the Nafion-coated device gives no response while the PEI-coated device shows a significant conductance decrease. These results undoubtedly demonstrate that multiplexed nanotube sensor arrays are promising for detecting multiple molecules in complex chemical environments.

Chen et al. (2007) reported a brand-new platform, which can serve as an e-nose with good discrimination factors, constructed with four different semiconducting nanostructured materials including In_2O_3 NWs, SnO_2 NWs, ZnO NWs, and SWCNTs.

The response of these sensors to hydrogen, ethanol, and NO_2 were measured at different concentrations and at both room temperature and 200°C.

Why is the integration of NWs with CNTs helpful in making e-nose? MOX NWs and SWCNTs have conspicuous differences in sensing mechanisms. In_2O_3, SnO_2, and ZnO NWs are N-type semiconductors, whereas SWCNTs are usually P-type semiconductors. Furthermore, the MOX NWs and nanotubes are believed to have different redox responses upon exposure to chemicals. This integration of both semiconducting MOX NWs and semiconducting SWCNT sensors therefore provides an important discriminatory factor for improved selectivity. In addition, the integrated micromachined-hotplate enables control of temperature individually and precisely for each sensor, with additional advantages such as ultralow power consumption (~60 mW) and short response time (20°C–300°C within 1 min). This temperature control works as the second discrimination factor. With these two discrimination factors, they achieved a "smell-print" library by detecting important industrial gases, such as hydrogen, ethanol, and nitrogen dioxide. In addition to this, they used principal component analysis (PCA) for data processing. Principal component analysis of the sensing results showed great discrimination of those three tested chemicals. Thorough analysis revealed clear improvements in selectivity by the integration of CNT sensors. This nanoelectronic nose approach has immense potential for detecting and discriminating between a wide variety of gases, including explosive ones and nerve agents.

8.10 CNT-, Nanowire- and Nanobelt-Based Chemical Nanosensors

An FET sensor has the structure of a common three-electrode transistor, where the source and drain electrodes bridge the semiconductor channel, and the gate electrode modulates the channel conductance. In the case of FET nanosensors, the semiconductor channel is made of a nanomaterial and is used as the "sensing" component of the device. Semiconductor channels can be fabricated using several nanomaterials, including CNTs and NWs. In order to provide selectivity toward a unique analyte, a specific recognition group (also called a receptor, ligand, or probe) is anchored to the surface of the semiconductor channel. This receptor is typically chosen to recognize its target molecule (also called the analyte) with a high degree of specificity and affinity.

8.10.1 CNT-Based ISFET for Nano pH Sensor

CNTs come in two different forms, namely MWCNTs and SWCNTs, which range in diameter from 1 to 10 nm and 10 to 50 nm, respectively. About 70%–80% of SWCNTs display semiconducting properties, whereas 70%–80% of MWCNTs show metallic properties.

How is a CNT handled to achieve accurate and exact positioning between source and drain? Dielectrophoresis (DEP) is a process through which neutral particles, such as CNTs, are translated through a suspending medium in a nonuniform electric field, which is generated between a pair of electrodes. DEP is used to separate, trap, and sort cells, bacteria, etc. The DEP technique is an accepted, preferred method for aligning CNTs. Dong and Wejinya (2009) and Dong et al. (2009) carried out alignment and testing of CNT *I–V* characteristics to verify whether they are metallic or semiconducting. CNTs with metallic properties serve as NWs, taking the place of the inversion layer in ion-sensitive field-effect transistor (ISFET), whereas the other kind of CNTs, with semiconducting properties, are used for the fabrication of nanotransistors. The microchip was originally made up of four layers from the bottom-up: silicon wafer, 300 Å of silicon dioxide, 200 Å of chromium, and 3000 Å of gold. It has triangular electrodes with an angle of 30° and electrode gaps ranging from 2 to 30 μm. The stock solution is a mixture of MWCNTs, surfactant, and deionized (DI) water. A sonication process is necessary to suspend CNTs uniformly. An AC power supply is used to produce the DEP forces. For this purpose, a function generator (electrical waveform generator) is used with an oscilloscope (electronic test instrument that allows observation of constantly varying signal voltages) to monitor the power function generated. For testing *I-V* characteristics, the gap distance is about 30 μm, and a micropipette (a very small pipette used in microinjection) is used to deposit a 1.5 μL CNT droplet from the solution onto the gap. The function generator is turned on and DEP forces are formed between the electrodes. After CNT alignment has been successfully performed, an NW consisting of MWCNTs is formed, connecting the source and drain electrodes. Its resistance measured by a multimeter, is ~1 kΩ. For *I-V* measurement, a DC power generator is used to supply DC potential and the corresponding current values are measured by the multimeter. SEM is adopted to examine how the CNT alignment is carried out.

Does the current sensing AFM help in CNT alignment and I-V measurement? Yes, the above process of CNT alignment and *I-V* measurement is also done by the current sensing AFM (CSAFM) function, using however many experiments related to electrical properties of nanostructures are performed, including impedance measurement and testing of *I-V* characteristics. Here an ultra-sharp AFM cantilever, coated with conductive film, probes the conductivity and topography (shapes and features) of the sample surface concurrently. A bias voltage is applied to the sample, with the cantilever being kept as virtual ground (a node of the circuit that is maintained at a steady reference potential, without being connected directly to the earth). During scanning, the tip force is held constant and the current is used to construct the conductivity image of the surface. From these experiments, a conclusion on whether these SWCNTs are semiconducting is established.

8.10.2 NW Nanosensor for pH Detection

Cui et al. (2001) prepared single-crystal boron-doped (P-type) SiNWs by a nanocluster-mediated vapor-liquid-solid growth method. Devices were fabricated by flow aligning SiNWs on oxidized silicon substrates and then making electrical contacts to the NW ends with electron-beam lithography. *How is the silicon dioxide surface modified to enable pH measurements?* The SiNW solid-state FET, the conductance of which is modulated by an applied gate, was transformed into a nanosensor by transforming the silicon oxide surface with 3-aminopropyltriethoxysilane (APTES) to provide a surface that undergoes protonation and deprotonation, so that changes in the surface charge can chemically gate the SiNW. Measurements of conductance as a function of time and solution pH demonstrated that the NW conductance increased stepwise with discrete changes in pH from 2 to 9 and that the conductance was constant for a given pH; the changes in conductance were also reversible for increasing and/or decreasing pH.

How are the results interpreted? These results are understood by considering the mixed-surface functionality of the modified SiNWs (Figure 8.12). Covalently linking APTES, (3-aminopropyl) triethoxysilane ($C_9H_{23}NO_3Si$), to SiNW oxide surface results in a surface terminating in both $-NH_2$ and –SiOH groups, which have different dissociation constants, pK_a. At low pH, the $-NH_2$ group is protonated to $-NH_3-$ and acts as a positive gate, which depletes hole carriers in the P-type SiNW and decreases the conductance. At high pH, –SiOH is deprotonated to –SiO–, which correspondingly causes an increase in conductance. The observed linear response is ascribed to an approximately linear change in the total surface charge density (*versus* pH) because of the united acid and base behavior of the two surface groups. To uphold this contention, Cui et al. (2001) also carried out pH-dependent measurements on unmodified (only-SiOH functionality) SiNWs. These conductance measurements showed a nonlinear pH dependence. The conductance change was small at low pH (2–6) but large over the high pH range (6–9). Notably, these pH measurements on unmodified SiNWs were in conformity with previous measurements of the pH-dependent surface charge.

Hsu et al. (2005) synthesized SiNWs on silicon substrates *via* a catalytic reaction in an N_2 atmosphere by thermal CVD at 955°C. The SiNWs have an average diameter of approximately 30–50 nm and a length of up to a few tens of micrometers. Silicon bulk materials and SiNWs were used alternatively as the sensing layer in an extended-gate FET configuration for

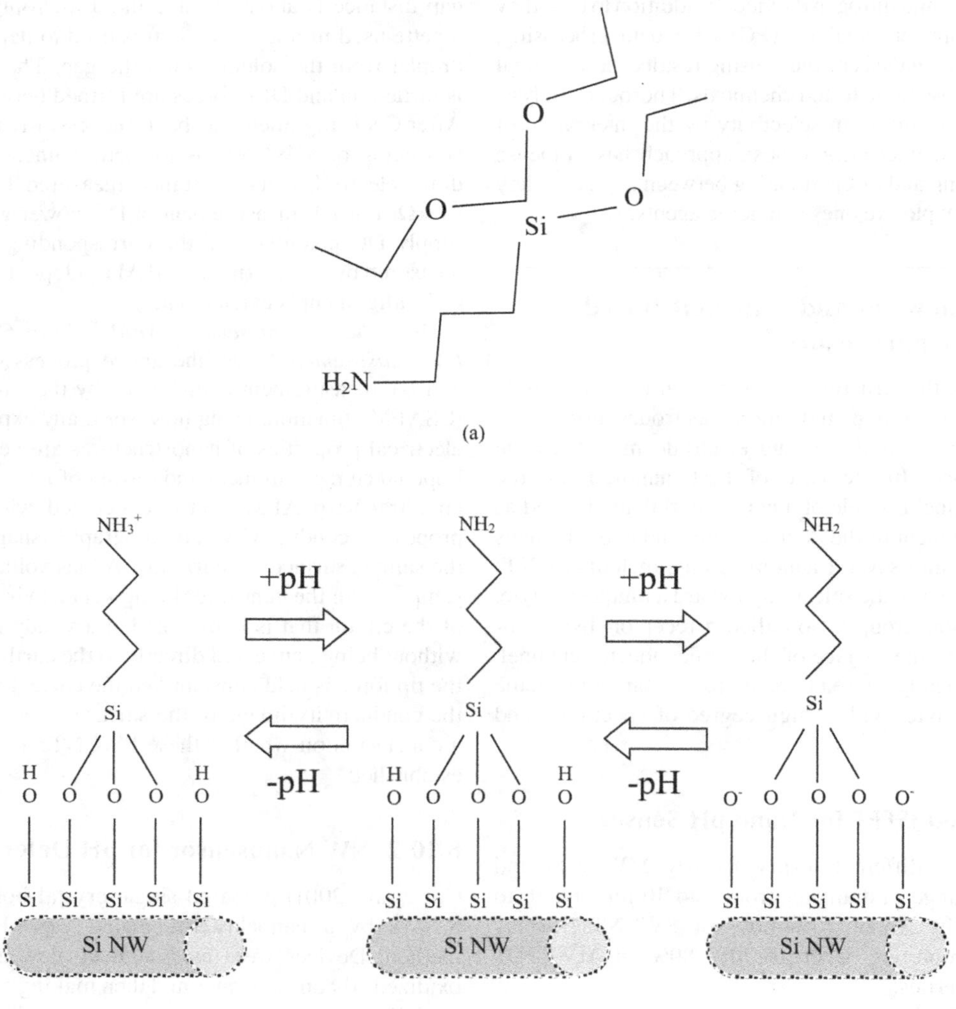

FIGURE 8.12 (a) Structural formula of 3-aminopropyltriethoxysilane (APTES). (b) APTES-modified SiNW surface, illustrating changes in the surface charge state with pH. (Cui, Y. et al., *Science*, 293, 1289, 2001.)

measuring solution pH. Experimental results showed that the pH sensitivity of silicon bulk materials was poor. However, good pH sensing properties of SiNWs, with a sensitivity of 58.3 mV pH^{-1}, was observed. Therefore, it was affirmed that the pH sensitivity of silicon bulk materials was greatly improved by downsizing them to the nanoscale. The sensitivities approach those of the pH sensors made of MOXs, such as SiO_2/Si_3N_4, SnO_2, and Ta_2O_8, which have sensitivities of 54, 58, and 59.2 mV pH^{-1}, respectively. The boron-doped SiNWs oxide surface resulted in a surface terminating in -SiOH groups, which would be deprotonated to $-SiO^-$, causing an increase in the conductance of the FET.

According to Hsu et al. (2005), Si-O asymmetric stretching bonds act as sites for H^+ ion adsorption. A higher Si–O asymmetric stretching bond density exists on the surface of the Si NWs sensing layer than on the Si bulk materials in the same magnitude sensing window, accounting for the improved pH sensitivity of NWs.

8.10.3 ZnS/Silica Nanocable FET pH Sensor

The NWs that have been extensively developed for sensing are those of silicon because of the vast knowledge that exists for the chemical modification of native Si oxide surfaces. *Excluding the CNT, is there any alternative material that has been used as an NW in FET?* Yes, the FET of the ZnS/silica (SiO_2) core/shell nanostructure is attractive to chemical sensors due to (i) the excellent and reproducible electronic characteristics of ZnS; (ii) the self-assembled silica shell is a natural insulator on the surface of the ZnS NW that serves as the protection layer against oxidation; and (iii) the enormous store of knowledge available about the chemical modification of silica surfaces.

He et al. (2007) demonstrated the reproducible transport characteristics of the ZnS/SiO_2 core/shell nanocable FETs, in which metal electrode and electrolyte solutions were used as a gate. The nanocable-based devices were used for label-free, real-time, and sensitive detection of chemical species. Amine- and hydroxyl (OH)-functionalized ZnS/silica nanocables exhibit linear pH-dependent conductance, which could be elucidated in terms of the changes in surface charge during protonation and deprotonation.

FET devices were fabricated using standard procedures with back-gate geometry. The nanocables were synthesized using vapor-liquid-solid growth. For the fabrication of nanocable FETs, the electrode pattern was designed to have a few parallel electrodes separated by 5–20 μm, employing standard photolithography and a lift-off process. The synthesized ZnS/silica nanocables were transferred from the Si substrate to pre-patterned Au/Ti electrodes by touching the nanocable sample with the electrodes. A single nanocable device could be achieved easily using this method. To firmly contact the Au/Ti metal electrodes and ZnS core of the nanocable, focused ion-beam (FIB) microscopy was employed to cut the nanocable at the two ends, so that the ZnS core was exposed, and, then, a Pt mixture was deposited at the ends to make contacts between the ZnS core and Au/Ti electrodes. The oxygen plasma treatment was carried out to get rid of the surface contamination of the ZnS/silica nanocable device before chemical sensing.

How is the ZnS/silica nanocable FET used as a chemical sensor? Chemical sensing in solution was carried out by monitoring electrical conductance through the surface-modified ZnS/silica nanocable device during additions of buffer solutions with different pH values. Here, the ZnS/silica nanocable-based FET, the conductance of which is modulated by an applied gate, is transformed into a chemical sensor by altering the surface of the silica shell with APTES to provide a surface that can undergo protonation and deprotonation, where changes in the surface charge are able to chemically gate the nanocable-based transistors. After the cleaning process in oxygen plasma to remove contaminants has been completed, ZnS/silica FETs were placed in the chamber under the saturation vapor pressure of a 1% ethanol solution of APTES overnight and then rinsed with ethanol thrice to achieve surface functionalization of nanocable FETs.

Measurements of conductance as a function of time and pH demonstrate that the NW conductance increases in steps corresponding to the pH values. The proposed mechanism for pH sensing is the same as described in Section 8.10.2.

To explore NW-based sensors for selective biological recognition, He et al. (2007) functionalized the silica shell of a nanocable functionalized with biotin, $C_{10}H_{16}N_2O_3S$. The conductance-time measurement showed that the conductance of the biotin-modified nanocable decreased rapidly and considerably to a constant value upon addition of the steptavidin solution. The mechanism is shown in Figure 8.13.

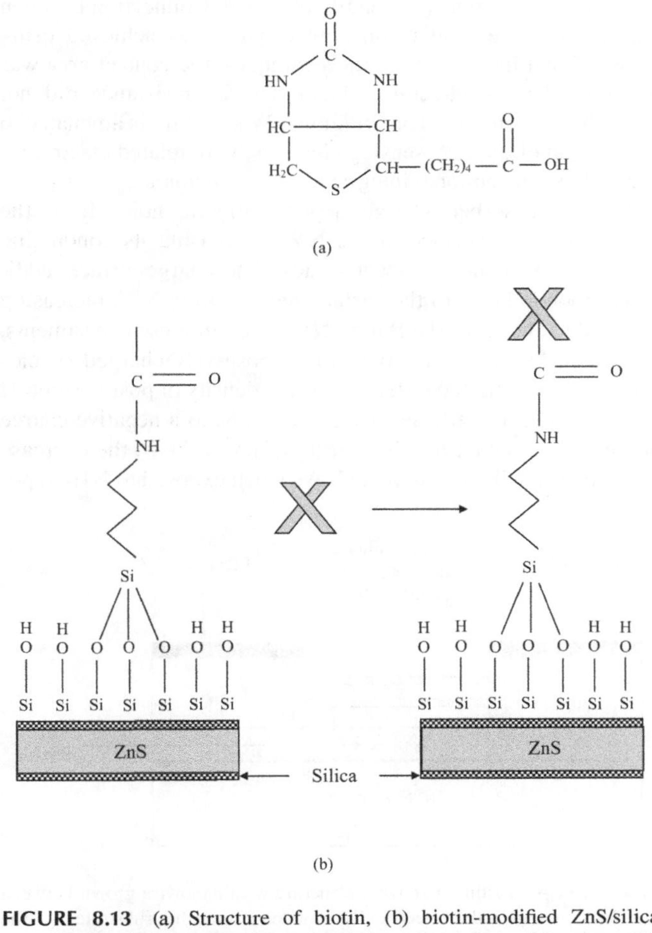

FIGURE 8.13 (a) Structure of biotin, (b) biotin-modified ZnS/silica nanocable (left) and binding of streptavidin to the surface of the nanocable (right). (He, Jr. H. et al., *J. Phys. Chem.* C, 111, 12152, 2007.)

8.10.4 Bridging Nanowire As Vapor Sensor

The high surface area and small volume of the NWs makes them especially attractive as effective field-effect sensing elements. In a field-effect sensor, charge from an adsorbed or nearby analyte induces compensating charge in the NW, modulating its conductance and, consequently, the current flowing between two electrodes. The device, therefore, acts as a chemically sensitive FET, with adsorbed charged species acting as the "gate." NW sensors can be very sensitive because a significant fraction of the carriers in the NW can be depleted so that the small number of carriers induced by the analyte changes the conductance significantly. *What is the problem in interfacing NWs with circuitry?* Using the NWs requires that they be connected to electrodes that can interface with other circuitry. The size mismatch between the nanoscale wires and conventional electrodes has impeded their adoption.

How does the bridging NW structure help in securing good electrical connection with microelectrodes? Kamins et al. (2006) demonstrated the use of the "bridging NW" structure (Figure 8.14) for sensing chemical species, such as NH_3 and HCl vapors. Metal-catalyzed, self-assembled silicon NWs were grown between two electrically isolated electrodes, so that intimate contact was made between the NWs and microscale electrodes during growth. Because the connection was between SiNWs and heavily doped Si electrodes of the same conductivity type, good electrical connection was readily obtained. Connection between the Si electrodes and metal contact pads was achieved using conventional integrated-circuit techniques; the contact area was large so that metal-semiconductor contact resistance did not limit the current flow. The bridging NW structure is amenable to integration of the NW sensing elements, with related electronics formed by conventional integrated-circuit technology.

Positive adsorbed charge depletes mobile holes from the region near the surface of the NW, decreasing its conducting area and conductance. Negative adsorbed charge attracts additional mobile holes to the surface region of the NW, increasing its conductance. For the P-type NWs used in these experiments, a current decrease corresponds to a positively charged species adsorbed onto the NWs, decreasing the density of positive mobile carriers (holes), while an increase pertains to a negative charge on the NWs, increasing the density of holes. Thus, the decrease in current flowing through the NWs when exposed to NH_3 vapor arises from positive surface charge, whereas the increase in current when exposed to HCl vapor is due to negative surface charge. The positive charge on the NW surface resulting from exposure to NH_3 vapor was undecidedly attributed to the reaction

$$NH_3 + H_2O \Leftrightarrow NH_4^+ + OH^- \tag{8.13}$$

whereas the negative charge from exposure to HCl vapor was attributed to Cl^-.

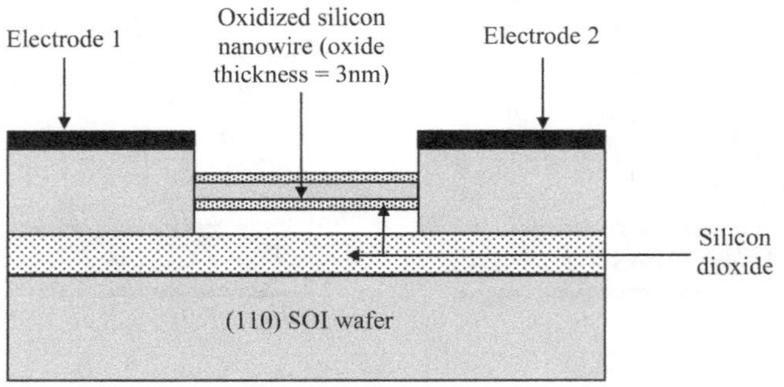

FIGURE 8.14 Briding nanowire structure with nanowire grown between two silicon electrodes electrically isolated from the substrate by the oxide of an SOI wafer. (Kamins, T. I. et al., *Nanotechnol.*, 17, S291, doi: 10.1088/0957-4484/17/11/S11, 2006.)

8.10.5 Palladium Functionalized Si NW H_2 Sensor

Chen et al. (2007) fabricated a hydrogen sensor by functionalizing SiNWs with Pd NPs. Recalling the Pd-functionalization process briefly (see Section 3.6.3), the oxide sheath on as-grown one-dimensional (1-D) SiNWs synthesized by thermal evaporation of high-purity SiO (99.9%) mixed with 1% Sn powder, was etched in a 5% HF aqueous solution, followed by immersion of the sample in a saturated palladium chloride ($PdCl_2$) solution. Pd^{2+} ions were reduced to Pd metal particles, which were deposited on the surface of SiNWs according to the chemical reaction: $Pd^{2+}+2Si\text{-}H \rightarrow Pd+H_2+2Si$.The Pd-modified SiNWs were dispersed on a silicon wafer with a 300-nm thick oxide layer. Gold electrodes were deposited on the sample by thermal evaporation through a shadow mask. *I–V* measurements were carried out with a two-probe measurement system.

The oxide-removed SiNWs showed N-type properties due to point defects and surface states in Si surfaces. The conductance of the Pd-coated SiNW was smaller than that of the pure SiNW due to electron depletion at the Pd and silicon interface. Exposure to 5% hydrogen increased the current signal of the sensor 20-fold. The effect of hydrogen on the conductivity was reversible. The response time was 3 seconds only, which is much faster than that of the macroscopic Pd wire sensor.

How does the H_2 gas sensor work? The mechanism of the gas sensor is expounded by electron depletion (or even inversion) at the Pd/Si interface and consequently reduction of electrical conductivity. At room temperature and atmospheric pressure, palladium absorbs up to 900 times of its own volume of hydrogen. Upon exposure to hydrogen, Pd reacts with hydrogen to form the hydride (PdH_x) rapidly. The Fermi level of PdH_x moves to a higher energy, which is above the dense region of the *d* bands; this brings up the Fermi level at the interface and results in the transition of electrons from Pd to SiNWs, returning them to a more N-type behavior. So, depletion width is a critical parameter that controls how significantly the Pd Fermi level shift can modulate the NW conductance. Due to the reversible reaction between Pd and hydrogen, PdH_x releases hydrogen quickly and leads to a decrease in conductance again. Compared with the back-gate modulation in FET with a partition of thick silicon oxide layer, the coated Pd NPs modulate the conductance of SiNWs directly and more effectively.

8.10.6 Polymer-Functionalized Piezoelectric-FET Humidity Nanosensor

Lao et al. (2007) demonstrated a humidity/chemical nanosensor based on piezoelectric field-effect transistor (PE-FET). *What is a PE-FET?* It is an FET structure with the gate insulator made of a

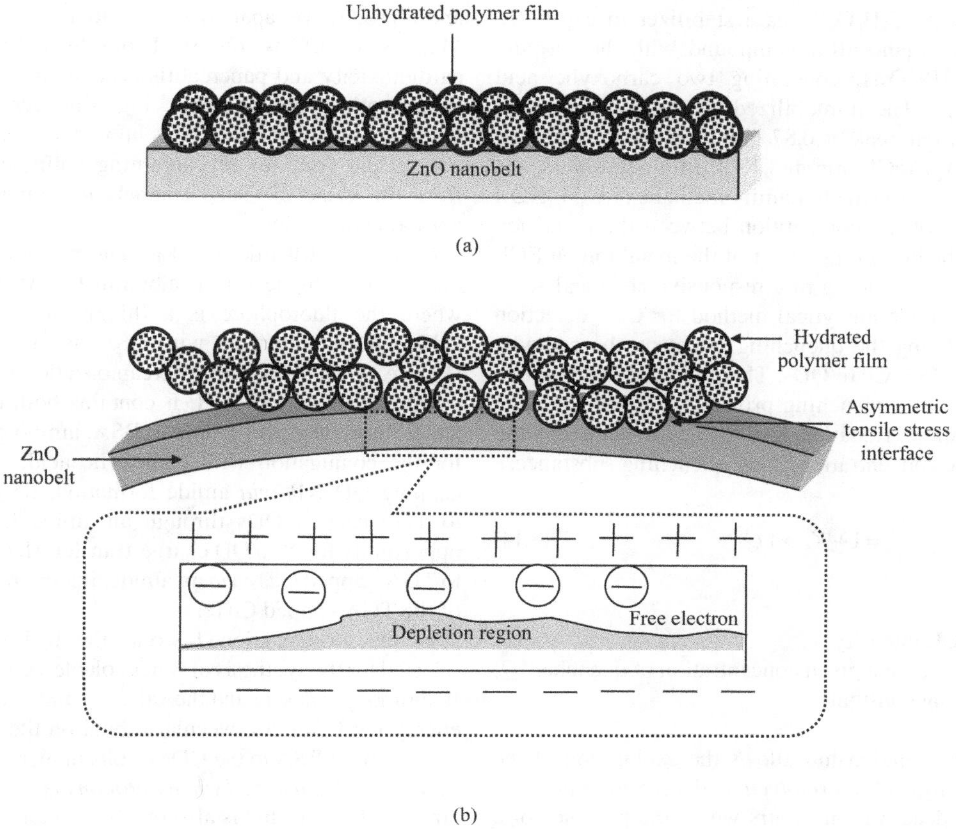

FIGURE 8.15 Polymer functionalized piezoelectric field-effect transistor (PE-FET: (a) Dehydrated polymer film lies straight on the zinc oxide nanobelt, causing no deformation of the belt. (b) Upon hydration, the polymer swells and expands in volume, causing bending of the ZnO belt with attendant piezoelectric field generation across the ZnO nanobelt. Positive and negative charges are produced on the outer and inner surfaces of ZnO nanobelt, respectively. The positively charged top surface attracts free electrons while the repulsive force of the negatively charged bottom surface creates a depletion region adjacent to it. Dehydration in a dry environment results in reverting of the sensor to condition (a). (Lao, C. S. et al., *Appl. Phys. Lett.*, 90, 262107–1, 2007.)

piezoelectric material that develops a voltage across it following deformation. The device (Figure 8.15) has a single-side-coated ZnO NB gate functionalized with multiple layers of polymers, using an electrostatic self-assembling process.

How is the stress produced in the ZnO NB? Polymers for functionalization of NB were anionically charged poly(*N*-isopropylacrylamide) (PNIPAM), $C_6H_{11}NO$, and poly(diallyldimethylammonium chloride) (PDADMAC), $(C_8H_{16}ClN)_n$. These polymers were closely packed before vapor exposure. Upon exposure to water vapor, the polymers underwent a hydration process. As a result, the volumes of these polymers were increased by several factors. Because the polymers were coated on only one side of the ZnO NB, the swelling of polymers generated an asymmetric tensile stress at the contact surface with ZnO NB. Consequently, the ZnO NB was bent, resulting in an asymmetric strain across the thickness of the NB. A bent ZnO NB could produce a piezoelectric potential across the NB due to the strain-induced piezoelectric effect. With the stretch and compression effects of a deformed ZnO NB, positively charged and negatively charged surfaces were produced at the outer and inner bending surfaces, respectively, of a ZnO NB. Consequently, a piezoelectric field was built across the ZnO NB.

Thus the operation of the PE-FET relies on the self-contraction/expansion of the polymer, which builds up a strain in the piezoelectric NB and induces a potential drop across the NB that serves as the gate voltage for controlling the current flowing through the NB. In return, the deformation of ZnO NB produces a piezoelectric field across the NB, which serves as the gate for controlling the flow of current along the NB.

The device is a useful component for nanopiezotronics.

8.11 Optochemical Nanosensors

8.11.1 Low-Potential Quantum Dot ECL Sensor for Metal Ion

What are the attractive features of quantum dot (QD)-based ECL analysis method? The QD-based electrochemiluminescent (ECL) analytical technique has developed rapidly in many fields because of advantages, such as low cost, high sensitivity and stability, unsophisticated instrumentation, and applicability to a wide range of analytes.

Cheng et al. (2010) constructed the ECL sensing system for low-potential detection of metal ions. They immobilized surface-unpassivated cadmium telluride (CdTe) QDs on a glassy carbon electrode. The surface-unpassivated CdTe QDs were prepared using meso-2,3-dimercaptosuccinic acid (DMSA), with the molecular formula $C_4H_6O_4S_2$, as a stabilizer to cap CdTe QDs. DMSA is an organosulfur compound with the formula $HO_2CCH(SH)CH(SH)CO_2H$, containing two carboxylic acid and two thiol groups. The immobilized QDs showed a strong cathodic ECL emission peak at 0.87 V with an onset potential at 0.64 V (*versus* Ag/AgCl/saturated KCl) in air-saturated, pH 9.0 HCl Tris [tris(hydroxymethyl)aminomethane]: $(C_4H_{11}NO_3)$ buffer. On the basis of the competition between the metal ion and the stabilizer, the quenching effect of the metal ion on ECL emission was observed, leading to a responsive chemical sensing application. A simple analytical method for Cu^{2+} detection was developed, utilizing the quenching effect on the cathodic ECL emission of DMSA-CdTe QDs. This followed the behavior of the fluorescence (FL) quenching principle, described by the Stern–Volmer equation (Equation 8.14), an expression relating FL quenching to the concentration of the quenching substance:

$$\frac{I_0}{I} = 1 + K_{sv} \times [Q] \tag{8.14}$$

where

I_0 is the initial ECL intensity

I is the ECL intensity at a given concentration of quencher $[Q]$

K_{sv} is the quenching constant

The Stern–Volmer relationship allows the exploration of the kinetics of a photophysical *intermolecular* deactivation process. An *intermolecular* deactivation occurs where the presence of a chemical species β accelerates the decay rate of a chemical α in its excited state.

The sensor showed a linear relationship between the quenching of ECL emission and the Cu^{2+} concentration in the range from 5.0 nM to 7.0 μM. The detection limit was 3.0 nM and the quenching constant was $4.6 \times 10^5\ M^{-1}$.

After the detection procedure, the ECL emission of QD-modified GCE was not recovered at the original intensity prior to the quenching of Cu^{2+}. This pointed toward a structure destruction of the QDs by Cu^{2+}. The competitive binding of Cu^{2+}, with a stronger metal-S interaction than the Cd-S bond, to the stabilizer led to a quenching effect on ECL emission. This was substantiated to be associated with the thermodynamic tendency for the formation of Cu-S on the DMSA-CdTe QD surface.

8.11.2 BSA-Activated CdTe QD Nanosensor for Sb^{3+} Ion

Why is it necessary to establish sensitive and accurate analytical methods for quantitative determination of antimony? It is necessary because antimony is a toxic element and the upper allowable limit of antimony in domestic water is 0.005 mg L^{-1}. Exposure to high levels of antimony has a variety of adverse health effects. Antimony toxicity takes place either through occupational exposure of personnel working in metal mining, smelting or refining industries, besides those engaged in antimony or antimony trioxide production; or during medical therapies administered to leishmaniasis and schistosomiasis patients (Sundarand Chakravarty 2010). Toxic effects to workers include respirational, cardiovascular, gastrointestinal, dermal, and reproductive, apart from genotoxicity, and carcinogenecity. Main side effects observed during antimony therapy are cardiotoxicity and pancreatitis necessitating critical monitoring of patient's health. Further, ingesting certain compounds of antimony has injurious effects upon body tissues and functions, e.g., acidic fruit juices containing antimony oxide, dissolved from the glaze of cheap enamelware containers, have caused antimony poisoning.

Ge et al. (2010) designed an antimony-selective nanosensor, based on a bovine serum albumin (BSA)-fluorophore system, where the fluorophore is a thioglycolic acid (TGA)-capped CdTe QD nanoparticle with BSA as an antimony conjugate (Figure 8.16). TGA or mercaptoacetic acid is the organic compound $HSCH_2COOH$. It contains both a thiol (mercaptan) and a carboxylic acid group. In BSA, amino groups are available for the conjugation with carboxylic acid, –COOH, for group capping QD NPs *via* amide formation. BSA is able to couple to TGA-capped QDs through an amide linkage, offering an opportunity for N to QD charge-transfer. Hence, BSA conjugates to TGA-capped CdTe *via* an amide link interacting with carboxyl of the TGA-capped CdTe.

In the aforementioned work, the hydrothermal route was selected for the synthesis of water-soluble TGA-capped CdTe QDs. If antimony is absent and the valence band of the QD is at a higher energy level than a molecular orbital on the ligand, the electron transfer from BSA to the QDs results in increased QD intensity

How is FL quenched when antimony is present? If the molecular orbital implicated is also involved in antimony binding, then, in the presence of antimony, the energy level is no longer available. Due to its lack of availability, electron migration is not permissible, and therefore it cannot occur. Thus the QD FL is

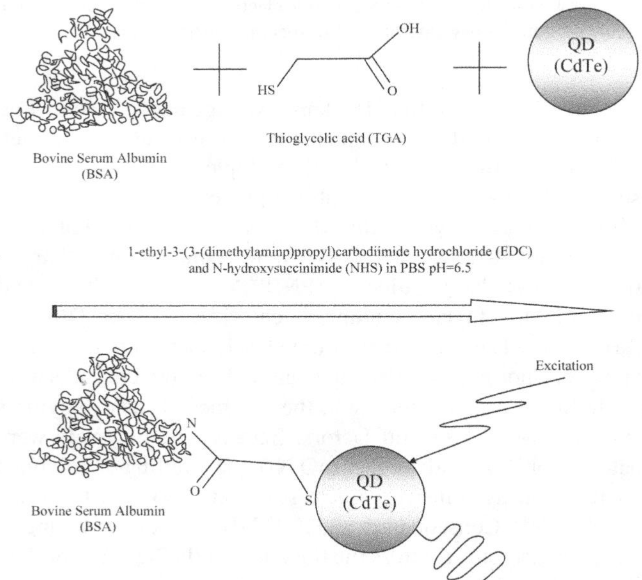

FIGURE 8.16 The covalent linking of BSA on the surface of TGA-capped QD nanoparticles by an EDC/NHS coupling reaction. QD modification utilizes the property that BSA has amino groups available for the conjugation with carboxylic acid group, capping QD nanoparticles *via* amide formation. The emission is switched on when the excitation light irradiates the TGA-QD-BSA conjugates. (Ge, S. et al., *Analyst*, 135, 111, 2010.)

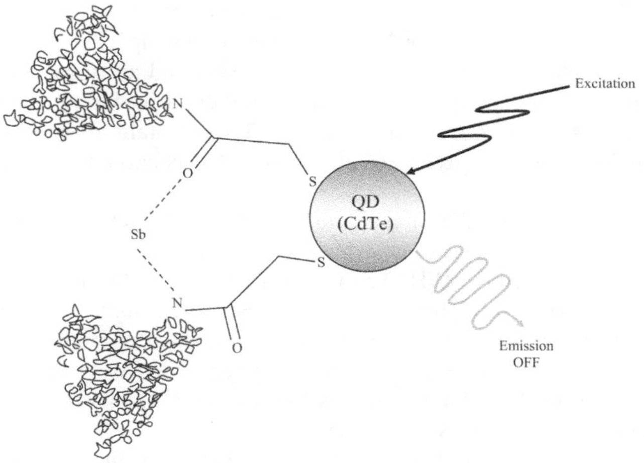

FIGURE 8.17 The quenching mechanism of TGA-CdTe-BSA-Sb. The quenching is indicated by representing the emission by a gray line, meaning that the emission is turned off. When an antimony ion enters the BSA, the lone electron pairs of nitrogen and oxygen atoms participate in the coordination, switching off the QD emission. Consequently, drastic quenching of the fluorescence intensity takes place enabling the detection of minute concentrations of Sb^{3+} ions. (Ge, S. et al., *Analyst*, 135, 111, 2010.)

quenched in the presence of antimony (Figure 8.17). When antimony ion enters the BSA, the lone pair electrons of the nitrogen and oxygen atom become involved in the coordination, switching off the QD emission and resulting in a dramatic quenching of the FL intensity. Extremely low concentrations of antimony ions thus become detectable.

A good linearity of the relationship between change in intensity (ΔI) and the concentration of antimony was observed in the range of 0.10–22.0 mg L^{-1}, with a correlation coefficient of 0.9984, and the detection limit was 2.94×10^{-8} g L^{-1}. This was estimated as the concentration of analyte that produced an analytical signal equaling three times the standard deviation of the background FL.

What is the effect of interfering ions? In a study of interferences, the antimony-sensitive TGA-QD-BSA sensor showed good selectivity. Even the presence of high concentrations of SO_4^{2-}, NO_3^-, Cl^-, F^-, K^+, Na^+, or NH_4^+ ions did not bring about any changes in the emission of the TGA-QD-BSA conjugates. This was probably due to poor complexation of the above ions with BSA, that do not change the quenching of the QD; a complexation reaction is a chemical reaction that takes place between a metal ion and a molecular or ionic entity.

Thus, interferences by metal ions in the detection of antimony were minimal. Therefore, a simple, fast, sensitive, and highly selective analysis for antimony was achieved. Furthermore, the sensor was applied successfully to the determination of antimony in real water samples with satisfactory results.

8.11.3 Functionalized CdSe/ZnS QD Nanosensor for Hg(II) Ion

What is the reason for the notoriety of mercury? Mercury is a widespread, dangerous global pollutant. Mercury poisoning, also known as *mercuralism*, is the phenomenon of intoxication by contact with mercury. Mercury poisoning from amalgam dental fillings is the root cause of a multitude of difficult-to-diagnose and often life-threatening diseases.

On what principles is the QD-based Hg nanosensor designed? The FL efficiency of QDs is sensitive to the presence and nature of adsorbates on the surface of QDs. Therefore, a chemical sensing system based on QDs can operate using FL changes induced by molecular recognition at the surface of QDs. L-carnitine (LC), an amine compund produced in the body from the amino acids lysine and methionine via a biosynthetic pathway, plays a vital role in metabolism of fatty acids. LC modifies QDs *via* chelation of carboxyl group with heavy metal ions; chelation is the formation or presence of two or more separate bindings between a polydentate (multiple bonded) ligand and a single central atom. LC ($C_7H_{15}NO_3$) is a quaternary ammonium compound, biosynthesized in the body from the amino acids lysine and methionine. It increases the use of fat as an energy source by transporting fatty acids into the mitochondria, where they are burned to release energy for body functions.

Li et al. (2008) prepared LC-capped CdSe/ZnS core/shell QDs (Figure 8.18), which were capable of selectively responding as selective fluorescent probe for the determination of mercury ions. Monodisperse CdSe QDs were synthesized and were tri-*n*-octylphosphine oxide- (TOPO-) protected. Then, the freshly washed CdSe/ZnS/QDs were dispersed in an ethanol solution of LC. Aggregated QDs were removed by centrifuging, and the target product of LC-capped QDs was separated. A centrifuge is a laboratory device that uses centrifugal force to separate the components of a complex mixture of fluids, liquids or gases, by spinning it at a high speed about a central axis in a container

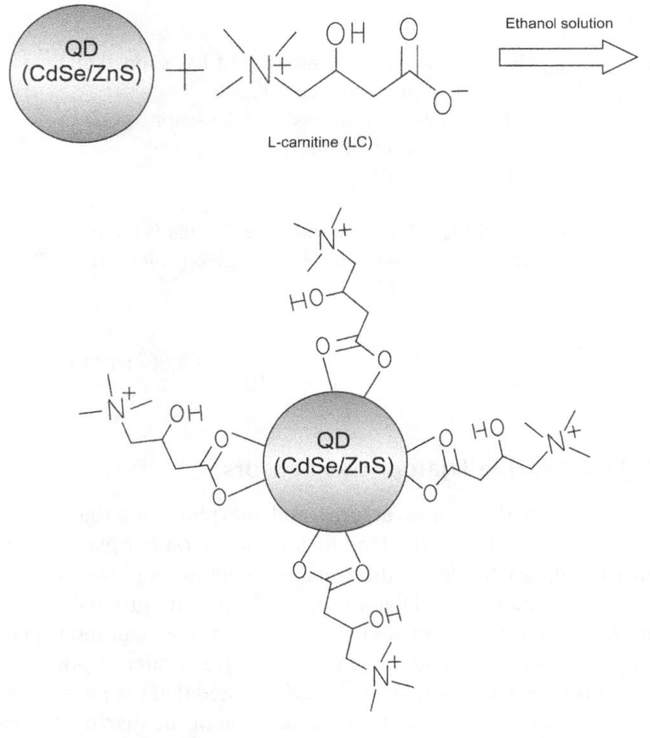

FIGURE 8.18 Synthesis of L-carnitine-capped QDs by the surface modification of QDs *via* a carboxyl anchor group of L-carnitine for the Hg(II) nanosensor. (Li, H. et al., *Microchim. Acta*, 160, 119, doi: 10.1007/s00604-007-0816-x, 2008.)

when the materials are spread apart according to their specific gravities, shapes, sizes, viscosities, and also rotor speed with the heavier materials pushed radially outwards and the less heavier ones migrating towards the axis.

Luminescence emission of the LC-capped QDs was highly sensitive to the presence of mercury ions in ethanol. The fluorescent emission of LC-capped QDs was increasingly quenched by adding mercury ions. The result indicated that the quenching metal ions remained adsorbed or linked to the QD NPs.

The finding that mercury ions quenched the FL intensity of LC-capped QDs in a concentration-dependent relationship, described by a Stern–Volmer-type equation, was applied to develop a method for the determination of mercury ions. The detection selectivity was checked by conducting the FL titration of LC-capped CdSe/ZnS/QDs with various metal ions. Only Hg(II) ions showed strong quenching of LC-capped QDs FL. The experimental data were understood in terms of a strong affinity to Hg(II) ions of the surface ligands of QDs. The interference of alkali, alkaline earth ions, Ni(II) ions, and Zn(II) ions was very weak. Thus, quenching of the luminescence emitted by the synthesized NPs allowed the detection of mercury concentration as low as 0.18 mM, affording a very sensitive detection system for this toxic chemical species.

Example 8.3

Mercury ions quench the FL emission of LC-capped QDs in accordance with the Stern–Volmer equation

$$\frac{I_{\max}}{I} = 1 + K_{sv} \times \left[\mathrm{Hg}^{2+}\right] \tag{8.15}$$

where

$I_{\max}$ is the luminescence intensity of LC-capped QD in a mercury ion-free solution

I is the luminescence intensity of LC-capped QD in a mercury ion solution

$K_{sv} = 1.167 \times 10^4\ \mathrm{M}^{-1}$

If the ratio $I_{\max}/I$ is observed to be 5, what is the mercury ion concentration $[\mathrm{Hg}^{2+}]$ in the given solution?

From Equation 8.15

$$\left[\mathrm{Hg}^{2+}\right] = \frac{(I_{\max}/I) - 1}{K_{sv}} = \frac{5-1}{1.167 \times 10^4} = 3.428 \times 10^{-4}\ \mathrm{M}$$

8.11.4 Marine Diatom Gas Sensors

Recognizing the nanostructures and morphologies that nature has optimized during the long history of life on our planet helps in framing a completely different approach to engineering systems at the nanoscale. Diatoms are an important group of algae, particularly microalgae, and are one of the most common types of phytoplankton found in the oceans, in freshwater, in soils and on damp surfaces. Stefano et al. (2005) noted that the photoluminescence emission from the silica skeleton of the marine diatom *Thalassiosira rotula* was strongly dependent on the surrounding environment. First, let us *see what electrophilic and nucleophilic substances are.* "Electrophilic" is the combination of two words (*electro* means "electron" and *phile* means "loving"). Therefore, electrophilic substances are electron lovers and are those substances that contain positive charges on them. Likewise, nucleophilic substances are nucleus lovers. They contain an atom or molecule with an excess of electrons and thus carry a negative charge.

Stefano et al. (2005) examined the effect of diatom exposure to photoluminescence-quenching gaseous substances: NO_2, and vapors of acetone (C_3H_6O), and ethanol (C_2H_6O). All these substances are electrophiles, so that they can attract some electrons from the silica skeleton of diatoms and quench its photoluminescence. The reverse effect was observed by exposing the diatoms to nucleophilic substances like xylene, C_8H_{10} [$C_6H_4(CH_3)_2$ or $C_6H_4C_2H_6$] and pyridine, C_5H_5N.

Thus, the researchers found that depending on their electronegativity and polarizing ability, some substances quench the luminescence, while others effectively enhance it. These phenomena allow reliable discrimination between different substances. This effect can be used for optical chemical sensing.

8.12 Discussion and Conclusions

Different types of gas nanosensors are listed in Table 8.1. These are grouped into three broad categories. The first group of nanosensors is the chemiresistor class, using metallic/MOX/CNT NPs. Other nanomaterials include polymer thin films, nanofibers, etc. The second class of nanosensors is the field-effect chemical sensors, using CNTs, Si NWs, or ZnS/silica nanocables. This includes both gas and pH sensors in liquid media. The third type of nanosensors consists of ion sensors, utilizing QD luminescence quenching. As demanded by the particular situation, the engineer chooses one from the various options available, e.g., if room temperature operation is essential, then obviously MOXs, working at high temperatures, are unsuitable. Similar criteria are applied to make the correct choices.

Review Exercises

8.1 What is a chemical nanosensor? How does it differ from any macroscopic chemical sensor? Elaborate the statement "Detection of chemical processes on a single-molecule scale is the ultimate goal of sensitive analytical assays."

8.2 Many chemical sensors operate through the variation of a surface parameter, like surface conductivity, in response to analyte concentration. Hence, the effective surface area of the device, i.e., the area actually interacting with the analyte, determines its sensitivity. How do nanomaterials provide an easy answer for increasing the surface area? Discuss the role played by different nanomaterials, like NPs, nanotubes, nanowires, etc., in chemical sensors.

8.3 Highlight the importance of gas sensing. What are the gases that commonly need to be detected? What roles

TABLE 8.1

Diversity of Chemical Sensors

Sl. No.	Name of the Nanosensor	Sensed Quantity/ Applications	Nanomaterial/ Nanostructure Used	Sensing Principle
1.	Metallic nanoparticle-based gas sensor	Gas	Metallic nanoparticle	Conductivity change (chemiresistor)
2.	Metal oxide (MOX) gas sensor	Gas	Metal oxide nanoparticle	Conductivity change (chemiresistor)
3.	Carbon nanotube (CNT) gas sensor	Gas	CNT	Conductivity change (chemiresistor)/ field-effect (FET)
4.	Porous silicon-based gas sensor	Gas	Nanoporous silicon	Conductivity change (chemiresistor)
5.	Thin organic polymer film-based gas sensor	Gas	Organic polymer	Conductivity change (chemiresistor)
6.	Electrospun polymer nanofibers as humidity sensor	Humidity	Polymer nanofiber	Conductivity change (chemiresistor)
7.	CNT-based ISFET for nano pH sensor	pH	CNT	Field effect
8.	NW nanosensor for pH detection	pH	SiNW	Field effect
9.	ZnS/silica nanocable FET pH sensor	pH	ZnS/silica nanocable	Field effect
10.	Bridging nanowire as vapor sensor	Vapor	Nano wire	
11.	Palladium functionalized Si NW H_2 sensor	H_2	Si NW	Field effect
12.	Polymer functionalized piezoelectric-FET humidity nanosensor	Humidity	Polymer	Effect of moisture- induced strain on piezoelectric gate
13.	Quantum dot ECL sensor	Metal ion	QD	Electrochemical luminescence
14.	BSA-activated CdTe QD nanosensor	Sb^{3+} ion	QD	Fluorescence quenching
15.	Functionalized CdSe/ZnS QD nanosensor	Hg(II) ion	QD	Luminescence quenching

do nanomaterials play in fabricating improved gas-sensing devices?

8.4 Explain the working of a palladium NW sensor. How do palladium NWs enhance the performance of hydrogen sensors?

8.5 Why do MOX-based gas sensors need high temperatures for their normal functioning? What will happen if the required temperature is not provided?

8.6 What different types of hotplates are used for constructing metal oxide gas sensors? Present a comparison between the common types of miniaturized hotplates.

8.7 What are the attractive features of CNTs for gas-sensing applications? Explain the two approaches that are commonly followed for fabricating CNT-based gas sensors.

8.8 Are gas sensors based on raw CNTs selective for gas detection? How are CNTs functionalized for their intended application?

8.9 Discuss, in view of the following two remarks, why the CNT-FET device presents serious impediments to commercialization: (i) Selective growth of metallic *versus* semiconducting CNTs is not possible with present technology. (ii) Furthermore, if an *in situ* CVD process is used in the CNT-FET fabrication sequence, it is an intricate process to make a single SWCNT grow horizontally, bridging a given distance between the source and the drain, so that one is forced to "pick and place" a semiconducting SWCNT from bulk samples.

8.10 DEP is a process through which neutral particles, such as CNTs, are translated through a suspending medium in a nonuniform electric field, that is generated between a pair of electrodes. Name a sensor that is fabricated using this technique.

8.11 Describe a nano-FET constructed with palladium coating on MWCNTs for hydrogen sensing.

8.12 What is a microgravimetric sensor? How does a CNT-based QCM sensor for ethanol work?

8.13 Describe a nanoporous silicon hydrogen sensor. Is it operated at room temperature? Why?

8.14 Explain the use of electrospun polymer nanofibers for humidity sensing. How does the modification of a gold electrode with PDDA before nanofiber deposition help in improving the sensor performance?

8.15 Describe the use of a platform, incorporating In_2O_3, SnO_2, and ZnO NWs, and SWCNTs, together with a micromachined hotplate for nano-electronic nose implementation.

8.16 How is a CNT-based ISFET used as a pH nanosensor? By what technique is the CNT aligned with contacting electrodes?

8.17 What surface-terminating functional groups are produced on a silicon NW by covalently linking APTES to it? Explain, with structural formulae, the pH sensing mechanism of an APTES-modified silicon NW.

8.18 A higher Si-O asymmetric stretching bond density exists on the surface of the SiNW sensing layer than on the Si bulk materials in the same magnitude sensing window. The Si-O asymmetric stretching bonds can act as sites for hydrogen ion adsorption. Present arguments to convince that the pH sensitivity of silicon bulk materials is greatly improved by downsizing the sensors to the nanoscale.

8.19 How does a ZnS/SiO_2 core/shell nanocable FET work as a pH sensor? With the help of structural formulae, illustrate the operation of a biotin-functionalized nanocable FET for detecting streptavidin.

8.20 How does a bridging silicon NW sensor with Si electrodes avoid the difficulties of establishing contacts with metal electrodes interfacing with signal processing circuitry?

8.21 Explain the construction and operating mechanism of a Pd-functionalized SiNW sensor for hydrogen.

8.22 How does a polymer-functionalized zinc oxide FET sense moisture? Explain, with diagrams, the mechanisms responsible for the generation of a humidity-dependent strain in the piezoelectric NB?

8.23 How is QD emission switched off in the antimony ion nanosensor in the presence of antimony? Illustrate your answer with diagrams.

8.24 How is mercury detected by analyte-induced changes in the photoluminescence of suitably modified QDs?

8.25 Explain the following statement: "NW sensors are highly sensitive because a significant fraction of the carriers in the NW can be depleted; therefore, a small number of carriers induced by the analyte alters the conductance of the NW appreciably."

REFERENCES

Cattanach, K., R. D. Kulkarni, M. Kozlov, and S. K. Manohar. 2006. Flexible carbon nanotube sensors for nerve agent simulants. *Nanotechnology* 17: 4123–4128, doi: 10.1088/0957-4484/17/16/022.

Chen, Z. H., J. S. Jie, L. B. Luo, H. Wang, C. S. Lee, and S. T. Lee. 2007. Applications of silicon nanowires functionalised with palladium nanoparticles in hydrogen sensors. *Nanotechnology* 18: 345502 (5 pp.), doi: 10.1088/0957-4484/18/34/345502.

Cheng, L., X. Liu, J. Lei, and H. Ju. 2010. Low-potential electrochemiluminescent sensing based on surface unpassivation of CdTe quantum dots and competition of analyte cation to stabilizer. *Analytical Chemistry* 82: 3359–3364.

Cui, Y., Q. Wei, H. Park, and C. M. Lieber. 2001. Nanowire nanosensors for highly sensitive and selective detection of biological and chemical species. *Science* 293(17): 1289–1292.

Dong, Z. and U. C. Wejinya. 2009. Design, fabrication and measurement of CNT based ISFET for NANO devices. *IEEE Nanotechnology Materials and Devices Conference*, June 2–5, 2009, Traverse City, MI. New York: IEEE, pp. 178–182.

Dong, Z., U. C. Wejinya, H. Yu, and I. H. Elhajj. 2009. Design, fabrication and testing of CNT based ISFET for NANO pH sensor application: A preliminary study. *IEEE/ASME International Conference on Advanced Intelligent Mechatronics*, Suntec Convention and Exhibition Center, Singapore, July 14–17, 2009. New York: IEEE, pp. 1556–1561.

Favier, F., E. C. Walter, M. P. Zach, T. Benter, and R. M. Penner. 2001. Hydrogen sensors and switches from electrodeposited palladium mesowire arrays. *Science* 293(5538): 2227–2231.

Fields, L. L., J. P. Zheng, Y. Cheng, and P. Xiong. 2006. Room temperature low-power hydrogen sensor based on a single tin dioxide nanobelt. *Applied Physics Letters* 88(26): 263102.

Francia, G. D., B. Alfano, and V. L. Ferrara. 2009. Conductometric gas nanosensors. *Journal of Sensors* 2009: 659275 (18 pp.), doi: 10.1155/2009/659275.

Ge, S., C. Zhang, Y. Zhu, J. Yu, and S. Zhang. 2010. BSA activated CdTe quantum dot nanosensor for antimony ion detection. *Analyst* 135: 111–115.

He, Jr. H., Y. Y. Zhang, J. Liu, D. Moore, G. Bao, and Z. L. Wang. 2007. ZnS/Silica nanocable field effect transistors as biological and chemical nanosensors. *Journal of Physical Chemistry C* 111(33): 12152–12156.

Hieu, N. V., L. T. B. Thuy, and N. D. Chien. 2008. Highly sensitive thin film NH_3 gas sensor operating at room temperature based on SnO_2/MWCNTs composite. *Sensors and Actuators. part B* 129: 888–895.

Hsu, J.-F., B.-R. Huang, C.-S. Huang, and H.-L. Chen. 2005. Silicon nanowires as pH sensor. *Japanese Journal of Applied Physics* 44(4B): 2626–2629.

Huang, X.-J. and Y.-K. Choi. 2007. Chemical sensors based on nanostructured materials. *Sensors and Actuators. part B* 122: 659–671.

Jun, J. H., J. Yun, K. Cho, I.-S. Hwang, J.-H. Lee, and S. Kim. 2009. Necked ZnO nanoparticle-based NO_2 sensors with high and fast response. *Sensors and Actuators B: Chemical* 140: 412–417.

Kamins, T. I., S. Sharma, A. A. Yasseri, Z. Li, and J. Straznicky. 2006. Metal-catalysed, bridging nanowires as vapour sensors and concept for their use in a sensor system. *Nanotechnology* 17: S291–S297, doi: 10.1088/0957-4484/17/11/S11.

Lao, C. S., Q. Kuang, Z. L. Wang, M.-C. Park, and Y. Deng. 2007. Polymer functionalized piezoelectric-FET as humidity/chemical nanosensors. *Applied Physics Letters* 90: 262107-1–262107-3.

Law, M., H. Kind, B. Messer, F. Kim, and P. Yang. 2002. Photochemical sensing of NO_2 with SnO_2 nanoribbon nanosensors at room temperature. *Angewandte Chemie International Edition* 41(13): 2405–2408.

Li, J., Y. Lu, Q. Ye, M. Cinke, J. Han, and M. Meyyappan. 2003. Carbon nanotube sensors for gas and organic vapor detection. *Nano Letters* 3(7): 929–933.

Li, J., Y. Lu, and M. Meyyappan. 2006. Nano chemical sensors with polymer-coated carbon nanotubes. *IEEE Sensors Journal* 6(5): 1047–1051.

Li, H., Y. Zhang, X. Wang, and Z. Gao. 2008. A luminescent nanosensor for Hg(II) based on functionalized CdSe/ZnS quantum dots. *Microchim Acta* 160: 119–123, doi: 10.1007/s00604-007-0816-x.

Li, P., Y. Li, B. Ying, and M. Yang. 2009. Electrospun nanofibers of polymer composite as a promising humidity sensitive material. *Sensors and Actuators B: Chemical* 141(2): 390–395.

Lu, Y., J. Li, J. Han, H.-T. Ng, C. Binder, C. Partridge, and M. Meyyappan. 2004. Room temperature methane detection using palladium loaded single-walled carbon nanotube sensors. *Chemical Physics Letters* 391(4–6): 344–348.

Lucci, M., P Regoliosi, A. Reale, A. Di Carlo, S. Orlanducci, E. Tamburri, M. L. Terranova, P. Lugli, C. Di Natale, A. D'Amico, and R. Paolesse. 2005. Gas sensing using single wall carbon nanotubes ordered with dielectrophoresis. *Sensors and Actuators. part B* 111–112: 181–186.

Lucci, M., A. Reale, A. Di Carlo, S. Orlanducci, E. Tamburri, M. L. Terranova, I. Davoli, C. Di Natale, A. D'Amico, and R. Paolesse. 2006. Optimization of a NOx gas sensor based on single walled carbon nanotubes. *Sensors and Actuators B: Chemical* 118(1–2): 226–231.

Luongo, K., A. Sine, and S. Bhansali. 2005. Development of a highly sensitive porous Si based hydrogen sensor using Pd nanostructures. *Sensors and Actuators. part B* 111–112: 125–139.

Ma, Y., W. L. Wang, K. J. Liao, and C. Y. Kong. 2002. Study on sensitivity of nano-grain ZnO gas sensors. *Journal of Wide Bandgap Materials* 10(2): 113–120.

Massera, E., I. Nasti, L. Quercia, I. Rea, and G. Di Francia. 2004. Improvement of stability and recovery time in porous silicon-based NO_2 sensor. *Sensors and Actuators. part B* 102(2): 195–197.

Nohria, R., R. K. Khillan, Y. Su, R. Dikshit, Y. Lvov, and K. Varahramyan. 2006. Humidity sensors based on polyaniline film deposited using layer-by-layer nano-assembly. *Sensors and Actuators. part B* 114: 218–222.

Padigi, S. K., R. K. K. Reddy, and S. Prasad. 2007. Carbon nanotube based aliphatic hydrocarbon sensor. *Biosensors and Bioelectronics* 22(6): 829–837.

Penza, M., G. Cassano, P Aversa, F. Antolini, A. Cusano, A. Cutolo, M. Giordano, and L. Nicolais. 2004. Alcohol detection using carbon nanotubes acoustic and optical sensors. *Applied Physics Letters* 85(12): 2379–2381.

du Plessis, M.. 2008. Nanoporous silicon explosive devices. *Materials Science and Engineering. part B* 147(2–3): 226–229.

Qi, P., O. Vermesh, M. Grecu, A. Javey, Q. Wang, H. Dai, S. Peng, and K. J. Cho. 2003. Toward large arrays of multiplex functionalized carbon nanotube sensors for highly sensitive and selective molecular detection. *Nano Letters* 3(3): 347–351.

Shen, Y., T. Yamazaki, Z. Liu, D. Meng, T. Kikuta, N. Nakatani, M. Saito, and M. Mori. 2009. Microstructure and H_2 gas sensing properties of undoped and Pd-doped SnO_2 nanowires. *Sensors and Actuators. part B* 135(2): 524–529.

Stefano, L. D., I. Rendina, M. D. Stefano, A. Bismuto, and P. Maddalena. 2005. Marine diatoms as optical chemical sensors. *Applied Physics Letters* 87: 233902-1–233902-2.

Suehiro, J., G. Zhou, H. Imakiire, W. Ding, and M. Hara. 2005. Controlled fabrication of carbon nanotube NO_2 gas sensor using dielectrophoretic impedance measurement. *Sensors and Actuators. part B* 108(1–2): 398–403.

Sundar, S. and J. Chakravarty. 2010. Antimony toxicity. *International Journal of Environmental Research and Public Health* 7(12): 4267–4277.

Walter, E. C., K. Ng, M. P. Zach, R. M. Penner, and F. Favier. 2002a. Electronic devices from electrodeposited metal nanowires. *Microelectronic Engineering* 61–62: 555–561.

Walter, E. C., F. Favier, and R. M. Penner. 2002b. Palladium mesowire arrays for fast hydrogen sensors and hydrogen actuated switches. *Analytical Chemistry* 74(7): 1546–1553.

Wong, Y. M., W. P. Kang, J. L. Davidson, A. Wisitsora-at, and K. L. Soh. 2003. A novel microelectronic gas sensor utilizing carbon nanotubes for hydrogen gas detection. *Sensors & Actuators. part B, Chemical* 93(1): 327–332.

Yamazoe, N. 2005. Toward innovations of gas sensor technology. *Sensors and Actuators. part B* 108: 2–14.

Yamazoe, N. and K. Shimanoe. 2008. Roles of shape and size of component crystals in semiconductor gas sensors: I. Response to oxygen. *Journal of the Electrochemical Society* 155(4): J85–J92.

Yamazoe, N. and K. Shimanoe. 2009. Receptor function and response of semiconductor gas sensor. *Journal of Sensors* 2009: 875704 (21 pp.), doi: 10.1155/2009/875704.

Zhang, W.-D. and W.-H. Zhang. 2009. Carbon nanotubes as active components for gas sensors. *Journal of Sensors* 2009: 160698 (16 pp.), doi: 10.1155/2009/160698.

Zhang, T., M. B. Nix, B.-Y. Yoo, M. A. Deshusses, and N. V. Myung. 2006a. Electrochemically functionalized single-walled carbon nanotube gas sensor. *Electroanalysis* 18(12): 1153–1158.

Zhang, B., R. Fu, M. Zhang, X. Dong, B. L. Wang, and C. U. Pittman. 2006b. Gas sensitive vapor grown carbon nanofiber/polystyrene sensors. *Materials Science Bulletin* 41: 553–562.

9

Nanobiosensors

9.1 Introduction

We have had the opportunity to speak about biomolecules and biological sensors several times in preceding chapters because, when we are dealing with physical or chemical sensors, biological phenomena invariably enter the discussion owing to their supreme importance, e.g., we treated protein, protease and maltose sensors in Sections 6.11.1–6.11.3, we described the role of sensors in DNA, protein, virus and protein activity analysis in Sections 7.5.1–7.5.4, and we discussed sensors and tumor marker detection in Section 7.8.5. The reason, already indicated in Section 8.1, is that, when a particular class of sensors is being considered, one cannot seal the borders totally against other classes. Therefore, this topic is not new and the objective here is to look at biological sensors more formally.

What is a biological sensor? Referring to Section 1.15, it is a device in which biomolecules and biological phenomena (Section 1.5) occupy the upfront, decisive positions, with physical (Section 1.3) and chemical (Section 1.4) effects being pushed to the background or into supporting roles. Ultimately, biomolecules are chemicals but essentially those linked to life processes.

The term "biosensor" once referred broadly to any device that responds to chemical species in biological samples or any device meant for biological applications or using biological components. But the definition now is more restricted and more sharply focused. A biosensor is a device for the detection of an analyte, that combines a biological component with a physicochemical detector. Then, by simply downscaling the biosensor, a nanobiosensor is a biosensor on the nanoscale size (Khanna 2008).

However, non-scalable parameters ought still to be remembered.

Figure 9.1 illustrates the components of a nanobiosensor, showing biorecognition, transduction, and interface elements. Molecular recognition is central to biosensing. A nanobiosensor is a simple analytical device having a biological, biologically derived, or biomimetic element which uses a nanomaterial property or nanoscale phenomenon to convert the recognition event from the analyte into a form suitable for physical or chemical transduction. *What then is the specialty about nanobiosensor?* It brings together the sensitivity and specificity of biological molecules with the versatility of physical/chemical transducers at the nano-level. The result is an investigative device whose distinctiveness lies in the fact that it unites the nanotechnological phenomena of the physical, chemical, and biological worlds.

What responsibilities are assigned to the various components of a biosensor? The *biological recognition element* or *bioreceptor* of a biosensor ought to be of biological origin, such as an enzyme, an antibody (Ab), an antigen (Ag), a cell, a tissue, a deoxyribonucleic acid (DNA) sequence. *Do all recognition elements have a biological background?* No, not all recognition agents in novel devices are of a biological origin, but sometimes they may be synthetically produced elements such as crown ethers, cryptands, calixarenes, or molecularly imprinted polymers. In a wider sense, more correctly, when loosely defined, these artificially produced elements may be accepted and included as components of biosensors as long as biological applications are intended. The recognition element is the component used to attach the target molecule, and must therefore be highly specific, stable under storage conditions, and immobilized.

The *physicochemical transducer* acts as an interface, measuring the physical change that occurs with the reaction at the bioreceptor, and then transforming that energy into a quantifiable electrical output. Finally, the signals from the transducer are fed to a *microprocessor*, where they are amplified and analyzed; the digitally dissected data is converted to concentration units and transferred to a *display* and/or *data storage device.*

Nanobiosensors are subdivided into five main categories, namely, electrochemical, field-effect, mechanical, optical, and magnetic types. Of these, the first three classes will be exhaustively treated here in Chapter 9. Optical nanobiosensors have already been described in Chapter 6, which deals with optical nanosensors, and only a cursory view will be presented here to add to what has already been included. Magnetic nanobiosensors have been dealt with at length in Chapter 7, on magnetic nanosensors, and therefore will not be covered here.

9.2 Nanoparticle-Based Electrochemical Biosensors

Because nanomaterials and biomolecules have identical dimensions, the combination of nanometer materials and biomolecules is of considerable importance in the fields of biotechnology and bioanalytical chemistry. Consequent to their large specific surface area and high surface free energy, there has been a growing interest in the study of electron-transfer (ET) between proteins and electrodes modified with nanomaterials (Ju et al. 2002; Wang et al. 2002a; 2002b) since the beginning of the 1990s. Studies have shown that nanomaterial films can offer not only a suitable platform on which to assemble protein molecules, but are also able to enhance the ET process between protein molecules and electrodes. Both these qualities have attracted the attention of researchers.

Which natural element seems to be of paramount importance here? Obviously, it is none other than gold, the ornamental metal in a very pure form. Yes, gold has a special role to play among the nanomaterials. Many synthetic procedures are found in the literature for controlling the size, monodispersion, morphology, and surface chemistry of gold nanoparticles (AuNPs). The easy

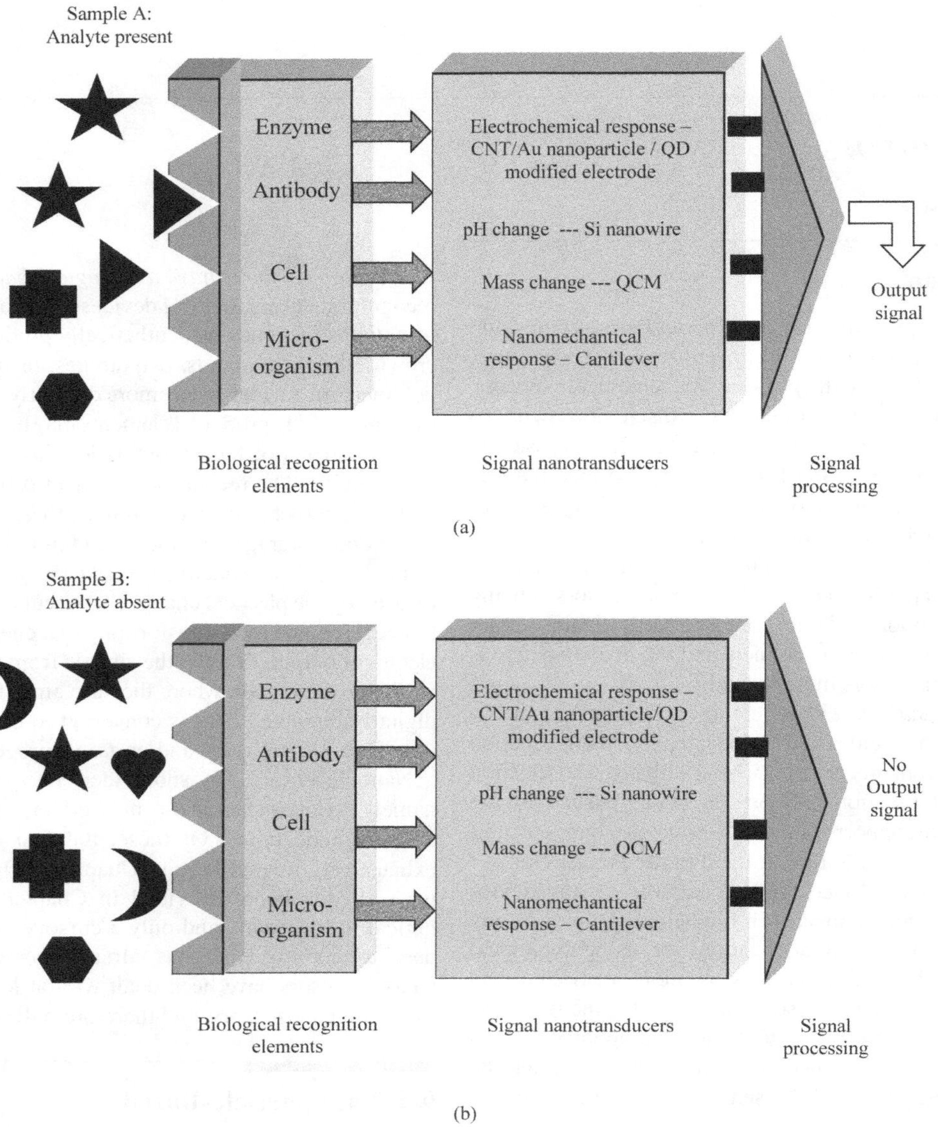

FIGURE 9.1 Nanobiosensor structure: (a) response of sample A containing the analyte and (b) response of sample B having no analyte. In (a), the analyte binds selectively to the bioreceptor, yielding a measurable response signal after processing.

modification of gold surface, using thiol-ended molecules makes then suitable for many different biological assemblies.

But gold is a noble metal and bulk gold is nonreactive, famous for retaining its properties and remaining unaffected by the environment. So, how do AuNPs behave differently? Yes, bulk gold is usually catalytically inert in chemical processes. On the other hand, nanometric gold particles have been found to be catalytically active. Gold nanoparticles (AuNPs) exhibit remarkable catalytic activity because of their enormous surface area, great availability of reactional sites, and interface-controlled characteristics.

Can metal NPs be used as electrodes? No, it is difficult to directly use metal NPs as electrodes. So, various approaches including self-assembly, grafting reaction electrophoresis etc., were developed to immobilize metal NPs onto the surfaces of supporting electrodes so that their useful properties could be utilized. Such electrodes are called *nanoparticle-modified electrodes. How are such electrodes prepared?* Gold NP-modified electrode surfaces are usually prepared in three ways (Yanez-Sedeno and Pingarron 2005): (i) by binding AuNPs with functional groups of self-assembled monolayers (SAMs); (ii) by direct deposition of NPs on the bulk electrode surfaces; and (iii) by incorporation of colloidal gold into the electrode by assimilating gold with other components in the composite electrode matrix.

Modification of surfaces of electrodes used in electrochemistry by decoration with AuNPs bestows on them unusual properties inherent in the nano-modifier Au particles. Several workers have investigated the electrocatalysis of biologically important small molecules, as well as macromolecules at gold-modified electrodes.

What electrochemical techniques have been used for biomolecule sensing? Various electrochemical techniques

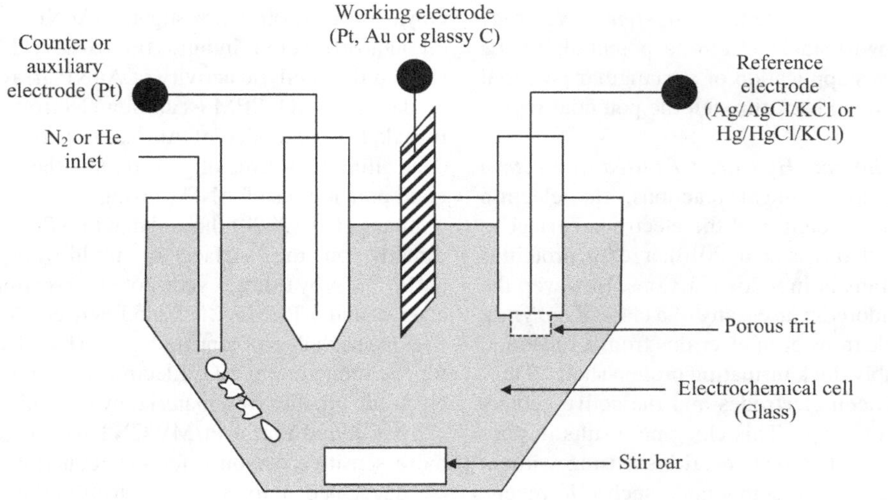

(a)

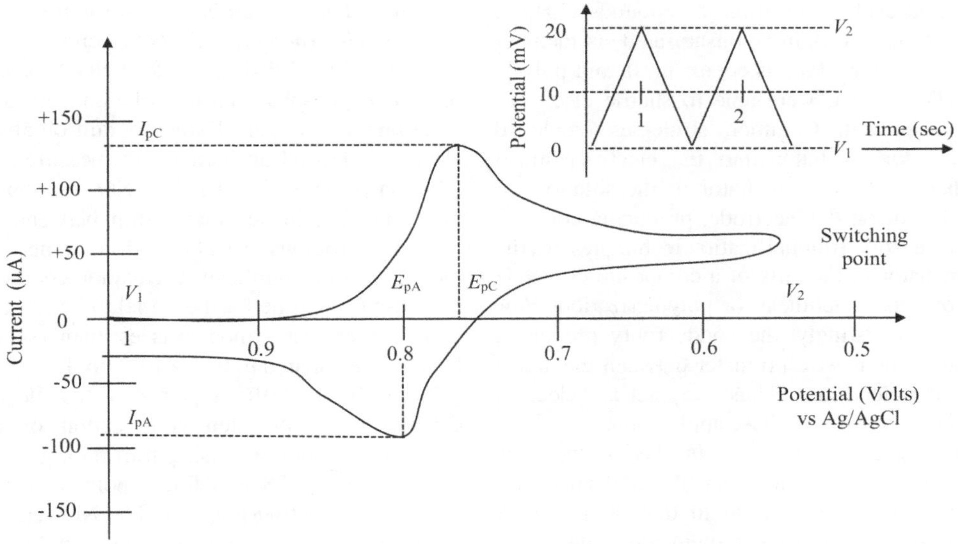

(b)

FIGURE 9.2 (a) An electrochemical cell used in voltammetry, consisting of a reference electrode, a working electrode, and a counter electrode, along with a nitrogen or helium inlet. (b) Typical cyclic voltammogram of an electrode at a specified sweep rate. I_{pC} and I_{pA} are the peak cathodic and anodic currents, respectively, and E_{pC} and E_{pA} the corresponding voltages, respectively. (The potential is applied between the reference electrode and the working electrode, and the current is measured between the working electrode and the counter electrode. The working electrode potential ramps linearly *versus* time between potentials V_1 and V_2, as shown in the inset. This ramping is known as the scan or sweep rate.)

have been used for sensing of biomolecules. Some common techniques are *voltammetry* (an analytical technique for the detection of minute quantities of substances by measuring the currents generated or the substances deposited on electrodes placed in electrolytic solutions, when known voltages are applied), *amperometry* (the measurement of current at a single applied potential), *potentiometry* (measurement of potential under the conditions of zero current flow), *impedometry* (analysis of a sample by impedance measurements), and *conductometry* (determination of the quantity of a material present in a mixture by measurement of its effect on the electrical conductivity of the mixture).

For enhanced current response, it is very important to develop a stable and highly target-specific interface by various surface modifications of conventional electrodes. *Electrocatalysis* is the speeding up of electrode kinetics by a material by minimization of overpotential, the superfluous potential over the equilibrium value that has to be applied to cause an electrodic reaction to take place at a given rate. An *electrocatalyst* is the material, normally employed on the surface or an electrode, which catalyzes a half-cell reaction.

Figure 9.2 summarizes basic information on voltammetry techniques. *Linear sweep voltammetry* (LSV) is a voltammetric method where the current at a working electrode is measured, while the potential between the working electrode and a reference electrode is swept linearly in time. The *working electrode* is the electrode on which the reaction is occurring. The working electrode is used with an auxiliary electrode and a reference electrode in a three-electrode system. A *reference electrode* is one which has a known, stable electrode potential. *Cyclic voltammetry* (CV) involves application of a triangular potential sweep, allowing one to sweep back through the potential region just covered.

What are electron mediators? How are NPs used as electron wires? In many bio-electrochemical reactions, the electron transfer between the redox-protein and the electrode surface is the key subject to be detected (Lia et al. 2010); a redox protein is one mediating electron transfer in redox reactions. However, the active centers of most oxidoreductases (any of a class of enzymes that catalyze the reversible transfer of electrons from a substrate) are enclosed in considerably thick insulating protein shells. Thus, the electron transfer between electrodes and the active centers is blocked by the protein casings. This clogging results in poor analytical performances of electrochemical biosensing without *electron-transfer mediators*. Some compounds, such as ferrocene (the organometallic compound: $Fe(C_5H_5)_2$) derivatives, quinones (either of two isomeric cyclic crystalline compounds $C_6H_4O_2$, having two carbonyl groups, CO, in an unsaturated six-member carbon ring; biologically important as coenzymes), and poly-2-aminoaniline (PANI) polymer, were able to shuttle electrons, giving clearance to current. Common strategies employed to incorporate the redox mediator into the electrochemical system involve either adding the mediator to the solution, or immobilizing it within or on the electrode, producing compact chemically modified sensors. Immobilization techniques are the dispersion of the mediator in the bulk of a composite electrode or its adsorption, covalent attachment, or polymerization. *How can NPs help here?* Interestingly, the conductivity properties of AuNPs can enhance the electron transfer between the active centers of proteins and electrodes. Thus, they act as "electron wires", favoring electron transfer in these applications.

How is the insulating effect of a protein shell reduced by AuNPs? Modification of electrode surfaces with AuNPs provides a sympathetic microenvironment similar to that of the redox proteins in native systems, giving the protein molecules more liberty in orientation, thereby reducing the insulating effect of the protein shell for direct electron transport through the conducting passageways or tunnels of gold nanocrystals. It is believed that the nanometric edges of gold particles penetrate slightly into the protein, thereby decreasing the distance between the electrode and the biomolecule redox sites for electron transfer.

9.2.1 Nitric Oxide Electrochemical Sensor

What is the utility of NO sensing? Nitric oxide (NO) plays a decisive role in physiology and pathology. NO has been implicated in the pathogenesis of several diseases; a deficiency of NO is partly responsible for some illnesses (hypertension, hyperglycemia, atherosclerosis, Parkinson's disease, Alzheimer's disease), while, conversely, excess NO may participate in others (arthritis, reperfusion injury, cancer). Thus, from biochemical as well as medical perspectives, it is vital to quantify the details of NO production in normal and abnormal tissues, including by direct measurements. Therefore, nanomaterial-based electrochemical sensors selective toward NO have fascinated scientists.

What efforts have been made for NO detection using NP-modified electrodes? Zhang and Oyama (2005) and Goyal et al. (2007) investigated AuNP arrays directly grown on nanostructured indium tin oxide (ITO) electrodes. They studied the catalytic activity of AuNP arrays toward the electro-oxidation of NO. SEM (scanning electron microscopy) imaging revealed that the size of Au nanospheres could be adjusted by controlling their time of growth, and the as-prepared arrays had good potential use for NO sensing.

Zhang et al. (2008) showed that Pt NPs can be electrodeposited directly on the surface of multi-walled carbon nanotube (MWCNTs) by using a cyclic potential scanning technique to form a novel nano-Pt/MWCNT-modified electrode; the as-prepared Pt-NP showed a potent and powerful electrocatalytic activity for the reduction of NO. Electrocatalysis is the enhancement of electrode kinetics by a material by minimizing the overpotential.

They found that a Pt/MWCNT-modified electrode elicited a more sensitive response for the reduction of NO, and it could be developed into a sensor to directly determine the NO concentration in aqueous solution.

How did they examine the response of nano-Pt/MWCNT-modified electrode toward NO? In order to inspect the response character of modified electrode to NO, they studied the detection of NO in phosphate buffer solution with a constant potential voltammetry. A typical amperogram obtained at the modified electrode showed an increase in measured currents with each addition of 4×10^{-7} mol L^{-1} of NO solution. According to this experiment, a linear relationship between peak currents and NO concentrations was obtained in a range of $4\times10^{-7}-1\times10^{-4}$ mol L^{-1} with a significant correlation coefficient, $r=0.997$. The response time (time for the signal to increase from 10% to 75%) in the amperometric mode was less than 5 seconds. The detection limit was estimated to be 1×10^{-7} mol L^{-1}.

Wang et al. (2010) reported a NO electrochemical sensor developed via one-step construction of a AuNPs-chitosan (CS) nanocomposite sensing film on a glassy carbon electrode (GCE) surface; CS is a linear polysaccharide widely used as an adhesive coating polymer. The NO sensor was prepared by single-step electrochemical deposition in solution containing $HAuCl_4$ and CS on pretreated GCE at an applied potential of –1.00 V for 60 seconds (GCE/CS-AuNPs). Because of the special electronic properties and excellent electrocatalytic ability of Au NPs, this method exhibited high sensitivity, broad linearity, and low detection limits. The anodic peak potential notably shifted negatively, compared with that at the control GCE. The high sensitivity and high stability of the method developed were coupled to a wide linear range from 3.60×10^{-8} to 4.32×10^{-5} M for the quantitative analysis of NO. The detection limit of 7.20 nM was much lower than the vast mainstream methods reported.

How was the sensor applied to biological samples? Fabricated GCE/CS-AuNPs were applied to NO monitoring from rat kidneys. No apparent NO current response was observed without kidney or L-Arg in the PBS (phosphate-buffered saline) containing trace O_2. Arginine, abbreviated as Arg, is an α-amino acid. The L-form is one of the 20 most common natural amino acids.

But a remarkable change of amperometric current could be detected after the addition of L-Arg into PBS containing a whole kidney, suggesting that NO was released from the kidney. The concentration of NO liberated from the rat kidney sample was calculated to be ~150 nM. The concentration of NO monitoring from the drug sample was calculated to be ~1.60 μM.

9.2.2 Determination of Dopamine, Uric Acid, and Ascorbic Acid

Why is it necessary to develop biosensors for these analytes? Because dopamine (DA), $C_6H_3(OH)_2$–OH_2–CH_2–NH_2, uric acid (UA), $C_5H_4N_4O_3$, and ascorbic acid (AA), $C_6H_8O_6$, are three important biomolecules, which are widely distributed in the body of many mammals, and exhibit vital physiological functions, such as message transfer (communication) in the brain and in defense of the body against disease. DA, one of the major catecholamines, belongs to the family of excitatory chemical neurotransmitters. It is a biogenic (produced by living organisms or biological processes) amine, synthesized in the hypothalamus (the part of the brain that lies below the thalamus), in the arcuate nucleus, the caudad (toward the tail or posterior end of the body), and various areas of the central and peripheral nervous systems. Its concentration is of great consequence in the function of central nervous, renal, hormonal, and cardiovascular systems. Because extreme abnormalities of DA level are warning signs of several diseases, such as schizophrenia and Parkinson's disease, the determination of the concentrations such compounds in real biological samples and the identification of changes in neurotransmission that correlate the behavioral states of animals are obvious targets in neurochemical studies, i.e., study of the chemical composition and processes of the nervous system and the effects of chemicals on it. Thus, it is important to develop sensitive, fast, and specific methods for the detection of DA, UA, and AA (Huang et al. 2008).

What are the main obstacles in the correct detection of DA? AA, DA, and UA generally coexist in the extracellular fluid of the central nervous system and serum, and their oxidation potentials are alike at most solid electrodes, which leads to overlapping signals that appear when one strives to detect these three compounds, when present together. The major predicament arises in the detection of DA in the presence of high concentrations of AA. As AA is oxidized at almost the same potential as DA on the uncovered or bare electrodes, the bare electrode very often is negatively affected by fouling effects due to accretion of the oxidized product on the electrode surface; hence, poor selectivity and sensitivity ensue. The homogeneous catalytic oxidation of AA by oxidized DA is another chief interference in the measurement of DA. In view of these two reasons it is very tricky to determine DA directly. In order to resolve these problems, some modified electrodes have been developed to determine DA.

Goyal et al. (2007) applied AuNP arrays for simultaneously detecting DA and serotonin, $C_{10}H_{12}N_2O$. Subsequently, Zhang and Oyama (2007) derivatized the AuNP arrays with 3-mercapto-propionic acid (MPA), whereby a 3D MPA monolayer was produced on the NP array. The 3D MPA monolayer showed stronger electrocatalytic activity toward DA and UA than the AuNP array on indium tin oxide (ITO).

Huang et al. (2008) deposited a class of novel nanocomposites, poly(3-methylthiophene) (P3MT)/gold nanoparticles (AuNPs) on the surface of a GCE. An important conducting polymer, 3-methylthiophene, C_5H_6S, is easily deposited on an electrode surface by an electrodeposition technique. The AuNPs were uniformly inserted into a P3MT layer, and formed a porous 3D structure.

The ability of an AuNPs-P3MT composite-modified electrode to analyze DA, UA, and AA at the same time was investigated. This modified electrode showed exceptional electrocatalytic activity toward the oxidation of AA, DA and UA; the overlapping anodic peaks of AA, DA, and UA were completely divided into three well-defined, contrasting voltammetric peaks. A supplementary study showed that there existed a linear relationship between the peak current and the concentration of DA in the range of 1.0×10^{-6}–3.5×10^{-5} mol L^{-1}, and UA in the range of 1.0×10^{-6}–3.2×10^{-5} mol L^{-1}. The detection limits were 2.4×10^{-7} mol L^{-1} for DA and 1.7×10^{-7} mol L^{-1} for UA. This proposed method was applied to the detection of real samples, obtaining satisfactory results.

Balal et al. (2009) constructed iron(III)-doped zeolite (one of a family of hydrous aluminum silicate minerals, the molecules of which enclose cations of sodium, potassium, calcium, strontium, or barium) NP-modified carbon paste electrode (CPE) for sub-micromolar determination of DA and tryptophan (Trp); Trp is a neutral amino acid, $C_{11}H_{12}N_2O_2$. Differential pulse voltammograms of DA and Trp were compared for a zeolite-modified carbon paste electrode (ZCME) and a bare CPE. This comparison was made with 30 µM DA and 50 µM Trp in a 0.1 M phosphate buffer at pH 5. The prepared modified electrode showed voltammetric responses with high sensitivity and stability for DA and Trp.

What are the underlying causes for improved detection capability of DA by NP-modified electrodes? The catalytic mechanism of AuNPs for the oxidation of biomolecules has been studied extensively. Generally, it was considered that the AuNP layer on the surface of an electrode decreases the overpotential needed, and enables faster ET kinetics, which renders the redox reaction kinetically viable, and the voltammetry appear to be reversible, e.g., with AuNPs on the surface of GCE, the DA oxidation process appeared to be more reversible and the peak shifted amply to be discernible from that of AA, the main interfering factor. The outstanding catalytic ability of the modified electrode was attributed to its special 3D nanostructure, which could accelerate electron transfer; this acceleration produced the observed difference.

9.2.3 Detection of CO

Catalytic oxidation of carbon monoxide has broad technological applications, including fuel-cell technology, air purification in gas products, long-duration space voyage, and CO conversion in automobile exhaust systems. In the experiments of Maye et al. (2000), catalytically active AuNP cores were covered with thiolate (any derivative of a thiol in which a metal atom replaces the hydrogen attached to sulfur; a thiol is a sulfur-containing organic compound having the general formula RSH, where R is another element or radical) monolayers resulting in a networked congregation of core/shell AuNPs on GC electrodes, involving exchange of 1,9 nonanedithiol (NDT), HS-$(CH_2)_9$–SH, with decanethiolate (DT)-capped AuNPs, cross-linking, nucleation and thin-film growth; decanethiol has the molecular formula HS-$(CH_2)_9$–CH_3. This assembly had a pronounced catalytic activity toward the oxidation of carbon monoxide, with impending sensing applications.

Geng and Lu (2007) studied AuNP on glassy carbon (GC) electrodes for electro-oxidation of carbon monoxide under basic conditions. Electrochemical cyclic voltammetric results illustrated that CO oxidation simultaneously took place in the anodic and cathodic sweeps during one cycle. Moreover, CO electro-oxidative activity was remarkably different in the anode for different-sized AuNPs. The ultrafine catalyst metal particles (2- and 6-nm Au) were found to be more effective, compared with the larger ones (12-, 24-, and 41-nm Au).

For a detailed coverage of CO nanosensors, the reader should see Section10.3.

9.2.4 Glucose Detection

Glucose oxidase (GO_x) is a structurally rigid glycoprotein (proteins that contain oligosaccharide chains (glycans) covalently attached to polypeptide side chains), having a molecular weight of 152,000–186,000 Da (unit of atomic mass roughly equivalent to the mass of a hydrogen atom, 1.67×10^{-24} g), and consisting of two identical polypeptide chains, each containing a flavin adenine dinucleotide (FAD) redox center, which are deeply embedded in the apoenzyme (enzyme that requires a cofactor but does not have one bound); the FAD is a redox cofactor involved in several important reactions in metabolism. Consequently, even if the enzyme approaches the electrode surface, the distance between either of its two FAD/FADH2 centers and the electrode surface stretches beyond the distance across which electrons are transferred at a measurable rate. Naturally, direct electron transfer from the enzyme to the electrode is not observed. In order to shuttle the electrons between the redox centers of GO_x and the electrode, ET mediators are often applied, which have been extensively used to fabricate glucose biosensors.

The bioactivity, stability, and quantity of the biological recognition elements immobilized on the electrode are important issues in bioelectrochemistry. Biological activity is a parameter expressing the effects of a molecule on living matter. *How does immobilization of biomolecules on naked surfaces of materials differ from that on NP surfaces?* The adsorption of biomolecules directly onto naked surfaces of bulk materials frequently results in their denaturation and subsequent loss of bioactivity. *Denaturation* is a process in which the folding structure of a protein is altered due to exposure to certain chemical or physical factors, e.g., heat, acid, solvents, causing it to become biologically inactive. The AuNPs offer excellent candidates for the immobilization platform. The adsorption of biomolecules onto the surfaces of AuNPs retains their bioactivity and stability because of the biocompatibility and the high surface free energy of AuNPs; *biocompatibility* is the property of not producing a toxic, injurious, or immunological response to living matter. As compared with flat gold surfaces, AuNPs have a much higher surface area, allowing loading of a larger amount of protein, and are potentially more sensitive. Thus, a number of laboratories have explored the contribution of AuNPs for biomolecular immobilization.

Xiao et al. (2003) reconstituted an apo-flavoenzyme, apo-glucose oxidase, on a 1.4-nm gold nanocrystal functionalized with the cofactor flavin adenine dinucleotide (FAD), a substance that must be associated with an enzyme for it to function. A flavoenzyme is an enzyme that possesses a flavin nucleotide as coenzyme; an apo-enzyme is the protein component of an enzyme, to which the coenzyme attaches to form an active enzyme. The electron transfer turnover rate of the reconstituted bioelectrocatalyst was ~5000 second^{-1}. The AuNP acted as an electron relay or "electrical nanoplug" for the alignment of the enzyme on the conductive support and for the electrical cabling of its redox active center. Figure 9.3 shows the procedure of assembling the AuNP-reconstituted GO_x electrode. For this purpose, AuNP (1.4-nm), with a single *N*-hydroxysuccinimide (NHS)-functionalization was reacted with *N*-6-(2-aminoethyl)-FAD (1) to yield the FAD-functionalized AuNP. FAD (1) is the structural gene for flavin adenine dinucleotide synthetase of a species of yeast called Saccharomyces cerevisiae. FAD1 gene has 1 transcript (RNA copy) and 305 orthologues (homologous gene sequenes); it is a member of 1 Ensembl protein family.

Apo-GO_x was then reconstituted with the FAD-functionalized AuNP. The AuNP-reconstituted GO_x was accumulated on an Au electrode with the FAD-functionalized AuNP. Thus, a strategy was formulated to electrically contact redox enzymes (e.g., GO_x) with their macroscopic environment by the reconstitution of an apoenzyme with an FAD-AuNP unit.

Andreescu and Luck (2008) developed a sensitive and reagent-less electrochemical glucose biosensor based on surface-immobilized periplasmic glucose receptors on AuNPs. The sensor was fabricated by immobilization of genetically engineered periplasmic glucose receptors to the AuNPs and showed selective detection of glucose in the micromolar concentration range, with a detection limit of 0.18 μM.

The analytical performances of several enzyme biosensor designs prepared by immobilization of glucose oxidase on different tailored AuNP-modified electrode surfaces were compared by Mena et al. (2005). Glucose oxidase (GO_x) and the redox mediator tetrathiafulvalene (an organosulfur compound with the formula $(H_2C_2S_2C)_2$) were co-immobilized in all cases by cross-linking with glutaraldehyde, $CH_2(CH_2CHO)_2$. Gold disk electrodes were modified with short-chain molecules, such as cysteamine (Cyst), $HSCH_2CH_2NH_2$, the simplest stable aminothiol, and 3-mercaptopropionic acid 3-mercaptopropionic acid (MPA). The 3-mercaptopropionic acid, $C_3H_6O_2S$, and its derivatives are versatile compounds, which are used as chain transfer and cross-linking agents in polymerizations. For Cyst-Au electrodes, the enzyme was immobilized on AuNPs, previously bound on Cyst SAMs, using two different configurations. One of them consisted of a Cyst SAM bound to the gold disk *via* the sulfur atom so that the colloidal gold (Au_{coll}) was linked to the -NH_2 moiety (a part or functional group of a molecule). The second one was realized by derivatizing the Cyst SAM with glutaraldehyde to form a thiol terminal group so that AuNPs were bound to the sulfur moiety. Moreover, biosensor configurations based on immobilization of the enzyme onto Cyst or MPA SAMs that were previously bound on electrodeposited AuNPs on GCEs were also evaluated and compared. It was concluded from this comparative investigation that colloidal gold bound to Cyst SAMs gave rise to high rates of enzyme immobilization, and the GO_x/Au_{coll}–Cyst–AuE design achieved a stability much greater than for the other configurations. Figure 9.4 presents the scheme for realizing this biosensor design.

The useful lifetime of one single GO_x/Au_{coll}-Cyst-AuE was verified by performing repetitive calibration graphs for glucose after storage of the biosensor in phosphate buffer, pH 7.4, at 4°C. The useful lifetime of GO_x/Au_{coll}–Cyst–AuE was 28 days, very much longer than that of the other GO_x biosensor designs. The

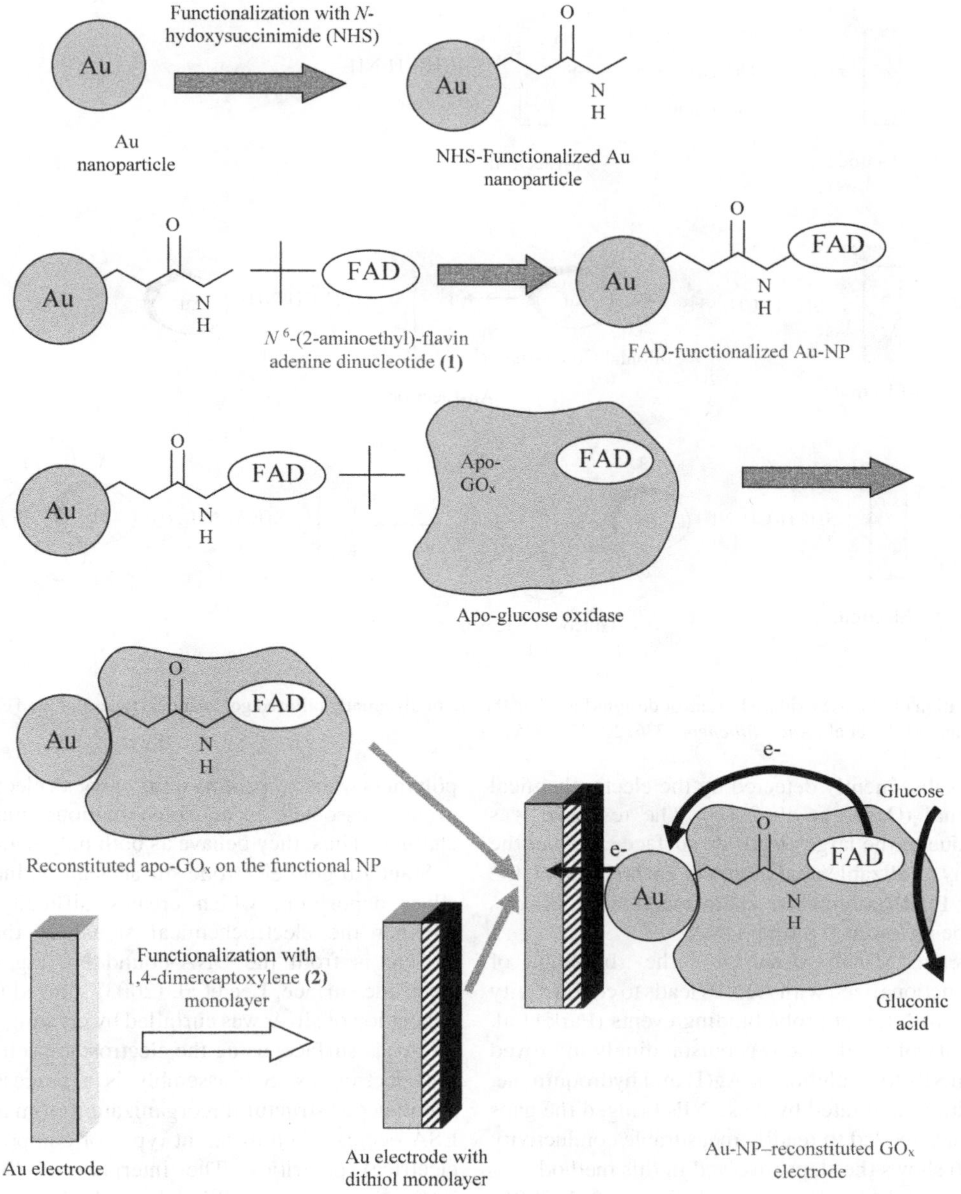

FIGURE 9.3 Steps in the assembly of AuNP-reconstituted GO_x electrode. (Xiao, Y., et al., *Science*, 299, 1877, 2003.)

same biosensor yielded slope mean values within the control limits for 28 days. One of the most claimed advantages of the biosensors based on the use of gold NPs is the stability that can be achieved, related to the possibility of enzyme immobilization without loss of bioactivity. This performance is markedly improved when compared with other GO_x biosensor designs.

9.2.5 Gold Nanoparticle DNA Biosensor

Figure 9.5 describes the rudimentary principle of a DNA biosensor. One way of fabricating this on an oxidized silicon surface is by modifying it with succinimidyl-4-(*p*-malemidophenyl)-butyrate (SMPB), $C_{18}H_{16}N_2O_6$, a heterobifunctional protein crosslinker reactive toward sulfhydryl and amino groups. Probe oligonucleotides (short oligomers of 2–20 nucleotides) are immobilized on the activated surface by spotting NaCl–phosphate buffer (pH = 7) solutions of the appropriate alkylthiol-modified oligonucleotide by manual pipetting. Single-stranded probe DNA is immobilized on the recognition surface. Target capture takes place to generate the recognition signal, which is transduced by electrochemical, mechanical, or optical means.

Colloidal Au provides a novel way to construct electrochemical DNA biosensors with high sensitivities for sequence-specific DNA detection (Drummond et al. 2003). Colloidal gold-based electrodes allow the electrode surface area, and consequently the amount of immobilized ssDNA, to be greatly enhanced. Colloidal gold nanoparticles have also been used to label DNA sequences for the enhancement of signal hybridization, e.g., using a sandwich-based design, the labeled target was captured by probe strands immobilized on a graphite pencil (GP) electrode, and

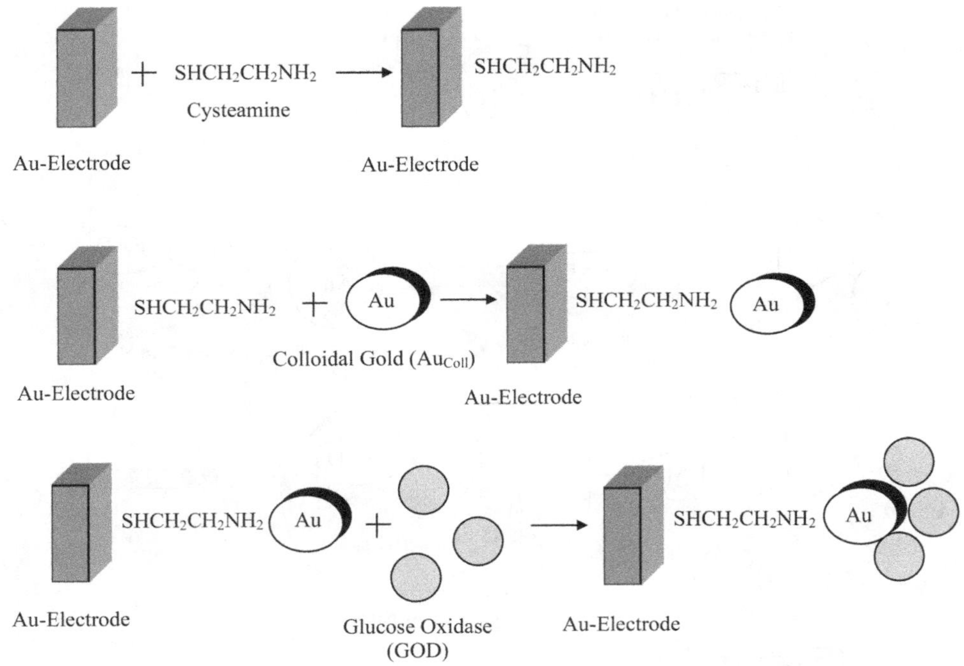

FIGURE 9.4 Preparation of glucose oxidase biosensor designs based on the use of differently tailored gold nanoparticle-modified electrode surfaces: GO_x/Au_{coll}–Cyst-AuE. (Mena, M. L., et al., *Anal. Biochem.*, 336, 20–27, 2005.)

hybridization was subsequently detected by the electrochemical gold-oxidation signal (Ozsoz et al. 2003). The response was greatly increased due to the large electrode surface area and the availability of many oxidizable gold atoms in each NP label, the detection limit for PCR (polymerase chain reaction) amplicons being reported to be as low as 0.8 fmol DNA.

In *conductometric DNA detection*, the binding of oligonucleotides functionalized with AuNPs leads to conductivity changes associated with target-probe binding events (Park et al. 2002). The sensitivity of the device was outstandingly improved by exposing the AuNPs to a solution of Ag(I) and hydroquinone, when silver deposition facilitated by these NPs bridged the gaps between the particles and led to readily measurable conductivity changes. Figure 9.6 shows the steps involved in this method.

In the *electrochemical stripping method* (Wang et al. 2001), after hybridization, the AuNPs were dissolved with HBr/Br_2 and related to DNA content. This was achieved by pre-concentration of gold(III) ions through electrochemical reduction and subsequent determination by anodic stripping voltammetry (ASV) (Figure 9.7a). To lower the detection limits, gold tracer "amplification" by silver deposition on the surface was applied (Figure 9.7b).

ASV works by electroplating certain metals in solution onto an electrode, thus pre-concentrating the metal. Then, the metals on the electrode are sequentially stripped off, generating an easily measurable current. The current (milliamps) is proportional to the amount of metal being stripped off. The potential (voltage, in millivolts) at which the metal is stripped off is characteristic for each metal, so that the metal can be identified as well as quantified.

Lee et al. (2003) reported an improved electrochemistry-based sequence-specified detection technique by modifying the electrode surface using polyelectrolytes. Polyelectrolytes are polymers whose repeating units carry an electrolyte group. These groups dissociate in aqueous solutions, making the polymers charged. Thus, they behave as both polymers and electrolytes.

Standard gold electrode surfaces are vulnerable to significant silver deposition, which creates difficulty in distinguishing whether the electrochemical signal of the silver-dissolution current is from the DNA-bound NP tag or the background electrode surface. Lee et al. (2003) showed that the background deposition of silver was curtailed by creating a positively charged electrode surface, using the electrostatic self-assembly (ESA) of polyelectrolytes. Self-assembly is a process occurring due to spontaneous structural reorganization from a disordered system. ESA occurs when different types of components have opposite electrical polarities. The interplay of Coulombic repulsive interactions between like-charged objects and the attractive interactions between unlike-charged ones results in the self-assembly of these objects into highly ordered, closed arrays.

Figure 9.8 shows the schematic representation of detection by electrochemical DNA-hybridization. The cleaned gold electrodes were immersed in an ethanolic solution of mercaptoundecanoic acid (MUDA), $HS(CH_2)_{10}CO_2H$ (2 mM), for 48 hours at room temperature, again rinsed with water, and then dried. The layer-by-layer (LBL) deposition of the polyelectrolytes by alternating the adsorption of the polycations (PAH: poly(allylamine hydrochloride)) and polyanions (PSS: poly(styrenesulfonate)) was performed. LBL is a simple, versatile, and inexpensive thin-film fabrication technique for multilayer formation, allowing precise thickness control at the nanoscale, in which the films are formed by depositing successive layers of oppositely charged materials, with wash steps in between. The outermost negatively charged PSS layer was used to bind the positively charged streptavidin (SA) or avidin (a tetrameric protein purified from the bacterium *Streptomyces avidinii*) between PSS and PAH.

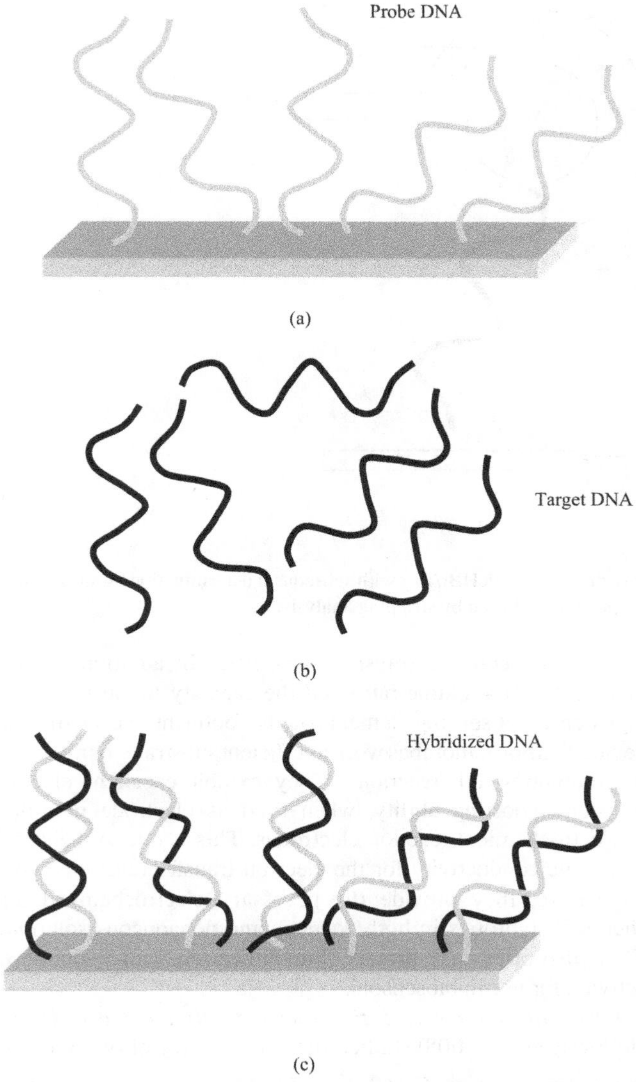

FIGURE 9.5 General DNA biosensor design: Probe DNA in (a) captures target DNA in (b) to form hybridized DNA in (c). The hybridization signal generated is transduced into an electronic signal.

The immobilization of the DNA probe was accomplished through avidin-biotin binding; biotin is a B vitamin, with formula $C_{10}H_{16}N_2O_3S$. The gold electrode was now prepared for hybridization with DNA analytes. The AuNP hybridization indicator was bound to the hybridized target by the interaction between the biotin group of the target and the SA group of the gold NR. Subsequent silver deposition on the AuNP resulted in discernible signal amplification. With the low background signal achieved by electrode-surface modification and selection, sensitive (high signal-to-noise ratio) electrochemical detection of the hybridization event, using the silver-enhanced AuNP approach, was demonstrated.

9.2.6 Monitoring Allergen-Antibody Reactions

Deriving the advantage of specific recognition between antibody (Ab) and antigen (Ag), through Ab–Ag interaction, that forms the basis for immune response to infectious disease-causing agents, immunoassays are widely applied in clinical practice. AuNPs have been used in these assays because of their large surface area and biocompatibility. They provide higher loading density and greater retention of immunoactivity. House dust mites cause heavy atopic diseases, such as asthma and dermatitis. Among allergens (examples of antigenic proteins) from *Dermatophagoides farinae*, Der f2 shows the highest positive rate for atopic patients. Huang et al. (2006) assembled recombinant dust mite allergen Der f2 molecules on a AuNP-modified GP carbon electrode and used electrochemical impedance spectroscopy (EIS) to monitor the interaction between the allergen and murine monoclonal antibody (MAb). To make an EIS measurement, a small-amplitude signal, usually a voltage between 5 and 50 mV, is applied to a specimen over a range of frequencies in the range 10^{-3}–10^5 Hz. The EIS instrument records the real (resistance) and imaginary (capacitance) components of the impedance response of the system.

The increase in Ab concentration was accompanied by deceleration of the interfacial electron transference of the redox probe, indicating the bonding of more Ab molecules to the immobilized allergen. Here, immobilization of a greater allergen density with higher immunoactivity retention was achieved due to the active sites of the Au colloidal monolayer. In EIS, the AuNPs served as electrical conduction paths.

9.2.7 Hepatitis B Immunosensor

The binding reaction of Ags at Ab-immobilized surfaces is often insufficient to produce a large signal change, e.g., for impedance measurements. A novel strategy used to overcome this shortcoming is the immobilization of Ab molecules *via* self-assembly on a Pt electrode modified with colloidal gold and polyvinyl butyral, $H_2(C_8H_{14}O_2)_n$ (PVB: a tough, clear, adhesive) as matrices, which combines the large specific surface area and high surface free energy of AuNPs with the entrapped effect of PVB (Tang et al. 2004). The novel immunosensors exhibited excellent electrochemical characteristics toward the Hepatitis B surface antigen, (HBsAg), the antigen most frequently used to screen for the presence of this infection, which appears in the serum from 4 to 12 weeks following infection. The method is applied generally to amplified assays of other biomolecules and other transduction modes, such as potentiometric or amperometric immunosensors. Another Hepatitis B nanosensor is described in Section 11.4.1.

9.2.8 Carcinoembryonic Antigen Detection

Most of the work involving immunosensors is related to the diagnosis and/or monitoring of human diseases. The carcinoembryonic Ag is a glycoprotein used as a tumor marker and has been frequently investigated in immunoreactions. It is present in fetal gastrointestinal tissue, but is generally absent from adult cells, with the exception of some carcinomas, which are malignancies that originate in the epithelial tissues.

A label-free amperometric immunosensor was fabricated by Ou et al. (2007) on a mercaptopropanesulfonic acid ($C_3H_8O_3S_2$)-modified gold electrode surface, based on LBL assembly of AuNPs, multi-walled carbon nanotubes-thionine (MWCNT-Thi) and -chitosan (Chit) $(AuNPs/MWCNT\text{-}Thi/Chit)_n$, and posterior anti-carcinoembryonic antigen (anti-CEA) immobilization *via*

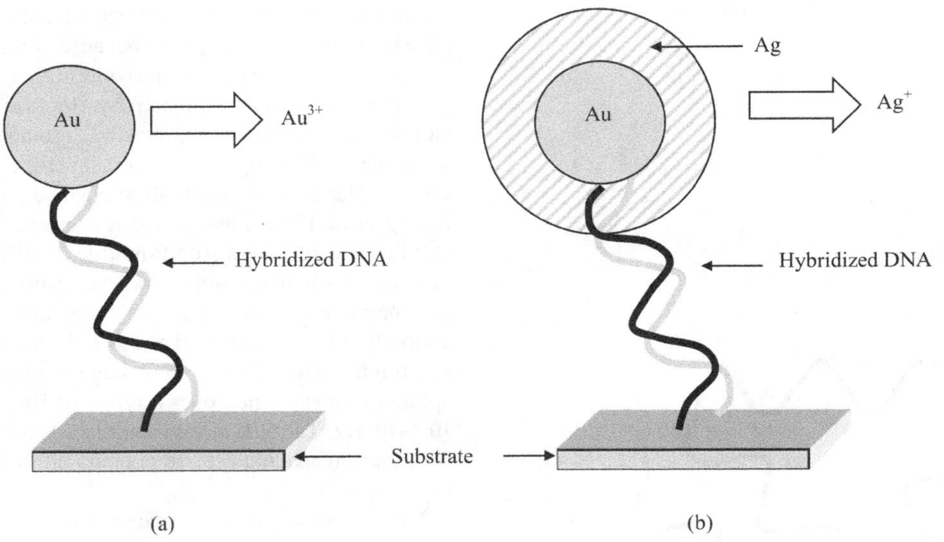

FIGURE 9.6 DNA detection by electrochemical stripping: (a) Au nanoparticles are dissolved with HBr/Br_2, with subsequent determination by anodic stripping voltammetry to relate to DNA. (b) Au nanoparticles are amplified by silver deposition followed by stripping analysis.

covalent bond; thionine has the formula Cl_2HSNaS. The detection was based on the variation of current responses before and after the immunoreaction. When the immobilized antibodies bound to the Ags, the Ag-Ab complex formed on the surface inhibited the ET. Since the film formed by LBL presents an electroactive profile, exhibiting a change in size or shape when stimulated by an electric field, the modified electrode amperometric signal was modulated by subsequent addition of target molecules. It was verified by a decrease in the amperometric signal as the concentration of Ag on the surface increased. Further examples of nanosensors for cancer detection are given in Section 11.3.

9.2.9 *Escherichia coli* Detection in Milk Samples

Lin et al. (2008) described the detection of *Escherichia coli* in milk samples. *E. coli* is a bacterium that is commonly found in the gut of humans and other warm-blooded animals, being one of the most frequent causes of many common bacterial infections, including cholecystitis, bacteremia, cholangitis, urinary tract infection (UTI), and traveler's diarrhea, and other clinical infections such as neonatal meningitis and pneumonia.

In this work, a disposable amperometric immunosensor was developed by using a screen-printed carbon electrode, modified with ferrocenedicarboxylic acid ($C_{11}H_{10}FeO_2$) and AuNPs, which lead to an increase in the amount of anti-*E. coli* immobilized on electrode surface, achieving high sensitivity. The anti-*E. coli* was immobilized *via* cross-linking by using glutaraldehyde; after adsorption of Ag, a second Ab, horseradish peroxidase (HRP)-labeled Ab (Ab conjugated to HRP) was immobilized. Section 10.10 lists several *E. coli* nanosensors.

9.3 CNT-Based Electrochemical Biosensors

In what ways do one-dimensional nanostructures aid in bioelectronic sensing? The importance of one-dimensional nanostructures, like CNTs, for bioelectronic detection stems from their electronic transport properties in addition to their high surface-to-volume ratio and the capacity to cluster a very high density of sensing elements on the footprint of a small array device. Carbon nanotubes work as efficient substrates or modifiers for promoting ET reactions. They exhibit excellent electron transfer-supporting ability, when used as electrode modifiers in electrochemical reaction electrodes. This is due to their high electronic conductivity for the electron transfer reactions. Over and above, they provide the necessary electrochemical and chemical stabilities in both aqueous and nonaqueous solutions. They also offer easy protein immobilization with retention of activity, for use in biosensors.

Where are the electroactive sites actually located in CNTs? Holloway et al. (2008) studied the voltammetry of two standard redox processes of $Fe(CN)_6^{4-}$ and $Ru(NH_3)_6^{3+}$ complex ions using three types of MWCNTs, with oxygenated edge-plane, edge-plane, or almost no edge-plane-like defects. It was found that the electrochemistry of CNTs was dominated by *the ends of the tubes*. The rate of electron transfer was faster in the case of *edge-plane defect sites*. This indicated that the electroactive sites on MWCNTs were located at the tube ends. It was also observed that aligned SWCNT-modified electrodes showed better electrochemical properties than randomly dispersed SWCNT-modified electrodes, further confirming that the electrochemistry of CNTs was dominated by the ends of the tubes. The length of the aligned CNTs also had a significant effect on the electron transfer rate. The electron transfer rate constant was inversely proportional to the mean length of the CNTs. The electron transfer rate at the CNT-modified electrode was further increased by dialyzing the CNTs after purification and shortening treatments. The adsorbed acid moieties during purification and the acid-treatment processes decreased the electrocatalytic activity of CNTs in electroanalysis.

How are CNT-modified electrodes made? CNT-biocomposite electrodes, consisting of a mixture of CNTs, binders, and biomolecules, are made by mixing carbon powder with different binders, such as mineral oil or bromoform ($CHBr_3$). *What are the*

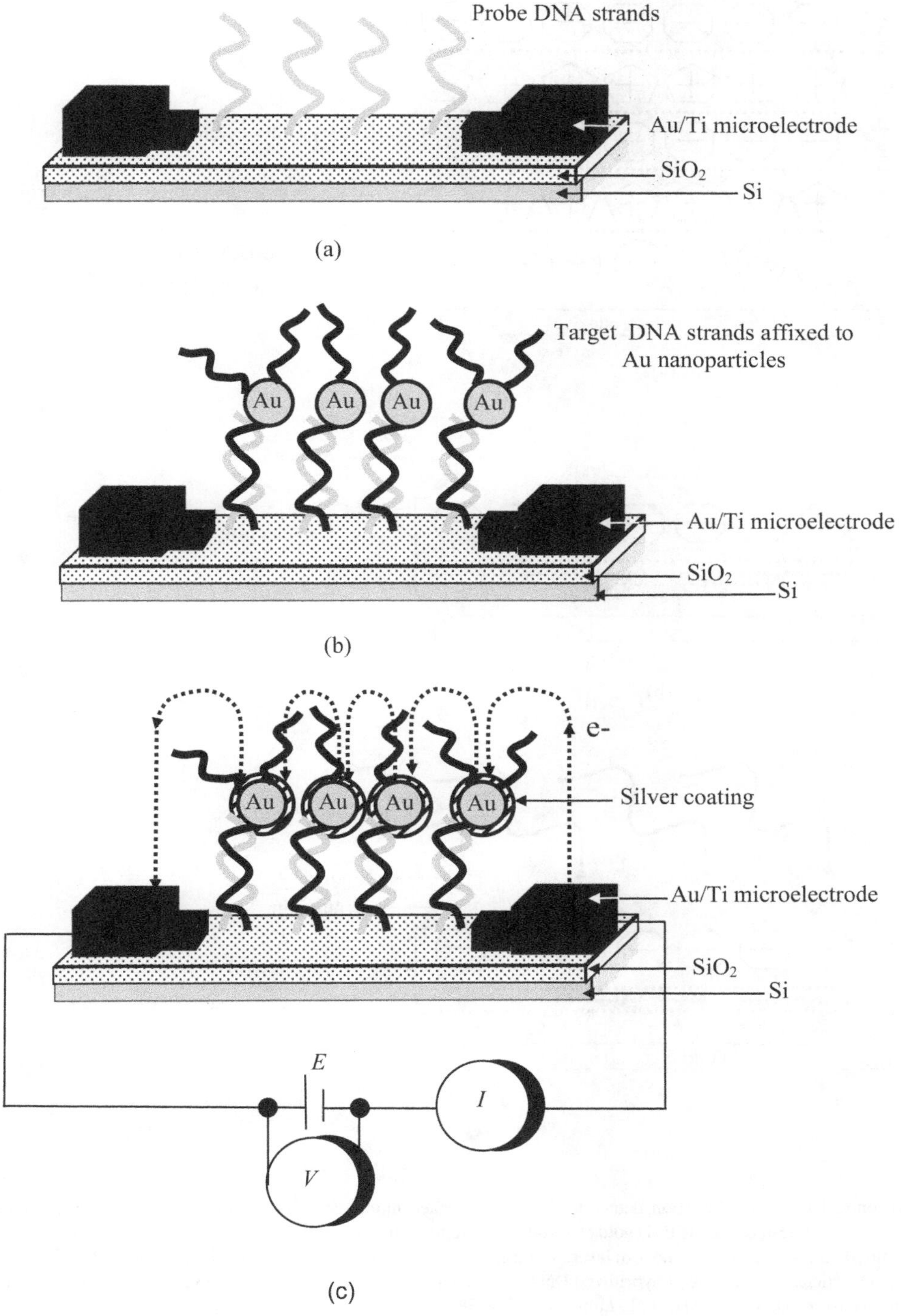

FIGURE 9.7 Conductivity DNA assay: (a) target DNA strand immobilization on activated silicon dioxide surface; (b) hybridization of Au nanoparticles tagged oligonucleotides with target DNA; (c) silver enhancement in $AgNO_3$ and hydroquinone solution, and resistance measurement across the electrode gap. (Wang, J. et al., *Langmuir*, 17, 5739, 2001.)

advantages of such electrodes? CNT-modified electrodes offer advantages such as high aspect-ratios of ~ 100–1000, nano sizes, specific catalytic activity together with sensitivity, selectivity, stability, and reliability.

How do CNTs help in communication between active sites of enzymes and the electrode? Establishing a satisfactory communication channel between the active site of the enzyme and the electrode surface is a formidable problem in amperometric enzyme electrodes. The enzyme cannot be oxidized or reduced at the electrode at any potential because it is embedded deep inside the shell. The ability of CNTs to reduce the distance between the redox site of the protein and the electrode helps in accelerating the overall rate of the reaction, the rate of electron transfer varying inversely with the exponential

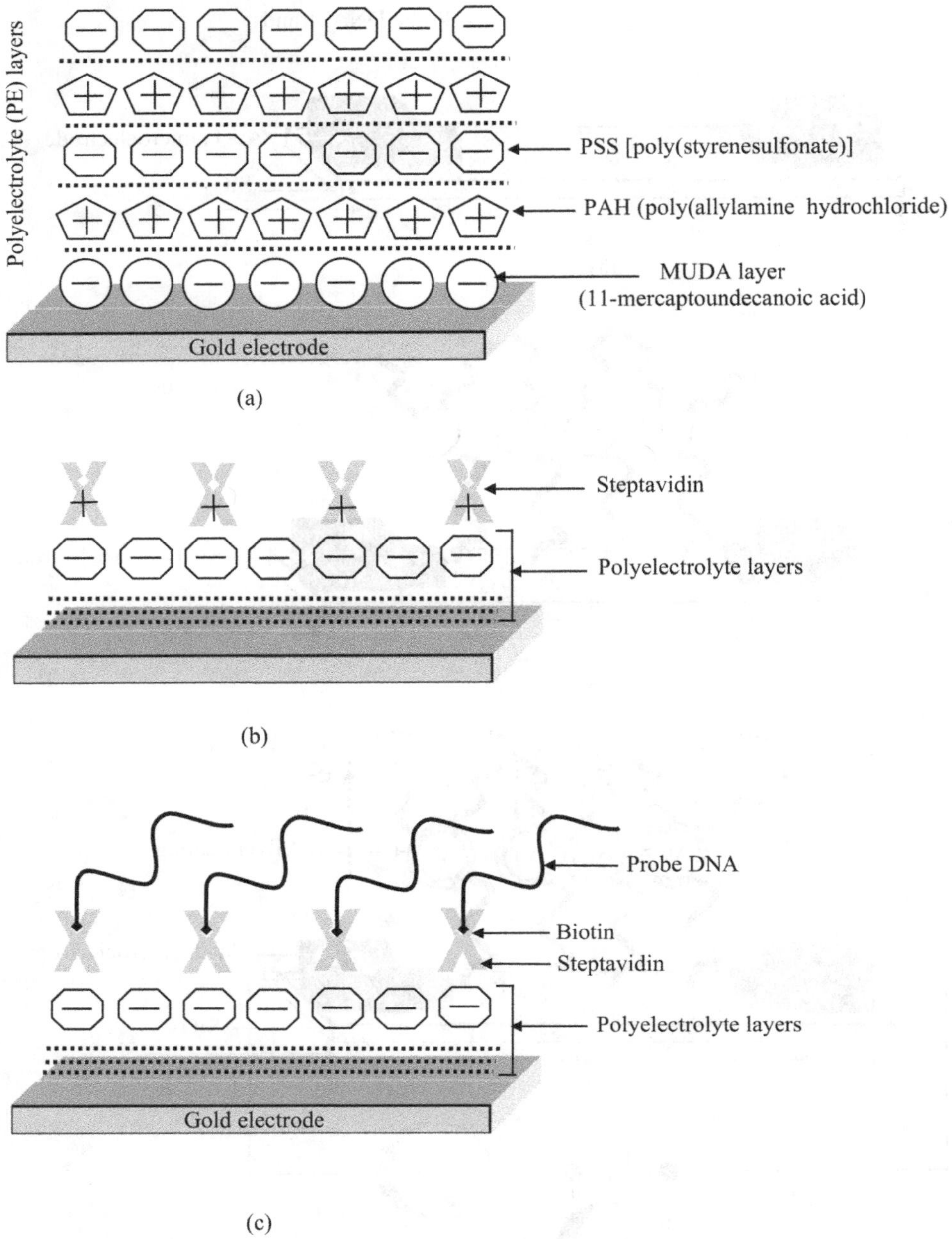

FIGURE 9.8 Electrochemical DNA hybridization detection, based on surface modification of gold electrode and silver-enhanced gold nanoparticle labeling: (a) making a polyelectrolyte-modified (PEM) gold electrode; (b) streptavidin immobilization over the last negatively charged polyelectrolyte layer; (c) probe DNA layer immobilization over the streptavidin layer; (d) target DNA and probe DNA hybridization; the target DNA is biotinylated, (e) binding of streptavidin with gold nanoparticles; (f) labeling of hybridized DNA by gold nanoparticles; (g) staining of gold nanoparticles with silver; and (h) measuring oxidative silver-dissolution current. (Lee, T. M.-H. et al., *Langmuir*, 19, 4338, 2003.)

separation between the redox center and the electrode. The small diameter and long length of CNTs allow them to be plugged into proteins with better electro-activity compared to other carbon-based electrodes. Patolsky et al. (2004) have reported the direct electrochemistry of enzymes at CNT-modified electrodes.

9.3.1 Oxidation of Dopamine

Sincere effort has been made in the development of a highly sensitive and selective method for the detection of DA, which is one of the important catecholamine ("fight or flight" family of hormones released by the adrenal glands in response to stress) neurotransmitters in the mammalian central nervous system. Abnormal concentrations of DA in body fluids influence the function of the central nervous system. *Why are conventional electrodes unable to detect DA?* Conventional electrodes are not suitable for the determination of DA due to the interference from AA and UA, which coexist in a real sample at 100 times higher concentrations than DA. These compounds can be easily oxidized at the similar potential as DA and thus always interfere with DA detection.

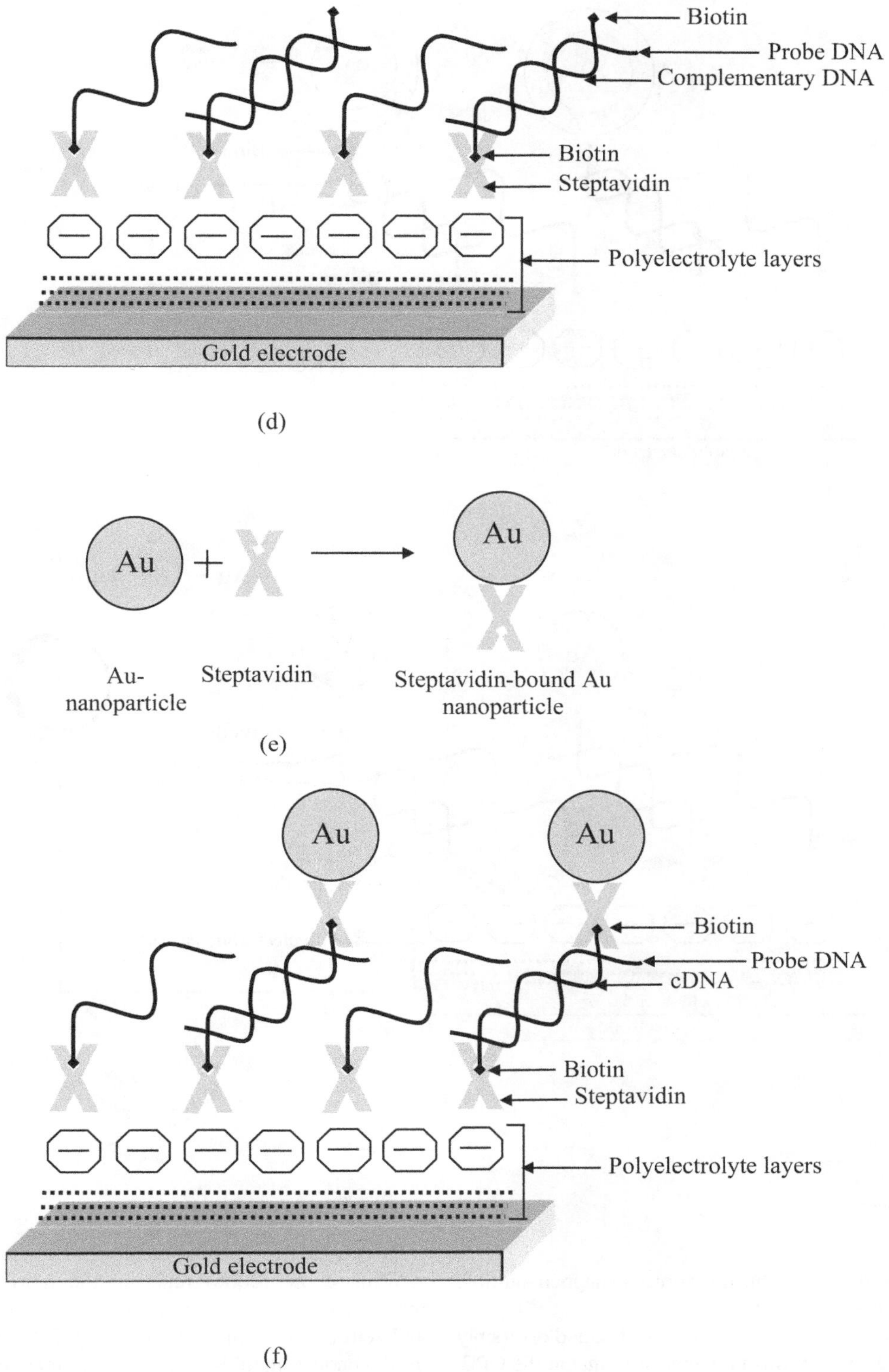

FIGURE 9.8 (Continued)

How do CNT-modified electrodes resolve this problem? CNT-modified electrodes are widely used for DA sensing. A CNT electrode was constructed by Britto et al. (1996) by mixing CNTs with bromoform and packing the paste inside a glass tube. Oxidation of DA was studied by CV with the constructed CNT-modified electrode. CV is a type of potentiodynamic electrochemical measurement, having a current peak on the forward scan and a second, inverted peak on the reverse scan, representing the opposite reaction (oxidation or reduction) to that observed on the forward scan. CV takes the experiment a step further than LSV, which ends when it reaches a set potential. When CV reaches a set potential, the potential ramp of the

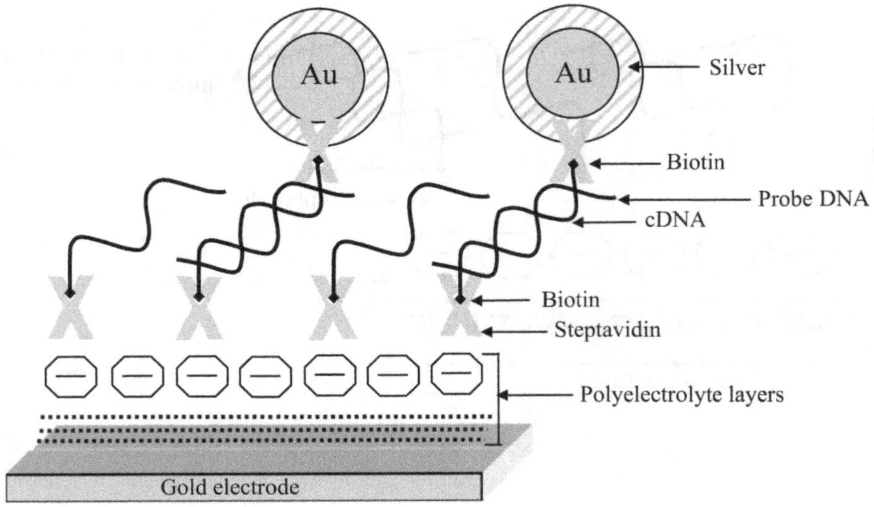

(g)

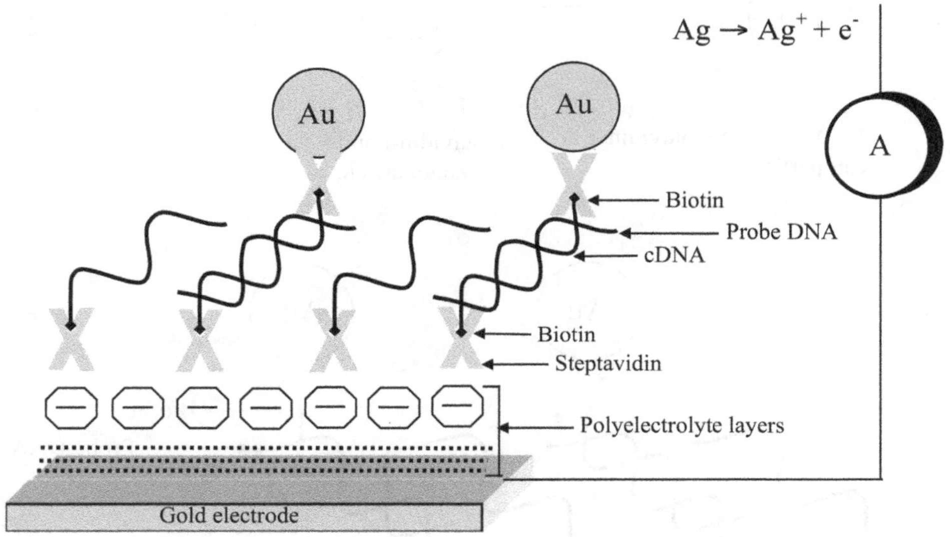

(h)

FIGURE 9.8 (Continued)

working electrode is inverted. This inversion can happen multiple times during a single experiment.

A peak potential separation of 30 mV was obtained reversibly at the CNT electrode, which was far superior to that at the CPE. Due to the catalytically active surface, background currents were larger in the former case. Voltammetric behavior of DA was unaffected by the treatment of a CNT electrode with homogenized brain tissue, suggesting that DA was oxidized mainly inside the tubes where the electrolytic product was stabilized. The increase in performance by the CNTs was attributed to the abundance of active sites on the tube surfaces and ends, their specific electronic structure, and the presence of edge-plane-like defects on the surfaces of CNTs.

Wang et al. (2006) reported the fabrication of a P3MT-modified GCE coated with Nafion/SWCNTs for highly selective and sensitive determination of DA. P3MT, C_5H_6S, is a widely used conducting polymer. The modified electrode enhanced the voltammetric signal of DA and effectively suppressed the interferences from AA and UA as well. A lower detection limit of 5.0 nM was achieved for DA. It was also successfully applied for the determination of DA in healthy human blood serum.

Goyal et al. (2008) used a fullerene-C60-coated gold electrode for the determination of DA in the presence of high concentrations of AA at physiological pH; fullerene is any molecule composed entirely of carbon, in the form of a hollow sphere, ellipsoid, or tube. Since the discovery of fullerenes in

1985, buckminsterfullerene (C60) has attracted large numbers of scientists due to its remarkable electrochemical properties. The modified electrode not only exhibited strong catalytic activity toward the oxidation of DA and AA, but also separated their voltammetric responses in the presence of each other, and thus DA was detected selectively in the presence of excess AA. Linear calibration curves for DA were obtained using square-wave voltammetry over the concentration range 1 nM–5.0 mM in 0.1 M phosphate buffer solution at pH 7.2, with a significant correlation coefficient of 0.9931 and the detection limit (3σ: an upper or lower control limit three standard deviations from the mean) was estimated to be 0.26×10^{-9} M.

The interference studies showed that the presence of physiologically common interferents (i.e., UA, citric acid, tartaric acid, glucose, and sodium chloride) negligibly affected the response of DA. On the other hand, the bare electrode was incapable of separating the responses of DA and AA. AA caused major interference in the voltammetric determination of DA. A somewhat broad oxidation peak was obtained, and the oxidation peak potentials of the analytes were indistinguishable.

9.3.2 Direct Electrochemistry or Electrocatalysis of Catalase

Wang et al. (2002a and b; 2005) and Li et al. (2003) prepared a CNT-modified electrode by casting an aliquot of a CNT suspension on the substrate surfaces. It was found that the as-prepared electrode facilitated the ET reactions of cytochrome *c* (cyt c), catalase (*Ct*), and DNA. Cyt *c* is defined as an electron transfer protein, having one or several heme *c* groups, bound to the protein. Heme is a prosthetic group (a tightly bound, specific non-polypeptide unit required for the biological function of some proteins) that consists of an iron atom contained in the center of a large heterocyclic organic ring called a *porphyrin*. *Ct* is an enzyme found in most plant and animal cells, that functions as an oxidative catalyst.

Cyclic voltammograms of *Ct* were recorded at a bare gold electrode and at a CNT-modified electrode. Within a typical potential window, no voltammetric response was observed for *Ct* at the bare gold electrode whereas a quasi-reversible redox process of *Ct* was obtained at the modified electrode, corresponding to the Fe(III)/Fe(II) redox center of the heme group of the *Ct* adsorbate. It was noted that the electron transfer rate of the *Ct* redox reaction was much faster at the SWCNT-modified electrode than at other electrodes based on carbonaceous materials (substances rich in carbon), like carbon soot and GP. Moreover, the catalytic activity of *Ct* was retained upon adsorption on the modified electrode.

9.3.3 CNT-Based Electrochemical DNA Biosensor

DNA analysis is helpful for understanding many diseases on a molecular level and promises new perspectives for medical diagnosis in the future. The DNA which carries the genetic information is a double helix molecule and the double helix is held together by two sets of forces: hydrogen bonding between complementary base pairs and base-stacking interactions.

What is the principle of electrochemical hybridization biosensors? In these biosensors, a single-stranded DNA probe is immobilized on the transducer surface. This converts the formation of a double-stranded DNA into the required electrical signal.

(i) The label-free approach works on the intrinsic redox-active properties (e.g., direct oxidation) of the DNA bases: guanine or adenine. The one-electron reduction potential vs normal hydrogen electrode (NHE) of the four DNA bases are: guanine 1.29V, adenine 1.42V , cytosine 1.6V and thymine 1.7V. So guanine is the most easily oxidized, and ease of oxidation decreases in the order of increasing potentials.

(ii) A redox-active-labeled system utilizes a redox-active label and there is a change in affinity of the redox molecule toward the probing single-stranded-DNA-modified interface, before and after exposure to the target DNA. In both the approaches, it is desirable that the current signal originating from the hybridization between the probe and the target be properly amplified.

How is this amplification done? In the label-free method, the guanine oxidation is amplified by using MWCNTs. Performing direct electrochemistry of guanine at a CNT-modified electrode yields a comparatively higher guanine oxidation current than an unmodified GCE, using CV. For the detection of DNA sequences related to the breast cancer *BRCA 1* gene, using CV analysis, the guanine oxidation current observed for MWCNT-modified electrode was 11 times that with an unmodified GCE (Wang et al. 2003a; 2003b). Kerman et al. (2004) employed sidewall and end-functionalized CNTs, providing a larger surface area for DNA immobilization and higher electron transfer efficiency to achieve further increase in guanine oxidation current. The detection limit for DNA was also lowered by this method.

9.3.4 Glucose Biosensor

What is the concept of molecular wiring? How are CNTs used for this purpose? Electron transfer in biological systems is one of the leading areas in the biochemical and biophysical sciences. In the past few years, there has been considerable interest in the direct electron transfer between redox proteins and electrode surfaces. Various immobilization strategies have been adopted to fabricate enzyme electrodes for biosensor applications. Evidently, the best strategy for successful enzyme biosensor fabrication is to devise a configuration by which electrons can directly transfer from the redox center of the enzyme to the underlying electrode. This has been accomplished in recent years using the idea of *molecular wiring*.

The resemblance in length scales between nanotubes and redox enzymes suggests the presence of interactions that may be encouraging for biosensor electrode applications. The strategy of physical adsorption or covalent immobilization of large biomolecules on the surface of immobilized carbon nanotubes may reflect a pathway through which direct electrical communication between electrodes and the active site of redox-active enzymes is achieved. The nanotubes are dispersed on the surface of a support electrode to form a haphazardly sprinkled array of high surface area, or else incorporated as a dispersion within a matrix to form a thin film.

How is molecular wiring done? Patolsky et al. (2004) reported the structural alignment of the enzyme glucose oxidase, GO_x, on electrodes by using SWCNTs as electrical connectors between the enzyme redox centers and the electrode. To achieve this, SWCNTs were purified and shortened by oxidizing in a mixture of sulfuric and nitric acids. The process yields SWCNTs with carboxyl groups at their ends and defect sites on sidewalls. The shortened SWCNTs were further purified by dialysis and filtering. The lengths of these SWCNTs were distributed over a broad range. So selection of nanotubes of given lengths was required. The differentiation and sorting of SWCNTs with respect to their lengths is known as length fractionalization. For this purpose, the SWCNT dispersion was loaded into a chromatographic column. The process gives groups of SWCNTs with narrow length distributions. Length-fractionalized SWCNTs of a particular average length, e.g., 25 nm, obtained after this process were coupled to the surface of an Au electrode on which a 2-thioethanol (C_2H_6OS)/cystamine (organic disulfide: $C_4H_{12}N_2S_2$) mixed monolayer (3:1 ratio) was assembled. As depicted in Figure 9.9, this coupling was done in the presence of the reagent 1-ethyl-3-(3-dimethylami-nopropyl)c arbodiimide hydrochloride (EDC or EDAC), a zero-length cross-linking agent used to couple carboxyl groups to primary amines. A densely packed, needle-like pattern of standing SWCNTs was obtained after 5 hours of coupling, leading to a preferred standing conformation (the spatial arrangement of atoms and chemical bonds in a molecule) of the SWCNTs onto the surface. The amino derivative of the FAD cofactor (1) was then coupled to the carboxyl groups, –COOH, at the free edges of the standing SWCNTs. After this step, apo-glucose oxidase (apo-GO_x) was reconstituted on the FAD units, linked to the extremities of the standing SWCNTs. The study demonstrated the aligned reconstitution of a redox flavoenzyme (GO_x) on the edge of carbon nanotubes that were linked to an electrode surface. The SWCNT acted as a nanoconnector that electrically contacted the active site of the enzyme and the electrode. Amperometric responses of Go_x-SWCNT electrodes were studied for different SWCNT lengths: 25 nm, 50 nm, 100 nm and 150 nm. It was found that the electrons were transported along distances greater than 150 nm, and the rate of electron transport was controlled by the length of the SWCNTs.

Lyons and Keeley (2008) studied carbon nanotube-based modified electrode biosensors. Among electrochemical biosensors based on enzyme attachment, the glucose oxidase (GO_x) assemblies are unquestionably the most deeply studied system. This enzyme has attractive characteristics, such as its well-known behavior, great stability, and robustness. Comparison of the voltammograms obtained using a bare, SWCNT-modified, and SWCNT/GO_x and SWCNT/GO_x/Nafion-modified GC working electrodes in a 50 mM phosphate buffer (pH = 7) revealed that a bare GC electrode exhibited a virtually flat and uninspired, featureless voltammetric response, whereas a pair of well-defined redox peaks were observed at the both the SWCNT/GO_x and SWCNT/GO_x/Nafion-modified electrodes. Nafion ($C_7HF_{13}O_5S{\cdot}C_2F_4$) is a commercial sulfonated tetrafluoroethylene-based fluoropolymer-copolymer.

The observed peaks were characteristic of those representing the redox behavior of an adsorbed species, that of the FAD. The summits were compellingly symmetrical and exhibited the characteristic bell shape expected for adsorbed redox species. Furthermore, it was noted that the addition of Nafion clearly increased the dispersion of the nanotube collections on the support electrode surface, which may then boost the effectiveness of the nanotube wiring to the entrenched flavin active site of the immobilized glucose oxidase, and hence result in more successful charge-transfer kinetics and potential directed turnover of the flavin group. Nafion-overcoated nanotube/GO_x composite-modified electrode was unmistakably superior to GC as an enzyme immobilization platform.

What explanations have been proposed for the improved response of modified electrodes? These are enumerated as follows:

1. Wang et al. (2003b) have reported that glucose oxidase adsorbs preferentially to edge-plane sites on nanotubes. It has been established that such sites contain a significant amount of oxygenated functionalities. These groups are created via the rupturing of carbon–carbon bonds at the nanotube ends and at defect sites, which may occur on the side walls. They render the hydrophilic (water loving: the tendency of a molecule to be dissolved by water) and ionic characters to the nanotubes, and it is believed that they are responsible for the "nesting" of the protein on the nanotube film. The nanotubes and enzyme molecules are of similar dimensions, which makes possible the adsorption of glucose oxidase without major loss of its biocatalytic shape, form, or function.
2. Lyons and Keeley (2008) suggested that nanotubes pierce the glycoprotein shell and gain access to the prosthetic group such that the electron tunneling distance is curtailed. Such access is generally not likely with conventional "smooth" electrodes, and notable unfolding of the protein shell can occur, resulting in the loss of biochemical activity. Furthermore, they proposed that the GO_x molecules are adsorbed in a pristine configuration, where the tunneling distance for electron transfer between the flavin sites and the nanotube strand is not too large, making the probability for the transition for electron transfer favorable. It is important to note that the carbon nanotube and the enzyme molecule share a similar length scale, so that the enzyme is able to adsorb onto the nanotube sidewall without losing its biologically active shape, form, and function.
3. Guiseppi-Elie et al. (2002) introduced the striking analogy of piercing a balloon with a long sharp needle, such that the balloon does not burst. Instead, by a gentle twisting action, the needle is made to enter the balloon without destruction. Similarly, it has been proposed that some of the nanoscale "dendrites" of CNT venture outward from the surface of a strand and act like bundled ultramicroelectrodes that are able to perforate the glycoprotein shell of glucose oxidase and gain access to the flavin prosthetic group, such that the electron tunneling distance is minimized and, consequently, the electron transfer probability is optimized. This degree of nanoscale electronic wiring and intimate access is not generally afforded with traditional smooth electrodes.

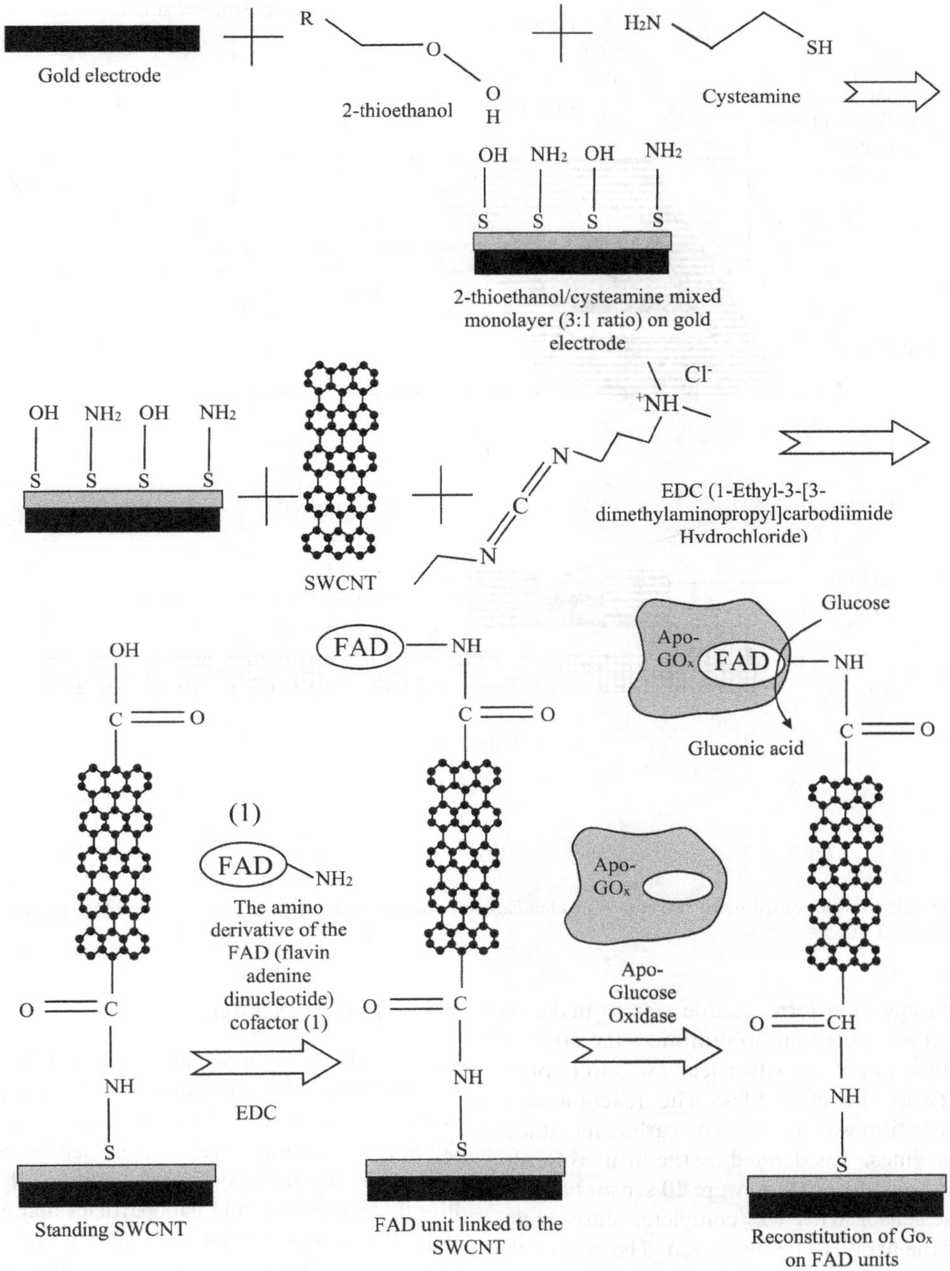

FIGURE 9.9 SWCNT electrically contacted glucose oxidase electrode. (Patolsky, F. et al., *Angew. Chem. Int. Ed.*, 43, 2113, 2004.)

9.3.5 Cholesterol Biosensor

Normal human blood serum contains less than 200 mg dL^{-1} cholesterol. A waxy, fat-like substance, that the body needs to function normally, it is naturally present in cell walls or membranes. Li et al. (2005) reported a carbon nanotube-modified biosensor for monitoring total cholesterol in blood; the schematic of the sensor is shown in Figure 9.10. They used carbon nanotubes to modify CPE, which promotes electron transfer and enhances the response current. This sensor consists of a carbon working electrode and a reference electrode screen-printed on a substrate of polycarbonate (a versatile, tough plastic: polymer containing carbonate groups (–O–(C=O)–O–), produced by the reaction of bisphenol A and phosgene). *What are the ingredients of the mixture applied on the electrodes?* Cholesterol esterase (an enzyme releasing cholesterol and a free fatty acid anion from cholesterol ester), cholesterol oxidase (an alcohol dehydrogenase/oxidase flavoprotein catalyzing the dehydrogenation of the C(3)-OH of cholesterol), peroxidase (HRP is a hydrogen-peroxide-reducing enzyme, occurring in animal and plant tissues, that catalyzes the dehydrogenation (oxidation) of various substances in the presence of hydrogen peroxide), and potassium ferrocyanide, $K_4[Fe(CN)_6]{\cdot}3H_2O$, were immobilized on the screen-printed carbon electrodes. MWCNTs were added to prompt electron transfer. The role of

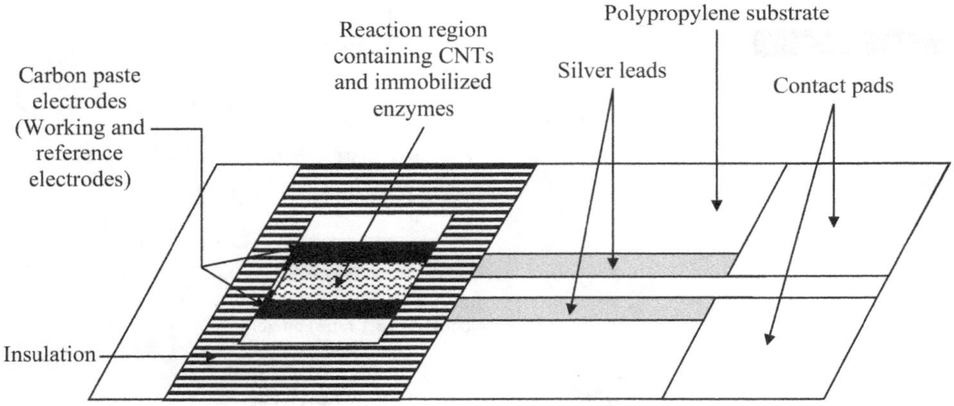

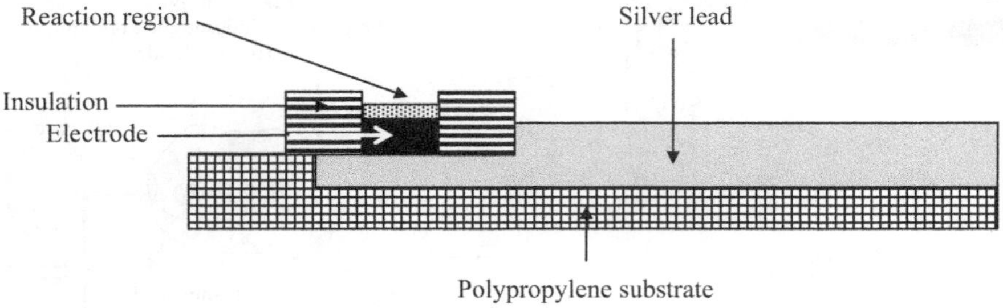

FIGURE 9.10 Cholesterol biosensor consisting of two screen-printed carbon paste electrodes: (a) the sensor strip and (b) its cross section. (Li, G. et al., *Biosens. Bioelectron.*, 20, 2140, 2005.)

the redox mediator potassium ferrocyanide was to make the reaction detectable at low potential, to minimize the effect of interferences. The function of the silver leads was to improve electric conductivity of the electrodes. The reaction area, where the carbon paste film was modified by carbon nanotubes and immobilized enzymes, was defined by the insulating film coated on the carbon paste film. There were 20 sensor bases on each plate. Once the sensor array was complete, sensor strips were excised out of the array, to be measured. The size of the reaction area was 2 mm × 2 mm while the sensor strip was 35 mm × 10 mm.

By how much do the responses of modified and unmodified electrodes differ? The responses to cholesterol were determined over the range of 100–400 mg dL^{-1}, using both the carbon nanotube-modified and unmodified sensors. An almost linear relationship between the cholesterol concentration and the response current of the carbon nanotube-modified electrodes was observed, whereas a curvilinear response was observed for unmodified electrodes. The average sensitivities were 0.0059 and 0.0032 μA mg^{-1} dL^{-1} for modified and unmodified electrodes, respectively. It can be seen that the carbon nanotubes promote the electron transfer and almost double the sensitivity. AA and UA at normal levels expected in blood had no effect on the response current. This indicates that these substances did not interfere during the total blood cholesterol measurements.

9.3.6 H_2O_2 Biosensor

Chen et al. (2007) have fabricated a H_2O_2 biosensor based on the immobilization of hemoglobin (Hb) on multi-wall carbon nanotubes and gold colloidal nanoparticles. Liu and Hu (2008) developed a similar analytical device by entrapping Hb in a composite electrodeposited CS-multi-wall carbon nanotubes film by assembling gold nanoparticles and hemoglobin, step by step. Hb is a kind of redox respiratory protein in red cells. It is the iron-containing oxygen-transport metalloprotein in the red blood cells of all vertebrates except fishes. CS is a form of a natural polysaccharide. AuNPs provide a mild microenvironment similar to that of redox proteins in native systems and give the protein molecules more freedom in orientation.

How is the hemoglobin-modified electrode prepared and what are its characteristics? The CNTs were dispersed in CS solution to obtain a homogeneous suspension. Then, the suspension was electrodeposited on to the surface of a gold electrode, and the modified electrode was transferred to AuNP solution for assembling. Finally, hemoglobin was adsorbed onto the AuNPs. The composite electrode showed excellent electrocatalysis to hydrogen peroxide and oxygen. The peak current was linearly proportional to H_2O_2 concentration in the range from 1×10^{-6} to 4.7×10^{-4} mol L^{-1} with a detection limit of 5.0×10^{-7} mol L^{-1}. This biosensor exhibited high stability, good reproducibility, and high selectivity. This is due to the synergistic action of both carbon nanotubes and AuNPs, since each of them has the ability to facilitate or promote electron transfer between the proteins and the electrode surface. The incorporation of CNTs and AuNPs in the film increased its electrocatalytic ability toward hydrogen peroxide.

9.4 Functionalization of CNTs for Biosensor Fabrication

There are two principal methods of CNT functionalization: (i) non-covalent and (ii) covalent. Non-covalent methods include encapsulation, physical and chemical adsorption, and hydrophobic interactions. Covalent immobilization of biomolecules on CNTs is implemented by oxidation of CNTs by sonicating (shearing open by high-frequency sound agitation, usually ultrasound) or refluxing (boiling a liquid in a vessel attached to a condenser so that the vapors continuously condense for re-boiling) in concentrated acid solution which results in the formation of carboxylic acid –COOH groups at the ends and sidewalls of the nanotubes. Carboxylic acid groups on the surfaces of CNTs react with amino functional groups (RNH_2 or R_2NH, or R_3N where R is a carbon-containing substituent) of biological receptors through carbodiimide (or methanediimine, a functional group of the formula RN=C=NR formed by dehydration of urea) coupling, e.g., the attachment of bovine serum albumin (BSA) protein (a serum albumin protein used as a carrier protein and as a stabilizing agent in enzymatic reactions) and glucose oxidase (GO_x) enzyme on the sidewalls of CNTs *via* amide (an organic compound that contains the functional group consisting of an acyl group (R–C=O) linked to a nitrogen atom (N)) linkages.

Most research studies on CNTs have employed the *droplet coating technique* for the preparation of CNT electrodes. Here, the CNTs are randomly aligned with respect to each other and are pointing in different directions, resulting in a tangle of nanotubes. An unknown spatial relationship exists between the biomolecules and the CNTs. But for biomolecules and CNTs to operate in harmony, attachment of biomolecules to CNTs needs cautious attention. In a CNT, there are two types of sites for immobilization of biomolecules. These attachment sites are on the tips and sidewalls of the nanotubes. Tip conjugation is favored in densely packed arrays because sidewalls provide limited accessibility. But a large surface area is available for conjugation on the sidewalls of CNTs because of their high aspect ratio. So, sidewalls cannot be ignored. In many applications, such as a CNT forming the channel of a MOSFET (metal-oxide-semiconductor field-effect transistor), sidewall conjugation is the sole possibility.

9.5 QD (Quantum Dot)-Based Electrochemical Biosensors

Electronic properties, together with biocompatibility, enable QDs to be used in electrochemical sensing.

9.5.1 Uric Acid Biosensor

Zhang et al. (2006b) performed UA ($C_5H_4N_4O_3$) biosensing with uricase-ZnS QD/L-Cyst ($HSCH_2CH_2NH_2$) assembly; uricase is a peroxisomal liver enzyme that catalyzes the oxidation of UA to allantoin during purine catabolism. Covalent binding of uricase and L-Cyst occur through the free carboxyl groups on the surface of ZnS QDs. Properties of carboxyl group-functionalized ZnS QDs, such as solubility, conductivity, and increased binding site number for increased enzyme loading resulted in improved electrocatalytic performance by the sensor.

9.5.2 Hydrogen Peroxide Biosensor

Where are H_2O_2 measurements required? The determination of hydrogen peroxide content is of great importance in many fields, including industrial environmental protection and clinical control.

Horseradish peroxidase (HRP) is one of the heme-containing redox enzymes, with a molecular weight of approximately 42,000 Da. The electron transfer converts Fe(III) in the heme of HRP to Fe(II). This process catalyzes some chemical reactions, such as the reduction of H_2O_2. Unfortunately, strong adsorption of HRP onto the electrode surface causes denaturation. On the other hand, the electrochemically active centers in HRP are always buried deeply in its extended three-dimensional structure, which makes direct electron transfer between HRP and the electrode surface very difficult.

Wang et al. (2009) immobilized HRP with lipophilic (fat-liking: the ability of a chemical compound to dissolve in fats, oils, lipids, and nonpolar solvents) CdSe/ZnS QDs on a GCE surface. It was found that HRP transfers electron directly onto the GC electrode only when the electrode was modified with QDs through vapor deposition. This may be due to the slow vapor deposition process of QDs, which results in a good combination of HRP and QDs, and HRP adjusts its molecule to an appropriate orientation for direct electron transfer. Based on the direct electrochemistry of HRP, Wang et al. (2009) fabricated an H_2O_2 biosensor. It was also found that the modified electrode could be used as a sensor for H_2O_2, and the linear range of detection was 5.0×10^{-6}–1.0×10^{-4} M, with a detection limit of 2.84×10^{-7} M. The sensor exhibited reproducibility and stability.

Compared with HRP, which is often used to construct hydrogen peroxide biosensors, Hb has many advantages, such as commercial availability, moderate cost, and its known and documented structure besides its intrinsic peroxidase activity; therefore, the use of Hb for the construction of hydrogen peroxide biosensors is possible.

Hb-on-nano cadmium sulfide (CdS)-modified electrode is used as an H_2O_2 biosensor. The forbidden bandwidth of nano CdS is 2.42 eV at room temperature, and its photovoltaic effect can be induced by visible light, generating oxidative cavities and reductive electrons. The electrons can transfer to the surface of nanoparticles rapidly because of the nanoscale effect of nano CdS, resulting in a higher charge-detaching efficiency, resulting in the enhanced electron transfer reactivity and catalytic activity. Hb is a typical heme protein. However, direct electron transfer between Hb and the electrode is not easy to achieve. So, various nanomaterials have been evaluated to obtain the satisfactory electrochemical response of the protein.

Zhou et al. (2005) immobilized Hb with CdS NPs on a pyrolytic graphite (PG) electrode. A cyclic voltammogram of a Hb/CdS NPs co-modified PG electrode in 0.1 M PBS at pH 6.0 showed a

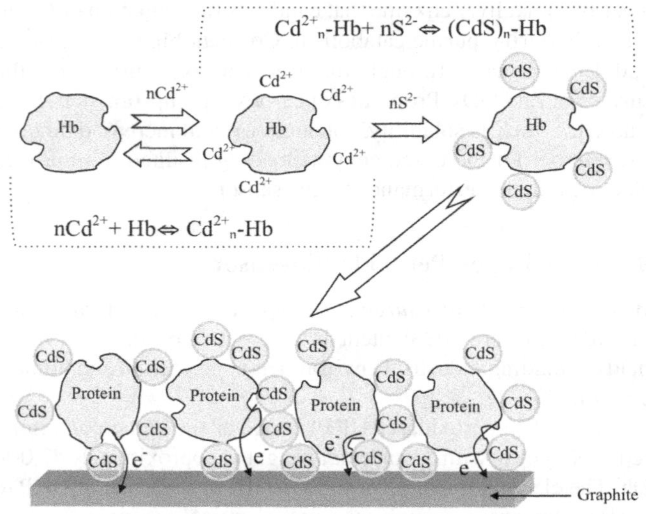

FIGURE 9.11 Process of preparing CdS-hemoglobin (Hb)/graphite electrode. (Xu, Y. et al., *J. Biol. Inorg. Chem.*, 12, 421, doi: 10.1007/s00775-006-0198-2, 2007.)

pair of pronounced redox peaks ascribed to Hb, with the anodic peak at –239 mV and the cathodic peak at –337 mV (*versus* SCE, saturated calomel electrode), respectively. The peak separation is about 98 mV. Hb cannot exhibit electrochemical response without the help of CdS NPs. If Hb alone is modified on the PG electrode surface, no wave can be observed. Even if the protein is immobilized onto the surface of CdS NPs, instead of being entrapped in the NPs, fine redox waves still cannot be obtained.

A linear relationship between the reductive peak current and the H_2O_2 concentration was obtained from 5.0×10^{-6} to 4.0×10^{-4} mol L^{-1}, on the basis of which an H_2O_2 biosensor might be developed.

Xu et al. (2007) proposed a hydrogen peroxide biosensor, based on the direct electrochemistry of hemoglobin modified with quantum dots. They described the direct electrochemical behavior of Hb modified with QDs (CdS) on a normal GP electrode. Figure 9.11 shows the modification of Hb with QDs. Because of the chemical interaction between CdS and Hb, the modified electrode is more stable. Hb modified with CdS showed direct electron transfer and the modified electrode was sensitive to hydrogen peroxide. The solution of Hb modified with CdS was deposited onto the surface of the GP electrode and dried in air. Then, Nafion was used to cover the Hb-CdS film as a binder to hold the film stably on the electrode surface. The Nafion/CdS-Hb/GP electrode can be used as a hydrogen peroxide biosensor because of its good peroxidase-like bioactivity. This biosensor showed an excellent response to the reduction of H_2O_2 without the aid of an electron mediator. The catalytic current sho36wed a linear dependence on the concentration of H_2O_2 in the range 5×10^{-7}–3×10^{-4} M, with a detection limit of 6×10^{-8} M.

9.5.3 CdS Nanoparticles Modified Electrode for Glucose Detection

Huang et al. (2005) studied the direct electrochemistry of glucose oxidase (GO_x) adsorbed on a CdS NP-modified PG electrode. PG is a unique form of GP, manufactured by decomposition of a hydrocarbon gas at a very high temperature in a vacuum furnace, resulting in an ultrapure product which is near the theoretical density and extremely anisotropic.

The enzyme showed significantly enhanced ET reactivity. GO_x adsorbed onto CdS NPs maintained its bioactivity and structure, and could electrocatalyze the reduction of dissolved oxygen, which resulted in a great increase in the reduction peak current. Upon addition of glucose, the reduction peak current decreased, which could be used for glucose detection. CdS NPs provided an ideal environment for retaining the enzyme activity as well as promoting the ET reactivity of GO_x. A mediator-free glucose biosensor was prepared. Considering its good stability and high selectivity, along with its low cost, this biosensor showed great promise for the rapid determination of glucose.

9.5.4 QD Light-Triggered Glucose Detection

Schubert et al. (2009) demonstrated the applicability of quantum dots for NADH sensing. NADH is the abbreviation for the reduced form of NAD (nicotinamide adenine dinucleotide), i.e., nicotinamide adenine dinucleotide: hydrogen reduced. It is a coenzyme that incorporates niacin and is involved in the Krebs cycle. NADH sensing is based on the immobilization of CdSe/ZnS nanocrystals on a gold electrode. The use of quantum dots on the electrode provides a photoswitchable interlayer, allowing the spatial readout of the sensor surface. Photogenerated excitons of semiconducting nanocrystals can electrically communicate with electrode surfaces and lead to an anodic or cathodic photocurrent. The current signal can be triggered by illumination of the quantum dot-modified electrode surface. Because of photoexcitation, electron–hole pairs are generated in the quantum dots. Excited conduction-band electrons of the quantum dots can be transferred to an electrode or to an electron acceptor in solution. Electrons can also be transferred from an electrode or a solubilized electron donor to valence-band holes in the quantum dots. Therefore, a quantum dot layer between the electrode and a redox system can be used for a light-triggered readout of the electron transfer reaction with the electrode.

The aforementioned light-induced carrier generation provides the basis for the combination of the CdSe/ZnS electrode with a NADH-producing enzyme reaction for the light-triggered detection of the corresponding substrate. Schubert et al. (2009) showed that glucose detection is possible with such an electrode system by photocurrent measurements. To prepare the electrodes (Figure 9.12), the quantum dots were first modified with a dithiol, 1,4-benzenedithiol (BDT), $C_6H_6S_2$, and then immobilized on the gold electrodes (Au-[QD-BDT]). A thiol is an organosulfur compound that contains a carbon-bonded sulfhydryl (-C-SH or R-SH) group (where R represents an alkane, alkene, or other carbon-containing moiety); a dithiol is a compound having two thiol groups. For immobilization of the CdSe/ZnS nanocrystals, their capping ligand, trioctylphosphine oxide (TOPO), $C_{24}H_{51}OP$, by synthesis was exchanged with BDT. TOPO is a tertiary alkylphosphine that can be used as an extraction or stabilizing agent. The use of a small dithiol provided the possibility of replacement of the original ligand TOPO, in a first step, as well as the strong coupling of the nanocrystals to the gold electrode surface *via* chemisorption in a second step (Au-[QD-BDT]). The bare gold electrode showed no photocurrent, but, after quantum

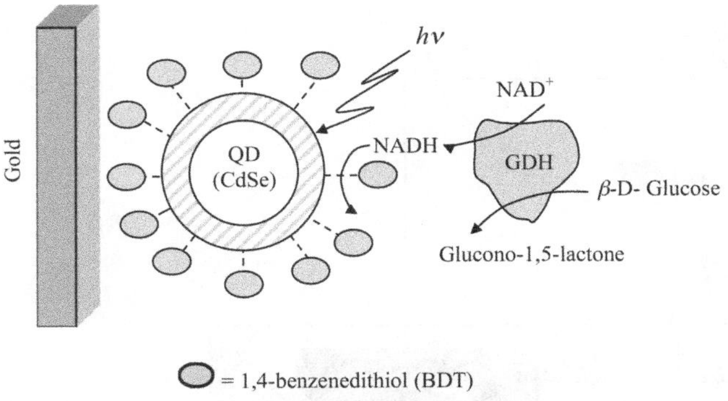

FIGURE 9.12 Glucose detection due to the catalytic production of NADH by the enzyme glucose dehydrogenase (GDH) in solution, performed at CdSe/ZnS quantum-dot-modified gold electrodes with QDs immobilized on gold *via* the ligand 1,4-benzenedithiol (Au-[QD-BDT]). (Schubert, K. et al., *Langmuir*, 26, 1395, doi: 10.1021/la902499e, 2009.)

dot immobilization, a negative photocurrent of approximately 10 nA was observed at an applied potential of +50 mV *versus* Ag/AgCl, 1 M KCl. The detection of NADH was possible in the range from 20 μM to 2 mM. The detection of glucose, using glucose dehydrogenase, was effectively demonstrated.

9.6 Nanotube and Nanowire-Based FET Nanobiosensors

Can a planar field-effect transistor (FET) be used for detecting biochemicals? Yes, its capability for this purpose has long been known (Agarwal et al. 2010). *What has limited its application?* Clearly, its lower sensitivity. Compared to the surface region of a planar device, due to the ultrahigh surface-to-volume ratio, carbon nanotubes (CNTs) and silicon nanowires (NWs) are ideal choices for sensors due to the introduced depletion/accumulation of charges near the surface, because of surface binding/adsorption of foreign molecules and species (Figure 9.13). The physical properties restraining sensor devices fabricated in planar semiconductors are readily improved by exploiting nanoscale FETs. Binding to the surface of a carbon nanotube (CNT) NW leads to depletion or accumulation of carriers in the "bulk" of the nanometer diameter structure (*versus* only the surface region of a planar device) and increases sensitivity to the extent that single-molecule detection is possible. Furthermore, the small size of CNT and NW building blocks and topical advances in assembly suggest that dense arrays of sensors could be prepared.

9.6.1 Nanotube *versus* Nanowire

CNTs could be considered an ideal material for sensing applications because: (i) every atom in an SWCNT is located on the surface, leading to extreme sensitivity to the proximal environment; (ii) surface-modified CNTs are biocompatible (nontoxic to living organisms, such as cells), thus providing the appropriate interface between biological entities and electronic circuits; and (iii) nanotubes are readily synthesized from inexpensive precursors, such as methane, or purchased from commercial sources at high purity.

Several properties of CNTs limit their development as nanosensors. Existing synthetic methods produce mixtures of metallic and semiconducting CNTs, which make systematic studies complicated, because metallic devices will not function as expected. The need to separate semiconducting nanotubes from metallic nanotubes and a nonuniform distribution of bandgaps leads to an inability to fine-tune the electronic properties. Flexible methods for the modification of CNT surfaces, which are required to prepare interfaces selective for binding a wide range of analytes, are not well established. Long-term, water-stable modification with bioreceptor molecules to develop specific and sensitive biosensors has proved to be a challenge.

NWs of semiconductors, such as Si, do not have these limitations, as they are always semiconducting, and the dopant type and concentration can be controlled, which enables the sensitivity to be tuned in the absence of an external gate. In addition, it should be possible to exploit the vast store of knowledge that exists for the chemical modification of oxide surfaces, e.g., from studies of silica and planar chemical sensors.

9.6.2 Functionalization of SiNWs

Two approaches for functionalizing the SiNWs are element-doping and surface chemical modification (Yang et al. 2006). Compared with the element-doping method, chemical modification is easily controlled and able to anchor various functional groups on the surface of SiNWs. For SiNWs, the linker molecule of choice depends on whether or not the wire has an oxide coating. Accordingly, the appropriate functionalization methods will be briefly described:

1. *Functionalization of SiNWs coated with a native oxide layer*: A range of linker molecules has been designed to bind to the native oxide coating on the SiNWs (Curreli et al. 2008). Among these, alkoxysilane derivatives (any alkoxy derivative of a silane) are the most widely used linkers. The Si-methoxide, CH_3SiO, or Si-ethoxide, C_2H_5OSi, reacts with the surface OH group, anchoring the linker molecule to the silicon oxide surface. Common linkers for Si/SiO_2NW functionalization are 3-(trimethoxy-silyl)propyl aldehyde (TMSPA) and aminopropyltrimethoxysilane (APTMS). Another popular linker molecule is 3-aminopropyl-triethoxysilane (APTES), $C_9H_{23}NO_3Si$. This reagent yields a surface coated with amino groups. These $-NH_2$ groups are activated toward bioconjugation (the process of coupling two biomolecules together in a covalent linkage) by using the appropriate coupling reagent.
2. *Functionalization of H-terminated SiNW surfaces*: The silicon oxide coatings are easily etched away by submerging the NWs in dilute HF. This replaces the native oxide layer with a hydride-terminated silicon surface. This hydride-terminated surface is found to be air-stable for several days. The Si-H bond is rapidly photodissociated with ultraviolet (UV) radiation to generate radical species on the Si surface. These radicals subsequently react with terminal olefin groups on linker molecules, thus

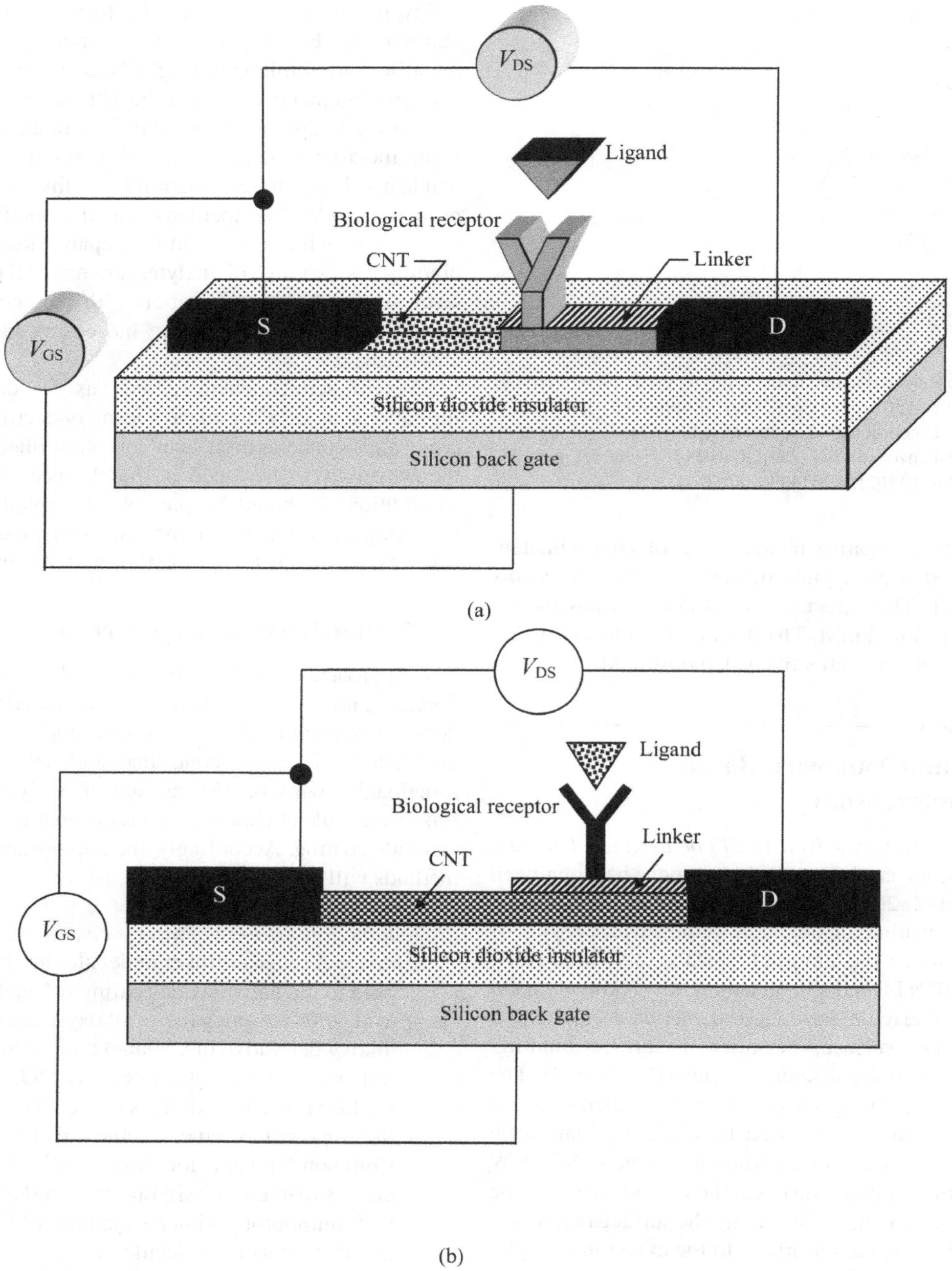

FIGURE 9.13 Biosensor based on a carbon nanotube field-effect transistor: (a) three-dimensional view and (b) cross section. The electrical current through the channel CNT is altered upon binding the ligand-receptor (because of the charge of the ligand) and is detected, by the change of drain–source current.

forming stable Si-C bonds at the Si surface. An olefin, olefine, or alkene, is an unsaturated chemical compound containing at least one carbon-to-carbon double bond. This photochemical hydrosilylation treatment selectively functionalizes the SiNWs but does not react with the underlying SiO_2 layer of the substrate. This photohydrosilylation treatment, carried out using an olefin derivative of an easily cleavable carbamate (organic compound derived from carbamic acid (NH_2COOH)), followed by deprotection, results in SiNWs coated with amino groups. The -NH_2 groups are then used to physically adsorb probe single-stranded (ssDNA) on the SiNWs, and to attach several biotin derivatives, antibodies, or probe single-stranded peptide nucleic acid (ssPNA).

Photohydrosilylation is the photochemically induced addition of a hydrosilane across the double bond of an alkene or alkyne, giving a more substituted silane. It is usually applied to cases where the hydrosilane is part of a silicon surface. The reaction uses UV irradiation and is initiated by an added photoactive compound, such as a platinum catalyst.

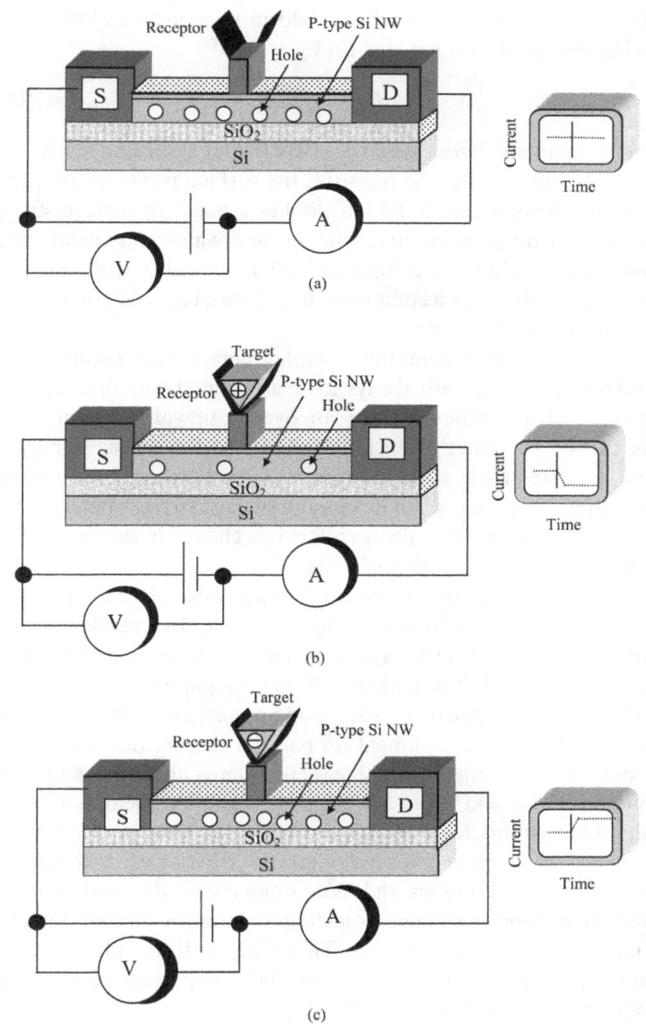

FIGURE 9.14 P-type oxidized silicon nanowire: (a) without any target molecule bound to the receptor site; (b) with a positively charged target molecule bound to the receptor site, causing hole depletion in NW and decreasing its conductance; and (c) with a negatively charged target molecule bound to the receptor site, causing hole accumulation in NW and increasing its conductance.

9.6.3 DNA and Protein Detection

If the bound analyte molecule carries a charge opposite to that on the main carriers in the FET, then charge carriers accumulate under the bound analyte, causing an increase in the device conductivity. This mechanism is shown in Figure 9.14a, where a negatively charged molecule, such as DNA, binds to the P-type NW, causing a gathering of hole carriers and resulting in an increase in conductivity. In contrast, analytes with molecular charges the same as that of the main carriers in the FET lead to depletion of main carriers beneath the bound analyte, causing a decrease in conductivity. The latter case is shown in Figure 9.14b, where a positively charged molecule, such as a protein below its isoelectric point (pI), depletes the carriers upon binding to the NWs. The isoelectric point is the pH at which a particular molecule or surface carries no net electrical charge.

A traditional approach to DNA detection utilizes a ssDNA probe (attached to the NW) that hybridizes with its complementary single strand. This strategy requires a minimum 10 mM of electrolyte concentration at room temperature to ensure robust hybridization. Operating at such high salt concentrations could significantly diminish the device sensitivity. Therefore, using a different probe/capture molecule such as the uncharged PNAs (peptide nucleic acids) offers several advantages over DNA probes. PNA-DNA hybridize at low electrolyte concentrations, conditions under which there is a long Debye length (the length at which mobile charge carriers screen out the external electric field) and the devices are expected to be particularly sensitive. It was found that immobilization of only the probe DNA produced a resistance increase of greater than 300%, whereas, for the PNA probe, the resistance of the device was unchanged. Upon hybridization with complementary DNA, a 14% conductance change was observed for the DNA probe in contrast to a 200% increase in resistance for the PNA-modified device.

Cancer-derived proteins or existing proteins present at abnormal concentration in tumors are biomarkers that are tracked to monitor the progress of cancer. Therefore, devices that quantify the level of biomarkers in serum or other human samples have potential applications in the diagnosis of cancer or other diseases.

The first example of a real-time, electrical detection of a protein from a solution using a FET nanosensor, reported by Cui et al. (2001), used the biotin-SA probe-receptor complex. The interaction of biotin and SA was monitored using P-type SiNWs, and binding was observed with a limit of detection (LOD) of 10 pM of SA, at $\lambda_D \sim 3.4$ nm. The conductance of the NW device increased rapidly in compliance with the gating effect of the negatively charged SA. Upon addition of pure buffer, the signal remained unwavering, because of the high binding affinity between biotin and SA. A system with lower binding affinity than biotin-SA, the biotin-antibiotin (a monoclonal Ab), was then preferred to prove reversible binding and regeneration of the sensing surface. As expected, the flow of fresh buffer rapidly detached the bound antibiotin and the device conductivity returned to baseline, suggestive of successful recovery of the sensor.

9.7 Cantilever-Based Nanobiosensors

Do these cantilevers have any relation to atomic force microscopy (AFM) cantilevers? Yes, microcantilevers are derived from the microfabricated cantilevers used in AFM and are based on the bending induced in the cantilever when a biomolecular interaction takes place on one of its surfaces. They translate the molecular recognition of biomolecules into nanomechanical motion (from a few nm to hundreds of nm). They are typically made of silicon/silicon nitride or polymer materials, with dimensions ranging from tens to hundreds of μm long, some tens of μm wide, and hundreds of nm thick (Carrascosa et al. 2006). Moreover, these devices can be fabricated in arrays comprising ten to thousands of microcantilevers. Hence, they are promising alternatives to current DNA and protein chips because they permit parallel, fast, real-time monitoring of thousands of analytes (e.g., proteins, pathogens, and DNA strands) without any need for labeling.

What are the capabilities of nanocantilevers? When fabricated at the nanoscale (nanocantilevers), the expected LODs are in the femtomole (fmol = 10^{-15} mol) to attomole (amol = 10^{-18} mol) range, with the astonishing possibility of detection at the single-molecule level in real time.

Microcantilevers themselves do not have any inherent chemical selectivity. *How, then, is an ordinary solid-state surface transformed into an intelligent sensor surface, recognizing and identifying complex biological systems?* The key to using microcantilevers for selective detection of molecules is the ability to functionalize one surface of the silicon microcantilever in such a way that a given target molecule will be preferentially bound to that surface upon its exposure. The sensitivity of detection is greatly improved by applying an appropriate coating to one cantilever surface. Selective biochemical sensors are prepared by coating or covalently binding a molecular recognition agent, i.e., a molecule or a polymer that has a strong and specific biomolecular interaction with a guest molecule, to the microcantilever surface (Hansen and Thundat 2005).

This strategy allows microcantilever sensors to measure extremely small changes due to molecular adsorption and, essentially for that reason, they are extremely sensitive biosensors; with the cantilever technique, it is possible to detect surface stress as small as about 10^{-4} N m^{-1}. Such measurement is also quantitative; it is related to the concentration of the analyte being detected. Nonetheless, it must be clearly borne in mind that the factors and the phenomena responsible for the surface-stress response during molecular recognition still remain ambiguous and confusing for biological adsorption, due to the complexity of the interactions involved.

What are the two main approaches for using cantilevers in biosensing?

1. By immobilizing a ligand on one side of the cantilever and placing it in contact with a receptor in solution, the cantilever bends in response to a change in surface stress generated by ligand-receptor binding; the greater the binding energy, the greater the bending. A ligand is an ion or molecule that binds to a central metal atom to form a coordination complex. The bending causes movement of the cantilever tip of the order of 1–100 nm.
2. Furthermore, the resonance frequency of a microcantilever varies sensitively with molecular adsorption. In the dynamic or resonance mode, cantilevers are excited close to their resonance frequency, which is typically in the kHz or even MHz range. When an additional mass is attached to the oscillating cantilever, its resonance frequency changes (adding a mass lowers the resonance frequency). This is not surprising since, at a first approximation, cantilevers behave like a harmonic oscillator, an ideal oscillating spring-mass system.

What is the explanation for bending and deflection of the cantilever? What are the compressive and tensile stresses? Changes in the environment around or directly on the surface of the cantilevers create a mechanical stress in the surface, which leads to an expansion or contraction of the cantilever surface. If this stress acts on only one side of the cantilever, then the asymmetrically stressed structure bends and the cantilever is deflected. For simple isotropic materials, the surface stress σ relates the reversible work dW of deforming a surface (elastically) to the change in surface area dA by

$$dW = \sigma \times dA \tag{9.1}$$

For spontaneous processes, i.e., $dW < 0$, and assuming a positive stress ($r > 0$), dA must be negative, the surface wants to contract, and the stress is said to be tensile. For a negative surface stress ($r < 0$), dA must be positive, the surface wants to expand, and the stress is said to be compressive. In a nutshell, expansion of a surface is defined as a compressive surface stress and contraction as a tensile surface stress.

Two general comments must be made here. Firstly, compressive and tensile stresses are always to be associated with one specific surface of a cantilever since an overall upward bending can be caused by either a tensile stress at the top surface or by a compressive stress at the bottom surface. Furthermore, it must be emphasized that, when discussing surface stresses detected by cantilever sensors, one always refers to a change in surface stress, but not to an absolute stress.

Why do scientists recommend using cantilever arrays for measurements instead of a single cantilever? Independent of the different modes, it is always advantageous to use several cantilevers in parallel. It is evident that one can get more information out of a single experiment when using more sensors in parallel. A more sophisticated argument for parallel sensing lies in the way biologists deal with the fact that their systems under investigation (biomolecules and cells) are not as general and ideal as typical physical systems (e.g., atoms and crystals). *How do biological systems differ from physical systems?* Biological systems are normally more complex and more unique than physical systems and their properties depend much more strongly on their history and their actual environment. Therefore, well thought-out control experiments constitute an especially important part of any experiment in the life sciences.

Cantilevers are temperature-sensitive sensors and might also respond to changes in buffer composition. In addition, several other molecules in the sample, and not only the target molecules, might interact with the sensor. One therefore needs a strategy to subtract these unwanted background signals from the "real" signal. In particular, first-time users want to obtain quick results from their samples, but are often not aware of the fact that signals from the environment in which the target molecules are presented can be much stronger than signals from the target molecules themselves. Therefore, researchers use arrays of cantilevers with several sensors in parallel: all cantilevers are physically identical and only differ in their surface coating, as shown for two cantilevers in Figure 9.15 to illustrate the concept. Their physical uniformity can be checked before an experiment by either applying a well-defined heat pulse and recording the thermal responses of the cantilevers or by measuring their individual resonance frequencies, which should be identical to within at least about 5%. Some cantilevers are then made sensitive for the target molecules whereas others act as a reference for nonspecific binding or for physical signals, such as temperature, refractive index changes, or different buffers. Experiments with complex biological samples are nowadays performed with cantilever arrays of up to eight cantilevers in parallel.

In contrast to competing technologies, such as surface plasmon resonance (SPR), which measures changes in dielectric constant,

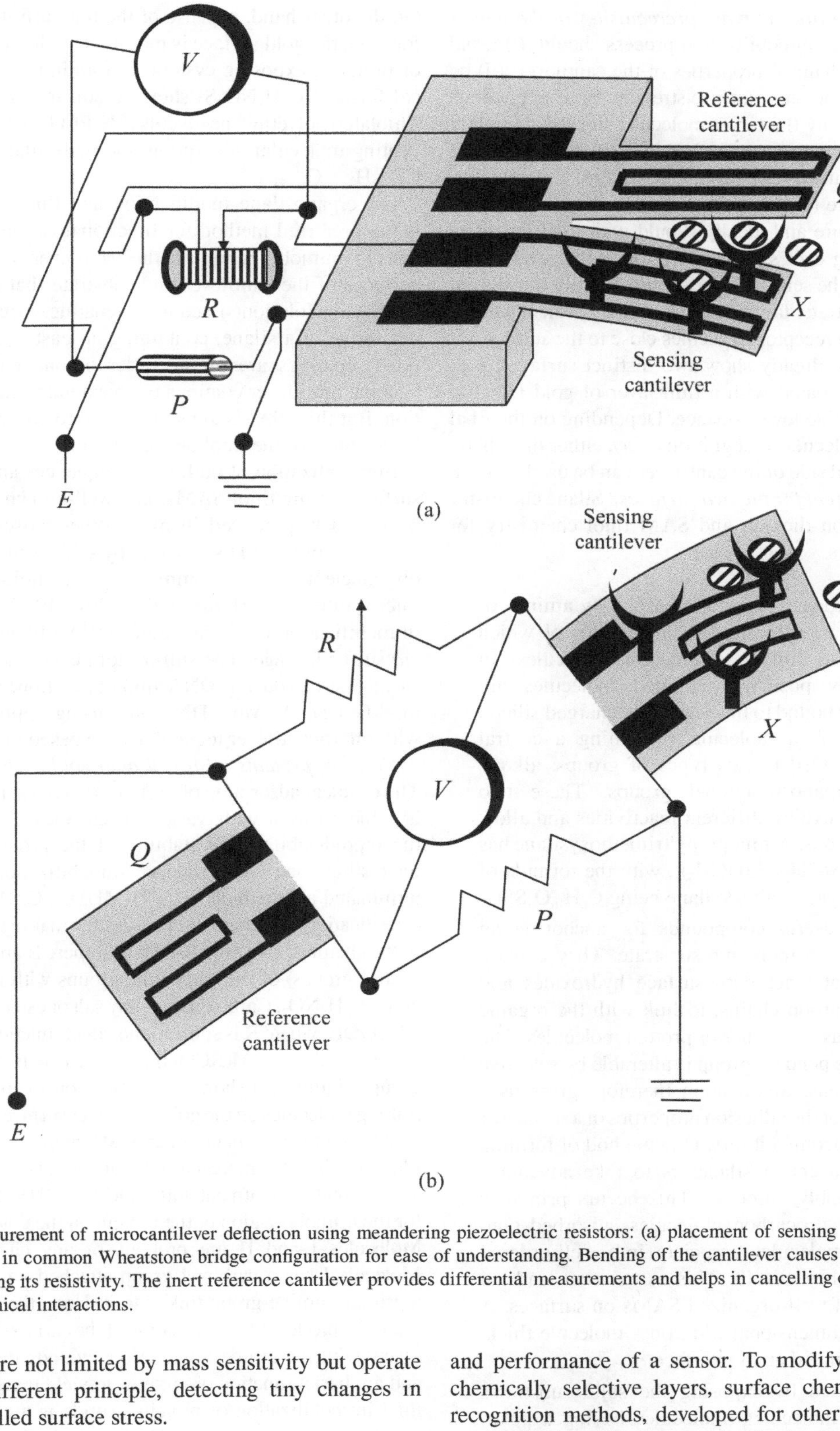

FIGURE 9.15 Measurement of microcantilever deflection using meandering piezoelectric resistors: (a) placement of sensing and reference cantilevers and (b) circuit drawn in common Wheatstone bridge configuration for ease of understanding. Bending of the cantilever causes elongation of the resistor on its surface, changing its resistivity. The inert reference cantilever provides differential measurements and helps in cancelling out the parasitic effects of temperature and chemical interactions.

cantilever arrays are not limited by mass sensitivity but operate *via* an entirely different principle, detecting tiny changes in in-plane forces, called surface stress.

9.7.1 Biofunctionalization of the Microcantilever Surface

Coating or functionalizing a sensor is a critical preparation step, because the recognition layer defines the application and performance of a sensor. To modify the cantilevers with chemically selective layers, surface chemistry and molecular recognition methods, developed for other sensors, are utilized, especially those for quartz crystal microbalance (QCM), surface acoustic wave devices (SAW), chemical field-effect transistors (chemFETs), and surface-enhanced Raman spectroscopy (SERS) sensors.

Many approaches can be used to immobilize the molecular recognition agents for the microcantilever sensor, depending upon the final application. *What are the main prerequisites of the immobilization process?* The immobilization process should: (i) avoid any change in the mechanical properties of the cantilever; (ii) be uniform, in order to generate a surface stress as large as possible; (iii) allow accessibility by the target molecule; (iv) should tightly attach biomolecules to the sensor surface but still be flexible and functional as in their natural environment; and (v) generate surface coatings which are reliable, robust against changes in buffer solution and temperature and ideally should withstand repetitive detection and cleaning cycles. Especially for cantilever sensors, interactions on top of the sensing layer should be fully transferred to the underlying substrate, favoring a dense and covalent surface functionalization, with receptor molecules close to the surface.

Usually, cantilevers already show two distinct surfaces, e.g., a gold upper surface coated with a thin layer of gold (20–100 nm) and a silicon dioxide lower surface. Depending on the final application for the molecular recognition assay, either the silicon dioxide side or the gold side of the cantilever can be used. *Are the coating methods different for the two surfaces?* Silane chemistry is employed for silicon dioxide and SAM thiol chemistry for gold-coated cantilevers.

1. The silicon dioxide surface is modified by amino- or mercaptosilane monolayers, the end groups of which are further cross linked to receptor molecules. In addition, highly positively charged molecules are electrostatically bound to the negatively charged silicon dioxide. Silane is a molecule containing a central silicon atom bonded to two types of groups: alkoxy groups and organo-functional groups. These two types of groups exhibit different reactivities and allow sequential reactions. Aminopropyltrimethoxysilane has the formula $H_2N(CH_2)_3Si(OCH_3)_3$, with the formula of 3-mercaptopropyltrimethoxysilane being $C_6H_{16}O_3SSi$.

   Silanes are useful compounds for anchoring an organic layer to an inorganic substrate. They contain Si-O bonds, that react with surface hydroxides and pendant hydrocarbon chains, to link with the organic overlayer, such as a polymer or protein molecules. The chemistry of the pendant group is alterable by selective chemical reactions. Silanization therefore gives us a technique to tailor the adhesion properties of a surface or to change its biocompatibility. One method of forming thin, uniform layers of silanes is to take advantage of a "self assembly" process. This occurs primarily in long-chain hydrocarbon molecules, adsorbed onto surfaces. The molecules associate into well-ordered domains after they adsorb. This association leads to the formation of well-organized SAMs on surfaces. A SAM is a two-dimensional film, one- molecule thick, covalently assembled at an interface.
2. SAMs are formed by exposing the solid substrates to amphiphilic molecules (molecules having a polar, water-soluble group attached to a nonpolar, water-insoluble hydrocarbon chain), with chemical groups having a strong affinity for the substrate. Alkanethiols absorb spontaneously from solutions on the Au surface. Gold is the most frequently used metal because it does not form a stable oxide layer under ambient conditions. On the other hand, because of the high affinity of sulfur groups for gold, the gold surface is modified by thiolabeled nucleic acids or proteins, exposing cysteines (α-amino acids with the chemical formula $C_3H_7NO_2S$) showing sulfur atoms at their surface. Thiolated polyethylene glycols (SH-PEG), act as inert layers, preventing molecular adsorption; the molecular formula of PEG is $C_{2n+2}H_{4n+6}O_{n+2}$.

For organosilane modification and thiol-SAMs, *dip coating* is the preferred method for functionalization to allow for high-density immobilization on the cantilever surface. All reactive surfaces of the cantilever and substrate that are exposed to the modifying solution(s) acquire a coating. Organosilane (organic derivative of a silane, containing at least one carbon-to-silicon bond) coatings are of the order of one monolayer thick but become multilayer coatings upon extended exposure to the solution. But thiol SAMs are self-limited to coverages of one monolayer or less of the thiol on a gold film.

Immobilization of both DNA sequences and proteins on gold surfaces, using thiol-SAMs, is a well-known chemistry and has been widely practiced in many other biosensing applications. Direct coupling of DNA probes by self-assembly of thiol-labeled oligonucleotides is a common, easy technique for gold-coated microcantilevers. Herne and Tarlov (1997) characterized the immobilization of single-strand DNA oligonucleotides on gold *via* sulfur linkage. The sulfur atom causes a marked change in surface stress during DNA immobilization, in contrast to non-modified DNA. Most DNA-biosensing applications performed with microcantilever technology are based on this strategy.

How are proteins adsorbed onto gold surfaces of cantilevers? The covalent adsorption of proteins on gold surfaces of cantilevers is achieved by a wide variety of chemical procedures, ensuring the reproducibility and stability of the protein coating. A very interesting one is covalent immobilization of carboxylate-terminated alkanethiols (e.g., 11-MUDA, $C_{11}H_{22}O_2S$) followed by esterification (the chemical process for making esters, compounds of the chemical structure R-COOR′, where R and R′ are either alkyl or aryl groups) of the carboxylic groups with EDC, $C_8H_{17}N_3$, and NHS, $C_4H_5NO_3$. Carboxylate is any salt or ester of a carboxylic acid (R-COOH, where R is some monovalent functional group) having a formula of the type $M(RCOO)_x$, where M is a metal and R an organic group. Figure 9.16 shows the deposition of mercapto-derivatized linking molecules on the gold cantilever surface; "mercapto" means containing the univalent radical –SH: *mercaptopurine.* Figure 9.17 illustrates how the mercapto-derivatized linking molecule forms an amide bond on treatment with EDC and NHS. The reaction is used for the immobilization of the protein on the gold-coated cantilever. Molecules I and II are proteins or any chemical compounds. Molecule I has a carboxyl group attached while molecule II has a primary amine group linked to it. Here, the intermediate active ester (the product of condensation of the carboxylic group and NHS) reacts with the amine (organic compounds that contain nitrogen and are basic) function of proteins to yield the amide bond and the final immobilization of protein. Amide is a chemical compound formed from an organic acid by the substitution of an amino (NH_2, NHR, or NR_2) group for the hydroxyl of a carboxyl (C**OOH**) group. An amide bond is a chemical bond formed between two molecules when the carboxyl group of one molecule reacts with the amine group of the other molecule, thereby releasing a molecule of water (H_2O).

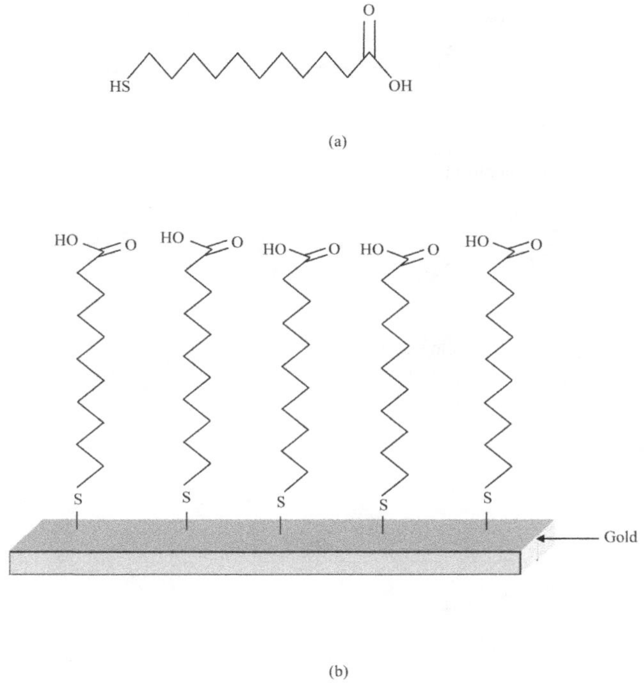

FIGURE 9.16 (a) 11-Mercaptoundecanoic acid, (b) SAM of mercaptoundecanoic acid on the gold surface.

Another option is to use cystamine modified with glutaraldehyde with the subsequent attachment of the protein through an amine group; cystamine is an organic disulfide. The immobilization of different bioreceptors on each cantilever of an array is a complicated task. There are many commercial platforms devoted to the specific functionalization of individual cantilevers.

Several approaches exist for exposing individual cantilevers of a cantilever array to specific biomolecules. The microfabricated cantilevers are incubated in individual glass capillaries or standard pipette tips filled with coating molecules. They are dipped in the channels of an open microfluidic network, or they are spotted with microliter drops of receptor molecules using an inkjet type of device. Special care is taken to provide a homogeneous coating at the hinge region, where the flexible cantilever beam is connected to the bulk silicon. Bending in this area influences the deflection of the free end of the cantilever much more strongly than bending somewhere in the middle or close to the end of the cantilever.

Biotin-streptavidin interactions are well-studied binding partners, that interact with very high affinity (Figure 9.18). Shu et al. (2007) investigated biotin-streptavidin binding interactions using microcantilever sensors. Three structurally different biotin-modified cantilever surfaces were produced as shown in Figure 9.19. The cantilever response to the binding of SA on these biotin-sensing monolayers was compared. The mechanical response of the cantilever strongly depends upon the nature of the biotin-modified surfaces: (i) biotin/PEG-coated microcantilevers did not bend upon the injection of SA; (ii) biotin-HPDP, (*N*-(6-(biotinamido)hexyl)-3'-(2'-pyridyldithio)-propionamide (a reversible biotinylation reagent)-coated microcantilevers bent downward; and (iii) biotin-SS-NHS (succinimidyl 2-(biotinamido)-ethyl-1,3′-dithiopropionate)-coated microcantilevers bent upward. Biotin-SS-NHS enables simple and efficient biotinylation of antibodies, proteins, and other primary amine-containing molecules.

9.7.2 Biosensing Applications

The first applications of cantilever sensors for biological systems were reported in 1996 with single cantilevers (Butt 1996). The first biosensing experiments with cantilever arrays were demonstrated in 2000, showing the proof of principle for DNA detection and the ability to identify single-base mismatches between sensing and target DNA oligonucleotides. This excellent example in the field of genomics involved the detection of a single-base mismatch with an LOD of 10 nM, reported by Fritz et al. (2000). For such detection, they used an array of two cantilevers with a control (noncomplementary) oligonucleotide in one of them and the DNA probe (complementary) immobilized in the other, giving hybridization-deflection signals as small as 10 and 16 nm for 12-mer and 16-mer DNA targets, respectively, and displaying a deflection noise of 0.5 nm. The length of the oligonucleotide is usually denoted by "mer," e.g., a fragment of 12 bases would be called a 12-oligomer nucleotide.

More detailed experiments include showing the dependence of cantilever bending on the length, grafting density, and orientation of DNA oligonucleotides on surfaces, and demonstrating the ability to detect different sequences in parallel and within a high background of nonspecific sequences.

Wee et al. (2005) have reported the detection of PSA (prostate-specific Ag: a protein manufactured exclusively by the prostate gland) and c-reactive proteins (CRP: one of the plasma proteins known as acute-phase proteins, produced by the liver and found in the blood), which is a specific marker of cardiac disease, by means of an electromechanical biosensor using self sensing piezoresistive microcantilevers.

In addition, a novel development for the detection of early osteosarcoma (malignant bone tumors common in children and young adults) has been described, sensing the interactions between antibodies against vimentin (any of a group of polypeptides that polymerize to form filaments in the cytoskeleton) and antigens with a single cantilever-based biosensor (Milburn et al. 2005), and supporting the idea that cantilever biosensors can provide a suitable platform for life sciences research.

The most elaborate experiments with nucleic acids report the detection of non-amplified RNA (ribonucleic acid) in total RNA from a cell, with a detection limit of 10 pM (Zhang et al. 2006a). These experiments were conducted within a high background of non-specific molecules and, for the lowest concentration, showed a differential deflection of 10 nm, or a surface stress of 1 mN m^{-1} (millinewtons per meter).

Example 9.1

The length (l) of a silicon cantilever is 20 μm and its thickness (t) is 300 nm. Upon antigen-antibody coupling, its deflection (Δz) was 7 nm. Find its radius of curvature (R), and also the surface stress (σ) developed in the cantilever. Young's modulus of silicon $E = 1.69 \times 10^{11}$ N m^{-2}, and Poisson's ratio (μ) = 0.25.

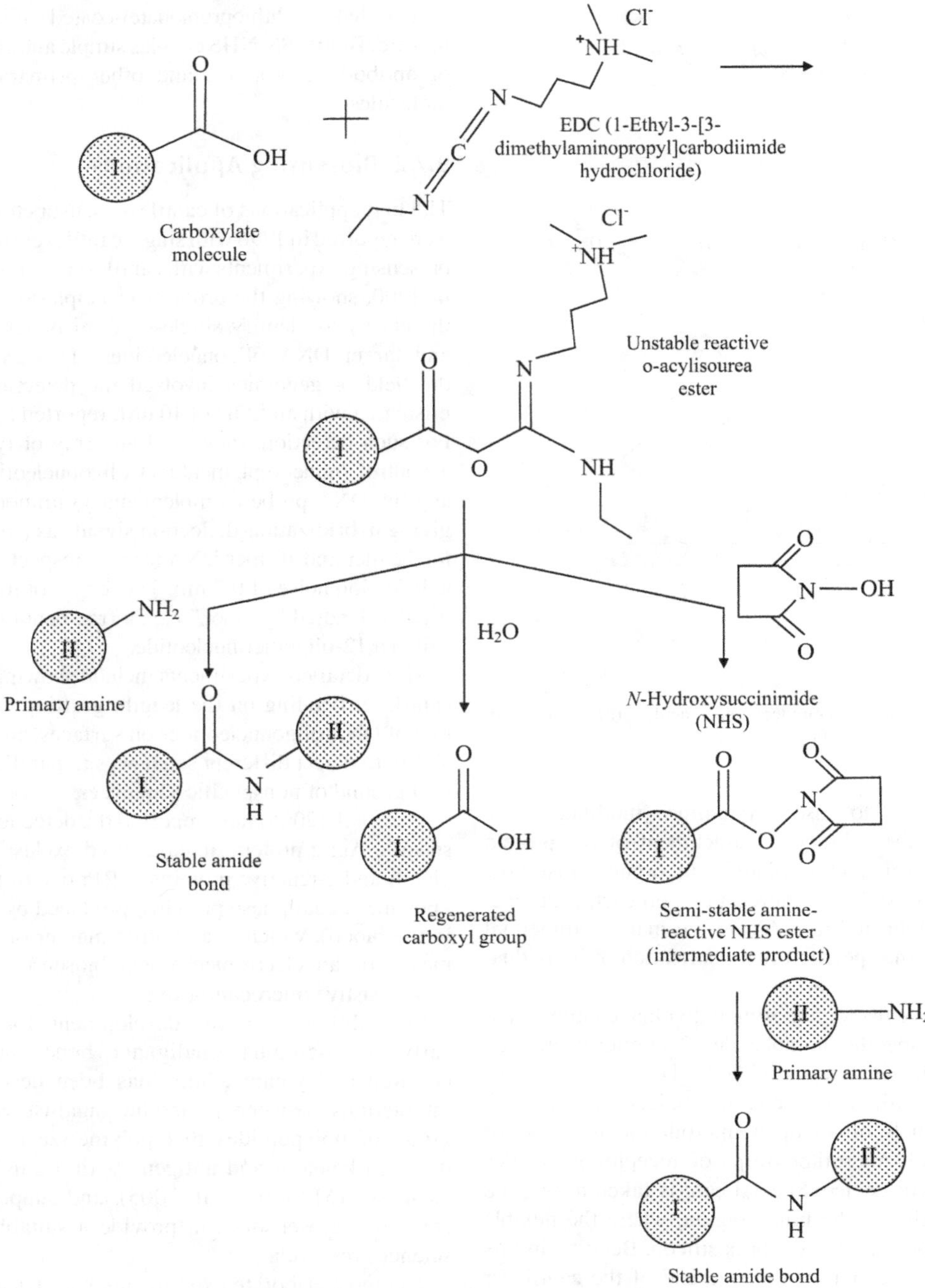

FIGURE 9.17 Esterification of the carboxylic groups with 1-ethyl-3-(3-dimethylaminopropyl)-carbodiimide (EDC) and *N*-hydroxysuccinimide (NHS). EDC is a water-soluble derivative of carbodiimide that catalyzes the formation of amide bonds between carboxylic acids or phosphates and amines by activating carboxyl or phosphate to form an *O*-urea derivative. This derivative reacts readily with nucleophiles. NHS is often used to assist carbodiimide coupling in the presence of EDC. The reaction includes the formation of the intermediate active ester (the product of condensation of the carboxylic group and NHS).

The radius of curvature is calculated from the cantilever length l and the sensor deflection Δz from the formula

$$R^{-1} \approx \frac{3\Delta z}{2l^2} = \frac{3\times 7\times 10^{-9}}{2\left(20\times 10^{-6}\right)^2} = 26.25 \text{ m}^{-1} \quad (9.2)$$

Hence $R = 1/26.25 = 3.81\times 10^{-2}$ m. The bending is related to the change of surface stress σ according to Stoney's formula, given in Equation 4.42

$$\sigma = \frac{Et^2}{6R(1-\mu)} = \frac{1.69\times 10^{11}\times\left(300\times 10^{-9}\right)^2}{6\times 3.81\times 10^{-2}\times(1-0.25)} \quad (9.3)$$

$$= 8.871\times 10^{-2} \text{ N m}$$

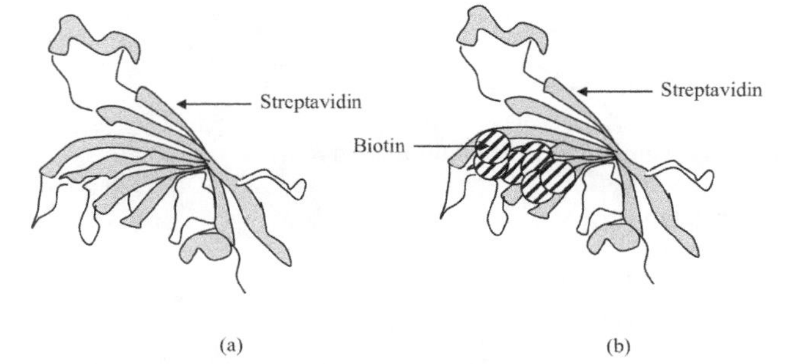

FIGURE 9.18 (a) Structure of streptavidin and (b) biotin molecules represented by spheres, attached to streptavidin.

Example 9.2

(a) A rectangular cantilever of silicon has dimensions 40 μm (length)×5 μm (width)×0.7 μm (thickness). In the dynamic mode, the cantilever was excited to its resonance frequency f_0. Determine this resonance frequency if the spring constant of the cantilever is 0.07 N m^{-1}, and the density of silicon is 2329 kg m^{-3}.

(b) Adsorption of biomolecules increased the mass of the cantilever by 100 pg and decreased the resonance frequency to f_1 by an amount $\Delta f = f_0 - f_1$ Find f_1 and Δf.

The resonance frequency f_0 of an oscillating cantilever is given by

$$f_0 = \left(\frac{1}{2\pi}\right)\sqrt{\frac{1}{m_{\text{eff}}}} \tag{9.4}$$

where
k is the spring constant
m_{eff} is the effective mass of the cantilever

The effective mass is used here for the following reason. All parts of the cantilever do not oscillate with the same amplitude. The largest deflection takes place near the free end with a decay to zero at the clamped end. Therefore, in the analytical model development, the cantilever is represented by a lumped mass model. The lumped mass m_{eff} is smaller than the real mass m, by a factor which depends on the geometry of the cantilever. A good rule of thumb says that the effective mass is one-third of the real mass. Gluing tips or mirrors on cantilevers adds their mass to the effective mass. In general, the effective mass is expressed as

$$m_{\text{eff}} = nm_b \tag{9.5}$$

where
n is the factor=0.24 for a rectangular cantilever
m_b is the mass of the cantilever beam given by

$$m_b = l \times w \times t \times \rho. \tag{9.6}$$

where
l, w, t are respectively the length, width, and thickness of the cantilever
ρ is the density of its material

Hence,

$$m_b = l \times w \times t \times \rho = 40\times10^{-6}\times5\times10^{-6}\times0.7\times10^{-6}\times2329$$
$$= 3.261\times10^{-13}\text{ kg} \tag{9.7}$$

from which $m_{\text{eff}} = 0.24 \times 3.261 \times 10^{-13}$ kg $= 7.826 \times 10^{-14}$ kg, and the resonance frequency is

$$f_0 = \left(\frac{1}{2\pi}\right)\sqrt{\frac{k}{m_{\text{eff}}}} = \frac{1}{2\times3.14}\times\sqrt{\frac{0.07}{7.826\times10^{-14}}}$$
$$= 1.506\times10^5\text{ Hz} \tag{9.8}$$

When adsorbates are uniformly deposited on the cantilever surface, the resultant mass change is

$$\Delta m = \frac{k}{4\pi^2 n}\left(\frac{1}{f_1^2} - \frac{1}{f_0^2}\right) \tag{9.9}$$

where f_1 and f_0 are the resonance frequencies after and before loading. Equation 9.9 yields

$$100\times10^{-12}\times10^{-3} = \frac{0.07}{4\times(3.14)^2\times0.24} \times\left(\frac{1}{f_1^2} - \frac{1}{(1.506\times10^5)^2}\right) \tag{9.10}$$

or

$$1\times10^{-13} = 7.395\times10^{-3}\times\left(\frac{1}{f_1^2} - 4.4091\times10^{-11}\right) \tag{9.11}$$

or

$$1\times10^{-13} = \frac{7.395\times10^{-3}}{f_1^2} - 3.261\times10^{-13} \tag{9.12}$$

or $f_1^2 = (7.395\times10^{-3})/(4.261\times10^{-13}) = 1.73551\times10^{10}$. Therefore $f_1 = 1.31739\times10^5$ Hz, and the frequency decreased by $(1.506-1.31739)\times10^5$ Hz $= 1.8861\times10^4$ Hz.

9.8 Optical Nanobiosensors

As these have already been described in the chapter on optical nanosensors (Sections 6.1, 6.3, 6.4, 6.5.3, and 6.11.1–6.11.4), only few additional examples will be presented here to supplement what has been discussed before.

9.8.1 Aptamers

What are aptamers? Aptamers (Apts) are single-stranded RNA or DNA sequences with ligand-binding ability, which have been developed to detect a variety of molecular targets including small molecules, nucleic acids, proteins, and cancer cells. *How are aptamers created?* They are created *in vitro* through systematic evolution of ligands by exponential enrichment for the recognition of target analytes with high affinity and specificity. *What is the relevance of aptamers for biosensors?* They have become

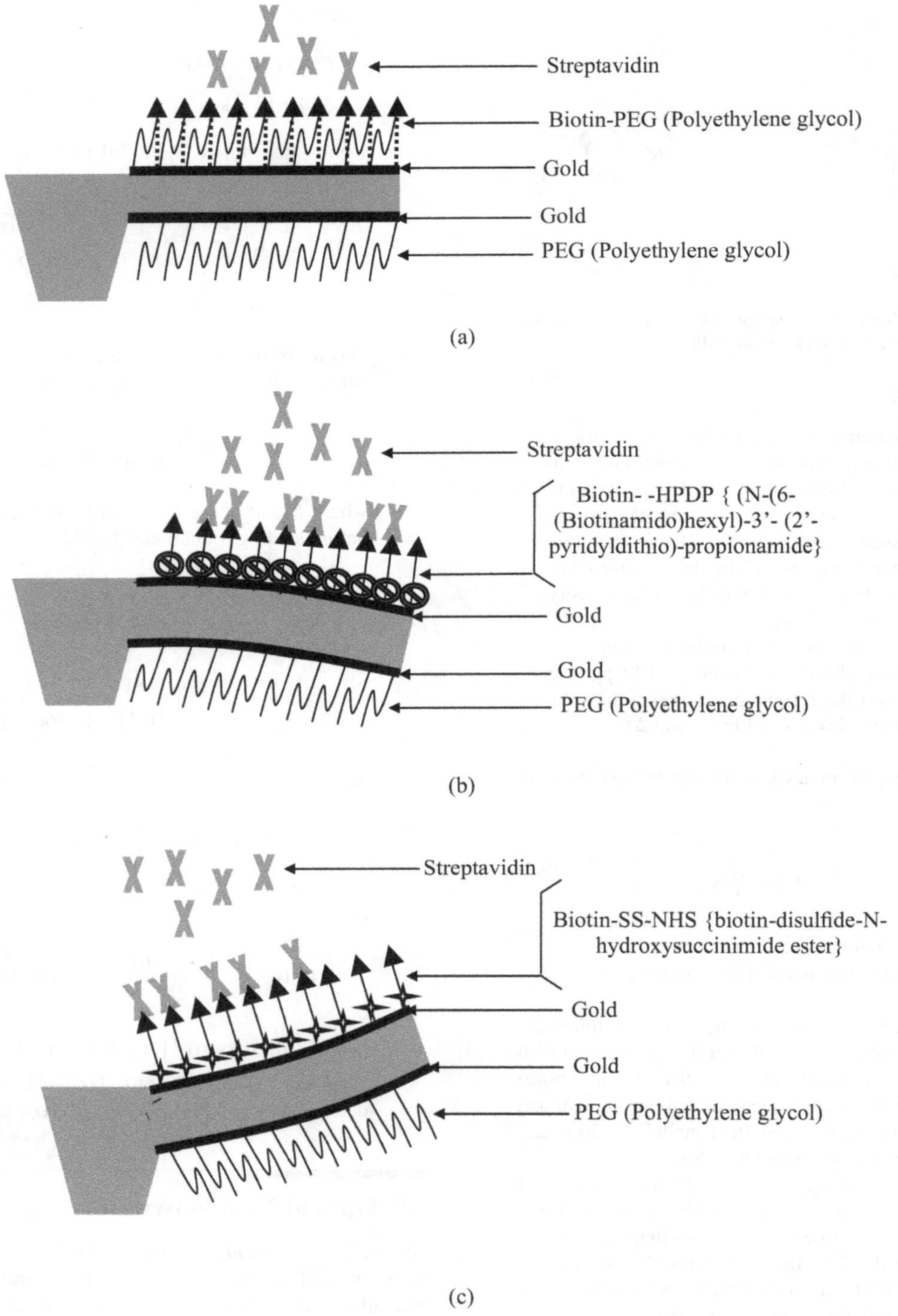

FIGURE 9.19 Difference in behavior of microcantilevers with various surface modifications on the upper gold surfaces: (a) biotin-PEG; (b) biotin-HPDP; and (c) biotin-SS-NHS. Lower gold surfaces of all the microcantilevers are coated with PEG, known to hinder nonspecific adsorption of proteins. (Shu, W. et al., *Biosens. Bioelectron.*, 22, 2003, 2007.)

important tools for molecular diagnostics and therapeutics in the biosensing of disease-related proteins. The aptamer biosensors offer detection limits down to the nanomolar (10^{-9} mol) level.

How can aptamers be compared with antibodies for developing biosensors? These artificial nucleic acid ligands appear as attractive alternatives to natural receptors, such as antibodies. They possess several advantages for biosensor design compared to antibodies: (i) due to their relative ease of isolation and modification, tailored binding affinity, and resistance against denaturation, specific aptamers can be selected *in vitro* for virtually any given target, ranging from small molecules to large proteins and even cells, paving the way to the development of a wide range of aptamer-based biosensors; (ii) unlike antibodies or enzymes, DNA aptamers are usually chemically stable and readily available commercially at low cost; and (iii) many aptamers undergo significant conformational changes upon target binding, offering high flexibility in biosensor development.

9.8.2 Aptamer-Modified Au Nanoparticles as a Colorimetric Adenosine Nanosensor

For what purposes is adenosine sensing necessary? Adenosine is an endogenous (growing or developing from within, e.g., endogenous cholesterol is cholesterol that is made inside the body and is not obtained in the diet) nucleoside (a structural subunit of nucleic acids: any of various compounds consisting of a sugar, usually ribose or deoxyribose, and a nitrogen base) interacting in various physiological processes which are also regulated by hormones and neurotransmitters (chemicals located and released in the brain to allow an impulse from one nerve cell to pass to another nerve cell). Adenosine modulates blood flow and neurotransmission and may be protective during pathological (concerned with disease) conditions such as ischemia (restricted blood flow within tissues) and stroke. Adenosine triphosphate (ATP), having the empirical formula $C_{10}H_{16}N_5O_{13}P_3$ and chemical formula $C_{10}H_8N_4O_2NH_2(OH)_2(PO_3H)_3H$, has important responsibilities in regulating cellular metabolism and biochemical pathways. ATP is a molecule found in any biological system, where it works as an energy carrier. Detecting ATP within cells can help observe energetic physiological processes, e.g., signal cascades, transport processes, etc. A signal cascade amplifies the initial signal caused by binding of the ligand to a receptor. ATP depletion is related to certain diseases, such as Parkinson's disease and ischemia (a restriction in blood supply). An adenosine sensor is needed for improved understanding of its physiological actions and the extent of receptor activation.

Chen et al. (2008) described a colorimetric sensing approach for the determination of ATP in urine samples (Figure 9.20). A colorimetric sensor measures the concentration of a known constituent of a solution by comparison with spectroscopic or visual standards. *What is the principle behind this approach?* In the absence of ATP, the color of the Apt-AuNP solution changed from wine-red to purple *via* self-induced aggregation. In the presence of ATP, binding of the analytes to the Apt-AuNPs induced folding of the aptamers on the AuNP surfaces into four-stranded quadruplex or tetraplex (fourfold) structures (G-quartet) and/or an increase in charge density. Consequently, the Apt-AuNP solution was wine-red in color when the analyte was present. G-quartets are atypical nucleic acid structures, consisting of a planar arrangement where each guanine is hydrogen bonded by Hoogsteen pairing to another guanine in the quartet.

9.8.3 Aptamer-Based Multicolor Fluorescent Gold Nanoprobe for Simultaneous Adenosine, Potassium Ion, and Cocaine Detection

This probe is useful for rapid screening of environmental or security-related molecular targets. Adenosine is a nucleoside, $C_{10}H_{13}N_5O_4$, composed of adenine linked to the sugar ribose, which is a structural component of nucleic acids. Cocaine ($C_{17}H_{21}NO_4$) is a highly addictive central nervous system stimulant that is snorted, injected, or smoked. Zhang et al. (2010) reported a gold nanoprobe for multiplex (two or more signals over a common channel) detection in a homogeneous solution, which was suitable for small molecular analytes and functioned in a mix-and-detect fashion. This nanoprobe (Figure 9.21) combined the high specificity of aptamers with the unique fluorescence quenching property of AuNPs; quenching is the suppression of fluorescence by absorption of the stimulating radiation. At the surface of AuNPs, a nanosurface energy-transfer (NSET) effect occurred, which led to long-range energy-transfer-based fluorescence quenching.

What is the strategy of their design? In their design, dyes were held in close proximity to the AuNP surface through hybridization of assembled single-stranded DNA (ssDNA) probes and dye-labeled aptamer sequences, causing efficient AuNP-quenched fluorescence (OFF state). In the presence of the targets, the aptamer-target binding separated the duplex (having two identical units), liberated the dye-labeled aptamer into solution, and restored the fluorescence.

Three 3′-thiolated DNA strands (P_A, P_K, and P_C,) probes for the anti-adenosine aptamer, the potassium-specific G-quartet, and the anti-cocaine aptamer, respectively, were mixed at equal molar ratios and co-assembled at the surface of AuNPs. The sequences of these three strands were complementary to those of the three aptamers (AA, the anti-adenosine aptamer; AK, the anti-potassium-specific G-quartet; and AC, the anti-cocaine aptamer). These three aptamers were labeled with different dyes at the 5′ end and then hybridized with their complementary sequences at the surface of AuNPs, which formed the multicolor gold nanoprobe. The choice of three dyes relied on the consideration of their spectral overlap, with the anti-adenosine aptamer labeled with R_{ox}, the G-quartet with fluorescein amidite (FAM), and the anti-cocaine aptamer with cyanine dye Cy5. In the OFF state, all fluorophores were in close proximity to AuNPs, which resulted in significant fluorescence quenching due to the NSET effect. Fluorescein amidite, abbreviated FAM, is an important dye used in molecular biology, frequently in the creation of probes to detect the presence of specific nucleic acid sequences. Cyanine dye is any of a class of dyes containing a –CH=group, linking two heterocyclic rings containing nitrogen.

Regarding the NSET effect, Jennings et al. (2006a; 2006b) reported the first successful application of a dipole surface-type energy transfer from a molecular dipole to a nanometal surface that more than doubled the Förster radius (22 nm) in place of 10 nm for traditional Förster radius value. Furthermore, it followed $1/R^4$ distance dependence. Because a 1.4 nm gold particle is below the normal mean free-path of an electron in gold at normal temperatures, the only scattering event an electron will sense is that of the surface potential. In other words, it is expected that the electrons spend all of their time at the surface of the particle. If a common fluorophore, such as fluorescein ($C_{20}H_{12}O_5$), is placed near such a particle, then the fluorescence quantum efficiency of the dye begins to decrease with a $1/R^4$ distance dependence. The basis of NSET arises from the damping of the fluorophore's oscillating dipole by the free electrons of gold metal as a through-space mechanism. Because the electrons of the Au particle are homogeneously oriented, the constraint on dipole-dipole coupling has been greatly relaxed and thus gives rise to energy-transfer efficiency at much greater distances.

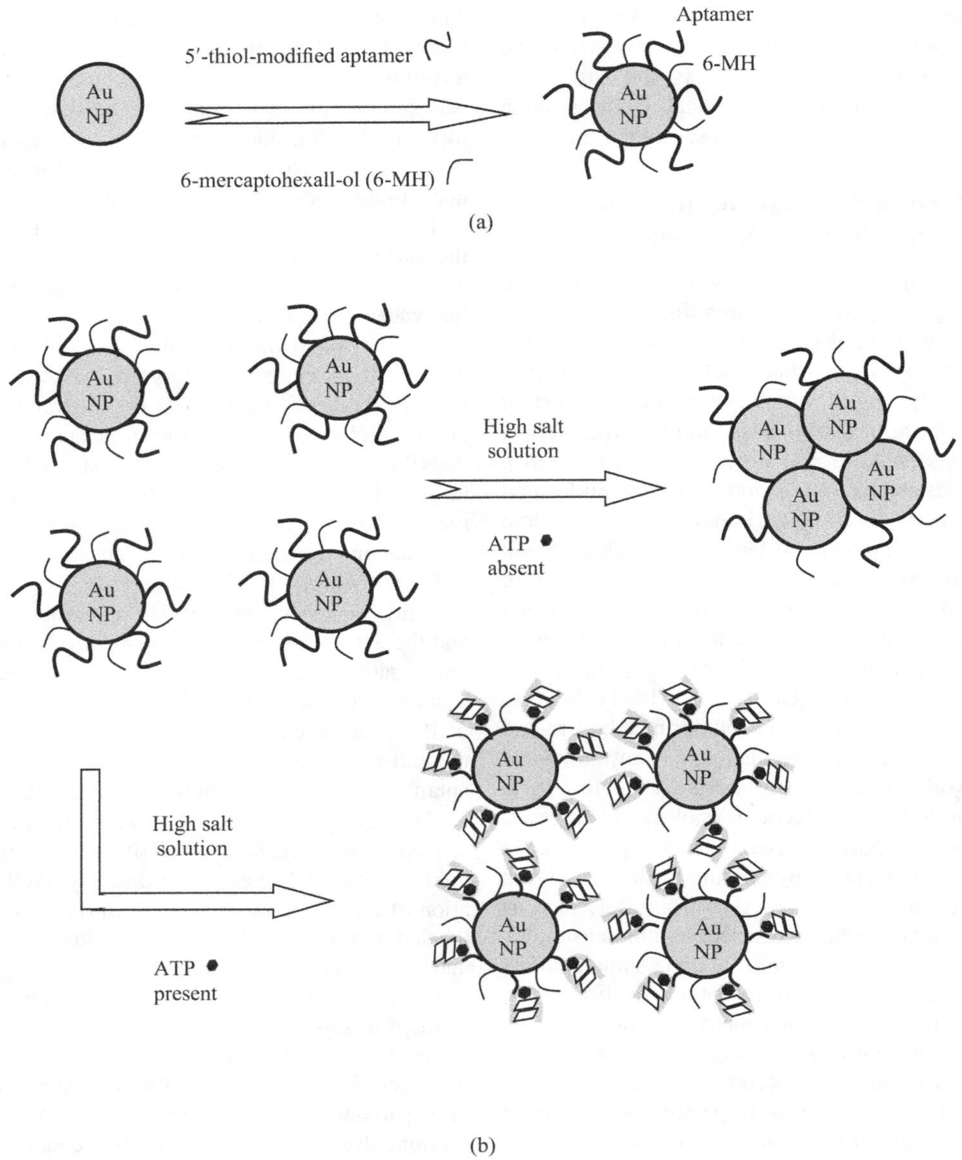

FIGURE 9.20 (a) Synthesis of aptamer-Au nanoparticles. (b) Sensing mechanism of Apt-AuNPs for the colorimetric determination of ATP: the solution containing no ATP changes color from wine-red to purple due to aggregation, whereas the one containing ATP retains its original red-wine color. (Chen, S.-J. et al., *Biosens. Bioelectron.*, 23, 1749, 2008.)

In the aforementioned experiments, the specific binding of individual aptamers with their specific targets separated the aptamer from the AuNP surface, thus leading to fluorescence recovery that provided quantitative measurement of the analyte concentration. This gold nanoprobe was highly selective since each analyte displaced only the specific aptamer, thus leading to the corresponding dye emission.

9.8.4 Aptamer-Capped QD as a Thrombin Nanosensor

For what purpose is thrombin detection required? Thrombosis is an important pathophysiological component of many cardiovascular diseases. Pathophysiology is concerned with the study of the changes of normal mechanical, physical, and biochemical functions, either caused by a disease, or resulting from an abnormal syndrome. Thrombin is a trypsin-like serine protease (an enzyme that cuts certain peptide bonds in other proteins), playing a central role in thrombosis formation (of a blood clot inside a blood vessel, obstructing the flow of blood through the circulatory system). This role is carried out through the conversion of the soluble plasma protein fibrinogen into insoluble sticky strands of fibrin. The fibrin clot stops bleeding. Detection and imaging of thrombin activity is thus of considerable biomedical interest. Thrombin is also known as coagulation factor II (F2).

Choi et al. (2006) described the application of aptamer-capped quantum dots for thrombin detection (Figure 9.22). A 15-mer thrombin-binding aptamer (TBA) and lead acetate, $Pb(CH_3COO)_2$, were mixed in TAE buffer (pH 8), followed by the injection of sodium sulfide, $Na_2S{\cdot}9H_2O$, with vigorous stirring, to

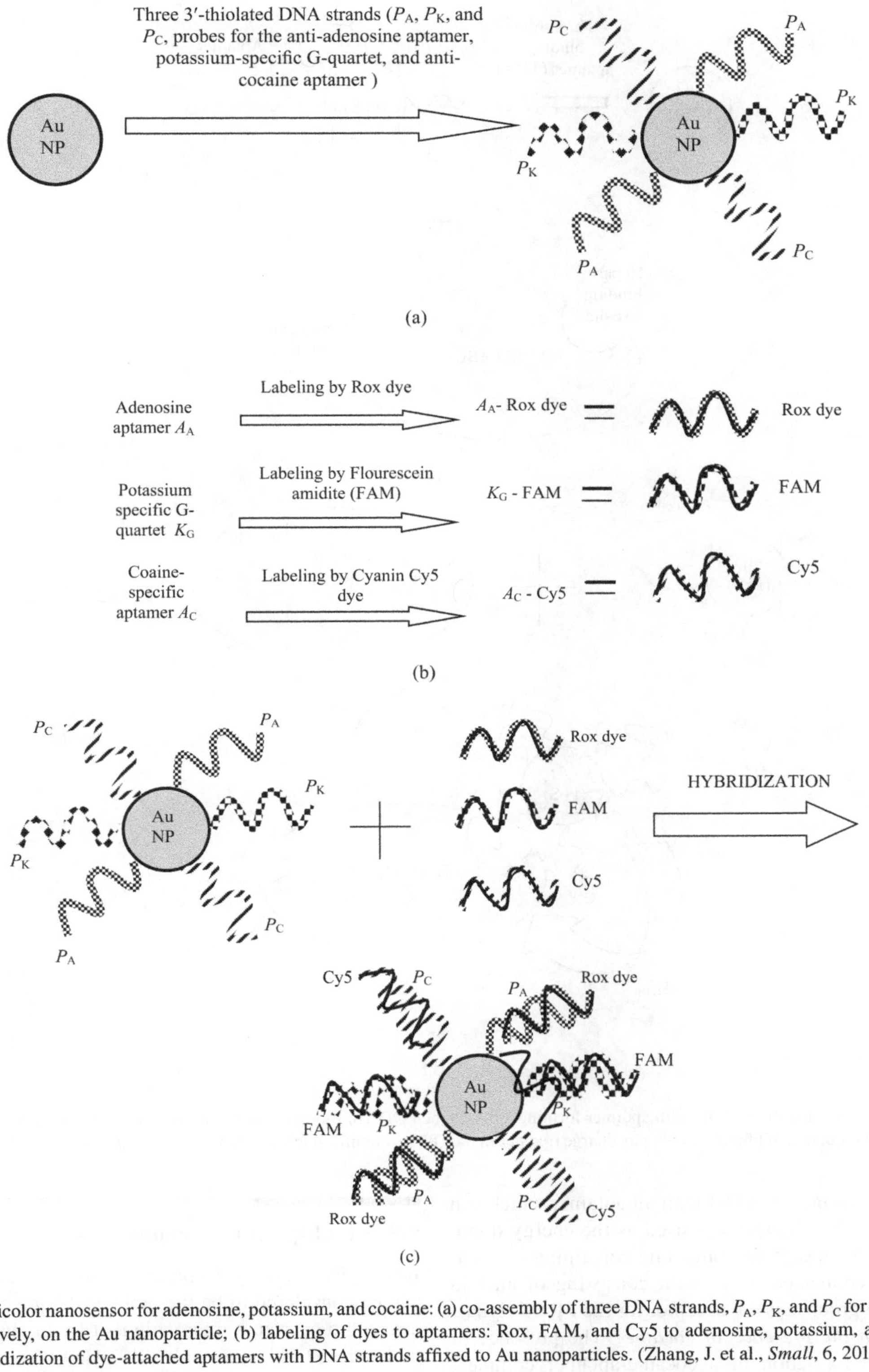

FIGURE 9.21 Multicolor nanosensor for adenosine, potassium, and cocaine: (a) co-assembly of three DNA strands, P_A, P_K, and P_C for adenosine, potassium, and cocaine, respectively, on the Au nanoparticle; (b) labeling of dyes to aptamers: Rox, FAM, and Cy5 to adenosine, potassium, and cocaine aptamers, respectively (c) Hybridization of dye-attached aptamers with DNA strands affixed to Au nanoparticles. (Zhang, J. et al., *Small*, 6, 201, 2010.)

produce TBA-capped QDs which were stable for several months and had diameters of 3–6 nm.

How is thrombin detected? When the aptamer-functionalized QD binds to its target (thrombin), there is a highly selective quenching of the photoluminescence, which was ascribed to the charge transfer from functional groups on thrombin to QD. This interaction was highly selective resulting in an NP optical probe for thrombin.

9.8.5 QD Aptameric Cocaine Nanosensor

Why is cocaine detection necessary? Simple and sensitive detection of cocaine is crucial for law enforcement and clinical diagnostics.

Zhang and Johnson (2009) developed a single-QD aptameric sensor for cocaine (Figure 9.23). The sensor is formed by

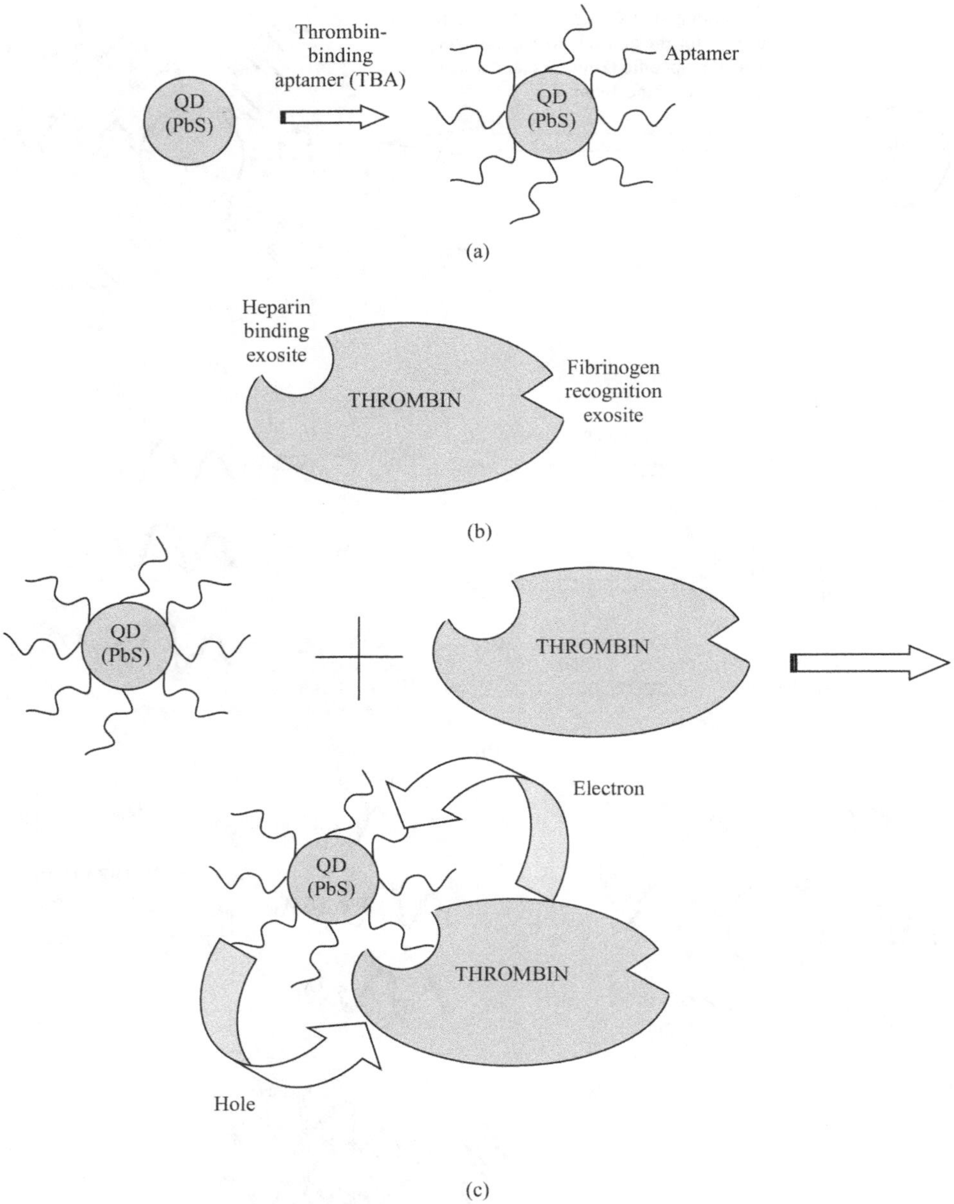

FIGURE 9.22 (a) Functionalization of QD with aptamer-forming TBA-capped QD. (b) Blood-clotting human α-thrombin. (c) Interaction of TBA-capped QD with thrombin and induction of photobleaching by charge transfer in the QD-thrombin complex. (Choi, J. H. et al., *J. Am. Chem. Soc.*, 128, 15584, 2006.)

functionalizing the surface of a QD with an aptamer which can recognize cocaine. The 605 QD was used as the energy donor and Cy5 (Cyanine 5: an organic compound containing two heterocyclic radicals connected by a chain consisting of an odd number of methine groups) as the acceptor for the FRET signal.

What happens in the presence of cocaine? The presence of cocaine led to the formation of a cocaine-aptamer complex, which made the Cy 5-labeled oligonucleotide dissociate from the aptamer and 605 QD. The decrease in the Cy 5 signal, due to the absence of FRET between 605 QD and Cy 5, signified that cocaine was present. *What are the possible applications of cocaine sensor?* This single QD-based aptameric sensor has potential applications in forensic analysis, environmental monitoring, and clinical diagnostics.

9.9 Biochips (or Microarrays)

What is the notion of a biochip? A biochip (biological microchip) can be considered to be the biological equivalent of a computer microchip or even a microchip that is able to interact with biomolecules (Tarakanov et al. 2010). However, instead of performing millions of mathematical operations, a biochip is intended to carry out thousands of biological interactions per minute, i.e., thousands of times faster in comparison with existing technologies. *How are biochips superior to traditional devices?* The main advantage of biochips over conventional analytical devices is the possibility of massive parallel analysis, while biochips are also smaller than conventional testing systems and highly economical in the use of specimens and reagents.

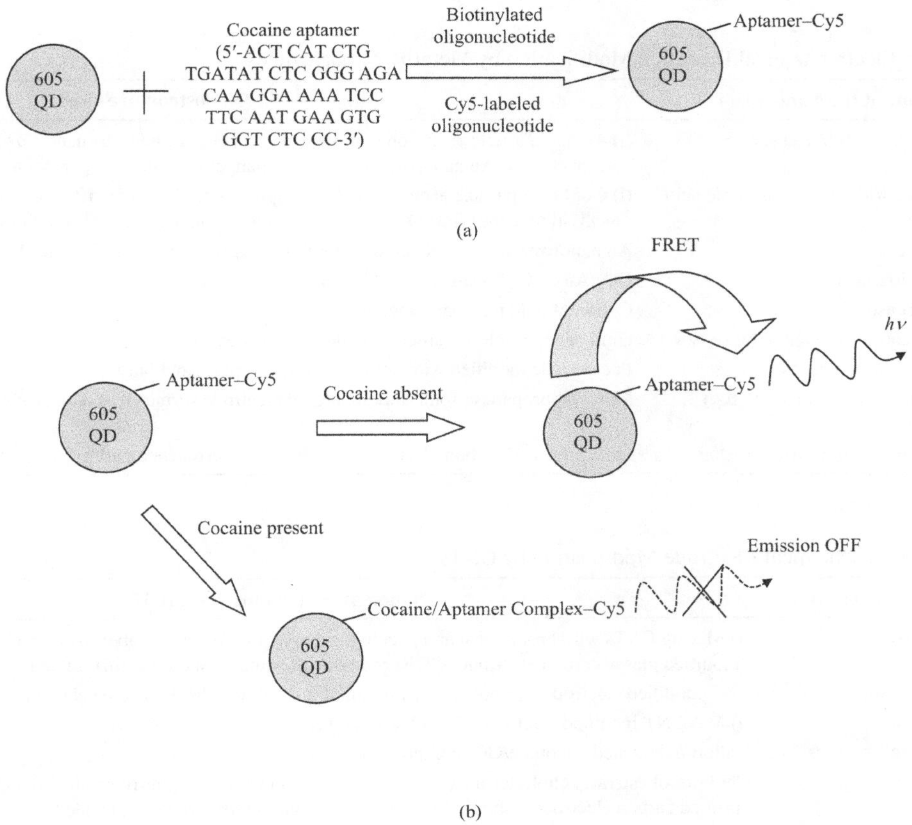

FIGURE 9.23 (a) Functionalization of 605 QD with cocaine-recognizing aptamer. (b) Detection of cocaine with QD-aptameric nanosensor: when cocaine is absent, fluorescence resonance energy transfer (FRET) between 605 QD and Cy5 dye take place, accompanied by emission of light, but, when cocaine is present, a complex structure is formed between cocaine and the aptamer quenching the emission. Abolishment of emission indicates the presence of cocaine. (Zhang, C.-Y. and Johnson, L.W., *Anal. Chem.*, 81, 3051, 2009.)

Some examples of biochips are given in this paragraph. The development of DNA chips has been one of the fastest-growing areas in DNA/RNA analysis. DNA chips rapidly decrease laboratory turnaround times so that results are available within 2–6 hours compared with perhaps 24 hours. The CNT-FET can be used as a label-free biosensor with a clear change of the CNT conductance in response to the ligand-binding event. CNT-based protein biochips are used for the simultaneous detection of several neurotransmitters, like DA, catecholamine, and serotonin, e.g., a multi-walled-CNT has been synthesized and used for sensitive DA detection (Wu et al. 2007). As previously stated, DA and serotonin are important neurotransmitters that interact in the brain. Whereas DA is easily detected with electrochemical sensors, the detection of serotonin is more difficult because reactive species formed after oxidation can absorb to the electrode, reducing sensitivity. The CNT-modified electrodes not only increase the sensitivity and reduce the fouling, but also allow monitoring of DA and serotonin changes simultaneously. These studies show that CNT-coated microelectrodes can be used with fast scanning techniques and are advantageous for *in-vivo* measurements of neurotransmitters.

9.10 Discussion and Conclusions

Tables 9.1 to 9.5 list the vast diversity of biosensor approaches. Electrochemical sensing has received considerable attention. Modified electrodes have been prepared by impregnating metallic NPs or CNTs, along with biomolecules. Immobilization of biomolecules on QDs has been successfully applied in several cases. Platforms like NW FETs and cantilevers have attracted great interest.

Biosensors are animated by the application of the target biomolecules on suitable platforms. Without these molecules, they are useless. As the effect of these biomolecules weakens, biosensors lose their senses and ultimately stop functioning. Therefore, biosensors need careful attention to be paid to storage and handling. They should also be used with caution bearing in mind their storage environment, their useful shelf lives and other limitations. Ultimately, they rely on living world molecules for their normal functioning.

Review Exercises

9.1 Define the term "biosensor." What are the three main components of a biosensor? Explain with the help of a diagram.

9.2 What types of chemical sensors are classified as biosensors? Justify the statement, "Biosensors constitute a subset of chemical sensors but are studied separately because of their extreme importance."

TABLE 9.1

Biosensors Based on Electrochemical Electrode Modification by Metallic Nanoparticles

Sl. No.	Name of the Nanosensor	Nanomaterial/Nanostructure Used
1.	Nitric oxide (NO) biosensor	(i) Au nanoparticle arrays on nanostructured ITO electrode; (ii) nano-Pt/MWNT modified electrode; (iii) Au nanoparticles-chitosan nanocomposite on a glassy carbon electrode
2.	Dopamine, uric acid and ascorbic acid biosensor	(i) Gold nanoparticle arrays with 3-mercapto-propionic acid (MPA) on ITO; (ii) poly(3-methylthiophene) (P3MT)/gold nanoparticles (AuNPs), on a glassy carbon electrode
3.	CO biosensor	Au nanoparticle cores covered with thiolate monolayers on GC electrodes
4.	Glucose biosensor	GO_x/Au_{coll}–Cyst–AuE
5.	DNA biosensor	Colloidal gold nanoparticles
6.	Allergen-antibody reaction biosensor	Gold nanoparticle-modified graphite carbon electrode
7.	Hepatitis B immunosensor	Pt electrode modified with colloidal gold and polyvinyl butyral
8.	Carcinoembryonic antigen (CEA) biosensor	Mercaptopropanesulfonic-modified gold electrode surface based on $(AuNPs/MWCNTThi/Chit)_n$
9.	*Escherichia coli* biosensor	Screen-printed carbon electrode-modified with ferrocenedicarboxylic acid and AuNPs

TABLE 9.2

Biosensors Based on Electrochemical Electrode Modification by CNTs

Sl. No.	Name of the Nanosensor	Nanomaterial/Nanostructure Used
1.	Dopamine biosensor	(i) Mixing CNTs with bromoform and packing the paste inside a glass tube; (ii) poly (3-methylthiophene) modified glassy carbon electrode (GCE) coated with Nafion/SWCNTs; (iii) fullerene-C60 coated gold electrode
2.	Catalase biosensor	CNT-modified electrode by applying an aliquot of a CNT suspension on the substrate surface
3.	DNA biosensor	(i) MWCNT-modified electrode; (ii) sidewall and end-functionalized CNTs
4.	Glucose biosensor	Nafion-overcoated nanotube/GO_x composite-modified electrode
5.	Cholesterol biosensor	Cholesterol esterase, cholesterol oxidase, peroxidase, and potassium ferrocyanide immobilized on the screen-printed carbon electrodes, and multi-walled carbon nanotubes (MWCNT) added
6.	H_2O_2 biosensor	(i) Immobilization of Hb on multi-wall carbon nanotubes and gold colloidal nanoparticles; (ii) entrapping Hb in a composite electrodeposited chitosan-multi-wall carbon nanotubes film by assembling gold nanoparticles and hemoglobin step by step

TABLE 9.3

Biosensors Based on QDs

Sl. No.	Name of the Nanosensor	Biomaterial Used
1.	Uric acid biosensor	Uricase-ZnS QD/L-cysteamine
2.	Hydrogen peroxide biosensor	(i) Horseradish peroxidase (HRP) with lipophilic CdSe/ZnS QDs on a glassy carbon electrode; (ii) Hb modified with QDs (CdS) at a normal graphite (GP) electrode
3.	Glucose biosensor	Glucose oxidase (GO_x) adsorbed on a CdS nanoparticles-modified pyrolytic graphite electrode
4.	H_2O_2 biosensor	Immobilization of Hb with cadmium sulfide (CdS) nanoparticles (NPs) on pyrolytic graphite (PG) electrode
5.	Light-activated glucose biosensor	Immobilization of CdSe/ZnS nanocrystals on gold electrode

TABLE 9.4

Biosensors Based on Nanowire FETs

Sl. No.	Name of the Nanosensor	Biomaterial/Nanostructure Used
1.	DNA biosensor	ssDNA probe
2.	Protein biosensor	Biotin-SA probe-receptor complex

9.3 What is a nanobiosensor? Bring out the implication of "nano" in this term.

9.4 Why do AuNPs behave as being "catalytically active"? Name two methods used to immobilize NPs on electrodes. Name three methods of preparing AuNP-modified electrodes.

9.5 Define the following terms: voltammetry, amperometry, potentiometry, impedometry, and conductometry.

9.6 What is meant by LSV and CV?

9.7 How do AuNPs help in reducing the insulating effect of the protein shell for the direct electron transfer to the electrode?

9.8 Mention some examples where NO sensing is required? Describe the electrochemical sensors for NO detection, using nanomaterials.

9.9 What human body abnormalities are detected by determination of DA? How does AA affect DA measurements? Describe the functioning of one AuNP-modified electrode for DA sensing.

9.10 How is direct electron transfer from the enzyme to the electrode inhibited in a glucose biosensor? How do AuNPs help in the immobilization of the enzyme?

TABLE 9.5

Biosensors Based on Cantilevers

Sl. No.	Name of the Nanosensor	Biomaterial/Nanostructure Used
1.	DNA biosensor	Two cantilevers with a control (noncomplementary) oligonucleotide on one of them and the DNA probe (complementary) immobilized on the other
2.	Osteosarcoma biosensor	Cantilever sensing the interactions between vimentin antibodies and antigens

Describe the main steps involved in the preparation of the structure GO_x/Au_{coll}–Cyst–AuE.

9.11 Describe, with the help of a diagram, the basic principle of a DNA biosensor. How is DNA detected by conductometric method? Illustrate your answer with a diagram. Briefly explain the electrochemical stripping method of DNA sensing.

9.12 How does the use of polyelectrolyte layers help in discriminating whether the electrochemical signal has originated from silver deposition on the background gold electrode surface or the AuNP label? Explain with diagrams the various steps of this scheme for enhanced detection of DNA hybridization.

9.13 How do CNTs function as electrode modifiers in biochemical reactions? Which sites on CNTs are more electroactive? Describe some strategies used with CNTs for DA measurements.

9.14 How are SWCNTs used for electrical connection between the enzyme redox centers and the electrode? Describe the assembly of the SWCNT electrically contacted glucose oxidase electrode. Highlight the role of edge plane sites on nanotubes in adsorption of glucose oxidase.

9.15 What is the upper limit of blood cholesterol in human beings? Describe a biosensor for monitoring total cholesterol in blood using cholesterol esterase, cholesterol oxidase, peroxidase, potassium ferrocyanide, and MWCNTs, immobilized on a CPE.

9.16 Name two methods for functionalizing CNTs for biosensors. What are the two types of sites on CNTs for immobilizing biomolecules?

9.17 How does Hb, modified with quantum dots, work as a hydrogen peroxide sensor? What are the useful features of this biosensor? What are the advantages of Hb over HRP?

9.18 How does a QD-based light-triggered glucose sensor operate?

9.19 Highlight the relative merits and demerits of CNTs and NWs for sensor fabrication. Outline the methods for functionalizing SiNWs with a native oxide layer and those with H-terminated silicon surfaces.

9.20 Describe a SiNW-based FET nanosensor for protein detection.

9.21 What constructional materials are used for fabricating microcantilevers? Do uncoated cantilevers exhibit chemical sensitivity? Point out some advantages of using several cantilevers in parallel.

9.22 What criteria are used for choosing an immobilization process for molecular recognition agents on the cantilever? How are silicon oxide and gold surfaces modified for attaching biomolecules on cantilever surfaces?

9.23 When biotin-containing materials are coated onto a cantilever surface and SA is applied to the surface, biotin-streptavidin binding takes place. In one case, the cantilever remained unaffected, whereas, in another situation, it bent downward, and in a still further case, it bent upward on applying SA. Name the three biotin-based compositions that were applied on the cantilever.

9.24 Elaborate the statement: "AuNPs act as electron transfer wires in many bioelectrochemical reactions."

9.25 How does the immobilization of biomolecules on naked gold surfaces differ from that on gold NPs?

9.26 Mention the advantages of aptamers, as compared with antibodies, for biosensing applications. Describe the functioning of a QD-based nanosensor for cocaine.

9.27 Describe the use of AuNPs to fabricate an aptamer-based multicolor nanoprobe for selective, multiplex detection of several analytes in a homogeneous solution. How does this probe utilize the highly specific binding abilities of aptamers with the ultra-high quenching ability of AuNPs?

9.28 Explain, through schematic diagrams, the interaction of TBA-Capped PbS QD with thrombin.

9.29 Describe a nanostructured electrode, based on CdSe/ZnS quantum dots, for the sensitive measurement of the enzyme cofactor NADH. Is glucose detection feasible with such an electrode system and photocurrent measurements?

9.30 Different kinds of NPs, and sometimes the same kind of nanoparticles, can play different roles in various biosensor systems. Elucidate this statement taking the example of "gold nanoparticles."

REFERENCES

Agarwal, A., V. K. Khanna, and C. Shekhar. 2010. CMOS compatible silicon nanowire arrays for CHEMFET and BioFET (PF-13). *International Conference on Quantum Effects in Solids of Today (I-ConQuEST)*, December 20-23, 2010, NPL, New Delhi, Souvenir & Abstracts, pp. 177–178. Available at: https://www.nplindia.in/sites/default/files/icnpweb.pdf

Andreescu, S. and L. A. Luck. 2008. Studies of the binding and signaling of surface-immobilized periplasmic glucose receptors on gold nanoparticles: A glucose biosensor application. *Analytical Biochemistry* 375(2): 282–290.

Balal, K., H. Mohammad, B. Bahareh, B. Ali, H. Maryamc, and Z. Mozhgand. 2009. Zeolite nanoparticle modified carbon paste electrode as a biosensor for simultaneous determination of dopamine and tryptophan. *Journal of the Chinese Chemical Society* 56: 789–796.

Britto, P. J., K. S. V. Santhanam, and P. M. Ajayan. 1996. Carbon nanotube electrode for oxidation of dopamine. *Bioelectrochemistry and Bioenergetics* 41: 121–125.

Butt, H.-J. 1996. A sensitive method to measure changes in the surface stress of solids. *Journal of Colloid and Interface Science* 180: 251–260.

Carrascosa, L. G., M. Moreno, M. Alvarez, and L. M. Lechuga. 2006. Nanomechanical biosensors: A new sensing tool. *Trends in Analytical Chemistry* 25(3): 196–206.

Chen, S., R. Yuan, Y. Chai, L. Zhang, N. Wang, and X. Li. 2007. Amperometric third-generation hydrogen peroxide biosensor based on the immobilization of hemoglobin on multiwall carbon nanotubes and gold colloidal nanoparticles. *Biosensors and Bioelectronics* 22: 1268–1274.

Chen, S.-J., Y.-F. Huang, C.-C. Huang, K.-H. Lee, Z.-H. Lin, and H.-T. Chang. 2008. Colorimetric determination of urinary adenosine using aptamer-modified gold nanoparticles. *Biosensors and Bioelectronics* 23: 1749–1753.

Choi, J. H., K. H. Chen, and M. S. Strano. 2006. Aptamer-capped nanocrystal quantum dots: A new method for label-free protein detection. *Journal of the American Chemical Society* 128: 15584–15585.

Cui, Y., Q. Q. Wei, H. K. Park, and C. M. Lieber. 2001. Nanowire nanosensors for highly sensitive and selective detection of biological and chemical species. *Science* 293: 1289–1292.

Curreli, M., R. Zhang, F. N. Ishikawa, H.-K. Chang, R. J. Cote, C. Zhou, and M. E. Thompson. 2008. Real-time, label-free detection of biological entities using nanowire-based FETs. *IEEE Transactions on Nanotechnology* 7(6): 651–667.

Drummond, T. G., M. G. Hill, and J. K. Barton. 2003. Electrochemical DNA sensors. *Nature Biotechnology* 21: 1192–1199.

Fritz, J., M. K. Baller, H. P. Lang, H. Rothuizen, P. Vettiger, E. Meyer, H.-J. Güntherodt, C. Gerber, and J. K. Gimzewski. 2000. Translating biomolecular recognition into nanomechanics. *Science* 288: 316–318.

Geng, D. S. and G. X. Lu. 2007. Size effect of gold nanoparticles on the electrocata-lytic oxidation of carbon monoxide in alkaline solution. *Journal of Nanoparticle Research* 9: 1145–1151, doi: 10.1007/s11051-007-9210-1.

Goyal, R. N., V. K. Gupta, M. Oyama, and N. Bachheti. 2007. Gold nanoparticles modified indium tin oxide electrode for the simultaneous determination of dopamine and serotonin: Application in pharmaceutical formulations and biological fluids. *Talanta* 72: 976–983.

Goyal, R. N., V. K. Gupta, N. Bachheti, and R. A. Sharma. 2008. Electrochemical sensor for the determination of dopamine in presence of high concentration of ascorbic acid using a Fullerene-C60 coated gold electrode. *Electroanalysis* 20(7): 757–764.

Guiseppi-Elie, A., C. Lei, and R. H. Baughman. 2002. Direct electron transfer of glucose oxidase on carbon nanotubes. *Nanotechnology* 13: 559–564.

Hansen, K. M. and T. Thundat. 2005. Microcantilever biosensors. *Methods* 37: 57–64.

Herne, T. M. and M. J. Tarlov. 1997. Characterization of DNA probes immobilized on gold surfaces. *Journal of the American Chemical Society* 119: 8916–8920.

Holloway, A. F., G. G. Wildgoose, R. G. Compton, L. Shao, and M. L. H. Green. 2008. The influence of edge-plane defects and oxygen-containing surface groups on the voltammetry of acid-treated, annealed and "super-annealed" multiwalled carbon nanotubes. *Journal of Solid State Electrochemistry* 12: 1337–1348.

Huang, Y., W. Zhang, H. Xiao, and G. Li. 2005. An electrochemical investigation of glucose oxidase at a CdS nanoparticles modified electrode. *Biosensors and Bioelectronics* 21: 817–821.

Huang, H. Z., Z. G. Liu, and X. R. Yang. 2006. Application of electrochemical impedance spectroscopy for monitoring allergen-antibody reactions using gold nanoparticle-based biomolecular immobilization method. *Analytical Biochemistry* 356: 208–214.

Huang, X., Y. Li, P. Wang, and L. Wang. 2008. Sensitive determination of dopamine and uric acid by the use of a glassy carbon electrode modified with poly(3-methyl-thiophene)/gold nanoparticle composites. *Analytical Sciences* 24: 1563–1568.

Jennings, T. L., M. P. Singh, and G. F. Strouse. 2006a. Fluorescent lifetime quenching near d = 1.5nm gold nanoparticles: Probing NSET validity. *Journal of the American Chemical Society* 128: 5462–5467.

Jennings, T. L., J. C. Schlatterer, M. P. Singh, N. L. Greenbaum, and G. F. Strouse. 2006b. NSET molecular beacon analysis of hammerhead RNA substrate binding and catalysis. *Nano Letters* 6(7): 1318–1324.

Ju, H. X., S. Q. Liu, B. X. Ge, F. Lisdat, and F. W. Scheller. 2002. Electrochemistry of cytochrome c immobilized on colloidal gold modified carbon paste electrodes and its electrocatalytic activity. *Electroanalysis* 14: 141–147.

Kerman, K., Y. Morita, Y. Takamura, M. Ozsoz, and E. Tamiya. 2004. DNA-directed attachment of carbon nanotubes for enhanced label-free electrochemical detection of DNA hybridisation. *Electroanalysis* 16(20): 1667.

Khanna, V. K. 2008. New-generation nano-engineered biosensors, enabling nanotechnologies and nanomaterials. *Sensor Review* 28(1): 39–45.

Lee, T. M.-H., L.-L. Li, and I.-M. Hsing. 2003. Enhanced electrochemical detection of DNA hybridization based on electrode-surface modification. *Eangmuir* 19: 4338–4343.

Li, N. Q., J. X. Wang, and M. X. Li. 2003. Electrochemistry at carbon nanotube electrodes. *Reviews in Analytical Chemistry* 22: 19–33.

Li, G., J. M. Liao, G. Q. Hu, N. Z. Ma, and P. J. Wu. 2005. Study of carbon nanotube modified biosensor for monitoring total cholesterol in blood. *Biosensors and Bioelectronics* 20: 2140–2144.

Lia, Y., H. J. Schluesenerb, and S. Xu. 2010. Gold nanoparticle-based biosensors. *Gold Bulletin* 43(1): 29–41.

Lin, Y. H., S. Chen, Y. Chuang, Y. Lu, T. Y. Shen, C. A. Chang, and C. S. Lin. 2008. Disposable amperometric immunosensing strips fabricated by Au nanoparticles-modified screen-printed carbon electrodes for the detection of foodborne pathogen *Escherichia coli* 0157:H7. *Biosensers and Bioelectronics* 23: 1832–1837.

Liu, C. and J. Hu. 2008. Direct electrochemistry of hemoglobin entrapped in composite electrodeposited chitosan-multiwall carbon nanotubes and nanogold particles membrane and its electrocatalytic application. *Electroanalysis* 20(10): 1067–1072.

Lyons, M. E. G. and G. P. Keeley. 2008. Carbon nanotube based modified electrode biosensors. Part 1. Electrochemical studies of the flavin group redox kinetics at SWCNT/glucose oxidase composite modified electrodes. *International Journal of Electrochemical Science* 3: 819–853.

Maye, M. M., Y. B. Lou, and C. J. Zong. 2000. Core-shell gold nanoparticle assembly as novel electrocatalyst of CO oxidation. *Eangmuir* 16: 7520–7523.

Mena, M. L., P. Yáñez-Sedeño, and J. M. Pingarrón. 2005 A comparison of different strategies for the construction of amperometric enzyme biosensors using gold nanoparticle-modified electrodes. *Analytical Biochemistry* 336: 20–27.

Milburn, C., J. Zhou, O. Bravo, C. Kumar, and W. O. Soboyejo. 2005. Sensing interactions between vimentin antibodies and antigens for early cancer detection. *Journal of Biomedical Nanotechnology* 1(1): 30–38.

Ou, C., R. Yuan, Y. Chai, M. Tang, R. Chai, and X. He. 2007. A novel amperometric immunosensor based on layer-by-layer assembly of gold nanoparticles-multi-walled carbon nanotubes-thionine multilayer films on polyelectrolyte surface. *Analytica Chimica Acta* 603(2): 205–213.

Ozsoz, M., A. Erdem, K. Kerman, D. Ozkan, B. Tugrul, N. Topcuoglu, H. Ekren, and M. Taylan. 2003. Electrochemical genosensor based on colloidal gold nanoparticles for the detection of Factor V Leiden mutation using disposable pencil graphite electrodes. *Analytical Chemistry* 75: 2181–2187.

Park, S. J., T. A. Taton, and C. A. Mirkin. 2002. Array-based electrical detection of DNA with nanoparticle probes. *Science* 295: 1503–1506.

Patolsky, F., Y. Weizmann, and I. Willner. 2004. Long-range electrical contacting of redox enzymes by SWCNT connectors. *Agnewandte Chemie International Edition* 43(16): 2113–2117.

Schubert, K., W. Khalid, Z. Yue, W. J. Parak, and F. Lisdat. 2009. Quantum-dot-modified electrode in combination with NADH-dependent dehydrogenase reactions for substrate analysis. *Langmuir* 26(2): 1395–1400, doi: 10.1021/la902499e.

Shu, W., E. D. Laue, and A. A. Seshia. 2007. Investigation of biotin streptavidin binding interactions using microcantilever sensors. *Biosensors and Bioelectronics* 22(9–10): 2003–2009.

Tang, D., R. Yuan, Y. Chai, J. Dai, X. Zhong, and Y. Liu. 2004. A novel immunosensor based on immobilization of hepatitis B surface antibody on platinum electrode modified colloidal gold and polyvinyl butyral as matrices via electrical impedance spectroscopy. *Bioelectrochemistry* 65: 15–22.

Tarakanov, A. O., L. B. Goncharova, and Y. A. Tarakanov. 2010. Carbon nanotubes towards medicinal biochips. *WIREs Nanomedicine and Nanobiotechnology* 2: 1–10.

Wang, J., R. Polsky, and D. Xu. 2001. Silver-enhanced colloidal gold electrochemical stripping detection of DNA hybridization. *Langmuir* 17: 5739–5741.

Wang, J. X., M. X. Li, Z. J. Shi, N. Q. Li, and Z. N. Gu. 2002a. Direct electrochemistry of cytochrome c at a glassy carbon electrode modified with single-wall carbon nanotubes. *Analytical Chemistry* 74: 1993–1997.

Wang, G., J. J. Xu, and H. Y. Chen. 2002b. Interfacing cytochrome c to electrodes with a DNA-carbon nanotube composite film. *Electrochemistry Communications* 4: 506–509.

Wang, J., A. N. Kawde, and M. Musameh. 2003a. Carbon-nanotube-modified glassy carbon electrodes for amplified label-free detection of DNA hybridisation. *Analyst* 128(7): 912.

Wang, J., M. Musameh, and Y. Lin. 2003b. Solubilization of carbon nanotubes by Nafion towards the preparation of amperometric biosensors. *Journal of the American Chemical Society* 125: 2408.

Wang, M. K., F. Zhao, Y. Liu, and S. J. Dong. 2005. Direct electrochemistry of microperoxidase at Pt microelectrodes modified with carbon nanotubes. *Biosensors and Bioelectronics* 21: 159–166.

Wang, H. S., T. H. Li, W. L. Jia, and H. Y. Xu. 2006. Highly selective and sensitive determination of dopamine using a Nafion/carbon nanotubes coated poly (3-methyl-thiophene) modified electrode. *Biosensors and Bioelectronics* 22: 664–669.

Wang, Z., Q. Xu, H.-Q. Wang, Q. Yang, J.-H. Yu, and Y.-D. Zhao. 2009. Hydrogen peroxide biosensor based on direct electron transfer of horseradish peroxidase with vapour deposited quantum dots. *Sensors and Actuators B: Chemical* 138(1): 278–282.

Wang, F., X. Deng, W. Wang, and Z. Chen. 2010. Nitric oxide measurement in biological and pharmaceutical samples by an electrochemical sensor. *Journal of Solid State Electrochemistry* 15(4): 829–836, doi: 10.1007/s10008-010-1157-y.

Wee, K. W., G. Y. Kang, J. Park, J. Y. Kang, D. S. Yoon, J. H. Park, and T. S. Kim. 2005. Novel electrical detection of label-free prostate specific antigen (PSA) using self-sensing piezoresistive microcantilever. *Biosensors and Bioelectronics* 20(10): 1932–1938.

Wu, W., H. Zhu, L. Fan, D. Liu, R. Rennenberg et al. 2007. Sensitive dopamine recognition by boronic acid functionalized multi-walled carbon nanotubes. *Chemical Communications* 23: 2345–2347.

Xiao, Y., F. Patolsky, E. Katz, J. F. Hainfeld, and I. Willner. 2003. Plugging into enzymes: Nanowiring of redox enzymes by a gold nanoparticle. *Science* 299: 1877–1881.

Xu, Y., J. Liang, C. Hu, F. Wang, S. Hu, and Z. He. 2007. A hydrogen peroxide biosensor based on the direct electrochemistry of hemoglobin modified with quantum dots. *Journal of Biological Inorganic Chemistry* 12: 421–427, doi: 10.1007/s00775-006-0198-2.

Yanez-Sedeno, P. and J. M. Pingarron. 2005. Gold nanoparticle-based electrochemical biosensors. *Analytical and Bioanalytical Chemistry* 382: 884–886.

Yang, K., H. Wang, K. Zou, and X. Zhang. 2006. Gold nanoparticle modified silicon nanowires as biosensors. *Nanotechnology* 17: S276–S279, doi: 10.1088/0957-4484/17/11/S08.

Zhang, C.-Y. and L.W. Johnson. 2009. Single quantum-dot-based aptameric nanosensor for cocaine. *Analytical Chemistry* 81: 3051–3055.

Zhang, J. D. and M. Oyama. 2005. Gold nanoparticle arrays directly grown on nano-structured indium tin oxide electrodes: Characterization and electroanalytical application. *Analytica Chimica Acta* 540: 299–306.

Zhang, J. D. and M. Oyama. 2007. Electrocatalytic activity of three-dimensional monolayer of 3-mercaptopropionic acid assembled on gold nanoparticle arrays. *Electrochemistry Communications* 9: 459–464.

Zhang, J., H. P. Lang, F. Huber, A. Bietsch, W. Grange, U. Certa, R. Mckendry, H.-J. Güntherodt, M. Hegner, and Ch. Gerber. 2006a. Rapid and label-free nanomechanical detection of biomarker transcripts in human RNA. *Nature Nanotechnology* 1: 214–220.

Zhang, F. F., C. X. Li, X. H. Li, X. L. Wang, Q. Wan, Y. Z. Xian, L. T. Jin, and K. Yamamoto. 2006b. ZnS quantum dots derived a reagentless uric acid biosensor. *Talanta* 68: 1353–1358.

Zhang, L., Z. Fang, G.-C. Zhao, and X.-W. Wei. 2008. Electrodeposited platinum nanoparticles on the multi-walled carbon nanotubes and its electrocatalytic for nitric oxide. *International Journal of Electrochemical Science* 3: 746–754.

Zhang, J., L. Wang, H. Zhang, F. Boey, S. Song, and C. Fan. 2010. Aptamer-based multicolor fluorescent gold nanoprobes for multiplex detection in homogeneous solution. *Small* 6(2): 201–204.

Zhou, H., X. Gan, T. Liu, Q. Yang, and G. Li. 2005. Effect of nano cadmium sulfide on the electron transfer reactivity and peroxidase activity of hemoglobin. *Journal of Biochemical and Biophysical Methods* 64: 38–45.

Part V

Emerging Applications of Nanosensors

10

Nanosensors for Societal Benefits

Hitherto, in this book, we strictly classified nanosensors into physical (Chapters 4–7), chemical and biological (Chapters 8–9) categories (Parts III–IV). But from here on, in Part V (Chapters 10–12), we focus on a pre-selected application and describe nanosensors from all categories that serve each particular application. This will enable us to get an overview of the vast store of innovative research papers falling under each application, giving a clear picture of the nanosensor landscape in the service of mankind. It is the applications that determine the utility of research.

Societal benefit areas are environmental fields, such as biodiversity and ecosystem, food and agriculture, public health, infrastructure and transport, energy and minerals, water resources, disaster resilience, and so forth. Air and water are the basic necessities of life, every human being needing pure air to breathe and uncontaminated water to drink. Nanosensors for air pollution and water quality monitoring will be presented. There will also be a change in our narrative style on nanosensors, from classification-based to application- and chronology-based approaches.

10.1 Air Pollutants

These are harmful substances released into the earth's atmosphere through industrial manufacturing plants, transport vehicles, and other agencies (Table 10.1). Such substances are detrimental to human health, the flora and fauna on the earth and to the planet as a whole.

10.2 Nanosensors for Particulate Matter Detection

10.2.1 Cantilever-Based Airborne Nanoparticle Detector (CANTOR)

The CANTOR works on the change in second mode resonance frequency after sampling of nanoparticles for 15 minutes (Wasisto et al. 2013; 2015). This portable system consists of two main sections: (i) a silicon piezoresistive cantilever (1250 μm × 26.5 μm × 39.3 μm) with a full square Wheatstone bridge at the clamped extremity; and (ii) a cylindrical-shaped electrostatic aerosol sampling unit for collection of engineered nanoparticles (Figure 10.1). The cantilever, operating at a second bending frequency f_2= 221502.36 Hz, has a quality factor of 1950 in air. It is fabricated by a bulk micromachining process, employing photolithography, diffusion, and deep reactive-ion etching facilities. The sampler has a tubular construction with a fan near the outlet for maintaining a laminar flow of nanoparticles over the cantilever. The increase in cantilever mass due to nanoparticle attachment lowers its resonance frequency.

Testing with 100-nm carbon-based nanoparticles (NPs) at a concentration of 6000 NPs/cm^3 revealed a mass sensitivity of 36.51 Hz/ng.

For CANTOR-1, a measurement precision of NPs with mass concentration of < 55% is achieved, with a limit of detection < 25 μg m^{-3}. CANTOR-1 has a cantilever with dimensions 2750 μm × 100 μm × 50 μm and frequency 9.4 kHz. Supporting modules, such as phase-locked loop (PLL)-based frequency tracking system, power supply and data acquisition control system, are included (Wasisto et al. 2015).

10.2.2 Nanomechanical Resonant Filter-Fiber

The resonance frequency response of a nanomechanical filter-fiber (3 μm width, 138 μm length and 220 nm thickness silicon nitride fiber, coated with aluminum film of thickness 50 nm) is determined when it is impacted with nanoparticles at a defined aerosol velocity (Schmid et al. 2013). Maximum frequency shift of a resonant filter-fiber is 1117 ± 403 Hz for 100 ± 8 nm silver particles.

10.2.3 Aerosol Sensing by Voltage Modulation

Figure 10.2(a) shows the organization of an aerosol-sensing chip (Zhang et al. 2017). The sensing chip consists of a microchannel with entrance/exit for aerosol particles passing through the chip. *En route*, the particles are electrically maneuvered. The path of the particles through the chip comprises three sections, in each of which there are two electrodes. These are the charging, precipitation, and measurement sections.

Particle charging is done by triggering a positive corona discharge between a planar electrode and a needle-shaped tungsten tip, by applying a high voltage. The accelerated electrons readily dislodge electrons from the air molecules, forming positive ions. The collisions of the particles with these ions make them charged.

When the charged particles and gaseous ions enter the precipitation compartment, a square voltage waveform of lower level V_1 and higher level V_2 is applied at a frequency of 0.1 Hz between the planar electrodes. At the lower level V_1, all the excess gas ions precipitate while the charged particles move onward. At the higher level V_2, the remaining excess gas ions and a proportion of the charged particles precipitate. Collection of the particles on the precipitation electrode takes place in accordance with their mobilities. The precipitation process is utilized to determine the average diameter of the particles.

In the measurement compartment, a voltage is applied to trap the particles on the planar measurement electrodes, and to export

TABLE 10.1
Major Air Pollutants: Their Sources and Effects

Sl. No.	Name of Pollutant	Sources	Effects on Humans and the Environment
1.	Particulate matter	Suspended dust, soot, pollen, mold, silica, asbestos particles in the air from buildings under construction, thermal power stations, etc. It is designated by particle size, such as PM2.5 for 2.5 μm particles and PM10 for 10 μm particles, expressed in μg/m³.	It leads to chronic respiratory tract problems, lung diseases, asthmatic attacks. It also decreases visibility through air and makes it hazy.
2.	Carbon monoxide	Burning of coal, oil, and other fuels; automobile exhausts; home heaters and industrial furnaces.	CO binds with hemoglobin, causing dizziness, tiredness, headaches, asphyxia, and death.
3.	Sulfur dioxide	Coal burning, petroleum extraction, paper manufacturing, metallic ore smelting, volcanoes.	It acts as a precursor to sulfuric acid formation and acid rain. It causes corrosion of metals. It produces chest tightness, breath shortness, and wheezing.
4.	Nitrogen dioxide	Fertilizer manufacturing, biomass or fuel burning, power plants, cars.	Forms nitric acid, irritates eyes, produces coughing and shortness of breath.
5.	Ground-level or tropospheric ozone, as distinguished from stratospheric ozone in the upper atmosphere which protects from ultraviolet radiation	Photochemical reactions in sunlight, in which nitrogen oxides, hydrocarbons and volatile organic compounds participate.	Causes respiratory problems; irritates eyes, nose, and throat.
6.	Volatile organic compounds (VOCs) e.g., benzene, toluene, and xylene (BTX), formaldehyde, etc.	Gasoline tanks, carburetors, microbial activities in sewage.	It creates carcinogenic and toxic effects.
7.	Ammonia	Livestock waste, human excreta, fertilizer plants, decaying organic matter.	Pungent odor, eye, nose, throat irritant.

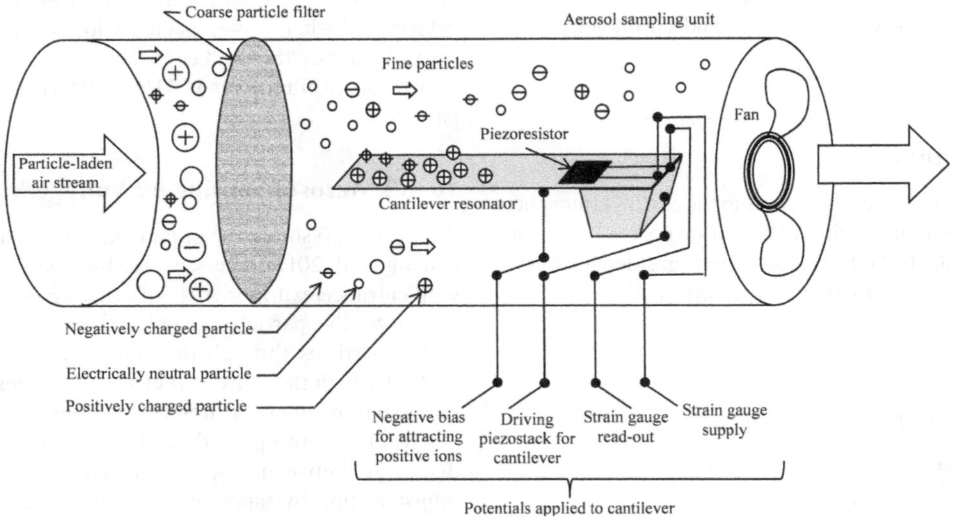

FIGURE 10.1 Nanoparticle sampling unit containing the cantilever (Wasisto et al. 2013, 2015).

two current levels, I_1 and I_2, pertaining to the voltage levels V_1, V_2. As a consequence, a sequence of currents related to the various measurement voltages are obtained for the multi-size particle distribution. These are labeled as I_{11}, I_{21}, I_{31}, I_{41} for I_1 and similarly, I_{12}, I_{22}, I_{32}, I_{42} for I_2. The output currents correspond to particle number concentrations in different ranges of particle sizes. The particle size distribution is estimated through a data fusion algorithm. Thus, particle size in a polydisperse distribution is found by modulating the voltage to control the rate of trapping of charged particles with resultant output current signal generation.

The chip size is 98 mm × 38 mm × 25 mm. Aerosol particles of sizes 30–500 nm are detectable in the concentration range 5×10^2 to 5×10^7 particles/cm³. The control circuitry Figure 10.2b) contains two microcontroller units, under the supervision of which, the corona discharge for particle charging works and the square voltage waveforms are supplied to the precipitation and measurement sections. Also included is a signal conditioning circuit for determining the very low output currents, which are in the femtoampere range (Zhang et al. 2017).

10.2.4 MEMS-Based Particle Detection System

Particle classification is implemented by a micromachined virtual impactor (Figure 10.3), which works by accelerating the

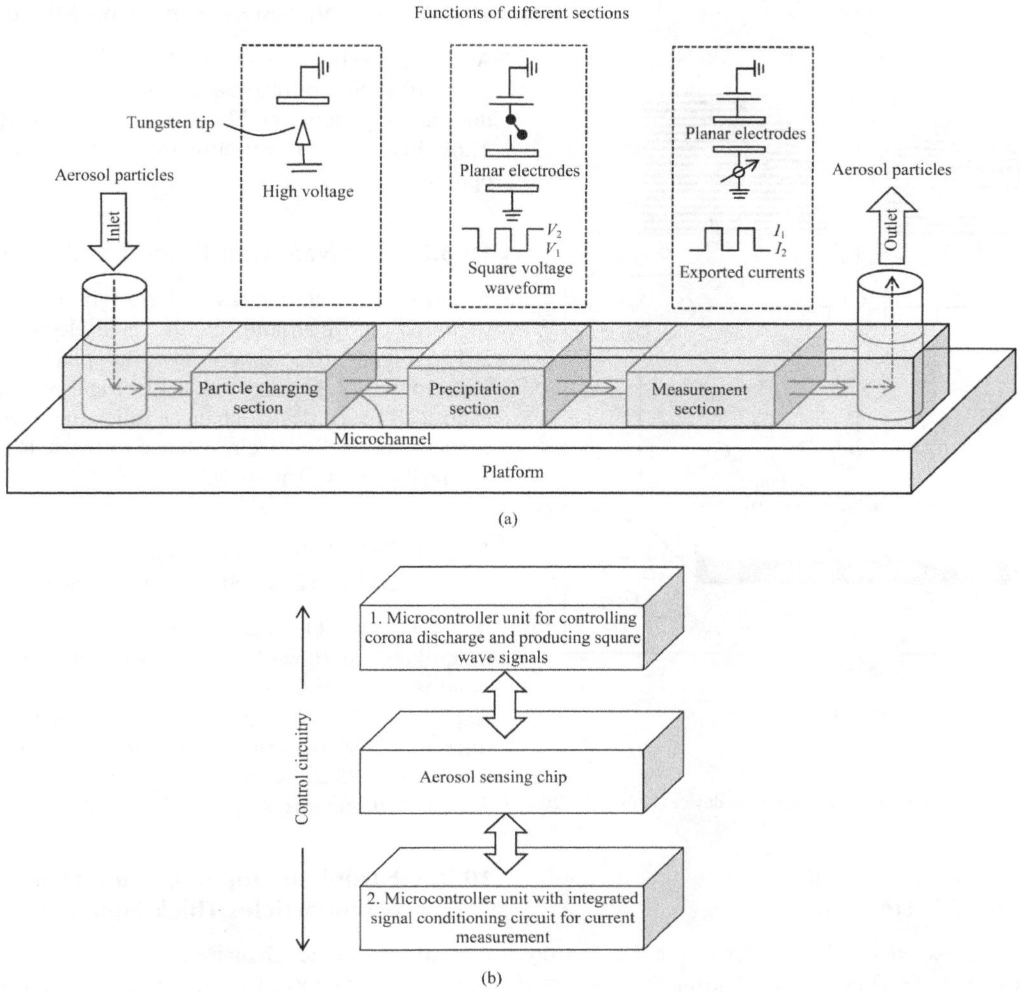

FIGURE 10.2 Aerosol sensing: (a) organization of the aerosol sensor chip and (b) subdivisions of the control circuit (Zhang et al. 2017).

particles through an injection nozzle and changing the direction of flow laterally after acceleration (Kim et al. 2018). The greater inertia or larger size of the heavy particles means that they are unable to change their directions and continue moving in the same direction along straight-line paths, while the lower inertia or smaller-sized light particles can follow the sharp change in direction and easily turn around to move in the side channel. So, the particle flow is subdivided into two orthogonal flows. The heavy particles, persisting on their journey straight ahead, strike an impaction plate coated with a greasy layer. They are coated on this plate and are later removed. Effectively, three subgroups of the particles are formed: a minor flow subgroup and two major flow subgroups, according to their aerodynamic diameters.

Sections for particle classification, charging, precipitation, and sensing are integrated into a single chip (Kim et al. 2018). The layout of the detection system is shown in Figure 10.4. The sensing section is realized with a metal filter, fabricated from sintered nickel. A precise electrometer is used to measure the current in the fA range due to the ultra-fine particles. The electrometer is a current-to-voltage converter. The voltage drop, produced by current flowing through a high-value feedback resistor (100 GΩ), is measured using an operational amplifier with a low-input bias current. The metal filter, along with the operational amplifier and high-value resistor, are all shielded against noise by enclosing them in an aluminum container kept at ground potential. The analog signals are converted into a digital format with a microcontroller. The high-voltage power supply also functions under its commands.

It is used for particles in the size range 20–300 nm. The range of concentrations is 320–10^6 particles/cm^3.

10.3 Nanosensors for Carbon Monoxide Detection

Carbon dioxide detection was considered in Section 9.2.3. Here, more nano devices for CO sensing will be described.

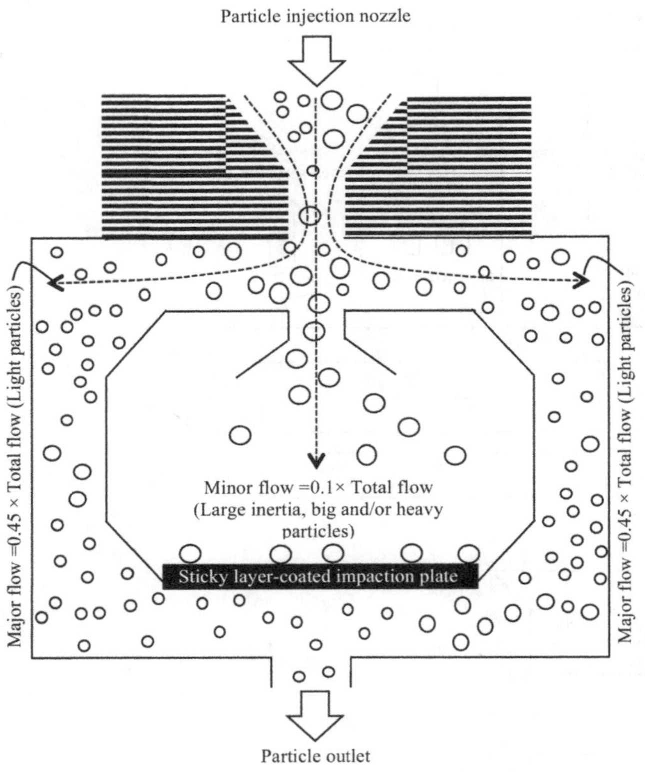

FIGURE 10.3 Virtual impactor particle classification device (Kim et al. 2018).

10.3.1 Au Nanoparticle-Based Miniature CO Detector

Au nanoparticles deposited on the surfaces of a CNT array catalyze the conversion of CO into CO_2, releasing free electrons which cause changes in resistivity of the CNT array (Bakhoum et al. 2013). The sensor is capable of detecting 100 ppm CO.

10.3.2 CuO Nanowire Sensor on Micro-Hotplate

A CuO nanowire is grown on a SiO_2-SiN_x membrane (Steinhauer et al. 2016). Sensor resistance change is measured after stabilizing the temperature at 325°C. Responses change from 6.4% to 27.6% in response to variation in CO concentration from 1 to 30 ppm.

10.3.3 ZnO Nanowall-Based Conductometric Sensor

A network of ZnO-intertwined 2D foils is created on Al film, sputtered onto platinum interdigitated electrodes (Bruno et al. 2017). Figure 10.5 shows how the platinum electrodes are covered with ZnO nano-walls. Electrical resistance of the ZnO nano-wall-based device changes following exposure to low CO concentrations. Relative resistance response is studied at 10, 25, 50 and 80 ppm CO at 400°C.

10.3.4 ZnO NPs-Loaded 3D Reduced Graphene Oxide (ZnO/3D-rGO) Sensor

A ZnO/3D-rGO sensor integrated on a microheater (microhotplate) shows high sensitivity together with linearity and rapid response and recovery times (Ha et al. 2018). The sensor response of 27.5%, response time of 14 seconds, and recovery time of 15 seconds to 1000 ppm CO at room temperature improves to 85.2% sensitivity, with response/recovery times of 7 seconds/9 seconds, respectively, at 200°C.

10.3.5 Europium-Doped Cerium Oxide Nanoparticles Thick-Film Sensor

A relative resistance change of $Ce_{1-(3/4)x}Eu_xO_2$ film is produced on exposure to CO (Ortega et al. 2019). Eu-doped ceria film shows a greater response than undoped ceria, with a 0.9s response time, as compared with 56s for the undoped film. Maximum response is achieved at 400°C.

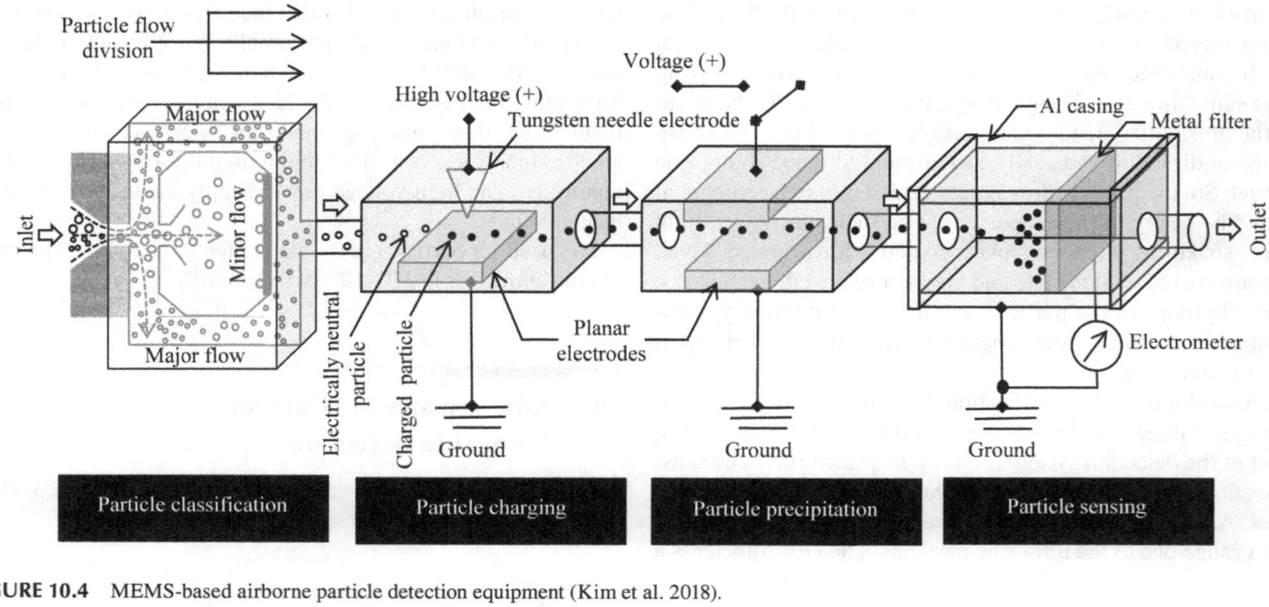

FIGURE 10.4 MEMS-based airborne particle detection equipment (Kim et al. 2018).

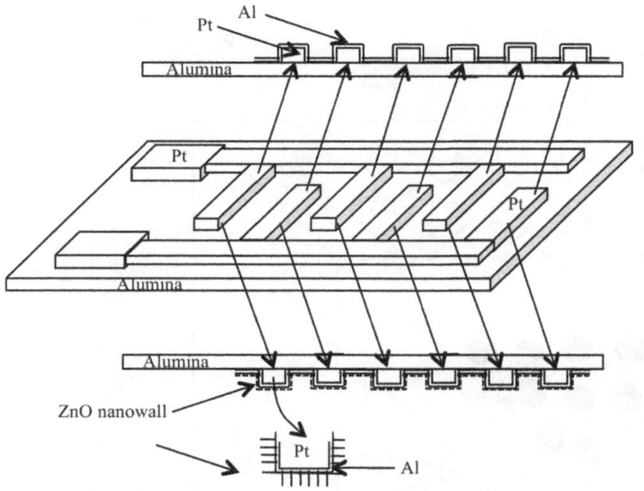

FIGURE 10.5 Aluminum deposition by magnetron sputtering and growth of ZnO nano-walls by chemical bath deposition (CBD) on the platinum interdigitated electrodes of the sensing platform (Bruno et al. 2017).

10.3.6 Pt-decorated SnO_2 Nanoparticles Sensor

Enhanced response of SnO_2 nanoparticles is achieved on decoration with platinum, along with reduction in operating temperature (Peng et al. 2019). A 1.5 mole % Pt-decorated SnO_2 shows response to CO up to 610.45 at 80°C, three times that of pristine SnO_2 sensor.

10.4 Nanosensors for Sulfur Dioxide Detection

10.4.1 Tungsten Oxide Nanostructures-Based Sensor

WO_3 inks are screen-printed on Au electrodes on an alumina substrate with a Pt heater (Boudiba et al. 2012). Response is determined from the resistance change in response to SO_2 exposure. Devices fabricated with WO_3 nanoplates and nanolamellae exhibited high sensitivity to 1–10 ppm SO_2 at 200–300°C, along with a short response time.

10.4.2 SnO_2 Thin-Film Sensor with Nanoclusters of Metal Oxide Modifiers/Catalysts

This sensor is based on the relative resistance change of a SnO_2 thin film with Pt interdigitated electrodes after incorporating catalysts like NiO, PdO, etc. (Tyagi et al. 2014). A maximum response of 56 is achieved with a SnO_2/NiO sensor at 180°C towards 500 ppm SO_2, with a response time of 80 seconds.

10.4.3 Fluorescence Nanoprobe

The nanoprobe works on fluorescence resonance energy-transfer (FRET). A reactive organic molecule (cyanine dye Cy) is functionalized on the surface of carbon nanodots (Sun et al. 2014). The bisulphite molecule has a reactive response to Cy. So, the energy-transfer path between Cy and the carbon nanodot is shut down in the presence of the bisulphite molecule in aqueous solution and thereby the fluorescence is enhanced. SO_2 gas in air is detected by assembling the probe on a test strip. The limit of detection is 1.8 μM and the relationship is linear.

10.4.4 Niobium-Loaded Tungsten Oxide Film Sensor

The response (resistance in dry air/resistance in analyzed gas) of sensing film, made from Nb-loaded hexagonal WO_3 nanorods on interdigitated Au electrodes on an alumina substrate, is measured (Kruefu et al. 2015). A 0.5 wt % Nb-loaded WO_3 nanorod film gives a response of ~10 with a response time of 6 seconds to 500 ppm SO_2 at 250°C.

10.4.5 Nickel Nanowall-Based Sensor

The relative resistance change in response to SO_2 concentration is measured (Hien 2018). Maximum response to SO_2 is 8.8% at 300°C, with the limit of detection being 1 ppm SO_2 with a response of 2.5%.

10.5 Nanosensors for Nitrogen Dioxide Detection

10.5.1 SnO_2 Nanoribbon Sensor

This sensor works by comparison of resistance of an SnO_2 nanoribbon deposited on prefabricated gold electrodes on an insulating substrate in pure air relative to an NO_2 environment (Law et al. 2002). Resistance of SnO_2 nanoribbon in the dark, under 254-nm UV or 365-nm UV is higher in 100 ppm NO_2 than the corresponding values in pure air.

10.5.2 Tris(hydroxymethyl) Aminomethane (THMA)-Capped ZnO Nanoparticle-Coated ZnO Nanowire Sensor

$\Delta R/R$ response of a ZnO nanowire film grown on alumina substrate, provided with Pt interdigitated electrodes and coated with THMA-functionalized nanoparticles, is determined in the presence of NO_2 (Waclawik et al. 2012). A meander Pt heater/temperature sensor is made on the back side. Response of the sensor coated with THMA-functionalized ZnO nanoparticles towards 2 ppm NO_2 is double that of the as-prepared ZnO nanowire sensor.

10.5.3 In_2O_3-Sensitized CuO-ZnO Nanoparticle Composite Film Sensor

Resistances of the film spin-coated on gold interdigitated electrodes are measured in the target gas and dry air (Li et al. 2018). A 1 wt% In_2O_3-decorated CuO-ZnO film shows a response of 82% to 100 ppm NO_2, with a response time of 7 seconds. The detection limit is 1 ppm with a response of 17% and a response time of 136 seconds.

10.5.4 UV-Activated, Pt-Decorated Single-Crystal ZnO Nanowire Sensor

This sensor relies on the resistance variation of ZnO nanowires loaded with Pt on interdigitated gold electrodes on a ceramic

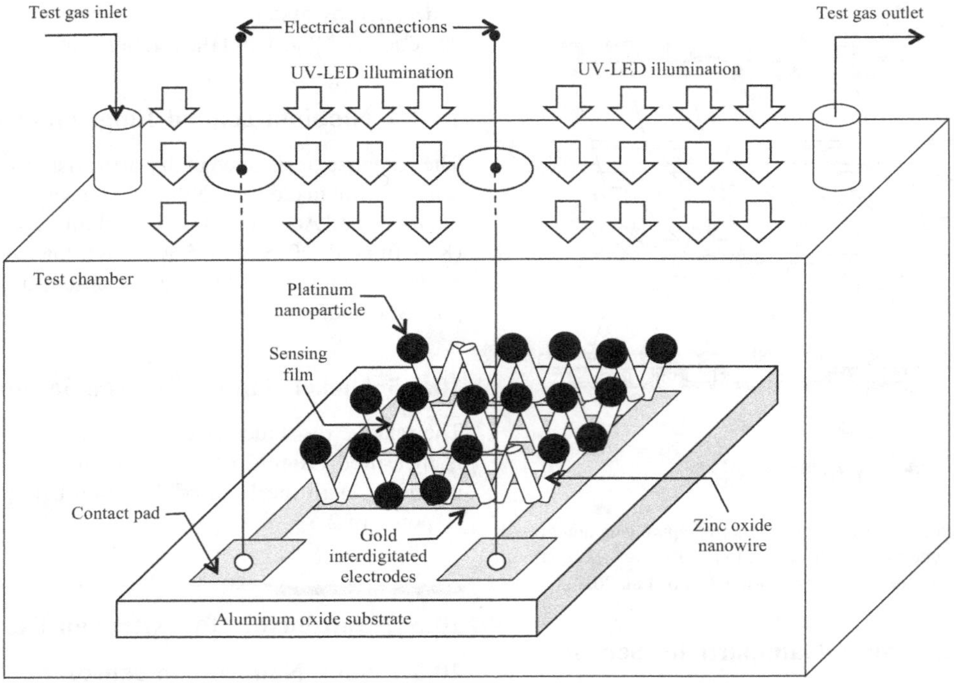

FIGURE 10.6 Photo-activated ZnO nanowire sensor with testing arrangement (Espid et al. 2019).

substrate (Espid et al. 2019) (Figure 10.6). Response toward 5 ppm NO_2 under 365-nm UV-LED is improved with ZnO nanowires and Pt-loaded nanowires as compared with ZnO nanoparticles.

10.6 Nanosensors for Ozone Detection

10.6.1 SnO_2/SWCNT Hybrid Thin-Film Sensor

This sensor determines ozone concentration from the changes in resistance of SnO_2/SWCNT thin film deposited by dip coating from SWCNTs dispersed in tin-based sol over interdigitated electrodes on silicon dioxide layer on silicon (Berger et al. 2011). It is capable of sensing 21.5 ppb ozone in air.

10.6.2 Nanocrystalline $SrTi_{1-x}Fe_xO_3$ (STF) Thin-Film Sensor

The sensor functions by measuring the change in resistance of a perovskite nanostructured $SrTi_{1-x}Fe_xO_3$ (STF) film synthesized on SiO_2/Si substrate by the polymeric precursor method (Masterlaro et al. 2013). Resistance of $x=0.075$ film changes by an order of magnitude on exposure to 600 ppb of ozone at 250°C; the response time is 2 minutes and the recovery time is <5 minutes.

10.6.3 ZnO Nanoparticle Sensor

It is fabricated by photolithography/laser ablation on a flexible substrate (Acautla et al. 2014). Normalized sensor response (resistance in target gas/resistance in dry air) of a ZnO nanoparticle thin film (200-nm thickness) deposited by drop coating on Ti/Au electrodes on a kapton polyimide flexible substrate is measured with respect to ozone concentration. The sensor is capable of 5 ppb to 300 ppb ozone detection at 300°C

10.6.4 Pd-Decorated MWCNT Sensor

Here, ozone is estimated from a resistance change of a (MWCNT-Pd) thin film drop-coated on interdigitated Pt electrodes on SiO_2 from a glycerol dispersion (Colindres et al. 2014). The best response is achieved at 120°C in the range 20–300 ppb, with a response time of 60 s.

10.6.5 UV-Illuminated ZnO Nanocrystal Sensor

This sensor works by $R_{ozone}/R_{dry\ air}$ measurement of ZnO nanoparticles spin-coated from colloidal solution on Ti/Pt interdigitated electrodes on a SiO_2/Si substrate (Bernardini et al. 2017). A concentration of 35 ppb ozone is detected by replacing thermal excitation with UV illumination.

10.7 Nanosensors for VOC Detection

10.7.1 Chemiresistive Sensor Using Gold Nanoparticles

The sensor is fabricated by inkjet printing of an Au nanoparticle suspension on spiral gold interdigitated electrodes (Garg et al. 2010). The Au nanoparticles are capped with trithiol ligands. On exposure to toluene, ethanol, dichloromethane, methanol and acetone, its resistance varies. The sensor is highly sensitive to toluene in the concentration range 300–1200 ppm. Response to other volatile organic compounds (VOCs) is moderate.

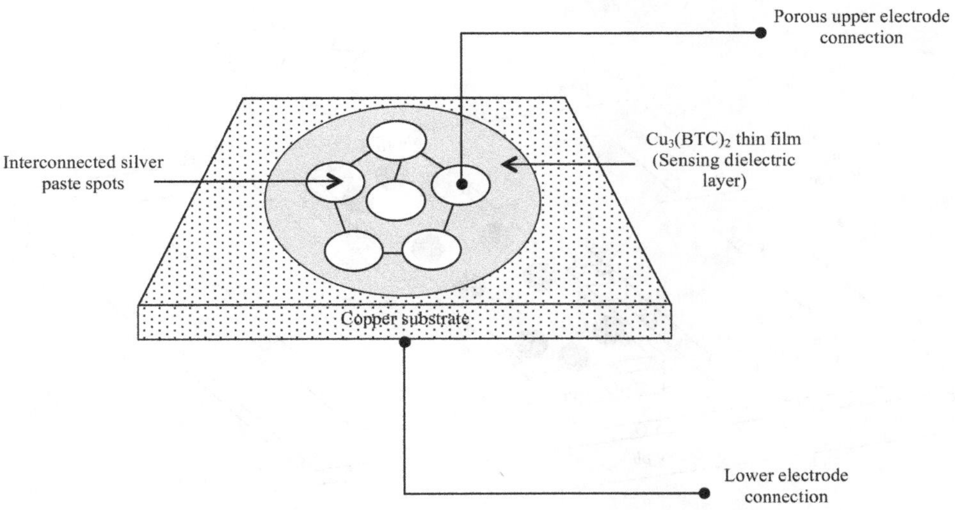

FIGURE 10.7 Capacitive sensor using MOF nanoparticles (Homayoonnia and Zeinali 2016).

10.7.2 Metal-Organic Framework (MOF) Nanoparticle-Based Capacitive Sensor

The lower electrode of the capacitor is made of copper, and the dielectric is an MOF with the chemical formula $Cu_3(BTC)_2$, where BTC=benzene tricarboxylate (Homayoonnia and Zeinali 2016). The upper electrode consists of interconnected silver spots (Figure 10.7). Variation of the capacitance of the sensor with analyte concentration is recorded. Acetone, methanol, ethanol, and isopropanol are easily detectable in the concentration range 250–1500 ppm.

10.7.3 Al-Doped ZnO Nanowire {(ZnO:Al)NW}Sensor

NWs are spread on SiO_2/Si wafer with pre-patterned Cr/Au pads for electrical contact and the

$$\text{Response } (R,\%) = \frac{\text{Current}_{\text{Gas}} - \text{Current}_{\text{Air}}}{\text{Current}_{\text{Air}}} \times 100 \quad (10.1)$$

is measured (Pauporté et al. 2018). Response to 50 ppm ethanol is 15%, to 50 ppm 2-propanol is 10%, and to 50 ppm *n*-butanol is 9%. Response time of individual nanowires (NWs) to VOCs is 10 s and recovery time is 30–40s.

10.7.4 Nickel-Doped Tin Oxide Nanoparticle (Ni-SnO_2 NP) Sensor for Formaldehyde

Ni-SnO_2 NPs are prepared by a one-step hydrothermal process (Hu et al. 2018). The sensor is fabricated by coating the Ni-SnO_2 NPs solution in ethanol on the outer surface of an alumina tube with two gold electrodes and Pt connection wires. The required electrical heating at temperatures 100–500°C is provided by a nichrome heating coil inserted inside the alumina tube. The sensor is heated at 180°C for one week to ensure long-term stability.

Among the different Ni-doping concentrations used, the 5 atomic percent (at%) Ni is adjudged to provide the most responsive sensor to HCHO, while the optimum operating temperature is 200°C. Resistance in air is high and decreases in ambient HCHO. The sensitivity to 50 ppm HCHO at 200°C is 104 as found from the ratio (resistance in air/resistance in HCHO). The limit of detection is 120 ppb. The response of 5 at% Ni-doped SnO_2 NPs is 10 times that of pure SnO_2 NPs.

10.7.5 Palladium Nanoparticle (PdNP)/Nickel Oxide (NiO) Thin- Film/Palladium (Pd) Thin-Film Sensor for Formaldehyde

The formaldehyde sensor is shown in Figure 10.8. A 100-nm thick SiO_2 film is deposited on a sapphire substrate by radio-frequency (RF) sputtering (Chen et al. 2018). The 10-nm Cr/15-nm Pt are thermally evaporated on a photoresist pattern, and Cr/Pt interdigitated electrodes are defined by lift-off photolithography. The 30-nm thick NiO sensing layer is deposited by RF sputtering and subjected to rapid thermal annealing (RTA). A 2-nm thick Pd thin film is deposited over the NiO layer by thermal evaporation. Finally, for PdNP deposition, the $PdCl_2$ solution is drop-coated on a Pd thin film and illuminated with UV for 5 hours. After annealing, the device is bonded to a metal can.

The optimal operating temperature is 250°C. The sensing response, *S*, defined as

$$S = \frac{\text{Resistance in HCHO ambience} - \text{Resistance in air}}{\text{Resistance in air}} \quad (10.2)$$

is 10.1 in 20 ppm HCHO/air gas. The limit of detection is 16 ppb.

10.7.6 Surface Acoustic Wave (SAW) Sensor With Polymer-Sensitive Film Containing Embedded Nanoparticles

A SAW device was described in Section 4.2. On a similar principle, a SAW delay line is fabricated on ST-X cut quartz substrate with gold interdigital transducers (Constantinoiu and Viespe 2019). The delay line is shown in Figure 10.9. The oscillation frequency is 69 MHz. The sensitive film consists of polyethyleneimine (PEI) and polydimethylsiloxane (PDMS) polymers, mixed with ZnO, TiO_2, or WO_3 nanoparticles.

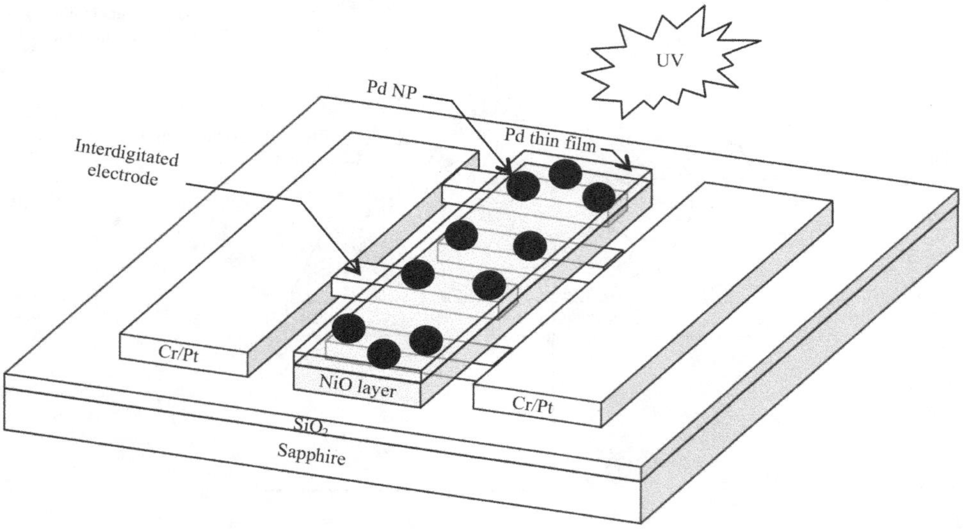

FIGURE 10.8 NiO thin-film sensor: (a) schematic layout and (b) operating mechanism (Chen et al. 2018).

The frequency shift of the sensor is associated with a concentration of the target gas. The PEI/WO_3 sensor has a limit of detection (LoD) = 6 ppm methanol, 15 ppm acetone and 9 ppm dichloromethane. For the PEI/ZnO sensor, the LoD=9 ppm toluene.

10.7.7 Resorcinol-Functionalized Gold Nanoparticle Colorimetric Probe for Formaldehyde Detection

Color change of the probe in line with formaldehyde concentration in solution or the gaseous phase is noted (Martínez-Aquino et al. 2019). The probe is synthesized as an aqueous red-wine-colored dispersion in the absence of formaldehyde. After 12–15 minutes of addition of 50 mM formaldehyde, the color changes to dark blue. Figure 10.10 depicts the effect of formaldehyde on the sensor.

LoD is 0.5 ppm for formaldehyde in an aqueous solution. Gaseous formaldehyde emissions are detected by placing the formaldehyde sources close to an open vial, containing the probe, inside a locked container for 15 hours, and recording the UV spectra.

10.8 Nanosensors for Ammonia Detection

In Section 8.5.5, we described an ammonia sensor using MWCNTs and tin oxide. More examples of ammonia nanosensors will be given here.

10.8.1 Polyaniline Nanoparticle Conductimetric Sensor

PANI NPs films are inkjet printed on interdigitated electrodes (Crowley and Killard 2008). The sensor is operated in conductometric mode. Sensitivity, linearity of response and recovery time are improved on operating at 80°C. The sensor has a logarithmic response to ammonia from 1 to 100 ppm.

10.8.2 MoO_3 Nanoparticle Gel-Coated Sensor

Sol-gel synthesized MoO_3 nanoparticles are spin-coated on Pt electrodes plated on an alumina substrate. Resistance response is measured with varying ammonia concentration (Gouma et al. 2010). Ammonia is detected up to 50 ppb at 500°C.

10.8.3 ZnO:Eu^{2+} Fluorescence Quenching Nanoparticle-Based Optical Sensor

The sensor is fabricated from ZnO:Eu^{2+} nanoparticles which are excited by near-UV 378 nm radiation to emit indigo colored light (Yang et al. 2018). Change in luminescence intensity of excitation spectrum is determined after exposing to ammonia. LoD is 20 ppm with linear characteristic from 0 to 80 ppm.

10.8.4 Pt Nanoparticle (Pt NP)-Decorated WO_3 Sensor

This has been described in Figure 10.11. On a sapphire substrate, 10-nm/15-nm Cr/Pt interdigitated electrodes are formed by vacuum evaporation and lift-off photolithography (Liu et al. 2019). Then 25-nm thick WO_3 film is deposited by RF sputtering. For Pt NPs, the $H_2PtCl_6(H_2O)_6$ solution in distilled water is dispersed on the WO_3 film, followed by UV illumination for reduction of $PtCl_6^{2-}$. The PtNPs are stabilized by annealing in nitrogen.

The optimal working temperature of the device is 250°C. At this temperature, the resistance of the device decreases from 73.9 MΩ in air to 2.67 MΩ in 1000 ppm NH_3/air, providing a response of

$$S = \frac{\text{Resistance in air} - \text{Resistance in gas}}{\text{Resistance in gas}} = \frac{73.9 - 2.67}{2.67} = 26.68 \tag{10.3}$$

At the same temperature, the response to 1 ppm NH_3/air is 3.73. The sensor owes its high sensitivity to the catalytic activity and larger surface area-to-volume ratio of PtNPs, as evidenced by

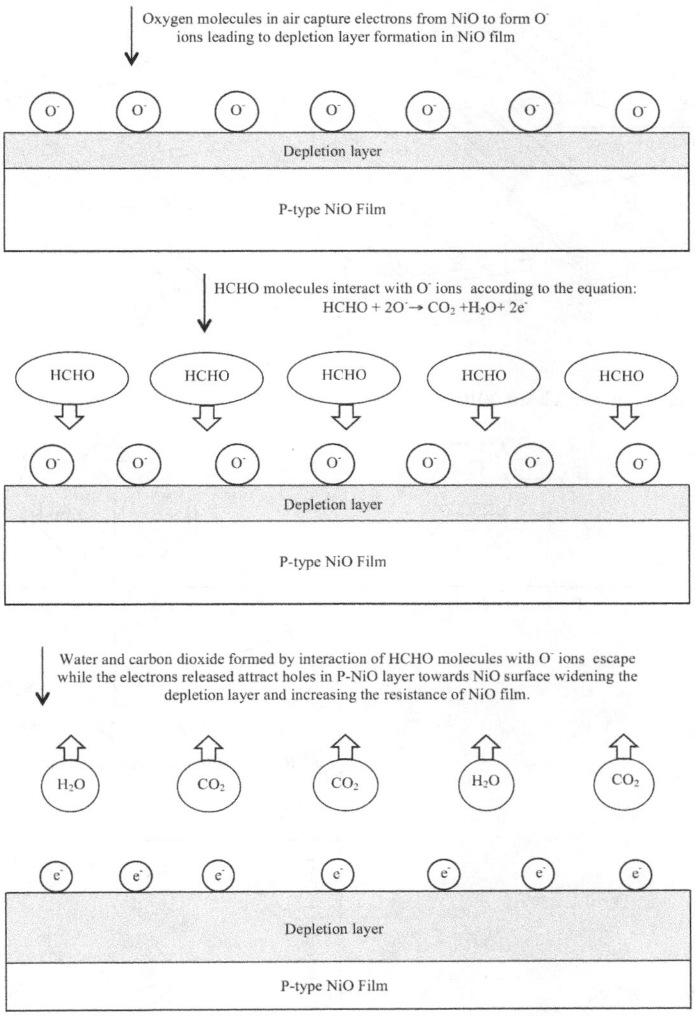

(b)

FIGURE 10.8 (Continued)

the superior performance of Pt NPs-decorated sensor than the pristine WO_3 film sensor.

The sensor shows high selectivity towards ammonia when tested against methanol, ethanol, and methane gases.

10.9 Water Pollutants

Water pollution is the contamination of natural sources of water, such as lakes, ponds, rivers, and oceans, by dumping of chemical substances, waste materials, or harmful microorganisms caused by human activities e.g., by industrial products or garbage. Table 10.2 offers a glimpse of these water contaminants.

10.10 Nanosensors for Detection of *Escherichia coli* 0157:H7

Escherichia coli 0157:H7 is a serotype of the *Escherichia coli* bacterium. *E. coli* infection can occur due to swallowing fouled water while swimming in a pond, drinking unpasteurized milk and fruit juices, or eating uncooked ground meat. It causes food poisoning and diarrhea. Moving ahead from preliminary discussion of *E. coli* detection in milk samples in Section 9.2.9, many recently developed *E. coli* nanosensors are explained in the ensuing subsections.

10.10.1 Magnetoelastic Sensor Amplified With Chitosan-Modified Fe_3O_4 Magnetic Nanoparticles (CMNPs)

Magnetoelastic sensors, made from amorphous ferromagnetic alloys of Fe, Ni, Mo, and B, vibrate mechanically under the influence of a time-varying magnetic field at a characteristic resonance frequency, producing a magnetic flux, which is remotely detected by a pick-up coil to provide wireless sensing. For sensor fabrication, 20-mm long magnetoelastic sensors are cut from the alloy: $Fe_{40}Ni_{38}Mo_4$ B_{18} (Lin et al. 2010). The uncoated sensor has a resonance frequency of 102.5 kHz in air. It is ultrasonically cleaned and coated with polyurethane.

The assay is carried out by adopting the sequence of steps shown in Figure 10.12. It is done by immersing the

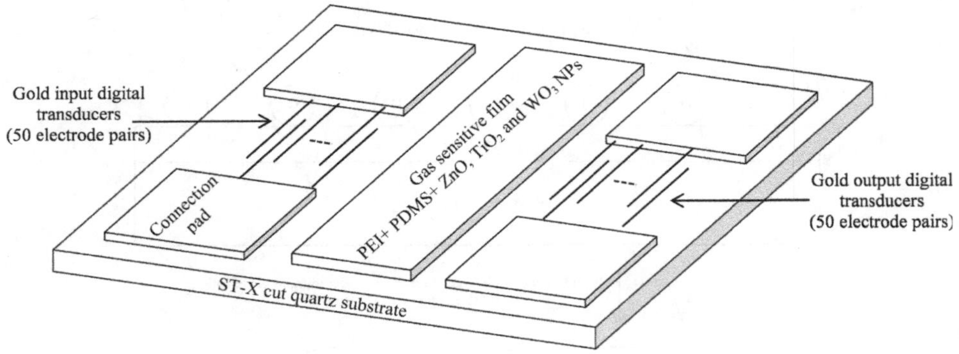

FIGURE 10.9 SAW delay line (Constantinoiu and Viespe 2019).

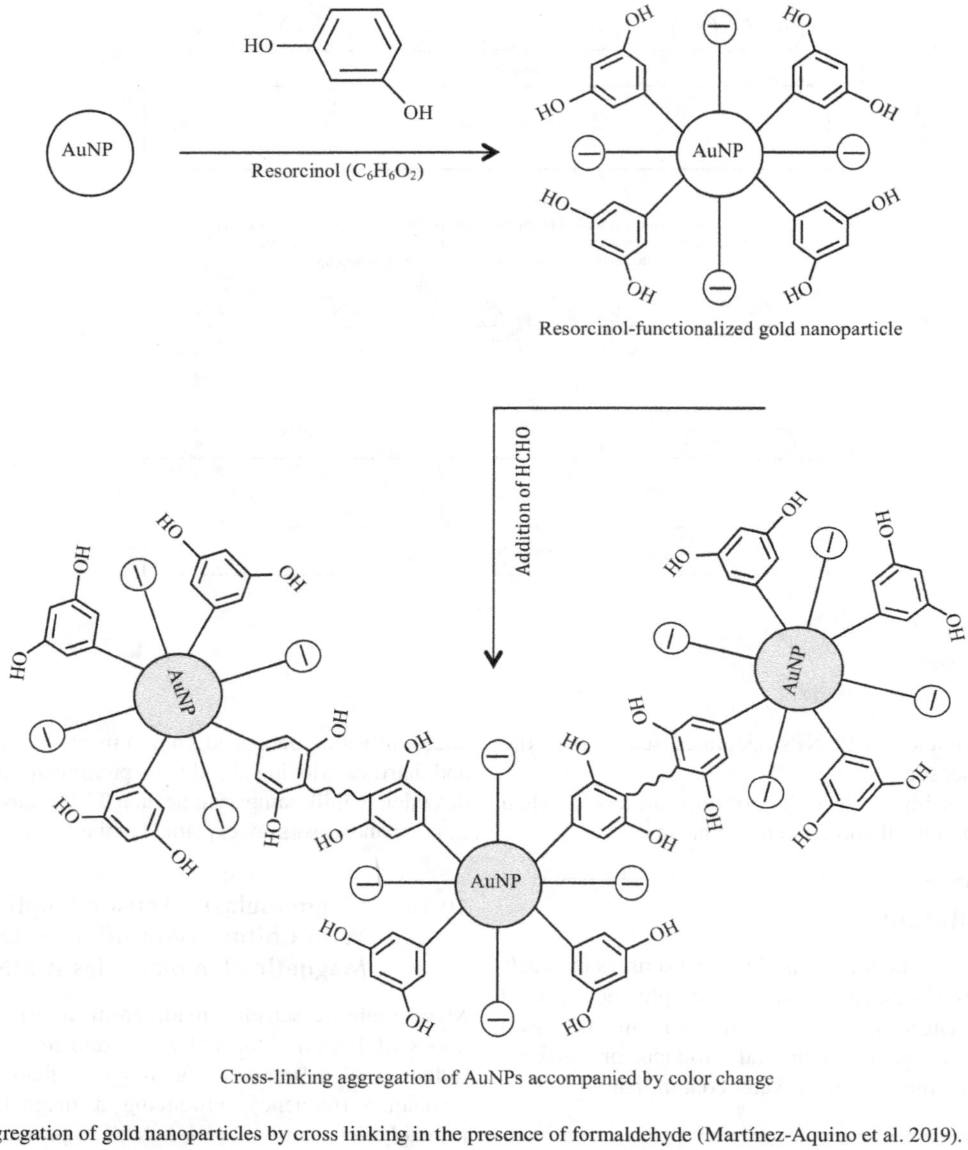

FIGURE 10.10 Aggregation of gold nanoparticles by cross linking in the presence of formaldehyde (Martínez-Aquino et al. 2019).

polyurethane-coated sensor in a cuvette containing *E. coli* and CMNPs in a phosphate buffer. At a suitable pH (5–6.5), the negatively charged *E. coli* O157:H7 is electrostatically attracted and adsorbed by the positively-charged CMNPs, and are then adsorbed on the polyurethane-coated magnetoelastic sensor under the action of magnetic flux to form an *E. coli*-CMNPs complex. Consequent upon the adsorption and loading with *E. coli*, the mass of the magnetoelastic sensor increases, causing a decrease in its resonance frequency. The shift in resonance frequency is proportional to the logarithm of *E. coli* concentration in the range from 10 cells/mL to 3.7×10^8 cells/mL. The detection limit is 10 cells/mL.

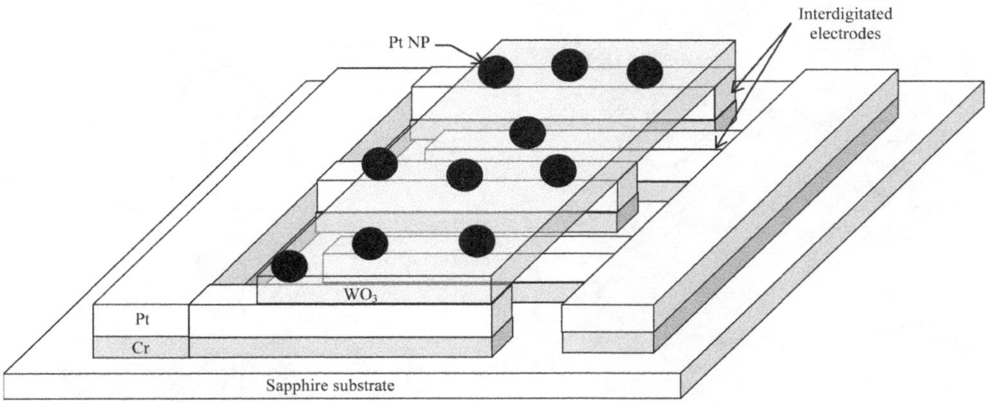

FIGURE 10.11 PtNPs-WO_3 sensor (Liu et al. 2019).

TABLE 10.2
Common Water Pollutants

Sl. No.	Name of Pollutant	Description
1.	Pathogens	Bacteria (coliform, *E. coli*, etc.), protozoa and viruses.
2.	Inorganic materials	Heavy metals such as mercury, lead, arsenic, copper, chromium, zinc, barium, etc.) due to leaching from wastes or industrial accidents.
3.	Organic materials	Methyl-*tert*-butyl ether (MTBE) causing leukemia, lymphoma, etc.
4.	Macroscopic items	Plastic waste, wooden or metal pieces, etc.

10.10.2 Mercaptoethylamine (MEA)-Modified Gold Nanoparticle Sensor

This sensor works on colorimetric principle (Figure 10.13). MEA is conjugated to gold nanoparticles, forming a MEA-NP sensor. MEA-NPs bind to *E. coli* 0157:H7 bacteria through electrostatic attraction between positively charged MEA and negatively charged *E.coli* 0157:H7, resulting in an intense color change from red to blue, which is easily differentiated by the unaided eye (Su et al. 2012).

The relationship between the absorption ratio (given by absorption at 625 nm/absorption at 520 nm) *versus E. coli* 0157:H7 concentration in the range 2.91×10^8 to 16×10^3 CFU/mL is linear.

10.10.3 Cysteine-Capped Gold Nanoparticle Sensor

Like the sensor of Section 10.11.2, the operating principle of this sensor is colorimetric. Electrostatic attraction between the positive potential of cysteine in CysAuNPs and the negative potential of *E. coli* 0157:H7 binds together the *E. coli* 0157:H7 and CysAuNPs (Raj et al. 2015). This binding causes a red shift in the plasmon absorption spectrum of CysAuNPs. Broadening of the spectrum takes place, with changes in *E. coli* 0157:H7 concentration. The solution exhibits a change in color from red to blue.

Limit of detection is 100 cells/mL and the sensor characteristic is linear in the range 1×10^{3}–4×10^3 cells/mL.

10.10.4 Three Nanoparticles-Based Biosensor (Iron Oxide, Gold, and Lead Sulfide)

This is an immuno biosensor with electrochemical measurements (Figure 10.14). Polyaniline (PANI)-coated iron (III) oxide magnetic nanoparticles (MNPs), gold nanoparticles (AuNPs) and lead sulfide nanoparticles (PbSNPs) are synthesized (Wang et al. 2015). MNPs are functionalized with monoclonal antibody (mAb) to *E. coli* O157:H7 forming mAb-MNP. AuNPs are conjugated with a polyclonal antibody (pAb) to *E. coli* to form pAb-AuNP. An oligonucleotide is added to the pAb-AuNP conjugate and the linkage between PbSNPs and oligonucleotides on pAb-AuNP is established, giving pAb-AuNP-oligo-PbSNP. When mAb-MNP is added to a sample solution containing the target *E. coli* antigen, the *E. coli* antigens are captured by mAb-MNP, resulting in *E.coli* antigen-mAb-MNP formation. After incubation and blocking with bovine serum albumin (BSA), the *E. coli* antigen-mAb-MNP are separated from the solution by attraction with a magnet. In combination with and after incubation with pAb-AuNP-oligo-PbSNP, the complex MNP-mAb-*E. coli* antigen-pAb-AuNP-oligo-PbSNP is obtained. This complex is magnetically separated and transferred to a screen-printed carbon electrode (SPCE). The PbSNPs are dissolved and deposited on SPCE and the signal of PbS is measured by square wave anodic stripping voltammetry (SWASV).

The roles played by the three nanoparticles are as follows: MNPs help to separate target *E. coli* bacteria from the sample, AuNPs serve as labels to the *E. coli* bacteria, and PbSNPs produce the electrochemical signal. The PbSNPs provide the signal amplification. Because each AuNP links to several PBSNPs, every binding event to target *E. coli* bacteria results in appreciable enhancement of the signal.

The lead signal is obtained on the square wave voltammetric sensorgram at −0.7 V. The biosensor can detect *E. coli* bacteria in the range 10^1–10^6 CFU/mL. The detection time is < 1 hour (Wang et al. 2015).

10.10.5 Magneto-Fluorescent Nanosensor (MfnS)

Magnetic relaxation and optical fluorescence principles are applied in this sensor (Figure 10.15). Monoclonal IgG1 antibodies

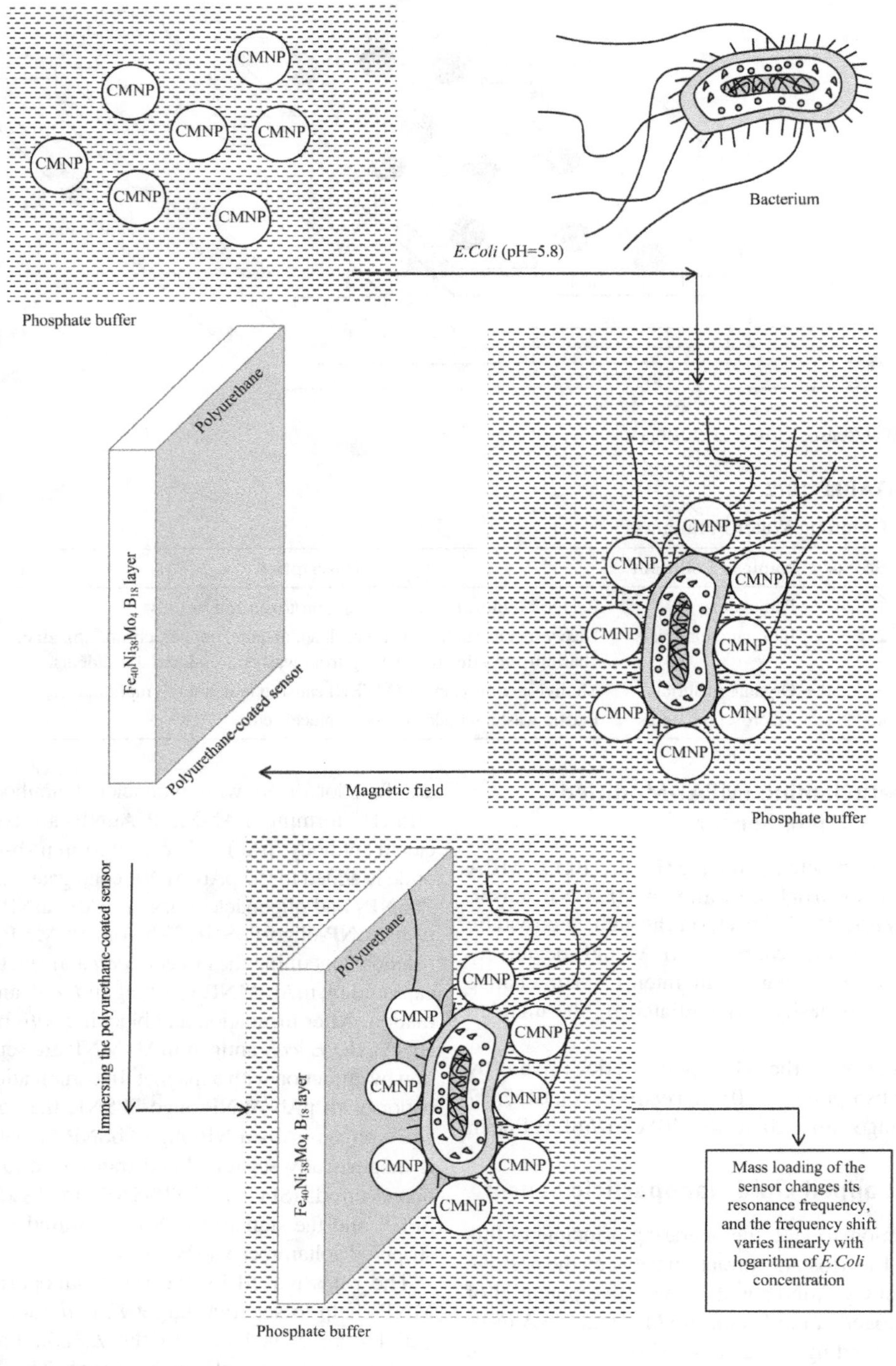

FIGURE 10.12 CMNPs-supported *E. coli* collection on the magnetoelastic sensor surface (Lin et al. 2010).

(Abs) specific to *E. coli* 0157:H7 are conjugated to poly(acrylic acid) (PAA)-coated iron oxide nanoparticles (IONPs) forming Ab-conjugated IONPs (Banerjee et al. 2016). Inside the PAA coatings of the IONPs, a lipophilic optical dye is encapsulated. Low concentrations of *E. coli* 0157:H7 are detected *via* an increase in magnetic relaxation time, whereas higher concentrations are measured from an increase in the intensity of fluorescence emission from the dye.

Magnetic relaxation and fluorescence modes work in complementary fashion to provide information from as low as 1 CFU (colony forming unit) of *E. coli* 0157:H7 to much higher CFU values.

Magnetic modality: Due to interaction between IgG1 antibodies (Abs) and bacterial epitopes, the magnetic nanosensors form clusters when placed in solution, whereby their interaction with water protons is inhibited, resulting in an increase in magnetic relaxation time T_2. At low bacterial concentrations, the magnetic nanosensor clustering increases, causing a larger change in T_2, i.e., a larger ΔT_2. At high bacterial concentrations, the magnetic nanosensors are dispersed far apart, producing smaller ΔT_2.

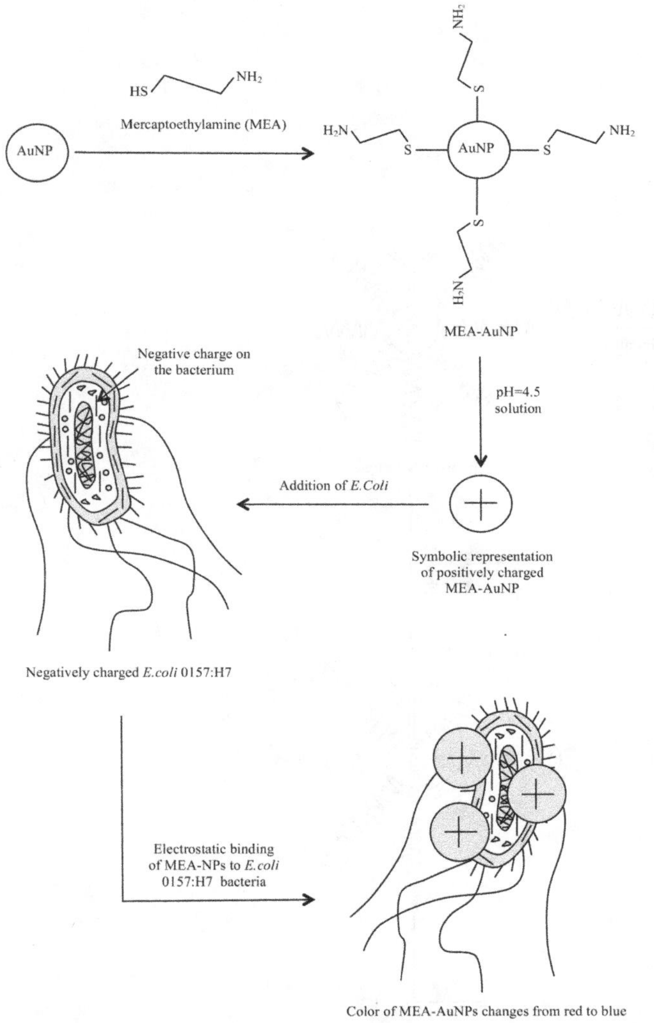

FIGURE 10.13 *E. coli* detection by a MEA-modified AuNP sensor (Su et al. 2012).

Optical fluorescence modality: The steps are: encapsulation of the lipophilic optical dye DiI in the PAA coating of the IgG1-conjugated IONP, purification by a magnetic column and dialysis, incubation with bacterial solution, centrifugation to remove unbound bacteria, resuspension, and fluorescence emission measurement. Emission intensity is related to bacterial concentration.

10.10.6 Signal-Off Impedimetric Nanosensor With a Sensitivity Enhancement by Captured Nanoparticles

Figure 10.16 shows the electrode modification procedure and binding of *E. coli* bacteria to the modified electrode. *E. coli* 0157:H7 antibody is covalently grafted on a self-assembled monolayer (SAM)-modified gold electrode (Wan et al. 2016). After the *E. coli* 0157:H7 bacteria are captured by the electrode, it is exposed to AuNPs. These AuNPs bind to the bacteria, serving as electron-transfer pathways across the SAM layer and thereby decreasing the electron-transfer resistance between the $[Fe(CN)_6]^{3-/4-}$ redox probe and the gold electrode surface, as measured by electrochemical impedance spectroscopy.

The limit of detection is 100 CFU/mL, and the dynamic range $= 3 \times 10^2 - 1 \times 10^5$ CFU/mL. A 10-fold signal amplification is obtained due to fastening of AuNPs.

10.10.7 An Impedimetric Biosensor for *E. coli* O157:H7 Based on the Use of Self-Assembled Gold Nanoparticles (AuNPs) and Protein G-Thiol (PrG-Thiol) Scaffold

Planar gold electrodes are modified with AuNPs through 1,6-hexathiol linker molecules and protein G (Lin et al. 2019). The sensor is characterized by cyclic voltammetry and electrochemical impedance spectroscopy.

The limit of detection is 48 CFU/mL and the dynamic range extends up to 10^7 CFU/mL.

10.10.8 Gold Nanoparticles Surface Plasmon Resonance (AuNP SPR) Chip

AuNPs are surface imprinted on the SPR chip (Figure 10.17). Interaction between negatively charged *E coli* cell wall components and amine-functionalized AuNPs is provided by adding positive Cu(II) ions during surface imprinting (Gur et al. 2019). Refractive index changes and resulting resonance frequency shifts are measured and reflect variations in *E. coli* concentration.

The method provides ultrasensitive detection, with a limit of detection of 1 CFU/mL.

10.10.9 Microfluidic Nanosensor Working on Aggregation of Gold Nanoparticles and Imaging by Smartphone

Figure 10.18 shows the working of the microfluidic biosensor. Immunomagnetic segregation of *E. coli* bacteria is carried out in the separation chamber, with the colorimetric estimation of aggregated AuNPs being done in the detection chamber of the microfluidic chip (Zheng et al. 2019). *E. coli* O157:H7 calls are mixed with: (i) polystyrene microspheres (PS) modified with detection antibodies (DAb) and catalase (Cat) enzyme molecules, and (ii) MNPs modified with capture antibodies (CAb), in the first mixing channel. After collection in the separation chamber, they are subjected to incubation. Magnetic nanoparticles (MNPs)-*E. coli* bacteria-polystyrene sphere (PS) or MNPs-*E. coli*-PS complexes are formed. The complexes are seized by the external magnetic field in the separation chamber and washed with PBST. Then, H_2O_2 is injected and catalyzed by catalase on the complexes. Finally, AuNPs, along with cross-linking agent, are injected for reaction with catalyzate in the second mixing channel. It is incubated in the detection chamber. Aggregation of AuNPs is activated by cross-linking, causing change in color from blue to red. The change in color is measured by an APP on the smartphone and the change is correlated to *E. coli* concentration.

E. coli concentration up to 50 CFU/ml is measured in 1 hour.

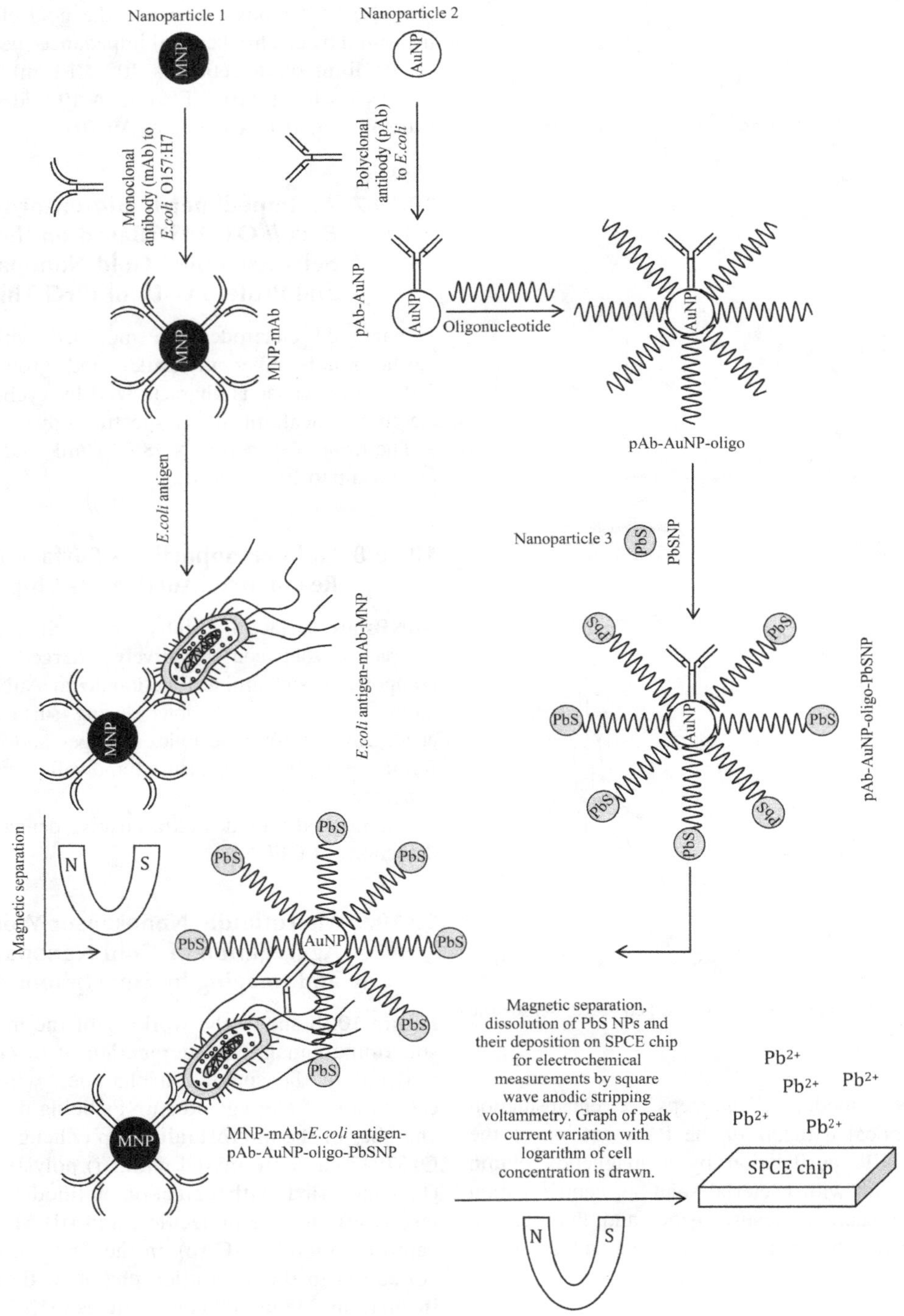

FIGURE 10.14 Biosensor using three types of nanoparticles (Wang et al. 2015).

10.11 Nanosensors for Detection of *Vibrio cholerae* and Cholera Toxin

Vibrio cholerae is the bacterium responsible for the intestinal disease cholera, with cholera toxin being the protein complex secreted by this bacterium. Cholera is a water- and food-borne disease, leading to severe watery diarrhea which may result in dehydration and death.

10.11.1 Lactose-Stabilized Gold Nanoparticles

A lactose derivative is self assembled on gold nanoparticles (Schofield et al. 2007). AuNPs (16-nm diameter) appear red in solution, with the surface plasmon absorption band centered around 520 nm. Cholera toxin B-subunit (CTB) binds to the lactose derivative, inducing aggregation of NPs, whereby the wavelength of the surface absorption band red shifts, acquiring deep purple color.

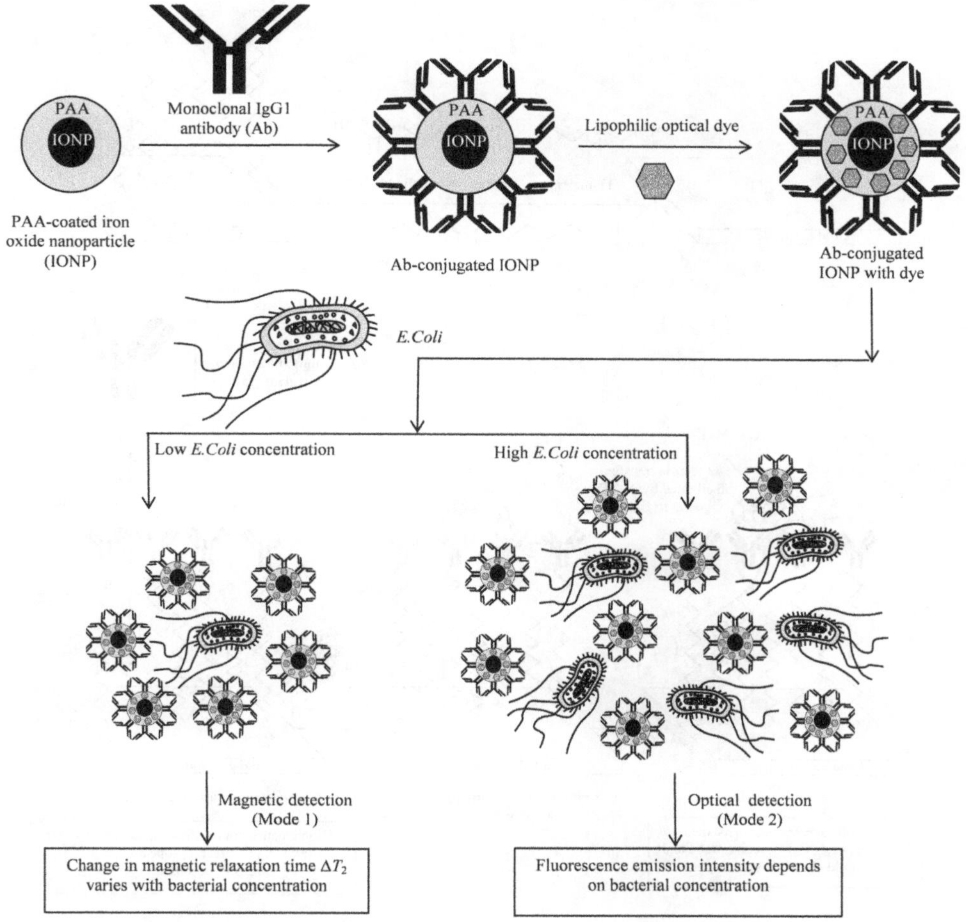

FIGURE 10.15 MfnS for dual-mode *E. coli* detection by magnetic resonance and fluorescence intensity (Banerjee et al. 2016).

The lowest measured value is 110 nM, whereas the theoretical detection limit is 54 nM, and the time of detection is ~ 10 min.

10.11.2 ssDNA/Nanostructured MgO (nMgO)/ Indium Tin Oxide (ITO) Bioelectrode

Vibrio cholerae detection is done through DNA hybridization, using fragmented target DNA (ftDNA) and differential pulse voltammetry (DPV) (Patel et al. 2013). ssDNA is immobilized on nMGO electrophoretically deposited on ITO substrate to form ssDNA/nMgO/ITO (Figure 10.19). Changing concentration of the ftDNA target is hybridized with ssDNA to study the response by DPV.

Peak current increases with rising ftDNA concentration from 100–500 ng/μL. Sensitivity is 16.8 nA/ng/cm^2. The limit of detection is 59.12 ng/μL, with a response time of 3 seconds.

10.11.3 Nanostructured MgO (nMgO) Photoluminescence Sensor

The principle applied is change in photoluminescence (PL) peak intensity in response to the concentration of complementary DNA (cDNA) from *Vibrio cholerae* samples (Patel et al. 2015). Probe DNA (pDNA) is conjugated with nMgO to form a pDNA-nMgO complex. The cDNA from *Vibrio cholerae* samples is hybridized with this complex and the resulting changes in PL intensity at 700-nm red emission are measured. The PL intensity increases with a rise in cDNA concentration.

10.11.4 Lyophilized Gold Nanoparticle/Polystyrene-Co-Acrylic Acid-Based Genosensor

V. cholerae detection is done electrochemically, according to the procedure described in Figure 10.20 by differential pulse anodic stripping voltammetry (DPASV) in acidic medium (HBr/Br_2) (Sheng et al. 2015). The procedure is subdivided into Parts A and B. For Part A, gold nanoparticles (AuNP)-polystyrene-co-acrylic acid (PSA)-avidin in lyophilized form are rehydrated and functionalized with the reporter probe to obtain the conjugate: AuNP-PSA-avidin-reporter probe. For the sandwich hybridization assay, the target DNA (PCR products amplified from genomic DNA (gDNA) of *V. cholerae*) is hybridized with the above conjugate, completing Part A.

Part B consists of a screen-printed electrode (SPE), modified with the capture probe. The hybridized solution made in Part A is pipetted onto SPE (Part B) for hybridization followed by DPASV.

The DPASV is carried out for different concentrations of the target DNA, where it is found that the assay compares well with the wet version, from 1 aM to 1 fM of linear target DNA. The limit of detection is 1 fM of PCR products amplified from gDNA of *V. cholerae*.

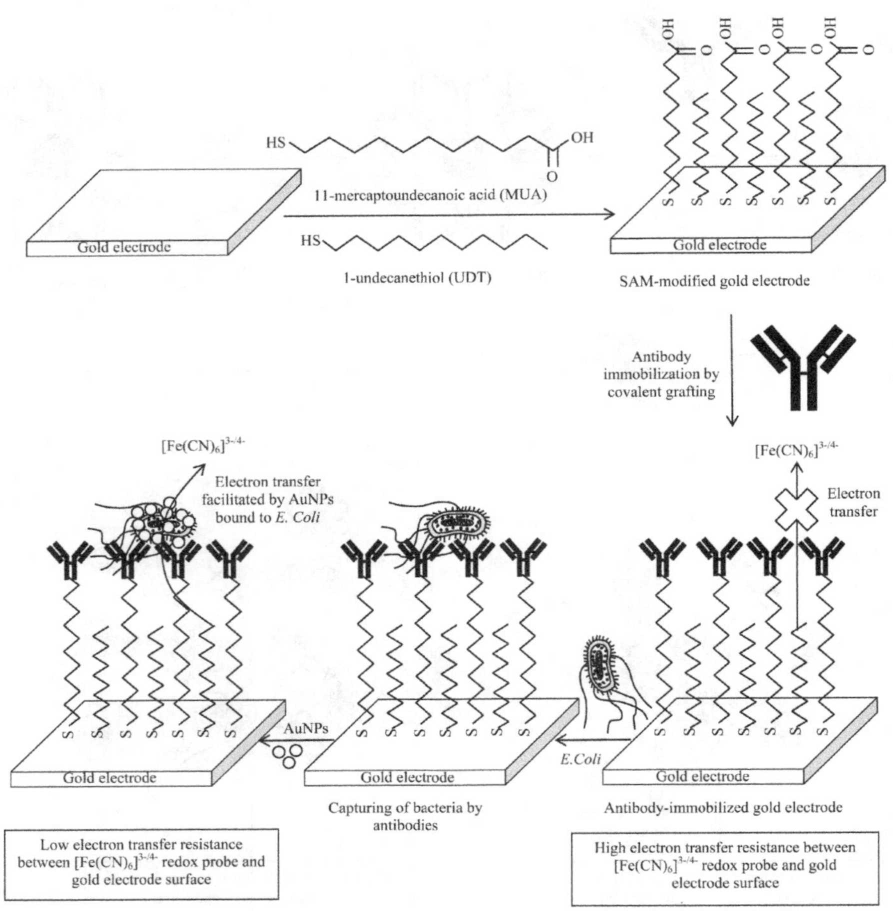

FIGURE 10.16 Gold nanoparticle-aided impedimetric biosensor (Wan et al. 2016).

10.11.5 Polystyrene-co-Acrylic Acid (PSA) Latex Nanospheres

A screen-printed electrode (SPE) is modified with polystyrene-co-acrylic acid (PSA) latex nanospheres and colloidal gold nanoparticles (AuNPs) (Rahman et al. 2017). Figure 10.21 shows the protocol followed in this biosensor. The capture DNA probe is immobilized on PSA latex-AuNPs. Then, the electrode is soaked in cDNA target of *V. cholerae* for partial hybridization, followed by conditioning with the reporter DNA probe for full hybridization. Differential pulse voltammetry is performed with anthraquinone monosulfonate (AQMS) redox species.

The current increases linearly with logarithm of *V. cholerae* cDNA concentration (μM) increases from 1.0×10^{-15} μM to 1.0×10^{-1} μM. The limit of detection is 1.0×10^{-15} μM $= 1.0 \times 10^{-21}$ M.

10.11.6 Graphene Nanosheet Bioelectrode with Lipid Film Containing Ganglioside GM1 Receptor of Cholera Toxin

Graphene nanosheets are deposited by pouring the graphene suspension on a Cu wire placed on a filter of glass fiber (Nikoleli et al. 2018). The stabilized lipid film is prepared by spreading the polymer mixture on the microfilter and irradiating it with UV radiation. The natural cholera toxin receptor ganglioside-monosialic acid GM1 is incorporated into this lipid membrane. The copper wire with the graphene nanosheets is packaged with the ganglioside GM1-immobilized lipid film stabilized on the microfilter to make the bioelectrode. The potential difference between the bioelectrode and the reference electrode is measured for varying concentrations of cholera toxin. The response is linear between 10 nM to 10 μM at pH 7.0.

The sensitivity is ~60 mV/decade ("factor of ten"), the limit of detection is 1 nM, and the response time is ~ 5 min.

10.12 Nanosensors for Detection of *Pseudomonas aeruginosa*

Commonly found in soil and water, this bacterium is known to infect surgical wounds, and urinary and respiratory tracts, particularly in healthcare settings.

10.12.1 Probe-Modified Magnetic Nanoparticles-Based Chemiluminescent Sensor

The procedure implemented for producing the chemiluminescence signal is shown in Figure 10.22. The *gyrB* gene of *Pseudomonas aeruginosa* is amplified by polymerase chain reaction (PCR) of biotin-deoxyuridine triphosphate (dUTP)-labeled DNA fragments bound to $Fe_3O_4@SiO_2$ magnetic nanoparticles (MNPs) (Tang et al. 2012, 2013). Fe_3O_4 MNPs are carboxyl-modified and conjugated

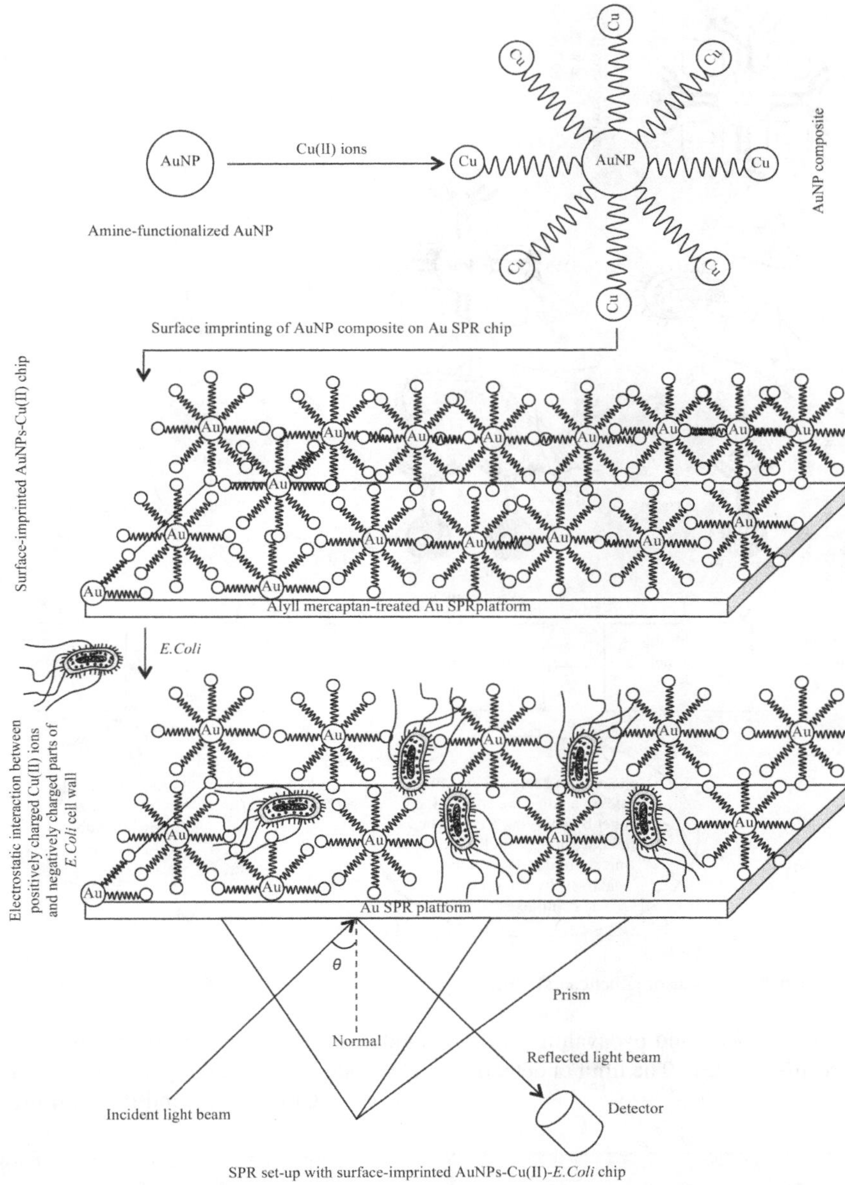

FIGURE 10.17 *E. coli* detection by a SPR nanosensor (Gur et al. 2019).

with amine-modified probes to yield probe-modified magnetic nanoparticles. These probes capture the aforesaid biotin-dUTP-labeled DNA fragments to form complexes which are bonded with streptavidin-modified alkaline phosphatase (SA-AP). AMPPD :3-(2′-spiroadamantyl)-4-methoxy-4-(3″-phosphoryloxy)-phenyl-1,2-dioxetane, the substrate for alkaline phosphatase (AP) enzyme, is added and the chemiluminsecence signal is recorded.

Chemiluminescent intensity increases from 3000 to 4800 as PCR product content rises from 7.5 to 30 fM. Limit of detection is 7.5 fM of *gyrB* fragments.

10.12.2 Reduced Graphene Electrode Decorated with Gold Nanoparticles (AuNPs)

Differential pulse voltammetry (DPV) is carried out, using a reduced graphene/AuNP graphite screen-printed electrode (Figure 10.23) as the working electrode, an Ag/AgCl reference electrode and a carbon counter electrode in $[Fe(CN)_6]^{3-/4-}$ redox mediator in the presence of varying concentrations of pyoverdine (PyoV: a fluorescent siderophore produced by *Ps. aeruginosa*) (Gandouzi et al. 2019). The current response is proportional to PyoV concentration from 0.5–100 μM.

Sensitivity is 0.14 μA/μM. Limit of detection is 66.90 nM.

10.12.3 Polyaniline(PANI)/Gold Nanoparticle (AuNP) Decorated Indium Tin Oxide (ITO) Electrode

Cyclic voltammetry (CV) is performed with a PANI/AuNP/ITO electrode as the working electrode, an Ag/AgCl reference electrode and a Pt counter electrode in a wide range of pyocyanin (a bio-marker for *Ps. aeruginosa*) concentrations from 1.9 to 238 μM (Elkhawaga et al. 2019). The redox current increases with rising pyocyanin concentration.

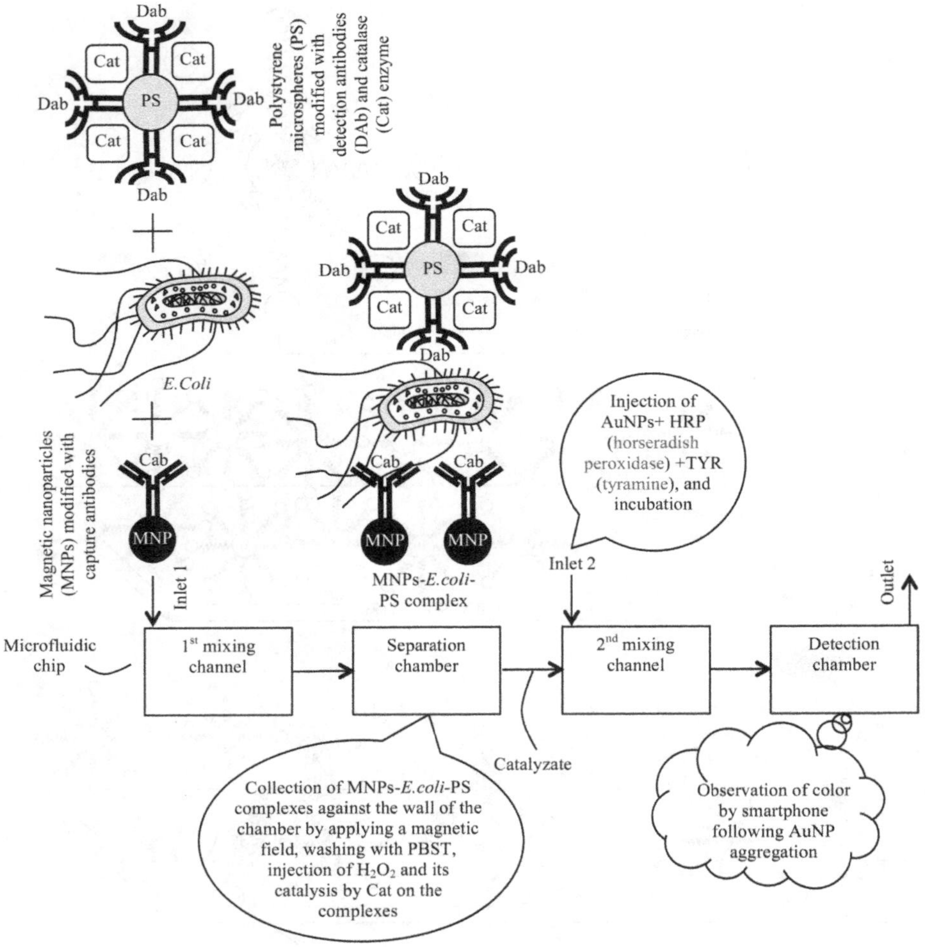

FIGURE 10.18 Microfluidic colorimetric biosensor (Zheng et al. 2019).

Relation between redox current peak and pyocyanin concentration in the range 1.9–25.62 μM is linear. The limit of detection is 500 nM.

10.13 Nanosensors for Detection of *Legionella pneumophila*

Prevalent in natural sources of water and often in manufactured water systems, *Legionella* is a bacterium which infects the lungs, causing severe pneumonia.

10.13.1 ZnO Nanorod (ZnO-NR) Matrix-Based Immunosensor

The immunosensing is done as presented in Figure 10.24. The ZnO-NR matrix is grown and patterned on a Ti/Au electrode on a glass substrate (Park et al. 2014). Primary antibodies are electrostatically immobilized onto the electrode. To these primary antibodies, different concentrations of the immune-dominant components of antigens of *Legionella pneumophila* (peptidoglycan-associated lipoproteins) are bound. In addition to these antigens, secondary antibodies conjugated with horseradish peroxidase enzyme are attached. Cyclic voltammetry (CV) is carried out in acetate buffer with tetramethylbenzidine (TMB) and H_2O_2. The three-electrode system is used in which ZnO-NR electrode is the working electrode. The reference electrode is an Ag/AgCl electrode and the working electrode is a platinum electrode.

The output peak reduction current logarithmically increases with the antigen concentration from 1–5000 pg/mL. The limit of detection is 1 pg/mL.

10.13.2 Azimuthally-Controlled Gold Grating-Coupling Surface Plasmon Resonance (GC-SPR) Platform

Reflectivity measurements are carried out on chromium/gold sinusoidal grating with a 500-nm pitch and a Cr(5 nm)/Au(40 nm) peak-to-valley amplitude (Figure 10.25) after functionalizing with the virostat antibody as bioreceptor (Meneghello et al. 2017). Bacterium concentrations used are 10^6, 10^4, 10^2, 10 CFU. Incident radiation wavelength is 600–800 nm, angle of incidence = 70°, and azimuthal orientation = 45°.

In the SPR data, three response regimes are noted. These are low-, intermediate-, and high- response regimes. In the low-response regime, the wavelength shift = 2 nm, corresponding to 10 CFU concentration. In the intermediate-response regime, the wavelength shift = 13 nm, corresponding to 10^2–10^4 CFU. The high-response regime pertains to a wavelength shift = 46 nm, and

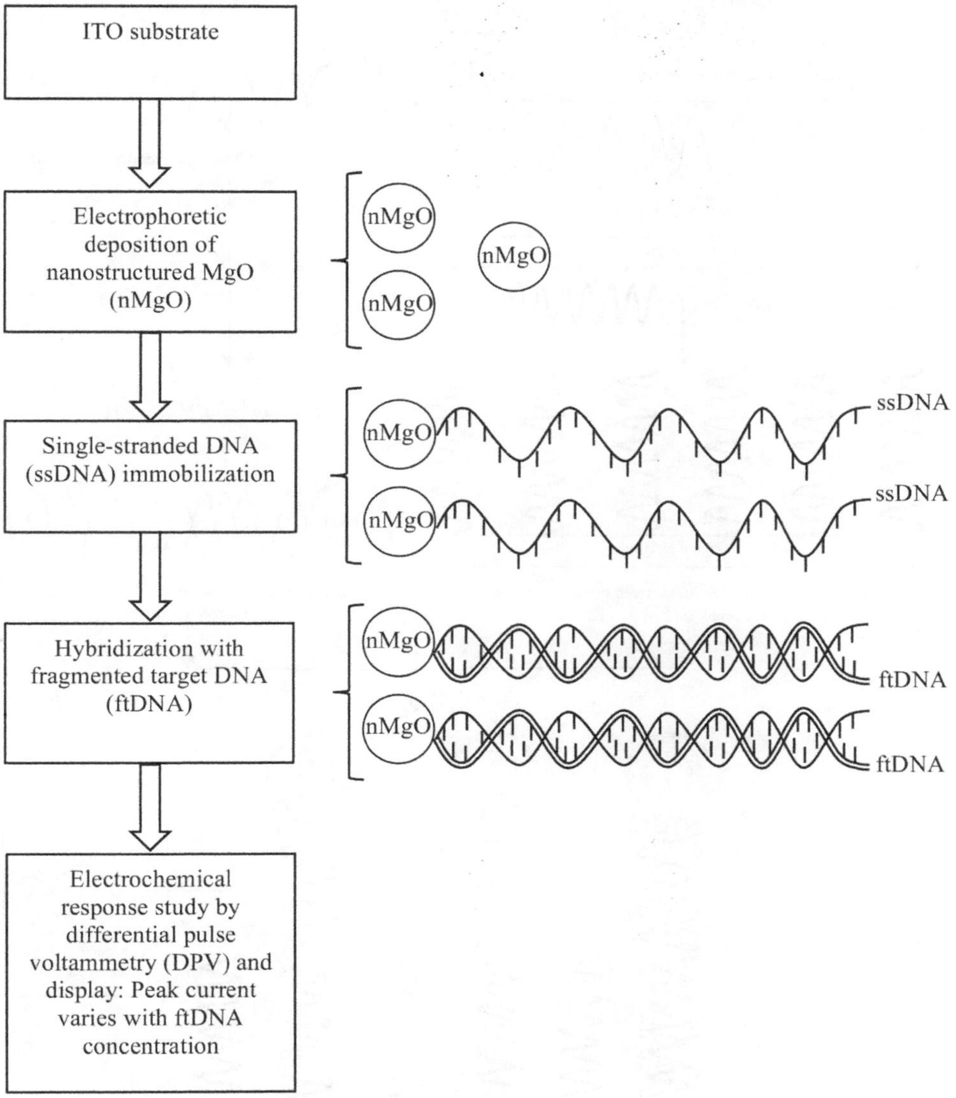

FIGURE 10.19 Flowchart of *Vibrio cholerae* detection by an ssDNA/nMgO/(ITO) bioelectrode (Patel et al. 2013).

the associated concentration is 10^6 CFU. The sensitivity of this system is three orders of magnitude higher than the sensitivity of the fluorescence method.

10.14 Nanosensors for Detection of Mercury Ions

In Section 8.11.3, a functionalized CdSe/ZnS QD nanosensor was described for detecting Hg(II) ions. Many newly developed nanosensors for this purpose will be described here.

10.14.1 Thymine Derivative (N-T) Decorated Gold Nanoparticle Sensor

This sensor utilizes the selective binding capability of N-T towards Hg^{2+} ions (Du and Peng 2015). N-T-decorated AuNPs exhibit changes in color from red to blue due to aggregation of AuNPs, caused by the presence of Hg^{2+} ions. These changes are discernible by the naked eye. Limit of detection is 0.8 nM.

10.14.2 Smartphone-Based Microwell Reader (MR S-phone) AuNP-Aptamer Colorimetric Sensor

The ambient-light sensor of a smartphone measures the color changes resulting from the interaction between the aptamer and the Hg^{2+} ions, leading to aggregation of AuNPs following treatment with NaCl (Xiao et al. 2016). Figure 10.26 illustrates the sensor operation. The readout is linear over the concentration range 1–32 ng/mL of Hg^{2+} ions. Limit of detection is 0.28 ng/mL.

10.14.3 Starch-Stabilized Silver Nanoparticle-Based Colorimetric Sensor

This sensor is used for Hg^{2+} ion detection in the presence of 0.005 mol/L HNO_3 (Figure 10.27). The parameter monitored is the absorbance strength of the localized surface plasmon resonance (LSPR) band of starch-coated AgNPs at 406 nm. It shows a proportionality relationship to the concentration of Hg^{2+}ions (Vasileva et al. 2017). In the Hg^{2+} ion concentration

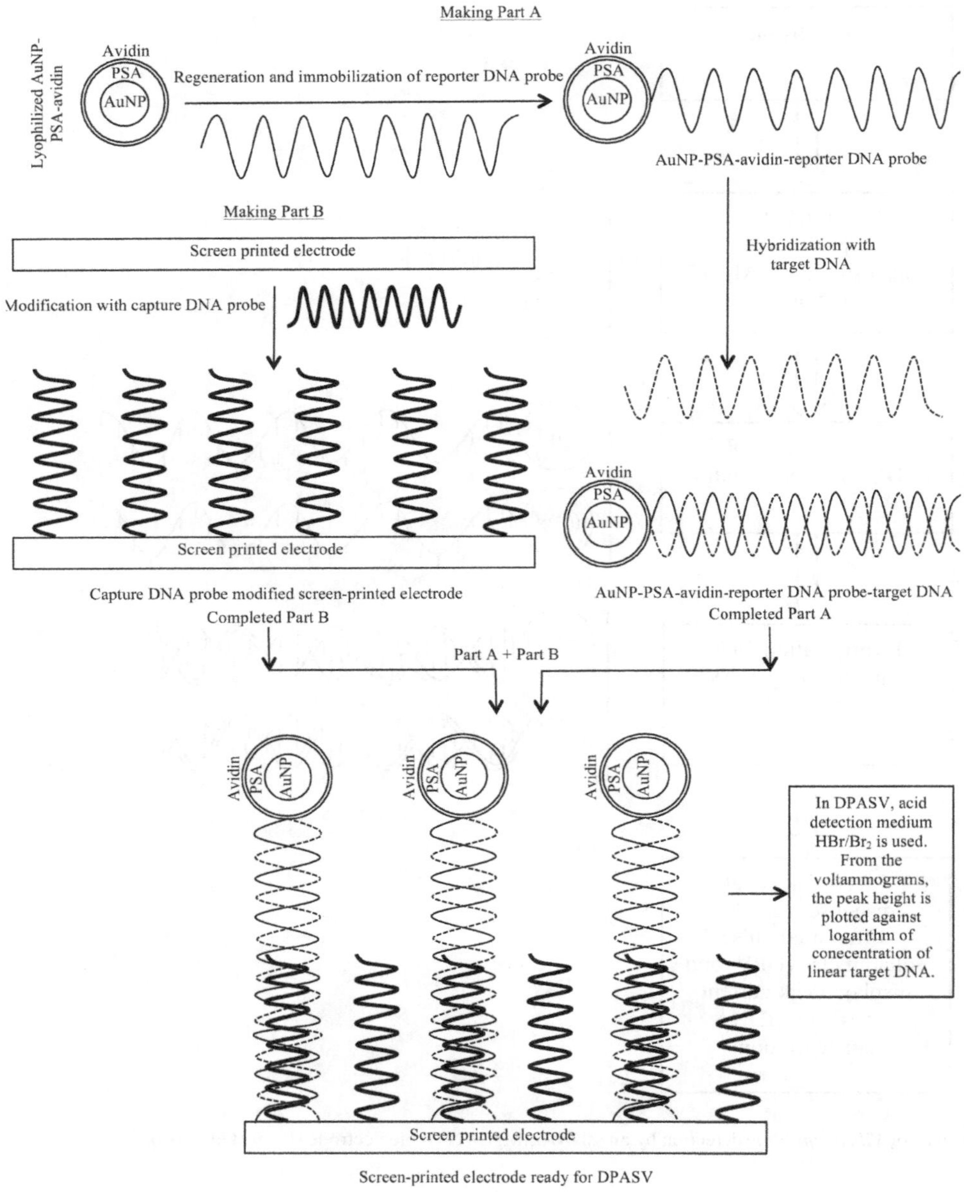

FIGURE 10.20 SPE preparation for AuNP/PSA genosensor (Sheng et al. 2015).

range 0–12.5 μg/L, the gradient is positive, whereas, in the concentration range 25–500 μg/L, it is negative. Ions such as Pb^{2+}, Na^+, K^+, etc. do not interfere in the assay. The limit of detection is 0.9 μg/L.

Aggregation of silver nanoparticles occurs *via* reduction of mercury ions and the amalgamation of freshly produced mercury atoms with silver atoms (Figure 10.27). This is accompanied by the diffusion of silver ions into the solution. Decrease in the surface charges of AgNPs in the presence of Hg^{2+} ions also causes their destabilization and aggregation.

10.14.4 Chitosan-Stabilized Silver Nanoparticle (Chi-AgNP)-Based Colorimetric Sensor

In the presence of Hg^{2+} ions, the color of Chi-AgNPs changes from brownish yellow to colorless. Other commonly present ions, such as Pb^{2+}, Cr^{3+}, Al^{3+}, etc., produce no such effects (Zarlaida et al. 2017). Limit of detection for the unaided eye is 1 μM.

10.15 Nanosensors for Detection of Lead Ions

10.15.1 Glutathione (GSH)-Stabilized Silver Nanoparticle (AgNP) Sensor

The color of the as-prepared GSH-AgNP colloidal solution is pale yellow (Anambiga et al. 2013). On mixing with Pb^{2+} ion concentrations from 10^{-3} to 10^{-9} M, the color changes to deep orange, due to aggregation of AgNPs arising from strong co-ordination bonding of Pb^{2+} ions with $-NH_2$ and –COOH groups of the glutathione modifier (Figure 10.28). The limit of detection is 10^{-9} M.

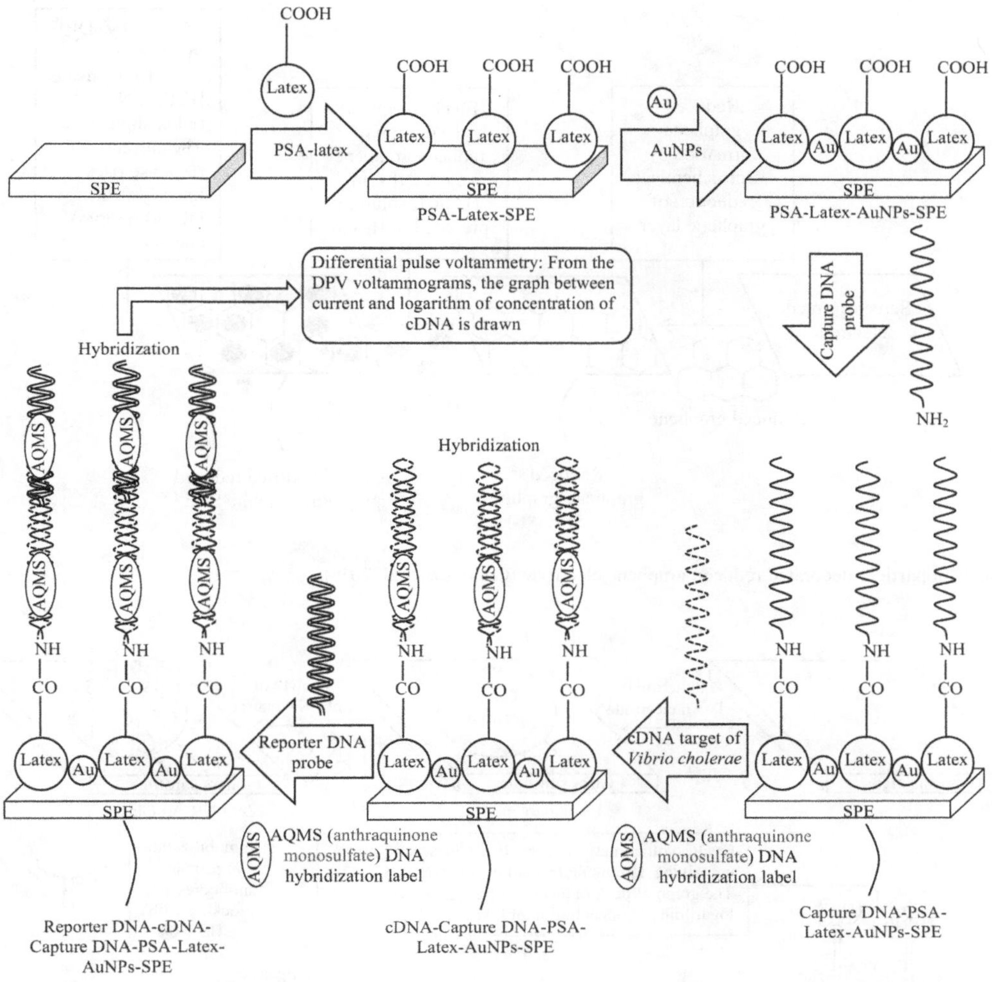

FIGURE 10.21 Biosensor using PSA-latex nanospheres and collidal gold nanoparticles (Rahman et al. 2017).

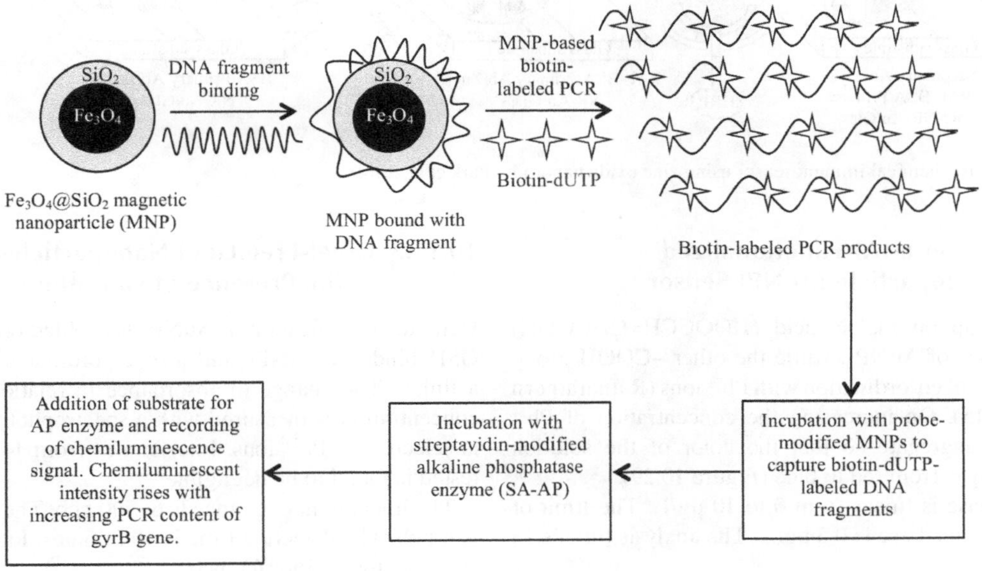

FIGURE 10.22 Chemiluminescent sensor signal generation steps (Tang et al. 2012, 2013).

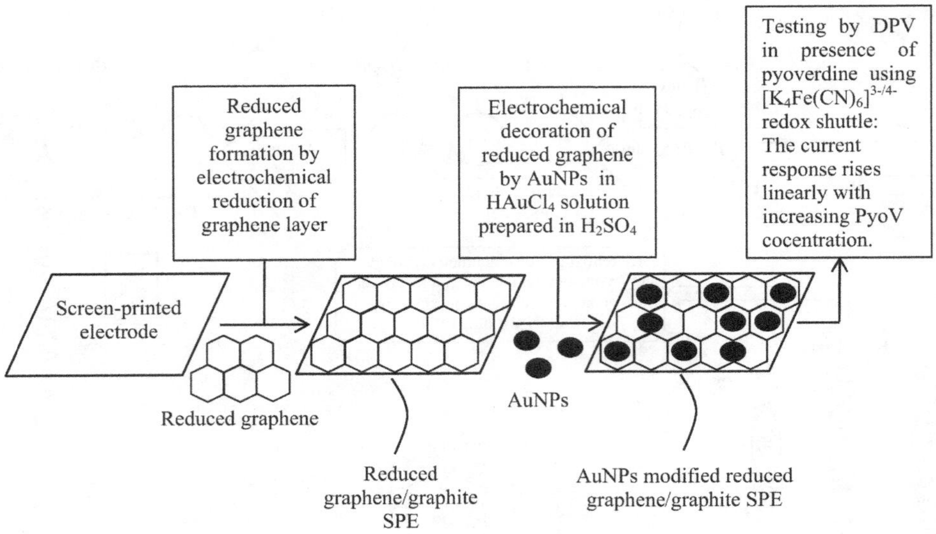

FIGURE 10.23 Gold nanoparticle-decorated reduced graphene electrode (Gandouzi et al. 2019).

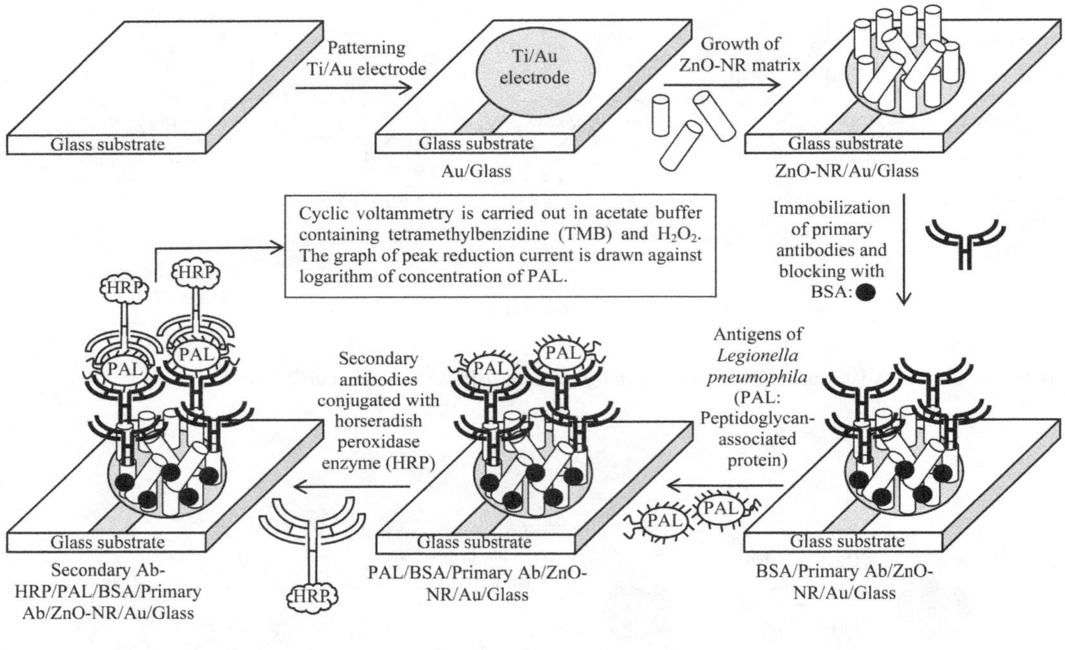

FIGURE 10.24 Electrochemical immunosensor using zinc oxide nanorods (Park et al. 2014).

10.15.2 Maleic acid (MA)-Functionalized Gold Nanoparticle (AuNP) Sensor

One –COOH group on maleic acid (HOOCCH=CHCOOH) modifies the surface of AuNPs, while the other –COOH group has a strong affinity of co-ordination with Pb^{2+} ions (Ratnarathorn and Dungchai 2015). On increasing the concentration of Pb^{2+} ions, the AuNPs aggregate so that the color of the solution progressively changes from red to blue (Figure 10.29).

The color response is linear from 0 to 10 μg/L. The limit of detection by the unaided eye is 0.5 μg/L. The analysis time is 15 minutes.

10.15.3 Label-Free Gold Nanoparticles (AuNPs) in the Presence of Glutathione (GSH)

Here, the modification of AuNPs is avoided (Zhong et al. 2015). GSH binds to AuNPs and also co-ordinates with Pb^{2+}ions as a linker. The change in absorbance in relationship to Pb^{2+} ion concentration is measured with a spectrophotometer. The assay is selective to Pb^{2+} ions. Interference from 14 other metal ions tested is found to be negligible.

The linear dynamic range is 6–500 ppb. The limit of detection is 6 ppb. The detection time is 10 minutes. Ions of other metals show negligible interference.

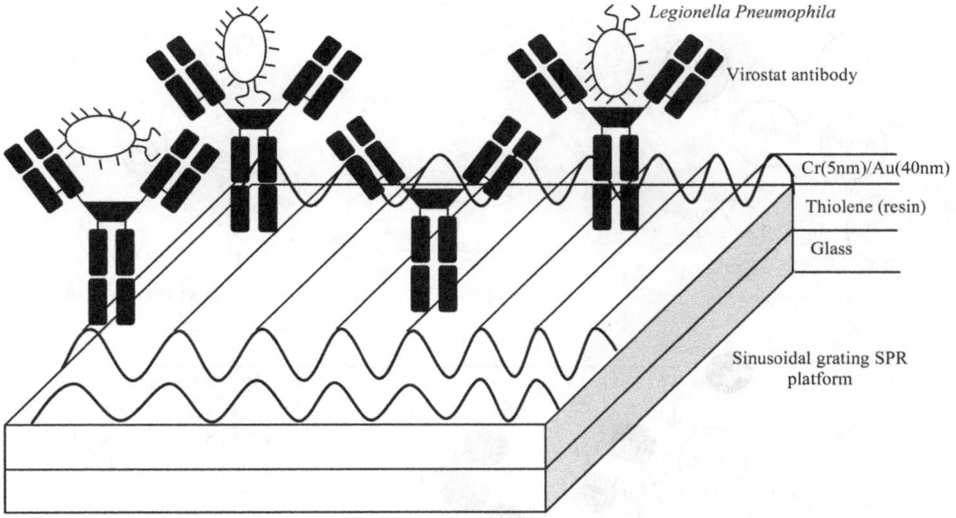

FIGURE 10.25 Nanosensor using sinusoidal grating SPR platform (Meneghello et al. 2017).

10.15.4 Gold Nanoparticles (AuNPs) Conjugated with Thioctic Acid (TA) and Fluorescent Dansyl Hydrazine (DNS) Molecules

The Color of the Au-TA-DNS solution changes from red to blue and the fluorescence intensity decreases upon addition of Pb^{2+}/Cu^{2+} ions (Nath et al. 2015). This is caused by metal ion-induced aggregation of nanoparticles. Figure 10.30 shows the sensor, which provides both colorimetric and fluorometric assessment. The UV-visible absorbance of AuNP-TA-DNS varies linearly with the concentration of Pb^{2+} and Cu^{2+} ions in the concentration range 0.001–10 ppm. The fluorescence spectrum of the sensor at 330 nm excitation wavelength also shifts with Pb^{2+}/Cu^{2+} ionic concentrations in this range.

The detection limit is < 10 ppb. The strong affinity of the sensor towards Pb^{2+}/Cu^{2+} ions makes it suitable as a turn-off sensor.

10.15.5 Valine-Capped Gold Nanoparticle Sensor

Well-dispersed gold nanoparticles are obtained due to the electrostatic repulsion between the $-COO^-$ groups on valine (Priyadarshini and Pradhan 2017). The valine-capped AuNPs (Figure 10.31) are highly selective in binding to Pb^{2+} ions in competition with other metal ions. Color changes are observed on adding a solution containing Pb^{2+} ions in the range 1–100 ppm. These changes arise from the analyte-induced aggregation of AuNPs. The limit of detection is 30.5 μM.

10.15.6 Polyvinyl Alcohol (PVA)-Stabilized Colloidal Silver Nanoparticles (Ag NPs) in the Presence of Dithizone

Dithizone is a ligand, which has a sulfide group that can be covalently bonded to AgNPs (Roto et al. 2019). It also has greater selectivity of binding with Pb^{2+}ions than with other ions at pH5–6. As the concentration of Pb^{2+} ions increases, from 0.5 to 10 μg/L, the localized surface plasmon resonance (LSPR) absorbance of AgNPs decreases linearly. The limit of detection is 0.64 μg/L.

10.15.7 Gold Nanoparticle (AuNP)-Graphene (GR)-Modified Glassy Carbon Electrode (GCE)

Using the AuNP-GR-GCE sensor, square wave anodic stripping voltammograms *versus* lead (II) ion concentration are recorded (Cheng et al. 2019).

In pH4.5 buffer solution, the peak current is proportional to Pb^{2+} ion concentration in the range 0.2–50 μg/L. The limit of detection is 0.34 μg/L.

10.16 Nanosensors for Detection of As(III) ions

10.16.1 Portable Surface-Enhanced Raman Spectroscopy (SERS) System

Figure 10.32 shows the implementation procedure of this arsenic ion sensor. The glass substrate is pre-treated with a self-assembled monolayer of mercaptosilane as linker molecules, followed by transferring and covalent attachment of a thiol-modified silver nanoparticle (40-nm) close-packed film to it *via* thiol linkage (Wang et al. 2011). A surface-enhanced Raman spectroscopy (SERS) module for direct measurements in water is constructed. A linear relationship of the relative intensity of arsenate peak *versus* arsenic concentration occurs in the range 1–500 ppb.

Detection of arsenic up to 1 ppb level can be demonstrated with this portable SERS system.

10.16.2 Surface Plasmon Resonance (SPR) Nanosensor

Figure 10.33 shows the measurement protocol. A 5-nm Ti/50-nm Au film is functionalized with *n*-alkanethiol (Salinas et al. 2014). By SPR in the Krestchmann configuration,

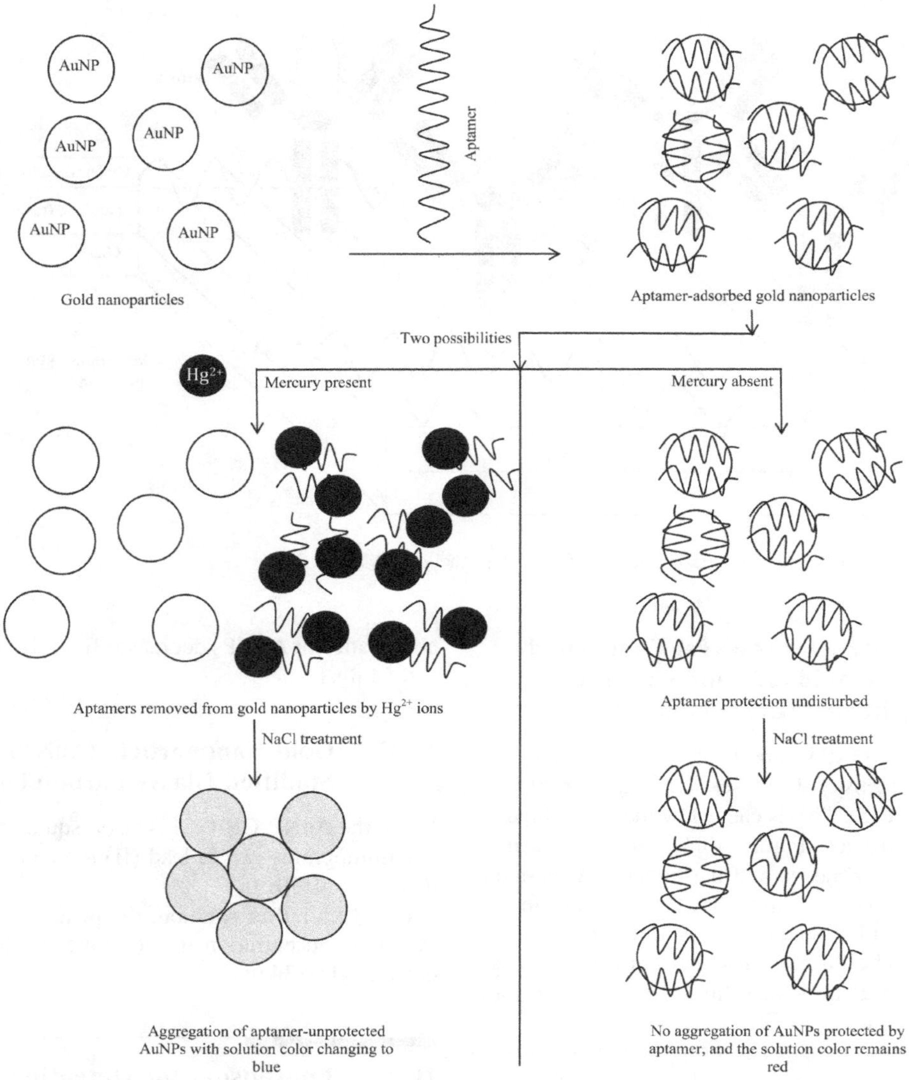

FIGURE 10.26 Mercury ion sensing by color changes of gold nanoparticles (Xiao et al. 2016).

transmittance is determined for different wavelengths for pristine gold substrate, gold substrate + dithiothreitol and gold substrate + dithiothreitol + arsenic.

Arsenic concentration can be determined down to 5 ppb.

10.16.3 FePt Bimetallic Nanoparticle (FePt-NP) Sensor

Electrochemically deposited bimetallic FePt, FePd, FeAu and AuPt nanoparticles on Si(100) substrate are used for As(III) detection by anodic stripping voltammetry (ASV) (Moghimi et al. 2015).

FePt NPs show the best performance, with a sensitivity~0.42 μA/ppb. The limit of detection is 0.8 ppb.

10.16.4 Gold Nanoparticles (AuNPs)-Modified Glassy Carbon Electrode (GCE) for Co-Detection of As(III) and Se(IV)

Cyclic voltammetry (CV) is used to modify GCE by AuNPs (Idris et al. 2017). Using it as the working electrode, square wave voltammetry is conducted for different concentrations of As(III) and Se(IV) between 0.01 and 12 ppm, with 0.1M H_2SO_4 as the supporting electrolyte.

The peak currents are proportional to concentrations of As(III) and Se(IV) ions. The limits of detection are 0.15 ppb for As(III) and 0.22 ppb for Se(IV) ions.

10.16.5 Silver Nanoparticle-Modified Gold Electrode

Silver nanoparticles, obtained by the chemical reduction of silver nitrate, are used to modify the gold electrode (Sonkoue et al. 2018). The electrode modification is done as shown in Figure 10.34. A well-defined reduction peak is observed in cyclic voltammograms. Linear sweep voltammetry is done in a 0.1 M HNO_3 solution for various arsenic concentrations.

The calibration curve is drawn for different arsenic concentrations in the range 0.05–0.2 μM. The limit of arsenic detection is 1.38×10^{-8}M.

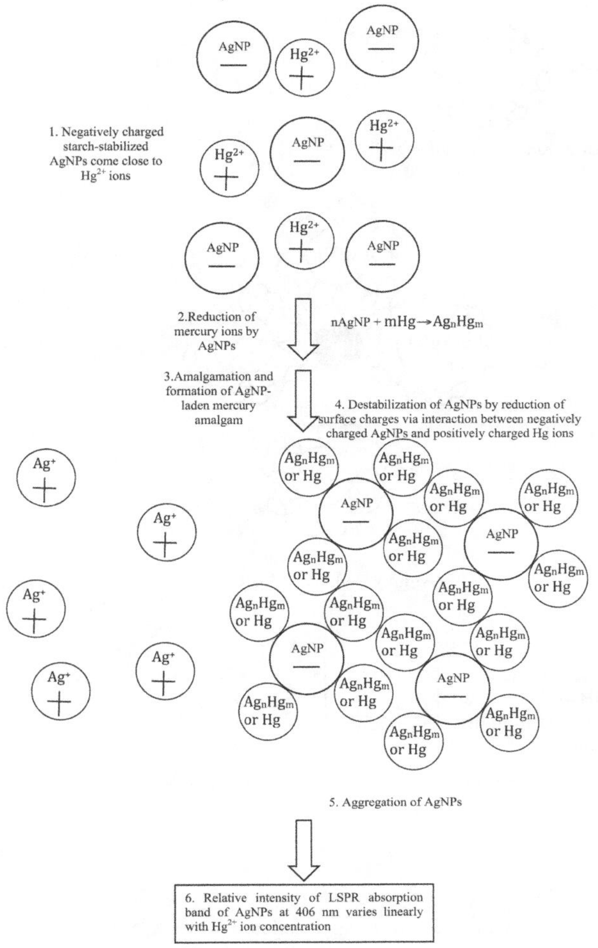

FIGURE 10.27 Aggregation-inducing interaction between starch-stabilized AgNPs and Hg^{2+} ions in solution (Vasileva et al. 2017).

10.16.6 Carbon Nanoparticle (CNP)/Gold Nanoparticle (AuNP)-Modified Glassy Carbon Electrode (GCE) Aptasensor

The aptasensor is shown in Figure 10.35. CNPs are drop-coated on GCE, followed by electrodeposition of AuNPs by cyclic voltammetry, producing a carbon-gold bi-nanoparticle platform (Mushiana et al. 2019).

Carbon nanoparticles provide signal enhancement, while gold nanoparticles help in immobilization of the aptamer by connecting the aptamer to the electrode *via* the Au-S bond. A 100-base thiolated Ars-3 aptamer is immobilized on the CNP/AuNP-modified GCE. Square wave voltammetry (SWV) is performed in a 0.1 mM $[Ru(NH_3)_6]^{3+}$ electrolyte solution after binding to various As(III) concentrations from 0.5 to 500 ppb. The peak current rises with As(III) concentration.

The limit of detection is 0092 ppb. The sensor is not susceptible to interference from Cu^{2+}, Hg^{2+}, or Cd^{2+} ions.

10.16.7 Gold Nanostructured Electrode on a Gold Foil (Au/GNE)

The Au/GNE is fabricated on a gold foil by electrochemical oxidation-reduction sweeping cycles in electrolyte solution (Babar et al. 2019). Arsenic detection is done by square wave anodic stripping voltammetry (SWASV), involving arsenic accumulation/deposition followed by stripping in 0.1 M HNO_3 electrolyte. The calibration curve is generated from the SWASV response for different concentrations of arsenic.

By visual analysis of calibration characteristics, the limit of detection is 0.1 ppb (1.3 nM). The sensitivity is 39.54 μA/ppb. The Au/GNE is selective for arsenic in the presence of several other metal ions, e.g. Hg^{2+}, Pb^{2+}, Cu^{2+}, etc.

10.16.8 Bimetallic Nanoparticle (NP) and [Bimetallic NP + Polyaniline (PANI)] Composite-Modified Screen-Printed Carbon Electrode (SPCE)

Square wave anodic stripping voltammetry (SWASV) is performed with solutions of different concentrations of arsenic in 0.1 M HCl, using AuPtNP/SPCE and AuPtNP/PANI/SPCE platforms. The results are applied to produce the calibration plot of the sensor (Melinte et al. 2019).

The range of detection is 0–200 nM and the limit of detection is 19.7 nM. The first platform is able to detect trace levels of As(III) with an analysis time 5 minutes. The second platform is used for higher As(III) concentrations, with an analysis time of 12 minutes.

10.16.9 Ranolazine (Rano)-Functionalized Copper Nanoparticles (CuNPs)

By the addition of As(III) ions to Rano-CuNPs, the ranolazine coated on the nanoparticle surface is removed, causing aggregation of the nanoparticles (Laghari et al. 2019). The color changes from brick red to dark green with increasing As(III) ion concentration (Figure 10.36).

The detection is linear in the range is 3×10^{-7}–8.3×10^{-6} M. The limit of detection is 1.6×10^{-8}M.

10.17 Nanosensors for Detection of Cr(VI) Ions

10.17.1 Colloidal Gold Nanoparticle (AuNP) Probe-Based Immunochromatographic Sensor

Immunoassay is performed with gold nanoparticles coated with monoclonal antibodies (mAbs) against Cr^{3+} and Cr^{6+} ions (Liu et al. 2012). Color changes are detected with the help of a portable lateral flow reader, the construction and operation of which are explained in Figure 10.37.

The linear range of detection is 5–80 ng/mL, with the visual lowest detection limit (LDL) being 50 ng/mL. The limit can reach 5 ng/mL with the strip reader. The assay can be carried out in 5 minutes with the test strip.

10.17.2 Amyloid-Fibril-Based Sensor

Amyloid fibrils are nanosized biomaterials, peptides, or protein aggregates. Amyloid fibrils of the hen lysozyme bind the Cr(VI) ions (Leung et al. 2013). When 1,5-diphenylcarbazide (DPC)/sulfuric acid is added to the solution, the absorbance of

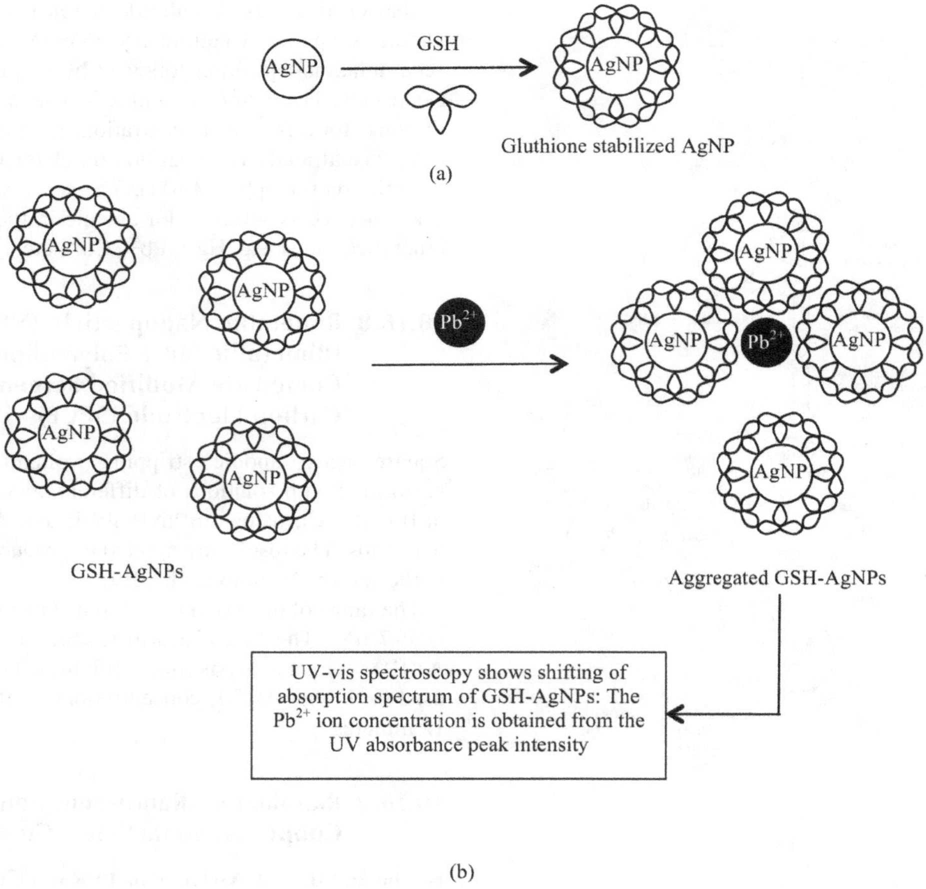

FIGURE 10.28 Pb^{2+} ion sensing using GSH-AgNPs: (a) stabilization of AgNPs with glutathione and (b) aggregation of GSH-AgNPs in presence of Pb^{2+}ions (Anambiga et al. 2013).

Cr(VI)-adsorbed fibrils varies linearly with Cr(VI) ion concentration in the range 0–0.75 ppm, serving as a colorimetric probe (Figure 10.38).

10.17.3 Gold Nanoparticle (AuNP)-Decorated Titanium Dioxide Nanotubes (TiO_2NTs) on a Ti Substrate

The Ti/TiO_2NTs/AuNPs electrode is the working electrode with the Pt foil counter electrode and Ag/AgCl reference electrode (Jin et al. 2014). The analytical performance of the working electrode against Cr(VI) is studied by amperometry by successively adding Cr(VI) ion solutions of different molarity in 0.1 M HCl, with the voltage adjusted to a constant value. The calibration plot is drawn from the observed current response.

The current response is linear in the concentration range 0.10–105 μM. Sensitivity is 6.91 μA/μM for Cr(VI). The limit of detection is 0.03 μM.

10.18 Nanosensors for Detection of Cd^{2+} ions

10.18.1 Gold Nanoparticle Amalgam (AuNPA)-Modified Screen-Printed Electrode (SPE)

Anodic stripping differential pulse voltammetry (ASDPV) is performed using AuNPA-modified disposable SPE (Zhang et al. 2010). A circular working electrode and a ring-shaped counter electrode are printed on a glass-fiber plate with carbon ink (Figure 10.39). Silver/silver chloride ink is used for printing the reference electrode. SPE modification is implemented through a two-step electrodeposition process, in which AuNPs are electrodeposited in the first step and mercury is electrodeposited in the second step.

The maximum electrochemical response is obtained with pH = 5.5 phosphate-buffered saline (PBS) containing 0.1M KCl as the supporting electrolyte. A distinct differential pulse voltammetry (DPV) peak is observed with the reduction current proportional to Cd^{2+} ion concentration in the range 8.4 ppb to 500 ppm. The limit of detection is 2.6 ppb. A portable detection system has been developed for on-site testing of tap water and river water specimens.

10.18.2 Turn-On Surface-Enhanced Raman Scattering (SERS) Sensor

Gold nanoparticles (41-nm) are encoded with a Raman-active dye and a Cd^{2+} chelating layer (a polymer ligand that reacts specifically with Cd^{2+} ions to form a complex), yielding SERS nanoparticles (Yin et al. 2011). Single SERS nanoparticles are optimized to remain spectrally silent. Figure 10.40 shows the SERS sensor.

On adding Cd^{2+} ions, interparticle self-aggregation occurs, turning on the SERS signal multiplied by a factor of 90.

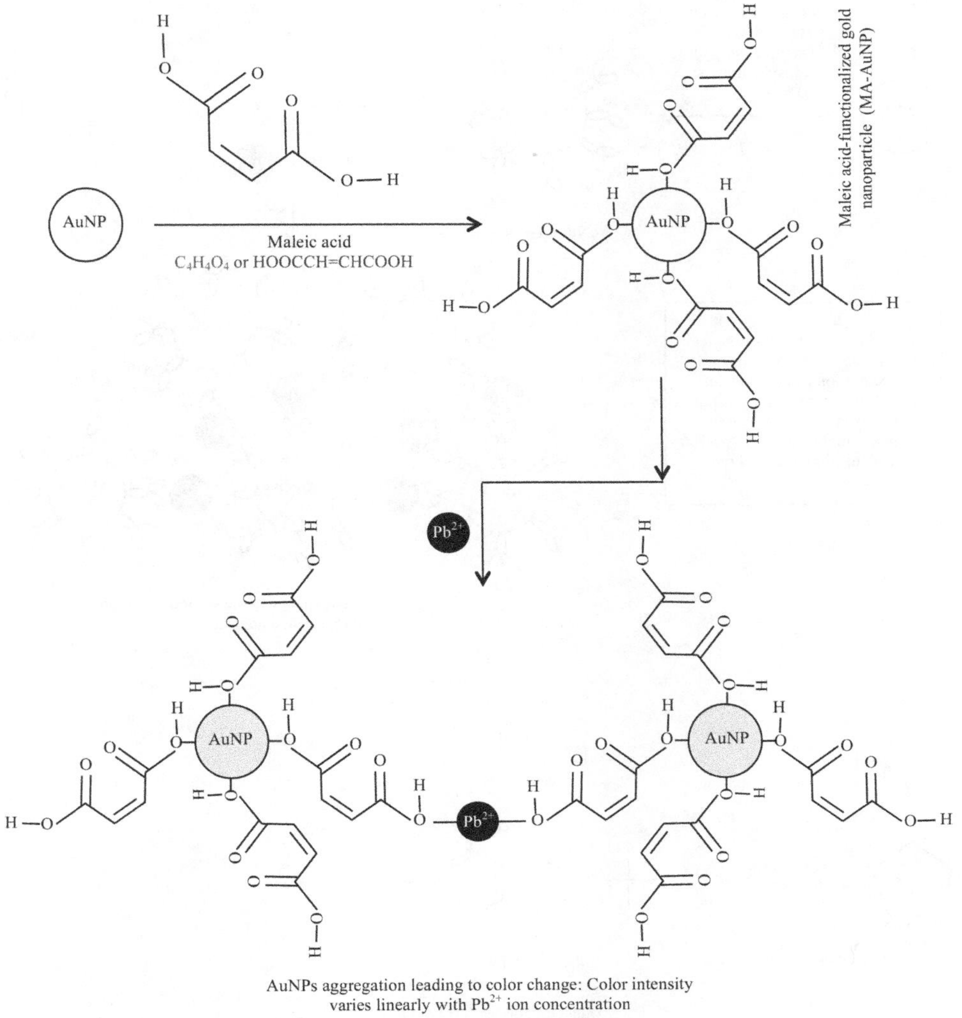

FIGURE 10.29 MA-AuNP sensor for Pb^{2+}ions (Ratnarathorn and Dungchai 2015).

10.18.3 Thioglycerol (TG)-Capped CdSe Quantum Dots (QDs)

The absorbance of TG-CdSe QDs increases and the fluorescence intensity decreases (due to quenching of emission) with increasing Cd^{2+} ion concentration, because cadmium cations form cadmium oxide on the surface of QDs by interacting with hydroxyl groups of TG (Brahim et al. 2015). The mechanism of interaction of TG-capped CDSe QDs with Cd^{2+} ions is shown in Figure 10.41.

The absorbance of TG-CdSe QDs at 427 nm is proportional to Cd^{2+} ion concentration in the range 5–22 μM. The slope is 0.0278/μM and the limit of detection is 0.32 μM.

10.18.4 CdTe Quantum (CdTe QD) Dot-Based Hybrid Probe

The hybrid probe-based sensor is shown in Figure 10.42. The hybrid probe consists of two quantum dots: (i) a red emissive QD (rQD) embedded in a silica sphere (reference QD); and (ii) a green-emissive QD (gQD), the variable fluorescence QD, covalently attached to the surface of the silica sphere (Qian et al. 2017).

First, 3-mercaptopropionic acid (MPA)-capped gQDs and -rQDs are prepared. Then, rQDs are loaded into silica spheres and the silica surfaces are modified with amino groups to form amino-functionalized rQDs@SiO_2 nanospheres, from which rQDs@SiO_2@gQDs hybrid nanospheres are formed.

Then, 1,10-phenanthroline (Phen) is complexed to the probe. As a result of this complexation, the intense fluorescence of gQDs is quenched but that of the silica shell-protected rQDs remains constant. The quenching reaches 97.79% when the Phen concentration is 20 μM.

Using a single excitation wavelength, the nanohybrid probe shows dual well-resolved emission peaks at the wavelengths of 525 (gQD) and 635 nm (rQD). When Cd^{2+} ions are added, the Phen ligands are detached. Depending on the Cd^{2+} ion concentration, and hence the number of detached ligands, the fluorescence quenching of the gQD decreases. The fluorescence of gQD recovers accordingly. The ratio of fluorescence signals of the variable fluorescence gQD to the constant fluorescence rQD, serving as the internal reference signal, is a measure of Cd^{2+} ion concentration.

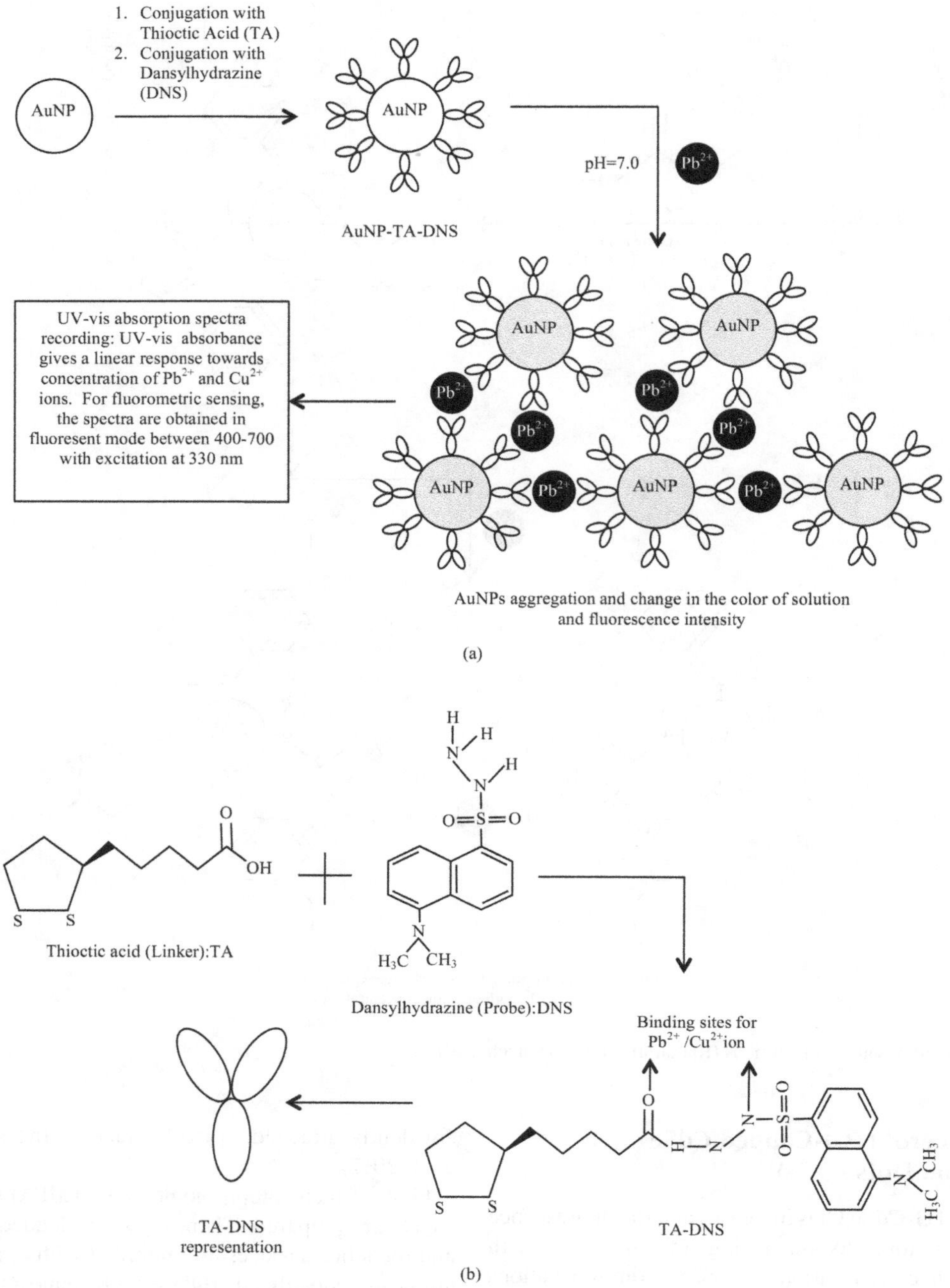

FIGURE 10.30 AuNPs-TA-DNS sensor: (a) AuNPs modification and their aggregation induced by Pb^{2+}ions, and (b) illustrating the formation of TA-DNS (Nath et al. 2015).

The ratio

$$r = \left(\frac{\text{Intensity}_{\text{Green}}}{\text{Intensity}_{\text{Red}}}\right)_{\text{Cd present}} \Bigg/ \left(\frac{\text{Intensity}_{\text{Green}}}{\text{Intensity}_{\text{Red}}}\right)_{\text{Cd absent}} \tag{10.4}$$

is proportional to Cd^{2+} ion concentration in the range 0.5 nM–2 μM. The limit of detection is 0.17 nM. The naked eye can detect Cd^{2+} ions in the range 2 nM–1 μM. The sensor can unambiguously differentiate between Cd^{2+} and Zn^{2+} ions, and can be applied to tap water and rice samples.

10.18.5 Aptamer-Functionalized Gold Nanoparticle (AuNP) Sensor

The aptamers make the AuNPs less susceptible to aggregation (Gan and Wang 2020). Addition of Cd^{2+} ions causes specific interaction between the Cd^{2+} ions and the aptamers. As a consequence, the free aptamer percentage decreases. With a decrease in free aptamer content, AuNPs become prone to aggregation. The change in color produced due to AuNP aggregation is picked up by a smartphone-based colorimetric system (SBCS). Figure 10.43 displays the response of the sensor to different samples.

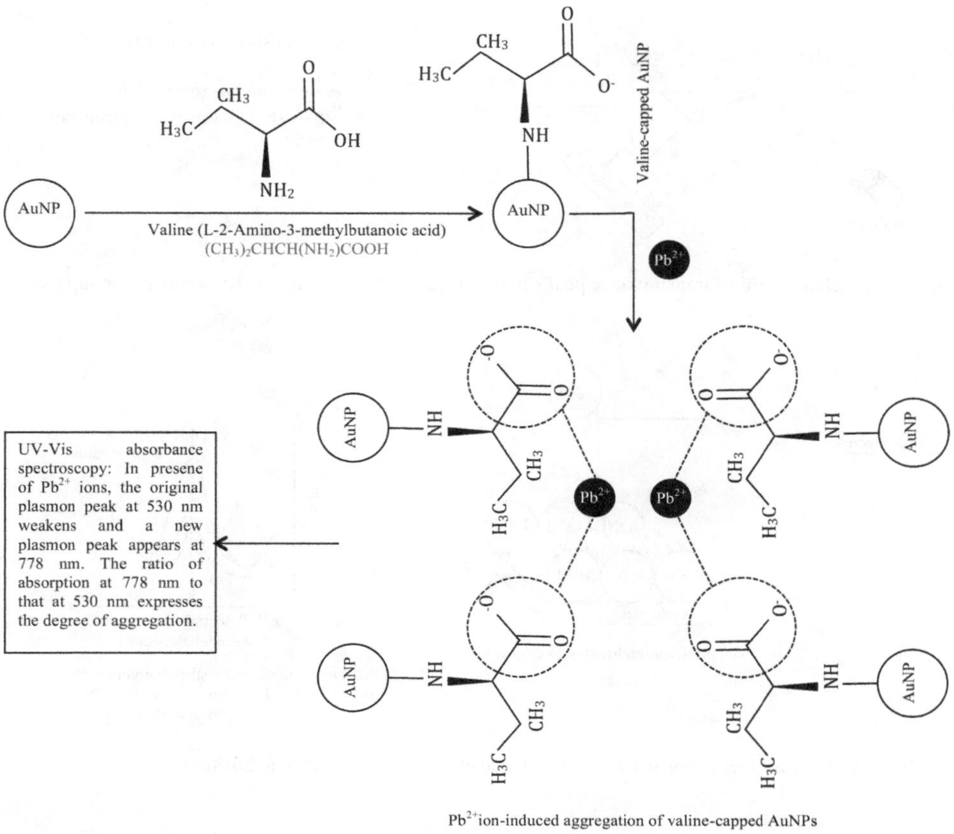

FIGURE 10.31 Pb^{2+}ion sensing by valine-capped AuNPs (Priyadarshini and Pradhan 2017).

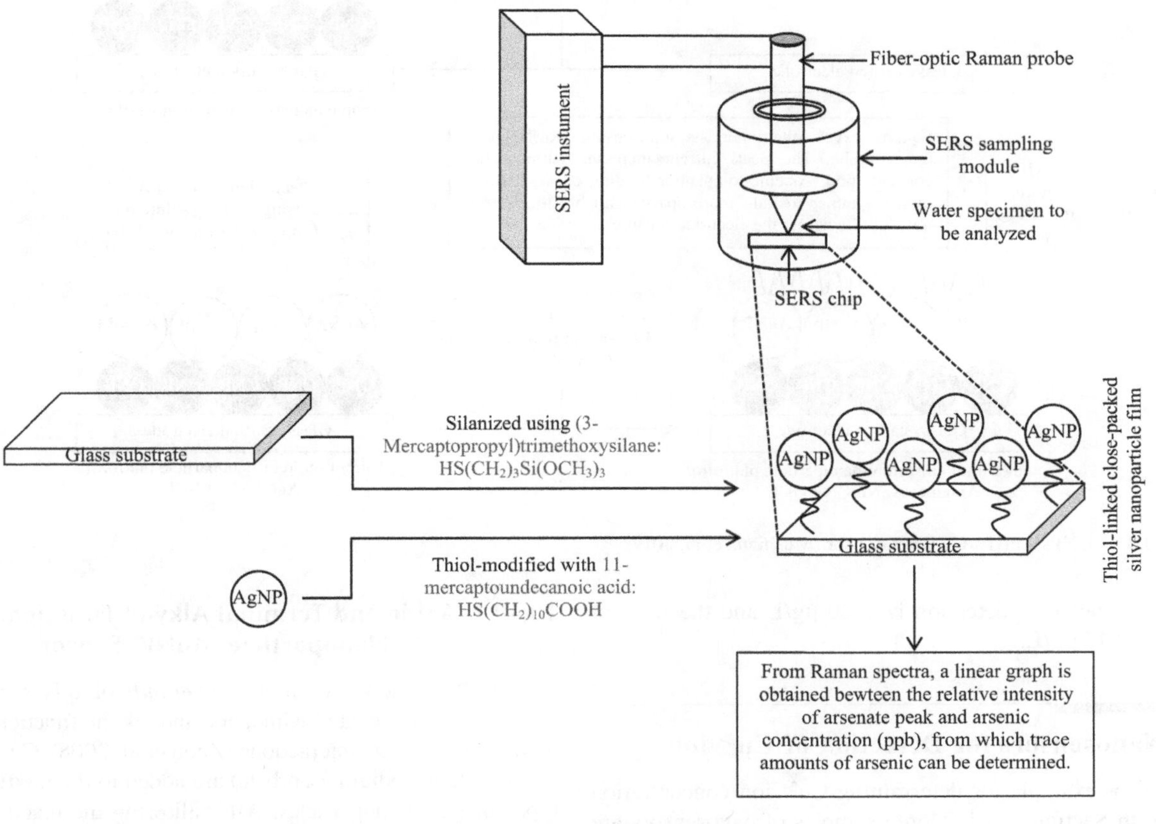

FIGURE 10.32 On-site arsenic sensing in groundwater by the SERS technique (Wang et al. 2011).

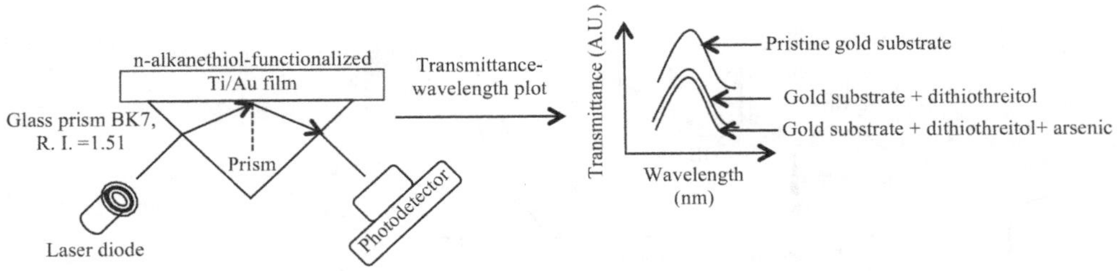

FIGURE 10.33 Measurement of relative shift of transmittance peaks in the presence of arsenic by the Krestchmann set-up (Salinas et al. 2014).

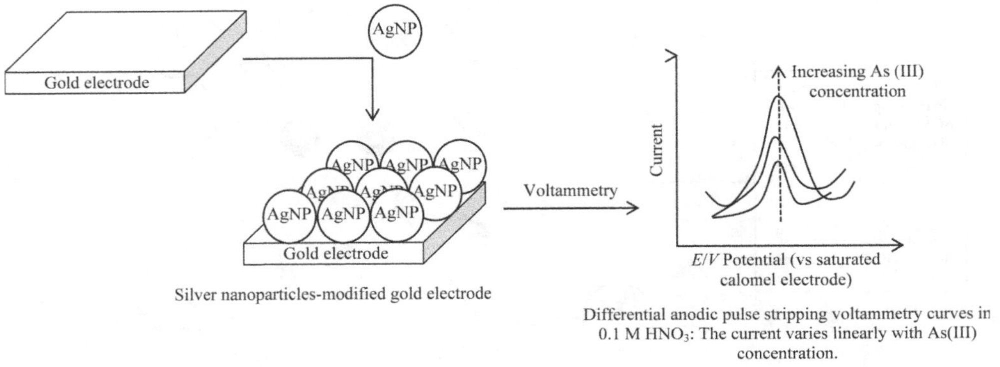

FIGURE 10.34 Arsenic(III) sensing by a silver nanoparticle-modified gold electrode (Sonkoue et al. 2018).

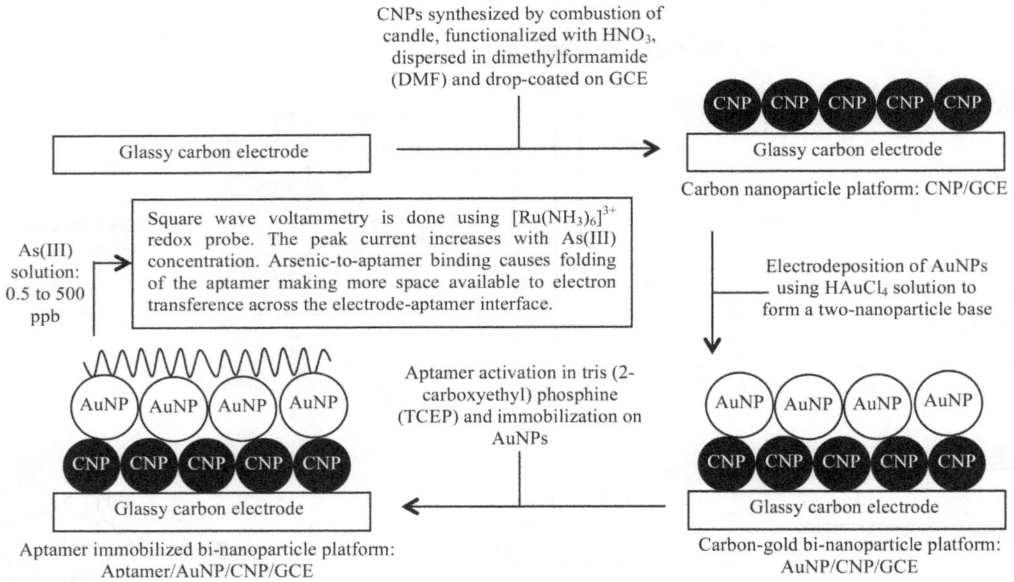

FIGURE 10.35 CNPs/AuNPs/GCE aptasensor (Mushiana et al. 2019).

The linear range of detection is 2–20 μg/L and the limit of detection is 1.12 μg/L.

10.19 Nanosensors for Detection of Cu^{2+} ions

A nanosensor example for determining Cu^{2+} ion concentration was given in Section 8.11.1. More examples of nanosensors are presented here.

10.19.1 Azide and Terminal Alkyne-Functionalized Gold Nanoparticle (AuNP) Sensor

Figure 10.44 shows the *modus operandi* of this sensor. Gold nanoparticles are coated with azide and alkyne functional groups by ligand-exchange interactions (Zhou et al. 2008). Cu^{2+} ions and the reductant (sodium ascorbate) are added to the mixture of two types of gold nanoparticles. After allowing the reaction to take place overnight, color changes caused by Cu^{2+}ions are non-distinct below a concentration of 20 μM. But the color changes and

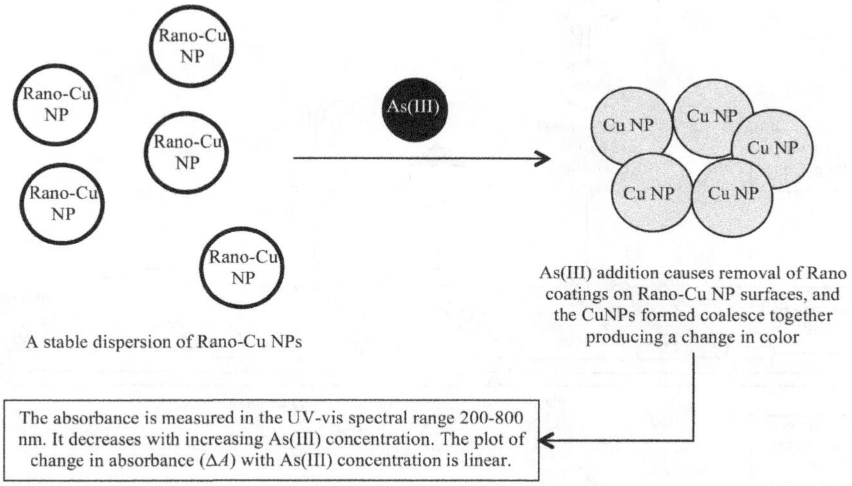

FIGURE 10.36 Arsenic sensing by aggregation of Rano-CuNPs (Laghari et al. 2019).

precipitate formation become obvious at concentrations > 50 μM. The minimum eye-detectable concentration of Cu^{2+} ions is 50 μM.

It may be noted that azide-alkyne cycloaddition used in this analysis fulfills many of the prerequisites of click chemistry, a class of high-yielding biocompatible small-molecule reactions that are simple to perform in benign solvents and produce minimal and inoffensive byproducts that can be removed easily without chromatography. Hence this sensing method is known as Cu^{2+} ion sensing by click chemistry.

10.19.2 Cadmium Sulfide Nanoparticle (CdS NP)-Gold Quantum Dot (Au QD) Sensor

The paste of CdS NP-Au QD is cast on an indium tin oxide (ITO) electrode to form CdS NP-Au QD/ITO electrode (Ibrahim et al. 2016). The modified electrode is shown in Figure 10.45. Photocurrent responses are measured for different Cu^{2+} ion concentrations, using the CdS NP-Au QD/ITO electrode as the working electrode, a Pt auxiliary electrode and an Ag/AgCl reference electrode, with 0.1 M KCl-containing 0.5 M triethanolamine (TEA) electrolyte under illumination from a 150-W halogen lamp as the source of light, turned on and off.

The calibration graph of change in current (ΔI) with respect to Cu^{2+} ion concentration is linear in the concentration range 0.5–120 nM. The limit of detection is 6.73 nM.

When illuminated, CdS captures photons, causing electron excitation from valence to conduction band, with holes left behind in the valence band. The electrons activate the reduction of Cu^{2+} ions to Cu^{+}, while holes are forged by TEA in the electrolyte. Cu_xS is formed by competitive displacement of Cd^{2+} by Cu^{2+}, providing a lower energy level. Electron–hole recombination centers are created, with a resulting decrease of photocurrent density in the presence of Cu^{2+}.

10.19.3 Multiple Antibiotic Resistance Regulator (MarR)-Functionalized Gold Nanoparticle (AuNP) Sensor

Colorimetric assay is performed using MarR (a regulatory protein from *Escherichia coli*) as a highly selective biorecognition element for Cu^{2+} ions (Wang et al. 2017). MarR-coated gold nanoparticles (MarR-AuNPs) undergo aggregation upon addition of Cu^{2+}ions. This nanoparticle grouping changes the color of the solution from red to purple. Figure 10.46 shows the coating of gold nanoparticles with MarR and nanoparticle aggregation in the presence of Cu^{2+} ions.

The limit of detection by the naked eye is 1 μM. For a UV-vis spectrometer, the limit is 405 nM and, by the smartphone method, it is 61 nM. The time required for detection is 5 minutes. Absorbance measurements with UV-vis spectrometer are done between 450 and800 nm. In the absence of Cu^{2+} ions, the surface plasmon absorption peak is located around 520 nm. On adding Cu^{2+}ions, it shifts to 600 nm. The absorbance ratio, = absorbance at 600 nm/absorbance at 520 nm = A_{600}/A_{520}, is plotted with respect to Cu^{2+} ion concentration. The smartphone recorded color image is processed for red-green-blue (RGB) values.

10.19.4 Casein Peptide-Functionalized Silver Nanoparticle (AgNP) Sensor

Cu^{2+} ions interact with the peptide ligands on the surfaces of AgNPs (Ghodake et al. 2018). This interaction leads to aggregation of AgNPs. So, the color of casein peptide-capped AgNPs changes from yellow to red when Cu^{2+} ions are added. The change in absorbance is measured by UV-vis spectroscopy for different Cu^{2+}ion concentrations.

The response is linear in the concentration range 0.08–1.44 μM. The limit of detection is 0.16 μM.

10.20 Nanosensors for Detection of Pesticides

10.20.1 DDT (Dichlorodiphenyltrichloroethane)

DDT is an organochlorine pesticide. It is detected by a gold nanoparticle (AuNP)-based dipstick competitive assay (Lisa et al. 2009). The assay consists of the following steps (Figure 10.47):

(i) Optimal concentration (1 μg) of DDA (1,1,1- trichloro-2,2-*bis*(chlorophenyl) acetic acid)-BSA (bovine serum albumin) conjugate (antigen) is immobilized on nitrocellulose (NC) membrane-containing strip. The DDA-BSA conjugate (antigen)- coated NC strip is reactant A.

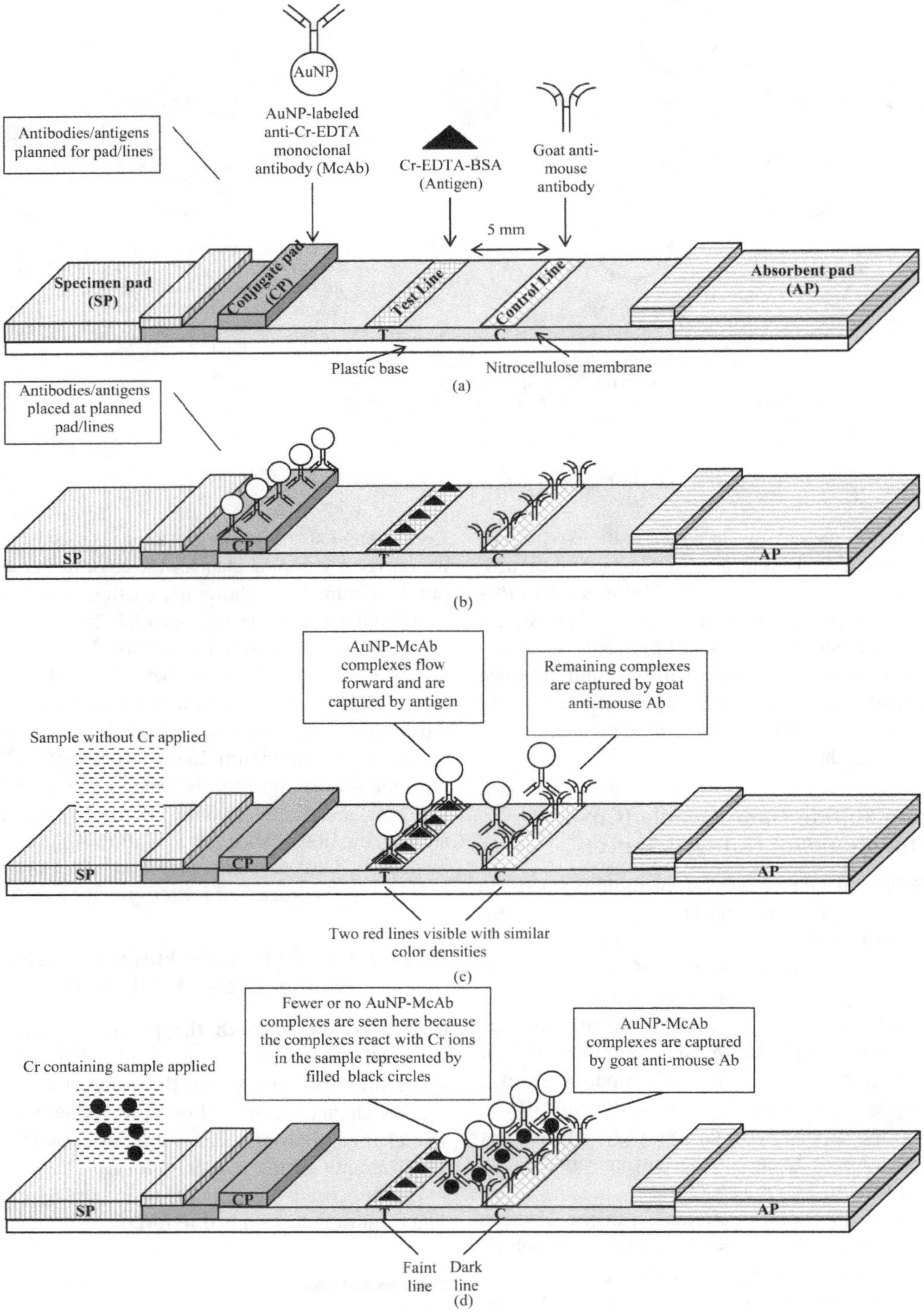

FIGURE 10.37 Cr(VI) sensing by immunochromatographic assay; (a) schematic layout of the lateral flow reader and its constituent parts, (b) placement of antibodies and antigens on the reader, (c) injecting a Cr ion-negative sample on the reader, and (d) flowing a Cr ion-positive sample on it (Liu et al. 2012).

(ii) Gold nanoparticles are conjugated with different concentrations of anti-DDT antibodies to form anti-DDT antibodies/AuNPs. This is reactant B.

(iii) Different free DDT concentrations are taken: 0.7–1000 ng/mL (Reactant C).

(iv) Reactant A is dipped in B and the consequent intense red color is the zero-DDT control.

(v) Reactant A is dipped in reactants (B+C) and the color development on NC membrane (reactant A) is observed. The competitive bindings, (B+C) and (B+A), provide detection of free DDT.

Depending on the concentration of reactant C (free DDT), the binding of reactant B (anti-DDT antibodies/AuNPs) to A (DDA-BSA conjugate antigen) varies. The intensity of color on the NC

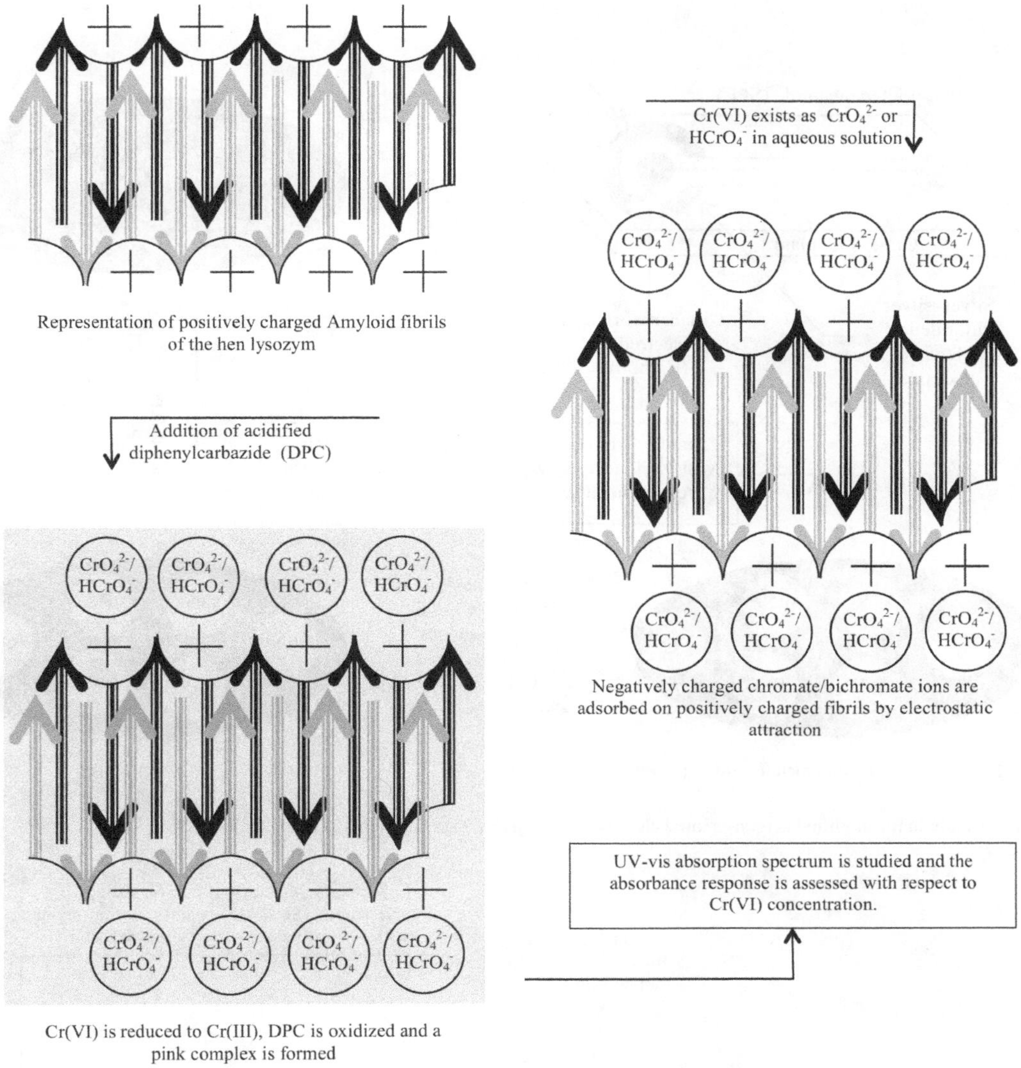

FIGURE 10.38 Colorimetric nanosensor for Cr(VI) ions using amyloid fibrils (Leung et al. 2013).

membrane is inversely proportional to the concentration of free DDT. The color intensity is maximum at zero DDT concentration. The limit of detection is 27 ng/mL.

10.20.2 2,4-Dichlorophenoxyacetic Acid (2,4-D)

This is a systemic herbicide for killing broadleaf weeds. A gold nanoparticle-catalyzed chemiluminescence (CL) sensor is used for its detection (Boro et al. 2011). A microtiter plate is coated with 2,4-D-BSA conjugate, followed by blocking of non-specific sites by skim milk in PBS (Figure 10.48). A competitive inhibition assay is performed. Then, 2,4-D analyte solutions at concentrations of 0–2000 ng/mL, pre-incubated with IgG 2,4-D antibody/colloidal gold (CG) nanoparticle conjugate, are added to the well, allowing free antibody binding with the 2,4-D-BSA conjugate. The chemiluminescence signal generated at 425 nm by CG after the addition of luminol and $AgNO_3$ is measured.

The linear range of detection is 0–100 ng/mL. The detection limit is 3 ng/mL.

10.20.3 Carbofuran (CBF)

It is a broad-spectrum anticholinesterase carbamate pesticide.

10.20.3.1 Amperometric Immunosensor

The amperometric immunosensor is based on cyclic voltammetry (CV) and electrochemical impedance spectroscopy (EIS) in a background solution of 5 mM $[Fe(CN)_6]^{3-/4-}$ containing 0.1 M KCl (Zhu et al. 2013). The GCE is modified with multi-walled carbon nanotubes (MWCNTs)/graphene sheets (GS)-polyethyleneimine polymer (PEI)-gold (Au) nanocomposites to form GS-PEI-Au/MWCNT/GCE (Figure 10.49). This electrode is further modified with gold nanoparticles (AuNPs)-antibody (Ab) conjugate and blocked with bovine serum albumin (BSA) forming BSA/AuNP-Ab/PEI-Au/MWCNT/GCE.

CV responses of the immunosensor to different concentrations of CBF are recorded. The change in current (ΔI) is plotted against the logarithm of the CBF concentration; the graph is linear in the range 0.5–500 ng/mL. The current decreases as CBF

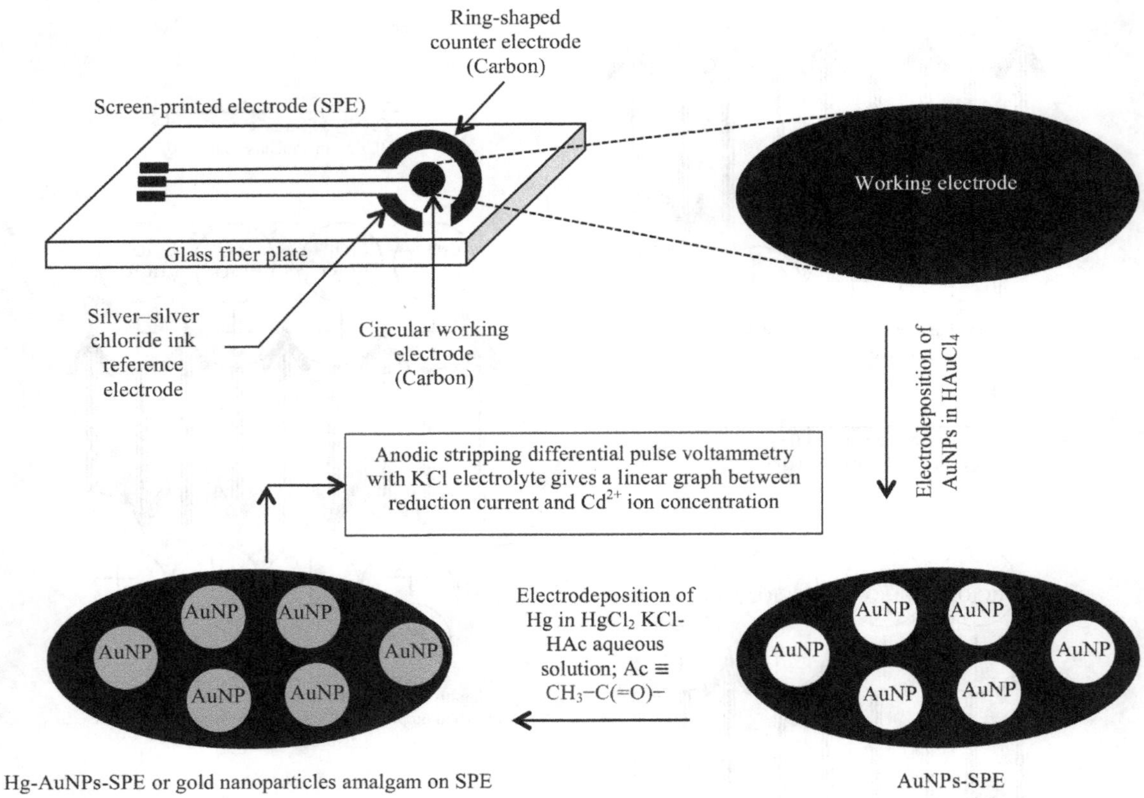

FIGURE 10.39 AuNPs amalgam-modified screen-printed electrode for ASDPV measurements (Zhang et al. 2010).

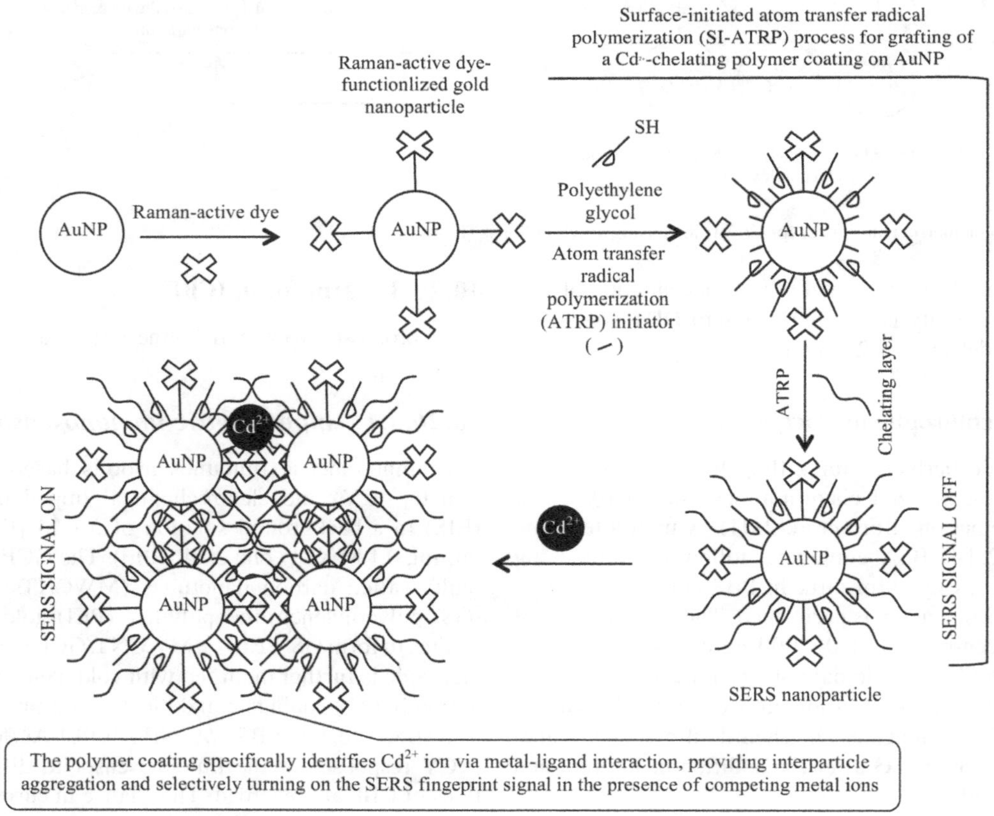

FIGURE 10.40 Turn-on SERS sensor (Yin et al. 2011).

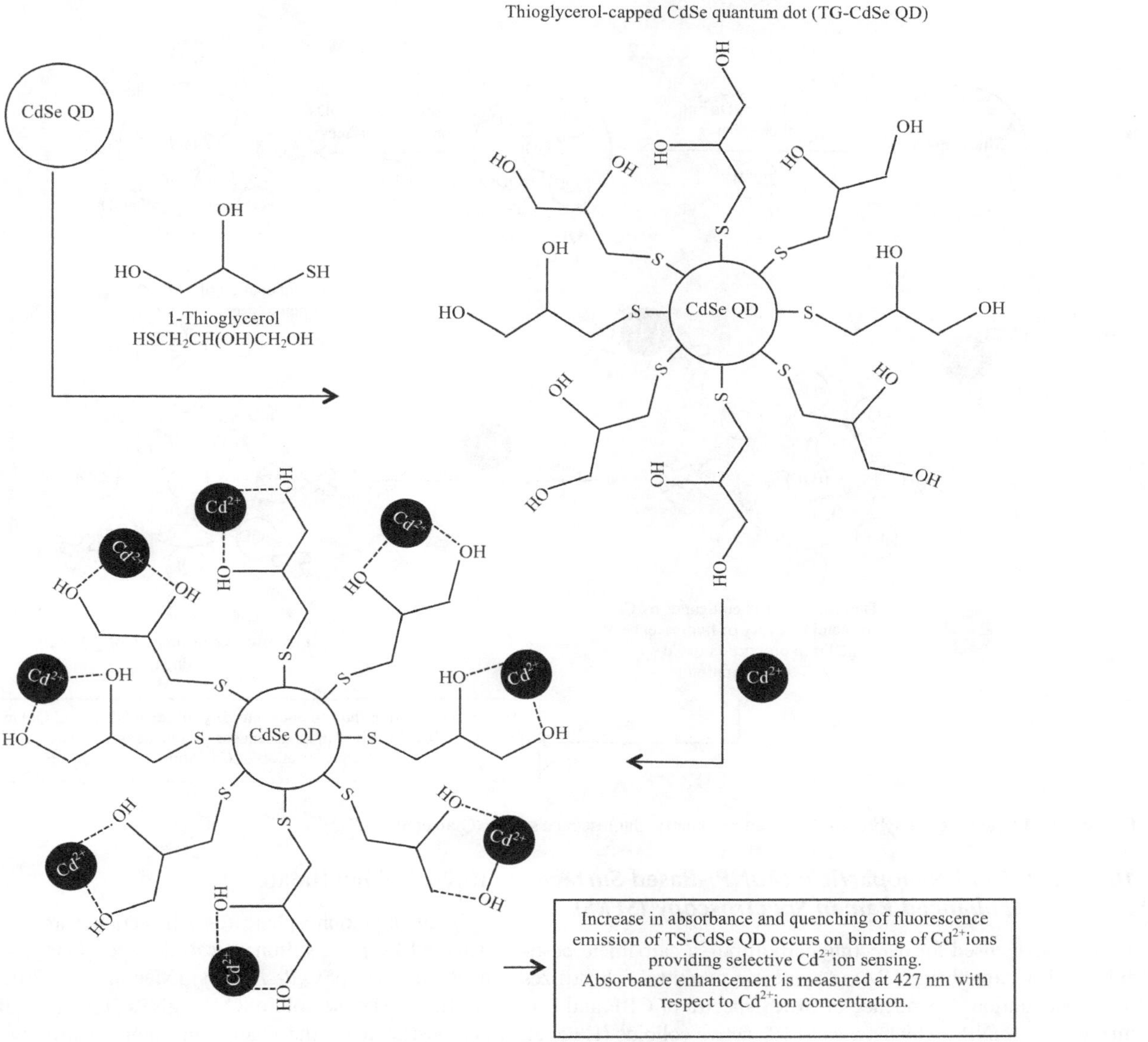

FIGURE 10.41 TG capping of CDSe QDs and operation of TG-CdSe QDs as Cd^{2+}ion sensor (Brahim et al. 2015).

concentration increases because binding of more CBF with antibody raises the barrier to electron transfer. The detection limit is 0.03 ng/mL.

10.20.3.2 Molecularly Imprinted Polymer (MIP)-Reduced Graphene Oxide and Gold Nanoparticle (rGO@AuNP)-Modified Glassy Carbon Electrode (GCE)

The CBF-sensing procedure is presented in Figure 10.50. For preparing the rGO@AuNP-coated GCE, the rGO@AuNPs composite is ultrasonically dispersed in Nafion and ethanol, dripped onto the GCE and air-dried (Tan et al. 2015). The MIP is prepared on the rGO@AuNP/GCE surface to form MIP/rGO@AuNP/GCE, using the CBF template molecule with methyl acrylic acid (MAA) as the functional monomer. The cross-linker is ethylene glycol maleic rosinate acrylate (EGMRA). The template is removed to get the imprinted rGO@AuNPs electrode.

Cyclic voltammetry (CV), electrochemical impedance spectroscopy (EIS) and differential pulse voltammetry (DPV) are performed using hexaferrocyanate as the probe in a solution of $K_3[Fe(CN)_6]$, containing 0.1 mol/L KCl.

An important observation from the DPV responses is that the CBF concentration can be determined from the peak current. Particularly in the range from 5×10^{-8} to 2×10^{-5} mol/L, the peak current and concentration of CBF are directly proportional to each other. The detection limit is adjudged to be 2×10^{-8} mol/L. Furthermore, selectivity to CBF sensing was investigated by measuring the current responses of CBF carbaryl, metrolcarb, etc., at the same concentration (2×10^{-5} mol/L) of these analytes. Carbaryl and metrolcarb are compounds which are structurally related analogs of CBF. The stronger response towards CBF than other anlytes confirmed selectivity of the sensor for recognizing CBF.

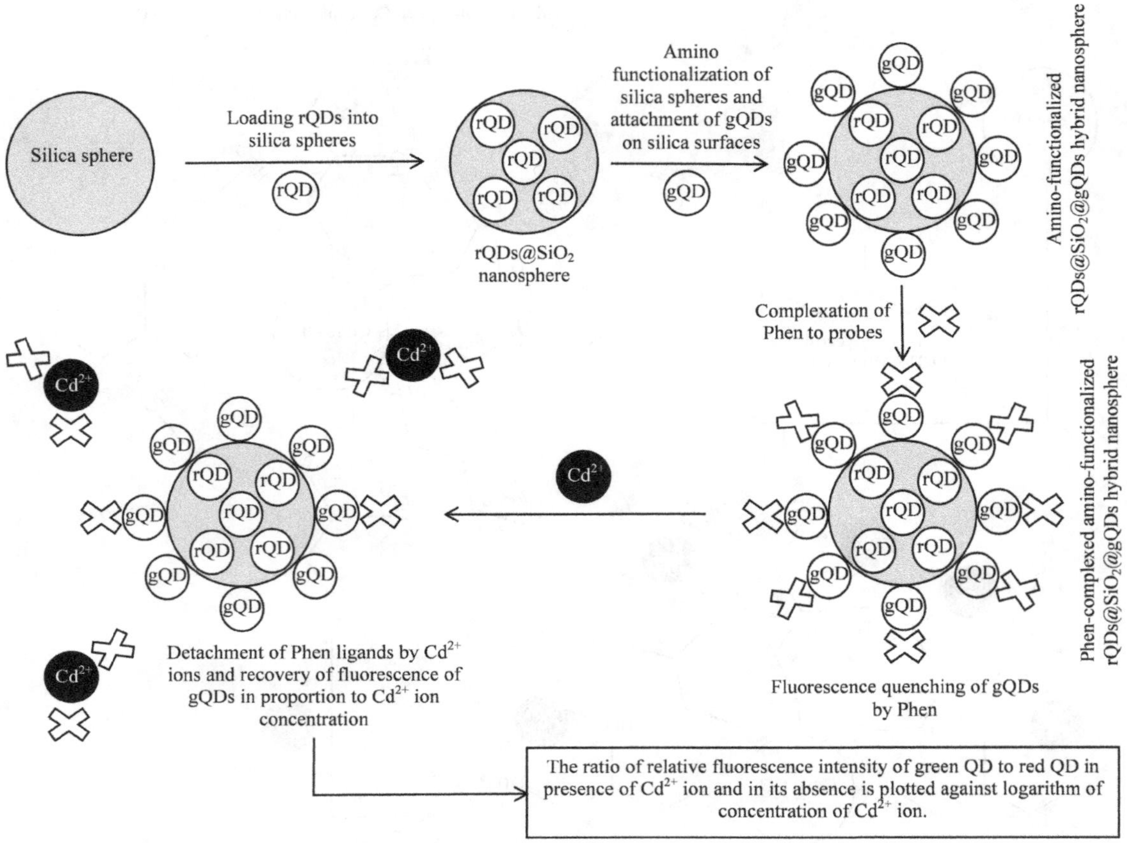

FIGURE 10.42 rQD-gQD hybrid-probe-based ratiometric fluorescence sensor (Qian et al. 2017).

10.20.3.3 Gold Nanoparticle (AuNP)-Based Surface Enhanced Raman Spectroscopy (SERS)

This sensor is used for detecting carbofuran, a carbamate pesticide, and deltamethrin (DM), a synthetic pyrethroid. It utilizes the concentration-dependence of SERS spectra of CBF and DM mixed with AuNPs, which act as SERS-active colloids (He et al. 2019). CBF in soil is qualitatively detected by characteristic peaks at 1000 cm^{-1} and 1299 cm^{-1} while DM detection in soil is detected through peaks at 560 cm^{-1} and 1000 cm^{-1}.

SERS intensity varies linearly with a concentration of CBF and DM in the range 0.01–10 mg/L. The limit of detection of CBF residue in soil is 0.053 mg/kg and that of DM is 0.056 mg/kg.

10.20.4 Methomyl

This is a carbamate pesticide. It is detected with the help of a glassy carbon electrode (GCE), modified with a composite made from Fe_3O_4 and Ag nanoparticles (Fe_3O_4/Ag composite) (Gai et al. 2019). The nanoparticles composite is attached to the GCE, using chitosan (CS) as linker molecules (Figure 10.51). The cyclic voltammetric response of Fe_3O_4/Ag-GCE to different concentrations of methomyl is determined. During the experiment, 0.2 M, pH = 6.9 phosphate-buffered saline (PBS) is used as supporting electrolyte.

The vital parameter to be observed is reduction peak current, which varies linearly with the concentration of methomyl. The linear relation is restricted to the range 3.47×10^{-5}–3.47×10^{-4} mol/L. The limit of detection is 2.97×10^{-5} mol/L.

10.20.5 Dimethoate

It is an organophosphate (OP) insecticide and acaricide. It is detected by *para*-sulfonato-calix[4]resorcinarene-capped silver nanoparticles (pSC_4R- AgNPs) (Menon et al. 2013). Selectivity of the response of pSC_4R-AgNPs towards dimethoate is demonstrated by the change in color on mixing with various pesticides, when it is found that the color changes from yellow to red only with dimethoate (Figure 10.52). A red shift takes place in wavelength and the surface plasmon band is broadened with dimethoate, but not with any other pesticide.

For 0.15 mM pSC_4R- AgNPs, the ratio of absorbance at 519 nm (A_{519}) to absorbance at 420 nm (A_{420}) increases linearly with dimethoate concentration from 10×10^{-8} M to 100×10^{-8} M. The limit of detection is 80 nM.

10.20.6 Atrazine

It is a widely used triazine-based herbicide applied to prevent pre-and post-emergence of broadleaf and grassy weeds.

10.20.6.1 Gold Nanoparticle (AuNP)-Modified Gold (Au) Electrode

Figure 10.53 shows the atrazine immunosensor. The Au electrode is immersed in 3-mercaptopropionic acid (MPA) solution for 2 h and mercaptoethylamine (MEA) solution for 1 h (Liu et al. 2014). Then, it is incubated with AuNP suspension for 2 h. AuNPs/Au electrode is coated with anti-atrazine monoclonal antibodies

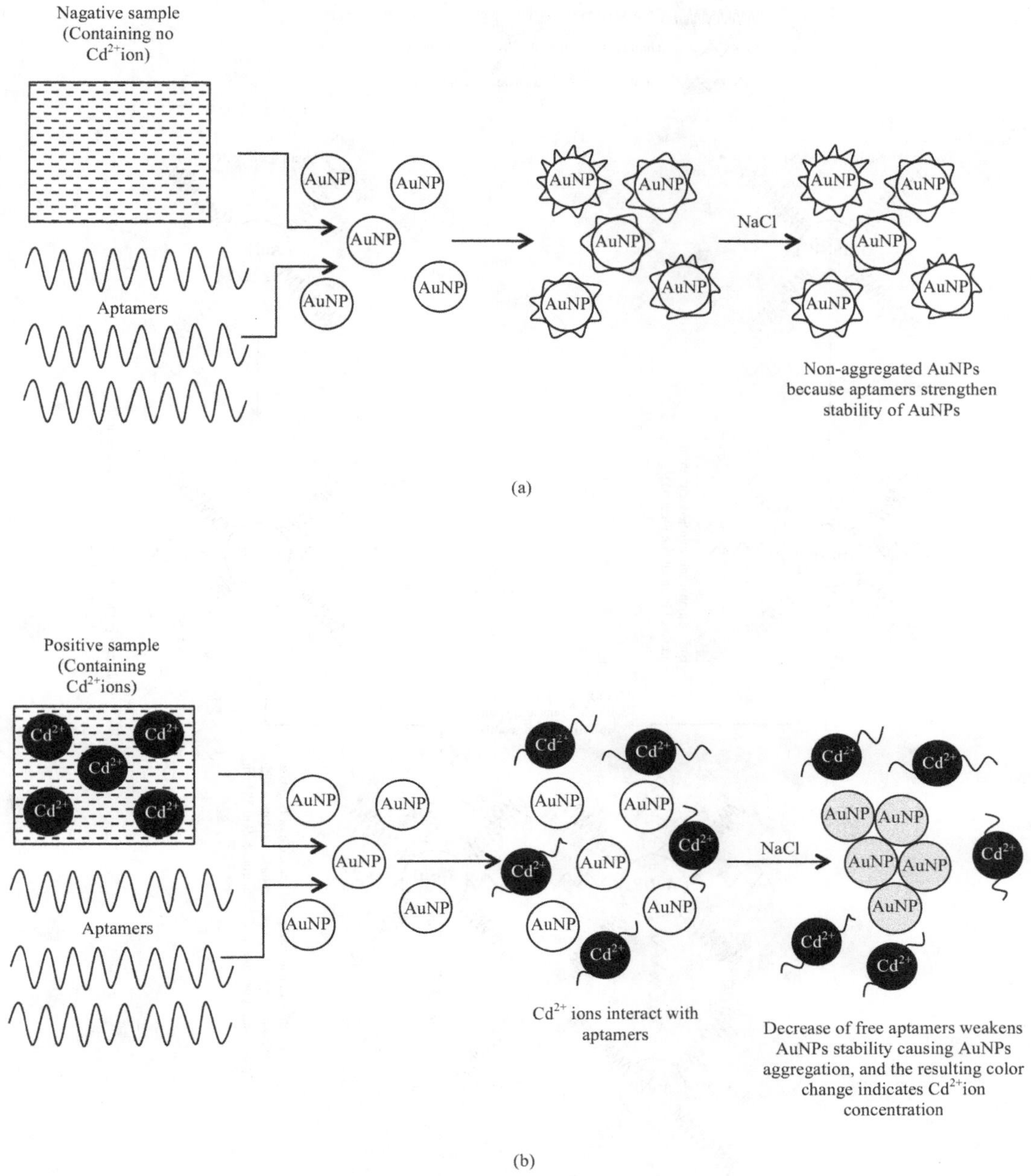

FIGURE 10.43 Combined use of aptamers and gold nanoparticles for detecting cadmium ions: (a) negative sample and (b) positive sample (Gan and Wang 2020).

(anti-atrazine) and the left-over sites are blocked with bovine serum albumin (BSA) to form: BSA/anti-atrazine/AuNPs/Au. Differential pulse cyclic voltammetry (DPV) is done with ferricyanide redox indicator. The peak current decreases with increasing atrazine concentration, due to blocking of electrons shuttling between the redox center of anti-atrazine and the electrode.

The current response with respect to atrazine concentration is linear in the range 0.05–0.5 ng/mL. The detection limit is 0.016 ng/mL.

10.20.6.2 Cysteamine (Cys)-Functionalized Gold Nanoparticles (AuNPs)

Cys-AuNPs act as a recognition probe for atrazine at pH = 3.8 (Liu et al. 2016). The mercapto group (–SH) of Cys binds with AuNPs to form Au-S bonds. The amino functional groups ($-NH_2$) in Cys provide stability to Cys-AuNPs (Figure 10.54a). The Cys-AuNPs colloidal solution has a wine-red color. In pH 3.8 buffer, the $-NH_2$ groups on the surfaces of Cys-AuNPs become positively charged and hydrogen bonding occurs between $-NH_2$ converted into $-NH_3^+$ and atrazine, causing aggregation of cysteamine-AuNPs. The aggregation of Cys-AuNPs changes the color of the solution from red to blue or purple (Figure 10.54a). The color change can be examined by the naked eye or UV-vis spectrophotometry. Absorption spectra are recorded at 400–750 nm.

The change of absorbance between the blank and the specimen, as measured at 523 nm

$$\Delta A_{523} = A_{\text{Blank}} - A_{\text{Specimen}} \tag{10.5}$$

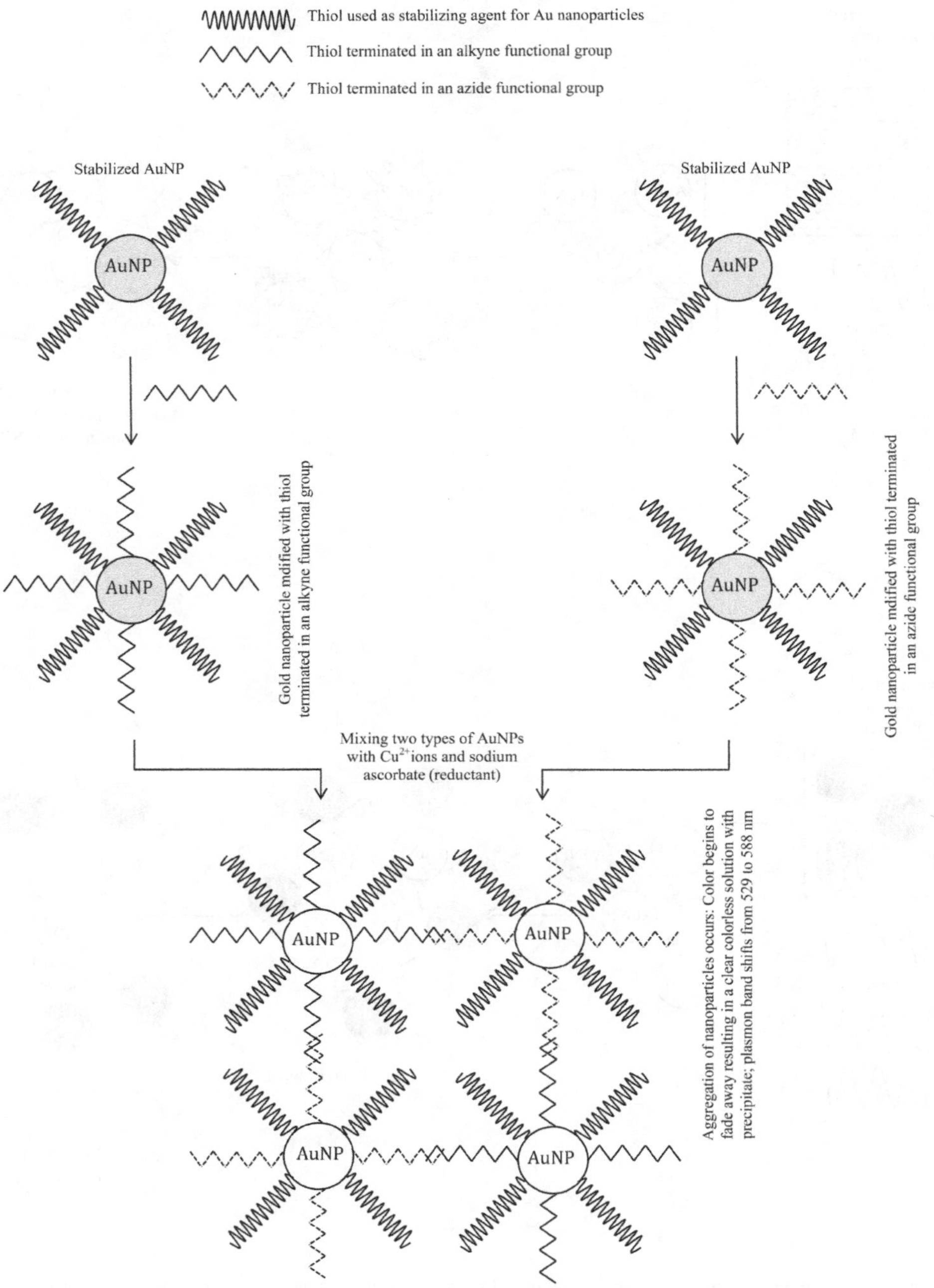

FIGURE 10.44 Use of copper ion-click chemistry between azide- and alkyne-functionalized gold nanoparticles for Cu^{2+} ion sensing (Zhou et al. 2008).

is proportional to atrazine concentration in the range 0.033–6.67 μg/g. The detection limit is 0.0165 μg/g.

10.20.6.3 Nitrogen-Doped Carbon Quantum Dot-Based Luminescent Probe

The carbon quantum dot is amine-functionalized (Mohapatra et al. 2018). When atrazine is added, hydrogen bonding takes place between the surface amino groups of the carbon quantum dot and atrazine. Photoluminescence spectra are recorded. The fluorescence emission intensity at 490 nm turns on with increasing atrazine concentration and increases because hydrogen bonding between the carbon quantum dot and atrazine induces aggregation.

The sensor works linearly in the concentration range 5pM–7 nM. The detection limit is 3 pM.

10.20.7 Paraoxon-Ethyl

It is an acetycholinesterase (AChE)-inhibiting organophosphate (OP) insecticide. It is detected by AChE/gold nanoparticle (AuNP)-Polypyrrole (Ppy)-reduced graphene oxide (rGO) nanocomposite/glassy carbon electrode (GCE) biosensor (Yang et al. 2014). Figure 10.55 shows the sensing protocol.

The current decreases with increase in concentration of paraoxon-ethyl from 1 nM to 5 μM. The current falls because

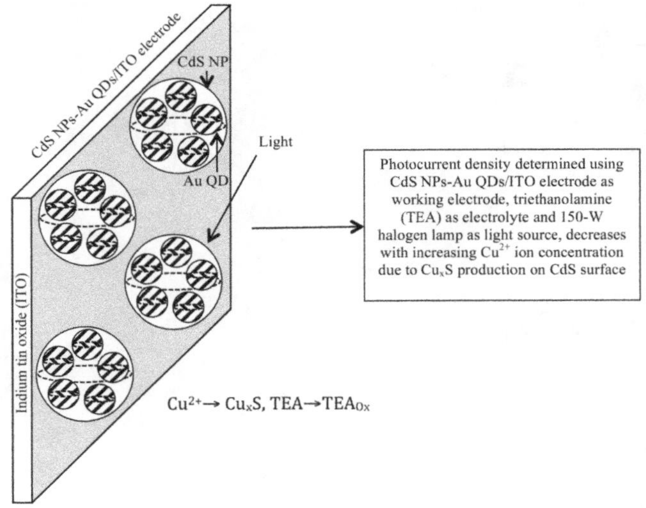

FIGURE 10.45 Gold quantum dots loaded on cadmium sulfide nanoparticles on ITO electrode for Cu^{2+} ion sensing (Ibrahim et al. 2016).

paraoxon-ethyl is an AChE inhibitor, i.e., it reduces the enzymatic activity. The limit of detection is 0.5 nM.

10.20.8 Acetamiprid

It is a neonicotinoid insecticide. It is detected by the biosensor based on acetamiprid-specific S-18 ssDNA aptamer-controlled inhibition of gold nanozyme (nanoparticle-based artificial enzyme) activity (Weerathunge et al. 2014). Inherent peroxidase-like activity of gold nanoparticles (AuNPs) is masked/inhibited by conjugation with the acetamiprid-specific S-18 ssDNA aptamer. In the presence of acetamiprid, the aptamer suffers structural changes accompanied by desorption from the surface of AuNPs. This desorption restores the peroxidase-like activity of AuNPs thereby oxidizing the colorless tetramethylbenzidine (TMB) to a purplish-blue compound in accordance with the desorption of the aptamer and hence acetamiprid concentration. The sensing scheme is displayed in Figure 10.56.

Relative peroxidase-like activity of the AuNPs-S-18 conjugate, with respect to acetamiprid concentration, is either visualized from color change or quantified by UV-vis spectroscopy. It is linear in the range 0.1–10 ppm. Up to 0.1 ppm, acetamiprid is detectable in a 10-minute assay.

10.20.9 Hexachlorobenzene (HCB), Perchlorobenzene

It is a lipotropic organochlorine fungicide. It is detected by the tin oxide nanoparticles (SnO_2 NPs): Graphene nanopellets (GNP) hybrid/silver-decorated optical fiber platform (Sharma et al. 2017). A 40-nm thick silver film is deposited over the unclad portion of a 15-cm long plastic-clad fiber, followed by SnO_2 NP/GNP nanohybrid coating over the silver film (Figure 10.57). After cleaving both ends of the optical fiber to ensure guidance of maximum light, the fiber is mounted inside a glass flow cell. One end of the fiber is illuminated with a polychromatic tungsten halogen lamp. At the opposite end of the fiber, a spectrometer is placed. HCB solutions of different concentrations are introduced in the flow cell. These solutions interact with the fiber optic probe. GNP/SnO_2 acts as recognition element for HCB. It is the sensing layer. GNP interacts with HCB causing its dechlorination. The effective refractive index of GNP/SnO_2 is governed by this interaction and hence varies with HCB concentration.

The peak absorbance wavelengths determined from the absorbance spectra recorded for different HCB concentrations show a red positive shift in the range $0–10^{-2}$ g/L. The calibration curve is drawn between peak absorption wavelength and HCB concentration. The red shift in peak wavelength is 43 nm for the range $10^{-12}–10^{-2}$ g/L, whereas the same is 74 nm in the range $0–10^{-2}$ g/L. The sensitivity is high in the lower concentration range but decreases at high concentrations because of the finite number of active sites in the SnO_2 NP/GNP hybrid.

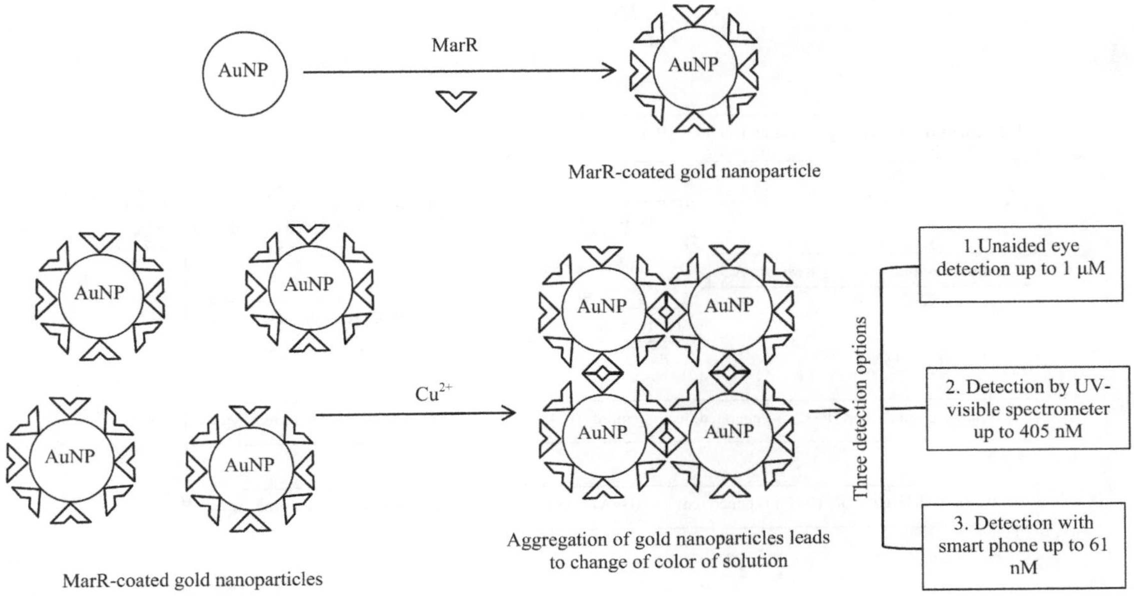

FIGURE 10.46 Colorimetric assay using MarR-coated gold nanoparticles (Wang et al. 2017).

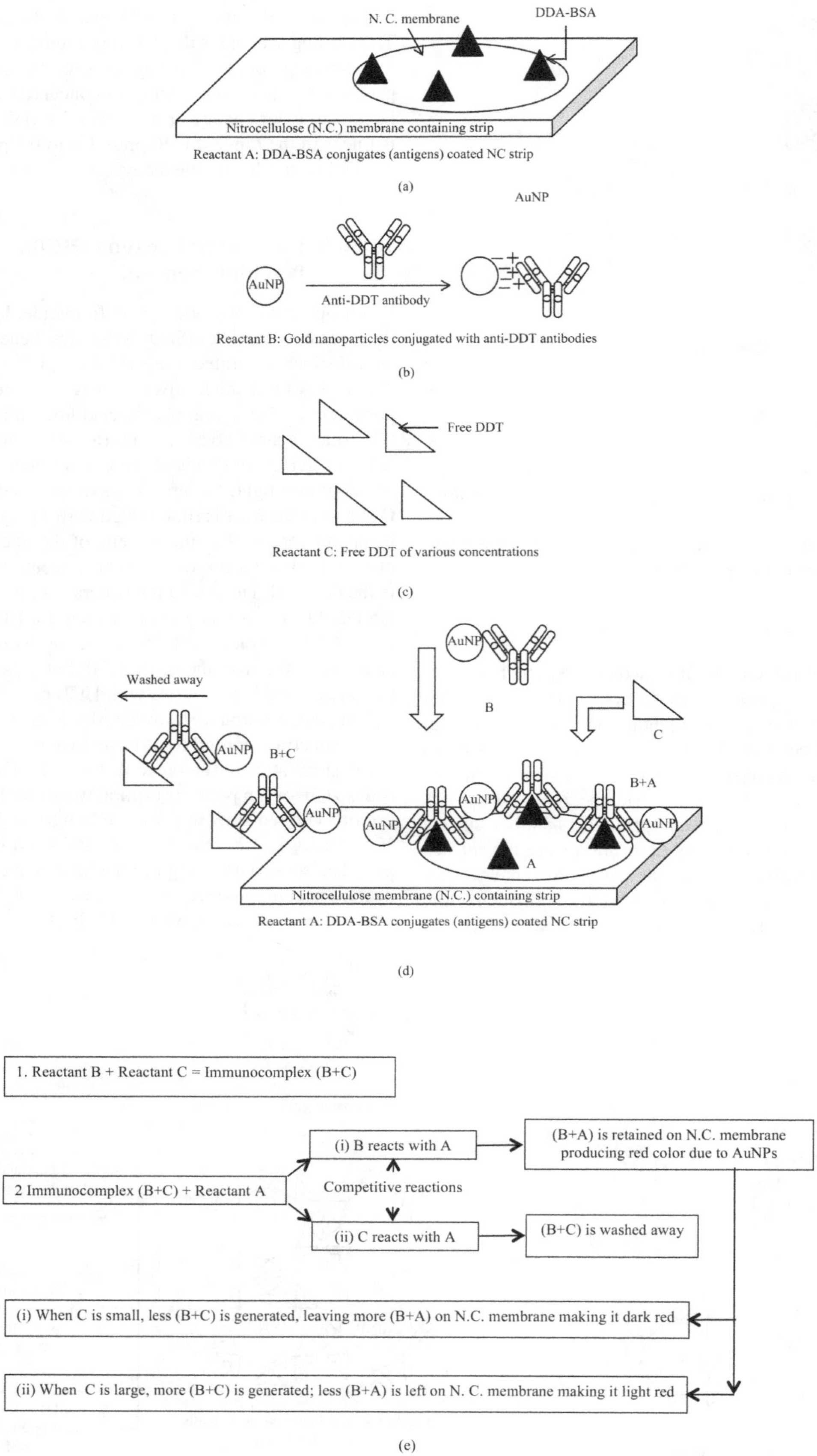

FIGURE 10.47 Competitive assay for DDT detection: (a) reactant A, (b) reactant B, (c) reactant C, (d) reactants B and C applied to reactant A, and (e) scheme of the assay (Lisa et al. 2009).

10.20.10 Malathion (MLT): Diethyl 2-[(dimethoxyphosphorothioyl)sulfanyl]butanedioate

It is a non-systemic organophosphorus insecticide. It is detected by a gold nanoparticle- (AuNPs)-chitosan (CS)-ionic liquid (IL/pencil graphite electrode (PGE) (Bolat and Abaci 2018). The pencil graphite electrode (PGE) is coated with a uniform chitosan (CS)-ionic liquid, IL: (1-butyl-3-methylimidazolium tetrafluorophosphate, ($BMIMPF_6$) hybrid nanocomposite to form CS-IL/PGE. Electrodeposition of gold nanoparticles (AuNPs) produces AuNP-CS-IL/PGE. Cyclic voltammetry is done in 0.1M KCl containing 5 mM $[K_4Fe(CN)_6]^{3-/4-}$. As square wave voltammetry (SWV) provides high sensitivity, SWV voltammograms are recorded for different concentrations of MLT.

The peak MLT reduction currents, measured with respect to the Ag/AgCl electrode for various MLT concentrations, reveal two linear ranges: one in the range 0.89–5.94 nM with greater sensitivity, with a steep slope arising from a larger number of available binding sites; and the other in the range 5.94–44.96 nM with lower sensitivity and a shallow gradient due to restricted number of binding sites on the surface of the electrode. The limit of detection is 0.68 nM.

10.20.11 Dithiocarbamate (DTC) Pesticide Group

This group consists of ziram: zinc dimethyldithiocarbamate, a fungicide; zineb: zinc ethylene bis(dithiocarbamate), a fungicide; and maneb: manganese ethylene-1, 2-bisdithiocarbamate,

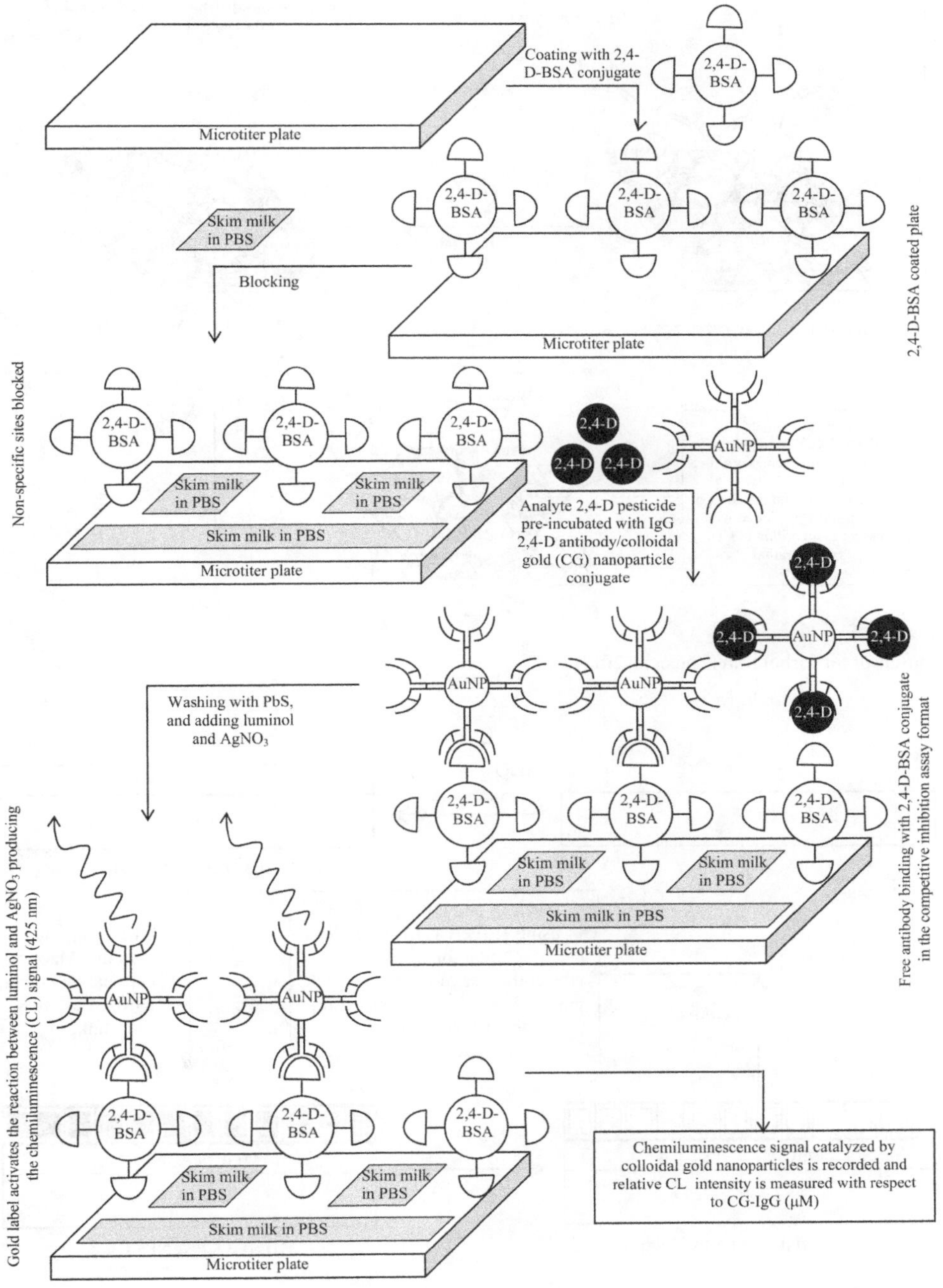

FIGURE 10.48 Colloidal gold nanoparticles catalyzed chemiluminescence immunosensor for detecting 2,4-dichlorophenoxyacetic acid (Boro et al. 2011).

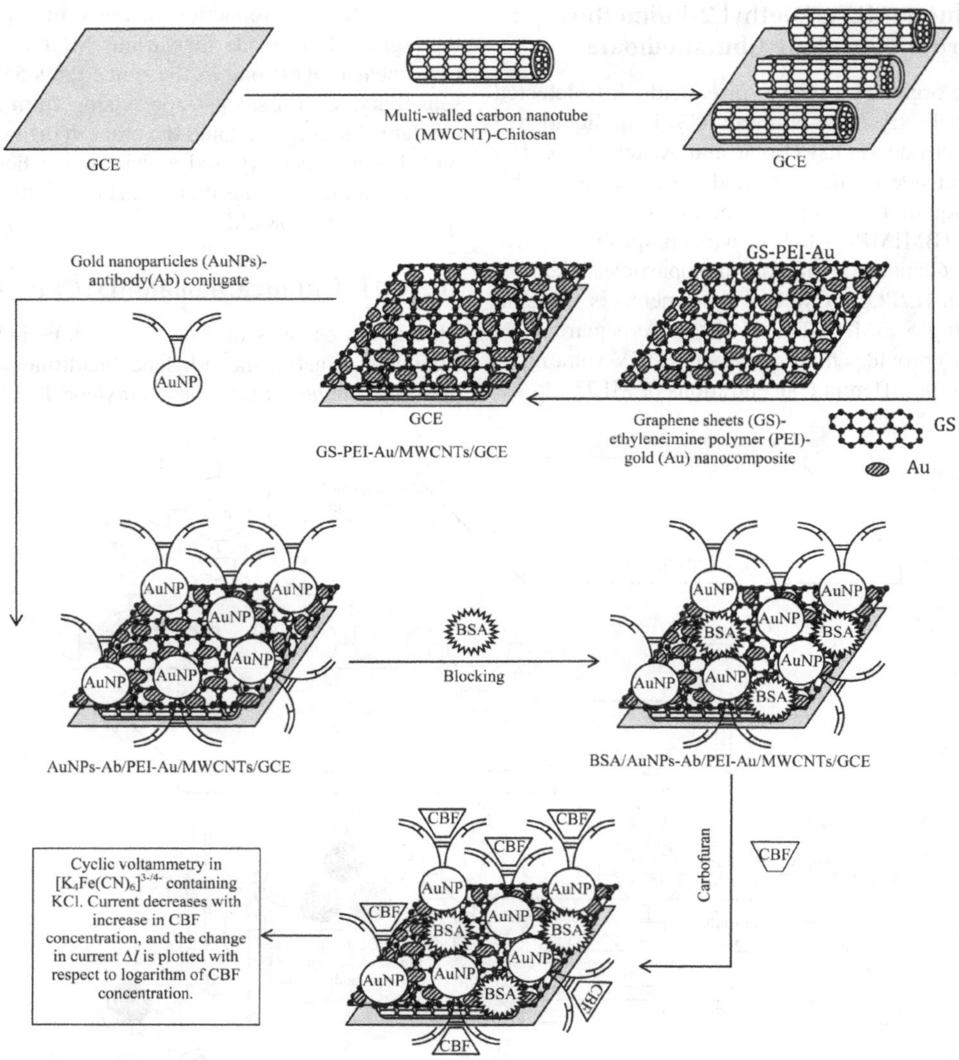

FIGURE 10.49 Immunosensor for carbofuran (Zhu et al. 2013).

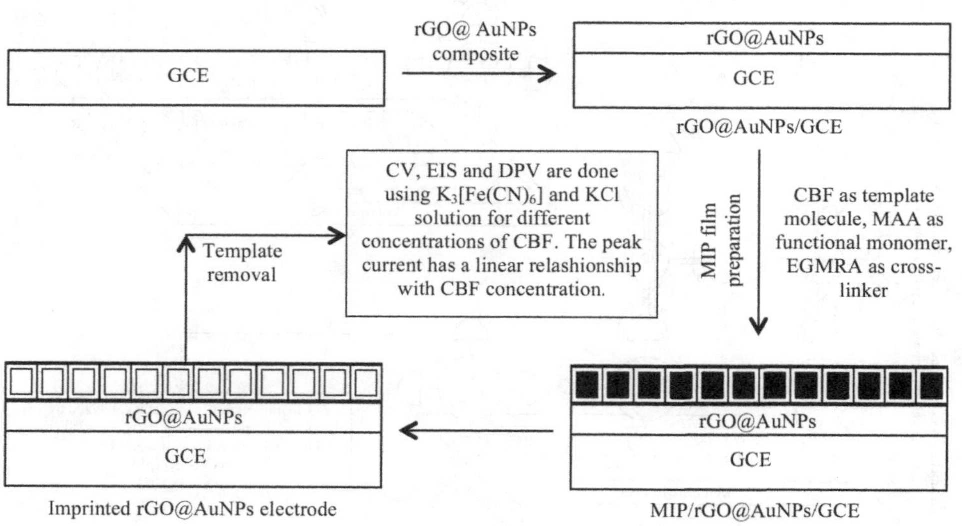

FIGURE 10.50 MIP-rGO-AuNP-modified electrode for CBF sensing (Tan et al. 2015).

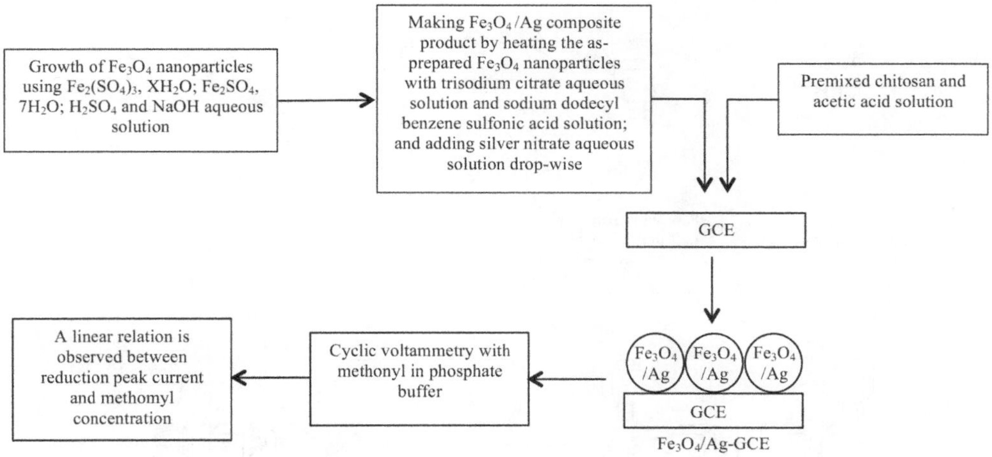

FIGURE 10.51 Methonyl detection with a Fe_3O_4/Ag-GCE electrochemical sensor (Gai et al. 2019).

a fungicide. It is detected by cetyltrimethylammonium bromide (CTAB)-coated copper nanoparticles (CTAB-CuNPs) (Ghoto et al. 2019). CTAB is used as a capping agent for CuNPs due to its amphiphilic character, originating from its hydrophobic and hydrophilic parts, the former producing a positive charge on the surface of CuNPs and the latter providing steric hinderance to prevent coagulation and oxidation of CuNPs.

The absorption intensity at 490 nm increases linearly with the concentrations of the three pesticides. The colorimetric detection range is 97.8–489.3 ng/mL for ziram; it is 8.8–44.1 ng/mL for zineb, and 8.4–42.4 ng/mL for maneb. The corresponding limits of detection are 150.9 ng/mL, 0.7 ng/mL, and 6.1 ng/mL, respectively.

10.21 Discussion and Conclusions

10.21.1 Particulate Matter

Nanosensors for particulate matter exploit either the frequency shift associated with mass increment of resonators or rely on successive processes of particle classification, charging and subsequent measurement by voltage modulation (Table 10.3). Accurate current determination is required.

10.21.2 Gases

CO nanosensors mostly work on the chemiresistive principle, either at room temperature or at elevated temperatures (Table 10.4). They are embellished with Au or cerium oxide nanoparticles, CuO nanowires or ZnO nanowalls for increasing sensitivity.

For SO_2 sensing, resistance variation of various nanostructures, such as tungsten oxide thin film, Ni nano-walls and nano-clustered tin oxide film, with gas concentration is exploited (Table 10.5). An optical nanosensor works on the FRET principle.

Resistance variations of SnO_2 nanoribbons, ZnO nanowires, and Cu-ZnO nanoparticle composite film are exploited for nitrogen dioxide sensing (Table 10.6).

Ozone estimation relies on resistance changes in SnO_2/SWCNTs, nanocrystalline $SrTi_{1-x}Fe_xO_3$, ZnO nanoparticle/ZnO nanocrystal, and MWCNTs-Pd thin films (Table 10.7).

Sensors for VOCs mainly work on resistance and capacitance changes. A colorimetric probe and a SAW sensor are also studied (Table 10.8).

Conductivity changes of PANI NPs film, MoO_3 nanoparticles film, and PtNPs covered WO_3 film are employed for ammonia sensing (Table 10.9). Also, quenching of fluorescence of ZnO. Eu^{2+} NPs is utilized.

10.21.3 Pathogens

Mechanical, fluorescence, surface plasmon resonance, electrochemical, and magnetic approaches have been utilized for detecting *E. coli* bacteria, using Au and magnetic nanoparticles (Table 10.10).

Nanosensors for *V. cholerae* are designed around DPV, anodic stripping voltammetry, and potentiometry. A colorimetric approach is also used (Table 10.11).

Nanosensors for *Ps. aeruginosa* are developed using the chemiluminescence phenomenon and the electrochemical principle (Table 10.12).

Optical and electrochemical principles are used for sensing *L. pneumophila* (Table 10.13).

10.21.4 Metals

Nanosensors for Hg^{2+} ions are mainly based on the colorimetric principle (Table 10.14).

A colorimetric approach is mostly used for lead ion sensing; one electrochemical sensor is also included (Table 10.15).

For arsenic (III) sensing, voltammetry in various forms, such as anodic stripping, linear sweep, square wave etc., has been applied, in addition to optical methods, like SERS, SPR, and colorimetry (Table 10.16).

Nanosensors for Cr(VI) ions are colorimetry- and electrochemistry-based (Table 10.17).

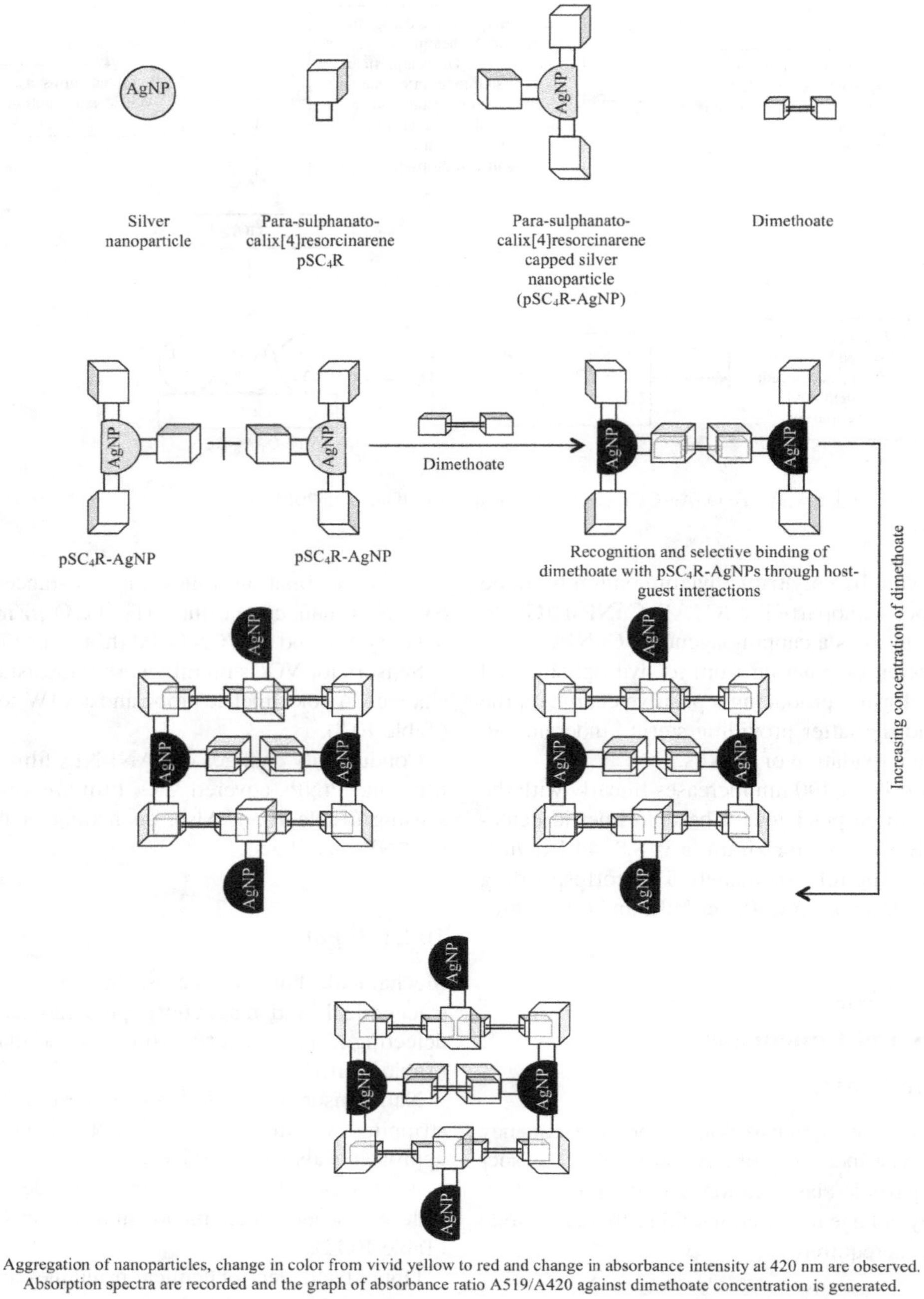

FIGURE 10.52 Formation of dimethoate inclusion complex with *para*-sulfonato-calix[4]resorcinarene-capped silver nanoparticles (Menon et al. 2013).

Cd^{2+} ion detection utilizes voltammetry, SERS, fluorescence, and colorimetric approaches (Table 10.18).

Copper ions are detected by colorimetric and photoelectrochemical principles (Table 10.19).

10.21.5 Pesticides

Pesticide detection relies on optical techniques, such as colorimetry, surface plasmon resonance, fluorescence, chemiluminescence, and SERS (Table 10.20). Apart from optical principles, other techniques rest on the pedestal of electrochemistry.

Review Exercises

10.1 How is particulate matter in air designated by particle size? What are its harmful effects on human beings and the environment?

10.2 From where does carbon monoxide come in the atmosphere? How does it cause death?

10.3 Which industries create sulfur dioxide and nitrogen dioxide pollution in the atmosphere? What are the corrosive acids associated with these gases?

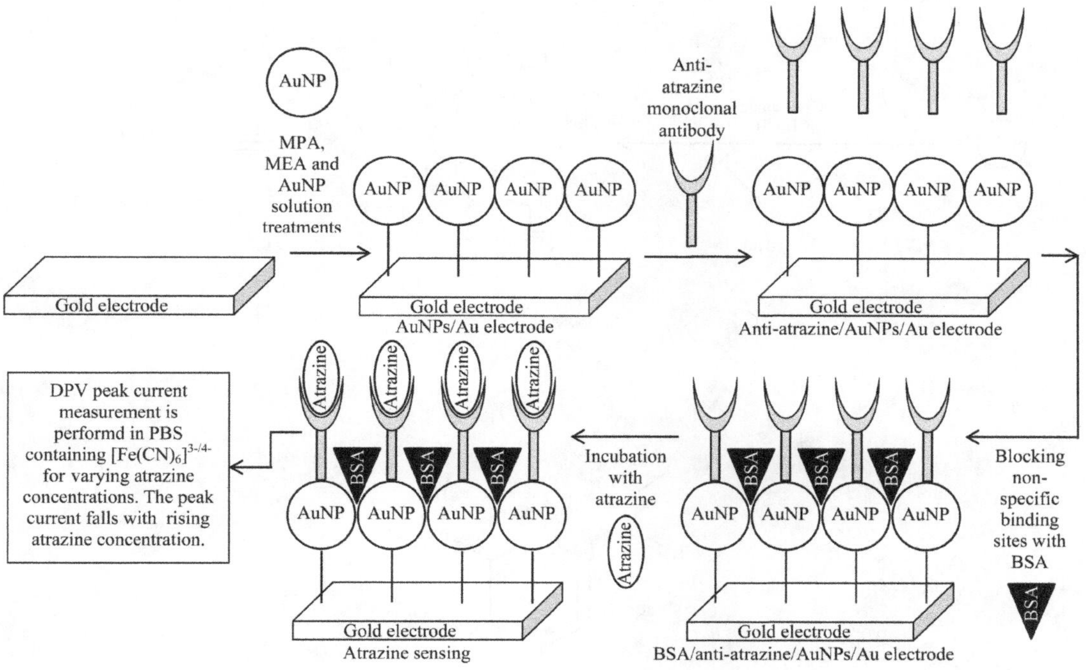

FIGURE 10.53 Gold nanoparticle-gold electrode immunosensor for atrazine (Liu et al. 2014).

10.4 What are the sources of ground-level ozone? What problems does ozone pollution bring with it?

10.5 Name a few organic gaseous pollutants which have carcinogenic effects. What are their sources?

10.6 Name a gas having a pungent odor, which is emitted from fertilizer plants.

10.7 What are the two main constituent sections of a cantilever-based airborne nanoparticle detector (CANTOR)? How does the attachment of nanoparticles to the cantilever impact its resonance frequency?

10.8 Explain the construction of the aerosol sensor chip with the help of a diagram. Name the three sections comprising the aerosol sensor chip and elucidate their roles. How is voltage modulation applied to determine the particle number concentrations?

10.9 How does a virtual impactor particle classification device work? Explain with a diagram.

10.10 Draw the layout diagram and explain the operation of a MEMS-based airborne particle detection equipment. What is the typical range of current magnitude? How is this current measured?

10.11 Name three nanosensors for carbon monoxide: one nanoparticle-based, one nanowire-based and one nano-wall-based. Describe how they work.

10.12 How is a fluorescence nanoprobe used to detect sulfur dioxide? Of what material are the nanodots used in this sensor made?

10.13 Name two metal oxides whose nanostructures are used in sensors for detecting sulfur dioxide.

10.14 Describe two sensors which use carbon nanotubes for detecting ozone. Name an ozone sensor in which thermal excitation is replaced by illumination with ultraviolet radiation.

10.15 Describe a capacitive sensor for VOC detection. Draw the diagram.

10.16 Draw the constructional sketch of a NiO thin-film sensor for detecting formaldehyde. Explain, giving relevant diagrams, its sensing mechanism.

10.17 How is a SAW delay line used for VOC detection?

10.18 Explain the use of a colorimetric probe for detecting formaldehyde in its liquid and gaseous forms.

10.19 Explain, with a diagram, the working of a platinum nanoparticle-decorated WO_3 sensor for ammonia detection.

10.20 Describe a nanoparticle-based ammonia sensor which works by fluorescence quenching on exposure to ammonia.

10.21 How are the *E. coli* collected on the surface of a magnetoelastic sensor? What is the function of chitosan nanoparticles in this sensor? How is the resonance frequency of the sensor affected by *E. coli* loading?

10.22 How are color changes produced on binding of *E. coli* with MEA-conjugated gold nanoparticles? Can cysteamine-capped gold nanoparticles be used to obtain a color response?

10.23 Explain the operation of an *E. coli* sensor, which uses three types of nanoparticles: iron oxide, gold and lead sulfide, clearly pointing out the roles of the three nanoparticle types.

10.24 How do magnetic relaxation and fluorescence modes work co-operatively to realize a nanosensor for *E. coli*, with the capability of measuring at low and high concentrations?

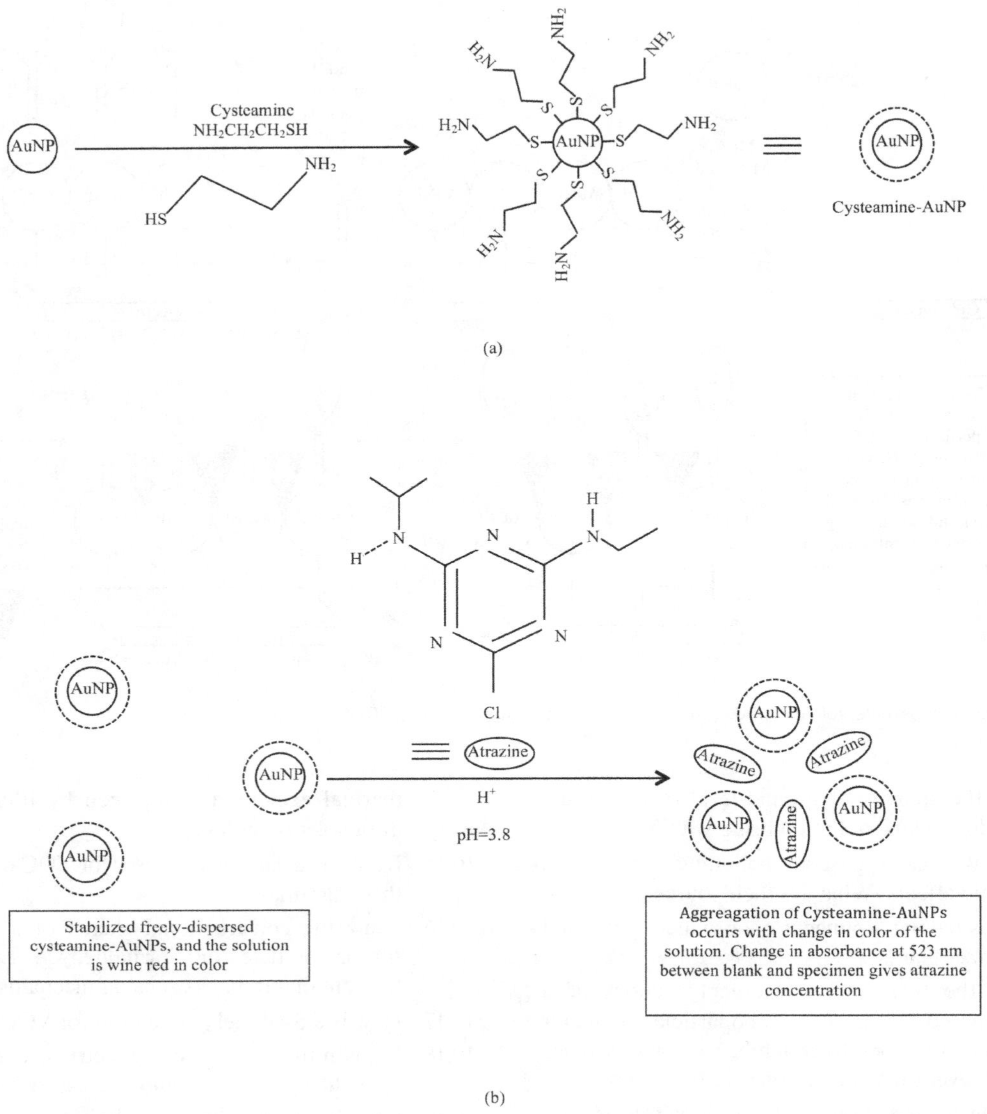

FIGURE 10.54 Cysteamine-gold nanoparticle sensor for atrazine: (a) stabilization of gold nanoparticles by cysteamine and (b) atrazine-induced aggregation of nanoparticles in acidic pH medium, resulting in color change (Liu et al. 2016).

10.25 How does a signal-off impedimetric nanosensor for *E. coli* work? Explain with diagrams.

10.26 How is a surface plasmon resonance sensor for *E. coli* made by incorporating positively charged Cu(II) ions during gold nanoparticle imprinting?

10.27 Describe, with a flow diagram, the operation of a microfluidic colorimetric biosensor for *E. coli.*

10.28 Explain the use of lactose-stabilized gold nanoparticles for *E. coli* detection from the shift in surface plasmon absorption band.

10.29 How does the ssDNA/nMgO)/ITO bioelectrode sense *Vibrio cholerae*?

10.30 How do we detect *V. cholerae* from changes in photoluminescence peak intensity of a nanostructured MgO sensor?

10.31 Describe a lyophilized gold nanoparticle/poly (styrene-co-acrylic acid)-based genosensor for *V. cholerae.*

10.32 How are *V. cholerae* sensed using a screen-printed electrode (SPE) modified with polystyrene-co-acrylic acid latex nanospheres and colloidal gold nanoparticles?

10.33 Explain the measurement of cholera toxin by a bioelectrode, using natural cholera toxin receptor ganglioside GM1 incorporated into a lipid membrane.

10.34 Describe sequentially the steps leading to production of the chemiluminescence signal in a probe-modified magnetic nanoparticle-based sensor for *Ps. aeruginosa.*

10.35 What is the analytical principle of the gold nanoparticle-decorated- reduced graphene oxide electrode for detecting PyoV to detect *Ps. aeruginosa*?

10.36 Name a biomarker for *Ps. aeruginosa* and describe its detection by PANI/gold nanoparticle-decorated ITO electrode.

10.37 Explain the use of an electrochemical immunosensor, using zinc oxide nanorods for measuring *Legionella pneumophila*?

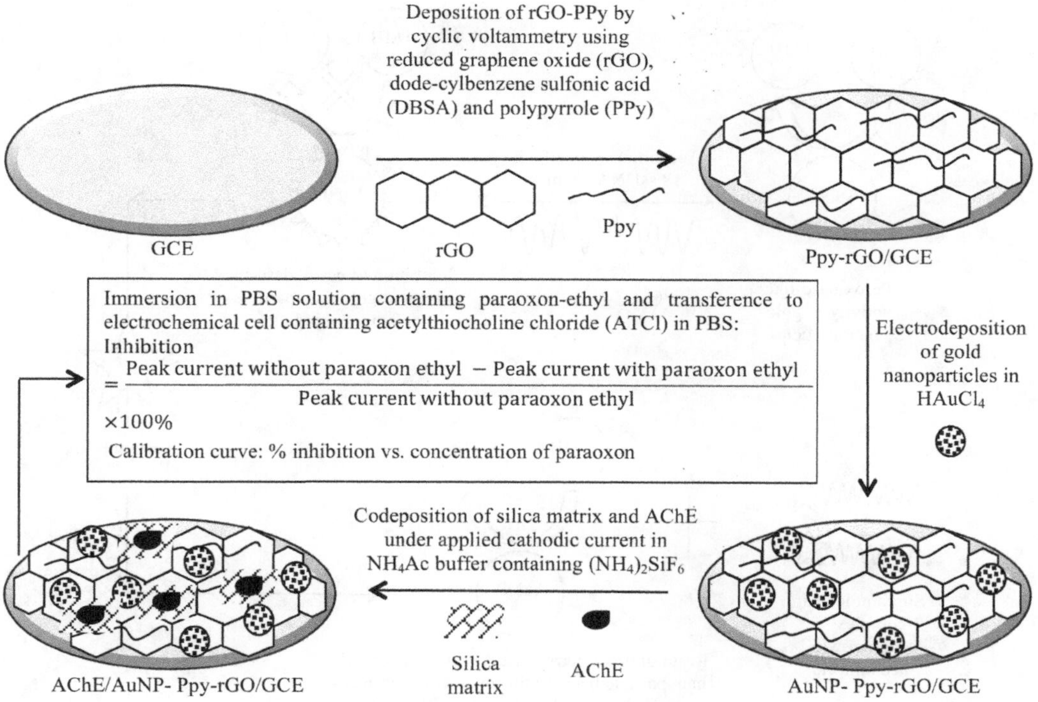

FIGURE 10.55 AChE biosensor for amperometric detection of paraoxon-ethyl (Yang et al. 2014).

10.38 For which bacterium is the virostat antibody a bioreceptor? How is the concentration of these bacteria measured by an azimuthally controlled gold grating-coupling surface plasmon resonance platform? What are the three response regimes of this platform?

10.39 How are mercury ions detected by thymine-derivative decorated gold nanoparticles?

10.40 Explain the use of aptamer-protected gold nanoparticles for Hg^{2+} ion sensing.

10.41 Explain, with a diagram, the interaction of starch-stabilized silver nanoparticles with Hg^{2+} ions, and their use for Hg^{2+} ion sensing.

10.42 Describe the use of glutathione (GSH)-stabilized silver nanoparticles as a sensor for Pb^{2+} ions.

10.43 Why do gold nanoparticles aggregate in the presence of Pb^{2+} ions in a maleic acid-functionalized gold nanoparticle sensor?

10.44 How are surface-unmodified (unlabeled) gold nanoparticles used for sensing Pb^{2+} ions?

10.45 Why does the UV-visible absorbance of thioctic acid- and dansyl hydrazine- conjugated gold nanoparticles vary in the presence of Pb^{2+} and Cu^{2+} ions?

10.46 Name the metal ion in the presence of which the color of valine-capped gold nanoparticles changes? What is the reason of this change in color?

10.47 Describe a silver nanoparticle-localized surface plasmon resonance absorbance sensor for Pb^{2+} ions.

10.48 Explain the use of gold nanoparticle-graphene modified glassy carbon electrode for lead ion sensing.

10.49 Describe the portable on-site arsenic testing system for ground water, employing SERS technique.

10.50 How is arsenic (III) detected by a surface plasmon resonance sensor? What is the lowest arsenic concentration determined by this method?

10.51 Which of the following bimetallic nanoparticles give the best performance for As(III) sensing: FePt, FePd, FeAu or AuPt nanoparticles?

10.52 Can we use a silver nanoparticle-modified gold electrode for As(III) sensing? How?

10.53 Explain the principle of carbon nanoparticles/gold nanoparticles modified glassy carbon electrode aptasensor for As(III) ions.

10.54 Describe the use of gold nanostructured electrode on gold foil for As(III) ion sensing.

10.55 How do the AuPtNP/SPCE and AuPtNP/PANI/SPCE platforms differ in their responses towards As(III) ions?

10.56 How do ranolazine-functionalized copper nanoparticles contribute to As(III) ion sensing?

10.57 Explain, with a diagram, the working of the immunochromatographic assay for Cr(VI) ions.

10.58 What are amyloid fibrils? Explain their use in making a colorimetric sensor for Cr(VI) ions.

10.59 Describe the use of gold nanoparticles decorated titanium dioxide nanotubes on Ti substrate as an amperometric sensor for Cr(VI) ions.

10.60 How is a screen-printed electrode modified with gold nanoparticles amalgam for sensing Cd^{2+} ions?

10.61 How are gold nanoparticles modified to build a SERS sensor for Cd^{2+} ion detection?

10.62 What property of thioglycerol-capped CdSe quantum dots helps in the detection of Cd^{2+} ions?

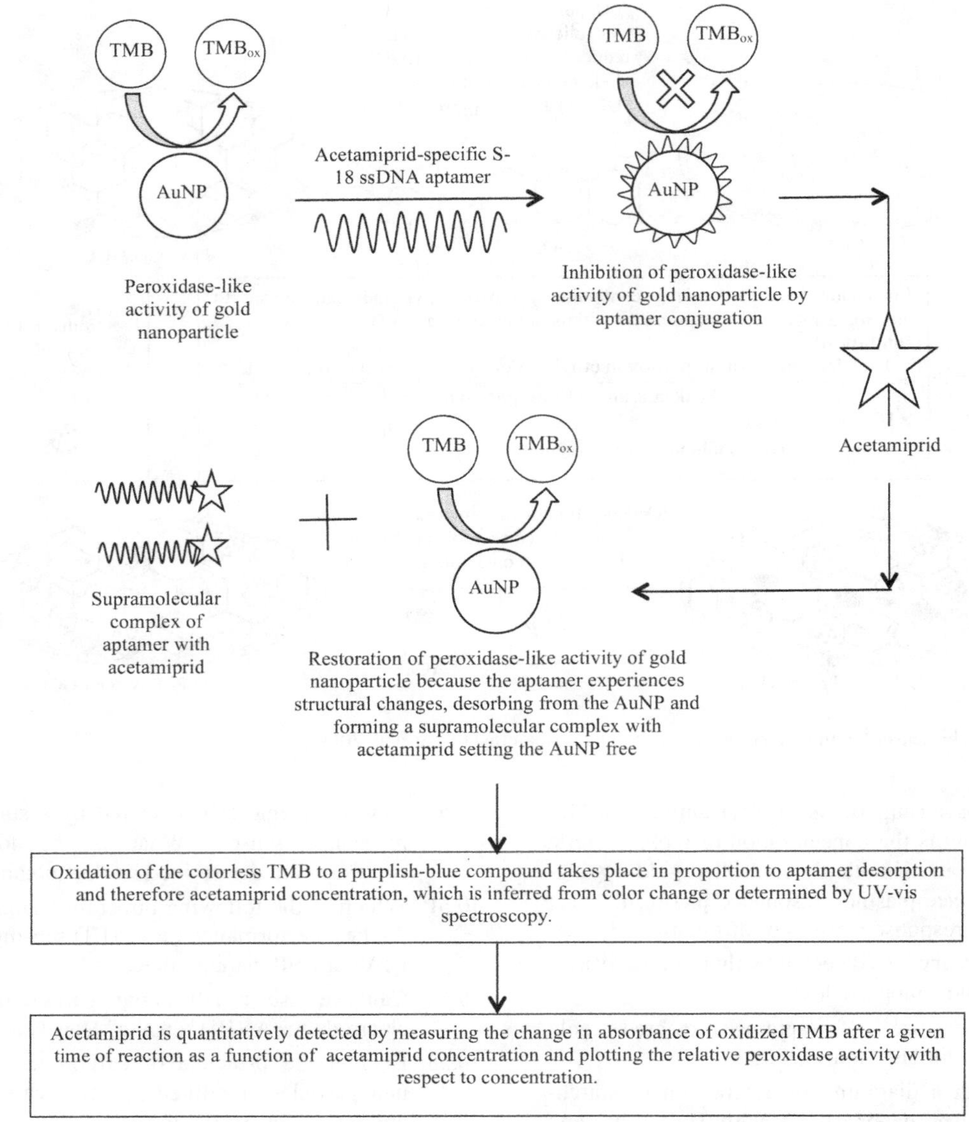

FIGURE 10.56 Acetaprimid sensor using aptamer-controlled gold nanoparticle enzymatic activity inhibition (Weerathunge et al. 2014).

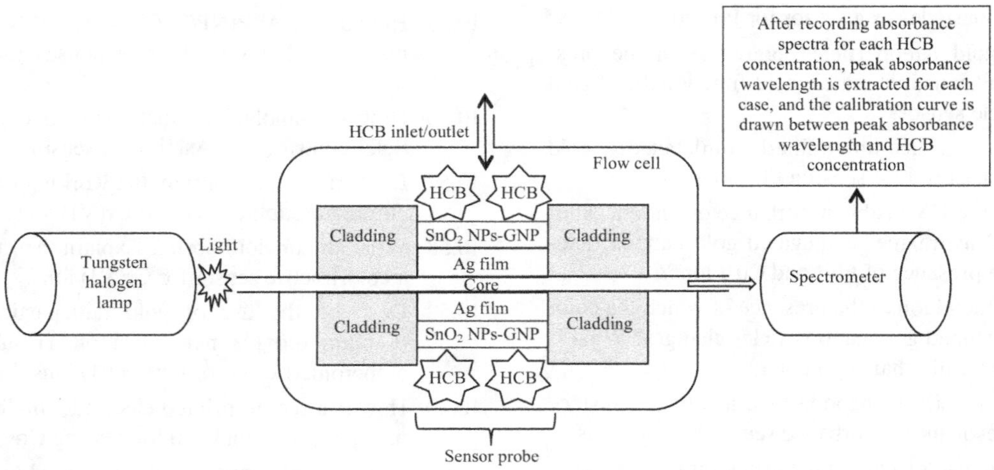

FIGURE 10.57 SnO_2 NP-GNP optical fiber probe for HCB sensing (Sharma et al. 2017).

TABLE 10.3

Particulate Matter Detection

Sl. No.	Nanosensor/System	Principle of Detection	Reference(s)
1.	CANTOR	Second-mode resonant frequency change.	Wasisto et al. (2013; 2015)
2.	Resonant filter-fiber	Resonant frequency response.	Schmid et al. (2013)
3.	Aerosol sensor	Voltage modulation to control charged particle trapping rate.	Zhang et al. (2017)
4.	MEMS chip	Voltage-based and precision electrometry.	Kim et al. (2018)

TABLE 10.4

Carbon Monoxide Detection

Sl. No.	Nanosensor	Principle of detection	Reference(s)
1.	Au nanoparticles-based CO detector	Changes in resistivity of the CNT array.	Bakhoum et al. (2013)
2.	CuO nanowire sensor	Resistance change at 325°C.	Steinhauer et al. (2016)
3.	ZnO nano-walls sensor	Change in electrical resistance.	Bruno et al. (2017)
4	ZnO/3D-rGO sensor	ZnO/3D-rGO integrated on microheater.	Ha et al. (2018)
5.	Eu-doped cerium oxide nanoparticle thick-film sensor	Relative resistance change.	Ortega et al. (2019)
6.	Pt-decorated SnO_2 nanoparticle sensor	Enhanced response of SnO_2 nanoparticles decorated with platinum.	Peng et al. (2019)

TABLE 10.5

Sulfur Dioxide Detection

Sl. No.	Nanosensor	Principle of detection	Reference(s)
1.	Tungsten oxide nanostructures	Resistance change.	Boudiba et al. (2012)
2.	SnO_2 thin-film sensor with nanoclusters	Resistance change.	Tyagi et al. (2014)
3.	Fluorescence nanoprobe	Fluorescence resonance energy transfer (FRET).	Sun et al. (2014)
4.	Niobium-loaded tungsten oxide film	Resistance change.	Kruefu et al. (2015)
5.	Nickel nano-walls	Resistance change.	Hien (2018)

TABLE 10.6

Nitrogen Dioxide Detection

Sl. No.	Nanosensor	Principle of Detection	Reference(s)
1.	SnO_2 nanoribbon	Resistance comparison.	Law et al. (2002)
2.	THMA-capped ZnO nanoparticle-coated ZnO nanowire	$\Delta R/R$ response.	Waclawik et al. (2012)
3.	In_2O_3-sensitized CuO-ZnO nanoparticle composite film	Resistance measurements.	Li et al. (2018)
4.	UV-activated, Pt-decorated, single-crystal ZnO nanowire	Resistance variation.	Espid et al. (2019)

TABLE 10.7

Ozone Detection

Sl. No.	Nanosensor	Principle of Detection	Reference(s)
1.	SnO_2/SWCNT thin-film sensor	Change in resistance.	Berger et al. (2011)
2.	Nanocrystalline $SrTi_{1-x}Fe_xO_3$ thin- film sensor	Resistance change.	Masterlaro et al. (2013)
3.	ZnO nanoparticle thin-film sensor	Normalized resistance response.	Acautla et al. (2014)
4.	Pd-decorated MWCNTs	Resistance change.	Colindres et al. (2014)
5.	UV-illuminated ZnO nanocrystals	$R_{ozone}/R_{dry\ air}$ measurement.	Bernardini et al. (2017)

TABLE 10.8

VOC Detection

Sl. No.	Nanosensor	Principle of Detection	Reference(s)
1.	Chemiresistive sensor using gold nanoparticles	Resistance change on exposure to toluene, ethanol, dichloromethane, methanol, and acetone.	Garg et al. (2010)
2.	MOF nanoparticles-based capacitive sensor	Variation of capacitance.	Homayoonnia and Zeinali (2016)
3.	Al-doped ZnO nanowire	Current changes.	Pauporté et al. (2018)
4.	Ni-SnO_2NPs for formaldehyde	Resistance decrease in HCHO ambience.	Hu et al. (2018)
5.	PdNPs/NiO thin-film/Pd thin-film for formaldehyde	Resistance change.	Chen et al. (2018)
6.	SAW sensor with polymer- sensitive film	Frequency shift.	Constantinoiu and Viespe (2019)
7.	Resorcinol-functionalized gold nanoparticle probe for formaldehyde	Color change of the probe.	Aquino et al. (2019)

TABLE 10.9

Ammonia Detection

Sl. No.	Nanosensor/System	Principle of Detection	Reference(s)
1.	Polyaniline nanoparticles	Conductometric mode operation.	Crowley and Killard (2008)
2.	MoO_3 nanoparticles gel-coated sensor	Resistance response.	Gouma et al. (2010
3.	ZnO.Eu^{2+}NPs fluorescence quenching	Change in luminescence intensity.	Yang et al. (2018)
4.	Pt NPs decorated WO_3 sensor	Relative resistance change in air and target gas.	Liu et al. (2019)

TABLE 10.10

Detection of *Escherichia coli* 0157:H7

Sl. No.	Nanosensor/System	Principle of Detection	Reference(s)
1.	Magnetoelastic sensor	Resonance frequency diminution.	Lin et al. (2010)
2.	MEA-modified gold nanoparticle	Colorimetric.	Su et al. (2012)
3.	Cysteine-capped gold nanoparticle	Colorimetric.	Raj et al. (2015)
4.	Iron oxide, gold, and lead sulfide nanoparticle-based biosensor	Electrochemical immunobiosensor.	Wang et al. (2015)
5.	Magneto-fluorescent nanosensor (MfnS)	Magnetic relaxation time and intensity of fluorescence emission measurements.	Banerjee et al. (2016)
6.	Signal-off impedimetric nanosensor	Electrochemical.	Wan et al. (2016)
7.	Impedimetric biosensor	Electrochemical.	Lin et al. (2019)
8.	AuNP SPR chip	Surface plasmon resonance.	Gur et al. (2019)
9.	Microfluidic nanosensor	Colorimetric.	Zheng et al. (2019)

TABLE 10.11

Detection of *Vibrio cholerae* and Cholera Toxin

Sl. No.	Nanosensor/System	Principle of Detection	Reference(s)
1.	Lactose-stabilized AuNPs	Colorimetric.	Schofield et al. (2007)
2.	ssDNA/nMgO/ITO bioelectrode	DNA hybridization and DPV response.	Patel et al. (2013)
3.	nMgO photoluminescence sensor	Change in PL peak intensity.	Patel et al. (2015)
4.	Lyophilized Au nanoparticle/polystyrene-co-acrylic acid-based genosensor	Differential pulse anodic stripping voltammetry.	Sheng et al. (2015)
5.	Polystyrene-co-acrylic acid (PSA) latex nanospheres	Differential pulse voltammetry.	Rahman et al. (2017)
6.	Graphene nanosheet bioelectrode	Potentiometric principle.	Nikoleli et al. (2018)

TABLE 10.12

Detection of *Pseudomonas aeruginosa*

Sl. No.	Nanosensor/System	Principle of Detection	Reference(s)
1.	Probe-modified magnetic nanoparticle-based chemiluminescent sensor	Chemiluminescence.	Tang et al. (2012, 2013)
2.	Reduced graphene electrode decorated with AuNPs	Differential pulse voltammetry.	Gandouzi et al. (2019)
3.	PANI/AuNPs decorated ITO electrode	Cyclic voltammetry.	Elkhawaga et al. (2019)

TABLE 10.13

Detection of *Legionella pneumophila*

Sl. No.	Nanosensor/System	Principle of Detection	Reference(s)
1.	ZnO-NR matrix-based immunosensor	Electrochemical.	Park et al. (2014)
2.	GC-SPR platform	Surface plasmon resonance.	Meneghello et al. (2017)

TABLE 10.14

Detection of Mercury(II) Ions

Sl. No.	Nanosensor/System	Principle of Detection	Reference(s)
1.	N-T decorated gold nanoparticle sensor	Colorimetric.	Du and Peng (2015)
2.	MR S-phone AuNP-aptamer sensor	Colorimetric.	Xiao et al. (2016)
3.	Starch-stabilized silver nanoparticle-based sensor	Colorimetric.	Vasileva et al. (2017)
4.	Chi-AgNP-based sensor	Colorimetric.	Zarlaida et al. (2017)

TABLE 10.15

Detection of Lead(II) Ions

Sl. No.	Nanosensor/System	Principle of Detection	Reference(s)
1.	GSH-stabilized AgNP sensor	Colorimetric.	Anambiga et al. (2013)
2.	MA-functionalized AuNPs sensor	Colorimetric.	Ratnarathorn and Dungchai (2015)
3.	Label-free AuNPs in the presence of GSH	Colorimetric.	Zhong et al. (2015)
4.	AuNPs conjugated with TA and DNS molecules	Colorimetric and fluorescent quenching principles.	Nath et al. (2015)
5.	Valine-capped gold nanoparticle sensor	Colorimetric.	Priyadarshini and Pradhan (2017)
6.	PVA-stabilized AgNPs in the presence of dithizone	Colorimetric.	Roto et al. (2019)
7.	AuNP-GR-modified GCE	Electrochemical principle.	Cheng et al. (2019)

TABLE 10.16

Detection of As(III) Ions

Sl. No.	Nanosensor/System	Principle of Detection	Reference(s)
1.	SERS system	SERS.	Wang et al. (2011)
2.	SPR nanosensor	SPR.	Salinas et al. (2014)
3.	FePtNP sensor	Anodic stripping voltammetry.	Moghimi et al. (2015)
4.	AuNP-modified GCE for As(III) and Se(IV)	Square wave voltammetry.	Idris et al. (2017)
5.	Ag nanoparticle-modified Au electrode	Linear sweep voltammetry.	Sonkoue et al. (2018)
6.	CNP/AuNP-modified GCE aptasensor	Square wave voltammetry.	Mushiana et al. (2019)
7.	Au/GNE sensor	Anodic stripping voltammetry SWASV.	Babar et al. (2019)
8.	Bimetallic NPs and bimetallic NPs+PANI composite-modified SPCE	Square wave anodic stripping voltammetry.	Melinte et al. (2019)
9.	Rano-functionalized CuNPs	Colorimetric.	Laghari et al. (2019)

TABLE 10.17

Detection of Cr(VI) Ions

Sl. No.	Nanosensor/System	Principle of Detection	Reference(s)
1.	AuNP probe-based immunochromatographic sensor	Immunosensing + colorimetry.	Liu et al. (2012)
2.	Amyloid-fibril-based sensor	Colorimetric principle.	Leung et al. (2013)
3.	AuNP decorated TiO_2NTs on Ti substrate	Electrochemical principle: Amperometry.	Jin et al. (2014)

TABLE 10.18

Detection of Cd^{2+} Ions

Sl. No.	Nanosensor/System	Principle of detection	Reference(s)
1.	AuNPA-modified SPE	Anodic stripping differential pulse voltammetry.	Zhang et al. (2010)
2.	Turn-on SERS sensor	SERS.	Yin et al. (2011)
3.	TG-capped CdSe QDs	Fluorescence intensity quenching.	Brahim et al. (2015)
4.	CdTe QD-based hybrid probe	Ratiometric fluorescence.	Qian et al. (2017)
5.	Aptamer-functionalized AuNP sensor	Colorimetric.	Gan and Wang (2020)

TABLE 10.19

Detection of Cu^{2+} Ions

Sl. No.	Nanosensor/System	Principle of Detection	Reference(s)
1.	Azide- and terminal alkyne-functionalized AuNP sensor	Color changes.	Zhou et al. (2008)
2.	CdSNP-Au QD sensor	Photoelectrochemical principle.	Ibrahim et al. (2016)
3.	MarR-functionalized AuNP sensor	Colorimetric assay.	Wang et al. (2017)
4.	Casein peptide-functionalized AgNP sensor	Color/absorbance changes.	Ghodake et al. (2018)

TABLE 10.20

Detection of Pesticides

Sl. No.	Pesticide	Nanosensor/System	Principle of Detection	Reference(s)
1	DDT	AuNP-based dipstick competitive assay	Colorimetric immunoassay.	Lisa et al. (2009)
2	2,4-dichlorophenoxyacetic acid (2,4-D)	AuNP-catalyzed chemiluminescence sensor	Chemiluminescence-based competitive inhibition immunoassay.	Boro et al. (2011)
3.	Carbofuran	Amperometric immunosensor	Cyclic voltammetry and electrochemical impedance spectroscopy.	Zhu et al. (2013)
		MIP-rGO@AuNP-modified GCE	Electrochemical principle.	
		AuNP-based SERS	SERS spectra.	
4.	Methomyl	Fe_3O_4/AgNP-modified GCE	Cyclic voltammetry.	Gai et al. (2019)
5.	Dimethoate	pSC_4R- AgNP	Colorimetric.	Menon et al. (2013)
6	Atrazine	AuNPs-modified Au electrode	Electrochemical immunosensor.	Liu et al. (2014)
		Cys-functionalized AuNPs	Colorimetric assay.	Liu et al. (2016)
		Nitrogen-doped carbon QD-based luminescent probe	Fluorescence.	Mohapatra et al. (2018)
7.	Paraoxon-ethyl	AChE/AuNP-Py- oGO/GCE biosensor	Amperometry.	Yang et al. (2014)
8.	Acetamiprid	Biosensor based on acetamiprid-specific S-18 ssDNA aptamer-controlled inhibition of Au nanozyme activity	Colorimetric biosensor.	Weerathunge et al. (2014)
9.	Hexachlorobenzene	SnO_2 NP: GNP hybrid/silver-decorated optical fiber platform	Surface plasmon resonance.	Sharma et al. (2017)
10.	Malathion (MLT)	AuNP-chitosan CS-IL/PGE	Non-enzymatic electrochemical sensing.	Bolat and Abaci (2018)
11.	Dithiocarbamates (DTCs) pesticide group	CTAB-CuNPs	Colorimetric probe.	Ghoto et al. (2019)

10.63 How is a hybrid probe made from red emissive and green emissive quantum dots used in detecting Cd^{2+} ions by ratiometric fluorescence measurements?

10.64 What is the idea behind using aptamer functionalized gold nanoparticles for Cd^{2+} ion sensing?

10.65 How does copper ion-click chemistry between azide and alkyne functionalized gold nanoparticles enable Cu^{2+} ion sensing?

10.66 How are Cu^{2+}ions detected from photocurrent responses using Au quantum dots loaded on CdS nanoparticles on ITO electrode? What is the role of optical illumination in this sensor?

10.67 What is the multiple antibiotic resistance regulator (MarR)? How do MarR-coated gold nanoparticles help in detecting Cu^{2+} ions?

10.68 Explain the root cause of peptide functionalization of silver nanoparticles for Cu^{2+} ion sensing.

10.69 Explain, with diagrams, the gold nanoparticle-based dipstick competitive assay for DDT pesticide detection.

10.70 How is the chemiluminescnce signal obtained in the gold nanoparticle-catalyzed immunosensor for 2,4-D herbicide?

10.71 Explain, with suitable diagrams, the main steps involved in the modification of the glassy carbon electrode for fabricating an amperometric immunosensor for carbofuran.

10.72 Describe the construction and operation of a glassy carbon electrode modified with molecular imprinted polymer and nanostructures for carbofuran sensing.

10.73 How is gold nanoparticle-based SERS spectroscopy used for detecting carbofuran and deltamethrin?

10.74 Explain the electrochemical detection of methomyl using Fe_3O_4/Ag nanoparticles composite-modified glassy carbon electrode.

10.75 Name the insecticide toward which *para*-sulfonato-calix[4]resorcinarene capped silver nanoparticles show selectivity in response. How is this selective response observed and used for insecticide sensing?

10.76 How is a gold electrode modified to make an electrochemical sensor for atrazine?

10.77 Describe the use of cysteamine-functionalized gold nanoparticles for sensing atrazine.

10.78 Name the herbicide which causes changes in fluorescence emission intensity of nitrogen-doped carbon quantum dots.

10.79 Explain the working of the AChE biosensor for amperometrically detecting paraoxon-ethyl.

10.80 How is acetamiprid-specific S-18 ssDNA aptamer utilized to make a nanosensor for acetamiprid, in which peroxidase-like activity of gold nanoparticles is restored by acetamiprid?

10.81 Describe the sensing of hexachlorobenzene using tin oxide nanoparticles: graphene nanopellets hybrid/silver-decorated optical fiber platform.

10.82 Explain the detection of malathion, using a gold nanoparticle-chitosan-ionic liquid/pencil graphite electrode.

10.83 What is the pesticides group detected using cetyltrimethylammonium bromide-coated copper nanoparticles? Name the pesticides in this group.

REFERENCES

Acuautla M., S. Bernardini, L. Gallais, T. Fiorido, L. Patout, and M. Bendahan. 2014. Ozone flexible sensors fabricated by photolithography and laser ablation processes based on ZnO nanoparticles. *Sensors and Actuators B: Chemical* 203: 602–611.

Anambiga I. V., V. Suganthan, N. Arunai Nambi Raj, G. Buvaneswari, and T. S. Sampath Kumar. 2013. Colorimetric detection of lead ions using glutathione stabilized silver nanoparticles. *International Journal of Scientific & Engineering Research* 4(5): 710–715.

Babar N.-U.-A., K. S. Joya, M. A. Tayyab, M. N. Ashiq, and M. Sohail. 2019. Highly sensitive and selective detection of arsenic using electrogenerated nanotextured gold assemblage. *ACS Omega* 4: 13645–13657.

Bakhoum E. G and M. H. M. Cheng. 2013. Miniature carbon monoxide detector based on nanotechnology. *IEEE Transactions on Instrumentation and Measurement* 62(1): 240–245.

Banerjee T., S. Sulthana, T. Shelby, B. Heckert, J. Jewell, K.Woody, V. Karimnia, J. McAfee, and S. Santra. 2016. Multiparametric magneto-fluorescent nanosensors for the ultrasensitive detection of Escherichia Coli O157:H7. *ACS Infectious Diseases* 2: 667–673.

Berger F., B. Ghaddab, J. B. Sanchez, and C. Mavon. 2011. Development of an ozone high sensitive sensor working at ambient temperature. *Journal of Physics: Conference Series* 307: 012054 (pp. 1–3).

Bernardini S., M. H. Benchekroun, T. Fiorido, K. Aguir, M. Bendahan, S. B. Dkhil, M. Gaceur, J. Ackermann, O. Margeat, and C. Videlot-Ackermann. 2017. Ozone sensors working at room temperature using zinc oxide nanocrystals annealed at low temperature. *Proceedings* 1: 423 (pp. 1–5).

Bolat G. and S. Abaci. 2018. Non-enzymatic electrochemical sensing of malathion pesticide in tomato and apple samples based on gold nanoparticles-chitosan-ionic liquid hybrid nanocomposite. *Sensors* 18: 773 (16 pp.).

Boro R. C., J. Kaushal, Y. Nangia, N. Wangoo, A. Bhasin, and C. R. Suri. 2011. Gold nanoparticles catalyzed chemiluminescence immunoassay for detection of herbicide 2,4-dichlorophenoxyacetic acid. *Analyst* 136: 2125–2130.

Boudiba A., C. Zhang, C. Bittencourt, P. Umek, M.-G. Olivier, R. Snyders, M. Debliquy. 2012. SO_2 gas sensors based on WO_3 nanostructures with different morphologies. *Procedia Engineering* 47: 1033–1036.

Brahim N. B., N. B. H. Mohamed, M. Echabaane, M. Haouari, R. B. Chaâbane, M. Negrerie, H. B. Ouada. 2015. Thioglycerol-functionalized CdSe quantum dots detecting cadmium ions. *Sensors and Actuators B: Chemical* 220: 1346–1353.

Bruno E., V. Strano, S. Mirabella, N. Donato, S. G. Leonardi, and G. Neri. 2017. Comparison of the sensing properties of ZnO nanowalls-based sensors toward low concentrations of CO and NO_2. *Chemosensors* 5: 20 (pp. 1–9).

Chen H.-I., C.-Y. Hsiao, W.-C. Chen, C.-H. Chang, I-P. Liu, T.-C. Chou, and W.-C. Liu. 2018. Formaldehyde sensing characteristics of a NiO-based sensor decorated with Pd nanoparticles and a Pd thin film. *IEEE Transactions on Electron Devices* 65(5): 1956–1961.

Cheng Y., F. Sun, J. Lee, T. Shi, T. Wang, and Y. Li. 2019. Gold-nanoparticles-graphene modified glassy carbon electrode for trace detection of lead ions. *E3S Web of Conferences* 78: 03007 (pp. 1–5).

Colindres S. C., K. Aguir, F. C. Sodi, L. V. Vargas, J. A. M. Salazar, and V. G. Febles. 2014. Ozone sensing based on palladium decorated carbon nanotubes. *Sensors* 14: 6806–6818.

Constantinoiu I. and C. Viespe. 2019. Detection of volatile organic compounds using surface acoustic wave sensor based on nanoparticles incorporated in polymer. *Coatings* 9: 373 (pp. 1–9).

Crowley K. and A. J. Killard. 2008. Fabrication of an ammonia gas sensor using inkjet-printed polyaniline nanoparticles. *Talanta* 77(2): 710–717.

Du J. and X. Peng. 2015. Gold nanoparticle-based colorimetric detection of mercury ion via coordination chemistry. *Sensors and Actuators B: Chemical* 212: 481–486.

Elkhawaga A. A., M. M. Khalifa, O. El-badawy, M. A. Hassan, and W. A. El-Said. 2019. Rapid and highly sensitive detection of pyocyanin biomarker in different *Pseudomonas aeruginosa* infections using gold nanoparticles modified sensor. *PLOS ONE* 14(7): e0216438 (pp. 1–16), https://doi.org/10.1371/journal.pone.0216438.

Espid E., B. Adeli, and F. Taghipour. 2019. Enhanced gas sensing performance of photo-activated, Pt-decorated, single-crystal ZnO nanowires. *Journal of The Electrochemical Society* 166 (5): H3223–H3230.

Gai K., H. Qi, L. X. li, and X. Liu. 2019. Detection of residual methomyl in vegetables with an electrochemical sensor based on a glassy carbon electrode modified with Fe_3O_4/Ag composite. *International Journal of Electrochemical Science* 14: 1283–1292.

Gan Y. and P. Wang. 2020. In-situ detection of cadmium with aptamer functionalized gold nanoparticles based on smartphone-based colorimetric system. *Talanta* 208: 120231.

Gandouzi I., M. Tertis, A. Cernat, D. Saidane-Mosbahi, A. Ilea, and C. Cristea. 2019. A nanocomposite based on reduced graphene and gold nanoparticles for highly sensitive electrochemical detection of *Pseudomonas aeruginosa* through its virulence factors. *Materials* 12: 1180 (13 pp.).

Garg N., A. Mohanty, N. Lazarus, L. Schultz, T. R. Rozzi, S. Santhanam, L. Weiss, J. L. Snyder, G. K. Fedder, and R. Jin. 2010. Robust gold nanoparticles stabilized by trithiol for application in chemiresistive sensors. *Nanotechnology* 21: 405501 (6 pp.).

Ghodake G. S., S. K. Shinde, R. G. Saratale, A. A. Kadam, G. D. Saratale, A. Syed, F. Ameen, and D.-Y. Kim. 2018. Colorimetric detection of Cu^{2+} based on the formation of peptide–copper complexes on silver nanoparticle surfaces. *Beilstein Journal of Nanotechnology* 9: 1414–1422.

Ghoto S. A., M. Y. Khuhawar, T. M. Jahangir, and J. U. D. Mangi. 2019. Applications of copper nanoparticles for colorimetric detection of dithiocarbamate pesticides. *Journal of Nanostructure in Chemistry* 9: 77–93.

Gouma P., K. Kalyanasundaram, X. Yun, M. Stanaćević, and L. Wang. 2010. Nanosensor and breath analyzer for ammonia detection in exhaled human breath. *IEEE Sensors Journal* 10(1): 49–53.

Gür S. D., M. Bakhshpour, and A. Denizli. 2019. Selective detection of *Escherichia coli* caused UTIs with surface imprinted plasmonic nanoscale sensor. *Materials Science & Engineering. part C* 104: 109869 (pp. 1–7).

Ha N. H., D. D. Thinh, N. T. Huong., N. H. Phuong, P. D. Thach, and H. S. Hong. 2018. Fast response of carbon monoxide gas sensors using a highly porous network of ZnO nanoparticles decorated on 3D reduced graphene oxide. *Applied Surface Science* 434: 1048–1054.

He Y., S. Xiao, T. Dong, and P. Nie. 2019. Gold nanoparticles for qualitative detection of deltamethrin and carbofuran residues in soil by surface enhanced Raman scattering (SERS). *International Journal of Molecular Sciences* 20: 1731 (13 pp.).

Hien V. X. 2018. SO_2-sensing properties of NiO nanowalls synthesized by the reaction of Ni foil in NH_4OH solution. *Advances in Natural Sciences: Nanoscience and Nanotechnology (ANSN)* 9: 045013 (7 pp.).

Homayoonnia S. and S. Zeinali. 2016. Design and fabrication of capacitive nanosensor based on MOF nanoparticles as sensing layer for VOCs detection. *Sensors and Actuators B: Chemical* 237: 776–786.

Hu J., T. Wang, Y. Wang, D. Huang, G. He, Y. Han, N. Hu, Y. Su, Z. Zhou, Y. Zhang, and Z. Yang. 2018. Enhanced formaldehyde detection based on Ni doping of SnO_2 nanoparticles by one-step synthesis. *Sensors and Actuators B: Chemical* 263: 120–128.

Ibrahim I., H. N. Lim, O. K. Abou-Zied, N. M. Huang, P. Estrela, and A. Pandikumar. 2016. Cadmium sulphide nanoparticles decorated with Au quantum dots as ultrasensitive photoelectrochemical sensor for selective detection of copper (II) ions. *Journal of Physical Chemistry C* 120(39): 22202–22214.

Idris A. O., N. Mabuba, and O. A. Arotiba. 2017. Electrochemical co-detection of arsenic and selenium on a glassy carbon electrode modified with gold nanoparticles. *International Journal of Electrochemical Science* 12: 10–21.

Jin W., G. Wu, and A. Chen. 2014. Sensitive and selective electrochemical detection of chromium(VI) based on gold nanoparticle-decorated titania nanotube arrays. *Analyst* 139: 235–241.

Kim H.-L., J. Han, S.-M. Lee, H.-B. Kwon, J. Hwang, and Y.-J. Kim. 2018. MEMS-based particle detection system for measuring airborne ultrafine particles. *Sensors and Actuators A: Physical* 283: 235–244.

Kruefu V., A. Wisitsoraat, and S. Phanichphant. 2015. Effects of niobium-loading on sulfur dioxide gas-sensing characteristics of hydrothermally prepared tungsten oxide thick film. *Journal of Nanomaterials* 2015: 820509 (pp. 1–8).

Laghari G. N., A. Nafady, S. I. Al-Saeedi, Sirajuddin, S. T. H. Sherazi, J. Nisar, M. R. Shah, M. I. Abro, M. Arain, and S. K. Bhargava. 2019. Ranolazine-functionalized copper nanoparticles as a colorimetric sensor for trace level detection of As^{3+}. *Nanomaterials* 9: 83 (pp. 1–12).

Law M., H. Kind, B. Messer, F. Kim, and P. Yang. 2002. Photochemical sensing of NO_2 with SnO_2 nanoribbon nanosensors at room temperature. *Angewandte Chemie International Edition* 41(13): 2405–2407.

Leung W.-H., L. Zou, W.-H. Lo, and P.-H. Chan. 2013. An amyloid-fibril-based colorimetric nanosensor for rapid and sensitive chromium(VI) detection. *ChemPlusChem* 78: 1440–1445.

Li T.-T., N. Bao, A.-F. Geng, H. Yu, Y. Yang, and X.-T. Dong. 2018. Study on room temperature gas-sensing performance of CuO film-decorated ordered porous ZnO composite by In_2O_3 sensitization. *Royal Society Open Science* **5**: 171788 (pp. 1–12).

Lin H., Q. Lu, S. Ge, Q. Cai, and C. A. Grimes. 2010. Detection of pathogen Escherichia coli O157:H7 with a wireless magnetoelastic-sensing device amplified by using chitosan-modified magnetic Fe_3O_4 nanoparticles. *Sensors and Actuators B: Chemical* 147: 343–349.

Lin D., R. G. Pillai, W. E. Lee, and A. B. Jemere. 2019. An impedimetric biosensor for E. coli O157:H7 based on the use of self-assembled gold nanoparticles and protein G. *Microchimica Acta* 186(3): 169.

Lisa M., R. S. Chouhan, A. C. Vinayaka, H. K. Manonmani, and M. S. Thakur. 2009. Gold nanoparticles based dipstick immunoassay for the rapid detection of dichlorodiphenyltrichloroethane: An organochlorine pesticide. *Biosensors and Bioelectronics* 25: 224–227.

Liu X., J.-J. Xiang, Y. Tang, X.-L. Zhang, Q.-Q. Fu, J.-H. Zou, and Y.H. Lin. 2012. Colloidal gold nanoparticle probe-based immunochromatographic assay for the rapid detection of chromium ions in water and serum samples. *Analytica Chimica Acta* 745C: 99–105.

Liu X., W.-J. Li, L. Li, Y. Yang, L.-G. Mao, and Z. Peng. 2014. A label-free electrochemical immunosensor based on gold nanoparticles for direct detection of atrazine. *Sensors and Actuators B: Chemical* 191: 408–414.

Liu G., S. Wang, X. Yang, T. Li, Y. She, J. Wang, P. Zou, F. Jin, M. Jin, and H. Shao. 2016. Colorimetric sensing of atrazine in rice samples using cysteamine functionalized gold nanoparticles after solid phase extraction. *Analytical Methods* **8**: 52–56.

Liu I.-P., C.-H. Chang, T. C. Chou, and K.-W. Lin. 2019. Ammonia sensing performance of a platinum nanoparticle-decorated tungsten trioxide gas sensor. *Sensors and Actuators B: Chemical* 291: 148–154.

Martínez-Aquino C., A. M. Costero, S. Gil, and P. Gaviña. 2019. Resorcinol functionalized gold nanoparticles for formaldehyde colorimetric detection. *Nanomaterials* 9: 302 (pp. 1–9).

Mastelaro V. R., S. C. Zílio, L. F. da Silva, P.I. Pelissari, M. I.B. Bernardi, J. Guerin, and K. Aguir. 2013. Ozone gas sensor based on nanocrystalline $SrTi_{1-x}Fe_xO_3$ thin films. *Sensors and Actuators B: Chemical* 81: 919–924.

Melinte G., O. Hosu, M. Lettieri, C. Cristea, and G. Marrazz. 2019. Electrochemical fingerprint of arsenic (III) by using hybrid nanocomposite-based platforms. *Sensors* 19: 1–13.

Meneghello A., A. Sonato, G. Ruffato, G. Zacco, and F. Romanato. 2017. A novel high sensitive surface plasmon resonance Legionella pneumophila sensing platform. *Sensors and Actuators B: Chemical* 250: 351–355.

Menon S. K., N. R. Modi, A. Pandya, and A. Lodha. 2013. Ultrasensitive and specific detection of dimethoate using a p-sulphonato-calix[4]resorcinarene functionalized silver nanoprobe in aqueous solution. *RSC Advances* 3: 10623–10627.

Moghimi N., M. Mohapatra, and K. T. Leung. 2015. Bimetallic nanoparticles for arsenic detection. *Analytical Chemistry* 87(11): 5546–5552.

Mohapatra S., M. K. Bera, and R. K. Das. 2018. Rapid "turn-on" detection of atrazine using highly luminescent N-doped carbon quantum dot. *Sensors and Actuators B: Chemical* 263: 459–468.

Mushiana T., N. Mabuba, A. O. Idris, G. M. Peleyeju, B. O. Orimolade, D. Nkosi, R. F. Ajayi, and O. A. Arotiba. 2019. An aptasensor for arsenic on a carbon-gold bi-nanoparticle platform. *Sensing and Bio-Sensing Research* 24: 100280 (pp. 1–5).

Nath P., R. K. Arun, and N. Chanda. 2015. Smart gold nanosensor for easy sensing of lead and copper ions in solution and using paper strips. *RSC Advances* 5: 69024–69031.

Nikoleli G. P., S. Karapetis, P. D. Nikolelis, C. G. Siontorou, and M.T. Nikolelis. 2018. A nanosensor for the rapid detection of cholera toxin using graphene electrodes with incorporated polymer lipid films. *Journal of Nanomedicine & Nanotechnology* 3(2): 000137 (8 pp.).

Ortega P. P., L. S. R. Rocha, J. A. Cortés, M. A. Ramirez, C. Buono, M. A. Ponce, and A. Z. Simões. 2019. Towards carbon monoxide sensors based on europium doped cerium dioxide. *Applied Surface Science* 464: 692–699.

Park J., X. You, Y. Jang, Y. Nam, M. J. Kim, N. K. Min, and J. J. Pak. 2014. ZnO nanorod matrix based electrochemical immunosensors for sensitivity enhanced detection of Legionella pneumophila. *Sensors and Actuators B: Chemical* 200: 173–180.

Patel M. K., M. A. Ali, V. V. Agrawal, Z. A. Ansari, S. G. Ansari, and B. D. Malhotra. 2013. Nanostructured magnesium oxide biosensing platform for cholera detection. *Applied Physics Letters* 102: 144106-1–144106-5.

Patel M. K., M. A. Ali, S. Krishnan, V. V. Agrawal, A. A. Al Kheraif, H. Fouad, Z.A. Ansari, S. G. Ansari, and B. D. Malhotra. 2015. A Label-free photoluminescence genosensor using nanostructured magnesium oxide for cholera detection. *Scientific Reports* 5: 17384 (pp. 1–8).

Pauporté T., O. Lupan, V. Postica, M. Hoppe, L. Chow, and R. Adelung. 2018. Al-Doped ZnO nanowires by electrochemical deposition for selective VOC nanosensor and nanophotodetector. *Physica Status Solidi A* 215: 1700824 (pp. 1–8).

Peng S., P. Hong, Y. Li, X. Xing, Y. Yang, Z. Wang, T. Zou, and Y. Wang. 2019. Pt decorated SnO_2 nanoparticles for high response CO gas sensor under the low operating temperature. *Journal of Materials Science: Materials in Electronics* 30: 3921–3932.

Priyadarshini E. and N. Pradhan. 2017. Metal-induced aggregation of valine capped gold nanoparticles: An efficient and rapid approach for colorimetric detection of Pb^{2+} ions. *Scientific Reports* 7: 9278 (pp. 1–8).

Qian J., K. Wang, C. Wang, C. Ren, Q. Liu, N. Hao, and K. Wang. 2017. Ratiometric fluorescence nanosensor for selective and visual detection of cadmium ions using quencher displacement-induced fluorescence recovery of CdTe quantum dots-based hybrid probe. *Sensors and Actuators B: Chemical* 241: 1153–1160.

Rahman M., L. Y. Heng, D. Futra, and T. L. Ling. 2017. Ultrasensitive biosensor for the detection of vibrio cholerae DNA with polystyrene-co-acrylic acid composite nanospheres. *Nanoscale Research Letters* 12: 474 (pp. 1–12).

Raj V., A. N. Vijayan, and K. Joseph. 2015. Cysteine capped gold nanoparticles for naked eye detection of E. coli bacteria in UTI patients. *Sensing and Bio-Sensing Research* 5: 33–36.

Ratnarathorn N. and W. Dungchai. 2015. Highly sensitive colorimetric detection of lead using maleic acid functionalized gold nanoparticles. *Talanta*, 132: 613–618.

Roto R., B. Mellisani, A. Kuncaka, M. Mudasir, and A. Suratman. 2019. Colorimetric sensing of Pb^{2+} ion by using Ag nanoparticles in the presence of dithizone. *Chemosensors* 7: 28 (pp. 1–12).

Salinas S., N. Mosquera, L. Yate, E. Coy, G. Yamhure, and E. González. 2014. Surface plasmon resonance nanosensor for the detection of arsenic in water. *Sensors & Transducers* 183(12): 97–102.

Schmid S., M. Kurek, J. Q. Adolphsen, and A. Boisen. 2013. Real-time single airborne nanoparticle detection with nanomechanical resonant filter-fiber. *Scientific Reports* 3: 128 (pp. 1–5).

Schofield C. L., R. A. Field, and D. A. Russell. 2007. Glyconanoparticles for the colorimetric detection of cholera toxin. *Analytical Chemistry* 79(4): 1356–1361.

Sharma S., S. P. Usha, A. M. Shrivastav, and B. D. Gupta. 2017. A novel method of SPR based SnO_2: GNP nano-hybrid decorated optical fiber platform for hexachlorobenzene sensing. *Sensors and Actuators B: Chemical* 246: 927–936.

Sheng L. P., B. Lertanantawong, L. S. Yin, M. Ravichandran, L. Yook Heng, W. Surareungchai. 2015. Electrochemical genosensor assay using lyophilized gold nano-particles/latex microsphere label for detection of Vibrio cholera. *Talanta* 139: 167–173.

Sonkoue B. M., P. M. S. Tchekwagep, C. P. Nanseu-Njiki, and E. Ngameni. 2018. Electrochemical determination of arsenic using silver nanoparticles. *Electroanalysis* 30(11): 2738–2743.

Steinhauer S., A. Chapelle, P. Menini, and M. Sowwan. 2016. Local CuO nanowire growth on microhotplates: In situ electrical measurements and gas sensing application. *ACS Sensors* 1(5): 503–507.

Su H., Q. Ma, K. Shang, T. Liu, H. Yin, and S. Ai. 2012. Gold nanoparticles as colorimetric sensor: A case study on E. coli O157:H7 as a model for Gram-negative bacteria. *Sensors and Actuators B: Chemical* 161: 298–303.

Sun M., H. Yu, K. Zhang, Y. Zhang, Y. Yan, D. Huang, and S. Wang. 2014. Determination of gaseous sulfur dioxide and its derivatives via fluorescence enhancement based on cyanine dye functionalized carbon nanodots. *Analytical Chemistry* 86: 9381–9385.

Tan X., Q. Hu, J. Wu, X. Li, P. Li, H. Yu, X. Li, and F. Lei. 2015. Electrochemical sensor based on molecularly imprinted polymer reduced graphene oxide and gold nanoparticles modified electrode for detection of carbofuran. *Sensors and Actuators B: Chemical* 220: 216–221.

Tang Y., Z. Li, N. He, L. Zhang, C. Ma, X. Li, C. Li, Z. Wang, Y. Deng, and L. He. 2012. Preparation of functional magnetic nanoparticles mediated with PEG 4000 and application in Pseudomonas aeruginosa rapid detection. *Journal of Biomedical Nanotechnology* 9: 1–6.

Tang Y., J. Zou, C. Ma, Z. Ali, Z. Li, X. Li, N. Ma, X. Mou, Y. Deng, L. Zhang, K. Li, G. Lu, H. Yang, and N. He. 2013. Highly sensitive and rapid detection of *Pseudomonas aeruginosa* based on magnetic enrichment and magnetic separation. *Theranostics* 3(2): 85–92.

Tyagi P., A. Sharma, M. Tomar, and V. Gupta. 2014. Efficient detection of SO_2 gas using SnO_2 based sensor loaded with metal oxide catalysts. *Procedia Engineering* 87: 1075–1078.

Vasileva P., T. Alexandrova, and I. Karadjova. 2017. Application of starch-stabilized silver nanoparticles as a colorimetric sensor for mercury(II) in 0.005 mol/L nitric acid. *Journal of Chemistry* 2017: 6897960 (9 pp.).

Waclawik E. R., J. Chang, A. Ponzoni, I. Concina, D. Zappa, E. Comini, N. Motta, G. Faglia, and G. Sberveglieri. 2012. Functionalised zinc oxide nanowire gas sensors: Enhanced NO_2 gas sensor response by chemical modification of nanowire surfaces. *Beilstein Journal of Nanotechnology* 3: 368–377.

Wan J., J. Ai, Y. Zhang, X. Geng, Q. Gao, and Z. Cheng. 2016. Signal-off impedimetric immunosensor for the detection of *Escherichia coli* O157:H7. *Scientific Reports* 6: 19806 (pp. 1–6).

Wang Y., G. Pan, and K. Q. Wang. 2011. The development of a portable arsenic detector based on surface-enhanced Raman spectroscopy for rapid on-site testing of groundwater. *NSTI-Nanotech* 3: 521–524.

Wang Y., P. A. Fewins, and E. C. Alocilja. 2015. Electrochemical immunosensor using nanoparticle-based signal enhancement for *Escherichia Coli* O157:H7 detection. *IEEE Sensors Journal* 15(8): 4692–4699.

Wang Y., L. Wang, Z. Su, J. Xue, J. Dong, C. Zhang, X. Hua, M. Wang, and F. Liu. 2017. Multipath colorimetric assay for copper (II) ions utilizing MarR functionalized gold nanoparticles. *Scientific Reports* 7: 41557 (9 pp.).

Wasisto H. S., S. Merzsch, A. Waag, E. Uhde, T. Salthammer, and E. Peiner. 2013. Portable cantilever-based airborne nanoparticle detector. *Sensors and Actuators B: Chemical* 187: 118–127.

Wasisto H. S., S. Merzsch, E. Uhde, A. Waag, and E. Peiner. 2015. Partially integrated cantilever-based airborne nanoparticle detector for continuous carbon aerosol mass concentration monitoring. *Journal of Sensors and Sensor Systems* 4: 111–123.

Weerathunge P., R. Ramanathan, R. Shukla, T. K. Sharma, and V. Bansal. 2014. Aptamer-controlled reversible inhibition of gold nanozyme activity for pesticide sensing. *Analytical Chemistry* 86: 11937–11941.

Xiao W., M. Xiao, Q. Fu, S. Yu, H. Shen, H. Bian, and Y. Tang. 2016. A portable smart-phone readout device for the detection of mercury contamination based on an aptamer-assay nanosensor. *Sensors* 16: 1–10.

Yang Y., A. M. Asiri, D. Du, and Y. Lin. 2014. Acetylcholinesterase biosensor based on a gold nanoparticle–polypyrrole–reduced graphene oxide nanocomposite modified electrode for the amperometric detection of organophosphorus pesticides. *Analyst* 139: 3055–3060.

Yang W., W. Feng, X. Yang, H. Chen, Y. He, D. Deng, and Z. Peng. 2018. Optical ammonia sensor based on ZnO:Eu^{2+}fluorescence quenching nanoparticles. *Zeitschrift für Naturforschung A* 73(9): 781–784.

Yin J., T. Wu, J. Song, Q. Zhang, S. Liu, R. Xu, and H. Duan. 2011. SERS-active nanoparticles for sensitive and selective detection of cadmium ion (Cd^{2+}). *Chemistry of Materials* 23: 4756–4764.

Zarlaida F., M. Adlim, M. Syukri Surbakti, A. F. Omar. 2017. Chitosan-stabilized silver nanoparticles for colorimetric assay of mercury (II) ions in aqueous system. *Materials Science and Engineering* 352: 012049 (pp. 1–5).

Zhang L., D.-W. Li, W. Song, L. Shi, Y. Li, and Y.-T. Long. 2010. High sensitive on-site cadmium sensor based on AuNPs amalgam modified screen-printed carbon electrodes. *IEEE Sensors Journal* 10(10): 1583–1588.

Zhang C., D. Wang, R. Zhu, W. Yang, and P. Jiang. 2017. A miniature aerosol sensor for detecting polydisperse airborne ultrafine particles. *Sensors* 17: 929 (pp. 1–11).

Zheng L., G. Cai, S. Wang, M. Liao, Y. Li, and J. Lin. 2019. A microfluidic colorimetric biosensor for rapid detection of Escherichia coli O157:H7 using gold nanoparticle aggregation and smart phone imaging. *Biosensors and Bioelectronics* 124–125: 143–149.

Zhong G., J. Liu, and X. Liu. 2015. A fast colorimetric assay for lead detection using label-free gold nanoparticles (AuNPs). *Micromachines* 6: 462–472.

Zhou Y., S. Wang, K. Zhang, and X. Jiang. 2008. Visual detection of copper (II) by azide- and alkyne-functionalized gold nanoparticles using click chemistry. *Angewandte Chemie International Edition* 47: 7454–7456.

Zhu Y., Y. Cao, X. Sun, and X. Wang. 2013. Amperometric immunosensor for carbofuran detection based on MWCNTs/GS-PEI-Au and AuNPs-antibody conjugate. *Sensors* 13: 5286–5301.

11

Nanosensors for Industrial Applications

The food industry is a collection and integration of diverse businesses, ranging from agriculture to food processing, that provide wholesome food to the world population. The healthcare industry delivers preventive, curative, and palliative treatments to people. The automotive industry is concerned with motor vehicles, whereas the aerospace industry deals with aircraft and space vehicles. The consumer industry sells goods directly to the consumer in contrast to the capital goods industry, which interacts with other companies, rather them dealing directly with consumers.

All these industries use different types of nanosensors, which are at the center of attention in this chapter. The food industry needs nanosensors to examine whether the food items are infected with any pathogenic bacteria or adulterated with toxins. The health industry requires nanosensors for disease diagnosis, mainly cancers of various types, and infectious diseases. Likewise, there is a huge demand for nanosensors by the automotive, aerospace, and consumer industries, mainly pressure and inertial sensors (accelerometers and gyroscopes), as well as magnetic field nanosensors.

11.1 Nanosensors for Detection of Food-Borne Pathogenic Bacteria

11.1.1 *Salmonella typhimurium*

This bacterium causes gastroenteritis in humans, leading to inflammation of the stomach and small intestine, with symptoms of vomiting, diarrhea and pain in the abdomen.

11.1.1.1 DNA Aptamers and Magnetic Nanoparticle (MNP)-Based Colorimetric Sensor

Bare magnetic nanoparticles (MNPs) promote the oxidation of 3,3′, 5,5′-tetramethylbenzidine (TMB) by peroxidase enzyme-like activity in the presence of hydrogen peroxide (H_2O_2), producing a blue-colored product (Park et al. 2015). The DNA aptamers considered here have a specific interaction with and high affinity for *Salmonella typhimurium.* These aptamers also influence the behavior of MNPs because they act as shields, inhibiting the catalytic peroxidase activity of MNPs. Therefore, attachment of aptamers to MNPs results in inhibition of enzymatic activity affecting the absorbance and color of the product. But, on adding a sample containing *S. typhimurium* to the DNA aptamers@ MNPs solution, the *S. typhimurium* detach the aptamers from the MNPs. By this action, the original intensity of absorbance at 650 nm and the blue appearance are recovered. The color change can be perceived by the unaided eye. Hence the extent of absorbance intensity/color recovery is a measure of the quantity of *S. typhimurium* present in the sample, easily allowing its detection.

MNPs are synthesized by the co-precipitation method, using $FeCl^{2+}$ and $FeCl^{3+}$ in water, adding NaOH until $pH = 10$ is reached, followed by sonication (in ultrapure water) and incubation for 35 minutes.

For the assay, *S. typhimurium* are added to a suspension containing DNA aptamers and magnetic nanoparticles, i.e., DNA aptamers@ MNPs solution, as shown in Figure 11.1. DNA aptamers leave the MNPs and become attached to *S. typhimurium.* After dilution in acetate buffer, the MNPs are incubated with white TMB powder dissolved in dimethyl sulfoxide (DMSO), and H_2O_2. The MNPs are removed with a magnet.

The color of the reaction product is carefully observed. *S. typhimurium* at concentrations of up to 7.5×10^5 CFU/mL are visibly detected; "CFU" are colony-forming units, representing individual bacteria. Typical time required for the assay is 10 min (Park et al. 2015).

11.1.1.2 Strip Sensor Using Gold Nanoparticle (AuNP)-Labeled Genus-Specific Anti-Lipopolysaccharide (LPS) Monoclonal Antibody (mAb)

Along with the LPS-mAb, the sensor also uses the *Salmonella* antigen of the LPS-bovine serum albumin (BSA) conjugate (Wang et al. 2016). The LPS-BSA *Salmonella* antigen is produced by oxidizing LPS with sodium periodate, then adding ethanediol and BSA, adjusting the pH at 9, allowing the reaction to continue for 24 hours, and then dialyzing the sample.

AuNPs (15 nm) are synthesized by the sodium citrate method. Mice are immunized with the immunogen produced by conjugating *S. typhimurium* LPS with BSA to obtain the *Salmonella* anti-LPS mAb. AuNPs are modified with *Salmonella* anti-LPS mAb, producing AuNP-labeled *Salmonella* anti-LPS mAb probes.

The immunochromatographic strip sensor contains a sample pad, a test line (T-line), a control line (C-line) and an absorption pad (Figure 11.2). The T-line is coated with LPS-BSA *Salmonella* antigen while the C-line is coated with goat anti-mouse IgG antibodies.

The sample under test is reacted with AuNP-labeled *Salmonella* anti-LPS mAb probes and loaded on the sample pad of the strip sensor. The solution flows from the sample pad towards the absorption pad. The colors of the T-line and C-line are observed after 10 min incubation with the naked eye.

In an analysis, a negative control sample is a particular sample with the special property that it is not affected by any variable

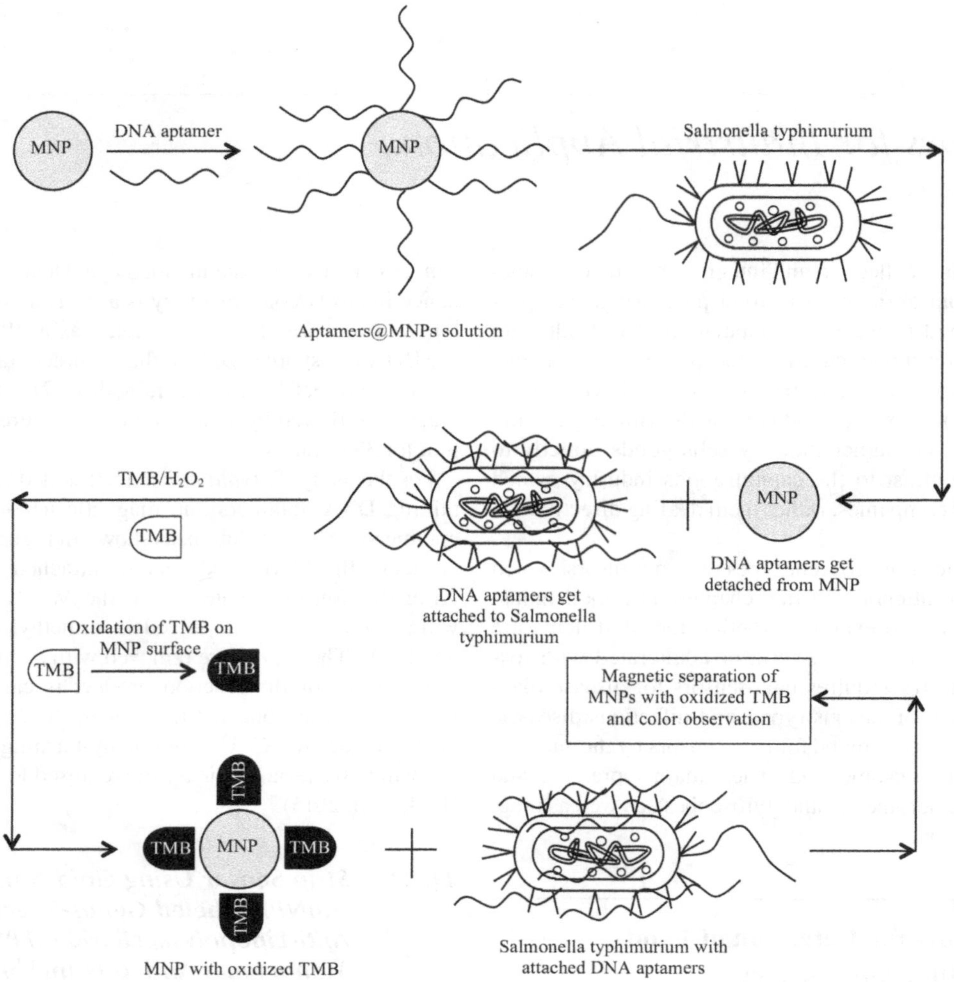

FIGURE 11.1 MNPs/aptamer-based sensor for *Salmonella typhimurium* (Park et al. 2015).

parameter in the experiment. It is treated in exactly the same way as other samples during the experiment. Here, the negative control sample is one not containing any *Salmonella*. When the negative control sample is loaded on the sample pad, the AuNPs labeled *Salmonella* anti-LPS mAb probes react with the LPS-BSA *Salmonella* antigen on the T-line and the remaining probes are captured by the anti-mouse IgG antibodies on the C-line. Both the T-line and the C-line show a red appearance due to the AuNPs present in them.

When the sample contains *Salmonella* (positive sample), these *Salmonella* samples will have already reacted with the AuNP-labeled *Salmonella* anti-LPS mAb probes. So, fewer *Salmonella* cells are left to be bound to the LPS-BSA *Salmonella* antigen on the T-line, making this line look faintly red. Weakening of the red color intensity on the T-line, relative to C-line, shows that the sample is positive for *Salmonella*. A negative correlation exists between the red color intensity on the T-line and the number of *Salmonella* cells in the sample.

Only 25 ng/mL of Ra mutant LPS from *Salmonella* is adequate for color inhibition on the T-line; this mutant strain produces a variant form of the LPS. The sensor can detect twelve Salmonella CFU. The sensitivity based on gray values is 10^3 CFU for all species of *Salmonella*, with the exception of *Salmonella thompson*, for which the lower limit is 10^4 CFU.

11.1.2 *Clostridium perfringens*

This is a rod-shaped, spore-forming Gram-positive anaerobic (growing without oxygen) bacterium, causing food poisoning.

The sensor used is ssDNA-immobilized cerium oxide (CeO_2) nanorods (NRs)/chitosan (CHIT) composite-modified glassy carbon electrode (GCE) (Qian et al. 2018). Crystalline CeO_2 NRs are produced by a modified hydrothermal method. In this method, aqueous $Ce(NO_3)_2.6H_2O$ solution is added dropwise to a NaOH solution in water, whereupon a white precipitate is formed. The next step involves heating the solution in an autoclave at 100°C for 24 hours. Sonication of the CeO_2 NRs is carried out in aqueous CHIT solution, containing acetic acid, to form a colloidal suspension. Drops of the CeO_2 NRs/CHIT composite are dropped onto GCE, and then air dried. Thus, CeO_2 NRs/CHIT composite/GCE is obtained (Figure 11.3). The ssDNA probe is pipetted over CeO_2 NRs/CHIT composite/GCE to form ssDNA/CeO_2 NRs/CHIT composite/GCE.

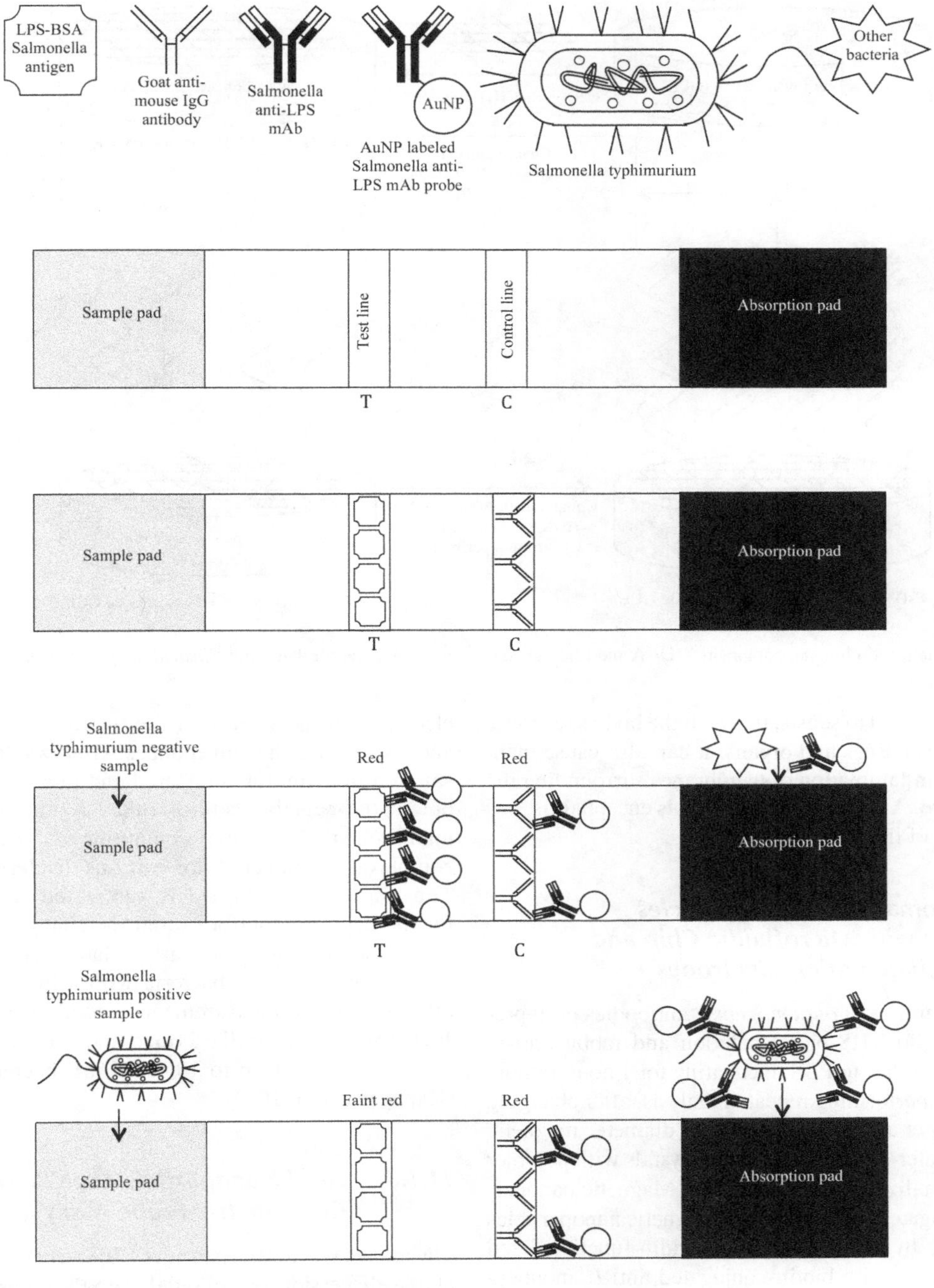

FIGURE 11.2 *Salmonella typhimurium* detection with immunochromatographic strip sensor (Wang et al. 2016).

Several concentrations of complementary target DNA (tDNA) sequence extracted from *Clostridium perfringens* are hybridized with ssDNA/CeO_2 NRs/CHIT composite/GCE forming dsDNA/CeO_2 NRs/CHIT composite/GCE. Measurements are performed in sodium phosphate buffer containing $[Fe(CN)_6]^{3-/4-}$ and KCl]. The working electrode is dsDNA/CeO_2 NRs/CHIT composite/GCE with a platinum counter electrode and a saturated calomel electrode.

From electrochemical impedance spectroscopy (EIS) measurements, it is found that the change in charge-transfer resistance (ΔR_{ct}) exhibits a linear relationship with the logarithm of tDNA concentrations ($\log C_{tDNA}$) in the range of concentrations 1×10^{-14}–1×10^{-7} mol/L. The limit of detection is 7.06×10^{-15} mol/L. Differential pulse voltammetry (DPV) measurements show that the current varies linearly with $\log C_{tDNA}$ in the range 1×10^{-14}–1×10^{-7} mol/L, with a detection limit of 1.95×10^{-15} mol/L. Satisfactory results are achieved by both techniques. The sensor is applied to detection of the DNA sequence of *C. perfringens* in dairy products (Qian et al. 2018).

11.1.3 *Listeria monocytogenes*

This is a facultatively anaerobic bacterium. It causes listeriosis, with symptoms of headache, stiff neck, and convulsions. This

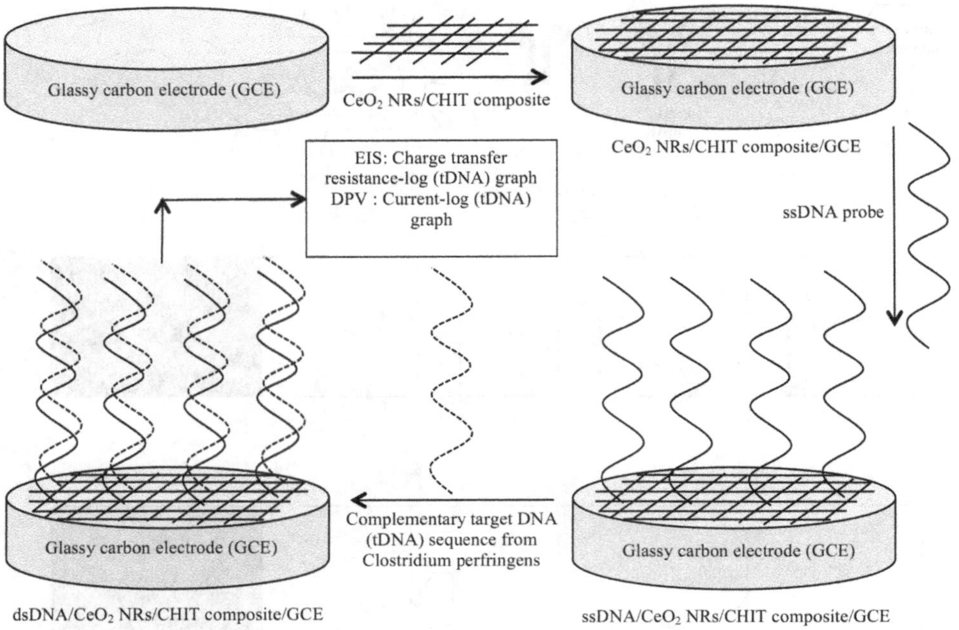

FIGURE 11.3 CeO_2 nanorod-chitosan composite/ssDNA-modified glassy carbon electrode for detecting *Clostridium perfringens* (Qian et al. 2018.)

disease condition can lead to sepsis, in which the body's response to infection damages tissues and organs. It can also cause meningitis, resulting in inflammation of membranes surrounding the brain and spinal cord. Another possible effect is encephalitis, i.e., acute inflammation of the brain.

11.1.3.1 Immunomagnetic Nanoparticles (IMNPs) with Microfluidic Chip and Interdigitated Microelectrodes

Biotin-conjugated anti-*L. monocytogenes* antibodies are prepared by adding sulfo-NHS-biotin solution and rabbit anti-*L. monocytogenes* antibodies to PBS, incubating for 1 hour, removing excess biotin by performing repeated dialysis, after changing the PBS (Kanayeva et al. 2012). The 30-nm diameter magnetic nanoparticles are water-soluble Fe_3O_4 nanocrystals with polymer coating and functionalized with streptavidin. Magnetic nanoparticle-antibody conjugates, called immunomagnetic nanoparticles (IMNPs), are made by mixing the streptavidin-functionalized magnetic nanoparticles with biotin-conjugated anti-*L.* monocytogenes antibodies in PBS, with continuous shaking for 2 hours, incubating in a magnetic field for 1 hour, washing with PBS and then re-suspending in PBS (Figure 11.4). The antibodies become attached to the magnetic nanoparticles *via* biotin-streptavidin bonds, to form IMNPs.

The microfluidic chip has a measurement chamber and a microfluidic channel. There are 25 pairs of interdigitated microelectrodes, each 10 μm wide, with a 10-μm microelectrode-to-microelectrode gap.

L. monocytogenes-IMP complexes are formed by mixing different concentrations of *L. monocytogenes* bacteria with IMPs. Now, the baseline signal is set by injecting mannitol into the microfluidic chip, waiting for 2 minutes for the stabilization of electrons on the surface of the interdigitated microelectrodes, and measuring the impedance. After water rinsing, the IMP solution in mannitol is injected and impedance is measured to obtain the negative control value. Again, after washing with mannitol, the solution containing *L. monocytogenes*-IMP complexes is injected for various bacterial concentrations. The impedance obtained is subtracted from the impedance for the negative control to find the change in impedance. The impedance change is linearly related with the concentration of *L. monocytogenes* bacteria within the range 10^3–10^7 CFU/mL. Impedance measurement is done in the frequency range 16.4–161 kHz. In milk, lettuce and ground beef samples, the sensor can detect up to 10^4 CFU/mL bacteria within 3 hours (Kanayeva et al. 2012).

11.1.3.2 Gold Nanoparticle (AuNP)/DNA Colorimetric Probe Assay

The assay (Wachiralurpan et al. 2018) works on: (i) maintenance of the dispersion of colloidal AuNPs following hybridization with the addition of the complementary target DNA (tDNA) and the optimum $MgSO_4$ salt concentration, showing red color (UV-vis absorption peak at 520–550 nm); and (ii) aggregation of AuNPs, because of the surface charge of AuNP/DNA probe becoming neutral in the absence of complementary tDNA, producing color changes from red to purple and then colorless (UV-vis absorption peak at 550–650 nm). Figure 11.5 shows the two contradictory situations of nanoparticle dispersion and their aggregation.

Colloidal AuNP solution (20-nm diameter) in citrate buffer is incubated with thiol-modified DNA to produce the AuNP/DNA probes. The red precipitate is dispersed in wash buffer containing phosphate buffer and sodium chloride. The optimum

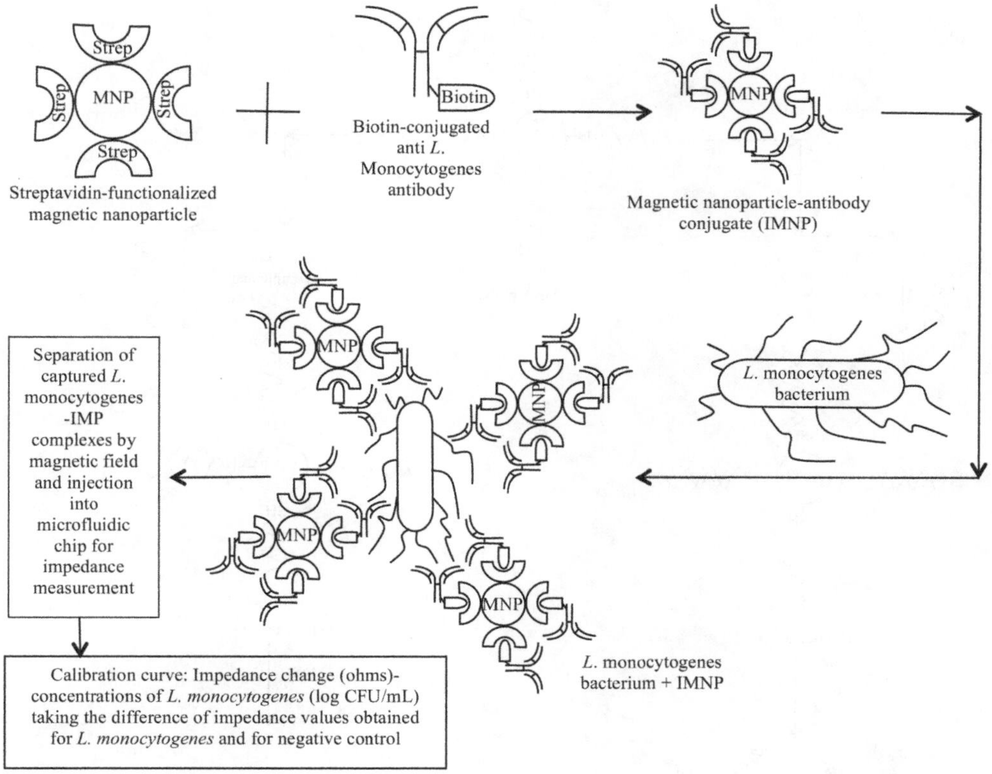

FIGURE 11.4 Immunomagnetic nanoparticles sensor for *Listeria monocytogenes* (Kanayeva et al. 2012).

concentration of $MgSO_4$ salt for visual detection is determined to be 45 mM. Test samples containing complementary tDNA are hybridized with AuNP/DNA probes, and the optimized $MgSO_4$ concentration is added. A UV-vis spectrophotometer is used for measurements. Using purified genomic DNA, the limit of detection is 800 fg. The detection limit using a pure bacterial culture is 2.82 CFU/mL. The assay can be used as a rapid screening test (Wachiralurpan et al. 2018).

11.1.4 *Campylobacter jejuni*

This is a helically shaped non-fermenting bacterium, which causes enteritis, the inflammation of the small intestine. Antibody-conjugated fluorescent dye doped-silica nanoparticles (FDS-NPs)-based assay is used for detection of the bacterium (Poonlapdecha et al. 2018). Spherically shaped FDS-NPs of 50-nm diameter are synthesized by mixing cyclohexane (oil), Triton X-100 (surfactant) and *n*-hexanol (co-surfactant) with Rubpye dye solution, then adding tetraethylorthosilicate (silica precursor) and ammonium hydroxide (initiator of the polymerization reaction) and mixing it for 24 hours. Precipitation with acetone yields FDS-NPs, which are washed with ethanol and DI water. After cross-linking the surfaces of FDS-NPs with 1-ethyl-3-(3-dimethylaminopropyl)carbodiimide hydrochloride (EDC.HCl) and *N*-hydroxysuccinimide (NHS), they are further cross-linked with glutaraldehyde and conjugated with anti-*C. jejuni* monoclonal antibody, blocked with BSA in Tris-HCl, and washed with PBS, to obtain anti-*C. jejuni* Ab-conjugated FDS-NPs (Figure 11.6). Then, the *C. jejuni* bacterial suspension is incubated with anti-*C. jejuni* Ab-conjugated FDS-NPS to form *C. jejuni* bacterium/anti-*C. jejuni* Ab-conjugated FDS-NPs complexes. These complexes are deposited on a slide and observed under a fluorescence optical microscope. The excitation wavelength used is 312 nm.

A positive result for *C. jejuni* bacteria is indicated by large fluorescent orange dots, together with moving bacterial cells. In a *C. jejuni* bacterium-negative result, only small fluorescent orange dots are seen, with the nanoparticles undergoing Brownian motion.

C. jejuni bacteria are detectable at concentrations up to 10^3 CFU/mL. At 10^3 CFU/mL concentration, no orange spot is seen. The assay time is 0.5 hours. The assay is used on chicken samples from slaughterhouses (Poonlapdecha et al. 2018).

11.1.5 *Yersinia enterocolitica*

This is a Gram-negative bacterium, causing yersiniosis (fever, pain in stomach and body, and diarrhea). The bacterium is detected by a graphene quantum dots (GQDs)-based immunosensor (Savas and Altintas 2019).

The working, counter, and reference electrodes are formed on a glass slide by electron beam evaporation of Ti (40 nm)/Au (200 nm), using a fine stainless-steel mask. GQDs (<5 nm) in distilled water are mixed with ethyl(dimethylaminopropyl) carbodiimide (EDC)/*N*-hydroxysuccinimide (NHS) and injected onto the Au working electrodes, followed by overnight incubation (Figure 11.7). On the GQDs-laminated Au electrodes, anti-*Yersinia enterocolitica* antibodies in sodium acetate (NaAc)

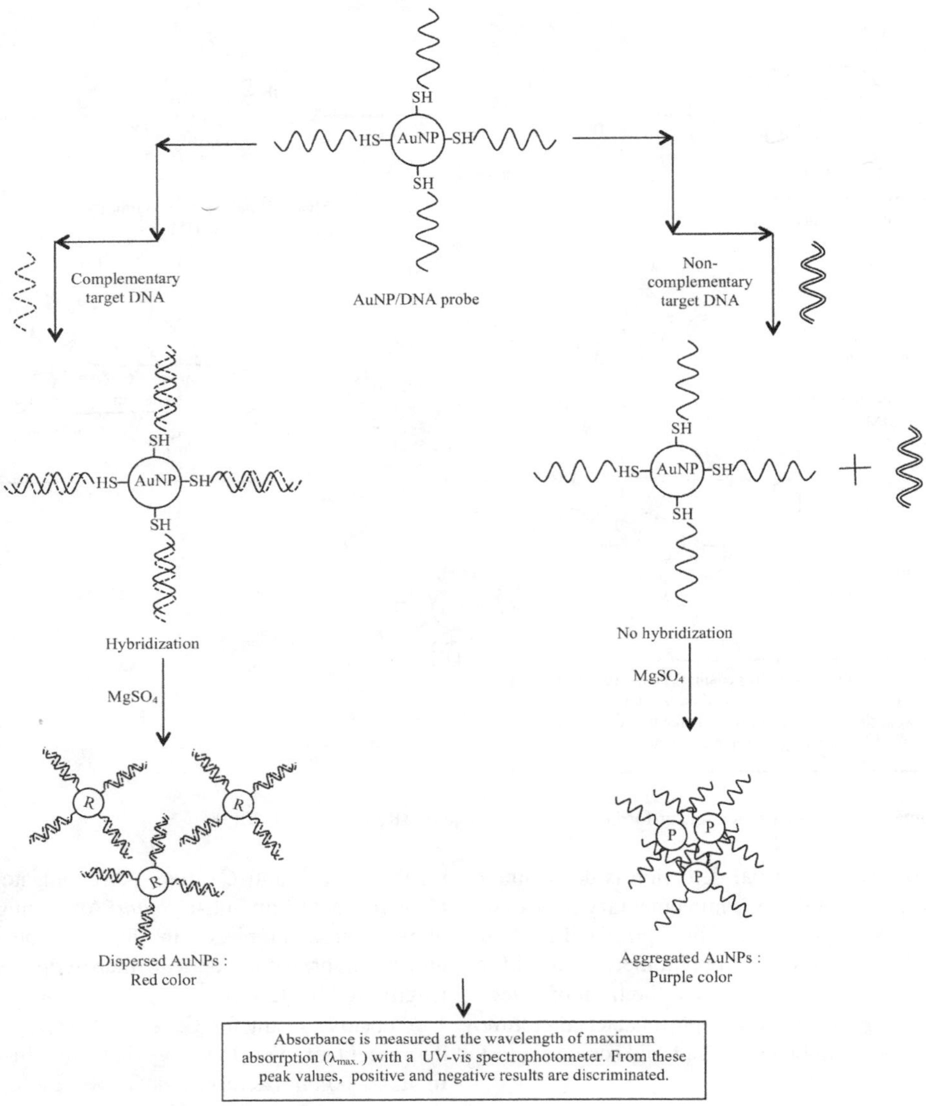

FIGURE 11.5 AuNP/DNA probe assay (Wachiralurpan et al. 2018).

buffer are deposited and incubated for 1 hour. Incubation with BSA and ethanolamine inactivates the remaining carboxyl groups, preventing non-specific binding interactions. Thus, we obtain GQD-laminated and covalently antibody-immobilized Au electrodes: Anti-*Y. enterocolitica* Abs/GQDs/Au/Glass slide.

Y. enterocolitica samples of different concentrations in PBS buffer are injected onto the surface of the Anti-*Y. enterocolitica* Abs/GQDs/Au/Glass slide, at a volume of 200 μL, incubated for 8 min, and then amperometric measurements are carried out at −0.2 V in the presence of 5 mM H_2O_2, using the three-electrode system.

Compared to the bare gold electrode, the GQDs-laminated gold electrode shows enhanced electrocatalytic property towards reduction of H_2O_2, due to the intimacy of electronic interactions between the GQDs and Au film, leading to improved electron transference. The detection of bacteria is based on the inhibition of electron transference by the antigen–antibody complex formation, so that the current decreases with an increase in bacterial concentration.

For the 5000 ppm GQDs electrode, a linear relationship is obtained between the sensor signal in mA and the *Y. enterocolitica* concentration 6.23×10^2–6.23×10^8 CFU/mL. Making electrodes with 5000 ppm GQD for milk samples and 50000 ppm GQD for human sera samples on the grounds of significant sensor signal generation at 6 CFU/mL in these samples, the limit of detection is ascertained as 5 CFU/mL for milk and 30 CFU/mL for human sera (Savas and Altintas 2019).

11.2 Nanosensors for Detection of Food-Borne Toxins

11.2.1 Botulinum Neurotoxin Serotype A (BoNT/A)

This is an acutely lethal neurotoxic protein with an intravenous/intramuscular lethal dose of 1.5 ng/kg.

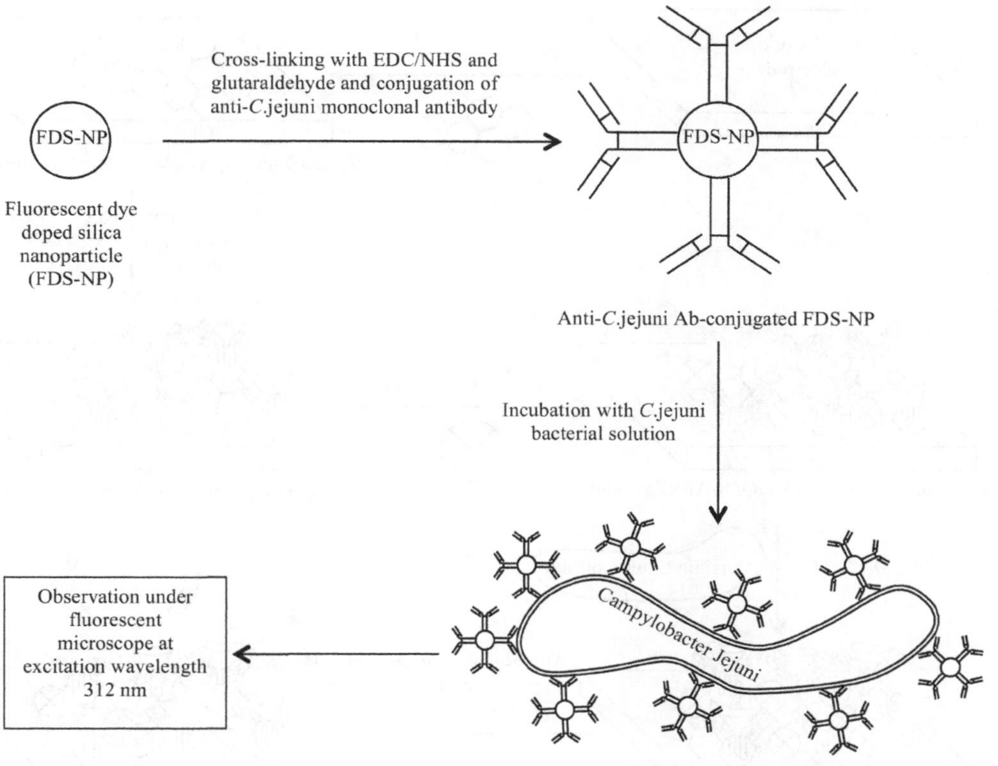

FIGURE 11.6 Sensor for *Campylobacter jejuni* using fluorescent dye-doped silica nanoparticles for signal reporting (Poonlapdecha et al. 2018).

11.2.1.1 Gold Nanodendrite (AuND)/Chitosan Nanoparticle (CSNP)-Modified Screen-Printed Carbon Electrode (SPCE) for Botulinum Neurotoxin Serotype A (BoNT/A)

The modified procedure of the working electrode is shown in Figure 11.8. AuNDs are electrochemically synthesized on the working electrode of the screen-printed carbon electrode by square-wave voltammetry, using an electrolytic solution containing chloroauric acid ($HAuCl_4$) and nicotinamide adenine dinucleotide (NAD^+) in H_2SO_4 (Sorouri et al. 2017). By immersion in CSNPs solution for 10 hours, self-assembled monolayers (SAMs) of CSNPs are formed. Glutaraldehyde, activated by mixing with 1-ethyl-3-(3-dimethylaminopropyl) carbodiimide (EDC) and *N*-hydroxy-succinimide (NHS) is deposited dropwise onto the electrode. Anti-BoNT/A polyclonal antibody is covalently immobilized on the AuNPs/CSNPs nanocomposite-covered SPCE by incubation for 1 hour, followed by rinsing in PBS to remove unbound antibodies, and blocking the remaining active sites with BSA.

Electrochemical measurements are performed using $K_3Fe(CN)_6/K_4Fe(CN)_6$ in PBS as a redox probe. The relative change in charge-transfer resistance R_{CT} is proportional to the BoNT/A concentration from 0.2 to 230 pg/mL. The limit of detection is 0.15 pg/mL. The electronic behavior of AuNDs, combined with the large surface area provided by CSNPs and the high specificity of anti-BoNT/A antibody yielded an effective platform for BoNT/A sensing (Sorouri et al. 2017).

11.2.1.2 Peptide-Functionalized Gold Nanoparticles (AuNPs)-Based Colorimetric Assay for Botulinum Serotype A Light Chain (BoLcA)

The different symbols used, and the two types of assays carried out, namely mixing and bridging types, are shown in Figure 11.9. In the absence of BoLcA, the designed biotinylated peptides 1 and 2 activate aggregation of AuNPs (Liu et al. 2014). As a result, the AuNP solution is aggregated. When BoLcA are present, they cleave the peptides, thereby preventing the aggregation of AuNPs. Hence, the AuNPs solution is dispersed. The color difference between aggregated and dispersed AuNPs is an indicator of BoLcA absence or presence.

Two types of assays, namely assay A and B, are conducted, assay A with Peptide 1 and assay B with peptide 2. The following reactants are prepared for the assay:

(i) Peptide 1 for the mixing assay, with biotin at one end: AuNP-Thiol (Anchor)-PEG (Spacer)-Peptide-Biotin. AuNPs functionalized with Peptide 1 are labelled as 1-AuNPs.
(ii) Peptide 2 for the bridging assay, with biotin at both ends: Biotin-Peptide-Biotin
(iii) Neutravidin-functionalized AuNPs are called Navi-AuNPs

Assay A: Various concentrations of BoLcA in HEPES buffer are incubated with 1-AuNPs for 20 hours and mixed with Navi-AuNPs. UV-vis spectra are recorded.

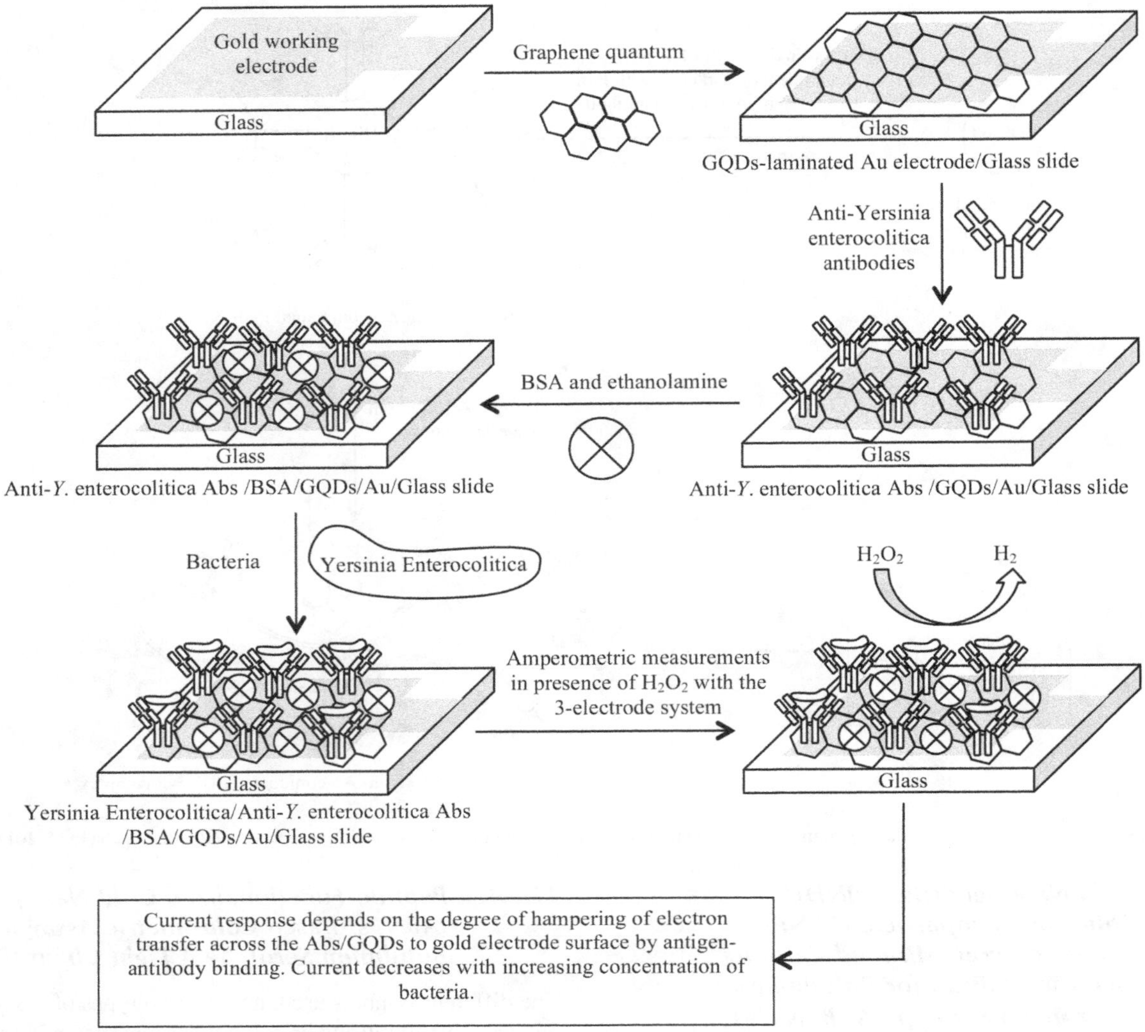

FIGURE 11.7 Detecting *Yersinia enterocolitica* with a graphene quantum dots-based immunosensor (Savas and Altintas 2019).

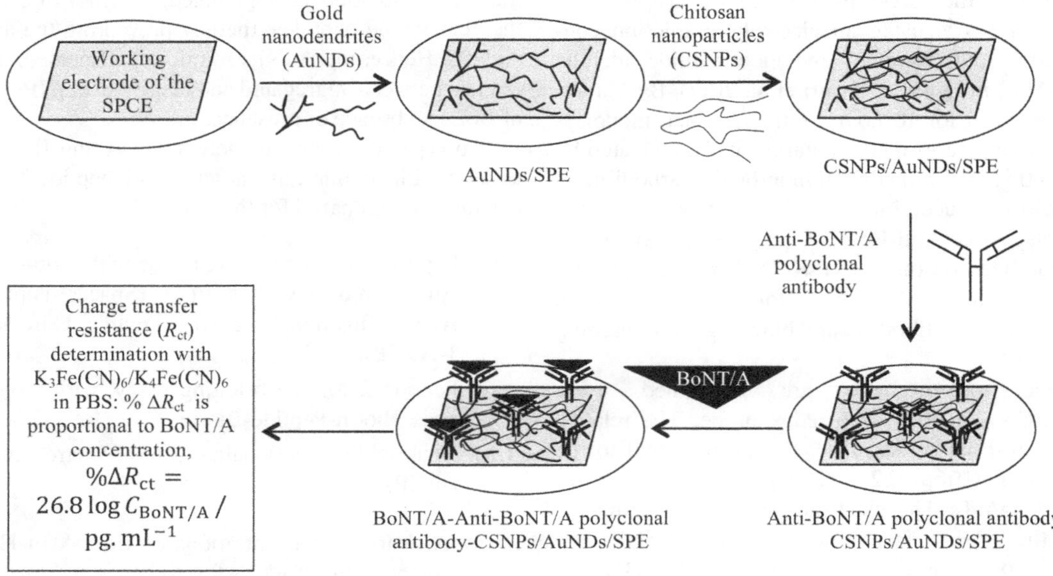

FIGURE 11.8 BoNT/A sensing with gold nanodendrites/chitosan nanoparticles modified screen-printed carbon electrode (Sorouri et al. 2017).

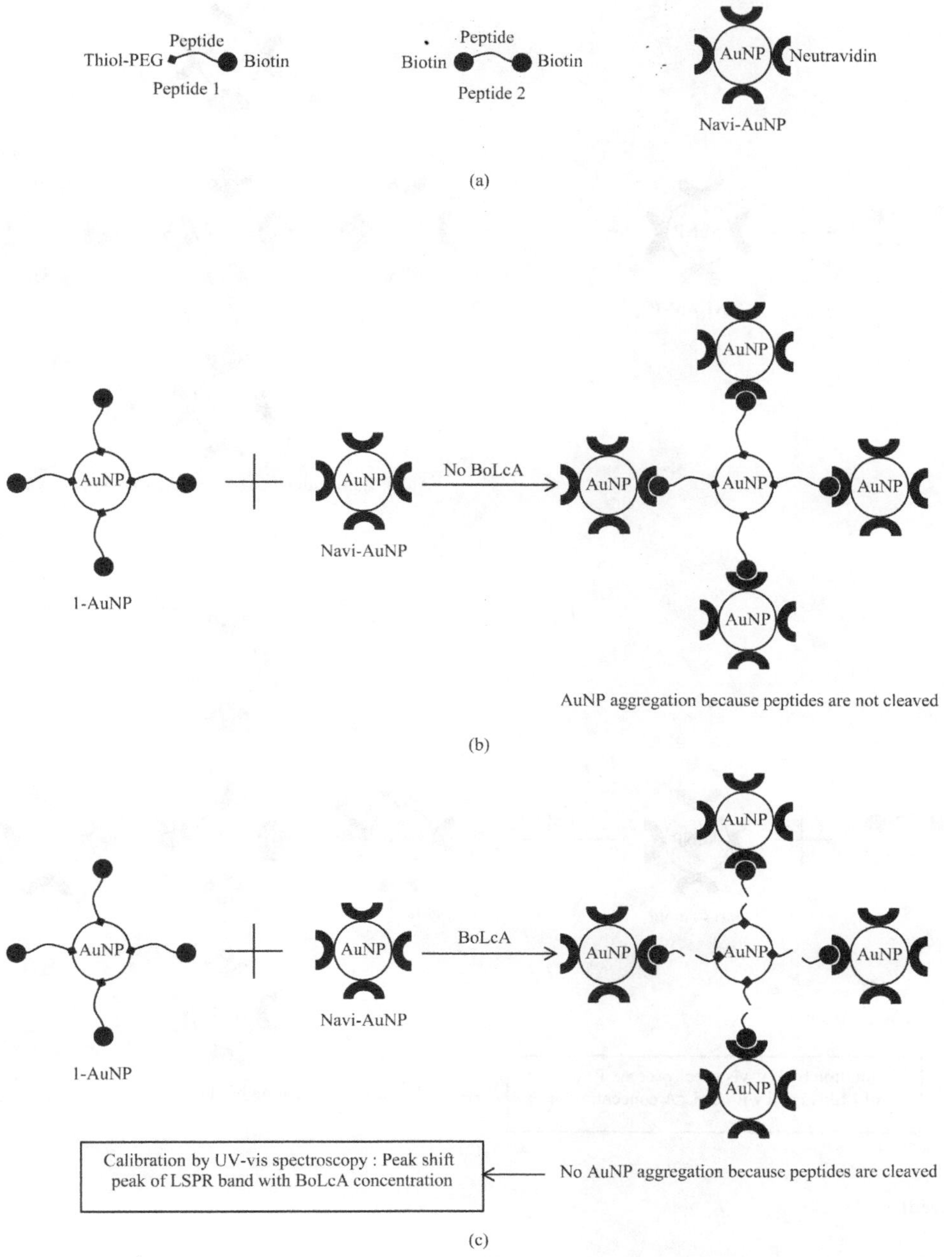

FIGURE 11.9 BoLcA light-chain sensing assay: (a) symbols, mixing assays without (b) and with (c) BoLcA, bridging assays without (d) and with (e) BoLcA (Liu et al. 2014).

Assay B: Different concentrations of BoLcA in HEPES buffer are incubated with peptide 2 for 3 hours and mixed with Navi-AuNPs. UV-vis spectra are obtained.

In both cases, the LSPR peak shifts are measured with respect to BoLcA concentration. The limit of detection is 5 nM for assay A and 0.1 nM for assay B. The limit is 50 times higher for assay B than for assay A. Furthermore, assay A suffers from the problem of a long assay time for producing sufficient response signal for detection. Assay B shortens the assay time (Liu et al. 2014).

11.2.2 Staphylococcal Enterotoxin B (SEB)

This toxin causes staphylococcal food poisoning, characterized by nausea, vomiting, and cramps in the stomach.

A gold-silver hybrid triangular nanoparticle (Au-AgNP)-based localized surface plasmon resonance biosensor is used (Zhu et al. 2009). The LSPR biosensor is shown in Figure 11.10. The Au-Ag NPs array is created on a glass substrate by nanosphere lithography, using self-assembly of nanospheres from the nanosphere solution. With evaporation of the solution, the capillary forces draw the nanospheres together into a hexagonal close-packed pattern. After depositing Ag and Au thin films, the nanosphere mask is removed by sonicating in ethanol, resulting in Au-Ag nanoparticles, with triangular footprints. The Au protective film on Ag nanoparticles isolates them from the atmosphere.

Dithiodiglycolic acid is reacted with the nanoparticle array, followed by activation of carboxyl groups on the acid with

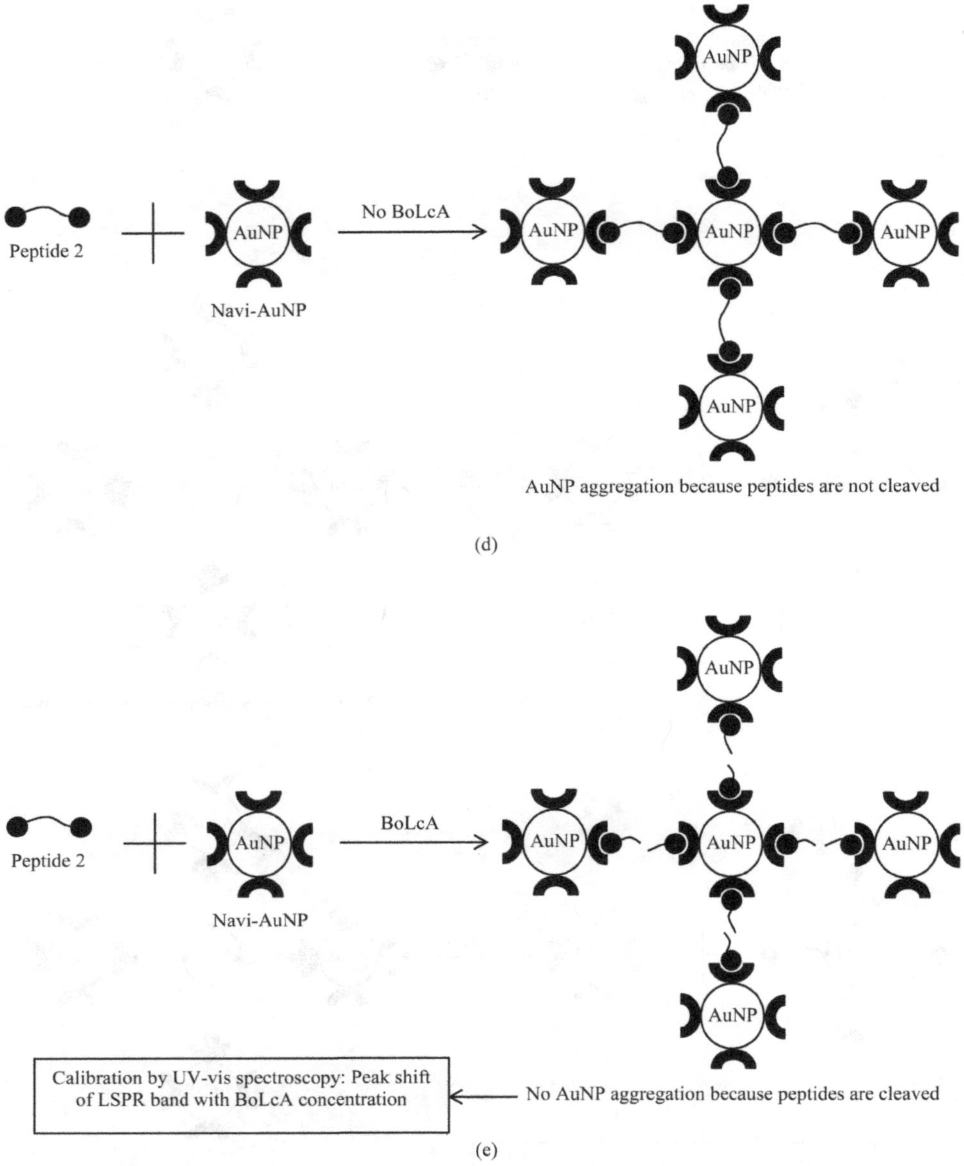

FIGURE 11.9 (Continued)

1-ethyl-3-(3-dimethylamino-propyl) carbodiimide-HCI (EDC. HCl) and sulfo-*N*-hydroxysuccinimide (S-NHS), rinsing with PBS buffer, drying with nitrogen, depositing anti-SEB monoclonal mouse IgG antibody solution in PBS buffer, allowing the chemical reaction between the alkaline aminophenol of IgG and the activated carboxyl, blocking the unwanted ester groups with ethanolamine aqueous solution and rinsing again with PBS.

SEB solutions of various concentrations in PBS buffer are mixed with the anti-SEB mouse monoclonal antibodies conjugated with Au-AgNPs. Binding of SEB target analyte with anti-SEB monoclonal mouse IgG antibodies alters the effective refractive index of the surrounding medium. UV-vis spectroscopy is applied for characterization. For a given SEB concentration, the extinction efficiency *versus* wavelength graph is plotted. This is the extinction spectrum of the localized surface plasmon resonance (LSPR). The peak wavelength of extinction spectrum is obtained. These measurements are carried out for SEB concentrations from 100 ng/mL to 10 μg/mL and the corresponding peak wavelengths are recorded. The correlation between SEB concentrations and the respective peak wavelengths helps in SEB level assessment (Zhu et al. 2009).

11.3 Nanosensors for Cancer Cell/Biomarker Detection

In Section 9.2.8, carcinoembryonic antigen detection was explained. Magnetonanosensors to monitor binding of several

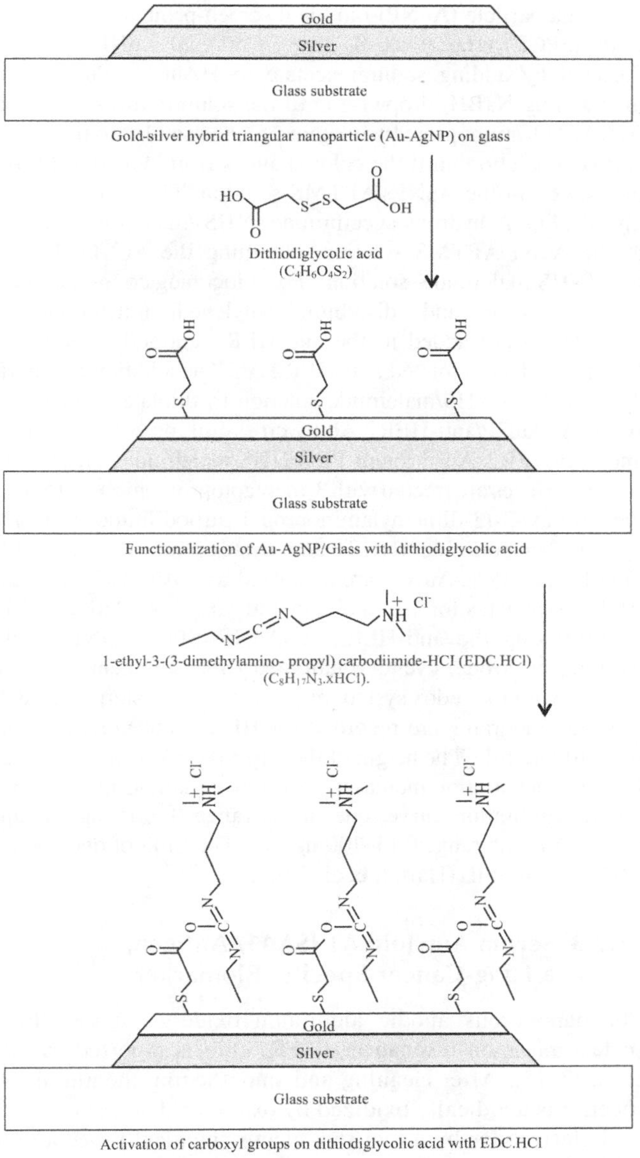

FIGURE 11.10 Au-AgNP-based LSPR biosensor (Zhu et al. 2009).

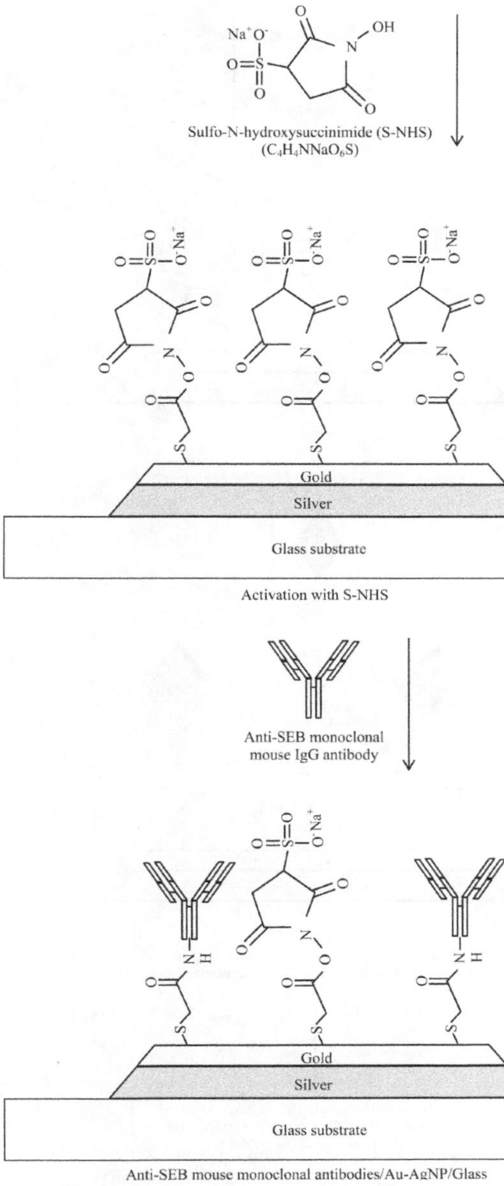

FIGURE 11.10 (Continued)

tumor markers in real time were discussed in Section 7.8.5. Nanosensors for diagnosing specific types of cancers will be presented here.

11.3.1 Breast Cancer Cell MCF-7

A human mucin-1 (MUC1) aptamer and folic acid (FA)-conjugated magnetic nanoparticles (MNPs)-based surface plasmon resonance (SPR) cytosensor is used (Chen et al. 2014). A sandwich SPR assay is organized by capturing MCF-7 breast cancer cells between two marker proteins: MUC1 aptamer and FA-conjugated MNPs (Figure 11.11). The MUC1 aptamer is a DNA chain which binds tightly with mucin-1, a highly glycosylated transmembrane protein overexpressed on the surfaces of breast cancer cells. FA is known to target cancer cells preferentially due to overexpression of a folate receptor on cancer cells. The presence of large, high refractive index MNPs in an evanescent field significantly influences the SPR response, thereby improving its sensitivity.

MNPs are synthesized by dissolving $FeCl_3.6H_2O$ and $FeSO_4.7H_2O$ in DI water, adding ammonia and sodium citrate, and raising the temperature to 90°C, when after a 30-minute incubation, a black precipitate is formed, which is washed with DI water and the MNPs are separated with the help of a magnet. The MNPs are dispersed in a mixture of 1-ethyl-3-(3-dimethy laminopropyl)-carbodiimide (EDC) and *N*-hydroxysuccinimide (NHS) in DI water, followed by the addition of FA solution in dimethylsulfoxide (DMSO) and the triethylamine catalyst. After pH adjustment to 9 and incubating for 4 hours, washing with DI water, and drying, MNP-FA conjugates are obtained.

MUC1 aptamer is immobilized on the gold disk for SPR studies by incubating with MUC1 aptamer solution in PBS buffer and rinsing with DI water.

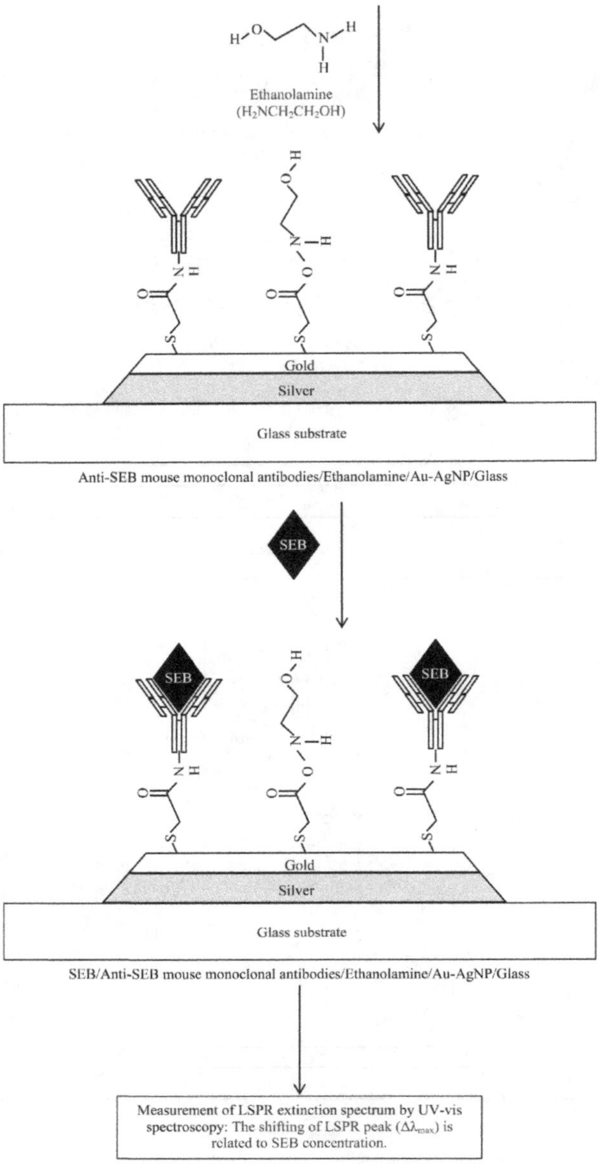

FIGURE 11.10 (Continued)

When the target breast cancer cell MCF-7 suspension is injected over the gold disk, the target cancer cells are captured by the aptamer. Non-specifically adsorbed cells are removed by PBS containing Tween-20. Then, on injection of MNP-FA solution and incubation, the MUC1 aptamer/Breast cancer cell MCF-7/MNP-FA sandwich structures are attached to the surface of the gold disk. SPR analysis is carried out. Similarly, SPR spectroscopy is repeated for various concentrations of breast cancer cells MCF-7.

The SPR angle increases linearly with the concentration of breast cancer cells MCF-7 in the range 500–10000 cells/mL. The limit of detection is marked at 500 cells/mL (Chen et al. 2014).

11.3.2 HER2, A Medical Sign of Breast Cancer

HER2 or Human Epidermal Growth Factor Receptor 2, the breast cancer biomarker, is detected by anti-HER2 antibody-conjugated gold nanoparticle (AuNP)-modified screen-printed carbon electrode (SPCE) (Hartati et al. 2018). Colloidal AuNPs are synthesized by adding sodium acetate to $HAuCl_4$ solution, and then adding $NaBH_4$ dropwise until the solution turns wine-red. APTMS (3-aminopropyl)trimethoxysilane is added to the AuNP colloidal solution until the color changes from wine-red to purple, to obtain the AuNPs/APTMS solution. Then, polyethylene glycol(PEG)/*N*-hydroxy succinimide (NHS)/maleimide is added to the AuNP/APTMS solution, obtaining the AuNP/APTMS/PEG/NHS/maleimide solution. For bioconjugate preparation, 2-iminithiolane and disodium ethylenediaminetetraacetate (Na_2EDTA) are added to the anti-HER2 antibody solution in PBS to produce thiolated anti-HER2 Ab. The addition of AuNP/APTMS/PEG/NHS/maleimide solution to thiolated anti-HER2 Ab gives AuNP/anti-HER2 Ab by cross-linking between AuNP and anti-HER2 Ab through PEG/NHS/maleimide. The SPCE-AuNP surfaces are treated with 3-mercaptopropionic acid (MPA), then 1-ethyl-3-(3-dimethylaminopropyl)carbodiimide (EDC)/*N*-hydroxy succinimide (NHS) and cysteamine (Figure 11.12). Finally, the SPCE-AuNPs are incubated with AuNPs/anti-HER2 Ab bioconjugates for 1 h to obtain anti-HER2 Ab/AuNP/SPCE-AuNP. Using the anti-HER2 Ab/AuNPs/SPCE-AuNP as the working electrode, cyclic voltammetry measurements are carried out with the redox system of $[Fe(CN)_6]^{3-/4-}$ solution in KCl.

Voltammograms are recorded for HER2 concentration from 0.15–100 ng/mL. The height of the current peak decreases as the HER2 concentration increases. Two-line equations are obtained for the calibration curve, one in the range 0–0.15 ng/mL and the other in the range 0.15–100 ng/mL. The limit of detection is 1.02×10^{-2} ng/mL (Hartati et al. 2018).

11.3.3 Serum Amyloid A1 (SAA1) Antigen, a Lung-Cancer-Specific Biomarker

The nanoporous anodic aluminum oxide (AAO)-localized surface plasmon resonance (LSPR) chip is reported by Lee et al. (2015). After cleaning and smoothening the aluminum sheet, it is anodically oxidized by oxalic acid and phosphoric acid electrolytes by a two-step electrochemical procedure to achieve a pore depth of 1 μm and different pore diameters of 15 nm to 95 nm (Figure 11.13). A Ni (5 nm)/Au (15 nm) film is formed over the AAO layer by electron-beam evaporation. The chip with the surface gold film is incubated in 11-mercapto-1-undecanol acid solution in ethanol for 24 h, when a self-assembled monolayer (SAM) is formed. The SAA1 antibody is bound using *N*-(3-dimethylaminopropyl)-*N*'-ethylcarbodiimide hydrochloride (EDC.HCl) and *N*-hydroxysuccinimide (NHS) in distilled water, mixed with dimethyl sulfoxide (DMSO), stirring for 1 hour and then removing non-specifically bound antibody with Tris-HCl. On the sensing membrane thus formed on the gold film on the surface of the AAO chip, various concentrations of the SAA1 antigen are applied. The shift in the reflection spectrum of the membrane is measured with a spectrometer.

The greatest sensitivity is obtained with the 95-nm pore diameter AAO chip. A linear relationship is found between the wavelength shift and the logarithm of SAA1 concentration in the range 10^{-16}–10^{-6} g/mL. From the measurements, the limit of detection is confirmed as 100 ag/mL (Lee et al. 2015).

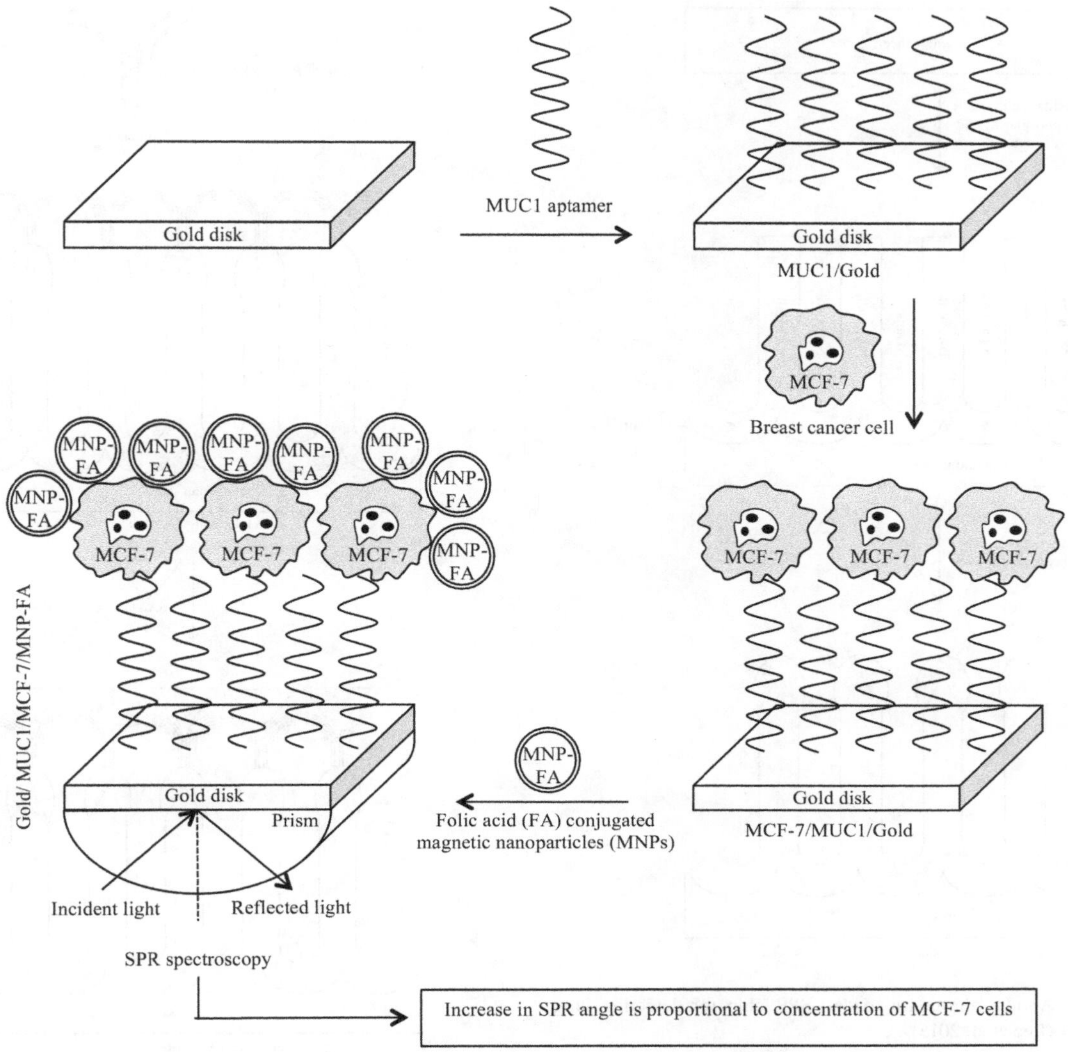

FIGURE 11.11 SPR breast cancer sensor using MUC1 aptamer and FA-conjugated MNPs (Chen et al. 2014).

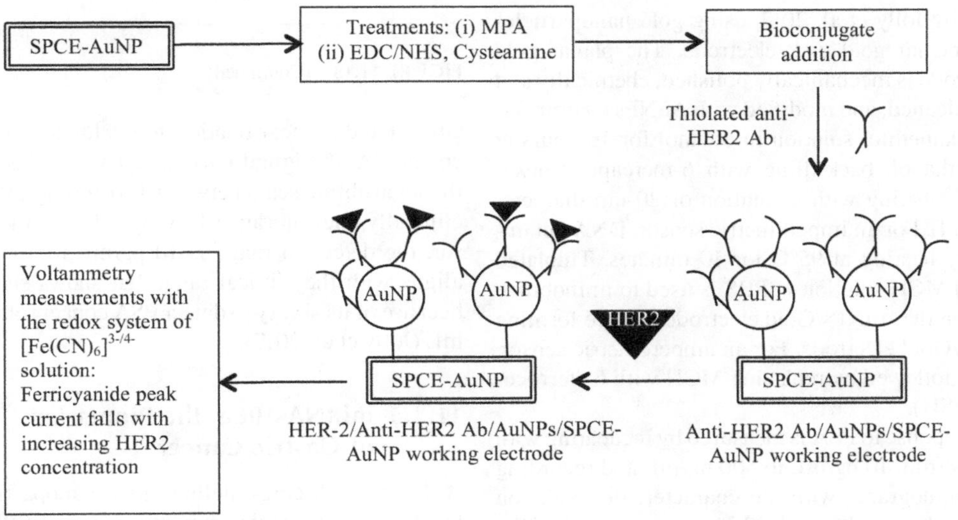

FIGURE 11.12 Anti-HER2 antibody-conjugated AuNP-modified SPCE for detecting breast cancer biomarker HER2 (Hartati et al. 2018).

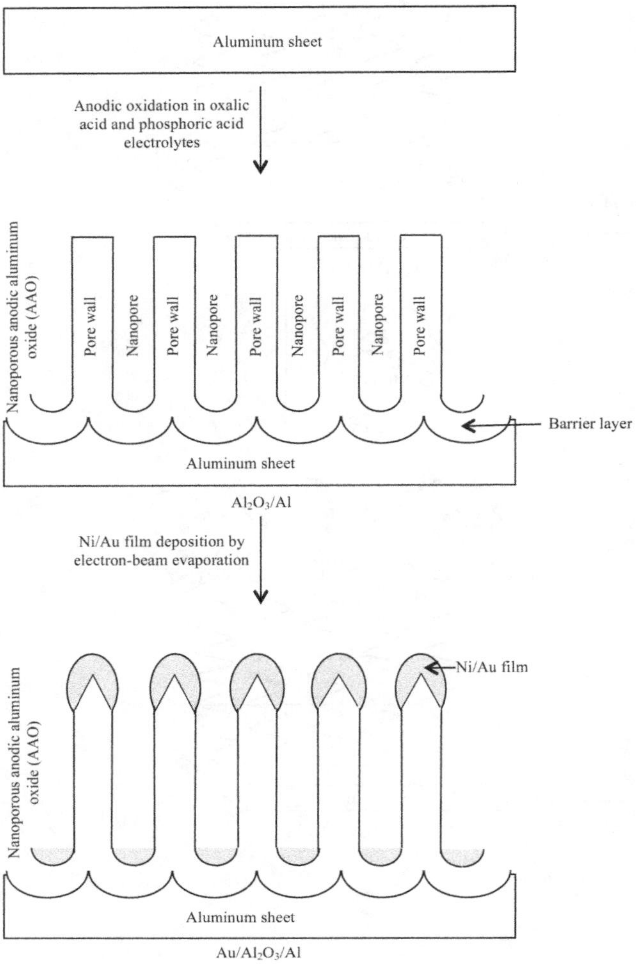

FIGURE 11.13 Anodic aluminum oxide-LSPR chip for detecting lung cancer biomarker (Lee et al. 2015).

11.3.4 Prostate-Specific Antigen (PSA), a Biomarker for Prostate Cancer

Dual-mode amperometric and impedimetric aptasensor platform is developed by Jolly et al. 2017, using gold nanoparticles (AuNP)-modified planar gold disk electrode. The planar gold disk working electrode is mechanically polished, chemically and electrochemically cleaned, and modified with AuNPs by immersing in 11-amino alkanethiol solution in ethanol for 16 hours at 4°C, washing in ethanol, backfilling with 6-mercapto-1-hexanol (MCH), and incubating with a solution of 20-nm diameter AuNPs (Figure 11.14). For an impedimetric sensor, DNA aptamers are activated by heating at 95°C for 10 minutes. Thiolated DNA aptamers and MCH solution in PBS is used to immobilize the DNA aptamer on the AuNPs/Gold electrode surface forming an aptamer/AuNPs/Gold electrode. For an amperometric sensor, the above steps are followed by replacing MCH with 6-(ferrocenyl)hexanethiol (FcSH).

Amperometric response to PSA is measured by incubating with PSA concentrations from 10 pg/mL to 500 ng/mL and recording square wave voltammograms with the characteristic oxidation peak of the FcSH redox marker at 0.27 V. An increase in PSA concentration is accompanied by a decrease in oxidation peak current and a linear relationship is found between the change in current (ΔI)/Original current (I_0) versus PSA concentration on the logarithmic scale between 1 to 100 ng/mL, which covers the clinically relevant range 1–10 ng/mL. For the impedimetric sensor, the detection range is 10 pg/mL to 10 ng/mL, which also aligns with the clinical range but shows superior performance because of its ability to detect PSA concentrations down to 10 pg/mL (Jolly et al. 2017).

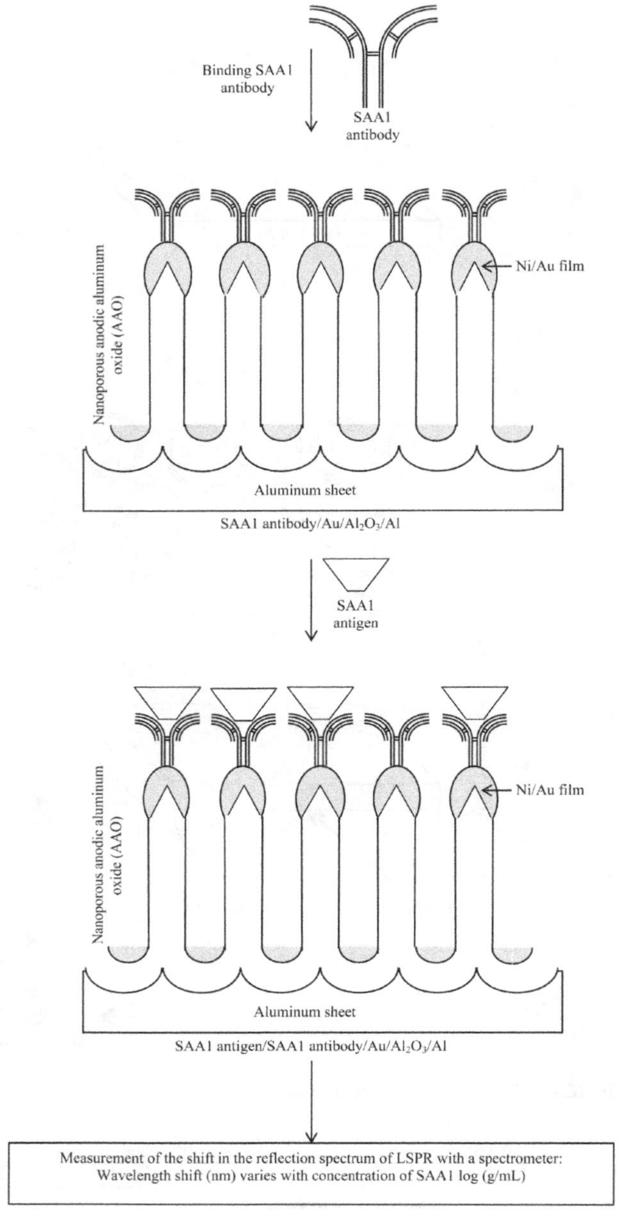

FIGURE 11.13 (Continued)

11.3.5 miRNA-106a, the Biomarker of Gastric Cancer

A hybridized core-satellite gold nanoparticle (AuNP)-based localized surface plasmon resonance (LSPRR) chip is fabricated and characterized by Park et al. (2018). Core-satellite

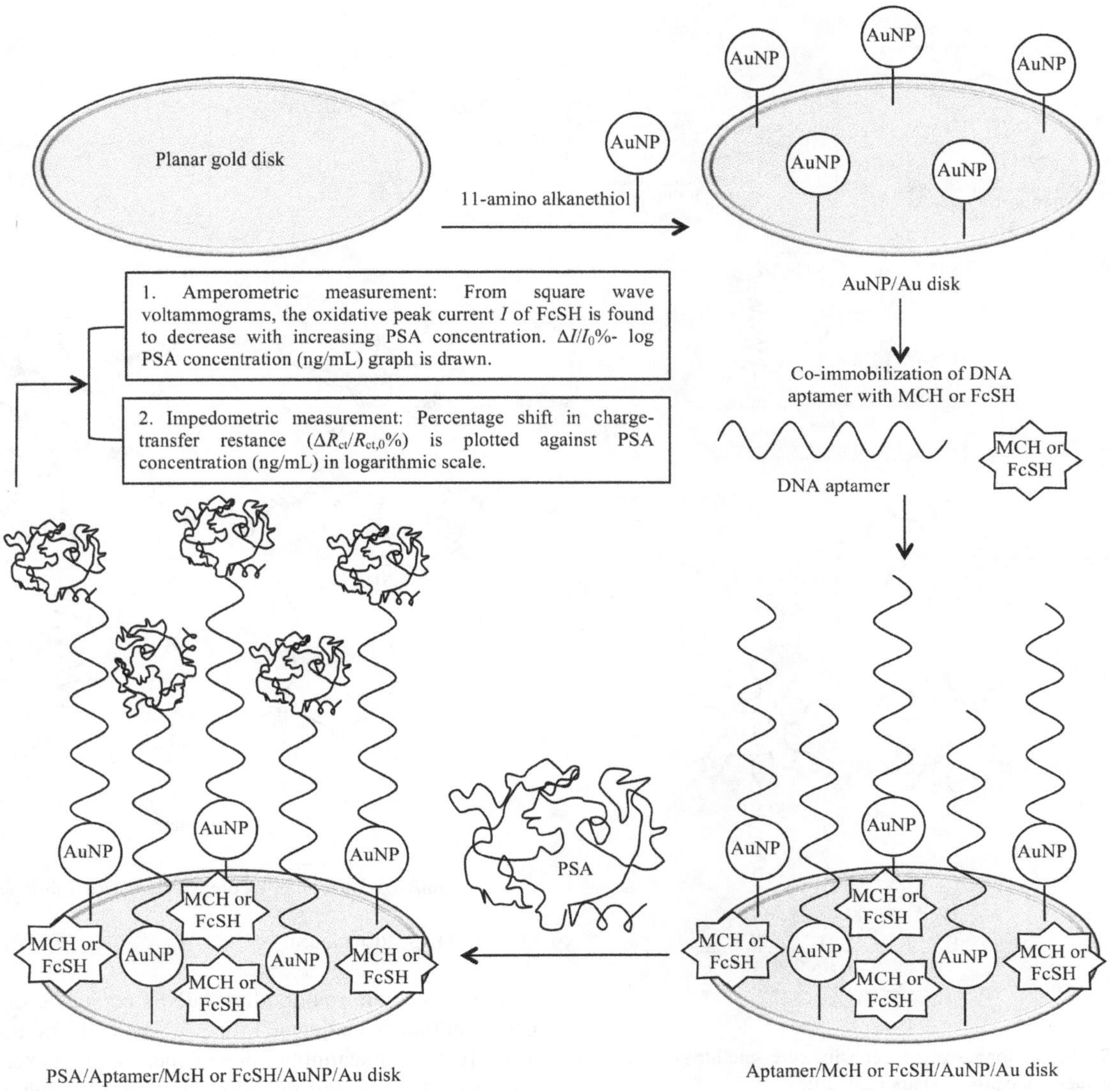

FIGURE 11.14 Prostate cancer biomarker detection using gold nanoparticle-aptamer-modified electrode (Jolly et al. 2017).

nanoparticles must be distinguished from core-shell nanoparticles. The core-shell nanoparticles are made of an inner core of a material A coated with an outer shell of material B. The core-satellite nanoparticles mentioned here comprise a core of 50-nm diameter AuNPs around which 30-nm diameter satellite AuNPs are distributed; notice the same material used for the core and the satellite nanoparticles.

A pre-modified mixed layer is formed to establish that a uniform green color, with a wavelength of 550 nm, is reproducibly obtained when 50-nm diameter core AuNPs are immobilized at an optimized density on a quartz substrate, using a mixed layer of (3-aminopropyl) triethoxysilane (APTES) and trimethoxy (octadecyl) silane (OTMS) in a mixing ratio of 3:1 APTES/OTMS. Following this predetermined trend, the AuNPs are biofunctionalized using oligonucleotides. Core AuNPs are functionalized with RNA probe 1, which is a 12-mer long oligonucleotide, having a sequence complementary to one part of the microRNA miRNA-106a. We obtain Probe1/Core AuNP/Quartz substrate (Figure 11.15). On this structure, the miRNA-106a is dispersed in concentrations from 1 pM to 10 μM. Since the RNA probe 1 on core AuNPs is complementary to one part of miRNA-106a, hybridization takes place between RNA probe 1 and one end of miRNA-106a forming miRNA-106a/probe1/core AuNP/quartz substrate. On this structure, 30-nm diameter satellite AuNPs, functionalized with RNA probe 2, are distributed. Like RNA probe 1, RNA probe 2 is also a 12-mer long oligonucleotide having a sequence complementary to another part of miRNA-106a, which is different from that of RNA probe 1. Hence, we have satellite AuNP/probe2. Therefore, on hybridization of RNA probe 2 with the complementary sequence on miRNA-106a, the opposite end of miRNA-106a gets attached to RNA probe 2 to form satellite AuNP/probe2/miRNA-106a/probe1/core AuNP/quartz substrate.

Colorimetric changes occur on the substrate in response to increasing concentrations of miRNA-106a. The spectral studies are performed using a spectrophotometer. The peak wavelength changes and the peak shift varies from ~ 1.4 nm at 1 pM to 28.33 nm at 10 μM (Park et al. 2018).

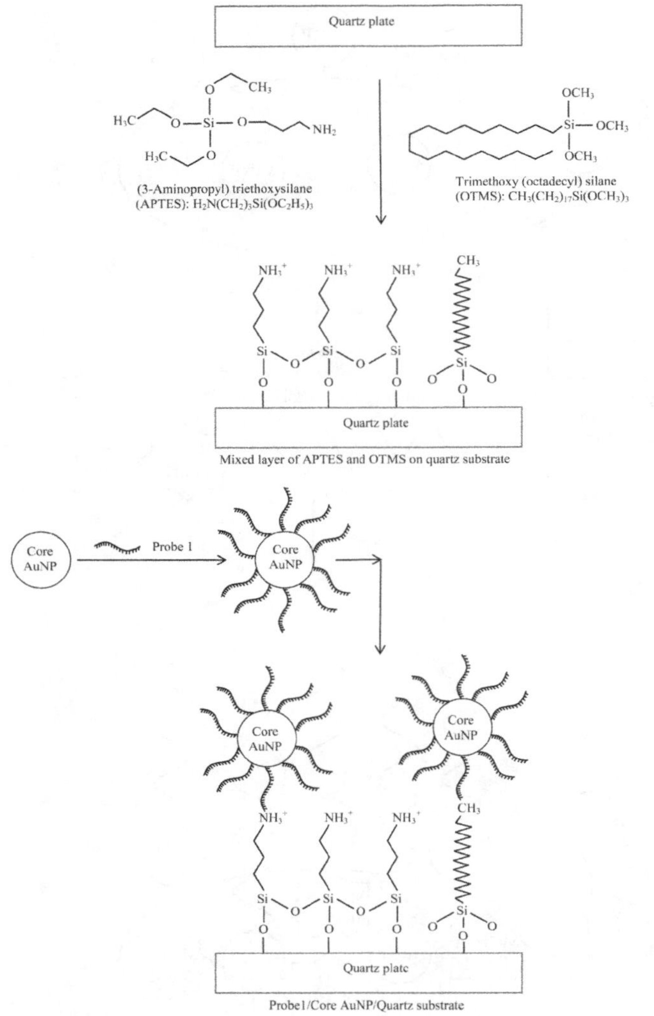

FIGURE 11.15 Sensor for gastric cancer using core-satellite gold nanoparticles and oligonucleotide probes (Park et al. 2018).

11.3.6 Colorectal Carcinoma Cell

Colorectal cancer cell imaging is performed using antibody-conjugated gold nanoparticles (Ab-AuNPs) as contrast agents (Lima et al. 2014). Colorectal cancer cells are specifically targeted for imaging by fluorescence confocal microscopy by the use of antibody/AuNP conjugates. The AuNPs are prepared by reduction of Au^{3+} with glycerol in an alkaline medium. Known quantities of polyvinylpyrrolidone (PVP) and gold chloride are dissolved in water to form an $AuCl_3$-PVP solution. Separately, specific amounts of glycerol and sodium hydroxide are dissolved in water, forming a glycerol-NAOH solution. The glycerol-NAOH solution is mixed with the $AuCl_3$-PVP solution. The mixture acquires a deep red color due to the AuNPs produced. Incubation is allowed to continue for 1 week. The pH is adjusted to 7 by adding HCl, and the colloidal solution is purified by dialysis. Anti-β-catenin and anti-E-cadherin antibodies are conjugated with AuNPs, using PBS/BSA as the medium, to yield antibodies+PBS/BSA+AuNPs (Figure 11.16). The conjugate solutions are allowed to rest for 1 hour before use.

Cancer cells HT29 (colorectal adenocarcinoma) are fixed with paraformaldehyde, permeabilized with Triton-X and incubated

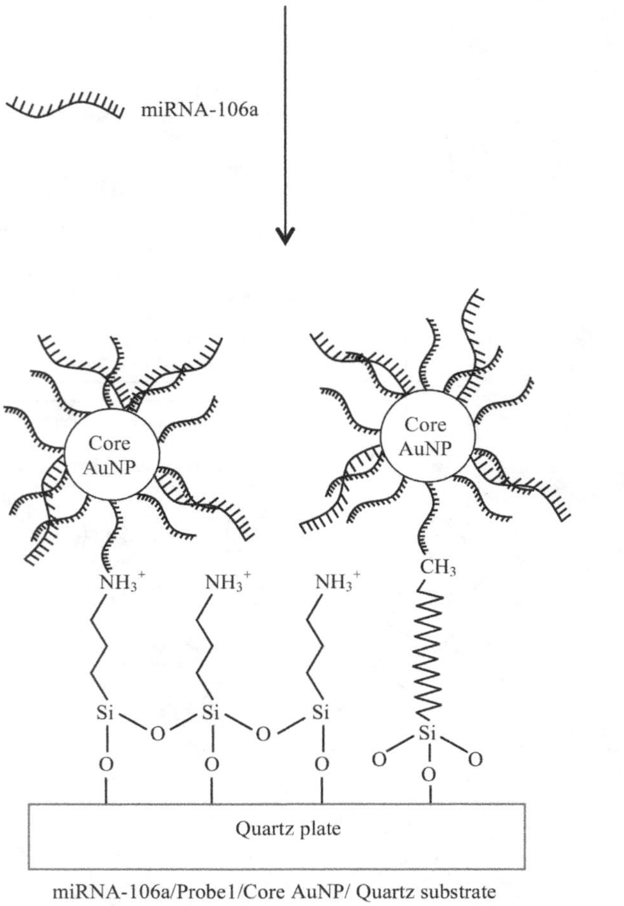

FIGURE 11.15 (Continued)

with anti-β-catenin antibodies-AuNPs conjugate or anti-E-cadherin antibodies conjugated for 1 h. The cells are observed with a confocal laser scanning microscope, using an excitation wavelength of 330 nm and recording emissions in the 700–900 nm wavelength range. Absorption of the antibodies onto the AuNPs slightly enhances the fluorescent emissions.

Identical images are obtained on labeling the cells with Alexa Fluor® 488 fluorescent dye. In the standard procedure, the antibody is incubated with Alexa Fluor® 488 dye for 24 h. The subsequent washing and fixation steps consume another 3 h, so that a total time of 27 h is necessary for cell imaging. The AuNP-based process needs 1 h for incubation. The total time of 1 h 10 min, including the 10-min rinsing step, is considerably shorter and labor-saving than for the standard procedure (Lima et al. 2014).

11.3.7 Cluster of Differentiation 10 (CD10) Antigen, the Common Acute Lymphoblastic Leukemia Antigen

These are detected by **a** quartz crystal microbalance (QCM) immunosensor with gold nanoparticle (AuNP) signal enhancers (Yan et al. 2015).

(i) The gold coating of the QCM crystal is cleaned, treated with 3-mercaptopropionic acid (MPA), and activated with EDC/NHS to produce the carboxyl groups

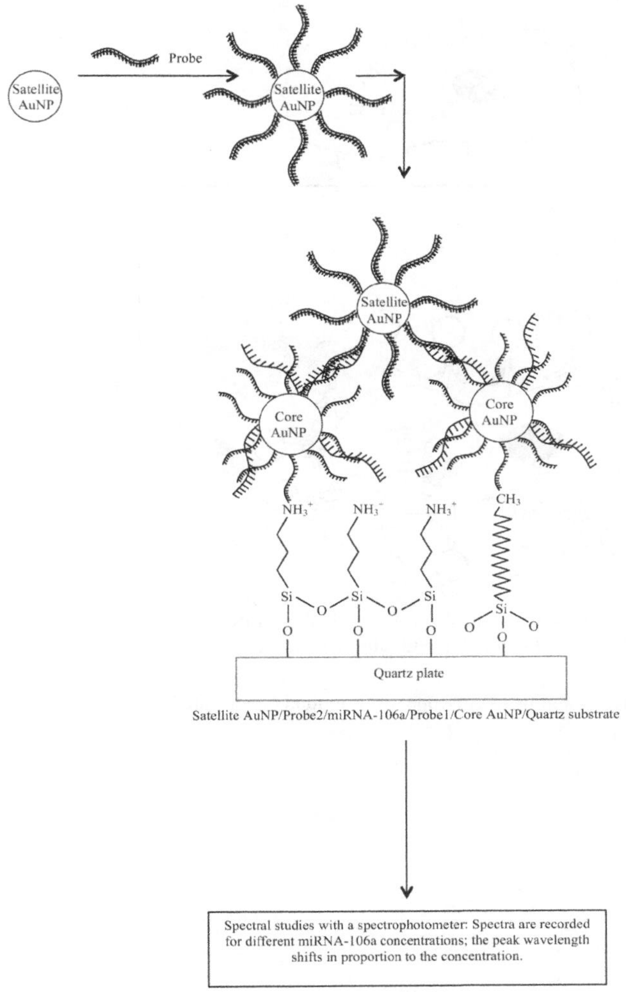

FIGURE 11.15 (Continued)

(Figure 11.17). The crystal is put into the reaction cell. Anti-CD-10 antibodies (labelled as Ab1) are immobilized onto the gold coating of the QCM crystal by crosslinking between the activated gold surface and the amino group, to obtain Ab1/gold coating. Let f_1 be the stable resonance frequency of the quartz crystal, with Ab1 antibodies attached to the gold coating.

(ii) Different concentrations of CD-10 antigens in PBS (1×10^{-8} to 1×10^{-11} M) are injected into the cell. After 1-h incubation, the CD-10 antigens bind with the antibodies Ab1 on the gold coating, forming CD-10 antigens/Ab1/gold coating.

(iii) AuNPs, prepared by reducing $HAuCl_4$ with sodium citrate, are capped with glutathione and then anti-CD-10 antibodies (labeled as Ab2) are attached to the AuNPs by forming amide bonds between carboxyl and amino groups, thereby generating AuNPs/Ab2.

(iv) Over the CD-10 antigens {step(ii)}, AuNPs/Ab2 in PBS are injected to form AuNP/Ab2/CD-10 antigen/Ab1/gold coating. Let the resonance frequency of the quartz crystal with all the bindings, including the CD-10 antigens and the AuNPs, be f_2.

Clearly, resonance frequency $f_2 <$ resonance frequency f_1, so that the frequency difference $\Delta f = f_2 - f_1$ is negative. As the frequency change is caused by the increase in mass of the crystal by all the bindings, excluding antibodies Ab1, it can be related to the quantity of CD-10 antigen. The sensor is calibrated by determining $-\Delta f$ for various concentrations of the CD-10 antigen. The shift in frequency varies linearly with CD-10 antigen concentration in the range 1×10^{-11} to 1×10^{-10} M. The limit of detection is 2.4×10^{-12} M (Yan et al. 2015).

11.4 Nanosensors for Detection of Infectious Disease Indicators

11.4.1 IgG Antibodies to Hepatitis B Surface Antigen (α-HbsAg IgG Antibodies)

HbsAg is a protein covering the external surface of the hepatitis B virus. The chronoamperometric mode gold nanoparticle electroactive label (AuNP)-assisted magnetosandwich assay is used for its detection (Escosura-Muňiz et al. 2010). The scheme of the assay is shown in Figure 11.18 and its description is described below:

(i) Tosylactivated magnetic beads (MBs) are incubated with HbsAg, separated with a magnet, suspended in PBS-BSA blocking buffer, and washed with PBS-Tween, forming MB/HbsAg.

(ii) The MB/HbsAg is incubated with human serum of post-infected patients forming the immunocomplex α-HbsAg IgG antibody/MB/HbsAg.

(iii) The 20-nm AuNPs, synthesized by reducing trichloroauric acid with trisodium acetate, are conjugated with goat polyclonal antibodies anti-human IgG (α-HIgG) followed by blocking with BSA, to obtain the α-HIgG antibody/AuNP conjugate.

(iv) α-HbsAg IgG antibody/MB/HbsAg is incubated with α-HIgG antibody/AuNP conjugate to form the complete magneto-immunosandwich: α-HbsAg IgG antibody/MB/HbsAg/α-HIgG antibody/AuNP conjugate.

The AuNPs are detected electrochemically through the measurement of their catalytic properties toward hydrogen evolution in an acidic medium. Chronoamperograms are recorded by placing a solution containing HCl and the magnetosandwich on the surface of the electrode, and thereafter keeping the working electrode at +1.35 V for 1 min, applying a negative potential of -1 V for 5 min, measuring the cathodic current and taking the absolute value of current at 200 s as the analytical signal. The procedure is repeated with sera samples having different α-HbsAg IgG antibody concentrations. The biosensor can detect 3 mIU/mL of α-HbsAg IgG antibodies in human serum, guaranteeing the detection of HbsAg IgG antibodies up to 10 mIU/mL (Escosura-Muňiz et al. 2010).

The unit mIU/mL stands for milli-international units per milliliter. The miU (milli-international unit) is 1/1000th of an international unit (IU), which is a unit for the amount of a substance (mass or volume). The mass or volume of a substance present in 1 IU varies with the substance and its biological activity. It is the amount agreed upon by scientists and doctors for easy comparison across substances. The mL (milli-liter) is unit of volume equal to 1/1000th of a liter. See also Section 9.2.7.

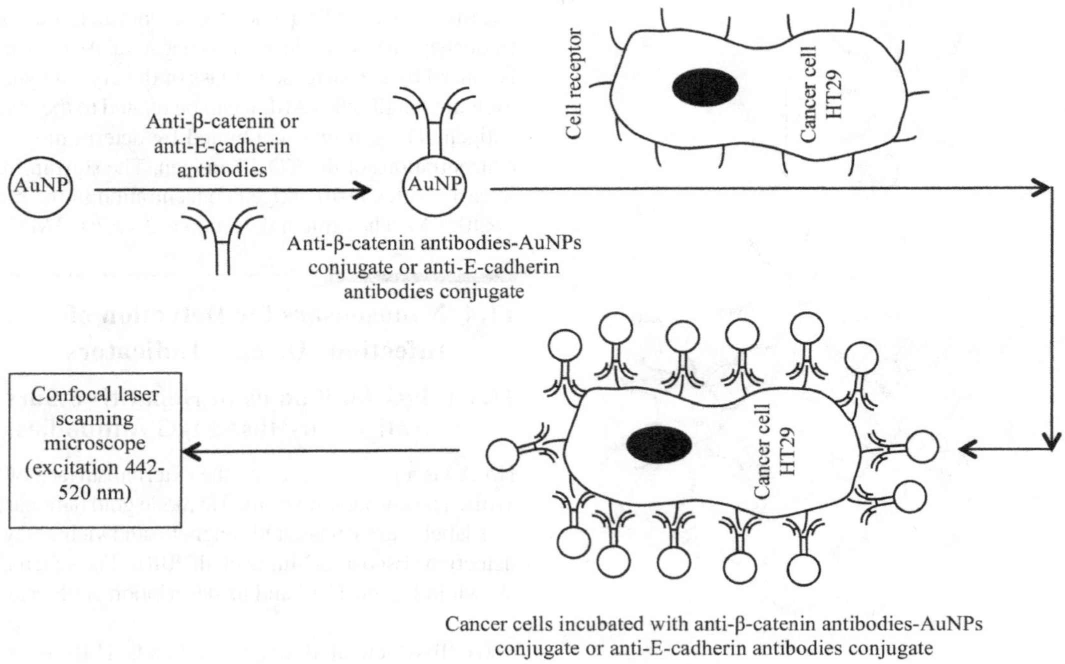

FIGURE 11.16 Preparation for imaging of colorectal cancer cells by confocal laser scanning microscope using antibody-conjugated gold nanoparticles as contrast agents (Lima et al. 2014).

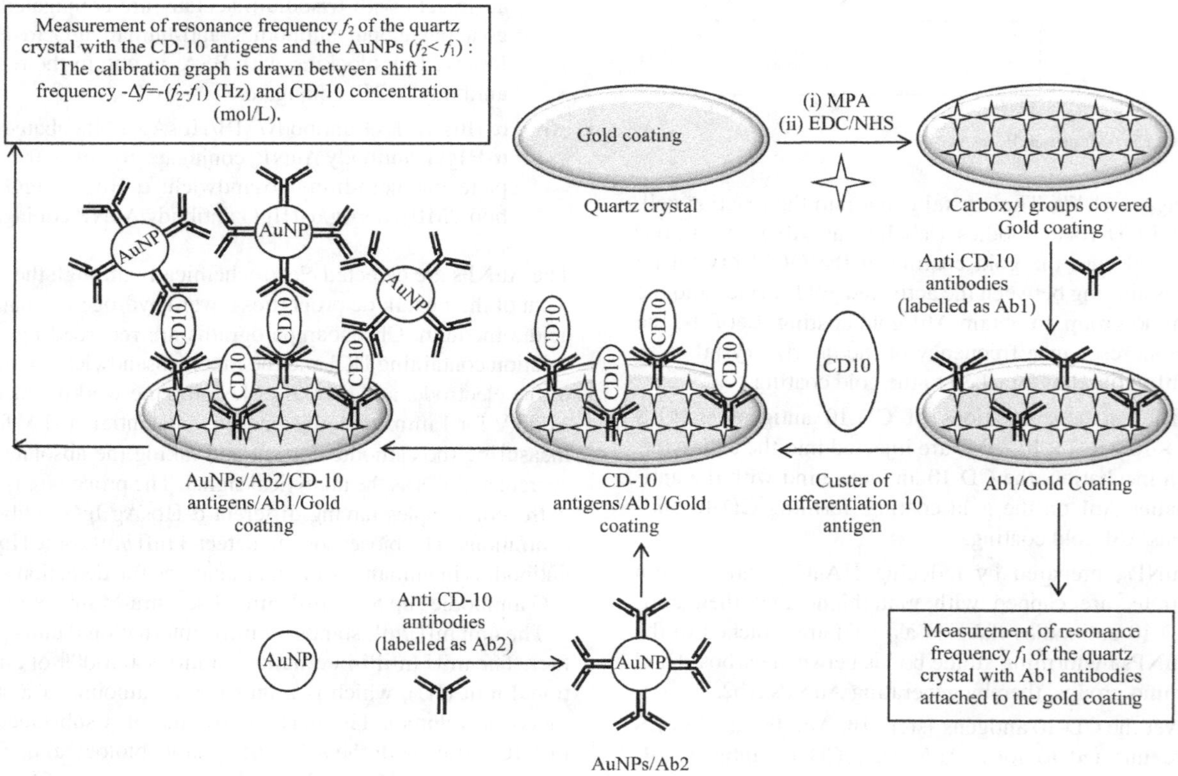

FIGURE 11.17 QCM sensor for CD-10 antigen, using gold nanoparticle-induced signal enhancement (Yan et al. 2015).

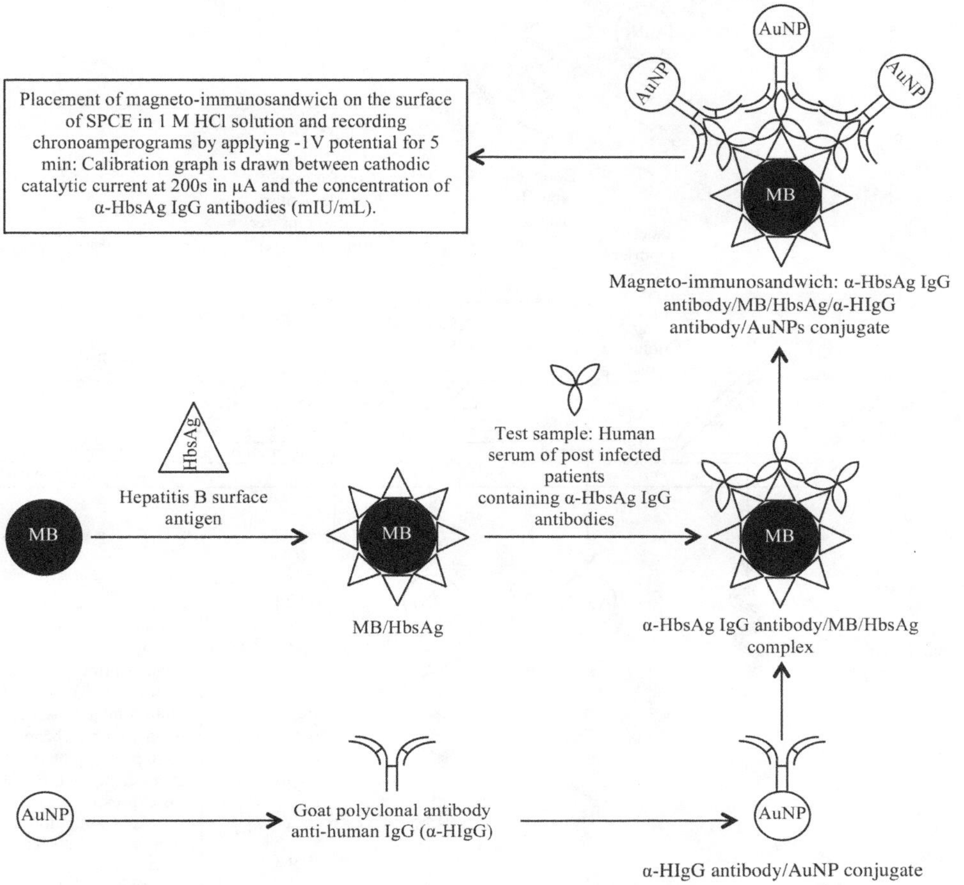

FIGURE 11.18 Making the magneto-immunosandwich for chronoamperometric sensor for detecting antibodies to hepatitis B surface antigen (Escosura-Muñiz et al. 2010).

11.4.2 Dengue-1 RNA

Dextrin-capped gold nanoparticle (AuNP)-based colorimetric lateral flow biosensor (LFB) is employed (Yrad et al. 2019). Three DNA probes are used (Figure 11.19):

(i) capture probe (dDNA) is dispersed on the test line of the nitrocellulose (NC) membrane to selectively capture the Dengue Type 1 virus (DENV-1);
(ii) control probe (cDNA) is designed as the complementary strand to the reporter probe, and dispersed on the control line of the membrane, 3 mm away from the test line; and
(iii) AuNP-labeled reporter probe (AuNP/rDNA), made from the 10-nm-sized AuNPs and prepared by heating $HAuCl_4$, sodium carbonate and dextrin at 150°C for 1 hour, and functionalized with thiolated rDNA in dithiothreitol (DTT), followed by purification, by centrifuging and washing with PBS.

The adhesive backing card carries the sample pad, the conjugate pad, the NC membrane with the test and control lines, and the absorbent pad. The conjugate pad is preloaded with AuNP/rDNA and therefore appears red. The test and control lines are invisible.

The assay starts by applying the sample, containing the target dengue-1 RNA (tRNA), to the sample pad along with the running buffer. The liquid flows by capillary action from the sample pad towards the absorbent pad. On the conjugate pad, the first hybridization takes place between the tRNA and AuNP/rDNA, forming the AuNP/rDNA-tRNA complex. Thus, the liquid is loaded with AuNPs.

Moving ahead, the AuNP/rDNA-tDNA complex reaches the test line, where it is captured by the dDNA and the second hybridization occurs, to form the sandwich: AuNP/rDNA-tRNA-dDNA. The AuNPs accumulated on the test line impart a red color to it, so that it looks like a red band if tRNA is present.

The remaining AuNP/rDNA moves onwards to the control line, where AuNP/rDNA-cDNA is formed by the third hybridization to give another red band at this line. The assay is completed in 20 minutes. If the sample does not contain tRNA, only the control line looks red, whereas the test line shows no red color.

The cut-off value of the sensor is 0.01 μM with a synthetic dengue-1 target, whereas it is 1.2×10^4 CFU/mL with pooled human sera (Yrad et al. 2019).

11.4.3 Japanese Encaphilitis Virus (JEV) Antigen

This is detected with a silver nanoparticle (AgNPs)-based sensing probe (Lim et al. 2017) by working through the steps given below:

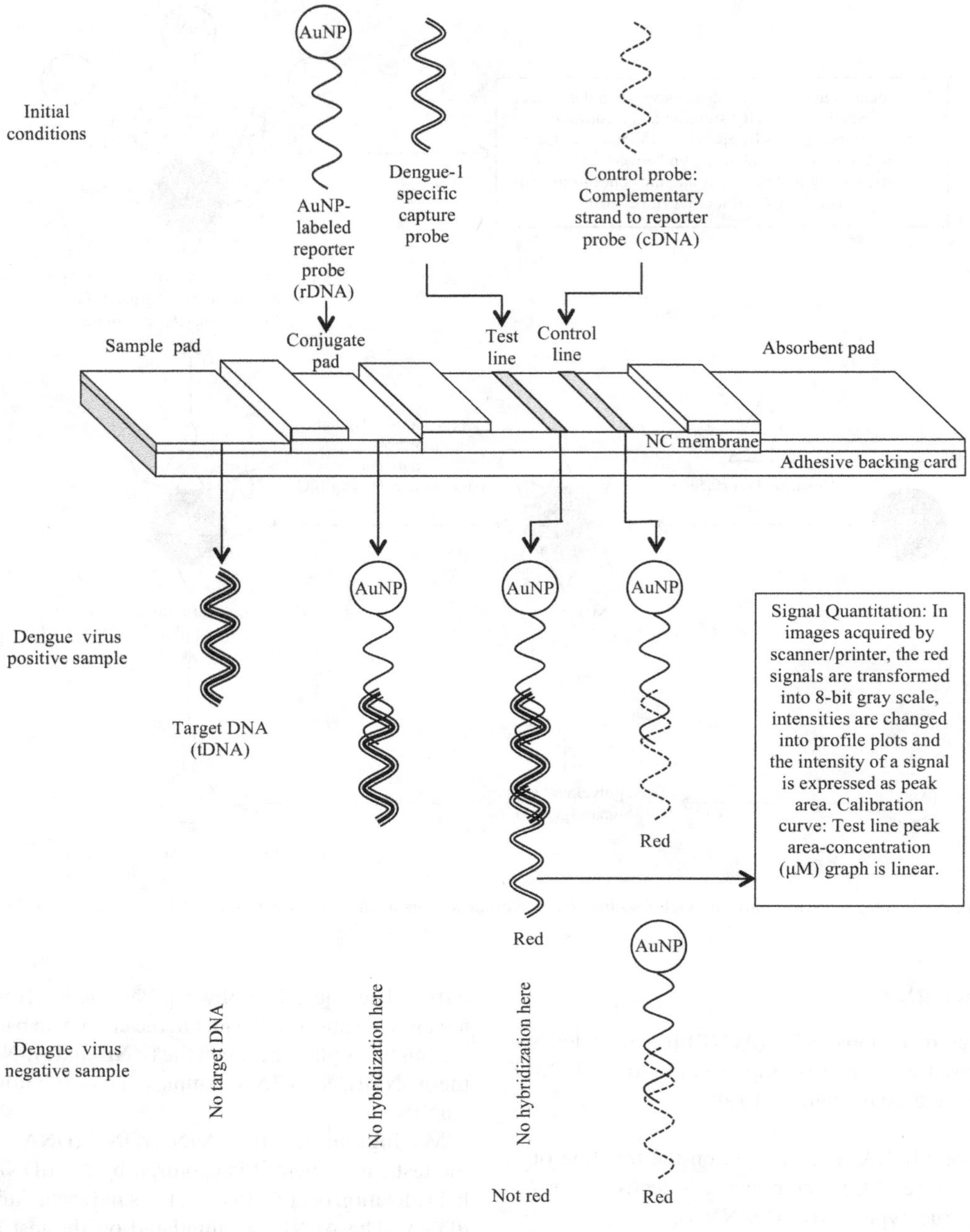

FIGURE 11.19 Lateral flow sensor for dengue virus, using gold nanoparticles (Yrad et al. 2019).

(i) AgNPs are prepared by adding trisodium citrate ($Na_3C_6H_5O_7$) dropwise to boiling silver nitrate ($AgNO_3$) solution. The appearance of a brownish yellow color indicates the formation of AgNPs. The spherical-shaped AgNPs have a diameter of 51.2 nm.

(ii) Prior to AgNP deposition on glass slides, the slides are cleaned and immersed in (3-aminopropyl) triethoxysilane (APTES) in ethanol for amine group functionalization by silanization, followed by rinsing in ethanol and heat treatment at 120°C for 30 minutes. Then the glass slides are immersed in the AgNP suspension overnight, and washed with ultrapure water, to obtain the AgNPs/glass slide as shown in Figure 11.20.

(iii) JEV antibodies are deposited onto the AgNP-modified glass slides, incubated for 2 h, rinsed with PBS, and treated with BSA to block the nonspecific binding sites. Ag NPs-based sensing probes are formed as:

JEV antibodies/BSA/AgNP/glass slide

(iv) Finally, the JEV antigen solution is poured over the Ag NPs-based sensing probes, forming:

JEV antigen/JEV antibodies/BSA/AgNP/glass slide

after incubation for 1 h.

The absorbance intensity of JEV-antigen-treated AgNP-based sensing probes is measured at 427 nm. A linear correlation is observed between the absorbance intensity and the logarithm of JEV antigen concentration in the range 14 to 100 ng/mL.

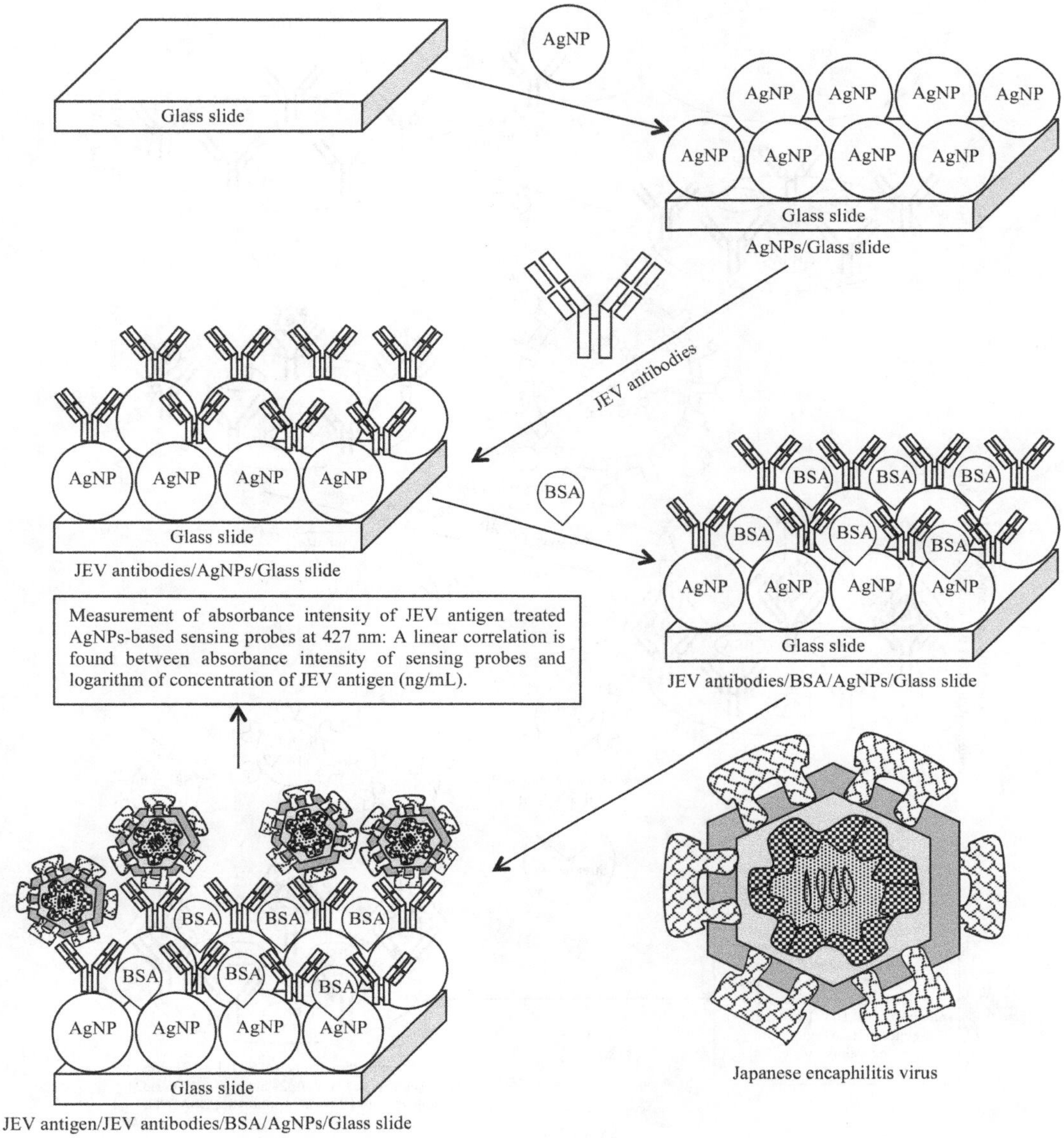

FIGURE 11.20 AgNP absorbance probe for Japanese encaphilitis virus (Lim et al. 2017).

The detection sensitivity is 0.0218 mL/ng. The limit of detection is 12.8 ng/mL, and the time required for the assay is 1 h (Lim et al. 2017).

11.4.4 HIV-1 p24 Antigen

This is a viral protein for screening/diagnosing HIV infections. Silica nanoparticle-based immunoassay (SNIA) is used for detecting its presence (Chunduri et al. 2017). Europium-doped silica nanoparticles (SiO_2: EuNPs) are synthesized by reaction between tetraethoxysilane (TEOS), ammonium hydroxide, europium chloride and ethanol for 24 h, washing with ethanol, centrifuging, and then annealing at 700°C for 2 h for Eu activation. Amino-functionalized europium-doped silica nanoparticles (SiO_2: Eu-NH_2 NPs) are formed by silanization with APTES. The SiO_2: Eu-NH_2 NPs are dispersed in DMF and added to succinic anhydride to get NPs with carboxylic groups. Streptavidin is conjugated to the carboxylic functionalized NPs using 1-ethyl-3-(3-dimethylaminopropyl) carbodiimide (EDC)/ sulpho *N*-hydroxysuccinimide(sulpho NHS) and incubation is carried out.

For the assay, the anti-HIV-1 p24 antibodies, diluted with carbonate-bicarbonate buffer, are coated on the fluorescence microplate well and washed with wash buffer, followed by blocking of the non-specific binding sites (Figure 11.21). Then, the samples containing different concentrations of HIV-1 p24 antigen are added and incubated. After washing with wash buffer, biotinylated HIV-1 p24 detector antibody is added and incubated, forming capture antibody-HIV-1 p24 antigen-biotinylated detector antibody complex, due to interaction between capture and detector antibodies. Finally, the streptavidin-conjugated SiO_2:EuNPs diluted in dimethyl sulfoxide (DMSO) are added and incubated to form SiO_2:EuNPs/streptavidin/biotin/detector antibody-HIV-1 p24 antigen-capture antibody.

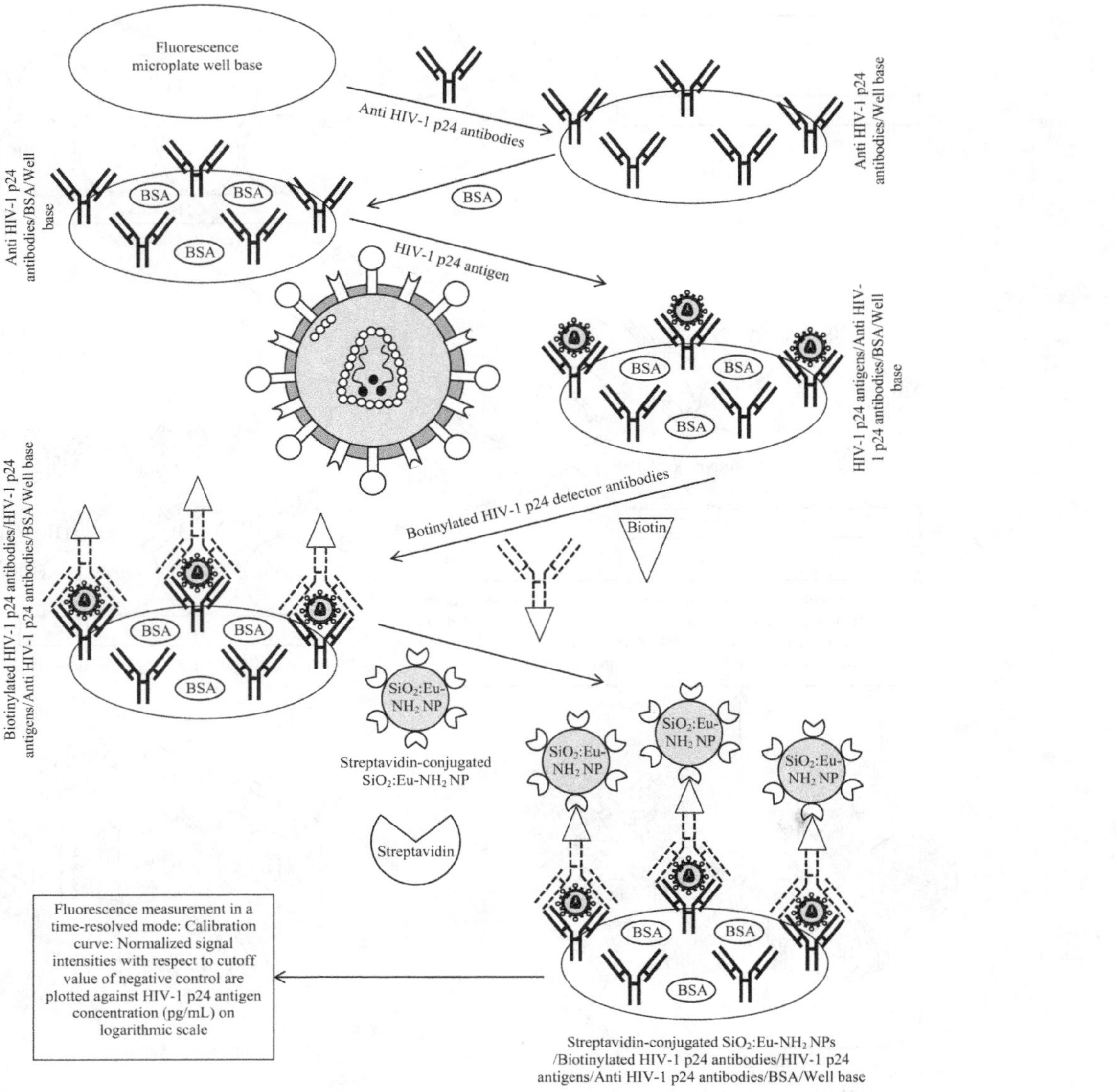

FIGURE 11.21 Fluorescence sensor for HIV virus detection, using silica nanoparticles (Chunduri et al. 2017).

Fluorescence measurement is carried out in a time-resolved mode. The excitation wavelength is 268 nm and the emission wavelength is 615 nm. The decay time and the measurement window are 400 μs each.

SNIA provides femtogram level sensitivity with linear detection ability of HIV-1 p24 antigen in the range 0.02–500 pg/mL, offering 1000 times improvement over the sensitivity of colorimetric ELISA (Chunduri et al. 2017).

11.4.5 Zika Virus (ZIKV)

This virus is detected by a platinum nanoparticle (PtNP)-based viral lysate assay on a paper microchip (Draz et al. 2018).

(i) The microchip is fabricated by screen-printing finger-like integrated electrodes of 2-mm width and with 2-mm spacing, using silver-graphene nanocomposite ink on cellulose paper substrate.

(ii) Magnetic beads (MBs) modified with anti-ZIKV mAbs, i.e.,

anti-ZIKV mAbs/MBs are incubated with ZIKV for 30 minutes obtaining:

ZIKV/anti-ZIKV mAbs/MBs as shown in Figure 11.22.

(iii) Citrate-capped PtNPs are treated with 3-(2-pyridyldithio)propionylhydrazide (PDPH) freshly reduced by tris(2-carboxyethyl) phosphine (TCEP). Before coupling the anti-Zika virus monoclonal antibodies (anti-ZIKV mAbs) to PtNPs, the antibodies are mixed with sodium metaperiodate and sodium acetate and incubated in the dark for 20 minutes. The oxidized antibodies are reacted with PtNPs for 1 hour to get:

anti-ZIKV mAbs/PtNP probes.

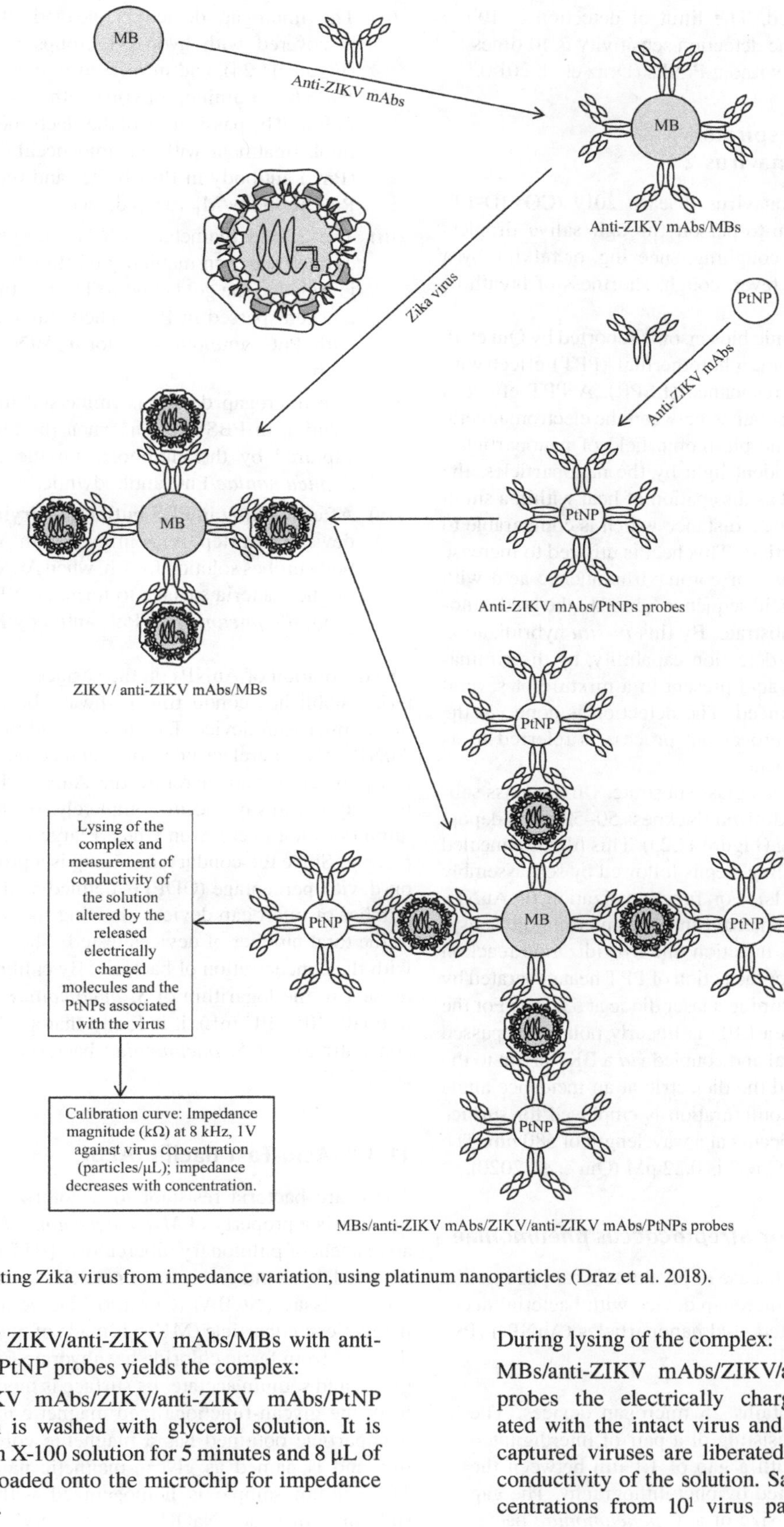

FIGURE 11.22 Detecting Zika virus from impedance variation, using platinum nanoparticles (Draz et al. 2018).

(iv) Incubation of ZIKV/anti-ZIKV mAbs/MBs with anti-ZIKV mAbs/PtNP probes yields the complex: MBs/anti-ZIKV mAbs/ZIKV/anti-ZIKV mAbs/PtNP probes which is washed with glycerol solution. It is lysed in Triton X-100 solution for 5 minutes and 8 μL of the lysate is loaded onto the microchip for impedance measurement.

During lysing of the complex: MBs/anti-ZIKV mAbs/ZIKV/anti-ZIKV mAbs/PtNP probes the electrically charged molecules associated with the intact virus and the PtNPs linked with captured viruses are liberated, altering the electrical conductivity of the solution. Samples with ZIKV concentrations from 10^1 virus particles/μL to 10^5 virus particles/μL are tested. The limit of detection is 10^1 virus particles/μL. The detection sensitivity is 10 times higher than achieved without PtNPs (Draz et al. 2018).

11.4.6 Severe Acute Respiratory Syndrome Coronavirus 2

This virus causes the coronavirus disease 2019 (COVID-19), which spreads from person-to-person through saliva droplets or nasal discharges during coughing, sneezing, or talking by a patient. Symptoms include fever, cough, shortness of breath or difficulty in breathing.

A dual-functional plasmonic biosensor is reported by Qiu et al. (2020). It combines plasmonic photothermal (PPT) effect with localized surface plasmon resonance (LSPR). A PPT effect is the production of heat by resonance between the electromagnetic field of a laser beam and the plasmonic field of nanoparticles. After the absorption of incident light by the nanoparticles, the absorbed energy is released as dissipation of heat within a small distance of the nanoparticles, a distance which is comparable to the diameter of the nanoparticle. This heat is utilized to increase the hybridization temperature of coronavirus nucleic acid with its complementary nucleic acid sequence bound to the gold nano-island formed on a glass substrate. By this *in-situ* hybridization promotion mechanism, the detection capability, by discrimination of the relevant nucleic acid present in a mixture of several genes, is significantly enhanced. The detection is done *via* the LSPR method, and the improvement process is referred to as thermoplasmonic enhancement.

The sensor is fabricated on a glass substrate. On the glass substrate, a thin gold film of optimized thickness 50–52 nm is deposited by magnetron sputtering (Figure 11.23). This film is annealed at 550°C for 3 h. The annealing step is followed by self-assembly of gold nano-islands (AuNIs). For functionalization of AuNIs, 0.1 nmol cDNA solution is injected into the microfluidic sensor chamber. After target DNA injection, the hybridization reaction is allowed to take place under the action of PPT heat generated by 532 mW optical power, by shining a laser diode at 532 nm. For the LSPR, the white light from a LED is linearly polarized, passed through a birefringent crystal and coupled *via* a BK7 prism to the interface between AuNI and the dielectric at an incidence angle of 72°. The Kretschmann configuration is employed for surface plasmon resonance, which occurs at a wavelength of 580 nm. The limit of detection of SARS-Cov-2 is 0.22 pM (Qiu et al. 2020).

11.4.7 *Pneumococcus* or *Streptococcus pneumoniae*

They cause pneumococcal disease or pneumococcal pneumonia. They are detected using a microgap device with bacterial decoration by antibody-conjugated gold nanoparticles (AuNPs) (Pyo et al. 2017).

(i) The sensor chip contains 25 microgap devices. The microgap device, consisting of a pair of interdigitated electrodes (IDEs), with a gap of 1.1 μm between the electrodes, is fabricated by photolithography. The gap is comparable to the size of a *S. peneumoniae* bacterium, which is 1–1.5 μm, the major axis length of a single cell, and 2–3 μm for diplococci.

(ii) The microgap device is cleaned, the SiO_2 surface is covered with hydroxyl groups in oxygen plasma (Figure 11.24), and immersed in methanol, containing 1% v/v (3-aminopropyl)trimethoxy-silane (APTMS) followed by passivation of the electrodes in 1-dodecanethiol, treatment with pneumococcal C-polysaccharide (PnC) antibody in PBS buffer and then BSA. We get: PnC antibody/Microgap device.

(iii) The 25-nm diameter AuNPs are synthesized by the citrate reduction method, and thioglycolic acid (TGA) is added to the gold colloid. The mixture is centrifuged and redispersed in PBS. Then, AuNPs are conjugated with PnC antibodies to form AuNP@PnC antibody probes.

(iv) The microgap device is immersed in *S. pneumoniae* solution in PBS for 1 h, when the target bacteria are captured by the antibodies on the device, forming: *S. pneumoniae*/PnC antibody/microgap device.

(v) After washing in PBS buffer and drying, the microgap device from step (iv) is immersed in AuNP@PnC antibody probes solution for 1 h, when AuNPs are decorated on the bacteria surface to form: AuNP@PnC antibody probes/*S. pneumoniae*/PnC antibody/Microgap device.

The decoration of AuNPs on the surface of *S. pneumoniae* bacteria establishes conducting pathways between the electrodes of the microgap device. Due to this bridging of the gap by the AuNPs, the interelectrode conductance depends on the bacterial population and therefore the AuNPs. However, the bridging may not always occur completely, so that the conductance jump does not occur even when a large number of bacteria are present. Since the conductance jump is a probabilistic event, the on-device percentage (ODP) is defined as the ratio between the number of microgap devices showing the jump in conductance to the total number of devices tested. The ODP value increases with the concentration of bacteria. By calibrating the ODP with respect to the logarithm of *S. pneumoniae* bacteria concentration ($10–10^8$ CFU/mL), it is found that a minimum measurable concentration of *S. pneumoniae* bacteria is 10 CFU/mL (Pyo et al. 2017).

11.4.8 Acid-Fast Bacilli (AFB)

These are bacteria resistant to decolorization by acids. Acid fastness is a property of *Mycobacterium tuberculosis*, the causative agent of pulmonary tuberculosis (PTB). These bacteria are detected by a magnetic nanoparticle-based colorimetric biosensing assay (NCBA) (Gordillo-Marroquín et al. 2018). The magnetic nanoparticle (MNP) is made of magnetite (Fe_3O_4) synthesized from ferric chloride hexahydrate $FeCl_3.6H_2O$, ethylene glycol and sodium acetate. Its surface is modified with glycan to form the glycan-functionalized magnetic nanoparticle. So, the nanoparticle obtained has a magnetite core with glycan coating, and is named as glycan-magnetic nanoparticle (GMNP). The sputum sample is homogenized with a 1:1 mixture of sodium hydroxide (NaOH) and *N*-acetyl-L-cysteine (NALC). The GMNP is mixed with the sputum sample, and the mixture is incubated for 10 min (Figure 11.25). A complex is formed

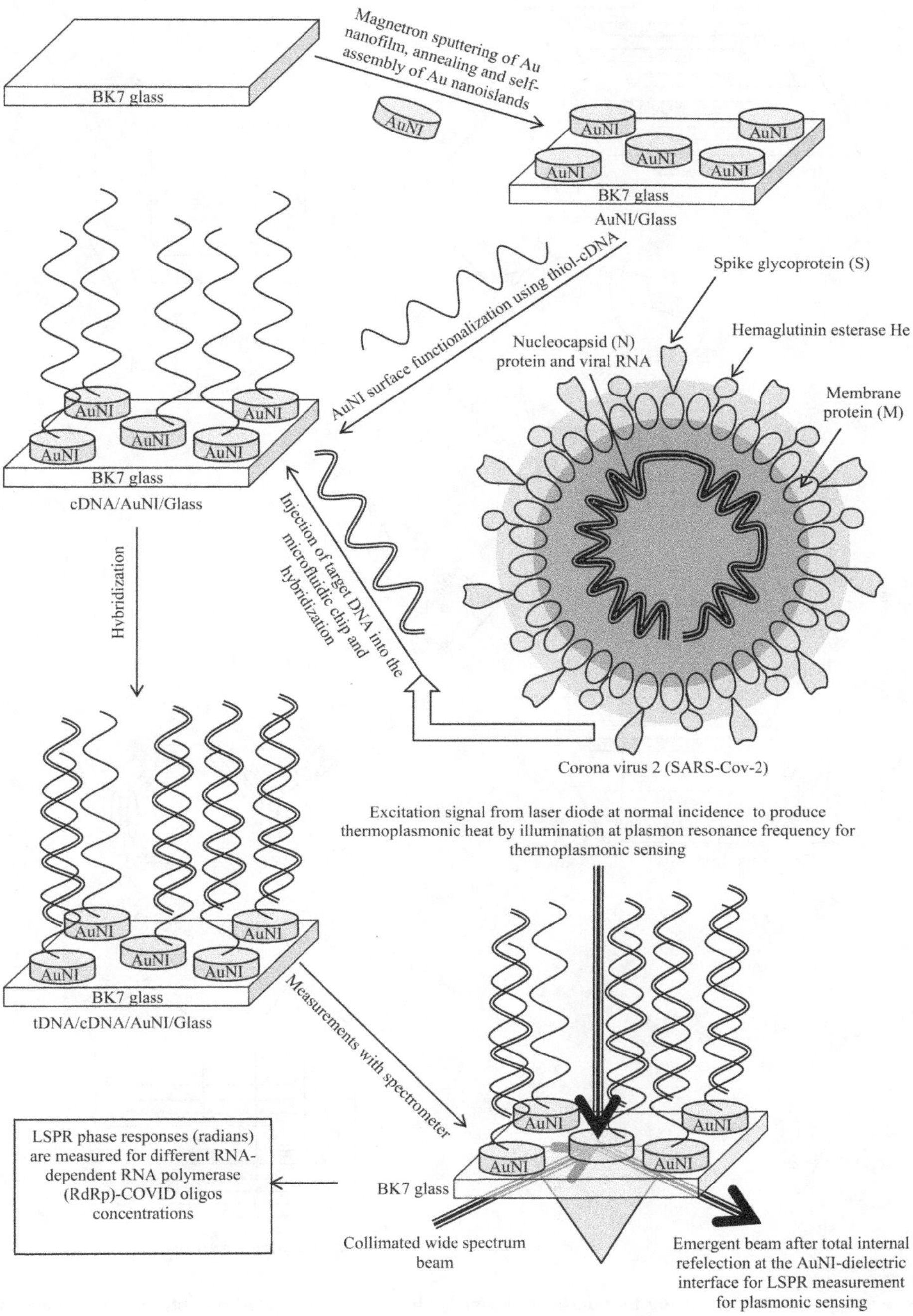

FIGURE 11.23 Photothermal effect-improved LSPR device for SARS-Cov-2 sensing (Qiu et al. 2020).

between GMNP and the acid-fast bacilli (AFB), if present in the sputum sample. As a result, the GMNP-AFB complexes are formed. The GMNP tube is placed in a magnetic rack when GMNP-AFB complexes are separated, and the supernatant is discarded. The GMNP-AFB complexes are washed and resuspended in PBS. They are kept on a slide and stained. The AFB are read and counted.

GMNPs help in extracting and concentrating the AFBs. Presence of the mycobacteria is indicated by red clumps surrounded by brown nanoparticles. It is a low-cost rapid test requiring < 20 min (Gordillo-Marroquín et al. 2018).

11.4.9 *Streptococcus pyogenes* Single-Stranded Genomic-DNA (*S. pyogenes* ssg-DNA)

Streptococcus pyogenes is the bacterium causing human rheumatic heart disease. It is detected using a single-walled carbon nanotubes (SWCNTs)-based chemiresistive genosensor (Singh et al. 2014).

(i) SWCNTs are ultrasonically dispersed in *N*, *N′* dimethyl formamide (DMF), aligned by AC electrophoresis across a pair of gold electrodes at a distance of 3 μm, and annealed at 300°C for 1 h in a reducing

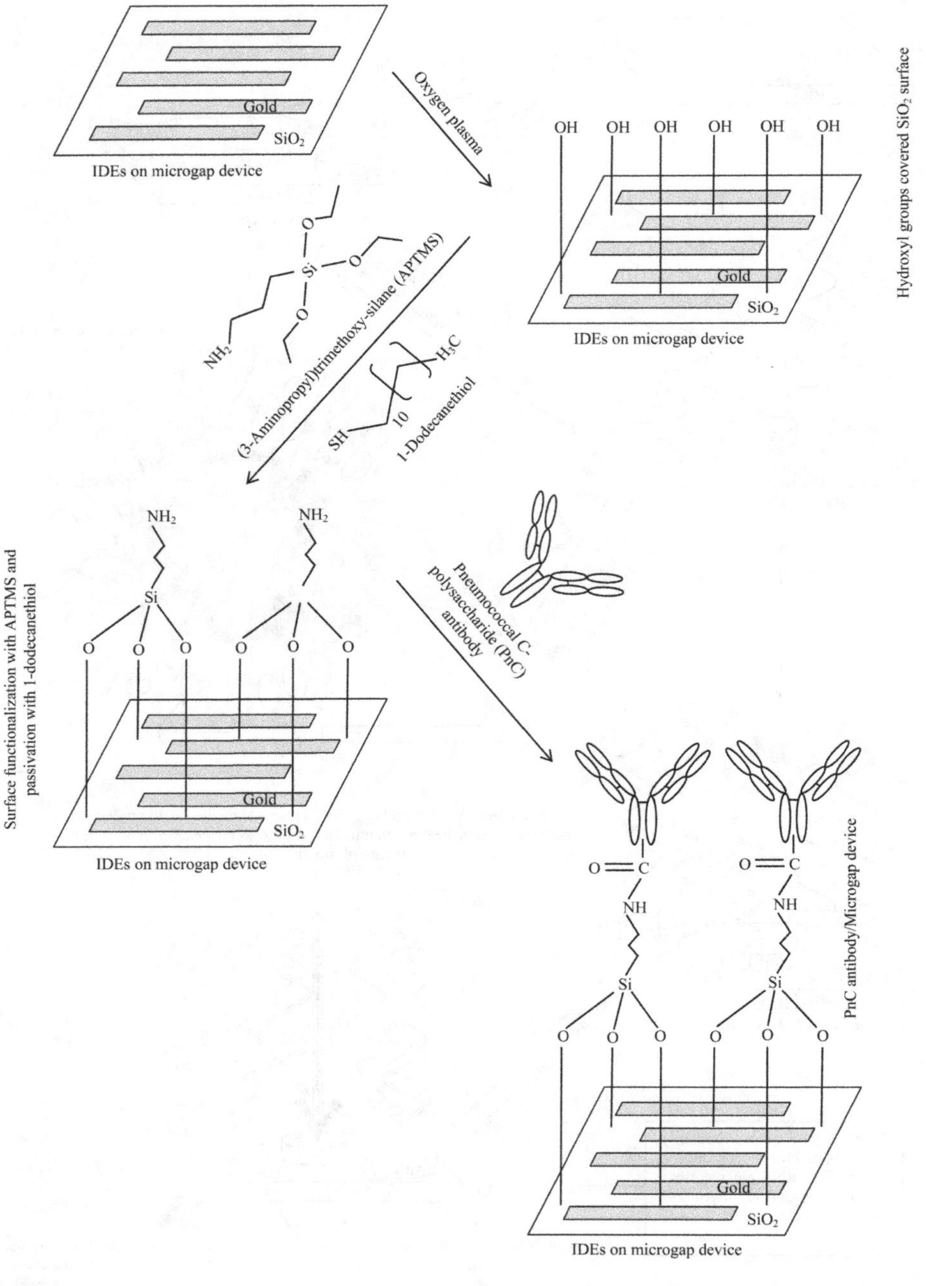

FIGURE 11.24 *Streptococcus pneumoniae* sensing by microgap device with bacteria decorated on antibody-conjugated gold nanoparticles (Pyo et al. 2017).

atmosphere, followed by incubation in 6-mercapto-1-hexanol (MCH) in DMF for 1 h to block non-specific sites (Figure 11.26).

(ii) After incubation of the aligned SWCNTs with the molecular bi-linker 1-pyrenemethylamine (PMA) for 2 h, washing with DMF, and drying, *S. pyogenes mga* gene-specific 24-mer ssDNA probe is covalently immobilized on the SWCNTs by incubation in TE buffer for 20 h. Non-specific binding sites are blocked with BSA. We get: 24-mer ssDNA probe/SWCNTs.

(iii) Different concentrations of ssg-DNA in TE buffer are dispersed on 24-mer ssDNA probe/SWCNTs and washed off after incubation for 2 min. Hybridization takes place between the target ssg-DNA and its complementary 24-mer ssDNA probe of *S. pyogenes* to give: ssg-DNA/24-mer ssDNA probe/SWCNTs.

The current-voltage characteristic of SWCNTs is measured in dry condition from −0.5 V to +0.5 V. With a rise in the concentration of ssg-DNA, the current decreases because the resistance increases. The normalized resistance bears a linear relationship

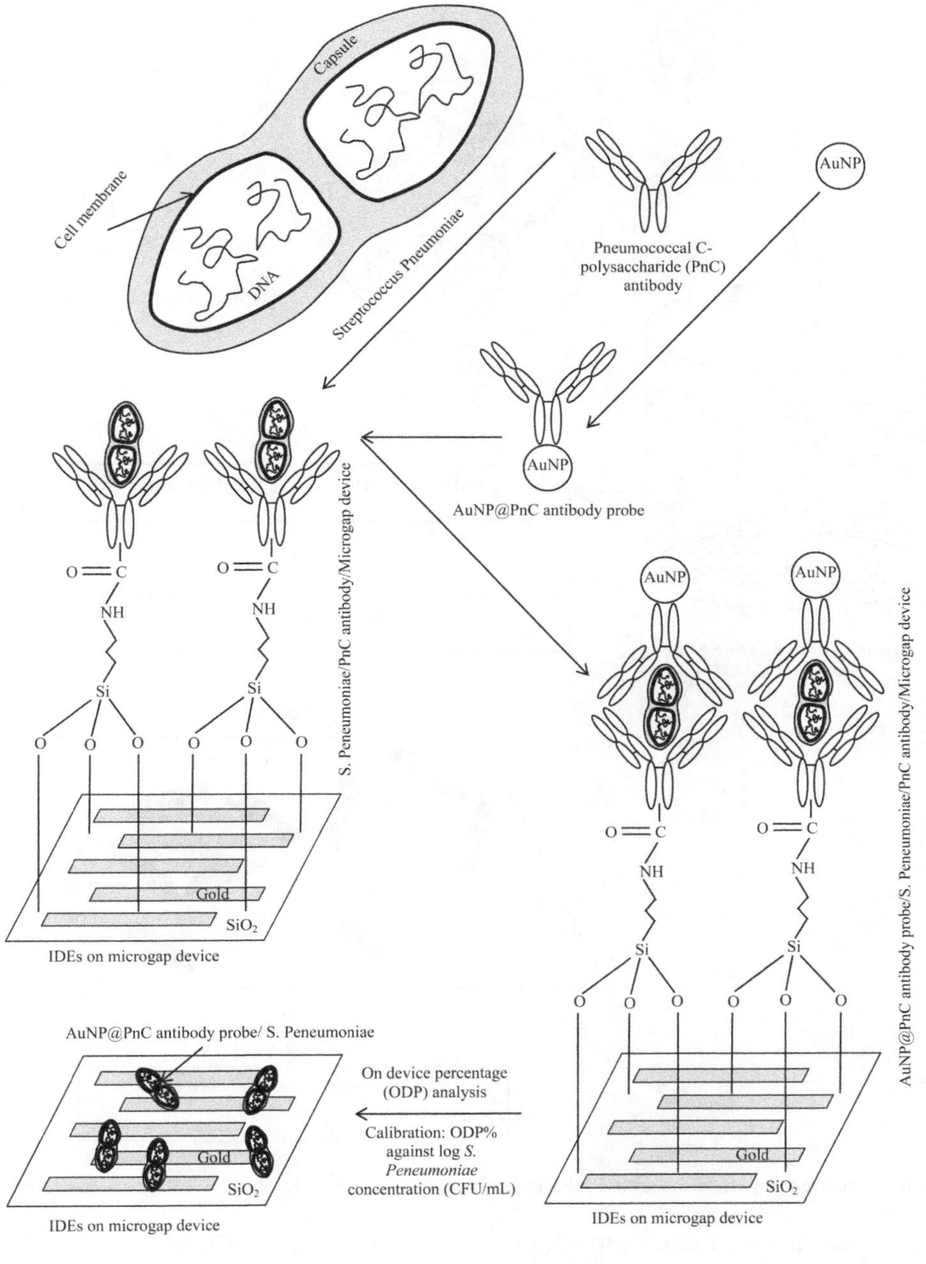

FIGURE 11.24 (Continued)

with the ssG-DNA concentration between 1–1000 ng/mL. By these measurements, the limit of detection is determined as 0.16 ng/mL. Selectivity of the sensor is demonstrated by its negligible response to a non-complementary DNA, the *Staphylococcus aureus* ssg-DNA (Singh et al. 2014).

11.4.10 *Plasmodium falciparum* Heat-Shock Protein 70 (PfHsp70)

Hsp70 is one of the two prominent proteins expressed by *Plasmodium falciparum*, the protozoan malaria-causing parasite. Gold nanoparticle (AuNP)-based competitive fluorescence immunoassay is used for detecting this parasite (Guirgis et al. 2012). The symbols used and the two cases described below are shown in Figure 11.27. The immunoassay progresses as follows:

(i) The citrate capping of AuNPs, synthesized by the citrate method, is exchanged with 11-mercaptoundecanoic acid (MUA) by incubation with MUA to form: AuNP-MUA.
(ii) AuNP-MUA is conjugated with anti-PfHsp70 monoclonal antibody (2E6) by: (a) covalent bonding using EDC/NHS or (b) electrostatic adsorption without EDC/NHS cross-linkage; approach (b) is found to provide higher response activity and linearity than approach (a). After BSA blocking of non-specific sites, we get: BSA-blocked-AuNP-MUA-2E6 conjugates.

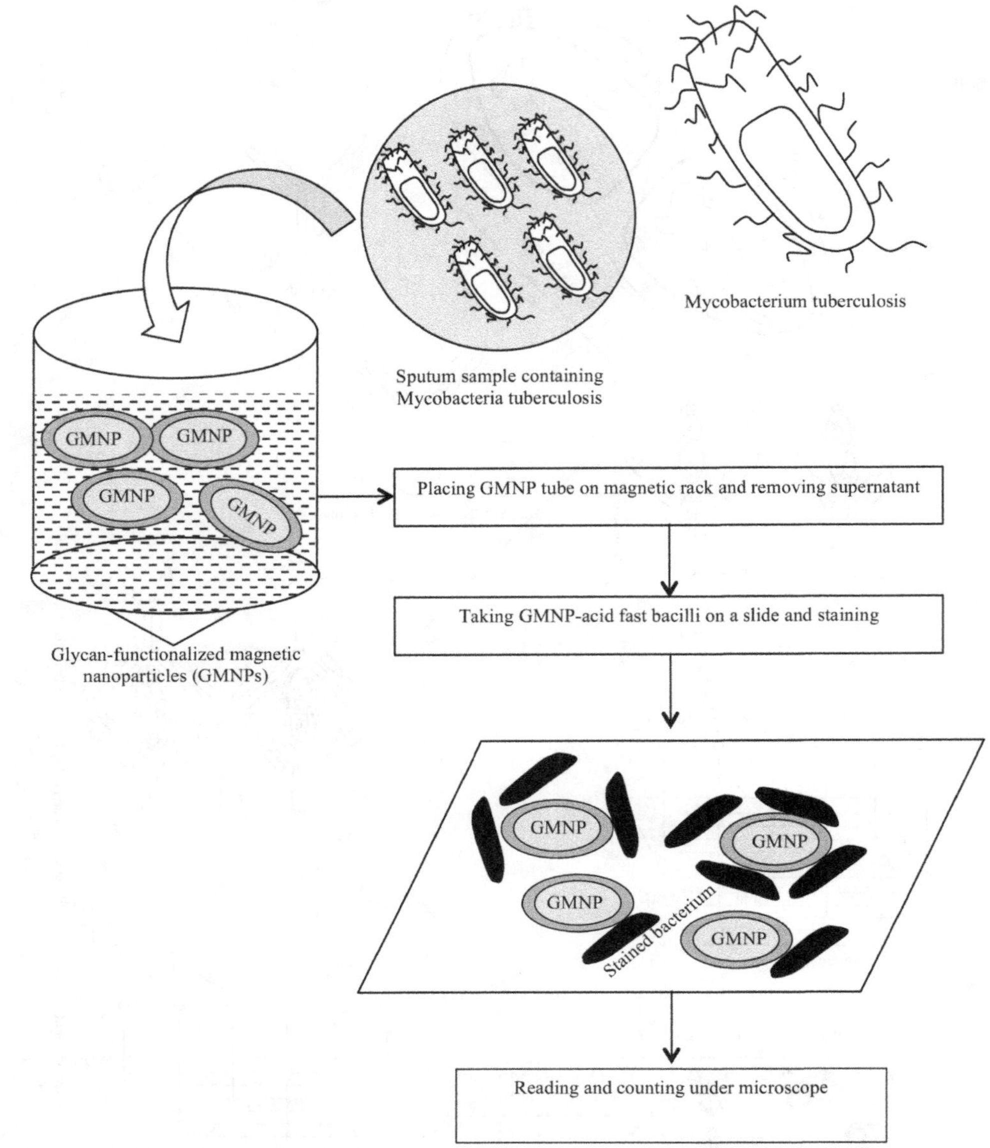

FIGURE 11.25 Glycan-functionalized magnetic nanoparticle colorimetric biosensing assay for *Mycobacterium tuberculosis* (Gordillo-Marroquín et al. 2018).

(iii) The PfHsp70 antigen is incubated with Cy3B (Cy3B-*N*-hydroxisuccinimidine monoester) fluorescent dye to get: Cy3B-labeled PfHsp70 antigen.

(iv) The fluorometric experiment is performed with the solution mixture: PfHsp70 antigen, BSA-blocked-AuNP-MUA-2E6 conjugates, and Cy3B-labeled PfHsp70 antigen, after 3 h of incubation.

Two cases are of interest:

Case I: BSA-blocked-AuNP-MUA-2E6 conjugates are mixed with Cy3B-labeled PfHsp70 antigen (PfHsp70 antigen absent): The optical density spectrum of BSA-blocked-AuNP-MUA-2E6 conjugates, centered at 520 nm, overlaps with the photoluminescence spectrum of Cy3B-labeled PfHsp70 antigen with an emission maximum at 580 nm. Hence, the photoluminescence intensity of Cy3B-labeled PfHsp70 antigen is low when BSA-blocked-AuNP-MUA-2E6 conjugates are present due to quenching of intensity by the BSA-blocked-AuNP-MUA-2E6 conjugates.

Case II: PfHsp70 antigen is added to BSA-blocked-AuNP-MUA-2E6 conjugates mixed with Cy3B-labeled PfHsp70 antigen. Competition occurs between the PfHsp70 antigen and Cy3B-labeled PfHsp70 antigen to bind with the 2E6 antibodies on the BSA-blocked-AuNP-MUA-2E6 conjugates. Therefore, a greater quantity of free Cy3B-labeled PfHsp70 antigen remains in solution leading to an increase in fluorescence intensity. The fluorescence increase (FI) is defined as

$$\mathrm{FI} = \left\{1 - \frac{\mathrm{Peak} - \mathrm{PI}(\mathrm{Blank})}{\mathrm{Peak} - \mathrm{PI}(\mathrm{PfHsp70\ antigen})}\right\} \times 100\% \quad (11.1)$$

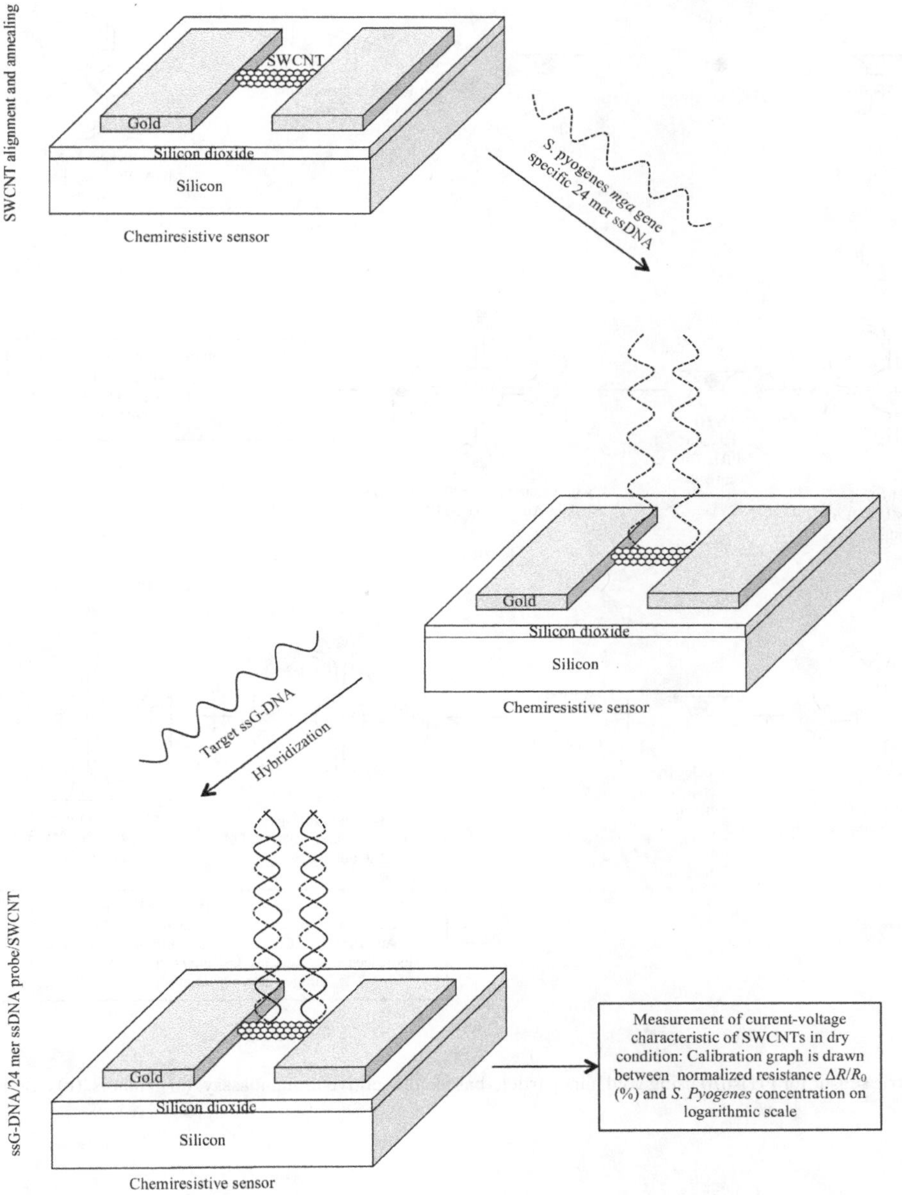

FIGURE 11.26 SWCNTs chemiresistive sensor for *Streptococcus pyogenes* bacterium (Singh et al. 2014).

where Peak-PI (PfHsp70 antigen) is the photoluminescence intensity with antigen and Peak-PI(Blank) is the same without antigen, both at 574 nm. Using electrostatically conjugated antibodies, the FI varies linearly with PfHsp70 antigen concentration from 8.2 to 23.8 μg/mL. The limit of detection is 2.4 μg/mL and the limit of quantification is 7.3 μg/mL (Guirgis et al. 2012).

11.5 Nanosensors for Automotive, Aerospace, and Consumer Applications

In continuation with the membrane-based CNT electromechanical pressure sensor (Section 4.12), and nanogap Pirani gauge (Section 5.17), we present more examples of pressure sensors here.

11.5.1 Strain/Pressure Sensors

11.5.1.1 Polymer-Metallic Nanoparticles Composite Pressure Sensor

This is a resistive-type pressure sensor, consisting of a sealed chamber containing an elastic polymer medium, such as a polymer network liquid crystal (PNLC), containing a dispersion of metallic nanoparticles, e.g., silver nanoparticles (AgNPs) (Hashimura et al. 2014). Electrical connections to the polymer medium are made through two electrodes (Figure 11.28). Electric current I flows through quantum-mechanical tunneling between the metallic nanoparticles at a mean distance d apart, as given by the equation

$$I = KeV\exp\left(-\frac{2\sqrt{2m\Phi}}{\hbar}d\right) \qquad (11.2)$$

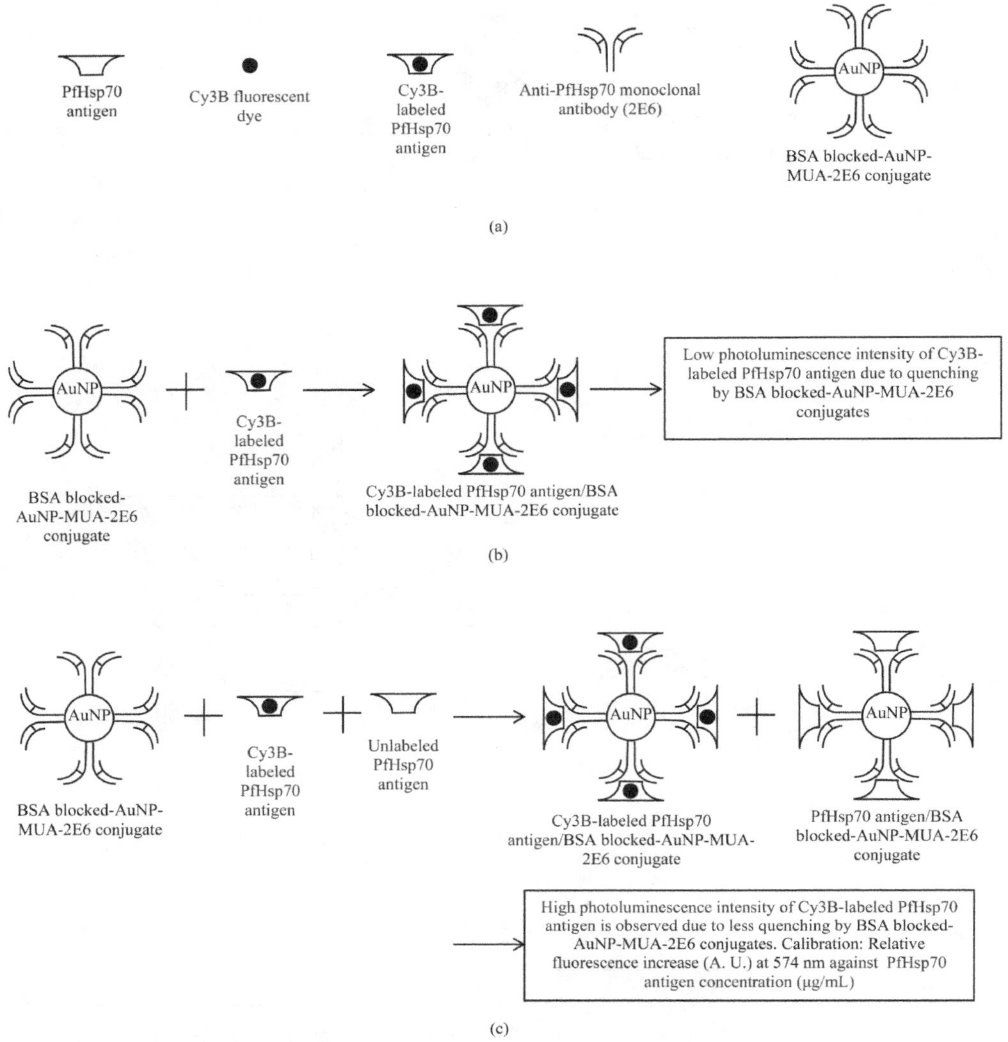

FIGURE 11.27 Fluorometric sensor for PfHsp70 using gold nanoparticle-based competitive immunoassay: (a) symbols, (b) case I, and (c) case II (Guirgis et al. 2012).

where K is a constant. The symbol e stands for the electronic charge, V is the applied bias, m is the mass of the electron, Φ denotes the work function of the metal from which NPs are made, e.g. −4 eV for AgNPs, and $\hbar$ is the reduced Planck's constant. The sensor finds applications in touchscreen displays.

When pressure is applied on the top surface of the sensor, the metallic nanoparticle-to-metallic nanoparticle distance decreases, resulting in an increase in tunneling current. Conversely on releasing the applied pressure the inter-nanoparticle distance increases, so that the tunneling current declines.

The metallic nanoparticle concentration in the polymer is determined by an important parameter, called the percolation threshold. The conductivity of a polymer-metallic nanoparticle composite, consisting of nanoparticle filler in the polymer matrix, is determined by the volume ratio of the filler. At a low volume ratio, the metallic nanoparticles are unable to aid in conduction so that the conductivity of the composite is very low, ~10^{-15} Ω-cm. With increasing volume ratio of filler, a stage is reached when the nanoparticles form tunneling-assisted linkages, thereby setting up a conducting path through the polymer matrix. The necessary filler content is referred to as the percolation threshold. Close to the percolation threshold, the conductivity of the composite is highly sensitive to filler content. Hence the interparticle distance plays a decisive role. A small change in filler content produces several orders of magnitude variation in this distance. But as we recede away from the threshold, this sensitivity is lost. So, the pressure sensor is designed to operate nearby the percolation threshold to maximize the sensitivity (Hashimura et al. 2014).

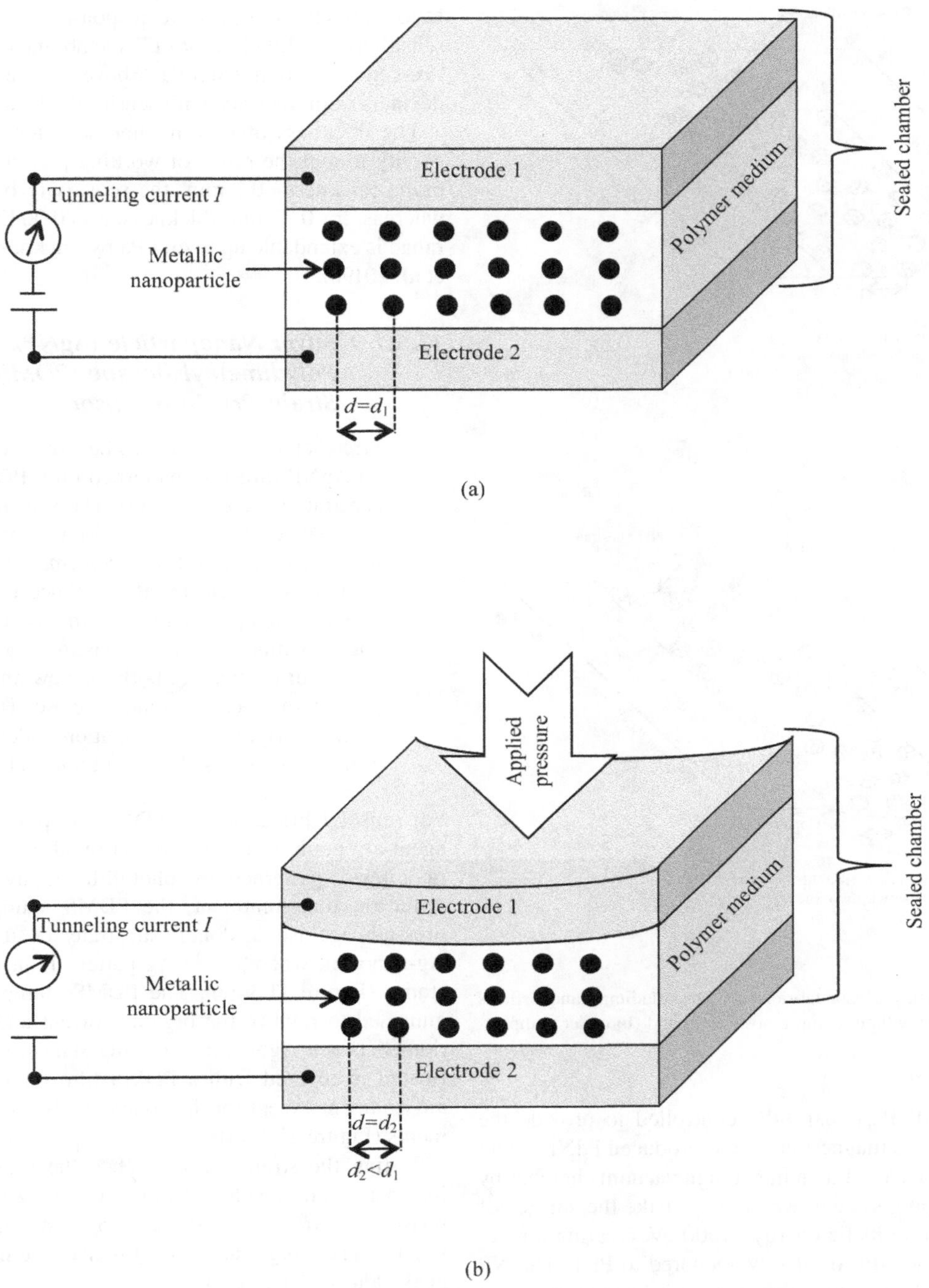

FIGURE 11.28 Pressure sensor using metallic nanoparticles suspended in a polymer medium: (a) without pressure and (b) under applied pressure (Hashimura et al. 2014).

11.5.1.2 Percolative Pd Nanoparticle (PdNP) Array-Based Pressure Sensor

Conventional piezoresistive MEMS sensors use doped silicon membranes in which carrier mobility varies with pressure. The metal nanoparticle sensor also works on the piezoresistive effect, but its operation is fundamentally governed by quantum-mechanical tunneling in percolation-based nanostructures, translating the pressure exerted on an elastic membrane into changes in tunneling conductance across NP percolating networks. It utilizes the ultra-sensitivity of tunneling conductance among nanoparticle arrays to changes in inter-particle separation by the pressure applied on a flexible membrane.

The sensor consists of Ag interdigitated (IDE) electrodes defined on a PET membrane (substrate) by shadow-mask evaporation, over which palladium NPs (PdNPs) are deposited from a magnetron plasma gas aggregation cluster source (Chen et al. 2019b). Figure 11.29 shows the membrane, the electrodes, and the nanoparticles. The distance between the IDEs is 15 μm with intervening resistance > $10^{10}\Omega$. PdNPs are preferred over AuNPs, being less coalescence-prone, and hence are capable of achieving greater stability.

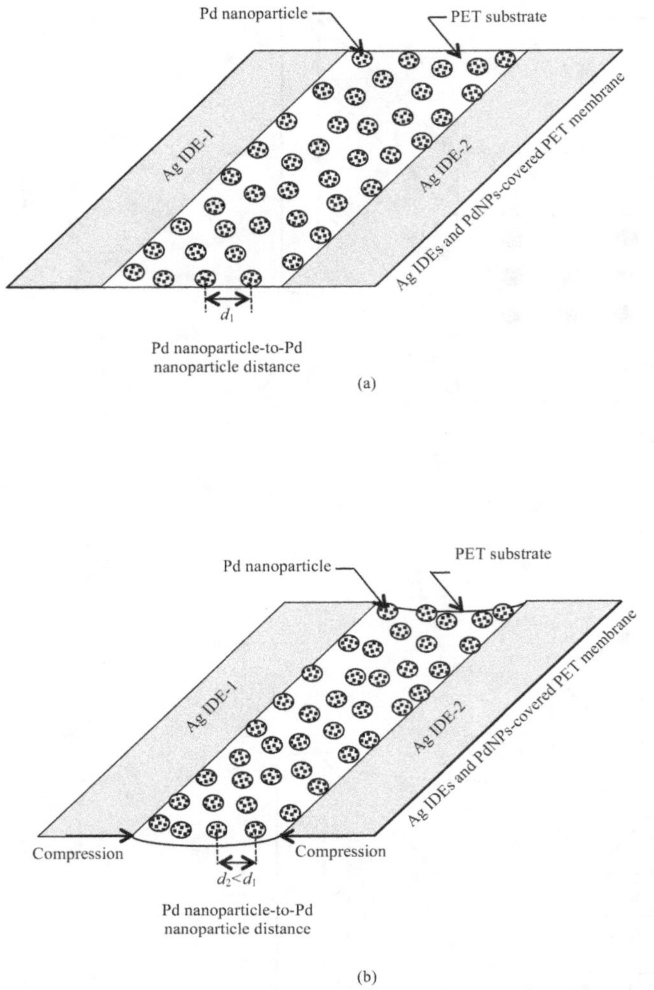

FIGURE 11.29 Pressure sensor fabricated with palladium nanoparticles on a PET substrate: (a) without compressive strain and (b) under compression (Chen et al. 2019b).

Deposition of PdNPs is carefully controlled to provide the desired operation. The magnetron source-produced PdNPs in an argon stream at 80 Pa are drawn into a high vacuum chamber by a differential pumping system, where they strike the surface of PET substrate with a kinetic energy ~1000 eV, creating a reactive site at which the NPs are firmly anchored to PET. The NP deposition is controlled by monitoring the conductance between the IDEs in real time by measuring the current at a fixed bias. The proximity to the percolation threshold is revealed from the evolution of conductance with deposition time.

The sensor is applied as a barometric altimeter with an altitude resolution of 1 m. Potential applications include automotive and consumer electronics.

If G_0 is the conductance in the absence of an applied differential pressure ΔP, and G is the conductance when ΔP is applied, the sensitivity S is obtained from the gradient of the pressure-response curve:

$$S = \frac{G - G_0}{\Delta P} = \frac{\Delta G}{\Delta P} \tag{11.3}$$

Below 60 kPa, the pressure response curve is linear, and the sensitivity is 0.129/kPa for PET membrane thickness = 0.05 mm. Pressure resolution is 0.5 Pa. Above 60 kPa, the pressure graph deviates from linearity, with sensitivity decreasing to 0.049/kPa.

The thickness of PET membrane determines both the sensitivity to and the range of working pressures. For PET membrane thickness = 0.1 mm, the sensitivity becomes 0.0247/kPa whereas, for 0.25 mm thickness, it is 0.0042/kPa. The working range is extendable up to 40 kPa by thickness adjustment (Chen et al. 2019b).

11.5.1.3 Silver Nanoparticle (AgNP)/ Polydimethylsiloxane (PDMS) Strain/Pressure Sensor

(a) Stretchable strain sensor: The strain-sensing element is an AgNP thin film patterned on a PDMS elastomeric substrate (Lee et al. 2014). The sensing mechanism is the change in electrical resistance of the AgNP thin film when subjected to a deformation. When a strain is applied, the electrical resistance increases because of the development of new microcracks in the film and widening of already existing microcracks. When the strain is released, both the new and the previously present microcracks tend to close. Hence, resistance changes originate from creation/widening and closure of microcracks, as shown in Figure 11.30(a).

For sensor fabrication, the PDMS pre-polymer and the curing agent are poured into the silicon mold with the SU-8 negative photoresist patterned by photolithography (Figure 11.30(b)). After thermal treatment, the PDMS stamp is detached and pressed against a donor substrate coated with ion-based Ag-nano-ink whereby the Ag pattern is printed on the PDMS stamp (Figure 11.30(c)). The PDMS stamp is withdrawn and annealed to remove the organic solvent and establish connections between AgNPs by sintering. The strain sensor thus fabricated is covered with a PDMS film for protection from the environment except for the contact pads for electrical measurements (Figure 11.30(d)).

Testing the strain sensor: When the sensor is subjected to increasing strain values from ε=0% to 20%, the normalized resistance ($\Delta R/R_0$) is 0.06 at ε=5% and increases to 0.41 at ε=20%. The gauge factor (GF) is maximum at ε=20% and the peak value of GF is 2.05.

When the strain sensor is tied on the top surface of the wrist band and the wrist is bent downward, the induced strain is tensile, and the resistance of the sensor increases. On upward bending of the wrist, the strain becomes compressive and the sensor resistance decreases. The ability to sense tensile and compressive strains makes the sensor useful in detecting wrist motion. In this way, the sensor can be applied to real-time, parallel sensing of multiple human motions over a broad range of strain magnitudes in wearable electronics systems.

(b) Pressure sensor: The pressure sensor (Figure 11.30(e)-(f)) consists of a 100–500 μm thick PDMS membrane as a deformable diaphragm and a spiral AgNP thin film as the piezoresistive element (Lee et al. 2014).

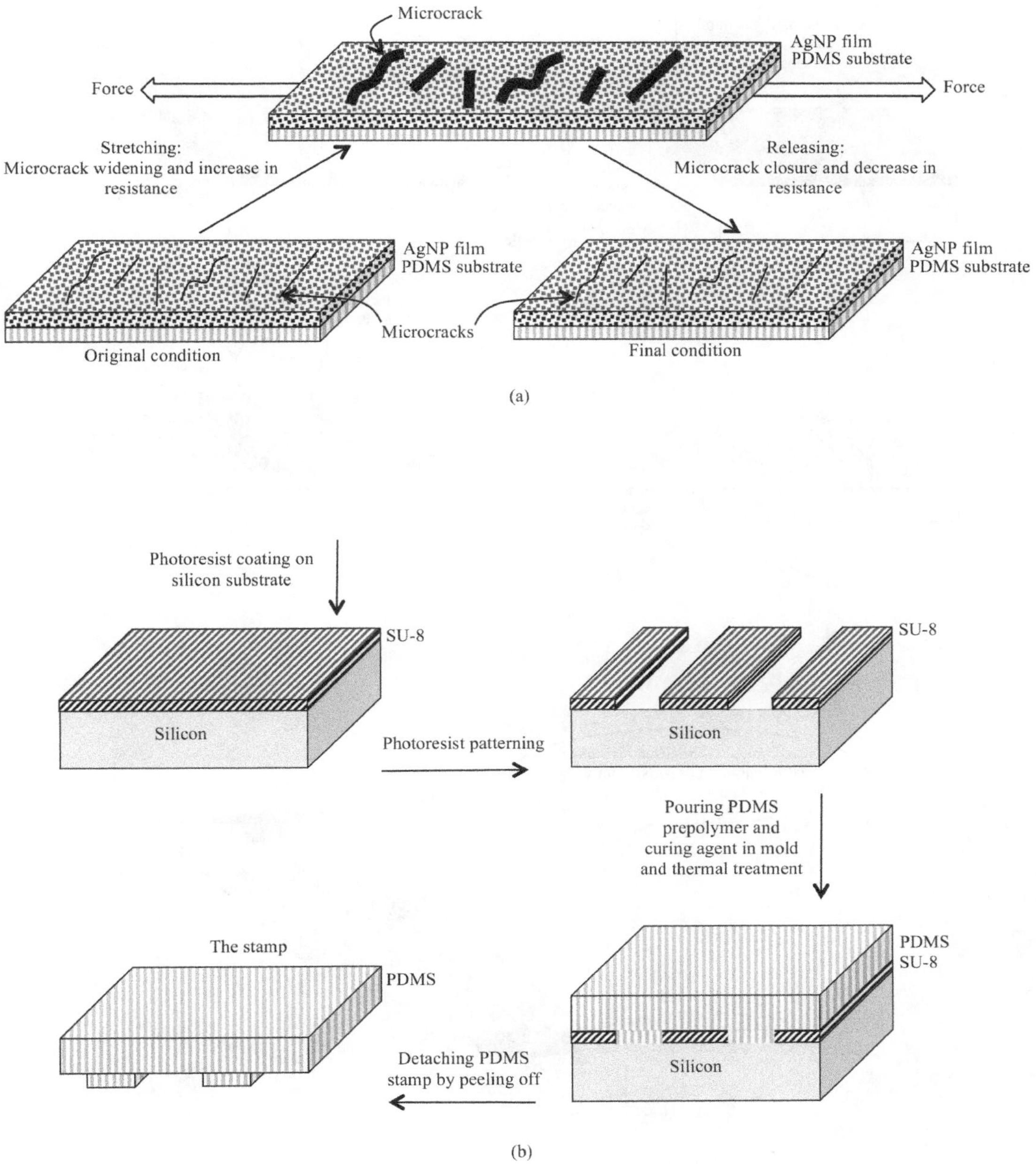

FIGURE 11.30 Strain/pressure sensor: (a) principle of the sensor, (b) fabrication of PDMS stamp, (c) use of stamp to print Ag nano-ink pattern, (d) packaged strain sensor, (e) pressure sensor without applied pressure and (f) pressure sensor with membrane subjected to a downward pressure (Lee et al. 2014).

The response and recovery times of the sensor pronounce how quickly it detects the signal and how quickly it relapses to its original state. The response time of the pressure sensor is 1 second, and its recovery time is 0.5 s. The sensitivity, given by the normalized resistance ratio ($\Delta R/R_0$), is 0.35. A comparison of sensitivities may be done between sensors made with 100 μm and 500 μm diaphragm thicknesses. This difference is expressed in terms of the normalized resistance ($\Delta R/R_0$) values for the two sensors at a fixed pressure of 3 kPa. For a sensor with a 100-μm thick diaphragm, $\Delta R/R_0$ is 28 times higher than for a sensor with a 500-μm thick diaphragm.

A possible application of this sensor is measurement of blood pressure on a stent and catheter.

11.5.1.4 Polyacrylamide (PAAm)/Gold nanoparticle (AuNP) Pressure Sensor

The stress-responsive sensor consists of a PAAm film embedded with AuNPs (Figure 11.31). The AuNPs, obtained by reduction of the $HAuCl_4$ salt, are sterically stabilized by poly(vinyl pyrrolidone), PVP (Topcu and Demir 2019). The dispersion of AuNPs has a deep red color. With introduction of PAAm, the PAAm/AuNP film acquires a saturated blue appearance due to aggregation of AuNPs.

The working range extends up to 160 MPa. On application of pressure, the blue color turns to ruby red depending upon the magnitude of pressure. These vivid color changes occur due to

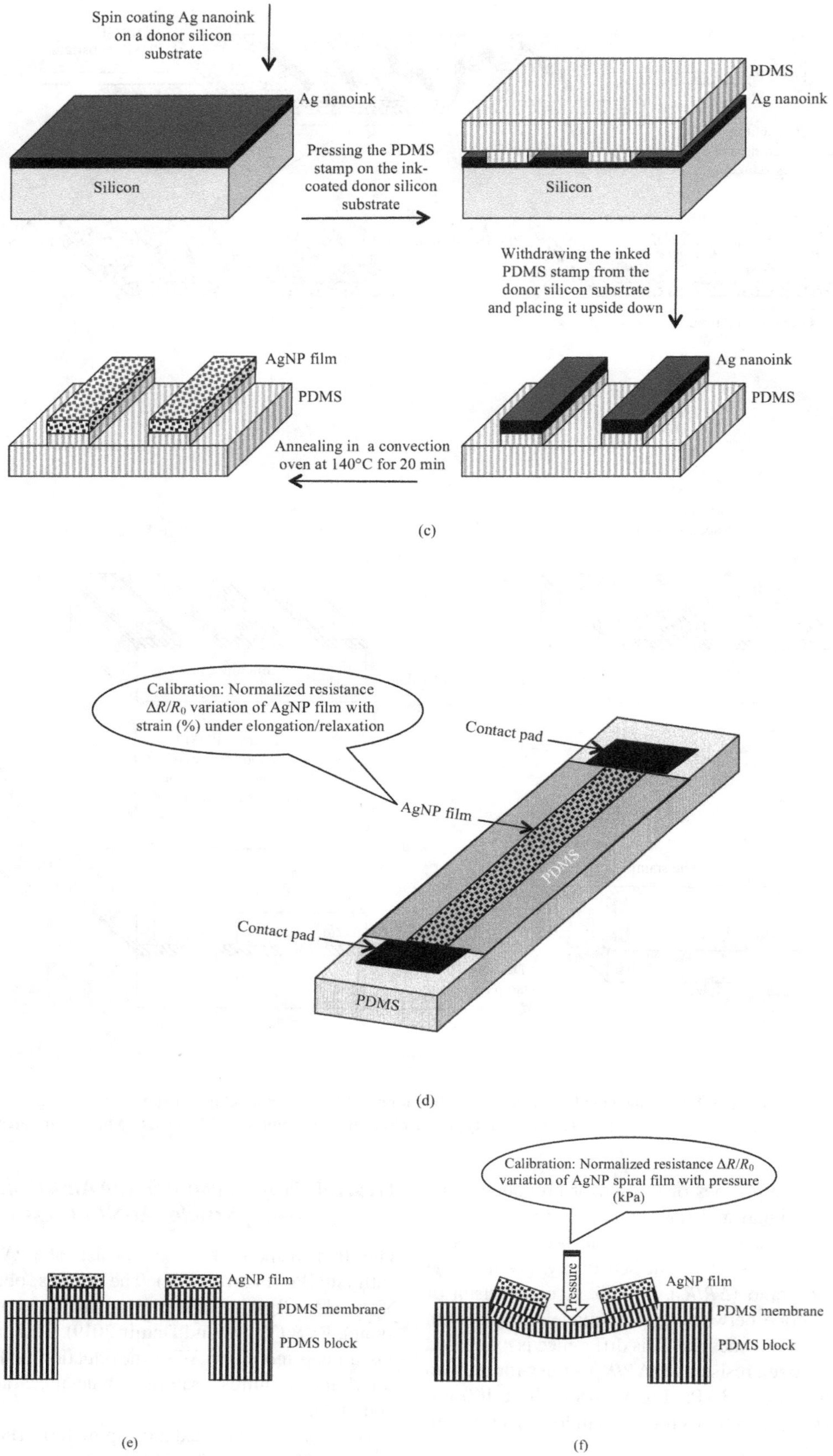

FIGURE 11.30 (Continued)

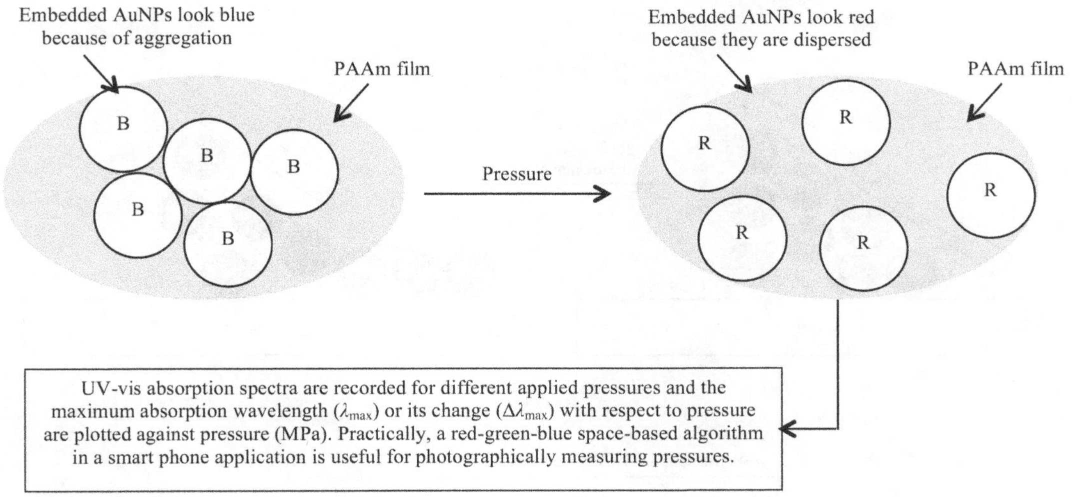

FIGURE 11.31 PAAm/AuNPs pressure sensor (Topcu and Demir 2019).

disassembly of Au nanoparticle clusters. At a high temperature, the viscosity of PAAm decreases, slightly influencing the color changes. Smartphone-aided assessment of pressure is carried out using a specially designed simple red-green-blue (RGB) space-based algorithm for color quantification. Interesting applications are soft robotics, electronic skin, and touchscreen displays (Topcu and Demir 2019).

11.5.2 Acoustic Vibration Sensor

A gold nanoparticle (AuNP) monolayer-based acoustic vibration sensor is used (Zhang et al. 2014; Yi et al. 2015). The active unit is one strip of AuNP monolayer deposited on a PET substrate. For high sensitivity, the nanoparticle diameter of 100 nm is chosen, noting that the sensitivity improves by 15 times when the diameter increases from 15 to 97 nm. The AuNPs are synthesized by seed-mediated growth from gold (III) chloride trihydrate ($HAuCl_4$) and capped with cetyltrimethylammonium bromide (CTAB) surfactant to prevent aggregation. A uniform monolayer of AuNPs is formed on the PET substrate by convective assembly at the confined contact angle, which involves nanoparticle convection and self-assembly, prompted by evaporation of solvent near the interface between the colloidal gold and the supporting substrate (Zhang et al. 2014). Figure 11.32 (a) shows the self-assembly of AuNPs. Au electrodes are deposited 90-μm apart on the AuNP monolayer. Figure 11.32 (b) shows the arrangement for studying the sensor response.

Dynamic response of the sensor is characterized by alternately loading and unloading with mechanical strains at frequencies in the range 1–20000 Hz and measuring the resistance of the monolayer film. Ultrafast dynamic response of the resistance to external stimulus demonstrates the capability of reliable detection of acoustic vibrations by this sensor. Applications include ultrasensitive pressure sensors and safe-ingress sentinel systems (Zhang et al. 2014; Yi et al. 2015).

11.5.3 Acceleration Sensor

Following the tunnel-effect accelerometer (Section 4.13), the NEMS accelerometer (Section 4.14), the silicon nanowire accelerometer (Section 4.15), and the thermal convective accelerometer with a CNT-sensing element (Section 5.14), we consider here an ultra-long vertically aligned barium titanate ($BaTiO_3$) nanowire (NW) arrays-based accelerometer reported by Koka and Sodano (2013). The sensor utilizes the piezoelectric effect to detect acceleration from mechanical vibration. Figure 11.33 shows the construction of the sensor. $BaTiO_3$ NWs (45 μm long) are grown by a two-step hydrothermal process: (i) sodium titanate NW arrays are grown on a Ti substrate by a hydrothermal reaction with Ti; and (ii) the sodium titanate NW arrays thus formed are immersed in barium hydroxide octahydrate solution to perform another hydrothermal growth in which barium ions diffuse into the sodium titanate NWs to yield barium titanate NWs.

The barium titanate NW arrays are stripped from the Ti substrate and transferred to a borosilicate glass substrate, to which they are bonded by silver epoxy. The silver film acts as the bottom electrode. The top electrode is a solder foil fixed on the upper surface of the NWs. The solder foil, weighing 16 mg, serves as the proof mass. For polling the $BaTiO_3$ NW arrays, an electric field of 75 kV/cm is applied between the silver film and solder foil electrodes for 12 h.

After fixing the base of the sensor on a permanent magnet shaker, the shaker is turned on when the inertia from the top electrode exerts a time-varying compressive and tensile stress on the $BaTiO_3$ NWs. The frequency, direction, and amplitude of this stress determine the piezoelectric potential produced across the NWs. So, the stress on the NWs varies as the acceleration of the base of the sensor. Therefore, the piezoelectric voltage generated across the NWs is proportional to the acceleration of the base of the sensor.

The sensitivity is 50 mV/g, which is much higher than that of the ZnO NW sensor (2.5 mV/g). The operating bandwidth extends up to 10 kHz (Koka and Sodano 2013).

11.5.4 Orientation, Angular Rate, or Angle Sensors

11.5.4.1 CNT Field-Emission Nano Gyroscope

The CNT gyroscope works on the Coriolis effect (Yang et al. 2012). When the CNT vibrates along the *X*-direction, rotation

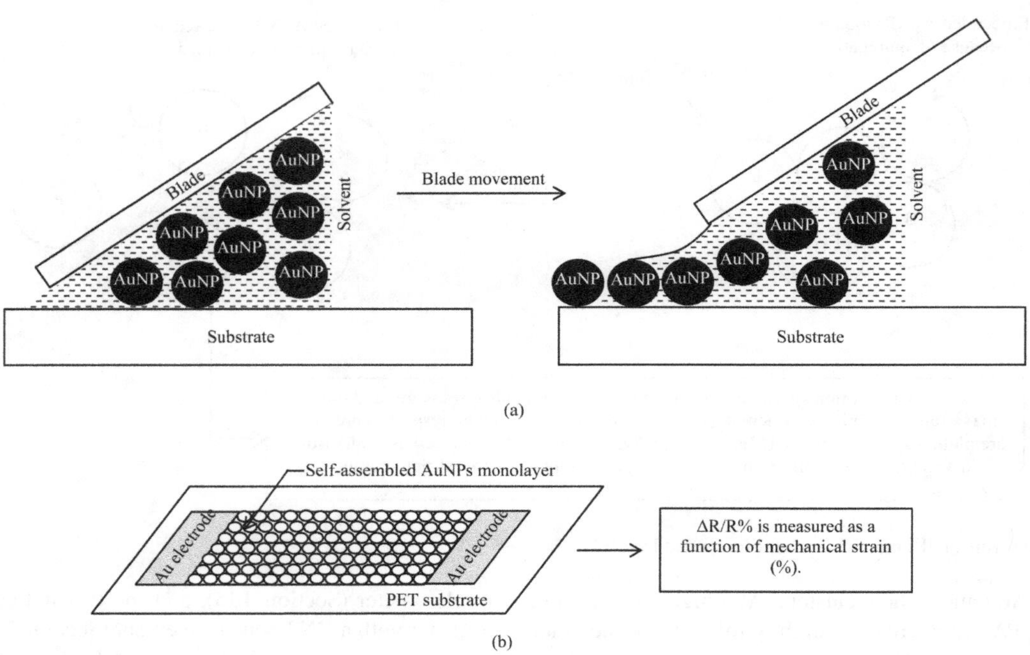

FIGURE 11.32 (a) Contact angle triggered self-assembly of gold nanoparticles. The assembly is stimulated on or inhibited off by varying the contact angle across a threshold level of 4.2°. (b) Arrangement of the active unit of the strain gauge (Yi et al. 2015; Zhang et al. 2014).

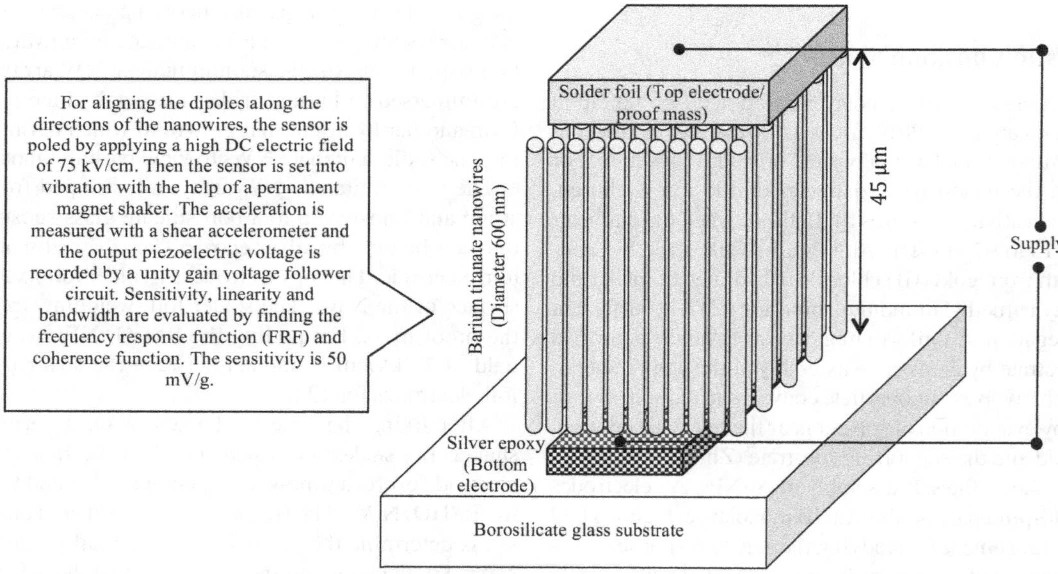

FIGURE 11.33 Barium titanate nanowire acceleration sensor (Koka and Sodano 2013).

around the *Z*-direction leads to acceleration and hence the Coriolis force in the *Y*-direction.

It is a vibratory gyroscope in which a vibrating CNT is used to determine the angular rate. The field emission property of CNT is utilized for measuring frequency and amplitude of the vibration. Changes occurring in vibrational parameters due to rotation help in deduction of the angular rate. Field emission, also called cold emission, is actually high-field electron emission, entailing the extraction of electrons from a surface by application of a large electrostatic field. The phenomenon facilitating this extraction is tunneling of electrons through the surface potential barrier. CNTs are bestowed with exceptional physical and chemical properties. Possession of these properties favors field emission from CNTs. The CNTs are mechanically strong. They also display high chemical stability. In addition, they have nanoscale diameters and long lengths, resulting in high width-to-height or aspect ratios. The CNT gyroscope offers advantages of low energy consumption. It is also capable of high resolution and can provide a large scale of measurement.

A 1-μm long CNT is fixed to the microcantilever of AFM (Figure 11.34). An AC voltage of 2 V applied between the CNT and an electrode positioned close to and facing the CNT tip forces the CNT into vibration at its resonance frequency; a DC voltage < 100 V is applied for field emission. The natural frequency of CNT vibration is determined by its inner/outer diameter and length. A CNT with a resonance frequency of 1 MHz and a quality factor of 400 is tested at a rotational speed of 20 rad/second (Yang et al. 2013). The gyroscope is highly sensitive to rotational speeds up to 100 rad/s.

11.5.4.2 Magnetic Nanoparticles-Based Gyroscopic Sensor

The sensor consists of a container with a liquid, such as water or toluene, in which magnetic nanoparticles~10 nm diameter are dispersed (Krug and Asumadu 2018). The stochastic Brownian motion of the particles due to bombardments of surrounding molecules of the liquid maintains the nanoparticle distribution and prevents them from settling to the bottom of the container due to gravity.

Each nanoparticle acts as a miniature gyroscope. These miniature gyroscopes act together to constitute a single gyroscope, which can be used to monitor the angular deviation of an aeroplane or spacecraft.

The working of the gyroscope can be understood from Figure 11.35. An alternating current is supplied from a power source to the Helmholtz conductive coils wound around magnetic cores. This current generates a rotating magnetic field, which sets the nanoparticles into rotatory motion along axes in consonance with the rotating magnetic field, and at the same frequency as the current in the coils. An alternative arrangement to Helmholtz coils consists of two straight insulated conductors, oriented perpendicular to each other and carrying 90° out-of-phase alternating currents.

When an external force acts on the system to produce angular velocity or acceleration, the coils, and hence the rotating magnetic field, undergo angular movement or tilting. The force causes changes in the axes of rotation of the magnetic nanoparticles. The shift in the axes of rotation of magnetic particles around their previous positions is called precession. The rotating magnetic field engenders a force, which has a tendency to prevent the precession and maintain the axes of rotation the particles at a constant orientation, with respect to the axis of the system. The resulting changes in the rotating magnetic field are detected in the form of small changes in the input alternating currents in the conductive coils. These changes are correlated with angular movement or orientation of the coils. The input currents are measured by current sensors. A processor connected to the current sensors determines the angular movement.

In the initial stage, the rotational axes of magnetic nanoparticles are misaligned with the axis of rotation of magnetic field. Owing to the loss of magnetic nanoparticle spin torque, the input current diminishes. Afterwards, the rotational axes of magnetic nanoparticles regain alignment with the axis of rotation of the magnetic field and the input current increases until equilibrium is attained. Thus, the changes in input current last as long as precession of magnetic nanoparticle motion occurs and cease when the precession stops. The greater the extent of precession, the greater are the input current variations.

The frequency of oscillation and viscosity/composition of the solution are controlling parameters to optimize gyroscope performance (Krug and Asumadu 2018).

11.5.5 Ultrasound Sensor

An optically activated/read nano ultrasonic transducer was developed on the basis of the production of acoustic waves following absorption of light by a material (Smith et al. 2015). It consists of an optically transparent film sandwiched between partially transparent metal films, such as the structure: gold film (thickness 30 nm)-indium tin oxide film (ITO, thickness 150 nm)-gold

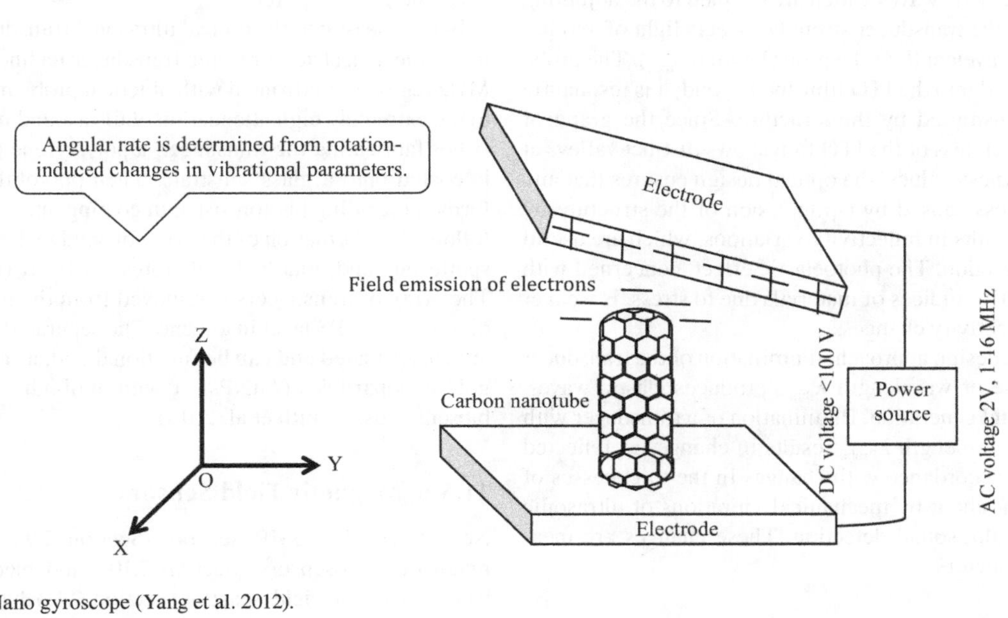

FIGURE 11.34 Nano gyroscope (Yang et al. 2012).

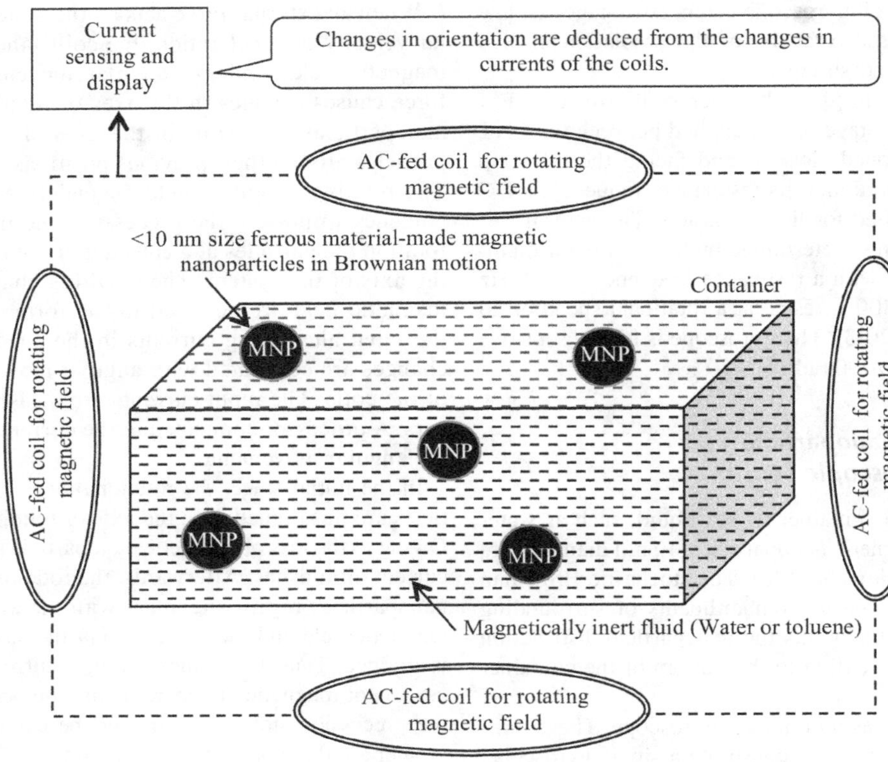

FIGURE 11.35 Conceptual diagram of magnetic nanoparticle gyroscope (Krug and Asumadu 2018).

film (thickness 30 nm), resembling a Fabry-Perot interferometer (Figure 11.36a). The transducer is made in a range of sizes with lateral widths from 5 μm to 15 μm. It works as an optical and ultrasonic resonator at the same time. Because it uses short laser pulses for ultrasound generation and detection, it falls under the purview of picosecond laser ultrasonics.

Optical design considerations for this transducer are: (i) the transducer strongly absorbs light of one wavelength, the wavelength of the pump beam (λ_{Pump}), so that the transducer efficiently produces thermoelastic waves which are coupled to the adjoining medium; and (ii) the transducer strongly reflects light of another wavelength, the wavelength of the probe beam (λ_{Probe}). The probe beam is transmitted into the ITO film and beyond; it is resonantly reflected and transmitted by the structure. Since the graph of reflectivity vs. thickness of the ITO film shows distinct valleys at certain film thickness values, the optical design ensures that any change in thickness caused by compression of the structure by an elastic wave results in reflectivity variations, which are useful for ultrasonic detection. The photoelastic effect, concerned with changes in refractive indices of materials due to stress, is weaker here than the reflectivity changes.

With the above design approach, illumination of the transducer with a laser pulse of wavelength λ_{Pump} produces elastic waves, which is ultrasound generation. Illumination of a transducer with a laser pulse of wavelength λ_{Probe} results in changes in reflected light intensity in accordance with changes in the thicknesses of the films, brought about by mechanical vibrations of ultrasonic waves, allowing ultrasound detection. These changes are measured by photodetectors.

The frequency of ultrasound produced by the photoacoustic transducer is determined by the optical absorption in the structure and the range of frequencies in the pump laser pulse beam. Ultrasound frequencies up to 100 GHz are achieved. The lateral size of the ultrasound beam produced is dictated by the size of the optical spot. As a result, the lateral resolution of ultrasound is limited, despite the fact that the wavelength of ultrasound is << wavelength of light waves. Additionally, with the efficiency of optical generation and detection being low, a larger amount of laser energy is required.

In comparison to the optical ultrasonic transducer, the conventional piezoelectric ultrasonic transducer technology used in the MHz region is confronted with practical problems, primarily due to the extremely high attenuation of ultrasound in the GHz range.

For fabricating the transducer, a polystyrene (PS) buffer layer is coated onto a glass substrate. Then photolithography is performed including photoresist spin coating and pattern definition, followed by formation of the array of sandwich structures by DC sputtering, and, finally, by photoresist lift-off (Figure 11.36(b)). The array of transducers is removed from the glass substrate by dissolving the PS layer in toluene. The separated transducer chips are encapsulated and can be functionalized, as needed, e.g., with gold nanoparticles (AuNPs) or with antibodies for chemical or biosensor use (Smith et al. 2015).

11.5.6 Magnetic Field Sensor

Needle-type SV-GMR sensors (Section 7.9), superconductive magnetic nanosensors (Section 7.10), and electron tunneling-based magnetic field sensors (Section 7.11) have already been

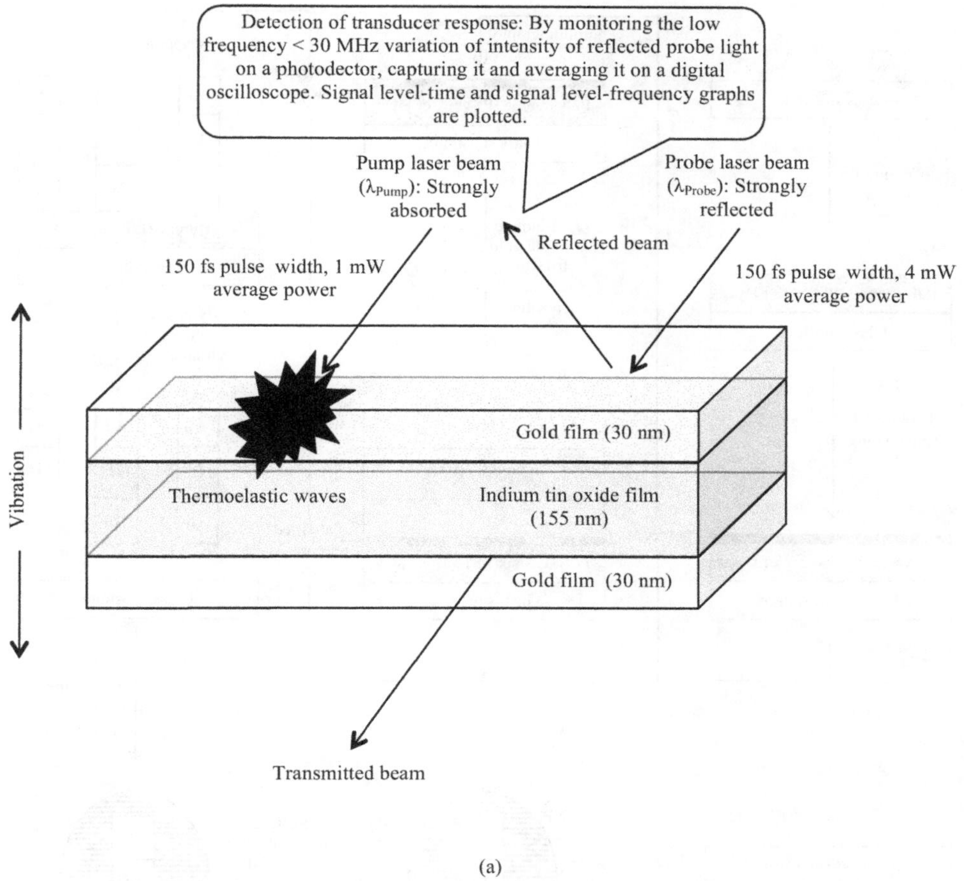

FIGURE 11.36 Optically excited ultrasonic transducer: (a) principle and (b) fabrication process flowchart (Smith et al. 2015).

described. Here, some other types of magnetic field sensors are examined.

11.5.6.1 Nanoparticle Core-Based Fluxgate Magnetometer

The use of magnetic nanomaterial to build the core of a fluxgate magnetometer provides high magnetization and negligible hysteresis. By provision of high permeability and a hysteresis-free core, the power losses, and therefore the power consumption, are significantly reduced. Moreover, higher bandwidths are obtained, owing to the lowering of eddy current losses by the high-resistivity nanopowder.

The fluxgate magnetometer is a high-precision magnetic field sensor, capable of measuring weak magnetic field strengths, e.g., the small variations in the Earth's magnetic field (Kennedy et al. 2014). The capability of the flux-gate magnetometer is limited by the properties of its magnetic core material. For the magnetometer core, iron oxide nanopowder is synthesized in an oxygen-rich low-pressure environment by the arc-discharge method. After sieving the powder to remove large-sized Fe residues, it is pressed into pellets. The magnetometer made with a nanoparticle core is shown in Figure 11.37.

The saturation magnetization of the nanopowder is 68 emu/g and the hysteresis is 2 mT. The sensitivity of the single-core magnetometer increases with frequency up to 30 kHz but decreases at higher frequencies due to the impedance of the coils and eddy current losses. This sensitivity value is adequate to monitor the variations in the terrestrial magnetic field. The three-axis magnetometer module gives identical and repeatable results at given locations when walking around a room, showing its suitability for magnetic field mapping in indoor locations (Kennedy et al. 2014).

11.5.6.2 Magnetic Nanoparticle (MNP)-Functionalized Magnetometer

The sensor works on the refractive index (RI) changes induced in a distribution of MNPs dispersed in a liquid near the gold film of a surface plasmon resonance (SPR) device (Chen et al. 2019a). It consists of a gold film-coated few-mode side-polished fiber immersed in a suspension of MNPs in a liquid, called a magnetic fluid (MF) or ferrofluid, which itself is enclosed inside a capillary tube (Figure 11.38). The carrier liquid is water and the Fe_3O_4 nanoparticles have an average diameter of 10 nm. The anisotropic aggregation of MNPs is influenced by the intensity and orientation of an externally applied magnetic field. In turn, the change in anisotropic MNP aggregation alters the refractive index adjoining the SPR device. The SPR device measures the changes in surface plasmon wavelength resulting from the new refractive index, due to modified direction-dependent clustering of MNPs. These wavelength changes are finally used to determine both the magnetic field magnitude and orientation.

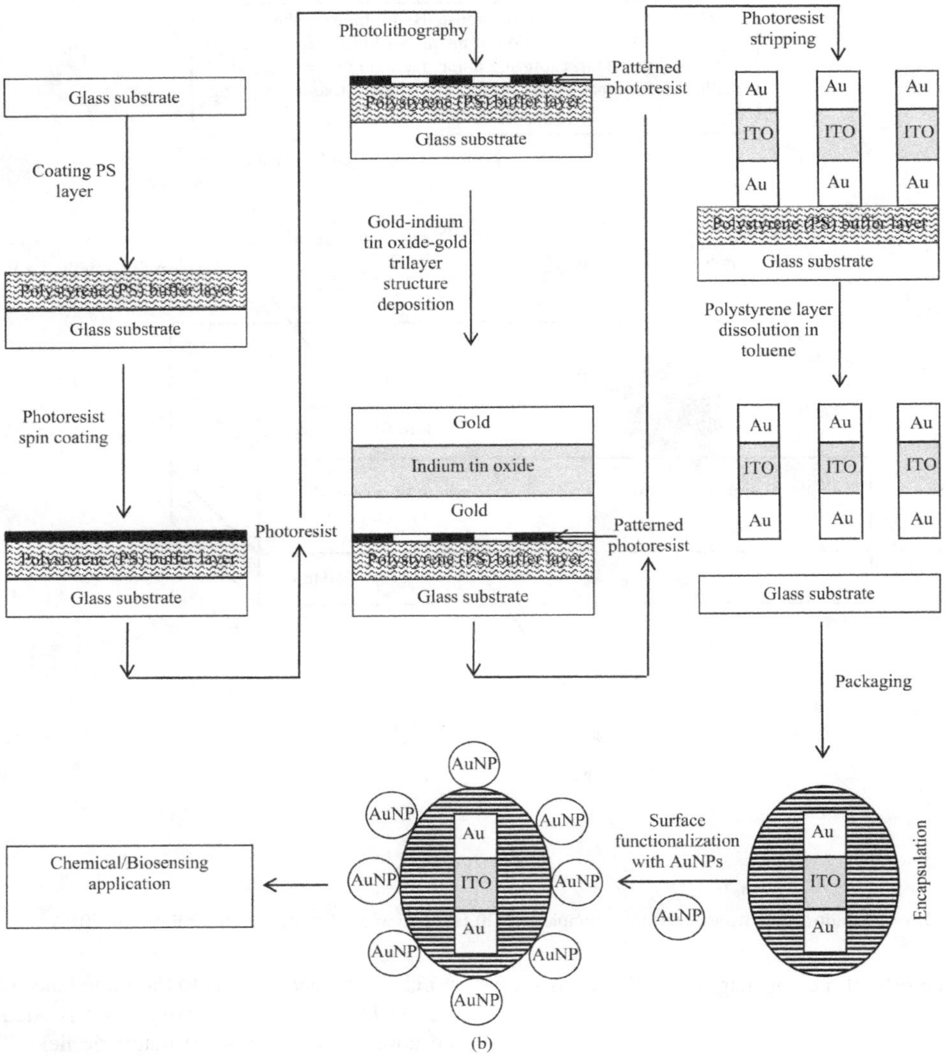

FIGURE 11.36 (Continued)

Knowledge of magnitude and direction yields the magnetic field vector; hence it is called a vector magnetometer. Applications include determination of weak magnetic field vectors. Such measurements are useful in military, industrial, and power transmission sectors.

The combination of MF with SPR technologies benefits from the high refractive index sensitivity of SPR, ~ 10^4 nm per refractive index unit (RIU) around RI = 1.33 (refractive index of water). The sensitivity of the magnetic field intensity is 0.0692 nm/Oe while that of the magnetic field orientation is −11.917 nm/degree.

To fabricate the sensor, a piece of few-mode fiber (FMF) is coarsely polished by a wheel polishing machine followed by fine polishing. A 5-nm thick chromium film and then a 50-nm thick Au film are deposited on the FMF. The side-polished section of FMF is inserted in a capillary tube, the capillary tube is embedded in a holder with a rectangular groove, the magnetic fluid is drawn into the capillary tube by capillary action, and finally the capillary tube is sealed with UV glue at both ends. The experimental setup consists of a tungsten halogen lamp, coupled into the fiber. Transmission spectra recording is done with a spectrometer. An electromagnet is used for producing the magnetic field (Chen et al. 2019a).

11.6 Discussion and Conclusions

11.6.1 Pathogens

Colorimetric, fluorescence and electrochemical methods are used for detection of pathogenic organisms (Table 11.1).

11.6.2 Toxins

Toxins are sensed by impedimetric, colorimetric and LSPR principles (Table 11.2).

11.6.3 Cancer

SPR/LSPR, cell imaging, QCM, and electrochemical methods are exploited for revealing cancers (Table 11.3).

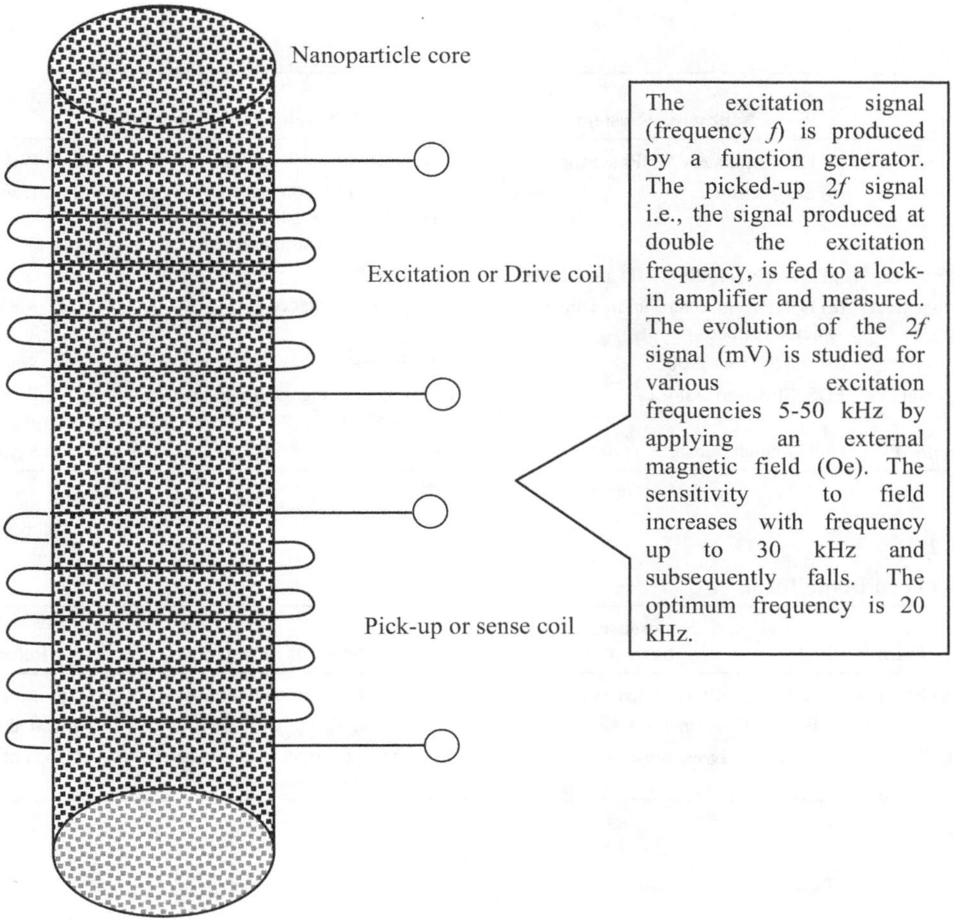

Nanoparticle core
Excitation or Drive coil
Pick-up or sense coil
The excitation signal (frequency *f*) is produced by a function generator. The picked-up 2*f* signal i.e., the signal produced at double the excitation frequency, is fed to a lock-in amplifier and measured. The evolution of the 2*f* signal (mV) is studied for various excitation frequencies 5-50 kHz by applying an external magnetic field (Oe). The sensitivity to field increases with frequency up to 30 kHz and subsequently falls. The optimum frequency is 20 kHz.

FIGURE 11.37 Fluxgate magnetometer with a nanoparticle core.

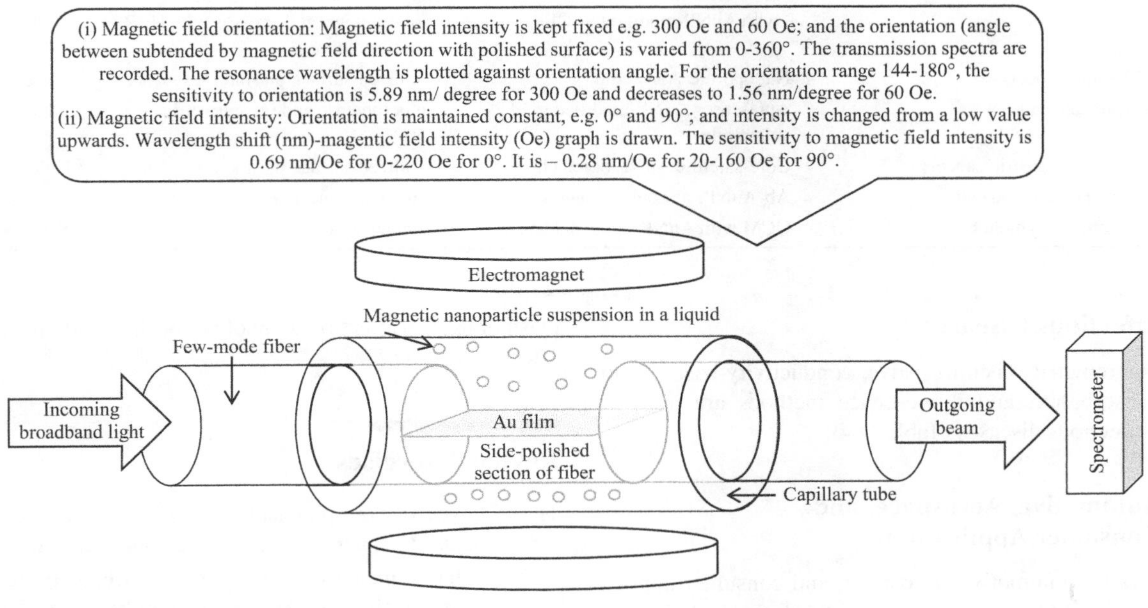

FIGURE 11.38 Magnetometer using ferrofluid coupled with surface plasmon resonance (Chen et al. 2019a).

TABLE 11.1

Detection of Food-Borne Pathogens

Sl. No.	Bacterium	Nanosensor/System	Principle of Detection	Reference(s)
1.	*Salmonella typhimurium*	(a) DNA Aptamers-MNPs sensor	Color recovery	Park et al. (2015)
		(b) Strip sensor	Negative correlation between the intensity of red color on the T-line and the amount of the *Salmonella* strain present	Wang et al. (2016)
2.	*Clostridium perfringens*	ssDNA-CeO_2 (NRs)/CHIT modified GCE	EIS and DPV	Qian et al. (2018)
3.	*Listeria monocytogenes* *L. monocytogenes*	(a) IMNPs with microfluidic chip and microelectrodes	Change in impedance	Kanayeva et al. (2012)
		(b) AuNP/DNA probe	Colorimetric	Wachiralurpan et al. (2018)
4.	*Campylobacter jejuni*	FDS-NP-based assay	Fluorescent orange dots and moving bacterial cells	Poonlapdecha et al. (2018)
5.	*Yersinia enterocolitica*	GQD immunosensor	Amperometry	Savas and Altintas (2019)

TABLE 11.2

Detection of Food-Borne Toxins

Sl. No.	Toxin	Nanosensor/ System	Principle of Detection	Reference(s)
1.	(a) BoNT/A	AuND/CSNP-Modified SPCE	Impedimetric immunosensor	Sorouri et al. (2017)
	(b) BoLcA	Peptide functionalized AuNP-based assay	Colorimetric	Liu et al. (2014)
2.	SEB	Au-AgNP-based biosensor	Localized surface plasmon resonance atmosphere	Zhu et al. (2009)

TABLE 11.3

Detection of Cancer

Sl. No.	Type of cancer	Nanosensor/ System	Principle of Detection	Reference(s)
1.	Breast Cancer Cell MCF-7	MUC1 aptamer and FA- conjugated MNP-based cytosensor (SPR)	Surface plasmon resonance	Chen et al. (2014)
2.	HER2 (breast cancer)	Anti-HER2 antibodies conjugated with AuNP-modified SPCE	Height of the current peak in voltammogram decreases with HER2 concentration	Hartati et al. (2018)
3.	SAA1 (lung cancer)	Nanoporous AAO chip	Localized surface plasmon resonance	Lee et al. (2015)
4.	PSA (prostate cancer)	Aptasensor AuNP-modified gold electrode	Amperometric and impedimetric	Jolly et al. (2017)
5.	miRNA-106a (gastric cancer)	Core-satellite AuNP-based chip	Localized surface plasmon resonance	Park et al. (2018)
6	Colorectal carcinoma cell	Ab-AuNPs as contrast agents	Colorectal cancer cell imaging	Lima et al. (2014)
7.	CD10 antigen (lymphoblastic leukemia)	QCM with AuNP signal enhancers	Immunosensor	Yan et al. (2015)

11.6.4 Infectious Diseases

Chronoamperometric, chemiresistive, conductivity-based, colorimetric, absorbance, and fluorescence methods are used for detecting infectious diseases (Table 11.4).

11.6.5 Automotive, Aerospace, and Consumer Applications

Nanosensors for automotive, aerospace, and consumer applications are fabricated by exploiting a variety of physical phenomena, including resistance changes, piezoelectric, colorimetric, plasmonic, Coriolis force, photo-acoustic, and magneto-optic effects (Table 11.5).

Review Exercises

11.1 Give two examples each of sensors needed by the food, healthcare, automotive, and aerospace industries.

11.2 What is meant by the statement, "Magnetic nanoparticles display catalytic peroxidase enzyme-like activity"? How do DNA aptamers interact with *Salmonella*

TABLE 11.4

Detection of Infectious Diseases

Sl. No.	Parameter	Nanosensor/ System	Principle of Detection	Reference(s)
1.	IgG Antibodies to hepatitis B surface antigen	AuNP electroactive label-assisted assay	Chronoamperometry	Escosura-Muñiz et al. (2010)
2.	Dengue-1 RNA	AuNP-based lateral flow biosensor	Colorimetric	Yrad et al. (2019)
3.	Japanese Encaphilitis Virus (JEV) Antigen	AgNP-based probe	Absorbance intensity variation with the logarithm of JEV antigen concentration	Lim et al. (2017)
4.	HIV-1 p24 antigen	Silica nanoparticle-based immunoassay	Fluorescence measurement	Chunduri et al. (2017)
5.	Zika Virus (ZIKV)	PtNP-based assay on paper microchip	Changes in the electrical conductivity of the solution	Draz et al. (2018)
6.	Novel coronavirus	Dual functional plasmonic biosensor	Combination of plasmonic photothermal effect with LSPR	Qiu et al. (2020)
7.	*Streptococcus pneumoniae*	Microgap device using AuNPs	Bridging of the gap by the AuNPs changes the interelectrode conductance according to the bacterial population	Pyo et al. (2017)
8.	Acid-Fast Bacilli	GMNP-based biosensing assay	Colorimetric	Gordillo-Marroquín et al. (2018)
9.	*Streptococcus pyogenes* ssg-DNA	SWCNT-based genosensor	Chemiresistive	Singh et al. (2014)
10.	*Plasmodium falciparum* heat- shock protein 70 (PfHsp70)	AuNP-based immunoassay	Competitive fluorescence	Guirgis et al. (2012)

TABLE 11.5

Devices for Automotive, Aerospace, and Consumer Applications

Sl. No.	Parameter	Nanosensor/ System	Principle of Detection	Reference(s)
1.	Strain/pressure	(a) Polymer-metallic nanoparticles composite sensor	Resistive-type pressure sensor	Hashimura et al. (2014)
		(b) Percolative PdNP array-based sensor	Quantum-mechanical tunneling in percolation-based nanostructures	Chen et al. (2019b)
		(c) AgNP/PDMS sensor	Resistance changes from formation/broadening and closing of microcracks	Lee et al. (2014)
		(d) PAAm/AuNP sensor	Colorimetric and plasmonic principles	Topcu and Demir (2019)
2.	Acoustic vibration	AuNP monolayer-based sensor	Change in electrical resistance of an Au monolayer with acoustic vibrations	Yi et al. (2015),Zhang et al. (2014)
3.	Acceleration	$BaTiO_3$ NW arrays-based accelerometer	Piezoelectric effect	Koka and Sodano (2013)
4.	Orientation, angular rate, or angle	(a) CNT field-emission gyroscope	Coriolis effect	Yang et al. (2012)
		(b) Magnetic nanoparticle-based gyroscopic sensor	Angular movement or orientation of the coils is extracted from the variations in the rotating magnetic field detected as small changes in the input alternating currents in the conductive coils	Krug and Asumadu (2018)
5.	Ultrasound	Optically activated/read ultrasonic transducer	Photoacoustic or optoacoustic effect	Smith et al. (2015)
6.	Magnetic field	(a) Nanoparticle core-based fluxgate magnetometer	Gating of the flux by the core in and out of the sense coil	Kennedy et al. (2014)
		(b) MNPs-functionalized magnetometer	The magneto-optic effect	Chen et al. (2019a)

typhimurium? How do they influence the behavior of magnetic nanoparticles? How are the above properties utilized to build a nanosensor for *S. typhimurium*? Explain the working of this nanosensor with a diagram.

11.3 What are the names of the different pads and lines on the immunochromatographic strip sensor for *Salmonella typhimurium*? Explain their function. How does one infer whether the test sample contains *Salmonella* strains (positive sample) or the reverse (negative sample)?

11.4 How is a glassy carbon electrode modified to build a sensor for detecting *Clostridium perfringens*? What is the electrical parameter correlated with *C. perfringens* bacteria concentration?

11.5 Describe the use of immunomagnetic nanoparticles to design a sensor relating the *Listeria monocytogenes* concentration with impedance changes?

11.6 In the gold nanoparticle/DNA probe assay for *Listeria monocytogenes*, why do the gold nanoparticles remain dispersed in the presence of the target DNA but aggregate in its absence. Explain diagrammatically.

11.7 How are *Campylobacter jejuni* bacteria detected using antibody-conjugated fluorescent dye-doped silica nanoparticles? What is the difference observed between positive and negative results?

11.8 How does the inhibition of electron transference by the formation of antigen-antibody complex help in detecting *Yersinia enterocolitica* in a graphene quantum dots-based immunosensor?

11.9 Describe the modification of screen-printed carbon electrode with gold nanodendrites and chitosan nanoparticles to fabricate a sensor for Botulinum Neurotoxin Serotype A.

11.10 Explain, with diagrams, the mixing and bridging assays for Serotype A Light Chain (BoLcA), using peptide-functionalized gold nanoparticles.

11.11 In the sensor for breast cancer cell MCF-7, two vital components are human mucin-1 (MUC1) aptamer and folic acid-conjugated magnetic nanoparticles. What are the roles played by these components?

11.12 Name a biomarker for breast cancer. Describe how this biomarker is measured by an antibody- conjugated gold nanoparticles-modified screen-printed carbon electrode.

11.13 Name a biomarker for lung cancer. Explain, with diagrams, the working of a sensor, using nanoporous anodic aluminum oxide for detecting this biomarker.

11.14 What is the biomarker for prostate cancer? How is this biomarker detected by an aptasensor platform? Distinguish between amperometric and impedimetric modes of this platform.

11.15 What is the biomarker for gastric cancer? Explain its detection by a localized surface plasmon resonance chip.

11.16 Describe the preparatory steps for the imaging of colorectal cancer cells by confocal laser scanning microscope? What nanoparticles are used as contrast agents for imaging, and how is this done?

11.17 How do gold nanoparticles enhance the detection signal in a quartz crystal microbalance for ascertaining the common acute lymphoblastic leukemia antigen?

11.18 Explain with the help of diagrams how a magneto-immunosandwich is made for detecting antibodies to hepatitis B surface antigen in a chronoamperometric sensor?

11.19 What are the three probes made for the dengue-1 RNA sensor, using a lateral flow sensor? What are the nanoparticles used in this sensor? Will the sensor work without the nanoparticles?

11.20 How is an absorbance probe designed for sensing Japanese encephalitis virus, using silver nanoparticles?

11.21 How is the HIV-1 p24 antigen detected by a fluorescence measurement? Which nanoparticles are used in this measurement?

11.22 How are platinum nanoparticles used in making a sensor for Zika virus, in which electrical conductivity changes indicate the virus concentration?

11.23 What is meant by plasmonic photothermal effect? How is this phenomenon utilized to improve the detection sensitivity of novel corona virus by localized surface plasmon resonance? Describe the operation of the nanosensor with a diagram.

11.24 How does decoration of *Streptococcus pneumoniae* bacteria with gold nanoparticles help in their detection by a microgap device? Explain the details of the bacterial decoration process.

11.25 How are magnetic nanoparticles used for making a sensor for *Mycobacterium tuberculosis*? What are the core and coatings of the magnetic nanoparticles made from?

11.26 Which bacterium causes rheumatic heart disease? Describe the chemiresistive sensor using SWCNTs for detecting this bacterium?

11.27 In what way is the competition between the PfHsp70 antigen and Cy3B-labeled PfHsp70 antigen for binding with 2E6 antibodies related to fluorescence intensity variations in the sensor for the malarial parasite?

11.28 Write the equation connecting the tunneling electric current *I* flowing between the metallic nanoparticles at a mean distance *d* apart? How is variation of this current applied to pressure sensing in a polymer-metallic nanoparticles composite device?

11.29 What is meant by a percolation threshold? Explain its significance in the context of a polymer-metallic nanoparticles composite pressure sensor.

11.30 Describe the construction and operation of a percolative Pd nanoparticles array-based pressure sensor. Is it different from a conventional MEMS piezoresistive pressure sensor? If yes, explain the difference? What parameter determines the sensitivity and working range of this sensor? Can the working range be extended?

11.31 How is a silicon mold used to make a PDMS stamp? Describe the coating procedure of PDMS with silver nanoparticle thin film? How is this Ag nano-ink film on PDMS used for measuring pressure? Explain the construction of strain sensor and pressure sensor using this film. Use diagrams to illustrate your description.

11.32 How do color changes take place in a polyacrylamide/gold nanoparticles pressure sensor? Mention some applications of this type of colorimetric sensor.

11.33 How is a monolayer of gold nanoparticles deposited on a PET substrate for fabricating an acoustic vibration sensor? What is the nanoparticle deposition method called? What is the critical parameter responsible for triggering or inhibiting the nanoparticle self-assembly?

11.34 Describe, with a diagram, the fabrication of a barium titanate array accelerometer. What electrical parameter is the indicator of acceleration changes?

11.35 Name the physical effect on which a CNT gyroscope works. Why is it called a vibratory gyroscope? What specific property of CNT is utilized in this gyroscope? What are the typical DC and AC voltages used in gyroscope operation? Give a few advantages of a CNT gyroscope.

11.36 What types of conductive coil arrangements are used to generate rotating magnetic fields in a magnetic nanoparticle gyroscope? How is the rotational motion of nanoparticles affected by angular acceleration, and how are these effects used to determine the angular movements?

11.37 An optically excited ultrasonic transducer uses two laser beams of different wavelengths, one for ultrasound generation and the other for ultrasound detection. What are these laser beams called? How do they perform their generation/detection tasks?

11.38 How does the use of nanomaterial for making the core of flux-gate magnetometer improve its performance? What material is used for making the core?

11.39 How does a magnetic field produce refractive index changes? Describe how the magnetic field-induced refractive index changes are exploited to develop a magnetic nanoparticles functionalized magnetometer. What makes it a vector magnetometer?

11.40 Explain the magnetic-*cum*-optical principle, construction, and working of a magnetic nanoparticles-functionalized magnetometer?

REFERENCES

Chen H., Y. Hou, Z. Ye, H. Wang, K. Koh, Z. Shen, and Y. Shu. 2014. Label-free surface plasmon resonance cytosensor for breast cancer cell detection based on nano-conjugation of monodisperse magnetic nanoparticle and folic acid. *Sensors an Actuators B: Chemical* 201: 433–438.

Chen Y., W. Sun, Y. Zhang, G. Liu, Y. Luo, J. Dong, Y. Zhong, W. Zhu, J. Yu, and Z. Chen. 2019a. Magnetic nanoparticles functionalized few-mode-fiber-based plasmonic vector magnetometer. *Nanomaterials* 9: 785 (pp. 1–12).

Chen M., W. Luo, Z. Xu, X. Zhang, B. Xie, G. Wang, and M. Han. 2019b. An ultrahigh resolution pressure sensor based on percolative metal nanoparticle arrays. *Nature Communications* 10: 4024 (pp. 1–9).

Chunduri L. A. A., A. Kurdekar, M. K. Haleyurgirisetty, E. P. Bulagonda, V. Kamisetti, and I. K. Hewlett. 2017. Femtogram level sensitivity achieved by surface engineered silica nanoparticles in the early detection of HIV infection. *Scientific Reports* 7: 7149 (pp. 1–10).

Draz M. S., M. Venkataramani, H. Lakshminarayanan, E. Saygili, M. Moazeni, A. Vasan, Y. Li, X. Sun, S. Hua, X. G. Yu, and H. Shafiee. 2018. Nanoparticle-enhanced electrical detection of Zika virus on paper microchip. *Nanoscale* 10(25): 11841–11849.

Escosura-Muñiz A. D. L., M. M.-D. Costa, C. Sánchez-Espinel, B. Díaz-Freitas, J. Fernández-Suarez, Á. González-Fernández, and A. Merkoci. 2010. Gold nanoparticle-based electrochemical magnetoimmunosensor for rapid detection of anti-hepatitis B virus antibodies in human serum. *Biosensors and Bioelectronics* 26: 1710–1714.

Gordillo-Marroquín C., A. Gómez-Velasco, H. J. Sánchez-Pérez, K. Pryg, J. Shinners, N. Murray, S. G. Muñoz-Jiménez, A. Bencomo-Alerm, A. Gómez-Bustamante, L Jonapá-Gómez, N.Enríquez-Ríos, M. Martín, N. Romero-Sandoval, and E. C. Alocilja. 2018. Magnetic nanoparticle-based biosensing assay quantitatively enhances acid-fast bacilli count in paucibacillary pulmonary tuberculosis. *Biosensors* 8: 128 (pp. 1–13).

Guirgis B. S. S., C. S. E. Cunha, I. Gomes, M. L. Cavadas, I. Silva, G. Doria, G. L. Blatch, P. V. Baptista, E. Pereira, H. M. E. Azzazy, M. M. Mota, M. Prudêncio, and R. Franco. 2012. Gold nanoparticle-based fluorescence immunoassay for malaria antigen detection. *Analytical and Bioanalytical Chemistry* 402: 1019–1027.

Hartati Y. W., D. Nurdjanah, S. Wyantuti, A. Anggraeni, and S. Gaffar. 2018. Gold nanoparticles modified screen-printed immunosensor for cancer biomarker HER2 determination based on anti HER2 bioconjugates. In The 3rd International Seminar on Chemistry, Green Chemistry and its Role for Sustainability, Surabaya, Indonesia, 18–19 July, 2018, AIP Conference Proceedings 2049, AIP Publishing, NY, USA, 020051-1–020051-8.

Hashimura A., L. Tang, and A. T. Voutsas. 2014. Metallic nanoparticle pressure sensor, Patent No. US008,669,952B2, March 11, 13pp.

Jolly P., P. Zhurauski, J. L. Hammond, A. Miodek, S. Liébana, T. Bertok, J. Tkác, and P. Estrela. 2017. Self-assembled gold nanoparticles for impedimetric and amperometric detection of a prostate cancer biomarker. *Sensors and Actuators B: Chemical* 251: 637–643.

Kanayeva D. A., R. Wang, D. Rhoads, G. F. Erf, M. F. Slavik, S. Tung, and Y. Li. 2012. Efficient separation and sensitive detection of *Listeria Monocytogenes* using an impedance immunosensor based on magnetic nanoparticles, a microfluidic chip, and an interdigitated microelectrode. *Journal of Food Protection* 75(11): 1951–1959.

Kennedy J., J. Leveneur, J. Turner, J. Futter, and G. V. M. Williams. 2014. Applications of nanoparticle-based fluxgate magnetometers for positioning and location. In IEEE Sensors Applications Symposium (SAS), Queenstown, New Zealand, 18–20 February 2014. New York: IEEE, pp. 228–232.

Koka A. and H. A. Sodano. 2013. High-sensitivity accelerometer composed of ultra-long vertically aligned barium titanate nanowire arrays. *Nature Communications* 4: 2682 (pp. 1–10).

Krug B. G. and J. A. Asumadu. 2018. Magnetic nanoparticle-based gyroscopic sensor, International Publication No. WO 2018/023033 Al, Publication date: 01 February, 2018, 64 pp.

Lee J., S. Kim, J. Lee, D.Yang, B. C. Park, S. Ryu, and I. Park. 2014. A stretchable strain sensor based on a metal nanoparticle thin film for human motion detection. *Nanoscale* 6: 11932–11939.

Lee J.-S., S.-W. Kim, E.-Y. Jang, B.-H. Kang, S.-W. Lee, G. Sai-Anand, S.-H. Lee, D.-H. Kwon, and S.-W. Kang. 2015. Rapid and sensitive detection of lung cancer biomarker using nanoporous biosensor based on localized surface plasmon resonance coupled with interferometry. *Journal of Nanomaterials* 2015: 183438 (11 pp.).

Lim L. S., S. F. Chin, S. C. Pang, M. S. H. Sum, and D. Perera. 2017. A novel silver nanoparticles-based sensing probe for the detection of Japanese encephalitis virus antigen. *Sains Malaysiana* 46(12): 2447–2454.

Lima K. M. G., R. F. Araújo Junior, A. A. Araujo, A. L. C. S. Leitão Oliveira, and L. H. S. Gasparotto. 2014. Environmentally compatible bioconjugated gold nanoparticles as efficient contrast agents for colorectal cancer cell imaging. *Sensors and Actuators B: Chemical* 196: 306–313.

Liu, X., Y. Wang, P. Chen, Y. Wang, J. Zhang, D. Aili, and B. Liedberg. 2014. Biofunctionalized gold nanoparticles for colorimetric sensing of Botulinum Neurotoxin A light chain. *Analytical Chemistry* 86(5): 2345–2352.

Park J. Y., H. Y. Jeong, M. I. Kim, and T. J. Park. 2015. Colorimetric detection system for *Salmonella Typhimurium* based on peroxidase-like activity of magnetic nanoparticles with DNA aptamers. *Journal of Nanomaterials* 2015: 527126 (9 pp.).

Park S.-H., J. Lee, and J.-S. Yeo. 2018. On-chip plasmonic detection of microRNA-106a in gastric cancer using hybridized gold nanoparticles. *Sensors and Actuators B: Chemical* 262: 703–709.

Poonlapdecha W., Y. Seetang-Nun, K. Tuitemwong, and P. Tuitemwong. 2018. Validation of a rapid visual screening of Campylobacter jejuni in chicken using antibody-conjugated fluorescent dye-doped silica nanoparticle reporters. *Journal of Nanomaterials* 2018: 4571345 (10 pp.).

Pyo H., C. Y. Lee, D. Kim, G. Kim, S. Lee, and W. S. Yun. 2017. Electrical detection of Pneumococcus through the nanoparticle decoration method. *Sensors* 17: 2012 (pp. 1–8).

Qian X., Q. Qu, L. Li, X. Ran, L. Zuo, R. Huang, and Q. Wang. 2018. Ultrasensitive electrochemical detection of *Clostridium perfringens* DNA based morphology-dependent DNA adsorption properties of CeO_2 nanorods in dairy products. *Sensors* 18: 1878 (pp. 1–15).

Qiu G., Z. Gai, Y. Tao, J. Schmitt, G. A. Kullak-Ublick, and J. Wang. 2020. Dual-functional plasmonic photothermal biosensors for highly accurate severe acute respiratory syndrome Coronavirus 2 detection. *ACS Nano* 14(5): 5268–5277 (10 pp.), https://doi.org/10.1021/acsnano.0c02439.

Savas S. and Z. Altintas. 2019. Graphene quantum dots as nanozymes for electrochemical sensing of *Yersinia enterocolitica* in milk and human serum. *Materials* 12: 2189 (pp. 1–14).

Singh S., A. Kumar, S. Khare, A. Mulchandani, and Rajesh. 2014. Single-walled carbon nanotubes based chemiresistive genosensor for label-free detection of human rheumatic heart disease. *Applied Physics Letters* 105: 213701-1–213701-4.

Smith R. J., F. P. Cota, L. Marques, X. Chen, A. Arca, K. Webb, J. Aylott, M. G. Somekh, and M. Clark. 2015. Optically excited nanoscale ultrasonic transducers. *Journal of the Acoustical Society of America* 137(1): 219–227.

Sorouri R., H. Bagheri, A. Afkhami, and J. Salimian. 2017. Fabrication of a novel highly sensitive and selective immunosensor for Botulinum Neurotoxin Serotype A based on an effective platform of electrosynthesized gold nanodendrites/chitosan nanoparticles. *Sensors* 17: 1074 (pp. 1–13).

Topcu G. and M. M.Demir. 2019. Colorimetric and plasmonic pressure sensors based on polyacrylamide/Au nanoparticles. *Sensors and Actuators A: Physical* 295: 503–511.

Wachiralurpan S., T. Sriyapai, S. Areekit, P. Sriyapai, S. Augkarawaritsawong, S. Santiwatanakul, and K. Chansiri. 2018. Rapid colorimetric assay for detection of *Listeria monocytogenes* in food samples using lamp formation of DNA concatemers and gold nanoparticle-DNA probe complex. *Frontiers in Chemistry* 6: 90 (pp. 1–9).

Wang W., L. Liu, S. Song, L. Xu, H. Kuang, J. Zhu, and C. Xu. 2016. Gold nanoparticle-based strip sensor for multiple detection of twelve Salmonella strains with a genus-specific lipopolysaccharide antibody. *Science China Materials* 59(8): 665–674.

Yan Z., M. Yang, Z. Wang, F. Zhang, J. Xia, G. Shi, L. Xia, Y. Li, Y. Xia, and L. Xia. 2015. A label-free immunosensor for detecting common acute lymphoblastic leukemia antigen (CD10) based on gold nanoparticles by quartz crystal microbalance. *Sensors and Actuators B: Chemical* 210: 248–253.

Yang Z., M. Nakajima, Y. Shen, and T. Fukuda. 2012. Nano-gyroscope device using field emission of isolated carbon nanotube. In International Symposium on Micro-NanoMechatronics and Human Science (MHS), Nagoya, Japan, 4–7 November 2012, IEEE, NY, USA, pp. 256–261.

Yang Z., M. Nakajima, Y. Shen, P. Wang, C. Ru,Y. Zhang, L. Sun, and T. Fukuda. 2013. Test of a CNT gyroscope based on field emission. In The 7th IEEE International Conference on Nano/Molecular Medicine and Engineering, Phuket, Thailand, 10–13 November 2013. New York: IEEE, pp. 59–62.

Yi L., W. Jiao, K. Wu, L. Qian, X. Yu, Q. Xia, K. Mao, S. Yuan, S. Wang, and Y. Jiang. 2015. Nanoparticle monolayer-based flexible strain gauge with ultrafast dynamic response for acoustic vibration detection. *Nano Research* 8: 2978–2987.

Yrad F. M., J. M. Castañares, and E. C. Alocilja. 2019. Visual detection of dengue-1 RNA using gold nanoparticle-based lateral flow biosensor. *Diagnostics* 9: 74 (pp. 1–14).

Zhang C., J. Li, S. Yang, W. Jiao, S. Xiao, M. Zou, S. Yuan, F. Xiao, S. Wang, and L. Qian. 2014. Closely packed nanoparticle monolayer as a strain gauge fabricated by convective assembly at a confined angle. *Nano Research* 7(6): 824–834.

Zhu S., C. L. Du, and Y. Fu. 2009. Localized surface plasmon resonance-based hybrid Au–Ag nanoparticles for detection of Staphylococcus aureus enterotoxin B. *Optical Materials* 31: 1608–1613.

12

Nanosensors for Homeland Security

12.1 Necessity of Nanosensors for Trace Explosive Detection

Due to the vast variety of material options exercised in making explosives, the low vapor pressure of several explosive materials precluding their easy detection, and the copious trickeries by which explosives can be concealed, there is a desperate need for high sensitivity/selectivity nanosensors to detect explosives in minuscule quantities. Table 12.1 presents a few commonly used explosives (Senesac and Thundat 2008).

12.2 2,4,6-Trinitrotoluene (TNT) Nanosensors

12.2.1 Curcumin Nanomaterials Surface Energy Transfer (NSET) Probe

Curcumin (diferuloylmethane) is a bright yellow substance obtained from *Curcuma longa* plants. It is the ingredient responsible for the yellowish-golden color of turmeric. It is an anti-inflammatory agent.

Tumeric-extracted curcumin nanoparticles or nanocurcumin undergo reactions with trace quantities of TNT (Pandya et al. 2012):

(i) Absorption mode: changes in the absorption spectra are observed in the presence of TNT. The π-donor–acceptor interactions take place between the π-electron surplus nanocurcumin and the electron-lacking TNT causing aggregation of the nanocurcumin. The higher the concentration, the larger the size of the nanocurcumin aggregates (Figure 12.1). Due to this aggregation, the plasmon band of the nanocurcumin shifts to longer wavelengths. The yellow color of nanocurcumin transforms to a red color. This red shift arises from the change in the local refractive index of curcumin nanoparticles.

(ii) Fluorescence mode: better TNT detection is achieved using the fluorescence method as compared to absorption. The effect of TNT concentration on curcumin nanoparticle aggregation is investigated by studying the changes in fluorescence emission intensity at 670 nm. The fluorescence emission intensity varies linearly with a TNT concentration in the range 1–100 nM. Other nitrocompounds, such as HMX, RDX, PETN, etc., are unable to affect the fluorescence emission intensity, which shows that the nanocurcumin probe provides selective detection of TNT (Pandya et al. 2012).

12.2.2 Amine-Functionalized Silica Nanoparticles (SiO_2-NH_2) Colorimetric Sensor

SiO_2 nanoparticles are amine functionalized using aminopropyltriethoxysilane (APTES) (Idros et al. 2015). The functionalized nanoparticles are self-assembled on quartz substrates to form a thin film. In the presence of TNT, the primary amine attached to the silica nanoparticles captures TNT to form Meisenheimer amine–TNT complexes (Figure 12.2), the resonance-stabilized cationic reactive intermediates formed in electrophilic aromatic substitution.

To begin, the SiO_2-NH_2 nanoparticle film is green. As the concentration of TNT molecules increases, the formation of an increasing number of SiO_2-NH_2–TNT complexes changes the color of the film from green to red. This color change takes place due to the increase in the average distance between the nanoparticles because of the formation of a red-colored compound mediated by a strong acceptor–donor interaction between the amine group on the silica nanoparticle and the nitro group of TNT. It is measured in terms of normalized reflection intensity at the 400–800 nm wavelength using a highly reflective silver mirror as the baseline:

$$\text{Normalized reflection intensity} = \left(\frac{\text{Reflection intensity of the film}}{\text{Reflection intensity of the mirror}} \Big/ \frac{\text{Integration time for the film}}{\text{Integration time for the mirror}}\right) \times 100\%$$

The method is applied to a wide range of TNT concentrations from 10^{-12} to 10^{-4} molar (Idros et al. 2017).

12.2.3 Amine-Modified Gold@Silver Nanoparticles-Based Colorimetric Paper Sensor

A paper sensor is fabricated using hydrophilic amine-protected Ag@Au nanoparticles (Arshad et al. 2019). The -NH_2-Au@Ag nanoprobe binds selectively with TNT through Meisenheimer complex formation (Figure 12.3). The nanosensor works on the colorimetric principle. Non-occurrence of this type of interaction with other nitroaromatics imparts selectivity to the nanoprobe toward TNT detection. The linear range of detection is 0–20 μg/mL. The limit of detection (LOD) is 0.35 μg/mL. The non-requirement of expensive instrumentation makes the method cost-effective (Arshad et al. 2019).

TABLE 12.1
Some Explosives

Sl. No.	Name of Explosive	Formula	Molecular Weight (g/mol)	Smell, Color	Vapor Pressure (Torr) at 293 K
1.	2,4,6-Trinitrotoluene (TNT)	$C_7H_5N_3O_6$	227.13	Odorless yellow-colored solid	4.8×10^{-6}
2.	Tetryl (tetranitro-N-methylamine)	$C_7H_5N_5O_8$	287.15	Odorless crystalline solid, yellow in color	3.7×10^{-10}
3.	Picric acid (PA)	$C_6H_3N_3O_7$	229.1	Odorless crystalline solid, pale yellow in color	3.1×10^{-8}
4.	1,3,5-Trinitro-1, 3,5-triazacyclohexane (RDX: Royal Demolition eXplosive)	$C_3H_6N_6O_6$	222.12	Odorless white-colored solid	8.3×10^{-10}

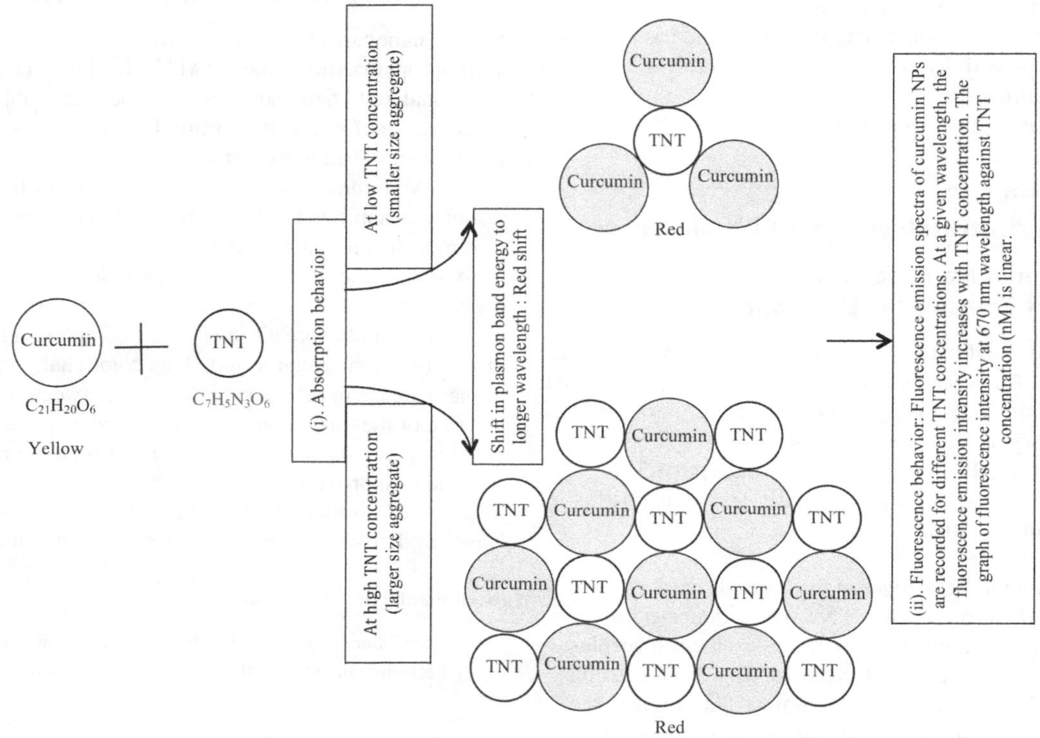

FIGURE 12.1 Effect of TNT on nanocurcumin nanoparticles fluorescence probe: absorption and fluorescence behaviors. (Pandya et al. 2012.)

12.2.4 Polyethylenimine (PEI)-Capped Downconverting β-$NaYF_4$:Gd^{3+},Tb^{3+}@PEI Nanophosphor Luminescence Sensor

Downconversion is a process in which a high-energy photon splits up into a pair of photons of lower energy. A downconversion material can convert a high-energy photon into low-energy photons.

$NaYF_4$ nanophosphors prepared using the hydrothermal method are added to a solution containing $Y(NO_3)_2$, $Gd(NO_3)_2$, $Tb(NO_3)_2$, and PEI (Malik et al. 2019). After reaction at 180°C for 24 h and cooling to room temperature, green luminescent β-$NaYF_4$:Gd^{3+},Tb^{3+}@PEI nanophosphors are obtained (Figure 12.4). PEI functionalizes the surfaces of nanophosphors with amino groups. These nanophosphors emit sharp emission peaks at several wavelengths. Among these peaks, the one at 544 nm is drastically quenched in contact with TNT.

The sensor works on fluorescence quenching due to the formation of a Meisenheimer complex by the charge-transfer mechanism between the electron-rich amino groups on the surface of nanophosphors and the electron-deficient nitro groups on the TNT in aqueous medium. In aqueous solution, the color changes from white to brownish-yellow. This color change is visually detected.

When the photoluminescence (PL) intensity is observed at 544 nm green light, it is found that the intensity is quenched due to Meisenheimer complex formation between nanophosphors and TNT. The PL intensity is quenched by the luminescence resonance energy transfer (LRET) from nanophosphors to the Meisenheimer complex. The range of linear detection is 0.1–300 μM. The limit of detection is 119.9 nM (27.2 ppb). Selectivity with respect to pesticides, amino acids, etc., is confirmed from the insignificant intensity changes caused by these analytes (Malik et al. 2019).

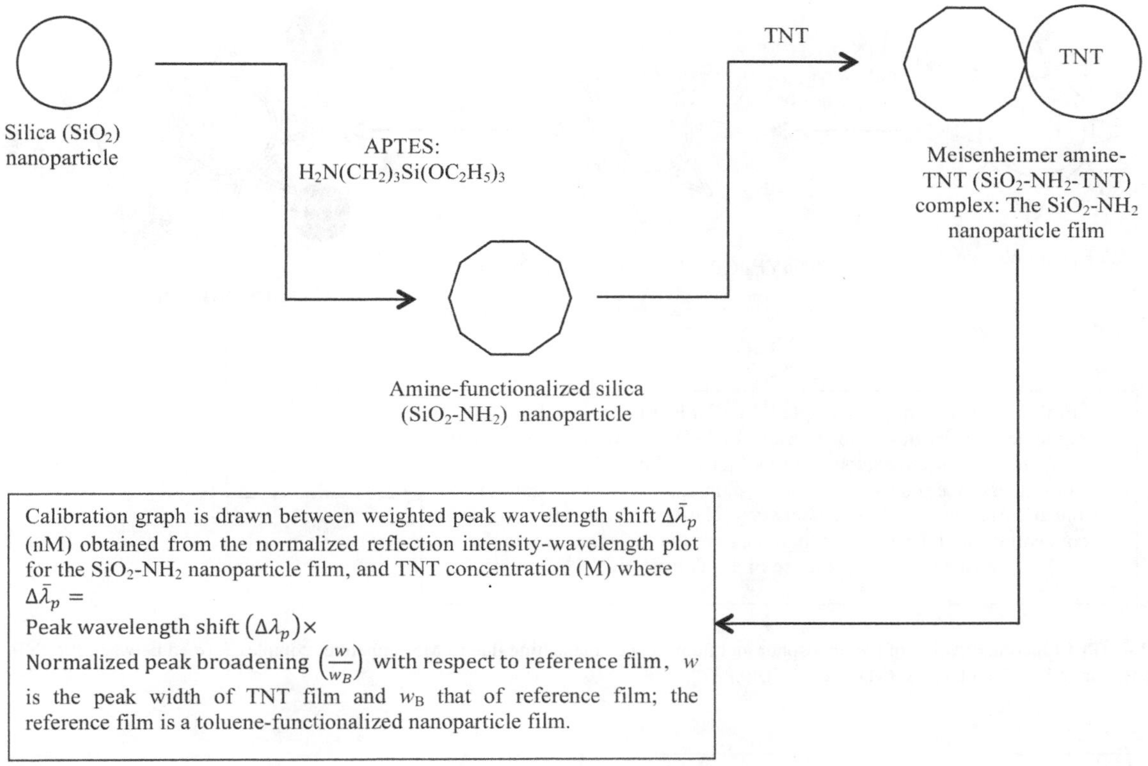

FIGURE 12.2 Formation of amine-functionalized SiO_2 nanoparticle and Meisenheimer amine–TNT complex. (Idros et al. 2015.)

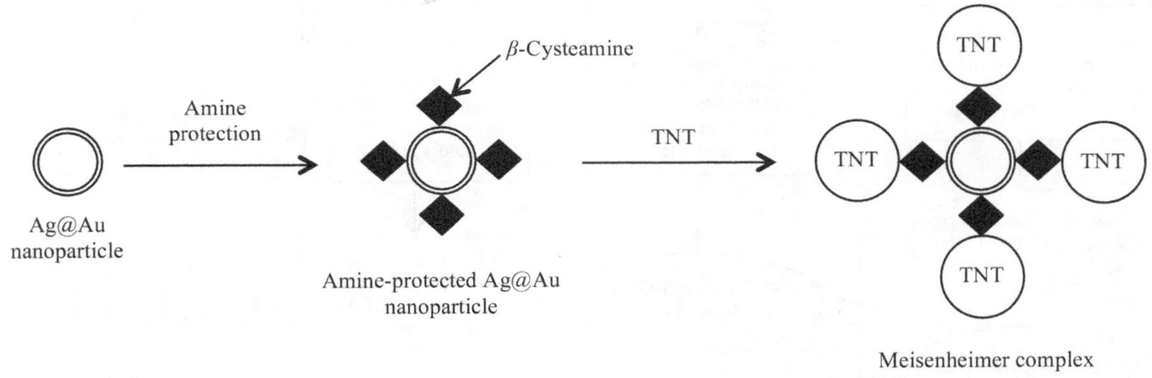

FIGURE 12.3 Amine protection of Ag@Au nanoparticle and binding of -NH_2-Au@Ag nanoprobe with TNT. (Arshad et al. 2019.)

12.2.5 Janus Amine-Modified Upconverting $NaYF_4$:Yb^{3+}/Er^{3+} Nanoparticle (UCNP) Micromotor-Based On-Off Luminescence Sensor

Micromotors are very small particles that propel themselves in a particular direction autonomously when dispersed in a solution. Upconverting nanoparticles are nanoscale particles that emit a photon of higher energy by absorbing two or more lower-energy photons.

This sensor works on the luminescence quenching of Janus amine–modified upconverting nanoparticle micromotors by TNT through a combination of their two properties, viz., their self-propelled motion in H_2O_2 solution and the chemical recognition of TNT by amino groups (Yuan et al. 2019). The motion of UCNPs greatly increases the chances of their collision with TNT molecules and consequently the reaction rate. Figure 12.5 shows a flowchart explaining the fabrication and operation of this sensor.

Janus amine-UCNP capsule motors are fabricated as follows: (i) $NaYF_4$:Yb^{3+}/Er^{3+} UCNPs are synthesized; (ii) for amine modification, they are treated with poly(acrylic acid) (PAA) and then 3-aminopropyltriethoxysilane to get amine group–functionalized UCNPs; (iii) eight layers of poly(allylamine hydrochloride) (PAH)/poly(acrylic acid) are deposited on silica particles by layer-by-layer (LbL) assembly; (iv) the amine group–functionalized UCNPs of step (ii) are mixed with $(PAH/PAA)_8$-coated silica particles, and thus UCNP-coated silica particles are formed; (v) a monolayer of UCNP-coated silica particles is deposited on a glass substrate; (v) 20 nm-thick platinum (Pt) is

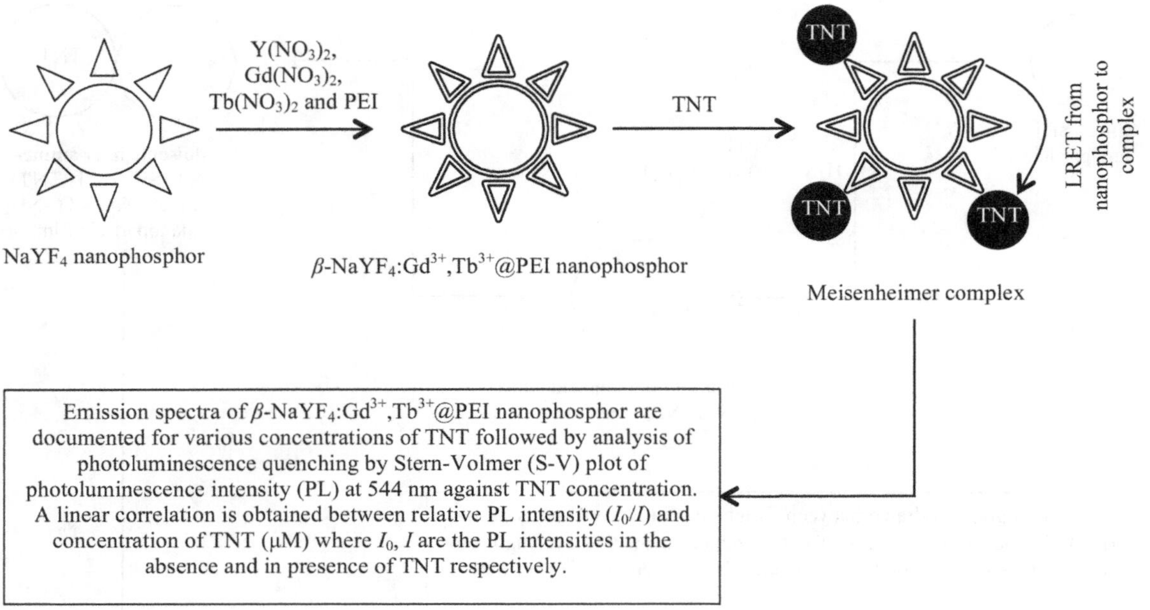

FIGURE 12.4 PEI functionalization of nanophosphor and fluorescence quenching due to Meisenheimer complex formed between β-$NaYF_4$:Gd^{3+},Tb^{3+}@PEI nanophosphor and TNT molecules. (Malik et al. 2019.)

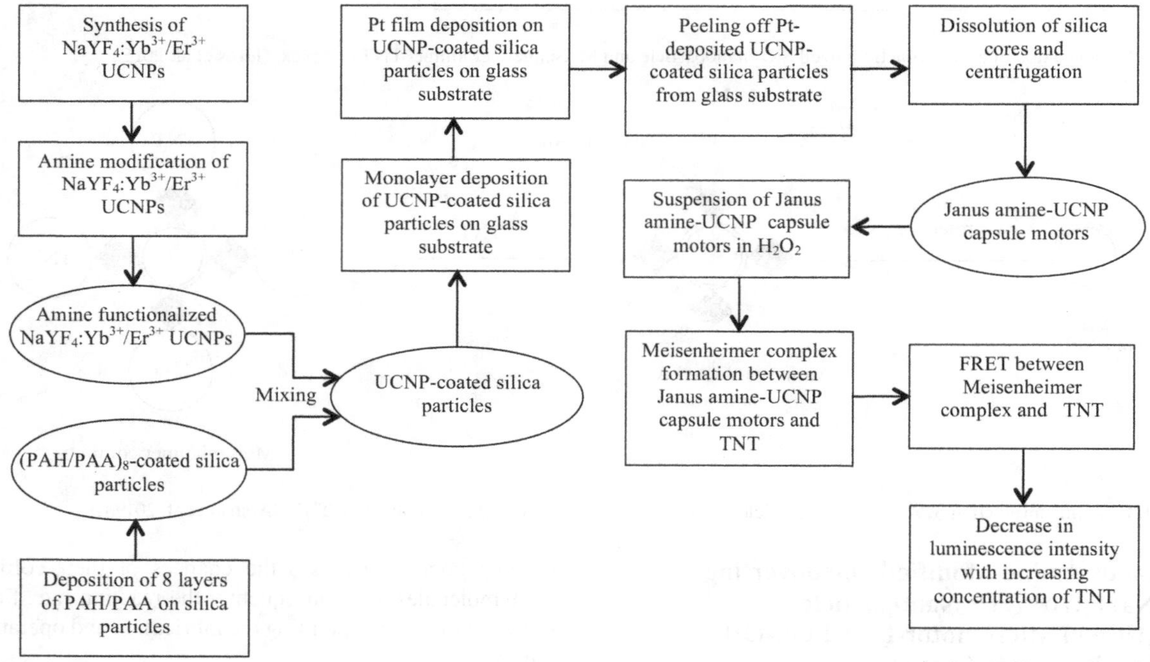

FIGURE 12.5 Flow diagram of fabrication and use of Janus amine–modified $NaYF_4$:Yb^{3+}/Er^{3+} UCNP micromotor–based luminescence sensor. (Yuan et al. 2019.)

deposited over the surface of the particles; (vi) the particles are peeled off the glass substrate; (vii) after dissolution of the silica cores in hydrofluoric acid (HF), Janus amine-UCNP capsule motors are obtained by centrifugation and washing in water.

When taken in a H_2O_2 solution, the Janus amine-UCNP capsule motors are able to swim at a speed of 110 μm/s. The force for this swimming is provided by the oxygen bubbles released through the catalytic decomposition of H_2O_2 on the Pt side. The locomotion capability increases the chances of their coming in contact with TNT molecules. Then, the amino groups on the surfaces of the PAA chains swing into action. These groups chemically recognize the TNT molecules. Together with the TNT molecules, they form Meisenheimer complexes. This complex absorbs strongly in the emission spectrum of the UCNPs. Fluorescence resonance energy transfer (FRET) takes place between the excited UCNPs and the complex. Using the FRET mechanism, the luminescence intensity of the Janus amine-UCNP capsule motors decreases. The decrease in luminescence intensity indicates the presence of TNT.

The luminescence intensity decreases linearly with the natural logarithm of TNT concentration. The limit of detection is 2.4 ng/mL. The Janus amine-UCNP capsule motors display superior performance in the form of a "3.5 times better detection limit" than static UCNPs, i.e., those without motor action (Yuan et al. 2019).

12.2.6 $AgInS_2$ (AIS) Quantum Dot (QD) Fluorometric Probe

AIS quantum dots are synthesized by a thermal decomposition reaction using $Ag(NO_3)_2$, $In(NO_3)_2$, and diethyldithiocarbamate (DDTC) (Baca et al. 2017). The QDs are stabilized by the ligand dodecylamine. No special ligand is required for specifically binding TNT molecules. The emission intensity vs. the wavelength measurements on AIS QDs exposed to various TNT concentrations using acetone as a solvent show that the emission intensity is increasingly quenched as the TNT concentration rises. The intensity changes are ascribed to the increasing aggregation of QDs with TNT concentration. A TNT concentration as low as 6 μM is detectable.

An SU-8 photoresist pattern defined on a filter paper is soaked in an AIS QDs solution in chloroform when the QDs are deposited on the clear spaces in the pattern. The QDs on the filter paper are used for the visual detection of TNT (Baca et al. 2017).

12.2.7 TNT Recognition Peptide Single-Walled Carbon Nanotubes (SWCNTs) Hybrid Anchored Surface Plasmon Resonance (SPR) Chip

The surface of a 50 nm-thick gold layer on an SPR chip is modified with aminopropyltriethoxysilane, as shown in Figure 12.6, and immersed in an SWCNTs solution to obtain an SWCNTs-SPR chip (Wang et al. 2018). Then, the TNT recognition peptide *TNTHCDR3* is non-covalently immobilized on the surface of SWCNTs via π-π interaction. In the TNT recognition peptide, three types of aromatic amino acids actively participate in binding to TNT molecules: tyrosine, tryptophan, and phenylalanine.

Real-time sensorgrams of the sensor exposed to different TNT concentrations in the range 0–100 ppm are taken. The response (resonance unit: RU) is highly linear from 0.8 to 12.5 ppm. The limit of detection is 772 ppb (3.4 μM). Selectivity is shown by the fact that TNT analogues, such as dinitrotoluene (DNT), RDX, etc., at 100 ppm TNT had an imperceptible effect on the sensor as compared to TNT. The sensor response (RU) to 100 ppm TNT decreased to 84% after 7 days, 83% after 21 days, and 70.8% after 28 days showing that the reduction in peptide activity with storage time and hence long-term stability is satisfactory (Wang et al. 2018).

12.2.8 Non-Imprinted and Molecularly Imprinted Bis-Aniline–Cross-Linked Gold Nanoparticles (AuNPs) Composite/Gold Layer for Surface Plasmon Resonance, and Related Sensors

Thioaniline-functionalized AuNPs (3.5 nm diameter) prepared by coating AuNPs with a (thioaniline + mercaptoethane

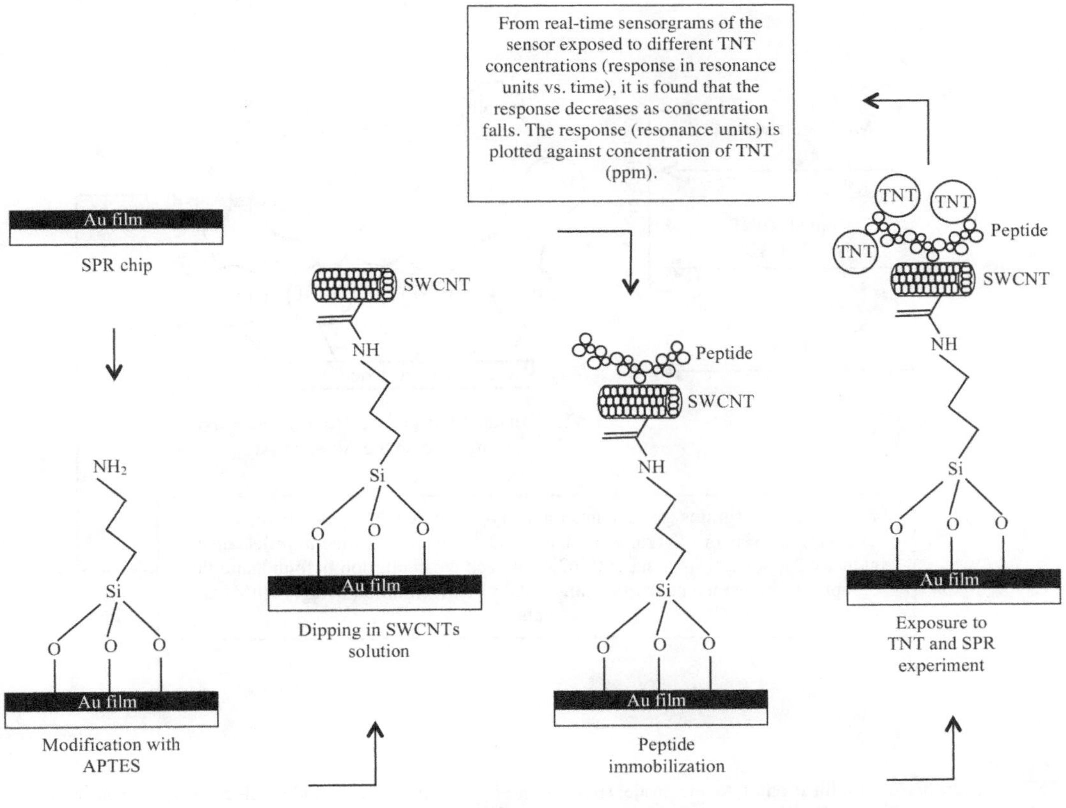

FIGURE 12.6 Preparation and use of TNT recognition peptide-SWCNTs hybrid anchored SPR chip. (Wang et al. 2018.)

sulfonate) mixed monolayer are electropolymerized on a thioaniline monolayer–modified gold electrode (50 nm gold film–coated glass plate) to form a bisaniline–cross-linked AuNPs composite on the Au electrode (Riskin et al. 2009). The gold electrode modification is shown in Figure 12.7(a). Gold nanoparticles functionalized with a mixed monolayer, as shown in Figure 12.7(b), are electropolymerized on the modified gold electrode of Figure 12.7(a). The modified gold electrode together with the functionalized AuNPs is used for TNT sensing, as shown in Figure 12.7(c). The linkage of TNT molecules to the bisaniline bridging units through π-donor–acceptor complex formation combined with the coupling of the localized plasmon of AuNPs and the surface plasmon wave of the Au surface provides amplified TNT detection up to a lower limit of 10 pM.

A further amplification of the response toward TNT is produced by the molecular imprinting of recognition sites for the TNT analyte on the bisaniline–cross-linked AuNPs composite of the above non-imprinted sensor. To achieve this goal of fabricating an ultrasensitive TNT sensor, the electropolymerization of a thioaniline monolayer–modified

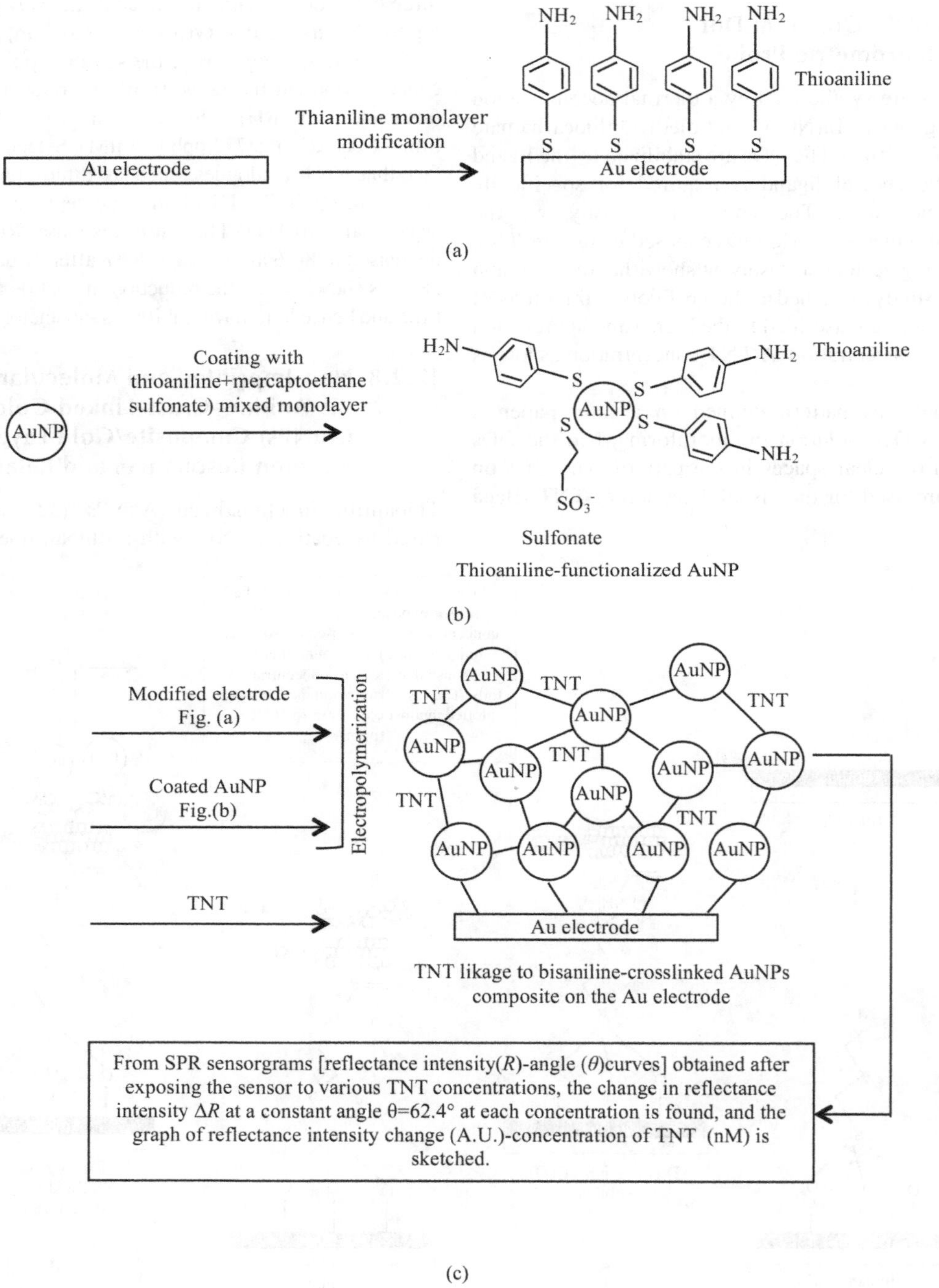

FIGURE 12.7 TNT nanosensor: (a) modification of Au electrode; (b) coating of gold nanoparticle; and (c) electropolymerization of (b) over (a) followed by TNT exposure. (Riskin et al. 2009.)

gold electrode is performed in the presence of picric acid (PA), which is the analogue of TNT (Figure 12.8). Then, π-donor–acceptor complexes are formed between the picric acid and the thioaniline-modified AuNPs. Removal of the picric acid imprint molecules leaves behind molecular contours in which the π-donor sites are optimally positioned for grabbing TNT molecules. Thus, thioaniline-modified AuNPs act as electropolymerizable nano-dimensional units for imprinted site production. The SPR shift response of the imprinted sensor at 1 pM TNT is the same as was seen for a non-imprinted sensor at 200 nM TNT, meaning that the sensitivity is drastically improved.

The calibration characteristics of the non-imprinted and imprinted sensor matrices are drawn, representing the changes in reflectance intensities with respect to TNT concentration. For the non-imprinted matrix, the change in reflectance commences at 10 pM and saturates at 1 μM. For the imprinted matrix, the reflectance calibration curve entails two steps. In the first step, the change in reflectance begins at 10 fM and levels off at 5 pM. These reflectance changes arise from the attachment of TNT to imprinted π-donor sites. In the second step, the reflectance change starts at 100 pM and becomes constant at 1 μM. The reflectance changes in this range originate from the association of TNT with non-imprinted sites.

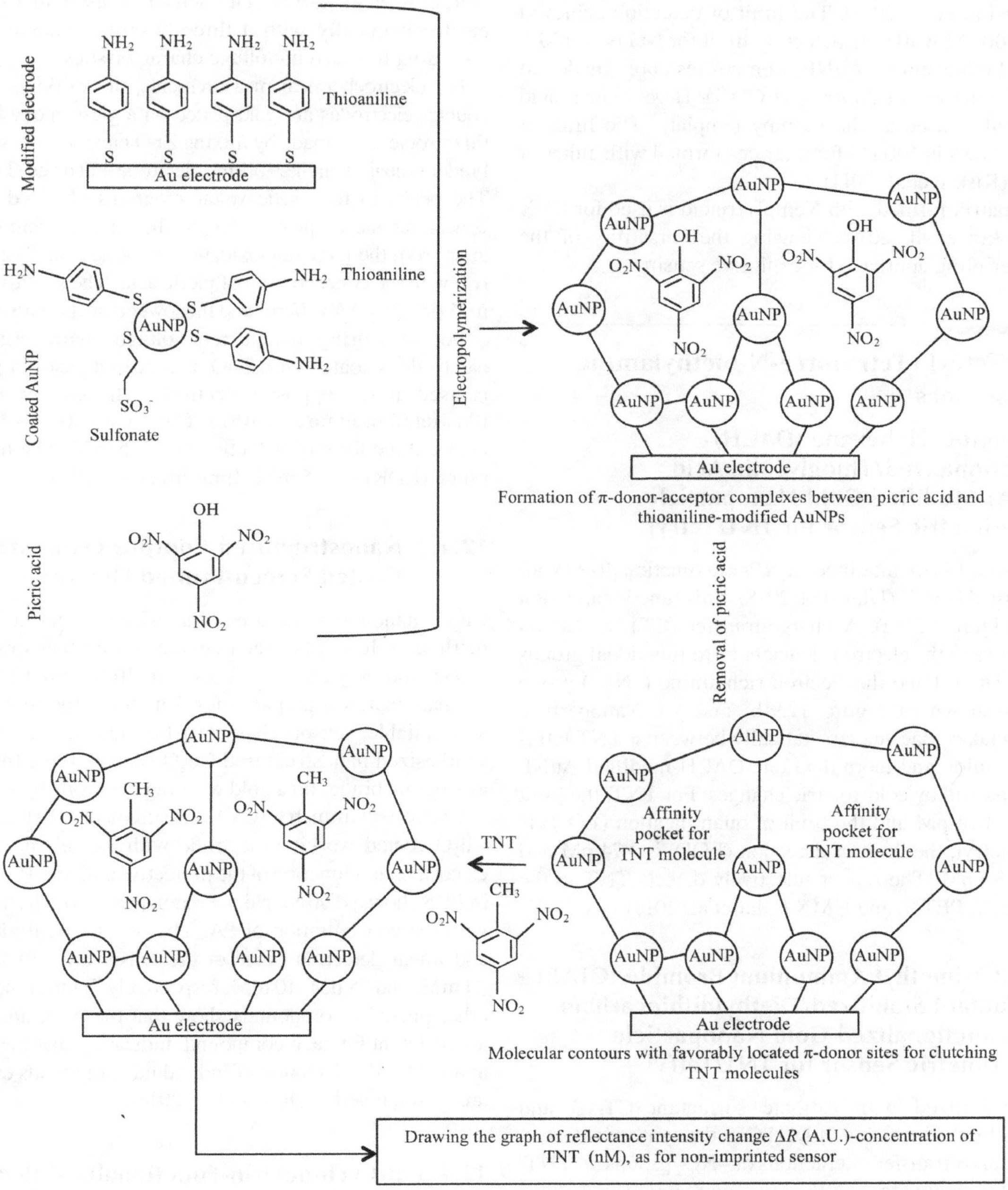

FIGURE 12.8 Implementation and working of molecular imprinted TNT sensor: electropolymerization of (thioaniline+mercaptoethane sulfonate)-coated gold nanoparticles on thioaniline-modified gold electrode in the presence of picric acid, taking out picric acid and the association of TNT molecules. (Riskin et al. 2009.)

The selectivity of the sensor toward TNT becomes evident on comparing the responses to TNT and DNT. While reflectance changes are detectable at 10 fM TNT concentration, they are apparent only at 50 pM DNT. Moreover, the saturation values are much higher for TNT than DNT. The sensitivity of the imprinted matrix for TNT is 1000 times that for DNT (Riskin et al. 2009).

In further work, a dummy template of citric acid is used during the electropolymerization of thioaniline-functionalized AuNPs on a thioaniline-modified Au surface to prepare a molecular imprinted AuNPs composite, and the subsequent removal of the imprinting template from the AuNPs matrix yields the imprinted AuNPs composite for specifically analyzing pentaerythritol tetranitrate (PETN) and nitroglycerin (NG) by surface plasmon resonance (Riskin et al. 2011). The limit of detection achieved for PETN is 200 fM while the detection limit for NG is 20 pM.

Similarly, high-affinity AuNP composites are made to determine ethylene glycol dinitrate (EGDN). Here, maleic acid or fumaric acid is used as the dummy template. The limit of detection for EGDN is 400 fM for matrices formed with either of the two acids (Riskin et al. 2011).

A sensing matrix formed with Kemp's triacid is used for RDX detection (Riskin et al. 2010), showing the versatility of the molecular imprinting approach for explosive sensing.

12.3 TNT/Tetryl (Tetranitro-N-methylamine) Nanosensors

12.3.1 Diaminocyclohexane (DACH)-Functionalized/Thioglycolic Acid (TGA)-Modified Gold Nanoparticle Colorimetric Sensor for TNT/Tetryl

Negatively charged TGA-modified AuNPs are functionalized with positively charged DACH (Ular et al. 2018). This functionalization is shown in Figure 12.9(a). A charge-transfer (CT) interaction takes place between the electron-deficient nitro functional groups ($-NO_2$) of TNT/tetryl and the electron-rich amino ($-NH_2$) group of DACH, as shown in Figure 12.9(b) and (c). Nanoparticle agglomeration takes place electrostatically between a TNT/tetryl Meisenheimer anion and more than one DACH-modified AuNP cation, accompanied by colorimetric changes. For TNT, the limit of detection is 1.76 pM and the limit of quantification (LOQ) is 5.87 pM. For tetryl, the limit of detection (LOD) is 1.74 pM and the LOQ is 5.80 pM. The sensor selectively detects TNT in the presence of RDX, PETN, and HMX (Ular et al. 2018).

12.3.2 Cetyl Trimethyl Ammonium Bromide (CTAB) Surfactant Stabilized/Diethyldithiocarbamate-Functionalized Gold Nanoparticle Colorimetric Sensor for TNT/Tetryl

AuNPs are stabilized with cationic surfactant CTAB and functionalized by the inclusion of DDTC in the synthesis (Ozcan et al. 2019). Charge-transfer interactions of $-NO_2$ groups on TNT/tetryl with DDTC bound to AuNPs cause aggregation of AuNPs, which results in color changes depending on the concentration of the TNT/tetryl. For TNT, the absorbance is measured at 534 nm and the limit of detection is 8 mg/L (3.52×10^{-5} mol/L). For tetryl, the absorbance is recorded at 458 nm and the detection limit is 0.8 mg/L (2.78×10^{-6} mol/L) (Ozcan et al. 2019).

12.4 Picric Acid Nanosensors

12.4.1 Zinc Oxide (ZnO) Nanopeanuts–Modified Screen-Printed Electrode (SPE)

White ZnO nanopeanuts are synthesized via a 7 h hydrothermal reaction at 170°C between zinc nitrate hexahydrate and polyethylene glycol (PEG) solutions in water with the pH maintained at 10.15 by the addition of ammonium hydroxide (Ibrahim et al. 2017). The sensor is used in two ways, viz., electrochemically with a three-electrode system or by simply recording the current-voltage characteristics.

For electrochemical measurements, the working, reference, and counter electrodes are gold plated on a glass epoxy board. A thick thixotropic paste made by mixing ZnO nanopeanuts with an organic binder is coated on the working electrode and dried (Figure 12.10(a)). The peaks in the cyclic voltammograms obtained with different concentrations of picric acid in phosphate-buffered saline (PBS) arise from the reduction/oxidation of picric acid. The peak currents rise with the concentration of picric acid. The sensitivity is 0.12 μA/mM or 9.23 μA/mM/cm^2 and the lower detection limit is 7.8 μM.

For measuring the current-voltage characteristics, a silver electrode is coated with ZnO nanopeanut paste. A platinum wire is used as the opposite electrode. The sensing mechanism is illustrated in Figure 12.10(b). The sensitivity is 493.64 μA/mM/cm^2 and the limit of detection is 0.125 mM. The linear dynamic range (LDR) is 1–5 mM (Ibrahim et al. 2017).

12.4.2 Nanostructured Cuprous Oxide (Cu_2O)-Coated Screen-Printed Electrode

Cu_2O nanomaterials are synthesized using a hydrothermal method at 160°C between copper nitrate trihydrate solution in water, and ethylene glycol for 8 h, 10 h, and 12 h from which the nanomaterial prepared for 8 h shows the most reproducible and reliable sensor characteristics (James et al. 2018). The synthesized nanostructured Cu_2O is coated in a thick paste with an organic binder on a gold working electrode by screen printing.

Cyclic voltammetry (CV) measurements performed using the Cu_2O-coated working electrode with Au reference and counter electrodes as a function of the protective antigen (PA) concentration in PBS show a distinct pair of redox peaks which shifts in position with the concentration of PA. The sensitivity, limit of detection, and linear detection range of the sensor are 130.4 μA/mM/cm^2, 39 mM, and 78 μM–10 mM, respectively. Interference studies with other phenolic compounds show that the peak anodic potentials are different for each compound, indicating that the sensor can be applied for the detection of individual compounds even when they are mixed together (James et al. 2018).

12.4.3 β-Cyclodextrin-Functionalized Reduced Graphene Oxide (rGO) Sensor

Interdigitated Ti (10 nm)/gold (20 nm) electrodes are defined at 20 μm gaps by electron beam lithography on an SiO_2-coated Si

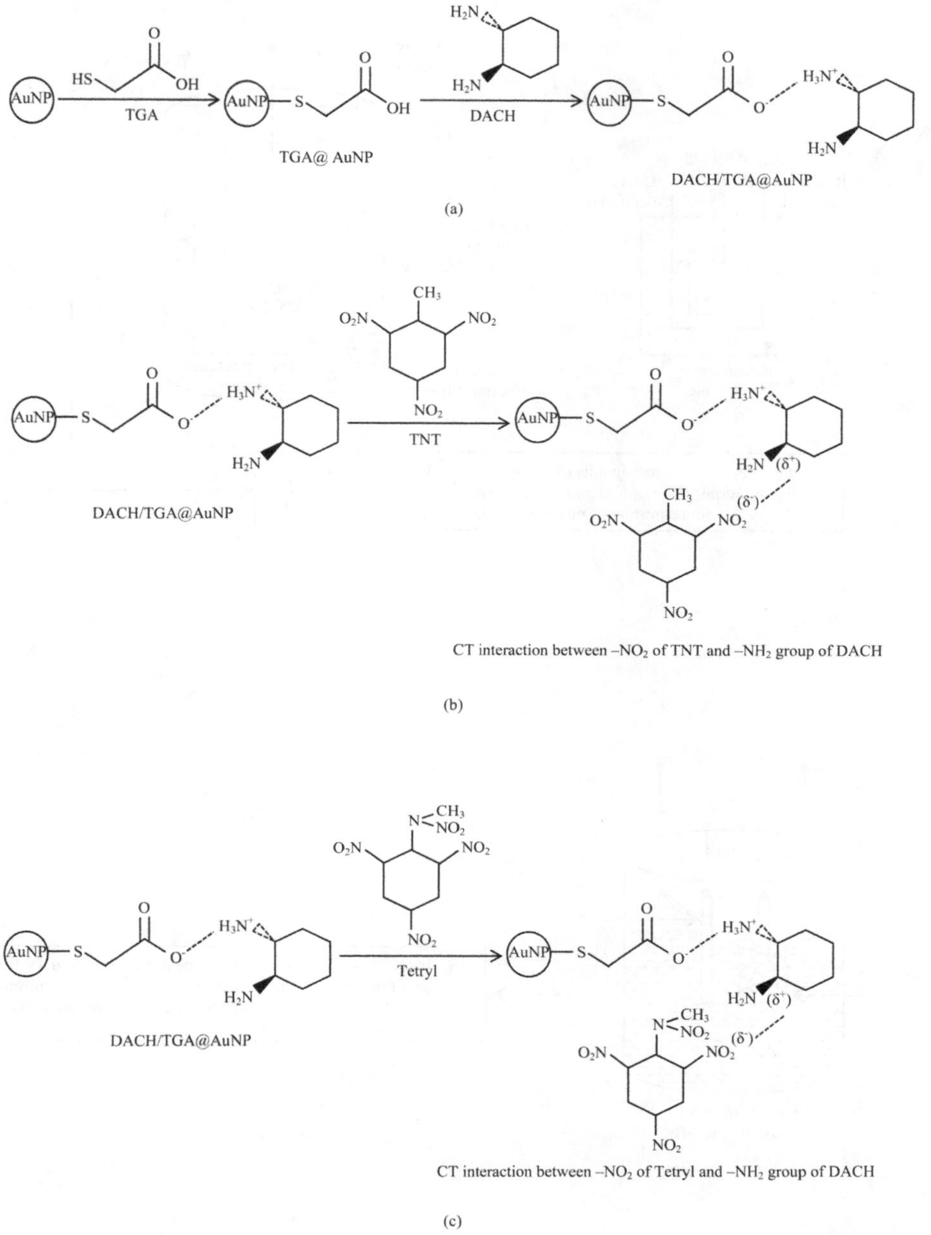

FIGURE 12.9 TGA/DACH-AuNP-based TNT/tetryl nanosensor: (a) AuNP treatment with TGA/DACH and its usage for (b) TNT detection and (c) tetryl detection. (Ular et al. 2018.)

substrate (Huang et al. 2014). After coating with graphene oxide dispersion and the formation of a homogeneous film by self-assembly, two gold wires are welded to the gold electrodes by silver paste. The GO-coated device is immersed in KH_2PO_4 solution and a cathodic potential is applied by a potentiostat when GO is electrochemically reduced to rGO. Then, the device is dipped in separately synthesized 1-pyrenebutyl-amino-β-cyclodextrin (PyCD) solution for 12 h and rinsed with deionized (DI) water. It is modified with PyCD via π-π interactions (Figure 12.11).

The β-cyclodextrin derivative molecule has a hydrophobic inner cavity but its outer surface is hydrophilic. The cavity size matches the size of the benzene molecule. This unique structure helps it to recognize PA. The $-NO_2$ groups of PA are electron-withdrawing groups whereas the lone electron pairs on the oxygen atoms in the –OH groups of β-cyclodextrin are electron-rich sites. The electron-withdrawing groups adsorb on the electron-rich sites. Further, the hydrophobic cavity of β-cyclodextrin catches the $-NO_2$ groups to form an inclusion complex.

The current response of the sensor increases with a rise in PA concentration. When the PA concentration is low, the change in current, ΔI, is 0.0095 μA. The response time is 5 s. In the PA concentration range 5–215 μM, the current changes linearly. The sensitivity is 0.00613 μA/μM and the limit of detection is 0.54 μM. The PyCD-rGO sensor has a larger response and a broader linear detection range than the bare rGO sensor (Huang et al. 2014).

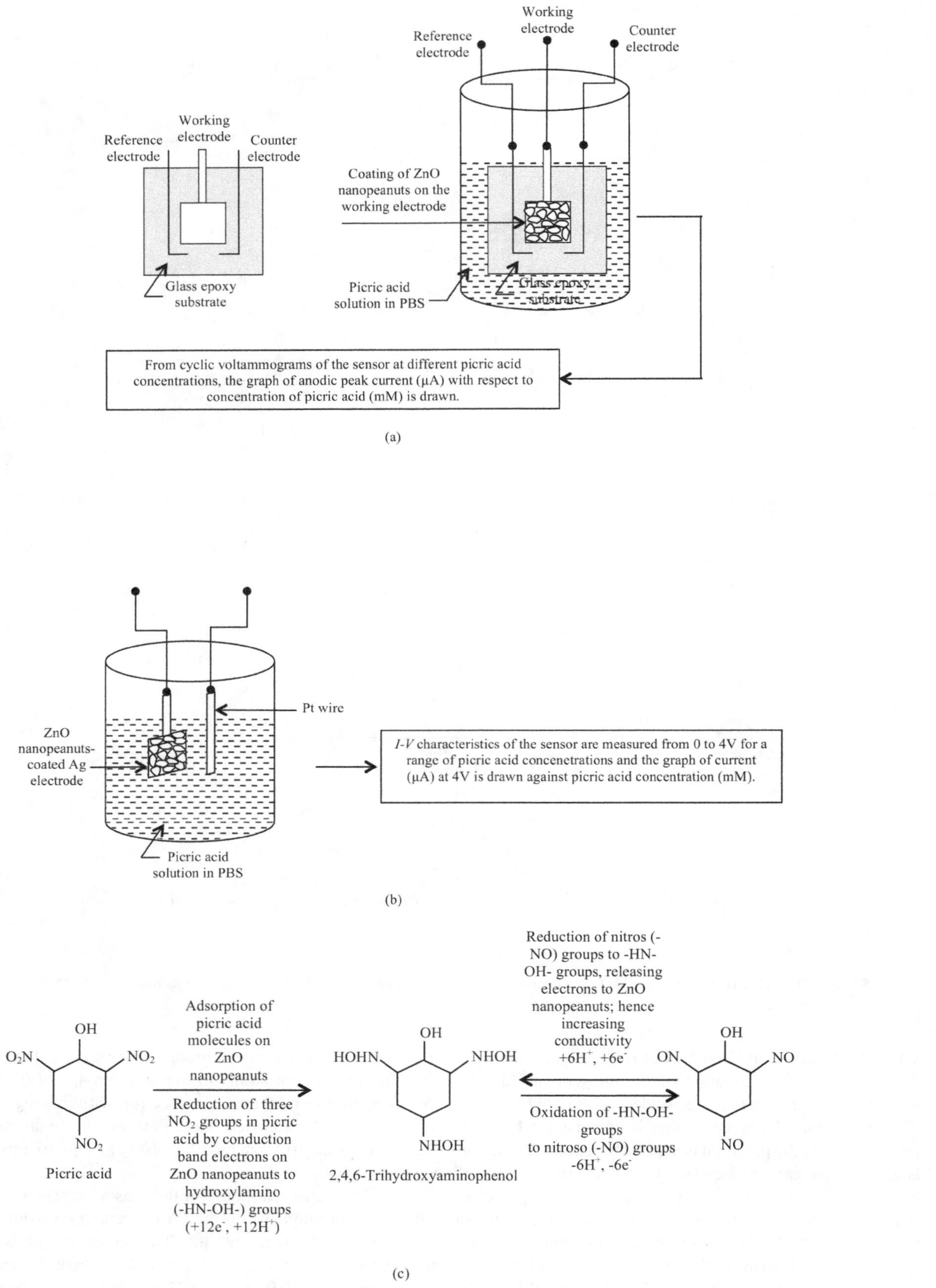

FIGURE 12.10 ZnO peanuts–based picric acid nanosensor: (a) three-electrode system for electrochemical sensor with a working electrode coated with ZnO peanuts; (b) I-V sensor using ZnO nanopeanuts–modified silver electrode; and (c) sensing mechanism of the I-V sensor. (Ibrahim et al. 2017.)

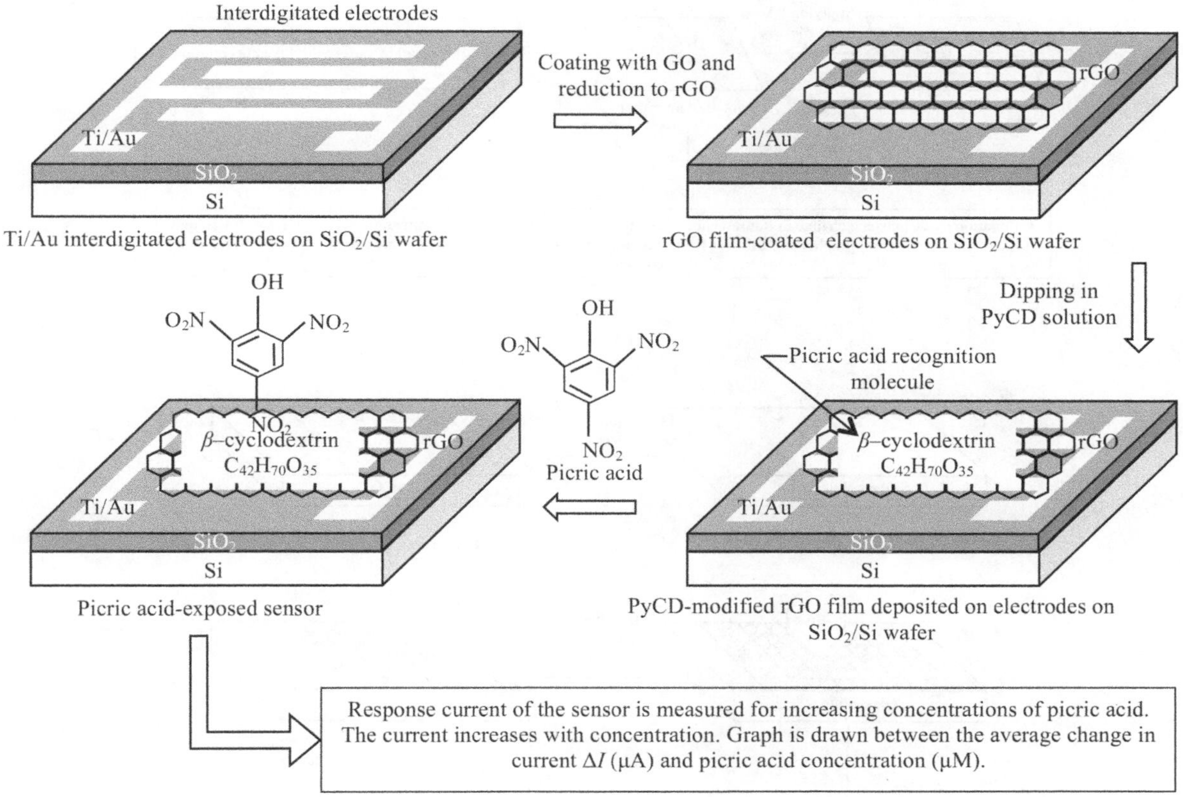

FIGURE 12.11 PyCD-rGO nanosensor for picric acid. (Huang et al. 2014.)

12.4.4 Conjugated Polymer Nanoparticles (CPNPs) Fluorescence/Current Response Sensor

Poly(3,3′-((2-phenyl-9H-fluorene-9,9-diyl)bis-(hexane-6,1-diyl)) bis(1-methyl-1H-imidazol-3-ium) bromide) (PFMI) is synthesized by the Suzuki coupling reaction (Malik et al. 2015). PFMI nanoparticles (PFMI-NPs) are formed by the re-precipitation method. The sensor is used in two detection modes:

(i) Contact mode detection: for this mode, portable fluorescent paper strips are made by immersing Whatman filter paper in a solution of PFMI-NPs in methanol and drying the solvent (Figure 12.12(a)). PA solution (10 μL) is applied on the paper for PA detection. Under 365 nm ultraviolet (UV) excitation, dark spots of different strengths are observed depending on the PA concentration. The limit of detection is 22.9 fg/cm^2.

 Fluorescence spectra are recorded for solutions of PFMI-NPs with different concentrations of PA in water by excitation at 380 nm. The sensor is calibrated by measuring the changes in the fluorescence intensity with PA concentration. The limit of detection is 30.9 pM (7.07 ppt) for PA in an aqueous solution.

(ii) Vapor phase detection: to detect PA in vapor form, a two-terminal device is made on a microscopic glass slide by depositing 150 nm-thick aluminum electrodes through a mask by thermal evaporation, drop casting 10 μL of PFMI-NPs solution in water over the space between the electrodes, and drying the solvent (Figure 12.12(b)). When PA vapors of different concentrations (0.2–1 ppb) are injected into the testing chamber and the output current of the two-terminal device is measured, it is found that the current increases with PA concentration. Negligible current changes are noticed with vapors of other nitroaromatics showing the selectivity of the response toward PA (Malik et al. 2015).

Sensing mechanism: the PA recognition sites in PFMI-NPs are the imidazolium groups attached on the side chains of PFMI. The imidazolium cation is $C_3H_5N_2^+$. The acidic phenolic –OH groups in PA readily deprotonate in aqueous solution. So, the negatively charged picrate ions $(O_2N)_3C_6H_2O^-$ are electrostatically attracted to the positively charged PFMI-NPs. High detection sensitivity is attributed to electrostatic interaction and the PET/RET mechanisms involved (Figure 12.12(c)).

12.4.5 Surface-Enhanced Raman Scattering (SERS) Using Hydrophobic Silver Nanopillar Substrates

The SERS substrate is a silver-on-silicon substrate (Hakonen et al. 2017). The surface of the base substrate is covered with silicon nanopillars formed by reactive ion etching (RIE), as shown in Figure 12.13. Using evaporation, a 225 nm-thick silver film is deposited on this substrate, producing a forest of flexible silver nanopillars. Due to the hydrophobic nature of the substrate, small

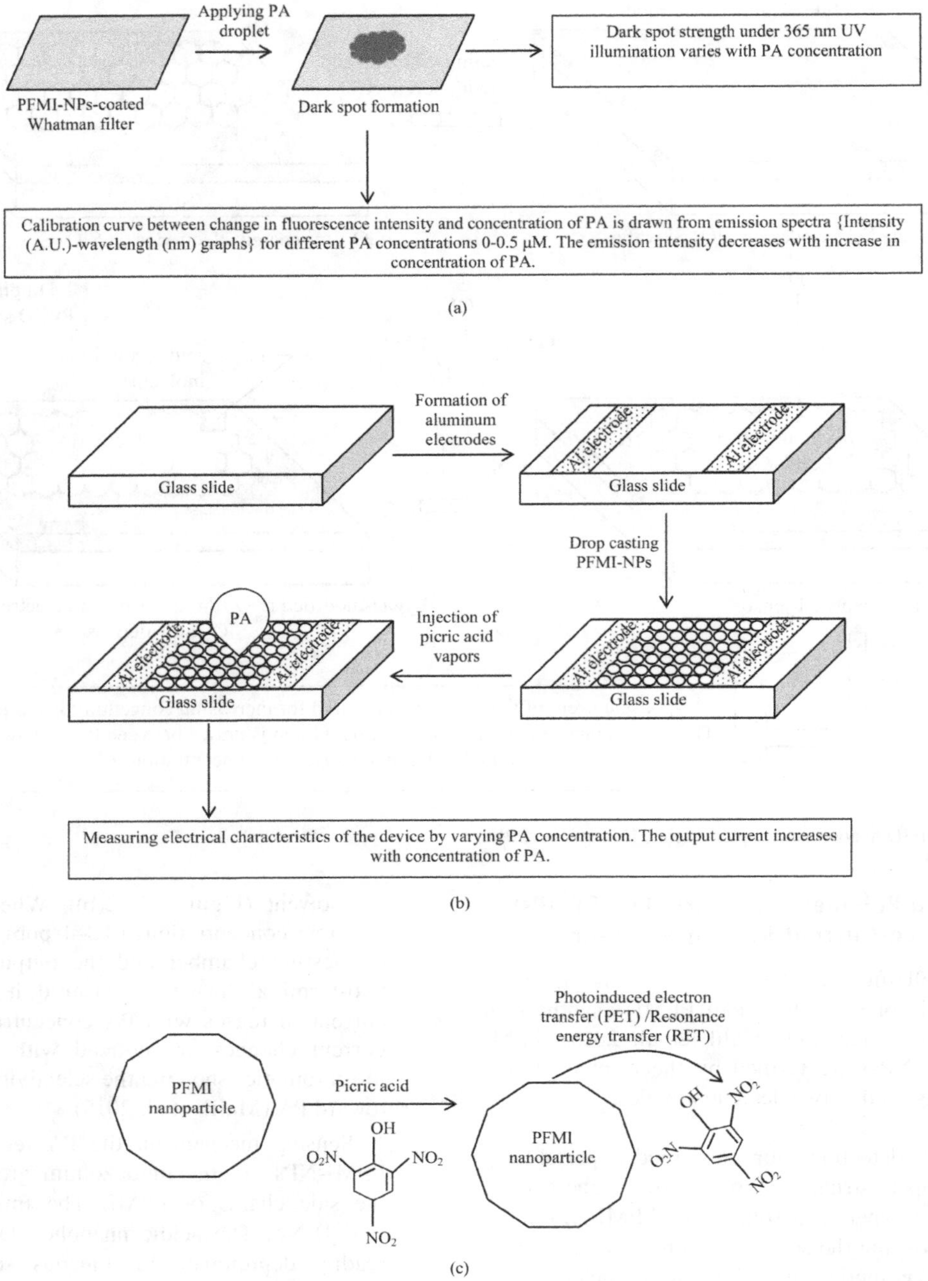

FIGURE 12.12 PFMI nanoparticle sensor for picric acid: (a) contact mode detection; (b) vapor phase detection; and (c) the underlying PET/RET phenomena. (Malik et al. 2015.)

aqueous droplets are easily formed on the substrate. They adhere to the substrate with large contact angles 120°–140°. The intensity vs. Raman shift (cm^{-1}) SERS spectra are measured with a hand-held Raman instrument. The peak heights are found to vary with PA concentration. The sensor is calibrated by plotting the peak height at 820 cm^{-1} for PA concentrations of 0–20 ppb. The lowest measured PA concentration is 0.04 ppb. The limit of detection is 18 ppt and the limit of quantification is 0.06 ppb. Rapid adhesion of the droplets together with hand-held measurements are advantageous features enabling PA direct measurements on-site outside sophisticated laboratories (Hakonen et al. 20 17).

12.5 Nanosensors for 1,3,5-Trinitro-1,3,5-Triazacyclohexane (RDX) and Other Explosives

12.5.1 Gold Nanoparticles Substrate for RDX (Cyclotrimethylenetrinitramine) Detection by SERS

Gold nanoparticles are synthesized by seed-mediated growth in which $HAuCl_4$ is reduced with borohydride in the presence of

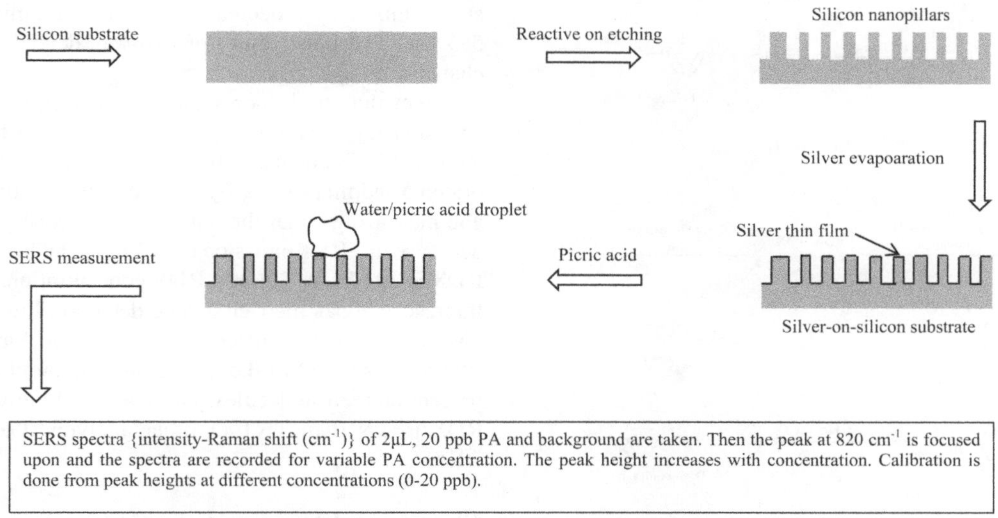
Silicon substrate
Reactive on etching
Silicon nanopillars
Silver evapoaration
Water/picric acid droplet
Silver thin film
SERS measurement
Picric acid
Silver-on-silicon substrate
SERS spectra {intensity-Raman shift (cm^{-1})} of 2μL, 20 ppb PA and background are taken. Then the peak at 820 cm^{-1} is focused upon and the spectra are recorded for variable PA concentration. The peak height increases with concentration. Calibration is done from peak heights at different concentrations (0-20 ppb).

FIGURE 12.13 SERS silver nanopillar substrate–based picric acid sensor. (Hakonen et al. 2017.)

trisodium citrate (Hatab et al. 2010). The citrate acts as a capping agent in the solution. The average diameter of AuNPs is 90–100 nm and the NPs are spherical to diamond shaped. Specimens for SERS analysis are made by mixing an AuNPs suspension with standard RDX solutions of varying concentrations, 10^{-4} to 10^{-6} M. Droplets of the AuNPs-RDX suspension are placed on a glass slide and dried in air.

The Raman band for RDX is strongest at 874 nm^{-1} and the intensity increases with RDX concentration. A calibration graph is obtained by integrating the characteristic peak area at 874 nm^{-1} and plotting the band area centered at 873 nm^{-1} with respect to the RDX concentration. The graph is linear in the range 1×10^{-6} M to 5×10^{-5} M. But when the RDX concentration is $>5 \times 10^{-5}$ M, the graph becomes nonlinear. The limit of detection is 1×10^{-6} M or 0.22 mg/L (Hatab et al. 2010).

12.5.2 4-Aminothiophenol (4-ATP)-Functionalized Gold Nanoparticle Colorimetric Sensor for RDX (Cyclotrimethylenetrinitramine)/HMX (Octahydro-1,3,5,7-Tetran itro-1,3,5,7-Tetrazocine)

Gold nanoparticles are synthesized by the classical citrate reduction method (Üzer et al. 2013). After pH adjustment of an AuNP solution at 3, the acidic solution is mixed with 4-ATP and stabilized at room temperature for 48 h to obtain 4-ATP-AuNPs (Figure 2.14(a)). For the assay, first the hydrolysis of RDX/HMX is carried out forming nitrite (NO_2^-) (Figure 12.14(b)). The diazonium salt formed in Figure 12.14(c) couples with naphthyl-ethylene diamine (NED) in Figure 12.14(d). The method is called the 4-ATP-AuNPs + NED method because the hydroxylate formed from RDX/HMX is spectrophotometrically examined using naphthyl-ethylene diamine as a coupling agent to yield azo dye. Briefly, the steps are: hydroxylate (Figure 12.14(b)) +HCl + 4-ATP-AuNPs (Figure 12.14(a)) +HPO4 (for pH = 2) +NED followed by an absorbance measurement 30 min after adding NED.

RDX and HMX are demarcated on the basis of their differential kinetics in hydrolysis. After room temperature hydrolysis in (sodium carbonate + sodium hydroxide) solution, only RDX is hydrolyzed to give a sufficient colorimetric response in the neutralized solution; HMX gives no response. The molar absorptivity for RDX at 565 nm is 17.6×10^3 L/mol/cm. The limit of detection for RDX is 0.55 μg/mL.

Contrarily, using a 60°C hot water bath hydrolysis, adequate colorimetric responses are recorded for both RDX and HMX. The molar absorptivity for RDX is 32.8×10^3 L/mol/cm and its limit of detection is 0.20 μg/mL. The molar absorptivity and detection limit for HMX are 37.1×10^3 L/mol/cm and 0.24 μg/mL, respectively (Üzer et al. 2013).

12.5.3 Cadmium Sulfide-Diphenylamine (CdS QD-DPA) FRET-Based Fluorescence Sensor for RDX (Cyclotrimethylenetrinitramine)/PETN (Pentaerythritol Tetranitrate)

This is a sensor that works on fluorescence quenching and its restoration in the presence of an analyte (Ganiga and Cyriac 2015). In this FRET sensor (Figure 12.15), DPA is the donor and CdS QD is the acceptor. Resonance energy transfer takes place from DPA to CdS QD. As a result, the intrinsic fluorescence of DPA is quenched. DPA is chosen because of its high selectivity to nitroamines and nitroesters. Furthermore, the amine group of DPA binds with the carboxylate group of 3-mercaptopropionic acid (MPA) in MPA-capped CdS QDs, strengthening the donor–acceptor interaction. Lastly, the DPA (donor) has a high fluorescence emission at 330–430 nm while the absorption wavelength of the CdS QD (acceptor) is 350–440 nm. Consequently, the fluorescence of DPA is quenched by CdS QD.

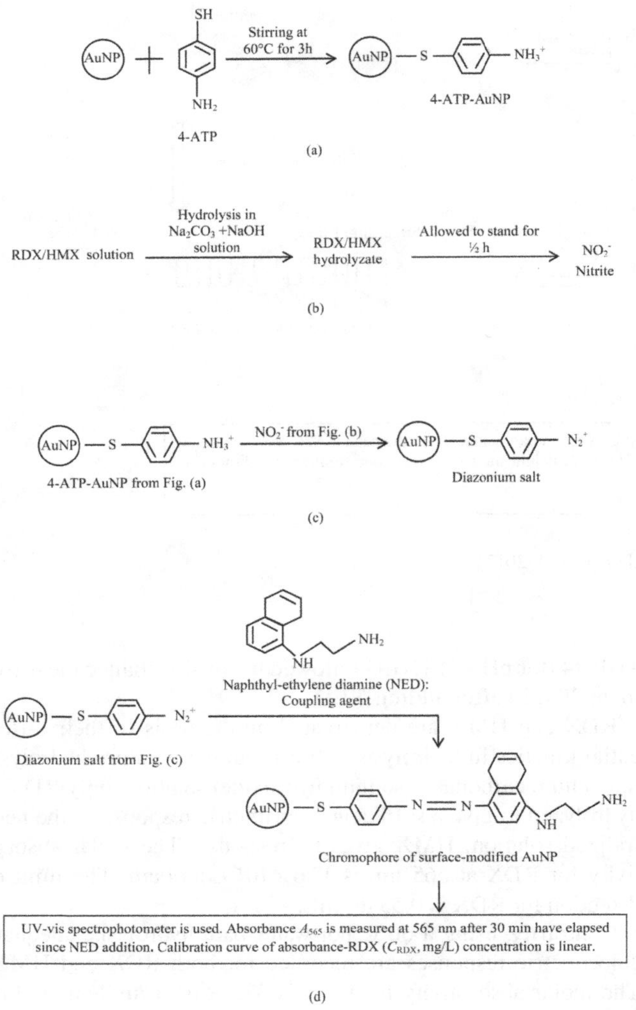

FIGURE 12.14 The 4-ATP-functionalized AuNP colorimetric sensor for RDX/HMX (a) functionalization of gold nanoparticles with 4-ATP; (b) nitrite production from the hydrolysis of RDX/HMX; (c) formation of diazonium salt by reaction between 4-ATP-AuNP from (a) and nitrite from (b); (d) coupling of diazonium salt with NED to form the chromophore of a surface-modified AuNP. (Üzer et al. 2013.)

CdS QDs coated with 3-mercaptopropionic acid are prepared by mixing $CdCl_2$, H_2O with thiourea in DI water, adding MPA to the mixture, adjusting the pH value at 10, maintaining in an autoclave at 100°C for 2 h, and cooling to room temperature. Adequate FRET emission is observed on adding 250 µL of 100 nM DPA solution in ethanol to 3 mL of CdS solution, which is referred to as the "CdS QD-DPA sensor."

When RDX or PETN molecules act on a CdS QD-DPA sensor, they break the interaction between CdS QD and DPA, thereby restoring the quenched fluorescence of DPA. This restoration, or turning on of fluorescence, occurs at 355 nm. Another approach to look at RDX/PETN sensing is turning off the resonance energy transfer responsible for fluorescence quenching. The turning-off wavelength is 585 nm. Thus, the RDX/PETN explosive detection can be pursued in two ways, either from fluorescence evolution at 355 nm or by a decrease in the FRET emission intensity at 585 nm. The opportunity of two-way detection minimizes the chances of false positives.

The evolution of fluorescence or the diminution of the FRET emission intensity takes place in proportion to the amount of RDX or PETN and hence their concentrations. The sensor is calibrated by adding successively increasing amounts of RDX/PETN and measuring either the fluorescence recovery at 355 nm or the decrease in FRET emission at 585 nm. The limit of detection for RDX is 20 nM. At 200 nM RDX concentration, 60–65% of DPA fluorescence has recovered. The detection limit for PETN is 10 nM (3200 pg). The difference in detection limits for RDX and PETN is ascribed to the nature and the number of nitrogroups present on their molecules. The sensor is highly selective toward RDX/PETN than TNT or other analytes (Ganiga and Cyriac 2015).

12.5.4 Gold Nanoparticles/Nitroenergetic Memory-Poly(Carbazole-Aniline) P(Cz-*co*-ANI) Film-Modified Glassy Carbon Electrode (GCE) for RDX (Cyclotrimethylenetrinitramine), TNT (2,4,6-Trinitrotoluene), DNT (2,4-Dinitrotoluene), and HMX (Octahydro-1,3,5,7-Tetranitro-1,3,5,7-Tetrazocine) Detection

These are electrochemical sensors that use square wave voltammetry (SWV) with tetra-n-butylammonium bromide (TBABr) as the supporting electrolyte and acetonitrile (ACN) as the solvent (Sağlam et al. 2018). For selective detection, separate sensor electrodes are prepared for each explosive. The GCE modification procedure is shown in Figure 12.16 taking RDX as an example.

The general procedure for electrode preparation entails the following steps: (i) the glassy carbon electrode is coated with the template of the relevant energetic material and carbazole-aniline monomers, e.g., RDX template-GCE/P(Cz-*co*-ANI), TNT template-GCE/P(Cz-*co*-ANI), DNT template-GCE/P(Cz-*co*-ANI), HMX template-GCE/P(Cz-*co*-ANI) using RDX, TNT, DNT, and HMX as template molecules; P(Cz-*co*-ANI) stands for poly(carbazole-aniline) copolymer film. The coating method used is electrochemical polymerization. (ii) The modified surfaces of electrodes, imprinted for molecular recognition of different explosives, are functionalized by electrodeposition of AuNPs. (iii) The template is extracted to get RDX memory-GCE/P(Cz-*co*-ANI)-AuNPs, TNT memory-GCE/P(Cz-*co*-ANI)-AuNPs, DNT memory-GCE/P(Cz-*co*-ANI)-AuNPs, HMX memory-GCE/P(Cz-*co*-ANI)-AuNPs.

From voltammetric measurements, the reduction potential of RDX is –1.03 V, that of TNT and DNT is –1.0 V, and that of HMX is –1.05 V. The calibrations of the four electrodes for TDX, TNT, DNT, and HMX are carried out at the respective reduction potentials. It is found that the currents are proportional to concentrations of analytes. For RDX, the linear range of detection is 50–1000 µg/L and the limit of detection is 10 µg/L. For TNT, the linear range of detection is 100–1000 µg/L and the limit of detection is 25 µg/L. These parameters for DNT are 100–1000 µg/L and 30 µg/L while the parameters for HMX are 50–1000 µg/L and 10 µg/L (Sağlam et al. 2018).

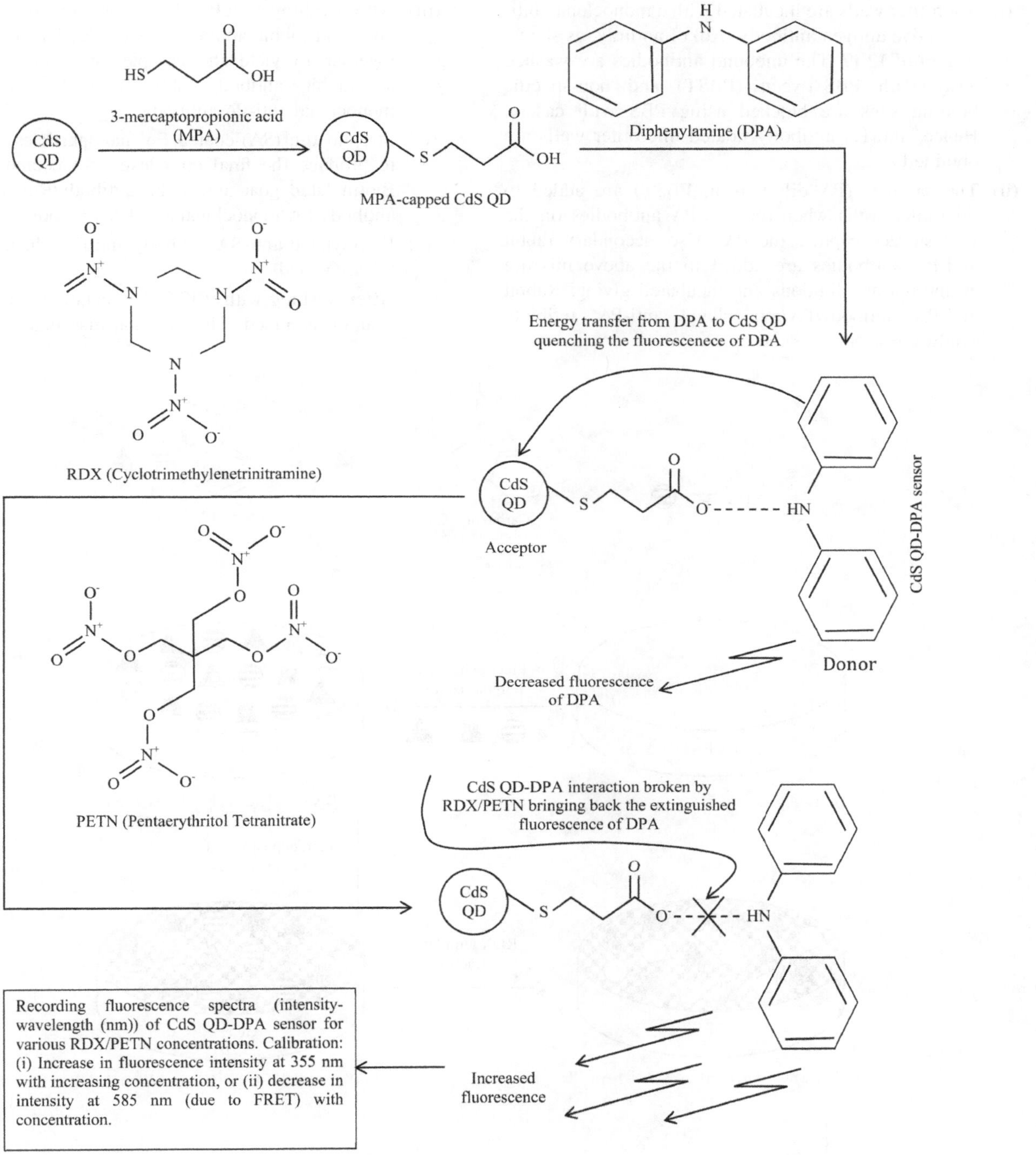

FIGURE 12.15 CdS QD-DPA nanosensor for RDX/PETN. (Ganiga and Cyriac 2015.)

12.6 Nanosensor Requirements for Detection of Biothreat Agents

While explosives destroy practically everything, viz., buildings, installations, and human populations, biothreat agents such as those given in Table 12.2 exclusively target human beings leaving the infrastructure unharmed.

12.7 Anthrax Spore Nanosensors

12.7.1 Europium Nanoparticle (Eu^+ NP) Fluorescence Immunoassay (ENIA) for *Bacillus anthracis* Protective Antigen

The *Bacillus anthracis* protective antigen is determined through the following steps (Tang et al. 2009):

(i) Microtiter wells are incubated with a monoclonal anti-protective antigen antibody (Ab) of anthrax, as shown in Figure 12.17. The unbound antibodies are washed away with PBS-Tween (PBST) and non-specific binding sites are blocked using PBS with casein. Hence, anti-PA antibody–coated microtiter wells are obtained.

(ii) The samples (PA dilutions in PBST) are added to microtiter wells when the anti-PA antibodies on the well surface capture the PA. Also, secondary rabbit anti-PA antibodies are added to the above mixture at appropriate dilutions and incubated giving: Rabbit anti-PA antibody/PA/monoclonal anti-PA antibody sandwiches.

(iii) After washing with PBST, suitably diluted biotinylated goat anti-rabbit antibodies are added and incubated together to yield the complexes: Biotinylated goat anti-rabbit antibody/Rabbit anti-PA antibody/PA/ monoclonal anti-PA antibody.

(iv) Streptavidin (SA)-coated Eu^+ nanoparticles are added to produce the final complexes: SA-coated Eu^+ NP/ Biotinylated goat anti-rabbit antibody/Rabbit anti-PA antibody/PA/monoclonal anti-PA antibody.

(v) Biotinylated anti-SA antibody and Eu^+ chelate-labeled SA are added.

(vi) After washing with PBST, the chelating enhancement solution is added. This solution dissociates Eu^+ from

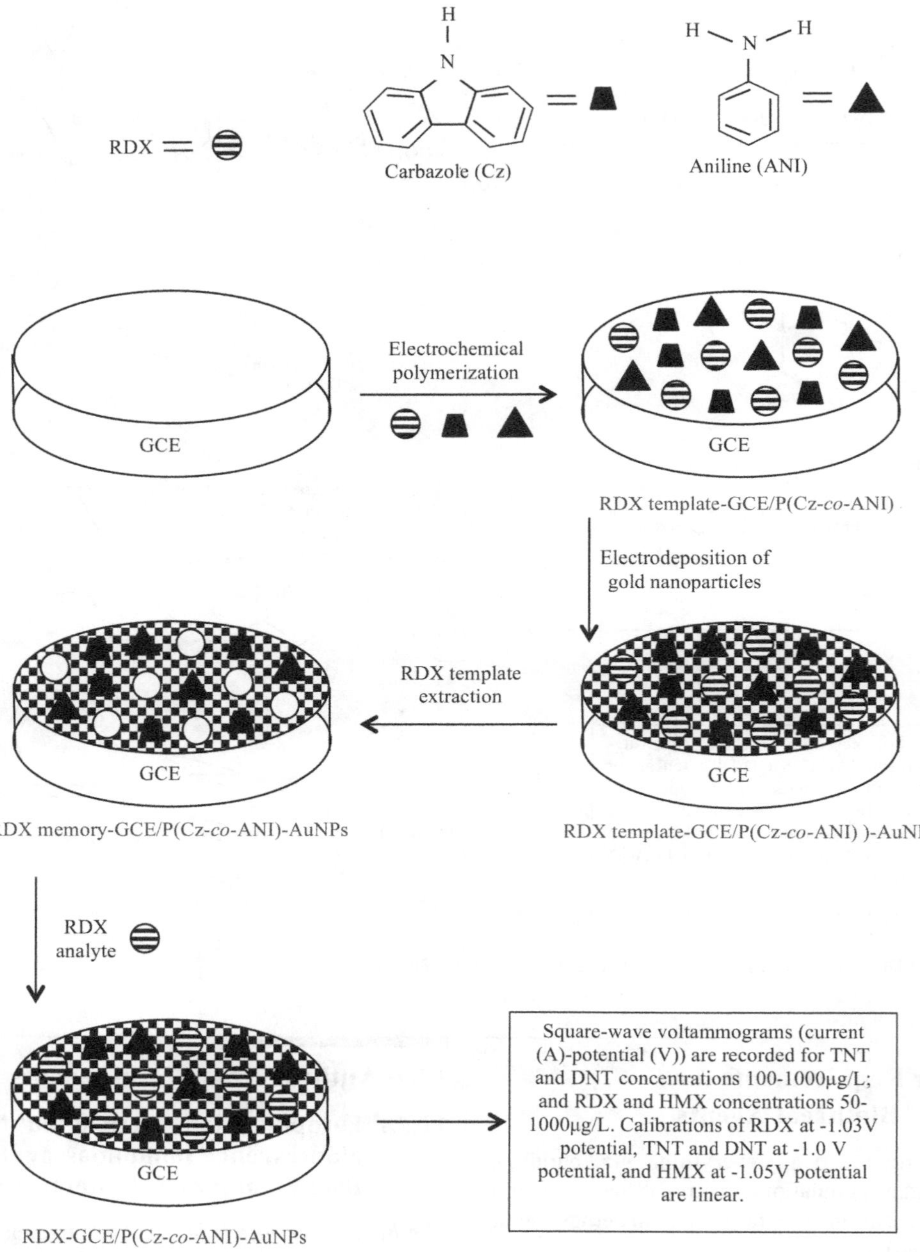

FIGURE 12.16 Modification of glassy carbon electrode for RDX sensing. (Sağlam et al. 2018.)

TABLE 12.2
Some Biothreat Agents

Sl. No.	Name of Biothreat Agent	Infection/Exposure and Effects
1.	Anthrax spores	Infection occurs by inhalation of spores or through contacting infected animals. Symptoms include difficulty in breathing, skin ulcers, etc.
2.	Plague bacterium (*Yersinia pestis*)	Infection occurs through Oriental rat flea. Symptoms: fever, headache, etc.
3.	*Francisella tularensis* bacterium	Causes tularemia (rabbit fever). Symptoms: fever, skin ulcers.
4.	Brucellosis bacterium	Spread through contaminated food such as raw meat; also through air or contact with open wound. Symptoms: similar to flu, appetite loss, backache, etc.
5.	Smallpox virus (Variola)	Spreads via blood products, coughs/sneezes, handshakes, or touching contaminated objects. Symptoms: High fever, vomiting, skin rash on face, forearms, torso, etc.
6.	Ebola virus	Spreads through blood products, saliva or contact with objects contaminated with body fluids. Causes hemorrhagic fever, body aches, diarrhea.
7.	Ricin toxin	A poisonous lectin found in the seeds of castor oil plant. Affects by inhalation, ingestion, skin or eye exposure.
8.	Staphylococcal Enterotoxin B (SEB) toxin	Toxic by inhalation or ingestion. Poisoning is through an aerosol or in food, water, etc. Causes fever, chills, nausea, vomiting, diarrhea, etc.
9.	Aflatoxin	Exposure to aflatoxin occurs by eating aflatoxin-contaminated plant products such as peanuts or by consuming meat, eggs, or milk products from animals that have eaten aflatoxin-tainted fodder. It has carcinogenic effects causing liver damage and liver cancer.

the solid phase attached to the antibodies. Thus, the released Eu^+ forms a homogeneous and high fluorescence solution. Hence, the Eu^+ NPs enhance the intensity of the fluorescence signal.

(vii) The fluorescence signal is recorded by a fluorometer.

The fluorescence signal intensity varies linearly with the concentration of *B. anthracis* protective antigen in the detection range 0.01–100 ng/mL. The sensitivity is 100 times that of the enzyme-linked immunosorbent assay (ELISA) (Tang et al. 2009).

12.7.2 Gold Nanoparticle–Amplified DNA Probe-Functionalized Quartz Crystal Microbalance (QCM) Biosensor for *B. Anthracis* at Gene Level

A DNA probe 1 is developed for *B. anthracis* detection (Hao et al. 2011). The DNA probe is designed to recognize *B. anthracis* on the gene level toward its specific sequences on two sites: 168 base pair (bp) fragment of Ba813 on the *B. anthracis* chromosome and the 340 bp fragment of the protective antigen (*pag*) gene on the *B. anthracis* plasmid PXO1. Based on this development, the thiol DNA probe 1 is immobilized on the gold surface of the quartz crystal microbalance wafer through self-assembly via gold-sulfur (Au-S) bonding (Figure 12.18). The active sites on the Au surface of QCM are blocked by washing with 6-mercaptohexan-1-ol (6-MHO) to prevent non-specific adsorption during the ensuing test. Thus, the QCM biosensor is formed.

Asymmetric PCR amplified target ssDNA sequences from *B. anthracis* are made to interact with the DNA probes on the QCM surface. Hybridization takes place between the DNA probe 1 and the target ssDNA. The hybridization results in the formation of DNA probe 1–Target ssDNA complexes. As a consequence, the mass of the QCM biosensor increases. The mass increment affects the resonance frequency of QCM. The resonance frequency decreases. It is necessary to amplify the mass increase so that the resonance frequency diminishes by a larger amount. With this intention, complementary DNA (cDNA) probes 2 to the target DNA functionalized with gold nanoparticles are kept ready. These cDNA probe 2-AuNPs are bound to the opposite ends of the target ssDNA. The binding completes the structures: DNA probel-Target ssDNA-cDNA probe2-AuNPs. The additional increase in the mass of the QCM biosensor and the extra resonance frequency of the QCM diminution due to this mass loading combine together with the initial mass/frequency effects due to target ssDNA alone to produce an enhanced overall effect, thereby helping in *B. anthracis* DNA detection and improving the sensitivity of the biosensor.

The limit of detection is 3.5×10^2 CFU/mL of *B. anthracis* vegetative cells. The typical detection time is 30 min. The sensor easily identifies and distinguishes *B. anthracis* DNA from a similar associated species *B. thuringiensis* (Hao et al. 2011).

12.8 Rapid Screening Lateral Flow Plague Bacterium (*Yersinia pestis*) Nanosensor

The detection of sylvatic plague is based on the detection of specific antibodies to the fraction 1 (F1) capsular antigen and the recombinant virulence-associated V-antigen (LcrV protein) found in plague pathogen *Yersinia pestis*, the causative agent of plague. It is done using a lateral flow cassette (Abbott et al. 2014).

On the opposite ends of the lateral flow cassette are the sample pad and the absorbent pad (Figure 12.19). The analyte solution is applied to the sample pad. The conjugate pad follows the sample pad. This pad is produced by impregnating a Whatman pad with F1/CMLNPS or V/CMLNPS conjugate and Biotin/CMLNPS conjugate; CMLNPS = carboxylate-modified latex nanoparticles. These conjugates are prepared as follows: (1-ethyl-3-(3-dimeth ylaminopropyl)carbodiimide) (EDC)/(N-hydroxysuccinimide) (NHS) coupling is used to attach purified F1- and V-antigen proteins with blue CMLNPS to produce the conjugate:

F1-antigen/CMLNPS and V-antigen/CMLNPS

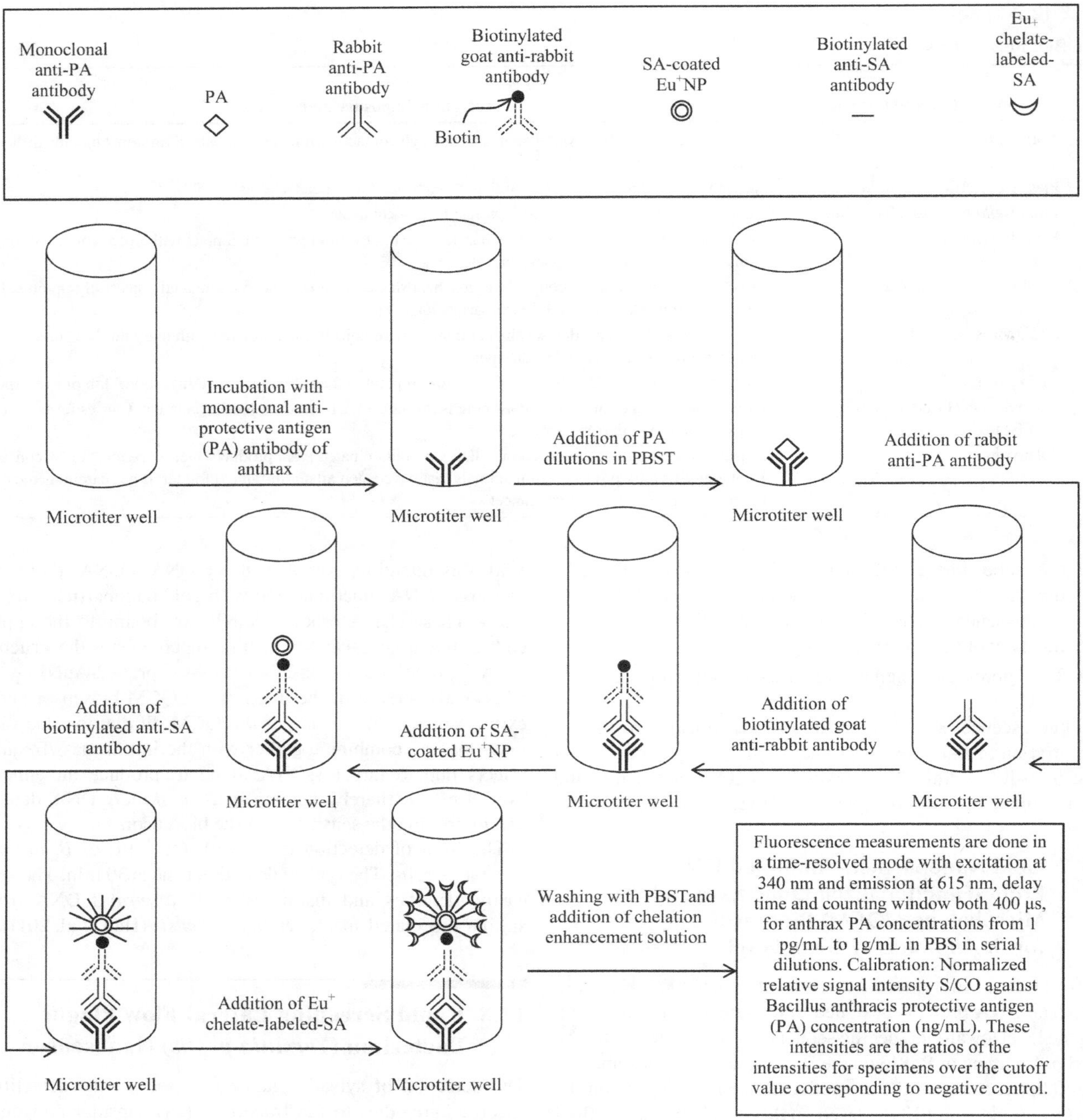

FIGURE 12.17 Eu^+ NP immunoassay for detecting the *Bacillus anthracis* protective antigen. (Tang et al. 2009.)

Biotin, bovine serum albumin (BSA), and red CMLNPS are combined to produce the conjugate:

Biotin/CMLNPS

After drying at 37°C, the pads are stored in a desiccator.

A short distance away from the conjugate pad is a line called the test line. Further away and before the absorbent pad is another line called the control line. Protein G is deposited on the test line, while the control line is coated with streptavidin. Again, drying at 37°C is followed by storage in a desiccator. Finally, the strips are cut into 4 mm-wide pieces and assembled in plastic packages.

During testing, the serum or eluate is added to the sample well and buffer solution is poured in the buffer well. The solution flows over the sample pad and through the nitrocellulose membrane of the lateral flow cassette to reach the conjugate pad. If any *Yersinia pestis* antibody, either F1-antibody or V-antibody, is present in the sample, it binds with the corresponding antigen-containing nanoparticle: F1-antigen/CMLNPS or V-antigen/CMLNPS in the conjugate pad, to produce the relevant complex:

F1-antibody/F1-antigen/CMLNPS

V-antibody/V-antigen/CMLNPS

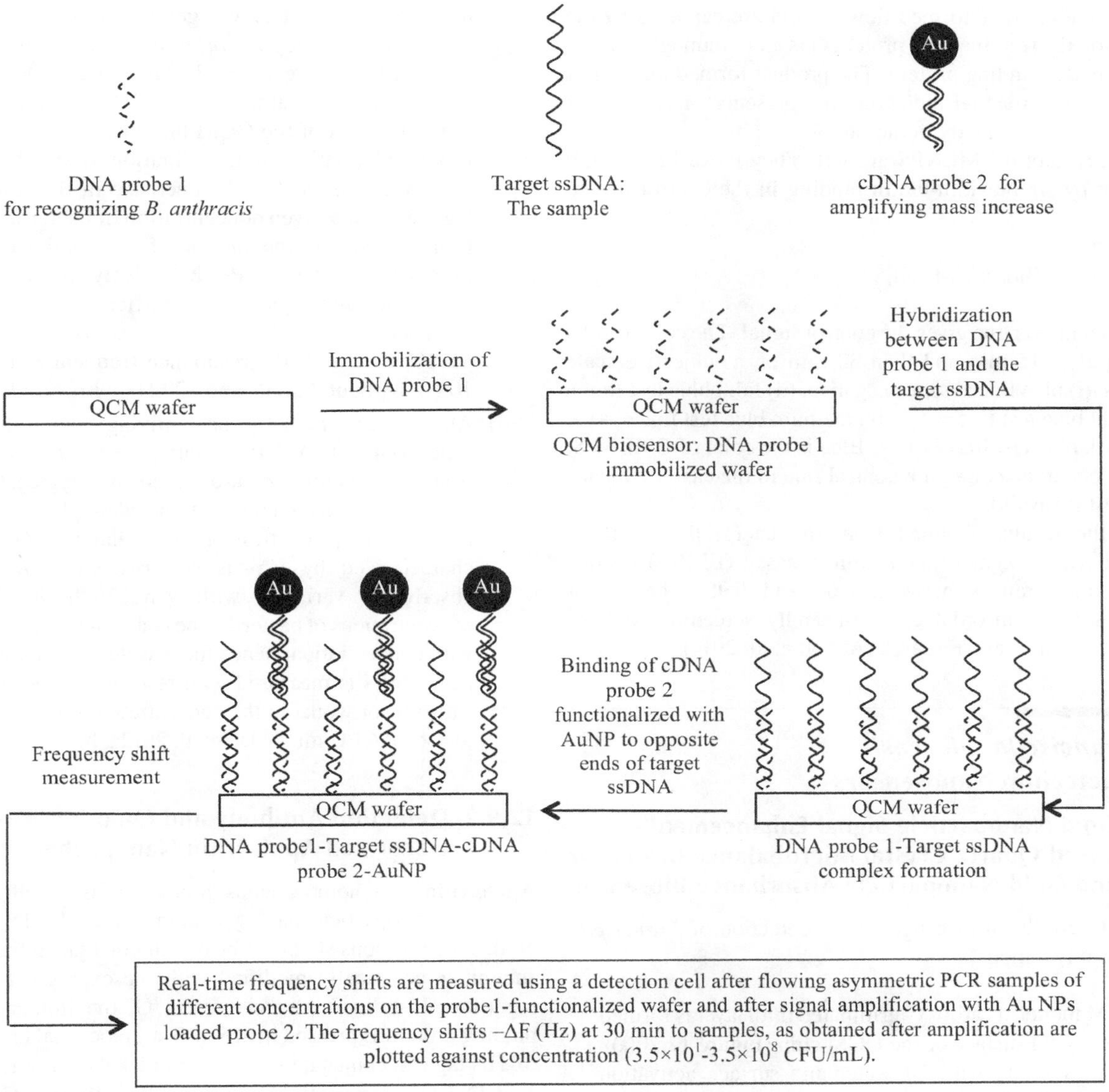

FIGURE 12.18 QCM biosensor for *B. anthracis* at gene level. (Hao et al. 2011.)

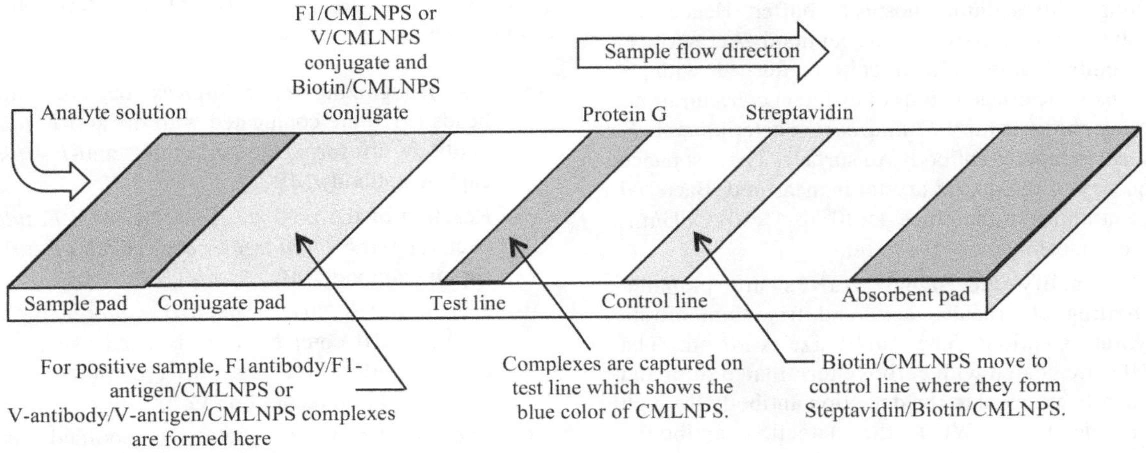

FIGURE 12.19 Lateral flow cassette for detecting plague. (Abbott et al. 2014.)

The complexes thus formed flow on and are captured by the protein G on the test line. The protein G is an immunoglobulin G (IgG) antibody–binding protein. The product formed on the test line gives the test signal indicating the presence or absence of F1-antibody or V-antibody in the sample.

Now, only Biotin/CMLNPS are left. They move further and are caught by streptavidin-biotin binding in the control line to form:

Streptavidin/Biotin/CMLNPS

The product in this line gives the control signal. The color results are noted after 15 min and then 30 min on a subjective scale defined as: (i) blue line absent = negative, (ii) faint blue test line = +, (iii) light blue test line = ++, (iv) medium blue test line = +++, and (v) dark blue test line = ++++. Blue is the color of CMLNPS. The non-appearance of a pink control line in the cassette means that the test is invalid.

When the results obtained from the lateral flow test are compared with enzyme-linked immunoassay (ELISA) results, the correlation ranges between 0.68 and 0.98. The sensor provides a rapid, affordable, user-friendly screening tool for plague surveillance and research (Abbott et al. 2014).

12.9 *Francisella tularensis* Bacterium Nanosensors

12.9.1 Gold Nanoparticle Signal Enhancement–Based Quartz Crystal Microbalance Biosensor and Gold Nanoparticle Absorbance Biosensor

Kleo et al. (2012a, b) investigated the detection of *Francisella tularensis* in two models:

(i) QCM model: a carboxy-terminated thiol layer is formed on the gold surface of the QCM chip (Figure 12.20(a)). After washing with DI water and surface activation using carbodiimide crosslinker chemistry, the chip is incubated with anti-*F. tularensis* antibody solution in sodium phosphate buffer followed by incubation in ethanolamine solution for surface saturation and flushing with sodium phosphate buffer. Hence, we get anti-*F. tularensis* capture antibody/Au surface. The antibody-immobilized chip is flushed with *F. tularensis* bacteria solutions of different concentrations forming the structure: *F. tularensis* bacterium/anti-*F. tularensis* capture antibody/Au surface. The resonance frequency of the quartz crystal is measured. Bacterial concentrations in the range 4×10^9 to 4×10^3 CFU/mL are detectable in about 15–20 min.

To amplify the signal, AuNPs are prepared by boiling chloroauric acid and trisodium citrate dihydrate solutions. The AuNP size is 20 nm. The AuNPs are coated with carboxy-terminated thiol and covalently attached to the detection antibody through an amide bond. When the detection antibody–functionalized AuNPs are added to the structure: *F. tularensis* bacterium-anti-*F. tularensis* capture antibody/Au surface, we get the structure: AuNP/detection antibody-*F. tularensis* bacterium/anti-*F. tularensis* capture antibody/Au surface. Due to the extra mass of the attached AuNP/detection antibody, the total mass of the QCM biosensor is substantially increased relative to the situation when this extra mass was absent. So, the decrease in the resonance frequency at a given concentration of bacteria is more pronounced with the antibody-functionalized AuNPs than without these NPs. Effectively, the change in the resonance frequency at a particular concentration is raised, e.g., at the *F. tularensis* concentration of 4×10^8 CFU/mL, the resonance frequency change is 10 Hz without AuNPs and ~27 Hz with AuNPs.

(ii) Absorbance model: on mixing an antibody-functionalized AuNPs solution with *F. tularensis* bacteria solution (Figure 12.20(b)), aggregation of gold nanoparticles and bacteria takes place affecting the optical properties of the solution, which are characterized by UV-vis spectroscopy showing the absorbance variation with wavelength at different concentrations of bacteria. The peaks of the spectra shift with concentration; hence the wavelength by which the peak shifts is measured with respect to concentration. This sensor works in the concentration range of 4×10^5 to 4×10^8 CFU/mL (Kleo et al. 2012a, b).

12.9.2 Detection Antibody and Quantum Dots Decorated Apoferritin Nanoprobe

Apoferritin is a homogeneous protein. It is a self-assembled spherical nanoparticle having a diameter of 10–15 nm (Kim et al. 2015). It consists of 24 heavy and/or light subunits. Each subunit is genetically modified with the expression of 6x-His tag (hexa histidine-tag) and protein G, the immunoglobulin-binding protein expressed in group G streptococcal bacteria. The 6x-His tag can conjugate with several nickel (Ni)-nitrilotriacetic acid (NTA)-functionalized quantum dots. Protein G can bind with the fragment crystallizable (Fc) region of the antibody of *F. tularensis* bacteria. It bestows directionality on the anti-*F. tularensis* antibody by binding exclusively to the Fc region.

The detection scheme shown in Figure 12.21 consists of the following steps:

(i) For recognizing *F. tularensis* bacteria, magnetic beads (MB) are conjugated with the anti-*F. tularensis* antibody to form the structure: anti-*F. tularensis* capture antibody/MB.

(ii) Reaction of the product of step (i) with *F. tularensis* bacteria gives: *F. tularensis* bacteria/anti-*F. tularensis* capture antibody/MB.

(iii) *F. tularensis* bacteria/anti-*F. tularensis* capture antibody/MB complexes are isolated using a magnet and washed in 1X Tris-Buffered Saline with 0.1% Tween® 20 Detergent (TBST).

(iv) The apoferritin is genetically modified and self-assembled in bacterial cells. The cells are lysed and apoferritin is purified.

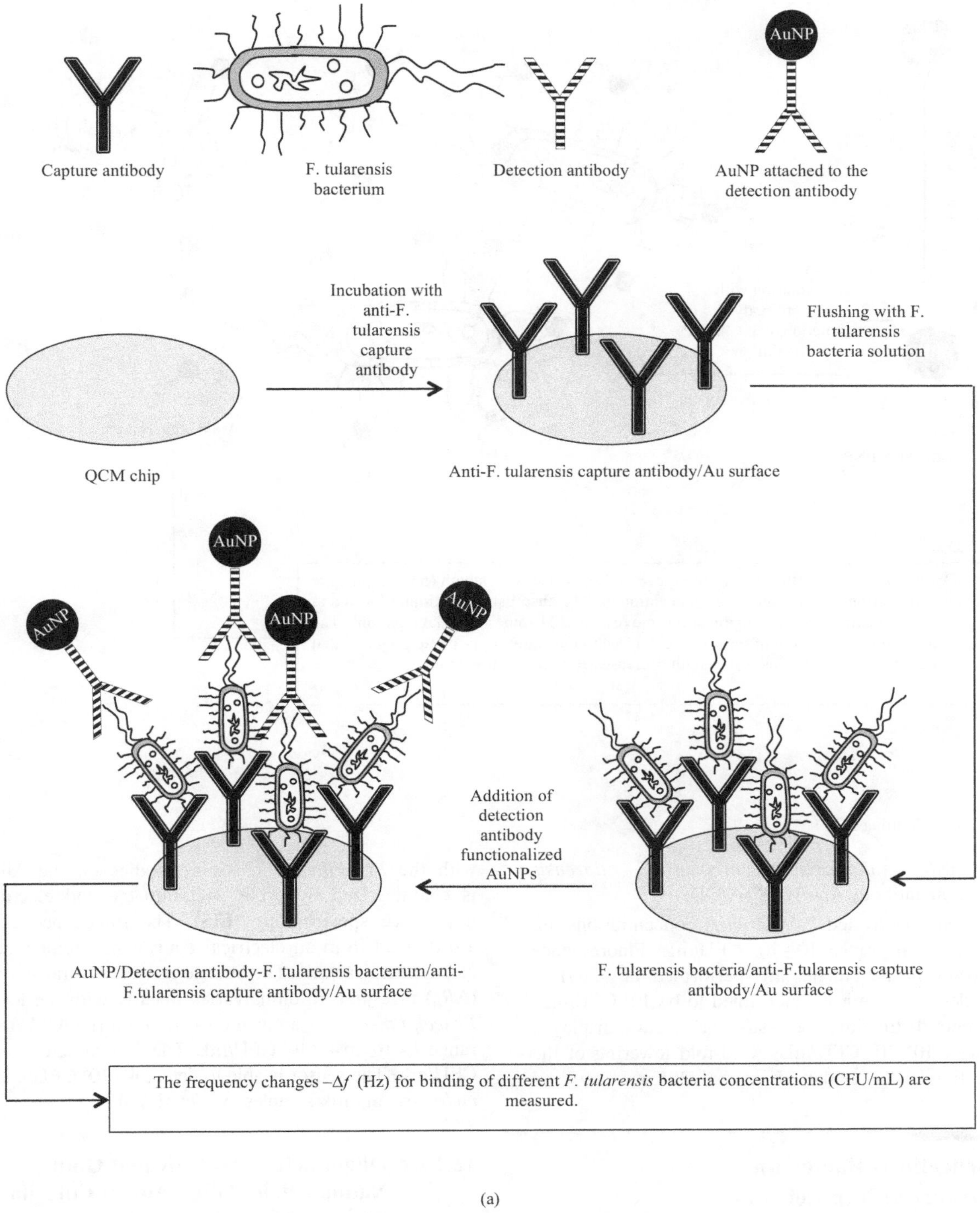

FIGURE 12.20 QCM biosensor for *Francisella tularensis* bacterium: (a) QCM model and (b) absorbance model. (Kleo et al. 2012a, b.)

(v) The quantum dots are water solubilized and modified using myristoyl-2-hydroxy-sn-glycero-3-phosphocholine (MHPC), polyethylene glycol, and Ni-NTA (Ni^{2+} ion coupled to nitrilotriacetic acid) to get Ni-NTA functionalized quantum dots Ni-NTA/QD.

(vi) The genetically modified apoferritin constructs (Apo) are mixed with Ni-NTA/QD in Dulbecco's phosphate-buffered saline (DPBS) and incubated. By conjugation of QDs with apoferritin, we obtain the complex: Apo/Ni-NTA/QD.

(vii) The Apo/Ni-NTA/QD complexes are conjugated with anti-*F. tularensis* antibody to form the complex: anti-*F. tularensis* detection antibody/Apo/Ni-NTA/QD.

(viii) Sandwich targeting of antigen is achieved by mixing the products formed in step (iii) with those of step (vii). The final product is: MB/anti-*F. tularensis* capture

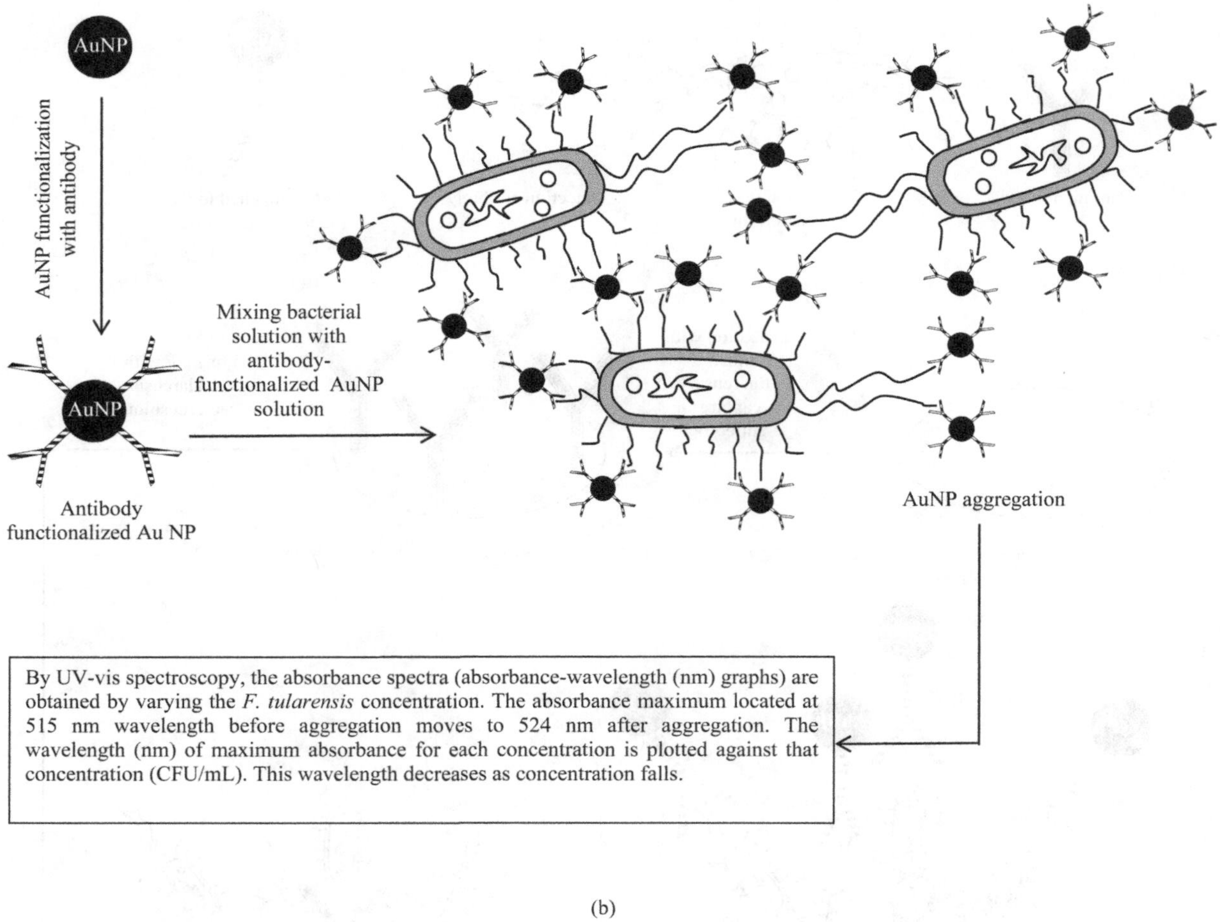

FIGURE 12.20 (Continued)

antibody/*F. tularensis* bacteria/anti-*F. tularensis* detection antibody/Apo/Ni-NTA/QD.

(ix) Different inactivated *F. tularensis* concentrations are chosen in the range 10^3–10^7 CFU/mL. Fluorescence intensity measurements are performed at 620 nm. The detection limit is determined to be 10^4 CFU/mL. In contradistinction, the traditional assays display a limit of 10^5–10^7 CFU/mL. A 10-fold lowering of the detection level is evident (Kim et al. 2015).

12.10 Brucellosis Bacterium (*Brucella*) Nanosensors

12.10.1 Gold Nanoparticle–Modified Disposable Screen-Printed Carbon Electrode (SPCE) Immunosensor for *Brucella melitensis*

An AuNPs-functionalized SPCE consists of an AuNP/carbon electrode, a carbon counter electrode, and an Ag reference electrode (Wu et al. 2013). Figure 12.22 shows the AuNP/carbon electrode. This electrode is immobilized with *Brucella melitensis* antibody to form: anti-*Brucella melitensis* antibody/AuNP/carbon electrode. After BSA treatment, the interaction of the different concentrations of *Brucella melitensis* antigens with the *Brucella melitensis* antibodies on the AuNP/SPCE is characterized by cyclic voltammetry and electrochemical impedance spectroscopy (EIS). The impedance components are derived from an electrical equivalent circuit model of the electrochemical cell. The change in electron-transfer resistance (ΔR_{et}) measured in ohms varies linearly with the logarithm of *Brucella melitensis* antigen concentration [log(CFU/mL)] in the range 4×10^4 to 4×10^6 CFU/mL. The limit of detection is 1×10^4 CFU/mL. The sensor is able to detect 4×10^5 CFU/mL *Brucella melitensis* in milk samples (Wu et al. 2013).

12.10.2 Oligonucleotide-Activated Gold Nanoparticle (Oligo-AuNP) Colorimetric Probe for *Brucella Abortus*

The designed DNA probe is specific to the BCSP31 gene of *Brucella* (Pal et al. 2017). The colloidal AuNPs solution is prepared by citrate reduction of gold(III) chloride hydrate. The AuNPs are functionalized by conjugating with a thiol-modified oligonucleotide probe to give oligo-AuNP. The assay is validated by observing the difference in behavior on interaction of the oligo-AuNP probe with specific and non-specific target DNA (Figure 12.23). Specific target DNA is the complementary DNA, e.g., the genomic DNA from *Brucella abortus*. Non-specific target DNA is the non-complementary DNA, e.g., *E. coli* DNA.

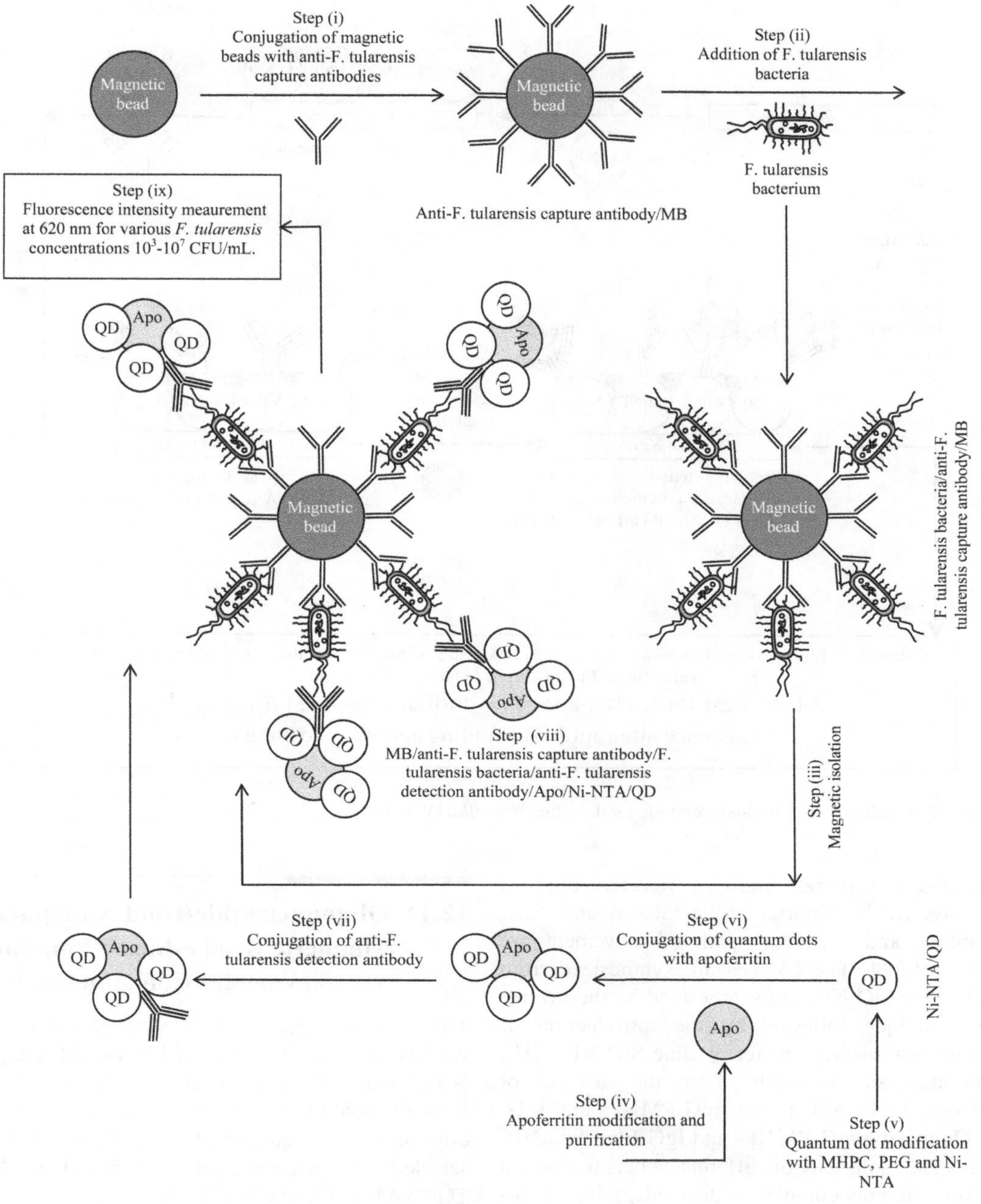

FIGURE 12.21 Detecting *Francisella tularensis* bacterium using a detection antibody and quantum dots decorated apoferritin nanoprobes and magnetic beads with capture antibodies. (Kim et al. 2015.)

Specific target gDNA is injected into the oligo-AuNP solution, kept at 60°C for 15 min and HCl is added. The color change is visually observed and also through absorbance measurements. Here, the target DNA undergoes hybridization with the oligo-AuNP probe. The aggregation of AuNPs is limited and the solution of unaggregated AuNPs remains red. So, a positive sample (hybridized) does not cause any change in color. The UV-vis spectrum shows an absorbance peak at 520 nm.

Next, non-specific target DNA is injected into the oligo-AuNP solution, kept at 60°C for 15 min and HCl is added. In this case, there is no hybridization reaction between the target DNA and the oligo-AuNP probe. So, AuNPs are aggregated and the color of the solution changes to purple indicating that a negative control sample (unhybridized) produces a color change. The UV-vis spectrum shows a red spectral shift with an absorbance peak above 560 nm.

The sensitivity is 5 ng/μL. The visual detection limit is 10^3 CFU/mL *Brucella* organism. The assay is completed within 30 min (Pal et al. 2017).

12.10.3 Colored Silica Nanoparticles Colorimetric Immunoassay for *Brucella abortus*

This sensor uses three probes: paramagnetic nanoparticles (PMNPs) as capture probes, colored silica (SiO_2NPs)

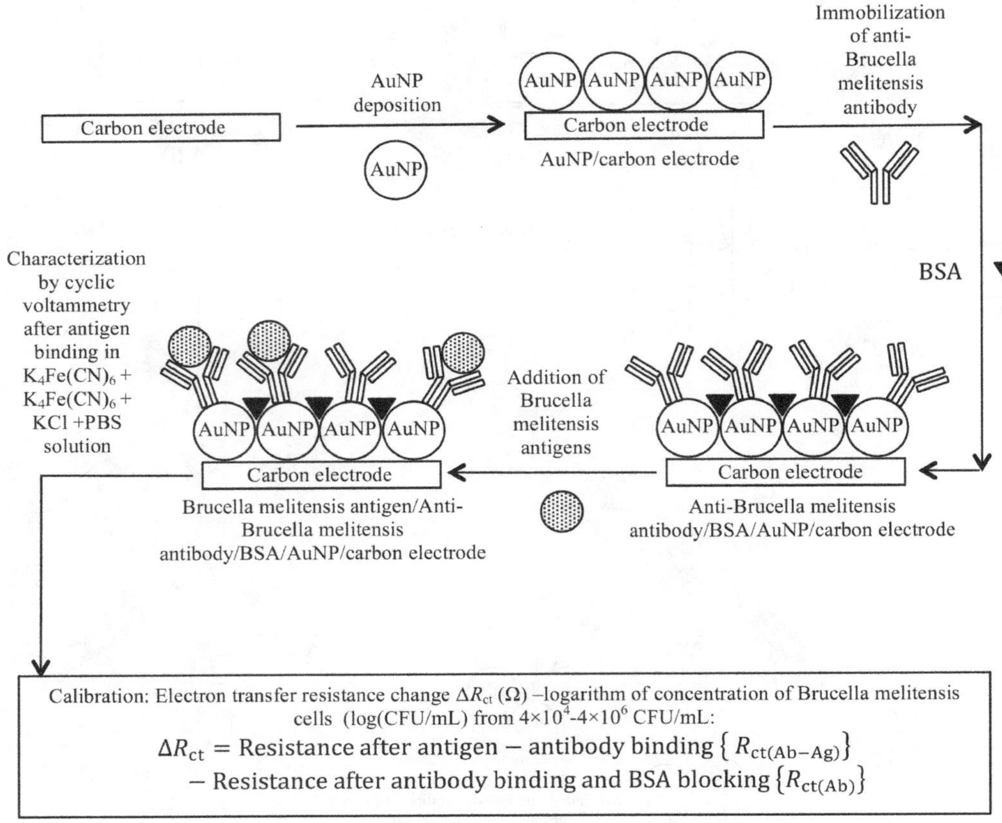

FIGURE 12.22 AuNP-modified screen-printed electrode for detecting *Brucella*. (Wu et al. 2013.)

nanoparticles as detection probes, and C. I. Reactive Blue 21, a polar organic dye as the reporter probe (Shams and Zarif 2019). The symbols and the assay path of movement are depicted in Figure 12.24. The PMNPs are synthesized from Fe_3O_4 and Fe_2O_3. Blue SiO_2NPs are synthesized by the inverse microemulsion technique followed by the introduction of amino groups on their surfaces to form: Blue-SiO_2NPs-NH_2. Polyclonal antibodies are immobilized on the surfaces of PMNPs and Blue-SiO_2NPs-NH_2 to get: IgG-PMNPs and IgG-Blue-SiO_2NPs. On mixing IgG-PMNPs and IgG-Blue-SiO_2NPs with *Brucella abortus* solutions of different concentrations, a sandwich structure immunocomplex is formed giving: Blue-SiO_2NPs-IgG-*Brucella abortus*-IgG-PMNPs. These complexes are separated with the help of a magnet and washed several times to remove unbound IgG. The blue dye is released from the SiO_2NPs using an aqueous solution of NaOH.

The absorbance of the blue dye is measured with a spectrophotometer at a wavelength of 670 nm. The sensor shows a wide dynamic range: 1.5×10^3 to 1.5×10^8 CFU/mL. The relationship between the absorbance and the logarithm of the *Brucella abortus* concentration is linear. The limit of detection is 450 CFU/mL. The assay shows no deterioration in performance after repetition up to 120 days, showing the stability of the conjugated nanoparticles. In a mixture of different bacteria, the assay shows the highest absorbance for *Brucella abortus* and *Brucella melitensis*, confirming its specificity (Shams and Zarif 2019).

12.11 Oligonucleotide/Gold Nanoparticles/Magnetic Beads–Based Smallpox Virus (Variola) Colorimetric Sensor

This is a non-aggregated AuNPs colorimetric assay utilizing AuNPs and MBs (Liu et al. 2012). The advantage of using MBs is that magnetic nanoparticles (MNPs) can be easily isolated from the solution by an externally applied magnetic field. An oligonucleotide sequence related to the smallpox variola virus is selected as the target sequence: VV 5′AGTTGTAACGGAAGA–TGCAATAGTAATCAG 3′.

The assay uses two specially designed probes: the AuNP reporter probe and the MB capture probe. Figure 12.25 shows the course of action for the assay, described as follows:

(i) AuNP reporter probe: to make this probe, PEG-thiol-modified oligonucleotides are taken. This oligonucleotide sequence is partially complementary to the target sequence: 5′AGTTGTAACGGAAGA 3′.

The 30 nm AuNPs are loaded with the reporter probe by treating the citrate-stabilized AuNPs with the solution containing PEG-thiol-modified oligonucleotides to get: AuNP-PEG-thiol-modified oligonucleotide.

(ii) MB capture probe: this probe is made by conjugating carboxylated magnetic beads with

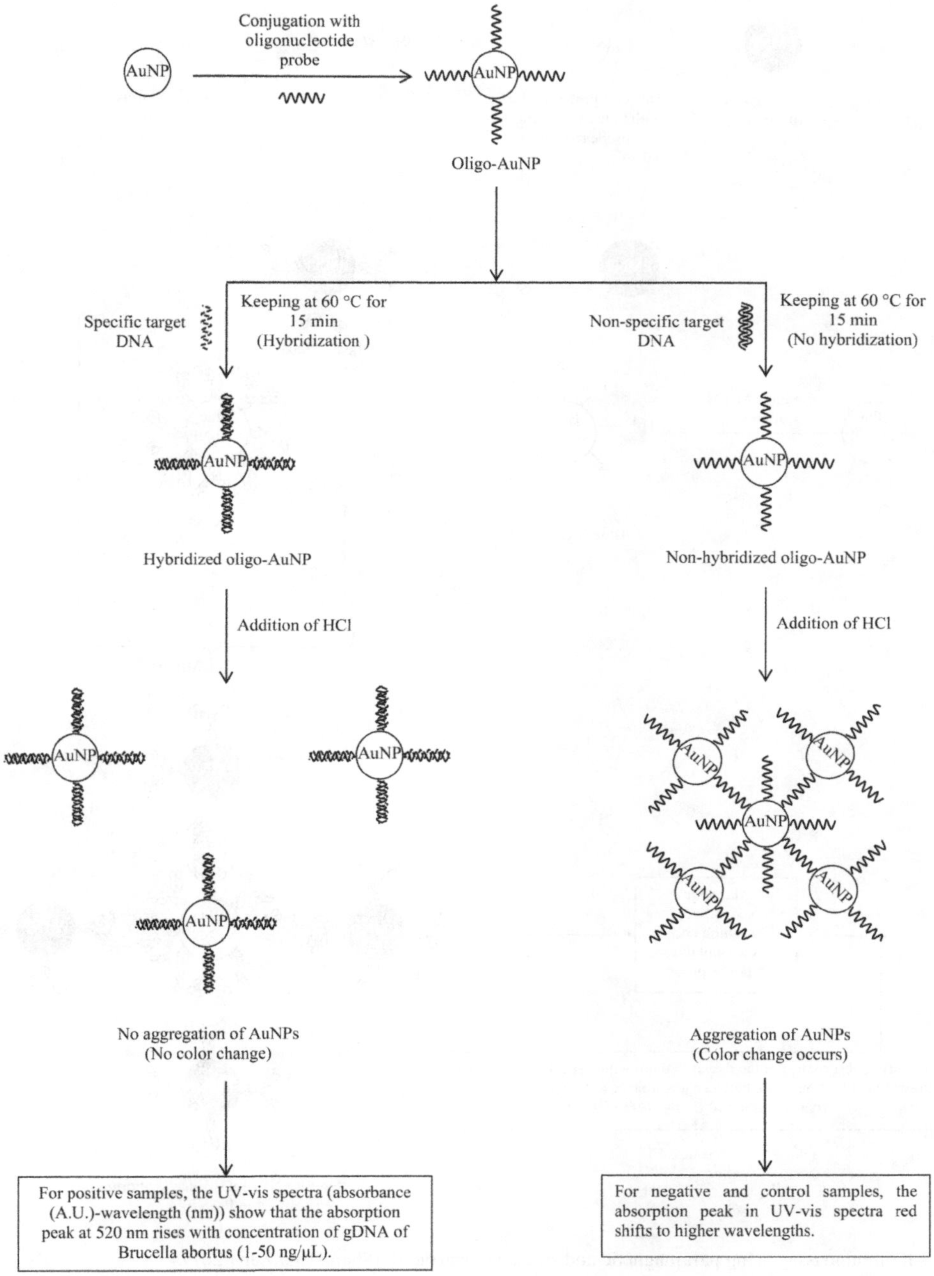

FIGURE 12.23 AuNP non-aggregation/aggregation-based probe for specific/non-specific target DNA of *Brucella*. (Pal et al. 2017.)

amine-modified oligonucleotides. These oligonucleotides match one 15-mer portion of the target sequence: 5′AGTTGTAACGGAAGA3′.

The sequence taken here differs from the previous one for AuNPs. The coupling is done with the crosslinker: (1-ethyl-3-(3-dimethylaminopropyl)carbodiimide hydrochloride)(EDC.HCl)/(N-hydroxysuccinimide).

The assay is started by adding the sample solution containing the target DNA to the MB probe solution followed by gentle vortexing. Then, AuNP probes are added to the mixture with a gentle whirlpool and hybridization is allowed to take place. After hybridization is completed, the MB probes/target DNA/linked AuNP probes are pulled to the wall of the tube and removed from the solution leaving the supernatant containing the remaining AuNPs. This supernatant is closely examined for color changes. With an increasing concentration of target DNA, the color of the supernatant changes from deep red to light red and the color difference is easily discernible with the naked eye. UV-vis spectroscopy is used for precise measurements. The peaks in the absorbance vs. wavelength curves need to be focused on. Absorbance at 528 nm has a linear relationship with the concentration of target DNA in the range 4–400 fmol. The limit of

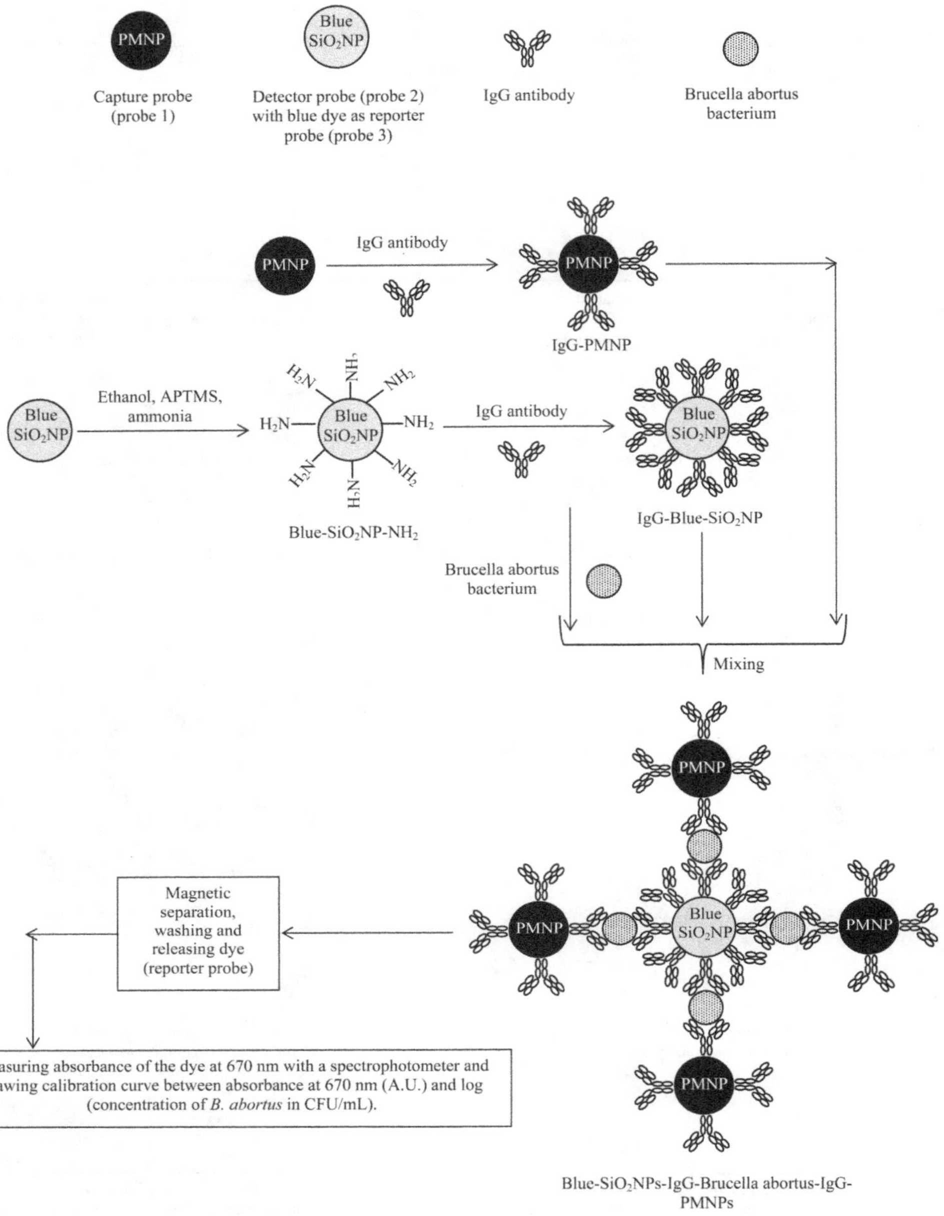

FIGURE 12.24 Sandwich immunoassay using paramagnetic and silica nanoparticles. (Shams and Zarif 2019.)

detection is 4 fmol. The assay is highly selective in providing accurate detection of target DNA in a mixture of target DNA with uncorrelated DNA (Liu et al. 2012).

12.12 Ebola Virus (EBOV) Nanosensors

12.12.1 Reduced Graphene Oxide–Based Field Effect Transistor (FET)

In this transistor, reduced graphene oxide is used as the conducting channel (Chen et al. 2017). Initially, gold interdigitated electrodes are formed on an SiO_2/Si substrate. The surface of the electrodes is modified with cysteamine. The surface-modified electrodes are immersed in an aqueous suspension of graphene oxide (Figure 12.26). The dried electrodes are annealed in argon ambience at 400°C for 10 min. During annealing, graphene oxide is thermally converted to reduced graphene oxide. Thus, the channel of the FET biosensor is formed. Over the rGO layer, a 3 nm-thick Al_2O_3 film is formed by atomic layer deposition. It is used as both a passivant and a gate insulator. Gold nanoparticles are sprayed on the Al_2O_3 film by sputtering. The AuNPs-sprayed Al_2O_3 film is subjected to cysteamine and glutaraldehyde treatments. Ebola antibodies are conjugated with AuNPs by incubation. Thus, the dielectric overlying the channel of rGO FET carries anti-Ebola antibodies as capture probes for the Ebola virus antigen.

To test the sensor, the baseline current I_0 is adjusted by pipetting 0.01 × PBS. The signal is allowed to settle down. After signal

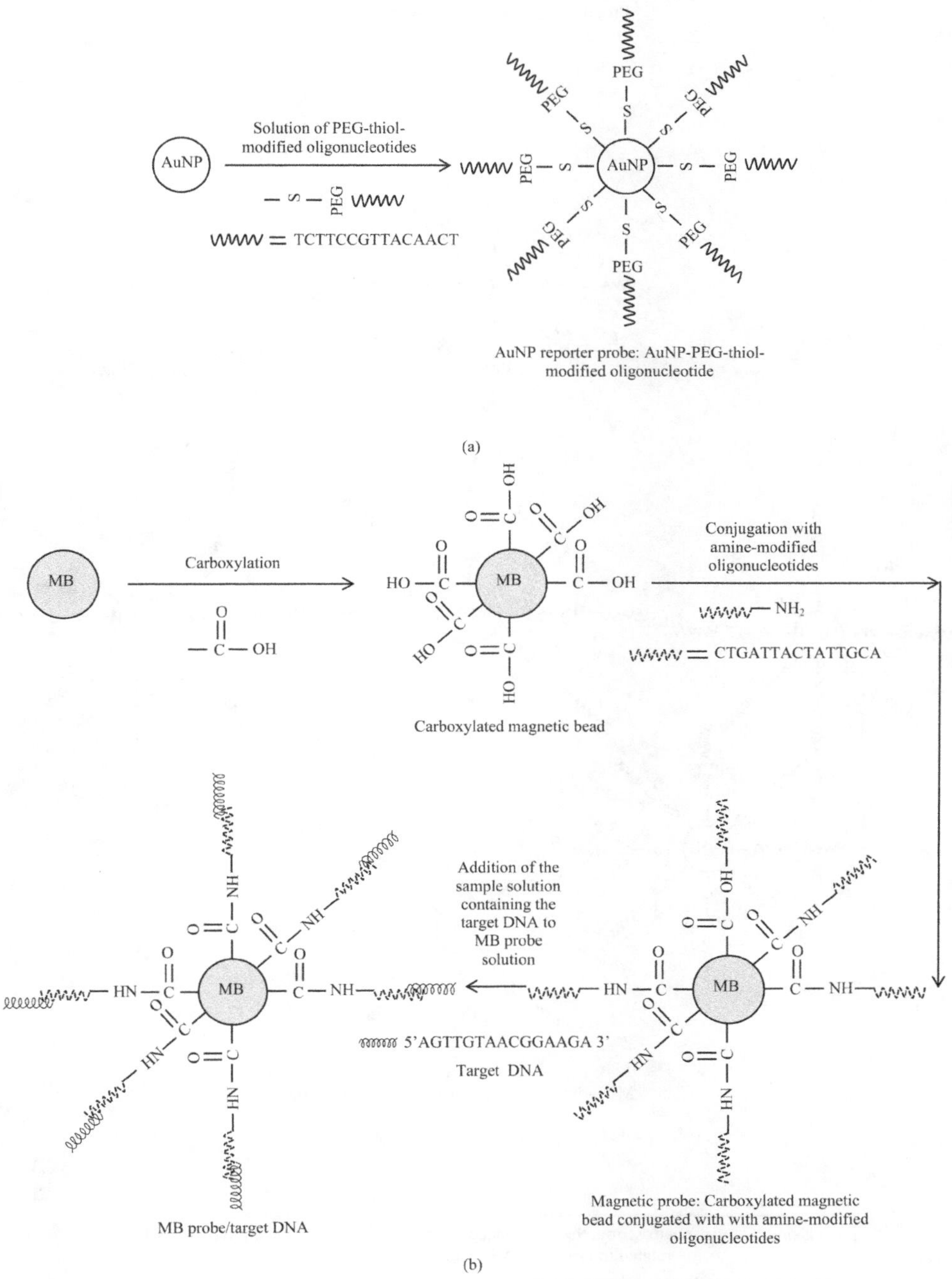

FIGURE 12.25 Scheme of non-aggregated AuNPs colorimetric assay using magnetic beads: (a) making an AuNP reporter probe, (b) making an MB capture probe and adding sample solution to it, and (c) adding the AuNP reporter probe of (a) to the output of (b). (Liu et al. 2012.)

stabilization, Ebola virus antigen solutions having concentrations from 1 to 444 ng/mL are taken. These solutions are made in 0.1 × PBS. They are pipetted on the dielectric surface overlying the FET channel. A definition of sensitivity for this sensor is introduced. It is defined as the ratio = change in the drain current ΔI_D divided by the baseline current I_0, i.e., $\Delta I_D/I_0$. At high concentrations of the antigen, the sensor response is different from that at low concentrations because the current change decreases. This is expected because the number of antibody probes immobilized on the FET channel is limited. Owing to the smaller number of free probes available, the current variation saturates. In other words, the occupation of a large number of probes leads to a decline in the change in current. This accounts for the nonlinear variation of sensitivity with concentration. The limit of detection is estimated as 1 ng/mL (Chen et al. 2017).

12.12.2 Bio-Memristor for Ebola VP40 Matrix Protein Detection

Nanowire field-effect transistors have a memristor property. When the source-drain voltage is swept between negative and positive values, a current minimum appears in the graph of the drain current

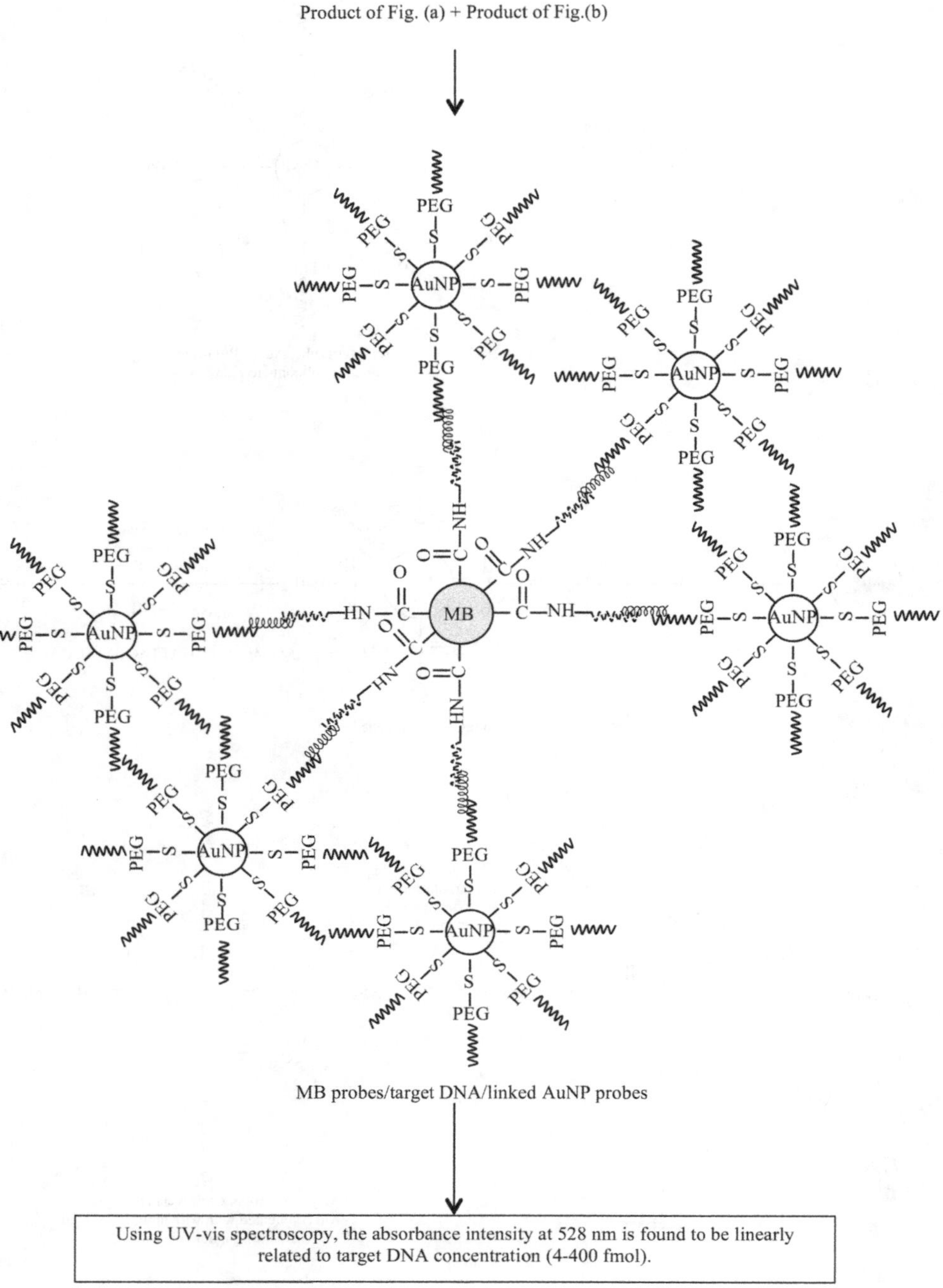

(c)

FIGURE 12.25 (Continued)

(Y-axis) vs. the source-drain voltage (X-axis). If charged molecules are present in a liquid environment in proximity to the FET channel, the position of this current minimum shifts to a different location in the X-direction, creating a voltage gap V_{GAP} = source-drain voltage difference between the former and the latter V_{SD} values of the current minimum. This gap can also be produced by applying a bias to a nearby reference electrode, mimicking the effect produced by charged species in the neighborhood of the channel.

Ibarlucea et al. (2017, 2018) fabricated a FET biosensor for the Ebola virus. The FET uses a honeycomb pattern of Si nanowires (Figure 12.27(a)). It is modified with antibodies to capture the VP40 matrix protein of Ebola (Figure 12.27(b)). When the experiment is started, a V_{GAP}= 1.5 V is opened by applying a bias to the reference electrode (Figure 12.27(c)). In the presence of the VP40 matrix protein of the Ebola virus (Figure 12.27(d)), the initially opened V_{GAP} is altered depending on the protein concentration (Figure 12.27(e)). By adjusting the reference electrode voltage, the gap is restored to its initial value. The necessary voltage applied for restoring the gap to its initial value is a measure of the VP40 matrix protein concentration. By this method, it is possible to detect femtomolar concentrations of the Ebola protein.

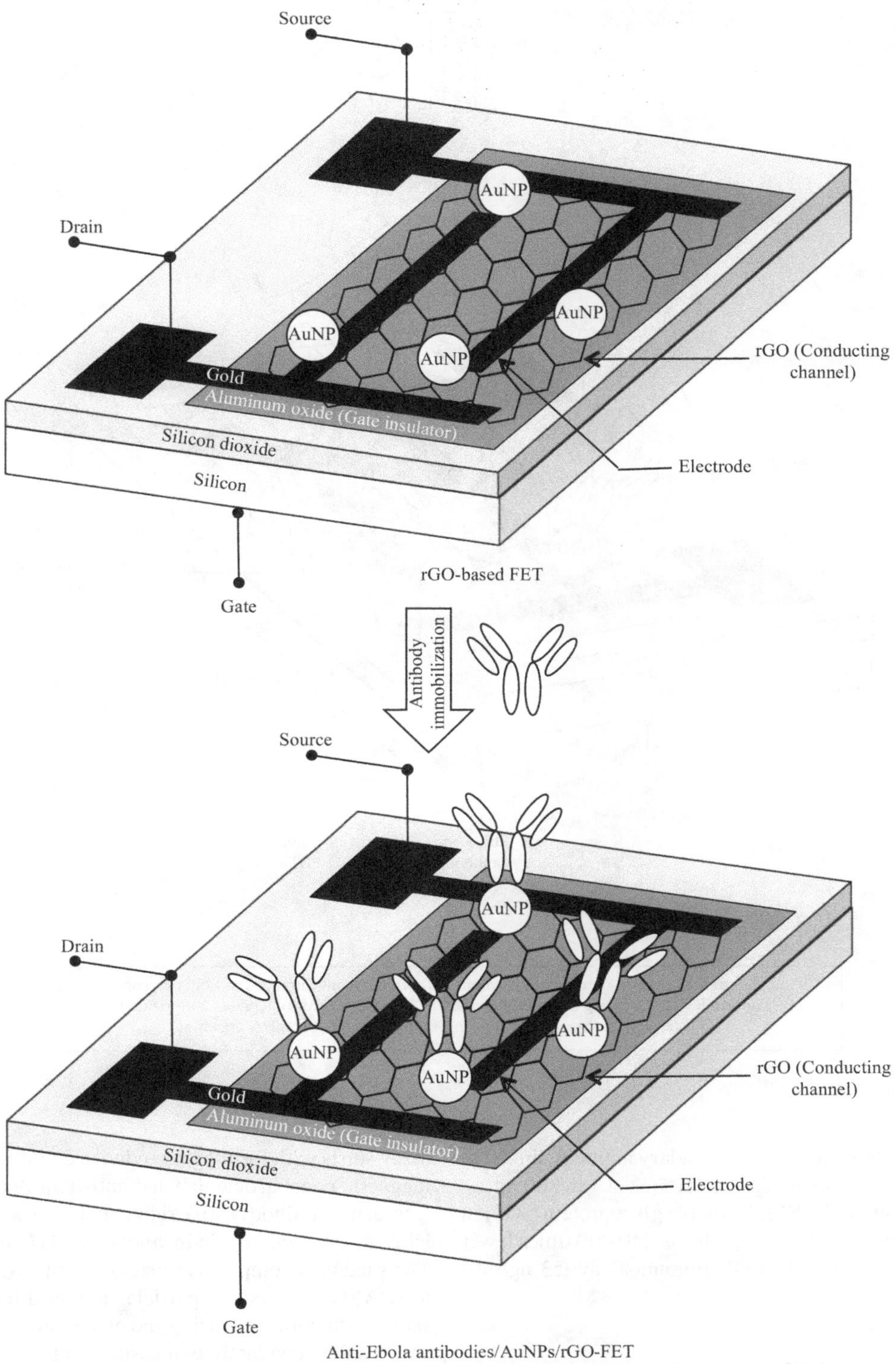

FIGURE 12.26 rGO-conducting channel-based FET biosensor. (Chen et al. 2017.)

12.12.3 3-D Plasmonic Nanoantenna Sensor

Here, a 3-D nanoantenna array is used as an immunoassay platform in place of the conventional flat gold substrate (Zang et al. 2019). The array consists of silicon dioxide nanopillars (Figure 12.28). Each nanopillar has randomly oriented gold nanoparticles on its sidewalls. These are called nanodots. Adjoining the bottom of each nanopillar is a gold backplane while the top of the nanopillar is capped by a gold nanodisk. This arrangement essentially serves as a nanoplasmonic cavity array which considerably enhances the fluorescent signal relative to traditional substrates.

The assay is carried out by fixing the protein A/G on the substrate via the 3,3 dithiobis(sulfosuccinimidylpropionate) (DTSSP) cross-linking reagent. Then, the capture antibody is immobilized. A sandwich assay is carried out with the capture antibody on one side of the Ebola virus antigen. On the other side, the detection

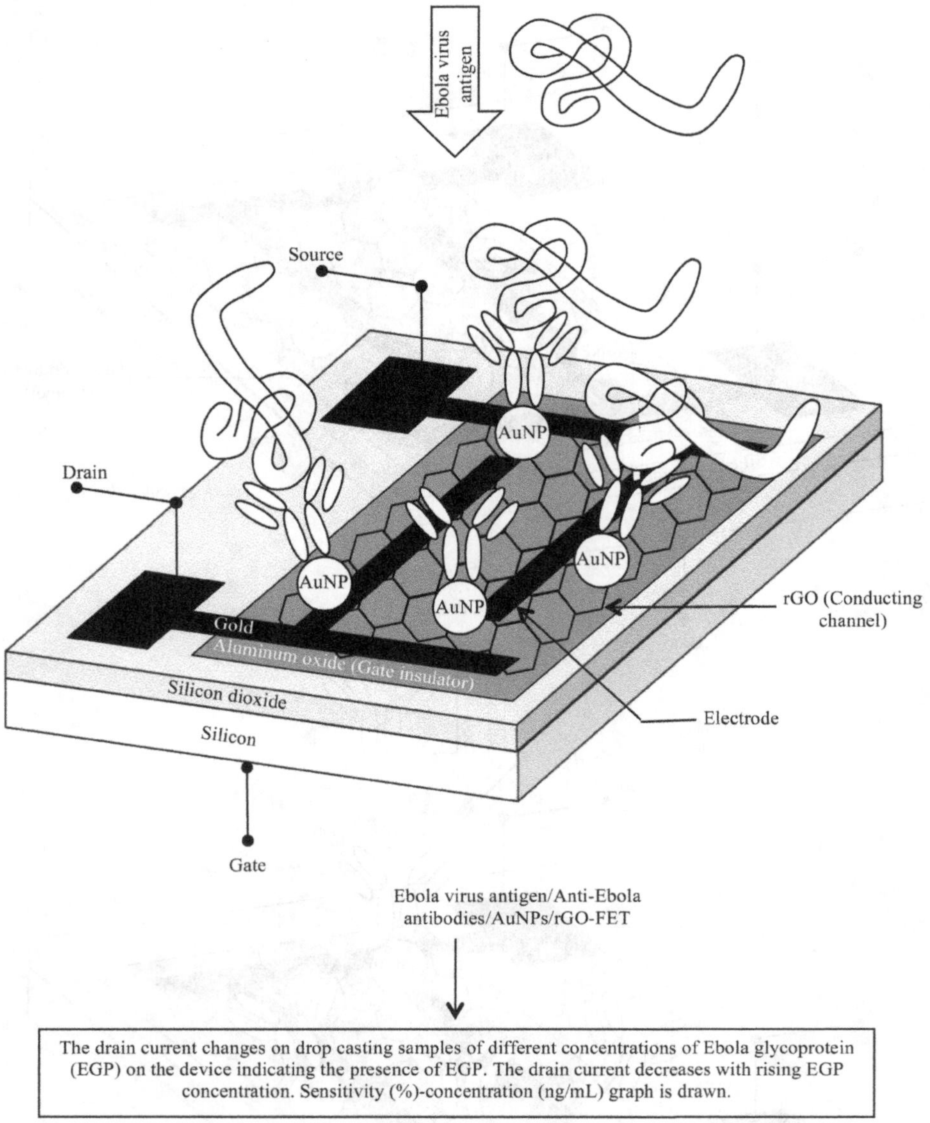

FIGURE 12.26 (Continued)

antibody is attached together with a secondary antibody linked to IRDye800cw with excitation/emission maxima at 785/800 nm.

The detection limit for the EBOV soluble glycoprotein (sGP) in human serum samples is 220 fg/mL. This is 240,000 times lower than that of the present rapid EBOV immunoassay (53 ng/mL) (Zang et al. 2019).

12.13 Ricin Toxin Nanosensors

12.13.1 Silver Enhancement Immunoassay with Interdigitated Array Microelectrodes (IDAMs)

The separation and enrichment properties of magnetic nanoparticles are combined together with the catalytic properties of gold nanoparticles in promoting a reduction in silver to design a sensitive immunoassay for detecting ricin (Zhuang et al. 2010). The assay starts with forming a sandwich complex consisting of: (i) a magnetic nanoparticle–labeled anti-ricin A chain antibody 6A6 (capturing antibody), (ii) ricin, and (iii) a gold nanoparticle–labeled anti-ricin B chain antibody 7G7 (detecting antibody). The sandwich complex is separated by a magnet and transferred to IDAMs. Here, silver particles are used to bridge the gaps in IDAMs, thereby enhancing the electrical signal, which is easily perceived by conductivity measurements.

For making component (i) of the sandwich immunocomplex, amino group–modified Fe_3O_4 magnetic nanoparticles are coated with a silica shell and the amino group–modified silica-coated Fe_3O_4 magnetic nanoparticles are carboxy modified. These carboxy group–modified MNPs are conjugated with anti-ricin A chain antibodies 6A6 (Figure 12.29(a)). Antibody-labeled MNPs are separated using a magnet and washed with PBS. An antibody-labeled MNP dispersion in PBS is made. For carrying out the assay, a series of ricin dilutions (component (ii)) are added to component (i) followed by incubation. The MNPs are separated

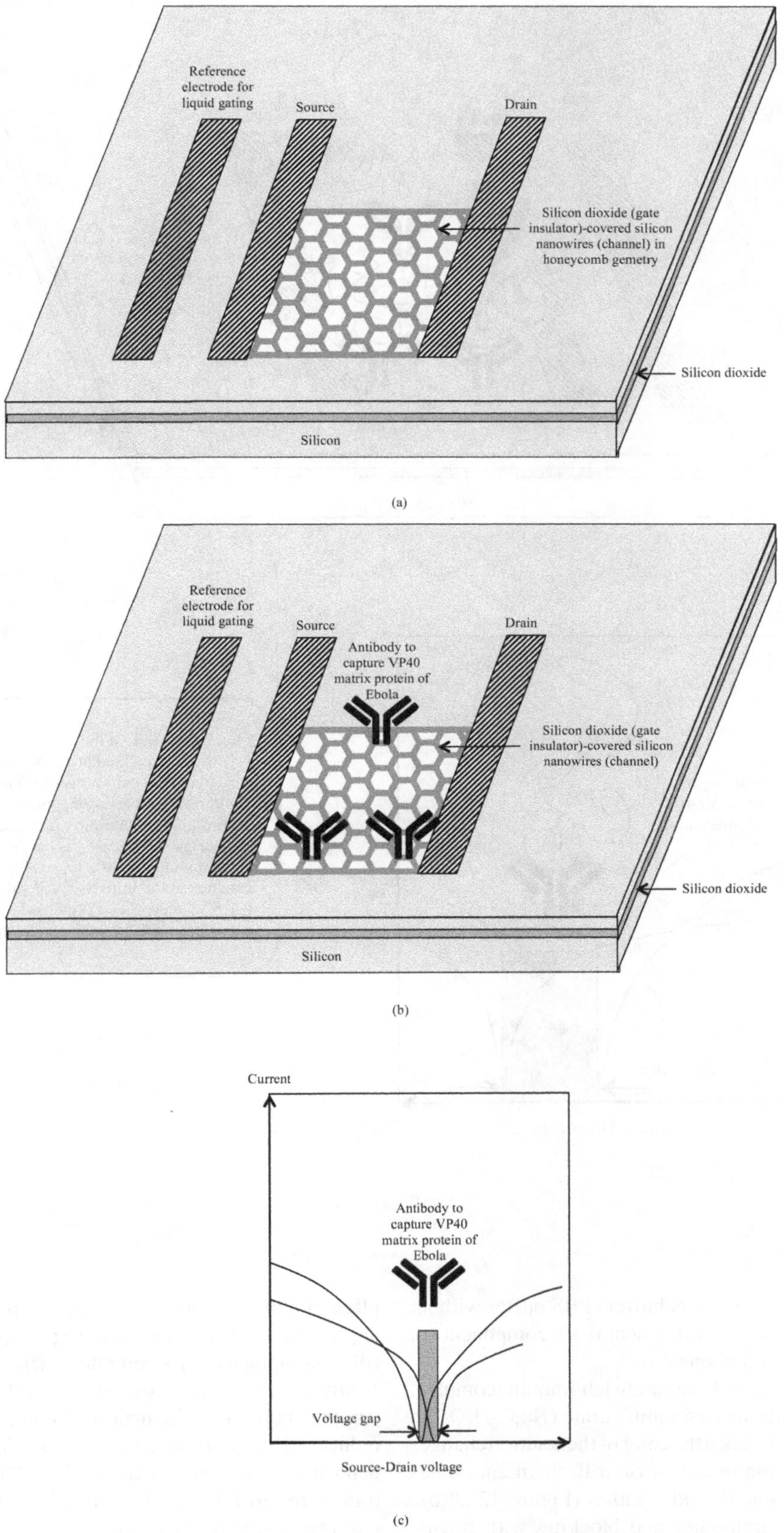

FIGURE 12.27 Silicon nanowire FET biosensor for the Ebola virus: (a) the sensing platform; (b) and (c) the platform with antibody immobilization, and its current-voltage characteristics; (d) and (e) the platform with antibody-Ebola virus, and its current-voltage characteristics, with calibration procedure in memristor mode. (Ibarlucea et al. 2017, 2018.)

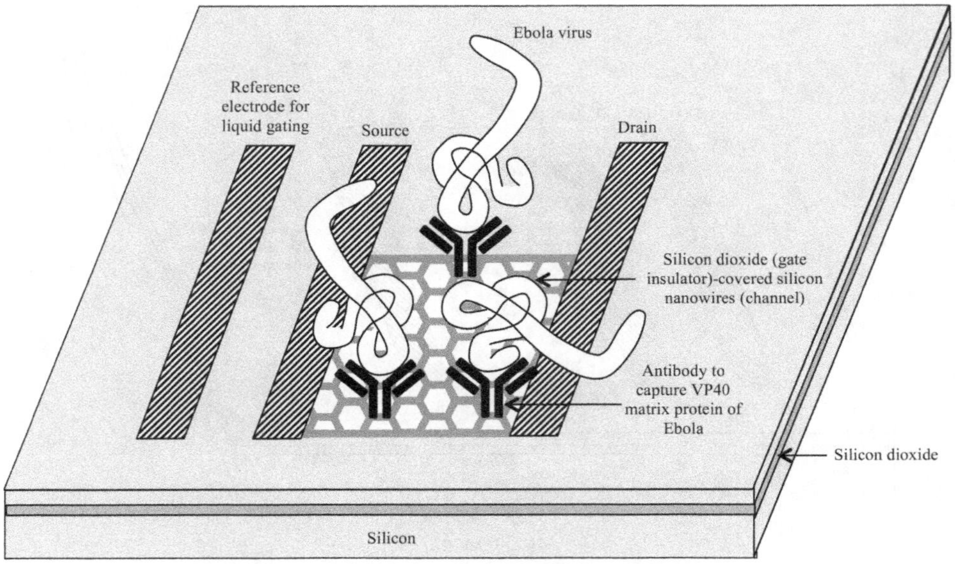

(d)

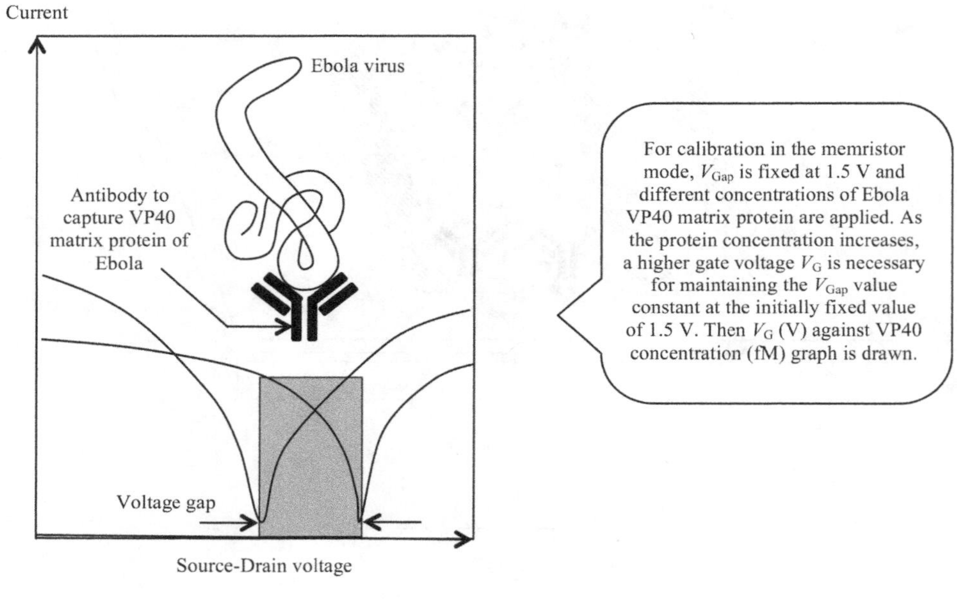

(e)

FIGURE 12.27 (Continued)

with a magnet and washed with PBN buffer (PBS buffer without NaCl). The resulting product is: component (ii) + component (i), which will be treated with component (iii).

For making component (iii) of the sandwich immunocomplex, AuNPs are prepared by mixing trisodium citrate ($Na_3C_6H_5O_7$) to boiling $HAuCl_4$ and stirring, when the color of the solution changes gradually from gray to wine red. Anti-ricin B chain antibodies 7G7 are added to the colloidal gold solution (Figure 12.29(b)). After conjugation of the antibodies and blocking with bovine serum albumin, the probes are suspended in PBS.

Component (iii) is added to the product: component (ii) + component (i) as obtained earlier, when the reaction takes place (Figure 12.29(c)). The reaction products are separated by applying a magnetic field, as done previously. We get: component (iii) + component (ii) + component (i).

Silver enhancement solution is added to component (iii) + component (ii) + component (i) (Figure 12.29(d)). Hydroquinone reduces silver ions to silver atoms. The reduced silver atoms deposit on the surfaces of AuNPs. The reaction products are transferred to IDAMs. The electrical signal is recorded with a multimeter (Figure 12.29(e)).

The resistance remains low and decreases with increasing ricin concentration in the range 10^{-11} M to 10^{-9} M. However, at a concentration of $<10^{-11}$ M, the resistance rises sharply. The limit

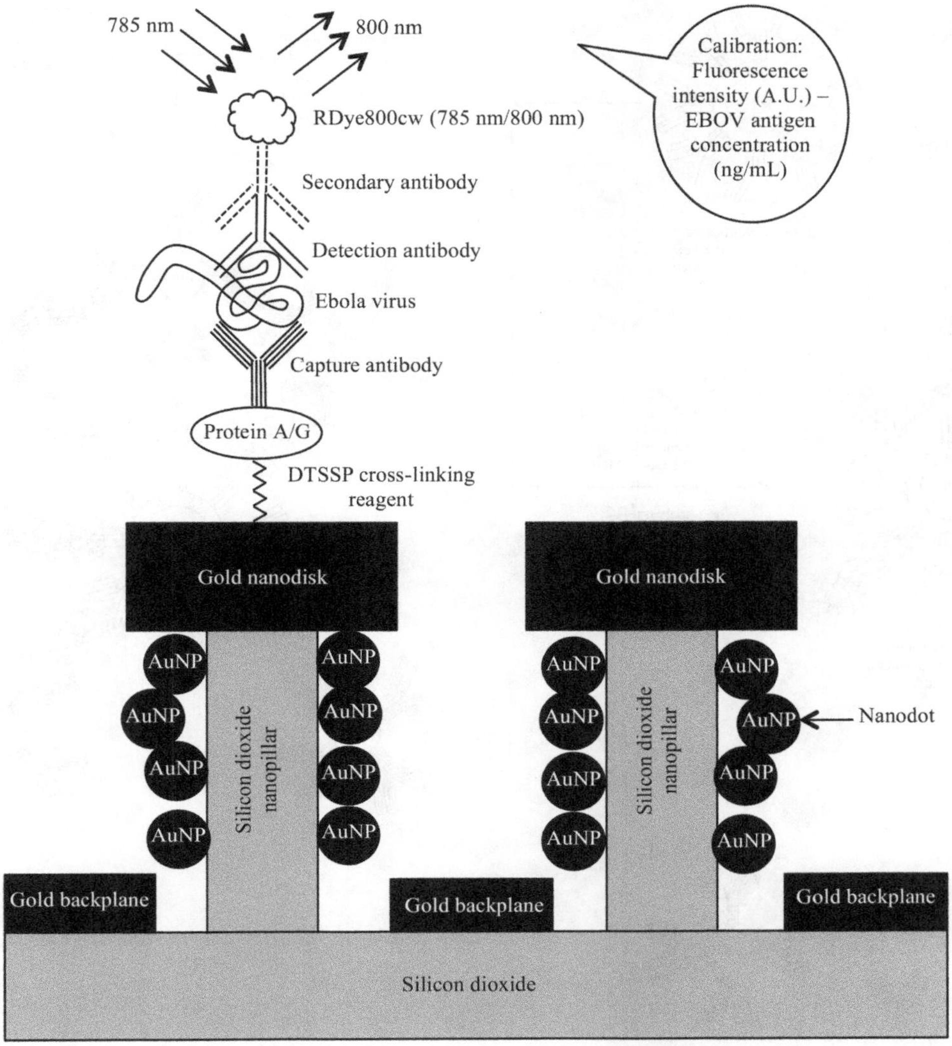

FIGURE 12.28 Nanoantenna biosensor for the Ebola virus. (Zang et al. 2019.)

of detection is 10 pM. The assay is completed in 1 h as opposed to 3 h required for ELISA. Also, the assay is five times more sensitive than ELISA (Zhuang et al. 2010).

12.13.2 Modified Bio-Barcode Assay (BCA)

The bio-barcode assay uses two probes (Hill and Mirkin 2006), viz., a magnetic microparticle (MMP) carrying a recognition agent for the target; and a gold nanoparticle carrying a different recognition agent and several thiolated single-stranded oligonucleotides. The sandwich structures formed by the reaction between the probes and the target are isolated by a magnet. The barcode strands are released in dithiothreitol (DTT) solution at high temperature and identified scanometrically.

The difficulties faced in this time-consuming assay are avoided in the assay designed and demonstrated by Yin et al. (2014). The MMP probe is prepared by binding a monoclonal anti-ricin A antibody to toluenesulfonyl-functionalized MMP (Figure 12.30). For AuNP probe preparation, the AuNP suspension is incubated with an anti-ricin A chain antibody and modified with specifically designed and synthesized thiolated single-stranded signal DNA. Unbound DNA is removed by centrifuging in PBS and is then resuspended in PBS. Ricin solutions of different concentrations are added to the MMP solution followed by washing in PBS and incubation with AuNP probes. The AuNP–MMP complex is separated magnetically and thoroughly washed in PBS. The signal DNA is then detected by polymerase chain reaction (PCR) or molecular photocopying, which is a DNA amplification technique employed to make numerous copies of a DNA segment providing an exponential multiplication of DNA sequences.

The limit of detection is 10^{-2} fg/mL. As a 10^{6} fg/mL ricin A chain is detected by ELISA, the assay of Yin et al. (2014) is 8 orders of magnitude more sensitive than ELISA. It is also 2 orders of magnitude superior to BCA.

12.13.3 Electroluminescence Immunosensor

This sensor uses two probes (Mu et al. 2016): (i) the magnetic-capturing probe consisting of a *Staphylococcus* protein A (SPA)-coated gold-magnetic nanoparticle immobilized with an anti-ricin polyclonal antibody (pcAb); and (ii) the electroluminescent (ECL) probe, which is a phage-displayed

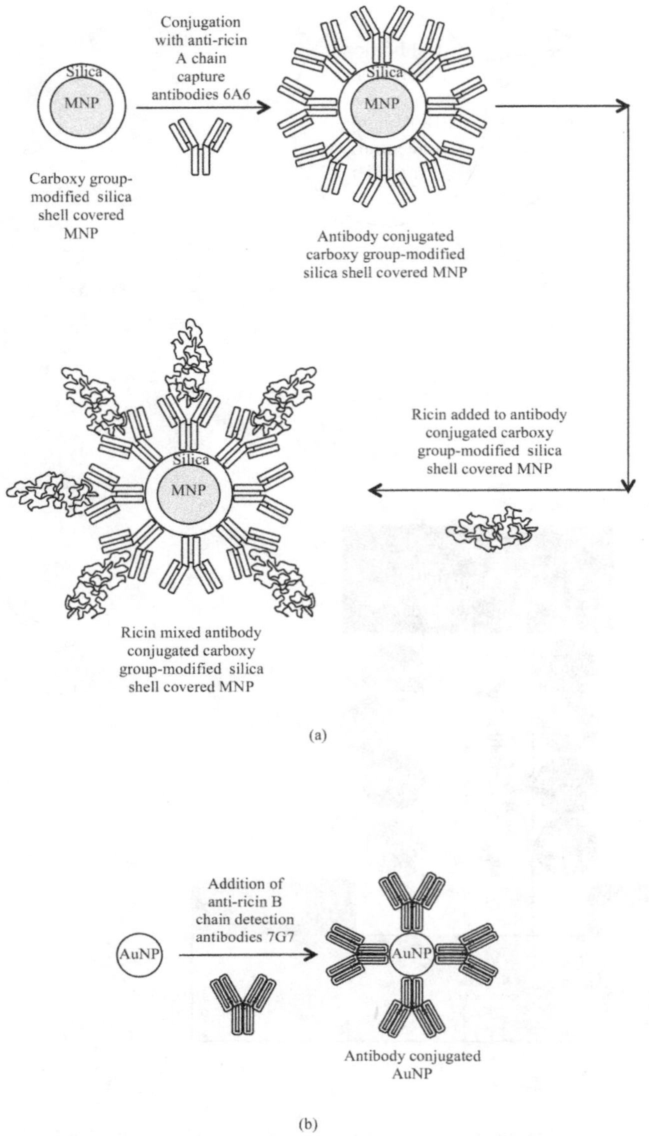

FIGURE 12.29 Immunoassay using magnetic and gold nanoparticles with interdigitated array microelectrodes: (a) exposing modified magnetic nanoparticles to ricin antibodies; (b) modifying gold nanoparticles for assay; (c) mixing magnetic nanoparticles from (a) with gold nanoparticles from (b); (d) adding silver enhancement solution to the reaction product of (d); and transferring the reaction products of (d) to IDAMs for resistance measurements (e) (Zhuang et al. 2010).

anti-ricin antibody labeled with tris(2,2′-bipyridyl)ruthenium(II) fluorophore $Ru(bpy)_3^{2+}$. Figure 12.31 shows the plan of action of an electroluminescent sensor.

Probe (i): for the preparation of gold-magnetic nanoparticles, superparamagnetic Fe_3O_4 nanoparticles made by the microwave co-precipitation method are dissolved in water and ultrasonicated followed by the addition of $HAuCl_4.4H_2O$ and then hydroxylamine. Following magnetic separation of the gold-magnetic nanoparticles, the HCl solution is added to remove the uncoated Fe_3O_4 nanoparticles. SPA is physically adsorbed on the surfaces of the gold-magnetic nanoparticles (Figure 12.31(a)). Then, an anti-ricin polyclonal antibody is added to get probe (i).

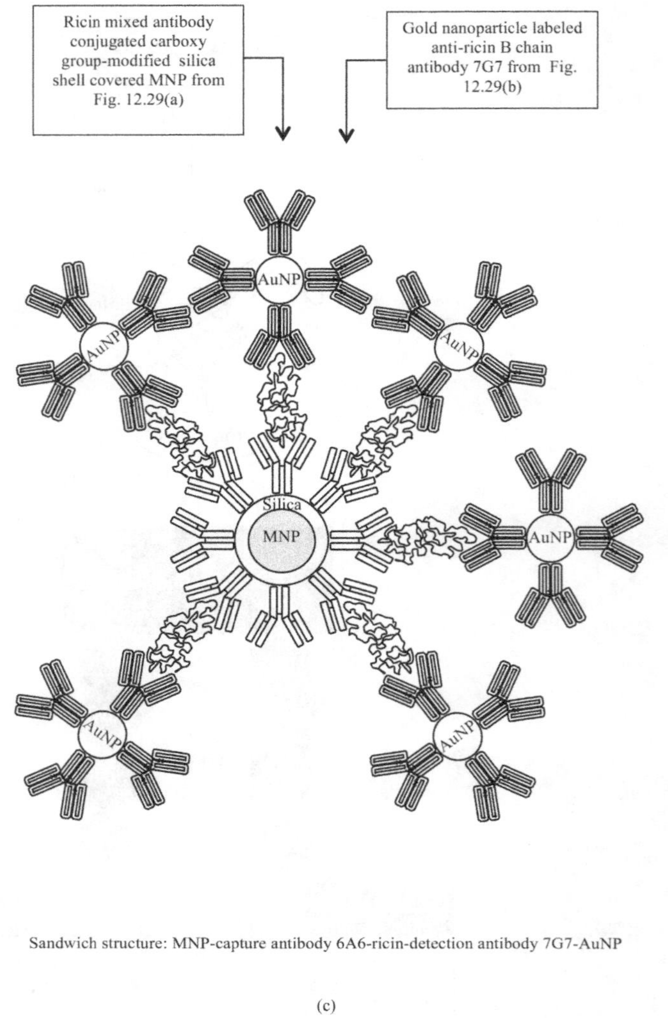

FIGURE 12.29 (Continued)

The assay is carried out by mixing probe (i) with ricin samples of different concentrations while stirring, followed by washing with PBS buffer. We get: Probe 1 + Ricin.

Probe (ii): a phage display antibody is an antibody made by the antibody phage display (APD) technique, a method for the production of recombinant antibodies of desired specificities based on the genetic engineering of bacteriophages, the viruses infecting bacterial cells. $Ru(bpy)_3^{2+}$-N-hydroxysuccinimide ester and N, N′-dimethylformamide (DMF) are added to a phage-displayed anti-ricin polyclonal antibody mixed with carbonate buffer (Figure 12.31(b)). The reaction is allowed to take place away from light. After removing the unbound $Ru(bpy)_3^{2+}$, a suspension of ECL probes in PBS buffer is made, giving probe (ii).

Probe (ii) is added with stirring to Probe 1 + Ricin as obtained above, followed by washing to get the SPA-gold-magnetic nanoparticles/anti-ricin pcAb/ricin/phage-displayed anti-ricin Ab/$Ru(bpy)_3^{2+}$complex.

The complex obtained is transferred to a detector cell by injection (Figure 12.31(c)). It must be deposited on the surface of the working electrode. The assistance of a magnet is required for this purpose. The potential difference between the working electrode and the reference electrode is adjusted at 1.25 V. This

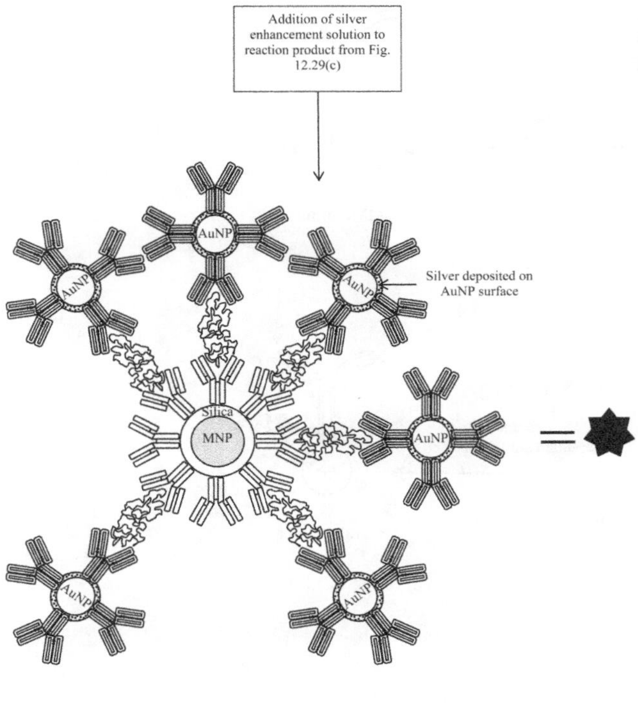

Silver-enhanced sandwich structure: MNP-capture antibody 6A6-ricin-detection antibody 7G7-AuNP-Ag

(d)

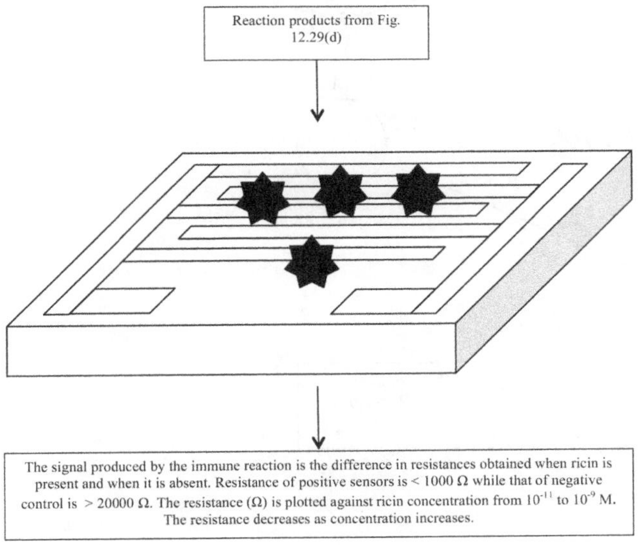

(e)

FIGURE 12.29 (Continued)

step is required to activate the electroluminescence reaction. The reaction occurs between the $Ru(bpy)_3^{2+}$-labeled phage-displayed anti-ricin antibody (Ab) and tripropylamine (TPA) in the solution. Photons are generated in this reaction. The photons produced are detected using a photomultiplier tube. On completion of the reaction, the magnet on the electrode is withdrawn. To restore the ECL signal to its baseline level, the detector cell is thoroughly cleaned.

A rectilinear correlation is observed between the ECL intensity and the logarithm of ricin concentration in the range 0.0001–200 µg/L. The limit of detection is 0.0001 µg/L, which is 2500 times lower than that achieved with ELISA. Gold-magnetic nanoparticles lower the LOD 3-fold. SPA also lowers it 3-fold while the $Ru(bpy)_3^{2+}$-labeled phage-displayed anti-ricin antibody lowers it 20-fold so that the integrated amplifying effect is $3 \times 3 \times 20 = 180$; the lowering factors are reckoned with respect to ECL sensors without one of the three effects (Mu et al. 2016).

12.14 Staphylococcal Enterotoxin B (SEB) Toxin Nanosensors

12.14.1 SEB Detection Through Hydrogen Evolution Inhibition by Enzymatic Deposition of Metallic Copper on Platinum Nanoparticles (PtNPs)-Modified Glassy Carbon Electrode

The detection is carried out by pursuing the following steps (Figure 12.32):

(i) Randomly distributed PtNPs are electrodeposited on a GCE in H_3PtCl_6 solution in HCl containing KCl using the amperometric method at a fixed potential of −0.5 V with respect to the Ag/AgCl electrode (Sharma et al. 2014). As a result, a PtNPs-coated GCE is obtained (Figure 12.32(a)).

(ii) (a) Rabbit anti-SEB polyclonal antibodies (primary antibodies) are physically adsorbed on the surface of a screen-printed electrode, and non-specific binding sites are blocked with a BSA solution in Tris-HCL (Figure 12.32(b)).

(b) SEB toxin solutions of different concentrations are incubated with an anti-SEB primary antibodies-immobilized SPE to form a sandwich immunocomplex by antigen–antibody interaction.

(c) Mice anti-SEB monoclonal antibodies (secondary antibodies) and then alkaline phosphatase (ALP) enzyme conjugated anti-mice immunoglobulin G antibodies are added to the immunocomplex-modified SPE, followed by incubation and washing in PBS.

(iii) Electrodes prepared in steps (i) and (iii) are immersed in glycin-NaOH buffer solution (Figure 12.32(c)). A fixed quantity of copper deposition solution is added and incubation is carried out. Under the bio-catalytic activity of the enzyme ALP, the hydrolysis of ascorbic acid-2 phosphatase (AAP) occurs producing ascorbic acid and causing a reduction in copper ions in the solution to metallic copper, which is deposited on the PtNPs-modified GCE.

(iv) The washed GCE is dipped in HCl solution and examined by linear sweep voltammetric measurements. The potential at which hydrogen is evolved is measured at 100 µA current. This potential shifts leftward with an increasing concentration of SEB. The measurements are done in the concentration range of 1 ng/mL tol µg/mL. The limit of detection is 1 ng/mL (Sharma et al. 2014).

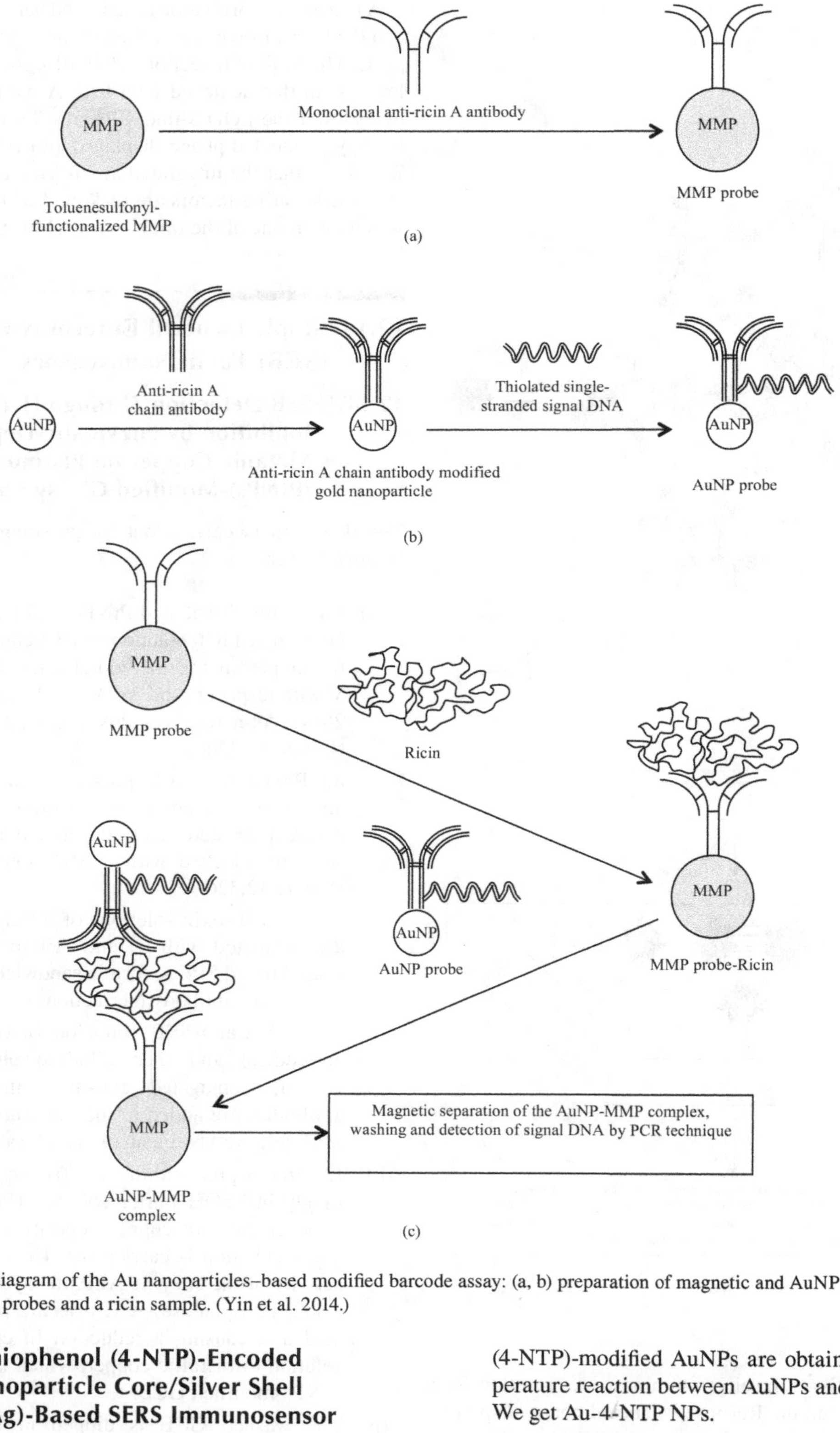

FIGURE 12.30 Flow diagram of the Au nanoparticles–based modified barcode assay: (a, b) preparation of magnetic and AuNP probes, and (c) carrying out the assay using these probes and a ricin sample. (Yin et al. 2014.)

12.14.2 4-Nitrothiophenol (4-NTP)-Encoded Gold Nanoparticle Core/Silver Shell (AuNP@Ag)-Based SERS Immunosensor

The SERS-based immunoassay is designed by Wang et al. (2016). It consists of the following steps (Figure 12.33):

(i) AuNPs (18 nm diameter) are prepared by boiling the $HAuCl_4$ solution and adding trisodium citrate to the solution. The solution turns red. Then, 4-nitrothiophenol (4-NTP)-modified AuNPs are obtained by room temperature reaction between AuNPs and aqueous 4-NTP. We get Au-4-NTP NPs.

(ii) The Au-4-NTP is mixed with polyvinylpyrrolidone, silver nitrate, and ascorbic acid to get Au-4-NTP@Ag NPs (Figure 12.33(a)). Silver shell thickness is optimized at 6.6 nm for best SERS intensity. The Au-4-NTP@Ag NPs are modified with monoclonal antibody mAb 1B3 (detector antibody). This gives: mAb 1B3 labeled-Au-4-NTP@Ag NPs.

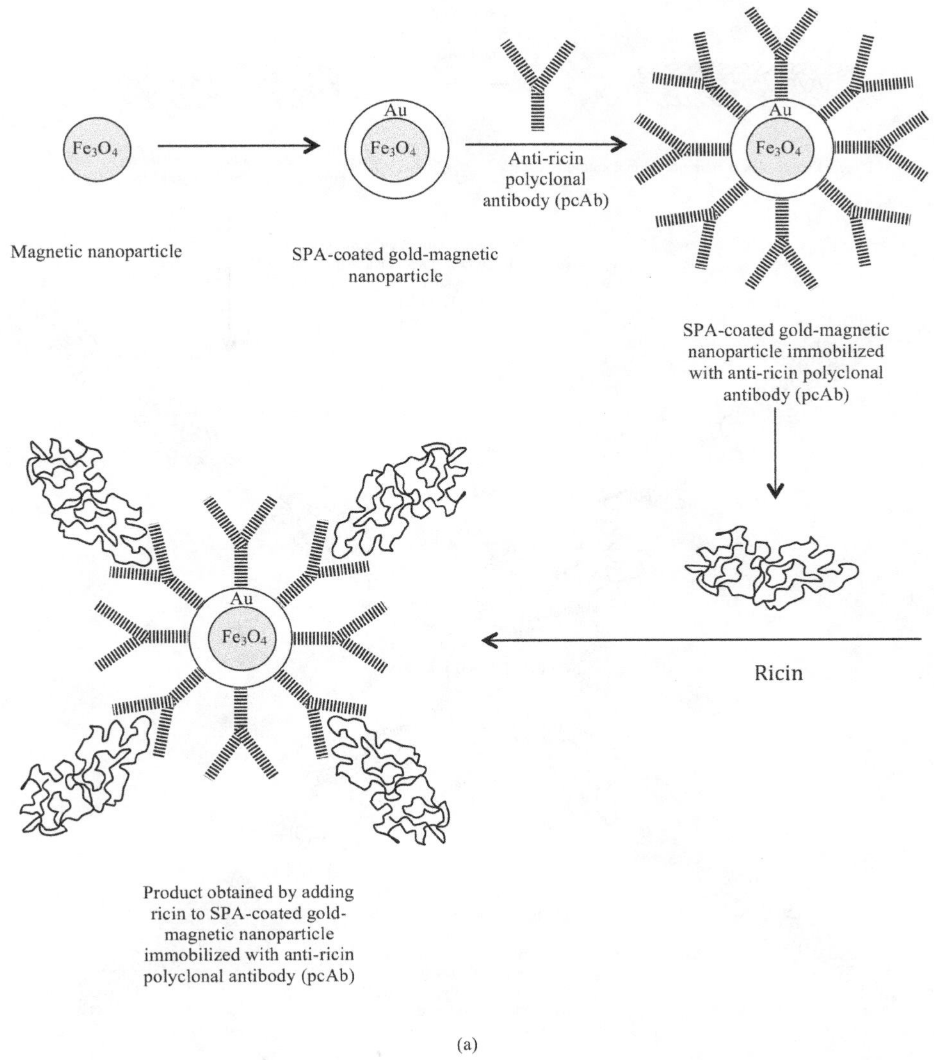

FIGURE 12.31 Electroluminescence biosensor using gold-magnetic nanoparticles and phage display antibodies, showing: (a) ricin added to an SPA-coated gold-magnetic nanoparticle immobilized with an anti-ricin polyclonal antibody (pcAb); (b) phage-displayed anti-ricin Ab/Ru(bpy)$_3^{2+}$ mixed with the complex produced in (a); and (c) electroluminescence signal generation from the complex produced in (b). (Mu et al. 2016.)

(iii) The microplate for the immunoassay is coated with monoclonal antibody mAb 4F2 (capture antibody) in sodium carbonate buffer, incubated and washed in (PBS, Tween) buffer, treated with blocking buffer (carbonate-bicarbonate buffer containing gelatin), incubated, and washed. We get: mAb 4F2-microplate (Figure 12.33(b)).

(iv) An SEB sample is added to the washed microplate followed by incubation (Figure 12.33(b)). Then, mAb-labeled Au-4-NTP@Ag NPs from step (ii) are applied on the microplate prepared in step (iii) (Figure 12.33(c)). After incubation and washing, we get the immunocomplex: microplate-mAb 4F2-SEB-mA b 1B3 labeled-Au-4-NTP@Ag NPs. The microplate is washed and SERS spectra are recorded for different concentrations of SEB in a Raman spectrometer.

The limit of detection is 1.3 pg/mL. The 4-NTP significantly enhances the signal between the AuNPs and the Ag shell. It also permits discrimination of the signal between the specimen and the microplate (Wang et al. 2016).

12.14.3 Aptamer Recognition Element and Gold Nanoparticle Color Indicator–Based Assay

The assay uses a single-stranded SEB-binding aptamer (SEB2) with a random coil structure and gold nanoparticles (Mondal et al. 2018). Two cases are distinguished (Figure 12.34):

Case (a): without SEB: the single-stranded SEB-binding aptamer (SEB2) is added to the microtiter plate. Then, a citrate-protected AuNPs solution is added and incubated. When a high concentration of NaCl salt solution is added, the AuNPs remain dispersed. The color of the solution remains unchanged.

Case (b): with SEB: the single-stranded SEB-binding aptamer (SEB2) along with SEB solution in PBS is

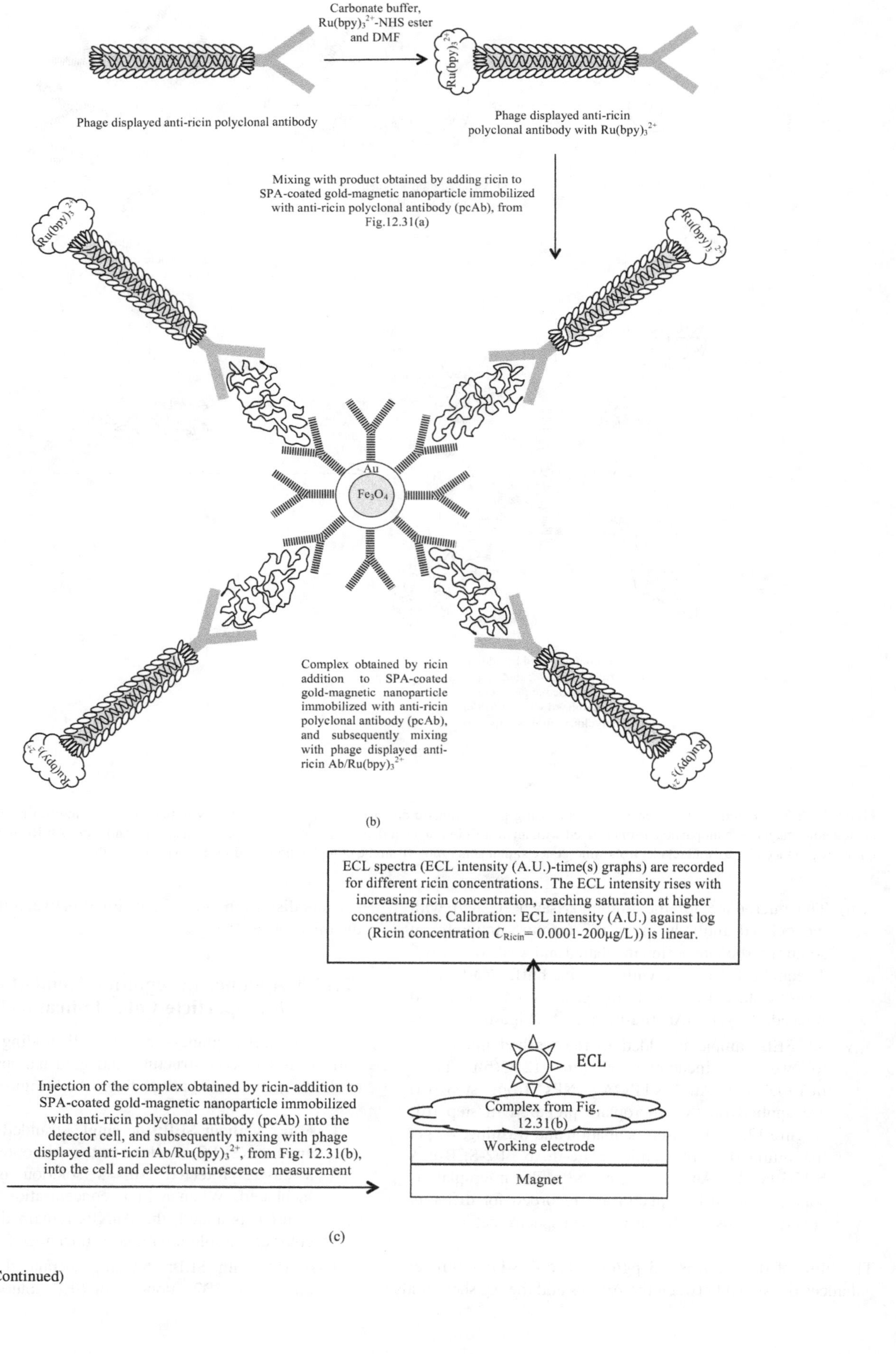

FIGURE 12.31 (Continued)

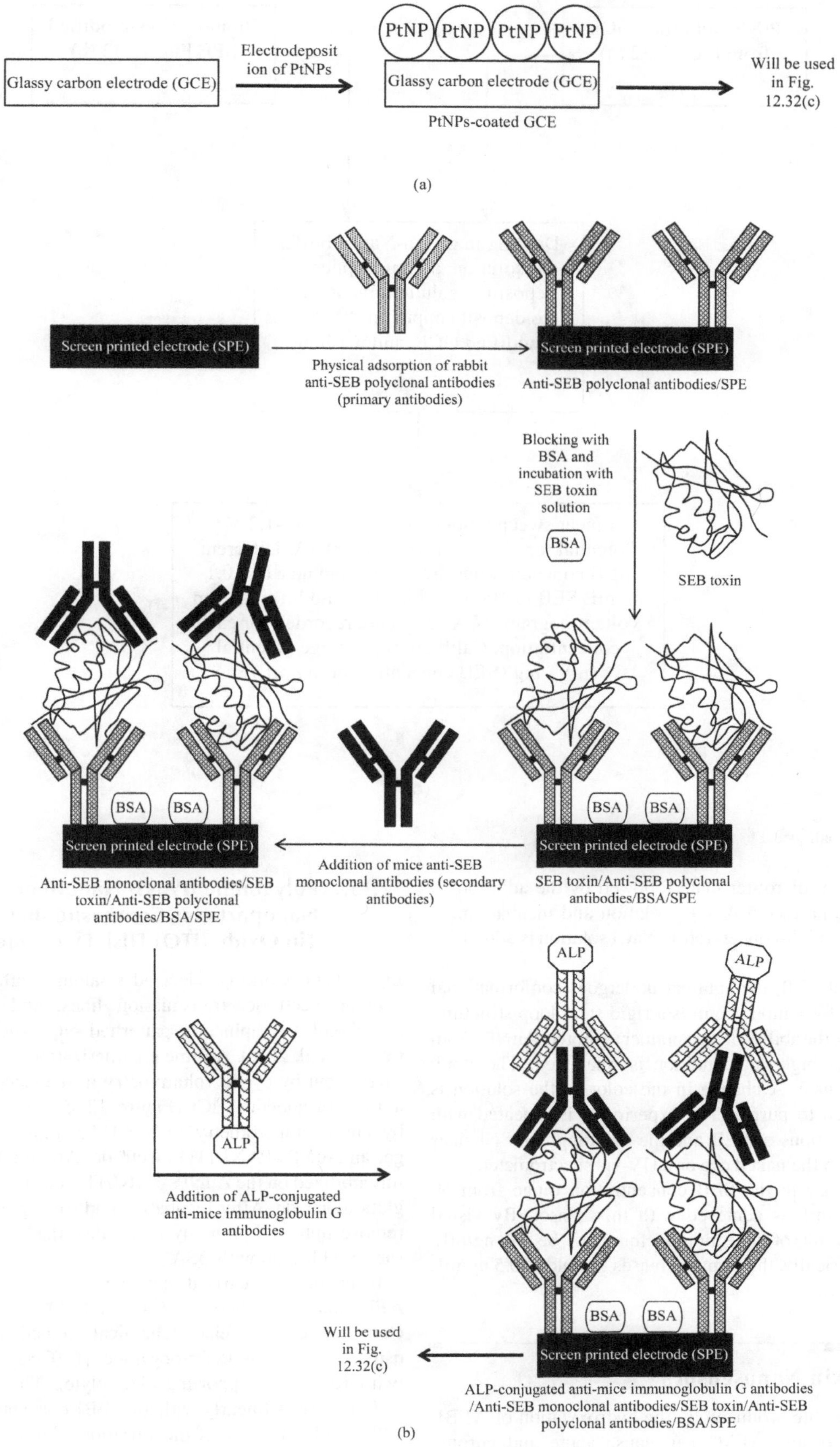

FIGURE 12.32 Use of PtNPs-modified GCE for SEB sensing: (a) modification of GCE by PtNPs, (b) modification of SPE by biomolecules, and (c) joint experiment with electrodes made in parts (a) and (b) of the diagram. (Sharma et al. 2014.)

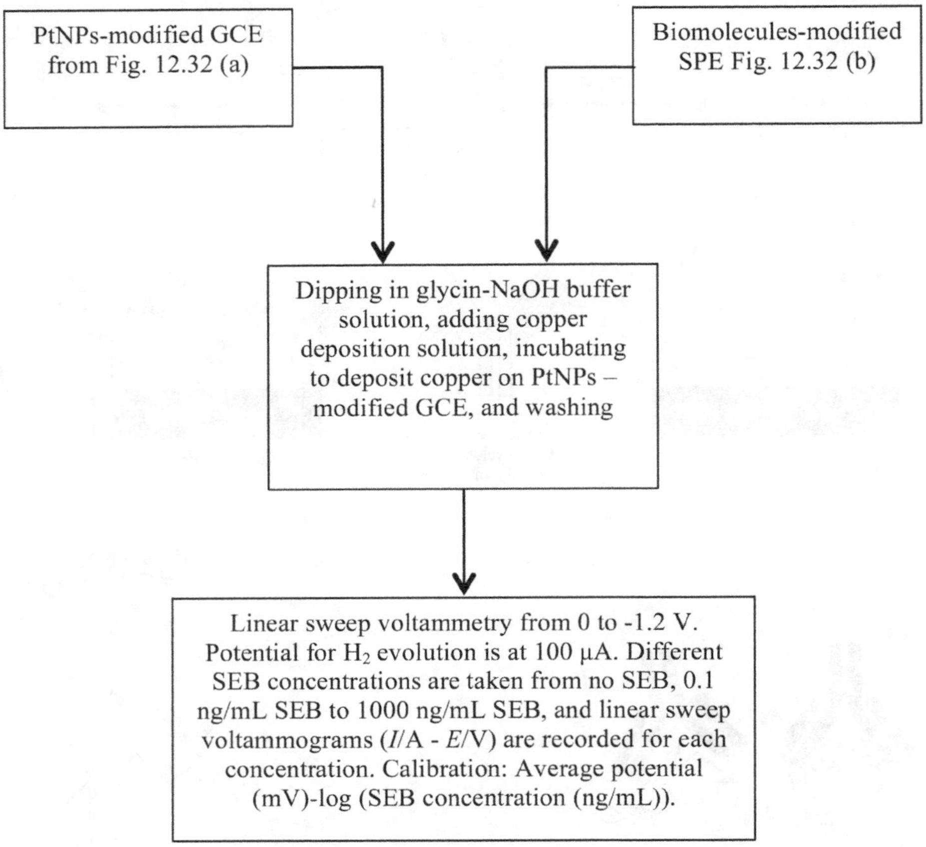

(c)

FIGURE 12.32 (Continued)

added to the microtiter plate followed by the addition of a citrate-protected AuNPs solution and incubation. As before, a high-concentration NaCl solution is added.

In the presence of SEB, the aptamer undergoes conformational changes. The conformation acquires a rigid stem-loop structure. As a consequence, the ability of the aptamer to protect AuNPs from aggregation under high-salt condition is weakened. Therefore, the AuNPs aggregate. A change in the color of the solution is observed from red to purple. The experiment is repeated with different concentrations of SEB and the resulting color changes are monitored with the naked eye or a UV-vis spectrometer.

A linear response to the SEB concentration range from 50 µg/mL to 0.5 ng/mL is achieved with this sensor. By visual inspection, the limit of detection is found to be 50 ng/mL. Spectrophotometrically, the limit decreases 10-fold to 0.5 ng/mL (Mondal et al. 2018).

12.15 Aflatoxin Nanosensors

AFM1 is a metabolite produced by the hydroxylation of AFB1. Both toxins AFB1 and AFM1 can cause acute and chronic mycotoxicosis.

12.15.1 Polyaniline (PANI) Nanofibers–Gold Nanoparticles Composite–Based Indium Tin Oxide (ITO) Disk Electrode for AFB1

Eight ITO working electrodes along with a single large counter electrode, transmission lines, and connection pads are photolithographically patterned on a slide glass substrate (Yagati et al. 2018). Aniline polymerization on ITO electrodes is carried out by cyclic voltammetry in the electrolyte containing aniline in aqueous HCl (Figure 12.35). AuNPs are deposited by chronoamperometry in the $HAuCl_4.3H_2O$ electrolyte. We get an AuNPs-PANI/ITO electrode. An anti-AFB1 antibody is immobilized on the AuNPs-PANI/ITO electrode by linking with glutaraldehyde. After incubation and rinsing with PBS buffer to remove unbound antibody molecules, the non-specific binding sites are blocked with BSA.

Incubation is carried out with known concentrations of AFB1 antigen of between 0.1 and 1000 ng/mL. The sensor is characterized by electrochemical impedance spectroscopy using a ferricyanide/ferrocyanide $[Fe(CN)_6]^{3-/4-}$ redox probe (with KCl as a supporting electrolyte). The impedance |Z| at 1 Hz increases linearly with an AFB1 concentration from 0.1 to 100 ng/mL. Because of the variation of impedance with AFB1 antigen concentration, it is an impedimetric biosensor. The limit

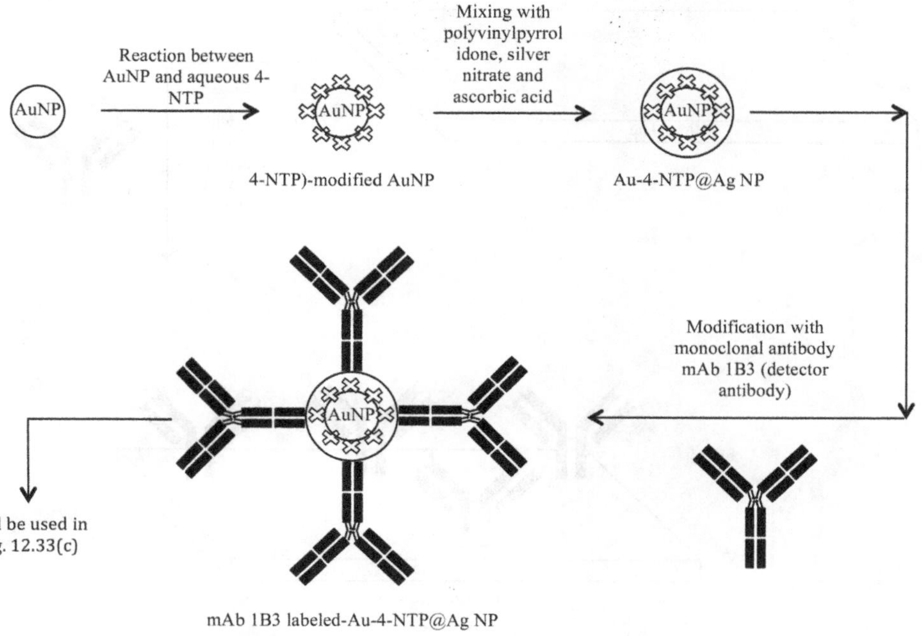

(a)

FIGURE 12.33 AuNP@Ag SERS immunosensor (a) making mAb-labeled Au-4-NTP@Ag NPs; (b) preparing mAb 4F2-microplate with SEB toxin; and (c) mixing products from (a) and (b), and recording of spectra. (Wang et al. 2016.)

of detection is 0.05 ng/mL. The impedance at 1 Hz is 14 times lower than that of a bare electrode because of the easier electronic and ionic transport through AuNPs-PANI. Selectivity is proven against the mycotoxin Ochratoxin A (OTA) (Yagati et al. 2018).

12.15.2 Gold Nanodots (AuNDs)/Reduced Graphene Oxide Nanosheets/Indium Tin Oxide Substrate for Raman Spectroscopy and Electrochemical Measurements for AFB1

(i) An Au-nanodots/rGO-nanosheets/ITO substrate is synthesized using the layer-by-layer electrodeposition technique (Althagafi et al. 2019). A thin film of rGO is electrochemically deposited on ITO-coated glass substrates by applying a negative potential to a GO solution in sodium sulfate (Figure 12.36). This gives rGO-nanosheets/ITO substrate, which is electrochemically decorated with Au nanostructures by applying a negative potential to the $HAuCl_4$ solution against an Ag/AgCl reference electrode. Thus, we get Au-nanodots/rGO-nanosheets/ITO substrate.

(ii) On an Au-nanodots/rGO-nanosheets/ITO substrate, the anti-AFB1 antibody is self-assembled to get anti-AFB1 antibody/Au-nanodots/rGO-nanosheets/ITO substrate.

(iii) Raman spectroscopy: AFB1 antigens are bound to the substrate obtained in step (ii). From Raman spectra, the intensity of the Raman band at 1461 cm^{-1} is found to vary linearly with an AFB1 concentration in the range 1 pg/mL to 5 ng/mL. The limit of detection is 8.1 pg/mL

(iv) Electrochemical response: after binding AFB1 antigens to an anti-AFB1 antibody/Au-nanodots/rGO-nanosheets/ITO electrode, the peak value of the oxidation current shows a linear relationship with the logarithm of the AFB1 antigen concentration in the range 1 pg/mL to 100 ng/mL. The detection limit is 6.9 pg/mL (Althagafi et al. 2019).

12.15.3 AFM1 Aptamer–Triggered and DNA-Fueled Signal-On Fluorescence Sensor for AFM1

This aptasensor performs AFM1 detection through fluorescence intensity of single-stranded DNA (Zhang et al. 2019). It works on the target-initiated DNA strand displacement scheme. The released DNA strand triggers an enzyme-based signal amplification recycle. This recycle produces colossal G-quadruplex structures for enhancing the signal. These structures associate with the fluorescent dye to produce the fluorescent signal whereby AFM1 detection is possible to very low levels.

(i) A biotin-modified AFM1 aptamer is mixed with streptavidin-immobilized magnetic nanobeads in Tris buffer, and incubated and washed in Tris buffer to remove the uncombined aptamer in a magnetic field (Figure 12.37). We thus get aptamer-coated magnetic nanobeads.

(ii) Complementary single-stranded DNA (C-strand DNA) in Tris buffer is mixed with the aptamer-coated magnetic nanobeads of step (i). As a result, double-stranded (aptamer/C duplex)-coated magnetic nanobeads are

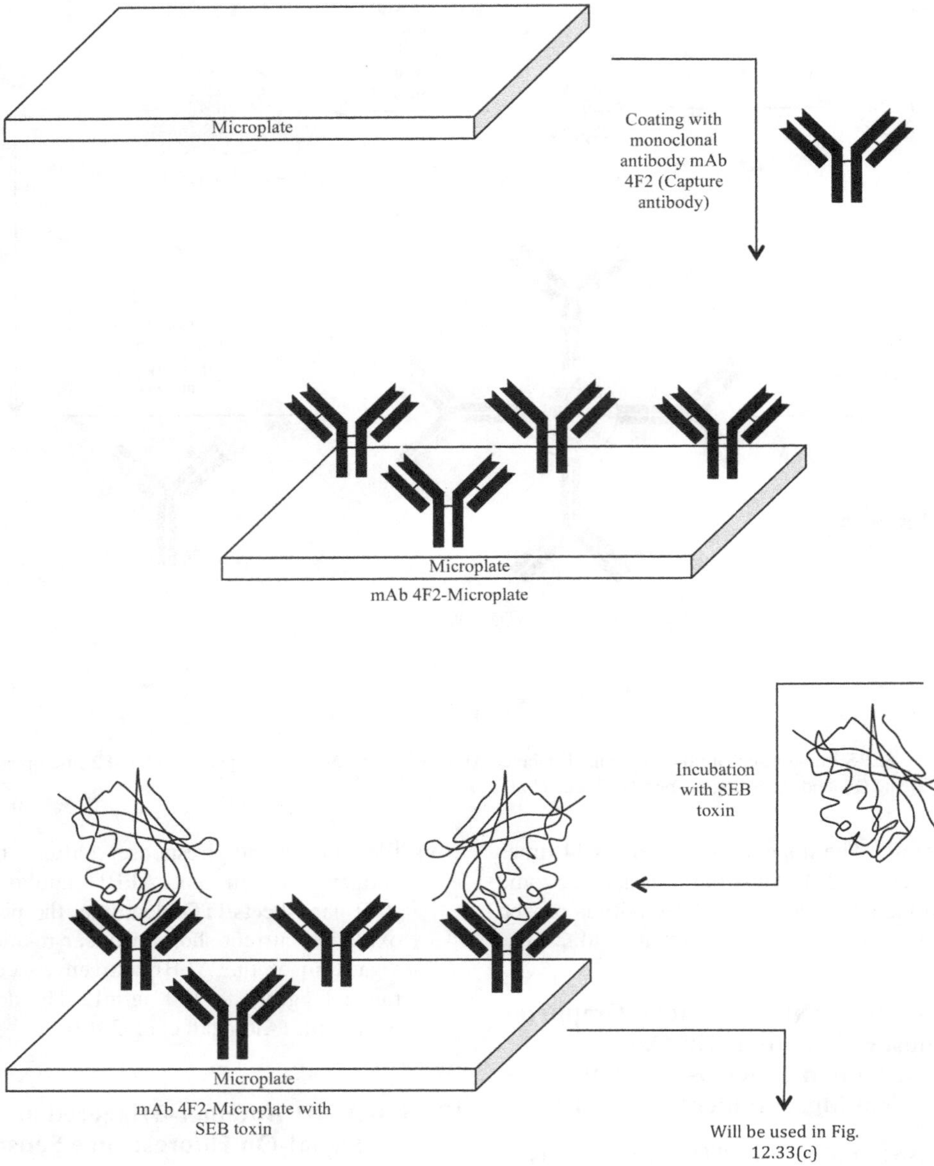

(b)

FIGURE 12.33 (Continued)

formed, which are washed in Tris buffer in a magnetic field to get rid of the uncombined C-strand DNA pieces.

(iii) AFM1 solutions of different concentrations are incubated with double-stranded (aptamer/C duplex)-coated magnetic nanobeads. AFM1 dissociates the aptamer/C-duplex. The C-strand DNA is released in the supernatant, which is removed with the help of a magnet.

(iv) Single-stranded template DNA (T-strand DNA) and guanine-rich DNA (G-strand DNA) in Tris buffer are preheated, and incubated after cooling for hybridization. Since the T-strand DNA partly complements the G-strand DNA and partly complements the C-strand DNA, it can hybridize both with the G-strand DNA and the C-strand DNA. Here, hybridization takes place between the G-strand DNA and the T-strand DNA to generate double-stranded DNA (G/T duplex); the C-strand complementarity will be exploited in step (v).

(v) The supernatant containing the C-strand DNA is added to the double-stranded (G/T duplex) from step (iii) along with Exonuclease III (ExoIII) enzyme. The C-strand DNA is able to recognize its complementary sequence in the T-strand DNA and hybridization sets in. A new duplex DNA is created. Under the influence of the enzyme, G-strand DNA and C-strand DNA are liberated. The G-strand DNA forms a quadruplex structure. The C-strand DNA identifies and hybridizes with the new G/T duplex DNA. It initiates a new amplification recycle.

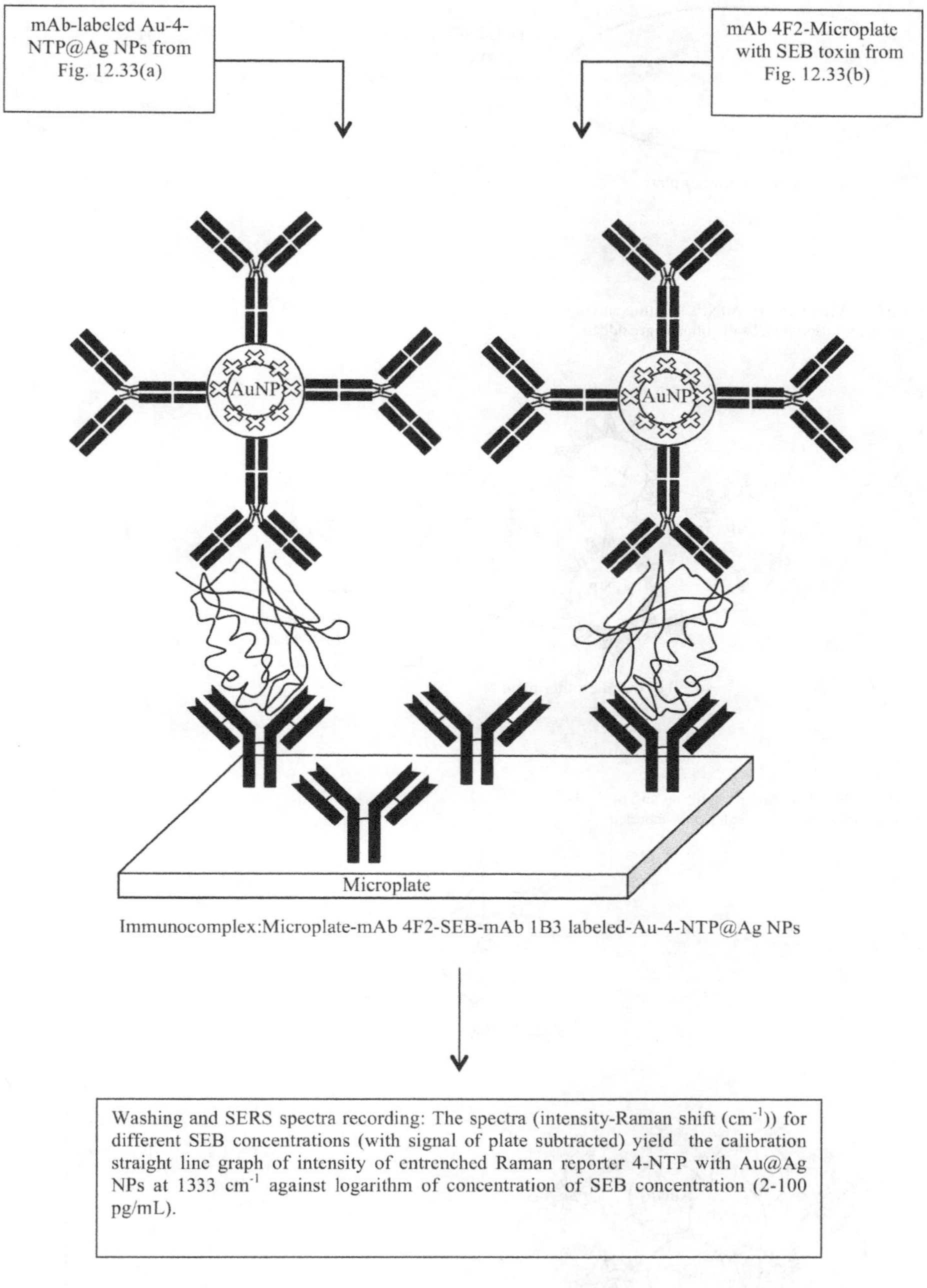
mAb-labeled Au-4-NTP@Ag NPs from Fig. 12.33(a)
mAb 4F2-Microplate with SEB toxin from Fig. 12.33(b)
AuNP
AuNP
Microplate
Immunocomplex:Microplate-mAb 4F2-SEB-mAb 1B3 labeled-Au-4-NTP@Ag NPs
Washing and SERS spectra recording: The spectra (intensity-Raman shift (cm^{-1})) for different SEB concentrations (with signal of plate subtracted) yield the calibration straight line graph of intensity of entrenched Raman reporter 4-NTP with Au@Ag NPs at 1333 cm^{-1} against logarithm of concentration of SEB concentration (2-100 pg/mL).

(c)

FIGURE 12.33 (Continued)

(vi) N-methyl mesoporphyrin IX (NMM), a fluorescent binder specific to G-quadruplex DNA, is added and incubation is carried out. The fluorescence intensity of the resultant product is measured.

(vi) The sensor determines the AFM1 concentration from the fluorescence intensity. The fluorescence intensity increases with the AFM1 concentration from 0 to 20 ng/mL. At 20 mg/mL, it attains saturation. Practically, the limit of detection is 9.73 ng/kg. Detection is not interfered with by toxins such as AFB1, AFM2, and Ochratoxin A (Zhang et al. 2019).

12.16 Discussion and Conclusions

12.16.1 Nanosensors for Explosives

12.16.1.1 TNT

All eight nanosensors for TNT detection considered here were optical. Luminescence intensity changes and color variations of nanoparticles are the most commonly employed strategies. The surface plasmon resonance approach is also used. The amino group has been a popular moiety utilized for binding with TNT

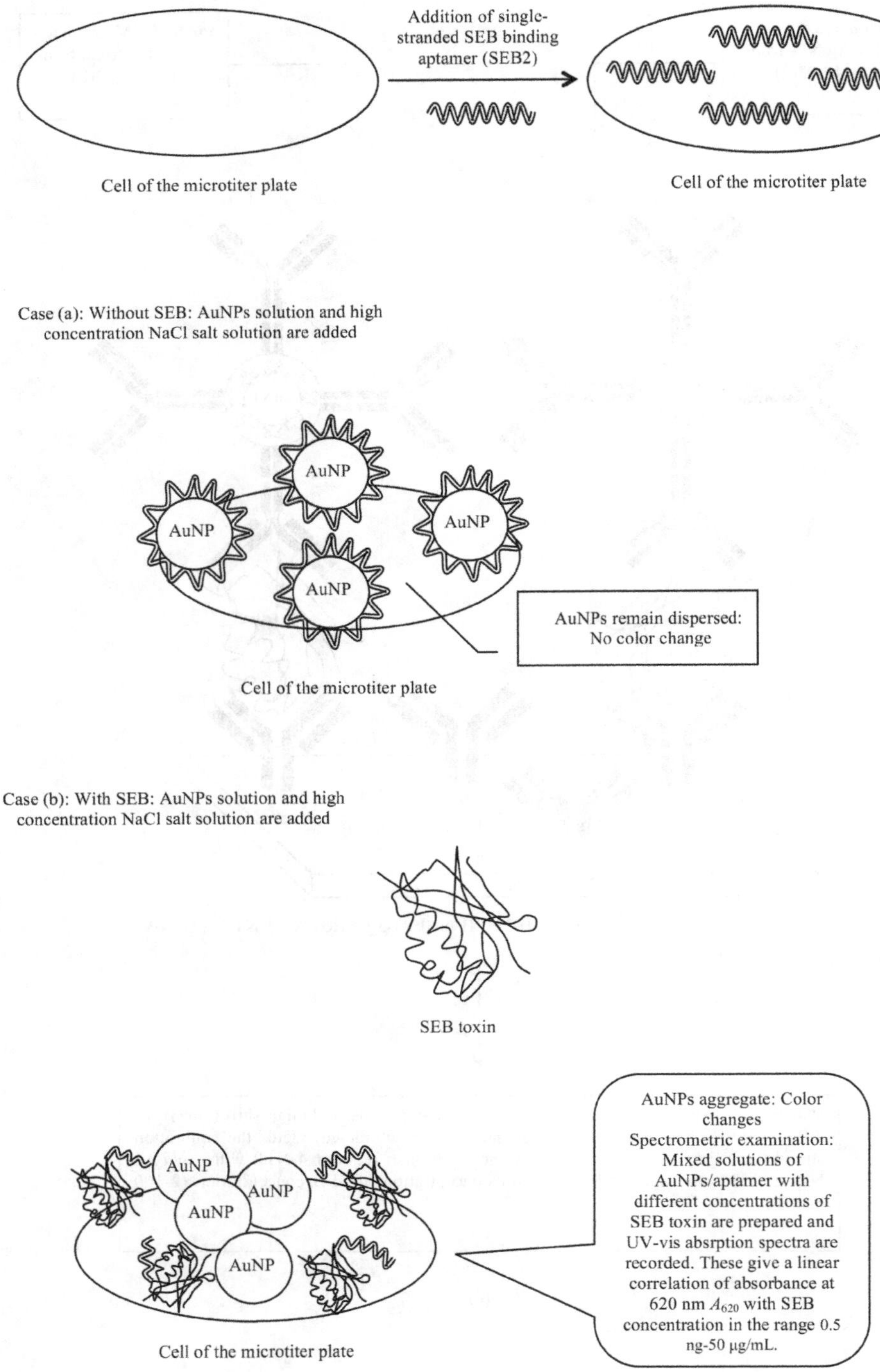

FIGURE 12.34 SEB detection assay using gold nanoparticles and aptamer. (Mondal et al. 2018.)

molecules. A TNT recognition peptide is also used, e.g., in the surface plasmon resonance sensor. However, functionalization with a specific TNT-binding ligand may not always be necessary, as exemplified by the $AgInS_2$ quantum dot probe. Table 12.3 summarizes these nanosensors.

12.16.1.2 TNT:Tetryl

Two types of colorimetric nanosensors for TNT/tetryl detection were described (Table 12.4).

12.16.1.3 Picric Acid

Among the five nanosensors for PA attempted (Table 12.5), there are two types of electrochemical nanosensors, one of which utilizes ZnO peanuts to modify the screen-printed electrode and the other uses nanostructured Cu_2O on this electrode. A β-cyclodextrin-functionalized reduced graphene oxide nanosensor works on the changes in current with PA adsorption. A fluorescence sensor is designed to work for PA detection in three platforms: as a nanoparticle dispersion in aqueous solution, in

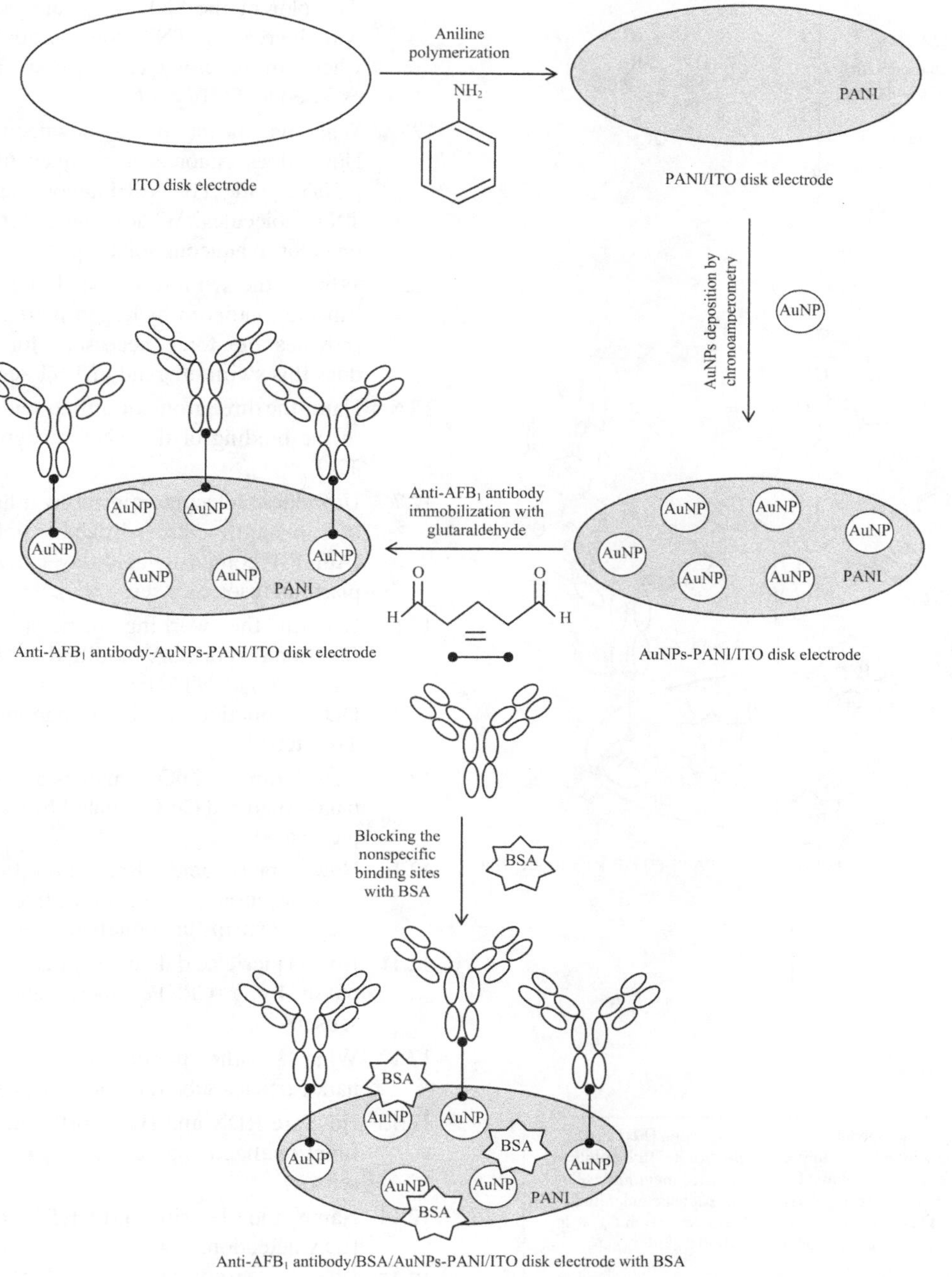

FIGURE 12.35 Preparation of an ITO working electrode for characterization by EIS. (Yagati et al. 2018.)

contact mode as a fluorescent paper strip, and for vapor detection as a two-terminal device. Silver nanopillar substrates are made for PA detection by surface-enhanced Raman scattering.

12.16.1.4 RDX/Other Explosives

Different detection schemes have been tried such as Raman scattering, colorimetry, fluorescence or FRET emissions, and molecular recognition–based voltammetric methods (Table 12.6). RDX is detected by specially designed gold nanoparticles–modified glass slides as SERS substrates. The RDX/HMX nanosensor distinguishes between the two explosives from their differential kinetics in hydrolysis to provide colorimetric responses for the target analytes. The quantum dot RDX/PETN fluorescence sensor produces responses from the recovery of fluorescence or a decrease in FRET emission. The RDX/TNT/DNT/HMX electrochemical sensor uses highly selective glassy carbon electrodes whose surfaces are imprinted with templates of target molecules for their accurate recognition.

12.16.2 Nanosensors for Biothreat Agents

Nanosensors on different biothreat agents are summarized in Tables 12.7–12.15.

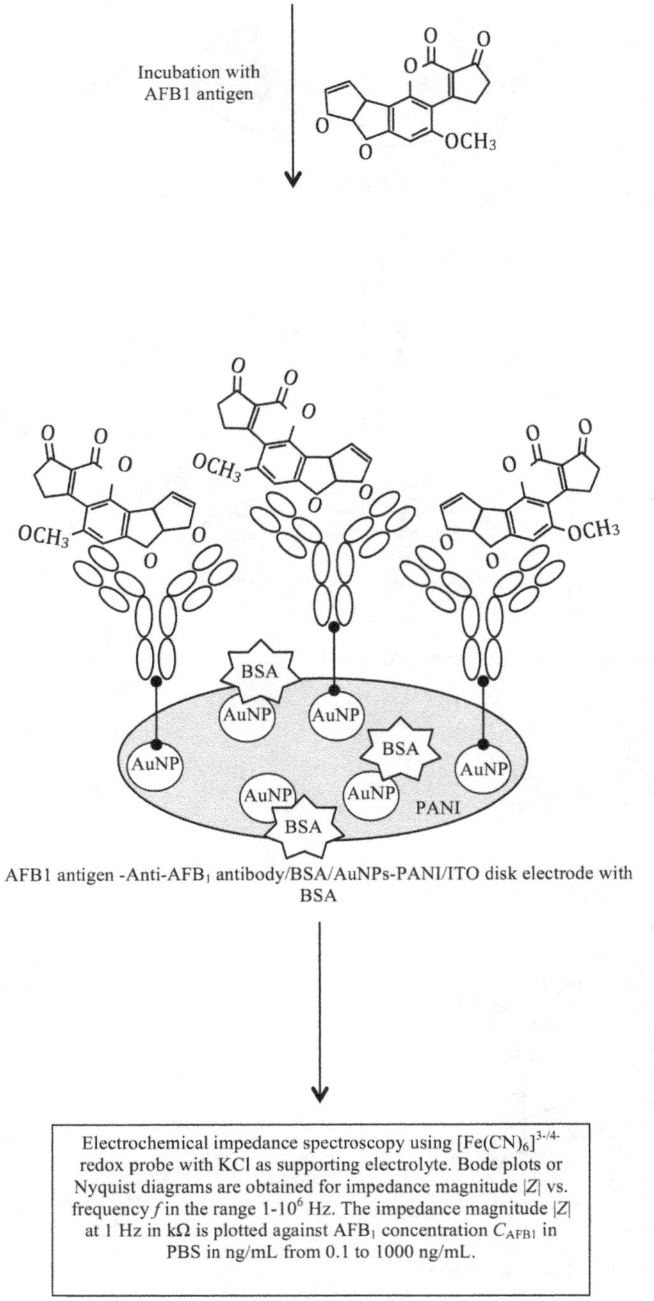

FIGURE 12.35 (Continued)

Review Exercises

12.1 Write the chemical formulae of TNT, tetryl, picric acid, and RDX. Which of these explosives exerts the lowest vapor pressure?

12.2 Name the natural source of curcumin. How are curcumin nanoparticles used in the detection of TNT? Do RDX, HMX, or PETX interact with curcumin?

12.3 Why is it necessary to functionalize silica nanoparticles with amine for TNT sensing? How does the color of the SiO_2-NH_2 nanoparticle film change with increasing TNT concentration? What happens when amine-protected Ag@Au nanoparticles are exposed to TNT?

12.4 What is meant by a downconversion material? How does fluorescence quenching occur when β-$NaYF_4$:Gd^{3+},Tb^{3+}@PEI nanophosphors interact with TNT molecules? What is the effect of this quenching on color in aqueous solution?

12.5 What is the swimming speed of Janus amine-UCNP capsule motors in hydrogen peroxide solution? What provides the force necessary for swimming? How does this swimming aid in TNT sensing?

12.6 Name the three aromatic amino acids which participate in the binding of the TNT recognition peptide with TNT.

12.7 How does molecular imprinting influence the response of bis-aniline-cross-linked gold nanoparticles (AuNPs) composite/gold layer for a TNT surface plasmon resonance sensor?

12.8 Explain the working principles of: (a) DACH-functionalized/TGA-modified Au nanoparticle sensor for TNT/tetryl and (b) CTAB-stabilized/DDTC-functionalized Au nanoparticle sensor for TNT/tetryl.

12.9 How are ZnO nanopeanuts–modified and nanostructured Cu_2O–coated SPEs used for detecting picric acid?

12.10 How is picric acid adsorbed on β-cyclodextrin? How does picric acid adsorption affect the response of a β-cyclodextrin-functionalized rGO sensor?

12.11 How is picric acid detected in contact mode and vapor phase by a CPNPs fluorescence/current response sensor?

12.12 What sensing parameter is utilized by a gold nanoparticles substrate for RDX detection by SERS?

12.13 How are RDX and HMX differentiated by a 4-ATP-functionalized gold nanoparticle colorimetric sensor?

12.14 Name and describe a FRET-based nanosensor for RDX detection.

12.15 How are RDX, DNT, and HMX concentrations electrochemically determined by modified glassy carbon electrodes? Describe the electrode modification procedures for the three explosives and measurement methods.

12.16 Give two examples each of bacteria, viruses, and toxins used as biothreat agents.

12.17 Explain the function of europium in the Eu^+ nanoparticle fluorescence immunoassay for the *Bacillus anthracis* protective antigen.

12.18 How is a quartz crystal microbalance used to detect *B. anthracis* at gene level?

12.19 Describe in detail the lateral flow test for plague bacterium. How are the results displayed on a subjective scale?

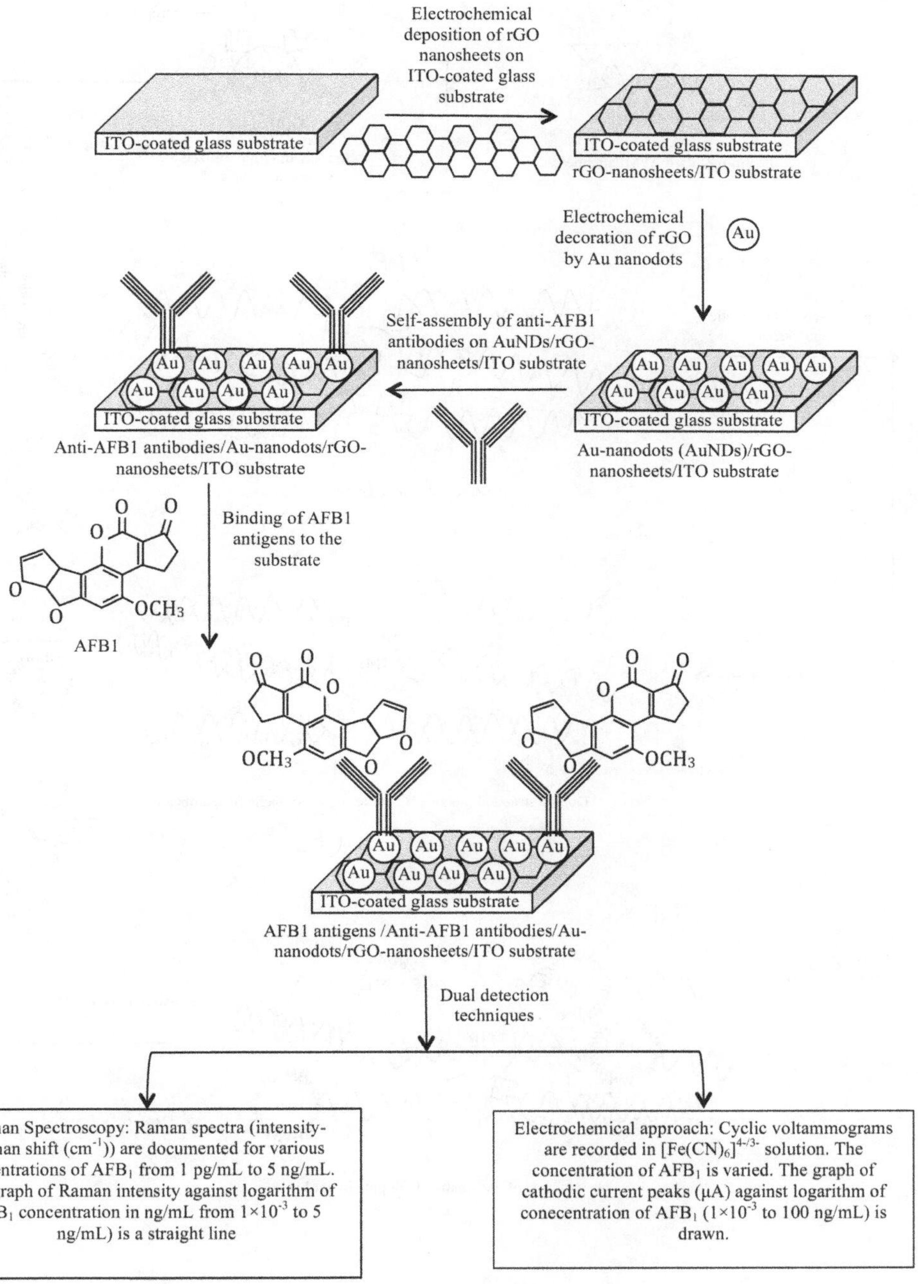

FIGURE 12.36 Substrate surface modification for Raman spectroscopy/electrochemical measurements. (Althagafi et al. 2019.)

12.20 How does an Au nanoparticle signal enhancement–based QCM biosensor measure *Francisella tularensis* bacterium concentration?

12.21 Discuss the use of apoferritin for constructing a nanosensor for *Francisella tularensis* bacterium.

12.22 How is an Au nanoparticle–modified screen-printed carbon electrode used for sensing *Francisella tularensis* bacterium by antibody–antigen binding?

12.23 How is an oligonucleotide-Au nanoparticle probe used to sense *Brucella abortus* by color changes?

12.24 How is a colorimetric immunoassay for *Brucella abortus* performed with silica nanoparticles?

12.25 Describe a nanosensor for detecting the smallpox virus.

12.26 How does an rGO-based FET detect the Ebola virus?

12.27 How is the voltage gap created in the presence of the VP40 matrix protein of the Ebola virus utilized for measuring the Ebola protein concentration in the sample?

12.28 What is the reason for using a 3-D nanoantenna array instead of a conventional substrate for Ebola virus detection?

12.29 Justify the use of magnetic nanoparticles in silver enhancement immunoassay for ricin.

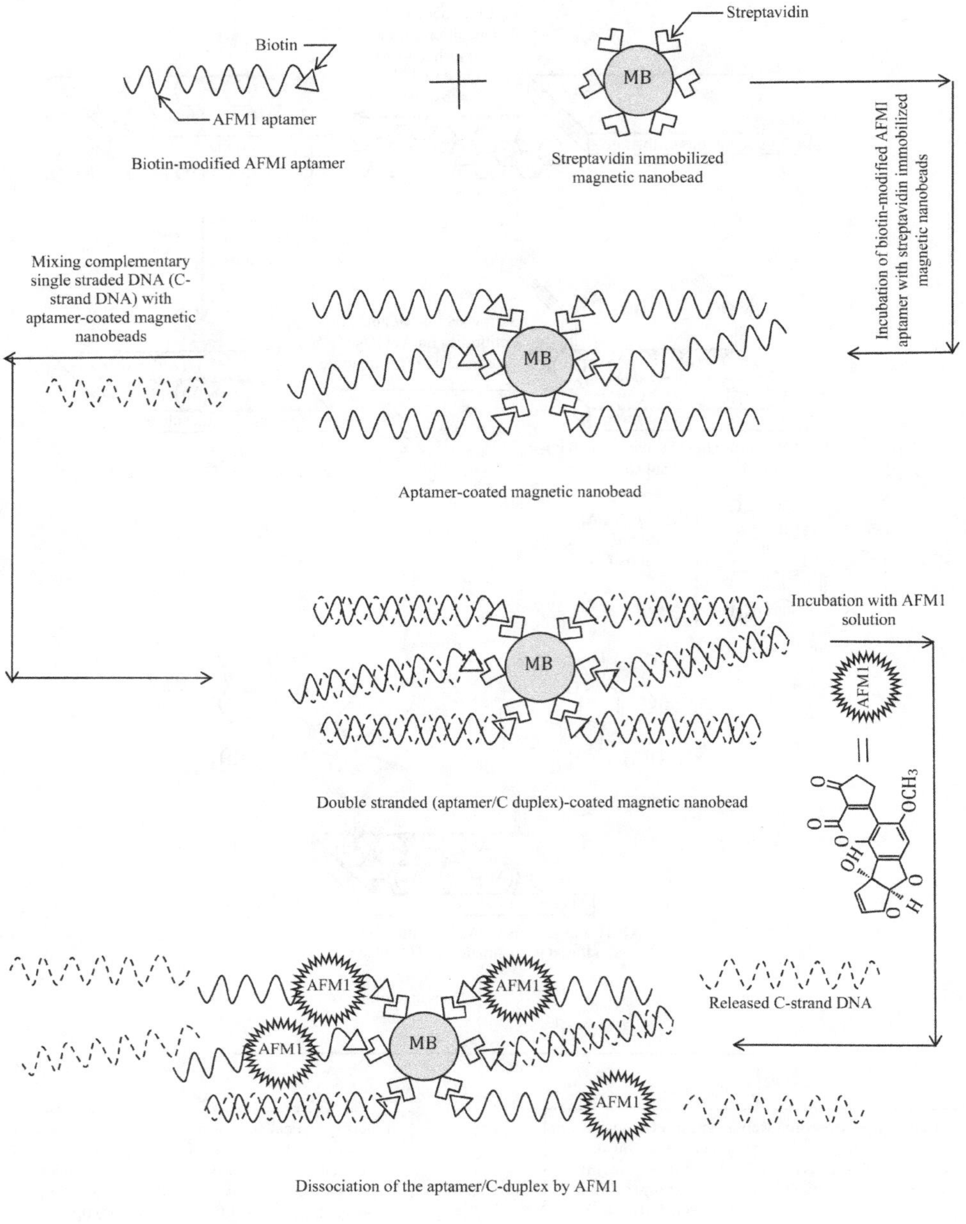

FIGURE 12.37 AFM1 fluorescence aptasensor using magnetic nanobeads: (a) making aptamer-coated magnetic nanobeads and treating with AFM1; (b) hybridization of T- and G-strand DNAs; and (c) ExoIII-actuated interaction between G/T duplex (from (b)) and released C-strand DNA (from (a)). (Zhang et al. 2019.)

12.30 Explain ricin detection by modified bar-code assay.

12.31 What is a phage display antibody? How is an electroluminescent probe for ricin detection built using this antibody?

12.32 Explain the sequence of steps involved in SEB detection by Pt nanoparticles–modified glassy carbon electrode.

12.33 How is the immunocomplex for SERS spectra recording obtained in a 4-NTP-encoded Au nanoparticle@Ag–based SERS sensor for SEB?

12.34 How does the aptamer act as a recognition element in the Au nanoparticle color indicator–based assay for SEB?

12.35 How is AFB1 electrochemically detected using a PANI nanofibers–Au nanoparticles composite–based ITO electrode?

12.36 Explain the Raman spectroscopy and electrochemical modes of using Au-nanodots/rGO-nanosheets/ITO substrate for AFB1 detection.

12.37 Describe the use of the target-initiated DNA strand displacement scheme for building an AFM1 aptamer-triggered and DNA-fueled sensor for detecting AFM1.

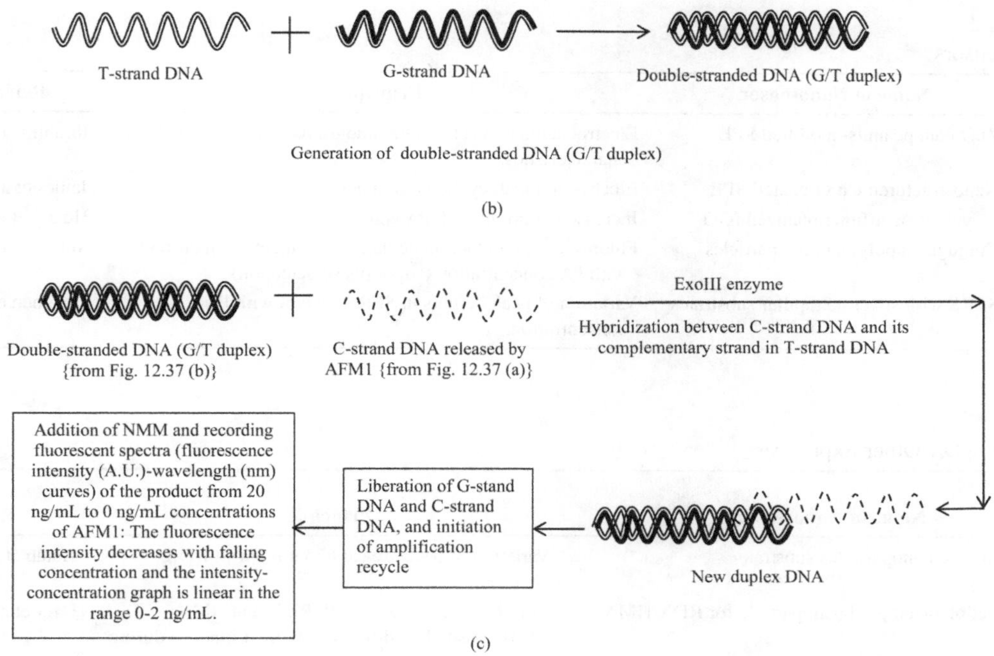

FIGURE 12.37 (Continued)

TABLE 12.3

TNT Nanosensors

Sl. No.	Name of Nanosensor	Principle	References
1.	Curcumin nanoparticles	Nanoparticle aggregation on reaction with TNT, and color shift from yellow toward red.	Pandya et al. (2012)
2.	Amine-functionalized silica nanoparticles	Color change from green to red with increasing TNT concentration.	Idros et al. (2017)
3.	Amine-modified gold@silver nanoparticles	Colorimetric.	Arshad et al. (2019)
4.	PEI-capped β-$NaYF_4$:Gd^{3+},Tb^{3+}@PEI nanophosphor	Photoluminescence intensity quenching by luminescence resonance energy transfer (LRET) from nanophosphor to the Meisenheimer complex formed between nanophosphor and TNT.	Malik et al. (2019)
5.	$NaYF_4$:Yb^{3+}/Er^{3+} nanoparticle	Luminescence quenching by TNT.	Yuan et al. (2019)
6.	$AgInS_2$(AIS) quantum dot	Emission intensity quenching with rise in TNT concentration.	Baca et al. (2017)
7.	TNT recognition peptide-SWCNTs hybrid anchored SPR chip	Surface plasmon resonance: sensorgrams of the sensor are recorded at different TNT concentrations.	Wang et al. (2018)
8.	Bis-aniline-cross-linked gold nanoparticles composite/gold layer	Surface plasmon resonance.	Riskin et al. (2009)

TABLE 12.4

TNT/Tetryl Nanosensors

Sl. No.	Name of Nanosensor	Principle	References
1.	DACH-functionalized/TGA-modified Au nanoparticle	Nanoparticle agglomeration and color change.	Ular et al. (2018)
2.	CTAB stabilized/(DDTC)-functionalized Au nanoparticle	Nanoparticle aggregation resulting in color change.	Ozcan et al. (2019)

TABLE 12.5

PA Nanosensors

Sl. No.	Name of Nanosensor	Principle	References
1.	ZnO nanopeanuts–modified SPE	Electrochemical: cyclic voltammograms and current-voltage characteristics.	Ibrahim et al. (2017)
2.	Nanostructured Cu_2O-coated SPE	Electrochemical: cyclic voltammetry.	James et al. (2018)
3.	β-cyclodextrin-functionalized rGO	Increase in current with PA concentration.	Huang et al. (2014)
4.	Conjugated polymer nanoparticles	Fluorescence (contact mode detection), increase in current with PA concentration (vapor phase detection).	Malik et al. (2015)
5.	SERS with silver nanopillar substrates	Variation of peak heights in SERS spectra with PA concentration.	Hakonen et al. (2017)

TABLE 12.6

Nanosensors for RDX/Other Explosives

Sl. No.	Name of Nanosensor	Principle	References
1.	SERS with gold nanoparticles substrate	Variation of peak area at 873 nm^{-1} with RDX concentration.	Hatab et al. (2010)
2.	4-ATP functionalized gold nanoparticle for RDX/HMX	Colorimetric response with RDX and HMX distinguished by difference in their kinetics during hydrolysis.	Üzer et al. (2013)
3.	CdS QD-DPA sensor for RDX/PETN	FRET (Förster or fluorescence resonance energy transfer).	Ganiga and Cyriac (2015)
4.	Gold nanoparticles/P(Cz-*co*-ANI) film-modified GCE for RDX TNT, DNT, and HMX	Electrochemical using square wave voltammetry.	Sağlam et al. (2018)

TABLE 12.7

Nanosensors for Anthrax Spores

Sl. No.	Name of Nanosensor	Principle	References
1.	Eu^+ nanoparticle (NP)	Fluorescence immunoassay.	Tang et al. (2009)
2.	Gold nanoparticle–amplified DNA probe-functionalized QCM	Measurement of increase in mass arising from hybridization of ssDNA from *B. anthracis* with DNA probe on QCM surface.	Hao et al. (2011)

TABLE 12.8

Plague Bacterium Nanosensor

Sl. No.	Name of Nanosensor	Principle	References
1.	Lateral flow test using F1- and V-antigen proteins fixed to blue carboxylate–modified latex nanoparticles on Whatman pad	Color results are noted on a subjective scale after binding of *Yersinia pestis* F1- or V-antibody, if present in the sample.	Abbott et al. (2014)

TABLE 12.9

Francisella tularensis Bacterium Detection

Sl. No.	Name of Nanosensor	Principle	References
1.	Gold nanoparticle sensors	QCM and absorbance models.	Kleo et al. (2012a, b)
2.	Apoferritin nanoprobe	Change in fluorescence intensity by interaction between anti-*F. tularensis* detection antibody/apoferritin/Ni-NTA/QD complex with *F. tularensis* bacteria/anti-*F. tularensis* capture antibody/magnetic bead.	Kim et al. (2015)

TABLE 12.10
Brucella Nanosensors

Sl. No.	Name of Nanosensor	Principle	References
1.	Gold nanoparticle–modified SPCE	Change in electron-transfer resistance with *Brucella melitensis* antigen concentration.	Wu et al. (2013)
2.	Oligonucleotide-activated gold nanoparticle	Colorimetric: target DNA hybridizes with oligo-AuNP probe whereby Au nanoparticle aggregation is restricted and therefore the color of Au nanoparticle solution remains red. In the absence of target DNA, there is no hybridization leading to nanoparticle aggregation and color change.	Pal et al. (2017)
3.	Silica nanoparticles	Colorimetric immunoassay.	Shams and Zarif (2019)

TABLE 12.11
Variola Nanosensor

Sl. No.	Name of Nanosensor	Principle	Reference
1.	Gold nanoparticles/magnetic beads	Target DNA is mixed with oligonucleotide-conjugated magnetic bead reporter probes followed by the addition of gold nanoparticle reporter probes modified with oligonucleotides partially complementary to target DNA. The hybridization product is taken out and supernatant solution is examined for color changes. Its absorbance changes with target DNA concentration.	Liu et al. (2012)

TABLE 12.12
Ebola Nanosensors

Sl. No.	Name of Nanosensor	Principle	References
1.	rGO-based FET	Change in drain current by antigen–antibody binding on the gate dielectric surface.	Chen et al. (2017)
2.	Bio-memristor	Voltage applied to a reference electrode to restore a predefined voltage gap to initial value.	Ibarlucea et al. (2017, 2018)
3.	Plasmonic nanoantenna	Sandwich immunoassay on a 3-D nanoantenna array.	Zang et al. (2019)

TABLE 12.13
Ricin Nanosensors

Sl. No.	Name of Nanosensor	Principle	References
1.	Magnetic nanoparticles/gold nanoparticles–based assay	Sandwich immunoassay with magnetic separation and silver enhancement.	Zhuang et al. (2010)
2.	Magnetic microparticles/gold nanoparticles–based assay	Immunoassay with magnetic separation and detection of signal DNA by amplification using polymerase chain reaction.	Yin et al. (2014)
3.	Gold-magnetic nanoparticles–based assay	Immunoassay followed by deposition of the immunocomplex on working electrode with a magnet. Electroluminescence reaction is carried out; emitted photons are detected by a photomultiplier and ECL intensity is correlated with ricin concentration.	Mu et al. (2016)

TABLE 12.14
SEB Nanosensors

Sl. No.	Name of Nanosensor	Principle	References
1.	Pt nanoparticles–modified GCE	Sandwich immunocomplexes are formed by antigen–antibody interactions on Pt nanoparticles on GCE. Copper is enzymatically deposited on Pt nanoparticles. Using voltammetry, potential of hydrogen evolution is measured. This potential is related to SEB concentration.	Sharma et al. (2014)
2.	Gold nanoparticles/silver shell–based sensor	SERS spectra of immunocomplex formed in the assay are recorded.	Wang et al. (2016)
3.	Gold nanoparticle–based assay	Colorimetric using aptamer as a recognition element. Conformational changes in aptamer with SEB binding degrade its capability to prevent gold nanoparticle aggregation producing color change in the gold nanoparticle solution.	Mondal et al. (2018)

TABLE 12.15
Aflatoxin Nanosensors

Sl. No.	Name of Nanosensor	Principle	References
1.	PANI nanofibers–gold nanoparticles composite–based electrode	Electrochemical impedance spectroscopy is done after immunoreaction on the electrode. Impedance varies with AFB1 concentration.	Yagati et al. (2018)
2.	Gold nanodots/rGO-nanosheets–modified substrate/electrode	After immunoreaction on the substrate/electrode, AFB1 is measured in two ways: by Raman spectroscopy and by electrochemical method.	Althagafi et al. (2019)
3.	Magnetic nanobeads–based aptasensor	Target-initiated DNA strand displacement occurs with liberated DNA strand promoting amplification of signal. In association with a fluorescent dye, a fluorescent signal is produced. The fluorescence intensity changes with AFM1 concentration.	Zhang et al. (2019)

REFERENCES

Abbott R. C., R. Hudak, R. Mondesire, L. A. Baeten, R. E. Russell, and T. E. Rocke. 2014. A rapid field test for sylvatic plague exposure in wild animals. *Journal of Wildlife Diseases* 50(2): 384–388.

Althagafi I. I., S. A. Ahmed, and W. A. El-Said. 2019. Fabrication of gold/graphene nanostructures modified ITO electrode as highly sensitive electrochemical detection of Aflatoxin B1. *PLoS ONE* 14(1): e0210652. HYPERLINK "https://doi.org/10.1371/journal.pone.0210652" https://doi.org/10.1371/journal.pone.0210652.

Arshad A., H. Wang, X. Bai, R. Jiang, S. Xu, and L. Wang. 2019. Colorimetric paper sensor for sensitive detection of explosive nitroaromatics based on Au@Ag nanoparticles. *Spectrochimica Acta. Part A: Molecular and Biomolecular Spectroscopy* 206: 16–22.

Baca A. J., H. A. Meylemans, L. Baldwin, L. R. Cambrea, J. Feng, Y. Yin, and M. J. Roberts. 2017. $AgInS_2$ quantum dots for the detection of trinitrotoluene. *Nanotechnology* 28: 015501 (8pp.).

Chen Y., R. Ren, H. Pu, X. Guo, J. Chang, G. Zhou, S. Mao, M. Kron, and J. Chen. 2017. Field-effect transistor biosensor for rapid detection of Ebola antigen. *Scientific Reports* 7: 10974 (pp. 1–8).

Ganiga M. and J. Cyriac. 2015. Detection of PETN and RDX using a FRET-based fluorescence sensor system. *Analytical Methods* 7: 5412–5418.

Hakonen A., F. C.Wang, P. O. Andersson, H. Wingfors, T. Rindzevicius, M. S. Schmidt, V. R. Soma, S. Xu, Y.Q. Li, A. Boisen, and H. A. Wu. 2017. Hand-held femtogram detection of hazardous picric acid with hydrophobic Ag nanopillar SERS substrates and mechanism of elasto-capillarity. *ACS Sensors* 2: 198–202.

Hao R.-Z., H.-B. Song, G.-M. Zuo, R.-F. Yang, H.-P. Wei, D.-B. Wang, Z.-Q. Cui, Z. P. Zhang, Z.-X. Cheng, and X.-E. Zhang. 2011. DNA probe functionalized QCM biosensor based on gold nanoparticle amplification for *Bacillus anthracis* detection. *Biosensors and Bioelectronics* 26: 3398–3404.

Hatab N. A., G. Eres, P. B. Hatzinger, and B. Gu. 2010. Detection and analysis of cyclotrimethylenetrinitramine (RDX) in environmental samples by surface-enhanced Raman spectroscopy. *Journal of Raman Spectroscopy* 41: 1131–1136.

Hill H. D. and C. A. Mirkin. 2006. The bio-barcode assay for the detection of protein and nucleic acid targets using DTT-induced ligand exchange. *Nature Protocols* 1(1): 324–336.

Huang J., L. Wang, C. Shi, Y. Dai, C. Gu, and J. Liu. 2014. Selective detection of picric acid using functionalized reduced graphene oxide sensor device. *Sensors and Actuators B: Chemical* 196: 567–573.

Ibarlucea B., T. F. Akbar, K. Kim, T. Rim, C.-K. Baek, A. Ascoli, R. Tetzlaff, L. Baraban, and G. Cuniberti. 2017. Ebola biosensing with a gate controlled memristor mode. In Trends in Nanotechnology, TNT2017, Dresden, Germany, June 5–9, Phantoms Foundation, Madrid, Spain, p. 63, Available at: http://www.tntconf.org/2017/TNT2017_abstractsbook.pdf

Ibarlucea B., T. F. Akbar, K. Kim, T. Rim, C.-K. Baek, A. Ascoli, R. Tetzlaff, L. Baraban, and G. Cuniberti. 2018. Ultrasensitive detection of Ebola matrix protein in a memristor mode. *Nano Research* 11: 1057–1068.

Ibrahim A. A., P. Tiwari, M. S. Al-Assiri, A. E. Al-Salami, A. Umar, R. Kumar, S. H. Kim, Z. A. Ansari, and S. Baskoutas. 2017. A highly-sensitive picric acid chemical sensor based on ZnO nanopeanuts. *Materials* 10: 795 (15pp.).

Idros N., M. Y. Ho, M. Pivnenko, M. M. Qasim, H. Xu, Z. Gu, and D. Chu. 2015. Colorimetric-based detection of TNT explosives using functionalized silica nanoparticles. *Sensors* 15: 12891–12905.

Idros N., M. Y. Ho, M. Pivnenko, M. M. Qasim, H. Xu, Z. Gu, and D. Chu. 2017. Colorimetric-based detection of TNT explosives using functionalized silica nanoparticles. *Procedia Technology* 27(2017): 312–314.

James S., B. Chishti, S. A. Ansari, O. Y. Alothman, H. Fouad, Z. A. Ansari, and S. G. Ansari. 2018. Nanostructured cuprous-oxide-based screen-printed electrode for electrochemical sensing of picric acid. *Journal of Electronic Materials* 47: 7505–7513.

Kim J.-E., Y. Seo, Y. Jeong, M. P. Hwang, J. Hwang, J. Choo, J. W. Hong, J. H. Jeon, G.-E. Rhie, and J. Choi. 2015. A novel nanoprobe for the sensitive detection of *Francisella tularensis*. *Journal of Hazardous Materials* 298: 188–194.

Kleo K., C. Nietzold, R. Grunow, and F. Lisdat. 2012a. Fast detection of pathogenic bacteria by using different sensor techniques. IMCS 2012-The 14th International Meeting on Chemical Sensors, 20–23 May, Nürnberg/Nuremberg, Germany, Copyright© (2012) by AMA Service GmbH, Wunstorf, Germany, Printed by Curran Associates, Inc.(2014), NY, USA, pp. 834–836. Available at: http://toc.proceedings.com /20476webtoc.pdf

Kleo K., D. Schäfer, S. Klar, D. Jacob, R. Grunow, and F. Lisdat. 2012b. Immunodetection of inactivated *Francisella tularensis* bacteria by using a quartz crystal microbalance with dissipation monitoring. *Analytical and Bioanalytical Chemistry* 404(3): 843–851.

Liu Y., Z. Wu, G. Zhou, Z. He, X. Zhou, A. Shen, and J. Hu. 2012. Simple, rapid, homogeneous oligonucleotides colorimetric detection based on non-aggregated gold nanoparticles. *Chemical Communications* 48: 3164–3166.

Malik A. H., S. Hussain, A. Kalita, and P. K. Iyer. 2015. Conjugated polymer nanoparticles for the amplified detection of nitro-explosive picric acid on multiple platforms. *ACS Applied Materials & Interfaces* 7: 26968–26976.

Malik M., P. Padhye, and P. Poddar. 2019. Downconversion luminescence-based nanosensor for label-free detection of explosives. *ACS Omega* 4: 4259–4268.

Mondal B., S. Ramlal, P. S. Lavu, N. Bhavanashri, and J. Kingston. 2018. Highly sensitive colorimetric biosensor for Staphylococcal Enterotoxin B by a label-free aptamer and gold nanoparticles. *Frontiers in Microbiology* 9: 179 (pp. 1–8).

Mu X., Z. Tong, Q. Huang, B. Liu, Z. Liu, L. Hao, H. Dong, J. Zhang, and C. Gao. 2016. An electrochemiluminescence immunosensor based on gold-magnetic nanoparticles and phage displayed antibodies. *Sensors* 16: 308 (22 pp.).

Özcan C., A. Üzer, S. Durmazel, and R. Apak. 2019. Colorimetric sensing of nitroaromatic energetic materials using surfactant-stabilized and dithiocarbamate-functionalized gold nanoparticles. *Analytical Letters* 52(17): 2794–2808.

Pal D., N. Boby, S. Kumar, G. Kaur, S. A. Ali, J. Reboud, S. Shrivastava, P. K. Gupta, J. M. Cooper, and P. Chaudhuri. 2017. Visual detection of *Brucella* in bovine biological samples using DNA-activated gold nanoparticles. *PloS ONE* 12(7): e0180919.

Pandya A., H. Goswami, A. Lodha, and S. K. Menon. 2012. A novel nanoaggregation detection technique of TNT using selective and ultrasensitive nanocurcumin as a probe. *Analyst* 137: 1771–1774.

Riskin M., R. Tel-Vered, O. Lioubashevski, and I. Willne. 2009. Ultrasensitive surface plasmon resonance detection of trinitrotoluene by a bis-aniline-cross-linked au nanoparticles composite. *Journal of the American Chemical Society* 131: 7368–7378.

Riskin M., R. Tel-Vered, and I. Willner. 2010. Imprinted Au-nanoparticle composites for the ultrasensitive surface plasmon resonance detection of hexahydro-1,3,5-trinitro-1,3,5-triazine (RDX). *Advanced Materials* 22(12): 1387–1391.

Riskin M., Y. Ben-Amram, R. Tel-Vered, V. Chegel, J. Almog, and I. Willner. 2011. Molecularly imprinted Au nanoparticles composites on Au surfaces for the surface plasmon resonance detection of pentaerythritol tetranitrate, nitroglycerin, and ethylene glycol dinitrate. *Analytical Chemistry* 83(8): 3082–3088.

Sağlam Ş., A. Üzer, E. Erçağ, and R. Apak. 2018. Electrochemical determination of TNT, DNT, RDX, and HMX with gold nanoparticles/poly(carbazole-aniline) film-modified glassy carbon sensor electrodes imprinted for molecular recognition of nitroaromatics and nitramines. *Analytical Chemistry* 90(12): 7364–7370.

Senesac L. and T. G. Thundat. 2008. Nanosensors for trace explosive detection. *Materials Today* 11(3): 28–36.

Shams A. and B. R. Zarif. 2019. Designing an immunosensor for detection of *Brucella abortus* based on colored silica nanoparticles. *Artificial Cells, Nanomedicine and Biotechnology* 47(1): 2562–2568.

Sharma A., V. K. Rao, D. V. Kamboj, and R. Jain. 2014. Electrochemical immunosensor for Staphylococcal Enterotoxin B (SEB) based on platinum nanoparticles-modified electrode using hydrogen evolution inhibition approach. *Electroanalysis* 26(11): 2320–2327.

Tang S., M. Moayeri, Z. Chen, H. Harma, J. Zhao, H. Hu, R. H. Purcell, S. H. Leppla, and I. K. Hewlett. 2009. Detection of anthrax toxin by an ultrasensitive immunoassay using europium nanoparticles. *Clinical and Vaccine Immunology* 16(3): 408–413.

Ular N., A. Üzer, S. Durmazel, E. Erçağ, and R. Apak. 2018. Diami nocyclohexane-functionalized/thioglycolic acid-modified gold nanoparticle-based colorimetric sensing of trinitrotoluene and tetryl. *ACS Sensors* 3(11): 2335–2342.

Üzer A., Z. Can, I. Akın, E. Erçağ, and R. Apak. 2013. 4-Aminothiophenol functionalized gold nanoparticle-based colorimetric sensor for the determination of nitramine energetic materials. *Analytical Chemistry* 86: 351–356.

Wang W., W. Wang, L. Liu, L. Xu, H. Kuang, J. Zhu, and C. Xu. 2016. Nanoshell-enhanced Raman spectroscopy on a microplate for Staphylococcal Enterotoxin B sensing. *ACS Applied Materials & Interfaces* 8: 15591–15597.

Wang J., S. Du, T. Onodera, R. Yatabe, M. Tanaka, M. Okochi, and K. Toko. 2018. An SPR sensor chip based on peptide-modified single-walled carbon nanotubes with enhanced sensitivity and selectivity in the detection of 2,4,6-trinitrotoluene explosives. *Sensors* 18: 4461 (8pp.).

Wu H., Y. Zuo, C. Cui, W. Yang, H. Ma, and X. Wang. 2013. Rapid quantitative detection of *Brucella melitensis* by a label-free impedance immunosensor based on a gold nanoparticle-modified screen-printed carbon electrode. *Sensors* 13: 8551–8563.

Yagati A. K., S. G. Chavan, C. Baek, M.-H. Lee, and J. Min. 2018. Label-free impedance sensing of Aflatoxin B_1 with polyaniline nanofibers/Au nanoparticle electrode array. *Sensors* 18: 1320 (pp. 1–14).

Yin H.-Q., M.-X. Jia, L.-J. Shi, J. Liu, R. Wang, M.-M. Lv, Y.-Y. Ma, X. Zhao, and J.-G. Zhang. 2014. Evaluation of a novel ultrasensitive nanoparticle probe-based assay for ricin detection. *Journal of Immunotoxicology* 11(3): 291–295.

Yuan Y., C. Gao, D. Wang, C. Zhou, B. Zhu, and Q. He. 2019. Janus-micromotor-based on–off luminescence sensor for active TNT detection. *Beilstein Journal of Nanotechnology* 10: 1324–1331.

Zang F., Z. Su, L. Zhou, K. Konduru, G. Kaplan, and S. Y. Chou. 2019. Ultrasensitive Ebola virus antigen sensing via 3D nanoantenna arrays. *Advanced Materials* 31(30): 1902331.

Zhang F., L. Liu, S. Ni, J. Deng, G.-J. Liu, R. Middleton, D. W. Inglis, S. Wang, and G. Liu. 2019. Turn-on fluorescence aptasensor on magnetic nanobeads for Aflatoxin M1 detection based on an Exonuclease III-assisted signal amplification strategy. *Nanomaterials* 9: 104 (pp.1–13).

Zhuang J., T. Cheng, L. Gao, Y. Luo, Q. Ren, D. Lu, F. Tang, X. Ren, D. Yang, J. Feng, J. Zhu, and X. Yan. 2010. Silica coating magnetic nanoparticle-based silver enhancement immunoassay for rapid electrical detection of ricin toxin. *Toxicon* 55: 145–152.

Part VI

Powering, Networking, and Trends of Nanosensors

13

Nanogenerators and Self-Powered Nanosensors

Having learned the basics of nanosensors in Part I (Chapter 1), about nanomaterials and nanofabrication in Part II (Chapters 2 and 3), about physical nanosensors in Part III (Chapters 4–7), about chemical and biological nanosensors in Part IV (Chapters 8 and 9), and about the prolific applications of nanosensors in Part V (Chapters 10–12), it is now time to enquire about some issues and concerns in feeding power to nanosensors and interconnecting nanosensors to execute complicated jobs. Therefore, it is necessary to address such topics in Chapters 13 and 14 of this Part VI (Chapters 13–15) before summing up and wrapping up discussions in Chapter 15.

13.1 Devising Ways to Get Rid of Environment-Devastating Batteries

In Chapter 14, we will see that it is often necessary to connect a large number of nanosensors to gather information from a locality, and provide the designed nanosensor network with internet connectivity. When we think of powering individual nanosensors in such a network with separate batteries, we can foresee many problems. Firstly, the addition of a battery to a nanosensor increases its size, making it less portable. Secondly, batteries need to be frequently recharged. Thirdly, batteries may stop working and require repair or replacement. Fourthly, batteries use toxic chemicals which pollute the environment and thus are incompatible with eco-friendly green technology. The larger the number of nanosensors, the larger the number of batteries used for their operation and the more complicated the recycling of battery wastes. All these thoughts compel us to think in a new direction, namely, these nanosensors must be able to work without batteries in a sustainable, maintenance-free manner. The so-called batteryless nanosensors are said to be self-powered nanosensors. But energy must be supplied from somewhere? In place of batteries, the self-powered nanosensors receive energy from nanogenerators which are miniature, flexible, slim, lightweight, low-cost, and clean nanodevices that derive their energy from the environment of the nanosensor to act as sustainable power sources.

13.1.1 Vibration: The Abundant Energy Source in the Environment

To build a nanogenerator, we must search for an environmental energy source which is ubiquitous and easily accessible. One source which is omnipresent in the environment is mechanical energy in the form of petite or big vibrations. Vibrations have a close, infrangible association with a living environment, so much so that finding a vibration-free place or platform seems almost impracticable. A nanosensor placed on the human body is not stationary. A person breathing (inhalation and exhalation), stretching and relaxing muscles, any blood vessel contraction, walking or doing any work constantly produce vibrations in some form. For a nanosensor inside the body, blood circulation and body fluid movements cause vibrations. Most of the time, humans work or rest in a house or a building. Buildings constantly vibrate for a variety of reasons. Internal causes of vibration are elevators, pumps, fans, trollies, etc., and human activities such as shouting, eating, walking, jumping, dancing, singing, playing loud music (acoustic waves), and so forth. External causes include bells, buzzers, and alarm sounds; car horns; the traffic on a highway; the movement of an automobile; ground shocks due to trains passing on nearby railway tracks; the jerks produced by heavy machines such as cranes; construction works such as pile driving, hammering, cutting, and excavation in progress. The blowing of wind in the atmosphere can produce huge vibrations. During stormy weather, the shattering of doors and window panes and other structures in a building may produce vibrations. Particularly tall and slender buildings are prone to airflow or wind-induced vibrations. All these vibrational activities provide adequate energy which can be utilized by a nanogenerator.

13.1.2 Phenomena for Harvesting Vibrational Energy: Tribo- and Piezoelectricity

To harness the environmental energy, we must search for physical phenomena which can convert mechanical energy into electricity. One such versatile phenomenon that is widely used in powerhouses is electromagnetic induction, which is also suited to the large-scale generation of electricity. Two other noteworthy phenomena which immediately come to mind are triboelectricity or frictional electricity and piezoelectricity.

"Tribo" derived from the Greek *tribos* means to rub. Consider bringing together two objects in intimate contact and pulling them apart. By friction, both the objects are charged. Rubbing two materials induces equal and opposite charges in the rubbing material and the rubbed material. One of the two materials loses electrons to become positively charged. The other material gains the same number of electrons to acquire an equal amount of negative charge. This rubbing event goes by the name of the triboelectric charging effect or contact electrification. A common example is pulling a comb through dry hair where we find that the hair is attracted toward the comb. This is due to charging of both the hair and the comb. The opposite charges on the hair and the comb mutually attract each other.

Materials have been arranged in a series called a triboelectric series in descending order of their tendency to lose electrons. If we select any two materials in this series, the material high up in the series has a greater electron-losing tendency, i.e., the tendency to become positively charged. Any material lower down in the series has a lower electron-losing tendency. When rubbed with a higher placed material, it will become negatively charged because it will have a higher electron-gaining tendency. Glass is positioned higher in this list than polyester. So, rubbing a glass rod with polyester will create a positive charge on the glass and a negative charge on the polyester.

"Piezo" means to press, push, or squeeze. When we squeeze or compress a quartz crystal, a potential difference is produced between the opposite faces of the crystal. Electricity produced by applying mechanical stress to a material is called piezoelectricity; the material itself is termed piezoelectric. Piezoelectricity is a reversible effect because the application of a voltage to such a material gives rise to mechanical strain. Naturally occurring piezoelectric crystals are quartz (SiO_2), Rochelle salt (sodium potassium tartrate tetrahydrate), tourmaline (boron silicate compound with Al, Fe, Mg, Na, Li, or K), etc. Examples of man-made piezoelectric ceramics are zinc oxide (ZnO), lead titanate ($PbTiO_3$), barium titanate ($BaTiO_3$), lead zirconate titanate (PZT), lithium niobate ($LiNbO_3$), and sodium tungstate (Na_2WO_4), to mention a few examples.

13.1.3 Role of Nanotechnology in Energy Harvesting

Where are the nanoscale effects in triboelectricity or piezoelectricity? When we rub two materials, it is the nanoscale roughness of their surfaces that increases the surface area and hence promotes the generation of electricity. Sometimes, nanostructures are deliberately produced on the surfaces of the materials used for triboelectricity generation to intensify and make the process more effective. Similarly, the use of nanowires (NWs) or other nanoparticle (NP) structures of the piezoelectric material greatly enhances the piezoelectric effect and helps to obtain significant power. Thus, nanotechnology lies at the roots of both methods. They will not work to satisfaction without using nanotechnology; hence the term "nanogenerators."

A nanogenerator which works using the triboelectric effect is called a triboelectric nanogenerator or TENG. The nanogenerator which utilizes the piezoelectric effect is known as a piezoelectric nanogenerator or PENG. The TENG and PENG are inventions of the not too distant past. The TENG was invented by Fan et al. (2012) using kapton-polyethylene terephthalate (PET) sheets showing a peak voltage of 3.3 V, a current of 0.6 μA, and a power density of ~10.4 mW/cm^3; and PENG by Wang and Song (2006) using ZnO nanowires showing a power density of 10 pW/μm^2 for a nanowire density of 20/μm^2 at an efficiency of 17–30%.

13.1.4 Other Energy Sources: Do Not Overlook Light and Heat!

Light is a form of energy, which is propagated as electromagnetic waves at a tremendously high velocity through oscillations of electric and magnetic fields or as discrete particles called photons. Heat is the kinetic energy of the random motion of atoms and molecules in a material. Nanogenerators using light energy work on the photoelectric effect. Here, an electric current is produced by the ejection of electrons from the surface of a material when light falls on it. Those using heat energy work on the thermoelectric effect. In this case, power generation involves the conversion of a temperature difference between two dissimilar conductors or semiconductors into an electrical potential difference (the Seebeck effect). The related phenomena are known as photoelectricity and thermoelectricity.

13.2 Output Current of Tribo/Piezoelectric Nanogenerators as the Outcome of Second Term in Maxwell's Displacement Current

Maxwell's displacement current density, $\mathbf{J}_D$, is expressed in terms of electric displacement, **D**, as

$$\mathbf{J}_D = \frac{\partial \mathbf{D}}{\partial t} \tag{13.1}$$

where

$$\mathbf{D} = \varepsilon_0 \mathbf{E} + \mathbf{P} \tag{13.2}$$

with ε_0 the permittivity of free space, **E** the electric field, and **P** the polarization field. Hence,

$$\begin{aligned} \mathbf{J}_D &= \frac{\partial}{\partial t}\left(\varepsilon_0 \mathbf{E} + \mathbf{P}\right) = \varepsilon_0 \frac{\partial \mathbf{E}}{\partial t} + \frac{\partial \mathbf{P}}{\partial t} \\ &= \text{a component from time variation of} \\ &\quad \text{electric field} + \text{a component from time} \\ &\quad \text{variation of small motion of} \\ &\quad \text{charges bound in atoms} \\ &\quad \left(\text{dielectric polarization}\right) \end{aligned} \tag{13.3}$$

The first term of the displacement current density is responsible for the generation of electromagnetic waves serving as the starting point of wireless communications. The second term is ascribed to the electric current production caused by small displacements of charges in insulating and semiconducting media. It gives birth to tribo- and piezoelectric effects (Wang et al. 2017).

13.2.1 Principle of TENG

The basic TENG device consists of two dielectric layers of thicknesses d_1 and d_2, and relative permittivities ε_1 and ε_2, respectively (Figure 13.1). Each dielectric layer is covered on one side by a metal electrode. The two metal electrodes are joined through a load resistor R. The TENG device works by contacting and separation the dielectric layers.

(i) Contacting by applying a force: the dielectric surfaces (non-metal covered sides) are brought into contact when charge densities $-\sigma_{\text{Contact}}$ and $+\sigma_{\text{Contact}}$ are produced on the two surfaces. This takes place due to the gain and loss of electrons. After the transference of electrons, the electron donor dielectric has a deficiency of electrons while the electron acceptor dielectric has a surplus or excess of electrons. The charging mechanism

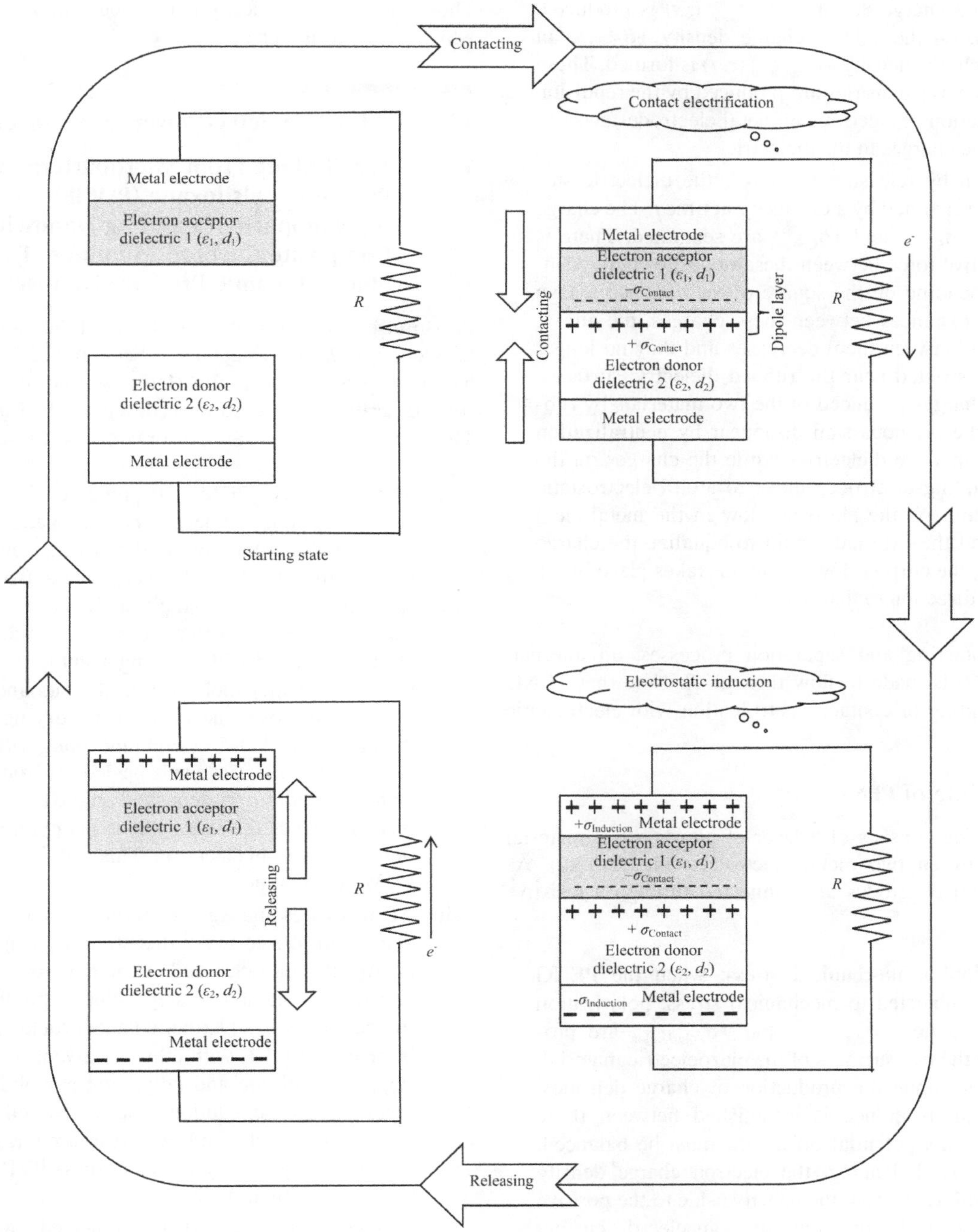

FIGURE 13.1 Working of the TENG device by contacting and separation.

is triboelectricity, i.e., the transfer of charge between two materials by friction or rubbing. These are less mobile charges because they are inside the dielectrics. Moreover, they are bound to each other by the mutual electrostatic attraction between opposite charges. As a consequence, the electrons remain localized near the interface of two dielectrics. A dipole layer is formed between the two dielectric surfaces due to an excess electron density on one side and a shortfall in electron density on the other side, producing an interfacial triboelectric potential.

Remember that the minimization of energy created by the triboelectric potential is mediated through the electron flow in the external circuit. Therefore, the load *R* experiences an electronic current flowing between the two electrodes. Inasmuch as mechanical energy becomes the cause of the production of an electric current, we can say that mechanical energy is converted into electrical energy, i.e., TENG operation has transpired.

By electrostatic induction, these charge densities create opposite charge densities on the respective metal electrodes. On the side of charge density $-\sigma_{Contact}$, an induced charge density $+\sigma_{Induction}$ (z,t) is produced. Similarly, on the side of charge density $+\sigma_{Contact}$, an induced charge density $-\sigma_{Induction}$ (z, t) is formed. These induced charge densities are produced by the repulsion and attraction of electrons in metal electrodes away or toward the charges in the dielectrics.

(ii) Separation by releasing the force: the dielectric surfaces are separated by a distance z at time t. The charge densities $-\sigma_{Contact}$ and $+\sigma_{Contact}$ are separated. There is an attractive force between these opposite charge densities depending on the square of the distance z. Due to larger distances between these charges, the attractive force between them decreases and they no longer remain restricted near the rubbed dielectric surfaces. So, the charges produced in the two materials by rubbing in the previous step disappear by neutralization within respective dielectrics while the charges on the corresponding electrodes move to attain electrostatic equilibrium. As the electrons flow in the metal electrodes and the external circuit to equalize the charge densities, the current flow through R takes place in the opposite direction to that in step (i).

Hence, by contacting and separation processes, an alternating current (AC) is made to flow through R. Note that TENG entails the coupling of contact electrification with electrostatic induction.

13.2.2 Principle of PENG

The PENG device consists of a layer of piezoelectric material sandwiched between two metal electrodes (Figure 13.2). As before, the metal electrodes are connected through a resistive load R.

(i) Application of mechanical stress: when the PENG device is subjected to mechanical stress, polarization charge densities $-\sigma_{Polarization}$ and $+\sigma_{Polarization}$ are produced on the two surfaces of the piezoelectric material. Consequent upon the production of charge densities, a potential difference is established between these surfaces. This potential difference must be balanced. For potential balancing, the electron charge density $-\sigma_{Flowing}$ migrates from the negative side to the positive side. By such electron migration, an electric current flows through the load R in the external circuit. The underpinning phenomenon responsible for current flow is the application of a mechanical stress on the PENG device, i.e., the PENG device works by transforming the mechanical energy into electrical energy. The current flows until the potential difference is neutralized, i.e., equilibrium is attained.

(ii) Release of mechanical stress: the stress release causes the PENG device to recover to its original shape. During this period, the developed piezoelectric potential becomes zero. The electrons accumulated on the two surfaces of the piezoelectric material therefore move in the opposite direction to the case when the stress was applied, leading to an oppositely directed current flow.

Thus, the PENG device produces current in two directions, which is an alternating current.

13.3 Triboelectricity-Powered Nanosensors

13.3.1 TENG Made From Micropatterned Polydimethylsiloxane (PDMS) Membrane/Ag Nanoparticles and Ag Nanowires Composite Covered Aluminum Foil as a Static/Dynamic Pressure Nanosensor

Pressure sensors are treated in Chapter 4: Section 4.12, Chapter 5: Section 5.17, and Chapter 11: Section 11.5.1; however, they have to be powered externally. Here, a self-powered pressure sensor is described. This sensor was developed by Lin et al. (2013b). The fabrication of this nanosensor consists of three parts:

(i) Micropatterned PDMS: to prepare this membrane (Figure 13.3(a)), the necessary micropattern is defined by photolithography on a silicon wafer and etched to the required depth by the wet process. Corresponding to the pattern, pyramidal pits are created on the silicon wafer. This silicon wafer is used as a mold. The PDMS elastomer mixed with a curing agent is spin-coated on the silicon wafer mold, filling the pits and forming a uniform film over the wafer. After drying and curing the PDMS, the PDMS membrane along with pyramidal projections on its surface is peeled off from the silicon wafer. A gold film is deposited on the underside of this membrane, i.e., the side without any pyramidal projections, to act as an electrode. This is the micropatterned PDMS membrane.

(ii) Ag nanowires and Ag nanoparticles composite covered aluminum foil: to make this structure (Figure13.3(b)), an Ag nanoparticles and Ag nanowires composite is prepared using a two-step solution-growth method. In the first step, platinum nanoparticles are formed by reducing $PtCl_2$ with ethylene glycol. In the second step, silver nitrate and poly(vinyl pyrrolidone) (PVP) are added to the solution followed by refluxing when silver nanoparticles and nanowires are formed by reaction with ethylene glycol. The excess PVP is removed by rinsing with acetone.

The aluminum foil is immersed in the silver nanoparticles and nanowires composite solution to fully adsorb the nanocomposite, and heated to evaporate the liquid.

(iii) Assembly of the TENG: structures produced in steps (i) and (ii) are bonded together using insulating tape with the pyramidal structures on the PDMS membrane facing the silver nanoparticles/silver nanowires covering the aluminum foil. As both the contacting surfaces have micro/nanostructures on them, the triboelectricity generation is considerably enhanced. Additionally, the pressure sensitivity of the device is improved. If these micro/nanostructures are not built, the triboelectric effect as well as the pressure sensitivity will be significantly reduced.

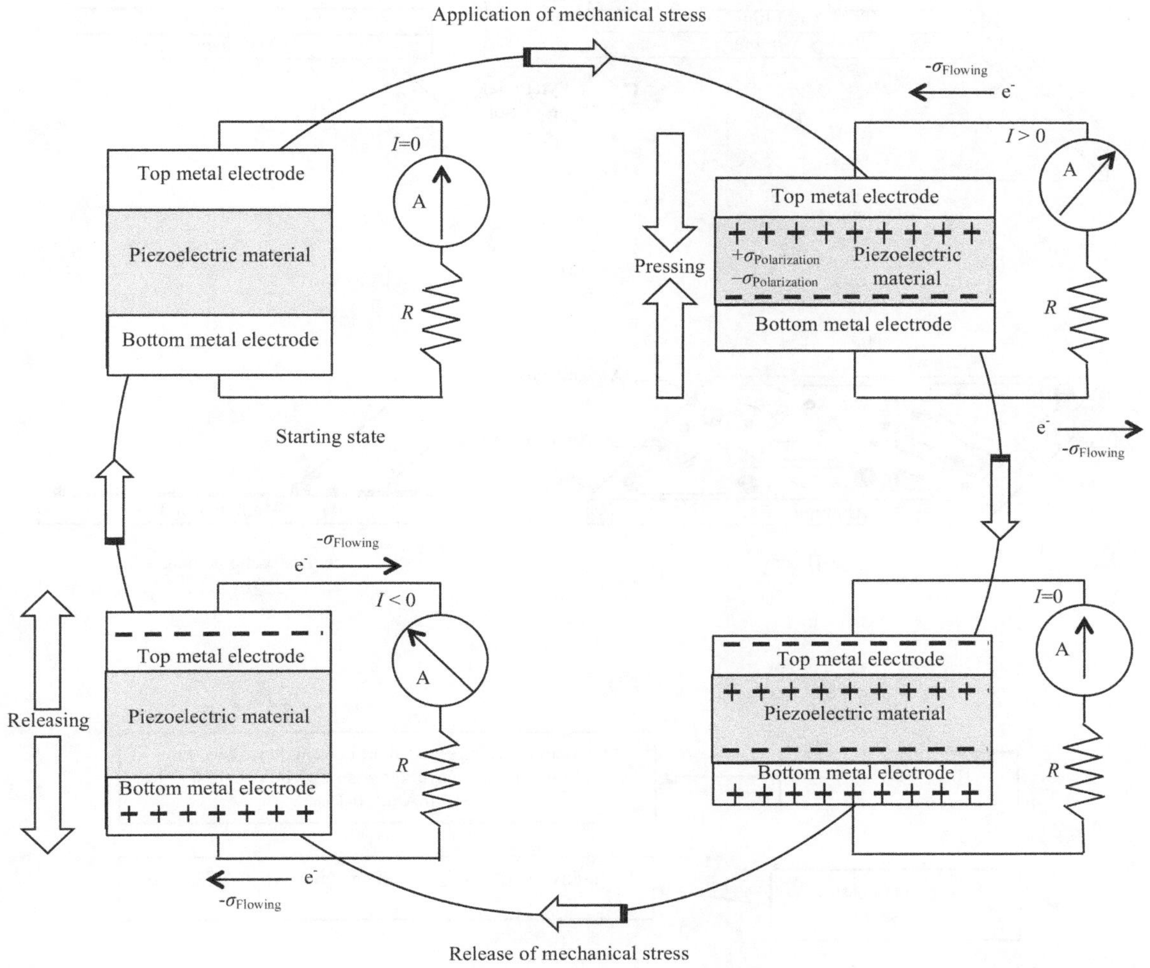

FIGURE 13.2 Working of the PENG device by stress application and stress release.

Working principle: the sensor is characterized by two main parameters:

(a) Open-circuit voltage V_{OC}: the Al foil and the PDMS differ in their capability to attract electrons. When the two layers are brought into contact by applying an external force and separated by a distance d in the vertical direction, the Al foil loses electrons to become positively charged while the PDMS layer gains electrons to become negatively charged. If $+Q$ and $-Q$ are the charges developed on the aluminum electrode and the gold electrode of the PDMS membrane, respectively, the charge:

$$Q = CV_{OC} \tag{13.4}$$

where C is the capacitance of the structure and V_{OC} is the open-circuit potential. The capacitance:

$$C = \frac{\varepsilon_0 A}{d} \tag{13.5}$$

where A is the surface area of the electrodes and ε_0 is the permittivity of free space. Combining Equations 13.4 and 13.5, we get

$$Q = \left(\frac{\varepsilon_0 A}{d}\right) V_{OC} \tag{13.6}$$

or

$$V_{OC} = \frac{Qd}{\varepsilon_0 A} \tag{13.7}$$

Thus, the open-circuit potential V_{OC} generated increases with the separation d between the electrodes. It is a static signal representative of the magnitude of the applied pressure.

(b) Short-circuit current density J_{SC}: repeated mechanical contact and separation of the two layers causes current flow in opposite directions in the external circuit, in one direction during contact and in the reverse direction during separation. The current moving back and forth is an alternating current. The short-circuit current density, J_{SC}, depends on the distance d and the loading rate of the external pressure ($\Delta d/\Delta t$). It is a dynamic signal representing the rate of change of the applied pressure.

Performance of the Sensor: by measuring the open-circuit voltage, the sensor is utilized for determining static pressure.

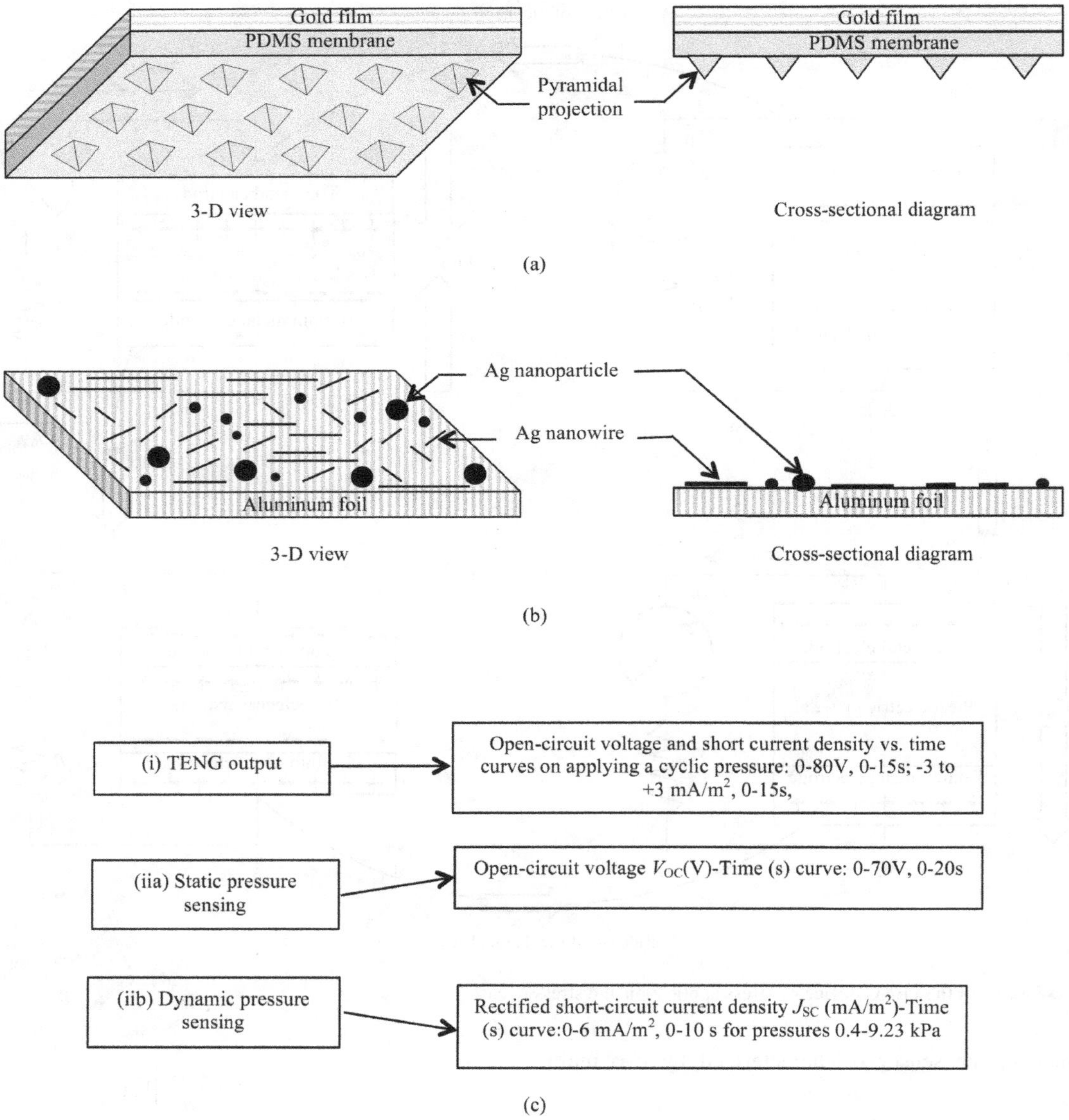

FIGURE 13.3 Parts of the TENG static/dynamic pressure nanosensor: (a) micropatterned PDMS, (b) aluminum foil covered with Ag nanowires and nanoparticles, and (c) performance characterization. (Lin et al. 2013b.)

A short-circuit current density measurement affords dynamic pressure sensing. The sensitivity is 0.31/kPa and the limit of detection is 2.1 Pa. The sensor responds in <5 ms. It has a stability of 30,000 cycles (Lin et al. 2013b).

13.3.2 Electrolytic Solution/Fluorinated Ethylene Propylene (FEP) Film TENG Nanosensor for pH Measurement

In distinction to pH sensors mentioned in Chapter 6: Section 6.7 and Chapter 8: Section 8.10.2, which need an external battery for driving, a self-powered device for pH measurement will be presented. This sensor is fabricated on a PET substrate (Wu et al. 2016).

On the PET substrate is a fluorinated ethylene propylene film which serves as the electrification layer (Figure 13.4(a)). FEP is hydrophobic in character. It is highly negative in a triboelectric series. Over the FEP film, four strip-shaped parallel copper electrodes are deposited through a mask by physical vapor deposition. Lead wires are attached to the electrodes.

Here, we see triboelectricity generation by friction between an FEP film and the solution in which it is dipped. It is a common belief that a TENG device is fabricated by rubbing two solid materials against each other. However, it is also possible to produce triboelectricity by rubbing a liquid against a solid and vice versa.

The sensor is immersed in solutions of different pH by mounting vertically on an electric motor. The motor moves in a direction perpendicular to the strip-shaped electrodes. Contact

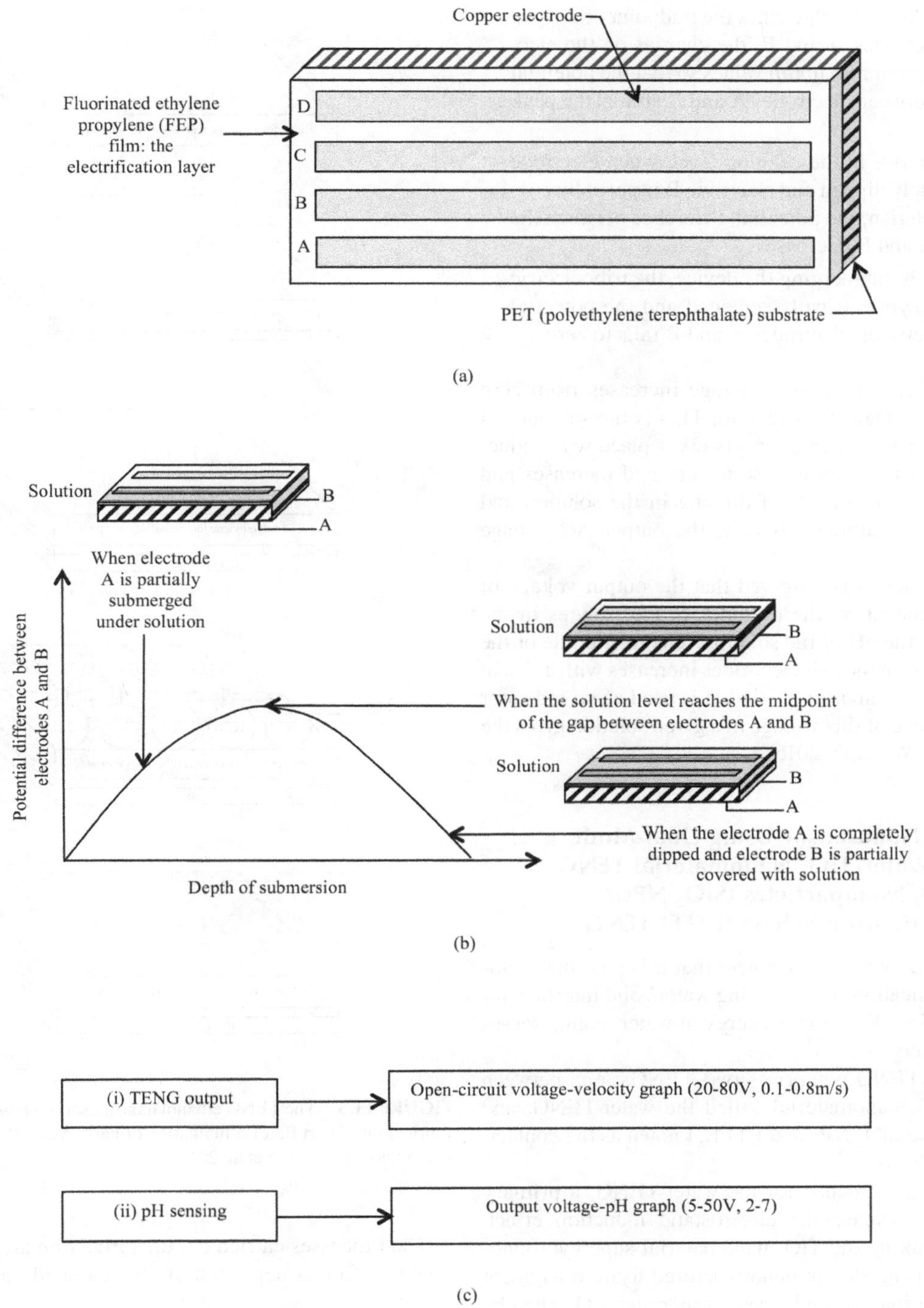

FIGURE 13.4 The TENG pH nanosensor: (a) structure, (b) voltage variation across electrodes A and B during submersion, and (c) performance characterization. (Wu et al. 2016.)

electrification takes place between the solution and the FEP film. As a result, nanogenerator action is observed. The area of the sensor submerged varies cyclically. For that reason, an alternating current flows between the electrodes.

(i) Considering two electrodes A and B, and the situation when electrode A is partially submerged in an aqueous solution (a buffer solution, suppose), coverage of an increasing portion of electrode A with liquid solution causes the appearance of a rising number of negative triboelectric charges on the FEP film. These negative charges on the FEP film attract the positive ions in the solution, setting up an electrostatic double layer at the solution/FEP film interface. The charge distribution on the FEP film surface being asymmetric, a potential difference is produced between electrodes A and B (Figure 13.4(b)). Consequently, an electron current flows from electrode B to electrode A.

(ii) When the solution level reaches the midpoint of the gap between electrodes A and B, the charges on the electrodes reach their maximum values so that the potential difference between electrodes A and B attains the peak value.
(iii) With further rise in the solution level, when electrode A is completely dipped and electrode B is partially covered with solution, the potential difference between the electrodes A and B decreases.
(iv) On completely submerging the device, the triboelectric charges are symmetrically screened and the potential difference between electrodes A and B falls to zero.

Thus, in steps (i)–(iv) the output voltage increases from zero to a peak value and again falls to zero. This is the submersion sequence. A similar sequence of events takes place with reduction of the solution level giving rise to repeated increases and decreases in voltage. The cycle of dipping in the solution and coming out from it continues as does the output AC voltage variations.

During these cycles, it is observed that the output voltage of the sensor is dependent on the presence of electrolytes in the solution and hence the pH of the solution. The magnitude of the output voltage between the two electrodes increases with a rise in the pH of the solution, and this variation is used as an indicator of the pH. Knowledge of this voltage in a given solution gives the pH of the solution (Wu et al. 2016).

13.3.3 Ethanol Nanosensor Using Dual-Mode TENG: Water/TiO_2 Nanomaterial TENG and SiO_2 Nanoparticles (SiO_2 NPs)/Polytetrafluoroethylene (PTFE) TENG

As in Section 13.3.2, we again see here that it is possible to utilize contact electrification at the flowing water/solid interface for electricity production. The kinetic energy of water motion serves as an additional energy source.

The dual-mode TENG comprises two TENGs: one between water and the TiO_2 nanomaterial called the water-TENG, and another between the SiO_2 NPs and PTFE, known as the contact-TENG (Lin et al. 2014).

Water-TENG Requirements: for the water-TENG, a primary requirement is to increase the electrostatic induction effect. This is done by making the TiO_2 nanomaterial superhydrophobic. Further, TiO_2 is needed in nanostructured form to augment the contact area for interaction between water and TiO_2, thereby enhancing the triboelectric effect.

Contact-TENG Requirements: for the contact-TENG, SiO_2 NPs are chosen because they act as a highly positive charging material and PTFE is selected because it is a highly negative charging material. So, there is a large difference between the quantities of tribocharges generated on SiO_2 NPs and PTFE, yielding a high output voltage. Here also, SiO_2 NPs are used instead of bulk SiO_2 to increase the interaction between SiO_2 and PTFE through the increased area provided by SiO_2 NPs.

Sensor Fabrication: to fabricate the sensor, a PET film is used to make Part A and a poly(methyl methacrylate) (PMMA) sheet is taken to make Part B (Figure 13.5).

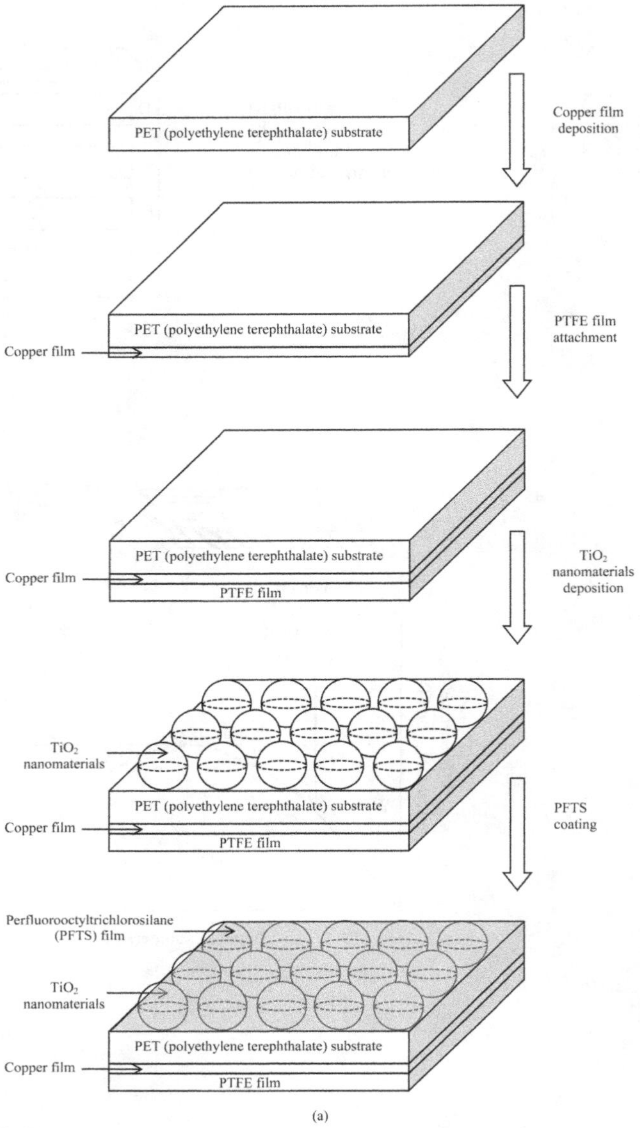

FIGURE 13.5 The TENG ethanol nanosensor: (a) fabrication of Part A, (b) fabrication of Part B, (c) combination of Parts A and B, and (d) performance characterization. (Lin et al. 2014.)

The processes carried out on a PET film are (Figure 13.5(a)): a copper film is deposited on the underside of the PET film to form:

PET film/Cu film

The PTFE film is attached to the Cu-film side of the PET film/Cu film structure to form:

PET film/Cu film/PTFE film

Superhydrophobic TiO_2 nanomaterials are grown over the PET film to obtain:

TiO_2 nanomaterials/PET film/Cu film/PTFE film

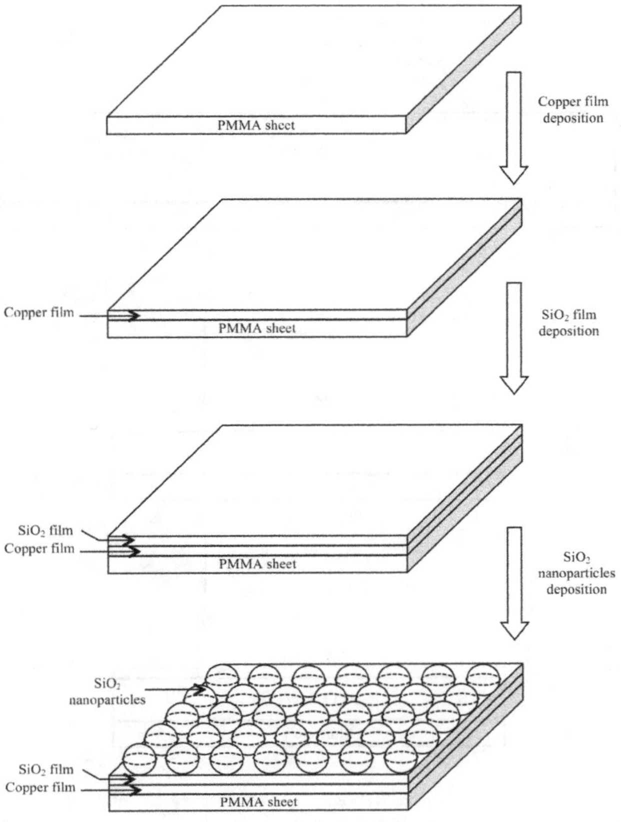

(b)

FIGURE 13.5 (Continued)

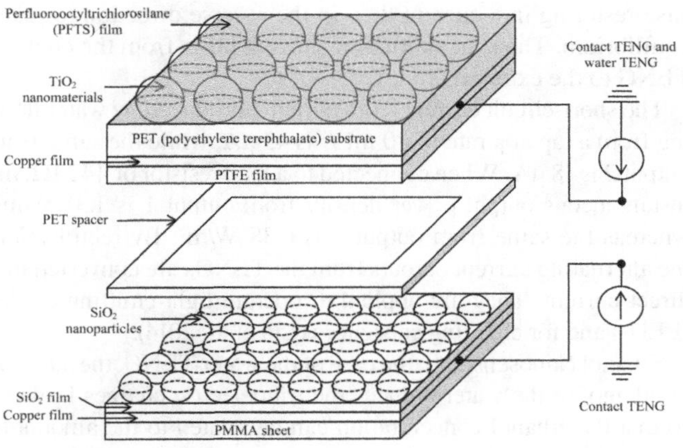

(c)

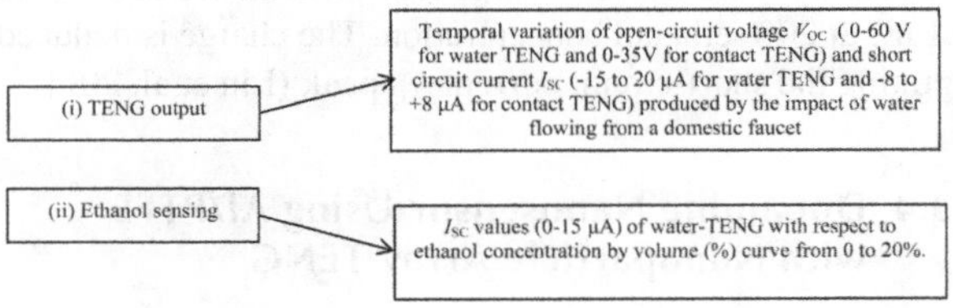

(d)

FIGURE 13.5 (Continued)

Perfluorooctyltrichlorosilane (PFTS) is coated over TiO_2 nanomaterials to provide superhydrophobicity. Thus, we get:

PFTS-coated TiO_2 nanomaterials/PET film/Cu film/PTFE film

This is Part A.

The processes carried out on a PMMA sheet are (Figure 13.5(b)): a copper film is deposited on the upper surface of a PMMA sheet forming:

Cu film/PMMA sheet

The SiO_2 thin film is deposited on the copper film of the Cu film/PMMA sheet to form:

SiO_2 film/Cu film/PMMA sheet

SiO_2 NPs are self-assembled over the SiO_2 film to obtain:

SiO_2 NPs/SiO_2 film/Cu film/PMMA sheet

This is Part B.

Parts A and B are combined using a PET film as a spacer to maintain the distance between Parts A and B (Figure 13.5(c)). The integrated device has a layered composition:

PFTS-coated TiO_2 nanomaterials/PET film/Cu film/PTFE film/PET film spacer/SiO_2 NPs/SiO_2 film/Cu film/PMMA sheet

Electric power generated by the water-TENG and contact-TENG is collected through the output 1 and output 2 terminals. Output 1 contains the power produced by both TENGs, the water-TENG and the contact-TENG, while output 2 contains the power generated by the contact-TENG only.

Water-TENG Operation: the operation of the water-TENG starts when a charged water drop (positive, suppose) contacts the TiO_2 film. Then, electrons flow from the ground to neutralize the positive charges until equilibrium is reached. When the water drop leaves the TiO_2 film surface, negative charges are left on the copper film. So, the electron flow from the copper film to the ground restores equilibrium. Thus, the water-TENG causes an alternating current to flow in the external circuit.

Contact-TENG Operation: the contact-TENG functions using the kinetic energy of water flow. The impact of water droplets on the PMMA provides the energy for pressing the surface of SiO_2 NPs against the PTFE surface. When a water droplet falls on the TiO_2 film, the surface PTFE film contacts the SiO_2 NPs surface. Both surfaces are charged. Potential differences are created between SiO_2 NPs with respect to the ground and the PTFE film relative to the ground. Equilibrium is attained by the electron flow between the ground and the upper copper electrode and between the lower copper electrode and the ground. As soon as the water droplet leaves the TiO_2 film, the surfaces of SiO_2 NPs and PTFE are separated. So, the potential differences between SiO_2 NPs with respect to the ground and the PTFE film relative to the ground are set up in the opposite direction to the preceding case resulting in a current flow in the reverse direction to reach equilibrium. Thus, an alternating current flows from the contact-TENG to the external circuit.

The short-circuit current from output 1 produced by water flowing from a tap at a rate of 40 mL/s is 43 μA, while the same from output 2 is 18 μA. When connected to a load resistor of 44 MΩ, the instantaneous output power density from output 1 is 1.31 W/m^2, whereas the same from output 2 is 0.38 W/m^2. By rectification, the alternating current outputs from the TENGs are converted into direct current. They are applied to operate light-emitting diodes (LEDs) and for charging capacitors (Lin et al. 2014).

Ethanol Nanosensor: in the case of the water-TENG, the addition of ethanol to the water reduces the triboelectric charges in water so that the ethanol concentration can be related to the amount of charge detected. Hence, the ethanol concentration is found from the quantity of charge. This makes the water-TENG a self-powered nanosensor for ethanol. The charge falls from 1.5 nC at 0% ethanol to 0.1 nC at 20% ethanol concentration. The charge is deduced by integrating the short-circuit current I_{SC} peak (Lin et al. 2014).

13.3.4 Dopamine Nanosensor Using Al/PTFE with Nanoparticle Array TENG

We spoke about dopamine sensing in Chapter 9: Section 9.2.2 using an externally powered device. We shall now describe a self-powered nanosensor for dopamine. This sensor consists of two parts which are connected together by four springs mounted at the corners to maintain a gap of 15 mm between the two parts (Jie et al. 2015).

It is fabricated by making the upper and lower parts separately, and then assembling them together (Figure 13.6):

Upper Part: on a PMMA substrate, an aluminum film is deposited through a rectangular opening in a mask to form the structure (Figure 13.6(a)):

Al film/PMMA substrate

Ease of handling and exceptional impact strength along with light weight are the advantages favoring the choice of PMMA as a substrate material.

Lower Part: electron beam evaporation is used to deposit a 50 nm-thick film of aluminum on the underside of a 0.15 mm-thick PTFE layer (Figure 13.6(b)). This gives:

PTFE layer/Al film

A 50 nm-thick gold film is deposited as a mask on the surface of the PTFE film. A nanoparticle array is etched on the surface of the PTFE film by inductively coupled plasma reactive ion etching (RIE). The etching is carried out in a mixed environment of argon, oxygen, and tetrafluoromethane (CF_4). The average diameter of PTFE nanoparticles is 300 nm. We get:

PTFE NPs/PTFE layer/Al film

The etching of PTFE nanoparticles on the surface of a PTFE layer serves a two-fold purpose. Firstly, it is aimed to improve its performance for triboelectricity generation by increasing the surface roughness as well as the surface area; and secondly, it is intended to enhance the ability of the PTFE film to selectively capture dopamine in an alkaline solution.

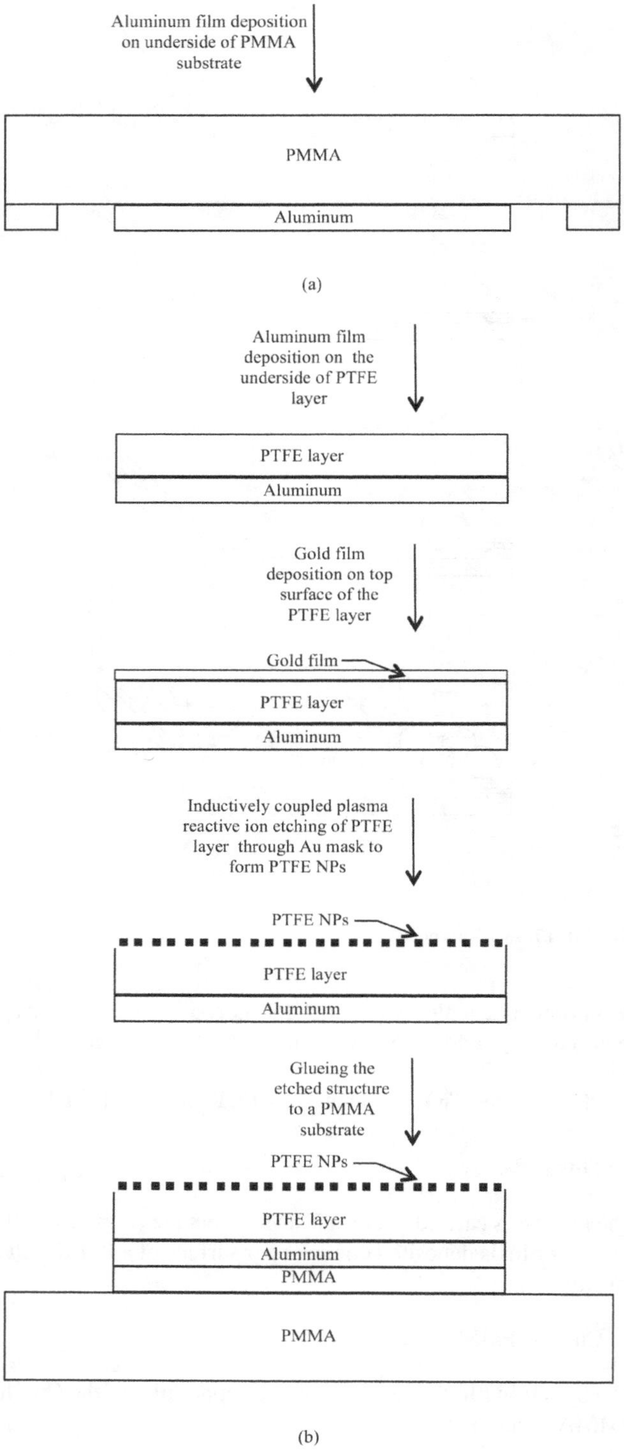

FIGURE 13.6 Fabrication of the TENG dopamine nanosensor: (a) upper part, (b) lower part, (c) assembly of two parts, (d) 3-D drawing, and (e) performance characterization. (Jie et al. 2015.)

Glueing the PTFE NPs/PTFE layer/Al film structure to a PMMA substrate gives

PTFE NPs/PTFE layer/Al film/PMMA substrate

which is the lower part.

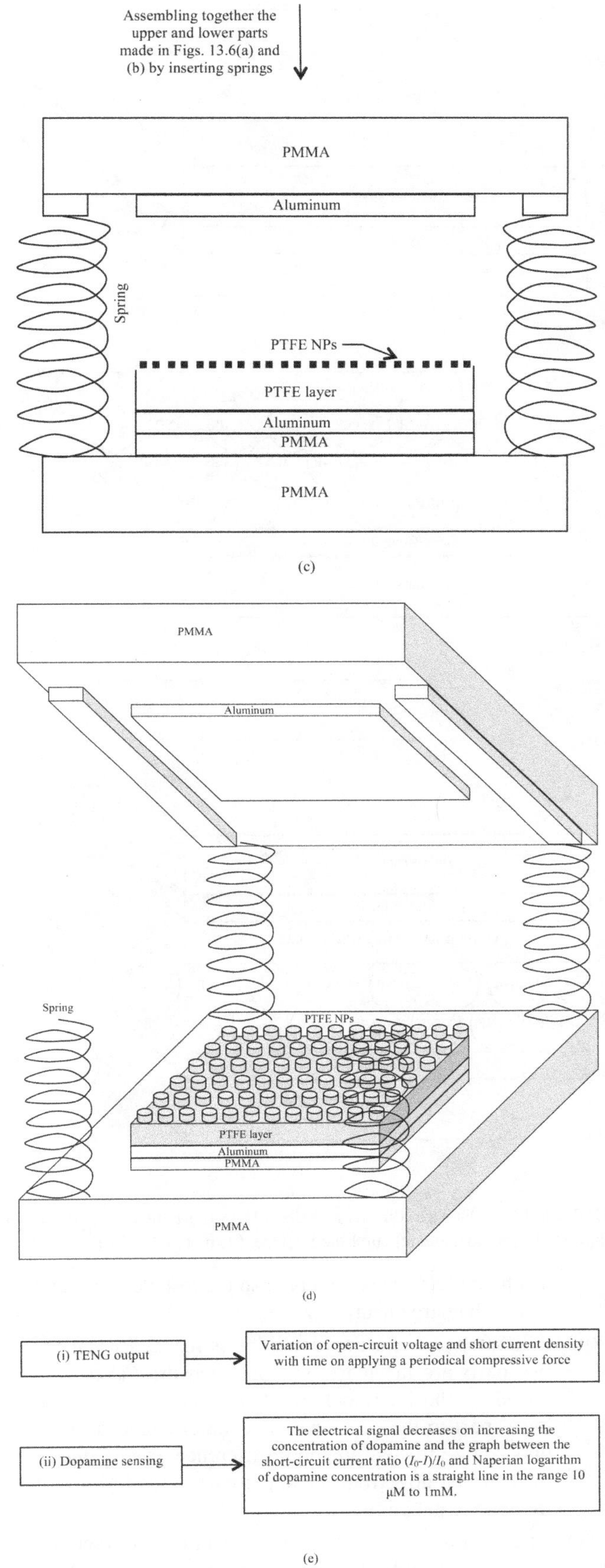

FIGURE 13.6 (Continued)

The upper and the lower parts are assembled together by inserting the springs between them to keep them apart (Figure 13.6(c)).

TENG Working Mechanism: to begin, there are no triboelectric charges on the aluminum film or the PTFE layer. When the Al film is brought into contact with the PTFE layer by applying an external force, positive charges are produced in the Al film and negative charges are generated in the PTFE layer with the transfer of electrons from the Al film to the PTFE layer. On removing the external force, the Al film and the PTFE layer move apart due to the recoiling of the spring. The positive and negative triboelectric charges are no longer coincident on the same plane. This creates an inner dipole moment which impels electron movement from the PTFE layer to the Al film, and a current flows in the external circuit between the top Al film and the bottom Al film underneath the PTFE layer. When the starting position is reached, the electrons have neutralized all the positive triboelectric charges on the Al film. Again, by application of an external force, the Al film comes in contact with the PTFE layer. Electron transference takes place from the Al film to the PTFE layer. A new cycle commences. An electrical potential difference is produced between the Al film and the PTFE layer. This potential difference appears between the top Al film and the Al film underneath the PTFE layer. Thus, contacting and separation of the Al film and the PTFE layer leads to an alternating current flow in the external circuit. It is the reason behind the current flow, which has therefore a triboelectric origin.

The TENG can be considered a capacitor. If the lower plate is taken to be at zero potential, the potential of the upper plate in an open-circuit condition is given by Equation 13.7. The potential difference increases with the increase in the distance d between the lower and the upper plates. This distance is maximum in the starting position of the TENG when the two plates are kept farthest apart by the force exerted by the spring. The open-circuit potential, V_{OC}, is at its peak value in this condition. If the two plates are connected by a wire, a short-circuit current, I_{SC}, flows in the external circuit. The optimal conditions for maximizing V_{OC} and I_{SC} are the application of a 60 N periodical compressive force at 1 Hz when $V_{OC} = 116$ V and $I_{SC} = 33$ μA.

Dopamine Sensing: the PTFE film is dopamine-sensitive. The PTFE film surface is soaked with buffer solution of tris (hydroxymethyl) aminomethane (Tris) at an alkaline pH=8.5. This tris buffer solution contains pre-mixed dopamine at various concentration levels. After soaking, the PTFE layer is washed with water and dried. The short-circuit I_{SC} depends on the time allowed for the interaction of the dopamine (DA) with the PTFE via oxidative self-polymerization. This time is optimized at 90 min with 500 μM dopamine. Catechol in dopamine is oxidized to dopamine-quinone. This oxidation is induced by alkaline pH and is accompanied by dopamine-quinone cross-linking.

A linear relationship is obtained between the current ratio $(I_0 - I)/I$ and the dopamine concentration in the range 10 μM to 1 mM. The limit of detection is 0.5 μM (Jie et al. 2015).

13.3.5 Mercury Ion Nanosensor Using Au Film with Au Nanoparticles/PDMS TENG

Several mercury ion nanosensors fed by an external power source were treated in Chapter 8: Section 8.11.3 and Chapter 10: Section

10.14. We now study a device in the self-powered nanosensor format.

The TENG works through interaction between the gold nanoparticles–covered gold film and a PDMS film because gold and PDMS differ widely in their capabilities for electron attraction and retention (Lin et al. 2013a). Mercury sensing takes place through its recognition by 3-mercaptopropionic acid (3-MPA) and binding with gold nanoparticles, thereby providing a large surface area-to-volume ratio. From the surface area viewpoint, the gold nanoparticles are also beneficial for increasing the electrical output of the triboelectric effect. Also, steady gaps can be sustained between the two rubbing surfaces under zero-strain condition.

Fabrication of the TENG sensor is shown in Figure 13.7:

(i) Upper Part: it consists of a glass substrate on the underside of which is a gold film and underneath the gold film is a polydimethylsiloxane film leading to the structure:
Glass substrate/gold film/PDMS film
as shown in Figure 13.7(a). It is made by laminating the gold film between the glass substrate and the PDMS film.

(ii) Lower Part: it consists of a glass substrate on which a gold film is deposited (Figure 13.7(b)). Gold nanoparticles of sizes 13, 32, and 56 nm, synthesized by reducing Au^{3+} ions with sodium citrate, are assembled on the gold film; 1,3 propanedithiol linker molecules are used. To ensure selectivity toward mercury detection by the gold nanoparticles, 3-MPA is self-assembled on the nanoparticle surfaces by Au-S coupling. Recognition of Hg^{2+} ions by 3-MPA urges them to bind with the gold nanoparticles.

Figure 13.7(c) shows the complete device made by bringing the upper and lower parts together.

TENG Operation: Figure 13.8 shows how the TENG moves through different stages:

(i) Initially when the upper and lower parts are far apart (Figure 13.8(a)), there is no transference of charge from one part to the other, and hence there is no potential difference or current flow between the two parts.

(ii) On bringing the upper part in contact with the lower part (Figure 13.8(b)), electrons are lost by the AuNPs–covered gold film of the lower part and gained by the PDMS film of the upper part. The lower part is positively charged and the upper part is negatively charged. However, with the two parts in contact, there is no potential difference between them. If the two parts are connected by a wire, no current flows in the external circuit.

(iii) On withdrawing the upper part from the lower part (Figure 13.8(c)), a potential difference is produced between the two parts depending on the distance between them, with the lower part at a positive potential and the upper part at a negative potential. Under this potential difference, electrons flow from the gold electrode of the upper part having an excess of electrons to the gold electrode of the lower part which has a deficit of electrons, to replenish the lost electrons and establish charge neutrality.

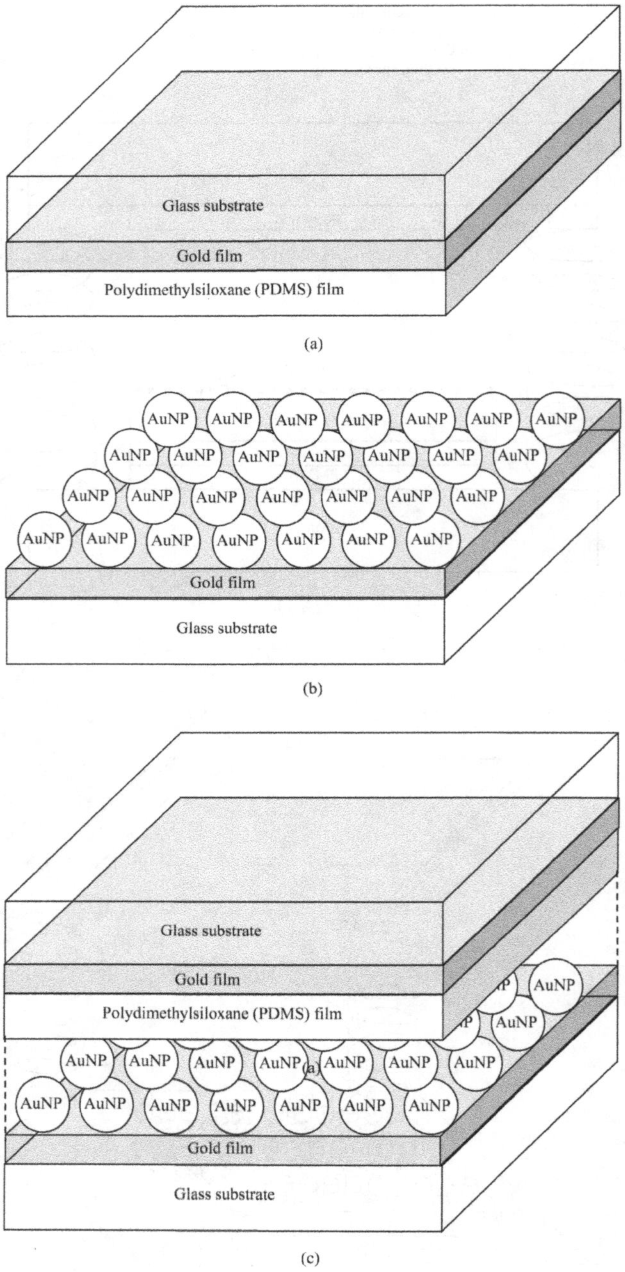

FIGURE 13.7 Fabrication of the TENG Hg^{2+} ion nanosensor: (a) upper part, (b) lower part, and (c) combined device. (Lin et al. 2013a.)

(iv) Ultimately, the starting condition is reached when the two parts are far apart. The potential difference produced by the internal flow of the electrons from the lower part to the upper part is counterbalanced by an oppositely directed potential difference arising due to the electron flow from the upper part to the lower part in the external circuit.

(v) Once again bringing the two parts in contact together (Figure 13.8(d)) leads to the disappearance of the internal potential difference, keeping the lower part positive

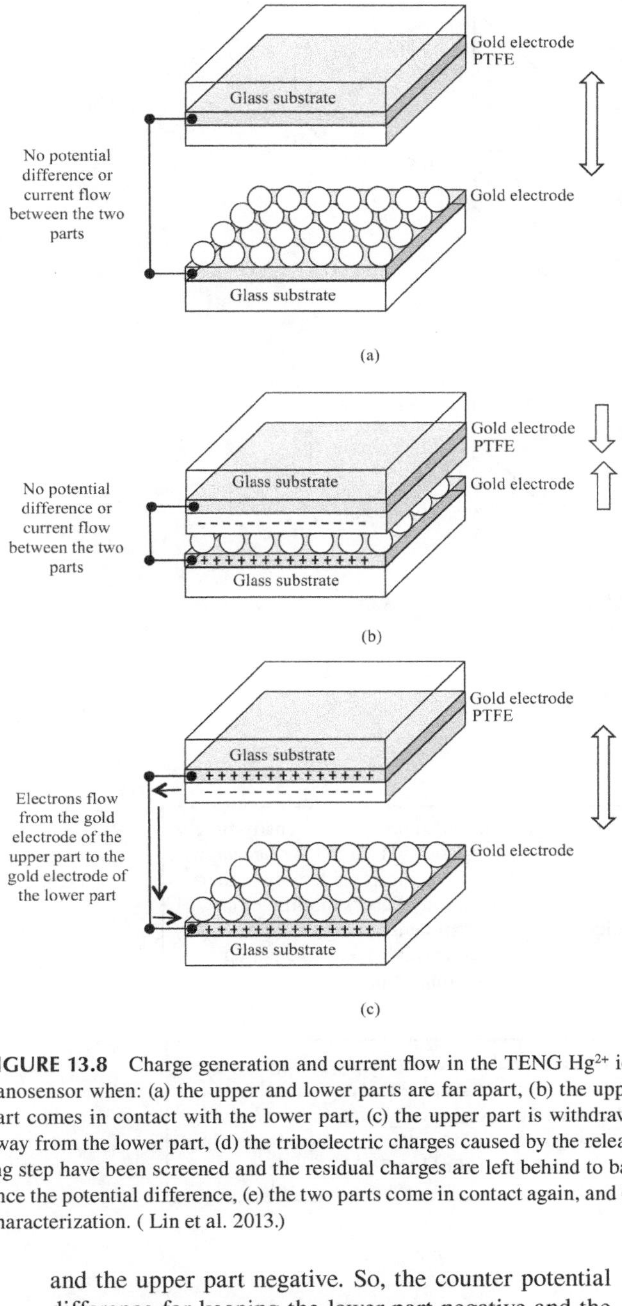

FIGURE 13.8 Charge generation and current flow in the TENG Hg^{2+} ion nanosensor when: (a) the upper and lower parts are far apart, (b) the upper part comes in contact with the lower part, (c) the upper part is withdrawn away from the lower part, (d) the triboelectric charges caused by the releasing step have been screened and the residual charges are left behind to balance the potential difference, (e) the two parts come in contact again, and (f) characterization. (Lin et al. 2013.)

and the upper part negative. So, the counter potential difference for keeping the lower part negative and the upper part positive must also vanish. This happens when the electrons flow back in the external circuit from the gold electrode of the lower part to the gold electrode of the upper part (Figure 13.8(e)).

(vi) The directions of the current flow in steps (iii) and (v) are opposite. Hence, an alternating current flows in the external circuit.

(vii) During contact of the two parts, the electrons are once again transferred from the AuNPs–covered gold film of the lower part to the PDMS film of the upper part, and the above cycle of events is repeated.

TENG Performance: important characterizing parameters for TENG operation are open-circuit voltage V_{OC} (when the two parts are far apart) and short-circuit current density J_{SC} (when the two parts are in contact) (Figure 13.8(f)). For the TENG made with 56 nm AuNPs, the V_{OC}= 105 V and the peak value of J_{SC}= 63 μA/cm^2. The electrical output of a TENG made without AuNPs is less than that of a TENG made with 13 nm AuNPs while the electrical output of a TENG made with 13 nm AuNPs is less than that of a TENG with 32 nm AuNPs. Further, the electrical output of a TENG made with 32 nm AuNPs is less than that of a TENG with 56 nm AuNPs. Also, a TENG made with 56 nm AuNPs has 5.09 times higher V_{OC} and 6.8 times higher J_{SC} than a TENG without AuNPs.

Hg^{2+} Ion Sensing: the AuNPs-covered Au film is soaked with solutions of different concentrations of Hg^{2+} ions in tris-borate buffer. A reaction is allowed to take place for an optimized time of 1 h. After washing with water and drying, the J_{SC} of the TENG is measured. The J_{SC} is found to decrease depending on the concentration of Hg^{2+} ions due to the modification of the triboelectric behavior of the AuNPs–covered gold film after interaction with Hg^{2+}ions. The short-circuit current ratio $(I_0 - I)/I_0$ varies linearly with the Hg^{2+}ion concentration in the range 100 nM to 5 μM. The limit of detection is 30 nM. Due to the high power density ~6.89 mW/cm^2 generated by the TENG, an LED can be used as an indicator with this triboelectric nanosensor in place of an electrometer (Lin et al. 2013a).

13.4 Piezoelectricity-Powered Nanosensors

13.4.1 ZnO Nanowire PENG as a Pressure/Speed Nanosensor

The TENG pressure sensor was described in Section 13.3.1. The PENG is also utilized to make a self-powered pressure sensor. The ensuing discourse is devoted to this nanosensor.

If we look at a tire of a vehicle moving on the road, we find that the portion of the tire pressing against the road flattens a little. With the rotation of the wheel, the flattened portion loses contact with the road and another portion comes in contact with the road. Now this portion in contact with the road flattens a little while the portion, which was previously flattened, regains its shape. Thus, successive portions of the tire undergo flattening and recovery as the vehicle advances. In other words, the portion of the tire of a vehicle experiences a change in shape at the positions at which it comes in contact with the road or loses contact from it. These repetitive shape changes associated with a portion of the tire contacting with the road or losing contact from it are used as triggers for activating the TENG (Hu et al. 2011).

TENG Structure and Fabrication: the TENG has a layered structure consisting of several layers (Figure 13.9). It is fabricated on a polyester substrate. After deposition of a thin Cr-adhesion layer in the central region on both sides, thin ZnO seed layers are deposited on the two sides of the polyester substrate. Then, ZnO nanowires are sequentially grown on the two sides employing a low-temperature hydrothermal growth method using zinc nitrate hexahydrate and hexamethylenetetramine (HMTA). The Cr-adhesion layer–coated substrate floats on the nutrient solution with one surface contacting the solution. The ZnO nanowires are grown on the surface in contact

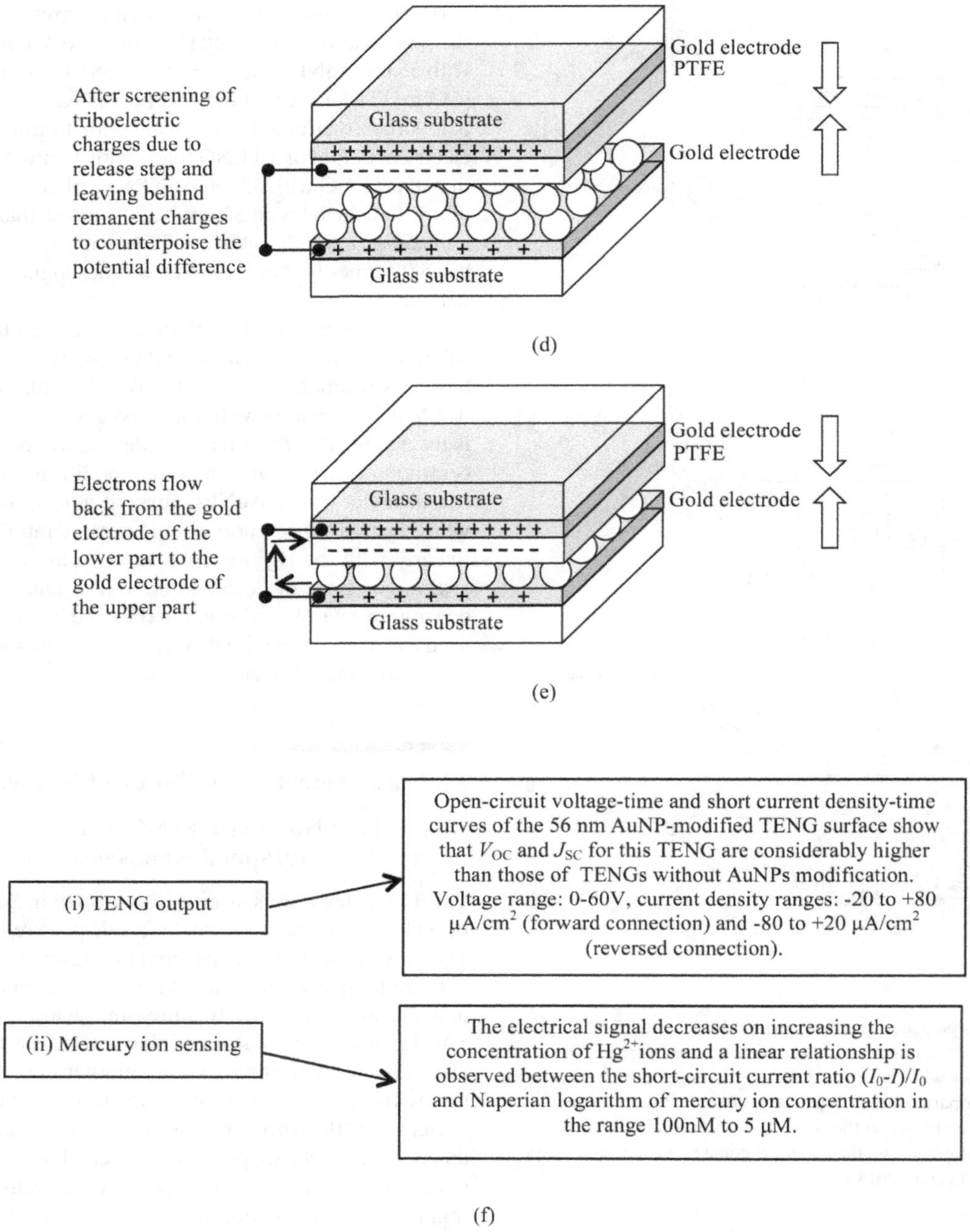

FIGURE 13.8 (Continued)

with the solution in a convection oven at 95°C. Then, the ZnO nanowires are similarly grown on the opposite surface. The ZnO nanowires have a diameter of 150 nm and are 2 μm long. Using the spin-coating method, poly(methyl methacrylate) layers are deposited on both sides. Finally, Cr/Au contact electrodes are deposited on the two surfaces of the structure. Following the attachment of the connection leads, the device is protected by coating with PDMS.

TENG Parameters: the flexible nanogenerator cum nanosensor is attached to the inner surface of a tire using adhesive tape. The output voltage is 1.5 V, the output current is 20 nA, and the maximum power density is 70 μW/cm³ of the ZnO volume, which can be used to light a liquid-crystal display (LCD) screen.

Pressure/Speed Sensing: the deformation of the tire is correlated with the tire pressure. The lower the pressure, the flatter the tire. Therefore, lowering the tire pressure results in more flatness; hence, a larger deformation and therefore a larger voltage output. A larger rotational speed of the tire increases its straining rate, which also translates to a higher output voltage. Thus, a self-powered tire pressure sensor/speed detector is realized (Hu et al. 2011).

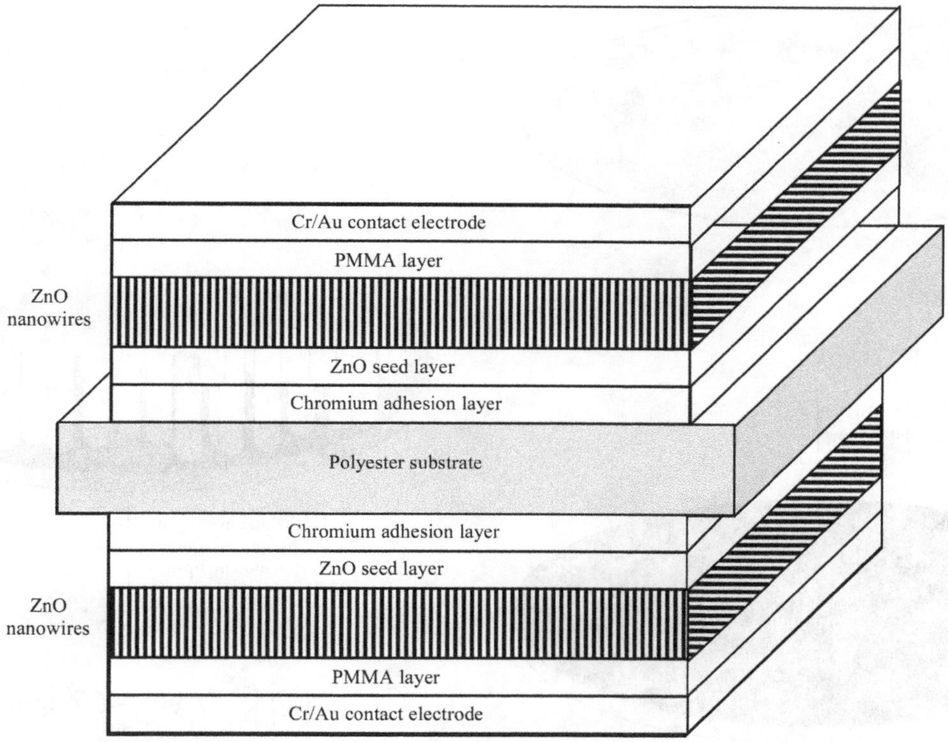

(a)

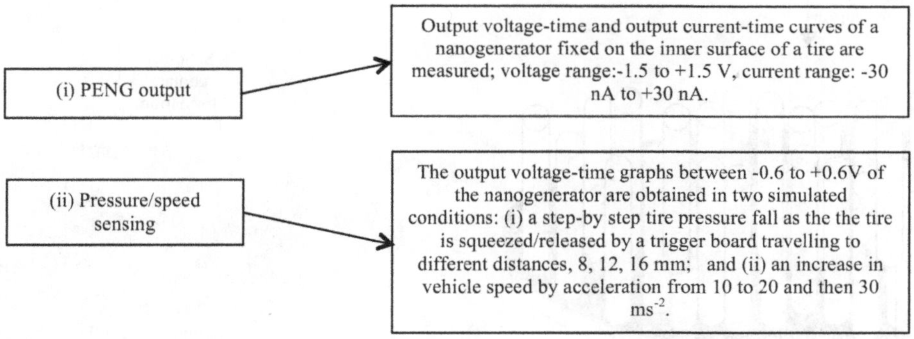

(b)

FIGURE 13.9 The PENG pressure/speed nanosensor: (a) constituent layers and (b) characterization. (Hu et al. 2011.)

13.4.2 UV and pH Nanosensors with ZnO Nanowire PENG

pH measurement was discussed in Chapter 6: Section 6.7, Chapter 8: Section 8.10.2, and Section 13.3.2. In Chapter 6: Section 6.13, we talked about a resistive UV nanosensor using crossed ZnO nanorods. The operation of a batteryless device for UV/pH estimation will be explained in this subsection.

A vertically integrated nanowire generator (VING) made from ZnO nanowires is connected to a ZnO nanowire pH or UV sensor to construct a self-powered pH or UV nanosensor system (Xu et al. 2010).

The fabrication and application of a VING or a PENG sensor for UV/pH sensing is shown in Figure 13.10.

VING Fabrication and Operation: an Si(100) wafer is coated with a Ti/Au layer by magnetron plasma sputtering (Figure 13.10(a)). The Ti film is an adhesion promoter for the deposition of subsequent films. The gold film is an intermediate layer for assisting ZnO nanowire growth. Over the gold film,

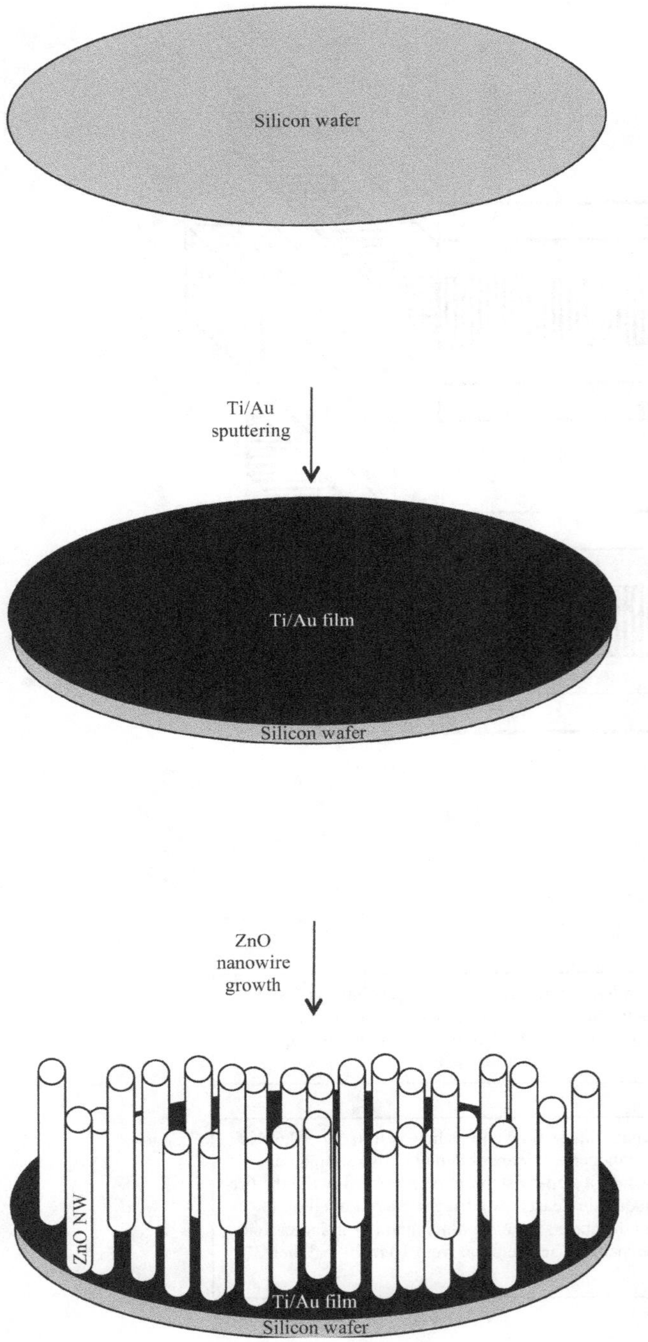

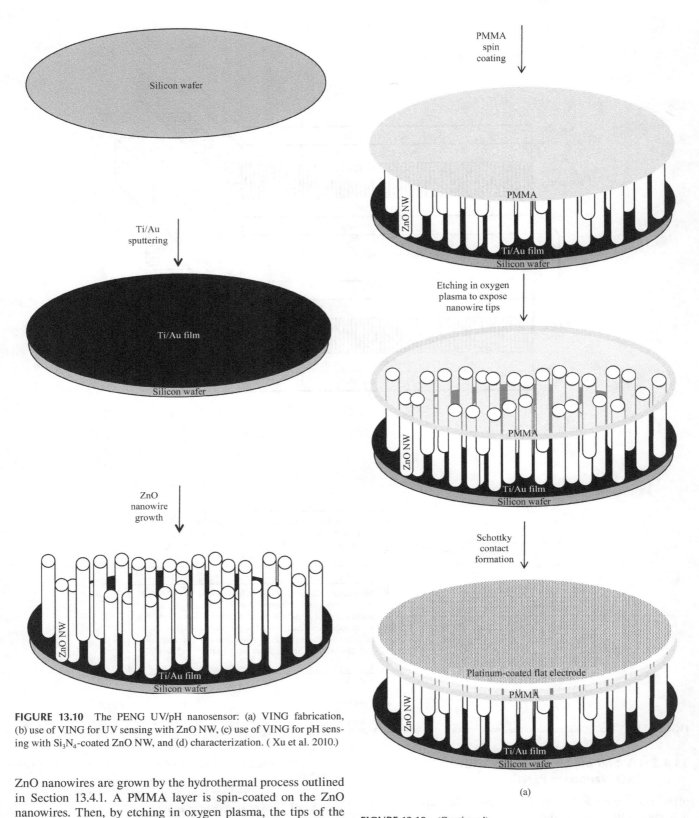

FIGURE 13.10 The PENG UV/pH nanosensor: (a) VING fabrication, (b) use of VING for UV sensing with ZnO NW, (c) use of VING for pH sensing with Si_3N_4-coated ZnO NW, and (d) characterization. (Xu et al. 2010.)

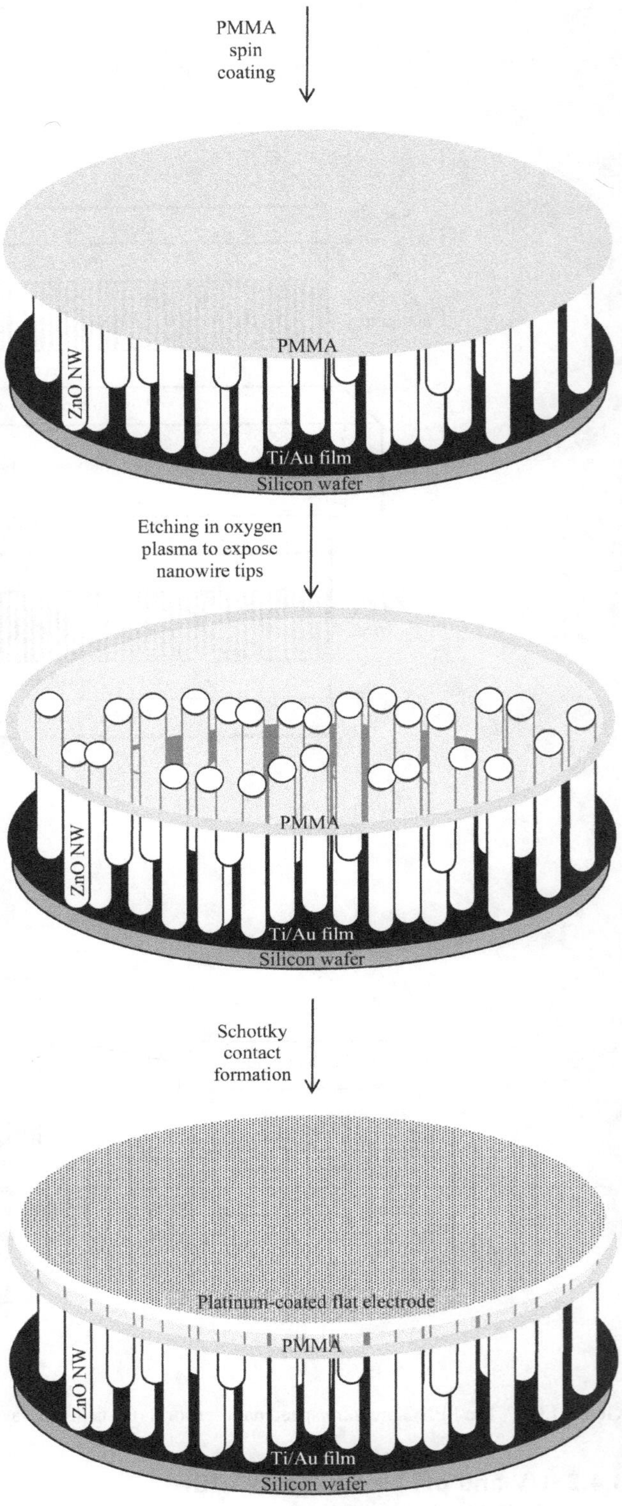

(a)

FIGURE 13.10 (Continued)

ZnO nanowires are grown by the hydrothermal process outlined in Section 13.4.1. A PMMA layer is spin-coated on the ZnO nanowires. Then, by etching in oxygen plasma, the tips of the nanowires are exposed while the bottoms and lower regions of the nanowires remain covered by the PMMA to provide ruggedness to the structure. A Schottky contact to the ZnO nanowires is formed with a platinum-coated flat electrode. On compressing the ZnO nanowires from the top surface by a uniaxially applied force, the crystallographically aligned nanowires are strained. Consequently, a piezoelectric voltage is produced along the *c*-axis-oriented growth direction of the nanowires.

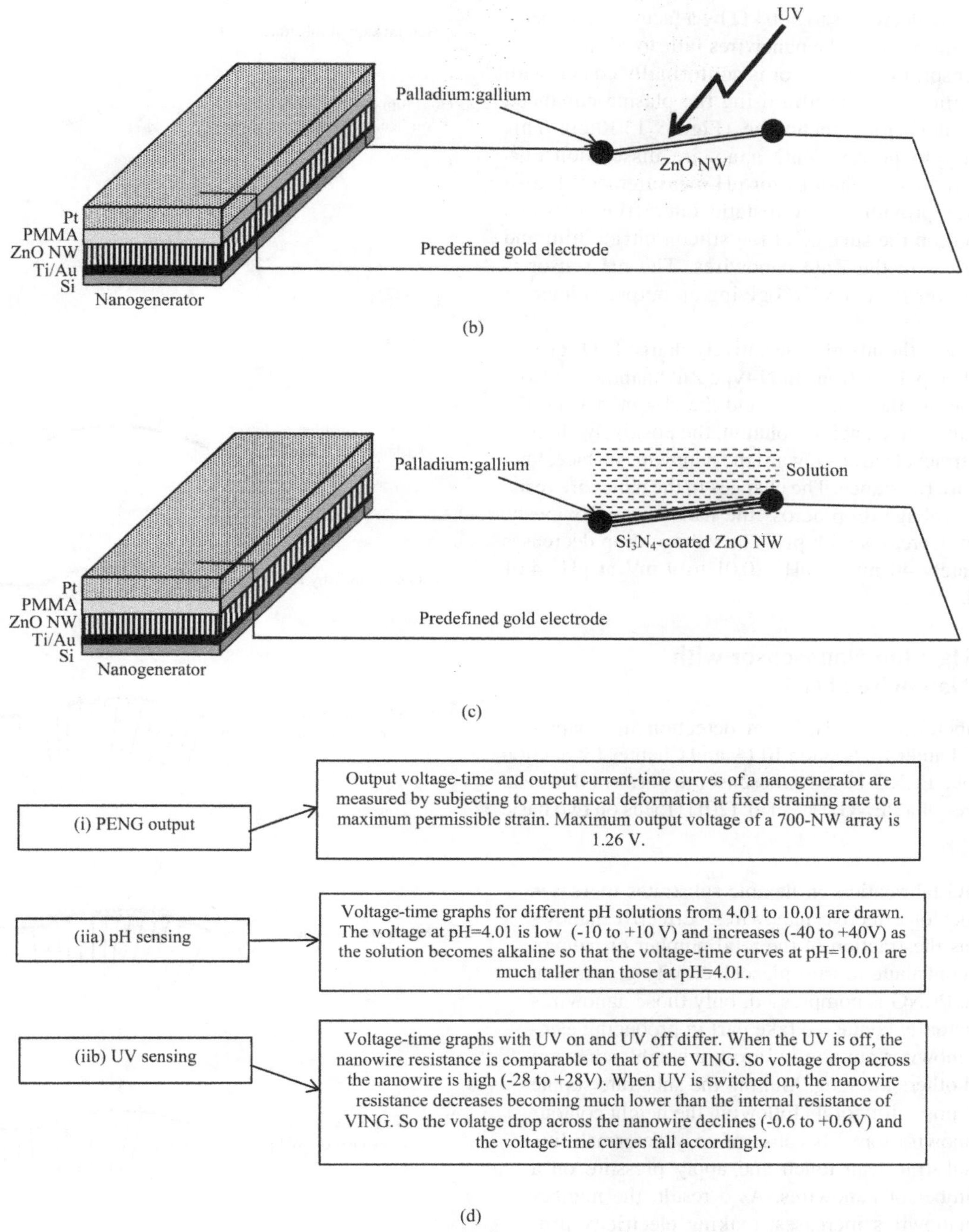

FIGURE 13.10 (Continued)

A series connection of three VINGs with output voltages of 0.08, 0.09, and 0.096 V leads to a total output voltage of 0.243 V. A parallel connection of three VINGS carrying current densities of 6, 3.9, and 8.9 nA/cm^n provides an aggregate output current density of 18 nA/cm^2. The maximum achieved power density is 2.7 mW/cm^3.

UV and pH Sensor Fabrication and Operation: the ZnO nanowires for making these sensors are formed by physical vapor deposition of zinc oxide. The zinc oxide source is taken in powdered form for vaporization. In these sensors, the ZnO nanowires are laid across predefined gold electrodes. Palladium:gallium is deposited by a focused ion beam microscope to fix the ends of the ZnO nanowires and for making ohmic contacts with the nanowires. The structure is used as a UV sensor (Figure 13.10(b)).

Without illumination with UV, the ZnO nanowires have a high resistance of ~10 MΩ. On exposure to UV radiation, the nanowire resistance decreases to 500 kΩ by a factor of 20; therefore, the voltage drop across the nanowires falls to 25 mV.

For the pH sensor, the UV sensor is conformally coated with a 10 nm-thick silicon nitride film using the plasma-enhanced chemical vapor deposition technique (Figure 13.10(c)). This film is necessary to prevent ZnO nanowire dissolution during immersion in buffer solutions for pH measurement. It also serves as a layer providing electrostatic interaction between charges adsorbed on the surface of the silicon nitride film and the charge carriers in the ZnO nanowires. The pH sensor is supplied with power from a VING giving an output voltage of 40 mV.

In a basic solution, the adsorbed negatively charged $-O^-$ groups on silicon nitride repel electrons in N-type ZnO nanowires from the surface depleting the nanowires and thereby increasing the nanowire resistance. In an acidic solution, the positively charged $-OH_2^+$ groups attract electrons toward the nanowire surface, lowering the nanowire resistance. The changes in the nanowire resistance affect the voltage drop across the nanowires. Obviously, the voltage drop decreases with pH. The voltage drop decreases from approximately 40 mV at pH = 10.01 to 9 mV at pH = 4.01 (Xu et al. 2010).

13.4.3 CNT Hg^{2+} Ion Nanosensor with ZnO Nanowire PENG

Further to deliberations on Hg^{2+} ion detection in Chapter 8: Section 8.11.3; Chapter 10: Section 10.14; and Chapter 13: Section 13.3.5 we look at a PENG-based nanosensor to perform this task.

Salient features of a PENG: Lee et al. (2011) focus on two basic issues:

(i) The PENG fabrication on flexible substrates increases the number of effective nanowires. Effective nanowires means the fraction of the total number of nanowires that contribute toward piezoelectricity generation. When the PENG is compressed, only those nanowires that are actually contacted take part in producing electricity. Nanowires are non-uniform in height; some are short and others are long. Making the substrate supple raises the possibility of its following the height contour of the nanowire forest because of easy deformability. So, the substrate can touch and apply pressure on a larger number of nanowires. As a result, the number of such nanowires increases, making electricity production more efficient. On these arguments, 50.8 μm-thick kapton (polyimide) substrates are used for PENG fabrication.

The PENG fabrication on a kapton substrate along with the SWCNT sensor and the two-loop circuit used is shown in Figure 13.11. On the bottom kapton substrate is a 50 nm-thick gold film deposited by radio-frequency (RF) sputtering through a shadow mask (Figure 13.11(a)). On the top kapton substrate, an indium tin oxide (ITO) film is deposited by RF sputtering for electrical contact. Also, a ZnO seed layer is sputtered for ZnO nanowire growth (Figure

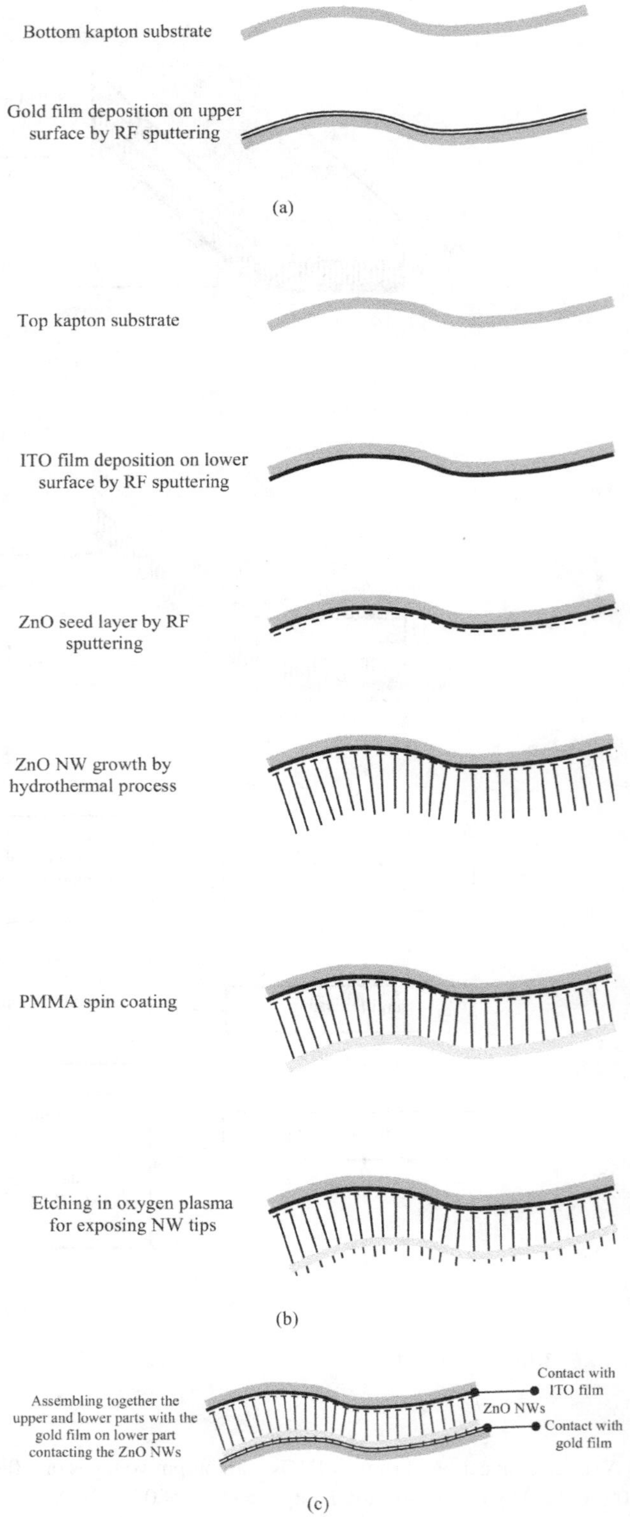

FIGURE 13.11 Layers in the construction of the PENG on a flexible substrate for mercury ion sensor: (a) upper part of the PENG, (b) lower part of the PENG, and (c) full assembly of the PENG, (d) SWCNT sensor used with this PENG, (e) two-loop circuit for charging and sensing using this PENG, and (f) characterization. (Lee et al. 2011.)

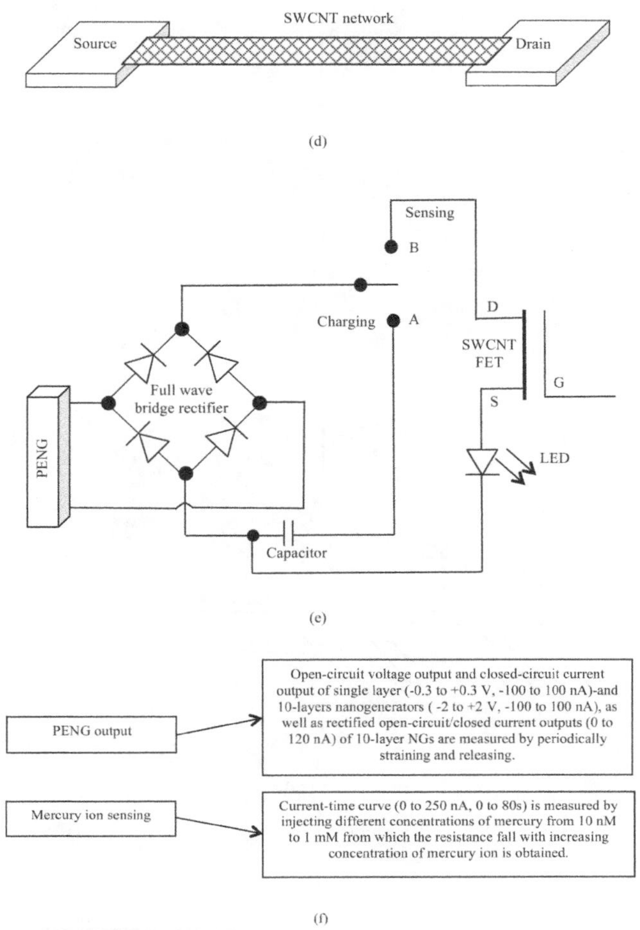

FIGURE 13.11 (Continued)

13.11(b)). The ZnO nanowire array is hydrothermally grown, as previously described in Section 13.4.1. A PMMA layer is spin-coated on the ZnO nanowires to make them sturdy and robust. The tips of the nanowires are uncovered by oxygen plasma etching. The two substrates are contacted and assembled together with the gold film touching the ZnO nanowires (Figure 13.11(c)). The gold film serves as one electrode and the ITO film as the other electrode.

(ii) Nanowires with a wurtzite crystal structure grow along or parallel to the *c*-axis, enabling the integration of several PENGs serially or in parallel along the *c*-axis. A 10-layer PENG is therefore used. The output voltage of a single PENG is 0.35 V and the output current density is 125 nA/cm^2. The open-circuit voltage of the 10-layer PENG is 2.1 V and the closed-circuit current is 105 nA. The peak power density is 0.3 mW/cm^3.

A bridge rectifier circuit with four diodes is used. For charge storage, a supercapacitor is fabricated. In this supercapacitor, the electrodes are made of ITO-coated glass substrates on which multi-walled carbon nanotubes (MWCNTs) are uniformly dispersed by spray coating. A polyvinyl alcohol (PVA)/phosphoric acid (H_3PO_4) polymer solution acts as the electrolyte. A charge of 3.6×10^{-4} C is stored after charging for 1 h at a vibration frequency of 1 Hz with a displacement of 1.2 cm.

Single-Walled Carbon Nanotube (SWCNT) Field-Effect Transistor (FET) Sensor: a low density SWCNT network–based FET working in enhancement mode is fabricated (Figure 13.11(d)). The sensor is characterized by measuring the drain-source current for different concentrations of Hg^{2+} ions in water droplets placed on the gate of the sensor. The sensor is calibrated in terms of resistance variation with a sequential Hg^{2+} ion injection. The resistance change measured from 1 nM to 1 mM Hg^{2+} ion concentration becomes significant at 10 nM and decreases further up to 1 mM.

Two-Loop Circuit Design: the circuit consists of two loops (Figure 13.11(e)). One loop A contains the PENG along with the rectifying circuit and the storage capacitor. After sufficiently charging the capacitor, the power is fed to the other loop B for sensor operation. In loop B, an LED is included which emits light in proportion to the Hg^{2+} ion concentration. The light from the LED becomes noticeable at 10 nM concentration and becomes brighter as the concentration rises to 1 mM (Lee et al. 2011).

13.4.4 Smelling Electronic Skin (e-Skin) with ZnO Nanowire PENG

The smelling e-skin is intended for monitoring the mining environment in real time (He et al. 2018). The e-skin is fabricated on a 150 μm-thick kapton substrate. The PENG function is realized through pressure-induced piezoelectricity generation by ZnO NWs while the smelling function is implemented by making a matrix of four gas sensors on the e-skin. These sensors are designed for measuring the relative humidity (RH), ethanol vapor, hydrogen sulfide, and methane gases. ZnO nanowires serve as the basis for all the functions of the device. Thus, four types of ZnO nanowire–based gas sensing units are formed on the kapton substrate. These are: bare ZnO NWs for relative humidity, Pd/ZnO NWs for ethanol vapor, CuO/ZnO NWs for H_2S, and TiO_2/ZnO NWs for CH_4. Vertically aligned NW arrays are grown using a hydrothermal method on a titanium foil and smeared from this foil on defined areas of the kapton substrate (Figure 13.12).

Nanowire Synthesis: (i) ZnO NW Array: the Ti foil is immersed in a zinc nitrate solution in water, ammonia is added and the reaction is allowed to take place at 83°C for 12 h. (ii) Pd/ZnO NW Array: the ZnO NWs/Ti foil is loaded with Pd nanoparticles by transferring it to an aqueous solution of $PdCl_2$ containing a small quantity of HCl and NaOH, and keeping it at 180°C for 30 min. (iii) CuO/ZnO NW Array: the ZnO/Ti foil is decorated with CuO nanocones by transferring it to an aqueous solution of copper nitrate trihydrate, maintaining it at 60°C for 4 h, then cooling, washing with deionized (DI) water, drying, and annealing at 300°C for 2 h. (iv) TiO_2/ZnO NW Array: the ZnO/Ti foil is loaded with TiO_2 by immersion in a mixed solution of tetrabutyl titanate in ethanol and ethanol/DI water for 5 min, drying, and annealing at 500°C for 1 h.

Sensor Fabrication: four kinds of gas sensing units are smeared on the kapton substrate. This smearing is done on specifically

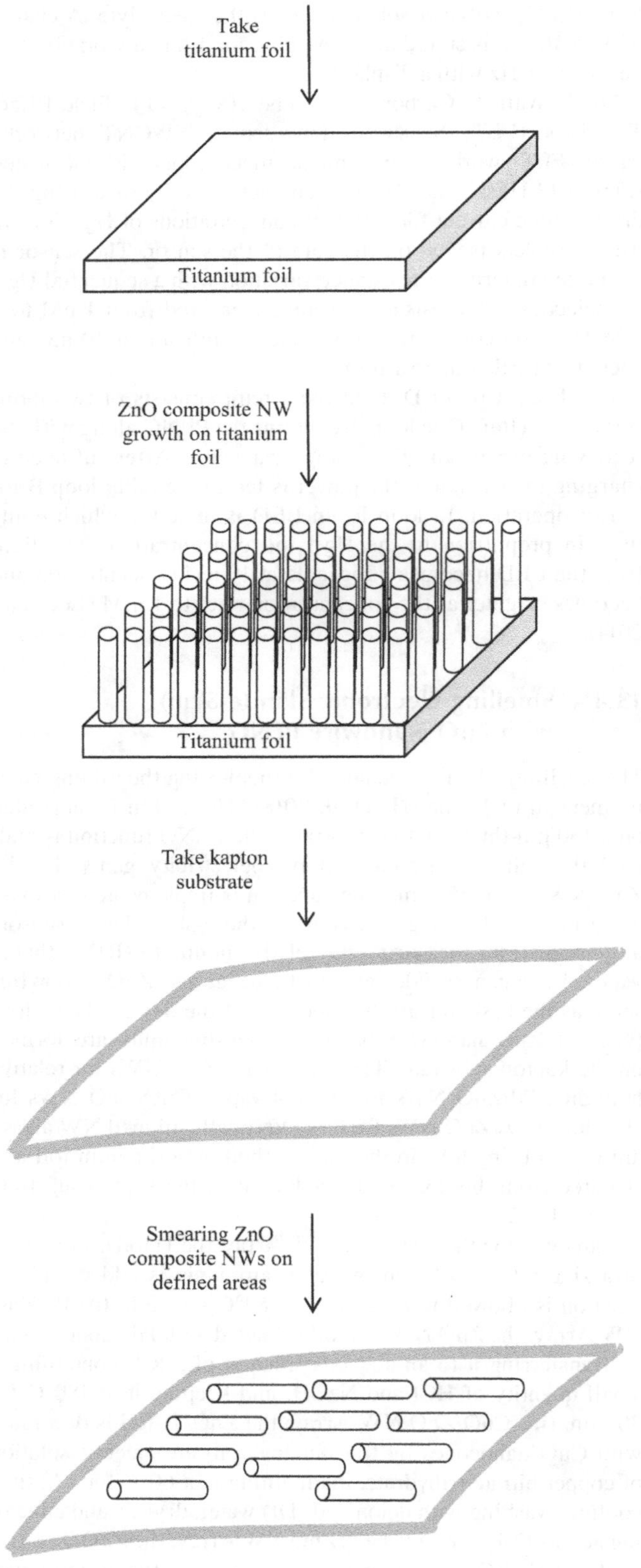

FIGURE 13.12 Smelling e-skin: (a) fabrication and (b) characterization. (He et al. 2018.)

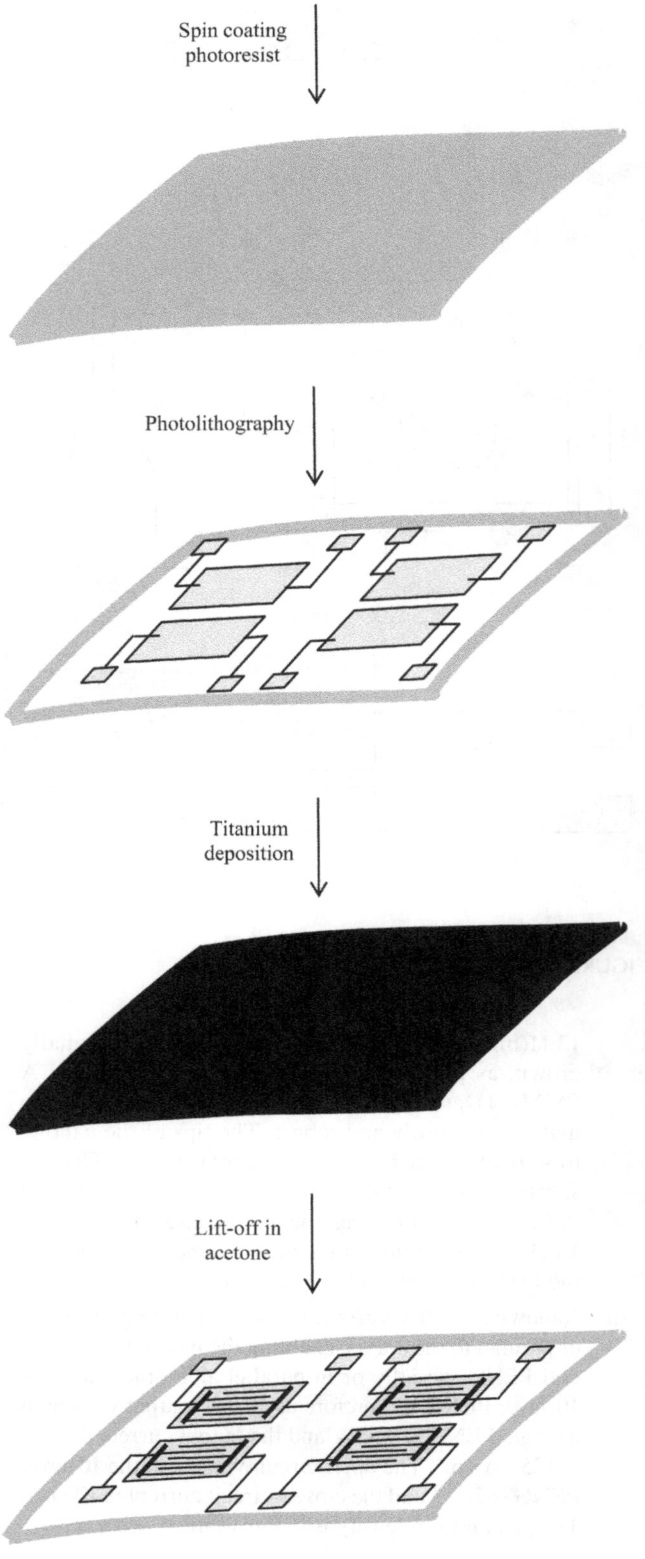

(a)

FIGURE 13.12 (Continued)

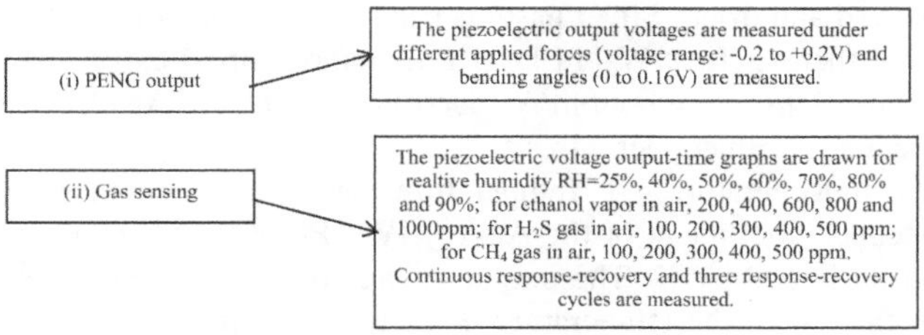

(b)

FIGURE 13.12 (Continued)

defined regions, which are different for each type of nanowire composite. ZnO NW composites are dried, a cross-finger electrode pattern is defined in photoresist over the NW areas, and a titanium film is deposited by electron beam evaporation followed by removal of unwanted Ti and photoresist in acetone.

e-Skin Performance: the e-skin is fixed on a part of a miner's body to monitor the environment in a mine. It is driven by body motions. Under the action of a compressive force, the output voltage of the relative humidity sensor decreases with increasing moisture level, e.g., it decreases from 0.098 V at 25% RH to 0.012 V at 90% RH. The response time is 52 s and the recovery time is 413 s. Similarly, on applying a compressive force, the output voltage of the ethanol vapor sensor decreases as the ethanol vapor concentration increases, e.g., it decreases from 0.094 V to 0.045 V as the ethanol vapor concentration rises from 200 ppm to 1000 ppm. The response and recovery times are 47 s and 92 s, respectively. The identical behavior of a decrease in the output voltage with an increase in the H_2S concentration is observed for the H_2S sensor on application of a compressive force. The output voltage decreases from 0.065 V to 0.017 V as the H_2S concentration upturns from 100 ppm to 500 ppm. The response time is 67 s and the recovery time is 106 s. However, the output voltage of the methane sensor is found to increase with increasing methane concentration when the sensor is subjected to a compressive force. The output voltage increases from 0.108 V to 0.165 V as the CH_4 gas concentration upsurges from 100 ppm to 500 ppm. The response time is 172 s and the recovery time is 68 s. The gas sensing readings are slightly influenced by the deformation of the e-skin under different applied forces and bending angles (He et al. 2018).

13.5 Miscellaneous Powered Nanosensors

13.5.1 Photovoltaic Effect-Powered H_2S Nanosensor Using P-SWCNTs/N-Si Heterojunction

This device utilizes the energy received from visible light to drive the sensor (Liu et al. 2017). It consists of a P–N heterojunction between the P-type single-walled carbon nanotubes and the N-type silicon. The difference in the Fermi levels of the P-SWCNTs and the N-Si leads to a built-in electric field (BEF) at the junction. When light falls on the device, photocarriers are produced near the junction. The BEF helps to separate the photocarriers, thereby generating a current flow to drive the sensor. Thus, the sensor is able to work under light illumination. When the light is switched off, it stops working.

Starting with an SiO_2-coated N-Si wafer, photolithography is performed to create the pattern defining the areas where silicon is to be exposed by removing the oxide using reactive ion etching (Figure 13.13). After oxide etching, photolithography and sputtering processes are applied to form Ti/Au electrodes. A homogeneous dispersion of SWCNTs in dimethylformamide (DMF) is prepared after purifying them by annealing at 300°C and treating with HCl and H_2O_2. This dispersion is drop-casted over the Ti/Au electrodes and intervening gaps. The residual solvent is removed and ohmic contacts are formed between the SWCNTs-Ti/Au and silicon-Ti/Au by annealing at 300°C in argon ambience. Connection pad X is connected to silicon-Ti/Au while connection pad Y is connected to SWCNTs-Ti/Au. Thus, these two pads serve as the connections to the P-SWCNTs/N-Si heterojunction diode.

In the dark, the sensor shows rectifying characteristics because it is a P–N junction. When illuminated with light, the sensor exhibits a photocurrent flow and the photogenerated current density is 0.81 mA/cm^2 at zero bias.

H_2S Sensing: the open-circuit voltage under visible light of wavelength 600 nm and power density 1.8 mW/cm^2 at zero bias decreases after exposure to H_2S gas. If V_0 is the V_{OC} in dry air and V_g is the V_{OC} in the target gas, then sensitivity S is

$$S = \frac{V_g - V_0}{V_0} \tag{13.8}$$

The sensitivity varies linearly with the H_2S concentration in the range 100–800 ppb. The limit of detection is 100 ppb. The sensitivity of a P-SWCNTs/N-Si heterojunction diode sensor is 4.5 times higher than that of the sensor using SWCNTs only, in which the resistance of SWCNTs is measured as a function of the H_2S concentration at 2 V (Liu et al. 2017).

13.5.2 Thermoelectricity-Powered Temperature Nanosensor Using Ag_2Te Nanowires/Poly (3,4-ethylenedioxythiophene):Poly(styrene sulfonate) (PEDOT:PSS) Composite

Several temperature nanosensors were presented in Chapter 5. These nanosensors are fed power externally. A self-powered temperature sensor will now be described. This sensor is developed by Jao et al. (2017). Figure 13.14 presents the structural diagram of the sensor showing the constituent layers.

To fabricate this sensor, Ag_2Te NWs are synthesized by a two-step chemical method:

(i) Synthesis of Tellurium Nanowires: this is done by a solution-phase chemical reduction approach using a tellurium dioxide solution in hydrazine containing sodium dodecyl sulfate (SDS). After a reaction time of 2 h, the Te NWs have a diameter of 18 nm and are 820 nm long.
(ii) Synthesis of Silver Telluride Nanowires: the Te NWs are redispersed in SDS solution and silver nitrate solution is added. The tellurium NWs are converted into Ag_2Te NWs through a redox reaction between silver ions and tellurium atoms. The Ag_2Te NWs have an average diameter of 21 nm.

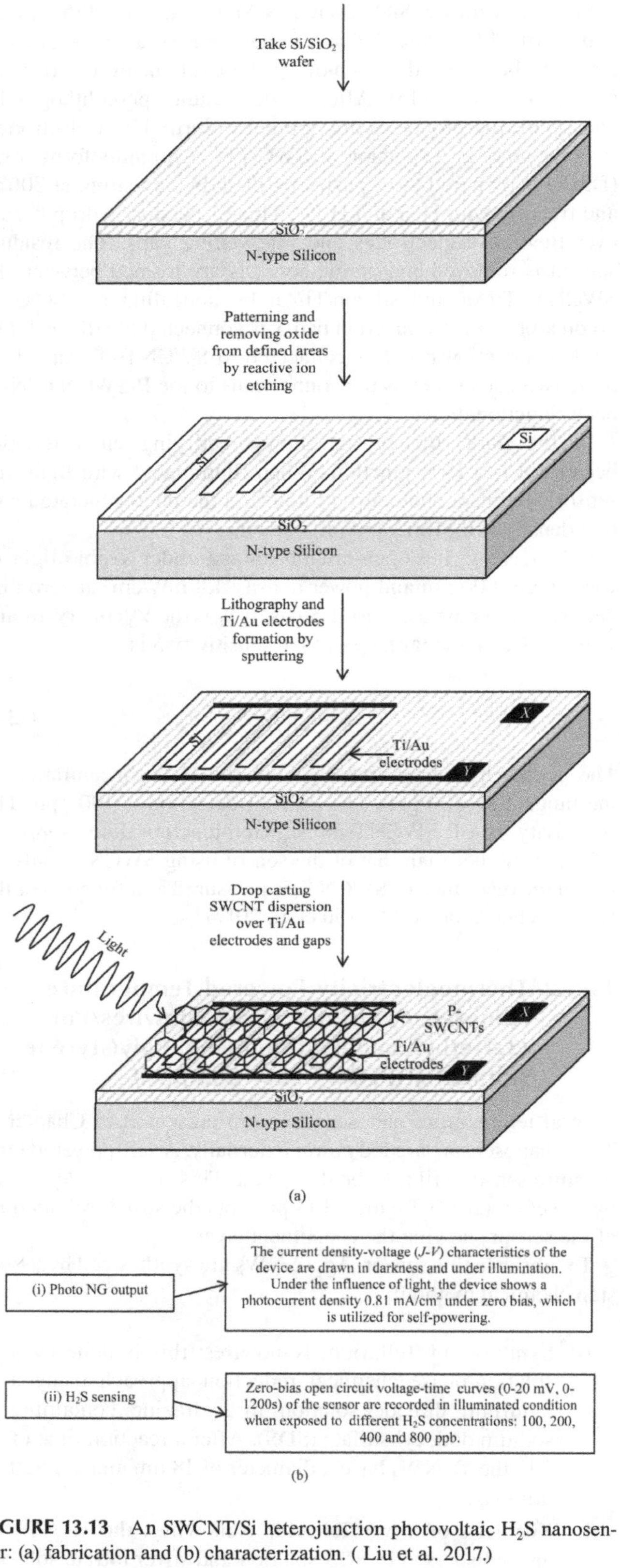

FIGURE 13.13 An SWCNT/Si heterojunction photovoltaic H_2S nanosensor: (a) fabrication and (b) characterization. (Liu et al. 2017.)

Sensor Fabrication: on a polyethylene terephthalate substrate, aluminum is deposited to serve as the bottom electrode. Over this bottom Al electrode, the Ag_2Te NWs nanocomposite with poly(3,4-ethylenedioxythiophene):poly(styrenesulfonate) is poured and dried. Silver paste is used to make the top electrode over the Ag_2Te NWs–PEDOT:PSS nanocomposite. Electrical connections to the bottom Al electrode and the top Ag electrode are made with copper wires.

Nanogenerator and Temperature Sensor Performance: the Seebeck coefficient of the Ag_2Te NWs–PEDOT:PSS nanocomposite is 100 μV/K. The output voltage of the device increases linearly with the temperature. It increases from −0.21 mV at a temperature difference $\Delta T = -12°C$ to 1 mV at $\Delta T = 50°C$. The sensor is fixed to a smart mug and the temperature of the water in the mug is measured from its output voltage (Jao et al. 2017).

13.6 Discussion and Conclusions

Often, it is necessary to arrange a large number of nanosensors in a network to collect information distributed in a region of space. Powering individual nanosensors with separate batteries is riddled with several problems such as an increase in the system size, recharging/repairing/replacing issues with batteries, and the disturbance of the ecological balance by battery-associated toxic substances. So, the idea of self-powered nanosensors emerges, prompting the search for energy sources from which environmental energy can be reaped. Vibrational energy is identified as the most universally available resource to be pooled, ranging from subtle sources such as human breathing and blood circulation to moderate sources such as the vibrations of buildings and more intense sources, e.g., automobile motion on jerky roads and heavy machinery shocks in factories.

A major chunk of work on self-powered nanosensors has concentrated on two phenomena, namely, the generation of electricity by friction between materials (triboelectricity) and by applying pressure on certain materials (piezoelectricity). In both cases, nanotechnology plays a deterministic role. Triboelectricity generation is significantly improved by controlling the roughness of the rubbed materials at the nano level. Similarly, piezoelectric output increases when the pressed material is shaped in the form of nanowires. So, the term 'nanogenerators' is coined for these energy sources. Besides these two sources, nanosensors can be powered using optical and thermal effects.

The equation for displacement current density in Maxwell's electromagnetic theory contains two terms: a component related to the temporal variation of the electric field and another component arising from the temporal variation of the dielectric polarization field. It is the second component which is responsible for tribo- and piezoelectric phenomena.

The TENG works by the collaborative action of two familiar phenomena. In a TENG, contact electrification and electrostatic induction mutually co-operate. During contact of metal-covered dielectrics, electrons flow across the interface of the dielectric materials, thereby creating a potential difference across the interface. During the separation of the materials, electrostatic induction takes over. Using the induction mechanism, opposite charges are induced on the metal electrodes. These induced charges being opposite to the inducing charges cause a reversal of the direction of the current flow in the external circuit with regard to that during contacting.

The PENG operation consists of repeated steps of applying and releasing mechanical stress. The repeated stress application/withdrawal operation leads to oppositely directed current flows in the two steps.

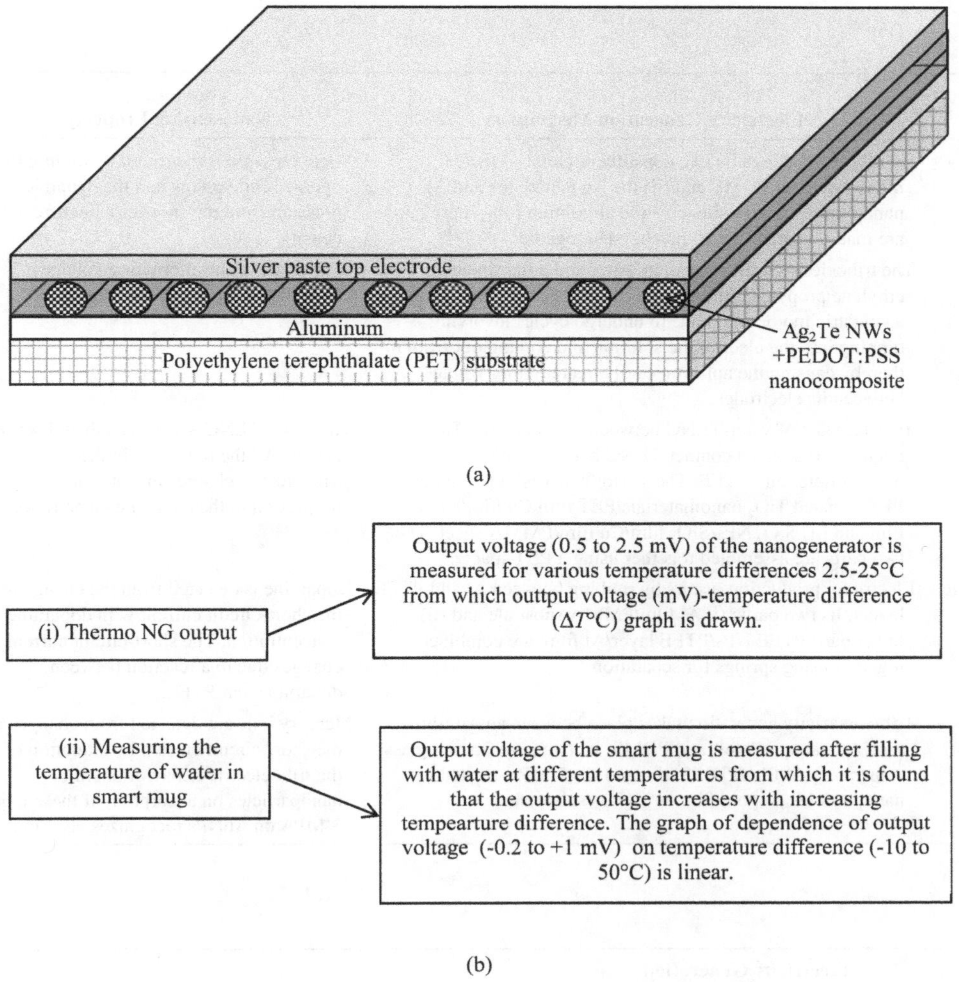

FIGURE 13.14 An Ag_2Te NWs–based thermoelectric nanogenerator: (a) structural schematic diagram and (b) characterization. (Jao et al. 2017.)

Five types of TENG-based nanosensors were described. These were for pressure, pH, ethanol, dopamine, and mercury ions. Their main characteristics are given in Table 13.1.

The operation of four types of PENG-based nanosensors was explained. These were for pressure/speed, UV/pH, Hg^{2+} ion, and a smelling electronic skin. These nanosensors use ZnO nanowires for electricity production. The main features of these nanosensors are given in Table 13.2.

Photoelectricity- and thermoelectricity-powered nanosensors are described in Table 13.3.

Review Exercises

13.1. Give four reasons to justify the elimination of batteries from a nanosensor network.

13.2. Name a source of energy which is ubiquitously available in the environment.

13.3. Cite four examples of vibratory motion in the human body which can be exploited for electricity generation.

13.4. Give examples of vibrations in buildings or automobiles moving on roads which can be used for electricity generation.

13.5. Name two phenomena which are most commonly used for powering nanosensors.

13.6. Why is hair attracted to a comb when it is pulled through the hair? What is this phenomenon of electricity generation called?

13.7. Name two phenomena by whose collaborative action electricity is produced in a TENG.

13.8. What is meant by a triboelectric series of materials? In a triboelectric series, material *X* is placed at a higher position than material *Y*. What do the relative positions of materials *X*, *Y* signify about their behavior when they are rubbed against each other?

13.9. Is piezoelectricity a reversible phenomenon? What is the composition of Rochelle salt? By what name is the lead, zirconium, and titanium-based piezoelectric material called?

TABLE 13.1
TENG Nanosensors

Sl. No.	Nanosensor Name	Electricity Generation Mechanism	Nanosensing Property	References
1.	TENG-based pressure nanosensor	By rubbing between its two constituent parts: (i) the micropatterned PDMS and (ii) the Ag nanowires and Ag nanoparticles composite–covered aluminum foil. These parts are made separately and then bonded together.	Static pressure is measured by finding the open-circuit voltage and the dynamic pressure from the short-circuit current density.	Lin et al. (2013b)
2.	TENG pH nanosensor	The triboelectric effect between water and a fluorinated ethylene propylene film is utilized. The sensor is mounted on an electric motor and made to undergo cycles involving dipping into the electrolytic solution and coming out from it, thereby causing the appearance of an alternating voltage between its electrodes.	Magnitude of the alternating voltage depends on the pH of the solution.	Wu et al. (2016)
3.	TENG ethanol nanosensor	Two TENGs: A water-TENG between water and the TiO_2 nanomaterial, and a contact-TENG between SiO_2 nanoparticles and PTFE. The sensor consists of two parts: (i) PFTS-coated TiO_2 nanomaterials/PET film/Cu film/PTFE film and (ii) SiO_2 NPs/SiO_2 film/Cu film/PMMA sheet. The two parts are assembled together using a PET spacer.	The water-TENG acts as an ethanol sensor because of the reduction in the triboelectric charges in water in proportion to the ethanol concentration.	Lin et al. (2014)
4.	TENG dopamine nanosensor	Electricity production occurs by rubbing between Al and PTFE layers. Its two parts: (i) Al film/PMMA substrate and (ii) lower part: PTFE NPs/PTFE layer/Al film, are combined together using springs for separation.	Dopamine is detected from the change in the short-circuit current with dopamine concentration. The short-circuit current changes due to a reaction between dopamine and PTFE.	Jie et al. (2015)
5.	TENG mercury ion nanosensor	Triboelectricity generation takes place between an Au film with Au nanoparticles, and PDMS. Its two parts are: (i) glass substrate/gold film/PDMS film and (ii) 3-MPA-treated Au nanoparticles on gold film on a glass substrate.	Mercury ions are detected from changes in the short-circuit current by alteration of the triboelectric behavior of Au nanoparticles on adsorption of these ions. 3-MPA on AuNPs recognizes Hg^{2+} ions.	Lin et al. (2013a)

TABLE 13.2
PENG Nanosensors

Sl. No.	Nanosensor Name	Electricity Generation Mechanism	Nanosensing Property	References
1.	PENG pressure/speed nanosensor	Piezoelectricity from ZnO nanowires	The nanosensor is used to monitor the pressure/speed by fixing to the tire of a vehicle. A larger voltage output is obtained when the tire pressure decreases as well as when the vehicle speed increases. So pressure/speed are correlated with voltage output.	Hu et al. (2011)
2.	PENG UV/pH nanosensor	ZnO nanowires	The nanosensor works on the decrease in nanowire resistance on UV exposure. For pH sensing, it is coated with a silicon nitride film which adsorbs negatively charged $-O^-$ groups in an alkaline solution and positively charged $-OH_2^+$ groups in an acidic solution. The resulting repulsion/attraction of electrons toward the nanowire surface changes the nanowire resistance and hence the voltage drop across the nanowires which is used for pH measurement.	Xu et al. (2010)
3.	PENG mercury ion nanosensor	ZnO nanowires	The nanosensor has an SWCNT network–based FET operating in enhancement mode. Mercury sensing is done through the variation of its drain-source current with mercury ion concentration.	Lee et al. (2011)
4.	PENG smelling electronic skin	ZnO nanowires	The skin uses four nanosensors: (i) bare ZnO NWs for moisture, (ii) Pd/ZnO NWs for ethanol vapor, (iii) CuO/ZnO NWs for hydrogen sulfide, and (iv) TiO_2/ZnO NWs for methane.	He et al. (2018)

13.10. Explain the central key role played by nanotechnology in harvesting energy by tribo- and piezoelectric phenomena. Why are electricity generators based on these phenomena named nanogenerators?

13.11. How is the thermoelectric effect used for electricity generation?

13.12. Write down the equation of displacement current density and explain the physical significance of electric and polarization field terms.

13.13. Describe the flow of an electric current in the external circuit from a triboelectric nanogenerator when the two layers are: (a) contacted by applying a force and (b) separated by releasing the force.

13.14. Explain the flow of an electric current in the external circuit from a piezoelectric nanogenerator when: (a) a mechanical stress is applied and (b) the stress is released.

13.15. In the TENG pressure nanosensor, why and how is the PDMS layer micropatterned? Why and how is

TABLE 13.3
Photo- and Thermoelectricity-Based Nanosensors

Sl. No.	Nanosensor Name	Electricity Generation Mechanism	Nanosensing Property	References
1.	Photoelectricity-powered H_2S sensor	Photocarriers produced by shining a light on the nanosensor comprising a P–N heterojunction between the P-type single-walled carbon nanotubes (SWCNTs) and the N-type silicon, provides the energy for driving the nanosensor.	The variation of the open-circuit voltage of a P–N heterojunction between the P-type (SWCNTs) and the N-type Si is the nanosensing property used for H_2S gas detection.	Liu et al. (2017)
2.	Thermoelectricity-powered temperature nanosensor	The voltage generated by the Seebeck effect across an Ag_2Te NWs/PEDOT:PSS composite provides the driving energy.	Decrease of the Seebeck voltage with temperature is the nanosensing property.	Jao et al. (2017)

the aluminum film covered with silver nanowires and nanoparticles? In what ways do its open-circuit voltage and short-circuit current density vary with static or dynamic pressure?

13.16. How does the TENG pH nanosensor produce the necessary driving voltage for its operation? How does this output voltage depend on the pH of the solution?

13.17. Can the flowing water/solid interface be used for triboelectricity generation? How is ethanol concentration measured by the TENG ethanol nanosensor?

13.18. Describe how triboelectricity is generated and how dopamine is detected by a TENG dopamine nanosensor.

13.19. How does the TENG Hg^{2+} ion nanosensor produce electricity? How does it measure mercury ion concentration?

13.20. How is the PENG pressure/speed nanosensor fabricated? What parameters are correlated with tire pressure and speed?

13.21. What phenomena are exploited to detect UV and pH by a PENG UV/pH nanosensor? Explain its working mechanism.

13.22. How is the two-loop circuit design used for electricity generation and nanosensing by a CNT Hg^{2+} ion nanosensor?

13.23. What and how are the gases detected by the smelling electronic skin? How does it produce electricity?

13.24. Describe the construction and operation of a photovoltaic effect–powered H_2S nanosensor.

13.25. By what phenomenon is electricity produced from an Ag_2Te NWs/PEDOT:PSS composite? How does the voltage produced vary with temperature?

REFERENCES

Fan F. R., Z. Q. Tian, and Z. L. Wang. 2012. Flexible triboelectric generator. *Nano Energy* 1: 328–334.

He H., C. Dong, Y. Fu, W. Han, T. Zhao, L. Xing, and X. Xue. 2018. Self-powered smelling electronic-skin based on the piezo-gas-sensor matrix for real-time monitoring the mining environment. *Sensors and Actuators B: Chemical* 267: 392–402.

Hu Y., C. Xu, Y. Zhang, L. Lin, R. L. Snyder, and Z. L. Wang. 2011. A nanogenerator for energy harvesting from a rotating tire and its application as a self-powered pressure/speed sensor. *Advanced Materials* 23(35): 4068–4071.

Jao Y.-T., Y.-C. Li, Y. Xie, and Z.-H. Lin. 2017. A self-powered temperature sensor based on silver telluride nanowires. *ECS Journal of Solid State Science and Technology* 6(3): N3055–N3057.

Jie Y., N. Wang, X. Cao, Y. Xu, T. Li, X. Zhang, and Z. L. Wang. 2015. Self-powered triboelectric nanosensor with poly(tetrafluoroethylene) nanoparticle arrays for dopamine detection. *ACS Nano* 9(8): 8376–8383.

Lee M., J. Bae, J. Lee, C.-S. Lee, S. Hong, and Z. L. Wang. 2011. Self-powered environmental sensor system driven by nanogenerators. *Energy and Environmental Science* 4: 3359–3363.

Lin Z.-H., G. Zhu, Y. S. Zhou, Y. Yang, P. Bai, J. Chen, and Z. L. Wang. 2013a. A self-powered triboelectric nanosensor for mercury ion detection. *Angewandte Chemie* 52: 1–6.

Lin L., Y. Xie, S. Wang, W. Wu, S. Niu, X. Wen, and Z. L. Wang. 2013b. Triboelectric active sensor array for self-powered static and dynamic pressure detection and tactile imaging. *ACS Nano* 7(9): 8266–8274.

Lin Z.-H., G. Cheng, W. Wu, K. C. Pradel, and Z. L. Wang. 2014. Dual-mode triboelectric nanogenerator for harvesting water energy and as a self-powered ethanol nanosensor. *ACS Nano* 8(6): 6440–6448.

Liu L., G. H. Li, Y. Wang, Y. Y. Wang, T. Li, T. Zhang, and S. J. Qin. 2017. A photovoltaic self-powered gas sensor based on a single-walled carbon nanotube/Si heterojunction. *Nanoscale* 9: 18579–18583.

Wang Z. L. 2017. On Maxwell's displacement current for energy and sensors: The origin of nanogenerators. *Materials Today* 20(2): 74–82.

Wang Z. L. and J. Song. 2006. Piezoelectric nanogenerators based on zinc oxide nanowire arrays. *Science* 312: 242–246.

Wu Y., Y. Su, J. Bai, G. Zhu, X. Zhang, Z. Li, Y. Xiang, and J. Shi. 2016. A self-powered triboelectric nanosensor for pH detection. *Journal of Nanomaterials* 2016:5121572 (6 pp.).

Xu S., Y. Qin, C. Xu, Y. Wei, R. Yang, and Z. L. Wang. 2010. Self-powered nanowire devices. *Nature Nanotechnology* 5: 366–373.

14

Wireless Nanosensor Networks and IoNT

Pursuant to our brief discussion on wireless nanosensor networks (WNNs) in Chapter 8: Section 3.8.4, we take up the subject in greater depth in this chapter.

14.1 Evolution of Wireless Nanosensor Concept

A single nanosensor can provide data about its immediate neighborhood within its spatial range of detection. Moreover, it can measure only one or a few interdependent parameters. So, the capability of an individual nanosensor is restricted to a near-distance range and to monitoring one or a few variables. Sometimes, measurements are performed over a larger region of space. The acquisition of knowledge of multiple independent parameters may be indispensable in carrying out complex operations. Then, the capabilities of a solo nanosensor need to be supplemented by resorting to a collection of nanosensors working in a co-operative manner. Obviously, the nanosensors in this collection must be linked through a communication network for the distribution of information and its access by the outside world. Effectively, a network of nanosensors is formed.

Let us illustrate the visualized scenario with two simple examples. The first example we cite is a human intrabody network. Wearable nanosensors are fastened onto the body of a patient to record physiological parameters such as body temperature, heart rate, and blood pressure, including electrocardiogram (ECG) and electroencephalogram (EEG). The collected data are transmitted in real time via the patient's mobile phone to a remote healthcare provider. This kind of network is designed to provide continuous 24×7 health monitoring of the patient. Another example that can be contemplated is an interconnected office network where nanosensors are fixed to the professional and personal belongings of an individual, e.g., pens, folders, and bags. The intent is to constantly update the individual about their whereabouts and safety.

If the above networks are created without using wired connections, they are referred to as wireless nanosensor networks. Thus, a wireless nanosensor network is a group of nanosensors dispersed over a region of space in which each nanosensor is dedicated to the measurement of a specific assigned parameter and the different nanosensors can communicate among themselves. Hence, the local data can be shared in the network, stored at a central location, and conveyed via a gateway to the internet. In this way, the wireless nanosensor network can provide information about a larger region and can accomplish challenging tasks that are outside the competency of a single nanosensor.

The concept of wireless nanosensor networks is in an early research and development stage because many ideas about wireless sensor networks are not directly applicable at the nanoscale (Lee et al. 2015). This chapter will attempt to envision the various possible ways in which such networks are likely to progress and materialize.

14.2 Promising Communication Approaches for Nanonetworking

There are two principal approaches for communication among nanosensors (Table 14.1):

(i) Molecular or chemical: by using molecules as information carriers (Malak and Akan 2012).
(ii) Electromagnetic: by encoding information in a radio frequency (RF) carrier wave at the transmitter and extracting it from the RF carrier at the receiver.

The salient features of molecular and electromagnetic nanonetworking techniques are elaborated in the ensuing sections.

14.3 Molecular Communication (MC)

14.3.1 A Common Natural Phenomenon

Although molecular communication appears unprecedented in the electronics world, it is a routine phenomenon in biological systems, e.g., intra-cellular, inter-cellular, and inter-organ communication occurs in plants via vesicle transport, neurotransmitters, and hormones, respectively (Hiyama et al. 2005). Insects use pheromones, chemical secretions that affect members of the same species and trigger a social response, providing long-distance signaling. Molecular communication is therefore a paradigm inspired by biological systems; hence it is called a bio-inspired paradigm. It is useful for communication over short distances of a few tens of microns up to 1 m.

14.3.2 Steps in Molecular Communication

In molecular communication, messages are digitally encoded at the transmitter, e.g., in the form of the presence or absence of a molecule. They are transmitted through air or water and intercepted without any antennae tuned to the wavelength of the signal. On reaching the receiver, the molecules trigger a chemical reaction which recreates the information. The exchange of information by molecular communication consists of five steps (Figure 14.1). These steps are (Llatser et al. 2012; Wang et al. 2017):

(i) Coding (the production of information molecules having special physical and/or chemical properties).
(ii) Transmission (the release of the generated information molecules into the fluidic medium).
(iii) Propagation (the movement of the information molecules in the medium).

TABLE 14.1

Comparison between Molecular and Electromagnetic Communication

Sl. No.	Feature	Molecular Communication	Electromagnetic Communication
1.	Carrier of information	Molecule	Electromagnetic wave
2.	Type of information carried	Chemical state or phenomena	Textual, audio, video
3.	Type of signal	Chemical	Electrical and/or optical
4.	Device used for communication	Bio-nanomachine	Electronic/photonic device
5.	Reception of information	By chemical reaction	By demodulation of carrier wave
6.	Medium of transmission	Fluid (air or water)	Vacuum, air, or conducting cable
7.	Distance of transmission	nm to m	m to km
8.	Speed of transmission	nms^{-1} or $\mu m\ s^{-1}$	Velocity of light
9.	Accuracy of transmission	Stochastic	Highly accurate
10.	Source of noise	By colliding molecules, particles in the path, or chemical reaction in the transport medium	Interfering signal or electromagnetic field
11.	Energy consumption	Low	High

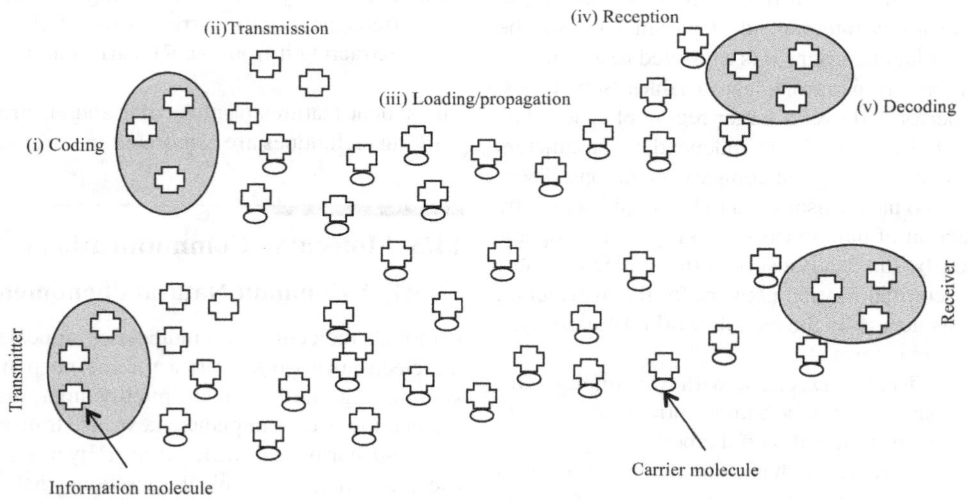

FIGURE 14.1 Sequential steps in molecular communication.

(iv) Reception (the detection of the information molecules by the receiver).

(v) Decoding (the reconstruction of the arriving information).

Steps (i) and (v) of molecular communication are implemented by one of the following methods (Wang et al. 2017):

(a) Coding/decoding based on the type of molecule: for coding, different types of molecules are used to denote different information bits. Information is decoded from the type of reaction between the molecules and the respective receptor.

(b) Coding/decoding based on the structure of the molecule: here, the sequence of DNA nucleotides determines the coding. The method provides large throughput and is invulnerable to noise.

(c) Coding/decoding based on the concentration of molecules: in this scheme, molecules of different concentrations are used to represent bits 0 and 1. It is the analogue of amplitude modulation.

(d) Coding/decoding based on the rate of concentration, $dC(t)/dt$, of molecules: in this coding scheme, different molecular concentration rates are used to signify the bits 0 and 1. It is the analogue of frequency modulation.

(e) Coding/decoding based on the release time of molecules: here, the molecular release time is used as the coding parameter. It is similar to phase modulation.

Steps (ii)–(iv) of molecular communication come under information propagation techniques (Noel 2013), which have two sub-categories (Table 14.2):

Sub-category 1: passive transport mechanisms, which do not require an energy input:

(a) Free diffusion: this process is initiated by the transmitter within a single cell, among different cells, or between two organisms by inundating the environment with randomly moving molecules, some of which are able to reach the receiver where they bind with its surface.

(b) Diffusion across gap junctions: a gap junction is an intercellular conductive channel (Figure 14.2). The signaling

TABLE 14.2
Passive and Active Transport Mechanisms

Sl. No.	Feature	Passive Mechanism	Active Mechanism
1.	Type of motion	Diffusion of molecules	Directional motion of molecules
2.	Number of information molecules necessary	Large	Small
3.	Facilities used	None	Transport, guiding, and interface molecules
4.	Driving energy	No chemical energy input. Driven by environmental parameters.	Chemical energy required
5.	Application environment	Dynamic and unpredictable	Large molecules
6.	Propagation distance	Small	Large
7.	Propagation delay	High	Low

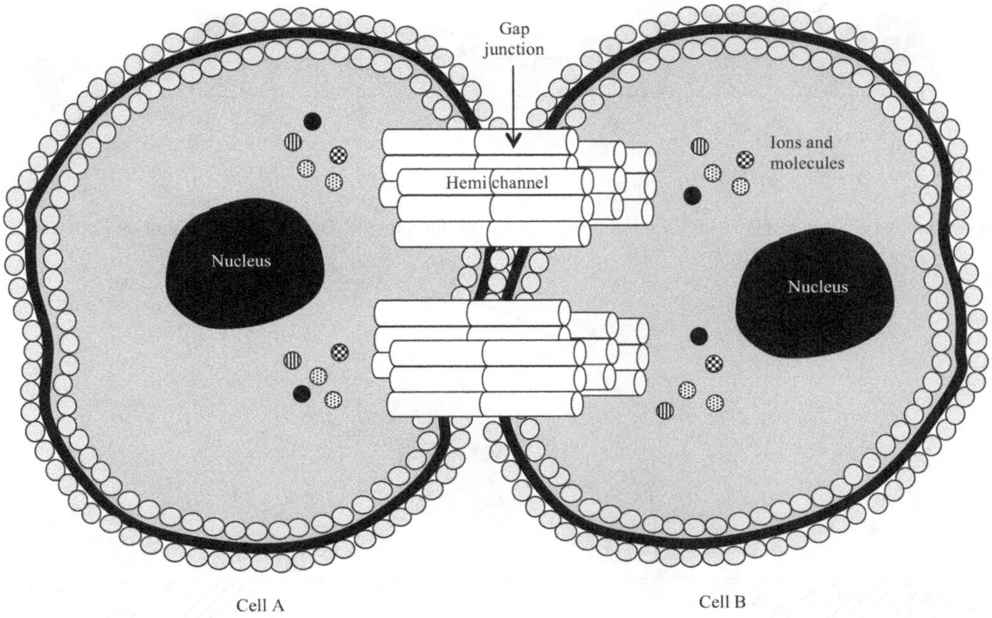

FIGURE 14.2 Communication between two cells A and B through gap junctions.

molecules produced by the transmitter diffuse to the adjoining cell. Here, they are amplified by a regenerative feedback mechanism for traveling longer distances with the intermediary cells acting as relays.

Sub-category 2: active transport mechanisms, which consume cellular energy:

(a) Molecular motors: these motors are miniscule protein machines that harness the chemical energy released by the hydrolysis of adenosine triphosphate (ATP) to perform mechanical work for the intracellular trafficking of large molecules. The information is loaded into a container that travels along a rail and is opened at the receiver (Figure 14.3).

(b) Bacterial flagellar motors (BFM): the flagella are long, thin, helical appendages to a bacterial cell that enable their movement through their habitat. The transmitter delivers the DNA information to the bacteria, which propels toward the receiver giving it the DNA information upon contact. Figure 14.4 shows the flagellar motor.

14.3.3 Advantages of MC

It is practically viable with a reasonable chance of success. It is dimensionally suitable for nanonetworks. Its bio-compatibility makes it safe for living systems. Its biochemical reactions show high energy efficiency and have been perfected by billions of years of the evolution of life on the planet.

14.3.4 Difficulties of MC

It suffers from stochasticity or randomness and fragility. Random molecular motion makes it susceptible to noise. Biological processes are sensitive to pH and temperature effects. The propagation delay is relatively long when compared with the speed of light. The propagation distance is also too short.

14.4 Electromagnetic Communication (EMC)

As the name implies, it is a wireless communication paradigm using electromagnetic waves. It has received impetus from the

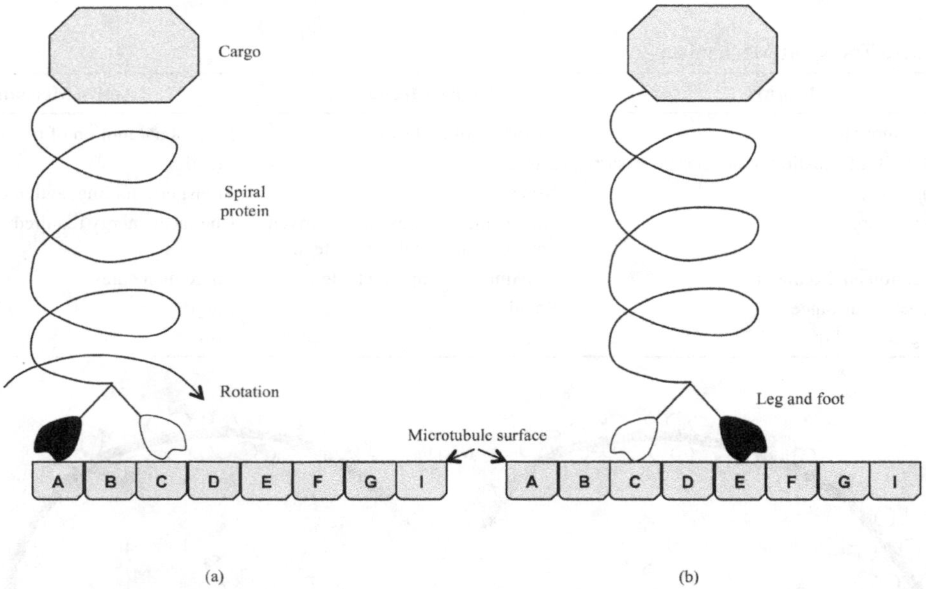

FIGURE 14.3 Rotary locomotory mechanism of kinesin spiral protein motor on microtubule: (a) initial position and (b) after rotation.

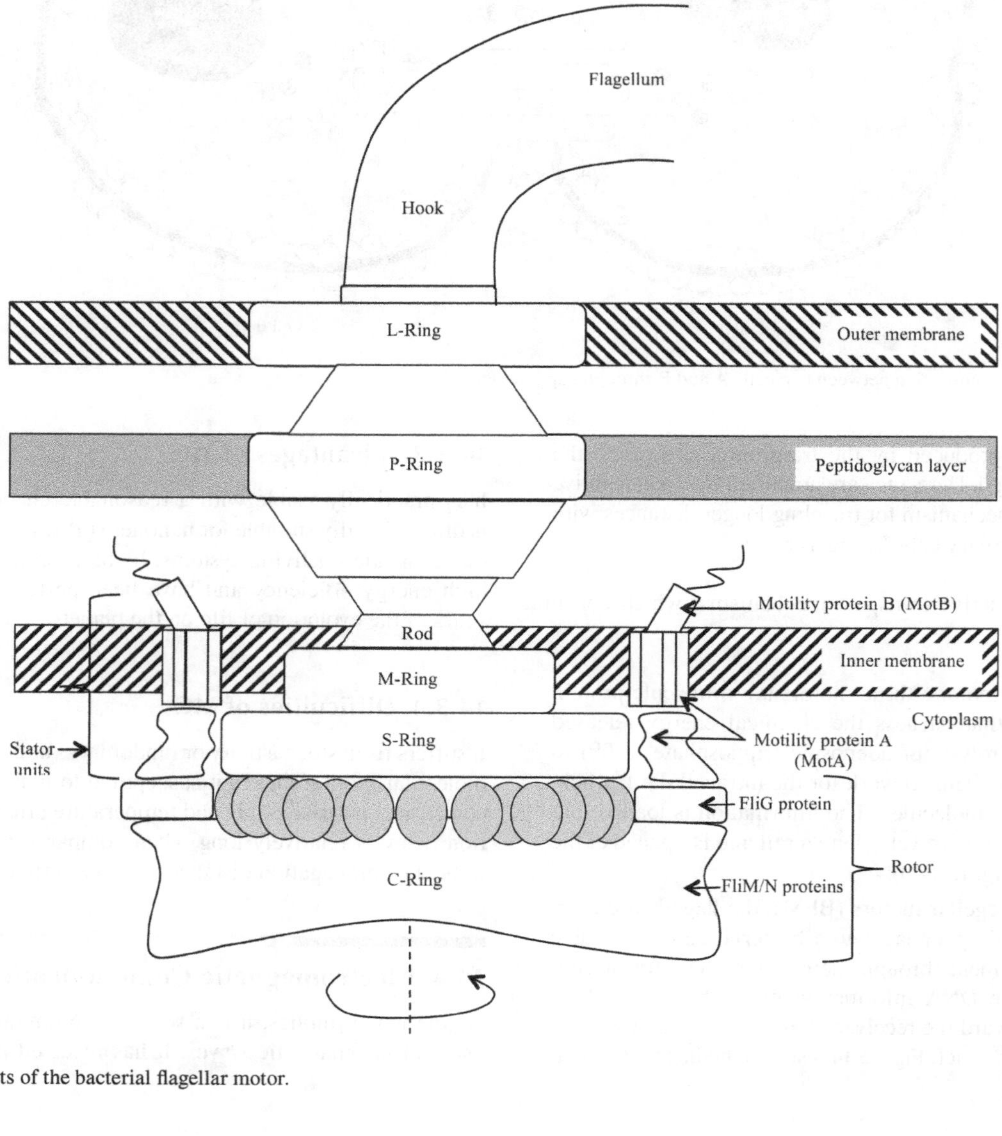

FIGURE 14.4 Parts of the bacterial flagellar motor.

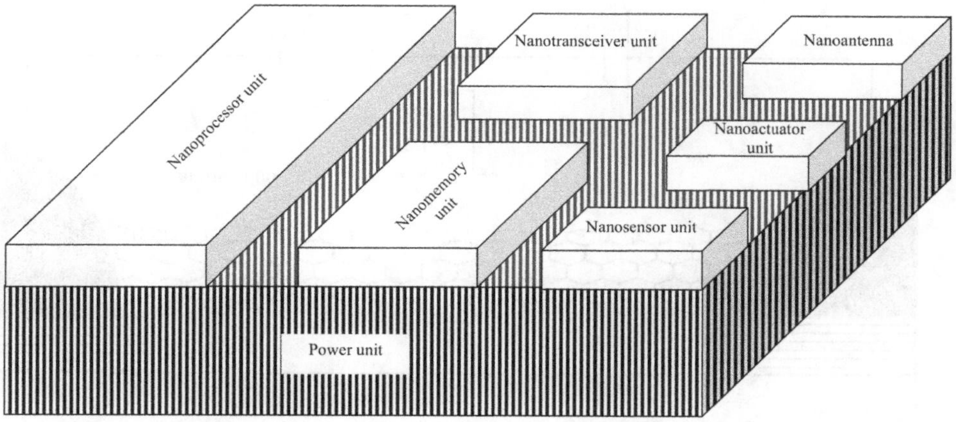

FIGURE 14.5 Visualized integrated nanosensor module.

recent advances in the fields of nanostructured materials and carbon nanoelectronics, comprising devices made from structural elements such as graphene (mono-atomic layer thick hexagonal carbon lattice with honeycomb appearance), graphene nanoribbons (GNRs) (narrow graphene sheets <50 nm wide), and carbon nanotubes (graphene sheets rolled into a cylindrical form). Henceforth, in this chapter we will focus on electromagnetic communication because molecular communication is still in its early infancy and these ideas may take longer to concretize.

14.5 Envisaged Electromagnetic Integrated Nanosensor Module

Although the nanosensor is the focal component of interest, an integrated nanosensor module is imagined by Akyildiz and Jornet (2010a) as a basic structural unit of the wireless nanosensor network. This module contains several components, as shown in Figure 14.5. Such a module becomes necessary because the powering, data processing, and communication requirements for nanosensors cannot be met by methods developed for bulky microscopic sensors. Therefore, novel techniques for managing these requirements have to be devised.

14.5.1 Nanosensor Unit

The WNN idea is methodically and meticulously formulated by Akyildiz and Jornet (2010a) considering three versions of a carbon nanotube field-effect transistor (CNT FET) as a physical nanosensor, as a chemical nanosensor, and as a bio-nanosensor. CNTs can be used as nanosensors in multiple ways. The conductance variations of CNTs with gaseous species or biomolecules can also be exploited for sensing.

The principles of a CNT nanosensor are shown in Figure 14.6. Here, the physical CNT sensor (Figure 14.6a) works by exploiting the change in conductance of the device when the CNT is subjected to a bending or deforming force. The chemical CNT sensor (Figure 14.6b) functions as a chemi-resistor through the variation of resistance in proportion to the concentration of gas adsorbed on the surface of the CNT. The CNT FET bio-nanosensor (Figure 14.6c) registers a change in the threshold voltage of an FET device when an antigen binds to an antibody tied to the CNT surface or a complementary DNA molecule gets attached to a DNA molecule fixed to the CNT surface or a protein molecule adheres to a properly functionalized CNT surface.

Any of the three versions of CNT sensor is a complex integrated device possessing elementary data processing and actuation capabilities. Therefore, it has accessory devices affixed to it, which are described in the following subsections.

14.5.2 Nanoactuation Unit

An actuator is a mechanical device. It converts the control signal into some kind of motion to open a valve or to operate a pump/motor. A physical nanoactuator works on the principle of nanoelectromechanical systems (NEMS), e.g., an applied voltage bends a CNT to produce motion. Nanotweezers and nanoscissors can be built on similar approaches. Bio-nanosensors are made with nanoshells containing drug molecules as nanoactuators. The surfaces of these nanoshells are chemically modified using antibodies or DNA strands by which they get attached to target cells. The nanoshells are melted by external or local irradiation whereby the medicine is delivered to the relevant cells. Nanoactuators consisting of nanoheaters are used to selectively kill cancerous cells.

14.5.3 Power Unit

Lithium nanobatteries providing a high-power density, an adequate lifetime, and controlled discharging/charging rates have been developed as power sources for WNNs. They use nanoporous aluminum oxide filled with polyethylene oxide (PEO) and lithium triflate (LiTf) electrolyte and are sealed with a cathode material. However, their deployment is hampered by the fact that the nanobattery needs periodic recharging to replenish lost charge. Consequently, nano energy harvesters have been built (Canovas-Carrasco et al. 2018). They scavenge mechanical energy from human muscular motion, vibrational energy from moving automobiles or the slight shaking of buildings due to heavy machinery in operation, or hydraulic

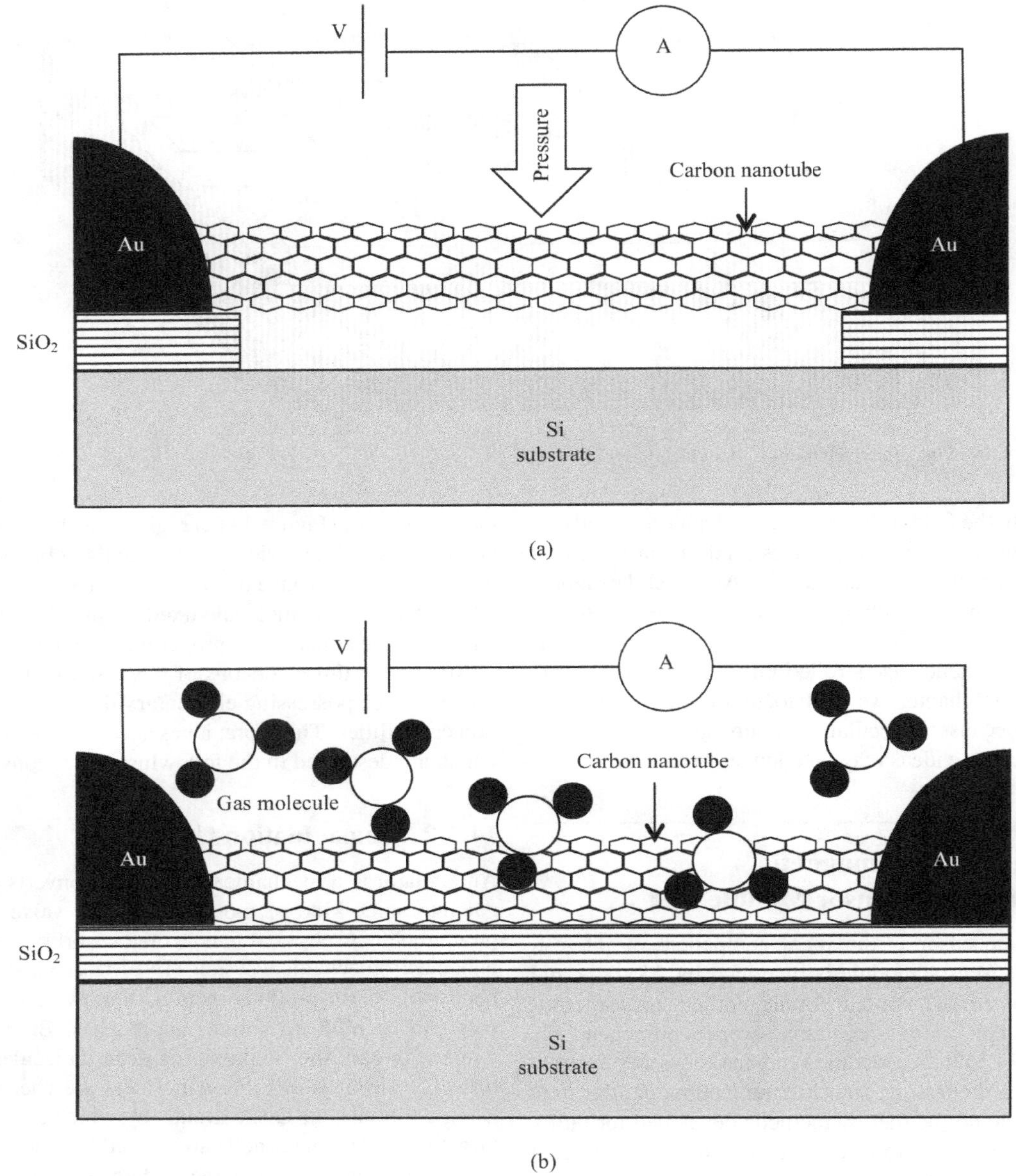

FIGURE 14.6 CNT-based sensors: (a) physical, (b) chemical, and (c) biological.

energy from moving fluids such as blood circulation. The salvaged energy is transformed into electricity by the piezoelectric effect in ZnO nanowires. The nanowires are electrically polarized when mechanically stressed. The electricity generated is either immediately used or saved for future use by charging the nanobattery.

A proposal is to convert the energy of electromagnetic waves in the environment into electricity. This can be done by using a NEMS resonator in which vibratory motion is activated and reinforced by the electromagnetic waves in air. The induced vibrations, in turn, are used to generate piezoelectricity, e.g., by ZnO nanowires. An alternative suggestion is based on the fact that a long antenna located near a powerful radio transmitter can recoup sufficient electromagnetic energy to provide electricity for lighting an electric bulb. On this principle, a rectifying antenna, known as the rectenna, has been designed by combining an antenna with a rectifying diode to obtain direct current from electromagnetic waves. Even though there is no strong transmitter nearby, the energy thus garnered is enough to power the nanosensor module.

Besides nanobatteries, supercapacitors are competitive power sources. Supercapacitors are capacitors exhibiting very high capacitance values due to electrolytic double-layer capacitance and pseudocapacitance from the Faradaic electron charge transfer of reversible redox reactions, electro-adsorption, or intercalation processes on electrode surfaces. The tussle for

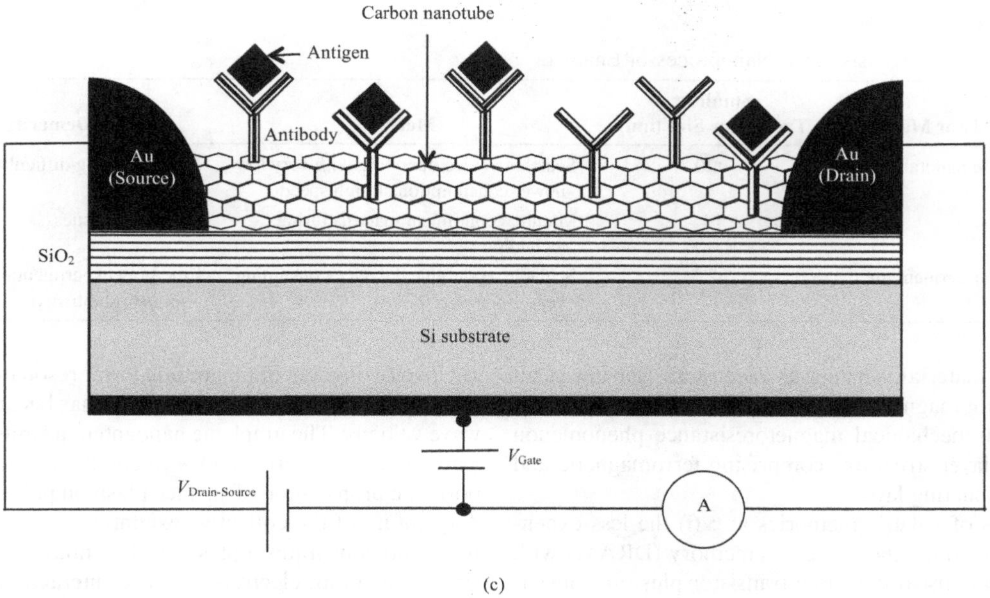

FIGURE 14.6 (Continued)

TABLE 14.3

Nanobatteries and Supercapacitors as Power Sources

Sl. No.	Power Source	Advantages	Disadvantages
1.	Nanobattery	Provides high energy density	Uses toxic chemicals Prone to degradation Mechanically inferior Viability not clear
2.	Supercapacitor	Absence of toxic chemicals Less prone to degradation Mechanically rugged Feasible	Provides lower energy density

supremacy between nanobatteries and supercapacitors is clear from Table 14.3.

14.5.4 Nanoprocessor Unit

The nanoprocessor is the nanodevice, which bears the responsibility for driving the hardware in the module with the exclusion of the nanobattery and the nano energy harvester. Therefore, it must contain the required number of structural units, which are the transistor elements. The size of the transistor and its power consumption play decisive roles in selecting the transistor technology. The nanoscale processor can be enabled with CNT, SiGe, or Si transistors. CNT FETs made from hollow cylindrical sheets of carbon atoms look promising for nanoprocessor fabrication due to their high energy efficiencies, but currently the technology is at the experimental stage. Aggressive downscaling of silicon metal-oxide-semiconductor field-effect transistors (MOSFETs) leads to undesirable short-channel effects, namely, drain-induced barrier lowering, velocity saturation, mobility degradation by surface scattering, impact ionization, hot carrier injection, etc. The industry has come up with novel partially depleted and fully depleted silicon-on-insulator (SOI) MOSFETs and also three-dimensional structures such as fin field-effect transistors (FinFETs) as prospective solutions. However, these non-planar structures are not favored from an energy-saving perspective. Germanium-strained silicon technology enables high-speed P-type metal-oxide-semiconductor (PMOS) FETs, thereby improving CMOS digital circuit performance. Table 14.4 lists the advantages/disadvantages of CNT, Si, and SiGe devices following Canovas-Carrasco et al. (2014).

The upcoming graphene transistors are even faster switching devices than the foregoing options, working on the near ballistic transport of electrons, thereby maintaining thermal noise due to inelastic electron scattering at very low levels. They are theoretically projected to operate up to a few hundred terahertz.

14.5.5 Nanomemory Unit

The intricacy and the extent of the programming code are essentially governed by the accessible memory space. Important non-volatile memories include the NAND and NOR gate flash memories in which the memory cell is a MOSFET with a floating gate surrounded by a dielectric for retaining charge, along with a capacitively coupled control gate. Mention may also be made of the racetrack or domain-wall memory that works by storing oppositely oriented magnetic regions in the nanowires

TABLE 14.4
CNT, SiGe, and Si Transistors as Nanoprocessor Enablers

Sl. No.	Transistor Material	Smallest Transistor Size (nm)	Merits	Demerits
1.	Carbon nanotubes	<20	Scalable device operating with very low power consumption at high speed	Manufacturing difficulties
2.	Silicon	14	Mature, economical manufacturing technology	Scalability issues
3.	Silicon-germanium	7	Scalable, economical device consuming very low power	Emerging experimental technology, not yet established

of ferromagnetic material, serving as racetracks (Parkin et al. 2008); and the giant magnetoresistance (GMR) memory that utilizes the quantum-mechanical magnetoresistance phenomenon in thin film multilayer structures comprising ferromagnetic and non-magnetic conducting layers.

The main types of volatile memories are: (i) the less expensive but slower dynamic random access memory (DRAM) with each memory cell consisting of one transistor plus one capacitor, and requiring periodic refreshing; and (ii) the expensive but faster static RAM (SRAM) that uses bistable latching circuits or flip-flops, and needs no refreshing. The advanced RAM or A-RAM is the DRAM version that uses single transistor cells without separate capacitors. Here, the capacitor function is realized from the floating body of the silicon-on-insulator transistor. The phase-change RAM (PRAM) stores data by switching the phase of chalcogenide glass between amorphous and crystalline states with the help of a titanium heating element.

Choosing from the memory types described, the NOR flash memory appears suitable from the viewpoint of its shorter read access times than the NAND counterpart, in addition to its satisfactory bit density and moderate energy requirement during reading. For similar reasons, the A-RAM appears to be a suitable technological choice.

In this context, an atomic memory design is also promulgated. In this design, the storage of a single bit is indicated by the presence or absence of an atom. Bennewitz et al. (2002) created a self-assembled memory structure by depositing gold on the (111) surface of a silicon wafer. The deposition of 0.4 monolayers of gold at 700°C was followed by annealing at 850°C. The writing operation of the memory involves the removal of silicon atoms from a nearly filled lattice. For writing, a nano tip is lowered toward the concerned silicon atom, generally by 0.6 nm for 30 ms. No voltage is applied during this period. Alternatively, the tip hovers about the silicon atom with a –4 V pulse applied for 30 ms. The readout is done by a one-dimensional line scan along the track defined by gold, when distinct peaks are observed for any extra silicon atom prominently projecting outside the noise level.

14.5.6 Nanoantenna

Metallic antennas are inappropriate for nano transceivers. The incongruity stems from the fact that their resonant frequencies rise to hundreds of terahertz at the nanoscale, leading to excessive channel weakening and hence shorter transmission distance. To avoid channel attenuation, it is desirable to work at lower frequencies in the order of units of terahertz, which is possible by selecting a new material.

Up to two orders of magnitude lower resonant frequencies can be obtained with graphene nanoantennas because of the smaller wave velocity. The graphene nanoantenna emits electromagnetic waves in the terahertz band with good efficiency. Graphene supports the propagation of surface plasmon polariton (SPP) waves. A quantum of the collective excitation of delocalized electrons in a solid constitutes a plasmon. Essentially, plasmons are density waves in an electron gas. The interaction and blending of electromagnetic waves with an excitation carrying an electric or magnetic dipole forms a quasiparticle called a polariton. The surface plasmon polariton is a combined excitation made up of a surface plasmon and a photon, i.e., an excitation resulting from the coupling of light energy with collective oscillations of electrons. It propagates as a wave along the surface of the solid in a way similar to the guiding of light by an optical fiber. A notable advantage of graphene is that the surface plasmon polariton waves moving on it exhibit a low ohmic loss. Further, they can be easily tuned.

CNT nanoantennas show similar properties to graphene. Graphene nanoribbons show slightly lower resonant frequency than CNT. The resonant frequency depends on the dimensional parameters such as the length and width of the GNR. The size and dielectric constant of the substrate also affect the resonant frequency. The wider the GNR, the higher the resonant frequency. Clearly, a narrow GNR is preferred. A larger-size substrate provides a higher absorption cross section (the percentage of incident power absorbed by the antenna) leading to improved performance. For a larger dielectric constant substrate, the maximum value of the absorption cross section is transferred to lower frequencies. At the same time, its value decreases, thereby adversely impacting the absorption efficiency. Thus, silicon dioxide with a dielectric constant of 3.9 shows a larger increase in the absorption coefficient than silicon with a dielectric constant of 11.7 albeit at higher frequencies.

14.5.7 Nano Transceiver

The layout and planning of the nano transceiver is shown in Figure 14.7. It consists of two blocks: the transmitter and the receiver (Jornet and Akyildiz 2014a). The transmitter contains three sub-blocks: the electric signal generator, the plasmonic nanotransmitter, and the nanoantenna. Likewise, the receiver is composed of three sub-blocks: the nanoantenna, the plasmonic nanoreceiver, and the electric signal detector.

At the transmitter, the electric signal generator produces the carrier wave, which modulates the information to be sent. The main component of the plasmonic nanotransmitter

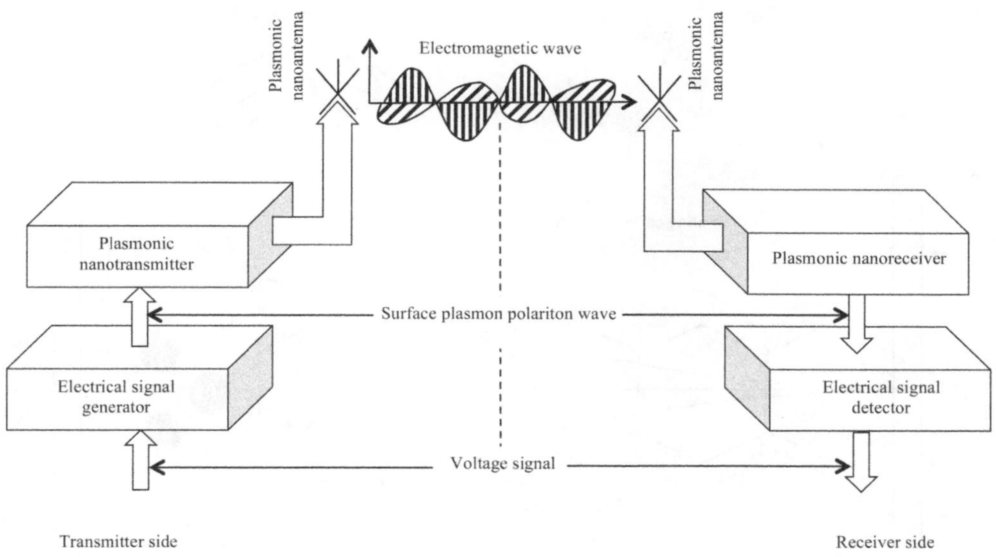

FIGURE 14.7 The nano transceiver concept.

is a high-electron-mobility transistor (HEMT) device. This HEMT device is made from a III-V semiconductor. It has a graphene-based gate. The application of a positive voltage across the source and drain terminals of the HEMT causes the acceleration of the electrons in the HEMT channel. The electrons traversing the channel excite a plasma wave. For a channel length of ~100 nm, the resonant frequency of the plasma wave in the two-dimensional electron gas of the HEMT channel lies in the terahertz band. This plasma wave is not radiated as such, but is instead used for the induction of an SPP wave at the interface with the graphene layer. From here on, the SPP wave reaches the graphene nanoantenna. The excitation of the nanoantenna launches the electromagnetic wave.

On the receiver side, the electromagnetic waves striking the graphene nanoantenna inject the SPP waves on the graphene-based gate of the HEMT. Consequently, the plasma waves are excited in the two-dimensional electron gas (2DEG) of the HEMT channel. The resulting electron motion produces a potential difference across the source and drain terminals of the HEMT. The electrical signal is detected and demodulated.

Use of the same nanomaterial (graphene) in the HEMT gate and the nanoantenna facilitates their coupling by avoiding losses caused by impedance mismatch.

14.5.8 Alternative Nanotube Electromechanical Nano Transceiver

The CNT transceiver is partially mechanical in operation in contrast to the conventional radio transceiver, which is purely electrical.

(i) CNT radio transmitter: a single CNT can be used to realize all four functions of a radio transmitter, namely, oscillation, modulation, amplification, and radiation by antenna (Weldon et al. 2008). Figure 14.8 shows how a CNT is used for the transmission of radio signals.

A CNT with one end connected to a power supply and the other end free starts vibrating. Therefore, self-oscillations are induced in the CNT clamped at one end by applying a direct current (DC) voltage to it. These self-oscillations depend on the field emission from the CNT to a counter electrode. The frequency of the oscillation of the CNT is its mechanical resonant frequency. Further, the application of a DC voltage to the CNT engenders the concentration of charges at its tip. Therefore, when a CNT is set into oscillation, electromagnetic waves are radiated in space. Additionally, when a DC voltage is applied to the V_{Tension} electrode near the oscillating CNT, the CNT bends. This bending of the CNT from its previous position alters its resonant frequency. In this way, the resonant frequency of the CNT is varied in accordance with the voltage of the information signal, thereby achieving frequency modulation.

(ii) CNT radio receiver: the incoming electromagnetic waves from the CNT transmitter impinge upon the CNT in the receiver (Rutherglen and Burke 2007; Jensen et al. 2007). They induce vibrations in the CNT, which are significant only when the frequency of the electromagnetic waves equals the natural vibration frequency of the CNT. A DC voltage is applied to the CNT with respect to a nearby counter electrode. Consequently, a field emission current is generated. Due to the irradiation by electromagnetic waves, there are nonlinearities in the field emission current replicating the received signal. Following these nonlinearities, the input radio signal is demodulated. Moreover, the radio signal is amplified because the field emission current arises from the applied battery voltage instead of the electromagnetic waves. Thus, the CNT is able to implement all four functions of the receiver, viz., antenna, tuning, demodulation, and amplification. It may also be noted that a 1 μm-long CNT has a resonant frequency of a few hundred megahertz.

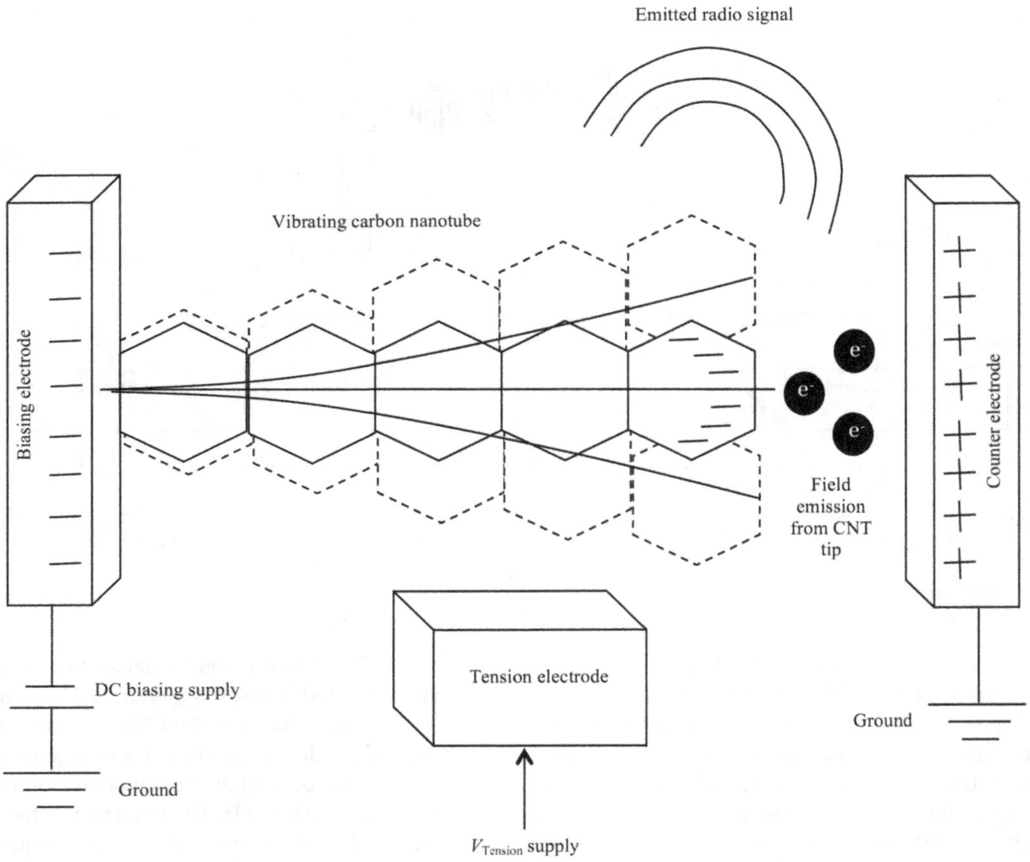

FIGURE 14.8 Nanomechanical CNT radio transmitter.

14.6 WNNs Formation Using EMC Nanosensor Modules: The WNN Architecture

After laying down the capabilities of the nanosensor module, we will now inquire how such nanosensor modules are arranged to form a WNN. To explain the WNN formation, we return to the examples of the human intrabody network and the interconnected office network mentioned in Section 14.1.

Figure 14.9 shows the WNN human intrabody network. We find that this network contains several distinct components. Our sight first falls on the nansosensor modules corresponding to signals from ECG, EEG, blood pressure, and other nanosensors. These modules, represented by circular dots, are placed at designated locations on the body, known as the nanonodes. The output signals from ECG, EEG, blood pressure, and other nanosensor modules are generated at the nanonodes. Each set of nanonodes has a nanorouter to which the signals are fed. The nanorouter is represented by a small upward pointing triangle with a cross. Thus, there is one nanorouter for ECG signals, a second nanorouter for EEG signals, a third one for blood pressure signals, and similarly there are nanorouters for other signals. All the nanorouters are connected through nano links to a central nano–micro interface, shown by a big multiple-lined rounded rectangle. The nano links are shown by dotted lines. The output of the nano–micro interface is the input to the gateway, which is here a mobile telephone. Note that a micro link is indicated by a full line joining the micro–nano interface with the gateway. Finally, through another micro link (full line), the gateway sends the signal to the internet. There is another micro link (full line) between the internet and the healthcare provider. So, the signals produced on the patient's body ultimately reach the healthcare provider to initiate the necessary corrective action for medical treatment by analyzing the status of the patient's health.

A WNN interoffice network is shown in Figure 14.10. Here also we can recognize the nanosensor modules represented by circular dots on the pens, file folders, and books of the officer. These are the nanonodes of this network. Sensing signals recorded at the nanonodes pass through nano links (dotted lines) to reach the nanorouters (upright cones with crosses). As before, we have one nanorouter each for the signals from the pens, folders, and books. The three nanorouters are joined by micro links (full lines) to the micro–nano interface (big multiple-lined rounded rectangle). Another micro link (full line) sends the signal to the gateway, here a modem from where it is conveyed to the internet by another micro link (full line).

From a consideration of Figures 14.9 and 14.10, and the related discussion, it can be concluded that the information is carried along the pathway:

Nanonode (Nanosensor module) → Nanorouter → Micro–nano interface → Gateway → Internet

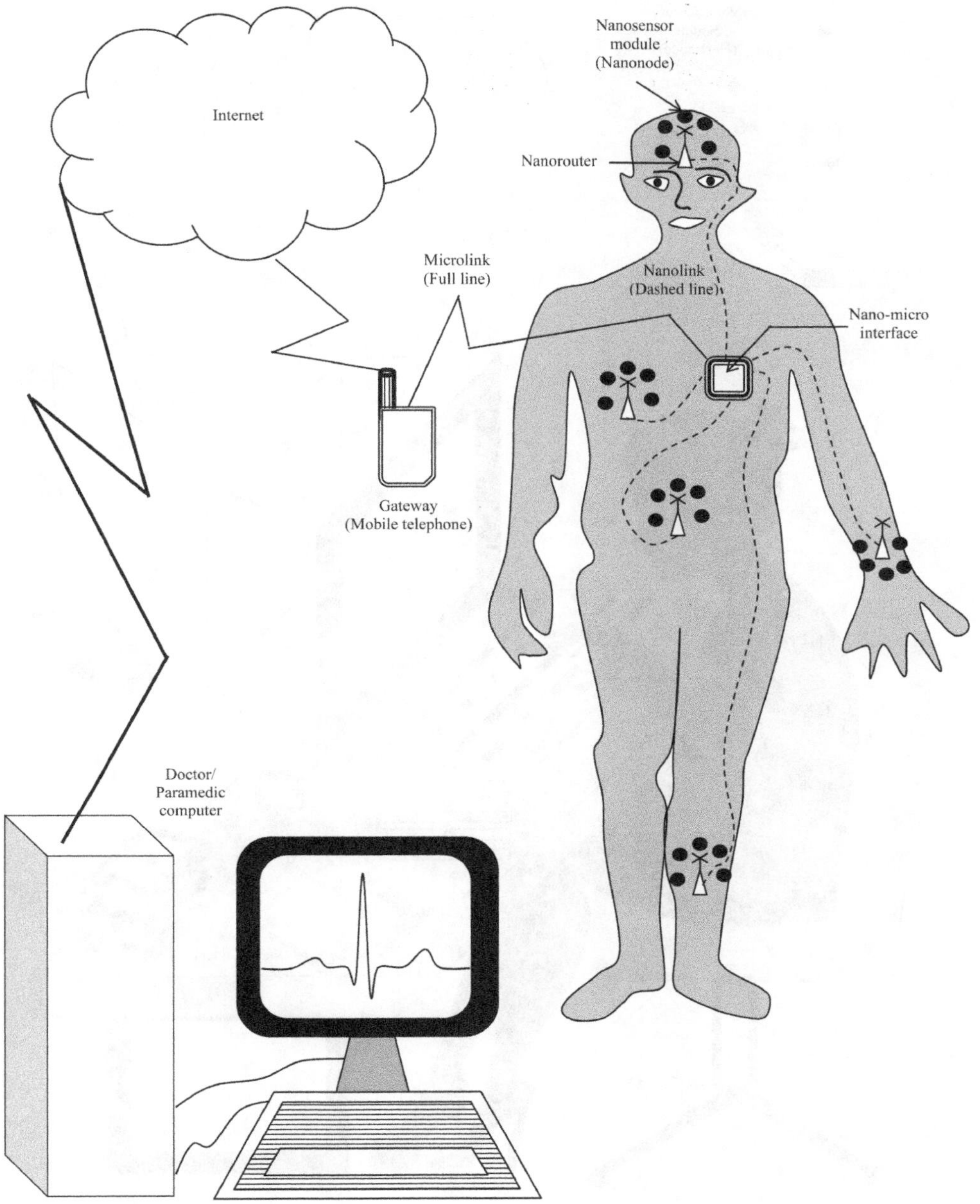

FIGURE 14.9 Layout and components of human intrabody nanosensor network.

so that the architecture of the WNN consists of the following chief components:

(i) Nanonode: it is the most basic component of the WNN, housing the nanosensor and its accessories, namely, the nanoactuator, nanogenerator, energy harvester, nano-processor, nanomemory, nanoantenna, and the nano transceiver. Hence, it has all the facilities for operating independently and executing elementary computation as well as signal transmission and reception. It is the nanosensor module of Section 14.5. Nonetheless, we must keep in mind that it has limitations to its memory capacity and transmission distance. Both these parameters are kept to the bare minimum needed because the power consumption of the nanosensor module must be kept as low as possible.

(ii) Nanorouters: these are much more resourceful than the nano modules in computational complexity. They can gather information sent by the various nanonodes. They can also transmit simple controlling commands to the nanonodes, such as on/off and sleep. However, their enhanced capabilities may add to their size and weight, which is obviously a restraint on making them more elaborate and multifarious.

(iii) Micro–nano interface: this is the bridge between the nano and microworlds. It receives a signal in the nanoscale and converts it into a form compatible to microscale. It can also do the reverse. It is thus

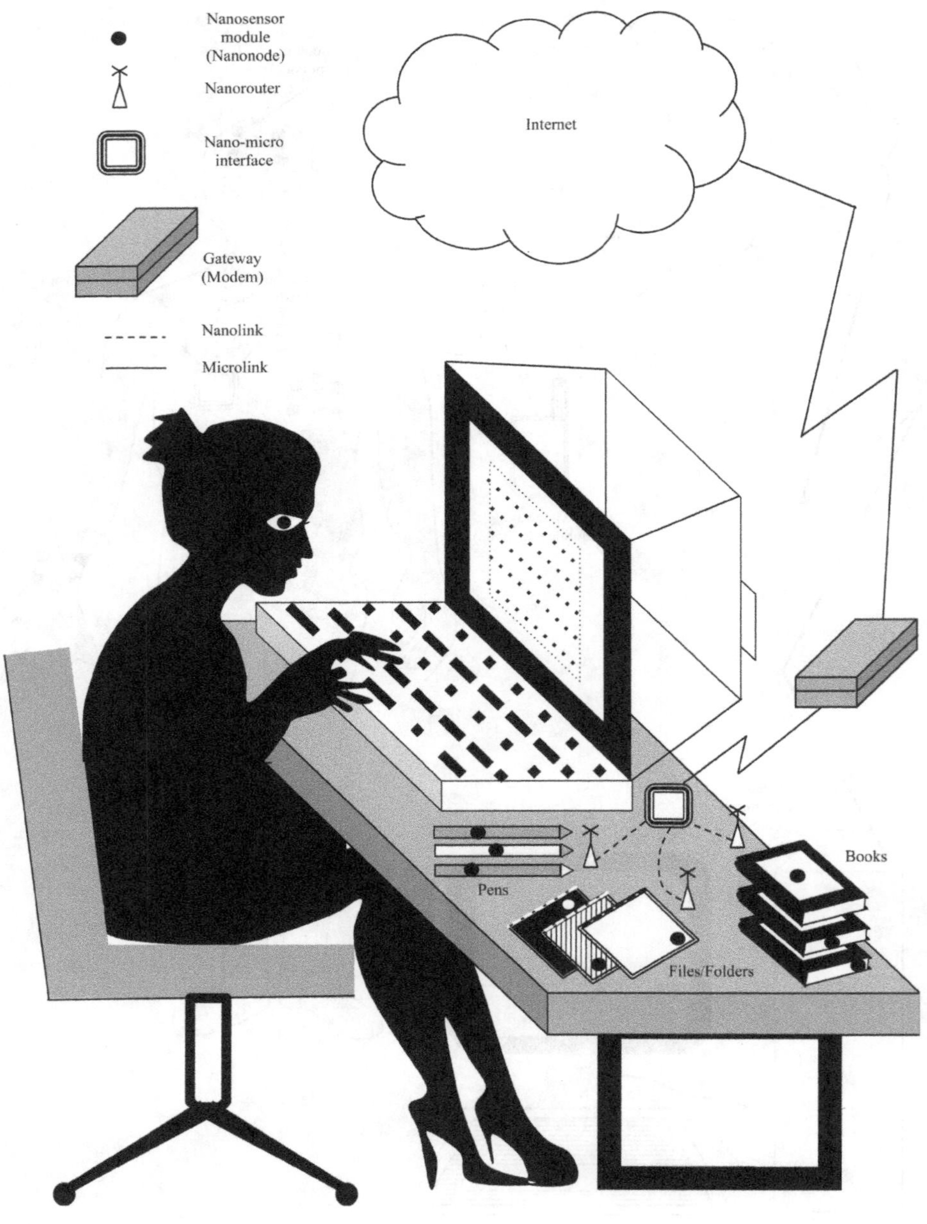

FIGURE 14.10 Architecture of interoffice nanosensor network.

conceived as a hybrid device. On the one side, it can work by a nano communication method. On the opposite side, it is capable of working in the conventional micro communication model.

(iv) Gateway: the gateway couples the information from the WNN to the internet, enabling its function to be controlled from a large distance, e.g., by a doctor sitting in a clinic or by an officer on holiday at his/her home. So, it is crucial to the communication of the WNN with the outside world, both from the WNN to the outside and vice versa.

(v) Nano link: it is the short-distance communication between the nanonodes and the nanorouter. It is taken care of by the nano transceiver embedded in the nanosensor module. It works by surface polariton–based communication, as explained in Section 14.5.7.

(vi) Micro link: it is the long-distance communication between the nanorouter and the micro–nano interface and also beyond the micro–nano interface. It adopts the classical communication techniques.

Looking at the hierarchical structure of a WNN, it is possible to assign an address to each nanonode, e.g., the nanonode 4 feeding the nanorouter 3 and connected via the micro–nano interface 2 to the gateway 1 is stipulated by a unique address {G1.I2.R3.N4}.

14.7 Frequency Bands of Electromagnetic WNN Operation

As we see, there are two links in the WNN architecture: the nano link and the micro link. The GNR nanoantenna approach mentioned in Section 14.5.7 is applicable to the terahertz operation. The electromechanical nano transceiver (Section 14.5.8) can be used in the megahertz band. However, it transpires that the mechanical method of electromagnetic wave generation has low energy efficiency for the nanoscale device. So, the terahertz band operation looks favorable at the nanoscale, i.e., for the nano-link communication between nanosensor modules. The megahertz band could be used for the micro-link communication from the nanorouter to the gateway and then to the internet.

The selected frequency range (0.1–10 THz) and the corresponding wavelength range (3 mm–30 μm) are known as the terahertz gap. It is a frequency/wavelength range, which has hitherto received much less attention in comparison to the frequency bands above it (far infrared) and below it (microwaves). Even if it has been explored and modeled to some extent, the preliminary studies have been made on devices stationed at distances of several meters. Differences between physical mechanisms and characteristics at nano- and microscales have not been accounted for. Nonetheless, its non-ionizing nature makes its usage near humans and inside the human body very safe.

14.7.1 THz Channel Model for Intrabody WNNs

Elayan et al. (2017) have developed an analytical THz channel model for communication at the nanoscale. The attenuation factor due to the total path loss is the product of three factors contributed by spreading loss, molecular absorption loss, and scattering loss. The spreading loss arises from the expansion of the electromagnetic wave during propagation through the medium. Molecular absorption loss occurs when the molecules of the medium are excited by the THz waves. In the molecules, the atoms are moving periodically while the molecule is undergoing translational and rotational motions. During the vibration induced by molecular excitation, a part of the electromagnetic energy is converted into the internal kinetic energy of the excited molecules. This part of the energy is lost from the communication aspect, and signifies the molecular absorption loss. The resonant frequency differs for the various types of molecules. At each resonance, the absorption is not limited to a single frequency but takes place over a stretch of frequencies, making the THz channel highly discriminative to frequency. Further, during its passage through the human body, the electromagnetic wave suffers deflection from cells, organelles, protein molecules, and other composites, which together give rise to the scattering loss.

When the model is applied to determine the three types of losses, it is found that the spreading loss rises with the increase in distance and frequency; the molecular absorption loss is relatively higher in blood molecules than in skin and other human tissues revealing their more absorptive nature; and the scattering effects are negligible, noting firstly that the scattering is more pronounced for wavelengths which are smaller than the dimensions of the scattering particle, and secondly that the radii of the various body particles such as red blood cells (4×10^{-6} m) and water particles (1.4×10^{-10} m) are very small.

14.7.2 Channel Capacity for WNNs

Jornet and Akyildiz (2010) proposed a terahertz propagation model taking the spreading and molecular absorption losses into account. Keeping in view the transmission capability of ultra-short pulses by graphene-based nanoelectronics, they envision a communication scheme for WNNs, which works by exchanging pulses of <1 ps in length. The channel capacity is determined as a function of the distance for different power spectral densities and for different molecular compositions of the medium. It is found that exceedingly high channel capacities of approximately a few terabits per second are achievable within distances up to a few tens of mm. This huge channel capacity supported by the terahertz channel can be utilized by the nanosensor modules in the WNN by sending information about every individual molecule of the medium, leading to the reporting of a very large number of events. Moreover, efficient multiple access schemes may be allowed in the WNN.

14.7.3 Multi-Path Fading

It is a phenomenon in wireless telecommunication in which the transmitted signal is subdivided and sent to the receiver antenna through two or more paths. The signals reaching from different paths are out of phase, and may undergo constructive/destructive interferences. As a consequence of phase shifting and interference effects, the received signal is weakened. Materials in the human skin and items in an office differ in roughness. They change the amplitude and phase of the reflected signal. Therefore, the reflection coefficients of objects in the imagined scenarios must be measured. Accurate fading models should be developed using these data and applied to include their effects.

14.8 Modulation Techniques for Electromagnetic WNNs

As ultra-short pulse-based techniques have been suggested for electromagnetic WNNs, it is evident that the applicable modulation methods could be chosen from among those used in the existing pulse-based communication systems, viz., pulse-amplitude modulation (PAM), in which the amplitude of the pulse is varied according to the instantaneous amplitude of the information signal; pulse-position modulation (PPM), in which the position of the pulse is changed according to the amplitude of the information signal keeping the amplitude and width of the pulse fixed; and pulse-width modulation (PWM) in which the time duration of the ON phase and hence the duty cycle is changed according to the amplitude of the signal. Other methods are pulse rate modulation (PRM), in which the rate of the pulses is altered, and communication through silence, in which the time between pulses is the variable parameter. These methods are not useful for nanosensor modules. Coding the information in the shape of the pulse is not advisable because of the frequency selective nature of the channel. Nor is coding through temporal pulse positions advocated. Furthermore, the pulse width is fixed if we wish to reside in the terahertz band.

14.8.1 Time Spread On-Off Keying (TS-OOK) Modulation Scheme

Jornet (2014) and Jornet and Akyildiz (2014b) proposed a modulation scheme based on the transmission of a 100 fm-long pulse to represent logic state 1 and staying silent, i.e., transmitting no pulse for the logic 0 state (Figure 14.11). This scheme of the

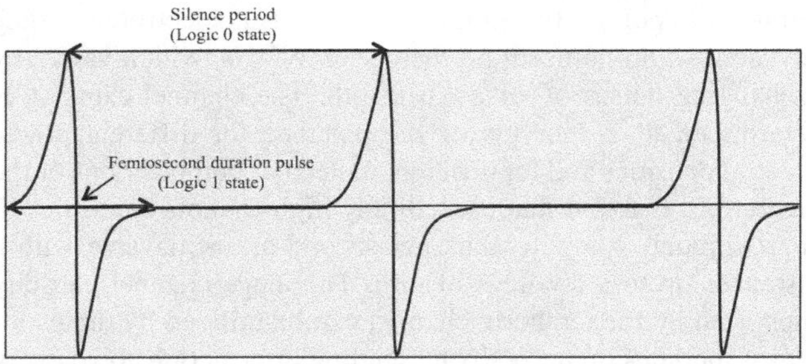

FIGURE 14.11 TS-OOK modulation scheme.

transmission of a sequence of pulses with intervening randomly chosen constant silence intervals is known as time spread on-off keying. In this scheme, the time duration between symbols is much greater than the duration of the symbol itself for three reasons: (i) pulse emission in bursts is not possible due to technological constraints; (ii) the longer inter-pulse separation gives sufficient time for the vibrating molecules in the channel to relax; and (iii) during the relaxation interval, a nanosensor module either remains lazy or receives information from other modules, enabling simpler multiple access rules. From an examination of the information rates obtained in single-user and multi-user circumstances, it is found that the adoption of this scheme makes it possible to support a large number of nanosensor modules, which are transmitting at rates up to terabits per second at the same time.

14.8.2 Symbol Rate Hopping (SRH)-TSOOK Modulation Scheme

Mabed (2017) showed that the TS-OOK is inadaptive to variations in traffic load. A dynamic TS-OOK modulation scheme is put forward. It is called symbol rate hopping TS-OOK. Here, the interval between two successive pulses of a transmission is regularly altered following a sequence which is pseudo-random in nature allowing the dissemination of produced interference over different communications and providing similar access conditions to all simultaneous communications. Using probabilistic analysis, the performance of this scheme is shown to be superior to TS-OOK when the number of active nodes is very large. For 300 active nodes, the throughput of communication with TS-OOK decreases to $<10^5$ frames per second. In the case of SRH-TSOOK, it becomes stable at 20^7 frames per second for every active communication.

14.9 Channel Sharing Protocol in WNN

Multiple access control (MAC) protocol allows several users to share the allotted spectrum effectively by coordinating access to the link. Carrier-sense multiple access control (CSMAC) is a protocol in which a node confirms the absence of traffic on a transmission medium prior to sending information on it. But in pulse-based transmission, there is no carrier signal. So, CSMAC cannot be used. Notwithstanding the inapplicability of CSMAC, we note that extremely short pulses are used for information conveyance and the time gap between consecutive pulses is extremely long. So, the chances of a collision between pulse streams from nanosensor modules attempting access at the same time are fairly remote. Further, it is expected that a given nanosensor module will transmit pulses when necessary. By detection of this first pulse, the remaining nanosensor modules will know when the next pulse is due, presuming that the time interval between pulses is constant. They will then transmit a pulse, keeping the arrival of the next pulse in mind.

14.10 Information Routing in WNNs

Information flow between nanosensor modules will take place by deciding the routes, which is called information routing.

14.10.1 Multi-Hop Routing

The scarce resources available to nanosensor modules will limit information transmission to their closest neighbors. When sending information to distant nanosensor modules, a multi-hop mechanism will be used in which the nanosenor modules on the route will act as relay centers. Remembering that the transceiver consumes a lot of power and long-distance transmission requires more power, multi-hop routing ensures higher energy efficiency than single-hop routing.

14.10.2 Sensing-Aware Information Routing: The Cross-Layer Protocol

The nanosensor module has the ability to know the information it is routing. So a WNN placed in a CO_2 environment will know through its nanosensor modules that it is placed in a CO_2 ambient environment. The GNR nanoantenna of this WNN will also be affected by the presence of CO_2 because of the adsorption of CO_2 on its surface. As a result, its resonant frequency will change. If the nanoantenna has been designed with the effect of the adsorption of CO_2 in mind, it will be able to adapt its resonant frequency to the sensed environment and radiate effectively. Otherwise, its radiation efficiency will be degraded, which is a type of power loss. Thus, a cross-layer routing protocol is advantageous for WNN in which the interdependence between different layers can be addressed. The cross-layer protocol allows the interaction, coordination, and optimization of protocols after crossing different layers of the stack, i.e., it permits co-operation and the sharing of information about the network status between protocols of different layers, thereby enabling the choice of the best route from energy consumption and performance perspectives.

14.11 Failure Mechanisms and Reliability Issues of WNNs

Two types of failure mechanism are possible:

(i) Individual nanosensor module failure: physically, the carbon-based networks are very strong as can be surmised from the mechanical strength of CNTs, but their degradation on exposure to chemical and biochemical environments may cause detrimental changes in their behavior, which are likely to make them function in an undesirable way.

(ii) Upsetting of the network by transitory molecular outbursts: a transient molecular disturbance may dislodge many nanosensor modules from their locations and may also impair the functioning of some of the devices. However, designing networks with a larger number of nanosensor modules and incorporating built-in adaptive controls in the network may avoid any calamities resulting from this disturbance. A routing protocol designed to take care of such a mishap can come to the rescue. However, a larger number of nanosensor modules may introduce fresh problems for multiple access control, which must be kept in mind.

14.12 Internet of Nano Things (IoNT): The Nanomachine

The IoNT is a network formed by combining the network of nanoscale devices, objects, or organisms having unique identifiers, with the existing telecommunications networks, computer and cellular networks, and wired or unwired networks (Akyildiz and Jornet 2010b). The nanoscale device in the definition of the IoNT is actually a nanomachine, which is defined as a natural entity or an artificial mechanical or electromechanical contrivance with nanoscale dimensions and constructed by nanotechnologies. The nanomachines of the IoNT are endowed with rudimentary actuation, computing, control, and communication capabilities besides the power unit. They function like the nanosensor modules described in Section 14.5. Due to the embedding of nanosensor modules in the IoNT, most of the WNN concepts apply to the IoNT and can be understood on similar lines.

By connecting nanosensor modules to the internet, knowledge sharing is made possible and many real-time applications are enabled by the IoNT; however, the drawback is that the nanodevices are exposed to attacks from malicious unauthenticated users (Atlam et al. 2018). The confidentiality and integrity of messages exchanged must be preserved and network intrusion must be prevented. Unfortunately, due to the limited computational capabilities of nanodevices and the large number of nanosensor modules to be protected, sophisticated security algorithms cannot be used, and new algorithms must be designed specially suited to the nanoscale networks.

Regarding the nanomachines in the IoNT, they are largely hypothetical and currently in the research and development stage. An example of such nanomachines is respirocyte (Freitas 1988), which is proposed as an artificial red blood cell carrying 236 times more oxygen per unit volume than a natural blood cell. Respirocytes can be injected into the human body as post-accident life-saving aids. Another example is synthetic platelets, which have been demonstrated to reduce the bleeding time by one-half through intravenous injection in a rat model of major trauma (Betram et al. 2009). They are based on nanoparticles functionalized with arginylglycylaspartic (arg-gly-asp) acid.

A bio-hybrid nanoengine has been reported by Valero et al. (2018). It consists of a catalytic stator, which rotates a DNA wheel in one direction. The power is derived from the hydrolysis of nucleotide triphosphate (NTP).

14.13 Discussion and Conclusions

The limited capabilities of a single nanosensor were pointed out and the motivation for a nanosensor network was explained by emphasizing the need to reinforce these inadequate capabilities. Following Akyildiz and Jornet (2010a), examples of a human intrabody network and an interoffice network were quoted. Speaking of networking, the first topic of discussion is the communication approach. In this context, two approaches viz., molecular and electromagnetic have great scope.

The molecular approach is greatly exploited by nature but much work must be done to apply it to wireless nanosensor networks. The coding/decoding, transmission, propagation, and reception of signals in molecular communication are very interesting and are thus garnering increasing attention from scientists.

Moving toward electromagnetic communication, it is appreciated that graphene, GNRs, and CNTs are destined to play key roles in WNNs. Presenting the ideas of CNT FET–based physical, chemical, and bio-nanosensors, the integrated nanosensor module was introduced as the fundamental structural unit of a WNN to be placed at its nodes. This nanosensor module is embellished with many basic capabilities although this is done at a very low level, keeping in sight the extremely small energy and space at its disposal. The capabilities invoked are nanoactuation with NEMS-based actuators, powering with lithium nanobatteries rejuvenated with nano energy harvesters, signal processing with CNT-FET nanoprocessors, data storage using atomic memory, signal radiating/intercepting antenna with graphene nanoantenna supporting SPP waves and a HEMT or CNT-based transceiver.

The human intrabody and the interoffice networks were revisited to sketch the WNNs to achieve the anticipated goals of these networks. From the drawings of these networks, it was unequivocally found that the WNNs have a well-defined hierarchical architecture, which is formed by the proper organization of a few basic elements. These elements were identified as nanonodes, nanorouters, micro–nano interfaces, and the gateway. At the bottom of this hierarchy lie the nanonodes where the nanosensor modules are placed. A little higher up are the nanorouters which receive signals from the nanonodes. Still higher up lie the micro–nano interfaces to which the nanorouters feed the signals. Topmost in this order is the gateway, which is the recipient of the signals from the micro–nano interfaces. It is the doorway of the WNN to the internet.

Climbing up the hierarchical ladder, it was also noticed that two types of communication links are used to join the network elements, one is a nano link connecting the nanonodes with the nanorouters and the other is a micro link connecting the nanorouters with the micro–nano interface and up to the gateway for internet access. The short-distance nano-link and the long-distance micro-link signals use different frequency bands for electromagnetic wave propagation. The micro link works at megahertz frequencies and uses established communication networks. For the nano link, the services of the terahertz band are solicited. This band sandwiched between the microwave and far-infrared regions has only recently caught the attention of researchers because of practical difficulties in electronic communication at its lower end and photonic communication at its upper extremity.

Terahertz channel modeling has been attempted but only for distances of the order of a few meters. Very limited efforts have been made at the nanoscale. Rigorous models for nanoscale communication are therefore the need of the hour. Nevertheless, the insufficient models available have yielded interesting results regarding the spreading and molecular absorption losses suffered by terahertz waves as they travel through air, water, or the human body.

When ultra-short pulses are used for communication, channel capacities in the range of terabits per second become available. Multi-path fading in WNNs is a subject of careful study.

Coming to modulation techniques for WNNs, it soon becomes clear that conventional methods based on amplitude, position, or pulse width cannot be applied to WNNs. One proposal is for time spread on-off keying in which a logic high state is registered by sending a 100 fs-long pulse, and a logic low state by sending no pulse. Any two pulses are interleaved with a long silence period, which is kept constant. Another approach, named the symbol rate hopping TS-OOK method, is proposed to overcome the inadaptability drawback of TS-OOK. The relatively long gap between pulses in TS-OOK helps to prevent collision between messages, which is essential because the carrier-sense MAC protocol becomes redundant in the absence of the carrier signal.

Regarding information routing in WNNs, a multi-hop mechanism is resorted to because of the relatively little power that nanosensor modules can expend. Further, a cross-layer routing protocol is recommended if the WNN is to derive the benefit of sensing the environment and adjusting to it.

Among the failure mechanisms of WNNs, it is obvious that CNT-based nanosensors are physically very rugged, but similar remarks do not hold following their exposure to chemical and biological surroundings. Perturbations in WNNs due to molecular outbursts can be dealt with by increasing the number of nanosensor modules in the network in a restricted way.

Finally, the WNNs pave the way toward the internet of nano things, just as the internet of things follows wireless sensor networks. Future nanomachines will bring forth many groundbreaking inventions in this area, making possible what are so far deemed incredible.

Review Exercises

14.1 In what respects are the capabilities of a single nanosensor limited? Why is it necessary to form a network of nanosensors?

14.2 Define a wireless nanosensor network. Give two examples.

14.3 What are the two possible communication approaches which can be applied to construct a wireless nanosensor network? Make a comparison of their main features.

14.4 Why is molecular communication called a bio-inspired paradigm? What is the spatial range over which it can be used? Mention its advantages and disadvantages.

14.5 Describe the main steps in molecular communication. Illustrate with a diagram.

14.6 How are the coding and decoding processes implemented in molecular communication?

14.7 What are the passive and active transport mechanisms used in molecular communication? Cite two examples of each mechanism. Briefly describe them.

14.8 Describe the main components of the integrated nanosensor module imagined for making wireless nanosensor networks.

14.9 Point out the chief merits and demerits of supercapacitors in competition with nanobatteries as sources of power in a wireless nanosensor network.

14.10 Name the three types of transistors that enable nanoprocessors for WNNs. Prepare a comparative chart mentioning their relative sizes, advantages, and shortfalls.

14.11 What is meant by atomic memory? How is a self-assembled memory structure produced by depositing gold on silicon?

14.12 Why are metallic antennas unsuitable for nano transceivers? Give an example of a material that efficiently emits electromagnetic waves at terahertz frequencies.

14.13 What is the difference between surface plasmon and surface plasmon polariton waves? Does graphene support the propagation of SPP waves?

14.14 Draw a block diagram and explain the working principle of a nano transceiver.

14.15 How is the resonant frequency of a CNT varied according to the information signal voltage? How is the signal demodulated in a CNT radio receiver?

14.16 Draw a diagram illustrating the layout and components of a human intrabody network. Identify the components of a wireless nanosensor network on this diagram.

14.17 For what communication links, micro or nano links, do the megahertz and terahertz frequency bands look favorable? Why?

14.18 Name the three contributing factors to the attenuation of waves in the terahertz channel model.

14.19 What is the reason for the inability to use the carrier-sense multiple access control (CSMAC) protocol in pulse-based transmission?

14.20 Why is a multi-hop mechanism favored for sending information to remote nanosensor nodes?

14.21 Explain the two main failure mechanisms imagined for wireless nanosensor networks.

14.22 What is meant by the internet of nano things? How is it related to a wireless nanosensor network?

REFERENCES

Akyildiz I. F. and J. M. Jornet. 2010a. Electromagnetic wireless nanosensor networks. *Nano Communication Networks* 1(1): 3–19.

Akyildiz I. F. and J. M. Jornet. 2010b. The internet of nano-things. *IEEE Wireless Communications* 17(6): 58–63.

Atlam H. F., R. J. Walters, and G. B. Wills. 2018. Internet of nano things: Security issues and applications. In Proceedings of the 2018 2nd International Conference on Cloud and Big Data Computing (ICCBDC 2018), 3–5 August, Barcelona, Spain. New York: ACM, 7 pp.

Bennewitz R., J. N. Crain, A. Kirakosian, J. L. Lin, J. L. McChesney, D. Y. Petrovykh, and F. J. Himpsel. 2002. Atomic scale memory at a silicon surface. *Nanotechnology* 13(4): 499–502.

Bertram J. P., C. A. Williams, R. Robinson, S. S. Segal, N. T. Flynn, and E. B. Lavik. 2009. Intravenous hemostat: Nanotechnology to halt bleeding. *Science Translational Medicine* 1(11): 11ra22–11ra22.

Canovas-Carrasco S., A.-J. Garcia-Sanchez, F. Garcia-Sanchez, and J. Garcia-Haro. 2014. Conceptual design of a nano-networking device. *Sensors* 16(12): 2104 (27 pp.).

Canovas-Carrasco S., A.-J. Garcia-Sanchez, and J. Garcia-Haro. 2018. On the nature of energy-feasible wireless nanosensor networks. *Sensors* 18: 1356 (13 pp.).

Elayan H., R. M. Shubair, J. M. Jornet, and P. Johari. 2017. Terahertz channel model and link budget analysis for intrabody nanoscale communication. *IEEE Transactions on Nanobioscience* 16(6): 491–503.

Freitas R. A., Jr. 1988. Exploratory design in medical nanotechnology: A mechanical artificial red cell. *Artificial Cells Blood Substitutes and Biotechnology* 26(4): 411–430.

Hiyama S., Y. Moritani, T. Suda, R. Egashira, A. Enomoto, M. Moore, and T. Nakano. 2005. Molecular communication. In Technical Proceedings of the 2005 NSTI Nanotechnology Conference and Trade Show, NSTI-Nanotech. 2005, *TechConnect Briefs* 3: 391–394.

Jensen K., J. Weldon, H. Garcia, and A. Zettl. 2007. Nanotube radio. *Nano Letters* 7(11): 3508–3511.

Jornet J. M. 2014. Low-weight error-prevention codes for electromagnetic nanonetworks in the terahertz band. *Nano Communication Networks* 5: 35–44.

Jornet J. M. and I. F. Akyildiz. 2010. Channel capacity of electromagnetic nanonetworks in the terahertz band. In IEEE International Conference on Communications, Cape Town, South Africa, 23–27 May. New York: IEEE, 6 pp.

Jornet J. M. and I. F. Akyildiz. 2014a. Femtosecond-long pulse-based modulation for terahertz band communication in nanonetworks. *IEEE Transactions on Communications* 62(5): 1742–1754.

Jornet J. M. and I. F. Akyildiz. 2014b. Graphene-based plasmonic nano-transceiver for terahertz band communication. In The 8th European Conference on Antennas and Propagation (EuCAP 2014), The Hague, Netherlands, 6–11 April, IEEE, NY, USA, pp. 504–508.

Lee S. J., C. Jung, K. Choi, and S. Kim. 2015. Design of wireless nanosensor networks for intrabody application. *International Journal of Distributed Sensor Networks.* 2015: 176761 (12 pp.).

Llatser I., A. Cabellos-Aparicio, and E. Alarcón. 2012. Networking challenges and principles in diffusion-based molecular communication. *IEEE Wireless Communications* 19(5): 36–41.

Mabed H. 2017. Enhanced spread in time on-off keying technique for dense terahertz nanonetworks. In IEEE Symposium on Computers and Communications (ISCC), Heraklion, Greece, 3–6 July. New York: IEEE, 7 pp.

Malak D. and O. B. Akan. 2012. Molecular communication nanonetworks inside human body. *Nano Communication Networks* 3: 19–35.

Noel A. 2013. Introduction to molecular communication, Digital Communication Guest Lecture, University of Erlangen-Nürnberg, November 5, 2013, https://warwick.ac.uk/fac/sci/eng/staff/ajgn/research/2013_07_mc_intro.pdf.

Parkin S. S. P., M. Hayashi, and L. Thomas. 2008. Magnetic domain-wall racetrack memory. *Science* 320: 190–194.

Rutherglen C. and P. Burke. 2007. Carbon nanotube radio. *Nanoletters* 7(11): 3296–3299.

Valero J., N. Pal, S. Dhakal, N. G. Walter, and M. Famulok. 2018. A bio-hybrid DNA rotor-stator nanoengine that moves along predefined tracks. *Nature Nanotechnology* 13(6): 496–503.

Wang J., B. Yin, and M. Peng. 2017. Diffusion-based molecular communication: Principle, key technologies, and challenges. *China Communications*, IEEE 14(2): 1–18.

Weldon J., K. Jensen, and A. Zettl. 2008. Nanomechanical radio transmitter. *Physica Status Solidi (b)* 245(10): 2323–2325.

15

Overview and Future Trends of Nanosensors

15.1 Introduction

The progression from macro- to nanosensors has been a very long and interesting evolution (Khanna 2008, 2009). Several nanomaterials, notably metallic nanoparticles (NPs), carbon nanotubes (CNTs), and quantum dots (QDs), together with enabling nanotechnologies, duly supported by microtechnologies have been exploited to realize nanosensors.

15.1.1 Interfacing Nanosensors with Human Beings

The core idea is to understand that a nanosensor is not necessarily a device merely reduced in size to a few nanometers, but a device that makes use of the unique properties of nanomaterials and NPs to detect and measure new types of events and entities in the nanoscale. Nanotechnology received a boost from the renewed interest in colloidal science combined with the introduction of a new generation of analytical tools such as the scanning tunneling microscope (STM) and the atomic force microscope (AFM). Deliberate manipulation of nanostructures by electron-beam lithography and molecular beam epitaxy enabled the observation of novel nanoscale phenomena.

An optical fiber pierces the cell wall non-destructively to peep into what is happening inside the cell. The active portion of the sensor penetrating the cell wall has to be of nano dimensions to avoid damage to the cell, but if its connection to the outside world is of that scale, no one will be able to see or operate it. The interface between human beings and nanosensors is of utmost importance. Ultimately, whether it is through an instrument or some improvisation, the nanosensor has to be connected to the macroworld.

Of serious concern are the compatibility issues at the borderline of human beings and nanosensors. After all, the nanosensor has to work in this macroworld and human beings are neither microscopic nor nanoscopic creatures although our structural units are of those scales. *Could we manipulate NPs if necessary tools were not developed?* No, we can neither see nor work with nano-objects without aiding tools. All nanoscale sensors have to be operated by our macroscopic-level capabilities using convenient aids. Man is a tool-using animal.

15.1.2 Three Main Types of Nanosensors

Nanosensors belong to one of three types: physical, chemical, or biological. Physical nanosensors are used to measure magnitudes such as mass, pressure, force, or displacement. Chemical nanosensors are employed to measure parameters such as the concentration of a given gas, the presence of a specific type of molecule, or the molecular composition of a substance. Biological nanosensors are used to monitor biomolecular processes such as antibody/antigen interactions, DNA interactions, enzymatic interactions, or cellular communication processes.

15.1.3 Using the Response Properties of the Same Nanomaterial in Different Types of Nanosensors

Can the same nanomaterial act as the central element of different types of nanosensors? Yes, the same nanomaterial, for example, CNT, can be used as the key element of physical, chemical, and biological nanosensors (Akyildiz and Jornet 2010); CNT-based sensors have been described in almost all chapters. A CNT can be used to build a field-effect transistor (FET) whose on/off threshold changes on a local deformation of the tube serving as a force sensor (mechanical sensor). CNTs are used to sense chemical compounds. When used as a gas sensor (chemical sensor), the presence of a specific type of gas molecule changes the threshold voltage of the transistor. The CNT biological sensors work in a similar way to chemical nanosensors; however, in this case, the change in the electronic properties of, for example, a CNT-based FET, is induced by biomolecules, by either a protein or any other chemical composite that binds itself to the functionalized nanotube, a specific antigen that binds itself to an antibody glued to the nanotube, or a single-stranded DNA chain that binds itself to another DNA chain which has been attached to the nanotube. But the most important aspect of the investigation of a variety of sensors is the three S's, that is, sensitivity, selectivity, and stability.

15.1.4 Nanosensor Science, Engineering, and Technology: Three Interrelated Disciplines

If science comprises the rules of the game, then engineering is actually playing the game and technology contains the tips that help us win the game. In this background, there are three important disciplines in reference to nanosensors: nanosensor science, nanosensor engineering, and nanosensor technology. The first discipline involves appreciating the nano-level phenomena and the laws of nature that can be applied in designing various nanosensors. This is the field of nanosensor scientists. The second discipline deals with the application of these laws by nanosensor engineers to construct nanosensors according to the desired application. The third discipline constitutes the skills acquired for the successful fabrication of a vast variety of nanosensors described in the preceding pages of this book. This is the intellectual property of the nanosensor technologists who have been granted patents for their inventions.

15.1.5 Scope of the Chapter

This chapter will provide a summary of the present status of the main types of nanosensors: physical, chemical, and biological, which have been described in this book, pointing out their current capabilities and limitations, and highlighting the research problems that need to be addressed in this area.

15.2 Scanning Tunneling Microscope

In a scanning tunneling microscope (Giessibl 2003), a pointed tip is brought near an electrically conducting surface biased at a voltage, V_t. When the separation between the tip and the surface is very small, typically a few atomic diameters, electrons are transported by tunneling and a current, I_t, flows between them. The strong distance dependence of the tunneling current lays the foundation for the atomic resolution capability of the STM. However, the two objects involved in tunneling must be conducting or at least semiconducting but not insulating, imposing a restriction on the nature of the specimen that can be examined.

15.3 Atomic Force Microscope

The main component of an AFM (Giessibl 2003) and its major instrumental difference from an STM is the spring that is used for sensing the force between the tip and the sample. *What requirements should be met by the force sensor to sense normal tip–sample forces?* It must be rigid in two axes and relatively soft in the third axis. *Which device realizes this property?* It is fulfilled by a cantilever beam, thus cantilever geometry is typically used for force detection. Access to the nanoworld is gained by means of a local probe operating with *surface contact* or in *near-field mode* at a few angstroms sample–tip distance only. The first AFMs were mostly operated in the *static contact mode*, and for this mode, the restriction on the stiffness of the cantilever is that it should be less than the interatomic spring constants of atoms in a solid, which is $k \leq 10$ Newton meter^{1} (N m^{-1}). In *dynamic atomic force microscopy*, k values higher than hundreds of N m^{-1} assist in reducing noise and increasing operational stability. The Q factor depends on the damping mechanisms present in the cantilever. For micromachined cantilevers operated in air, Q is principally limited by the viscous drag of air and characteristically amounts to a few hundred. In vacuum, internal and surface effects in the cantilever material are responsible for damping and Q values of hundreds of thousands are commonplace. Huge progress has been made in fabricating attonewton-sensitive cantilevers from both silicon nitride and silicon.

15.4 Mechanical Nanosensors

The ability to measure and quantify the motion of an object is a basic sense required in advanced control systems. Mechanical nanosensors are of fundamental scientific interest. They have been employed in applications including ultrasensitive displacement, acceleration, force, and resonant mass sensing of chemical and biological species.

How was the tunneling phenomenon applied in developing mechanical or other nanosensors? The tunneling phenomenon was used in various nanosensors for displacement, acceleration, and magnetic field. Miniature high-sensitivity accelerometers have been developed using a high-resolution displacement transducer based on electron tunneling. In a microelectromechanical systems (MEMS) technology–based tunneling accelerometer, the tunneling current between the tip and the counter electrode is maintained constant through a feedback system. The electrical force used to hold the current constant is measured and converted into an acceleration measurement. Magnetic fields are measured through an indirect application of tunneling.

How have MEMS/nanoelectromechanical systems (NEMS) technologies contributed toward nanosensors? With microelectronics technology pushing deep into the sub-micrometer size regime, a concerted effort surfaced to realize even smaller electromechanical systems: nanoelectromechanical systems. NEMS convert electrical current into mechanical motion on a nanoscale and vice versa (Ekinci 2005). They can be viewed as the descendants of MEMS, which operate at the micron scale and are found in commercial applications. Improved performance is expected from NEMS devices due to their small sizes and higher eigen frequencies.

What are the important NEMS devices used as nanosensors? NEMS devices include cantilevered and doubly clamped beam resonators operated in their fundamental flexural resonant modes, much like simple tuning forks. In this size regime, NEMS offer access to GHz frequencies, with quality factors in excess of 10^4. In general, the smaller the device, the more susceptible are its physical properties to perturbation by external influences. This enhanced sensitivity of NEMS is opening a variety of unprecedented opportunities for applications such as mass spectrometry. The vibrational frequency of a NEMS resonator is an exquisitely sensitive function of its total mass. Small variations in mass, such as those resulting from adsorbed addenda, can measurably alter its resonant frequency. Theoretical calculations for physically realizable devices indicate that NEMS mass sensitivity below a single Dalton (1 Da = AMU) is achievable.

What construction materials have been used in NEMS resonators? The NEMS resonating structures currently used are based on single-crystal Si fabricated from silicon-on-insulator substrates, because the mechanical and electrical properties of these systems are well known and the processing techniques are well developed. The combination of electron-beam lithography and Si micromachining techniques has made it possible to fabricate submicron mechanical structures from single crystal substrates. Efforts have been made to utilize other materials with a particularly high Young's modulus, such as aluminum nitride, silicon carbide, and ultra-nanocrystalline diamond.

What are the chief resonant mass nanosensors? Resonant mass nanosensors are already employed in many assorted fields of science and technology. These devices operate by providing a frequency shift that is directly proportional to the inertial mass of the analyte molecules deposited upon them. Among the most sensitive realizations are those based on the acoustic vibratory modes of crystals, thin films, and micron-sized cantilevers. Quartz crystal microbalances (QCM) utilize the characteristic of frequency variation with mass loading of the resonator. A QCM sensor can detect up to 10^{-13} g, and is used to probe nanometer-thick film loads. QCM is capable of measuring mass changes as small as a fraction of a monolayer or a single layer of atoms.

How have MEMS/NEMS-based balances been used? MEMS/NEMS with various designs and approaches have been used as balances to detect the mass of the species attached to the NEMS structure with mass sensitivities ranging from picograms (10^{-12} g) to zeptograms (10^{-21} g). Self-sensing cantilevers with metallic piezoresistors, having dimensions approaching the mean free path at atmospheric pressure, maintain high resonance quality factors in ambient conditions. This enables measurements in air at room temperature, with an unmatched mass resolution of less than 1 ag (10^{-18} g).

How has a single-electron transistor (SET) been applied in displacement sensing? The single-electron transistor has gained prominence for its exceptional sensitivity to charge. The most outstanding property of SET is the possibility to switch the device from the insulating to the conducting state by adding only one electron to the gate electrode. The elegant charge sensitivity of SET at cryogenic temperatures is exploited to measure displacement by capacitively coupling it to the mechanical resonator (Knobel and Cleland 2003). The detection of the piezoelectrically induced charge in a mechanical resonator with a SET has been shown to be a prime candidate for nearly quantum-limited displacement sensing. If instead, the resonator is fabricated from a piezoelectric material, such as GaAs, AlGaAs, or AIN, and the SET is configured to sense the piezoelectric voltage developed when the beam flexes, then a considerably higher displacement sensitivity can be achieved.

What are the applications of CNTs for determining mechanical parameters? CNTs act as sensing materials in force, pressure, flow, mass, position, stress, and strain sensors. (i) A tunable doubly clamped CNT electromechanical oscillator executes guitar string–like oscillation modes. Its resonance frequency can be widely tuned and the devices can be used to *transduce very small forces.* (ii) *Atomic-scale mass sensing* has been accomplished using doubly clamped suspended CNT nanomechanical resonators, in which their SET properties allow self-detection of the nanotube vibration. The detection of shifts in the resonance frequency of the nanotubes has been applied to sense and determine the inertial mass of atoms as well as the mass of the nanotube. (iii) The band structure of a CNT is drastically altered by *mechanical strain* leading to a strain nanosensor. (iv) The rate of *fluid flow* around a CNT sensor affects the heat transfer between the CNT sensor and the environment. In turn, this heat transfer affects the power required to heat the CNT sensor to a specific temperature; hence, a CNT sensor carefully biased to self-heating is used for fluid-flow rate sensors (Chow et al. 2010). (v) CNT transistors, incorporated into microfluidic channels, locally sense the change in the electrostatic potential induced by the flow of an ionic solution (Bourlon et al. 2007).

Example 15.1

A cantilever having a spring constant of 0.01 N m^{-1} and a resonance frequency of 1 MHz vibrates with an amplitude of 3×10^{-6} m at 27°C. The quality factor is 10^4 and the bandwidth is 1 Hz. Calculate: (a) the minimum detectable frequency change; (b) the minimum detectable force; and (c) the mass resolution achievable with this structure.

(a) The frequency shift is given by

$$\delta\omega = \left(\frac{1}{A}\right)\left(\frac{\omega_0 k_B T B}{kQ}\right)^{0.5} \tag{15.1}$$

where

A is the amplitude of the oscillation
ω_0 is the resonance frequency
k_B is the Boltzmann constant
T is the temperature
B is the bandwidth
k is the spring constant of the cantilever
Q is the quality factor of the cantilever

Here, $A=3\times10^{-6}$ m, $\omega_0=2\pi\times10^6$ Hz$=6.28\times10^6$ Hz, $k_B=1.381\times10^{-23}$ J K^{-1}, $T=27+273=300$ K, $B=1$ Hz, $k=0.01$ N m^{-1}, and $Q=10^4$.

$$\delta\omega = \left(\frac{1}{3\times10^{-6}}\right)\left(\frac{6.28\times10^6\times1.381\times10^{-23}\times300\times1}{0.01\times10^4}\right)^{0.5} = 5.3767\times10^{-3}\ \text{s}^{-1} \tag{15.2}$$

The minimum detectable angular frequency shift is 5.3767×10^{-3} Hz.

(b) The minimum force that can be detected by a cantilever is limited by thermomechanical fluctuations to

$$F_{\text{minimum}} = \left(\frac{2kk_B TB}{\pi Q f_0}\right)^{0.5} \tag{15.3}$$

where

k_B is Boltzmann's constant$=1.381\times10^{-23}$J K^{-1}
T is the temperature$=300$ K
B is the detection bandwidth$=1$ Hz
$k=0.01$ N m^{-1} is the cantilever spring constant
$f_0=10^6$ Hz is the resonance frequency
$Q=10^4$ is the quality factor

Inputting the values in Equation 15.3, we get

$$F_{\text{minimum}} = \left(\frac{2kk_B TB}{\pi Q f_0}\right)^{0.5} = \left(\frac{2\times0.01\times1.381\times10^{-23}\times300\times1}{3.14\times10^4\times10^6}\right)^{0.5} = 5.137\times10^{-17} = 51.37\,\text{aN} \tag{15.4}$$

(c) The expected mass resolution is

$$\delta m = k\left(\frac{1}{\omega_1^2}-\frac{1}{\omega_2^2}\right) \tag{15.5}$$

where

ω_1 is the initial resonance frequency of the cantilever = 6.28×10^6 Hz

ω_2 is the final resonance frequency of the cantilever = $6.28 \times 10^6 + 6.28 \times 5.3767 \times 10^{-3} = 6.28 \times 1.0000000053767 \times 10^6$ Hz

Therefore,

$$\delta m = k\left(\frac{1}{\omega_1^2} - \frac{1}{\omega_2^2}\right) = \frac{0.01}{(6.28)^2}\left\{\frac{1}{(1\times10^6)^2} - \frac{1}{(1.0000000053767\times10^6)^2}\right\}$$

$$= \frac{0.01}{(6.28)^2}\left(1\times10^{-12} - 9.999999892466\times10^{-13}\right)$$

$$= 2.54\times10^{-4}\times1.07534\times10^{-20}\ \text{kg} \tag{15.6}$$

$$= 2.731\times10^{-24}\ \text{kg}$$

Example 15.2

An Al/Al_2O_3/Al SET is fabricated. For optimal performance of the SET, the total island capacitance must be minimized, that is, the area of the tunnel junctions must be as small as possible. Taking the total central island capacitance as 200×10^{-18} F, what is the operating temperature of the SET device? ($k_B = 1.381 \times 10^{-23}$ J K^{-1}).

For a SET operation based on a Coulomb blockade, the relation:

$$k_B T \ll \frac{q^2}{2C} \tag{15.7}$$

must be satisfied, that is,

$$T \ll \frac{q^2}{2k_BC} = \frac{\left(1.6\times10^{-19}\right)^2}{2\times1.381\times10^{-23}\times200\times10^{-18}} = 4.634\ \text{K} \tag{15.8}$$

Therefore, to ensure reliable operation, the temperature must be <1 K.

Example 15.3

Find the resonance frequency of a doubly clamped beam resonator made of gallium arsenide having the following dimensions: length = 8 μm, width = 600 nm, and thickness = 259 nm, given that the Young's modulus (E_{GaAs}) of gallium arsenide is 85.5 GPa and its density (ρ_{GaAS}) is 5317 kg m^{-3}. If E_{Si} = 185 GPa and ρ_{Si} = 2328 kg m^{-3}, will the resonance frequency of a silicon beam be lower or higher than that for gallium arsenide?

What is the ratio of the resonance frequencies for the two materials? The fundamental resonance frequency, *f*, of a doubly clamped beam of length, *L*, and thickness, *t*, varies linearly with the geometric factor t/L^2 according to the simple relation:

$$f = 1.03\sqrt{\frac{E}{\rho}}\left(\frac{t}{L^2}\right) \tag{15.9}$$

where

E is Young's modulus
ρ is the mass density

Note that the beam width (*w*) does not enter the calculations.

1 Pa = 1 N m^{-2}. Hence, E_{GaAs} = 85.5 GPa = 85.5×10^9 N m^{-2} = 8.55×10^{10} N m^{-2} and E_{Si} = 185 GPa = 185×10^9 N m^{-2} = 1.85×10^{11} N m^{-2}. From Equation 15.9, the resonance frequency of a silicon beam is

$$f_{Si} = 1.03\sqrt{\frac{E_{Si}}{\rho_{Si}}}\left(\frac{t}{L^2}\right) = 1.03\sqrt{\frac{1.85\times10^{11}}{2328}}\left\{\frac{\left(259\times10^{-9}\right)}{\left(8\times10^{-6}\right)^2}\right\}$$

$$= 1.03\times8.9144\times10^3\times4.047\times10^3 = 3.716\times10^7 = 3.72\times10^7\ \text{Hz} \tag{15.10}$$

while that of a GaAs beam is

$$f_{Si} = 1.03\sqrt{\frac{E_{GaAs}}{\rho_{GaAs}}}\left(\frac{t}{L^2}\right) = 1.03\sqrt{\frac{8.55\times10^{10}}{5317}}\left\{\frac{\left(259\times10^{-9}\right)}{\left(8\times10^{-6}\right)^2}\right\}$$

$$= 1.03\times4.01\times10^3\times4.047\times10^3 = 1.672\times10^7\ \text{Hz} \tag{15.11}$$

Silicon gives a higher resonance frequency than gallium arsenide because of its larger (Young's modulus/density) ratio. The ratio of the resonance frequencies for the two materials is given by

$$\frac{f_{Si}}{f_{GaAs}} = \frac{1.03\left(t/L^2\right)\sqrt{E_{Si}/\rho_{Si}}}{1.03\left(t/L^2\right)\sqrt{E_{GaAs}/\rho_{GaAs}}}$$

$$= \frac{\sqrt{E_{Si}/\rho_{Si}}}{\sqrt{E_{GaAs}/\rho_{GaAs}}} = \frac{8.9144\times10^3}{\left(4.01\times10^3\right)} = 2.223 \tag{15.12}$$

which is the same as given by the ratio of Equations 15.10 and 15.11, that is, $3.716\times10^7/(1.672\times10^7) = 2.222$.

15.5 Thermal Nanosensors

A temperature nanosensor was made by electron-beam chemical vapor deposition (EB-CVD) using Pt source gas material. A thermal nanosensor with a 3-D nanostructure was made by fluidized bed chemical vapor deposition (FIB-CVD) using $W(CO)_6$ source gas material (Ozasa et al. 2004).

How have CNTs been exploited as temperature sensors? (i) CNTs grown on nickel film deposited on a float glass substrate serve as low-temperature sensors. (ii) A laterally grown CNT between two electrodes is used as a nano-temperature sensor. *Have carbon nanowires (NWs) been used as temperature sensors?* Yes, temperature sensors have been developed using an array of carbon NWs written by a Ga^+-focused ion beam. The NW technology eliminates the difficulty in the selective growth or placement of a CNT. *How is a CNT-based infrared sensor constructed?* The infrared photoresponse in the electrical conductivity of single-walled carbon nanotubes (SWCNTs) is greatly enhanced by embedding SWCNTs in an electrically and thermally insulating polymer matrix, enabling the construction of an infrared sensor.

What is the status of silicon nanowires (Si NWs) as thermal sensors? Silicon NWs are employed as nano-temperature sensors in two configurations: (i) a resistance temperature detector (RTD) and (ii) a diode temperature detector (DTD). Si NWs reap the benefits of nanoelectronics technology with regard to geometrical patterning by nanolithographic techniques with the attendant precise location advantage.

How are NPs utilized as temperature sensors? (i) A ratiometric fluorescent NP has been reported for temperature sensing using an alkoxysilanized dye as a reference. Er^{3+}-doped $BaTiO_3$ nanocrystals are suitable for use as far-infrared (FIR)-based temperature nanosensors. (ii) Gd_2O_3: Er^{3+}/Yb^{3+} nanocrystalline phosphor has been used as a temperature sensor.

What are the promises of bolometers with reference to a superconducting transition edge nanosensor? Since their original invention by Samuel P. Langley in 1878, bolometers have gone a long way in improving the sensitivity and expanding the frequency range, from x-rays and optical/ultraviolet (UV) radiation to sub-millimeter waves. The latter range contains approximately half the total luminosity of the universe and 98% of all the photons emitted since the Big Bang started the universe. Because the performance of ground-based terahertz (THz) telescopes is greatly limited by the strong absorption of THz radiation in the terrestrial atmosphere, the development of space-based THz telescopes will be crucial in better understanding the evolution of the universe. Although new detector concepts are now coming into play, bolometers still have great potential to achieve the most challenging goals. The notion of improving the sensitivity of bolometers by employing hot-electron effects at ultralow temperatures has been enthusiastically pursued (Wei et al. 2008). The hot-electron direct detector (HEDD) element is a superconducting transition edge nanosensor made from a thin Ti film with a superconducting transition temperature of $T_C \sim 0.2 - 0.4$ K. The current leads to the nanosensor are fabricated from Nb films with $T_C \sim 8.5$ K; a large superconducting gap in Nb blocks the outdiffusion of "hot electrons" to the current leads. The nanostructure is fabricated on a silicon substrate using electron-beam lithography and electron-gun deposition of Ti and Nb. The detector has a low noise equivalent power (NEP) $= 3 \times 10^{-19}$ W Hz$^{-1/2}$ at 0.3 K, and is capable of reckoning THz photons.

15.6 Optical Nanosensors

What is the main refractive index change strategy used in optical nanosensors? The localized surface plasmon resonance (LSPR) nanosensors induce small local refractive index changes at the surface of metallic NPs and function by transducing small changes in the refractive index near the metallic surface into a measurable wavelength shift response (Barbillon et al. 2007). *What are the prospects of surface-enhanced Raman scattering (SERS)-based nanosensing?* Compared with the refractive index–based detection schemes, SERS is a vibrational spectroscopic method that yields unique vibrational signatures for small molecule analytes, as well as quantitative information (Kneipp et al. 2010). The SERS signal transduction mechanism has many characteristics that can be utilized in biosensing applications: sensitivity, selectivity, low laser power, and no interference from water molecules. Examples are the trace analysis of DNA, bacteria, glucose, living cells, and the posttranslational modification of proteins, enzyme, chemical warfare agents, and CNTs. A miniaturized, inexpensive, and portable Raman instrument makes the technique practical for trace analysis in clinics, field, and urban settings.

Nanobiotechnology is a new word to describe the use of nanotechnology in biological systems. *How does the colloidal gold nanobiosensor work?* A colorimetric gold NP sensor for probing biomolecular interactions works on the change in the absorbance spectrum of a self-assembled monolayer of colloidal gold on glass, as a function of biomolecular binding to the surface of the immobilized colloids.

What are the capabilities of optical fiber nanobiosensors? Fiber-optic nanobiosensors are an inimitable class of biosensors that enable analytical measurements in individual living cells and the probing of individual chemical species in specific locations within a cell. Fiber-optic nanobiosensors consisting of antibodies, as biorecognition molecules, coupled to an optical transducer element, have been developed and used to detect biochemical targets, benzopyrene tetrol (BPT), and benzo[a]pyrene (BaP), inside single cells.

How are optical fibers utilized for optochemical sensing applications? Standard silica optical fibers (SOFs) coated with SWCNTs Langmuir–Blodgett multilayers are used as volatile organic compound (VOC) sensors based on light reflectometry, that is, through film reflectance–induced changes. Metal oxides (MOX) are another interesting class of materials widely exploited in gas sensing applications. The term "reflectometric configuration" is generally used to refer to the standard extrinsic Fabry–Perot (SEFP) configuration, which relies on reflectance measurements with low finesse and an extrinsic Fabry–Perot interferometer, realized by depositing a flat thin sensitive layer onto the distal end of properly cut optical fibers. Nano-coated, long-period fiber gratings, modified Fabry–Perot interferometers involving near field effect, and photonic band-gap modifications in hollow-core optical fibers have been proposed, respectively, to provide the best sensing performance.

How does the fluorescence-based pH nanosensor operate? A ratiometric pH nanosensor contains a pH-sensitive fluorescent dye as well as a pH-insensitive reference dye, embedded in a polymer matrix. A charged fluorescent molecule, for example, a reporter dye, is trapped and concentrated in solution at the nanopipette tip. The dye is excited by focusing a laser beam using far-field optics at the 100 nm inner diameter pipette tip, producing a local nanosensor where the fluorescence is dependent on the analyte concentration in the bath. SNARF-l-dextran, a negatively charged ratiometric pH-sensitive fluorophore, is used to demonstrate the concept of the nanopipette as a pH sensor.

What are the vital attributes of photonic explorers for bioanalysis with biologically localized embedding (PEBBLES)? PEBBLEs are nanometer-sized spherical optical sensing devices made from biologically inert polymers and are capable of *in vitro* measurements (Sumner et al. 2006). These sensors are fabricated in a microemulsion and consist of fluorescent indicators entrapped in a polyacrylamide matrix. The sensing mechanism involves the permeation of analytes into the NP matrix and their selective interaction with sensing components, resulting in signal changes. The fluorescent indicators and reference dyes are loaded into the polymeric NP core or its surface-coated layer and there is no interaction between the dyes. Such NP sensors have been developed for H^+, Ca^{2+}, and Mg^{2+} ions, OH radicals, small molecules such as O_2, etc.

What is the electrochemiluminescent (ECL) nanosensing approach? In the electrochemiluminescent nanosensor, ECL light is initiated by the gold nano-ring electrode in the presence of a co-reactant biospecies, the reduced form of nicotinamide adenine dinucleotide represented as NADH, meaning nicotinamide adenine dinucleotide + hydrogen (H) (Chovin et al. 2006).

How does the UV nanosensor work? Naturally self-assembled crossed ZnO nanorods exhibit a response of ~15 mA W^{-1} for UV light (361 nm) under 1 V bias (Chai et al. 2009).

What is the impact of QD nanosensors in the sensing arena? Compared to conventional organic dye molecules, fluorescent semiconductor nanocrystals (QDs) have several promising advantages. A few examples are given as follows: (i) quantitative maltose sensing by QD-FRET has shown the *modus operandi* by which QDs might play a role in enzyme assays; (ii) several studies have exploited QD-FRET for imaging the activity of proteases; (iii) individually, FRET-based QD bioprobes are able to detect the actions of protease, deoxyribonuclease, and DNA polymerase, or changes in pH; additionally, (iv) two such QD-mounted biosensors were excited at a single wavelength and were shown to operate simultaneously and independently of each other in the same sample solution. Thus, multiplex detections of the action of the protease, trypsin, were possible in the presence of deoxyribonuclease.

Example 15.4

Surface plasmon resonance measures the change in the index refraction of the fluid medium near the sensor surface. For an incident light of wavelength 635 nm, what is the difference in the wavelength of the emergent beams from two protein binding layers of refractive indices 1.5 and 1.515?

If the velocity of light in a medium is v and the same in vacuum is c, the relationship between the wavelength $\lambda = v/f$ in a medium and the wavelength $\lambda_0 = c/f$ in vacuum is

$$\frac{\lambda}{\lambda_0} = \frac{v/f}{c/f} = \frac{v}{c} \tag{15.13}$$

The frequencies cancel because frequency does not change as light moves from one medium to another. Because the refractive index n is defined as the ratio of the velocity of light in a vacuum to that in a medium, Equation 15.13 reduces to

$$\frac{\lambda}{\lambda_0} = \frac{1}{n} \tag{15.14}$$

Hence,

$$\lambda = \frac{\lambda_0}{n} \tag{15.15}$$

The refractive index n in this equation measured with respect to vacuum is called the *absolute refractive index*. In practice, air makes little difference to the refraction of light with an absolute refractive index of 1.0008, so the value of the absolute refractive index can be used assuming the incident light is in air. So, Equation 15.15 can be written for the first protein binding layer as

$$\lambda_{\text{protein1}} = \frac{\lambda_{\text{air}}}{n_{\text{protein1}}} = \frac{635}{1.5} = 423.33 \text{ nm} \tag{15.16}$$

For the second protein layer, Equation 15.15 is rewritten as

$$\lambda_{\text{protein2}} = \frac{\lambda_{\text{air}}}{n_{\text{protein2}}} = \frac{635}{1.515} = 419.14 \text{ nm} \tag{15.17}$$

Thus, the required difference in wavelengths is

$$\Delta\lambda = \lambda_{\text{protein1}} - \lambda_{\text{protein2}} = 423.33 - 419.14 = 4.19 \text{ nm} \tag{15.18}$$

15.7 Magnetic Nanosensors

How has the discovery of the giant magnetoresistance (GMR) phenomenon profoundly influenced our lives? The 2007 Nobel Prize in Physics is the global recognition of the rapid development of giant magnetoresistance, from both the physics and the engineering points of view. The use of GMR can be regarded as one of the first major applications of nanotechnology (Freitas et al. 2007). GMR garnered prolific interest from researchers in fundamental physics as well as from industry. Applications of this phenomenon have revolutionized techniques for retrieving data from hard disks. Behind the utilization of GMR structures as read heads for massive storage magnetic hard disks, breakthrough applications as solid-state magnetic nanobiosensors have emerged. The GMR nanosensor chip searches for up to 64 different proteins concomitantly and has been shown to be effective in the early detection of tumors in mice, suggesting that it may open the door to much earlier detection of even the most elusive cancers in humans.

What is the significance of tunneling magnetoresistance (TMR)? Magnetic tunnel junction (MTJ) sensors, which use the tunneling magnetoresistance effect, are quickly becoming the technology of choice for many magnetic sensor applications.

How do magnetic NPs act as magnetic relaxation switches (MRSws)? Magnetic NPs serve as MRSws, switching from a dispersed to a clustered state, or the reverse, due to the presence of molecular targets, with changes in the spin–spin relaxation time of water (*T*2), which can be easily detected by nuclear magnetic resonance (NMR) or magnetic resonance imaging (MRI) instrumentation (Perez et al. 2004). MRSw assays are designed to cause self-assembly of magnetic NPs upon the addition of molecular targets (forward switching, decreasing *T*2) or the disassembly of preformed clusters by enzymatic cleavage or competitive binding (reverse switching, increasing *T*2).

Forward MRSw assays are based on cross-linking magnetic NPs into clusters using molecular target bridges. Thus, they are preferably suited for detecting small molecule analytes such as drugs, metabolites, oligonucleotides, and proteins because the short cross-links guarantee that the magnetic NPs are placed in close enough proximity to encourage relaxation switching. In addition, there is no need for time-consuming separation or capture strategies because MRSw assays can be performed in turbid solutions such as blood; the removal of unbound magnetic NPs is not required. Interestingly, increasing the valency of a target by coupling to a protein or microparticle carrier has been shown to improve detection sensitivity by a factor greater than the increase in valency itself. This result is due to the evidence of higher valency targets being more effective in promoting NP clustering.

For *reverse MRSw assays*, NP clusters are first formed analogously to a forward assay before the addition of an enzyme that cleaves the molecular bridges at specific sites or a competitive binding molecule that destabilizes the cross-links.

How has magnetic NP labeling been used with cell surface markers? Magnetic NPs are used to tag cell surface markers so as to impart a magnetic moment to the cell that is proportional to the number of NPs bound. However, this method requires the removal of excess magnetic NPs preceding the measurement of *T*2 relaxation time. But this is readily accomplished by centrifugation or filtration. The magnetic tagging concept has been commonly employed for high-contrast MR imaging.

How has DNA hybridization been detected by magnetic nanosensors? Superparamagnetic crystalline Fe_2O_3/Fe_3O_4 particles, caged in epichlorohydrin cross-linked dextran, have been used as nanosensors in DNA hybridization experiments. The underlying principle of this approach is that the presence of a particular DNA sequence brings the particles coated with the complementary probes in close proximity, leading to an amply enhanced spin–spin relaxation of the resulting clusters.

How has peroxidase enzyme activity been sensed? Magnetic NP conjugates act as sensitive and selective nanosensors for peroxidase enzyme activity. The immobilization of dopamine or serotonin to the magnetic NPs has allowed for sensitive and direct detection by MRI of horseradish peroxidase and myeloperoxidase activity, respectively.

Example 15.5

The longitudinal/transverse relaxivity r_1 or r_2 of a maghemite contrast agent is the increase in the longitudinal/transverse relaxation rate observed in an aqueous solution when the concentration in an active compound (such as iron or gadolinium) is increased by 1 mM unit. These quantities allow the comparison between different contrast agents and reflect their efficiency: the higher the relaxivities are, the smaller the quantity of agent to be injected into the patient. AMI-25 (Endorem), a small particle of iron oxide (SPIO > 40 nm) has $r_1 = 9.95\ s^{-1}\ mM^{-1}$, $r_2 = 158\ s^{-1}\ mM^{-1}$ with an r_2/r_1 ratio = 15.88, while NC100150 (Clariscan), an ultrasmall particle of iron oxide (USPIO < 40 nm) has $r_1 = 24.0\ s^{-1}\ mM^{-1}$, $r_2 = 36.4\ s^{-1}\ mM^{-1}$ with an r_2/r_1 ratio = 1.52. For what applications are these NPs suitable and why?

It is obvious that the transverse relaxivity of superparamagnetic contrast agents is far greater than their longitudinal relaxivity. For this reason, they are used mainly for *T*2-weighted imaging. SPIOs are characterized by a high r_2/r_1 ratio. However, USPIOs present a lower r_2/r_1 ratio and are therefore better suited for *T1*-weighted imaging.

15.8 Chemical Nanosensors

Chemical sensors have been developed over decades to detect gases and vapors at various concentration levels for deployment in a wide range of industrial applications. *In what ways was the hydrogen nanosensor better than the corresponding macrosensor?* A hydrogen sensor fabricated using Si NWs modified by palladium (Pd) NPs showed a superior sensitivity to hydrogen and a faster response time than the macroscopic Pd wire hydrogen sensor (Chen et al. 2007).

How are CNTs used in gas detection? A sensor platform consisting of an interdigitated electrode (IDE) pattern has been fabricated for sensing gas and organic vapors. Purified SWCNTs in the form of a network laid on an IDE by solution casting serve as the sensor material. The electrical conductivity of the SWCNT network changes reproducibly upon exposure to various gases and vapors. Selectivity to specific gases, e.g., chlorine and hydrochloric acid vapor, was achieved by coating the SWCNTs with polymers such as chloro-sulfonated polyethylene and hydroxypropyl cellulose (Li et al. 2006).

What is the role of metal oxide NPs in gas sensors? A semiconductor gas sensor made with a porous assembly of tiny crystals of an N-type metal oxide semiconductor, typically SnO_2, In_2O_3, or WO_3, often loaded with a small amount of a foreign substance (noble metals or their oxides) called a *sensitizer*, possesses an electrical resistance. When operated at adequate temperatures in air (200°C–500°C), the resistor changes its resistance sharply on contact with a small concentration of a reducing gas or an oxidizing gas, enabling us to know the gas concentration from the resistance change. Enhancement in the selectivity and the overall efficiency of the sensors is achieved by tailoring the size, structure, and shape of the NPs.

What are the physicochemical models of gas sensors? Gas elements are prepared in polycrystalline form, and in the process, form large numbers of grain boundaries and necks. The model that combines the neck and the grain boundary mechanisms illustrates the effects of neck and grain boundaries on sensitivity. The gas sensitivity increases with decreasing grain size. The non-agglomerated necking of the NPs induced by heat treatment significantly enhances the gas sensing characteristics of the NP-based gas sensors. A model of the receptor function and the response of small nanosize semiconductor crystals was formulated by using the chemical parameters of the *gases side*, such as partial pressure, adsorption constant, and rate constant, and the physical parameters of the *semiconductor side*, for example, shape and size, donor density, and Debye length.

How are CNTs used in ion-sensitive field-effect transistors (ISFETs)? A CNT-based nano ISFET was fabricated by testing the CNT *I-V* characteristic to verify if the CNT is metallic or semiconducting. The metallic CNTs are used as NWs connecting the source and the drain, while the semiconducting ones are used for nanotransistors. *What are the other possible pH nanosensors?* Si NWs have been used as a sensing layer in an extended-gate field-effect transistor (EGFET) for the measurement of solution pH. Compound semiconductor/isolator (ZnS/silica) core/shell nanocables have been used to fabricate single NW–based FETs. Following chemical modification, amine- and oxide-functionalized nanocables exhibit linear pH-dependent conductance.

How is the moisture content of air detected by a nanosensor? By coating one side of the surface of a ZnO nanobelt with multilayer polymers using an electrostatic self-assembling process, a humidity/chemical nanosensor based on the piezoelectric field-effect transistor (PE-FET) principle was demonstrated (Lao et al. 2007).

How do quantum dots (QDs) help in measuring the concentrations of toxic ions such as antimony and mercury? A bovine serum albumin (BSA)-activated CdTe QD nanosensor for antimony was reported. When an antimony ion enters the BSA, the QD emission is switched off (Ge et al. 2010). Luminescent and stable CdSe/ZnS core/shell QDs capped with L-carnitine were prepared for optical determination of mercury ions in ethanol via analyte-induced changes in the photoluminescence of the QDs (Li et al. 2008).

15.9 Nanobiosensors

What makes QCM a useful tool in biosensing? Utilizing its ability to simultaneously detect mass and viscoelastic property changes, the QCM with dissipation monitoring (QCM-D) is the supreme tool to study biological interactions. Oligonucleotide immobilization is followed with the QCM system by measuring the frequency shifts of the crystal, which was 126 ± 12 Hz under optimal condition for a mass deposition of 380 ng cm^{-2} in a study carried out by Duman et al. 2003 for developing a nucleic acid sensor. This frequency shift is sufficient for sensitive measurements.

What is the role of gold NPs in electrochemical nanosensors? Electrochemical sensing (ES) techniques are proliferating components in various fields in which an accurate, low-cost, fast, and online measuring system is indispensable. Electrochemical biosensors created by coupling biological recognition elements with electrochemical transducers based on or modified with gold NPs have carved a niche for themselves. A few reasons explaining why gold NPs have aroused interest are as follows: (i) AuNPs provide a stable surface for the immobilization of biomolecules with the retention of their biological activities, probably due to enhanced orientational freedom. This perception is extremely useful when preparing biosensors. (ii) Moreover, gold NPs allow direct electron transport between redox proteins and bulk electrode materials. Thus, they permit ES to be performed without the need for electron transfer mediators. (iii) The desirable characteristics of gold NPs, such as their high surface-to-volume ratio, their large surface energy, their capability to decrease the distance between proteins and metal particles, and their ability to serve as electron-conducting pathways between prosthetic groups and the electrode surface, facilitate electron transfer between redox proteins and the electrode surface. (iv) The routing of electrons from redox proteins, particularly redox enzymes, to electrodes has been the subject of wide-ranging research. It is possible to electrically contact redox enzymes (GO_x) with their macroscopic environment by reconstituting an apo-flavoenzyme, apoglucose oxidase, on a 1.4 nm gold nanocrystal functionalized with the cofactor flavin adenine dinucleotide.

How is DNA detection through AuNP labels enhanced? Gold NP labels, coupled with signal amplification by the silver-enhancement technique, is a suitable choice of detection scheme for enhanced electrochemical detection of DNA.

Has a comparison of gold NP–decorated glucose biosensor electrodes been made? Which AuNP-tinted glucose electrode shows superior response? Comparing different amperometric glucose biosensor electrodes modified with gold NPs, the GO_x/Au_{coll}–Cyst–AuE electrode design showed a sensitivity for glucose determination higher than that achieved with GO_x/Cyst-AuE and GO_x/Au_{coll}–Cyst/Cyst–AuE and similar to that achieved with GO_x/MPA–AuE. Moreover, the useful lifetime of one single GO_x/Au_{coll}–Cyst–AuE was 28 days, remarkably longer than that of the other GO_x biosensor designs (Mena et al. 2005).

How are CNTs helpful in biosensing? For nanobiosensing applications, CNTs have multiple advantages such as a small size with a larger surface area, excellent electron transfer promoting ability when used as an electrodes modifier in electrochemical reactions, and easy protein immobilization with the retention of its activity for potential biosensors. CNTs play an important role in the performance of electrochemical biosensors, immunosensors, and DNA biosensors.

Has CNT been used as a nanoconnector? Yes, aligned reconstitution of a redox flavoenzyme (glucose oxidase) has been done on the edge of CNTs that are linked to an electrode surface. The SWCNT acts as a nanoconnector that electrically contacts the active site of the enzyme and the electrode. The electrons are transported along distances greater than 150 nm and the rate of electron transport is controlled by the length of the SWCNTs.

What is the function of Hb in a modified electrode for H_2O_2? Hb can transfer electrons directly to a normal graphite (GP) electrode when it is modified with QDs (CdS). The Nafion/CdS-Hb/GP electrode is used as a hydrogen peroxide biosensor because of its good bioactivity.

High-impact diseases, including cancer, cardiovascular disease, and neurological disease, are exigent to diagnose without supplementing clinical evaluation with laboratory testing. *How does a saliva test help in cancer detection?* A sensor array chip has been developed for the direct electrochemical detection of the cancer markers (RNA and protein) from saliva associated with oral cancer (Gau and Wong 2007). The sensor assay system relies on the efficient binding of target RNA molecules or proteins onto the sensor surface.

What is the contribution of aptamers in nanobiosensors? With the increasing application of nucleic acid aptamers as a new class of molecular recognition probe in bioanalysis and biosensors, the development of general and simple signaling strategies to transduce aptamer–target binding events to detectable signals is demanding more attention. A new signaling method based on aptamers and a DNA molecular light switching complex, $[Ru(phen)_2(dppz)]^{2+}$, was developed for sensitive protein detection. Aptamer-conjugated QDs are able to specifically target U251 human glioblastoma cells with potential applications such as cancer targeting and molecular imaging. By functionalizing the surface of a QD with aptamers which recognize cocaine, and taking advantage of single-molecule detection and fluorescence resonance energy transfer between 605QD and Cy5 and Iowa Black RQ, a single QD–based aptameric sensor was fabricated that was capable of sensing the presence of cocaine through both signal-off and signal-on modes (Zhang and Johnson 2009).

Example 15.6

Nair and Alam (2006) showed that the average response settling time (t_s) and the minimum detectable concentration (ρ_0) for nanobiosensors and nanochemical sensors obey the relationship:

$$\rho_0 \times t_s^{M_D} \sim k_D \tag{15.19}$$

where M_D and k_D are dimensionality-dependent constants for one-, two-, and three-dimensional nanosensors. Given that $M_D = 0.5$ for planar ISFET and unity for cylindrical NW and nanosphere. Also, $k_D = N_s\sqrt{(2/D)}$ for planar ISFET where N_s is the minimum number of molecules to be captured for detection and D is the diffusion coefficient of biological or chemical target molecules in the solution; k_D is $(N_s a_0/D)$ for cylindrical NW where a_0 is the radius of NW. For 12 base pair (bp) DNA, $D = 4.9 \times 10^{-6}/(\text{bp})^{0.72}$ cm^{-2} s^{-1} at 23°C, $N_s = 10$ μm^{-2}, and $a_0 = 30$ nm. Note that $N_s = 10$ μm^{-2} corresponds to two conjugations on a 1 μm long, 30 nm radius NW. If an experimenter waits for 5 min before recording observations, calculate the minimum detectable concentration of DNA for the planar ISFET and wire nanostructures. Also determine the ratio r_{DL} of detectable concentrations for the two cases. Here,

$$D = 4.9\times10^{-6}/(\text{bp})^{0.72}\ \text{cm}^2\ \text{s}^{-1}$$

$$= 4.9\times10^{-6}/(12)^{0.72}\ \text{cm}^2\ \text{s}^{-1} = 8.1881\times10^{-7}\ \text{cm}^2\ \text{s}^{-1}$$

$$= 8.1881\times10^{-7}\times(10^{-2}\times10^{-2})\text{m}^2\ \text{s}^{-1} = 8.1881\times10^{-11}\ \text{m}^2\ \text{s}^{-1}$$

$$N_s = 10\,\mu\text{m}^{-2} = 10\times10^6\times10^6 = 10^{13}\ \text{m}^{-2}$$

$$a_0 = 30\ \text{nm} = 30\times10^{-9}\text{m} = 3\times10^{-8}\text{m}$$

The detection limit (ρ_0) of a planar ISFET sensor is

$$\rho_0 = \frac{k_D}{t_s^{M_D}} = \frac{N_s\sqrt{2/D}}{t_s^{0.5}} = \frac{N_s\sqrt{2/D}}{(300)^{0.5}} = \frac{10^{13}\times\sqrt{2/(8.1881\times10^{-11})}}{17.321}$$

$$= \frac{1.5629\times10^{18}}{17.321}\ \text{molecules m}^{-3} = 9.02315\times10^{16}\ \text{molecules m}^{-3}$$

$$= 9.02315\times10^{10}\ \text{molecules cm}^{-3} \tag{15.20}$$

Now, 6.022×10^{23} molecules = 1 mol:

$$1\ \text{molecule} = 1/(6.022\times10^{23})\ \text{mole} = 1.661\times10^{-24}\ \text{mol}$$

$$9.02315\times10^{10}\ \text{molecules} = 9.02315\times10^{10}\times1.661\times10^{-24}\ \text{mol}$$
$$= 1.498\times10^{-13}\ \text{mol}$$

Hence, concentration = 1.498×10^{-13} mol cc^{-1} = 1.498×10^{-13} mol/10^{-3} L = 1.498×10^{-10} mol L^{-1} = 0.1498 nM.

The detection limit (ρ_0) of a cylindrical NW sensor is

$$\rho_0 = \frac{k_D}{t_s^{M_D}} = \frac{N_s a_0/D}{t_s^1} = \frac{(10^{13}\times3\times10^{-8})/(8.1881\times10^{-11})}{300}$$

$$= \frac{3.664\times10^{15}}{300}$$

$$= 1.2213\times10^{13}\ \text{molecules m}^{-3}$$

$$= 1.2213\times10^{7}\ \text{molecules cm}^{-3} \tag{15.21}$$

As before,

$$1.2213\times10^7\ \text{molecules} = 1.2213\times10^7\times1.661\times10^{-24}\ \text{mol}$$

$$= 2.0286\times10^{-17}\text{mol}$$

Hence, concentration = 2.0286×10^{-17} mol cc^{-1} = 2.0286×10^{-17} mol/10^{-3} L = 2.0286×10^{-14} mol L^{-1} = 20.286 fM.

The ratio defined as r_{DL} = detection limit of planar ISFET sensor/detection limit of NW sensor = 0.1498 nM/20.286 fM = $0.1498 \times 10^{-9}/20.286 \times 10^{-15} = 7.384 \times 10^3$.

Example 15.7

In a QD fluorescence quenching analysis, the fluorescence extinction coefficient (K) was 2×10^5 M^{-1}. If the analyte concentration was 10^{-5} M, by what percentage might the fluorescence intensity have fallen?

The extinction of fluorescence takes place according to the Stern–Volmer equation expressed as

$$\frac{I_{max}}{I} = 1 + K\times[X] = 1 + 2\times10^5\times10^{-5} = 3 \tag{15.22}$$

Therefore, $I/I_{max} = 1/3 = (1/3) \times 100\% = 33.33\%$.

15.10 Nanosensor Fabrication Aspects

Looking at the enabling technologies of nanosensors, the obvious question arises: *by what approach have nanosensors been hitherto fabricated*? Although there has been momentous progress in the fabrication of nanostructure-based nanosensors, most of these functional devices and integrated systems are fabricated by a *top-down approach* using a combination of lithography, etching, and deposition. Microcontact printing, imprint lithography, or direct-write dip-pen nanolithography are different nanofabrication techniques currently used to fabricate components with at least one of their dimensions in a scale below 100 nm. Despite several technological and physical limitations, the evolution of classical lithography techniques and other non-standard procedures have been used in different forms to realize nanoscale components with atomic precision.

As already known, the technique for these top-down approaches is complicated, time-consuming, and expensive, which limits their practical applications to a large extent. An appreciation of the kinetics and the thermodynamics of nanostructured materials synthesized by the bottom-up and top-down approaches and their subsequent integration into sensors is an interesting research area.

What are the essential principles of a bottom-up approach? In a bottom-up approach, the focus is on building smaller components into more complex assemblies. *Molecular manufacturing*, that is, the process of assembling nanodevices molecule by molecule, exemplifies a bottom-up approach. Currently, the self-assembly of nanocomponents by *DNA scaffolding* is one of the most promising techniques. It is believed that although top-down integration techniques will predominate at least for one more decade, novel bottom-up procedures, such as the evolution of DNA scaffolding, will be the way to obtain integrated nanodevices with higher complexity. *What is DNA scaffolding?* DNA molecules self-assemble in solution. This occurs via a reaction between a long single strand of viral DNA and a mixture of different short synthetic DNA strands. These short segments act as staples effectively folding the viral DNA into required two-dimensional shapes through complementary base-pair binding. Thus, DNA nanostructures such as squares, triangles, and stars are prepared measuring 100–150 nm on an edge and as thick as the DNA double helix width. Techniques have been developed to orient and position self-assembled DNA shapes and patterns, or "DNA origami," on surfaces that are well suited to semiconductor manufacturing equipment. These correctly positioned DNA nanostructures serve as scaffolds or miniature circuit boards for the precise assembly of computer chip components.

What are the issues regarding NPs? Although some studies of the influence of the shape, size, and distribution of NPs on sensing behavior, response, and recovery are reported, more work in this field is necessary. Studies on the binding of molecules to sensor elements, surface functionalization, and the kinetics of adsorption and desorption of molecules will help in resolving many issues.

What are the problems related to CNTs? The "rediscovery" of CNTs in 1991 was perhaps the most noteworthy event in the short history of nanotechnology, stimulating research into all aspects of CNTs from their manufacture to end use. The number of publications and patents on CNT synthesis is rapidly increasing and deluging the research journals. Still, many challenges remain that need to be addressed (Shanov et al. 2006). *What are these hurdles?* One of them is the production of large-scale and low-cost SWCNTs and multi-wall carbon nanotubes (MWCNT). Currently, research into nanotubes and their applications is hampered by the lack of a suitable technique for manufacturing them in large quantities, which is defined as 10,000 ton per plant per year. Concerning CNT synthesis via fluidized bed chemical vapor deposition, a survey shows that no methodical study of the key parameters has been undertaken (See and Harris 2007). There is no clear understanding of the influence of the key variables (e.g., temperature, pressure, and carbon source) on CNT properties (e.g., CNT diameter, length, and morphology), which shows that further research is necessary to optimize this process and, eventually, comprehend the science behind CNT growth using this technique. Another field of interest comprises the pursuit of controlled CNT growth in terms of selective deposition, orientation, and preselected metallic or semiconducting properties. Some of the long-standing problems in the nanotube area are due to the following reasons: (i) lack of control in the synthesis and chemical processing of SWCNTs; (ii) in chirality control that determines whether a nanotube is metallic or semiconducting; (iii) in diameter control determining the bandgap of a semiconducting SWCNT; and (iv) in the placement and orientation control on large substrates needed for scalable production of nanotube electronics and other devices (Joselevich et al. 2008).

To summarize, currently, there are four main challenges in the field of nanotube synthesis: (i) *mass production*, that is, the development of low-cost, large-scale processes for the synthesis of high-quality nanotubes, including SWCNTs; (ii) *selective production*, that is, control over the structure and electronic properties of the manufactured nanotubes; (iii) *organization*, that is, control over the location and orientation of the produced nanotubes on a flat substrate; and (iv) *mechanism*, that is, the development of a systematic understanding of the processes of nanotube growth. Our understanding of the CNT growth mechanism has been evolving, but more consideration is still required to explain the variety of the observed growth features and experimental results.

15.11 *In Vivo* Nanosensor Problems

One of the key advantages of the small size of nanosensors is their *in vivo* use. *What are the critical issues concerning nanobiosensors to be inserted inside the human body?* For implanted sensors, the major bottleneck is *biofouling* (Drake et al. 2007). This occurs due to the inflammatory response of the body toward any foreign material that is introduced into it. Various researchers have shown that this results in a decrease in the sensitivity of a sensor by almost 80% as compared to sensitivity during *in vitro* use. Various types of lipids, peptides, and proteins are actively adsorbed on the biosensor and diminish its functionality.

Research is still in the process of defining the various adsorbents and possible selective adsorption of an analyte. One of the approaches that scientists are currently exploring is modifying the surface properties by *biocompatible coatings. How do biocompatible coatings function?* These coatings inhibit the binding of non-specific elements and, at the same time, they do not affect the analyte. Different polymers are being used for these applications. These polymers should be biocompatible as well as non-immunogenic. Immunogenicity is the ability of a substance to provoke an immune response in the body.

What are the commonly used biocompatible polymers? The most widely used polymers for biocompatible coatings include polyethylene glycol (PEG), alkane-thiols, poly(vinyl alcohol), chitosan, and poly(acrylamide) for modifying the surface.

15.12 Molecularly Imprinted Polymers for Biosensors

How can the problem of the limited shelf life of biomolecules used in biosensors be solved? Biosensors utilize the advantage of the specific recognition properties of specialized biomacromolecules such as antibodies or enzymes. However, the latter tend to be unstable outside their natural environment. This results in an environmental intolerance and a short shelf life of the sensor. Thus, the biosensor becomes crippled after some period of time as the biomolecule loses its stability.

Synthetic biomimetic receptors such as molecularly imprinted polymers (MIPs) help to overcome these limitations. MIPs are capable of binding target molecules with affinities and selectivities akin to those of antibodies, enzymes, or hormone receptors, while being more stable, easier to prepare, less costly, and easier to integrate into standard industrial fabrication processes than the biomolecules. This is the same as aeroplanes built to mimic birds proved more efficient and powerful than the flying birds themselves, which were the genesis of the idea of a flying machine.

Bompart et al. (2010) combined the specific recognition properties of MIPs with noble metal nanocomposites for signal amplification and optical readout. Based on this combination mode, they fabricated a single-particle chemical nanosensor. They synthesized molecularly imprinted core–shell nanocomposite particles by precipitation and seeded emulsion polymerization. A layer of gold colloids was located between the polymeric core and the MIP shell. This layer resulted in signal enhancement when the particles were used as optical nanosensors. They described SERS measurements of the β-blocking drug (S)-propranolol adsorbed on to single Au-MIP core–shell composite particles of submicrometer size. By Raman measurements on single particles, the target molecule (S)-propranolol was detected after incubation at concentrations as small as 10^{-7} M.

15.13 Applications Perspectives of Nanosensors

Up to this point, we have described the detection mechanisms and the operation of diverse types of nanosensors according to the classification scheme of mechanical, thermal, optical, magnetic, and chemical sensors, and biosensors apart from treating the materials requirements and fabrication facilities for these nanosensors. Another way of looking at the field of nanosensors is from the applications perspective under which each application can be implemented by nanosensors from several categories. From this viewpoint, a relook at nanosensors from the applications aspect is a topic of unending research.

15.13.1 Nanosensors for Societal Benefits

Nanosensors of ever-increasing sensitivity and selectivity are required for air (CO, SO_2, NO_2, ozone, VOCs, ammonia, etc.) and water (pathogens, heavy metals, and pesticides) pollution monitoring.

15.13.2 Nanosensors for Industrial Applications

There is a persistent and ever-increasing demand for nanosensors for the agriculture and food industry, medicine and healthcare, and the automotive and aerospace industries. So, the reliability issues, aging characteristics, drift minimization and sensitivity improvement of these nanosensors will constitute the focal areas of research, and will always be centers of intense activity.

15.13.3 Nanosensors for Homeland Security

A key research area is likely to be nanosensor development for trace explosive sensing such as for TNT, PETN, picric acid, and RDX, and the detection of biological warfare agents, notably anthrax, influenza virus, tuberculosis, etc.

15.14 Interfacing Issues for Nanosensors: Power Consumption and Sample Delivery Problems

Both the continued downscaling of conventional semiconductor electronics and also the next-generation "molecular-based electronics" rely profoundly on a broad array of nanoscience and technology investigations that encompass areas such as physics, chemistry, biology, material science, along with the engineering sciences such as electronics and mechanics, and even the computer sciences. By their very nature, all these multidisciplinary nanoscience and technology endeavors must concern themselves with molecular-level processes. They must therefore incorporate methodologies for interfacing into the microscopic phenomenon.

The advantages of nanofabricated sensors over traditional sensors, namely, low power consumption, self-diagnosis, reliability, and cost reduction, have to be fully exploited by users. One of the most forgotten aspects in the nanoelectronics field is the *problem of wiring.* Nanosensors do not have the necessary efficient interfacing tools. An effective interface to the nanosensor is needed in order to extract the embedded signals. *How does one wire individual nanoelectronic devices within a nano-integrated circuit together? Furthermore, how does one extract and input information from such a circuit, that is, how does one allow it to communicate with the outside world?*

A nanoscale circuit or sensor array relinquishes its advantages of nanoscale integration if the wiring makes the final system just as heavy, bulky, or even larger than conventional systems. *What is the use if a nanosensor consumes pico or nanowatts of power while the power-hungry signal processing circuit uses milliwatts?* There will be a vast compatibility gap separating the two. For nanosensors, data acquisition and control system requirements need to be modified to be in consonance with the nanoscale sensor. Building congruous data processing and power systems is the need of the hour.

What in our imagination is the notion of a nanosensor in actual operation? For a device such as a nanoscale sensor array that could detect chemical and biological species as well as electrical potentials on tens of thousands of different channels, if at all possible, one would visualize a device integrated into a minuscule package that could be introduced into the human body, in the bloodstream or non-invasively in the brain. If physical wires are required to communicate with this sensor array, its connections severely impede the potential advantages of size and mobility In the end, one would like a system for wirelessly interfacing with the sensor array, both to extract information and provide power.

The developments of integrated circuits, which can detect, convert, process, and amplify minute signals, are required from the microelectronic/nanoelectronic community. As such, almost all the foundation work for the next generation of nanoelectronics is intrinsically defining new sensor modalities, while at the same time contributing to the advancement of the traditional capabilities (data and signal processing, computation, and communication) needed for the realization of intelligent sensors and integrated multifunctional sensor systems.

Scientists are looking forward to developing easy-to-operate systems with the ability to interface between nanoscale devices, microsystems, and macrosystems to supplement human analysis of data. The situation can only be managed by bringing novelty and through innovation of new systems. There is a demand for conceptualization and the generation of new nano ideas, their implementation, and taking them to the market place.

Apart from signal interfacing, what other problem arises at the nano–micro interface? Where microtechnology meets nanotechnology, not only do electrical interfacing problems arise, but also the delivery of microliter samples of analytes using microfluidic devices is no longer acceptable because nanoliters of samples are required for which suitable nanofluidic devices must be fabricated. Thus, the transition from the micro to the nanoworld must be envisaged in its entirety and this aspect of interfacing, notably, sample delivery, should not be ignored. The sample quantities used in microfluidics will inundate the nanosensor, which is neither necessary nor conducive to sensor functioning. A very similar situation arose when microsensors were developed. The obvious question is as follows: *if the sensor does not need microliters of blood sample, why should the same be withdrawn from a patient's body?*

15.15 Depletion-Mediated Piezoelectric Actuation for NEMS

What is the concept of depletion-mediated piezoelectric actuation? Masmanidis et al. (2007) investigated the use of a piezoelectric semiconductor in their nanomechanical device. The device consists of a 200 nm-thick epitaxially grown GaAs P-type/intrinsic/N-type (PIN) diode. An N-doped layer serves as the top electrode, and finally, the *charge-depleted high-resistance region* in the middle forms the piezoelectrically active layer. They used NEMS cantilevers to estimate and demonstrate the efficiency of depletion-mediated NEMS piezoelectric excitation. During a classic measurement, an alternating current (AC) signal applied across the PIN junction actuated the device at or near its resonance frequency, while the addition of a direct current (DC) voltage tuned the depletion region width. The nanoscale dimensions of the structure concentrated the electric field across a small width, which imparted excellent voltage sensitivity. Within the device, a transverse electric field, E_z, produced a longitudinal strain.

How does the traditional view change at the nanoscale? The traditional view calls for well-defined, alternating layers of electrodes and piezoelectric materials. However, this view crumbles in nanoscale devices made from semiconductors, where charge depletion smears the boundary between piezoelectrically inactive (electrically conducting) and active (electrically insulating) regions. This spreading effect is only significant when the total device thickness approaches the charge depletion width of the semiconductor. A bending moment was developed when the strain was asymmetrically distributed around the neutral axis of the beam. This bending moment resulted in mechanical resonance under a suitable range of driving frequencies.

The piezoelectric effect was capable of driving the device with AC signals as low as 5 mV before the inception of thermomechanical fluctuations corresponding to, roughly speaking, a single electronic charge on the cantilever itself. *What was the power consumption?* Assuming a maximum current flow of 1 nA at 5 mV of AC drive, the minimum required power consumption of this device approached 5 fW, with ~1 nW being more representative of typical operating conditions (an AC drive of 10 mV) during actuation. The reversibility of the piezoelectric phenomena offered the potential for ultrasensitive electrical measurement of nanomechanical motion.

Thus, they employed epitaxial piezoelectric semiconductors to obtain efficient and fully integrated NEMS actuation, which was based on the exploitation of the interaction between piezoelectric strain and built-in charge depletion. Other NEMS devices needed extrinsic methods of actuation, making the entire system bulky. In addition, they showed that the devices could be tuned in frequency using a DC bias, making them sensitive detectors of electrical potentials. The nanoscale dimensions of the structure concentrated the electric field across a small width, which imparted excellent voltage sensitivity. This tuning effect was linear, unlike other NEMS devices which are nonlinear in response. Finally, since all of the concepts presented by Masmanidis et al. (2007) were extendable to a wide variety of other materials beyond GaAs (such as AlN, SiC, or ZnO), enhanced electrical and mechanical properties are a natural and irresistible consequence of this actuation method.

To put it briefly, when talking about nanosensors, we should relinquish two basic notions that we shall ever use batteries as power sources or think of connecting nanosensors with each other or with other devices using wires, as we are conventionally trained to think from our knowledge of macro/microscopic devices.

15.16 Batteryless Nanosensors

Sustainable development has encouraged the quest for renewable energy sources such as wind, water, and solar energy, which are not depleted by usage. Batteryless nanosensors integrate nanogenerators with sensors and operate by harvesting energy from the environment, thereby avoiding the corrosive chemicals used in batteries. Vibration is an energy source, which is found in plenitudinous supply in everyday life, be it from music, loud sounds, engines, machinery, etc. But most of the energy harvesters derive energy by vibrating in one direction only. Their capability can be increased by simultaneously harvesting energy from multiple directions, e.g., by working with a 3-D TENG device by a hybridization of the contact-separation and sliding modes (Zhang et al. 2014).

15.17 Networking Nanosensors Wirelessly

Currently, electromagnetic communication is the established method for interconnecting nanosensensors with a communication system over the internet to support the development of the internet of nano things, impacting healthcare, agriculture, industry, etc. Considering the extremely large number of nanosensors in envisaged applications, which can transmit in a disorganized manner, collisions between symbols are highly likely, imposing restrictions on the rate of information transmission. Such situations are studied in the stochastic modeling of multi-user interference (Hossain et al. 2019).

Molecular communication is being developed as a viable alternative to electromagnetic communication. Security and privacy issues in molecular communication are being addressed. Existing cryptographic solutions are inappropriate for molecular systems. An example of an attack on a molecular communication channel could be thought of as a time disturbance of an *in vivo* nanosensor system in which the nanosensors are absorbed in the body after a stipulated time period to prevent immune responses (Loscri et al. 2014). A delay in the preset absorption time can result in triggering immune responses while causing absorption to occur before time, which will result in the failure of the nanosensors to accomplish their assigned tasks.

15.18 Discussion and Conclusions

Nanoscience and nanotechnology deal with the study and application of structures of matter with at least one dimension of the order of less than 100 nm. However, *properties related to low dimensions are more important than size.* Nanotechnology is based on the fact that some very small structures usually have new properties and behaviors that are not displayed by bulk matter with the same composition. With the advent of nanotechnology, research is underway to create miniaturized sensors. Miniaturized sensors can lead to reduced weight, batch fabrication, lower power consumption, and low cost.

Research activity in the areas related to nanosensors has seen phenomenal growth in the last 15 years. For well over a decade, nanoscale science and technology (nano-S&T)-related subjects have represented an exponentially growing portion of nearly all scientific and engineering disciplines. Indeed, the possibility of leveraging fundamental mechanisms at the nanoscale and molecular levels through a progressively increasing capability to understand, prescribe, and control all the basic properties (structural, chemical, mechanical, electronic, and photonic) of ultrasmall systems seems to offer fantastic and unlimited new possibilities for the future. In this book, an attempt was made to provide the most contemporary overview possible of nanosensors and their potential applications. Several nanostructures that are currently used in the development of nanosensors: NPs, nanotubes, nanorods, porous silicon, and self-assembled materials, were described, focusing on the type of nanomaterial used and the properties related to the particular nanostructure.

But all that has been said is just a drop in the ocean! Nanosensors promise to surpass microsensors; for example, going a step forward in microarray technology, nanoarrays are being developed based on the interaction between different types of receptors and ligands such as proteins or nucleic acids; approximately 400 nanoarray spots can be placed in the same area as a traditional microarray spot (Nanosensors in environmental analysis 2006: http://www.nanowerk.com/spotlight/spotid=366.php).

"Small is beautiful." Today, in looking at the range of applications that nanosensors find in varied fields of engineering and technology, we know that small is not only beautiful, but also powerful. Nanosensors represent a new frontier in technology, blurring the lines and defining a sort of conjugated system between physics, chemistry, biology, and materials. The essential and desirable features of nanosensors are summarized in the following lines:

All dimensions of nanosensors need not to be to nanoscale,
But they work only where "nanostate properties" prevail:
Quantum-mechanical tunneling, Quantum confinement effect,
Giant magnetoresistance, LSPR, SERS, QD-FRET,...
That is the nanosensor definition,
For everyone's attention!
What if all their dimensions be in nanorange?
Tinier and better but still nanosensor; status unchanged!
Sensitivities of nanosensors lie in the "nanoregime,"
The greater the sensitivity, the higher their esteem.
Their power consumption: nanowatts of power,
Lesser power-hungry devices are need of the hour.
They are batch-fabricated in a large foundry,
Still are low-priced, and almost free...
But interfacing them to the macroworld,
Is everyone's puzzle.
Building efficient nanogenerators is an arduous job,
... Behold, our hearts throb!
For the future, nanosensors have potentials immense,
Much remains unexplored, with lots of suspense,
Frenzy and excitement!
In nanosensor research and development.
Nanosensors: Physical, Chemical and Biological,
Will make us happy, healthy and comfortable.
Bridging ideas into practical realization,
Nanosensors will shrink instrumentation,
And write a new page!
Of glorious nanotechnology age!

The material presented in the chapters of this book was intended to serve only as a foundation course. Detailed information can be found in the excellent references provided at the end of each chapter, which the eager reader is recommended to kindly peruse in earnest.

Review Exercises

15.1 The cantilever structure has been widely used for fabricating nanosensors. Prepare a status review of the use of cantilevers in various types of nanosensors.

15.2 CNTs are sensitive to different physical and chemical properties. Compile a table illustrating the huge variety of nanosensors that utilize CNTs.

15.3 Metallic NPs have been used in nanosensors for performing various tasks. List and elucidate the different roles of NPs in building nanosensors.

15.4 QDs are used in optical nanosensors. They are also used in electrochemical nanosensors. Elucidate these two widely different applications of QDs as nanosensors.

15.5 Describe a model for the functioning of a gas sensor and discuss the influence of NP shape on sensitivity.

15.6 Explain how the gas concentration profile for the diffusion of a gas into a porous sensing material affects its sensing characteristics? Under steady-state conditions, the gas concentration inside the sensing layer decreases with depth, resulting in a diffusion-controlled gas concentration profile. What is the effect of such variation in the penetration of gas in the sensing layer, on the sensitivity of the sensor, and its response transients?

15.7 The Arrhenius equation describes the temperature dependence of the conductivity of a semiconductor. The conductivity is given by

$$\sigma = \sigma_0 \exp\left(-\frac{qV_s}{k_B T}\right) \quad (15.23)$$

where

σ_0 is a pre-exponential factor that includes the bulk intragranular conductance

k_B is Boltzmann's constant

T is the absolute temperature

qV_s is the potential energy barrier at the interface between two neighboring particles

The potential energy barrier is expressed as

$$qV_s = \frac{q^2 N_t^2}{2\varepsilon_0 \varepsilon_s N_d} \quad (15.24)$$

where

N_t is the surface density of adsorbed oxygen ions

ε_0, ε_s are the permittivity of free space and the semiconductor, respectively

N_d is the volumetric density of the electron donors

Apply Equations 10.23 and 10.24 to discuss the effect of particle size on the properties of a gas sensor, especially when the particle size is reduced to nanometers (see Zhang and Liu 2000).

15.8 An experiment is carried out with nanostructured silica shells of diatoms, a type of algae, as follows: laser beams are shone at the shells of *Thalassiosira rotula* in the presence of nitrous oxide, acetone, ethanol, air, xylene, and pyridine. The wavelengths of the light emitted by the shells are measured.

Are these wavelengths the same? Do the shells present slightly different colors depending on the surrounding gas? What is the application implied?

15.9 The pH sensitivity of silicon bulk materials is poor. However, a pH sensitivity of 58.3 mV pH^{-1} was observed for Si NW. Why?

15.10 Is GMR unique to multilayered structures? Can GMR occur in magnetically inhomogeneous media? (See Xiao et al. 1992.)

15.11 How are magnetic particles prevented from self-aggregation? What parameter changes when the particles assemble together to form nanoclusters? How is this property utilized in making a nanosensor?

15.12 Describe a magnetic nanosensor technology that is matrix insensitive, yet capable of rapid, multiplex protein detection with a resolution down to attomolar concentrations and extensive linear dynamic ranges.

15.13 Explain the term "biofouling" in reference to a biosensor. How can it be prevented?

15.14 What is the application of MIPs in biosensors? Give one example of their use.

15.15 The use of conventional approaches such as magnetomotive, electrostatic, and electrothermal techniques for NEMS actuation suffers from either low power efficiency, limited potential for integration, or poor nanoscale control over electromechanical coupling. Explain. How does the piezoelectric effect help in NEMS actuation?

15.16 How does the piezoelectric effect inherent to a material afford a highly efficient means of resonantly exciting NEMS devices with an alternating voltage? What is the concept of tunably coupled NEMS actuation?

15.17 Argue that nanosensors must be wireless or small battery-operated devices to utilize the real benefit of their small size.

REFERENCES

Akyildiz, I. F. and J. M. Jornet. 2010. Electromagnetic wireless nanosensor networks. *Nano Communication Networks* 1: 3–19.

Barbillon, G., J.-L. Bijeon, J. Plain, M. L. de la Chapelle, P.-M. Adam, and P. Royer. 2007. Biological and chemical gold nanosensors based on localized surface plasmon resonance. *Gold Bulletin* 40(3): 240–244.

Bompart, M., Y. D. Wilde, and K. Haupt. 2010. Chemical nanosensors based on composite molecularly imprinted polymer particles and surface-enhanced Raman scattering. *Advanced Materials* 22: 2343–2348.

Bourlon, B., J. Wong, C. Miko, L. Forró, and M. Bockrath. 2007. A nanoscale probe for fluidic and ionic transport. *Nature Nanotechnology* 2: 104–107.

Chai, G., O. Lupan, L. Chow, and H. Heinrich. 2009. Crossed zinc oxide nanorods for ultraviolet radiation detection. *Sensors and Actuators A* 150: 184–187.

Chen, Z. H., J. S. Jie, L. B. Luo, H. Wang, C. S. Lee, and S. T. Lee. 2007. Applications of silicon nanowires functionalised with palladium nanoparticles in hydrogen sensors. *Nanotechnology* 18: 345502 (5pp.), doi: 10.1088/0957-4484/18/34/345502.

Chovin, A., P. Garrigue, G. Pecastaings, H. Saadaoui, and N. Sojic. 2006. Development of an ordered microarray of electrochemiluminescent nanosensors. *Measurement Science and Technology* 17: 1211–1219.

Chow, W. W. Y., Y. Qu, W. J. Li, and S. C. H. Tung. 2010. Integrated SWCNT sensors in micro-wind tunnel for air-flow shear-stress measurement. *Microfluidics and Nanofluidics* 8: 631–640, doi: 10.1007/s10404-009-0495-5.

Drake, C., S. Deshpande, D. Bera, and S. Seal. 2007. Metallic nanostructured materials based sensors. *International Materials Reviews* 52(5): 289–317.

Duman, M., R. Saber, and E. Pi kin. 2003. A new approach for immobilization of oligonucleotides onto piezoelectric quartz crystal for preparation of a nucleic acid sensor for following hybridisation. *Biosensors and Bioelectronics* 18(11): 1355–1363.

Ekinci, K. L. 2005. Electromechanical transducers at the nanoscale: Actuation and sensing of motion in nanoelectromechanical systems (NEMS). *Small* 1(8–9): 786–797.

Freitas, P. P., R. Ferreira, S. Cardoso, and F. Cardoso. 2007. Magnetoresistive sensors. *Journal of Physics: Condensed Matter* 19: 165221, doi: 10.1088/0953-8984/19/16/165221.

Gau, V. and D. Wong. 2007. Oral fluid nanosensor test (OFNASET) with advanced electrochemical-based molecular analysis platform. *Annals of the New York Academy of Sciences* 1098: 401–410, doi: 10.1196/annals.1384.005.

Ge, S., C. Zhang, Y. Zhu, J. Yu, and S. Zhang. 2010. BSA activated CdTe quantum dot nanosensor for antimony ion detection. *Analyst* 135: 111–115.

Giessibl, F. J. 2003. Advances in atomic force microscopy. *Reviews of Modern Physics* 75: 949–983.

Hossain, Z., C. N. Mollica, J. F. Federici, and J. M. Jornet. 2019. Stochastic interference modeling and experimental validation for pulse-based terahertz communication. *IEEE Transactions on Wireless Communications* 18(8): 4103–4115.

Joselevich, E., H. Dai, J. Liu, K. Hata, and A. H. Windle. 2008. Carbon nanotube synthesis and organization. *Topics in Applied Physics* 111: 101–164.

Khanna, V. K. 2008. Nanoparticle-based sensors. *Defence Science Journal:* Special Issue on Nanomaterials: Science and Technology-II, 58(5): 608–616.

Khanna, V. K. 2009. Frontiers of nanosensor technology. *Sensors and Transducers Journal* 103(4): 1–16.

Kneipp, J., B. Wittig, H. Bohr, and K. Kneipp. 2010. Surface-enhanced Raman scattering: A new optical probe in molecular biophysics and biomedicine. *Theoretical Chemistry Accounts* 125: 319–327.

Knobel, R. G. and A. N. Cleland. 2003. Nanometre-scale displacement sensing using a single electron transistor. *Nature* 424: 291–293.

Lao, C. S., Q. Kuang, Z. L. Wang, M.-C. Park, and Y. Deng. 2007. Polymer functionalized piezoelectric-FET as humidity/chemical nanosensors. *Applied Physics Letters* 90: 262107-1–262107-3.

Li, J., Y. Lu, and M. Meyyappan. 2006. Nano chemical sensors with polymer-coated carbon nanotubes. *IEEE Sensors Journal* 6(5): 1047–1051.

Li, H., Y. Zhang, X. Wang, and Z. Gao. 2008. A luminescent nanosensor for Hg(II) based on functionalized CdSe/ZnS quantum dots. *Microchimica Acta* 160: 119–123, doi: 10.1007/s00604-007-0816-x.

Loscri, V., C. Marchal, N. Mitton, G. Fortino, and A. V. Vasilakos. 2014. Security and privacy in molecular communication and networking: Opportunities and challenges. IEEE Transactions on NanoBioscience 13(3): 198–207.

Masmanidis, S. C., R. B. Karabalin, I. D. Vlaminck, G. Borghs, M. R. Freeman, and M. L. Roukes. 2007. Multifunctional nanomechanical systems via tunably coupled piezoelectric actuation. *Science* 317: 780–782.

Mena, M. L., P. Yáñez-Sedeño, and J. M. Pingarrón. 2005. A comparison of different strategies for the construction of amperometric enzyme biosensors using gold nanoparticle-modified electrodes. *Analytical Biochemistry* 336: 20–27.

Nair, P. R. and M. A. Alam. 2006. Performance limits of nanobiosensors. *Applied Physics Letters* 88: 233120-1–233120-3.

Ozasa, A., R. Kometani, T. Morita, K. Kondo, K. Kanda, Y. Haruyama, J. Fujita, T. Kaito, and S. Matsui. 2004. Fabrication and evaluation of thermal nanosensor by focused-ion-beam chemical-vapor-deposition. In Digest of Papers: *Microprocesses and Nanotechnology Conference*, Osaka, Japan, October 27–29, 2004, 28P-7-15, IEEE, NY, USA, pp. 266–267.

Perez, J. M., L. Josephson, and R. Weissleder. 2004. Use of magnetic nanoparticles as nanosensors to probe for molecular interactions. *ChemBioChem* 5: 261–264, doi: 10.1002/cbic.200300730.

See, C. H. and A. T. Harris. 2007. A review of carbon nanotube synthesis via fluidized-bed chemical vapor deposition. *Industrial & Engineering Chemistry Research* 46(4): 997–1012, doi: 10.1021 /ie060955b.

Shanov, V., Y.-H. Yun, and M. J. Schulz. 2006. Synthesis and characterization of carbon nanotube materials (review). *Journal of the University of Chemical Technology and Metallurgy* 4(4): 377–439.

Sumner, J. P., N. M. Westerberg, A. K. Stoddard, C. A. Fierke, and R. Kopelman. 2006. Cu^{+}- and Cu^{2+}-sensitive PEBBLE fluorescent nanosensors using DsRed as the recognition element. *Sensors and Actuators B* 113: 760–767.

Wei, J., D. Olaya, B. S. Karasik, S. V. Pereverzev, A. V. Sergeev, and M. E. Gershenson. 2008. Ultrasensitive hot-electron nanobolometers for terahertz astrophysics. *Nature Nanotechnology* 3: 496–500.

Xiao, J. Q., J. S. Jiang, and C. L. Chien. 1992. Giant magnetoresistance in nonmultilayer magnetic systems. *Physical Review Letters* 68(25): 3749–3752.

Zhang, C.-Y. and L.W. Johnson. 2009. Single quantum-dot-based aptameric nanosensor for cocaine. *Analytical Chemistry* 81: 3051–3055.

Zhang, G. and M. Liu. 2000. Effect of particle size and dopant on properties of SnO_2-based gas sensors. *Sensors and Actuators B* 69: 144–152.

Zhang, H., Y. Yang, Y. Su, J. Chen, K. Adams, S. Lee, C. Hu, and Z. L. Wang. 2014. Triboelectric nanogenerator for harvesting vibration energy in full space and as self-powered acceleration sensor. *Advanced Functional Materials* 24(10): 1401–1407.

Dosen, G. S., Desmonce, D. Fera, and J. Sau. 20[illegible]. Metallic nanostructured materials based sensors. *International Materials Reviews* [illegible]–[illegible].

[illegible], M., R. [illegible] and [illegible]. 20[illegible]. A new approach for [illegible] of [illegible] nanotubes [illegible] electrode [illegible] crystal [illegible] microbalance sensor [illegible] *Sensors and Actuators B* [illegible].

[illegible], R. [illegible]. 20[illegible]. Electromechanical [illegible] of the nanoscale [illegible] and sensing [illegible] (NEMS) [illegible].

[illegible], P. [illegible], R. [illegible], and E. [illegible]. 20[illegible]. [illegible] *Journal of Physics: Condensed Matter* [illegible] doi: 10.1088/[illegible].

[illegible] and D. Wang. 20[illegible]. [illegible] with advanced electrochemical-based detection and analysis platform. *Proceedings of the National Academy of Sciences* [illegible] doi: 10.1073/pnas.[illegible].

[illegible], S., C. Zhang, Y. [illegible], and S. Zhang. 2010. [illegible] of the [illegible] nanocomposite [illegible]. *[illegible]* [illegible].

Giessibl, F. J. [illegible]. Advances in atomic force microscopy. *Reviews of Modern Physics* [illegible].

[illegible], Z., C. [illegible], J. [illegible], and J. [illegible]. 20[illegible]. [illegible] modeling and experimental validation [illegible] communication. *[illegible]* 18(2): 103–113.

[illegible], [illegible] Bai, [illegible], K. Hang, and [illegible]. 20[illegible]. [illegible] nanotube synthesis and applications. *Journal of Applied Physics* [illegible].

[illegible]. 20[illegible]. Nanoparticles and scales. *[illegible] Special Issue on Nanomaterials, Powder and [illegible]* [illegible].

K[illegible], [illegible]. [illegible] nanosecond [illegible]. *[illegible]* [illegible].

[illegible], [illegible], and K. [illegible]. 20[illegible]. [illegible] scanning [illegible] probe [illegible] *[illegible]* [illegible].

K[illegible], [illegible], and A. M. Cleland. 20[illegible]. Nanometre-scale displacement [illegible] single [illegible]. *Nature* 421: [illegible].

[illegible], [illegible], [illegible], [illegible], and Y. [illegible]. 200[illegible]. Polymer [illegible] microelectromechanical [illegible] *[illegible]* [illegible].

Li, [illegible], and M. Meyyappan. 200[illegible]. Nanoelectronics [illegible] based carbon nanotubes. *IEEE Sensors* [illegible].

Li, [illegible], X. Wang, and J. [illegible]. 2008. A [illegible] based on [illegible]. *[illegible]* 160: 115–123. doi: 10.1007/[illegible].

Lee, [illegible], R. Marchal, [illegible] Aimon, [illegible] Portier, and A. M. Ionescu. 20[illegible]. [illegible] and [illegible] of [illegible] molecular [illegible] and [illegible] opportunities and challenges. *IEEE Transactions on Nanotechnology* 17(2): [illegible]–207.

Masmanidis, S. C., R. B. [illegible], I. [illegible], C. [illegible], M. [illegible], and M. L. Roukes. 2007. Multifunctional nanomechanical systems via tunably coupled piezoelectric actuation. *Science* 317: 780–783.

[illegible], M. [illegible], [illegible], and [illegible]. 20[illegible]. A comparison of different strategies for the construction of amperometric enzyme biosensors using gold nanoparticle-modified electrodes. *Analytical Biochemistry* [illegible]: 20–27.

[illegible], [illegible] and [illegible]. 2006. [illegible] limits of nanomechanical sensors. *Applied Physics Letters* 88: [illegible].

Ono, [illegible], [illegible], [illegible], K. Kondo, K. Kaneko, Y. Hatakeyama, [illegible], and [illegible]. 2004. [illegible] and evaluation of thermal [illegible] by [illegible] chemical vapor deposition. In *Digest of Papers [illegible] Conference*, [illegible], Japan, October [illegible], 2004, [illegible], pp. 266–267.

[illegible], M. L., [illegible], and [illegible]. 20[illegible]. Use of [illegible] nanoparticles in [illegible] for [illegible]. *[illegible]* [illegible] doi: 10.1002/[illegible].

[illegible] and [illegible]. 2007. [illegible] of carbon nanotube synthesis via [illegible] chemical vapor deposition. *[illegible]* [illegible] doi: 10.1021/[illegible].

S[illegible], [illegible] and M. [illegible]. 2000. Synthesis and characterization of carbon nanotube materials (review). *Journal of the University of Chemical Technology and Metallurgy* [illegible].

[illegible], R. [illegible], [illegible], A. K. Stoddard, [illegible], and [illegible]. 20[illegible]. [illegible] sensitive [illegible] using [illegible] *Sensors and Actuators B* [illegible].

Vestal, [illegible], [illegible], S. [illegible], A. V. [illegible], and M. [illegible]. 20[illegible]. [illegible] for [illegible] anisotropy. *[illegible]* [illegible].

X[illegible], G., [illegible], and [illegible]. 1992. Giant magnetoresistance in [illegible] magnetic systems. *Physical Review Letters* 68(25): 3749–3752.

Zhang, [illegible] and [illegible] Johnson. 2009. Single quantum dot-based [illegible] nanosensor for [illegible] *[illegible] Chemistry* 81: [illegible].

Zhang, C. and M. Liu. 20[illegible]. Effect of particle size [illegible] of [illegible] SnO[illegible] based gas sensors. *Sensors and Actuators* [illegible]: 144–152.

Zhang, [illegible], Y. Yang, [illegible], K. [illegible], [illegible], and [illegible] Wang. 2014. Triboelectric nanogenerator for harvesting vibration energy in full space and as self-powered acceleration sensor. *Advanced Functional Materials* 24(10): [illegible]–1407.

Index